A Publication of the
National Research Council of Canada
Monograph Publishing Program

Alfalfa and Relatives
Evolution and Classification of *Medicago*

by ERNEST SMALL
Principal Research Scientist
National Environmental Program, Biodiversity Section
Research Branch, Agriculture and Agri-Food Canada
Ottawa, Ontario, Canada

NRC·CNRC
NRC Research Press
Ottawa, 2011

www.cabi.org

ISBN 978-0-660-19979-5
NRC 49740
Catalogue No.: NR84-6/2011E

Published exclusively in Canada and non-exclusively in other territories of the world (excluding Europe, Asia, Africa, the Middle East, and South America) by NRC Research Press as 978-0-660-19979-5
NRC No. 49740

Published exclusively in Europe, Asia, Africa, the Middle East, and South America, and non-exclusively in other territories of the world (excluding Australia, Canada, and New Zealand) by CAB International, with the ISBN 978 1 84593 750 8

CABI is a trading name of CAB International

CABI Head Office
Nosworthy Way
Wallingford
Oxfordshire OX10 8DE
UK
Tel: +44 (0)1491 832111
Fax: +44 (0)1491 833508
E-mail: cabi@cabi.org
Website: www.cabi.org

CABI North American Office
875 Massachusetts Avenue
7th Floor
Cambridge, MA 02139
USA
Tel: +1 617 395 4056
Fax: +1 617 354 6875
E-mail: cabi-nao@cabi.org

Library and Archives Canada Cataloguing in Publication

Small, Ernest, 1940-
Alfalfa and relatives: evolution and classification of *Medicago* / Ernest Small.

Includes bibliographical references and index.
Issued by: National Research Council Canada.
ISBN 978-0-660-19979-5

1. *Medicago*--Classification. 2. *Medicago*--Identification. 3. *Medicago*--Varieties. 4. Alfalfa.
I. National Research Council Canada II. Title.

SB205 M4 S53 2011 583′.74 C2011-980166-0

NRC Monograph Publishing Program
Editor: P.B. Cavers (University of Western Ontario)

Inquiries: Monograph Publishing Program, NRC Research Press, National Research Council of Canada, Ottawa, Ontario K1A 0R6, Canada. Web site: http://pubs.nrc-cnrc.gc.ca

Correct citation for this publication: Small, E. 2011. Alfalfa and relatives: evolution and classification of *Medicago*. NRC Research Press, Ottawa, Ontario, Canada. 727 pp.

Table of Contents

ACCOUNTS OF THE SPECIES

Note: The standard presentation format (described in Chapter 1) is followed, but the material is much more extensive than for the other 86 species of *Medicago* listed below, frequently providing data for each of the five infraspecific taxa, and also adding additional topics. To facilitate location of information, a detailed guide to this chapter follows.

THE OTHER SPECIES ([1]indicates a species of uncertain status)

Executive summary

The Eurasian genus *Medicago* is best known for alfalfa, *Medicago sativa*, known as lucern(e) in much of Europe and some other regions. Alfalfa is the world's most important livestock crop and one of the leading crops in most countries that have temperate climates. A dozen or so other species of *Medicago* have proven to be extremely useful pasture plants during the past century. Although best known as an animal feed, alfalfa is being developed as a source of human food extracts, pharmaceuticals, enzymes, industrial chemicals, and biofuels. Moreover, alfalfa is unsurpassed in its environmental benefits, and is increasingly being grown for ecological applications. Today, there is a growing need to produce cultivated varieties not just for livestock applications, but also for the other technological benefits mentioned above. Important in this regard are dozens of additional wild species of *Medicago*, with properties offering considerable potential for cultivar improvement by gene transfer. A substantial obstacle to the breeding of improved cultivars of alfalfa and the other cultivated species has been the lack of a comprehensive guide to all the species, subspecies, and varieties of *Medicago* and to their ecological and agronomic properties. This monograph provides an improved, comprehensive classification of all 87 *Medicago* species, as well as detailed information on the species useful for breeding improved cultivars. Notable achievements include:

(1) Adding to *Medicago* more than two dozen species that until recently almost universally were considered to belong to other genera.

(2) Providing a guide and extensive illustrations to identify the 100 or so species, subspecies, and varieties of *Medicago*.

(3) Summarizing key information on all the species based on 80 scientific papers produced by the author, a world expert with decades of research on the subject.

(4) Summarizing key information on all the species based on analyzing thousands of papers in the scientific literature.

(5) Reviewing the thousands of scientific names that have been confusingly applied to *Medicago* and accepting only those names that are necessary in accord with the *International Code of Botanical Nomenclature*.

(6) Presenting a simplified classification of the alfalfa species *Medicago sativa*, previously confusingly split into dozens of species and hundreds of forms.

(7) Providing extensive agriculturally and ecologically significant information for the species, to facilitate their development as new crops and/or new gene donors for the breeding of improved cultivars.

(8) Providing a user-friendly book. Clear language is employed to explain the extensive technical information included; the reader can quickly find details on a particular topic by examining the categories of information provided.

This comprehensive guide to one of the world's leading crops, and the numerous species allied to it, is an essential reference for a wide range of biological specialists, particularly taxonomists, agronomists, and ecologists.

Preface

The Eurasian genus *Medicago* is best known for the perennial species *M. sativa*. The cultigen, including its landraces and cultivars, is known as alfalfa in North America, South America, and Iberia, but is more commonly known as lucern(e) in South Africa, New Zealand, Australia, Oceania, and all European countries east of Spain. It is the world's oldest (Lesins and Lesins 1979), most important (e.g., Barnes et al. 1988), and most intensively studied livestock crop. In many countries it is one of the leading crops — indeed, it is the fourth most important crop in the United States. Alfalfa has aptly been crowned the "King of Fodders" and the "Queen of Forages" because it is considered unsurpassed in temperate regions as a productive, nutritious crop with numerous agricultural virtues. In fact, alfalfa is frequently the standard against which other forages are evaluated. While *M. sativa* has been cultivated for millennia, a dozen or so of the annual species of *Medicago* have proven useful as pasture components only during the 20th and 21st centuries. Several dozen forage cultivars of the annual species have been bred, and have proven to be extremely useful for dry, Mediterranean environments. All of the cultivated species of *Medicago* require additional agronomic improvement, and the germplasm for this purpose resides partly in the cultivars selected to date, but mostly in a vast array of wild populations of all the 87 species of *Medicago*. Moreover, many of the uncultivated species also have agricultural potential. The key to exploiting the wild resource is understanding the variation in the genus, and improving its taxonomy (classification and nomenclature).

Medicago has long been considered to be a classification nightmare. Shinners (1956) commented that "Few genera have accumulated a literature so abounding in misquotations, misinterpretations, and misapplications as *Medicago*." Bolton (1962) stated that "a reconciliation of conflicting views on nomenclature would be a monumental task… The taxonomic revision of the genus *Medicago* is long overdue. An authoritative treatment to include all the known species would be an invaluable guide to everybody interested in alfalfa... It would also bring some order to the confusion that now exists in the records of chromosome number, interspecific fertility, and agronomic performance."

A very disconcerting aspect of science is that studies of a given topic always seem to raise more questions than they answer, necessitating further research. For scientists, it is discouraging to contemplate that, just as they often correct the observations and/or interpretations made by past scientists, so are future scientists likely, with better evidence or insight, to make additional corrections. Constance (1964) described the work of taxonomists as an "unending synthesis". The situation was nicely summarized in the following quotation:

> The scientific man is always on the road, never at the journey's end... We necessarily work with incomplete materials and more or less inadequate tools. We have to build on foundations often poorly established, and no matter how clever or industrious we may be, posterity will have to revise and correct much of what we have done. [T.D.A. Cockerell 1935]

While the ultimate goal of taxonomists is to achieve a stable classification of species, with universally accepted names, this can be very difficult for groups that are inherently difficult to classify, as is the case with *Medicago*. Towards this goal, there have been several competent, systematic analyses of the genus, notably beginning with Urban (1873), and most recently by Lesins and Lesins (1979). The latter analysis is outmoded in several respects. In particular, many species formerly considered as belonging to the genus *Trigonella* are, in fact, undoubtedly species of *Medicago* (Small et al. 1987; Bena 1998*a*).

The taxonomic treatment of *M. sativa*, the most important species of *Medicago*, has been, in my opinion, intolerable. Several works detailing *Medicago* of the USSR (Grossheim 1945; Vassilczenko 1949; Lubenets 1972; Ivanov 1977, 1980) treat what is interpreted in this paper as a single species, *M. sativa*, as dozens of species and hundreds of infraspecific taxa. These works are invaluable as sources of information, but the classifications are based on (*a*) extreme taxonomic splitting, frequently recognizing landraces of alfalfa as separate species and (*b*) misinterpretation of geographically localized introgressed and hybrid wild alfalfa populations as genuine species. Unfortunately, dozens of species names that merely designate recombinant and minor forms of cultivated and wild alfalfas have been copied into reference works consulted by alfalfa researchers, giving the misleading impression that they represent species. This monograph is taxonomically conservative, i.e., recognizing as few taxa as seem necessary. In particular, all forms that are completely or substantially interfertile with *M. sativa* (when ploidy level is adjusted) are included in *M. sativa*, and only the main intraspecific lineages are afforded infraspecific recognition.

The present monograph has the advantage of including information generated not only by past monographers, but also by numerous other taxonomists. Nevertheless, no doubt there are mistakes of fact and interpretation that will require correction.

This monograph is notable for:

(1) Adding to *Medicago* more than two dozen species that until recently almost universally were considered to belong to other genera.

(2) Reducing to synonymy hundreds of variants of species recognized by past taxonomists on the basis of misunderstood natural variation patterns. The majority of the species treated here have previously received alternative systems of classification based on misunderstanding.

(3) Providing a comprehensive accounting of the thousands of names that have been applied to *Medicago*.

(4) Presenting a simplified classification of the alfalfa species *M. sativa*, which recognizes five subspecies, one of which has two varieties. These had been split into dozens of species by some taxonomists, with innumerable infraspecific taxa confusingly labelled by hundreds of scientific names.

(5) Combining classification analyses of *Medicago* with extensive analyses of evolutionary forces that were responsible for generating the variation observed.

(6) Providing extensive agriculturally and ecologically significant information for the species to facilitate practical application of the systematic information.

Regrettably, habitats occupied by *Medicago* species in Eurasia are extensively being eradicated, and a few of the annual and many of the perennial species are in danger of being eliminated from the wild. Moreover, there are very few, indeed often no, germplasm collections of many of these species. Hopefully the analysis of the species presented here will contribute to their conservation both in their natural habitats and as germplasm collections.

Acknowledgements and Dedication

Left: Brenda Brookes in an alfalfa leafcutter bee shelter in northern Alberta. Pollination by the alfalfa leafcutter bee (= "alfalfa leafcutting bee"; a female is shown at upper right) is responsible for most commercial alfalfa seed produced in North America. Although the alfalfa leafcutter bee does not produce honey, it is a greatly superior pollinator of alfalfa than the honey bee. Shelters such as this one, which are recommended for cold climates, are made of transparent plastic so that they act as greenhouses, providing warmth for the bees, which are reluctant to forage unless the temperature exceeds 20° C. Leafcutter bees are solitary, cavity-nesters, not social, hive-dwellers like honey bees. However, they are "gregarious" and can be maintained in large densities by providing "nesting boards" (shown inside the shelter) with small holes (typically 5–7 mm diameter, 65 mm deep) drilled closely together (see close-up at lower right). The females pack each hole with eggs placed on a mixture of pollen and nectar (which the larvae consume), with each egg and its ration of food separated by cut portions of leaves, forming a string of cells from the base of the hole to the exit (at lower right, note that several holes have been filled). The markings on the nesting boards are orientation guides that assist the bees in finding their own "apartment" (a shelter typically houses 20 000 – 60 000 female bees). The evolutionary relationship between the bee and the genus *Medicago* is discussed in Chapter 7.

This monograph is dedicated to Brenda Brookes, my technical assistant and frequent co-author over the many years I have studied *Medicago*. Her mastery of diverse techniques, ability to perceive obscure details, and accurate recording and analysis of mountains of data have been critical to the extensive new information recorded in this monograph.

Forty-nine of my 80 publications on *Medicago* and related subjects listed in the Literature Cited are co-authored; I am deeply grateful to my co-authors for their substantial contributions to my knowledge of *Medicago* species.

I particularly thank the following artists, who prepared illustrations for this monograph: David Beddoes, Brenda Brookes, Barry Flahey, Anne Hanes, Marcel Jomphe, Susan Rigby, and Lucy Yuzyk. The drawings they prepared are individually acknowledged in the figure captions.

Many individuals and institutions contributed photographs and drawings for this publication, as indicated in the legends accompanying many of the illustrations. Similarly, many experts provided information, and their contributions are acknowledged in the text.

The late K.A. Lesins and the late C.C. Heyn, who produced key comprehensive monographs on *Medicago*, generously shared their knowledge and provided seeds. Other experts on the taxonomy of *Medicago* who provided notable assistance include W. Greuter, P. Lassen, A. Huber-Morath, and R. Thanopoulos.

Herbarium loans (and the numbers of borrowed specimens), mostly of *Medicago* but also including allied genera, were provided by the following (standard herbarium abbreviations of Holmgren and Holmgren (1998): ALTA (86), B (118), BM (508), C (107), CAL (4), CDDA (5), E (494), FI (99), G (1182), GH + A (1451), HUJ (514), IBSC (9), K (1011), JEPS (24), KUN (9), KW (15), L (189), LE (73), LD (189), LY (1), MO (375), MPU (736), MW (1), NA (1109), NAS (8), NEBC (130), NY (125), P (257), PUN (2), S (179), SOM (1), TBI (1), UC (352), UPS (64), US (1372), USCH (60), W (1091), and WIR (325). The 12 276 specimens were borrowed between 1977 and 2010. In addition, essentially all the holdings of BM, CAN, HUJ, IZ, K, LE, LINN, P, US, and WIR, a total of about 70 000 collections, were studied during visits to these herbaria. About 10 000 vouchers are present in the herbarium of my institution (the Department of Agriculture (DAO)), based mostly on personal collections from Greece, Israel, Italy, Spain, and Turkey, and on material cultivated in Ottawa (the seeds obtained from a wide variety of sources), representing most of the taxa and their morphological and geographical variations. I owe a great debt to the curators and staff of the herbaria cited, both for loans and for assistance during my visits.

Seeds for experimental study were obtained from a variety of sources, including personal field collections. Over the years, I have particularly benefited from receiving thousands of accessions from the foremost two sources of germplasm for *Medicago*: the National Plant Germplasm System of the United States Department of Agriculture and the South Australian *Medicago* Genetic Resource Centre.

I am very grateful to four anonymous reviewers who provided extremely detailed critical reviews of the manuscript, and to K.P. Steele who reviewed Chapter 10. I also owe a great debt to the editorial staff of NRC Research Press for their competence, professionalism, and courtesy.

Color Illustrations

Color-1. A and B: Alfalfa leafcutter bees (*Megachile rotundata*) on flowers of alfalfa (*Medicago sativa* subsp. *sativa*). (A) Female. Note hairy scopa (pollen-collecting apparatus) on ventral abdomen. (B) Male. **(C, D) Rhizobial nodules on *Medicago murex*.** (C) and (D) courtesy of J. Howieson, Centre for Rhizobium Studies, Murdoch, Australia. Dr. Howieson noted "The *murex* nodulation is typical of that species when found on acid soils. I call it 'walnut' nodulation. My rationale for its development is that on the acid soils there are fewer opportunities for infection and hence the indeterminate nodules increase to a greater size."

Color-2. A–D: Flower color variation in *Medicago sativa*. (A) Flowering branch of *M. sativa* subsp. ×*varia*, showing a mixture of yellow flowers, purple flowers, and flowers variegated in color with yellow and purple. Source: Wikimedia Commons, photo by "Fornax" in Unterfranken, Germany in 2004. (B) Sulphur-yellow flowers, an occasional variation of *M. sativa* subsp. *falcata*. (C) Greenish flowers, common in F_1 hybrids between subspecies *sativa* and *falcata*. (D) White flowers, a rare mutation of *M. sativa*. (B–D) are from Barnes (1972).

Color-3 (Facing page). E–H: Two principal subspecies of *Medicago sativa* in flower. (E) Flowering branch of *Medicago sativa* subsp. *sativa*. Photo by S. Bauer, United States Department of Agriculture. (F) Field of *M. sativa* subsp. *sativa* in flower. Photo by G. DeLange. (G) Flowering branch of *Medicago sativa* subsp. *falcata*. Photo by K. Weller, United States Department of Agriculture. (H) *Medicago sativa* subsp. *falcata* in flower, growing on a farm in South Dakota. Photo by S. Bauer, United States Department of Agriculture.

E
F
G
H
MR. FALCATA

Color-4. *Medicago marina*, growing on the beach of El Alquián, Bay of Almería, Iberian Southeast, Spain. The substrate is based on sedimentary lands of Quaternary origin, with sand of volcanic origin. The species is frequent in the area, but urbanization of the coast threatens its future. Reproduced with the permission of the photographer, J.-J. Matamala García.

Color-5. (Facing page). The three arborescent species of *Medicago*. (A) *Medicago arborea*, photographed February 2001 near Saint-Raphael on the rocky coast of the French Riviera. Reproduced with the permission of the photographer, J. Reynaud, Laboratoire de Botanique, ISPB-Faculté de Pharmacie, Lyon, France. (B) *Medicago strasseri*, photographed 23 March 2008, Altres illes mediterrànies, by C.F, Paris, Sóller Botanic Garden, Spain, reproduced with permission. (C–F) *Medicago citrina*. (C) Plants growing on Grossa Island, Columbretes Archipelago, Spain. Photograph by M.B. Crespo, March 1999; supplied by A.I. Juan Gallardo and used with permission. (D) Branches with fruits and flowers. Photograph taken on Grossa Island, Columbretes Archipelago, Spain, by M.B. Crespo in April 1996; supplied by A.I. Juan Gallardo and used with permission. (E) Fruiting branches of a plant growing on Grossa Island, Columbretes Archipelago, Spain. Photograph by M.B. Crespo, May 1999; supplied by A.I. Juan Gallardo and used with permission. (F) Plants growing on Balearic Islands, Spain. Source: Universitat de les Illes Balears, http://herbarivirtual.uib.es, reproduced with the permission of J.R. Larrucea, Dept. Biologia, Universitat de les Illes Balears, Spain.

A
B
C
D
E
F

Color-6. *Medicago rhodopea*, a mountain-dwelling perennial species. Photo by K. Metodiev, 18 June 2004, Besaparski Hills, Pazardzhik District, Bulgaria. Used with permission.

Color-7. (Facing page). More mountain-dwelling perennial *Medicago* species. (A) Forage seed collection in Azerbaijan in 2004. Mountainous habitats such as shown here are commonly occupied by perennial *Medicago* species The collector shown is E. Hall from the Tasmanian Institute of Agricultural Research, who was accompanied by A. Humphries, a lucerne breeder at the South Australian Research and Development Institute (SARDI) and S. Hughes, curator of the Australian *Medicago* Genetic Resource Centre at SARDI. Photo provided by S. Hughes. (B) *Medicago prostrata* in edge of meadow behind Liechtenstein Castle, Mödling, Austria (southwest of Vienna), 13 June 2001. From the website of the Department of Botany, Naturhistorisches Museum Wien, Austria. Used with permission. (C) *Medicago ruthenica*. Photograph taken 20 June 2003 at Darhan, Mongolia (46°37′58″N, 109°24′38″E; altitude 1270 m), by S.-G. Li, University of Tsukuba, Ibaraki, Japan. Reproduced with the permission of Dr. Li. Note the stony soil frequently associated with the species.

A
B
C

Color-8. Some annual species. (A) Young fruit of *Medicago minima*. Photo courtesy of P. Busselen, Katholieke Universiteit Leuven, and reproduced with his permission. (B) *Medicago littoralis*. Notice the typical beach-sand habitat. Source: Wikimedia Commons. Photo by J. Tosti, April 2005, in Barcarès-Roussillon area, France. (C) *Medicago minima*. Photograph taken 20 June 2004. Source: German Wikimedia Commons. (D) *Medicago truncatula*, with flowers and a young fruit. Source: Forage Information System (2007), Oregon State University.

Color-9. (Facing page). Characteristic leaf markings. (E) *Medicago arabica*, showing central leaflet blotches. Photo courtesy of P. Busselen, Katholieke Universiteit Leuven, and reproduced with his permission. (F) Distinctive leaf marks of *Medicago littoralis* 'Angel' being grown for seed production in Bordertown, South Australia. These leaf marks are typical of 'Angel', a sulphonylurea herbicide residue tolerant medic, and its parent *M. littoralis* 'Herald', but they are not typical of the species. The marks are thought to trace to a *M. truncatula* donor parent used in a back-crossing program with 'Harbinger'. Photo taken September 2005 and supplied by C. Grant, SeedMark Pty Ltd. (G) Foliage of *Medicago murex*, showing characteristic white leaflet flecks that often are evident on living plants, but are rarely detectable on dried, herbarium specimens. This feature is absent in the very similar, closely related *M. sphaerocarpos*. Photographed by W. McDonald. Reproduced by permission of the copyright owner, the New South Wales (Australia) Department of Primary Industries.

E
F
G

Color-10. *Medicago laciniata*. Photograph taken 12 March 2004 in Eastern Province of Saudi Arabia by A. Alsirhan and reproduced with his permission. Note the desert habitat in which this species often occurs.

Color-11. Various *Medicago* species. (A, B) *Medicago coronata* in flower and in fruit, respectively. Photographed 25 February 2007 at Manfredonia, Italy, and reproduced here with the permission of the photographer, L. Rignanese. (C) *Medicago monspeliaca*, a prostrate plant with both flowers and fruits growing in an alpine region of France. This is the most common of 24 species with pulvinate (jointed) cotyledons that were transferred recently to the genus *Medicago*. Photo supplied through the courtesy of R. Doucet, Station Alpine Joseph Fourier, Université Joseph Fourier, Grenoble, France. (D) *Medicago lupulina*, in flower. Source: Forage Information System (2007), Oregon State University. (E, F). *Medicago hypogaea* growing in Israel. Notice the whitish "gynophores" that push the fertilized flowers underground where the fruits develop. Photos courtesy of A. Danin, The Hebrew University of Jerusalem (related to *Flora of Israel Online* at http://flora.huji.ac.il; current July 2010).

Color-12. *Medicago rugosa.* The species is shown growing in Malta (left: with immature fruit on 27 March 2007; right: with mature fruit on 08 May 2008). Photos courtesy of S. Mifsud, http://www.maltawildplants.com/FABC/Medicago_rugosa.php; current Aug. 2010.

Color-13. Annual *Medicago* ("medic") development in Australia. (A) Use of an annual, self-regenerating medic in crop rotation: sheep grazing on *Medicago truncatula* regenerating in wheat stubble in Mallala, South Australia. Photo taken 31 July 2007 and supplied by J. Howie, South Australian Research and Development Institute (SARDI). (B) A field of *Medicago polymorpha* 'Cavalier' grown for seed production in Freeling, South Australia. Photo taken 18 October 2005 and supplied by J. Howie, SARDI. (C) Small plot field evaluation of annual medics at Mallala, South Australia. Photo provided by S. Hughes.

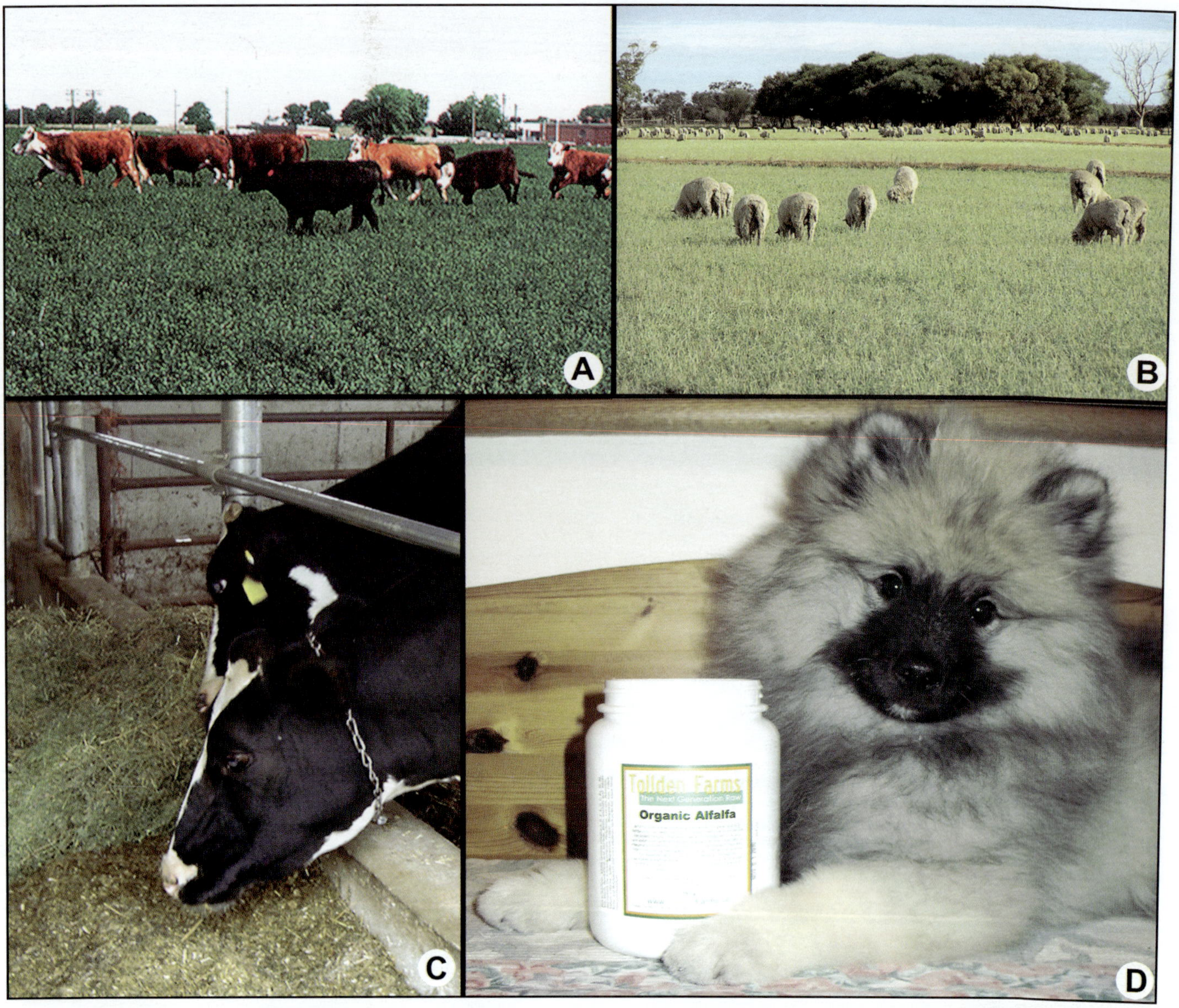

Color-14. Animals consuming alfalfa. (A) Cattle grazing in an alfalfa pasture. Photo courtesy of J. Caddel, Oklahoma State University. (B) Merino sheep grazing on alfalfa in Australia. Photo courtesy of I. Riley. (C) Dairy cows eating alfalfa fodder. Photo courtesy of G.R. Bauchan. (D) Alfalfa-based nutritional canine supplement and a 3-month-old Keeshond puppy. Photo by B. Brookes.

Color-15. Alfalfa harvest scenes. Photos courtesy of Forage Information System (2007), Oregon State University. This harvester cuts the plants, crimps the stems to speed drying, and forms the cut material into a row termed a windrow. After drying to the appropriate moisture content, the material can be collected for storage as silage, haylage, or dry hay.

Color-16. Alfalfa farm scenes. Top photo: Alfalfa field bordering farm buildings. Source: Forage Information System (2007), Oregon State University. Bottom photo: Loading rectangular bales of alfalfa hay onto a truck. Source: J. Caddel, Oklahoma State University.

1. Introduction

Because this monograph is intended for a diverse audience, an explanation of its content and purposes is in order. The following summarizes its organization and rationale.

First, the importance of the subject under discussion should be stated. The topic concerns the plant genus *Medicago*, which is economically important to society primarily because one of its perennial species is alfalfa (*M. sativa*). Alfalfa is the fourth most important crop in North America (after corn, soybean, and wheat, respectively), and the temperate world's most important forage crop. While alfalfa has been cultivated for millennia, a dozen annual species of *Medicago* have become significant pasture crops only in the past century. Several modern technologies are now being developed that have huge promise for extending the importance of alfalfa and its allied species well beyond their present primary use as a feed for livestock. For example, alfalfa has the potential of becoming the world's leading source of protein extract for nourishing humans, and the world's principal source of protein-based pharmaceuticals. Alfalfa is likely already the world's most environmentally friendly crop and has great potential for making agriculture more compatible with today's goal of minimizing negative ecological impacts. The economic values of *Medicago* are presented in the next chapter, "The Economic Importance of *Medicago*". The future development and value of the genus in general, and of alfalfa in particular, depend on the genes in wild *Medicago* species that can be used to breed superior crop cultivars. An assessment of available germplasm resources is given in the third chapter, "Genetic Resources".

The greatest explanatory theory in biology is evolution, and several evolutionary aspects critical to understanding the nature of the genus *Medicago* are dealt with in one of the book's two major sections, "The Evolution of *Medicago*". The section begins with an analysis of the biogeography of the species, starting first with an examination of its present geographical distribution. As will be seen, there are some interesting distribution patterns for species groups and even for plant characteristics. Key features of speciation in the genus *Medicago* — chromosomal evolution and the establishment of barriers to interbreeding — are reviewed next. This is followed by a detailed examination of adaptive character evolution among species, with emphasis on characters that have apparently been critical for the evolution of new species and, hence, have been used as diagnostic features for defining or identifying those species (i.e., the characters are not only ecologically important, but are also taxonomically important). As will be stressed, evolution of the genus *Medicago* has been guided to a remarkable degree by other organisms — principally herbivorous insects, pollinating bees, and symbiotic, nitrogen-fixing bacteria, less so mammals — and the nature of these associations is examined. It must be emphasized that this evolutionary information is not merely theoretical; an understanding of why and how the species have their present characteristics is key to utilizing them efficiently.

The second major section of the book is entitled "The Classification and Identification of *Medicago*". This section begins with a detailed examination of how the genus has been treated historically, followed by lines of evidence showing that *Medicago* is a considerably larger genus than traditionally defined. Utilizing diverse evidence, including recent examination of DNA sequence data, the genus is categorized into 14 taxonomic sections, several of which are subdivided into subsections, and a total of 87 species are recognized. Three species (*M. papillosa*, *M. prostrata*, and *M. suffruticosa*) are each divided into two subspecies. Two species (*M. rotata* and *M. truncatula*) are each divided into two varieties, and *M. sativa* is divided into five subspecies, one of which has two varieties. Dichotomous keys and detailed illustrations are provided to assist in identification of the taxa. This is followed by detailed taxonomic, ecological, and agricultural information on the 87 species, arranged according to the format described below.

PRESENTATION FORMAT FOR THE 87 SPECIES OF *MEDICAGO*

Full Scientific Name, Including Authorities

Authorities (authors) of names are given in standard, abbreviated form, following Brummitt and Powell (1992). The information can also be accessed at The International Plant Names Index: http://www.ipni.org/, current January 2010. The name is followed by an abbreviated citation of where the name was published. If this does not correspond to the basionym (i.e., the name on which the accepted name of the species is based, cf. Glossary), the publication that does give this information is indicated below under Synonymy/Homotypic. An explanation of the epithets is also given (e.g., for *M. arabica*, the epithet *arabica* is from the Greek for Arabian).

Subordinate Taxa

This gives information on subspecies, varieties, etc., of the species, including an identification key in cases where infraspecific groups are recognized.

Type Specimen

This gives the full citation of the type specimen, and in some cases a discussion of problems associated with the type. For the type specimens cited here, the convention of using an exclamation mark (!) to indicate that I have examined them is followed (an absence of "!" means I have not examined the type).

Throughout this monograph, standard herbarium abbreviations of Holmgren and Holmgren (1998) are followed (these are usually three or four uppercase letters).

Synonymy

Those unfamiliar with taxonomic treatments may be startled by the large numbers of old names sometimes associated with particular plant species. Since the starting point of plant nomenclature (Linnaeus' *Species Plantarum*, 1753), botanists operating with limited knowledge of the species (inevitable because often little information had been accrued), ignorant of the literature (which was much less available than today), or simply having different ideas of how plants should be classified, have been proposing new names to label variations allegedly new to science. Reviewing all these names in conjunction with an examination of the scientific merit of the plant groups distinguished by the names is essential, both to (1) ascertain whether any of the names actually designate populations genuinely worthy of recognition, and to (2) choose the names that must be adopted under the *International Code of Botanical Nomenclature* (McNeill et al. 2006; available at http://ibot.sav.sk/icbn/main.htm, current 1 January 2010). In the case of *Medicago*, the literature is greatly burdened by names proposed to designate variations that are quite unworthy of scientific recognition, and the ignorant use of these names is evident even in otherwise authoritative botanical and agricultural literature.

Homotypic

An abbreviated citation of junior synonyms that are based on the type specimen of the species (including some nomenclaturally problematic names) is given. Names are presented alphabetically based on the ultimate (i.e., last) epithet. Names employing the epithet are indented under it and sequenced by year of publication. Technical taxonomic phrases (e.g., nom. illegit.) are explained in the Glossary.

Heterotypic

In the same format as above, the abbreviated citation of junior synonyms that are not based on the type specimen of the species are given. Where types were located, they are cited. Types viewed by E. Small are indicated with the standard "!".

Notes regarding format of nomenclatural presentation

In the text (especially in synonymy), the abbreviation *M.* almost always stands for *Medicago*. If it refers to some other genus (*Medica*, *Melissitus*, etc.), the context makes this unambiguous.

In the Synonymy section above, homotypic synonyms (i.e., those based on the same type specimen) are presented in sequence of date. Since there are usually relatively few homotypic synonyms compared with heterotypic synonyms, this makes the history of nomenclatural changes easy to follow. By contrast, this time-honored convention is not followed for heterotypic synonyms (i.e., those based on different type specimens) because, in many cases, the number of synonyms is so large and the synonyms are so diverse that the nomenclature becomes overwhelming; see, for example, the treatments in Heyn (1963). Rather, heterotypic synonyms are presented in alphabetical sequence, arranged by final epithet. For each such epithet, the various names using the epithet in a different combination are indented under the original name, and are arranged in chronological sequence. This presentation makes it easy to follow the sequence of use of an epithet, and to understand how different authors have ranked and, in some cases, conceived the taxon in question. Where I have had doubts that a given epithet was in fact based on the same basionym (i.e., where two authors may have quite independently adopted the same epithet), the two names are simply presented in sequence, and one of the names not indented under the other.

Full scientific names with authorities are always presented with the correct citations, regardless of whether the author of the new name or combination correctly (by today's standards) cited the name. That is, if the original new scientific name or combination did not correctly cite the authorities for the name, the correction is made. This contrasts with Heyn (1963), for example, who presented synonymy using the exact names including authorities as they were originally published. This is a minor point since the synonyms that are given this way are not to be used for any purpose, but at least it does not present combinations that are now unacceptable.

Vernacular Names

English names are given in alphabetical sequence, with some explanation of the origin of the names. The French names are given similarly.

Geographical Distribution

Geographical information for the species was compiled from a wide variety of sources, including the International Legume Database & Information Service (http://www.ildis.org, current 1 January 2010). The presentation follows the conventions detailed in *World Geographic Scheme for Recording Plant Distributions* (Brummitt 2001), which has become the international standard for this. Accordingly, distributions are given in sequence for Africa, Asia, Australasia, Europe, Northern America, and Southern America (broad geographical categories that have been perhaps rather arbitrarily circumscribed). For each region, countries in which the species has been recorded are presented in alphabetical sequence, along with N to indicate "native", I to indicate "introduced", and U to indicate "uncertain status". In some cases, names of countries that no longer exist or have had their borders substantially revised are cited. This is because the information did not always allow a clearer specification, or because I simply lacked the patience to determine the exact distribution area from old collections (Yugoslavia being a particularly troublesome example). The presentation also includes a general statement describing the areas occupied.

Description

This includes a brief, standardized description of each species, as conventionally presented in taxonomic monographs, with size ranges given for parts of the plants. The format a(b–c)d means that most dimensions fall between "b" and "c", but that the number may be as small as "a" or as large as "d". For example, fruits with a diameter of (4)5–7(8) mm are mostly 5–7 mm wide, but occasionally have a diameter as small as 4 mm or as large as 8 mm. The descriptions are based both on information in the literature and on my examination of herbarium specimens. The descriptions apply to wild plants in their natural habitats, not to plants that are either cultivated or occur in unnaturally beneficial circumstances, as is sometimes the case with ruderal plants. As a rule, plants in excellent soil — with regular watering, good light, and protection from herbivores — grow much larger than wild plants, sometimes having vegetative organs that are much larger and even reproductive parts that are larger. The descriptions of plants in the last comprehensive monograph of *Medicago*, by Lesins and Lesins (1979), are based on greenhouse-grown plants, and usually exaggerate upper size ranges (sometimes greatly), especially for vegetative characteristics. The numerical ranges given by Lesins and Lesins should be employed simply as general information, and not used for identification purposes.

Identification Hints

This section provides information on how to distinguish the species under discussion from similar species that differ by characters that are sometimes difficult to evaluate. There is a particular risk that these species may be incorrectly identified by using the identification keys.

Habitat and Environmental Adaptation

Habitat: the settings in which the species usually occurs.

Phenology: information on the sequence of germination or resumption of growth in the spring, and timing of vegetative growth, flowering, and seed production.

Soil: information on soil texture, needs for elemental nutrients, and tolerance of pH and salinity.

Temperature: preferences for temperature during various phases of the life cycle, including frost and freeze tolerance.

Water: minimum requirement for rainfall, tolerance for drought, high humidity, and waterlogging.

Biotic Relationships

Pollination: degree of selfing and information on pollinators.

Dispersal: mode and adaptations for seed dispersal.

Herbivory: animals (pests, natural herbivores), especially insects that damage and limit growth and production.

Nodulation and mycorrhizae: information on rhizobial strains that fix nitrogen in root nodules for plant use and, where available, information on symbiotic fungi that infect the roots and benefit the plants.

Disease resistance: information on comparative resistance to harmful micro-organisms (fungi, bacteria, and viruses).

Economic History, Cultivars, and Uses

This gives a general statement assessing the contributions of the species to society.

Cultivars: brief review of cultivars and their comparative merits. The International Code of Nomenclature for Cultivated Plants (Brickell et al. 2009) requires cultivars to be cited by combining either the scientific name (of a genus or species) or an acceptable common name with the cultivar epithet in single quotations (e.g. *Medicago* 'Anik'; *Medicago sativa* 'Anik'; or alfalfa 'Anik'). However, the Code provides that cultivar epithets (in single quotations) may be employed alone as cultivar names when all of the cultivars mentioned belong to one genus, as is the case in this monograph.

Forage: uses of the species for forage (i.e., for grazing) in different geographical areas and in various agricultural circumstances.

Fodder: uses of the species for fodder (i.e., harvested for hay and other products) in different geographical areas and in various agricultural circumstances.

Green manure, rotation crop, ground cover: extent of such uses in different geographical areas.

Other uses: use for salad sprouts, ornamentals, plant products, etc.

Chromosome Number

This gives a completely referenced list of all reported chromosome numbers that were located.

Genetic Relationships and Gene Transferability

Intraspecific: information on barriers to hybridization within the species, where artificial hybridization of different populations has been attempted.

Interspecific: information on barriers to hybridization of the species with other species, where artificial hybridization of different populations has been attempted.

Exchange by advanced biotechnology: information on use of embryo rescue techniques and ovule–embryo culture to generate hybrids; generation of species bridges to transfer genes among species; generation of somatic hybrids (e.g., by electrofusion of protoplasts); genetic engineering.

Conservation Status

This gives information on the extent to which the survival of the species and its gene pools are threatened in native habitats, the extent to which conservation measures and legislation have been enacted, and the needs for the future.

Germplasm Sources

An indication is provided of the numbers of accessions of the species available from the major seed storage institutions of the world that have large collections of *Medicago* seeds.

Research Problems and Priorities

This section discusses specific botanical and agricultural research and development needs.

Representative Collections

This section cites exemplary specimens in the herbaria of the world, in all cases (with rare exceptions that are made clear) that I have personally examined. In the case of the type specimens cited here, I followed the convention of using an exclamation mark (!) to indicate that I have examined them (otherwise, I have not).

Illustrations

Every chapter dealing with a particular species presents plates illustrating vegetative, floral, and fruiting parts of the plant. The plants used as models for the line drawings are cited, and the artists are identified. For some species, photographs of type specimens, plants in their natural habitats, or plant parts are also presented.

ABOUT THE GUIDE TO SCIENTIFIC SYNONYMS

One achievement of this monograph is the evaluation of the thousands of scientific names for *Medicago* that have been published, clarifying their applicability and merit. A "Guide to Scientific Synonyms" is presented in Chapter 101 of this book. This list of species synonyms, in conjunction with the lists of infraspecific synonyms presented in the text of this monograph for every species recognized, is the first comprehensive guide to the *identity* of the thousands of names that have been recorded for the genus. Previously, for most of these names, only a few professional taxonomists were capable of understanding their meaning, and only by consulting scattered, specialized publications. It has not been possible to critically evaluate every name (most type specimens have not been examined, and indeed types for most of the taxa may not even exist), but generally enough information was found in the literature and original descriptions to assign names.

2. The Economic Importance of *Medicago*

For the 87 species of *Medicago* recognized in this monograph, extensive comments are presented in the chapters dealing with them, and the following information is concerned with reviewing their general uses and comparative values. As noted below, *Medicago* species are primarily used today for forage and fodder, but serve many other purposes, particularly as rotational crops and green manure.

DEFINITIONS OF FORAGE, FODDER, PASTURE, AND RELATED TERMS

The terms "forage", "fodder", and "pasture" are employed in different ways, according to different authors and indeed in different countries. *Medicago* species are primarily of value in agriculture as plants that are sown for the purpose of feeding animals, either as fodder (defined here as material that is harvested by humans prior to feeding) or forage (defined here as rooted plants consumed in the field by livestock). "Forage" is commonly used in a comprehensive sense that includes both forage and fodder as narrowly defined here (e.g., Allen et al. 1992). Thomas (1980) defined forage as "any plant material including herbage but excluding concentrates, used as feed for domestic herbivores". "Fodder" is also often similarly used in a comprehensive sense. The verb "graze", referring to the consumption of plants in the field by herbivores (Hodgson 1979), is equivalent to "foraging" as defined here. The term "browse" is often used to mean grazing on the twigs, shoots, and leaves of woody plants (three species of *Medicago* are woody: *M. arborea*, *M. citrina*, and *M. strasseri*), as opposed to grazing on "herbage" (entirely soft plant material), but this distinction is not made in this monograph. The noun "pasturage", although not a recommended term, is more or less equivalent to the noun "forage" as defined here (cf. Barnes et al. 1995). The essential difference between forage and fodder, as defined here, is simply whether the plant material is consumed in the field by livestock, or whether humans have harvested the material prior to consumption. In most cases, fodder will have been processed (at least dried and therefore no longer alive), but in the case of "green chop" (harvested material that is

Fig. 2-1. Alfalfa as the king of fodders and the queen of forages (see text; artist: B. Flahey).

chopped up fresh) the animals are fed still-living plant material. In most cases, forage is alive (or substantially so) while being consumed in the field, but often the dried-up remains of plants are consumed (in the case of annual *Medicago* species, the fruits attached to plants that have died often provide extremely nutritious forage). "Pastures" are grounds with vegetation used by grazing animals, especially livestock. Pastures can be natural (i.e., unaltered terrain furnishing grasses and other low-growing plants for grazing by domesticated herbivores), improved and/ or maintained, or created.

Medicago species are grown both in pure stands (normally for harvest as fodder), and in mixtures, typically with other legumes and grasses. Where a high degree of protection and care is afforded for mixed plantings, the material may either be grazed by livestock or harvested (as so-called "tame hay"), and the cultivated areas are often called "pastures", especially when enclosed, as opposed to "rangelands" (wildlands, whether public or privately owned, which are grazed by livestock but receive minimal or no care). *Medicago* species are also sown into low-maintenance, relatively natural wildlands to improve grazing (less commonly, the herbage is also harvested).

As searching Google will verify, alfalfa is known as the "Queen of Forages" and the "King of Fodders". "King of Forages" and "King of Forage Crops" are rarely used phrases, but when used they are also usually associated with alfalfa, although sometimes (in North America) with corn (maize). Similarly, "Queen of Fodders" and "Queen of Fodder Crops" are rarely used phrases that also refer to alfalfa. Alfalfa is not always the most important crop for feeding livestock. In some Western nations, cereals and oilseeds contribute substantially to the diet of livestock, sometimes more than alfalfa. Alfalfa is also exceeded in value for animals in some tropical countries by crops that are more adapted to equatorial climates. In general, however, alfalfa deserves its reputation as the world's leading livestock feed.

PERENNIAL *MEDICAGO* CROPS

Alfalfa

The species *Medicago sativa* (alfalfa) is usually grown as a perennial (rarely as an annual), mostly in mixed plantings in enclosed areas for use both as forage and fodder, but to a minor extent also to improve low-maintenance wildlands for grazing livestock.

Medicago sativa is not only the principal economic species of *Medicago*, it is arguably the world's most important temperate forage species, and indeed a considerable range of cultivars is available that enable the crop to be grown over a wide range of climates. In the United States, about 2.5% of agricultural land is represented by alfalfa, which generates about $7 billion annually (Şakiroğlu and Brummer 2007). When the value of animal products is added to the value of hay, the

Fig. 2-2. Alfalfa pageant. Queen Alfalfa is at the center, wearing a crown of alfalfa blossoms, and the girls seated at the sides carry signs such as LIME, INOCULATION, FERTILITY, and FIRM SEED BED, which indicate the factors making for good alfalfa growth. The pageant was called "The Coronation of Queen Alfalfa" and was performed at several high schools in Wisconsin in 1926. It was organized by L.F. Graber, chairman of the Department of Agronomy at the University of Wisconsin, who came to be known as "Mister Alfalfa". Photo (taken in 1926) from Graber (1976), courtesy of Department of Agronomy, University of Wisconsin-Madison.

dollar value of alfalfa in the USA alone reaches $145 billion annually (Bauchan and Greene 2000). Given that the United States grows about one-third of the estimated world acreage of 32 million ha (Michaud et al. 1988), a prorated annual world value of alfalfa would be about half a trillion dollars annually. However, given the undetermined but doubtless huge nitrogen-fixing and ecological values of alfalfa, it would not be surprising for the world value of alfalfa to exceed a trillion dollars annually.

In the USA, alfalfa seed is estimated to be worth more than $200 million annually and alfalfa sprouts about $250 million. Alfalfa also constitutes about one-third of the American honey crop, which has an annual production equal to about $150 million (Bauchan and Greene 2000). Bauchan and Greene stated that "alfalfa is the best animal feed for ruminant livestock and horses and it ranks fourth among all U.S. agricultural crops grown." Alfalfa is primarily an extraordinarily valuable food for all classes of livestock and is especially rich in vitamins, minerals, and protein. It can be grazed, albeit it may cause the digestive disorder "bloat" in ruminant animals. Some alfalfa is fed as green chop, but most is harvested and preserved as hay, compressed into cubes, or processed into silage, haylage, or baleage (see Chapter 100, Glossary). Alfalfa sprouts (germinated seeds allowed to grow into small seedlings) are a very popular human food. Recently, the prospect has arisen of producing large amounts of protein extracts from alfalfa, since alfalfa is the most efficient temperate crop for protein production. Alfalfa extracts also have tremendous potential for producing value-added goods such as biodegradable plastics and improved textiles. There is also great interest in using genetically transformed alfalfa for the production of pharmaceutical ingredients and industrial enzymes. Furthermore, alfalfa is being used as a model crop by plant molecular biologists because the plants are easily grown from single cells and are easily transformed genetically. Finally, the prospects of improving the crop through gene insertion are extremely promising.

Alfalfa is also of great interest today for its ecologically friendly properties. It has been known for millennia that the nitrogen-fixing abilities of the symbiotic bacteria housed in its root nodules can improve the soil. These bacteria can fix up to 560 kg of atmospheric nitrogen per hectare per year and, because of this, alfalfa is widely used in crop rotation. Alfalfa's extensive, perennial root system (which often includes a carrot-like taproot) also serves to reduce erosion by holding soil together, improving soil tilth and water infiltration, and benefiting soil ecology (including invertebrates and microorganisms). The fact that alfalfa requires minimal or no nitrogen fertilization makes it an energy-efficient crop. The perennial nature of the species means that it can be grown for several years without reseeding, again an energy-conserving feature that reduces tillage, and also serves to reduce soil erosion by maintaining a long-term, continuous cover. Alfalfa grows vigorously, which provides excellent weed control and reduces the need for herbicides. The biomass productivity of alfalfa is astonishing, with record yields in the order of 22 tons of dry matter per hectare without irrigation and 54 tonnes dry matter per hectare with irrigation. Such productivity is of great

Fig. 2-3. Alfalfa (*Medicago sativa*, discussed in detail in Chapter 12), is one of the world's most important crops, possibly contributing to agricultural products that exceed a trillion dollars (US) in value annually (see text; artist: B. Flahey).

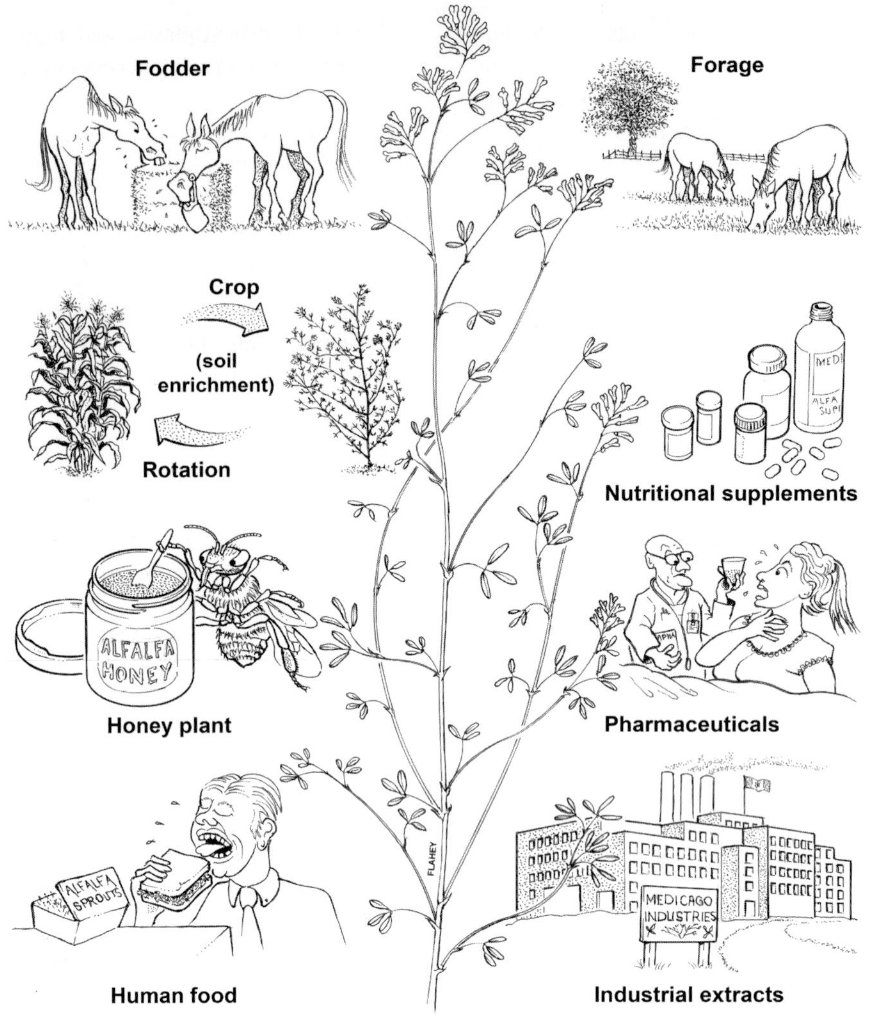

Fig. 2-4. Principal economic value of alfalfa (artist: B. Flahey).

value ecologically because it decreases agricultural pressure on the world's shrinking agricultural lands. Alfalfa typically contains 15% to 22% crude protein and can produce more than twice as much protein as soybean on a given area. This makes alfalfa the temperate world's most efficient source of protein production, a key feature for a planet with greatly diminished arable land, but with increasing protein requirements for livestock and people. The widespread use of alfalfa in rotation systems lessens the spread of diseases and insects among other crops used in rotation, which in turn lessens the need for environmentally damaging biocides. More recently, with growing concern about diminishing water supplies and increasing water pollution, growing alfalfa is viewed as one way to protect surface and ground water quality because it acts as a sponge for excess nitrates and water. Wild alfalfa (particularly *M. sativa* subsp. *falcata*) has proven to be an excellent plant for improving wildlands, not only to the benefit of livestock but also to wildlife, providing feed and protected habitats for deer, upland birds, rodents, and other animals. Experimentation is in progress to utilize alfalfa in additional, environmentally friendly ways, including generation of biofuels and as a fuel source for electricity.

One of the very few negative aspects of alfalfa is that the best productivity requires irrigation in a world in which the supply of water is becoming increasingly scarce. Alfalfa can deplete deep soil moisture reserves, affecting subsequent rotation crops such as small grains. However, breeding can improve the water efficiency of alfalfa, and there are many possibilities of using drought-resistant species as sources of genes for this purpose.

Alfalfa's Public Image Problem

The commercial value of *Medicago* is based principally on alfalfa (*M. sativa*). Unfortunately, the public views this crop as mundane and boring, and sometimes even considers it a mildly humorous subject. Why is this? Three rather uninspiring images are constantly conjured up when the average person hears the word alfalfa: alfalfa sprouts (nutritious but often associated with the granola and arugula crowd); actor Carl Switzer, memorable for his portrayal of the fictitious character "Alfalfa"; and a simplistic image of farming.

Fig. 2-5. "Alfalfa" eating alfalfa sprouts. Prepared by B. Brookes. Child performer Carl Switzer (1927–1959), nicknamed "Alfalfa", was a member of the "Our Gang" kids, stars of a series of hugely popular comedy movies of the 1930s (adapted as "The Little Rascals" on television in the 1950s). He was distinguished by his cowlicked hair, vacuous grin, and off-key singing. His movie career had dwindled by the early 1950s, although he became a semi-regular on Roy Rogers' television series. In the 1950s, he supported himself mostly as a hunting guide and bartender. At the age of 32 he was killed in a drunken brawl when he reportedly tried to collect a $50 debt with the aid of a knife. The killer pled self-defence and the death was ruled to be justifiable homicide.

Another example of how alfalfa is often not taken seriously is the Alfalfa Club of Washington, D.C., founded in 1913 by four men, over drinks, at the old Willard Hotel. The club was named for alfalfa because its "roots will allegedly do anything for a drink" (the root of alfalfa has the deserved reputation of growing very deep to obtain water). The club has no other purpose than to convene once a year in late January for an evening of food and cheap laughs, and has been described as a boozy, backslapping, black-tie bash with bawdy humor for the most important politicians (frequently including the President of the United States), Hollywood stars, captains of industry, and other VIPs. There are about 200 "Alfalfans" who are allowed to bring two guests each to the dinner. The press attends, but only in a cordoned off area in the lobby, because the event is officially off-the-record, although the content of the presentations is generally reported.

Unfortunately, the public has limited appreciation of the value of agriculture to their daily lives because farmers are becoming an increasingly smaller proportion of the population. Most alfalfa is consumed on-farm, so very few people are able to observe the tremendous quantities grown and used,

Fig. 2-6. "The Alfalfa Club" by B. Flahey.

and indeed it is difficult to determine its actual value. Moreover, except for alfalfa sprouts, alfalfa and other species of *Medicago* are mostly significant as contributors to animal products such as milk, cheese, butter, honey, leather, and wool clothing, and few people understand the vital role played by forage and fodder plants. Crops depend on political support for research and development, and inevitably support for alfalfa is difficult to obtain, because politicians are sometimes reluctant to back projects that are unfamiliar and difficult to explain to the general public. With obesity common today, dairy products are sometimes considered unhealthy, and this could be an obstacle to expanding this industry. However, as pointed out above and in other chapters, *Medicago* species have immense potential to greatly alleviate the planet's future needs for nutritious food, medicines, fuels, and industrial chemicals, and significantly contribute to the long-term sustainability of agricultural systems.

Other Perennial Species of *Medicago*

The only perennial species of *Medicago* grown extensively as a field crop is *M. sativa*, although several others are grown to a small extent. After *M. sativa*, *M. arborea* is the second most widely cultivated perennial species of *Medicago*. It is usually grown as an ornamental shrub, but has also proven useful when seeded into rangelands to improve livestock grazing or grown on steep slopes to stabilize the soil. However, *M. arborea* on the whole is used much less than some of the annual species of *Medicago*.

Medicago lupulina, although sometimes a biennial or perennial, mostly occurs as an annual. As discussed in the next section, *M. lupulina* and a dozen or so annual species of *Medicago* are grown primarily as crops that self-sow in an area for several years after initial seeding and are useful principally for improving low-maintenance pastures and wildlands for grazing livestock.

ANNUAL *MEDICAGO* CROPS: THE MEDICS

What are Medics?

Although there are a few annual cultivars of alfalfa (*M. sativa*), and this species is sometimes grown as an annual, the chief annual forage species of *Medicago* belong to other species. About 30 annual species of *Medicago* (all with coiled pods) are commonly called "medics" or "medicks". The principal species of these that have been used to breed cultivars are: *M. littoralis*, *M. murex*, *M. polymorpha*, *M. rugosa*, *M. scutellata*, *M. italica*, and *M. truncatula*.

The term medic is also traditionally applied to *M. lupulina*, which has obscurely coiled pods and is annual, biennial, or perennial; a few cultivars of this species are available. As noted in this monograph, 23 annual "medicagoid" species of *Trigonella* (none of these with coiled pods) were transferred to *Medicago* by Small et al. (1987), and while these (and another species now known to belong to the group) have not been called medics in the past and have not been the source of cultivars to date, the term may reasonably be expanded to cover them as well. The word medic is rarely applied to the perennial species of *Medicago*, and this usage is best avoided because the annual species generally have significantly different biological properties and agricultural applications. All the annual species of *Medicago* (including *M. lupulina*) are predominantly or entirely self-pollinating and set seed without pollinators. By contrast, perennial *Medicago* are weakly to strongly outcrossing. *M. sativa* in particular is generally strongly outcrossing and requires pollinators for seed set, although there are partially inbreeding wild populations and a very few strongly inbreeding cultivars.

The Mediterranean Adaptation of Commercial Medics

The Mediterranean Basin is the center of origin and primary region of diversity of almost all legumes used to feed livestock, including most of the annual species of *Medicago* currently used commercially (Sulas 2005). Such legumes have proven to be useful in other areas that have Mediterranean climates (i.e., mild wet winters followed by hot dry summers), including Cape Province of South Africa, California, central Chile, and southern Australia. The Mediterranean legumes have evolved distinctive adaptations suiting them to the climate and soils of the region, and it is critical to understand these limitations when adapting the plants for cultivation in new locations.

The annual medics with coiled pods are native primarily to semi-arid areas around the Mediterranean Sea, and are significant components of unimproved pastures in the region. In recent times, about a dozen of the weediest medics have become established throughout the world, mostly in areas with mild, rainy winters and alkaline soils. In such regions, medic seeds germinate at the start of fall or winter rains, grow vegetatively until early spring when they flower, produce seed by late spring or early summer, and die (their life cycles are often as short as 2 months, rarely as long as 6 months). Although they are often weedy, medics generally provide excellent pasturage where they are introduced. About a dozen or so annual species of *Medicago* (all with coiled pods and most among the weediest of species in the genus) are grown agriculturally. The comments below apply to them specifically.

The Role of Medics in Improving Natural Grazing Land

The most traditional method of raising animals is to allow grazing in permanent pastures that are typically not "improved" by manipulation, or only minimally so. Plant species that persist in pastures (either as annuals or perennials) must be capable of tolerating grazing over many years. In general, legume species are inferior to grasses as natural pasture species, likely because their greater palatability, digestibility, and nutritive value appeal to grazing animals. The basal meristem is a natural adaptation by grasses to grazing and, indeed, species of *Medicago* that are most able to withstand natural grazing are low-growing (or at least have low-growing ecotypes), thus maintaining their growing sites close to the ground, like grasses. Usually, *Medicago* species do not form extensive, pure stands in nature, but occur in scattered clumps or as scattered individuals, a strategy that is adaptive since herbivores are especially attracted to them. Although medics generally are limited in natural grazing lands, it is nevertheless possible to take advantage of their excellent nutritional contributions by seeding them to increase their frequency, and maintaining them by careful management of grazing pressure to allow the plants to reseed.

Regions of Medic Usage

Annual species of *Medicago* are grown as pasture legumes in their native Mediterranean region and elsewhere. Principal areas of planting and/or management include Australia, parts of South Africa, North Africa (including Algeria, Libya, Morocco, and Tunisia), the Middle East (including Iraq and Jordan), the USA (notably California, Oregon, and the southern states), parts of Europe (including France, Spain, and Portugal), and South America (notably Chile and Uruguay) (for reviews, see Puckridge and French 1983; Chatterton and Chatterton 1989; Masson and Gintzburger 2000; and Porqueddu 2001). Most areas where annual medics have proven successful have climates resembling the plants' Mediterranean homeland. These regions have wet, mild winters with rainfall typically 250–500 mm, falling mostly in the winter (often sporadically and variably), followed by very dry summers. They also have soils that are circumneutral to somewhat basic, with various textures, but tending to be coarse-grained and/or well-drained. In Mediterranean climates, the plants grow as winter annuals, germinating (in the Northern Hemisphere) from September to November in response to autumn rainfall and producing seed during spring and early summer, and reseeding spontaneously (self-seeding). Medics are high-yielding, under good conditions producing 6–10 t of dry matter per hectare per year (Brand et al. 1991*a*). In addition to supplying high quality winter pasture, the residues and seed pods can be utilized by sheep during the dry summer season in Mediterranean climates. An additional advantage of medics in many winter-mild areas is that they often grow in the winter and early spring when most grasses are dormant, and under these conditions, they sometimes outcompete local weeds.

Medics in West Asia

In West Asia, farming systems particularly produce cereals, mostly wheat and barley, as well as livestock products, mostly meat and milk from sheep and goats. Traditionally, a crop year was followed by a fallow year, largely to conserve moisture, but because of increasing land pressure, fallowing is becoming less common. Farmers often feed barley produced by continuous cropping to their livestock, which is unsustainable because the lands become exhausted. Grazing on common lands is widespread, and this is also unsustainable because of vegetation removal and consequent soil erosion. In either case, rotation with annual legumes, especially medics, is widely considered to be essential for improved sustainability (Cocks 1993). Most cultivars of annual *Medicago* have been bred in Australia and are poorly adapted to the countries of West Asia, especially lacking tolerance to frost (Cocks and Ehrman 1987). *Medicago rigiduloides* (often identified as "*M. rigidula*"), *M. noeana*, and *M. rotata* are considered to be better candidates for West Asia than the available Australian cultivars (Abd El Moneim and Cocks 1986).

Medics in North Africa

In dryland regions of North Africa, farming has long been based on rotation of crops with fallowing, the fallow period witnessing occupation of the land largely by weeds. Cereal production in the Mediterranean region (where rainfall averages 200–600 mm/year) traditionally has similarly used fallowing in alternate years to conserve crop-limiting soil moisture. Self-regenerating pasture medics were introduced to provide forage for livestock in the alternate year and to reduce the cereal's need for fertilizer. Azizi et al. (2004) pointed out that the Australian medic cultivars are not adapted to most of the agro-ecological zones of North Africa and West Asia because they were selected for the mild to cool winter environment prevailing in the Australian Mediterranean region, where frost is exceptional. Consequently, the Australian cultivars are ecologically well adapted to the low lands of North Africa (the coastal zones of Morocco, Tunisia, and Libya), but not the high plateau of North Africa and West Asia, where there is a significant occurrence of freezing days and lower winter temperatures. Azizi et al. (2004) suggested that by contrast to most of the Australian cultivars, *M. polymorpha* is preferable for the cooler climates of Africa, as well as such cool-tolerant species as *M. rigiduloides*, *M. noeana*, and *M. rotata*.

Medics in the European Mediterranean region

In the European Mediterranean region, annual, self-regenerating legumes are mostly used for pasture and rangeland improvement and reclamation of abandoned land (Piano 1987).

Medics in continental climates

The performance of annual medics in continental climates (especially in the USA) has generally not been as successful as in Mediterranean climates (Sims et al. 1991), with the exception of *M. lupulina*, a species that is adaptively distinct from the majority of annual species of *Medicago*. In New Mexico, Smith and Baltensperger (1984) found that alfalfa (*M. sativa*) outperformed *M. lupulina*, which in turn outperformed *M. truncatula* and *M. littoralis*. When annual medics are planted in the spring (for summer harvest) in regions with continental climates, they are being grown as "summer annuals" in contrast to their growth as "winter annuals" (with fall planting and spring harvest) in Mediterranean climates. In the Southern Great Plains, more than 5 t/ha of forage per year, with an average yearly seed yield of 570 kg/ha, is possible for a number of annual medics (Muir et al. 2005). There is interest in the United States in exploring the use of medics in sustainable agricultural systems, including for use as pasture, green manure, and cover crops.

The Australian Contribution to Medic Development

Australia has made extensive and invaluable contributions to medic technology and usage. Medics have been extensively utilized in Australia over the past century, and have been grown on about 40 million hectares, particularly in cereal–legume crop rotation systems in southern Australia where winter rainfall is 250–500 mm annually. Australian plant breeders have focused on earlier flowering, increased forage and seed production, tolerance to insects, and reduced pod spininess. Several dozen Australian cultivars are now available (in addition to cultivars produced elsewhere), almost all released in the last half of the 20th century. The use of ley and phase farming systems in Australia, described below, represents a major development in the employment of self-regenerating legumes, particularly *Medicago* species, and such usage has been adapted to Mediterranean climate regions outside of Australia.

Medics and the ley farming system in Australia

The traditional Mediterranean region rotation of cereal crops with a fallow period involving livestock production on medic-improved pastures is the basis for ley farming. Ley farming systems differ from permanent pastures in that there is a rotation between a crop (or several crops) that are grown as monocultures for a period, followed by a much less intensive management of the land during which the occupying plants self-regenerate for a period. The pasture (i.e., fallow) phase is typically used to produce livestock, and therefore represents a profitable strategy to diversify farm production.

The ley farming system of pasture–cereal rotations was developed intensively in Australia in the middle 1930s (Puckridge and French 1983) and has been adapted elsewhere, such as in South Africa starting in the middle 1960s (Brand et al. 1991*a*), and in North Africa and western Asia (Abd El Moneim and Cocks 1986). Contrarily, Cocks and Bennett (1999) stated that ley farming is not used in any other regions with Mediterranean-type climates. Ley farming is an integrated system of cereal and livestock production, common in the drier areas of southern and Western Australia, in which a planted, nitrogen-fixing annual legume pasture established for livestock grazing (mostly sheep in Australia) is rotated with a planted cereal crop. The legume self-regenerates to maintain the pasture for a number of years after the cereal is harvested. The ley system has been largely credited with the 50% increase in wheat yields in Australia between 1950 and 1980 (Donald 1981). Since the 1950s in Australia, annual species of *Medicago* have formed a major component of the pasture phase used to feed livestock, and are important to maintaining soil structure and fertility for the subsequent cereal phase. The rotation also provides a disease break for various cereal root pathogens. Economically, ley farming raises the stock-carrying capacity of land and tends to make farmers' income more stable. Medics have high levels of hardseededness (dormancy), enabling them to persist through cropping phases and to regenerate in subsequent years as pasture.

Annual Mediterranean species of *Medicago* germinate with the winter rainfall and grow, bloom, and produce seed during the Mediterranean winter and spring, thus avoiding the following long, hot, dry summer. Ley pastures are typically adapted to areas receiving 250–450 mm of winter rainfall. Medic pastures can produce 6–10 t of dry matter per hectare per year. Not only do medics supply green matter for livestock, but the fruits (burs or pods) can be consumed by sheep during the dry summer season. The forage nutritive value of medics may be greater if some herbage is removed during the growing season (Muir et al. 2005).

From 1995 to 2005, the role of annual medics in the Australian farming system declined (Nair et al. 2006*a*). This is attributed to a change in farming systems related to a decline in wool prices, and a need for perennial plants to restore hydrological stability. Legumes, including species of *Medicago*, are very sensitive to soil residues of sulfonylurea herbicides, which are widely used in the cereal phase of many Australian farming systems. Some of these herbicides are long-lived, and it has been hypothesized that their widespread persistence has been a contributing factor to decreased annual medic pasture productivity in recent decades. Cocks and Bennett (1999) summarized the problems as follows: "Unfortunately, ley farming has become unsustainable in southern Australia in four ways: it uses insufficient water resulting in rising water tables and dryland salinity; it results in soil acidification because of the leaching of excess nitrate and the removal of cations during cropping; an increased dependence on herbicides has resulted in herbicide resistant weeds; and finally there is widespread pasture deterioration due to many factors operating independently and together." Sulas (2005) pointed out that the problem is exacerbated by the high cost of seed production, the low persistence of annual medics in acid and sandy soils, the arrival of new insect pests and diseases, price problems due to changes in world trade, and the fact that there are potential sites of cultivation for which appropriate legumes have not yet been found.

Partly to address these problems, some new cultivars (of *M. littoralis* and *M. truncatula*) are being bred with good tolerance to sulfonylurea herbicide residues (Nair et al. 2006).

Medics and phase farming in Australia

In ley farming, the pasture species need to persist as dormant seeds in the soil during the period that the land is being used to cultivate other crops. By contrast, in "phase farming", such long persistence is not required because the pasture species are sown anew at the beginning of the pasture phase (the pasture phase lasts more than 1 year, so persistence is still required). This allows the crop phase to last for a longer period (Reeves and Ewing 1993).

Cocks and Bennett (1999) suggested that ley farming is likely to be replaced by phase farming, whereby the pasture period is lengthened. Sulas (2005) proposed that phase farming would involve continuous cropping for 4–6 years followed by a period of legume-dominated pastures for 2–3 years. Nichols et al. (2007) indicated that phase rotations involved long crop and pasture phases, and that while medics would continue to play an important role, additional legume species have also become important.

Medics as Smother Crops

"Smother crops" are grown to suppress weeds, a methodology considered desirable to reduce dependence on chemical weed control. A successful smother crop must compete strongly with weeds but minimally with the crop. Experimentation with annual species of *Medicago* as smother crops is in progress, but has produced less than desirable results in some cases. Sheaffer et al. (1991) found that several annual *Medicago* cultivars decreased weed yields, but also reduced soybean herbage and grain yields.

Summary of the Values of Medics

Medics grow rapidly, and some can produce considerable biomass, both pods and herbage providing excellent feed for livestock. In some cases a single harvest of medic species has produced more than 7 tonnes of dry matter per hectare in just over 2 months, with forage quality and crude protein content comparable to alfalfa. With effective *Rhizobium* inoculants, medics are known to produce up to 80 kg N/ha. Medics are mostly adapted to soils with a pH of 6.0 and above, but some species are more tolerant than alfalfa to soils with low pH when acid-tolerant *Rhizobium* is employed. Although they are annuals, the hard seeds invariably produced by medics germinate intermittently for years so that a crop often does not have to be reseeded for several years. Perhaps the chief disadvantage of medics is their lack of cold hardiness. While very valuable and of great potential, medics are far less economically important than alfalfa. Although conventional hybridization between the annual species of *Medicago* and alfalfa has proven to be almost impossible, research on transferring valuable medic traits to alfalfa by modern technologies is nevertheless under development.

Bauchan and Greene (2000) listed the following as established usages of medics:

1. Medics can be used as permanent pasture.
2. Medics can be used as a green manure crop because of their ability to fix nitrogen.
3. Medics can be used as a cover crop to prevent soil erosion.
4. Medics can be used as a companion crop to control weeds, add organic matter, lower soil temperatures, and improve fall residue after harvest. Various crops have been utilized with medics, including corn, soybeans, small grains, sunflower, squash, and some herbs.
5. Medics have been used in the alleyways of vineyards and fruit orchards to provide ground cover for soil erosion, providing an environment for beneficial insects, and a stable track for equipment.
6. The use of annual medics in wildlife habitats for deer, quail, and water fowl has been initiated.
7. Medics have potential usage in sustainable agriculture systems.

Clarkson (1977) listed the following positive and negative aspects of annual species of *Medicago* (medics) in Australia:

As a grazed protein supplement. Because they grow in winter when grasses are dormant and deficient in protein, medics can reduce winter weight losses of livestock, raise wool and beef production, and raise the fertility of animals mated in the spring.

To improve soil nitrogen fertility. Medics gradually improve soil fertility by fixing nitrogen from the atmosphere. Like other legumes, their roots develop nodules containing the *Rhizobium* bacteria, which do this important job. The added nitrogen promotes grass growth. In areas of adequate winter rainfall, the improved summer and winter pasture growth can support a higher overall stocking rate.

As a winter forage crop. Medics, like oats, may be grown as a winter grazing crop either on their own or with oats. They provide high quality winter grazing without lowering the soil nitrogen fertility, and can be valuable on soils where oats need nitrogen fertilizer.

As drought feed. The reserves of seed pods built up in favorable seasons are rich in protein and will sustain sheep for long periods. The pods are of little value to cattle because they cannot get at them; however, cattle can obtain useful material from dead medics for a short time after the plant dies.

For soil conservation. Their persistence, drought tolerance, and ability to colonize eroded areas make them useful pioneer plants. By providing nitrogen for grass growth, they may accelerate the restoration of protective ground cover.

The weaknesses of medics according to Clarkson (1977) include the following: production is highly variable from year to year and depends mainly on winter rain; production is reduced by competition from summer grasses; medics can cause bloat in cattle and are similar to alfalfa in this respect; the spiny pods of some medics reduce the value of wool by increasing the vegetable fault; and medics are not well adapted to strongly acid soils.

3. Genetic Resources

Fig. 3-1. An exceptionally tall *Medicago sativa* plant discovered in eastern Turkey during a germplasm collecting expedition (Small 1981*b*, 1982; photo by E. Small in 1981). Such vigorous, erect plants have potential for hay production.

THE IMPORTANCE OF GERMPLASM

Genetic erosion — the loss of genetic diversity of the world's biological resources — is a major threat to many of the crop and forage species that feed people and livestock. The diversity of wild species (and also of landraces) provides critical raw material for improvement of domesticated forms and selection of entirely new domesticates. Genetic erosion has been particularly pronounced in Asia generally, and in the Mediterranean Basin in particular (Prosperi et al. 1996), the principal regions inhabited by *Medicago* species.

Alfalfa (*Medicago sativa*) is one of the world's major crops, and more than a dozen annual species of *Medicago* are significant forages. Accordingly, the collection, characterization, and conservation of *Medicago* germplasm are critical for the future breeding and improvement of cultivars. In the individual treatments presented in this book for each species, information is given on the current state of germplasm availability, and also on the *in situ* conservation status of each species in its indigenous area. This chapter presents an overview of the state of genetic resources for *Medicago*.

Fortunately, *Medicago* seeds are long-lived. As pointed out in Chapter 7, a 200-year-old *M. polymorpha* seed extracted from Mexican adobe brick was still able to germinate. The predicted "safe storage time" (P_{10}), which is a 10% decrease in seed viability, has been suggested to be as high as 902 years for *Medicago* seeds (Stoyanova 2001). By comparison, Stoyanova reported that this value was only 31 years for *Lathyrus* seeds and only 27 years for *Vicia sativa*. However, Walters et al. (2005) reported that the time for germination to drop to 50% (P_{50}) is 93 years for *M. sativa*, 40 years for *Lathyrus odoratus,* and 71 years for *Vicia* species. Unfortunately, similar controlled studies of *Medicago* species other than *M. sativa* have not been reported.

The following table provides summary data on the numbers of seed accessions of *Medicago* species for three of the largest collections (VIR, GRIN, SARDI), and for a network of collections (SINGER) (see below for full explanations of these collections). For some species, the nomenclature used is somewhat ambiguous, but in almost all cases it has been possible to translate the names into those accepted as correct in this monograph. It should be kept in mind that there is considerable duplication of collections among institutions.

GERMPLASM HOLDINGS OF INSTITUTIONS WITH LARGE COLLECTIONS OF *MEDICAGO*

SINGER is the System-wide Information Network on Genetic Resources, the germplasm information exchange network of the Consultative Group on International Agricultural Research (CGIAR) and its partners. The website (http://www.singer.cgiar.org/) was consulted 15 June 2008. The data are based on combining the nomenclature under both *Medicago* and *Trigonella*.

VIR is the N.I. Vavilov Institute of Plant Industry, St. Petersburg, Russia (http://www.vir.nw.ru/; accessed June 2008). This is by far the largest and most important center of genetic resources in Russia and, unfortunately, since the demise of the USSR it has experienced funding difficulty.

GRIN is the Germplasm Resources Information Network of the United States Department of Agriculture (USDA). Data, including accessions unavailable for distribution, were

Fig. 3-2. Part of the seed storage facilities at the South Australia Research and Development Institute (SARDI). Photo provided by Steve Hughes.

provided by Stephanie L. Greene on 20 February 2008. Numbers for species include infraspecific groups. The collection was analyzed by Greene and Bauchan (2000) and Bauchan and Greene (2000, 2002). This is the world's most important research collection of *Medicago* germplasm, with exceptionally extensive on-line documentation of characteristics of many of the accessions (see http://www.ars-grin.gov/npgs/index.html; accessed 1 January 2010). It has excellent associated taxonomic information and free access to collections.

SARDI is the South Australia Research and Development Institute, run by the Government of South Australia. Data were provided by Steve Hughes on 15 July 2008. The South Australian pasture collection houses more than 44 000 seed samples of pasture species (*Medicago* and other forages) collected from more than 100 countries but, unfortunately, because of funding difficulties, access to this superb collection has recently required payment (http://www.seedhunter.com/article-so-what-about-pastures.html; accessed December 2008). The largest collections of the annual species with cultivars are found in SARDI.

Table 3-1. Number of accessions of *Medicago* species held by germplasm banks with the largest collections of the genus.

	Collection			
Species	SINGER	VIR	GRIN	SARDI
M. arabica	95		88	467
M. arborea	12		38	34
M. archiducis-nicolai				
*M. arenicola**				
*M. astroites**	35		2	13
*M. biflora**			1	3
M. blancheana[1]	154		22	95
M. bonarotiana[1]				
*M. brachycarpa**	3		5	6
M. cancellata	1	2	7	4
*M. carica**				
M. carstiensis	4	2	1	4
M. ciliaris[2]			73	158
M. citrina[3]				1
M. constricta	149		53	163
M. coronata	107		24	46
*M. crassipes**			1	1
M. cretacea		1	3	4
M. daghestanica			1	
M. disciformis	60		52	121
M. doliata	378		115	1008
M. edgeworthii	1		30	1
*M. fischeriana**	2		5	9
M. granadensis	25		13	23
*M. halophila**				
*M. heldreichii**				
M. heyniana	1		2	3
*M. huberi**				1
M. hybrida	2		1	
*M. hypogaea**				1
M. intertexta	218		20	263
*M. isthmocarpa**				
M. italica	193		93	853
M. laciniata	225		137	616
M. lanigera	1		1	2
M. laxispira				1
"*M. lesinsii*" (= *M. murex*)			10	5
M. littoralis	306		177	2852
M. littoralis × *truncatula*	10			
M. lupulina	167		264	362
M. marina	2		19	29

Table 3-1 (*Continued*).

Species	Collection			
	SINGER	VIR	GRIN	SARDI
*M. medicaginoides**	2		3	6
M. minima	514		336	671
*M. monantha**	275		78	159
*M. monspeliaca**	515		14	81
M. murex[4]	178		70	259
M. muricoleptis	12		9	9
M. noeana			19	64
M. orbicularis	772		343	1765
*M. orthoceras**	2		2	8
M. ovalis				
*M. pamphylica**				
M. papillosa			10	2
*M. persica**				
*M. phrygia**				28
M. pironae			4	3
M. platycarpos	1		16	6
M. plicata				1
*M. polyceratia**	21		10	13
M. polymorpha	1626		731	5121
M. popovii	1		1	1
M. praecox	96		24	76
M. prostrata		1	13	6
M. radiata	185		17	129
*M. retrorsa**				
M. rhodopea			2	2
*M. rhytidiocarpa**				
*M. rigida**				
M. rigidula[5]	1247		369	1327
M. rigiduloides			306	
*M. rostrata**				
M. rotata	280		26	177
M. rugosa	89		47	305
M. rupestris				
M. ruthenica			118	10
M. sativa	1148	2014	4167	3459
M. sativa subsp. *sativa*			3194	
M. sativa subsp. *caerulea*			89	

Table 3-1 (*Concluded*).

Species	Collection			
	SINGER	VIR	GRIN	SARDI
M. sativa subsp. *sativa* × subsp. *glomerata*			5	
M. sativa subsp. *glomerata*			7	
M. sativa subsp. *falcata*			514	
M. sativa subsp. *falcata* var. *viscosa*			22	
M. sativa subsp. × *varia*			327	
M. sauvagei	7		5	6
M. saxatilis			1	1
M. scutellata	153		76	441
M. secundiflora			2	2
M. shepardii	1		4	7
M. sinskiae				
M. soleirolii	6		13	22
M. sphaerocarpos[4]				572
M. strasseri				1
M. suffruticosa			18	16
M. suffruticosa subsp. *leiocarpa*			4	
M. syriaca				24
M. tenoreana	4		7	5
M. truncatula	906		339	5642
M. truncatula × *M. littoralis*[6]	38			
M. turbinata	217		97	255

Note: *As discussed in the text, the two dozen species indicated with an asterisk (*) were until recently almost universally considered to belong to genera other than *Medicago*.

[1]As discussed in the chapter on *M. bonarotiana*, many plants that are introgressants of *M. bonarotiana* and *M. rotata* have been traditionally placed in "*M. blancheana*", along with *M. bonarotiana*.

[2]As discussed in the chapter on *M. ciliaris*, this taxon is often considered to be a variety of *M. intertexta*.

[3]As discussed in the chapter on *M. citrina*, this taxon is often considered to be a variety of *M. arborea*.

[4]The name *M. murex* as used for current germplasm collections is often ambiguous, referring either to this species or to *M. sphaerocarpos*. Identifications are very likely correct in SARDI.

[5]The taxonomy of *M. rigidula* is problematical. Collections with this name may also be *M. rigiduloides*. The relationships with the rare species *M. sinskiae* and *M. syriaca* also are unclear.

[6]cf. *M. littoralis* × *truncatula*.

OTHER NOTABLE GERMPLASM COLLECTIONS AND DATABASES

The International Center for Agricultural Research in the Dry Areas (ICARDA), which has its main center in Alepo, Syria, maintains a collection of Mediterranean and temperate forage legumes. The largest collection is of *Medicago*, with about 8500 accessions (Bounejmate et al. 1999, Meglič et al. 2003).

The European Cooperative Programme for Crop Genetic Resources (ECPGR) maintains European, crop-specific databases, each managed by one of the participating institutes. The European perennial *Medicago* database is maintained by Groupe d'Étude et de Contrôle des Variétés et des Semences (GEVES). This organization coordinates the French national network on fodder-crop genetic resources with the Genetic Resources Board (BRG), the National Institute for Agronomic Research (INRA), and the Fodder Crops Breeders Association (ACVF). The *Medicago* database, representing 23 European gene banks, has passport data for about 3000 accessions, almost all of which are *M. sativa*. The data are available at http://www.ecpgr.cgiar.org/databases/Crops/Medicago.htm; accessed December 2008).

For many years, K. Lesins of the University of Alberta (see Lesins and Lesins 1979) collected *Medicago* seeds; his collection was given to the USDA (see GRIN) and shared with SARDI on joint collection trips. Similarly, C. Heyn conducted studies of *Medicago* (see Heyn 1969) at the Hebrew University of Jerusalem, and assembled a substantial collection of annual *Medicago*, much of it duplicated in the USDA collection.

GERMPLASM OF PERENNIAL *MEDICAGO*

Prosperi et al. (2001) presented an analysis of the conservation status of most of the perennial species of *Medicago*. Except for *M. sativa* subspecies *sativa* and *falcata*, they assessed most taxa as "endangered species, endemic or rare, few accessions in gene banks", and a few as "species with threatened diversity due to habitat erosion or occur in regions threatened by climate change, few available accessions".

Of the perennial species of *Medicago*, most germplasm collections are of *M. sativa*, and thousands of accessions are available. Germplasm acquisitions of *M. sativa* outnumber holdings of all other perennial *Medicago* species combined by more than 20 to 1. Although not reflected in Table 3-1, the infraspecific taxa of *M. sativa* recognized in this monograph are reasonably available, and so are good representations of landraces and cultivars. On the whole, germplasm of *M. sativa* for the improvement of alfalfa is not a particular concern. Dozens of collections of *M. arborea* are listed in the above table and, in fact, hundreds of accessions can be procured, suggesting that germplasm of this useful grazing and ornamental species is also reasonably available. However, the rare relative of *M. arborea*, *M. citrina*, and the very rare endemic relative *M. strasseri*, are poorly represented in gene banks. There are numerous collections of the annual–biennial–perennial *M. lupulina*. Collections of the common *M. marina* are moderately available. *Medicago ruthenica* has been well collected. Targeted expeditions by the USDA (see GRIN) have produced significant but small collections of *M. papillosa* and *M. platycarpos*. Other perennials are generally very poorly represented in gene banks, including *M. archiducis-nicolai*, *M. cancellata*, *M. carstiensis*, *M. cretacea*, *M. daghestanica*, *M. edgeworthii*, *M. hybrida*, *M. popovii*, *M. prostrata*, *M. rhodopea*, *M. rupestris*, *M. saxatilis*, and *M. suffruticosa*. Clearly, two factors have been responsible for the frequency of germplasm collections of perennial *Medicago* species: agricultural usage and the extent of geographical occurrence. All the perennial species widely used in agriculture (*M. sativa*, *M. lupulina*, and *M. arborea*) have been well collected. These species, along with *M. marina*, are widely distributed. The remaining species, which are only occasionally employed directly in agriculture and generally have narrow distribution ranges, have received limited attention, except for some targeted collection (largely by the USDA) of species for possible use in alfalfa breeding or for possible development of new forages. Many of the poorly collected perennials occur in mountainous areas, largely in Asia, and collection has been limited for geopolitical reasons. Climate change may represent a particular threat for mountain-dwelling, perennial species of *Medicago* since local circumstances may prevent migration to cooler areas necessary for long-term survival.

In situ conservation measures have been enacted for *M. citrina* more so than for any other species of *Medicago* (see Chapter 25). *Medicago hybrida* is protected in several areas of France. Also see the chapters on *M. rupestris* (Chapter 83) and *M. saxatilis* (Chapter 86) for examples of conservation analyses and recommended measures for preservation of natural populations in the Ukraine.

The perennial species of *Medicago* are of particular importance because many can be hybridized with *M. sativa* and, therefore, can serve as potential gene sources for cultivar improvement. These species are also themselves potential sources of new forages. Some key agriculturally important information about these species is presented in Table 3-2. This geographical and ecological information serves as a crude guide to agricultural potential.

Table 3-2. Synopsis of key agricultural information for perennial taxa of *Medicago*.

Taxon	Cultivars	Ecology	Chromosome number (2*n*)
M. arborea	A few forage cultivars; ornamental selections also available	Hills and rocky places, generally in foothills along shores; mainly on limestone and volcanic soils	32
M. citrina	None	Same as *M. arborea*	48
M. archiducis-nicolai	None	Valleys of rivers and on humusy soil on slopes in spruce forests; altitudes 3000–4000 m	16
M. cancellata	None	Infertile (often sandstone) dry, rocky or sandy soils of steppe and semi-desert areas	48
M. carstiensis	None	Rocky wooded areas; thrives in shade; uncommon. The only *Medicago* always with rhizomes	16
M. cretacea	None	Dry, calcareous (limestone) soils; lower mountain zone	16
M. daghestanica	None	Calcareous (limestone) taluses and rocky soils; submontane	16
M. edgeworthii	None	On mountains and stony pastures; frost resistant and tolerant of poor soils	16
M. hybrida	None	In grassy glades on limestone slopes; montane; adapted to high humidity of the Atlantic Ocean	16
M. marina	None	Almost always in loose sand very close to the sea	16
M. papillosa subsp. *macrocarpa*	None	Infrequent; dry, rocky alpine slopes, cliffs, talus slopes, and mountain pastures; soils of volcanic origin or calcareous	16, 32
M. papillosa subsp. *papillosa*	None	As above	16, 32
M. pironae	None	Rocky hillsides, submontane	16
M. platycarpos	None	Mountain regions, river valleys and meadows; moist soils, sometimes in open forests; found at higher latitudes towards the south; podzolic steppe soils, rocky soil, and sand; extends farther north (to 69E in the Yenisei River valley of Siberia) than any other species of *Medicago*	16
M. popovii	None	Clayey and stony slopes of upper tree–shrub zone, and in meadows in subalpine zone, extending upwards to 3500 m on southern slopes	16
M. prostrata subsp. *prostrata*	None	Rare; submontane, dry rocky limestone hillsides and grassy sites; sandy soils	16, 32
M. prostrata subsp. *pseudorupestris*	None	As above	?
M. rhodopea	None	Calcareous rocky sites in lower mountains	16
M. rupestris	None	Calcareous (limestone), rocky areas, lower mountains	16
M. ruthenica	None	Sunny hillsides, dry stony slopes, embankments, mixed grass steppes, and moist meadows, at times in large numbers; on sandy and stony soils, chiefly chernozem and chernozem-like steppe soils, including limestone	16
M. sativa subsp. *caerulea*	None, but germplasm in many cultivars	Dry to very dry areas, semi-deserts, dry slopes, sandy and loamy sands, rocky and clayey plains, in a variety of habitats	16
M. sativa subsp. *glomerata*	None, very rarely used in breeding	Submontane areas	16, 32
M. sativa subsp. *falcata* var. *falcata*	A few cultivars (e.g., 'Anik')	Boreal, dry steppes of Siberia, occasionally in moist areas; numerous ecotypes	16 (mostly), 32
M. sativa subsp. *falcata* var. *viscosa*	None	Particularly in submontane Caucasus	16 (mostly), 32

Table 3-2 (*Concluded*).

Taxon	Cultivars	Ecology	Chromosome number (2*n*)
M. sativa subsp. *sativa*	Includes most alfalfa cultivars	In a variety of generally open habitats with moderately fertile soil; numerous ecotypes	32
M. sativa subsp. *sativa* × subsp. *glomerata*	None	Hybrids particularly found in submontane Caucasus	32
M. sativa subsp. ×*varia*	Includes many alfalfa cultivars	In a variety of generally open habitats with moderately fertile soil; often weedy; numerous ecotypes	16, 32
M. saxatilis	None	Dry, rocky, calcareous (limestone), mid-mountain slopes	48
M. strasseri	None	Colonizes crevices of steep limestone cliffs in gorges	32
M. suffruticosa subsp. *leiocarpa*	None	Rocky places, montane; may be in moist soils; adapted to high humidity of Atlantic Ocean	16
M. suffruticosa subsp. *suffruticosa*	None	As above, allegedly at higher altitudes and in moister soils	16

Medicago species that have proven most useful for hay or grazing are all quite weedy, and it seems that weediness is a strong predictor of agronomic success. *Medicago sativa* is the only notably weedy perennial species of the genus, suggesting that the remaining perennials have limited potential. However, lack of experimental material of the wild perennials and ignorance of their properties have prevented assessment of their true potential. In any event, the perennials represent sources of genes for alfalfa improvement. *Medicago marina, M. arborea*, *M. citrina*, and *M. strasseri* are the best examples of *Medicago* perennials that are natural occupants of the dry Mediterranean climate. They are the only species that have perennial, aboveground shoots, not just perennial roots, crowns, or rhizomes, and are unsuitable for areas with low, freezing winter temperatures. Of these, *M. arborea* has some promise as a dryland range plant in Mediterranean areas. Fifteen species of perennial *Medicago* are refugial montane or submontane endemics: fourteen in Eurasia and one in northern Africa. Most of these occur regularly on dry, rocky, infertile, calcareous soils, and therefore have genes useful for Mediterranean rangelands. However, their very narrow distributions suggest that they possess corresponding narrow ecological tolerance ranges and, accordingly, most of these species would probably be directly usable only in pastures and rangelands that provide appropriate habitats.

"Tolerance to stress" is key to the agronomic value of a species, either for use directly as a crop or for gene transfer to a crop. Particularly important stresses that plants must endure include temperature, water availability, and edaphic factors (especially pH and ionic concentrations). From both an ecological and an agronomic perspective, tolerance of drought is the most important practical consideration. However, different factors contribute to drought tolerance in *Medicago*, and the issue is related to tolerance to soil infertility, soil salinity, and cold. Consequently, a wide variety of germplasm needs to be considered for improving adaptation through gene transfer.

GERMPLASM OF ANNUAL *MEDICAGO*

Because *M. truncatula* is an important source of cultivars and has been studied as a model legume for genome analysis, particularly large germplasm collections are available.

Small and Brookes (1984*c*) transferred *Factorovskya aschersoniana* to *M. hypogaea*, and Small et al. (1987) transferred 23 species of *Trigonella* (the so-called "medicagoid *Trigonella*") to *Medicago*. None of these annual species has been developed as an agricultural plant. Indeed, it appears that because these species have not been considered to belong to *Medicago* until recently, their collection as germplasm has been inhibited. For most of these species, seeds are completely unavailable or only a few accessions are available (notice in Table 3-1 that for the 24 species with an asterisk, there are good seed collections only for *M. monspeliaca* and *M. monantha*). *Medicago hypogaea* is a widespread plant, but very inconspicuous and quite ephemeral, and its underground fruits are difficult to acquire. In Israel, there are a few extremely small conservation areas where it is conserved *in situ*.

Most of the remaining annual species are spiny-fruited, or at least have populations with spiny fruits, an adaptation for dispersal that is usually associated with wide distribution. All the dozen or so annual species that are sown are widely distributed, and for each of these usually spiny-fruited, annual species, extensive seed collections are deposited in seed banks.

DEGRADATION OF THE ENVIRONMENT AND ASSOCIATED REDUCTION AND EXTERMINATION OF GERMPLASM RESOURCES *IN SITU*

Ideally, all germplasm would be maintained in nature, which is capable of conserving genes far more extensively than can germplasm banks. However, because of agricultural exploitation and urbanization, many of the world's species are threatened. In particular, many of the wild relatives of

Fig. 3-3. Photographs showing grazing pressure exerted on *Medicago* by sheep and goats in Turkey (Photos by E. Small in 1981). **Upper left:** Excavated, heavily grazed alfalfa plants, 26 km northeast of Van. **Upper right:** Sheep and goat trails on mountain slopes southeast of Sarikamiş, typical of a heavily grazed landscape. **Lower left:** Prostrate ecotype of alfalfa in dry clay soil, 10 km north of Dogubayazit. Grazing pressure favors the evolution of such low-growing forms. **Lower right:** Sheep and goats decimating plant cover near Marsden, Turkey. These grazers are allowed to roam freely and frequently eat all herbaceous vegetation down to ground level. The collection of germplasm from such heavily grazed areas is useful for finding genotypes that are resistant to continuous grazing pressure (pasture type), but less useful for finding genotypes suitable for the occasional harvest of large amounts of biomass (hay type) or for rotational grazing.

crop plants have undergone genetic erosion. The germplasm problem facing the majority of forage and fodder species is illustrated by Hardin's (1968) well known analysis of "The Tragedy of the Commons". Hardin discusses the global problem raised when human population growth causes the Earth's natural resources to be used excessively, and compares this with a pasture shared in common by local herders. He pointed out that each individual shepherd benefited by increasing the number of his own animals on the pasture. However, this inevitably results in every shepherd acting selfishly, and the pasture becoming so overgrazed and degraded that it can no longer be used by anyone. Unfortunately, numerous species of *Medicago* have had their natural ranges reduced by uncontrolled grazing. Many of the annual species have become weeds, distributed well beyond their natural ranges. However, all the perennials, as well as the narrowly distributed endemic annual species, are threatened in their indigenous areas. *Medicago arborea* is exceptional for the genus in that it is widely cultivated as an ornamental shrub so at least part of its germplasm base is being maintained as living, *ex situ* collections. For all other species of *Medicago*, seed collections are essential.

Very alarming destruction of much of the natural range of wild alfalfas has occurred in Eurasia as a result of poorly

Fig. 3-4. Graveyards in Turkey (photos by E. Small in 1981). Because they are often protected against grazing livestock, graveyards in much of the Middle East are refugia for *Medicago* species, and extremely important sources of germplasm.

controlled livestock grazing. In Turkey, for example, there are about two hoofed, domesticated grazers for every human, and this has resulted in severe pressure on wild alfalfa (Small 1981*b*, 1982). Most of the wild perennial species of *Medicago* have limited grazing resistance, certainly much less than many cultivars of domesticated alfalfa, which has been very highly selected for its tolerance to grazing (in some parts of the world, ten harvests per year are possible). It is critical that the habitats of the wild species be protected and, at the very least, efforts are necessary to collect and preserve germplasm of the uncommon species.

CORE COLLECTIONS

The large size of some genetic resource collections has led to concerns about their manageability. "Core collections" are subsamples of manageable size, defined by statistical procedures to be as representative as possible of the total genetic diversity in the whole collection, while minimizing allelic repetition (Brown 1989*a*, *b*). For the USDA gene bank (GRIN), core collections of *Medicago* have been made for the perennials (Basigalup et al. 1995) and for the annuals (Diwan et al. 1994, 1995). The annual *Medicago* core collection was grown at seven locations in the USA where the accessions were evaluated for 11 traits (days to flower, days to full pod stage, full pod stage, growth habit, biomass within species, biomass among species, pod production, pod spininess, plant height, plant width, and winter hardiness) with all evaluation data available in the GRIN system (http://www.ars-grin.org). For the SARDI collection, a core collection has been made for the annual *Medicago* (Skinner et al. 1999). Because *M. truncatula* is a model genome species it has received special attention. Core collections for *M. truncatula* have been defined for France (GEVES) by Ronfort et al. (2006) and for the GRIN collection by Han et al. (2007). A key problem with these core collections is their dependence on the taxonomic classifications upon which the accessions were identified. Several of the core collections concerned with *Medicago* are based on classifications that are inconsistent with those presented in this monograph. This could lead to under-representation of taxa when it has not been understood that an older name includes more than one species (e.g., "*M. murex*" is now understood to include two species, *M. murex* and *M. sphaerocarpos*). It could lead to a distorted representation of species for which a name has not accurately been interpreted to represent a given species (e.g., "*M. blancheana*" is a mixture of *M. bonarotiana* and hybrids between *M. bonarotiana* and *M. rotata*). As discussed in Chapter 12, many traditionally defined taxa (subspecies, varieties) of *M. sativa* are based on natural hybrids so core collections based on these may fail to emphasize the most different kinds of plant that are available.

4. Biogeography

Medicago is a genus of Eurasian origin and distribution, although some species have been introduced in many regions of the world. Only the (presumed or apparent) natural, indigenous distributions are considered in the following analyses, although defining the natural areas is problematical for those species of *Medicago* that are widely distributed. Determining the natural distribution of *Medicago sativa* is particularly difficult because cultivated forms (alfalfa) are sometimes indistinguishable from wild or ruderal forms, and for millennia humans have extended the distribution of this species.

In the following geographical analyses, annuals and perennials are considered separately. *Medicago lupulina* is encountered as an annual, biennial, and perennial. *Medicago brachycarpa* has been reported to occur sometimes as a perennial, sometimes as an annual. *Medicago edgeworthii* has the capacity to grow as an annual and a perennial in greenhouses, and may also do the same in nature. These three species, which exhibit flexibility in their longevity, are treated as annuals in the analyses presented below.

In the following geographical analysis, the 87 species of *Medicago* recognized in this monograph are split into three groups: strict perennials (21 species), annuals distinguished by cotyledons with "mobile" or "flex" joints at their bases (a group of 24 species, most of which have previously been referred to as the "medicagoid *Trigonella*"), and the remaining annuals (42 species). The evolutionary relationships of *Medicago* species and the organization of the species into natural groups (sections and subsections) are detailed in Chapter 10. The group of 24 annuals is almost certainly a natural phyletic group. The most important morphological character for taxonomic separation of these annuals is the presence of "pulvinate" cotyledons. Pulvini are joints at the base of leaves (usually leaf blades) that control leaf movement (for example, for orientation to light). *Medicago* leaves are trifoliolate, i.e., with three leaflets, and have pulvini (or "pulvinuli") at the base of each leaflet. However, the first seedling leaves (the cotyledons) lack pulvini in 12 of the sections of *Medicago* recognized in this monograph, but are present in *Medicago* sections *Lunatae* and *Buceras* (as noted above, all species in these sections are annuals, although two are sometimes also perennial) and also in the closely related genera *Trigonella* and *Melilotus*. While it seems clear that the 24 species with pulvinate cotyledons represent an isolated phyletic lineage, it is probable that the 21 perennial species collectively, and the remaining 42 annual species collectively, each contain several independent lineages. Nevertheless, it is useful to consider the non-pulvinate-cotyledon annuals and perennials as different classes for geographical examination because they are dramatically different ecologically. Accordingly, for purposes of biogeographical analysis, three groups are examined in detail: the non-pulvinate-cotyledon perennials, the non-pulvinate-cotyledon annuals, and the pulvinate-cotyledon annuals.

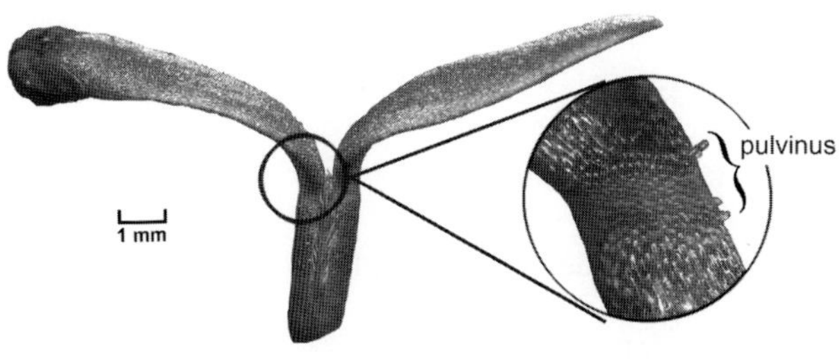

Fig. 4-1. A *Medicago edgeworthii* seedling, highlighting the cotyledonary pulvinus (flex joint) region. (Based on seed from the United States Department of Agriculture, accession code USDA, ACS19960, ACNO5.)

In the following analyses, approximate centers of distribution (to the nearest degree latitude and degree longitude) have been identified for all species. Also, the number of 5° latitude × 5° longitude quadrangles occupied has been found, as an indication of the relative areas occupied (compare with Holland (1978), who used 2.5° latitude × 2.5° longitude quadrangles). This was based on examining the presence in all quadrangles formed in the Old World, commencing east and west from the Prime Meridian, and north and south from the Equator. This areal measure is somewhat inaccurate for endemics, since merely straddling a meridian or parallel doubles the indicated area. In any event, narrow endemics are easily identified by occupation of one or two quadrangles. The areal analyses in the following are updated from Small and Lefkovitch (1986) and Small (2003), incorporating additional distributional data now available.

THE PERENNIAL SPECIES

The classification of the perennial species follows. As noted, the defined sections are distinguishable to an extent by their geographical distributions.

Classification of Perennial *Medicago* Species

A. Section *Dendrotelis*
  1. *M. arborea*
  2. *M. citrina*
  3. *M. strasseri*

B. Section *Medicago* subsection *Medicago*
  4. *M. cancellata*
  5. *M. cretacea*
  6. *M. daghestanica*
  7. *M. marina*
  8. *M. papillosa*
    a. *M. papillosa* subsp. *papillosa*
    b. *M. papillosa* subsp. *macrocarpa*
  9. *M. pironae*
  10. *M. prostrata*
    a. *M. prostrata* subsp. *prostrata*
    b. *M. prostrata* subsp. *pseudorupestris*
  11. *M. rhodopea*
  12. *M. rupestris*
  13. *M. sativa*
    a. *M. sativa* subsp. *sativa*
    b. *M. sativa* subsp. *caerulea*
    c. *M. sativa* subsp. *sativa* × subsp. *glomerata*
    d. *M. sativa* subsp. *glomerata*
    e. *M. sativa* subsp. *falcata* var. *falcata*
    f. *M. sativa* subsp. *falcata* var. *viscosa*
    g. *M. sativa* subsp. × *varia*
  14. *M. saxatilis*

C. Section *Medicago* subsection *Suffruticosae*
  15. *M. hybrida*
  16. *M. suffruticosa*
    a. *M. suffruticosa* subsp. *suffruticosa*
    b. *M. suffruticosa* subsp. *leiocarpa*

D. Section *Carstiensae*
  17. *M. carstiensis*

E. Section *Platycarpae*
  18. *M. archiducis-nicolai*
  19. *M. platycarpos*
  20. *M. popovii*
  21. *M. ruthenica*

The following table (4-1), figure (4-2), and distribution map (Fig. 4-3) show the centers of distribution of the 21 perennial species of *Medicago*, as well as the perimeter of their collective area of distribution (for more details, see Small 2003). Key geographical information for the species is also given in the table.

Table 4-1. Indigenous distributional information for the 21 perennial species of *Medicago*.

Taxon[1]	Distribution	Approximate center of the species	Relative area of the species[2]
M. arborea (1)	Coasts and islands of Mediterranean sea (mostly the northern area) and the Canary Islands	39°N, 10°E	13
M. citrina (2)	Narrow endemic: confined to a few Spanish islets close to Eivissa and Cabrera (Balearic Islands) and off the coast of Valencia Province (Columbretes Islets)	40°N, 1°E	2
M. archiducis-nicolai (18)	Central China (including northeast Tibet)	35°N, 103°E	6
M. cancellata (4)	North Caucasus and Caspian region (Ciscaucasia, Don Basin, Lower Volga)	45°N, 45°E	5
M. carstiensis (17)	Northeast Italy, southeast Austria, Yugoslavia, Albania, and Bulgaria	45°N, 19°E	7
M. cretacea (5)	Endemic to Crimea (Ukraine) and Transcaucasia (Russian Federation)	43°N, 39°E	5
M. daghestanica (6)	Endemic to Caucasus, west of Caspian Sea (Azerbaijan, Checheno-Ingushetia, Dagestan)	42°N, 47°E	1
M. edgeworthii	Himalayas: northwest Pakistan, northwest India, Kashmir, Nepal; China: Xizang (Gyangtse). Related forms in Afghanistan	33°N, 85°E	14
M. hybrida (15)	Southern France (Languedoc, Corbières, from eastern Pyrenées almost to the Rhone River)	43°N, 2°E	1
M. marina (7)	Mediterranean and Black Sea shores; Atlantic coasts of Iberia and France	38°N, 17°E	33
M. papillosa (8)	(See Fig. 62-6 for distribution map of subspecies)	40°N, 42°E	6
M. papillosa subsp. *macrocarpa*	Northeast Turkey (west of longitude 40°)		
M. papillosa subsp. *papillosa*	Northeast Turkey (east of longitude 40°) and in Transcaucasia (northern Armenia, southern Georgia)		

Table 4-1 (*Concluded*).

Taxon[1]	Distribution	Approximate center of the species	Relative area of the species[2]
M. pironae (9)	Endemic: Alps of northeast Italy, Austria, Yugoslavia	47°N, 15°E	2
M. platycarpos (20)	Widespread but discontinuous. In Europe: USSR (Bashkiria). In Asia: Russian Federation (Siberia); Kazakhstan (northeastern mountain regions); Mongolia (northern); China (western Manchuria and in the south to the Peking-P'o Hua Shan mountains)	53°N, 94°E	33
M. popovii (21)	Central Asian endemic (Kyrgyzstan, Tajikistan): mid-mountains of west Tian-Shan and west Pamiro-Alai	40°N, 70°E	6
M. prostrata (10)		45°N, 19°E	5
M. prostrata subsp. *prostrata*	Italy and Austria to Greece		
M. prostrata subsp. *pseudorupestris*	Albania, Macedonia, and Balkans		
M. rhodopea (11)	Endemic: southern Bulgaria	42°N, 25°E	2
M. rupestris (12)	Local endemic: lower Crimean Mountains (Ukraine)	45°N, 36°E	5
M. ruthenica (22)	Asia: Russian Federation (Siberia, Ussuri and Amur regions, Sakhalin); Mongolia; northeast China (Manchuria); North Korea; Japan?	47°N, 110°E	32
M. sativa (13)	(See Fig. 12-21 for distribution map of subspecies)	45°N, 45°E	137
M. sativa subsp. *caerulea*	Eastern Turkey, Iran, Russia		
M. sativa subsp. *glomerata*	Caucasus and adjacent Russia and in mountains of southern Europe		
M. sativa subsp. *falcata* var. *falcata*	Widespread in northern Eurasia and widely introduced elsewhere		
M. sativa subsp. *falcata* var. *viscosa*	Sympatric with subsp. *glomerata*		
M. sativa subsp. *sativa*	Native of Eurasia, includes most cultivated and weedy alfalfa plants		
M. sativa subsp. *sativa* × subsp. *glomerata*	Common hybrid swarms between subsp. *sativa* and subsp. *glomerata*; occurs within the range of subsp. *glomerata*		
M. sativa subsp. ×*varia*	Widespread in and outside of cultivation (the hybrid of subsp. *falcata* with either subsp. *sativa* or subsp. *caerulea*)		
M. saxatilis (14)	Local endemic: Ukraine (Crimean Mountains)	45°N, 35°E	4
M. strasseri (3)	Narrowly endemic to Crete	35°N, 25°E	2
M. suffruticosa (16)		38°N, 3°W	5
M. suffruticosa subsp. *leiocarpa*	Nature of geographical, ecological, and morphological separation from subsp. *suffruticosa* below, requires clarification		
M. suffruticosa subsp. *suffruticosa*	Southern France (east Pyrenées, Corbières, Languedoc), northeast Spain (east Pyrenées to Valencia), Morocco (Atlas Mountains)		

[1]Species numbers correspond to those in "Classification of perennial *Medicago* species" (see above) and the distribution map (Fig. 4-3).
[2]Number of 5° latitude × 5° longitude quadrangles occupied.

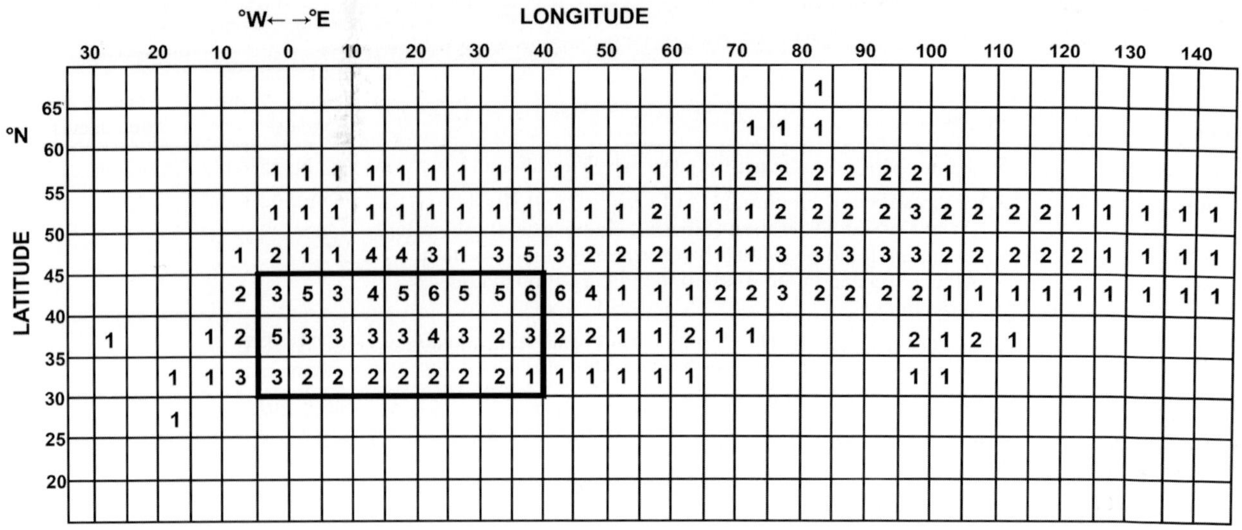

Fig. 4-2. The diversity of the strictly perennial species of *Medicago* in 5° latitude × 5° longitude quadrangles. Numbers of species in each sector are shown. The blocked area approximately includes the Mediterranean Circle countries.

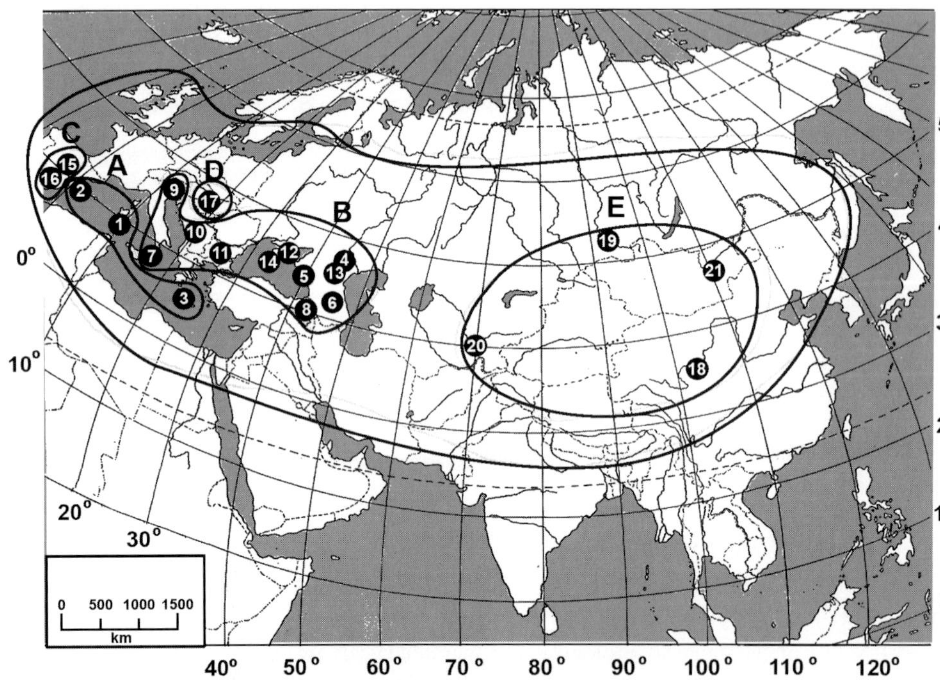

Fig. 4-3. Centers of distribution of the strictly perennial species of *Medicago*. The numbers correspond to the 21 species in "Classification of perennial *Medicago* species" (see above), the outermost perimeter indicates the natural distribution limits of the perennials, and the five perimeters labeled A–E within this correspond to the five taxonomic groups noted above.

Most of the perennial species of *Medicago* are very localized geographically and are legitimately represented by the centers of their distributions (note the relative distribution areas for the species in Table 4-1). *Medicago sativa* is the only very widespread species and its center is based on the presumed predomestication range. Notice that the five taxonomic groups recognized above (labeled A–E) are generally separable geographically (at least on the basis of centers of distribution). The historical biogeographical significance of this pattern is discussed below.

THE ANNUAL SPECIES WITHOUT PULVINATE (JOINTED) COTYLEDONS

The relative number of species in different areas is illustrated in Fig. 4-4 and the geographical information for the 42 annual species of *Medicago* lacking pulvinate (jointed) cotyledons is summarized in Table 4-2.

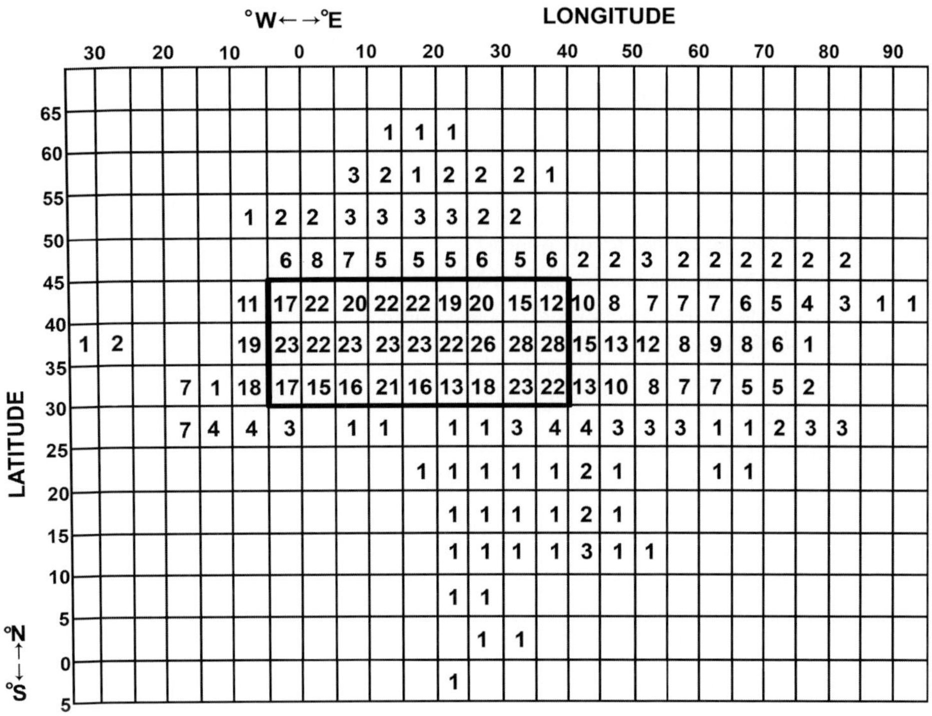

Fig. 4-4. Species diversity of the annual species of *Medicago* without pulvinate (jointed) cotyledons in 5° latitude × 5° longitude quadrangles. The numbers of species in each sector are shown. The blocked area approximately includes the Mediterranean Circle countries.

Table 4-2. Indigenous distributional information for the 42 annual species of *Medicago* without pulvinate (jointed) cotyledons (all have coiled pods albeit obscurely in a few of these).

Medicago	Distribution	Approximate center of the species	Relative area of the species[1]
M. arabica	Continental Europe, northern Africa, west Asia	42°N, 16°E	47
M. bonarotiana	Eastern Mediterranean	35°N, 35°E	4
M. ciliaris	Omni-Mediterranean	36°N, 9°E	30
M. constricta	Eastern Mediterranean	38°N, 37°E	15
M. coronata	Omni-Mediterranean	38°N, 26°E	29
M. disciformis	Omni-Mediterranean	39°N, 15°E	19
M. doliata	Omni-Mediterranean	38°N, 13°E	31
M. granadensis	Eastern Mediterranean	32°N, 36°E	7
M. heyniana	Southwestern Turkey, Crete, East Aegean Islands	37°N, 25°E	2
M. hypogaea	Southeastern Mediterranean	33°N, 31°E	7
M. intertexta	Western Mediterranean	37°N, 10°E	25
M. italica	Western and extreme eastern Mediterranean	38°N, 10°E	19
M. laciniata	Southern Mediterranean Circle, northern Africa	30°N, 35°E	70
M. lanigera	Asia northeast of Iran	34°N, 64°E	7
M. laxispira	Northern Iraq	36°N, 43°E	1
M. littoralis	Omni-Mediterranean and adjacent regions	38°N, 20°E	35
M. lupulina	Eurasia, North Africa	43°N, 31°E	93
M. minima	Eurasia, North Africa	38°N, 25°E	91
M. murex	Omni-Mediterranean	38°N, 10°E	24

Table 4-2 (*Concluded*).

Medicago	Distribution	Approximate center of the species	Relative area of the species[1]
M. muricoleptis	Italy and Greece	39°N, 15°E	5
M. noeana	Anatolia, Syria, and Iraq	38°N, 35°E	4
M. orbicularis	Omni-Mediterranean, Asia	38°N, 35°E	56
M. ovalis	Northwestern Africa, southern Spain	33°N, 5°W	6
M. plicata	South-central Anatolia	38°N, 35°E	2
M. polymorpha	Eurasia, North Africa	39°N, 30°E	80
M. praecox	Northern and eastern Mediterranean	40°N, 19°E	21
M. radiata	Eastern Mediterranean to central Asia	39°N, 50°E	22
M. rigidula	Omni-Mediterranean (North Africa and Europe) (see Chapter 78)	38°N, 10°E	ca. 25
M. rigiduloides	Asia (see Chapter 79)	38°N, 55°E	ca. 25
M. rotata	Eastern Mediterranean	36°N, 41°E	5
M. rugosa	Omni-Mediterranean	38°N, 15°E	21
M. sauvagei	Morocco	33°N, 8°W	1
M. scutellata	North Africa, southern Europe, eastern Mediterranean	40°N, 15°E	32
M. secundiflora	North Africa, southern Europe	36°N, 4°E	9
M. shepardii	South-central Anatolia	38°N, 35°E	1
M. sinskiae	Southwestern Turkmenistan; northern Iran?	37°N, 57°E	1 (2?)
M. soleirolii	Western Mediterranean	38°N, 10°E	6
M. sphaerocarpos	Omni-Mediterranean	38°N, 10°E	24
M. syriaca	Syria, Iran	36°N, 37°E	1
M. tenoreana	Northwestern Mediterranean	41°N, 12°E	8
M. truncatula	Omini-Mediterranean	37°N, 20°E	39
M. turbinata	Omni-Mediterranean	37°N, 13°E	27

Note: A question mark (?) indicates that the presence in a location needs confirmation.
[1]Number of 5° latitude × 5° longitude quadrangles occupied.

THE ANNUAL SPECIES WITH PULVINATE (JOINTED) COTYLEDONS

The relative number of species in different areas is illustrated in Fig. 4-5 and the geographical information for the 24 annual species of *Medicago* with pulvinate cotyledons is summarized in Tables 4-3 and 4-4.

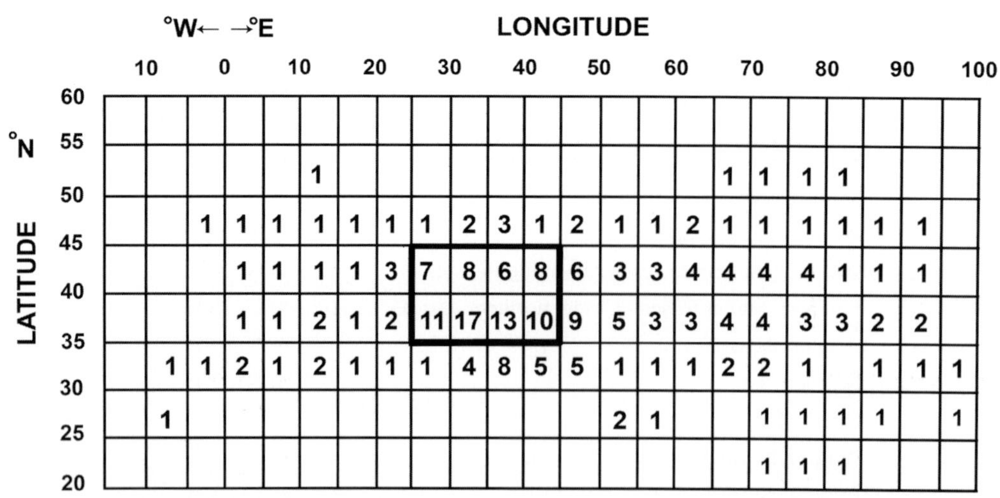

Fig. 4-5. The diversity of the annual species of *Medicago* with pulvinate (jointed) cotyledons, in 5° latitude × 5° longitude quadrangles. The numbers of species in each sector are shown. The blocked area approximately includes Anatolia.

Table 4-3. Countries (arranged from west to east) occupied by the annual species with pulvinate (jointed) cotyledons (*Medicago* sections *Buceras* and *Lunatae*).

Medicago species	North Africa	Europe	Anatolia	Iraq, Israel, Jordan, Lebanon, Qatar, Saudi Arabia, Syria, and/or United Arab Emirates	Iran	Armenia, Azerbaijan, Gruziya, Kazakhstan, Kyrgyzstan, Russia in Asia, Tadzhikistan, Turkmenistan, and/or Uzbekistan	Afghanistan, Bhutan, India, and/or Pakistan	Mongolia, China, and/or Nepal
M. arenicola			×					
M. astroites			×	×	×	×		
M. biflora			×	×	×	×		
M. brachycarpa			×	×		×		
M. carica			×					
M. crassipes			×	×	×			
M. edgeworthii							×	×
M. fischeriana		×	×	×	×	×		
M. halophila			×					
M. heldreichii			×					
M. huberi			×					
M. isthmocarpa			×					
M. medicaginoides		×	×		×	×		×
M. monantha			×	×	×	×	×	×
M. monspeliaca	×	×	×	×	×	×	×	
M. orthoceras		×	×	×	×	×	×	×
M. pamphylica			×					
M. persica				×	×			
M. phrygia			×	×	×			
M. polyceratia	×	×		×				
M. retrorsa							×	
M. rhytidiocarpa			×					
M. rigida			×					
M. rostrata			×					

Table 4-4. Indigenous distribution information for the 24 *Medicago* species with pulvinate cotyledons (*Medicago* sections *Buceras* and *Lunatae*).

Medicago	Distribution	Approximate center of the species	Relative area of the species[1]
M. arenicola	Known only from type. Antalya in Turkey	37°N, 31°E	1
M. astroites	Eastern Mediterranean to Iran	38°N, 40°E	10
M. biflora	Middle East, southern Transcaucasia	38°N, 37°E	10
M. brachycarpa	Middle East and Transcaucasia	37°N, 42°E	10
M. carica	Southwestern Anatolia, East Aegean Islands	37°N, 29°E	1
M. crassipes	Eastern Mediterranean	37°N, 39°E	10
M. edgeworthii	Himalayas: NW Pakistan, NW India, Kashmir, Nepal; China: Xizang (Gyangtse). Related forms in Afghanistan	33°N, 85°E	15
M. fischeriana	Southeastern Europe, southwestern Asia	42°N, 36°E	9
M. halophila	Southern Anatolia	37°N, 35°E	2

Table 4-4 (*Concluded*).

Medicago	Distribution	Approximate center of the species	Relative area of the species[1]
M. heldreichii	Southern Anatolia	31°N, 37°E	1
M. huberi	Southwestern Anatolia	38°N, 29°E	1
M. isthmocarpa	Local endemic of central Anatolia	38°N, 35°E	2
M. medicaginoides	Southeastern Europe to central Asia	39°N, 60°E	41
M. monantha	Asia, from Anatolia to China and India	33°N, 58°E	33
M. monspeliaca	Western Europe to central Asia	40°N, 25°E	53
M. orthoceras	Anatolia to western China	40°N, 55°E	22
M. pamphylica	Endemic to southern Anatolia	37°N, 31°E	1
M. persica	Southwestern Iran, adjacent Iraq	30°N, 50°E	2
M. phrygia	Anatolia to Iran	35°N, 40°E	15
M. polyceratia	Southwestern Europe, northwestern Africa	37°N, 7°E	7
M. retrorsa	Afghanistan	31°N, 69°E	1
M. rhytidiocarpa	Southern Anatolia	39°N, 35°E	1
M. rigida	Southern Anatolia	37°N, 35°E	2
M. rostrata	Central and southern Anatolia	38°N, 31°E	2

[1]Number of 5° latitude × 5° longitude quadrangles occupied.

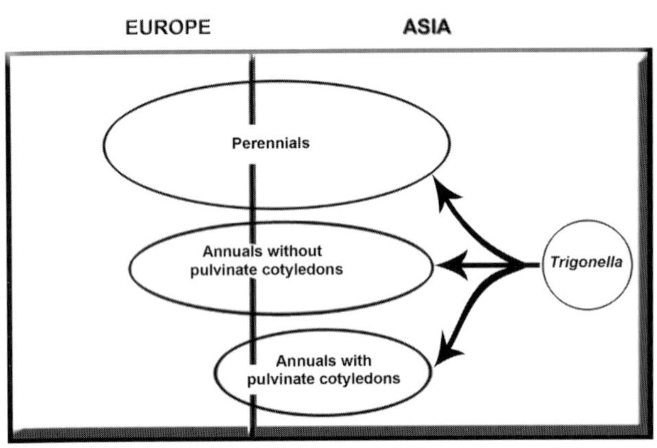

Fig. 4-6. A conception of biogeographical relationships among the perennials and the two annual groups of *Medicago*, and the genus *Trigonella*. The group of pulvinate-cotyledon annuals appears to be a phyletic lineage, but the remaining annuals and perennials are probably not. They are, however, ecologically and geographically differentiated, with the annuals tending to occupy drought-prone lowland areas, often in relatively southern areas, while the perennials tend to be in upland and mountainous areas, often in relatively northern areas.

As indicated above, 20 of the 24 species with pulvinate cotyledons occur in Anatolia; nine of these are extremely narrowly distributed endemics occurring only there, and one occurs only in Anatolia and in the adjacent Aegean Islands. One narrow endemic is found only in southwestern Iran and adjacent Iraq; another narrow endemic occurs only in Afghanistan.

BIOGEOGRAPHICAL RELATIONSHIPS AMONG THE PERENNIAL AND ANNUAL GROUPS

As reviewed by Cain (1944), one of the considerations used by biogeographers in assessing the likely center of origin of a genus is the postulate that the center should possess a greater variety of forms than may be expected in areas on the edge of the range. This assumes that only one center of origin exists, that polymorphism increases with time, and that the different evolved forms persist in the occupied region. As noted by Cain and others, there are difficulties with this interpretation, although it is not without some foundation. The frequency distributions of species of *Medicago* presented above indicate that it is most diverse in the area of the Mediterranean Sea, particularly around the northern and eastern shores.

The genus most closely related to *Medicago* is undoubtedly *Trigonella*, which is centered in central Asia. The group of annual species with pulvinate cotyledons, discussed above, is centered in Anatolia, and as the species in this group are distinctive from all other *Medicago*, and bear features indicating a relatively close relationship to *Trigonella*, it seems probable that the group is a biogeographical-phyletic offshoot of *Trigonella*. The group of perennial species tends to occur in upland and mountainous areas, often in relatively northern areas. The group of annual species lacking pulvinate cotyledons tends to occupy drought-prone lowland areas, often in relatively southern areas. It is apparent that most *Medicago* species ended up centered in the Mediterranean area, compared with the relatively predominantly Asian distribution of the likely (closest) apparent progenitor, *Trigonella* .

HISTORICAL BIOGEOGRAPHY

Most species of *Medicago* are native to the Mediterranean; therefore, it is important to understand the evolution, history, and ecology of the region. However, the wide, present-day geographical distribution of *Medicago* in Eurasia, coupled with the complex geological and climatological history of the continent, make it extremely difficult to trace the biogeographical history of the genus.

Naveh and Vernet (1991) provided the following information: "The landscapes of the Mediterranean Basin are relatively young geological systems: they gained their present geomorphological forms by violent uplift in the late Tertiary and early Quaternary periods... Final shaping of the landscapes took place in the Pleistocene in a high dynamic period of climatic fluctuation, and tectonic and volcanic activity. Site conditions, both physical and biological, diversified increasingly. From the Middle Pleistocene onwards, early Mediterranean peoples further diversified these landscapes on both a regional and local scale, thereby contributing to their present multi-dimensional heterogeneity."

The Mediterranean area was generally warm at the beginning of the Tertiary Period (about 55 million years ago) and by about 6 million years ago the Mediterranean basin was a hot desert. By the Pliocene Epoch (beginning about 6 million years ago) a cooling trend began. The Pleistocene Era (beginning almost 2 million years ago and ending 15 000 years ago) witnessed cycles of glaciation interspersed with warmer periods. In addition to these climatic events, beginning in the Tertiary, much of eastern Europe and part of Asia were covered by the huge lake Paratethys (of which the remnants are thought to be the modern Black and Caspian Seas). The Mediterranean Sea itself was closed off at the Gibraltar end from the Atlantic Ocean about 7 million years ago, drying up completely in the hot climate for periods and being periodically inundated by occasional openings to the Atlantic at the Gibraltar end.

Heyn (1963) concluded that "The Mediterranean region is considered by several authors to have provided a refuge for plants from adjacent regions, especially Central and Western Asia, in periods when climatic conditions worsened." Di Castri (1991) stated that "The Mediterranean Basin is at the crossroads biogeographically, having been covered by repeated waves of invasion during successive latitudinal shifts of vegetation in glacial and interglacial periods."

It seems likely that *Medicago* species evolved and migrated during the period that these climatic and geological events occurred, but there is little information available to link most of the species to the historical patterns. Lesins and Lesins (1979) had the view that the annual species of the genus originated from perennial species at the time when the Straight of Gibraltar was closed, as a result of successive flooding and drying of the Mediterranean Sea. Certainly it is obvious that most of the perennial species migrated to upland areas, particularly mountainous areas, to survive, while almost all the annuals became specialized Mediterranean-climate winter annuals.

In considering the distribution of the perennial species, it was pointed out above that the four species of section *Platycarpae* are the easternmost of the perennials (Fig. 4-3). These have been placed in the past in the genus *Trigonella*, and do seem to provide a link to this genus, the closest to *Medicago*. The perennial species of *Trigonella*, like those of *Medicago* section *Platycarpae*, tend to occur in the mountains of central Asia (Vassilczenko 1953), and probably represent the lineage from which *Medicago* arose or diverged. Viewed in this light, *Medicago* likely split from *Trigonella* somewhere in central Asia and migrated westward. Leaving aside the species of section *Platycarpae*, the remaining 17 species of perennial *Medicago* are centered in the north Mediterranean region, and very likely arose in these environs. Similarly, the 65 annual species of *Medicago* are distributed particularly around the Mediterranean Sea (the 24 species with noncoiled fruit particularly in Anatolia).

While the present distributions of *Medicago* species usually are difficult to explain precisely in terms of historical geological events, there are some apparent geographical relationships in some cases. For example, as pointed out by Lesins and Lesins (1979) "the occurrence of *M. suffruticosa* in both the Pyrenées Mountains of the Iberian Peninsula and in the Atlas Mountains of North Africa adds support to the view that these continents had been connected in the past."

Increasing aridity has undoubtedly been the principal evolutionary stimulus for Mediterranean annual plants. Ehrman and Cocks (1996) discussed the relationship of aridity gradients to reproductive patterns in annual species, including some species of *Medicago*, in western Asia. They noted that the highest development of variation patterns occurs under mesic conditions in coastal, Mediterranean areas with a high winter rainfall, a low incidence of frost, and a long growing season. Species occupying more arid areas tend to have high levels of seed dormancy, a high seed to pod ratio, restricted dispersal capability, and early flowering — all findings known to characterize desert annuals. Aridity has also been a major factor in the evolution of annual species of *Medicago*." As a rule, Mediterranean annuals are principally self-fertilized" (Rossiter 1966), a phenomenon that is explained as "fertility insurance" because the periodic droughts and other disastrous occurrences such as insect plagues require plants that can propagate when pollinators are scarce.

HUMAN INFLUENCES ON THE DISTRIBUTION OF *MEDICAGO*

Landscape alteration by humans has been responsible for alterations in the success (or failure) of numerous plant species, and this certainly applies to *Medicago*. Before the advance of civilization resulted in clearing, man-made fires, cropping, and grazing, the Mediterranean region was dominated by trees and perennial shrubs, along with perennial grasses. As a result of human activities, resistant shrubs and numerous annual plants have come to dominate much of the landscape, especially in regions where grazing can occur. The number of dominant genera of annuals is remarkably restricted, including mostly *Bromus*, *Festuca*, *Hordeum*, *Erodium*, *Trifolium*, and *Medicago* (Rossiter 1966). In the Old World, The creation of weedy habitats, particularly roadsides, cleared agricultural fields, and settlements, has favored the spread of many of the annual species with weedy tendencies. On the other hand, grazing on public lands, particularly by sheep, has often had a disastrous effect on perennial species, which by their nature are more subject to being eliminated than the annuals. As well, since there are many very narrowly distributed endemic species of *Medicago* (both annual and perennial), the continuing natural existence of some species is threatened.

Heyn (1963) emphasized that several of the annual species of *Medicago* with relatively soft-walled fruits possessing hooked spines have become widely distributed outside of their native, mostly Mediterranean home because of the extensive trade in raw wool not cleaned of the attached pods. Such weedy species have become particularly established in four regions of the world with "Mediterranean-type climates" (in addition to most of the Mediterranean Basin). These regions are California, central Chile, southern Africa, and southwestern and southern Australia. All five regions are located between parallels 30° to 40° North and South. "A Mediterranean climate can be very broadly defined as a transitional regime between temperate and dry tropical climates, with a concentration of rainfall in winter and the occurrence of a summer drought of variable length" (di Castri 1991). This type of climate has been firmly established only since the Pleistocene (Axelrod 1973). The vegetation of these regions is characteristically composed of dominating woody shrubs, with occasional woodlands and frequent open areas with herbs (both annual and perennial).

An interesting case of how the biogeographical success of a species became tied to humans is provided by *Medicago scutellata*. Olivieri et al. (1991) pointed out that this species used to be common in southern France until the 20th century because it was associated with vineyards and cereal rotation. However, during the 20th century, land use in the region changed and the species has become far less common.

Frequency of use is a principal determinant of the occurrence of cultivated species outside their natural areas of distribution. Among the perennial species, only *M. sativa* is widely and deliberately introduced, and indeed it is the only perennial *Medicago* (with the exception of *M. lupulina*, a species that is occasionally perennial) that is considered invasive. *Medicago arborea* is fairly widely introduced in the Old World, but its invasive potential is quite limited. Successful field crops frequently appear to have first been aggressive weeds of open, disturbed habitats before they were domesticated, and weediness does seem to be an excellent indicator of the potential value of plants as new crops (Small 1984*b*). Indeed, all the annual species, from which forage cultivars have been bred, are considered to be aggressive weeds in some countries.

RELATIVE FREQUENCY OF RECOGNITION OF NEW SPECIES OF *MEDICAGO* FROM 1753 TO 2000, IN RELATION TO THE SIZE OF THE GEOGRAPHICAL AREAS OCCUPIED

As indicated in Table 4-5, from the time of Linnaeus (1753) to 1950, there was a general decrease in the number of new species discovered and relatively smaller areas were occupied (as one would expect, because undiscovered species tend to have narrower distributions). The relatively large number of species (11) discovered from 1951 to 2000 reflects the recognition of a number of comparatively poorly studied species whose status requires additional study. Also, the relatively large area of the species discovered from 1951 to 2000 (compared with the first half of the century) is based in part on the splitting of each of two species of wide distribution (most of the new species discovered during this period are very narrowly distributed endemics).

Table 4-5. Dates of recognition of the 87 species as separate species (regardless of past nomenclature) and relative areas occupied.

Interval	Number of new species recognized	Relative area occupied[1]
1753–1800	26	39.8
1801–1850	28	12.3
1851–1900	17	4.5
1901–1950	5	4.2
1951–2000	11	5.5

[1]Mean number of 5° latitude × 5° longitude quadrangles occupied.

ADAPTIVE CHARACTER BIOGEOGRAPHY: THE EXAMPLE OF GLANDULAR HAIRS IN *MEDICAGO*

Chapter 7 discusses the adaptive nature of a number of characters in *Medicago*, and it is pointed out that there is some evidence that characters that have been used taxonomically reflect adaptation to local geographical circumstances. The best example of this is the remarkably localized presence of glandular hairs in the Caucasus region.

Plants of the *Medicago sativa* complex (see Chapter 12 on *M. sativa*) often have fruits covered with protective gland-tipped hairs, which occur less frequently on the vegetative parts of the plant. Several annual species of *Medicago* possess glandular trichomes on the vegetative parts and to a lesser extent on the fruits, and the glandular hairs of these species have been shown to be protective against certain stem- and leaf-eating insect pests (Small 1985*b*). In the perennial *M. sativa* (and indeed in several other perennial species of *Medicago*), glandular hairs, when present in large numbers, are almost invariably concentrated on the fruits, suggesting that they may be adaptive against fruit (seed)-eating insects. Such plants are especially found in the Caucasus region, and extend in a submontane belt over the Mediterranean Sea, with some occurring in the mountains of North Africa. Small (1985*b*) concluded that, at least in the Caucasus, the most likely insects responsible for this apparent adaptation are seed-eating weevils of the beetle genus *Tychius*. The taxonomy of forms of *M. sativa* with heavy concentrations of gland-tipped hairs has been quite problematical. Small (1986*a*) recognized one main taxon (with both diploids and tetraploids): *M. sativa* subsp. *glomerata*. Natural hybridization of subsp. *glomerata* with subsp. *sativa* is common. A form of subsp. *falcata* in the Caucasus (subsp. *falcata* var. *viscosa*) is also characterized by high concentrations of gland-tipped hairs. With careful examination, a few gland-tipped hairs can be found on most plants of the *M. sativa* complex, and it seems obvious that increased density of glandular hairs has been selected in regions where they are especially needed. A similar geographic concentration of gland-tipped hairs occurs in *M. papillosa* of Armenia, where dense coverings of gland-tipped hairs are more common in the western part of its distribution range (Small 1986*b*). There is also evidence of this phenomenon in other perennial species, but this has not been as well documented.

RELATIONSHIPS AMONG GEOGRAPHY, MORPHOLOGY, AND INTERFERTILITY IN *MEDICAGO*

The problems of interpreting ancient dispersal patterns in angiosperm plants have been extensively discussed (see for example Cain 1944; Camp 1947; Good 1974; and Croizat et al. 1974). Numerous factors determine the geographical relationships among species, and several considerations serve to influence the distribution ranges of species in ways that cannot, or at least cannot easily, be predicted. After long periods, geographical relationships are confounded by migrations due to cataclysmic changes in climate and landscape. Natural, long-distance dispersal may occur. As well, many events have altered the distribution of most organisms.

In the case of *Medicago*, Small and Lefkovitch (1986) examined the relationships among geography, morphology, and interfertility based on the perennial species and the annuals with coiled pods. It was found that, based on geographical centers, the mean inter-species distance between nearest neighbors (i.e., the mean of the distances among every pair of species based on smallest morphological distance) was 2800 km. This extremely large distance indicates that considerable migration of species has occurred, probably over a long period of time, making the detection of relationships difficult. Nevertheless, Small and Lefkovitch (1986) were able to demonstrate small but statistically significant relationships between geographical distance and morphological similarity (i.e., the closer species of *Medicago* are geographically, the more similar they are), and between each of these and their ability to hybridize (i.e., species that are closer geographically or more similar are more likely to be able to hybridize).

A positive association among species of *Medicago* between geographical distance and similarity, as well as between geographical distance and intersterility, is consistent with two of the three main models of speciation (see for example Lewis 1966; White 1978; and Wiley 1981). The most commonly accepted model is the "slow" one of "ecogeographic extension", which is the gradual divergence of populations as a result of occupation of increasingly different adaptive zones. This model implies that there is an intimate association between morphological divergence and intersterility, with both becoming established gradually and concomitantly. Both processes are viewed as consequences of increasing geographical isolation. Thus, the postulate of an association between morphological divergence and geographical separation is applicable up to the point of the evolution of complete genetic isolation between populations (species) and beyond. Another commonly accepted model of speciation holds that physiological barriers to interbreeding are established relatively suddenly in peripheral, more or less temporarily (spatially) isolated populations. It has been reasoned that, under such circumstances, marked genetic drift, drastic selection, and perhaps even accelerated mutation may result in the rapid generation of new "species". The two preceding models of rapid and slow speciation are not mutually exclusive, and both may be expected to lead to a positive correlation between contiguity, morphological similarity, and interfertility so that the existence of such a condition does not identify which of the processes may have occurred. A third model embraces paradigms of sympatric speciation that are more controversial and less widely held than the preceding models. By definition, the sympatric establishment of breeding, and possibly also morphological divergence, does not involve change in geographical relationships, although the first two processes may still be expected to develop concomitantly. In any event, Small and Lefkovitch (1986) found no support for this model with respect to *Medicago*. Yet a fourth model is intrinsic in "Jordan's Law", which, as stated by Cain (1944, p. 265) holds that "given any species in any region, the nearest related species is not likely to be found in the same region, nor in a remote region, but in a neighboring district separated from the first by a barrier of some sort or at least by a belt of country the breadth of which gives the effect of a barrier." Once again, with respect to *Medicago*, Small and Lefkovitch (1986) found no support for this model.

5. Chromosome Evolution

K.A. Lesins and his colleagues, particularly C.B. Gillies, published dozens of research papers (see Literature Cited) clarifying chromosome evolution in *Medicago*. Lesins and Gillies (1972) presented somatic idiograms for the chromosome complements of 24 species, as well as meiotic (pachytene) idiograms for several species. These authors reported that chromosome morphology differed appreciably within, as well as among, species. Karyotypic evolution among the four species of section *Spirocarpos* subsection *Intertextae* was examined by Schlarbaum et al. (1984). Extensive examination of the karyotypes of the subspecies of *M. sativa* has been conducted by Bauchan and Campbell (1994), Bauchan and Hossain (1997*a, b*; 1998*a, b*; 2001*a*), and Bauchan et al. (2002, 2003). Bauchan (2009) reviewed the cytogenetics of *M. sativa* and related species. Chromosome differences among species of *Medicago* are critical to limiting crossing and hence gene exchange by conventional hybridization. In some cases, clarification of chromosome differences assists in determining relationships among species. Meiotic chromosome pairing in hybrids can reveal the degree of relationship between species, but because relatively few species of *Medicago* have proven to be crossable (see Chapter 6), such information has been of limited value to date. However, *M. sativa* has been hybridized with several related species and chromosomal relationships among these have received study (e.g., Stanford et al. 1972; McCoy and Bingham 1988).

CHROMOSOME NUMBER IN *MEDICAGO*

Chromosome increase by polyploidy (both autopolyploidy and allopolyploidy) and decrease by aneuploidy are known, or at least hypothesized to occur, in *Medicago*. Supernumerary chromosomes have been identified (Hossain and Bauchan 1999), but confusion with satellites is suspected of being more significant in the production of erroneous counts (Bauchan and Elgin 1984). Translocations have probably been an important mechanism of chromosomal change in *Medicago*, but there has been limited study to date. The following comprehensive listing of chromosome numbers in *Medicago* (Table 5-1) is intended to summarize knowledge accumulated to date and to stress that, for some species, several reported chromosome number determinations are suspicious and require confirmation.

Table 5-1. Frequency of chromosome numbers (2*n*) recorded in the literature for *Medicago* species.

	Chromosome number (2*n*)									
Species	14	16	17	18	24	28	30	32	44	48
M. arabica		20								
M. arborea								7		
M. archiducis-nicolai		2								
M. arenicola[1]										
M. astroites		1								
M. biflora[1]										
M. bonarotiana (+ "*M. blancheana*")		7								
M. brachycarpa		2								
M. cancellata										4
M. carica[1]										
M. carstiensis		4								
M. ciliaris		13	1	3						
M. citrina										4
M. constricta	4									
M. coronata	1	8								
M. crassipes[1]										
M. cretacea		1						1		
M. daghestanica		5								
M. disciformis		6								
M. doliata		11								

Table 5-1 (*Continued*).

Species	Chromosome number (2*n*)									
	14	16	17	18	24	28	30	32	44	48
M. edgeworthii		2								
M. fischeriana[1]										
M. granadensis		3	1	2						
M. halophila[1]										
M. heldreichii[1]										
M. heyniana		2								
M. huberi[1]										
M. hybrida		2								
M. hypogaea	2									
M. intertexta		20	2					1		
M. isthmocarpa[1]										
M. italica		18								
M. laciniata		18								
M. lanigera		3								
M. laxispira		1								
M. littoralis	1	19								
M. lupulina		72		2			1	12		
M. marina		15								
M. medicaginoides		2								
M. minima		35								
M. monantha							1		1	1
M. monspeliaca		14								
M. murex		17								
M. muricoleptis		3								
M. noeana		2								
M. orbicularis		36								
M. orthoceras									1	
M. ovalis								2		
M. pamphylica[1]										
M. papillosa		8						1		
subsp. *macrocarpa*		1						3		
subsp. *papillosa*		3								
M. persica[1]										
M. phrygia[1]										
M. pironae		4								
M. platycarpos		12								
M. plicata[1]										
M. polyceratia					1	10	1	1	1	
M. polymorpha	30	21								
M. popovii[1]										
M. praecox	5	1								
M. prostrata		3						3		
M. radiata		8					1			
M. retrorsa[1]										
M. rhodopea		4								
M. rhytidiocarpa[1]										
M. rigida		1								
M. rigidula	41	13								
M. rigiduloides	28									
M. rostrata[1]										
M. rotata		8	1	2						

Table 5-1 (*Concluded*).

Species	Chromosome number (2*n*)									
	14	16	17	18	24	28	30	32	44	48
M. rugosa							40	10		
M. rupestris		2								
M. ruthenica	1	7								
M. sativa										
subsp. *sativa*								64		
subsp. *caerulea*		20								
subsp. *glomerata*		5						9		
subsp. *falcata*		37	1					30		
subsp. *falcata* var. *viscosa*								2		
subsp. × *varia*		6						16		
M. sauvagei		2								
M. saxatilis										4
M. scutellata	1	4				3	75	21		
M. secundiflora		4								
M. shepardii		1								
M. sinskiae	1	1								
M. soleirolii		6								
M. sphaerocarpos	10									
M. strasseri								1		
M. suffruticosa		5								
M. syriaca		1								
M. tenoreana		7								
M. truncatula	2	13								
M. turbinata	2	16								

Note: Detailed references for all these counts are given in the individual chapters for the species. Somatic counts are reported, regardless of whether haploid numbers were originally examined (which was quite uncommon). The analysis is complete up to 2003, the latest date for which comprehensive surveys have been published (note online chromosome data of the International Organization of Plant Biosystematists at http://www.iopb.org and the Index to Plant Chromosome Numbers at the Missouri Botanical Garden at http://mobot.mobot.org/W3T/Search/ipcn.html).

[1]No chromosome counts were found in the literature.

The following problems need to be considered with respect to the reliability of chromosome number reports, as summarized in Table 5-1:

(1) Misidentification. Many species are very difficult to identify. It was not possible to check vouchers for the above table.

(2) Errors of technique. Most chromosome numbers reported for *Medicago* are based on somatic tissues. Somatic chromosomes are small in *Medicago;* for example, *M. sativa* chromosomes are 2–3 μm in length (Bauchan and Campbell 1994) and not easy to count. Satellite chromosomes may be miscounted as additional chromosomes. Deviating counts should be verified using chromosome counts of cells in the meiotic phase.

(3) Tendency to verify previous counts. The base number in *Medicago* is $x = n = 8$, and there are tetraploid and hexaploid populations. Accordingly, counts of $2n = 16$, 32, and 48 are expected, and it seems clear that some researchers have been biased in favor of these numbers.

The following generalizations are made with due respect to these potential sources of error.

Of the 87 species recognized here, chromosome numbers have been recorded in the literature for all but 17. All of these 17 are uncommon or indeed narrowly endemic species and were once accepted as species of *Trigonella*. Counts of $2n = 16$ that were listed for these species in Bauchan (2009) were included in error (G.R. Bauchan, personal communication).

Of the 70 species for which chromosome numbers have been reported, $2n = 16$ is the sole or predominant count in 50 of these and is the only recorded count in most of them. Clearly the base number is $n = 8$. Of these 50 species, the few reported counts other than $2n = 16$ should be viewed in the following species as requiring confirmation: *M. ciliaris*, *M. coronata*, *M. granadensis*, *M. intertexta*, *M. ruthenica*, *M. truncatula*, and *M. turbinate*. The basis of this recommendation is that the nonconforming counts appear to be statistical outliers. *Medicago truncatula* has been developed as a model legume plant for various genetic and developmental studies (see Chapter 97) and its genome has been mapped. In the light of this, it is disconcerting that there is uncertainty regarding the possibility that there are two chromosome levels.

A count of $2n = 14$ for *Medicago* presumably is the consequence of aneuploid reduction; in some cases this seems to have been based on chromosomal translocation involving two chromosomes and the subsequent loss of the resulting centric fragment (Quiros and Bauchan 1988). The following species have had counts of $2n = 14$ only: *M. constricta*, *M. hypogaea*, *M. rigiduloides*, and *M. sphaerocarpos*. The following species have predominantly had counts of $2n = 14$: *M. polymorpha*, *M. praecox*, *M. rigidula*, *M. sinskiae*, and *M. sphaerocarpos*. Many of the counts of $2n = 16$ in the latter five species could well be based on the biased tendency mentioned above to find this conventional number in *Medicago*, or a miscount involving the satellite chromosomes.

Polyploidy is relatively rare among the annual species of *Medicago*. *Medicago ovalis* is tetraploid ($2n = 32$). *Medicago rugosa* and *M. scutellata* have $2n = 30$ (the frequent counts of $2n = 32$ are probably erroneous). The $2n = 30$ species may have arisen by hybridization of a $2n = 14$ species and a $2n = 16$ species followed by the polyploidization of the hybrid. Alternatively, a pair of chromosomes could have been lost from a $2n = 32$ species (Bauchan and Elgin 1984). There are some diploid chromosome number reports for *M. scutellata*, but it is clear that most plants are tetraploid, and the diploid counts require study. Counts are very limited for *M. monantha* ($2n = 30, 44, 48$), *M. orthoceras* ($2n = 44$), and *M. polyceratia* ($2n = 24, 28, 30, 32, 44$), a trio of species that are difficult to distinguish and, in view of the odd and divergent chromosome counts reported, require additional study.

In several species there are both diploid ($2n = 16$) and tetraploid ($2n = 32$) populations. These include *M. cretacea* (requires verification, since only two counts are recorded), *M. lupulina* (the tetraploids are relatively uncommon), *M. papillosa*, *M. prostrata*, and *M. sativa*. Except for *M. lupulina*, which can be annual, biennial, or perennial, these species are all perennials. Given the close resemblance of the diploid and tetraploid phases of these species, the polyploids are probably predominantly autopolyploids.

Medicago cancellata and *M. saxatilis* are hexaploids ($2n = 48$) and both are perennials. Lesins (1970) speculated that *M. cancellata* arose as an autoalloploid involving hybridization of *M. rhodopea* and *M. sativa* (or a related species) and that *M. saxatilis* similarly arose as an autoalloploid involving hybridization of *M. rupestris* with *M. sativa* (or a related species).

The three very closely related species of woody shrubs in the genus *Medicago* are polyploids —*M. arborea* and *M. strasseri* have $2n = 32$ and *M. citrina* has $2n = 48$. *Medicago citrina* is likely a direct derivative of *M. arborea*; *M. arborea* and *M. citrina* represent a tetraploid–hexaploid complex and they have often been placed in the same species.

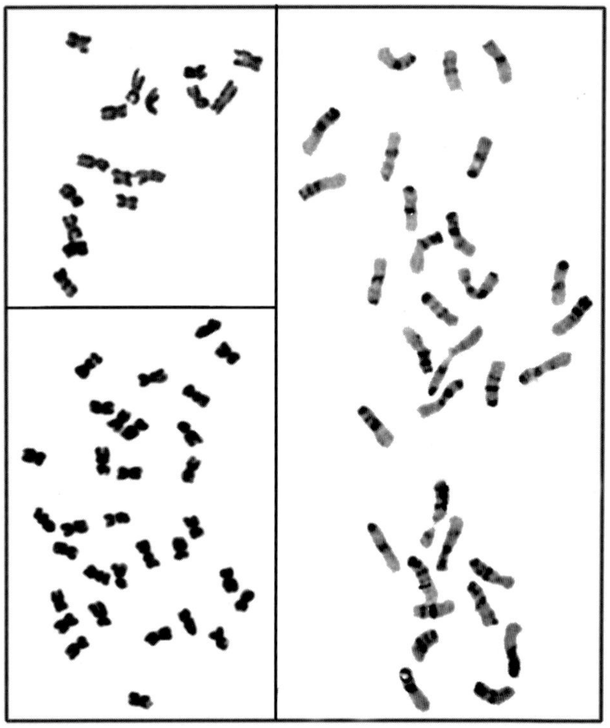

Fig. 5-1. Somatic chromosomes of *Medicago sativa*. **Right:** C-banded chromosomes of *M. sativa* subsp. *sativa* ($2n = 4x = 32$). **Lower left:** *M. sativa* subsp. *sativa* ($2n = 4x = 32$). **Upper left:** *M. sativa* subsp. *caerulea* ($2n = 4x = 16$). Photos from Bauchan (2009), courtesy of G.R. Bauchan.

6. Interbreeding Barriers

Information on reports to date of attempts to hybridize each of the 87 species of *Medicago* with others is summarized in the respective chapters in the section "Genetic Relationships and Gene Transferability". This chapter provides an overview.

Aside from the potential agricultural value of facilitating gene transfer among *Medicago* species, information on interfertility and/or intersterility is useful for assessing genetic and hence taxonomic relationships. The following table summarizes attempts at hybridization by conventional techniques that have been reported in the literature. The 87 species are numbered in the first column. The same numbers are used in the other columns, so the rows provide information, for each of the 87 species, about the results of attempts to hybridize it with other species. Also see the Note at the end of the table. Accounts of gene transfer by advanced technologies are given in the chapters dealing with each species.

Table 6-1. Summary of conventional hybridization reports for *Medicago*.

	Degree of success of hybridization			
	Unsuccessful hybridization	Sterile or substantially sterile hybrids	Partly fertile hybrids	Completely or substantially fertile hybrids
Strictly perennial species				
1. *M. arborea*		19		
2. *M. archiducis-nicolai*				
3. *M. cancellata*		18	19	
4. *M. carstiensis*	18			
5. *M. citrina*				
6. *M. cretacea*				
7. *M. daghestanica*	9, 18	11, 18		
8. *M. hybrida*	18	21		
9. *M. marina*	7, 18			
10. *M. papillosa*		18		
11. *M. pironae*	9, 18	7, 18		
12. *M. platycarpos*	18, 38, 43, 49			
13. *M. popovii*	17			
14. *M. prostrata*			18	
15. *M. rhodopea*		16, 18		
16. *M. rupestris*		15		
17. *M. ruthenica*	13, 18, 38, 43, 49			
18. *M. sativa*	4, 7, 8, 9, 11, 12, 13, 21, 27, 33, 38, 42, 43, 47, 70	1, 3, 7, 10, 11, 15, 19, 54	14	
19. *M. saxatilis*		18	3	
20. *M. strasseri*				
21. *M. suffruticosa*	18	8		
Annuals without pulvinate cotyledons				
22. *M. arabica*				
23. *M. bonarotiana*				51
24. *M. ciliaris*	29		32, 41	
25. *M. constricta*	33, 37, 59, 62			

Table 6-1 (*Continued*).

	Degree of success of hybridization			
	Unsuccessful hybridization	Sterile or substantially sterile hybrids	Partly fertile hybrids	Completely or substantially fertile hybrids
26. *M. coronata*	47			
27. *M. disciformis*	18, 61			
28. *M. doliata*	40, 62			
29. *M. granadensis*	24, 32, 41			
30. *M. heyniana*	43, 48			
31. *M. hypogaea*				
32. *M. intertexta*	29		24	
33. *M. italica*	18, 25, 40, 49, 56, 63	62	37	
34. *M. laciniata*		53		
35. *M. lanigera*				
36. *M. laxispira*				
37. *M. littoralis*	25, 40, 49, 63		33	
38. *M. lupulina*[1]	12, 17, 18, 55			
39. *M. minima*	62			
40. *M. murex*	28, 33, 37, 59, 62			
41. *M. muricoleptis*	29		24	
42. *M. noeana*	18, 52			
43. *M. orbicularis*	12, 18, 30, 17			
44. *M. ovalis*				
45. *M. plicata*				
46. *M. polymorpha*	47, 49			
47. *M. praecox*	18, 26, 46, 49			
48. *M. radiata*	30			
49. *M. rigidula*	12, 17, 33, 37, 46, 47, 62		50	
50. *M. rigiduloides*			49	
51. *M. rotata*				23
52. *M. rugosa*	42			
53. *M. sauvagei*	27	34		
54. *M. scutellata*	18	18		
55. *M. secundiflora*	38			
56. *M. shepardii*	33			
57. *M. sinskiae*				
58. *M. soleirolii*	33			
59. *M. sphaerocarpos*	25, 40			
60. *M. syriaca*				
61. *M. tenoreana*	27			
62. *M. truncatula*	25, 28, 39, 40, 49	33, 63	37	
63. *M. turbinate*	33, 37	62		
Annuals with pulvinate cotyledons (sections *Buceras* and *Lunatae*)				
64. *M. arenicola*				
65. *M. astroites*				

Table 6-1 (*Concluded*).

	Degree of success of hybridization			
	Unsuccessful hybridization	Sterile or substantially sterile hybrids	Partly fertile hybrids	Completely or substantially fertile hybrids
66. *M. biflora*				
67. *M. brachycarpa*[1]				
68. *M. carica*				
69. *M. crassipes*				
70. *M. edgeworthii*[1]	18			
71. *M. fischeriana*				
72. *M. halophila*				
73. *M. heldreichii*				
74. *M. huberi*				
75. *M. isthmocarpa*				
76. *M. medicaginoides*				
77. *M. monantha*				
78. *M. monspeliaca*				
79. *M. orthoceras*				
80. *M. pamphylica*				
81. *M. persica*				
82. *M. phrygia*				
83. *M. polyceratia*				
84. *M. retrorsa*				
85. *M. rhytidiocarpa*				
86. *M. rigida*				
87. *M. rostrata*				

Note: References to the literature are in the chapters dealing with each individual species. Tri-species bridges, questionable reports, and confirmatory reports have been excluded. Each attempted combination is listed twice, i.e., it is shown for each of the two species involved.

[1]Species that are both annual and perennial.

Most *Medicago* species are very strongly isolated by genetic and physiological barriers to interbreeding. As indicated in the above table, most attempts at conventional hybridization have been completely unsuccessful in generating hybrids; several have produced infertile hybrids, and only a few hybrids capable of setting seeds have been generated. Differences in chromosome number, particularly polyploidy (see Chapter 5 on chromosome evolution), account for some of the interbreeding barriers, but in most cases where chromosome numbers are the same between species, hybrids still cannot be generated. Where ploidy level has been adjusted (for example, doubled in a diploid to match a tetraploid species), generation of fertile hybrids is not usually improved (Lesins and Lesins 1979).

There have been numerous attempts to hybridize *Medicago* species with *M. sativa* because of the possibility of transferring genes to alfalfa. Conversely, there is only one reported attempt at hybridizing *M. sativa* with a member of the group of 24 species with pulvinate cotyledons (sections *Buceras* and *Lunatae*), a result in part of the lack of appreciation, until recently, that these species are members of the genus *Medicago*, not *Trigonella*.

Medicago sativa (discussed in Chapter 12) is recognized in this monograph as comprising the following groups:

Taxon	Ploidy
M. sativa subsp. *caerulea*	$2n = 16$
M. sativa subsp. *sativa*	$2n = 32$
M. sativa subsp. ×*varia*	$2n = 16, 32$
M. sativa subsp. *falcata* var. *falcata*	$2n = 16, 32$
M. sativa subsp. *falcata* var. *viscosa*	$2n = 16, 32$
M. sativa subsp. *glomerata*	$2n = 16, 32$

Fig. 6-1. Somatic chromosomes ($2n$ = 58) of a sterile hybrid between *Medicago sativa* ($2n$ = 32) and *M. scutellata* ($2n$ = 30), reported by Sangduen et al. (1982), the only known hybrid between an annual and a perennial species of *Medicago* produced by conventional hybridization. The hybrid was apparently the result of a combination of unreduced gametes but was a mixoploid, with shoots differing in chromosome number, and somatic chromosome number ranging from 30 to 64 in root tip cells. The hybrid was perennial but intermediate for most other morphological characteristics. Photo courtesy of N. Sangduen.

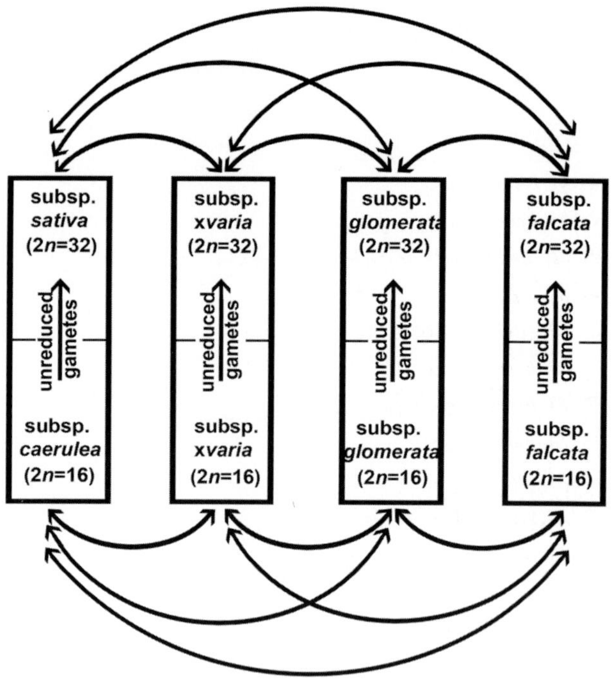

Fig. 6-2. Patterns of gene flow among the taxa and chromosome levels of the *Medicago sativa* complex. Unreduced gametes from the $2n$ = 16 taxa may also combine with gametes from all $2n$ = 32 taxa, but such gene flow is shown only within the four basic groups.

Hybridization occurs extensively in nature among the taxa of *M. sativa* and it appears that the complex constitutes a classical "polyploidy pillar complex" in which normal gametes can easily recombine among the four "pillars" of the taxa at the $2n = 16$ level, and similarly among the taxa at the $2n = 32$ level, with a low probability of genetic flow "upward" within the pillars via unreduced gametes and to the other taxa at the $2n = 32$ level (see Fig. 6-2). Occasionally, triploids may also be involved in gene transfer (Bingham and Saunders 1974). For breeding purposes, it is possible to transfer genes to alfalfa cultivars (almost all of which are tetraploid) from any of the taxa of *M. sativa* by conventional hybridization, adjusting ploidy level where necessary. See McCoy and Bingham (1991) for information on the use of unreduced gametes in the breeding of alfalfa.

7. Adaptations In *Medicago*

THE TAXONOMIC IMPORTANCE OF ADAPTATIONS IN *MEDICAGO*

All species are the result of natural selection and the principal selective forces acting on plants are related to climate, substrate (soil in the case of most plants, water or air for some other species), competition from other plants, depredations caused by other organisms (viruses, bacteria, fungi, animals), and symbiotic relationships with other species. The result is that plants acquire adaptive features, be they physiological, chemical, developmental, phenological, or morphological. Classification and identification of higher plants are based mostly on external morphology (notwithstanding the recent importance of molecular-based phylogenetic analysis), probably because vision is so important to humans and provides a wealth of easily obtainable data.

As a result of the studies reviewed in this monograph, it should be apparent that the majority of characters that have been used taxonomically to classify *Medicago* are the result of herbivory, i.e., they are adaptations to consumption by animals. Such adaptations are taxonomically important for distinguishing *Medicago* as a genus, for distinguishing related groups of species, for distinguishing infraspecific groups, and simply for clarifying infraspecific variation that does not merit taxonomic labels. Aside from the importance of these adaptations for systematic purposes, the adaptations are also of practical importance because the kinds of herbivory examined in this chapter are either extremely deleterious or extremely desirable from a human viewpoint, and comprehending the relationships is a key to improvement by breeding or management techniques.

Medicago is primarily from the Mediterranean region and the environment in which adaptations have evolved in the genus should be kept in mind. The climate north and east of the Mediterranean Basin, where most of the species of *Medicago* occur, is relatively continental, with occasional killing frosts. From a geological point of view, the Mediterranean Basin is extremely heterogeneous. However, although there are areas with acidic soils, the substrates are dominated by soils on calcareous rocks.

THE ECOLOGICAL IMPORTANCE OF ADAPTATIONS TO HERBIVOROUS ANIMALS

The consumption of plants by animals generally accounts for more than 10% of plant production in natural communities (Coley et al. 1985). There does not seem to be an analysis of the comparative degrees of herbivory affecting the different plant families, but it would not be surprising to find that legumes are especially attractive to herbivores because the symbiotic, nitrogen-fixing bacteria associated with their roots generate plant tissues that are especially rich in proteins (nitrogen, of course, is a key element of protein) and are, consequently, especially nutritious.

Adaptations to Insects

Small (1989*b*) analyzed character evolution among the 658 genera of Fabaceae and concluded that the genera appear to have evolved primarily in response to insects exerting pressures that resulted in new genus-defining, reproductive characters. As discussed below, this conclusion seems completely applicable to *Medicago*. On a world basis, insect pests are responsible for consuming about 13% of crop production (Cramer 1967). Most plant species are eaten by hundreds of insect species, and conversely most herbivorous insects are polyphagous. Both the animals and plants have evolved numerous adaptations to generalized feeding. These adaptations may involve chemistry, physiology, behavior, and morphology (for general discussions see Sorensen et al. 1988; Bernays 1991; Moore 1995; and Gatehouse 2002). Most investigations of insect resistance in *Medicago* have dealt with *M. sativa*. Numerous insects attack alfalfa and many have been studied in detail, but for the most part adaptations in alfalfa that are specific to given herbivorous insects have not been discovered. Moreover, while many alfalfa cultivars have been bred that are resistant to specific insect pests, the mechanism for their resistance is usually not clear. However, there are a few cases in which an adaptation allowing alfalfa and some other *Medicago* species to resist particular insect species has been discovered, as reviewed below, and these examples provide insight into character evolution in the genus.

Adaptations to Mammals

Grazing mammals may consume all aboveground parts of *Medicago* plants. Sheep and goats are particularly significant and are the chief natural mammalian herbivores of the genus. Characteristics that serve especially to protect perennial *Medicago* species against grazers are discussed in Chapter 12. These include low stature, tough stems, large stem-to-leaf ratio, small, hard leaves, and pubescence (several of these characteristics protect the plants against trampling as well as consumption). The annual species have also adopted these ways of defending themselves but, by virtue of being annuals, they put a much larger proportion of their energy budget into seed production and present less non-seed tissue to potential herbivores. Concomitantly, the annuals present palatable, non-seed tissues to herbivores for a much shorter period. Some of the perennial species and most of the annuals have indehiscent fruits that are tough and spiny, protecting the seeds (although goats are usually not dissuaded).

ADAPTATIONS IN *MEDICAGO*

1. Subterranean Organs

Radicles

The radicle (i.e., the primary root of the seedling) of *Medicago* species usually develops rapidly, which is consistent with the fact that most of the species grow in arid to semi-arid places that receive rainfall predominantly in one season. As described in Chapter 10, the relative length of the radicle and the cotyledons in the ungerminated seeds is a good diagnostic character for discriminating the taxonomic sections of *Medicago*. Curiously, the more "primitive" sections (i.e., those relatively similar to the most similar extant apparent ancestor, *Trigonella*) have seeds with longer radicles (typically three-quarters to as long as the length of the seed), compared with the more "advanced" sections (in which the radicle is about half the length of the seed). This pattern may be related to another phenomenon — the "advanced" sections of *Medicago* have indehiscent fruits that retain their seeds for a lengthy interval, sometimes for years. By contrast, the "primitive" sections have dehiscent fruits, which scatter the seeds as soon as they are ripe. Thus, the seeds in a given indehiscent fruit are confined to one spot, and because the seeds germinate over a long period, at least one seed will probably succeed eventually in producing a plant, regardless of water availability (in essence, the fruit has become the essential propagule). By contrast, seeds from the dehiscent fruits are under greater pressure to survive (i.e., they are independent of their siblings from the maternal fruit), and a longer radicle is a way of getting to water faster.

Roots

The annual species usually have a branching root system, mostly within the top 30 cm of soil, and this is reflective of habitats that, for the most part, are wet or damp early in the season and become progressively drier and hotter during the remainder of an often-short season. The exceptionally deep-penetrating root system of some perennial *Medicago* can take advantage of low water tables; consequently, these species are able to take advantage of winter rainfall stored in the soil in seasonally dry Mediterranean climates. Cultivars of *M. sativa* often have shallower root growth that maximizes productivity in an irrigated, upper soil layer, but wild perennial *Medicago* may have very deeply penetrating roots, sometimes with a main taproot, that permit survival in extremely dry circumstances. The root characteristics of *Medicago* species appear to reflect the principal function of obtaining water (and nutrients) from the soil, rather than adaptation to animals.

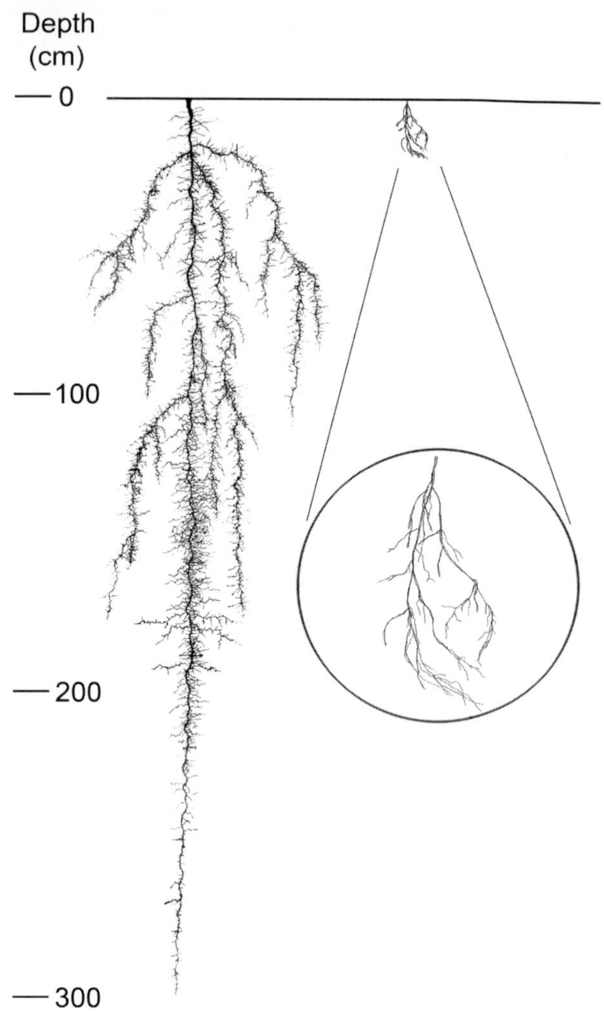

Fig. 7-1. A comparison of the root systems of perennial and annual species of *Medicago*. **Left:** The root of two-year-old *M. sativa*, based on fig. 102 of Weaver (1926). **Right:** The root of *M. bonarotiana* (from Fig. 19-1).

Rhizomes

Medicago carstiensis is the only species exclusively with rhizomes (see Figs. 23-1 and 23-3), a growth habit that protects the plants to some extent from grazing, and so may well represent an adaptation to herbivorous mammals. Either rhizomatous or "creeping-rooted" growth is characteristic of some populations of *M. sativa* subsp. *falcata* (see Chapter 12). These variants are native to cold-temperate regions, and it is possible that their ability to spread near the soil surface is advantageous in the spring when cold soils could significantly delay resumption of spring growth. It is therefore unclear whether a rhizomatous growth habit has been selected as an adaptation to reduce herbivory or as an adaptation to climate (or indeed to both factors). Rhizomes could also be a competitive and/or colonizing strategy against other species for access to space, or an adaptive strategy to avoid frost heaving injury common to taprooted plants.

Fig. 7-2. Growth habit adaptations to grazing mammals. The plants on the left are adapted to withstand grazing by exhibiting characteristics such as low or prostrate growth, numerous basal stems, very branched shoots (often with several slender stems arising from the base of the plant), very small (often slender) leaflets, and rhizomatous spread. These features make it more difficult for animals to inflict substantial or fatal damage to the entire plant. By contrast, the plants on the right are relatively susceptible to grazing because they grow upright, have large leaflets, and have single or few large succulent stems (artist: B. Flahey).

2. Stems

Hard versus succulent tissues

Several studies of *Medicago* species have indicated that tough tissues can represent an adaptation against insect herbivores. Sorensen et al. (1988) found that alfalfa cultivars that are resistant to the alfalfa weevil (*Hypera postica*) have heavy terminals and axillary branching, which seem responsible for the resistance. Brewer et al. (1986) found that the anatomical features of the stems in several experimental *Medicago* clones, mostly of alfalfa (*M. sativa*), contributed to resistance against the potato leafhopper (*Empoasca fabae*), one of the most injurious arthropod pests of alfalfa in North America. Stems with highly lignified tissues were more resistant to the leafhopper. A cylinder of hard tissues, made up of lignified xylem elements, phloem fibers, and interfascicular areas that surrounded the pith, was linked to reduced oviposition. Early lignification of tissues making up the cylinder appears to contribute to resistance by mechanically or chemically deterring or preventing feeding and oviposition. Many new alfalfa cultivars have very coarse stems compared with older ones such as Vernal and Saranac, which have relatively succulent stems (A. McElroy, personal communication). This could represent an adaptation to insects but, alternatively, coarse stems could facilitate growth or harvest, or may be linked to disease resistance. More succulent stems are preferred by grazing mammals and so are desirable agronomically, but the trade-off is that such stems may be more subject to insect damage.

Erect versus prostrate habit

Keeping a low profile is advantageous for plants, at least with respect to mammalian herbivores, simply because mammals preferentially tend to choose higher (more accessible) vegetation. Decumbent or prostrate growth habit in pasture plants in general, and in *Medicago* in particular, has been associated with resistance to grazing by mammals (Rossitor 1966; Small 1996. See Chapter 12 on *M. sativa* for more detail).

Numerous slim stems versus fewer thick stems

Producing a high number of narrow stems, rather than a smaller number of more robust stems, is likely advantageous because it increases the probability that more stems will escape consumption. Also, simply from a relative size perspective, stems that have a very narrow cross section are likely to be resistant to insects adapted to depositing eggs in large stems. Such a relationship was demonstrated by Small et al. (2007), who found that stem narrowness in hemp (*Cannabis sativa* L.) was negatively correlated with infestation by the European corn borer (*Ostrinia nubilalis*). Based on several experimental *Medicago* clones, Brewer et al. (1986*b*) found that stems with smaller cross sectional areas were resistant to the potato leafhopper (*Empoasca fabae*). Brummer and Bouton (1991) reported that thin stems were associated with persistence of alfalfa under regular mechanical harvests used to simulate grazing.

3. Leaves

Leaf size

There is considerable variation in leaf blade (or leaflet) size among species of *Medicago*. The leaves of *M. hypogaea* and the 24 species of sections *Buceras* and *Lunatae* tend to be quite small (they are all annuals, although two are sometimes also perennial). The leaves of the remaining annuals tend to be of moderate size (for the genus), but the leaves of *M. intertexta* and *M. ciliaris* are often relatively large. Generally, the leaves of the perennial species with annual, aboveground shoots are larger than those of the annuals. The leaves of section *Dendrotelis* (perennial shrubs, with leaves that can be perennial) tend to be relatively large. Givnish (1979, 1987) and Parkhurst and Loucks (1972) reviewed several hypotheses explaining the adaptive significance of leaf form and/or size in relation to environmental variables. As noted by Givnish, leaves tend to be small and thick in sunny or dry areas, and large and thin in shady or moist areas. The leaves also tend to be small on poor soils and larger on richer soils. These two trends do occur among the species of *Medicago,* at least when comparing the annuals with each other and the perennials with each other. Brown and Lawton (1991) noted that very small leaves probably reduce the feeding efficiency of insects and, conversely, larger leaflets tend to be more attractive to insect herbivores. Alfalfa cultivars usually have notably larger leaves (with wider leaflets) than most wild *M. sativa* plants, but this may be a result of selection for increased photosynthetic production.

Leaf marginal serrations

Givnish (1979) stated that an adequate explanation has not been advanced for the well-known observation that leaves with entire margins are frequent in the tropics and the arctic, whereas leaves without entire margins are frequent in the north temperate zone (cf. Baker-Brosh and Peet 1997). Indeed, the adaptive significance of the many known patterns of leaf margin is generally unclear. There are some differences in the extent and size of margin serrations among the species of *Medicago*, although these are generally not of value for identification purposes. Larger-leaved species tend to have more evident marginal serrations, suggesting that they are adaptive. In most *Medicago* species, approximately the terminal half of the leaflet is serrated, but there is a tendency in several species of sections *Buceras* and *Lunatae,* for about two-thirds of the leaflet to be serrated. The fact that *Medicago* species are exceptionally nutritious suggests the possibility that the saw-tooth appearance of the leaf edge plays a defensive role, as suggested in Fig. 7-3. It is not clear why the basal parts of *Medicago* leaflets (indeed in allied genera) have fewer serrations or none at all. There may be reasons other than insect resistance for the presence of marginal serrations on leaves; for example, they might act as "drip tips" for shedding raindrops (although most *Medicago* species are adapted to dry climates), which could explain the absence of basal serrations.

Leaf laciniation

Several species of *Medicago*, particularly *M. laciniata*, often have incised or laciniate leaves (this occurs regularly in a

Fig. 7-3. Possible anti-herbivore features of leaf margins. Laciniate leaves at the left and serrated leaves in the middle are illustrated as unattractive for reasons discussed in the text (artist: B. Flahey).

few other species, but rarely or extremely rarely in most other species of the genus). There is some evidence that such cut-up leaves mimic insect damage and so deter insects from visiting the presumably less attractive plant (Brown and Lawton 1991). If this seems far-fetched, note in the presentation of *M. rotata* (Chapter 81), which has a laciniate-leaved variety, var. *eliezeri*, that some specimens of *M. rotata* var. *rotata* (which has normal leaves) with leaves eaten by insects, resulting in a laciniate appearance, have been mislabeled as var. *eliezeri* in collections (so, at least humans have been fooled).

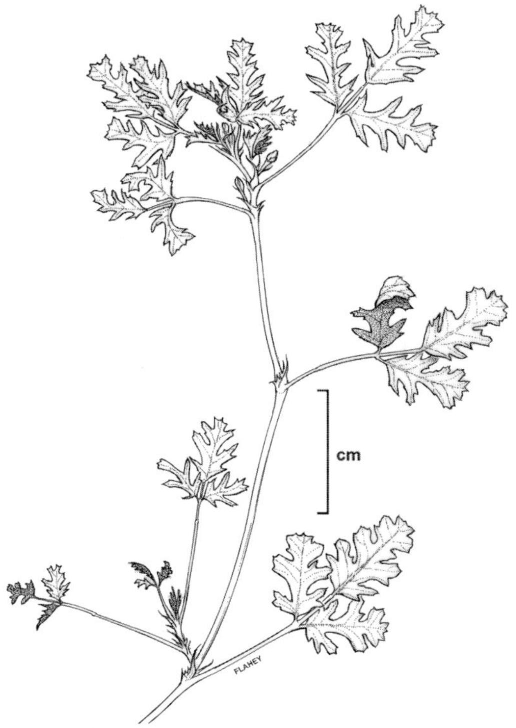

Fig. 7-4. Laciniate leaves of a plant of *Medicago polyceratia*. (Based on a plant grown in a greenhouse in Ottawa from seeds of United States Department of Agriculture, accession PI227395.85i. Artist: B. Flahey.)

Fig. 7-5. An interpretation of the function of leaflet blotches in some species of *Medicago*. Herbivorous insects are hypothesized to be deterred by eyespot-like blotches (artist: B. Flahey).

The frequency of alleles contributing to the development of laciniate leaves is likely affected by local herbivore pressure. Indeed, laciniation in species of *Medicago* is often relatively common in some particular areas of their distribution ranges. With respect to *M. monantha*, Halloran and Pennell (1981) noted that "one character exhibiting wide variation is the degree of indentation of the leaf, which varied from entire to deeply indented. The cut-leaf character occurred in this species at 550 m and above, indicating the possibility that the variation in this character is of adaptive significance or is closely linked to, or a pleiotropic expression of, genes conferring some measure of adaptability."

Leaf blotches

A large, usually anthocyanin-colored central blotch is a regular feature on the leaflets of some *Medicago* species (notably *M. arabica* and *M. intertexta*) and is an occasional feature of many other species. The function of these (if indeed they are adaptive) is unknown. There must be an energy cost to the plant in developing these blotches since they almost certainly interfere with photosynthesis. Given that insects are responsible for most of the visible characteristics of *Medicago* species, it may be that the blotches are protective against them. Eyespots (colored areas mimicking eyes), common on a variety of animals, are believed to be a means of deterring predators by arousing their instinctual fear of animals with large eyes (eyespots are also thought to confuse predators and deflect them from attacking vital regions of the body). The extent to which eyespots are adaptive in animals is unclear (see the review by Stevens 2005). The extent to which leaf blotches function as protection against herbivorous insects is unknown, but could be examined experimentally.

Leaf waxes

Bergman et al. (1991) found that the resistance of some alfalfa clones to the spotted alfalfa aphid *Therioaphis maculata* (Buckton) correlated with the quantity of epicuticular lipids on the leaves. They suggested that the aphids were stimulated to feed on older leaves by the lower amounts of lipids in these. The epicuticular wax layer of plants is known to modulate insect feeding and ovipositional behavior on some plants (references in Bergman et al. 1991).

4. Pubescence

Simple and glandular hairs

Epidermal pubescence is one of the most common plant adaptations against herbivorous insects. Trichomes (hairs) have been shown in numerous studies to be defensive against both adult and larval insect feeding and oviposition (Levin 1973). In *Medicago*, both simple hairs and gland-tipped hairs have been demonstrated to be defensive against a variety of insects (Small 1986; Small and Brookes 1986; Goertzen and Small 1993). Gland-tipped hairs are even present on the cotyledons near the seedling meristem, apparently to protect it (see Figs. 7-9 and 7-10). Gland-tipped hairs seem particularly useful as an adhesive that can immobilize small feeding insects (Ranger and Hower 2001; Fig. 7-11) and as a deterrent

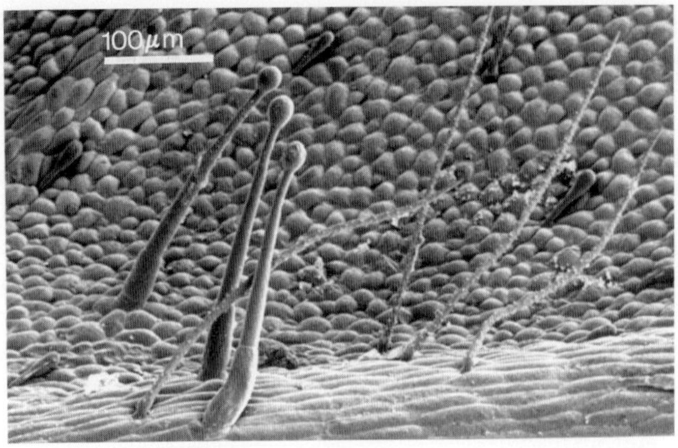

Fig. 7-6. Scanning electron micrograph of the abaxial leaf surface of *Medicago lupulina* showing midrib and lamina bearing gland-tipped trichomes (left) and simple trichomes on the midrib (right). (Illustration prepared by B. Brookes.)

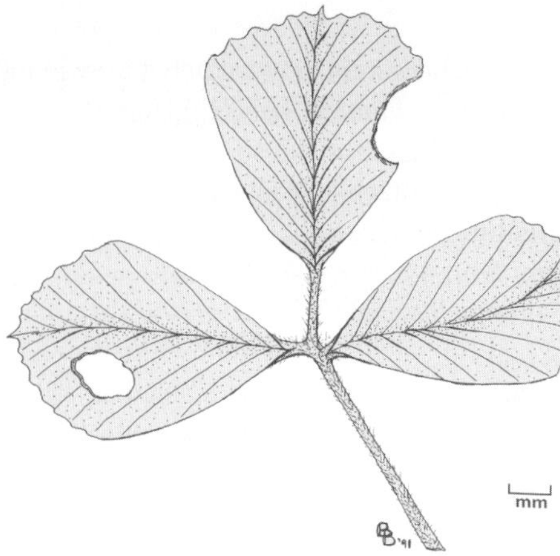

Fig. 7-7. Leaf of *Medicago lupulina* showing typical herbivore damage. An edge feeder has eaten a marginal portion of the middle leaflet and another herbivore has eaten a central portion of the lamina of a lateral leaflet. (Illustration prepared by B. Brookes.)

to oviposition. While gland-tipped hairs occur over many parts of the plant in several *Medicago* species, high concentrations of glands are normally found only in two lineages of the *M. sativa* complex (*M. sativa* subsp. *falcata* var. *viscosa* and *M. sativa* subsp. *glomerata*), predominantly on the flowers and fruits, and only in certain geographical regions. The Caucasus is the main region where the glands are found in very high concentrations. The most likely cause of this geographical pattern appears to be the sympatric seed-eating beetles of the genus *Tychius* (Small 1985*b*). Several of these beetles, especially *T. flavus* Beck. and *T. femoralis* Bris., are specialized alfalfa seed feeders (as larvae that develop inside the fruit), and their distribution ranges roughly coincide with the Eurasian distributions of glandular-podded wild alfalfas (Small 1985*b*). These beetles oviposit into the flowers and young fruits (Clarke and Burke 1977), precisely where the glands occur. Ironically, the beetles are devastating pests of the alfalfa cultivars without glands that are grown in Europe in close proximity to the wild, resistant plants that could have been used to breed resistant cultivars.

Danielson et al. (1989) examined the distribution of gland-tipped hairs on the stems and leaves of certain accessions of *M. prostrata*, *M. sativa* subsp. *glomerata* (identified as *M. glutinosa*), and (presumably) *M. sativa* subsp. *falcata* var. *viscosa* (identified by the ambiguous name *M. glandulosa*), all of which possessed appreciable densities of glandular hairs on their vegetative parts. These germplasm sources were used to develop potato leafhopper-resistant alfalfa varieties (Sulc et al. 2001, 2004. US Patent (1999) 5908974: Potato leafhopper-resistant alfalfa; http://www.patents.com/Potato-leafhopper-resistant-alfalfa/US5908974/en-US/; current January 2010).

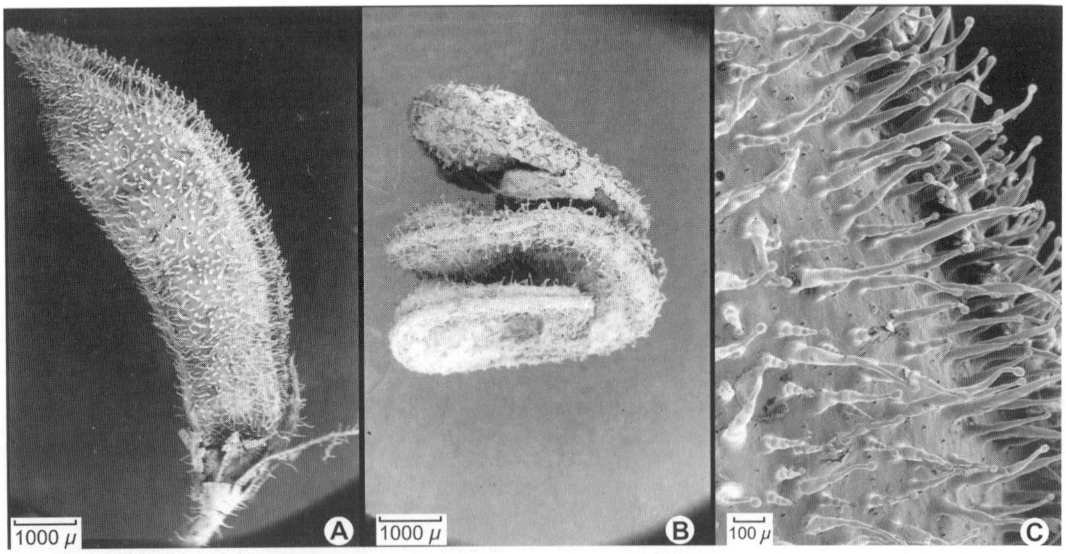

Fig. 7-8. Scanning electron photomicrographs of glandular-pubescent fruit of *Medicago sativa*. (A) Fruit of *Medicago sativa* subsp. *falcata* var. *viscosa* covered with gland-tipped trichomes. (C) Close-up of the trichome-covered surface (A and C based on E. Small M87, DAO). (B) *Medicago sativa* subsp. *glomerata* (based on E. Small M85, DAO). As discussed in the text, glandular trichomes on *M. sativa* are distributed in a distinctive geographical pattern (see Chapter 12). (Illustrations prepared by B. Brookes.)

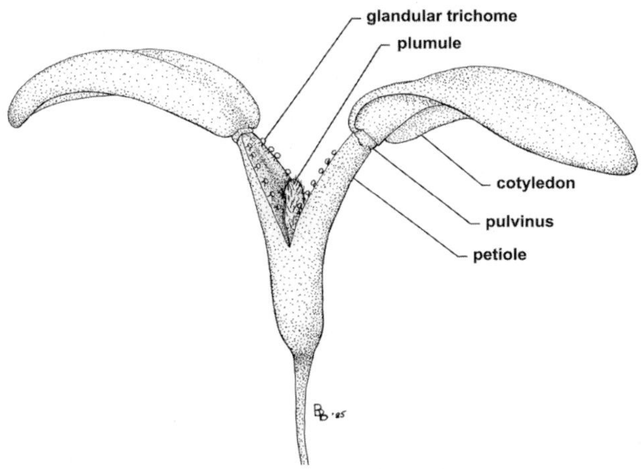

Fig. 7-9. Cotyledonary petiolar glandular trichomes in a *Medicago* seedling with pulvinate cotyledons (diagrammatic; see text for discussion about the presence of pulvini; artist: B. Brookes).

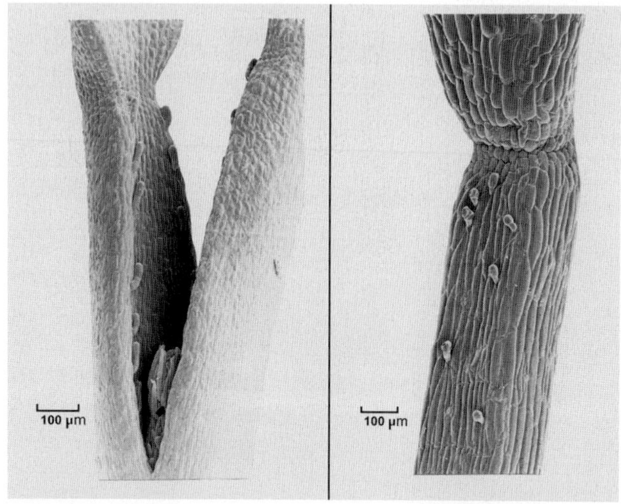

Fig. 7-10. Scanning electron micrographs of cotyledonary petioles of *Medicago monantha* showing glandular trichomes. (Illustrations prepared by B. Brookes.)

Othman et al. (1981) examined the distribution of gland-tipped hairs in four annual species of *Medicago* — *M. scutellata*, *M. rugosa*, *M. disciformis*, and *M. blancheana* (*M. bonarotiana* plus *M. bonarotiana* × *M. rotata*). They found that glandular hairs were present on all the vegetative and floral parts of the four species except on the cotyledons and petals. Each of the species had glandular hairs on the leaflet margins. The glandular hairs were distributed over the entire abaxial leaf surface of *M. scutellata* and *M. rugosa*, whereas they were confined to the midrib area on the other two species. The glandular hairs were densest on the peduncle, followed by the stem, petiole, seedpods, and leaflets.

In many plant species, epidermal pubescence serves to lower the rate of water loss, thereby conserving water. However, this is generally not postulated to be the principal function of glandular hairs and, in fact, Lenssen et al. (2001) found that the presence of glandular hairs on accessions of alfalfa has only a slight effect on water loss.

Kreitner and Sorensen (1979*a*, *b*, 1983) studied the structure of the glandular trichomes of several species of *Medicago* and found that, at maturity, the glands secreted what appeared to be anti-insect material that accumulated on the gland heads as well as on the stalks. Ranger and Hower (2001*b*) found that the glandular hairs of *Medicago* provide resistance to insects in two ways: (1) toxically, because the hairs exude a chemical that harms their rate of development and survival; and (2) mechanically, because the exudate traps nymphs and impairs oviposition.

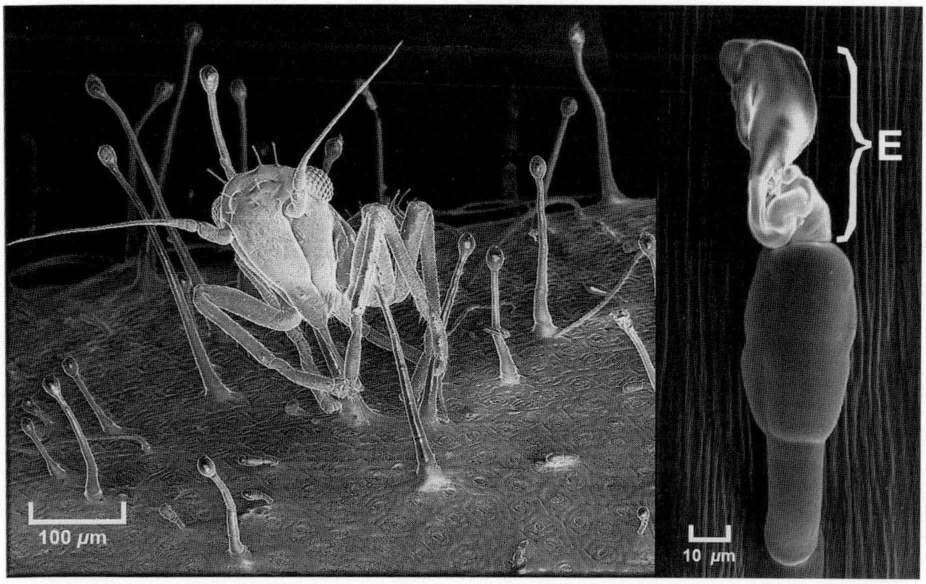

Fig. 7-11. Left: A first instar potato leafhopper trapped on the glandular trichomes of an alfalfa clone. **Right:** A glandular trichome which has released a sticky exudate (E). From Ranger and Hower (2001*a*); scanning electron micrographs courtesy of C.M. Ranger.

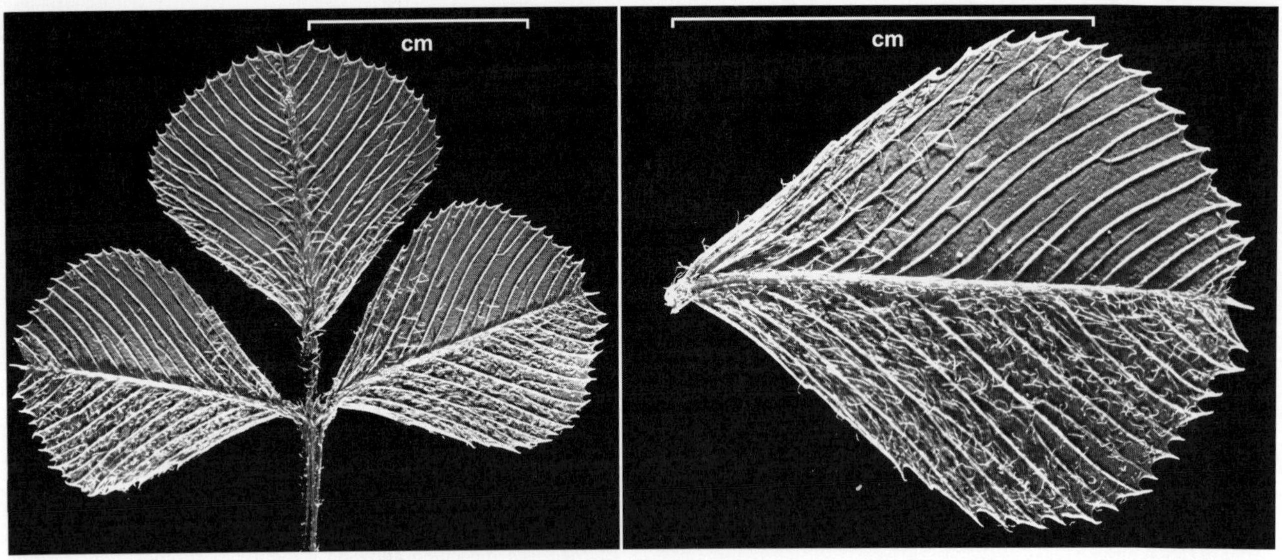

Fig. 7-12. Abaxial views of a leaf of *Medicago murex*. **Left:** Note the relatively high concentration of trichomes on the proximal portion of the lateral leaflets (the portion nearest the petiole base). **Right:** An enlargement of one of the lateral leaflets.

Fig. 7-13. A young leaf of *Medicago orbicularis*. Note how the leaflets are folded back from their midribs and adpressed, mainly exposing one side of the lateral leaflets.

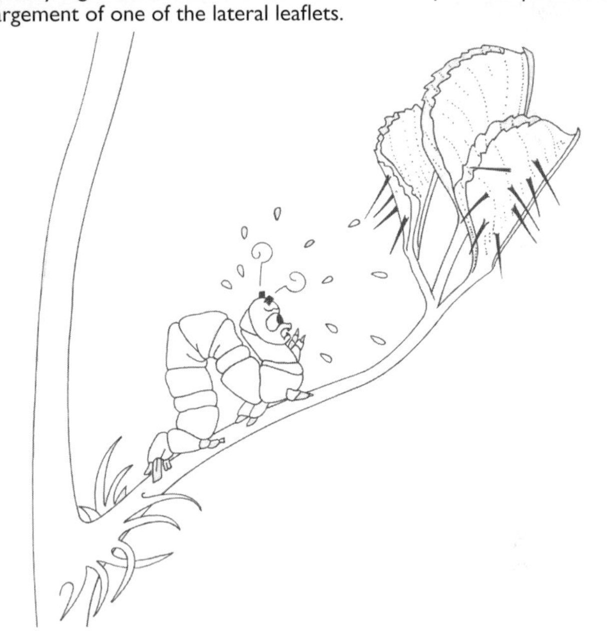

Fig. 7-14. An interpretation of how the pattern of pubescence on the trifoliolate leaves of *Medicago* might serve to discourage phytophagous insects (after Small et al. 1981c).

Pubescence pattern of trifoliolate leaves

Young and vigorous trifoliolate leaves of *Medicago* species fold elegantly at night to reduce their exposed surfaces by two-thirds while specifically presenting high concentrations of hairs on the exposed third of the surface, all of which can deter insects from feeding (Small et al. 1981*c*; Figs. 7-12, 7-13, and 7-14).

5. Flowers

Insect pollination represents a special kind of herbivory that is beneficial to plants. In the case of *Medicago*, the evolution of this symbiotic relationship is critical to understanding the adaptive nature of the plants.

Explosive versus passive tripping

Knuth (1908) described a variety of mechanical systems for effectuating pollination. Of these, the "simple valvular" arrangement, which characterizes most species of *Trigonella*, *Melilotus*, and *Trifolium* (close relatives of *Medicago*), and the explosive arrangement, which characterizes *Medicago*, are of particular relevance. In the simple valvular arrangement, the stamens and pistil project from the keel as long as the visiting pollinator continues to exert pressure. Once the visitor leaves they return to their original position, permitting repeated pollination. The mechanism in explosively tripping flowers, which trip only once, is complex. In brief, when a bee enters the throat of the *Medicago* flower to collect nectar, it must

Fig. 7-15. Alfalfa leafcutter bees (*Megachile rotundata*) on alfalfa flowers. A female is on the left, a male on the right. These bees are adapted to the explosive floral tripping mechanism of *Medicago* flowers and appear to have co-evolved with the genus.

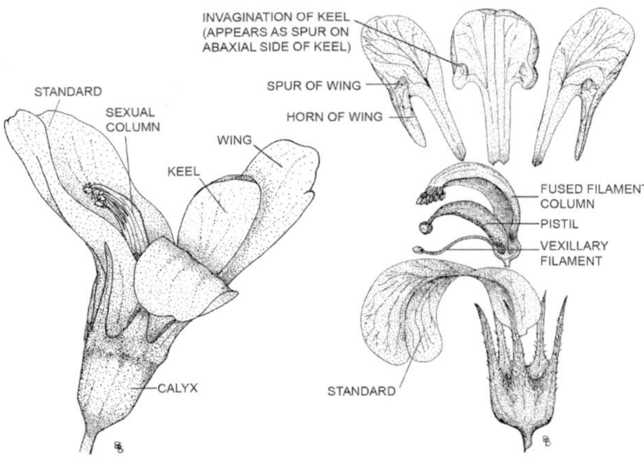

Fig. 7-16. Flower of alfalfa, undissected at the left and exploded at the right with the sexual column in the tripped position (artist: B. Brookes).

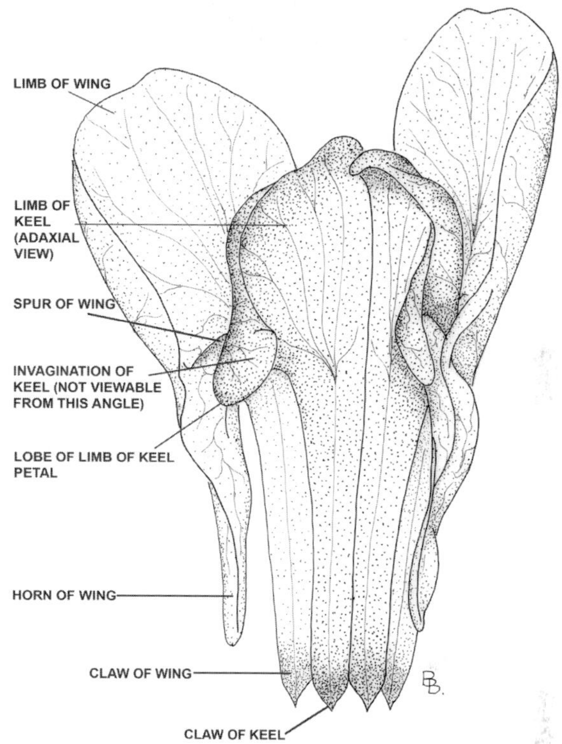

Fig. 7-17. Interconnection between keel and wing petals. The standard petal (not shown) is positioned in front of the keel in this view. A ventral (inner, adaxial) wing spur fits into a dorsal (outer, abaxial) keel invagination (cf. Fig. 7-17). Drawn from a tripped flower (i.e., adhesion tissue connecting keel lobes has separated; cf. Fig. 7-16). (Artist: B. Brookes.)

push aside special horns of the wing petals (the two outer or side petals) to reach the nectary at the base of the flower. In doing so, the sexual column is released and the stigma and 10 anthers slap explosively against the visitor, dusting it with pollen, and the stigma possibly picking up pollen that the bee previously collected from other flowers. This mechanism is explained in more detail below.

Explosive tripping in *Medicago* flowers in relation to the alfalfa leafcutter bee

The following comments about the explosive pollination mechanism in *Medicago* are based on *M. sativa* in particular because more publications have been produced about the pollination biology of alfalfa than for any other plant (McGregor 1976). As pointed out below, there is extensive evidence of coadaptation and perhaps also coevolution for pollination between the genus *Medicago* and the bee genus *Megachile*, and particularly between alfalfa and the alfalfa leafcutter bee (*Megachile rotundata* (Fabricius), also known as the alfalfa leafcutting bee and the lucerne bee, Fig. 7-15). *Medicago sativa* is basically an outcrossing species that requires cross-pollination, although some populations can set substantial amounts of seed by selfing. Most of the other perennial species are also at least partly outcrossing, but almost all the annual species appear to be strongly selfing, and so have substantially abandoned their relationships with pollinators. The floral structure of most of the annual species of *Medicago* is related to their mostly selfing nature. The annuals have flowers that may be closed (cleistogamous), although they are usually open (chasmogamous) and auto-tripping (Novosyelova 1998).

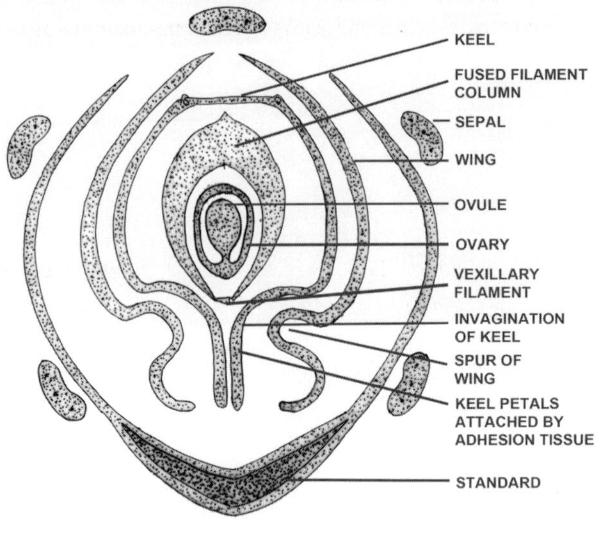

Fig. 7-18. A cross section of a flower of alfalfa (*Medicago sativa*) at the level of the juncture of the wing spurs and keel invaginations (artist: B. Brookes).

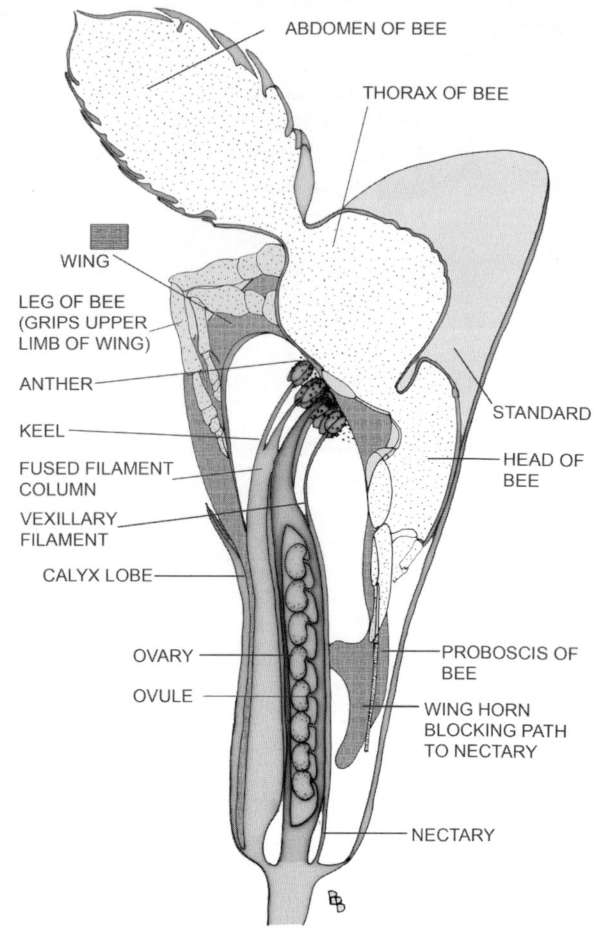

Fig. 7-19. A long section of an untripped alfalfa flower with a bee, *Megachile rotundata* (Fabricius), probing for the basal nectary (after Ivanochko 1979).

As detailed in Chapter 9, all species in the genus *Medicago* possess a complex syndrome of at least a dozen floral features designed for explosive cross-pollination. These floral features are in fact individual adaptations, but are functionally and developmentally correlated with each other so that they represent an adaptive syndrome. (For detailed information see Small 1987*c*, 1988, 1989*a*; Small and Jomphe 1989*b*; Small et al. 1981*b*; Small et al. 1987; for a detailed description of the mechanism in *M. sativa*, see Larkin and Graumann 1954.)

Papilionaceous flowers with nectaries at the base of the ovary or its stipe commonly have a free (vexillary) filament adjacent to the standard petal. The slits in the androecial column allow an insect's proboscis to reach the nectaries. In *Medicago,* the remaining fused nine filaments thicken progressively towards the keel side of the flower. The explosive tripping mechanism arises from the pressure exerted by elongated "expansion-restrained" cells of the keel side of the staminal column. This pressure results in a curving of the staminal tube, which arcs the sexual column towards the standard (Fig. 7-19).

Views of the flower of *M. sativa* are shown in Figs. 7-16 to 7-20. A complicated arrangement of the keel and wing petals acts to restrain the sexual column, preventing it from slapping against the standard petal. The free edges of the keel, especially the bottom of the limbs, possess "adhesion tissue". This adhesion tissue comprises an interlocking mesh of knobs and sockets, which grasp similar tissue. As indicated in the cross section of the flower (Fig. 7-18), the free edges of the keel petals cohere by this adhesion tissue, forming a tube around the sexual column, thereby restraining it. On each wing petal there is a "spur" (Figs. 7-16, 7-17, 7-18) that fits into an invagination of the keel petal at its area of maximum adhesion tissue (i.e., at the bottom of the keel limb). The wing spur and both sides of the keel invagination also possess adhesion tissue so that the wings and keel interlock strongly; Kreitner and Sorensen (1985) found that the interlocking is less strong in selfing species. The limb of the wing dilates into a "horn" basally from the wing spur (Fig. 7-16). The horns of the wings block the throat of the corolla, preventing a bee's proboscis from reaching the basal nectary (Fig. 7-19). When a bee of appropriate dimensions attempts to reach the nectary, it spreads the wing horns, which are the "trigger" of the explosive flowers. These act as levers, transmitting the spreading motion through to the keel petals (via the spur-invagination connection), resulting in pulling apart of the adhesion tissue holding the keel petals together, and a rapid release of the sexual column, which slaps against the ventral part of the insect. If the insect is carrying pollen from the flowers of another plant, the stigma may be cross-pollinated. The expansion-restrained cells of the fused filament column elongate quickly upon release of the column, "freezing" the sexual column in its new position and preventing it from returning to its original position for repeated tripping, as occurs in *Melilotus* and *Trigonella* s. str. (see Chapter 9).

The bee genus *Megachile* and its special relationship to *Medicago*

Species of the bee genus *Megachile* occur on all the major land masses of the world, with more than a hundred species in North America and a larger number in Eurasia. An analysis of 150 literature records of bee collections on alfalfa revealed that the Megachilidae was the most frequently collected family (aside from the social Apidae, which tend to visit all flowers) and that *Megachile* was the most frequently collected genus (Brookes 1992). The same observation was made by examining more than 500 bee collections in Canada (Brookes 1992). Megachilids are reputed to be efficient on alfalfa (Bohart 1958). This affinity of alfalfa flowers for *Megachile* species is suggestive of a well developed symbiosis between the bee genus and the plant species.

Of the perhaps 20 000 species of bees in the world, the honey bee and the alfalfa leafcutter bee may well be the only ones that have been truly "domesticated", i.e., selected genetically for increased usefulness to humans. The alfalfa leafcutter is second only to the honey bee in economic importance. It is a Palearctic species, occurring naturally from Europe and North Africa to Mongolia. It was inadvertently introduced from eastern Europe or western Asia to the east coast of North America about 1930. It is currently found throughout the northern United States and extends north to southern Ontario. The name leafcutter (or leafcutting) bee stems from the females' use of their serrated mandibles to cut round or oblong sections of leaves and petals. These pieces of plant material are shaped into small, thimble-like containers that are laid end to end in tunnels. Each cell is provisioned with an egg and a mixture of pollen and nectar. Although the bee collects alfalfa leaves if other vegetation is unavailable, it prefers softer plant materials, especially large petals of various other species. Many species of leafcutter bee occur in alfalfa fields. *Megachile rotundata* came to prominence simply by being noticed as an exceptionally efficient pollinator of alfalfa with a gregarious habit suiting it to domestication. The bee is now used to produce alfalfa seed in the western United States, western Canada, New Zealand, and in some South American and European countries.

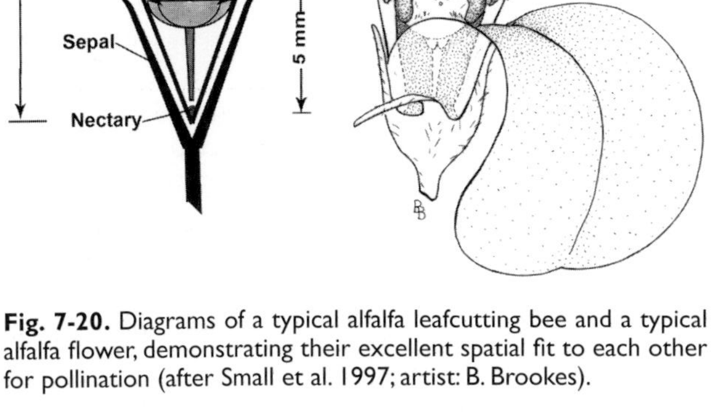

Fig. 7-20. Diagrams of a typical alfalfa leafcutting bee and a typical alfalfa flower, demonstrating their excellent spatial fit to each other for pollination (after Small et al. 1997; artist: B. Brookes).

The foraging period of the alfalfa leafcutter bee usually coincides with the blooming period of alfalfa in most climates (Bohart 1972), further suggesting a possibly intimate evolutionary relationship between the plant and the bee. Wild alfalfas lie within the Old World range of the alfalfa leafcutter bee. The bee is bivoltine in warm climates (Richards 1984), and since alfalfa flowering can extend over several months, the second generation of bees could take advantage of this.

The alfalfa leafcutter bee is "narrowly polylectic" (in the sense of Michener 1979), i.e., it collects pollen from unrelated plants of a relatively narrow range. It will visit flowers of only about 10% of showy-flowered plants (Small et al. 1994). It is an especially good pollinator of several species of the Fabaceae, but will visit a few species of many plant families. Experimental studies of its preferences (Small et al. 1997) have shown that it is attracted only to clusters of fairly small flowers with dimensions that allow the bee to readily reach nectaries. Indeed, the bee's dimensions appear specifically sized to fit the architecture of an alfalfa flower (Small et al. 1997).

Most bees dislike being slapped by alfalfa flowers and learn to avoid them. Indeed, most bees prefer flowers other than those of alfalfa (Brookes et al. 1994). By contrast, the alfalfa leafcutter bee seems instinctively programmed to accept the alfalfa flower. A characteristic stereotyped behavior of female alfalfa leafcutter bees has been observed on alfalfa flowers (Small 1996). The bee alights on the wing petals, her head oriented toward the throat of the flower. The flower is tripped by using a clawing motion of her front legs on the "thumb and pocket" region of the corolla and a simultaneous pressing down of the wing petals with the hind legs. After tripping, the sexual column slaps against the ventral surface of the bee, depositing pollen and pressing the standard away from the keel, thus opening the throat of the flower. The bee then inserts her proboscis down the throat of the flower and imbibes the nectar. During her floral visits the bee periodically pauses to groom pollen from her body onto the abdominal scopa. Newly emerged adults (that had never seen an alfalfa flower) when placed near alfalfa flowers have exhibited this precise sequence carried out by experienced adults. An anonymous reviewer of this monograph noted "I have also observed the bee to work the youngest flowers that have not reached the "explosive" stage. The sexual column is released but just barely; it is more of a passive contact with the body."

The alfalfa leafcutter bee's relatively narrow host preferences and stereotypical behavior on alfalfa suggest a possible coevolutionary relationship. "Coevolution" has been variously interpreted (see for example Janzen 1980; Thompson 1982; Futuyma and Slatkin 1983; and Thompson 1989). It involves reciprocally evolved adaptations between two species. "Diffuse coevolution" relaxes the criterion of interaction between only one pair of species to include other species (see for example Hougen-Eitzman and Rausher 1994). Because of the difficulty in evaluating historical events such as evolution, the intimacy and uniqueness of relationships are usually used to indicate whether a pair of species has coevolved. The innate behavior of the bee in triggering the explosive flower by a rather complex behavior likely means that the bee (or at least its close ancestors) evolved this behavior in response to the alfalfa flower (or at least its close ancestors). Explosive flowers are quite rare, and the bee's adaptation to the specific mechanism of the flower of the genus *Medicago* is unlikely to have evolved with respect to any other explosive flower.

It is interesting, however, that on related genera such as sweetclovers (*Melilotus*) and clovers (*Trifolium*) the bee does not exhibit the characteristic behavioral sequence it shows on the alfalfa flower, but rather proceeds directly to harvest nectar. Moreover, although the alfalfa leafcutter bee strongly prefers alfalfa over most plant species, it nevertheless prefers several species over alfalfa (Packer 1970; Small et al. 1997). Clearly, its flexible behavior allows it to exploit a variety of flowers despite its very limited preferences. Outcrossing *Medicago* species also exhibit flexibility and are commonly visited by many bee species, although of a smaller range than most plant species. It does seem that, while some coevolution has occurred, neither species has become dependent on the other.

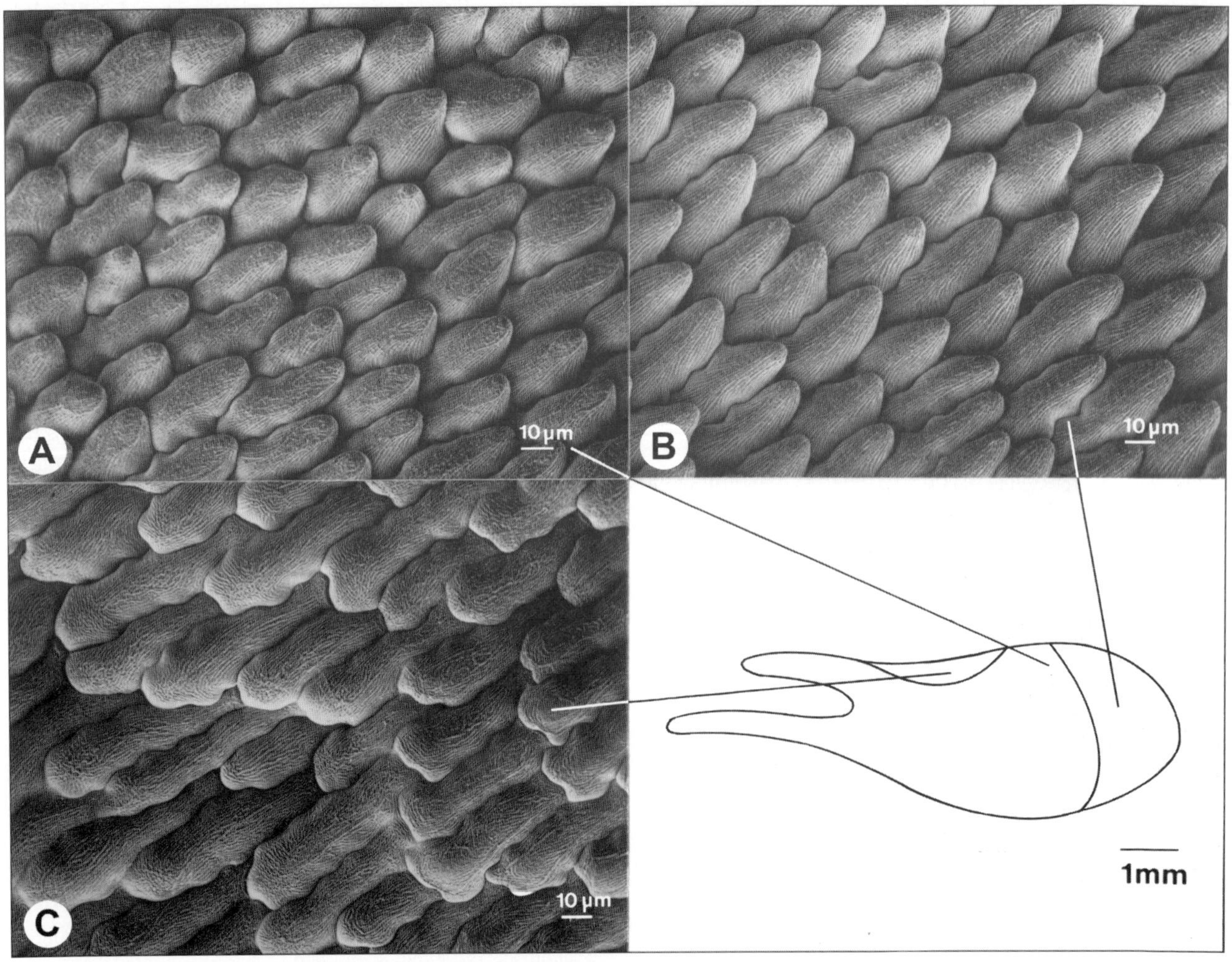

Fig. 7-21. Microsculptural surface pattern on a wing petal of alfalfa (*Medicago sativa*), a typical outcrossing species of the genus. The outline of a typical wing petal is shown: (A) region of indistinct microtexture; (B) region of apically-oriented papillate microtexture; (C) region of basally-oriented shingle-like microtexture. In the following three plates, compare the absence of well-defined texture in the corresponding areas of a selfing-cultivar of alfalfa, an obligatively selfing wild species of *Medicago*, and a species of *Trifolium*. (Scanning photomicrographs prepared by B. Brookes.)

Adaptations of wing petal surface topography as a landing platform for pollinators

One of the most interesting adaptations between alfalfa flowers and bees is an elaborate surface topography of the wing petals that appears to function as a friction-providing landing platform for the bees (Brookes and Small 1988; Fig. 7-21). Note that the three kinds of petals in a papilionaceous flower may have specialized functions in *Medicago* — the large standard petal serves to attract pollinators, the two fused keel petals surround and protect the sexual column, and the two wing petals serve as landing platforms for visiting bees as they forage for nectar at the base of the flower. In outcrossing species of the genus, the wings have surface textural patterns, based on overlapping epidermal cells, which would assist pollinators in gaining a stable foothold. Patterning is confined to the limb of the wing petal and is absent from the claw. At the far end of the limb, papillate cell extensions point toward the apex. Towards the base of the limb, on the side nearest to the standard petal, there is another distinctive region where cells appear to overlap in the direction of the base. Patterning on the remainder of the petal, if present, is much less obvious, so that the two regions are prominently different.

Alfalfa is primarily an outbreeder, but there are a few inbreeding cultivars, selected to obtain a seed crop in locations (especially in North America) where natural pollinators have decreased as a result of environmental destruction and/or the cost of providing cultured pollinators has become significant. Interestingly, in selfing cultivars of alfalfa such as 'Canauto', which have been bred to reduce the need for insect pollination,

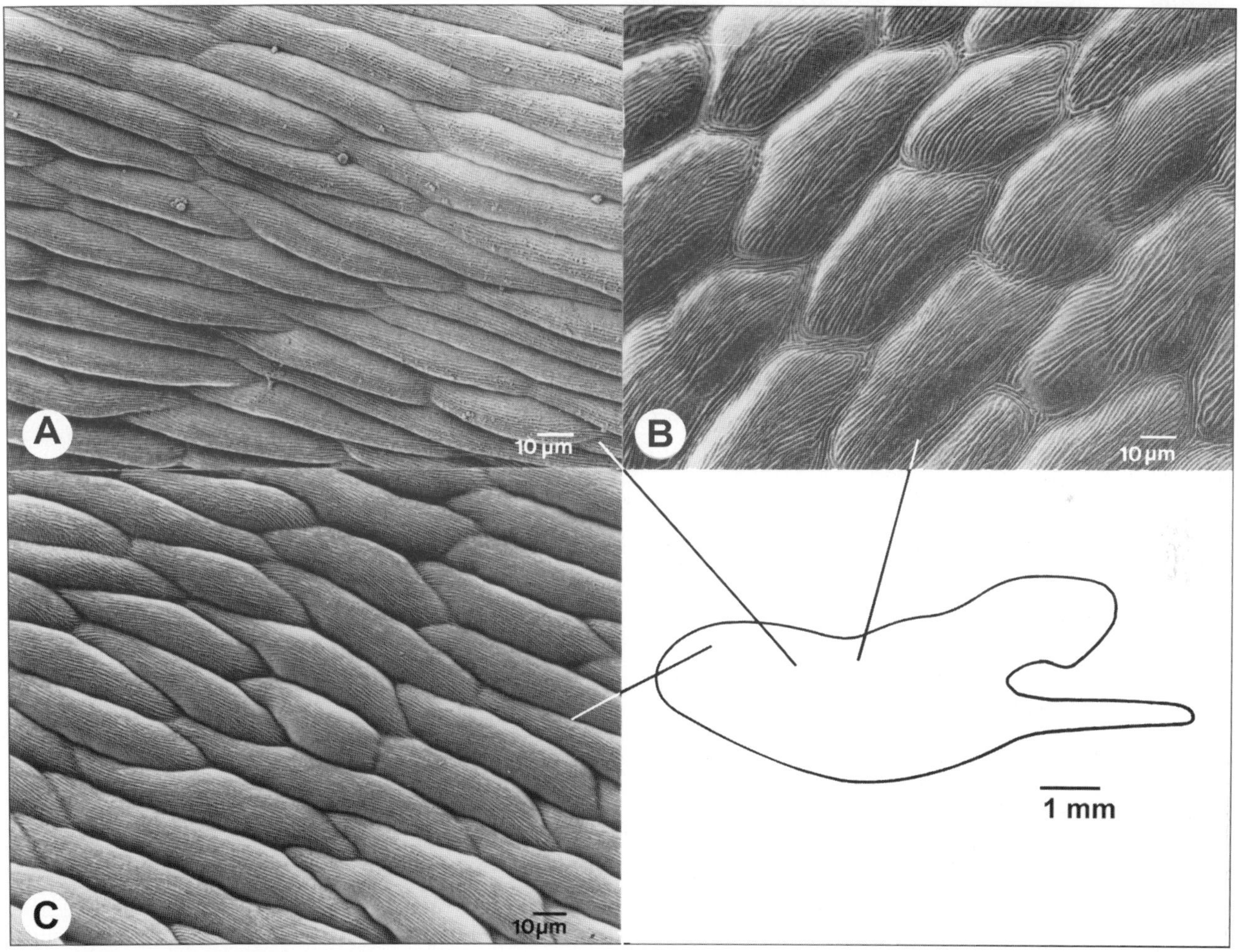

Fig. 7-22. Surface of a wing petal of 'Canauto', an inbreeding cultivar of alfalfa (*Medicago sativa*). Note how indistinct the microtexture is compared with the same regions (A, B, C) of the outcrossing alfalfa. (Scanning photomicrographs prepared by B. Brookes.)

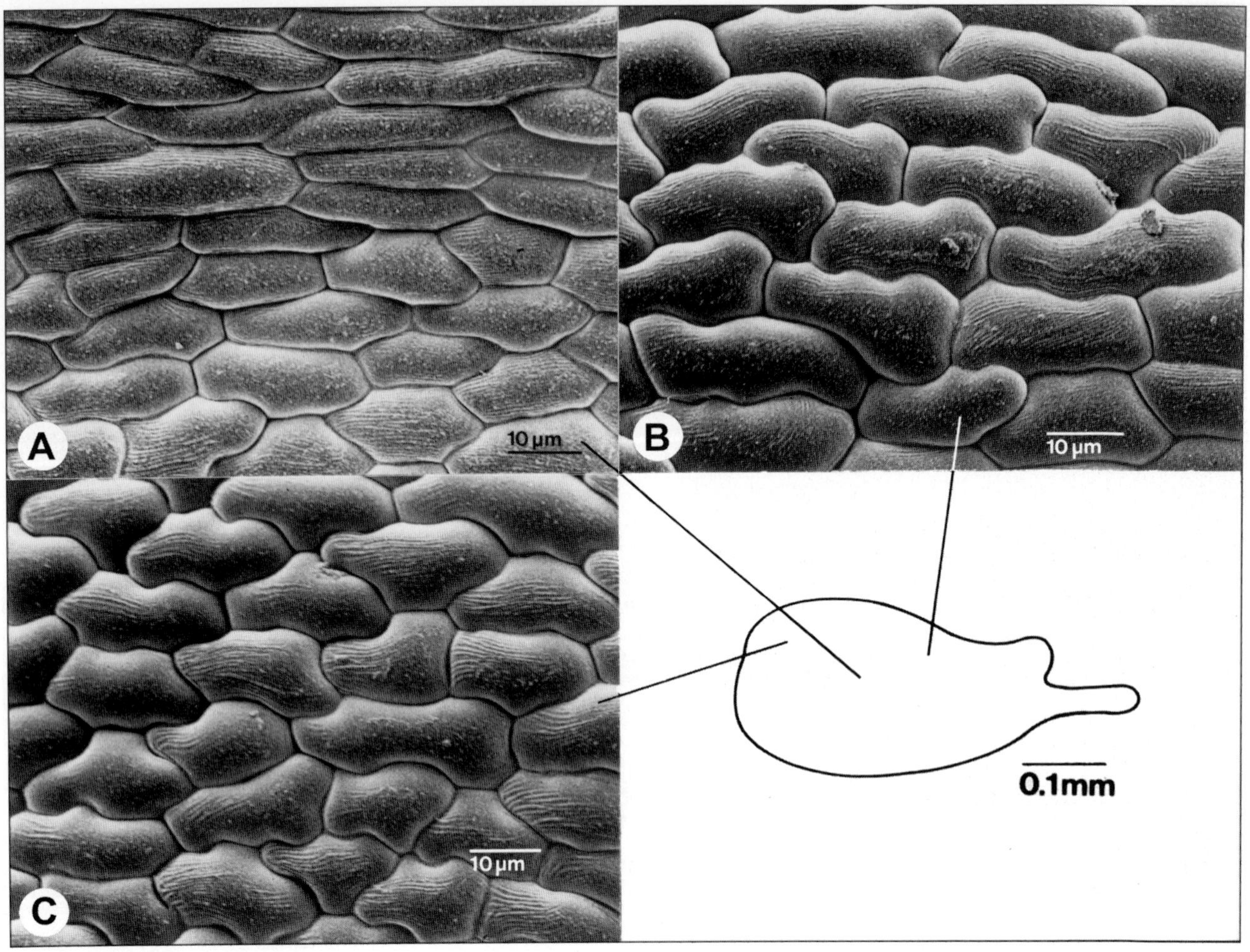

Fig. 7-23. Surface of a wing petal of the obligatively inbreeding *Medicago hypogaea*. Note the indistinct microtexture compared with the same regions (A, B, C) of the outcrossing alfalfa. (Scanning photomicrographs prepared by B. Brookes.)

the development of this anti-skid micropattern appears to be markedly reduced (Brookes and Small 1988; Fig. 7-22). This pattern has been lost presumably because the adaptive wing petal microtexture is no longer needed.

Similarly, there is no wing petal microtexture in the annual (inbreeding) species of *Medicago*, which is also consistent with their independence from pollinators. This parallels the rather pronounced degeneration of other floral features in inbreeding species of *Medicago* compared with that in the outbreeders (e.g., Small et al. 1981*b*; Small 1988). For example, in the obligate inbreeder *M. hypogaea* (Fig. 7-23), note the highly reduced size of the wing petal and the strikingly small proportion occupied by the horn (the "trigger" of the explosive pollination mechanism). The conclusion seems inescapable that there has been an evolutionary loss of petal microsculpturing because the advent of autogamy made it unnecessary to accommodate insect pollinators.

Brookes and Small (1988) reported that there is one further circumstance in which there is a lack of overlapping epidermal cell microscupturing — in papilionaceous flowers when they are in head-like inflorescences (note Fig. 7-24). The only such species in *Medicago* is *M. lupulina*, which in any event may lack microsculpturing because it is an inbreeder. The absence of texture on the wing petals when the flowers are in heads is probably because an anti-skid surface is not required as visiting bees can conveniently land on the crowded inflorescence.

As explained in Brookes and Small (1988), in addition to the theory that petal microtexture acts as an anti-skid surface for the "feet" of visiting pollinators, there are two alternative explanations: they could be tactile guides or light-orientation guides to the basal nectary. The anti-skid interpretation presented here is preferable since the direction of overlapping papillae is suited for this, while interpreting the distinct region at the base of the wing blade as an orientation guide (either as a kind of Braille or as a lighting guide) seems unnecessary as the flower is well supplied with color guides.

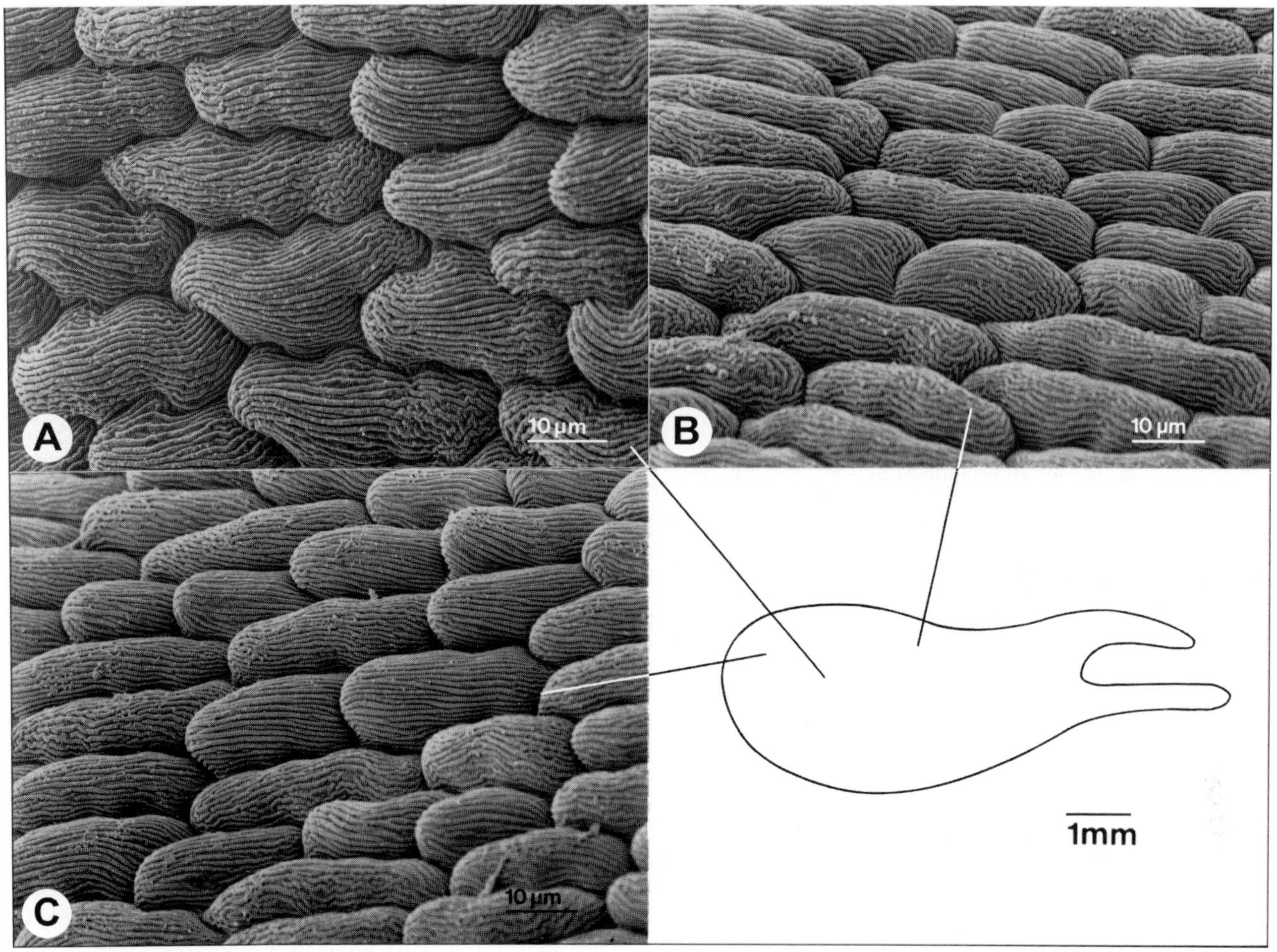

Fig. 7-24. Surface of a wing petal of *Trifolium medium* L., a species that has flowers in crowded heads. Note the indistinct microtexture compared with the same regions (A, B, C) of outcrossing alfalfa. (Scanning photomicrographs prepared by B. Brookes.)

6. Fruits

Several features of *Medicago* pods are significant adaptively as well as taxonomically. These are discussed below.

Fruit suture spines

Many *Medicago* species have coiled fruits with dorsal suture spines (this is the case for seven of the 21 exclusively perennial species and 33 of the 66 other species). All these species have indehiscent fruits. By contrast, none of the species with uncoiled fruits are spiny (although several have fruit sutures ornamented with wings or spinules) and the fruits of these species are more or less dehiscent. This suggests that the spines are present to protect the seeds, which are extremely nutritious for animals. Goats are generally not deterred by the spiny fruits of *Medicago*, but species with very hard spines (section *Spirocarpos* subsection *Pachyspirae*) are not preferred by most herbivores. The seeds remain in the indehiscent fruits for a long period (sometimes for years) before germinating and seem to require protection from being consumed. The majority of the annual species that are very widespread have spiny fruits, and this is clearly related to the ability of the spines to anchor the fruits to fur (the spines on the fruits of the annuals are often hooked at their tips, promoting such attachment). Lesins and Lesins (1979) suggested that the spines are a means of anchoring the indehiscent fruits in soil, as would be desirable on sloping land and in windy areas that have limited vegetation to stop the fruits from being blown. The perennial species with spines generally occupy hilly or mountainous areas, and some of the annuals are located in windy areas with limited vegetation, so at least in some species spines may serve to anchor fruits to the substrate. In species with spiny fruits, there are frequently plants with long spines, short spines, or very short spines (i.e. with tubercles), and these have in most cases (and very inadvisedly) been granted taxonomic status, since these characteristics are almost never correlated with other characteristics or with geography, and spine length is very variable in most species that possess fruit spines. It has been suggested that for *M. polymorpha* at least, forms without spines have been selected locally in some cases because this prevents them from becoming entangled in fur and carried away from a hospitable site (see Del Pozo et al. 2000, 2001, 2002).

Fruit spiralization

More than 50 species of *Medicago* have coiled fruits, once considered to be a defining characteristic of the genus. *Medicago sativa* is especially interesting since it is the only species with both coiled and uncoiled pods, with fruits ranging from the slightly curved fruit typical of *M. sativa* subsp. *falcata* (Fig. 7-25A) to the tightly coiled, snail-like fruit of *M. sativa* subsp. *sativa* (Fig. 7-25C). The degree of spiralization of the fruit has been shown to be strongly defensive against a specialized seed herbivore, the seed chalcid *Bruchophagus roddi* Gussakovsky (Hymenoptera: Eurytomidae) (Small and Brookes 1982, 1984*a*; Figs.7-25, 7-26). This insect is the world's most devastating pest of commercial alfalfa seed production, accounting for millions of dollars in damage each year and often virtually destroying seed crops. Pesticides have not provided a solution because of the need to avoid harming pollinators that are active on the plants for much of the time that the seeds are being formed. The discovery that pod coiling is very effective against the seed chalcid has provided a simple means of breeding resistance. *Bruchopophagus roddi* damages seeds by ovipositing through the fruit wall, generally near the placental suture, directly into young seeds. However, the insect's ovipositor is too short to reach the seeds in coiled fruits. This is because in coiled fruits the seeds are located at the inaccessible axis of the coil and the ovipositor is too short to penetrate through the dorsal suture area (the only exposed surface in tightly coiled pods, except for the ends of the pod) and still reach the seeds. Coiled fruits are occasionally encountered in other genera besides *Medicago*, and it would be interesting to examine whether coiling is of similar adaptive value in these other plants.

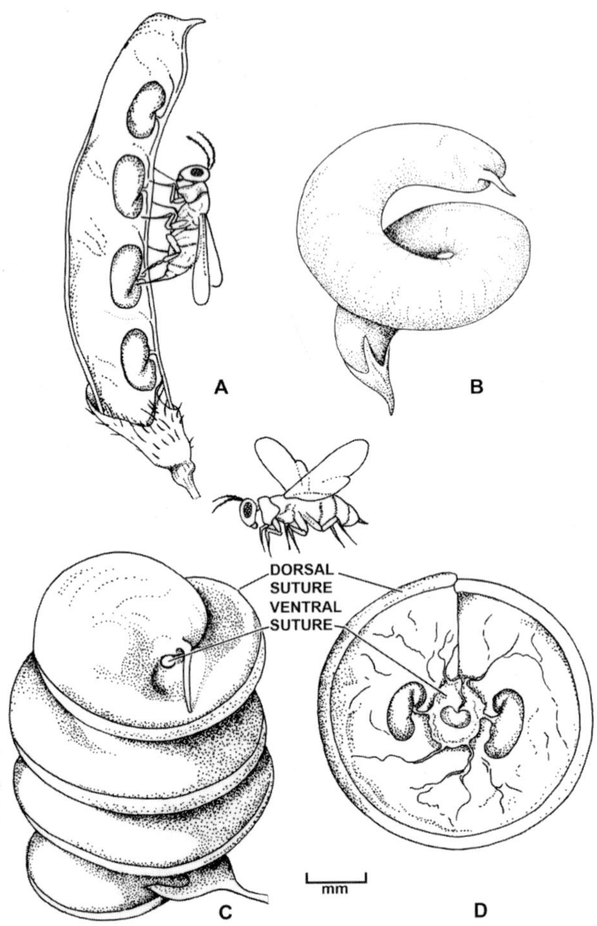

Fig. 7-25. Diagrams showing how fruit coiling in *Medicago sativa* provides a mechanical defense against seed chalcids (*Bruchophagus roddi*). (A) *Medicago sativa* subsp. *falcata* showing the relative accessibility of ovules to oviposition at the ventral suture. (B) *Medicago sativa* subsp. ×*varia* with 1.5 coils. (C) *M. sativa* subsp. *sativa* with 3.5 coils. (D) A cross section of (C) showing young seeds at the ventral suture, located at the axis of the pod. Note that in the highly coiled pod the ovules (other than those at the apex) are remote from, and hence out of range of, the chalcid's short ovipositor (after Small and Brookes 1982).

Between-seed mechanical separation within fruits

A problem with multi-seeded legume fruits is that once the fruit walls are penetrated by an insect it has access to all the seeds within the pod. At least this is the case if the fruit is like a pea pod. In *Medicago*, there are several fruit specializations that seem designed to prevent this (although experimentation to demonstrate evidence of effectiveness has not been attempted). An obvious mechanism is simply to have smaller pods, each with fewer seeds, and many species of *Medicago* are few-seeded (although only *M. lupulina* and *M. secundiflora* always have one-seeded fruits). Numerous species in section *Spirocarpos* and some in section *Medicago* have pads of spongy tissue separating the seeds. The pod walls of some of the flat-podded species are "shrink-wrapped" around the seeds so that they are effectively separated. A similar phenomenon exists in the two species of the linear-podded section *Isthmocarpae*. The only species in section *Pectinatae* has plicate fruits, i.e., the pod is pleated between the seeds, which is another clever way of preventing easy access to the seeds.

Flat pods

The species of several sections of *Medicago* (sections *Platycarpae* and *Lunatae* and the monotypic sections *Pectinatae* and *Ovales*) have quite flat fruits. Propagule flatness is frequently associated with wind distribution, and this may be the case for these species.

Fruit size

Large fruits tend to be relatively more attractive to herbivores than small fruits. This appears to be the case both for small and large herbivores. Small et al. (1989) noted that a bean weevil (*Bruchidius bimaculatus*) preferentially infested the larger fruits of *M. littoralis* (see Fig. 7-28).

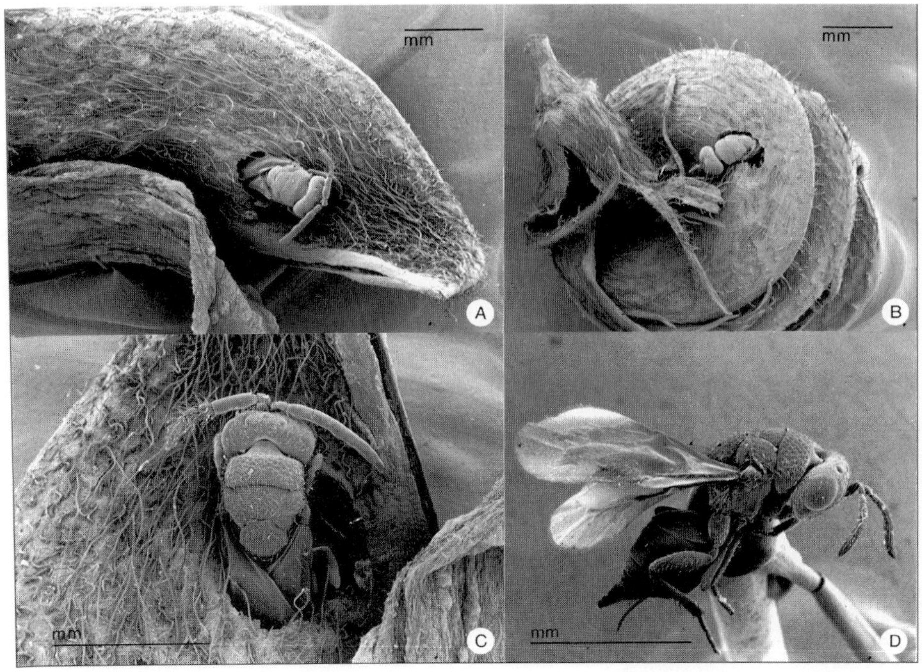

Fig. 7-26. Scanning electron photomicrographs of the alfalfa seed chalcid mounted on a pin (D), emerging from an uncoiled pod (A, C), and emerging from the end of a highly coiled pod (B) (after Small and Brookes 1984*a*).

Roggero and Porqueddu (1999) noted that "Small pods are less likely to be selected by grazing sheep in summer than large pods. Small pods also have the advantage that they are easily incorporated into the top soil, although they are subject to greater predation by ants... Among the morphological characteristics of the pod, the ones related to pod burial have an important evolutionary advantage in areas that have always been grazed...when tested in the field, pod burial by sheep trampling during summer grazing improved emergence in a genotype (of *M. murex*) characterized by lighter, more open pods that were more easily trampled than another with heavier, closed pods." Piano and Francis (1992) concluded that "a generalization which can be drawn from many ecogeographic surveys is that large podded and seeded species, such as *M. scutellata*, *M. ciliaris*, *M. intertexta*, and *M. blancheana*, are particularly rare in heavily grazed environments." They explained this on the basis that these species are relatively vulnerable to grazing, resulting in severe depletion of seed reserves in the soil, and hence jeopardizing long-term regeneration.

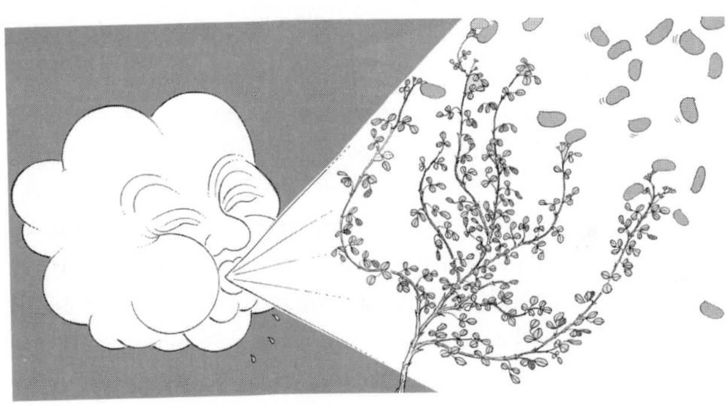

Fig. 7-27. Flat-podded fruits such as these may be adapted to distribution by wind (artist: B. Flahey).

Geocarpy

Medicago hypogaea is the only species in the genus that develops its seeds underground. According to Lev-Yadun (2000), there are only about 20 known plant species that are completely geocarpic (i.e., producing all their seeds underground) and about 50 that are amphicarpic (i.e., plants that produce some fruits underground and some fruits aerially). Lev-Yadun (2000) listed the following as habitat factors that favor the evolution of geocarpy: (*a*) maintaining occupation of a proven advantageous site in a heterogeneous environment; (*b*) overcoming uncertainties of survival through the growth season, related to threats such as fire, grazing, and drought; (*c*) protecting seeds from extreme fluctuations of microclimate at the soil surface.

7. Seeds

Seeds are of course essential to plant survival, but are also the most vulnerable phase of plant existence. As one would expect, despite their tiny size, they exhibit an extraordinary range of adaptations.

Seed size and number per pod

As in all plant species, a compromise is established between seed size (large seed reserves giving the seedling a greater ability to survive) and seed number (more, albeit

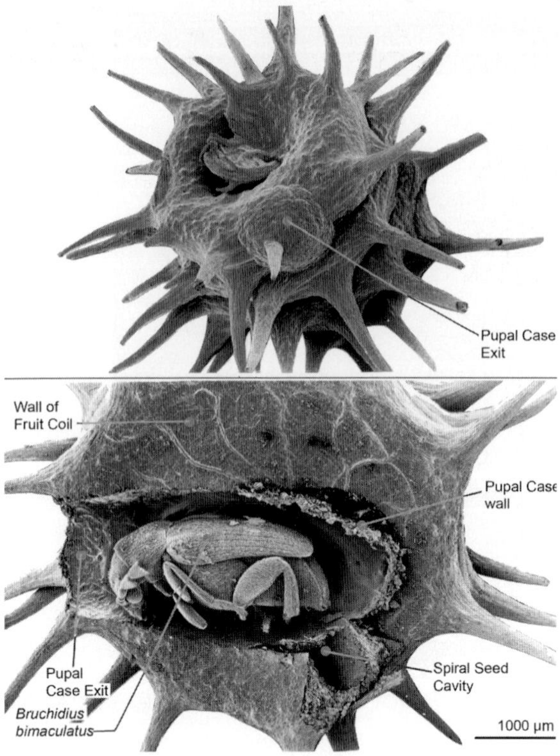

Fig. 7-28. Scanning electron micrographs of the spirally coiled, spiny fruits of *Medicago littoralis* (collection E. Small-Israel-97, DAO) infested by *Bruchidius bimaculatus*. **Upper panel:** Intact fruit in which a pupa has caused a characteristic, transparent surface swelling through which it will exit. **Lower panel:** Fruit split apart between two coils revealing an emergent bean weevil in its pupal case. (Prepared by B. Brookes. Based on Small et al. 1989.)

smaller, seeds giving the species a better chance of reaching suitable habitats). In *Medicago*, seed size is strongly influenced by pod position and plant maturity (Horovitz et al. 1976). Cocks (1990) found that in annual species of *Medicago* the seed in late-produced pods was only half as heavy as early-produced seeds.

Ehrman and Cocks (1988) observed that in annual *Medicago* (medics), relatively wet areas have a high frequency of species with large, soft seeds and few seeds per pod, whereas dry areas have a high frequency of species with small seeds, high levels of hardseedness and many seeds per pod (cf. Piano and Francis 1992). Small seeds are probably advantageous in that they require less water for germination because of their high surface-to-volume ratio and because they mature faster than large seeds, an important feature when moisture is limited (Piano and Francis 1992). However, there is often a cost to this — as seed size decreases so does seedling vigor (Twamley 1974).

Piano and Francis (1992) pointed out that medic species common on clay soils are generally larger-seeded than those that characteristically occupy sand. They suggested that "this characteristic may reflect some advantage in terms of plant emergence and root penetration in soils subject to compaction."

Seed longevity

Seeds of Fabaceae species are often quite long-lived and seeds of *Medicago* species are no exception. Taylor and Ewing (1988) examined the longevity of *M. polymorpha* and *M. truncatula* seeds buried at depths of 0, 2, 6 and 10 cm and observed that, except at the soil surface, burial depth had only a minor effect on seed survival. A seed of *M. polymorpha* (estimated to be 200 years of age) that was extracted from Mexican adobe brick was still able to germinate, and although the storage conditions were much drier than normally encountered, this is still an impressive longevity (Spira and Wagner 1983). For additional information on seed longevity, see Chapter 3 on "The Importance of Germplasm."

Seed dormancy

Most *Medicago* species, especially the annuals, produce "hard" seed, i.e., the seed coat is indeed hard (see, for example, Gachet and Elmir 1972; Ruiz and Devesa 1998). This is correlated with seed dormancy since the seed coat is hydrophobic and needs to be at least partially ruptured to allow imbibition of water and consequent germination (Crawford et al. 1989). Impermeability has been attributed to a variety of chemical deposits in, and anatomical features of, the seed coat (Russi et al. 1992). Increased permeability of the seed coat, or of particular parts such as the hilum, micropyle, and/or lens, seems to occur in many, perhaps most, annual species of *Medicago* in response to weathering by temperature fluctuations (Zeng et al. 2005). Light may also control germination in some species (Cocks and Bennett 1999). Ivory (1976) found that seed position in *M. scutellata* influenced the percentage of hard seed, this decreasing from the calyx end of the pod to the stigma end. In some species such as *M. monspeliaca*, strong scarification can result in 100% germination (Riasat et al. 2005). The same is true of *M. orbicularis* (Patanè and Gresta 2006). Additionally, the seeds of some *Medicago* species display degrees of physiological embryonic dormancy (after-ripening) for several months after the pods have abscised from the parent plant (Christiansen et al. 1996; Garcia et al. 2006), although Kernick (1978) stated that for *Medicago* "embryo dormancy does not play an important role." As long as the seed coat is intact the seeds can remain viable for many years, but tend to lose their viability rapidly once the seed coat is scarified. The rate of "softening" of hard seeds is species-specific, and there may be variation among ecotypes within species (Cocks 1993; Lloyd et al. 1997; Cocks and Bennett 1999). Moreover, environmental conditions may influence the development of seed dormancy. According to Kernick (1978), the proportion of hard seed depends on the stage of seed development when drying starts, and seeds that mature quickly because of dry weather have less hardseededness. Seed dormancy in wild plants is adaptive, enabling the seeds to remain quiescent until conditions become favorable for germination and growth.

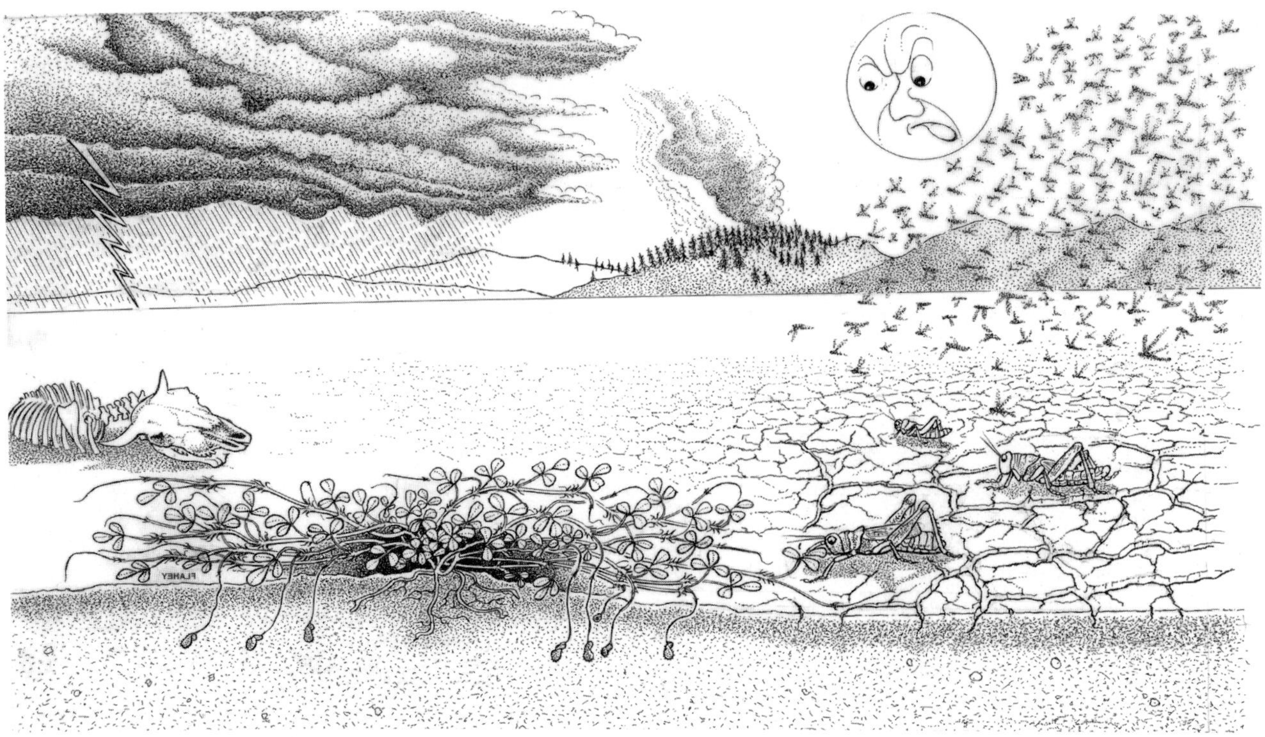

Fig. 7-29. An interpretation of *Medicago hypogaea* avoiding the many threats in its environment by passing most of its life cycle in the form of underground seeds (artist: B. Flahey).

Moreover, seed germination tends not only to be delayed, but also asynchronous, the erratic germination constituting an insurance that the plants produced from a given batch of seeds will not all be subject to extermination due to some sort of disaster. This is critical to annual plants in Mediterranean climates because of so-called "false breaks" — periods of rainfall followed by long droughts during which newly germinated seedlings are likely to die.

Sidhu and Cavers (1977) studied hardseededness in Canadian *M. lupulina*. They found that the seeds have an initial nondormant phase that depends on environment and parental genotype. Adaptively, the nondormant seeds can exploit favorable conditions immediately, establishing large populations rapidly in suitable habitats, whereas seeds that acquire dormancy are able to survive for many years in the soil, potentially exploiting suitable conditions for growth in the future.

Pavone and Reader (1982) studied seed bank dynamics of *M. lupulina* in an abandoned pasture in southern Canada. They observed that there were 600–2000 seeds/m^2 (a figure considered low for pasture species) and concluded that most seeds remain dormant for a period, with only 30% to 40% of the natural seed bank germinating in a given year.

Wagner and Spira (1994) studied germination, establishment, and survival in *M. polymorpha* in an open grassland in California. They found that seeds were dormant when produced in the spring and remained largely dormant after exposure to summer conditions. About 40% of the viable seeds in the seed bank were carried over from year to year. Seedling emergence was mostly simultaneous following the first substantial autumn rains. Survival to reproduction differed between years. Few individuals produced seeds prior to senescence in a dry year, whereas in a wet year most overwintering individuals successfully produced fruits.

Yao and Tan (2005) studied germination patterns in three annual species of *Medicago* (*M. medicaginoides*, *M. monantha*, and *M. orthoceras*) of the Zhungaer desert in Xinjiang province, China. Their observations demonstrated a pattern that is often present in ephemeral, drought-avoiding species that mature their fruits before the arrival of the arid season — seed germination occurred in the spring, summer, and autumn in all three species, but was most prevalent in the spring and autumn. They also found that seed germination time was very closely related to environmental factors such as temperature and precipitation.

Seed survival after ingestion by mammals

Sheep play a major role in medic seed distribution and germination. *Medicago* seeds are very nutritious, especially for sheep (Cocks 1988), and it appears that between 90% and 99% of the seeds of most medics consumed by sheep are digested (Carter 1980). Medic seeds ingested by sheep rarely survive, but the surviving seeds are of great significance to seed dispersal. Thomson et al. (1990) examined germination rates of eight species of annual *Medicago* after the seeds had passed through the guts of sheep and found that 2% to 20% of seeds could be recovered. These germinated only slightly better than seeds that

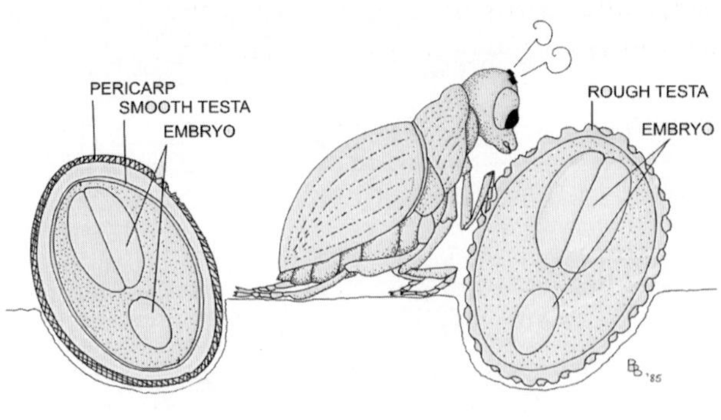

Fig. 7-30. A rationale for the presence of tubercle-protected seeds in dehiscent fruits and the absence of such protection in indehiscent fruit. **Left:** Smooth seed retained in indehiscent fruit is protected by the pericarp. **Right:** Seeds released from fruit may employ tubercles as protection against herbivores (artist: B. Brookes).

had not been fed to the sheep, indicating that scarification had been limited, so most of the seeds were still dormant. Relatively smaller seeds appear to have a higher survival rate following ingestion, although the fairly large-seeded *M. rotata* was found to have a relatively high survival rate (Thomson et al. 1990).

Ramos et al. (2006) employed both germination tests and viability tests (using tetrazolium), and found quite interesting results in an analysis of *M. polyceratia* seeds and two kinds of *M. sativa* seeds (from a wild and a domesticated variety, respectively) that had passed through sheep. About a third of the wild seeds (from *M. polyceratia* and wild *M. sativa*) were recovered intact and viable, but only about 2% of the domesticated *M. sativa* seeds survived passage through the sheep's guts. Some samples of the surviving seeds of *M. sativa* decreased in dormancy (i.e., a higher percentage germinated shortly thereafter), whereas the surviving seeds of *M. polyceratia* increased considerably in dormancy (i.e., a much smaller percentage germinated shortly thereafter, although the seeds remained viable.

The correlation of protective seed surface thickenings and fruit dehiscence

The majority of *Medicago* species lack rough-surfaced seeds although a few species do have them. Extra tissue is deposited on the surfaces of rough-surfaced seeds to make them rough (see Chapter 9), which requires that the plant expend energy, suggesting that there is an associated adaptive function.

Seed surface roughness (mostly tuberculate, sometimes rugose; see Chapter 9 on "Seed Characters" for definitions of these terms) probably has functional significance in *Medicago* and its nearest relatives. Almost all species of *Medicago* with rough seeds have fruits that dehisce at the sutures, or at least have fruits with pericarps that promptly decompose to release the seeds. However, smooth seeds mostly occur in fruits that are indehiscent with the pericarp providing substantial long-term covering (Small et al. 1990*a*). The surface roughenings may be protective against herbivorous insects and (or) increase friction and so help anchor the seeds in the soil. Puri (1979) hypothesized that surface bumps are a means of increasing seed surface area for imbibition of water, thereby contributing to rapid germination; however, there is no evidence that this is the case. Germination rate and longevity differences may exist between rough-surfaced and smooth-surfaced seeds in *Medicago*, perhaps related to microbial attack or freeze-thaw action, but these hypotheses remain to be explored.

Conjoined seeds: the dispersal of multi-seeded indehiscent fruit

Medicago species with dehiscent fruits typically scatter their seeds while the pods are still attached to the maternal plant. In some species (especially with flat pods), the pods fall off the plants and disintegrate in a short period, releasing the seeds. *Medicago lupulina* and *M. secundiflora* produce single-seeded fruits that are essentially achenes and are distributed singly (i.e., the fruits are equivalent to the seeds for practical purposes). The majority of *Medicago* species have coiled indehiscent fruits that fall off the plants at maturity and retain the seeds for periods (sometimes for years) before the fruit wall weathers and rots sufficiently to allow water to reach the seeds, resulting in germination (in a sense the fruits are "hard" like the seeds). (*Medicago heyniana* and *M. orbicularis* are exceptional for coiled-podded species in being more or less dehiscent.) The walls of the fruits of coiled-podded *Medicago* species differ greatly in their hardness and, therefore, in how long the fruits are retained before the walls disintegrate. The hardest fruits are produced by the species of section *Spirocarpos* subsection *Pachyspirae*. Because the seeds themselves are physiologically hard (i.e., dormant), only one or a few seeds within a given fruit will germinate at one time. Frequently, a seed germinates within a fruit with the shoot and root exiting but the remnants of the fruit remaining attached at the bottom of the plant (see Fig. 7-31). Sometimes several seeds will germinate simultaneously within a fruit, resulting in plants that are locked together like Siamese twins, which is an undesirable situation. However, this system is advantageous in that the fruit wall protects the seeds from herbivores until the seeds germinate. The system is

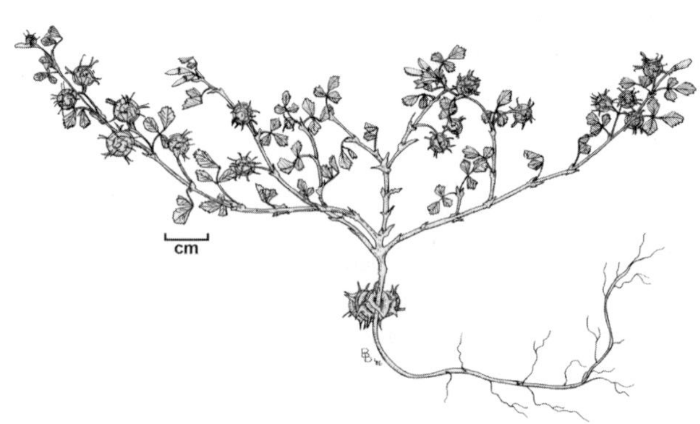

Fig. 7-31. *Medicago littoralis.*with a persistent fruit at the base of the plant. One of the seeds in the fruit gave rise to this plant. (Based on collection Phitos 8329, DAO. Artist: B. Brookes.)

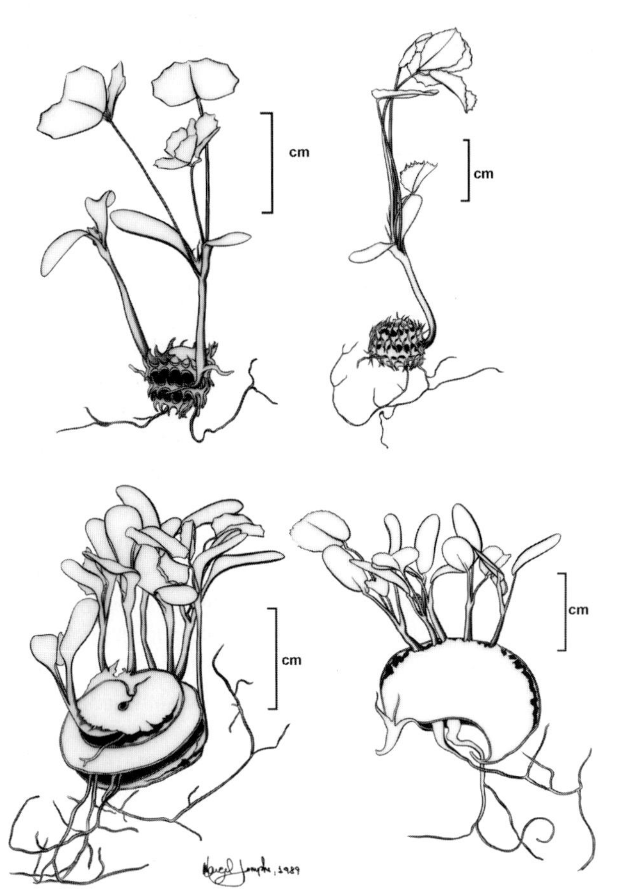

Fig. 7-32. *Medicago* fruits with emergent seedlings after 3 weeks in soil (as reported in Small et al. 1991). **Clockwise from upper left:** *M. laxispira* (based on E. Small H51); *M. granadensis* (based on E. Small M345); *M. biflora* (based on E. Small 157); *M. orbicularis* (based on E. Small M316). (Vouchers at DAO; artist: M. Jomphe.)

wasteful in that more seeds are deposited in one location than can successfully establish at one time, but since the seeds of a given fruit germinate irregularly, typically over several years, the net effect seems adaptive.

Small et al. (1991) reported a study of germination patterns when entire *Medicago* fruits are placed in soil and watered. As can be seen in the upper two figures in Fig. 7-32, only one or two seedlings emerge at a given time in hard-walled species, whereas in soft-walled species (the two lower figures), germination of many seeds can occur simultaneously. The lower two figures illustrate an experimental condition; normally the seeds of these species would be released quickly from the fruits and would scatter apart.

Small et al. (1991) compared the germination behavior of the two varieties of *M. truncatula*. As can be seen in Fig. 7-33, the seeds of variety *laxicycla*, an endemic of Libyan deserts, appear adapted to germinating rapidly and simultaneously on rare occasions when ample precipitation occurs (the plants subsequently grow and produce a new supply of fruits very rapidly). By comparison, only one seedling emerges at a given time from a fruit of variety *truncatula*. The very loose coiling of variety *laxicycla* is clearly an adaptation that facilitates water entry to the seeds and seedling egress from the fruit.

8. Protective Chemicals

Antiherbivorous toxins

In Chapter 9 it is pointed out that medicagenic acid glycosides, which are very uncommon saponins in plants, are characteristic of all species of *Medicago* and indeed serve to discriminate the genus from related genera. It seems reasonable to hypothesize that the ancestral species of the genus acquired the ability to synthesize these chemicals because they served a protective function. Saponins are known to be antinutritional to a wide variety of animals (see Chapter 12), and so the presence of medicagenic acid glycosides in *Medicago* is very likely an adaptation against herbivores.

Dyba et al. (2003) analyzed the saponin content of 15 *Medicago* species at four developmental stages: rosette, before flower formation, full bloom, and before seed harvest. They

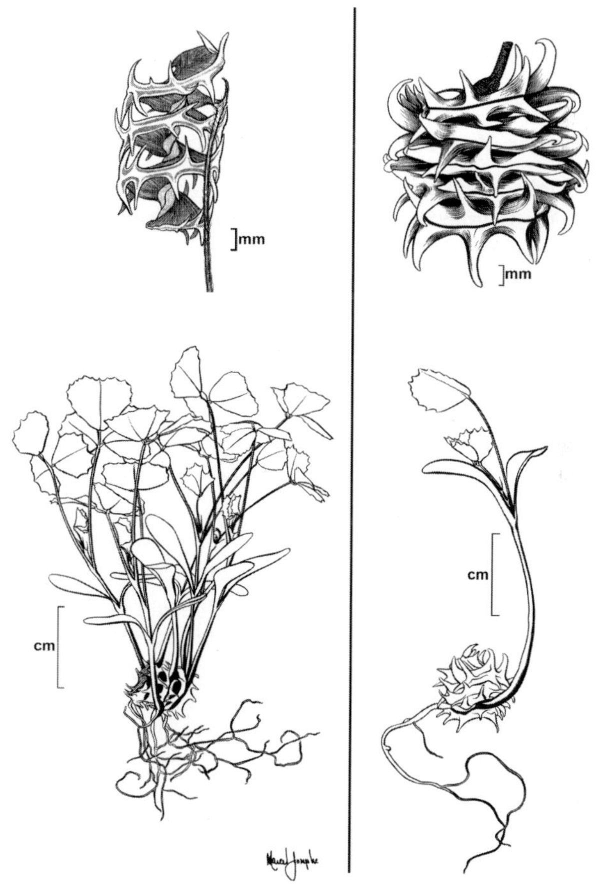

Fig. 7-33. A comparison of germination patterns in *Medicago truncatula* var. *truncatula* (on the right, based on E. Small IS10766, DAO) and *M. truncatula* var. *laxicycla* (on the left, based on E. Small M1353, DAO, as reported in Small et al. 1991). Above, fruits. Below, seedlings emerging from the fruits (many from var. *laxicycla*, just one from var. *truncatula*) after 3 weeks in damp soil. (Artist: M. Jomphe.)

found that saponin content was highest during the rosette stage and lowest before the seed harvest stage, suggesting that the presence of saponins protects young, growing parts of the plant.

Canavanine is a toxic, nonprotein amino acid present in the seeds of *Medicago* (for references, see Chapter 12). It is believed to protect seedlings against insects and even against competing weeds (Miersch et al. 1994). However, canavanine exemplifies the many chemicals that protect numerous species of the legume family and its presence is not reflective of adaptations unique to *Medicago*.

Calcium oxalate crystals are widespread in *Medicago*. Korth et al. (2006) demonstrated that in *M. truncatula* these are a defense against chewing insects.

Allelopathy

Chemicals that leach out of plants, particularly from the roots, may inhibit germination or growth of nearby plants and also other organisms inhabiting the nearby soil. Such allelopathy may be adaptive by preventing or lowering competition from other plant species. Medicagenic acid glycosides and canavanine (discussed above), as well as other constituents of *Medicago* species, are known to be allelopathic (Oleszek et al. 1992; Oleszek et al. 1999; Chon et al. 2002; Bousquet-Mélou et al. 2005).

9. Some Key Physiological Adaptations

Drought adaptation

Most of the annual species of *Medicago* are adapted to the winter-wet, summer-dry Mediterranean climate as drought evaders: their seeds germinate in late winter or early spring and the plants have matured new seeds by the time the upper levels of the soil, in which the roots occur, have dried up. In perennial *Medicago*, a comparable drought-evading strategy is summer dormancy, or at least a substantial lowering of summer growth. Very commonly, perennial *Medicago* drop a proportion of their leaves or young shoots in response to summer drought. Drought-adapted species of perennial *Medicago* are coarse with tough stems and small, narrow, firm leaflets that resist wilting and dissipate heat efficiently (for further information on the drought- and heat-adapting value of small, narrow leaves, see Small (1972), and references therein). The most universal mechanism helping perennial *Medicago* species adapt to drought is a deeply penetrating root system. Indeed, the biomass of the root system may exceed several times that of the mass of other vegetative organs (Ivanov 1980). Alfalfa's root system often penetrates loose soils to depths of 3–5 m, which is two to three times the rooting depth of cereal crops and many perennial grasses. Root depths of 6 m or more are common, with a reported record of 39 m (Sheaffer et al. 1988).

Essentially all the above mechanisms of drought adaptation in *Medicago*, including dormancy, lowered growth, abscission of leaves, sclerophyllous leaves and stems, dense pubescence, and marked allocation of biomass to the root, as well as physiological mechanisms that produce heat and drought resistance and increased hydrological suction potential, may be useful for maintaining or even increasing crop productivity in the face of drought.

The correlation between drought and cold resistance in perennial *Medicago*

Oakley and Garver (1917) proposed that the adaptations of alfalfa plants that make them resistant to cold are in considerable measure the same that enable the plants to endure or evade drought, although they provided little evidence to support this hypothesis. Ten of the 21 obligatively perennial species of *Medicago* are endemic to habitats that are both very cold in winter and very dry in the summer, and several of the remaining species often occupy such environments, so there is an obvious positive association between drought and cold resistance. Is this correlation causative or spurious? The deep roots of most perennial *Medicago* species, the most important means in the genus of surviving droughts, also serve as the critical energy storage organ for perennating overwinter and, because they are deeply buried, they evade the extreme cold of winter (although the critical crown of the plant is much less protected). Therefore, Oakley and Garver were at least partly correct in proposing that the same adaptations enable drought and cold evasion. "Bound water" was once commonly accepted as a measure by which alfalfa and other plants prevent freezing injury of cells, but the concept and its significance have been challenged (McKenzie et al. 1988). In any event, the decrease of "free water" content that occurs in alfalfa during cold hardening is intriguingly reminiscent of the same phenomenon observed during drought-hardening. It has been shown that desiccation resistance of alfalfa foliage and stems is strongly correlated among cultivars with a greater proportion of non-liquid protoplasmic water (Ivanov 1980). Once again, Oakley and Garver seem to have been at least partly correct in proposing that the same adaptations enable both drought and cold endurance. Indeed, it is now clear that cold and drought resistance in *Medicago* are intimately linked. Paquin (1986) showed that cold hardening alfalfa by exposing it to low temperatures simultaneously increases its drought resistance. Laberge et al. (1993) discussed chloroplast genes from *M. sativa* controlling responses to environmental stresses that simultaneously regulate both cold and drought resistance. Jung and Larson (1972) wrote with regard to alfalfa that "The mechanism(s) responsible for drought and cold tolerance is probably similar, since both stresses desiccate cells. Supporting evidence is as follows: (*i*) plants hardening to drought also become cold hardened and vice versa; (*ii*) drought and cold tolerance are both associated with small cell size; (*iii*) species, and occasionally varieties, have parallel rankings for drought and cold tolerance; (*iv*) changes in drought and cold tolerance are associated with plant development and growth; and (*v*) many physiological changes that occur during hardening to drought and cold are similar." Given the evidence for a close association between cold and drought resistance, it would be prudent to try to locate drought-resistant germplasm in accessions from cold climates. Finding cold-resistant germplasm in accessions from droughty Mediterranean climates, however, seems less likely — as pointed out elsewhere, the four perennial species of *Medicago* most adapted to the Mediterranean climate have very limited frost resistance.

Soil adaptation in *Medicago*

Perennial species of *Medicago* generally occur in circumneutral or calcareous soils that are mostly derived from limestone and volcanic substrates. Strongly acidic soils are not tolerated. The annual species of *Medicago* seem to have much the same soil associations (Ehrman and Cocks 1990), although a few of the annual species are relatively tolerant of acidic soils. Most of the perennial *Medicago* prefer loose, deep soils adapted to their spreading or deeply penetrating root system, but several species grow characteristically on stony soils.

The correlation between drought resistance and soil infertility tolerance in perennial *Medicago*

There is a little-appreciated similarity between many drought adaptations and adaptations to nutritionally deficient substrates, perhaps indicating a close physiological kinship between drought tolerance and tolerance to low nutrient availability (Small 1973). These adaptations include limited stature, reduction in internode length, perenniality, and leaf modifications such as small size, small cells, tissue compaction, thick cuticle, narrowed shape, glaucescence, pubescence, and the development of strengthening features. It is probably not coincidental that most perennial *Medicago* species have such features and often occur in soils that are deficient both in nutrients and water. For purposes of introducing new range plants to Mediterranean regions that are both dry and infertile, many of the perennial *Medicago* species would therefore be useful. For purposes of creating new *hay crop* domesticated forage species of perennial *Medicago*, however, it should be appreciated that plants adapted to infertile soils will have limited productivity. Domestication for higher productivity normally results in selecting plants that require fertile soils (Sinclair 1998).

Salt adaptation in *Medicago*

Increased salinity tolerance could adapt alfalfa to semi-arid areas that are seriously affected by salinity either because the lands are naturally saline or because irrigation has resulted in salinization. Saline soils are not only associated with droughty regions, but have been said to induce a water deficit and cause drought-like symptoms in plants (Bolton 1962). Moreover, salinity and drought may be synergistic, i.e., the harmful effects of salinity may be exaggerated in hot and dry conditions as opposed to cool and moist conditions (Bolton 1962). Alfalfa is considered only moderately resistant to salt. Wild forms of *M. sativa* subsp. *caerulea* from the shores of the Caspian Sea have been reported to be tolerant of saline soils and very resistant to drought (Bolton 1962). Rumbaugh and Pendery (1990) examined germination salt resistance of 761 United States Department of Agriculture alfalfa germplasm collections and found that accessions of subsp. *sativa* from the arid Indian and African sites were most resistant, whereas the northern subsp. *falcata* accessions were least resistant. Accessions of subsp. *caerulea* also had low salt resistance, which contradicts Bolton's observation above and is surprising given the saline habitats within the range of this subspecies. The sea beach species *M. marina* and several of the sea beach annual species (*M. littoralis* and *M. italica*) are likely resistant to salt — I have viewed *M. littoralis* half-immersed in sea water on the west coast of Italy. Transferring genes to alfalfa from these species may be a very useful breeding exercise; however, this would likely require non-sexual means since conventional sexual crossing does not seem possible.

8. Coevolution With Nitrogen-Fixing Bacteria

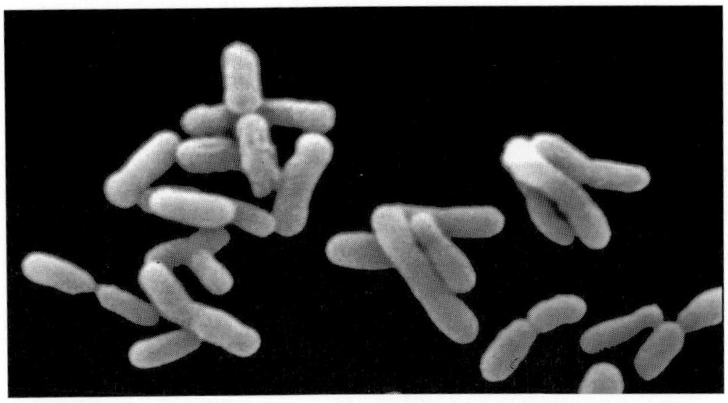

Fig. 8-1. A scanning electron micrograph showing the bacterium *Sinorhizobium meliloti* (courtesy of W. Margonlin and S.R. Long).

Fig. 8-2. Mature, nitrogen-fixing nodules on an alfalfa root. Photo courtesy of S.R. Long.

As pointed out in other chapters, the evolution of *Medicago* has been intimately connected with other organisms, particularly insect herbivores and pollinators. This chapter explores the extent to which much simpler organisms — nitrogen-fixing bacteria — may have played a role in the evolution of the genus. General reviews of the evolutionary aspects of the association between nitrogen-fixing bacteria and Fabaceae (the species referred to as "legumes" in the following) can be found in Howieson (1999); Gualtieri and Bisseling (2000); Hirsch et al. (2001); Sawada et al. (2003); Jones et al. (2007); Sprent (2007); and Sprent and James (2007).

The practical importance of the association between nitrogen-fixing bacteria and economic plants such as *Medicago* cannot be overestimated. Legume crops often fix as much as 200–300 kg of nitrogen per hectare, and total symbiotic nitrogen fixation has been estimated to be at least 70 million tonnes of nitrogen globally per year (Sessitisch et al. 2002). The world's consumption of chemically produced fertilizer nitrogen is of the order of 100 million tonnes annually. This has very significant environmental costs because of the pollution generated by run-off of fertilizer and because of the non-renewable energy resources used to produce, transport, and apply it. By contrast, biological nitrogen fixation is a critical component of sustainable agricultural systems and offers the producer more flexible management than fertilizer nitrogen because the pool of organic nitrogen becomes slowly available to non-legume species.

All organisms require nitrogen to build cellular constituents, including proteins and nucleic acids. Nitrogen forms about 79% of the earth's atmosphere, but this inert, free, molecular (dinitrogen) gaseous form cannot be metabolized by most species, which need it converted to mineral compounds such as ammonia (NH_3, NH_4^+) or nitrates (NO_3^-). The conversion may be done chemically (for commercial purposes) but, as pointed out above, much of it is carried out biologically through a process known as biological nitrogen fixation. Higher organisms (eukaryotes) lack the enzyme complex (nitrogenase) to effect the conversion; only some lower organisms (prokaryotes) have this capacity. Most biological nitrogen fixation is carried out by bacteria, many of which live independently of other organisms (so-called free-living), although many also live in close association with plants or other species (including termites and protozoa). Some so-called actinorhizal plant species, found in eight different families, are associated with the filamentous bacterium *Frankia*. However, unicellular nitrogen-fixing bacteria associated with plants are best

Fig. 8-3. The results of an experiment comparing the growth of alfalfa plants (left), inoculated with *Rhizobium*, with plants that have not received the bacterium. Photo by S. Bauer, United States Department of Agriculture.

known in association with special root structures (rhizobia, discussed below), in species of Fabaceae. (Exceptionally, the tropical tree *Parasponia* of the Ulmaceae also has the ability to form nodules with rhizobia.) Rhizobial bacteria live freely in the soil, but cannot fix nitrogen until they have invaded (colonized) the roots of a compatible legume. The interaction between a nitrogen-fixing bacterium and its host legume has been clarified recently. The bacterium synthesizes compounds called "nod" factors that initiate the nodulation process in the plant. Typically, the bacteria enter through tiny root hairs and proliferate within the cortex cells; this stimulates the legume root to form a tumor-like growth (nodule) on the root surface. Within the nodule, some of the bacteria develop into modified (bacteriod) forms that live in a symbiotic association with the plant (since the bacteriods do not reproduce, they sacrifice themselves altruistically for the benefit of the population). In this mutualism, the plant supplies some micronutrients and a carbon and energy source (carboxyhydrates, principally the dicarboxylic acids malate and succinate) to the bacteria, and the bacteria returns the favor by supplying NH_3 or amino acids to the plant. The bacteria also benefit by being protected within

Fig. 8-4. Nodules on the root system of *Medicago suffruticosa* subsp. *leiocarpa*, published in Rios et al. (1993). Reproduced with the permission of the authors (drawn by S. Rios).

the nodule, safe from competition and predation from other soil organisms.

All nitrogen-fixing species inducing rhizobial nodules were once placed in the bacterial genus *Rhizobium* (confusingly, the genus name is sometimes used as the singular of rhizobia, i.e., as a generic term inclusive of all rhizobial nitrogen-fixing bacteria). Recently, 76 species in 13 genera have been recognized, mostly in *Rhizobium*, *Mesorhizobium*, *Ensifer*, and *Bradyrhizobium* (Weir 2008).

Most, but not all, legumes have the capacity to enter into a mutualistic association with bacteria. Although different "species" of nitrogen-fixing bacteria are more or less specific to their host plants, they can often infect a range of plant species. Conversely, a single plant species can often be nodulated by different species of rhizobia. It is well known that some groups of legume species are particularly compatible with some strains of bacteria. In "cross-inoculation groups", certain rhizobial strains nodulate only certain legume species. For example, *Sinorhizobium meliloti* effectively nodulates species of *Medicago*, *Melilotus*, and *Trigonella*, whereas *Rhizobium leguminosarum* bv. (biovar.) *viciae* does so on *Pisum*, *Vicia*, *Lens*, and *Lathyrus*. Bena et al. (2005) showed that the genus *Medicago* includes at least five groups of species with different symbiotic specificities, including the following: (1) the closely related species pair *Medicago laciniata* and *M. sauvagei* interact efficiently with the specific ecotype of *Sinorhizobium meliloti* known as S. *meliloti* bv. *medicaginis*; (2) a group of *Medicago* species that are not closely related, including *M. rigiduloides*, *M. noeana*, and *M. radiata*, interacts with another ecotype of *S. meliloti*; most species interact well with either (3) *Sinorhizobium medicae* or (4) *S. medicae* and *S. melilotis* bv. *meliloti*; (5) *Rhizobium mongolense* (*Rhizobium gallicum* bv. *orientale*) is a specific symbiont of *Medicago ruthenica*.

Knowing the ability of specific bacterial strains to interact efficiently with given plant species is critical. However, such knowledge is sadly lacking at present. According to Brockwell and Bottomley (1995) "Preeminent rhizobial ecologists have suggested that 90% of all rhizobial inoculants applied are of no practical benefit to the productivity of legumes." The failure of inoculants could be because given microbial strains lack compatibility with given plants, or because introduced microbial strains simply cannot compete with local rhizobia, so determination of the cause of poor performance is complicated.

Coevolution has been defined as evolution in two or more evolutionary entities brought about by reciprocal selective effects between the entities (Ehrlich and Raven 1964; cf. discussion in Chapter 7). Coevolution may be associated with co-speciation, the joint speciation of two or more lineages that are ecologically associated. In the case of species of legumes and species of rhizobia, there has been interest in jointly using the molecular (nucleic acid) phylogenies of a group of related legumes and the group of rhizobia that are associated with them to examine whether co-speciation has occurred. On the whole, the few analyses conducted to date have produced only limited evidence of co-speciation between plants and bacteria (Bena et al. 2005). Indeed, one of the most thoroughly analyzed relationships is that between *Medicago* and its rhizobia, undertaken by Bena et al. (2005). These authors compared the phylogenies of 23 rhizobial strains (15 of *Sinorhizobium meliloti* and eight of *S. medicae*) isolated from *Medicago* species and previously established phylogenies of *Medicago* (see Chapter 10). They concluded that there is only weak evidence for co-divergence among the plants and the bacteria, and that specificity between the bacterial strains and *Medicago* species has evolved and de-evolved repeatedly. Some *Medicago* species proved to be compatible with both *Sinorhizobium meliloti* and *S. medicae*, while others were compatible with just one of the bacterial species; often, closely related *Medicago* species differed in this respect. This is dramatically illustrated by two very closely related species, *M. rigidula* and *M. rigiduloides*. As pointed out in the chapters dealing with these species, they contrast markedly in *Rhizobium* specificity. *Medicago rigiduloides* (a species that is largely Asian) was very compatible with Asian *Rhizobium* strains, but much less so with *Rhizobium* strains originating from Europe and Australia. Conversely, the European *Rhizobium* strains generally performed well with the European *M. rigidula*.

In the case of *Medicago*, there is evidence that in the distant past there was extensive genomic assortment or exchange among the species (Chapter 10). Obviously, this may have complicated the pattern of evolutionary relationships with symbiotic bacteria. Moreover, there is evidence that nitrogen-fixing bacteria associated with *Medicago* have also participated in gene exchange by lateral transfer (Bailly et al. 2006; Bailly et al. 2007). Thus, independent genetic assortment within *Medicago* and also among its symbionts may have limited the specificity of plant–microbe relationships.

Bena et al. (2005) examined the possibility that the distribution of bacterial strains limited the geographical spread of *Medicago* species. They noted that, with respect to the Mediterranean area, most of the widespread species of *Medicago* are able to establish efficient relationships with rhizobial groups that are similarly widespread around this region. However, many of the *Medicago* species that are confined to the eastern Mediterranean area are able to establish efficient relationships only with a group of rhizobial strains that are confined to the eastern region. Bena et al. concluded that "The lack of strains fixing nitrogen with these species in the western part of Europe strongly limits the geographical distribution area of such species and restricts them to the eastern part of the Mediterranean basin." The apparently most "primitive" lineages of *Medicago* (sections *Buceras* and *Lunatae*; see Chapter 10) are mostly found in Asia and the eastern Mediterranean (see Chapter 4), and presumably as the genus diversified westward, the species evolved compatibilities with the more western rhizobial strains. Interestingly, Bena et al. found that *M. monspeliaca*, the only "primitive" species that is extremely widespread, is compatible with *Sinorhizobium meliloti* type 3, which is widespread in western Mediterranean soils, at least partly explaining its considerable geographical spread.

The relative efficiency of nitrogen fixation depends on the particular partners involved and on environmental factors such as drought and temperature. Soil acidity appears to be a particularly important consideration, and there is evidence that some rhizobial strains, particularly those associated with *Medicago*, have evolved adaptations to acidic soils (Garau et al. 2005; Reeve et al. 2006). Because the productivity of legumes is intimately dependent on pairing with appropriate microsymbionts, the future commercial development of legumes in general, and of *Medicago* in particular, must involve clarification of the symbiotic relationships as well as appropriate development of technology to optimize the interaction.

9. Circumscription of the Genus *Medicago* and its Classification in the Fabaceae

ORIGIN OF THE LATIN NAME *MEDICAGO*

Linnaeus's name *Medicago* is apparently based on the Greek *medike*, from Medea (Media), the ancient empire from where it was supposed to have been derived. The Greek geographer and historian Strabo called alfalfa *mhdich*, a reference to Media (Russelle 2001). The root clearly persists in the Latin *medica*, the Old English *medick*, and the genus name *Medicago*. Grossheim (1945) wrote that *Medicago* was the name used by Pliny for alfalfa. Gledhill (1989) stated the name originated "from a Persian name for a grass."

TRIBAL POSITION IN THE FABACEAE

Traditionally, six principal genera (and sometimes segregates of these) have been placed in Fabaceae tribe Trifolieae. These are *Medicago*, *Trigonella*, *Melilotus*, *Trifolium*, *Ononis*, and *Parochetus*. *Ononis* has been considered to be marginally related to the other genera (Heyn 1981) and the inclusion of *Parochetus* has been considered to be doubtful (Small 1987*c*).

Schulz (1901) discovered a pattern of seed positioning in the fruit that he used to separate *Medicago*, *Melilotus*, and *Trigonella* from *Trifolium*, *Ononis*, and *Parochetus*. On this basis, he narrowed the circumscription of the Trifolieae to the latter three genera and placed the first three in a new tribe, the Trigonelleae (a viewpoint rarely followed today). Schulz noticed that in species he examined of *Medicago*, *Melilotus*, and *Trigonella* that the radicle, when travelling from base to tip, initially proceeds toward the bottom (pedicel or peduncle end) of the legume, but then arcs so that the tip points toward the stylar end of the legume, with a portion parallel to the ventral suture (as indicated by the arrows in Fig. 9-1A, C). Associated with this, the micropyle is located beneath the funiculus. By contrast, in species of *Trifolium*, *Ononis*, and *Parochetus* examined by Schulz, the radicle initially proceeds towards the apex of the fruit, parallel to the dorsal suture, then arcs towards the ventral suture, sometimes continuing to descend parallel to the ventral suture (Fig. 9-1B). Additionally, the micropyle is located above the funiculus. While there is no doubt that *Medicago*, *Melilotus*, and *Trigonella* do form a distinct lineage, Schulz's two classes of radicle orientation are not as well demarcated as he believed. In such species as *M. hypogaea*, *M. platycarpos*, *M. ruthenica*, *M. orbicularis*, *M. radiata*, *M. heyniana*, and *M. cretacea*, the radicle occurs at about a 90° angle to the pod's ventral suture, and in *M. carstiensis* it is at about a 45° angle.

As noted above, *Medicago* and its allies *Trigonella* and *Melilotus* have been considered to constitute tribe

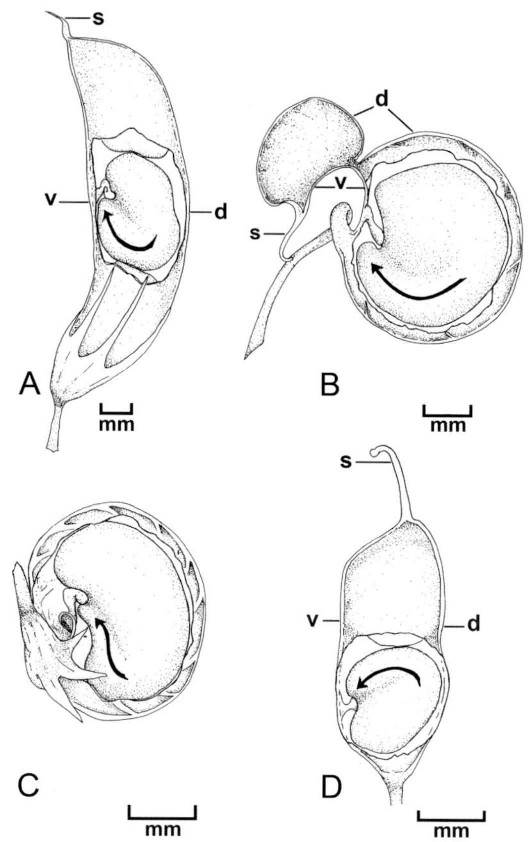

Fig. 9-1. Comparative orientation of seed and legume in some species of tribe Trifolieae. (A) *Medicago sativa* subsp. *falcata* (only one of several seeds is shown). (B) *M. lupulina*. (C) *M. hypogaea*. (D) *Trifolium pretense* L. The arrows show radicle direction; the letter "s" indicates style, "d" indicates dorsal suture, and "v" indicates ventral suture. As discussed in the text, Schulz (1901) thought radicle direction in relation to the ventral suture separated tribe Trifolieae into two subtribes, but this criterion is faulty. (Illustrations prepared by B. Brookes.)

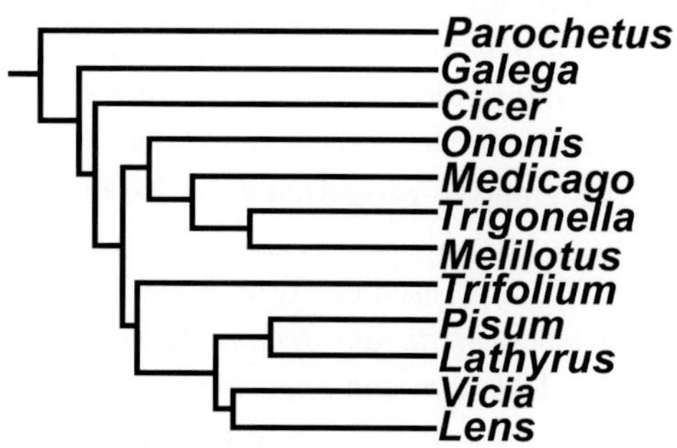

Fig. 9-2. A cladogram of plastid nucleic acid relationships among the genera discussed in this chapter, according to Wojciechowski (2003).

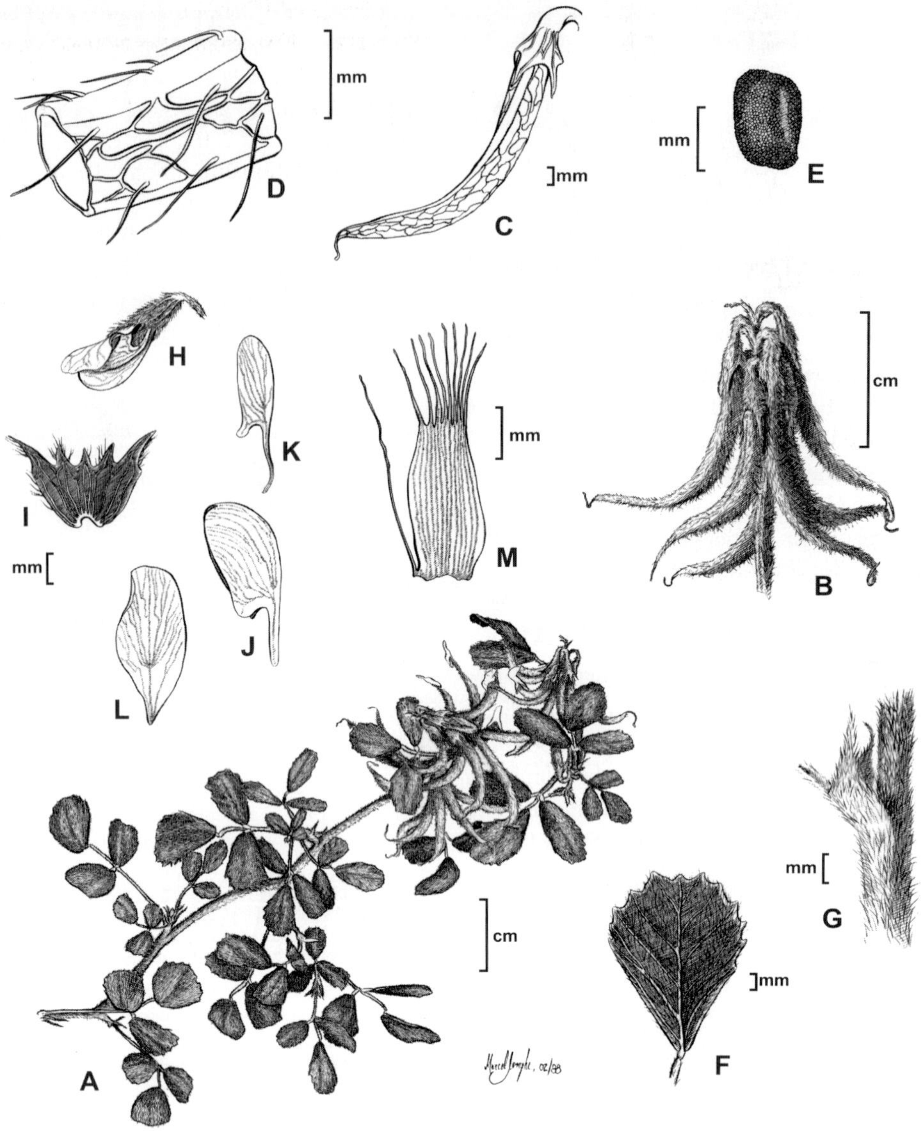

Fig. 9-3. An example of *Trigonella: T. sibthorpii* Boiss. (A) Fruiting branch. (B) Infructescence. (C) and (D) Fruit. (E) Seed. (F) Leaflet. (G) Junction of stem, petiole, and stipules. (H) Flower. (I) Calyx (slit between long lobes). (J) Keel. (K) Wing petal. (L) Standard petal. (M) Flattened staminal column (anthers removed). (After Danin and Small (1989); artist: M. Jomphe.)

Trigonelleae, in contradistinction to tribe Trifolieae, taken to include *Trifolium*, *Ononis*, and *Parochetus* (Schulz 1901). However, usually all these genera — sometimes with the omission of *Parochetus* or both *Ononis* and *Parochetus* — have been interpreted as comprising tribe Trifolieae. The correct name for what has been called Trigonelleae Schulz (if this group is recognized as a tribe of its own) is tribus Medicagineae Alef., Bot. Zeitung (Berlin) 17: 285 (1859). Alefeld (1859) used the form "Medicaginidae" and included the genera *Medicago*, *Trigonella*, *Grammocarpus* (= *Trigonella*), *Melilotus*, and *Pocockia* (= a mixture of the preceding). "*Hymenocarpos*" was also cited, evidently as a mistake for *Medicago radiata*. Although tribe Trifolieae is commonly credited to Bentham (1865), the correct citation (including the Medicagineae or not) is tribus Trifolieae Fries, Flora Scanica: 100 (1835).

Small (1987*c*) placed *Medicago*, *Trigonella*, and *Melilotus* together in a subtribe, as Trifolieae subtribe Trigonellinae E. Small subtribe nov. ≡ Tribus Papilionatae – Trigonelleae Schulz, Bot. Jahrb. 29: 679 (1901). Typus: *Trigonella* L. Small (1987*c*) provided the following description:

"Annual or perennial herbs, rarely shrubs, hairs mostly simple, microglands and septate hairs (glandular or not) sometimes present; leaves epulvinate, digitately trifoliolate (with some unifoliolate undeveloped leaves), stipules basally adnate to the petiole (but not completely encircling the stem); leaflets

Fig. 9-4. An example of *Melilotus*: *M. alba* L., from Hallier (1885, volume 23, plate 2360).

(often) pulvinulate, most at least distally marginally serrate, rarely (and intraspecifically) incised or partite; inflorescence axillary, bracteate; flowers (2)4–15(20) mm long, solitary or in umbellate, capitate, or spicate racemes; calyx of five equal or subequal lobes; corolla deciduous after fertilization (never marcescent); androecium free of corolla, diadelphous with free vexillary stamen; fruits one- to many-seeded, very polymorphic, including: short to long, straight, curved, coiled or twisted, terete or moniliform to flattened, sutures parallel or one or both curved or wavy, smooth, spiny, tuberculate, ciliate, winged or ridged; seeds smooth, tuberculate, verrucose, rugose, or papillose, radicle slightly less than half the length to slightly longer than the length of the cotyledons. Chromosome numbers $x = 8$, $2n$ usually 16, also 14, 18(?), 24, 28, 30, 32, 44, 48, supernumeraries sometimes present. Three conservatively defined, very closely related genera (*Medicago*, *Trigonella*, and *Melilotus*), of Eurasia and North Africa; very few species of *Trigonella* extending to Central and South Africa and to Australia."

Plastid inheritance in *Medicago* appears to be biparental (Keys et al. 1995), avoiding a complication in analysis of organisms in which it is maternal. Recent molecular (plastid nucleic acid) phylogenetic analysis (Wojciechowski et al. 2000; Steele and Wojciechowski 2003; Wojciechowski 2003) has confirmed the close relationship among *Medicago*, *Trigonella*, and *Melilotus*, and also the rather distant relationship of *Parochetus* from the other genera. However, this analysis has indicated that *Ononis* is much closer to this group than hitherto appreciated and that *Trifolium* is more distant than has been appreciated (indeed, it is more closely related to genera other than have traditionally been placed in tribe Trifolieae — i.e., to genera of tribe Vicieae). This interpretation of generic relationships suggests a redefinition of tribal composition is required and that nomenclaturally the tribal name Trifolieae may need to be abandoned. Before such action is taken, however, it is advisable to await additional studies of generic relationships.

GENERA THAT HAVE INCLUDED AT LEAST ONE SPECIES OF *MEDICAGO* IN THE PAST

Buceras Hall. ex All., Fl. Ped. 1: 313 (1785) (*Buceras* Haller, Hist. Stirp. Indig. Helvet. i. 164 (1768)); non *Buceras* P. Browne, Hist. Jamaic. t. 23 (1756) (Combretaceae).

Cochleata Riv. ex Rupp., Fl. Jen. ed. Hall. 264 (1745). (Pre-Linnaean.)

Crimaea (often spelled "*Crimea*") Vassilcz., Novosti Sist. Vyssh. Rast. 16: 131 (1979).

Diploprion Viv., Fl. Lib. Spec. 48. t. 19 (1824).

Factorovskya Eig, Zionist Org. Inst. Agric. & Nat. Hist. Bull. 6: 11 (1927).

Falcatula Brot., Phytogr. Lusitan. Select. (1800) 1: 160 (t. 65) (1801).

Kamiella Vassilcz., Novosti Sist. Vyssh. Rast. 16: 134 (1979).

Lotus L., Sp. Pl. 773–776 (1753).

Lupulina Noulet, Fl. Bassin Sous-Pyren. 156 (1837).

Lupularia Opiz Seznam 61 (1852). nom. nud. (Holub and Pouzar 1967).

Medica Hill in Scott, Suppl. Chamber's Cyclop. (1753), nom. inval.

Medica Miller, Gard. Dict. (Abridged), ed. 4 (1754), nom. illeg.; non *Medica* Druce in Rep. Bot. Exch. Cl. Brit. Isles, 3: 433 (1913), nec *Medica* Cothen., Disp. Veg. 7 (1790) (Bignoniaceae).

Medicula Medik., Vorles. Churpfälz. Phys.-Öcon. Ges. ii: 386 (1787).

Melilotus Mill. Gard. Dict. Abr. ed. 4, no. 2 (1754).

Melilotoides Heist. ex Fabr., Enum. Meth. Pl., ed. 2, 304 (1763).

Melisitus Medik. in Vorles., Churpf. Phys. Ges. 2: 383 (1787).

Nephromedia Kostel., Ind. Hort Prag. 92 (1844).

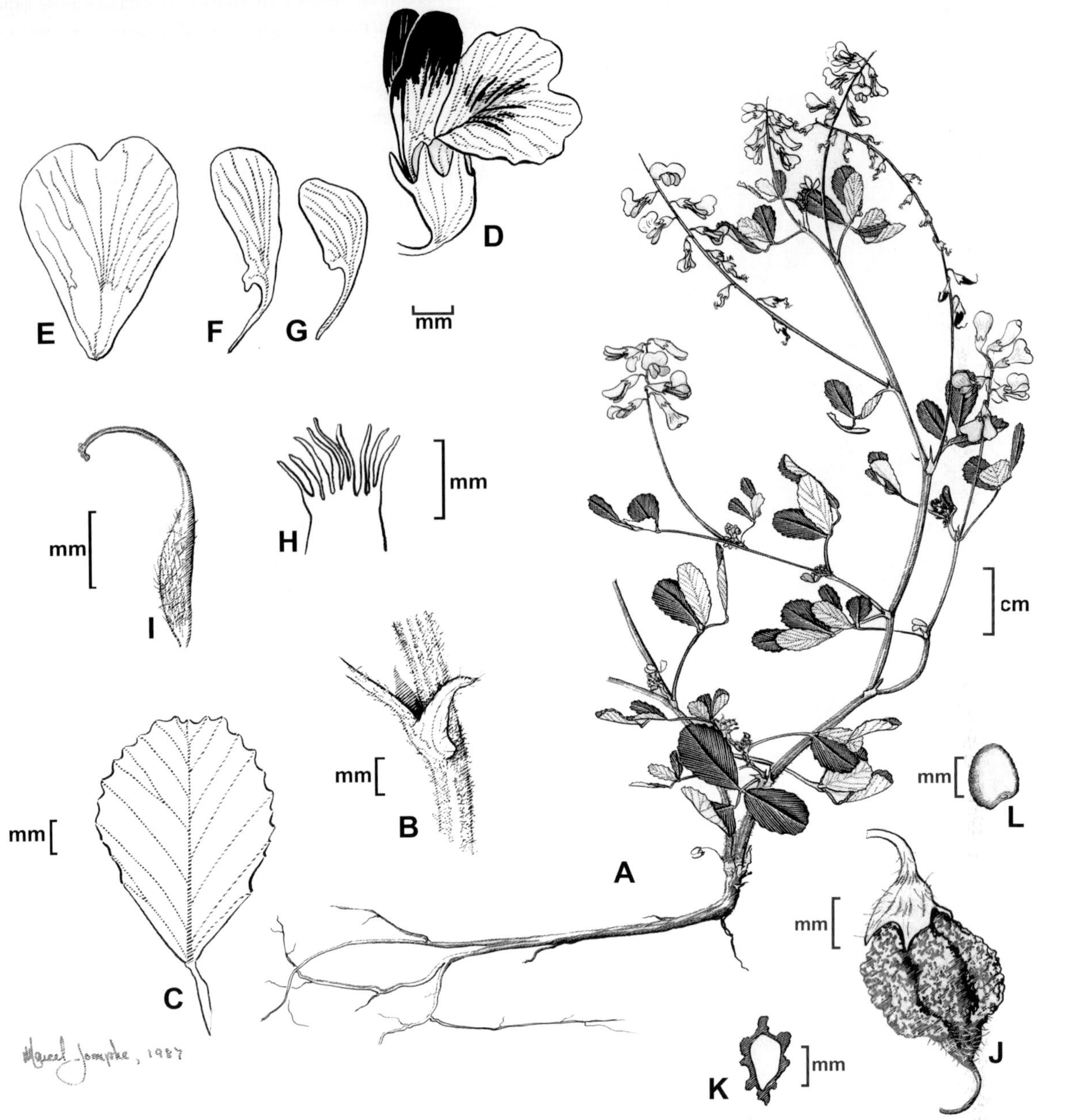

Fig. 9-5. An example of a species that has been controversially assigned either to *Melilotus* or *Trigonella*: *Melilotus bicolor* Boiss. & Balansa. (A) Habit. (B) The intersection of stipule, stem, and petiole. (C) Terminal leaflet. (D) Flower. (E) Standard petal. (F) Wing petal. (G) Keel petal. (H) A portion of the staminal column (with the anthers removed, free stamen not shown). (I) Pistil. (J) Fruit. (K) A cross section of the fruit. (L) Seed. (After Small and Jomphe (1989*a*); artist: M. Jomphe.)

Pocockia Ser. in DC., Prodr. 2: 185 (1825).

Pseudomelissitus Ovcz., M.R. Rasulova & G.K. Kinzikaeva in Fl. Tadzhikskoĭi SSR, 5: 328, 629 (1978).

Radiata Medikus, Philos. Bot. 1: 208 (1789).

Rhodusia Vassilcz., Novosti Sist. Vyssh. Rast. 9: 204 (1972).

Spirocarpus F.M. Opiz, Seznam Rost. Kvet. Cesk. 93 (1852). (See Holub and Pouzar (1967) regarding the spelling of *Spirocarpus*).

Trifolium L., Sp. Pl. 764-773 (1753).

Trigonella L., sp. Pl. 776-778 (1753).

Turukhania Vassilcz., Novosti Sist. Vyssh. Rast. 16: 132 (1979).

Fig. 9-6. An example of *Ononis: O. spinosa* L., from Hallier (1885, volume 23, plate 2325).

Fig. 9-7. *Parochetus communis* Buch.-Ham. ex D. Don, from Royle (1839, plate 35).

NOTABLE CONTRIBUTIONS TO THE CIRCUMSCRIPTION OF *MEDICAGO*

There has been uncertainty for centuries regarding how to delimit *Medicago* from the genera closest to it, particularly *Trigonella*. As reviewed below, my colleagues and I have documented several independent kinds of evidence in the literature since 1980 that make it clear that the genus is easily, clearly, and unambiguously separated from allied genera. As a practical matter, this means that about three dozen species predominantly placed in other genera (mostly *Trigonella*) in older publications need to be understood as species of *Medicago*. Recent literature and online databases have generally adopted the expanded circumscription of *Medicago*. A historical review is presented below regarding how the genus has been delimited by its most significant taxonomic investigators.

Aside from *Trigonella*, it is clear that the genus *Melilotus* is the next closest to *Medicago*. However, except for *M. lupulina* and *M. secundiflora*, which superficially appear to be close to *Melilotus*, taxonomists have not been troubled by the separation of *Medicago* from *Melilotus*. While it is not of concern to this monograph on *Medicago*, it may be noted that the separation of *Trigonella* and *Melilotus* remains controversial. Molecular analysis of some species of the two genera (Bena et al. 1998; Wojciechowski et al. 2000; Steele and Wojciechowski 2003; Wojciechowski 2003; cf. Small in press *a*, *b*) has indicated that the genera appear to overlap and should be united.

Linnaeus (1753) recognized nine species in the genus *Medicago*. His classification has been analyzed by Heyn (1959). Two of Linnaeus's species, *M. virginica* and *M. circinnata*, are now considered to be species of *Lespedeza* and *Hymenocarpos*, respectively. Linnaeus recognized *M. falcata* as a separate species, which has been reduced to a subspecies of *M. sativa* in this monograph (as well as by many others). Linnaeus recognized 13 varieties of *M. polymorpha*, 12 of which are recognized in this monograph as separate species, in addition to *M. polymorpha*. In total, therefore, Linnaeus recognized 19 of the species of *Medicago* accepted in this monograph as constituents (species or varieties) of *Medicago*. However, Linnaeus retained four species traditionally placed in *Trigonella* that are considered to be species of *Medicago* in this monograph (*M. ruthenica*, *M. platycarpos*, *M. polyceratia*, and *M. monspeliaca*). Heyn (1959) commented: "It seems obvious that Linnaeus classed his species with *Medicago* only on account of the shape of their pods ('Legumen compressum longum inflexum' – Gen. Pl. ed. 5: p. 339, 1754) regardless of other criteria such as characters of flowers or any vegetative characters." Linnaeus's characterization of *Medicago* was indefinite, but nevertheless the species included for the most part reflected clearly related taxa that botanists had previously recognized as belonging to a distinctive group. Moreover, Linnaeus did describe wing and keel petal connivency in *Medicago*, which reflect a syndrome of floral characters by which *Medicago* is best defined, as detailed below.

Fig. 9-8. Key researchers who have contributed to clarifying the taxonomy of *Medicago*. Carl Linnaeus (Carolus Linnaeus or Carl von Linné, 1707–1778), the "Father of Biological Nomenclature", was not a specialist on the genus, but his *Species Plantarum* (1753) represents the foundation of all higher plant taxonomy, including that of *Medicago* (source of illustration: Figuier 1867). The key monographs of the others are Urban (1873), Širjaev (1928–1934), Vassilczenko (1949), Heyn (1969), and Lesins and Lesins (1979). The sources of the remaining illustrations are: Ignatz Urban (1848–1931), photo courtesy of N. Kilian, Archives, Botanic Garden and Botanical Museum Berlin-Dahlem, Germany; Grigorii Ivanovitch Širjaev (1882–1954), photo courtesy of J. Danihelka, Masaryk University, Brno, Czech Republic (photo originally published in Vacek 1997); Ivan Tihonovich Vassilczenko (1903–1995), photo courtesy of I. Tatanov, Komarov Botanical Institute, St. Petersburg, Russia; Chaia Clara Heyn (1924–1998), photo by E. Small in 1985; Karlis Adolfs Lesins (1906–1982), photo courtesy of C.F. Quiros.

Seringe (1825) prepared a comprehensive review of *Medicago*. Urban (1873) commented that he did so "without critical analysis so that, far from shedding light on the species differentiation, he only added to the confusion" (translated from German). Like Linnaeus, Seringe included species in his concept of *Medicago* that are now universally accepted as members of other genera. Also like Linnaeus, Seringe's species of *Medicago* mostly have spiralled pods, whereas his species of *Trigonella* usually have straight pods. However, no single character was consistently found that separated the genera. Additional information on Seringe's system follows.

Koch (1837) distinguished *Medicago* from *Trigonella* by a straight ovary in early development in the latter, and a curved one in the former; however, subsequent researchers rejected this character. Nevertheless, Koch's criterion is of interest since a curved ovary reflects the development of fruit coiling, a characteristic that most succeeding students of *Medicago* employed, at least in part, to define the genus.

With growing frustration about how to distinguish *Medicago* from *Trigonella*, some botanists transferred species, which they considered to be not easily placed in either, to separate genera, especially on the basis of fruit characters. Medikus (in Vorles. Churpfälz. Phys.-Öcon. Ges. 2: 382. 1787; cf. Medikus 1789) assigned *Trifolium creticum* L. (= *Trigonella cretica* (L.) Boiss.) to a new genus, *Melisitus* (usually spelled *Melissitus*). Although this name is widely accepted in eastern Europe, according to Rudd in Farr et al. (1979), *Melilotoides* Heister ex Fabricius 1763 has priority (see Stafleu and Cowan 1976, pp. 810–811 for a discussion of the acceptability of Fabricius' generic names). Seringe (1825) placed the same species in his monotypic genus *Pocockia* (the latter therefore also illegitimate, although not infrequently encountered, for example in Boissier (1849)). Subsequently, other authors added a number of species with problematical affinities to this genus. Seringe's differentiation of the three genera was based mainly on fruit characters — *Medicago* having curving to spiralling, many-seeded pods, *Pocockia* having short, compressed, winged, samara-like pods, and *Trigonella* having cylindrical or compressed, straight, many-seeded pods.

Trautvetter (1841) published a "critical survey of the plant genera related to *Trifolium*" in which he defined *Medicago* as having many-seeded, wingless, almost always beakless pods, *Pocockia* as having few-seeded, membranous-winged pods, and *Trigonella* as having wingless, beaked pods. *Medicago* was delimited from *Melilotus* only by having three- to many-seeded pods, a criterion that is without merit. In 1841, Trautvetter recognized the unsatisfactory state of separating *Trigonella*: "I feel that this genus is one of the most unfortunate present classifications of all....a type of emergency genus — a genus to which anything is assigned which for one or the other reason does not quite fit into *Melilotus* or *Medicago* — which is also why the species currently classified as *Trigonella* do not have a single character in common" (translated from German).

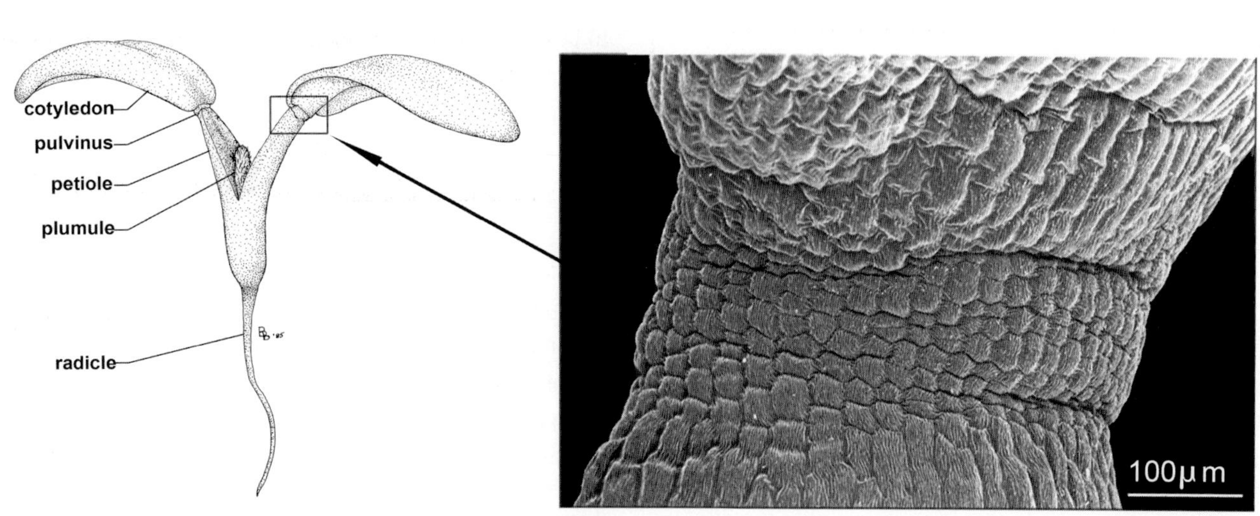

Fig. 9-9. Left: A diagram of the seedling of a plant with pulvinate cotyledons. **Right:** A scanning electron photomicrograph of the cotyledonary pulvinus of *Melilotus sulcatus* Desf. subsp. *segetalis* (Brot.) Maire. Cf. Fig. 9-10. (Illustrations prepared by B. Brookes.)

Bentham (1865) rejected the attempt to create a genus intermediate between *Medicago* and *Trigonella* and defined *Medicago* by spiral to falcate pods, and *Trigonella* by beaked, straight to falcate, sometimes flat and wide, sometimes linear pods.

Alefeld (1859) noticed the explosive tripping mechanism in some species of *Medicago* and suggested it might be a basis for delimiting the genus. However, he did not document his observations and his suggestion was not taken seriously for more than a century (see below).

Scofield (1908) simply emended *Medicago*, placing *M. sativa* and other species in *Medica*.

Urban (1873) was the first to prepare a carefully crafted, comprehensive analysis of the genus. He discovered a character — the presence of a swelling or "joint" (pulvinus) at the base of the cotyledons — that he found to be absent in every species of *Medicago* he examined, and present in species of the related genera *Trigonella* and *Melilotus* and many species of *Trifolium*. Stated otherwise, *Medicago* according to Urban is characterized as having epulvinate cotyledons (with the blades narrowed gradually into the petiole), whereas *Trigonella* has pulvinate cotyledons (with the blades demarcated from the petiole by means of a distinctive articulation). Using this character, Urban recognized 46 *Medicago* species, almost all of which are accepted in this monograph as species of *Medicago* and, in a postscript noted that, on the basis of his cotyledon joint character, he considered *Trigonella ovalis* to be a species of *Medicago* (it is recognized as *M. ovalis* in this monograph). All the species recognized within *Medicago* by Urban are assigned to *Medicago* in this monograph, although sometimes with different nomenclature and circumscriptions. Notably, Urban included in *Medicago* the following species considered to have controversial affinities: *M. cretacea*, *M. edgeworthii* (as *M. hybrida*), and *M. ovalis*, and he suggested that *M. platycarpos*, *M. ruthenica*, and *M. pubescens*

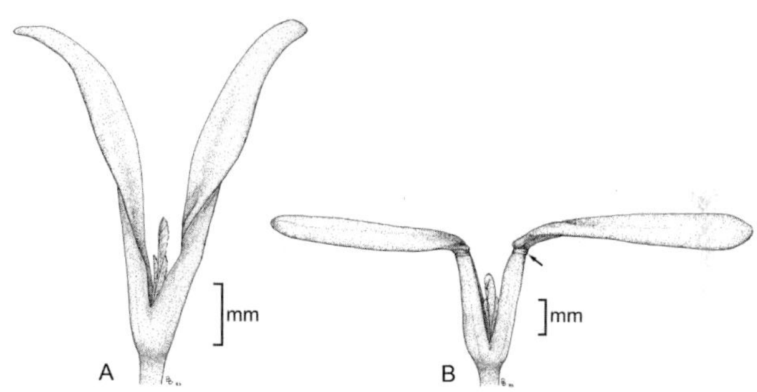

Fig. 9-10. The contrasting habits of seedlings with and without cotyledonary pulvini. (A) *Medicago sativa* L. subsp. *careulea* lacks pulvini and has erect cotyledons. (B) *Melilotus sulcatus* subsp. *segetalis* has cotyledons sharply flexed at the pulvini (note arrow). (Artist: B. Brookes.)

(not examined carefully by him because of limited material) also belong to the genus. As noted elsewhere, *M. edgeworthii* has pulvinate cotyledons (see Fig. 4-1), Urban apparently not having appreciated this.

In this monograph, certain species that Urban left in *Trigonella* (the "medicagoid *Trigonella*") are added to *Medicago* (as sections *Buceras* and *Lunatae*) although they possess cotyledonary pulvini (for reviews of pulvini, see Wilkinson 1983 and Gregory 1983). The justification for this is presented below, but it is opportune here to comment on the cotyledonary pulvinus feature. In Fig. 9-11, it is hypothesized that cotyledonary pulvini, an apparently "primitive" character (i.e., present in genera apparently close to ancestors), may have been lost in most species of *Medicago* because they acquired the character of coiled, indehiscent fruit. The proposed explanation is that pulvini result in cotyledons that are flexed at their petiole, making it difficult for them to escape from the cracks or holes that develop with aging and deterioration of the

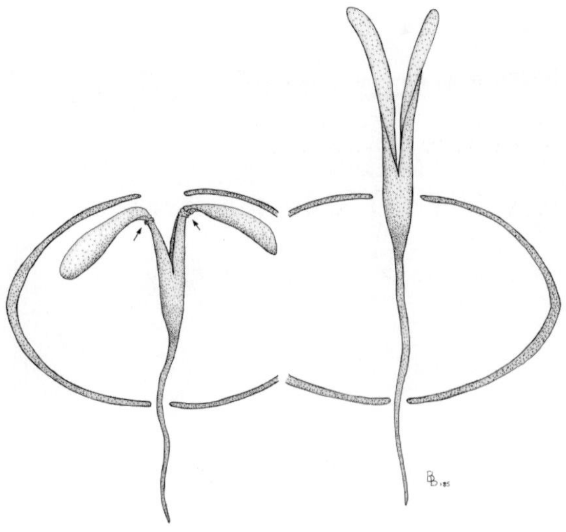

Fig. 9-11. A rationale for the evolutionary loss of cotyledonary pulvini in indehiscent fruits. **Left:** Cotyledons flexed at pulvini (arrows) are unable to escape through a small hole and remain trapped inside the fruit. **Right:** Cotyledons lacking pulvini are straight enough to be able to escape the fruit through a small hole. (Artist: B. Brookes.)

fruit. By contrast, *Medicago* species with cotyledonary pulvini (and indeed *Melilotus* and *Trigonella* species) have uncoiled fruit, and for the most part they are dehiscent or at least have fruit walls that are relatively weak. Small and Brookes (1984*c*) observed that in the pulvinate–cotyledon species of *Medicago*, there appears to be less flexing of the cotyledonary petiole than is true in *Trigonella* because the pulvini of the cotyledons seem to be less functional in the former. In this respect, the pulvinate–cotyledon species of *Medicago* show similarity to other *Medicago* species, which also exhibit comparatively limited flexing of the cotyledons.

Širjaev (1928–1934), the most authoritative monographer of the genus *Trigonella*, accepted Urban's criterion of cotyledonary joints as a means of defining *Medicago* and, accordingly, placed the following species in *Medicago* that at the time were often considered to belong to *Trigonella*: *M. cretacea*, *M. edgeworthii* (as *M. hybrida*), *M. ovalis*, *M. platycarpos*, *M. plicata*, *M. pubescens*, and *M. ruthenica*. Širjaev noted that species he placed in three of his sections of *Trigonella* (sections *Lunatae*, *Bucerates*, and *Isthmocarpae*), showed corolla features in common with species of *Medicago*, unlike all other species of *Trigonella*. However, these species have cotyledonary joints, and therefore were not assigned to *Medicago* by Širjaev. He used the term "medicagoid" to describe the *Medicago*-like flowers of these species. By extension, the *Trigonella* species with medicagoid flowers have been termed medicagoid *Trigonella*. It did occur to Širjaev, however, that the medicagoid *Trigonella* seemed to be transitional to *Medicago*, but he did not judge the relationship to be close enough to transfer the medicagoid *Trigonella* to *Medicago*.

Baum (1968, cf. Baum 1970) reviewed the generic treatment of *Medicago* and accepted Urban's criterion of absence of cotyledonary joints as a means of defining *Medicago*. Baum also confirmed that, as Širjaev had documented, floral features of the androecium were shared by *Medicago* and the medicagoid *Trigonella*. Baum thought that he could distinguish *Medicago* and the medicagoid *Trigonella* on the basis of floral features but, in a more extensive survey, Small et al. (1981*b*) found that they could not.

Heyn (1963), the monographer of several sections of the annual species of *Medicago*, and Lesins and Lesins (1979), who monographed what they thought was the entire genus, rejected the use of Urban's cotyledonary joint character as a suitable criterion for delimiting *Medicago* from *Trigonella* since, in their opinions, it placed species obviously similar to *Trigonella* in *Medicago*. Lesins and Lesins (1979) also rejected the use of floral characters as a means of defining *Medicago*.

I.T. Vassilczenko published very extensively on the classification of *Medicago* and *Trigonella*, but his work, marked by extraordinary taxonomic splitting, was not helpful in delimiting the two genera. His 1953 monograph on *Trigonella* included in the genus essentially all the species controversially included in *Medicago*, as well as the medicagoid *Trigonella* species. Moreover, he split several of these species into additional, poorly defined species. Vassilczenko (1953) stated that he had not examined the segregation of *Medicago* and *Trigonella* critically and, indeed, he did not even recognize the cohesiveness of the "medicagoid *Trigonella*". Moreover, additional work he carried out represents quite unjustified taxonomic inflation, as noted below. Vassilczenko (1972) transferred *M. arborea* to its own, new monotypic genus *Rhodusia* Vassilcz. In 1979, he also created three additional monotypic genera based on species of *Medicago*: *Crimea cretacea* (Bieb.) Vassilcz. (*M. cretacea*); *Turukhania platycarpa* (L.) Vassilcz. (*M. platycarpos*); and *Kamiella archiducis-nicolai* (Širj.) Vassilcz. (*M. archiducis-nicolai*). What had been called *Medicago gordejevi* Komarov and *Trigonella gordejevi* (Komarov) Grossh. was converted by Vassilczenko (1979) into still another monotypic genus, *Ursia* (with the one species *U. gordejevi* (Komarov) Vassilcz.). However, illustrative of Vassilczenko's lack of critical analysis, it is in fact a species of *Trifolium*, *T. gordejevi* (Kom.) Wei Zhi (Small 1987*b*).

Table 9-1 summarizes the formal transfer of 23 medicagoid species of *Trigonella* to *Medicago* by Small et al. (1987). These are now placed in *Medicago* sections *Buceras* and *Lunatae*. As noted elsewhere in this monograph, *M. edgeworthii* also belongs in section *Lunatae*.

Table 9-1. Transfers of medicagoid *Trigonella* to *Medicago*.

Trigonella	*Medicago*
T. arenicola Huber-Mor. in Notes Roy. Bot. Gard. Edinburgh 29: 322 (1969)	≡ *M. arenicola* (Huber-Mor.) E. Small, **comb. nov**.
T. astroites Fischer & C. Meyer, Index Sem. Horti Petrop. 1: 40 (1835)	≡ *M. astroites* (Fischer & C. Meyer) Trautv. in Bull. Sci. Acad. Imp. Sci. St. Petersb. 8: 271 (1841)
T. aurantiaca Boiss., Diagn. Pl. Orient. 1(2): 22 (1843), non *M. aurantiaca* Godr. 1853	*M. phrygia* (Boiss. & Balansa) E. Small, **comb. nov**., based on *T. phrygia* Boiss. & Bal. in Boiss., Diagn. Pl. Orient. 2(6): 45 (1859)
T. brachycarpa (Bieb.) Moris, Illustr. Rar. Stirp. Horti Taur.: 16 (1830)	≡ *M. brachycarpa* Bieb., Fl. Taur.-Cauc. 3: 517 (1819)
T. carica Huber-Mor. in Bauhinia 2: 299 (1965)	*M. carica* (Huber-Mor.) E. Small, **comb. nov**.
T. crassipes Boiss., Diagn. Pl. Orient. 1(2): 23 (1843)	≡ *M. crassipes* (Boiss.) E. Small, **comb. nov**.
T. fischeriana Ser. in DC., Prodr. 2: 183 (1825)	≡ *M. fischeriana* (Ser.) Trautv. in Bull. Sci. Acad. Imp. Sci. St. Petersb. 8: 271 (1841)
T. halophila Boiss., Diagn. Pl. Orient. 2(5): 78 (1856)	≡ *M. halophila* (Boiss.) E. Small, **comb. nov**.
T. isthmocarpa Boiss. & Balansa in Boiss., Diagn. Pl. Orient. 2(6): 44 (1859)	≡ *M. isthmocarpa* (Boiss. & Balansa) E. Small, **comb. nov**.
T. lunata Boiss., Diagn. 1(2): 19 (1843), non *M. lunata* Reichenb. 1832	*M. biflora* (Griseb.) E. Small, **comb. nov**. based on *T biflora* Griseb., Spicil. Fl. Rumel. 1: 41 (1843)
T. monantha C. Meyer, Verz. Pfl. Cauc.: 137 (1831)	≡ *M. monantha* (C. Meyer) Trautv. in Bull. Sci. Acad. Imp. Sci. St. Petersb. 8: 272 (1841). This includes *T. geminiflora* Bunge, *T. incisa* Benth., and *T. noeana* Boiss.
T. monspeliaca L., Sp. Pl.: 777 (1753)	≡ *M. monspeliaca* (L.) Trautv. in Bull. Sci. Acad. Imp. Sci. St. Petersb. 8: 272 (1841)
T. *orthoceras* Kar. & Kir. in Bull. Soc. Imp. Naturalistes Moscou 15: 399 (1841)	≡ *M. orthoceras* (Kar. & Kir.) Trautv. in Trudy Imp. S.-Petersburgsk. Bot. Sada 5: 421 (1877)
T. pamphylica Huber-Mor. & Širj. in Repert. Spec. Nov. Regni Veg. 323 (1938)	≡ *M. pamphylica* (Huber-Mor. & Širj.) E. Small, **comb. nov**.
T. persica Boiss., Diagn. Pl. Orient. 1(2): 22 (1843)	≡ *M. persica* (Boiss.) E. Small, **comb. nov**.
T. polycarpa Boiss. & Heldr. in Boiss., Diagn. Pl. Orient. 1(9): 18 (1849), non *M. polycarpa* Willd. ex Gren. & Godr. 1848	≡ *M. heldreichii* E. Small, **nom. nov**.
T. polyceratia L., Sp. Pl.: 777 (1753)	*M. polyceratia (L.)* Trautv. in Bull. Sci. Acad. Imp. Sci. St. Petersb. 8: 271 (1841)
T. retrorsa Boiss., Fl. Or. 2: 78 (1872)	≡ *M. retrorsa* (Boiss.) E. Small, **comb. nov**
T. rhytidiocarpa Boiss. & Balansa in Boiss., Diagn. Pl. Orient. 2(6): 44 (1859)	≡ *M. rhytidiocarpa* (Boiss. & Balansa) E. Small, **comb. nov**.
T. rigida Boiss. & Balansa in Boiss., Diagn. Pl. Orient. 2(5): 79 (1856)	≡ *M. rigida* (Boiss. & Balansa) E. Small, **comb. nov**.
T. rostrata (Boiss. & Balansa) Boiss., Fl. Or. 2: 89 (1872)	≡ *M. rostrata* (Boiss. & Balansa) E. Small, **comb. nov**. based on *Pocockia rostrata* Boiss. & Balansa in Boiss., Diagn. Pl. Orient. 2(6): 43 (1859)
T. sirjaevii Huber-Mor. in Repert. Spec. Nov. Regni. Veg. 46: 136 (1939), non *M. schirjaewii* Hand.-Mazz. 1938	*M. huberi* E. Small, **nom. nov.**
T. striata L. fil., Suppl.: 340 (1781), non *M. striata* Bast. 1814	*M. medicaginoides* (Retz.) E. Small, **comb. nov.** based on *Lotus medicaginoides* Retz., Obs. Bot. 2: 23 (1781). This includes *T. arcuata* C. Meyer, *T. cancellata* Pers., *T. striata* L. fil., and *T. tenuis* Bieb.

Note: Data are from Small et al. (1987). An ≡ indicates the names are based on the same type; **comb. nov.** and **nom. nov.** indicate that the new combination or new name, respectively, was applied in 1987.

Trigonella uncinata Banks & Sol. (= *T. brevidentata* Blakelock, see Townsend 1968), requires comment as it has been referred to *Trigonella* section *Buceras* (Townsend 1974; Rechinger 1984) which, if justified, would make it a pulvinate–cotyledon *Medicago* species. The types of both names were examined by Small et al. (1987), and both were found to have non-medicagoid flowers, and therefore should be excluded from *Trigonella* section *Buceras*, and consequently from *Medicago*.

THE MISLEADING NATURE OF FRUIT CHARACTERS IN DEFINING *MEDICAGO*

The taxonomic history of circumscribing *Medicago*, outlined above, provides an interesting example of taxonomic misperception. In determining relationships, plant taxonomists clearly pay particular attention to striking characteristics of species and, in *Medicago* and its relatives, the fruits have attracted the most interest. Fruits are essential for the positive identification of most species of *Medicago* and *Trigonella*, and therefore it is not surprising that pod morphology has dominated generic conceptions for centuries. Numerous authors (e.g., Bentham 1865; Vassilczenko 1953) relied on pod characters to distinguish *Medicago* from *Trigonella*. In the case of *Medicago*, the most striking feature of most of the species is the presence of pod coiling, which often makes the fruits look like snails. Early taxonomists had no difficulty in conceptualizing a genus with a nucleus of species with helicoid pods. However, this reliance on pod morphology prevented an accurate appreciation of the limits of *Medicago* for centuries, particularly with respect to how the genera *Medicago* and *Trigonella* should be separated. The most frequent pod pattern in *Medicago* is spiral coiling of the more or less compressed valves (55 of the 87 species recognized in this monograph have pods that develop at least one complete coil). By contrast, the pods of most species of *Trigonella* are usually cylindrical and are straight or only slightly curved (no species of *Trigonella* has pods that wind through a complete coil, although most show some pod curvature). Species with coiled pods have almost always been placed in *Medicago*, and related species with linear uncoiled pods have almost always been placed in *Trigonella*. However, the problem has been with species with other kinds of pods and such species have been variously placed at the whim of the individual botanist in either of the two genera or in additional genera created to solve the mystery of their appropriate placement. However, pods of species of the *Medicago–Trigonella* lineage have undergone remarkable morphological radiation in becoming adapted to different modes of dissemination (Heyn 1968) and in acquiring protective morphological adaptations against herbivores (see Chapter 7); so, the striking nature of fruit characters in some species can be an unreliable guide to their relationships. This is perhaps most apparent with respect to the enigmatic "*Factorovskya aschersoniana*", which uniquely in Trifolieae subtribe Trigonellinae possesses geocarpic fruits. The species was first assigned to *Trigonella* by Urban (1882), then reassigned by Eig (1926, 1927) to the monotypic genus *Factorovskya*, and finally Small and Brookes (1984*c*) recognized it as a species of *Medicago*, *M. hypogaea* E. Small. This species provides an instructive example of how taxonomists were misled by a particularly impressive characteristic (geocarpy), and so ignored fundamental similarities in numerous characters that place the species squarely in *Medicago*.

KEY RECENT EVIDENCE THAT SEPARATES *MEDICAGO* FROM ALLIED GENERA

As interpreted in this monograph, no single vegetative or fruit character cleanly separates *Medicago* from *Trigonella*, but each of a dozen or more floral characters does, and several other lines of evidence very strongly confirm the floral evidence. Below, the floral data are described in detail and the additional evidence is summarized.

1. Floral Characters

A syndrome of floral characteristics monothetically (i.e., uniquely) defines the genus *Medicago*. As noted above, there have been indications in the literature that the flowers of many species of *Medicago* and some traditionally included in *Trigonella* are characterized by a syndrome of floral features known to be associated in *M. sativa* flowers with an explosive pollination mechanism. Small et al. (1990*a*) circumscribed *Medicago* as including all the species having the explosively tripping pollination morphological syndrome. These floral features are in fact individual adaptations, but are functionally and developmentally correlated with each other so that they represent an adaptive syndrome (for detailed information see Small 1987*c*, 1988, 1989*a*; Small and Jomphe 1989*b*; Small et al. 1981*b*; Small et al. 1987; also see Chapter 7). In theory at least, it can be argued that since the character syndrome is clearly a collective set of adaptations to a well-defined mode of pollination, it could have arisen independently (by convergent evolution) in different lineages in response to pollinator pressures, and so should bear no more weight than a single character for purposes of assessing relationships. Indeed, it is clear that, in general among plants, given pollinator syndromes have evolved independently in unrelated lineages (Fenster et al. 2004). However, it seems unlikely that such an extensive, complex, and homogeneous set of characteristics could have arisen in parallel in *Medicago*, and as detailed below, independent evidence corroborates the validity of the floral syndrome as a means of circumscribing the genus *Medicago*.

A detailed analysis of the explosive pollination mechanism in *Medicago* is given in Chapter 7 (this section should be examined in conjunction with the following information on floral variation as the pollination mechanism is complex and the terminology of floral parts is extensive). The following diagram illustrates the differences in mode of pollination between *Medicago* and its nearest relatives, *Trigonella* and *Melilotus*.

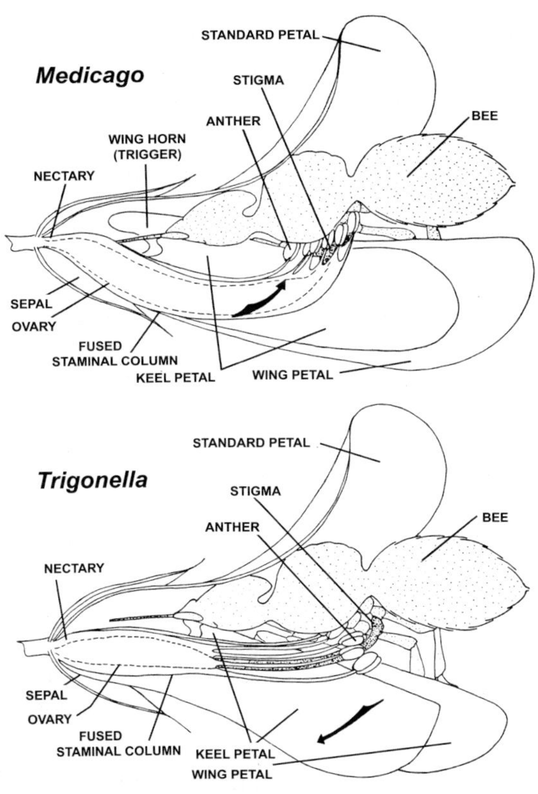

Fig. 9-12. Comparative pollinating arrangements of *Medicago* (explosive mechanism) and *Trigonella* (simple valvular mechanism). **Above:** A diagrammatic longitudinal section of an alfalfa flower showing a bee foraging for nectar at the base. The bee has tripped the flower resulting in an explosive release of the staminal column and pistil, which are slapping against the bee's thorax (arrow). **Below:** A diagrammatic longitudinal section of a typical *Trigonella* flower, showing a bee foraging for nectar at the base. The bee's weight has depressed the wing petals as well as the attached keel (the arrow indicates this movement), thus exposing the staminal column and pistil, which are passively contacting the bee's body. (Figures prepared by B. Brookes.)

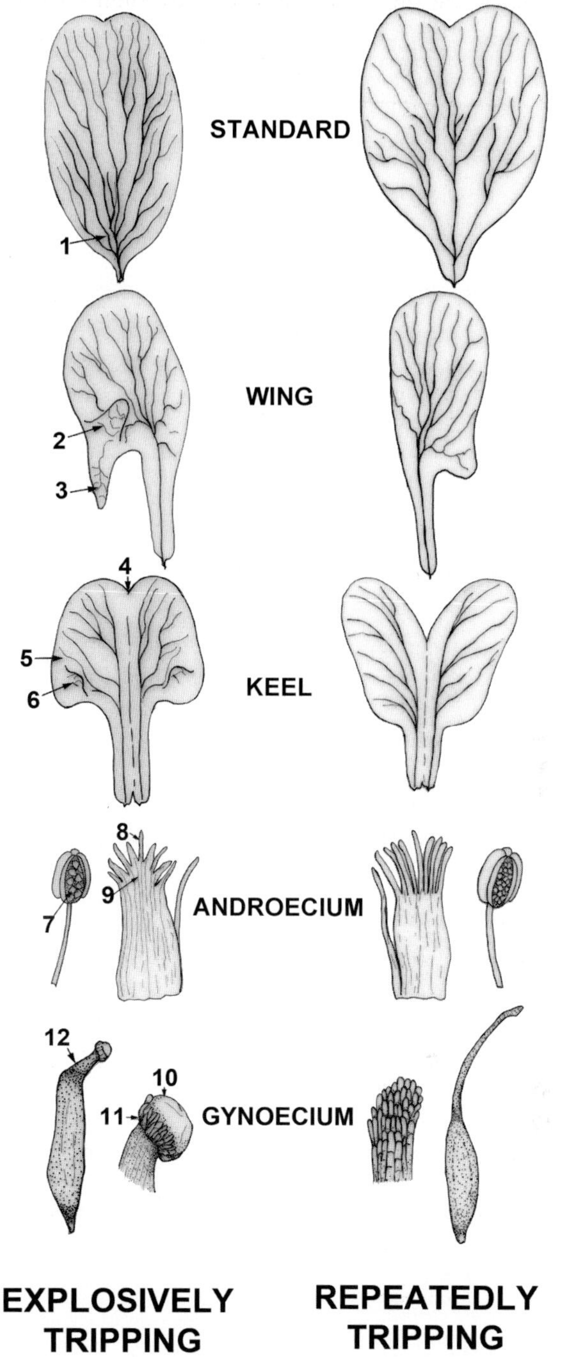

Fig. 9-13. General differences between the singly explosive pollinating mechanism of *Medicago*, shown on the left, and the repeating pollination mechanism of *Trigonella* and *Melilotus*, shown on the right. The 12 adaptive characters shown for explosively pollinated flowers are briefly discussed in numerical sequence in Table 9-2. (Figures prepared by B. Brookes.)

Table 9-2. The functions of 12 characters of the explosively tripping flowers shown in Fig. 9-13.

Character	Function
1. Basal venation of standard strongly branched.	Unclear. May be related to nectar guides associated with venation and with unique pollinators.
2. Wing petal prominently spurred.	Spur attaches to keel invagination as a component of tripping mechanism.
3. Wing horn large.	Wing horn functions as "trigger" of tripping mechanism and is large for good contact with pollinator.
4. Keel only slightly hooded.	Greater hooding of repeatedly tripping flowers protects dehiscing anthers (greater hooding is indicated by butterfly appearance of flattened keel).
5. Edges of keel possess detachable adherence tissue.	Allows temporary adherence of free edges of keel to restrain sexual column until it is explosively released.
6. Keel prominently invaginate.	Complementary to wing spur.
7. Pollen grains larger, fewer.	Related to larger stigma and more efficient pollen transfer.
8. Alternate filaments, thin (frequently).	Perhaps related to spreading anthers for maximum contact with pollinator.
9. Apex of anther column arched.	Results in simultaneous contact of all anthers with pollinator during explosive tripping.
10. Stigma fungiliform, large.	Facilitates reception of large pollen grains; withstands slapping.
11. Stigma with basal ring of hairs.	Prevents anthers at same height from transferring pollen to sister stigma. (Stigma and anthers must be at same height for simultaneous contact with pollinator.)
12. Style short, massive, in cross section it is thickest in the direction of movement (towards the standard petal).	Style must withstand slapping against pollinator and if long and thin it would be damaged. Like an axe handle, it is thickest in the direction of contact to withstand breaking.

The following examines in more detail key characteristics of the floral syndrome (summarized above) that characterizes the genus *Medicago*, based on example species of *Medicago* compared with species of *Trigonella* and *Melilotus*, including a number of species that have been alleged to represent intermediacy among these genera. As detailed in Chapter 7, while these characters are designed for explosive cross-pollination, they are also present in the self-pollinating species of the genus, albeit frequently in a reduced or relictual state. The following text and Figs. 9-14 to 9-21 have been adapted from Small et al. (1987).

Table 9-3. Species selected for comparing the syndrome of floral characteristics in *Medicago* with the floral characteristics of related genera.

Code	Species	Synonymy	Voucher (DAO)	Taxonomic Interest
1	*Medicago pironae*		M160	Outbreeder, traditionally placed in *Medicago*
2	*Medicago shepardii*		M340	Inbreeder, traditionally placed in *Medicago*
3	*Medicago lupulina*	*Melilotus lupulina* (L.) Trautv.	T172	Alleged intermediacy between *Medicago* and *Melilotus*
4	*Melilotus elegans* Salzm.		L77	Traditionally accepted as *Melilotus*
5	*Trigonella coerulea* (Desr.) Ser.	*Melilotus coerulea* Desr. In Lam.	T30	Alleged intermediacy between *Trigonella* and *Melilotus*
6	*Trigonella balansae* Boiss. & Reut.		T230	Traditionally accepted as *Trigonella* s. str.
7	*Medicago hypogaea*	*Factorovskya aschersoniana* (Urb.) Eig; *Trigonella aschersoniana* Urb.	F177	Enigmatic geocarpic species with debated affinities (Small & Brookes 1984)

Table 9-3 (*Concluded*).

Code	Species	Synonymy	Voucher (DAO)	Taxonomic Interest
8	*Medicago polyceratia*	*Trigonella polyceratia* L.	T163	A "medicagoid *Trigonella*" (now in *Medicago*)
9	*Medicago ruthenica*	*Trigonella ruthenica* L.	M1025	Alleged intermediacy between *Medicago* and *Trigonella*

A. Standard petals

The venation systems of the standard petals of *Medicago* are different from those in *Trigonella* and *Melilotus*. In *Medicago* standards, the veins branch abundantly near the bases of the claw. However, in *Trigonella* and *Melilotus* standards, one finds three, well-defined major basal veins at the claw that branch higher up than in *Medicago* standards. While one can often also perceive three clusters of veins in the standard of *Medicago*, branching is always more prominent near the bases than occurs in the standards of *Trigonella* and *Melilotus*. The functional significance of these differences is not clear in relation to the fact that *Medicago* flowers are explosively tripping, whereas those of *Trigonella* and *Melilotus* are not. Since nectar guides on the petals are often associated with venation, the differences may be related in some way to the spatial relationships of pollinator and flower.

B. Wing petals

The wing petals of the explosively tripping flowers of *Medicago* possess spurs, serving to interlock these petals with the keel more strongly than in the simple, valvular flowers of *Trigonella* and *Melilotus*. This strengthened interconnection between wing and keel petals is likely adaptive in efficiently transmitting the movement of the "trigger" of the wing petal to the keel. An evident morphological difference lies in the relative size of the horn compared with the remainder of the petal — it is smaller in the simple valvular flowers of *Trigonella* and *Melilotus* than in the explosively tripping flowers of *Medicago*. This has an obvious functional explanation, since relatively well-formed horns are needed in the explosively tripping flowers to serve as triggers.

C. Keel petals

The keels of *Medicago* are easily distinguished from those of *Trigonella* and *Melilotus* because they possess an invagination that serves to interlock them with the wing petals. Another difference lies in the much greater keel hooding that occurs over the sexual column in *Trigonella* and *Melilotus* than in *Medicago*. In Fig. 9-14, the keels have been flattened and laid back so that they are being viewed as if from the inside of the flower. The hood at the top was split open and parted, and in the keels in which the hood is well developed, the resulting flattened keel takes on the appearance of paired "wings", as is apparent for species 4, 5, and 6. It may be that in flowers that repeatedly trip (i.e., those of *Trigonella* and *Melilotus*), and so are functional for a long period, a hood is more desirable to protect the dehiscing anthers from the environment or from pollen collectors.

D. Androecia

Baum (1968) demonstrated that the staminal columns of *Medicago* and of medicagoid *Trigonella* (pulvinate–cotyledon *Medicago*) have an arched apex, whereas those of most *Trigonella* species have a relatively blunt apex (note in Fig. 9-15 that this contrast serves to differentiate most species of *Medicago* from most species of *Trigonella* and *Melilotus*). It may be that the arched apex is an adaptation to allow all anthers simultaneous contact with an insect visitor during explosive tripping. Because there is only one opportunity for pollen to be transferred to the insect, it is vital that all anthers strike the insect simultaneously. Because the stamens positioned on the keel side of the flower develop longer filaments, their anthers reach the pollinator at the same time as the anthers on the standard side of the flower (see Fig. 7-19).

There is some evidence that the fused staminal column has greater filament thickening in the explosively pollinating flowers of *Medicago* than in the simple, valvular flowers of *Trigonella* and *Melilotus* (Baum 1968). This is what one would expect, since the staminal column provides the tension responsible for tripping.

Flowers of Fabaceae section Trifolieae that trip repeatedly have different pollen–ovule relationships than the explosive flowers of *Medicago*. The latter have a lower pollen-to-ovule ratio, larger pollen grains, and somewhat more efficiency in pollen–ovule use (Small 1988).

E. Styles

Compared with the styles of *Medicago*, those of *Trigonella* and *Melilotus* are relatively slender and long. The relatively stout (breadth to length ratio) style of explosively tripping *Medicago* flowers seems to be an adaptation against breakage during the violent collision with visiting pollinators. Clearly, a long thin style would tend to deform or even break on strong collision. Moreover, the short styles usually assume an I-beam form (with the longer dimension oriented vertically), and such a shape is known to use materials efficiently to produce strength and dimensional stability that resist stresses.

A character widely considered as indicative of some sections of *Trigonella* is the presence of a beak on the fruit. Indeed, as noted in the above discussion, some past taxonomists have partly defined *Trigonella* by the presence of a beaked fruit. A "true beak" consists of a terminal, more or less elongated, more or less narrowed seedless part of the fruit. Its development probably depends mostly on differential growth of the top of the

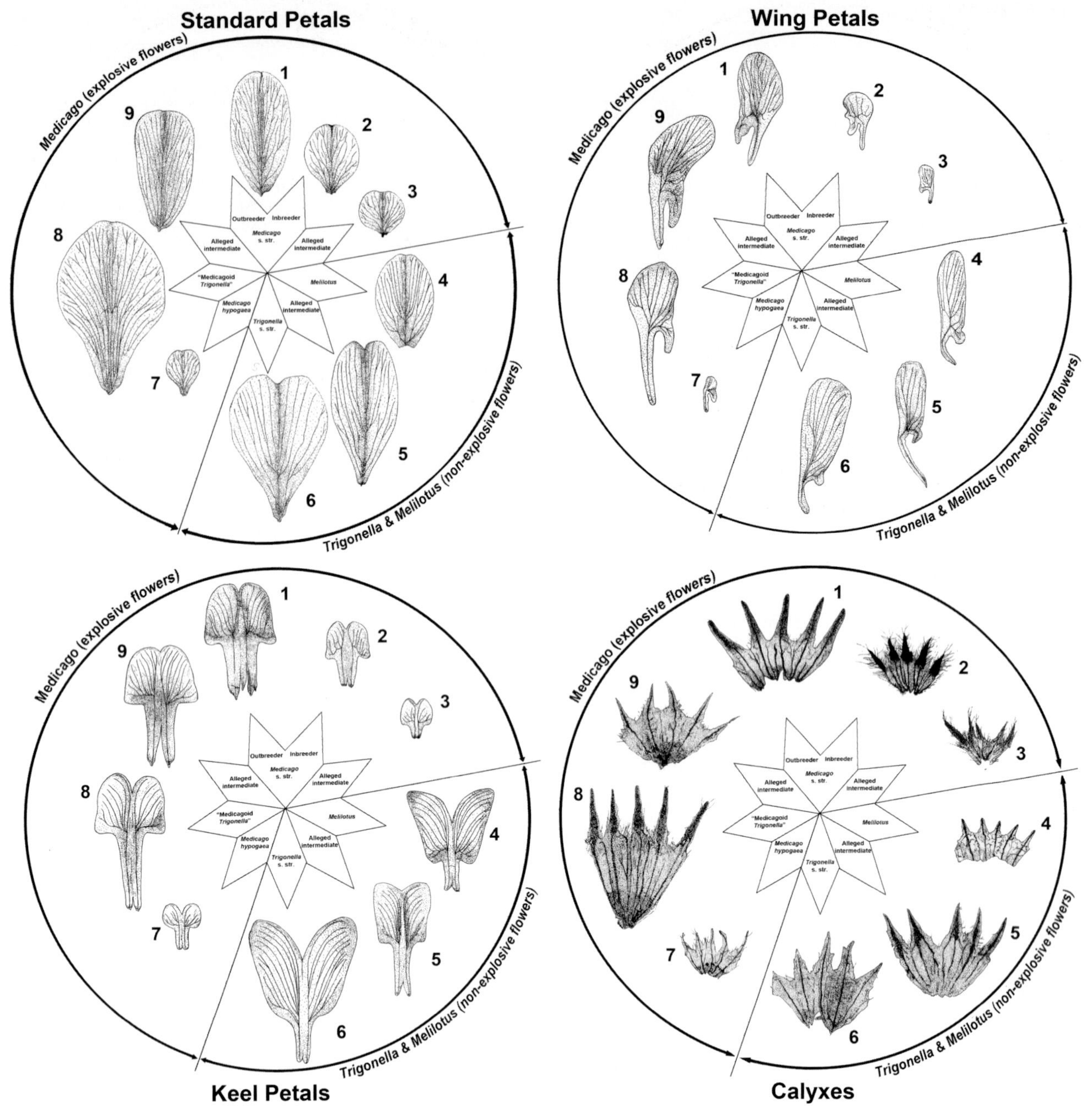

Fig. 9-14. Variation in petals and calyxes of *Medicago* and its closest relatives. The nine species examined are identified in Table 9-3 and the nine-pointed central part of the figure is expanded in the lower right side of Fig. 9-15. Note the different venation pattern in the standard petals of *Medicago* compared with the standard petals of *Trigonella* and *Melilotus*. Note the proportionately small size of the horn of *Medicago* wing petals (numbers 2, 3, and 7 are inbreeders, and inbreeders sometimes also show relatively small horns). The keels have split at the top during flattening; the two keel petals in their natural state are joined at the top notch to form a hood over the sexual column. The large notch in the non-explosively tripping species reflects a comparatively well-formed hood.

ovary (Širjaev 1928–1934 contended that the beak arises from the top of the ovary or from the style). The top of the ovary is relatively attenuated in many long-styled *Trigonella*, and this may predispose the species to develop a beak. The presence of a fruit beak in many *Trigonella* species might be an indirect consequence of its floral syndrome (i.e., the typically long style can be expected to produce a beak). The same situation prevails in the related genus *Melilotus* which, as noted above, should probably be combined with the genus *Trigonella*.

F. Stigmas

Stigma morphology of tribe Trifolieae was extensively examined by Small and Brookes (1983), who observed the following: The stigmas of *Medicago* are fungiliform, the base with a well-developed ring of long papillae and the apex apparently lacking papillar outgrowths. There are various stigma forms in *Trigonella* and *Melilotus*, such as pointed, sloping, etc., but none are strikingly fungiliform — the base always lacks a well-developed ring of long papillae and the apex has short

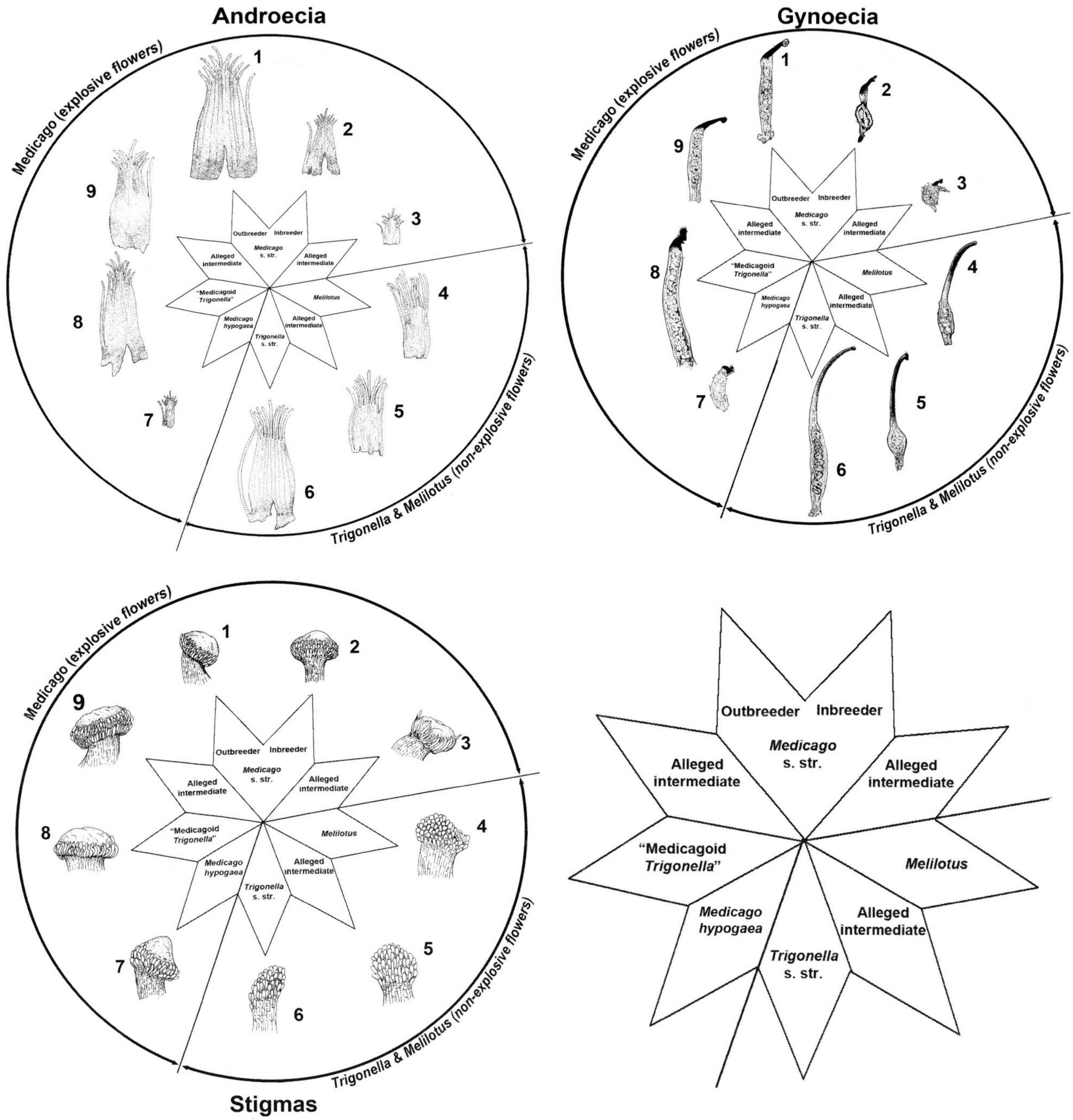

Fig. 9-15. Variation in the androecia, gynoecia, and stigmas of *Medicago* and its relatives. The nine species examined are identified in Table 9-3 and the nine-pointed central part of the figure is expanded in the lower right illustration. The androecia have been split at the vexillary filament (which is at the left or right) and flattened. Anthers are not shown. Note how the apex is arched in the explosively tripping species (i.e., the anthers remote from the vexillary stamen are born notably higher than those on filaments close to it). In the gynoecia, note the relatively short style necks of the explosively tripping species. Note the fungiliform shape of the stigma and the circumferential ring of papillae in the explosively tripping flowers.

papillae. Clearly, a blunt, mushroom-shaped stigma is better adapted than a pointed stigma to withstanding the shock of a violent collision with an insect, as occurs in *Medicago* flowers.

There is another possible advantage to possessing a mushroom-shaped stigma, which rests in the structure of the lateral part of the mushroom head. In *Medicago*, this lateral part comprises a circumferential ring of papillae to which pollen do not adhere (pollen are found on the top of the head, embedded in sticky exudate). This feature is characteristic of the explosively tripping flowers of *Medicago*, but is absent in related taxa (Small and Brookes 1983). In outbreeding species, a common mechanism for promoting outbreeding is the evolution of different height of anthers and stigma. However, in the explosively tripping flowers of *Medicago*, the stigma and anthers must be at the same height for simultaneous contact with a visiting bee. Therefore, the circumferential ring of papillae on the stigma seems to function to keep the anthers of a

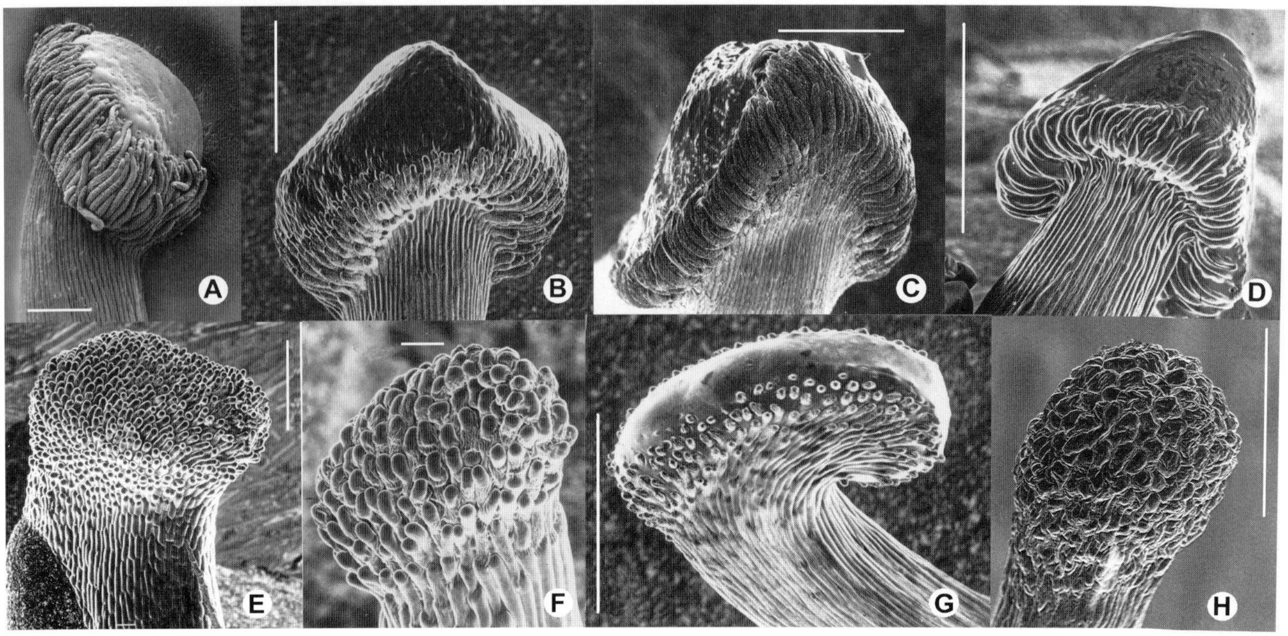

Fig. 9-16. Scanning electron micrographs showing representative stigmas of *Medicago* and related genera. (A) *Medicago sativa*; (B) *Medicago soleirolii*; (C) *Medicago biflora* (a species with pulvinate cotyledons); (D) *Medicago medicaginoides* (another pulvinate–cotyledon species of *Medicago*); (E) *Trigonella foenum-graecum*; (F) *Melilotus alba*; Desr.; (G) *Trifolium lagopus* Pourret ex Willd.; (H) *Ononis spinosa* L. The scales are 100 μm, except for (F) which is 10 μm. Photos reproduced from Small and Brookes (1983) and Brookes and Small (1988).

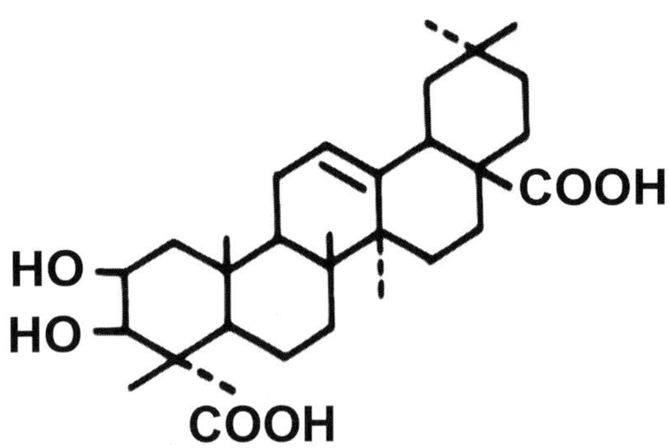

Fig. 9-17. Chemical structure of medicagenic acid.

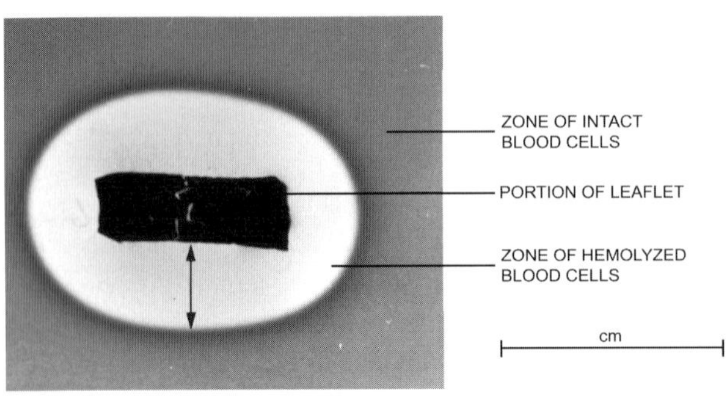

Fig. 9-18. An example of a hemolytic ring produced by a rectangle of alfalfa leaf on an emulsion of blood cells.

Medicago flower away from its stigma. In inbreeding *Medicago*, the circumferential ring of papillae is often reduced, the sticky exudate often extends farther down the stigma and, although the flower remains explosive, there is a notable tendency for pollen to be deposited on the stigma of the same flower from which they originate.

G. Calyxes

Asymmetrical development of the calyx, with a sepal lobe notably larger than the others, is fairly common in *Medicago*, *Trigonella*, and *Melilotus*. This is perhaps to be expected, given that the corolla is asymmetrical, with the standard petal larger than the others, and therefore perhaps benefiting from support provided by larger sepals. A more interesting pattern taxonomically is the presence of two large and three small sepal lobes. This is frequently the case in *Trigonella*, as well as in the unusual *Medicago hypogaea*, but is rare or absent in species of *Medicago* and *Melilotus*.

2. Saponin Chemistry

In this section, it is pointed out that saponin chemistry monothetically (i.e., uniquely) defines the genus *Medicago*: medicagenic acid glycosides in the leaves perfectly discriminate *Medicago sensu lato* from species in allied genera (Jurzysta et al. 1988).

Saponins are steroid or triterpenoid glycosides. Certain saponins can swell and rupture erythrocytes (red blood cells) and their presence can be measured quite simply by placing material in a stabilized emulsion of blood cells (note Fig. 9-18).

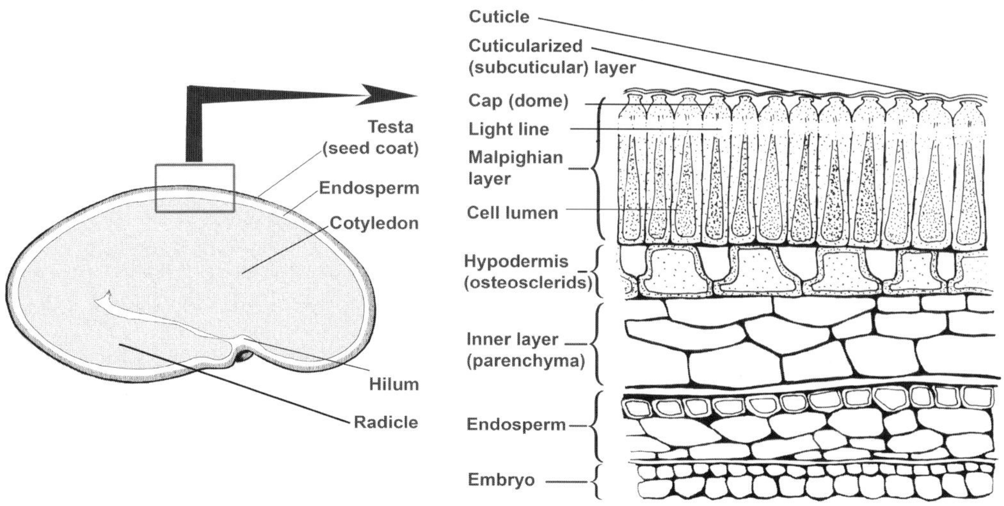

Fig. 9-19. Diagrams showing a longitudinal section of a *Medicago* seed (left) and a detailed section of the seed coat (right). (Illustrations prepared by B. Brookes.)

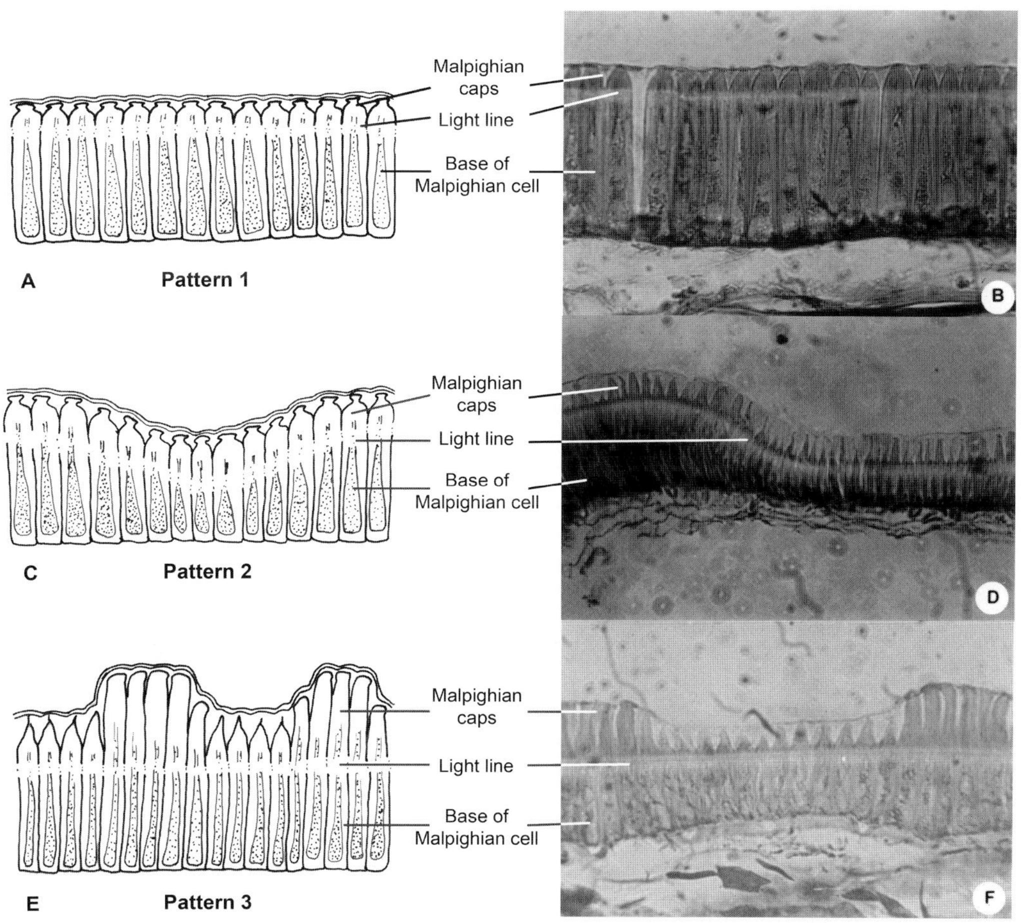

Fig. 9-20. Anatomical testa structure of the three common seed coat patterns in the legume tribe Trifolieae. The drawings on the left are diagrammatic representations of the pattern shown in the light microscope photomicrographs on the right. **Pattern 1:** (A) and (B) show the smooth seed of *Medicago arborea* (based on E. Small M129, DAO). **Pattern 2:** (C) and (D) show the rugose seed of *Medicago ruthenica* (based on E. Small M225, DAO). **Pattern 3:** (E) and (F) show the tuberculate seed of *Trigonella spinosa* (based on E. Small T16, DAO). (Illustrations prepared by B. Brookes.)

Notable of these are glycosides of medicagenic acid, a triterpenoid with an extremely limited distribution in the plant kingdom (Jurzysta et al. 1988). At least 14 medicagenic acid glycosides are known to exist in leaves of *Medicago sativa* (Jurzysta et al. 1988). Jurzysta et al. (1988) tested the content of medicagenic acid glycosides in 79 of the 87 species of *Medicago*, 32 of the species of *Trigonella* (more than half of those recognized in the genus), and all 19 of the species of *Melilotus*, as well as several species of *Trifolium*, *Ononis*, *Parochetus*, and *Lotus*. The results were strikingly clear: all species of *Medicago* produced a positive reaction for medicagenic acid glycosides, whereas none of the species of the other six genera produced a positive reaction.

Brenac and Sauvaire (1996) examined the composition of sterols and steroidal sapogenins in seven species representing *Medicago*, *Trigonella,* and *Melilotus*. They observed that *T. monspeliaca* L., one of the medicagoid species, had a sterolic composition indicating it was indeed a member of the genus *Medicago*.

3. DNA Analysis

By examining the ribosomal DNA sequences of 53 species of *Medicago*, 10 of the 23 "medicagoid *Trigonella*", 10 species of *Trigonella*, two species of *Melilotus*, and two species of *Trifolium*, Bena (2001) confirmed that the medicagoid species together with the other species of *Medicago* (including several species with controversial affinities) form a distinct lineage, separated from the other genera. As discussed in Chapter 10, several additional nucleotide studies confirm the collective distinctness of *Medicago* species from those of other genera.

4. Seed Characters

Small et al. (1990*a*; cf. Gupta 1991; Jha and Pal 1992) found that seed topography and the structure of the Malpighian layer of the seeds segregated the medicagoid *Trigonella* (pulvinate–cotyledon *Medicago*) from the remaining *Trigonella*, uniting them with species of *Medicago*, and discriminated most *Medicago* from most *Trigonella* and *Melilotus*. The following summarizes this information.

A diagrammatic section of the seed coat is illustrated in Fig. 9-19 (cf. Hamly 1932; Gunn 1981). The epidermis includes the Malpighian layer, a unicellular layer found in the seeds of various plant families, which is characteristic of many, but by no means all, genera of the Fabaceae (Fahn 1974). It often appears bicellular because of the presence of a refractory wall region, the light line (linea lucida). The light line is of unknown function

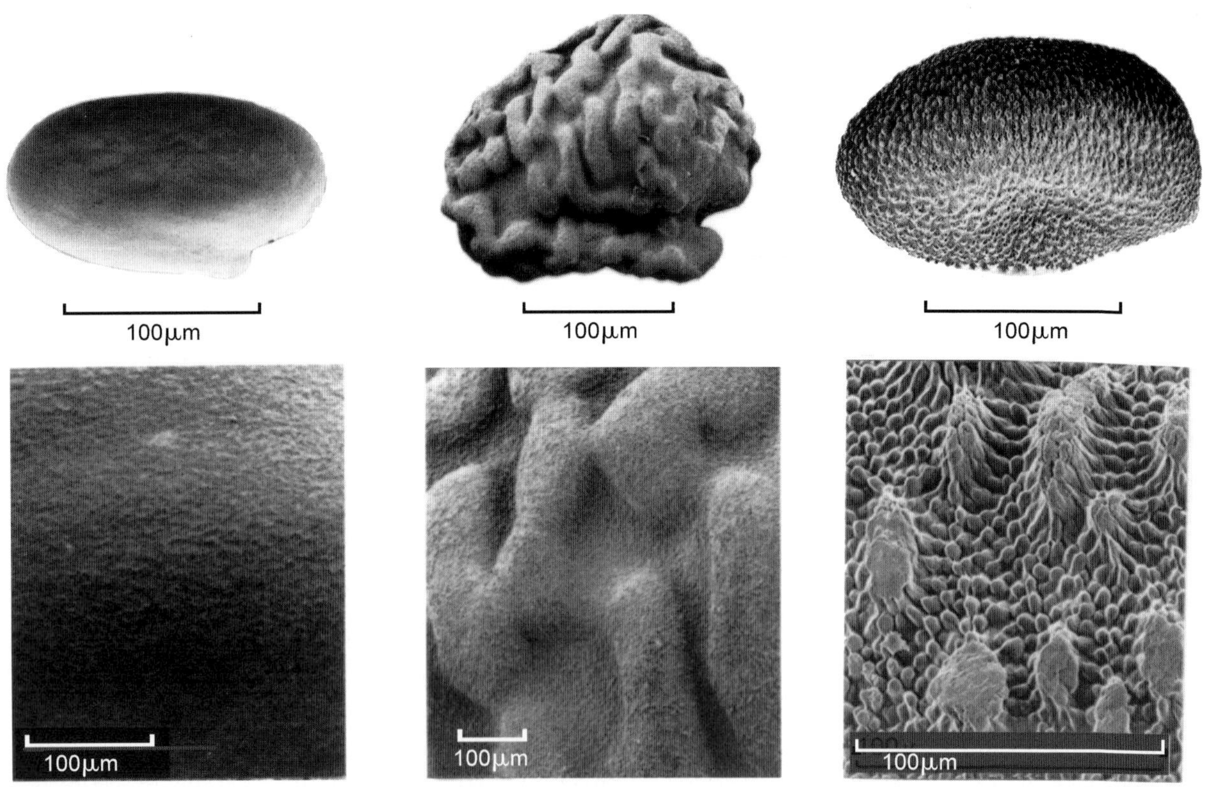

Fig. 9-21. Widespread patterns of seed coat structure in the legume tribe Trifolieae, showing (top row) scanning electron microscope photomicrographs of seeds and (bottom row) scanning electron microscope photomicrographs of seed surface. **Left column:** *Medicago edgeworthii* (based on E. Small M957, DAO), illustrating smooth seeds (Pattern 1). **Center column:** *Medicago heyniana* (based on E. Small M342, DAO), illustrating rugose seed (Pattern 2). **Right column:** *Trigonella sibthorpii* (based on E. Small T277, DAO) illustrating tuberculate seed (Pattern 3). (Illustrations prepared by B. Brookes.)

(Watson 1948), although it is thought to be related to the control of moisture into the seed (Esau 1977).

For ease of presentation, the portion of the Malpighian cell layer above the light line is designated *a* and that below the light line is designated *b*. There are three repetitive patterns, as discussed in the following. The sectioned seed appearance of the three patterns is shown in Fig. 9-20 and the surface appearance of the three patterns is shown in Fig. 9-21.

Pattern 1: The surface of the seed is smooth, with the length of *a* and *b* constant (Fig. 9-20A, B; Fig. 9-21, left column). This is the predominant pattern among species of *Medicago*. Pattern 1 also occurs in some species of *Trigonella* and *Melilotus*.

Pattern 2: The surface of the seed is rugose (usually wrinkled, the elevations always with rounded sides), with the length of *a* constant and the length of *b* varying directly with the height of the elevation (Fig. 9-20C, D; Fig. 9-21, center column). Pattern 2 appears to occur exclusively in *Medicago*, including in the pulvinate–cotyledon species.

Pattern 3: The surface of the seed is tuberculate ("warty", the elevations with markedly vertical sides), with the length of *a* varying directly with the height of the elevation and the length of *b* remaining constant (Fig. 9-20E, F; Fig. 9-21, right column). This is the most common pattern in *Trigonella*. Pattern 3 also occurs in *Melilotus* subgenus *Micromelilotus* Schulz but not in *Melilotus* subgenus *Melilotus*. *Medicago orbicularis* is the only species of *Medicago* to have Pattern 3. However, its tubercles are not as distinct as in most tubercular-seeded *Trigonella*.

In Chapter 7, it is pointed out that rough-surfaced seeds (patterns 2 and 3) strongly tend to occur in dehiscent fruits, and the ecological importance of seed surface roughness is discussed.

5. Phytoalexin Chemistry

Ingham (1979*a*, 1979*b*, 1981) examined the Trifolieae for the presence of wound-induced isoflavonoid phytoalexins, which commonly accumulate in the group when detached leaflets are inoculated with conidial suspensions of *Helminthosporium carbonum* Ullstrup. An examination of Ingham's data shows that while there are trends separating *Trifolium*, *Melilotus*, and *Medicago*, there are no consistent differences between *Medicago* and the species that were previously assigned to the medicagoid *Trigonella* and *Factorovskya*. Ingham's data suggest that the pulvinate–cotyledon *Medicago* bear a greater resemblance to *Medicago* than to the remainder of *Trigonella*.

6. Phenolic Chemistry

There are similarities in phenolic chemistry, suggesting that the pulvinate–cotyledon *Medicago* bear a greater resemblance to *Medicago* than to the remainder of *Trigonella* (Classen et al. 1982).

7. Pollen

Small et al. (1981*a*) compared the morphology of pollen grains of *Medicago*, *Trigonella*, and *Melilotus*. While no single character separated the genera, the pollen of species of *Medicago* proved to be substantially distinguishable from that of species of *Trigonella* and *Melilotus* when numerical discrimination techniques were used. As is evident in Table 9-4, there are very significant trends distinguishing *Medicago* from *Trigonella* and *Melilotus*.

Table 9-4. Morphological trends separating *Medicago* pollen from *Trigonella* and *Melilotus* pollen.

Character	*Medicago* (n = 55)	*Trigonella* (n = 34)	*Melilotus* (n = 18)
Presence of pyramidal pollen	In ca. 36% of species	Absent	Absent
Presence of pollen with more than, or less than, three colpi (elongated apertures)	In ca. 42% of species	In ca. 3% of species	In ca. 28% of species
Presence of prolate pollen only (i.e., with polar axis noticeably longer than equatorial axis)	In ca. 27% of species	Always present	Always present
Mean width more than 22.5 μm	In ca. 91% of species	In ca. 15% of species	Absent
Polar: equatorial index greater than 1.30	In ca. 6% of species	In ca. 80% of species	In ca. 80% of species
Pore width wider than colpus width	In ca. 9% of species	Always	Always
Presence of intra-colpus sculpturing	In ca. 18% of species	In ca. 90% of species	Always

Note: After Small et al. (1981*a*), which see for species surveyed and additional but weaker trends.

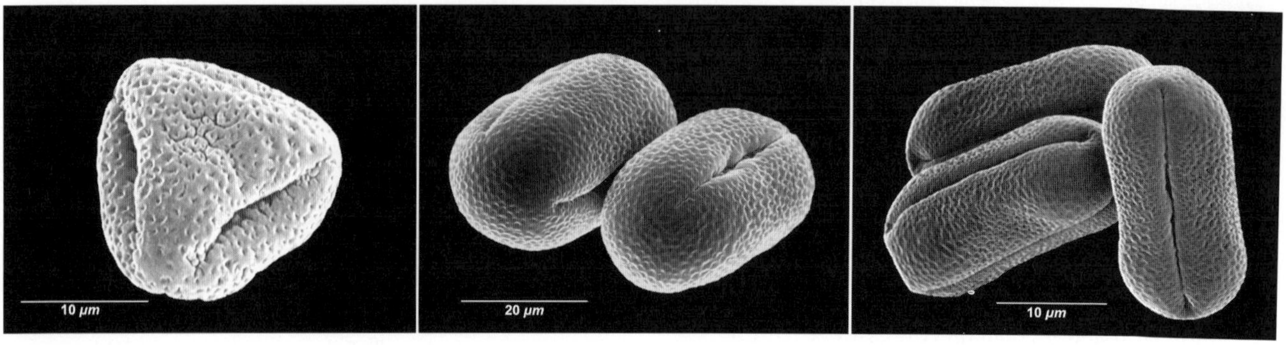

Fig. 9-22. Scanning electron micrographs of representative acetolyzed pollen grains of *Medicago*, *Trigonella*, and *Melilotus*. **Left:** Pyramidal grain of *Medicago hypogaea*. **Center:** Prolate grain of *Trigonella graeca* (Boiss. & Spruner) Boiss. **Right:** Prolate grain of *Melilotus dentatus* (Waldst. & Kit.) Pers. (Illustrations prepared by B. Brookes.)

TYPE OF *MEDICAGO* L.

Medicago L., Sp. Pl. 778 (1753)
M. sativa L. (typ. cons.)

For some time there was controversy over the choice of lectotype of *Medicago* L. Britton and Brown (1913) designated *M. sativa* L. as the lectotype and this decision became popularly accepted. However, Heyn (1959) pointed out that *Medicago* had been typified earlier by Scofield (1908). Scofield contended that the name *Medicago*, as used by Linnaeus, covered species assignable to two genera, *Medicago* and *Medica* (the latter listed in Farr et al. 1979 as an untypified name of P. Miller 1754). Scofield (1908) typified *Medicago* by *M. radiata* L. Heyn (1959) accepted Scofield's typification and, sympathetic to the viewpoint that *M. radiata* and *M. sativa* should not be considered congeneric, suggested that conservation of the name *Medicago* for *M. sativa* was appropriate. Stearn (1960) reviewed this situation and concluded that *M. radiata* was apparently the preferable choice of lectotype of *Medicago*. Ali (1968) thought that the viewpoint that there are debatable limits between the genera *Medicago* and *Trigonella*, with *M. radiata* being a transitional species, might result in complicated nomenclatural problems, and *ipso facto M. radiata* should not be considered the type of *Medicago*. Heyn (1969) rejected this contention. Baum (1970) argued that Scofield had used an incorrect and indeed illogical procedure in typifying *Medicago* on the basis of his own, comparatively narrow conception of the genus, rather than on Linnaeus's broader conception. This reasoning was rejected by Ali (1972). Farr et al. (1979) listed *M. radiata* as the type of *Medicago*. Soják and Pouzar (1984) proposed that the name *Medicago* be conserved with *M. sativa* as its type. The acceptance of this proposal made the above recondite perspectives moot.

DESCRIPTION OF THE GENUS *MEDICAGO*

The genus *Medicago* is defined here to include 87 species indigenous to Eurasia and North Africa:

Annual or perennial herbs, rarely shrubs; mostly slightly to moderately pubescent, sometimes with gland-tipped (occasionally septate) hairs; flowers in axillary racemes, heads or umbels, or sub-solitary, with an irreversible explosively tripping pollination mechanism (sometimes degenerate and not operational in some species, but nevertheless evidence of its morphological components remains); corolla mostly yellow, in some species violet, or intermediate or variegated, sometimes (intraspecifically) white; calyx teeth more or less sub-equal; vexillum (except when very small) with major basal vein usually with more than three branches near base; horn of wing petal large (approximately one-third length of wing limb); keel very strongly adherent to wing by a wing spur in a keel invagination; androecium with apex of fused portion of anther column arched, alternating free portions of filaments often relatively thick; stigma fungiliform; style always short; fruits very diverse: short to long, straight, curved or coiled, terete, compressed or flat, sometimes moniliform, sutures spiny, tuberculate, winged, fimbriate or without appendages; cotyledons with or (more frequently) lacking pulvini; seeds mostly smooth, sometime rugose or papillose.

MISASSIGNMENTS TO *MEDICAGO*

Various species have been placed in *Medicago* that in fact belong to other genera (indeed, they sometimes belong to families other than the Fabaceae). Figs. 9-23 and 9-24 illustrate two such species. A list of misassignments follows.

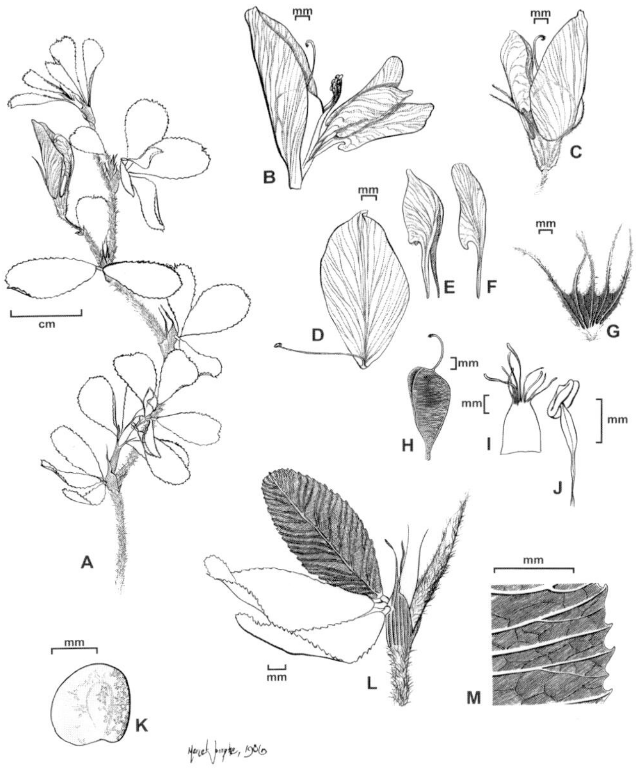

Fig. 9-23. *Trifolium gordejevi* (Kom.) Wei Zhi, previously erroneously assigned to *Medicago gordejevi* Kom. (A) Flowering branch. (B) Flower (floral parts teased apart). (C) Side view of flower. (D) Standard petal (free stamen at the base). (E) Keel. (F) Wing petal. (G) Calyx. (H) Pistil. (I) Staminal column (anthers and free stamen not shown). (J) Stamen (with narrow filament; these alternate with stamens with wider filaments). (K) Seed. (L) Trifoliolate leaf with stipules at the node. (M) Portion of leaflet margin. (After Small (1987*b*); artist: M. Jomphe.)

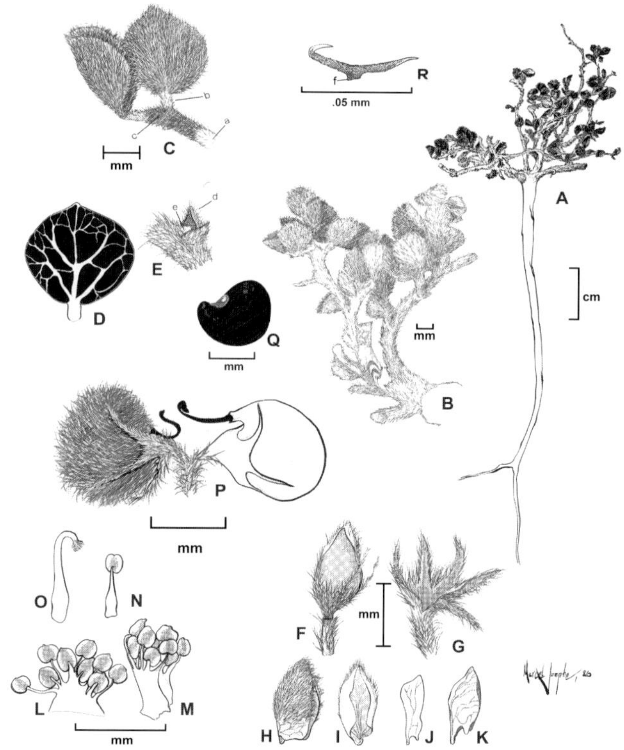

Fig. 9-24. *Indigofera nephrocarpa* Balf. f., previously erroneously recognized as *Medicago makranica* Heyn. (A) Habit. (B) Portion of shoot. (C) Leaf (a, pulvinus; b, pulvinule; c, extension of rachis, showing leaf is imparipinnate). (D) Adaxial view of leaflet with the hairs scraped away to reveal venation. (E) Node at stem the leaf removed (d, stipule; e, scar of petiole between stipules). (F) Flower bud (dissected to show G to O). (G) Calyx. (H) Side view of standard petal (note abaxial pubescence). (I) Adaxial view of standard petal. (J) Wing petal. (K) Keel. (L) Abaxial view of staminal column. (M) Side view of staminal column. (N) Free (vexillary) stamen. (O) Pistil. (P) Infructescence of two fruits. (Q) Seed. (R) Biramous hair (f, base). (Based on holotype of *Medicago makranica* Heyn. After Small et al. (1988); artist: M. Jomphe.)

SPECIES EXCLUDED FROM *MEDICAGO*

Medicago alba (Medik.) E.H.L. Krause in Sturm, Deutschl. Fl., ed. 2, 9: 128 (1901) ≡ *Melilotus alba* Medik.

M. altissima (Thuill.) E.H.L. Krause in Sturm, Deutschl. Fl., ed. 2, 9: 125 (1901) ≡ *Melilotus altissima* Thuill.

M. anguina (Delile) C. Romero Zarco, Lagascalia 18(2): 265 (1996) ≡ *Trigonella anguina* Delile.

M. besseriana (Ser.) E.H.L. Krause in Sturm, Deutschl. Fl. ed 2, 9: 122 (1901) ≡ *Trigonella besseriana* Ser.

M. circinnata L., Sp. Pl. 778 (1753) ≡ *Hymenocarpos circinnatus (L.)* Savi.

M. corniculata (L.) Trautv., Bull. Sc. Acad. Imp. Sci. St. Petersb. 8: 271 (1841) ≡ *Trigonella corniculata* L.

M. esculenta Rottler ex Willd., Enum. Hort. Ber. 799 (1809) = *Trigonella corniculata* L.

M. foenum-graecum (L.) E.H.L. Krause in Sturm, Fl. Deutschl., ed. 2, 9: 119 (1901) ≡ *Trigonella foenum-graecum* L.

M. gladiata (Steven ex M. Bieb.) E.H.L. Krause in Sturm, Deutschl. Fl., ed. 2, 9: 120 (1901) ≡ *Trigonella gladiata* Steven ex M. Bieb.

M. gordeievii (*gordejevi*) Komarov, Bot. Mater. Gerb. Glavn. Bot. Sada RSFSR 2: 133 (1921) ≡ *Trifolium gordeievii* (Kom.) Wei Zhi.

M. hamosa (L.) Trautv., Bull. Sc. Acad. Imp. Sci. St. Petersb. 8: 271 (1841) ≡ *Trigonella hamosa* L.

M. italica (L.) E.H.L. Krause in Sturm, Fl. Deutschl., ed. 2, 9: 125 (1901) ≡ *Melilotus italicus* (*italica* = orth. var.) (L.) Lam.

M. microcarpa Balbis in Bernh. (1840) (not *M. microcarpa* Urb.) = *Trifolium filiforme* L. (cited in Index Synonymique Francaise: http://www2.dijon.inra.fr/flore-france/me-mh.htm, current August 2008. Zohary and Heller (1984) referred to *T. filiforme* L. as a nom. ambig. and assigned it in part to synonymy of *T. dubium* Sibth., and also assignable to *T. micranthum* Viv.)

M. nummularia DC., Cat. Pl. Horti Monsp. 124 (1813) ≡ *Hymenocarpos circinnatus* (L.) Savi var. *nummularius* (DC.) Bornm.

M. officinalis (L.) E.H.L. Krause in Sturm, Deutschl. Fl., ed. 2, 9: 127 (1901) ≡ *Melilotus officinalis* (L.) Lam.

M. ornithopodioides (L.) Trautv., Bull. Acad. Sci. St. Petersb. 8: 271 (1841) ≡ *Trifolium ornithopodioides* (L.) Sm.

M. parviflora (Desf.) E.H.L. Krause in Sturm, Deutschl. Fl., ed. 2, 9: 125 (1901) = *Melilotus indica* (L.) All.

M. pungens Trautv., Bull. Acad. Sci. St. Petersb. 8: 271 (1841) = *Trigonella maritima* Delile.

M. sulcata (Desf.) E.H.L. Krause in Sturm, Deutschl. Fl., ed. 2, 9: 124 (1901) ≡ *Melilotus sulcata* Desf.

M. trigonella E.H.L. Krause in Sturm, Deutschl. Fl., ed. 2, 122 (1901) = *Trigonella laciniata* L.

M. virginica L., Spec. Plant. 778 (1753) ≡ *Lespedeza virginica* (L.) Britton

M. vassilczenkoi Vorosch., Fl. Sovetsk. Dal'n. Vost. 271 (1966) = *Trigonella schischkinii* Vassilcz.

KEY[1,2] TO GENERA OF FABACEAE, TRIBE TRIFOLIEAE

1. Leaves with three or more leaflets; when three leaflets are present, the three petiolules equally short-stalked; staminal column essentially free of corolla; petals often persistent in fruit .... subtribe Trifolinae: *Trifolium*
1. Leaves with three leaflets; the central petiolule of a leaf longer than the lateral petiolules; staminal column subtended by base of corolla; petals not persistent in fruit (subtribe Trigonellinae) 2
   2. Fruit coiled or not; flowers with explosive mechanism (not always evident); floral androecium with arched apex; style short compared with length of pistil .... *Medicago*
   2. Fruit never coiled; flowers lacking explosive mechanism; floral androecium with blunt apex; style long compared with length of pistil (*Trigonella* and *Melilotus*) 3
      3. Fruit nutlet-like, one-few-seeded, mainly indehiscent, often with sculptured surface; flowers in racemes .... *Melilotus*
      3. Fruit with a straight or rarely falcate legume, usually many-seeded and dehiscent; flowers single, or in heads or racemes .... *Trigonella*

[1]See comments in this chapter regarding questions about the genera that should be included in tribe Trifolieae, possible changes in nomenclature of the tribe that depend on which genera should be included, and the desirability of combining *Melilotus* and *Trigonella*.

[2]As pointed out above, the presence of medicagenic acid glycosides perfectly discriminates *Medicago* from allied genera, and can be easily tested by viewing the hemolytic effects of a very small sample of tissue on fresh blood. Indeed, within minutes I have confirmed the identity of vegetative collections more than a century old by viewing the reaction of a minute sample to fresh blood under a microscope (a more reliable test takes hours, using the method described). Of course, for those reluctant to obtain or contribute fresh blood, the above morphological key should suffice.

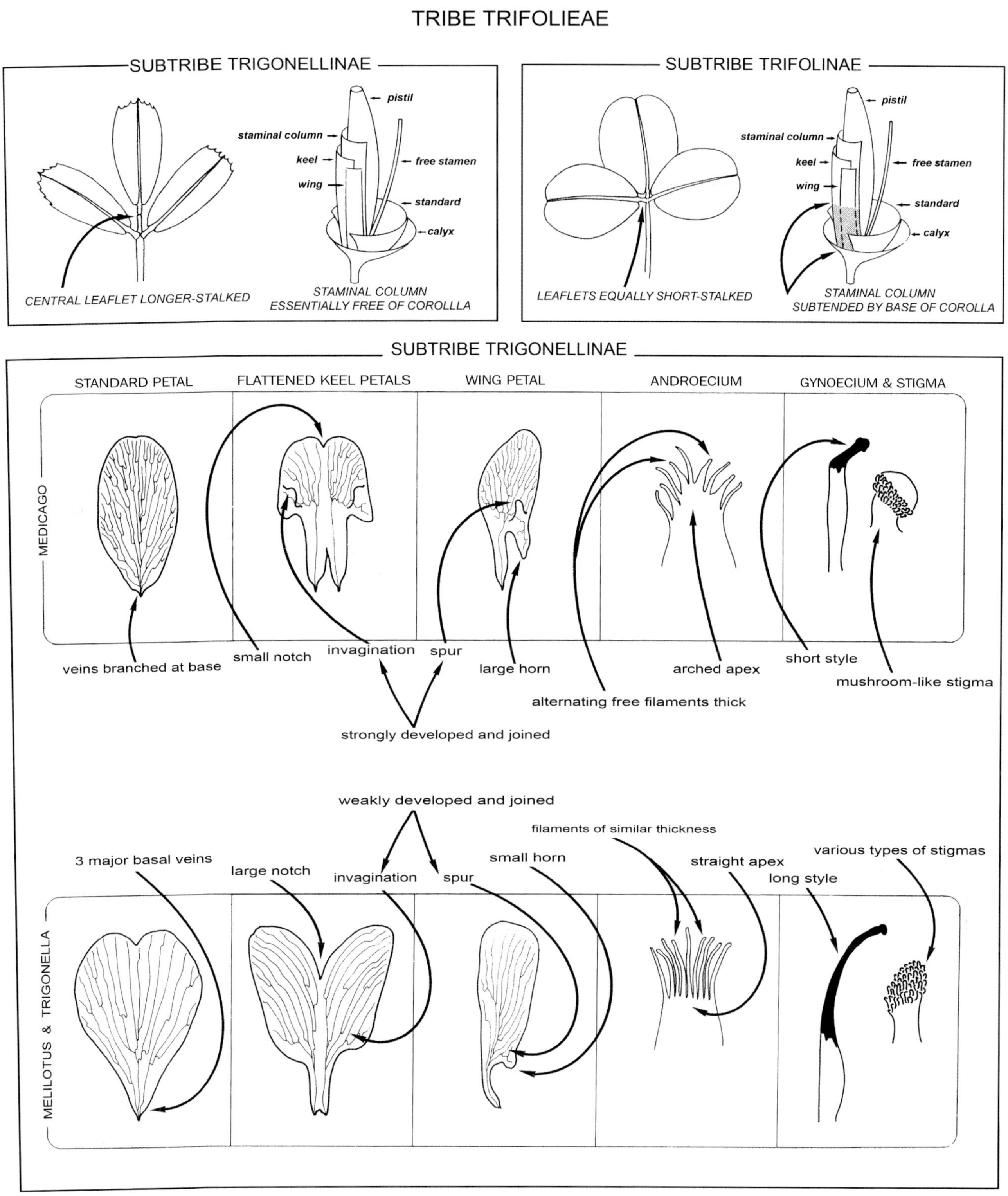

Fig. 9-25. Illustrations distinguishing *Medicago* (of Fabaceae tribe Trifolieae subtribe Trigonellinae) from other genera of tribe Trifolieae. (Artist: M. Jomphe.)

10. Phylogenetic Lineages and their Classification Into Sections and Subsections

As detailed below, several taxonomists utilizing classical morphological analysis have subdivided *Medicago* variously, sometimes agreeing on the species composition of particular groups and their ranking, but often not. Also, as detailed below, there have been several analyses of nucleotide sequences from both the plastid and nuclear genome in recent years, and these have shed new light on how the species of *Medicago* are related. Judged as a whole, the various molecular marker analyses conducted to date have confirmed the close relationships of some species that were previously deduced from consideration of their morphology, but have also shown that some species thought to be closely related are not, and that some species are more closely related than has been assumed on the basis of morphology. Consequently, the results of nucleotide sequence analyses have strongly suggested that some traditionally delimited taxonomic subdivisions are not justified and that the placement of some species in particular groups is invalid. Unfortunately, the phylogenetic analyses reviewed have not yet included all species of *Medicago* and different molecular markers often indicate different sets of relationships, and therefore suggest different taxonomic solutions. Accordingly, the relationships of many of the species and groups of species that have been examined remain ambiguous. This chapter attempts to account for the phylogenetic information that has been accumulated to date while changing traditional subdivision of *Medicago* only to the extent that now seems unavoidable.

KEY CLASSICAL MORPHOLOGICAL ANALYSES

Urban (1873), in the first competent, comprehensive taxonomic analysis of *Medicago* (almost all his species are accepted in this monograph), delimited *Medicago* into two *Hauptgruppen* (principal divisions or subgenera; these do not seem worthwhile) and ten sections (most of which are recognized in this monograph as sections or subsections, but not necessarily with the same species assignments).

Lesins and Lesins (1979), the second comprehensive systematic analysis of *Medicago*, divided it (often with no regard for nomenclatural requirements) into four subgenera: subgenus *Lupularia* (≡ section *Lupularia*, recognized below); subgenus *Orbicularia* (with section *Platycarpae* and five monotypic sections: *Carstienses*, *Orbiculares*, *Hymenocarpos*, *Heynianae*, and "*Cretaceae*" (*M. cretacea*), which, except for *Cretaceae*, are all accepted below; subgenus *Medicago* (including what is recognized in this monograph as sections *Medicago* and *Dendrotelis*, albeit split by Lesins and Lesins more extensively into subgroups); and subgenus *Spirocarpos* (recognized as section *Spirocarpos* in this monograph).

Heyn (1963) studied 27 annual species with coiled pods, which she recognized as section *Spirocarpos*. She divided it into four subsections and recognized *M. orbicularis* as an unrelated monotypic group.

None of the above works was based on an understanding that two dozen species, previously placed in *Trigonella*, properly (as recognized in this monograph: see Chapter 9) belong to *Medicago* (these are in sections *Geocarpa*, *Buceras*, and *Lunatae*).

As discussed in detail in Chapter 9, Urban (1873) employed the absence of a swelling or "joint" (pulvinus) at the base of the cotyledons as a criterion for circumscribing the genus *Medicago*. Cotyledonary pulvini are present in species of the related genera *Trigonella* and *Melilotus*. (which as noted in Chapter 9 probably need to be amalgamated into one genus). In this monograph, *M. hypogaea* (previously placed in other genera), and 24 species with cotyledonary pulvini, are assigned to *Medicago* (in sections *Buceras* and *Lunatae*). Cotyledonary pulvini are shared by these sections of *Medicago* and its sister group *Trigonella*. The retention of cotyledonary pulvini, a characteristic manifested at the embryonic/juvenile stage, appears to illustrate the much-debated evolutionary maxim "embryology recapitulates phylogeny" (Haeckel 1866).

Small (1987*c*) delimited *Medicago* into 10 sections, and subsequently Small and Jomphe (1989*b*) split one of these into three monotypic sections. The morphological relationships on which this delimitation was based are summarized in Figs. 10-1 and 10-2. Note that all the figures in this chapter and the text use the same number and/or letter codes for the infrageneric groups recognized.

NUCLEOTIDE SEQUENCE ANALYSES

Several phylogenetic analyses based on nucleotide sequence data are helpful in deciding on an appropriate infrageneric classification of *Medicago*. In particular, (1) Downie et al. (1998) presented a molecular phylogeny of *Medicago* based on 57 species, using sequences of the internal transcribed spacers (ITS) of nuclear ribosomal DNA ITS-1 and ITS-2; (2) Bena (2001) examined 61 species, also based on these two ribosomal internal transcribed spacers but additionally

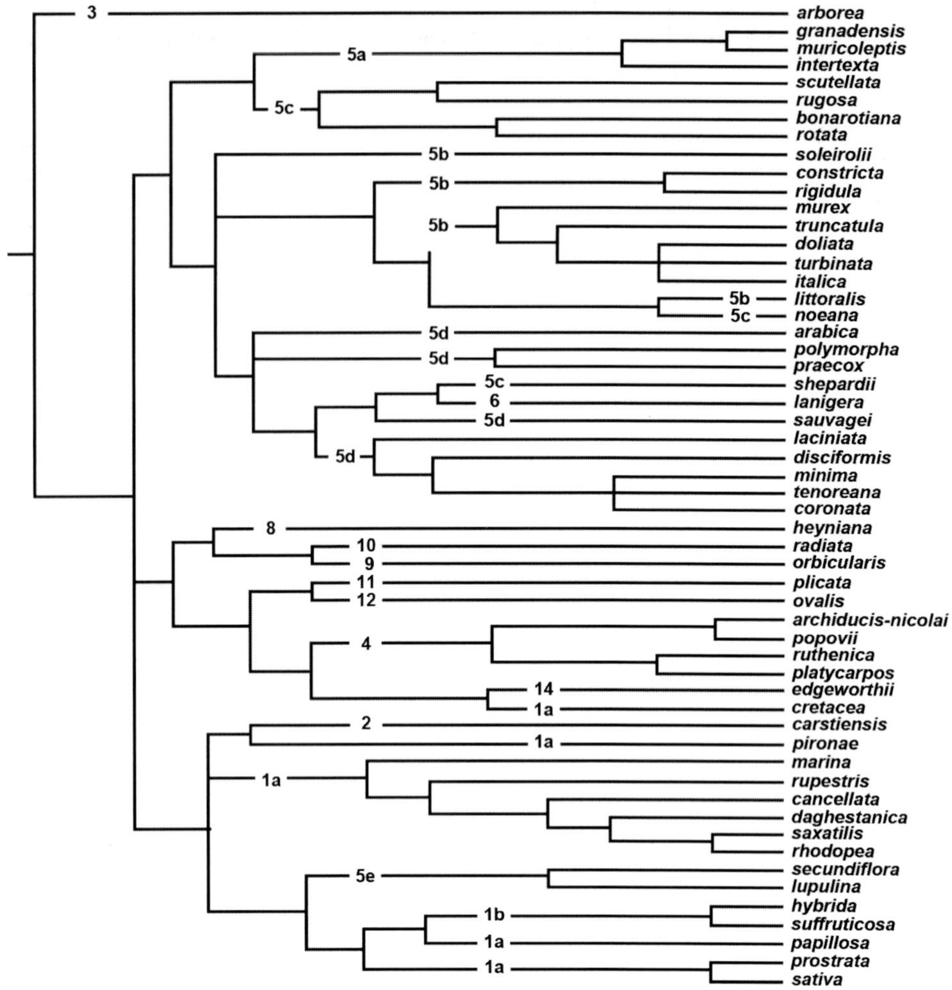

Fig. 10-1. Morphological relationships of species of *Medicago*, excepting most species of sections *Lunatae* (but including *M. edgeworthii*) and *Buceras*. Adapted from Small (1981*a*).

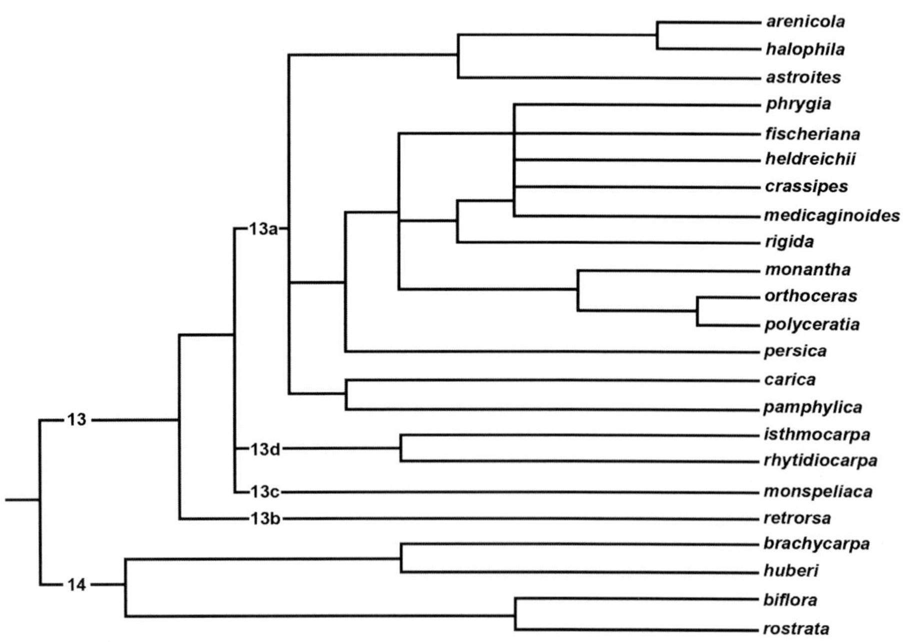

Fig. 10-2. Morphological relationships of species of *Medicago*, sections *Lunatae* and *Buceras*. Adapted from Small (1987*a*). *Medicago edgeworthii* was not appreciated to be a member of section *Lunatae* at the time of the study.

based on a part of the ribosomal external transcribed spacer (ETS); (3) Maureira-Butler et al. (2008) examined 56 species based on two nuclear genes (CNGC5 and β-*cop*) and a mitochondrial region (*rp*S14-*cob*); (4) Steele and Wojciechowski (2003) examined 19 of the species based on the plastid gene *trn*K/*mat*K; and (5) Steele et al. (2010) analyzed nucleotide sequence data from the plastid gene *trn*K/*mat*K (49 species) and the nuclear-encoded gene *GA3ox1* (45 species). Because the species sampled and molecular markers differed in these studies, it is useful to examine relationships on the basis of their collective observations. The following discussion first examines some general issues concerning the findings of these molecular studies, provides an infrageneric classification of *Medicago* that recognizes 14 sections (three of which are subdivided into subsections), and reviews the molecular studies to examine the degree of support or rejection for the infrageneric taxa.

INCONGRUENCE AMONG DEDUCED PHYLOGENETIC ANALYSES

Frequently, the different molecular markers used in the studies reviewed here have produced discordant results for some of the species. That is, the phylogenetic relationships observed among many of the *Medicago* species appear to differ notably, depending on which nucleotide sequence has been examined. Maureira-Butler et al. (2008) concluded that this incongruence is due to extensive hybridization among ancestral lineages of *Medicago* species (other possible explanations were discussed). They wrote that "if significant fractions of the genome track different histories, a single species tree, even one constructed from numerous genes, may be an unrealistic representation of the history of the species" and "given the extent of hybridization within *Medicago*, it is clear that a bifurcating topology is a grossly unrealistic representation of the origins of taxa in this genus." The hypothesis here is that hybridization occurred in the distant past, resulting in incorporation of portions of genomes among lineages; presumed allopolyploids, which by definition have a hybrid origin, were deliberately not included in the study of Maureira-Butler et al. (2008) to test this hypothesis. Such a scenario is discouraging from the point of view of ascertaining an unambiguous phylogeny, which could be used as a basis of infrageneric classification. Moreover, bifurcations or dichotomies are generally viewed as the basis of phylogeny-based, present-day biological classification, which is fundamentally inconsistent with the possibility that *Medicago* species are, in effect, multi-parental descendants. Despite these difficulties, as discussed below, the molecular sequence-based phylogenies of *Medicago* produced to date are useful in deciding whether certain lineages merit taxonomic recognition (whether they are derived from a single ancestral genome or are compounded of partial contributions from several divergent sources). It is important to note that the data of Maureira-Butler et al. indicated the presence of nine clades whose species composition was similar or identical regardless of the molecular marker used, and indeed there is appreciable congruence for certain species among many of the molecular studies reviewed. Accordingly, while there is disconcerting inconsistency among molecular markers in some respects, this has not prevented the elucidation of many relationships.

ANNUALS VERSUS PERENNIALS: ARE THEY MONOPHYLETIC GROUPS?

Virtually all taxonomists specializing on the genus *Medicago* have divided it into several perennial groups and several annual groups, although there has been disagreement about the composition and rank of these groups. Indeed, in the analysis of sections of *Medicago* presented below, each section is composed entirely of perennial species or entirely of annual species (although *M. lupulina*, *M. brachycarpa*, and *M. edgeworthii* are flexibly annual or perennial). Perennial *Medicago* have been interpreted as phylogenetically older, and annuals have been considered to be derivatives of the former (e.g., Heyn 1963; Lesins and Lesins 1979), but these interpretations are debatable and lacking in supporting evidence. With the exception of a few rather enigmatic annual species, the perennial groups and the annual groups have been considered to be well separated. However, Bena et al. (1998*a*) and Bena et al. (1998*b*), on the basis of phylogenetic analyses of nucleotide sequence data for *Medicago*, suggested that some of the annual species may be most closely related to some of the perennials, and vice versa. That is, perhaps some annual species arose independently from different perennial species, and vice versa. This may well be true (it is known that, under domestication, annual forms of the perennial *M. sativa* have been selected) and, indeed, prior to the use of molecular analysis, there was speculation that annuals and perennials had occasionally arisen from the other group. However, as discussed above, it appears that the results of phylogenetic analyses of nucleotide sequence data for *Medicago* may be confounded by extensive reticulate evolution (Maureira-Butler et al. 2008); if so, it may be that the data of Bena et al. (1998*a*) are insufficient to conclude that some annuals may be most closely related to some perennials, and vice versa.

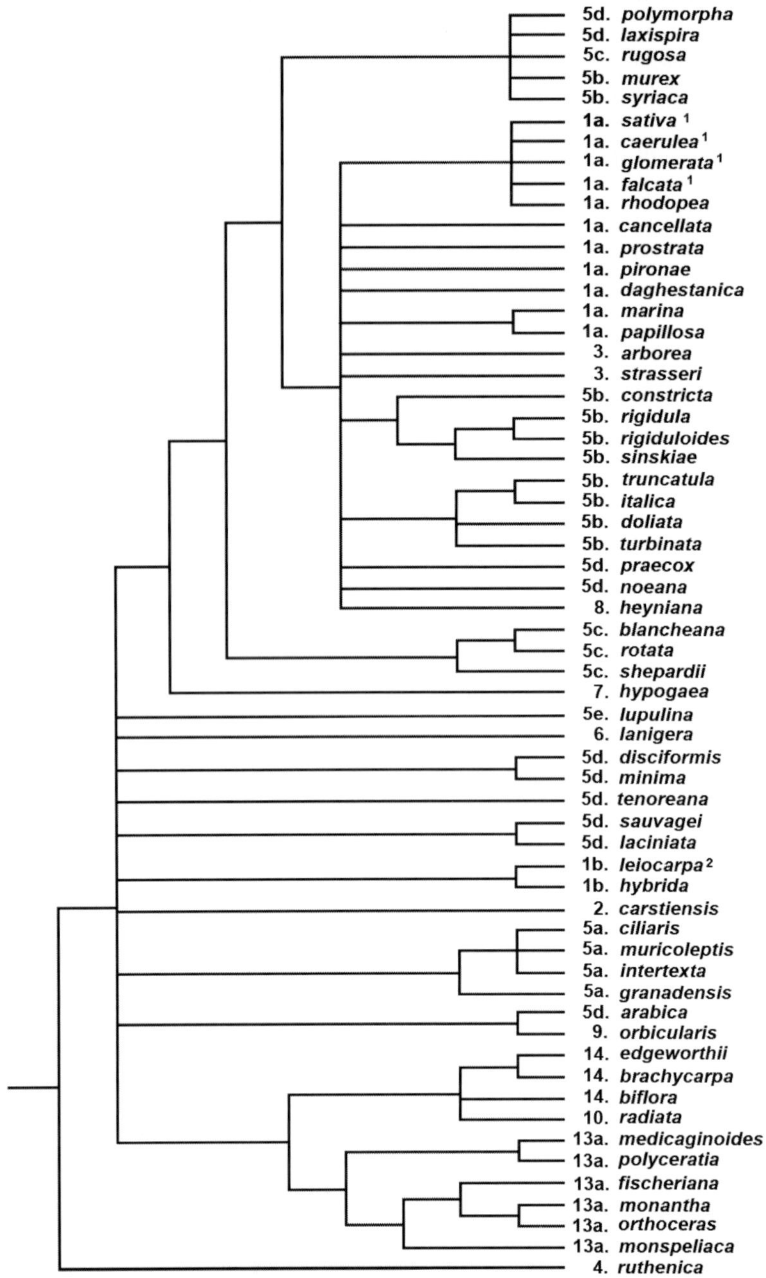

Fig. 10-3. Hypothesized relationships of species of *Medicago*, based on two ribosomal internal transcribed spacers (ITS-1 and ITS-2). Adapted from Downie et al. (1998). [1]Subspecies of *M. sativa*; [2]subspecies of *M. suffruticosa*.

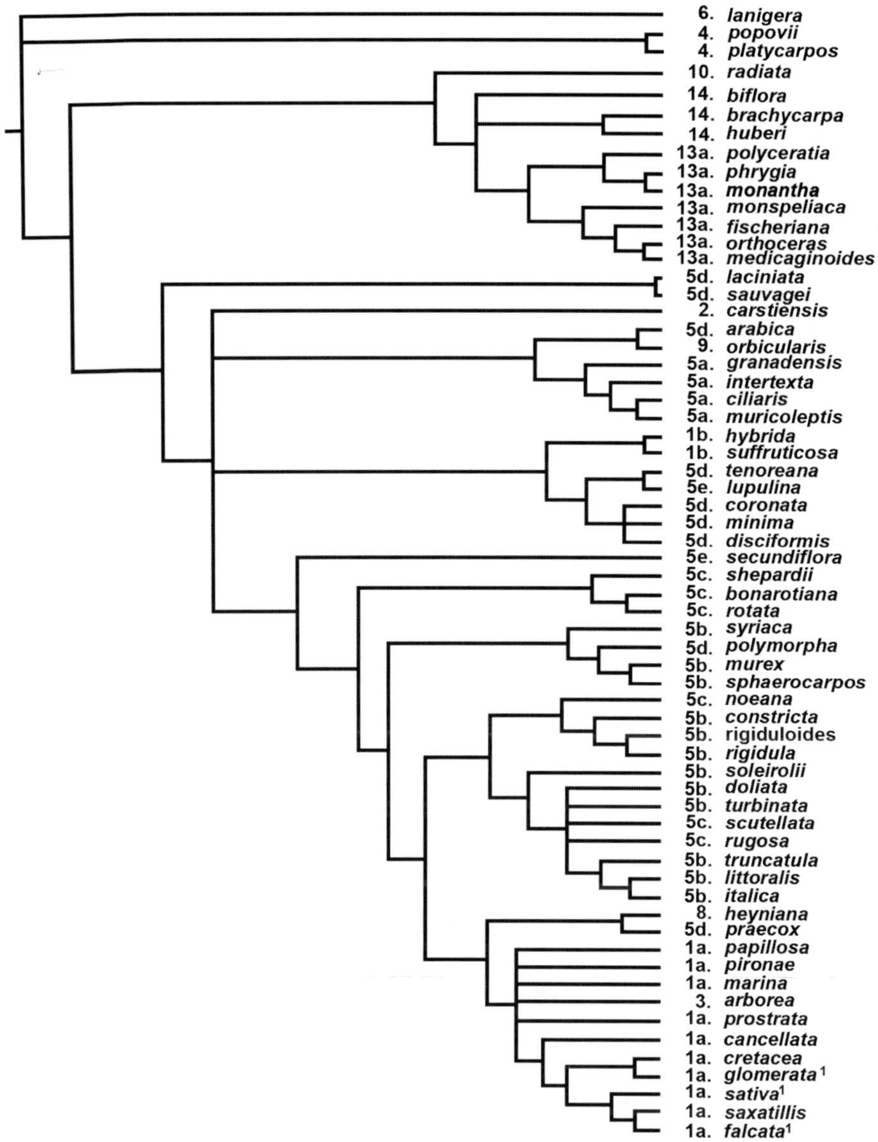

Fig. 10-4. Hypothesized relationships of species of *Medicago*, based on two ribosomal internal transcribed spacers (ITS-1 and ITS-2) and a part of the ribosomal external transcribed spacer (ETS). Adapted from Bena (2001). [1]Subspecies of *M. sativa*.

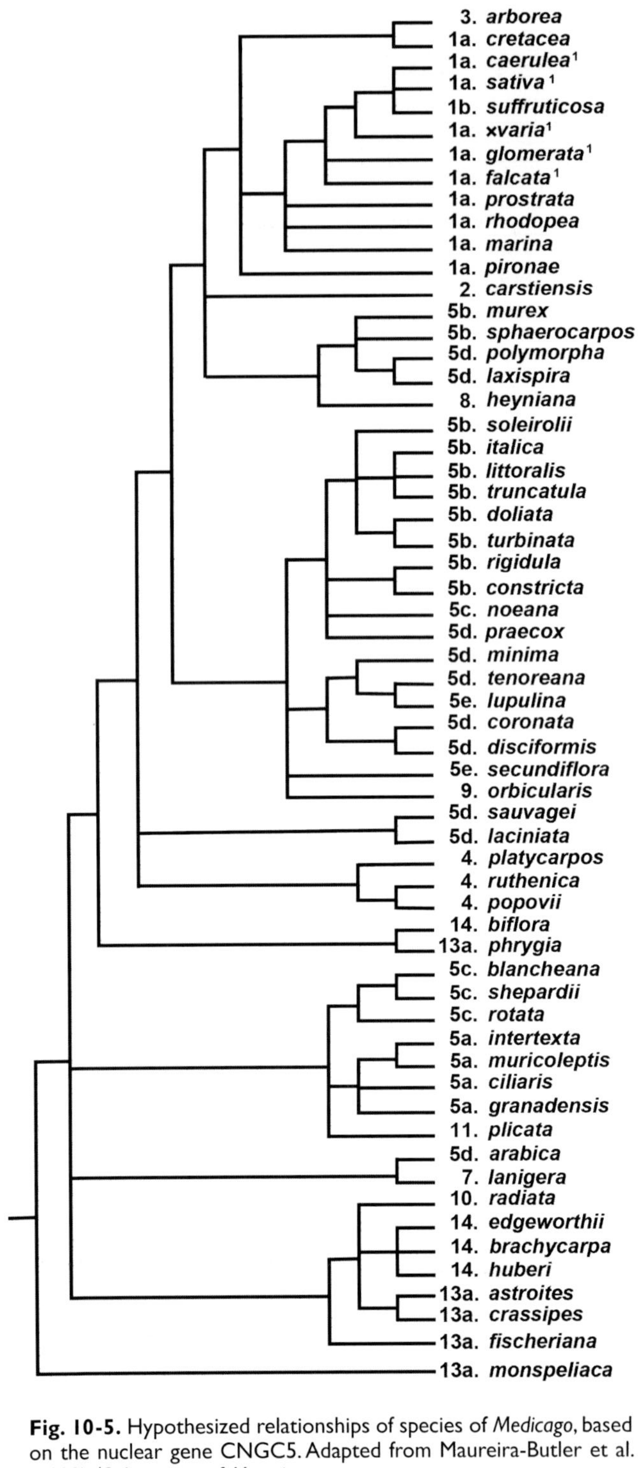

Fig. 10-5. Hypothesized relationships of species of *Medicago*, based on the nuclear gene CNGC5. Adapted from Maureira-Butler et al. (2008). [1]Subspecies of *M. sativa*.

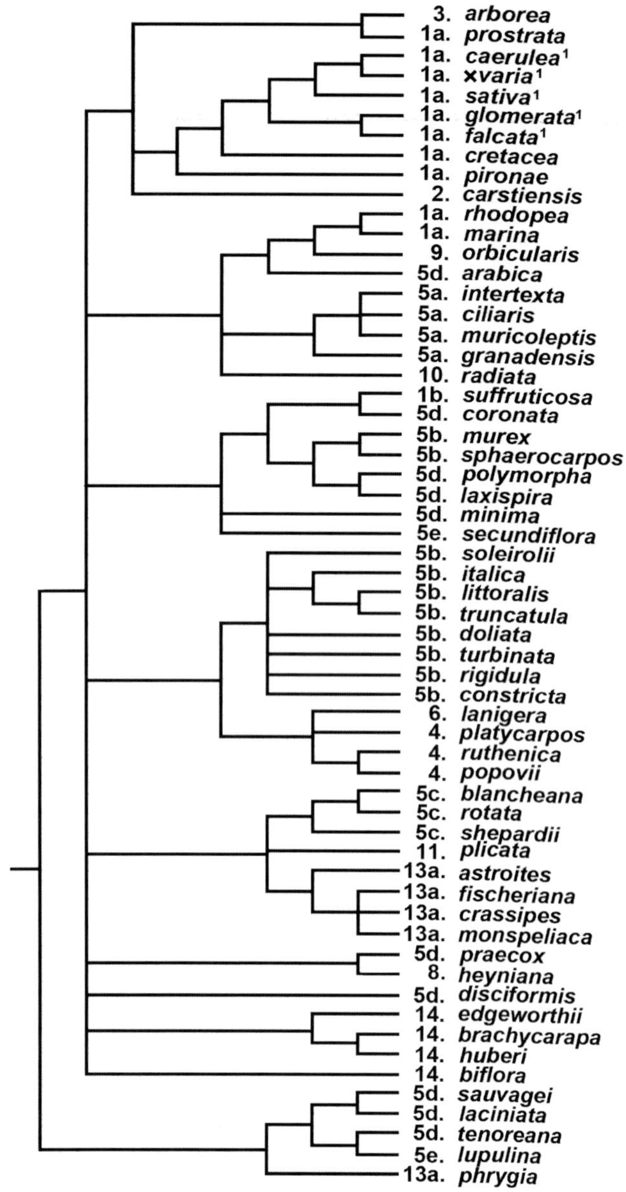

Fig. 10-6. Hypothesized relationships of species of *Medicago*, based on the nuclear gene β-*cop*. Adapted from Maureira-Butler et al. (2008). [1]Subspecies of *M. sativa*.

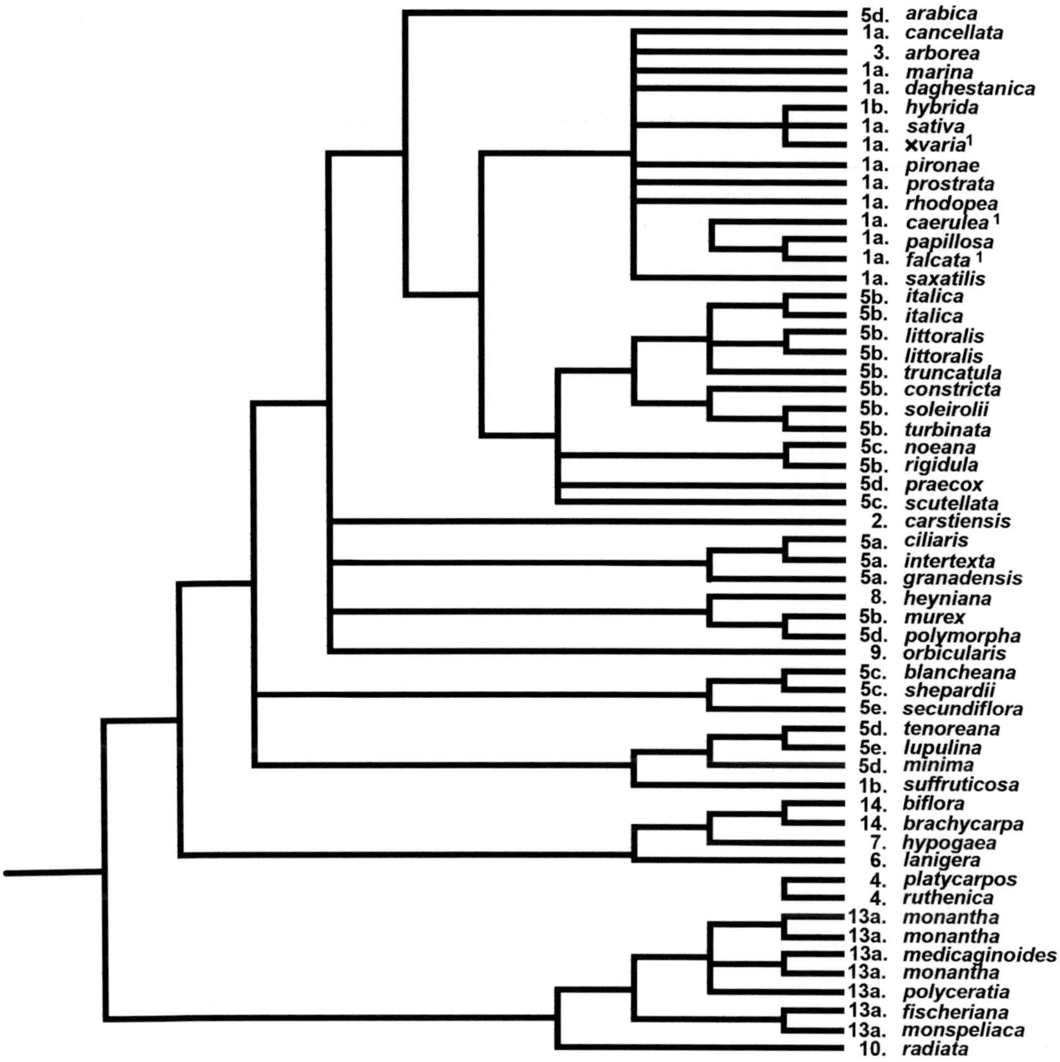

Fig. 10-7. Hypothesized relationships of species of *Medicago*, based on the plastid gene *trn*K/*mat*K. Adapted from Steele et al. (2010). [1]Subspecies of *M. sativa*.

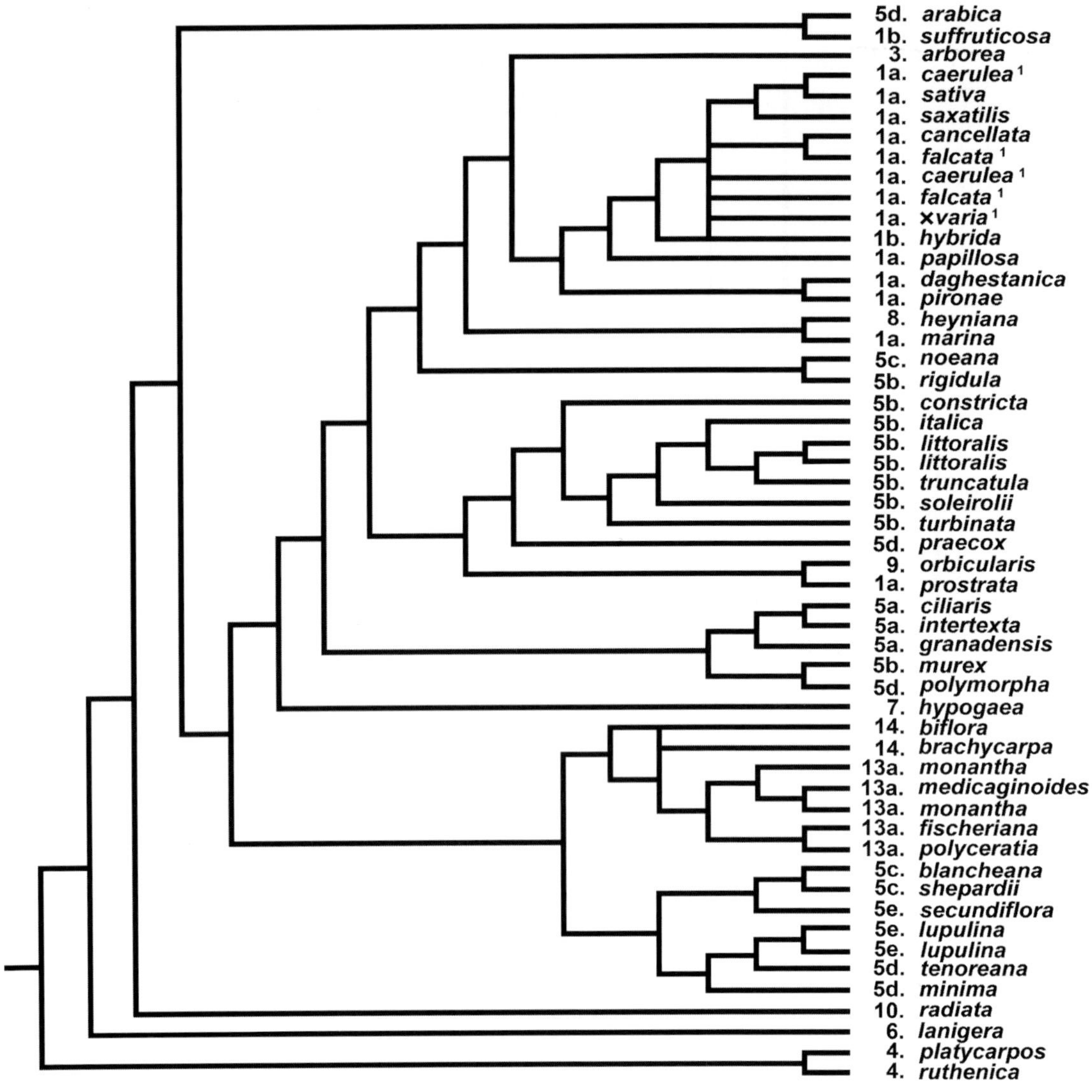

Fig. 10-8. Hypothesized relationships of species of *Medicago*, based on the nuclear gene *GA3ox1*. Adapted from Steele et al. (2010). [1]Subspecies of *M. sativa*.

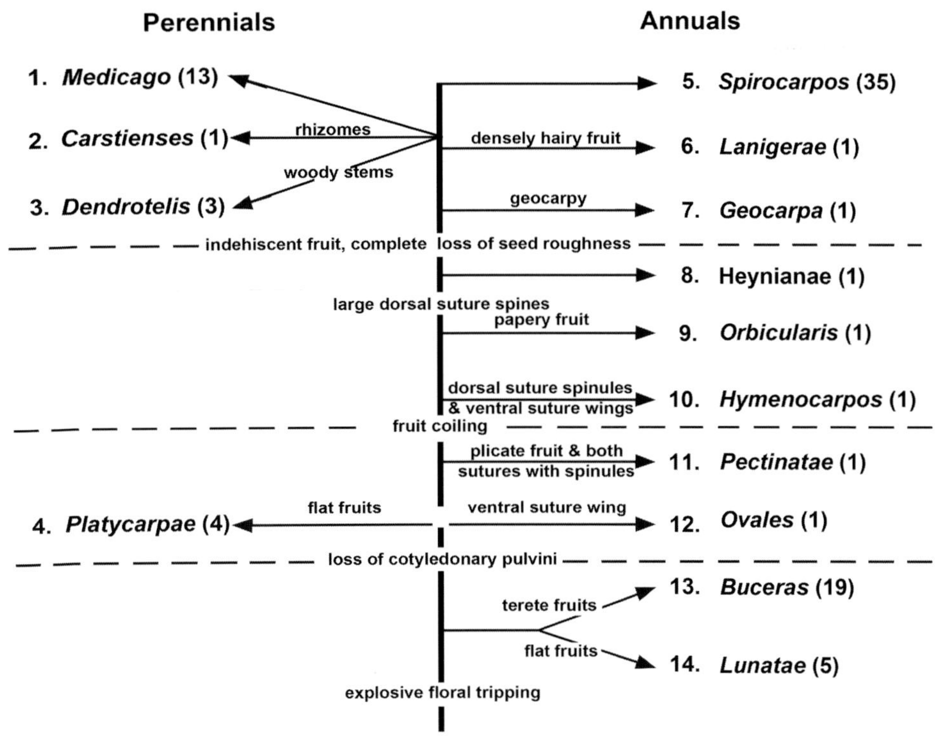

Fig. 10-9. Hypothesized evolutionary relationships based on a consideration of morphological and molecular data of the 14 sections of *Medicago* recognized in this monograph and described below. Key morphological changes (apomorphies) are indicated, although not all derived species have retained the changes. Sections higher up on the figure appear to be "advanced" (i.e., they show specialized traits not present in sections lower on the figure). Below the lowermost broken line the species tend to show apparently ancestral characteristics (i.e., features such as cotyledonary pulvini (always present in *Trigonella*, the sister group to *Medicago* in all analyses), lack of suture ornamentation (wings and fimbriation rarely occur), and seed roughness, which are absent or rare in most *Medicago* species. The numbers preceding the section name are the same as those used in all other figures in this chapter and in the section descriptions. Figures in parentheses show the number of species in the section. This simplified chain of groups is intended to summarize the most evident trends, not to indicate cladistic relationships.

KEY TO THE SECTIONS OF *MEDICAGO* PROVISIONALLY ACCEPTED IN THIS MONOGRAPH

1. Cotyledons epulvinate 2

  2. Fruit a 1-seeded nutlet .... 6. Section *Spirocarpos* (in part)

  2. Fruit 2-several-seeded (some 1-seeded pods may be present on the plant) 3

    3. Fruit flat, and with one or both sutures with spinules and/or a short wing, and seeds rough-surfaced 4

      4. Ventral fruit suture only ornamented (with a narrow wing or fringe) .... 12. Section *Ovales*

      4. Both fruit sutures ornamented with spinules or wings 5

        5. Fruits not plicate (pleated) .... 10. Section *Hymenocarpos*

        5. Fruits plicate (pleated) .... 11. Section *Pectinatae*

    3. Fruit lacking the combination of all three characteristics of the upper alternative 6

      6. Fruits coiled and seeds rough-surfaced (ridged, rugose, or papillate) 7

        7. Dorsal fruit suture spineless .... 9. Section *Orbiculares*

        7. Dorsal fruit suture spiny .... 8. Section *Heynianae*

      6. Fruit lacking the combination of both characteristics of the upper alternative 8

        8. Fruit flat, not coiled 9

9. Pod venation very distinct: veins parallel, straight, oblique near ventral suture, forming numerous roundly polygonal figures near dorsal suture (*M. hybrida* only) .... 1. Section *Medicago* (in part)
9. Pod venation not as described in upper alternative .... 4. Section *Platycarpae*
8. Fruit not flat, usually coiled (uncoiled and very rarely flat in *M. sativa* subsp. *falcata;* obscurely coiled in section *Geocarpa*) 10
10. Plants perennial 11
11. Branches ligneous .... 3. Section *Dendrotelis*
11. Branches herbaceous to lignescent 12
12. Usually 1–2 seeds per coil when fruit is coiled; rhizomes uncommon (principally in *M. sativa* subsp. *falcata*); radicle parallel to ventral fruit suture .... 1. Section *Medicago* (in part)
12. 4–5 seeds per fruit coil; the only section with species that always have rhizomes; radicle at 45° to ventral fruit suture .... 2. Section *Carstienses*
10. Plants annual 13
13. Fruit geocarpic .... 7. Section *Geocarpa*
13. Fruit aerial 14
14. Fruit densely covered with long, white, multicellular hairs .... 6. Section *Lanigerae*
14. Fruit not densely covered with long, white, multicellular hairs .... 5. Section *Spirocarpos* (in part)
1. Cotyledons pulvinate 15
15. Fruit flat, lunate, or hemicircular .... 14. Section *Lunatae*
15. Fruit terete to slightly flattened, linear (often curved) .... 13. Section *Buceras*

PROVISIONAL SECTIONS AND SUBSECTIONS OF *MEDICAGO*: NOMENCLATURE, DIAGNOSES, AND MERIT IN THE LIGHT OF NUCLEIC ACID SEQUENCE ANALYSES OF RELATIONSHIPS

1. *Medicago* sect. *Medicago*

Type: *M. sativa* L.; for information on conservation of the type and designation of the lectotype, see Chapter 12.

Synonyms:
Medicago sect. *Falcago* (Reichenb.) Gren. & Godr., Fl. Fr. 1: 38 (1949).

Diagnosis: Perennial herbs; pods straight, curved or coiled, spiny or spineless, when coiled usually 1–2(5) seeds per coil; seeds smooth; radicle half to two-thirds the length of the seed (usually slightly longer than half the length of seed); cotyledons epulvinate.

Phylogenetic relationships: Of the species in this section, the phylogenetic analyses of *Medicago* by Maureira-Butler et al. (2008) included *M. cretacea*, *M. marina*, *M. pironae*, *M. prostrata*, *M. rhodopea*, *M. sativa*, and *M. suffruticosa*. With the exception of *M. suffruticosa,* these species formed a monophyletic group, although the relationships within the group varied with the molecular marker used. The same result was evident in the analysis by Downie et al. (1998), which examined *M. daghestanica*, *M. marina*, *M. papillosa*, *M. pironae*, *M. prostrata*, *M. rhodopea*, and *M. sativa*. Of the species in this section, the phylogenetic analysis of *Medicago* by Bena (2001) examined *M. cancellata*, *M. cretacea*, *M. hybrida*, *M. marina*, *M. papillosa*, *M. pironae*, *M. prostrata*, *M. sativa*, *M. saxatilis,* and *M. suffruticosa*. *Medicago hybrida* and *M. suffruticosa* formed a group that was quite distant from the remaining species, which also formed a quite cohesive group. Steele et al. (2010) examined *M. cancellata*, *M. daghestanica*, *M. hybrida*, *M. marina*, *M. papillosa*, *M. pironae*, *M. prostrata*, *M. rhodopea*, *M. sativa*, and *M. saxatilis* (for *trn*K/*mat*K, and all but *M. rhodopea* for *GA3ox1*). The results of the analyses of a combined data set (*trn*K/*mat*K and *GA30x1* sequences) indicate a group, with all sampled species from sect. *Medicago* except *M. suffruticosa,* which is never found in a well-supported group with either molecular marker. Downie et al. (1998) found that *M. hybrida* and *M. suffruticosa* (subsp. *leiocarpa*) formed a distinct, well-supported group. Lesins and Lesins (1979) split *M. cretacea* as a monotypic section, *M. marina* as another monotypic section, and the pair of species *M. hybrida* and *M. suffruticosa* as another section. These three groups of species do have notable morphological distinctiveness, but on balance after reviewing the above studies, there is reasonable evidence for separateness only for the pair of species *M. hybrida* and *M. suffruticosa*, which are accordingly recognized here as a subsection of section *Medicago*.

Key to the subsections of section *Medicago*

Leaflets relatively narrow: length:width 3:2 or more .... 1a. Subsection *Medicago*

Leaflets relatively wide: length:width 2:1 or less .... 1b. Subsection *Suffruticosae*

1a. Subsection *Medicago*

DIAGNOSIS: Fruits coiled or (if uncoiled) terete, subcompressed, or flat, with or without dorsal suture spines; flowers yellow or violet; leaflets narrow.

SPECIES: *M. cancellata*, *M. cretacea*, *M. daghestanica*, *M. marina*, *M. papillosa*, *M. pironae*, *M. prostrata*, *M. rhodopea*. *M. rupestris*, *M. sativa*, *M. saxatilis*.

1b. Subsection *Suffruticosae* (Vassilcz.) E. Small, comb. nov.

TYPE: *M. suffruticosa* Ramond ex DC.

Synonyms:

Trigonella sect. *Suffruticosae* Vassilcz., Acta Inst. Bot. Komarovii, ser. 1/8: 74 (1949).

Medicago sect. *Suffruticosae* (Vassilcz.) Lesins & Lesins, Genus *Medicago* 134 (1979).

DIAGNOSIS: Fruits coiled or (if uncoiled) flat, without dorsal suture spines; flowers yellow; leaflets wide.

SPECIES: *M. hybrida*, *M. suffruticosa*.

2. *Medicago* sect. *Carstienses* Kozukharov, Bulg. Acad. Sci. Reports (Izvestiya) Bot. Inst. 15: 132 (1965; published as the linguistically incorrect "*Carstiensae*")

TYPE: *M. carstiensis* Wulfen

DIAGNOSIS: Perennial rhizomatous herbs; pods coiled, spiny, 4–5 seeds per coil; seeds smooth; radicle ca. two-thirds the length of the seed; cotyledons epulvinate.

SPECIES: *M. carstiensis*.

PHYLOGENETIC RELATIONSHIPS: None of the studies reviewed provides well-supported information on the relationships of this species.

3. *Medicago* sect. *Dendrotelis* (Vassilcz.) Lassen in Greuter et al., Willdenowia 12: 202 (1982)

TYPE: *M. arborea* L.

Synonyms:

Trigonella sect. *Dendrotelis* Vassilcz., Not. Syst. (Leningrad) (= Bot. Mater. Gerb. Bot. Inst. Komarova Akad. Nauk SSSR) 14: 229 (1951) (basionym; misprinted in Vassilczenko (1951) as *Trigonella arbosea*).

Rhodusia Vassilcz., Nov. Syst. Pl. Vasc. 9: 204 (1972).

Medicago sect. *Arboreae* K. Lesins & I. Lesins, Gen. Medic. 131 (1979) (nom. inval.).

DIAGNOSIS: Shrubs up to 4 m tall, developing annual stem rings of wood and bark; fruits with 0.5–3.0 coils, very rarely spiny; seeds smooth, radicle about half as long as seed; cotyledons epulvinate.

SPECIES: *M. arborea*, *M. citrina*, *M. strasseri*.

PHYLOGENETIC RELATIONSHIPS: Greuter et al. (1982) contended that the species of this section are unique by virtue of having keel petals as long as the standard petals (in most *Medicago* the standard petal is longer than the wing and keel petals) and in having a prominently stipitate fruit, but both characters are sometimes found in species of other sections. The perennial stems, characterized by annual rings of wood and bark produced by cambia, are unique. *Medicago marina* is the only other perennial *Medicago* that characteristically produces perennial aboveground branches, but these are not nearly as woody. The three species of this section are clearly closely related, and there is limited evidence that any other species is closely related. The study of Downie et al. (1998) did not indicate that *M. arborea* and *M. citrina* are closely related (although it is clear that the latter is a hexaploid relative of the former, which is tetraploid). The phylogenetic analysis of *Medicago* by Maureira-Butler et al. (2008) indicated that *M. arborea* is within a group that includes species of section *Medicago*. Steele et al. (2010) also found that *M. arborea* was included in a group with the species of section *Medicago*. This is somewhat consistent with a traditional view that the arborescent *M. arborea* is the sister group to at least some of the species of section *Medicago*. However, there is no support for the view that this section is sister to any other sections of the genus. As noted above, many authorities consider woodiness in plants to be an ancestral characteristic, although others hold that this character may be derived in particular groups. It has been contended that the woodiness of *M. arborea* reflects primitive (ancestral) status in the genus *Medicago* (Lesins and Lesins 1979), but there is no evidence of this.

4. *Medicago* sect. *Platycarpae* E. Small (Small 1987*c*, p. 180)

TYPE: *Trigonella platycarpos L.* ("*T. platycarpa*"; *Medicago platycarpos*)

Synonyms:

Medicago sect. *Platycarpae* Lesins & Lesins, Genus *Medicago* 74 (1979), nom. inval.

Trigonella sect. *Ellipticae* series *Platycarpae* Vassilcz., Acta Inst. Bot. Acad. Sci. URSS, Ser. 1(10): 168 (1959), nom. inval.

Turukhania Vassilcz., Novosti Sist. Vyssh. Rast. 16: 132 (1979; a monotypic genus based on *T. platycarpa*).

DIAGNOSIS: Perennial herbs; pods flat or compressed, not coiled, sutures lacking ornamentation (spines, spinules, tuberculae, or wings); seeds smooth or rugose, radicle almost equal to seed length; cotyledons epulvinate.

SPECIES: *M. archiducis-nicolai*, *M. platycarpos*, *M. popovii*, *M. ruthenica*.

PHYLOGENETIC RELATIONSHIPS: Maureira-Butler et al. (2008) included *M. platycarpos*, *M. popovii*, and *M. ruthenica* in their study. The latter three species formed a strongly supported clade. Bena (2001) noted that *M. platycarpos* was in a group with *M. popovii*. Steele and Wojciechowski (2003) and Steele et al. (2010) recorded a group of *M. platycarpos* and *M. ruthenica*.

5. *Medicago* sect. *Spirocarpos* Ser. in DC., Prodr. 2: 174 (1825)

TYPE: Lectotype, *M. polymorpha*, designated in Small (1987*c*)

DIAGNOSIS: Annual herbs; pods coiled, mostly with dorsal suture spines or tuberculae, mostly 1–2 seeds per coil; seeds smooth; radicle mostly slightly shorter than half the seed length (to two-thirds the length in some species); cotyledons epulvinate.

PHYLOGENETIC RELATIONSHIPS: Heyn (1963) divided section *Spirocarpos* into four subsections that can be separated by the following key. As noted below, Heyn's recognition of subsection *Intertextae* is extremely well supported by phylogenetic analyses of molecular data, her subsection *Pachyspirae* is relatively well supported, and her "subsection *Leptospirae*" (correctly, subsection *Spirocarpos*) and subsection *Rotatae* only weakly supported. *Medicago lupulina* and *M. secundiflora* are added here as subsection *Lupularia* because, as discussed below, both species seem closest to the species traditionally placed in section *Spirocarpos*, although they are not close to each other. The subsections presented below are in the main not supported by the molecular data reviewed, but because these data present many polytomies, they are insufficient as a basis for subcategorization of section *Spirocarpos*. The traditional morphological groupings are maintained pending more evidence because they are useful for identification.

Key to the subsections of *Medicago* section *Spirocarpos*

The following are as defined by Heyn (1963), except for the transfer of *M. soleirolii* from subsection *Rotatae* to subsection *Pachyspirae* (following Urban (1873) and Lesins and Lesins (1979)), the separation of *M. lanigera* as a section, the addition of new species to subsection *Pachyspirae*, and the addition of subsection *Lupularia*.

1. Fruit a 1-seeded nutlet .... 5e. Subsection *Lupularia*
1. Fruit with 2 or more seeds 2
 2. Seeds black or reddish-black; fruit spines with vein of coil face entering marginal spine distinctly on one side .... 5a. Subsection *Intertextae*
 2. Seeds yellow, brown, or yellow-brown; fruit spines (when present) with vein of coil face entering marginal spine directly 2
  3. Pods with soft walls at maturity (except in some *M. polymorpha*); spines (when present) with 2-rooted base .... 5d. Subsection *Spirocarpos*
  3. Pods with hard or tough walls at maturity; spines (when present) with conical base 3
   4. Fruits becoming extremely hard at maturity, often developing spongy tissue on coil faces and on spine bases; spines present in some populations of all species (except *M. soleirolii*) stocky and inflexible .... 5b. Subsection *Pachyspirae*
   4. Fruits hard at maturity or somewhat flexible; not usually developing spongy tissue on coil faces; spines absent (except short, thin spines in *M. rotata* and its hybrids) .... 5c. Subsection *Rotatae*

5a. Subsection *Intertextae* (Urb.) Heyn, Annual Species *Medicago* 26 (1963)

TYPE: *M. intertexta* (L.) Mill.

Synonym:

Medicago Sect. *Intertextae* Urb., Verh. Bot. Ver. Brandenb. 15: 36 (1873).

DIAGNOSIS: Mature fruit walls relatively soft at maturity and lacking spongy tissue; dorsal suture spines relatively thin and flexible and (more or less uniquely for this section) one root emerges diagonally from the coil surface; young pods emerge from the side of the calyx in a loose spiral and contracts subsequently; seeds dark red-brown to brownish black.

Species: *M. ciliaris*, *M. granadensis*, *M. intertexta*, *M. muricoleptis*.

Phylogenetic relationships: Downie et al. (1998), Bena (2001), Maureira-Butler et al. (2008), and Steele et al. (2010) observed these species formed a strongly supported group that never includes any other species. The species of this section are clearly morphologically distinctive from all other species of *Medicago*, and indeed some are capable of interbreeding. The very close relationships among the species of subsection *Intertextae* have been extensively documented (Lesins et al. 1971; Schlarbaum et al. 1984; Rose et al. 1988; Small et al. 1999).

5b. Subsection *Pachyspirae* (Urb.) Heyn, Annual Species *Medicago* 25 (1963)

Type: Not designated; given the uncertain composition and limits of this section, selection of a type seems inadvisable at this time.

Synonym:
Medicago sect. *Pachyspirae* Urb., Verh. Bot. Ver. Brandenb. 15: 36 (1873).

Diagnosis: Fruit walls becoming extremely hard at maturity (a very uncommon character outside this subsection) and a spongy tissue often developing on coil faces and on spine bases, obscuring venation (a very rare character outside this subsection, except in *M. polymorpha*); spines, if present on dorsal suture (may occur on fruits of all species except *M. soleirolii*), stocky, inflexible (mostly relatively thin and flexible in other sections); young pods in a contracted spiral concealed within the calyx; seeds usually yellow to brownish.

Species: *M. constricta*, *M. doliata*, *M. italica*, *M. littoralis*, *M. murex*, *M. rigidula*, *M. rigiduloides*, *M. sinskiae*, *M. soleirolii*, *M. sphaerocarpos*, *M. syriaca*, *M. truncatula*, *M. turbinata*.

Phylogenetic relationships: Bena (2001) examined all the above species except *M. sinskiae*. He observed that the species formed a group, albeit species from some other groups were also included. However, the closely related *M. murex* and *M. sphaerocarpos* form the sister group to a larger group including the other species, and the former two species are in a strongly supported group with *M. polymorpha* (a species that although in section *Spirocarpos* subsect. *Spirocarpos*, shares some characteristics of sect. *Spirocarpos* subsect. *Pachyspirae*). Maureira-Butler et al. (2008) examined all the above species (their "*M. murex*" is the very closely related *M. sphaerocarpos*; their "*M. lesinsii* is *M. murex*) except *M. rigiduloides* and *M. sinskiae*. Once again, most of the species are in the same group, but *M. sphaerocarpos* and *M. murex* formed a separate group with *M. polymorpha* (and the very closely related *M. laxispira*). Downie et al. (1998) also found two groups, one with *M. constricta*, *M. doliata*, *M. italica*, *M. rigidula*, *M. rigiduloides*, *M. sinskiae*, *M. truncatula*, and *M. turbinata*, the other with *M. murex* and *M. syriaca*. Steele et al. (2010) examined *M. constricta*, *M. italica*, *M. littoralis*, *M. murex*, *M. rigidula*, *M. soleirolii*, *M. truncatula*, and *M. turbinata*, and found for both genes examined that most of the species formed a group, but *M. rigidula* is the sister species to *M. noeana*, (currently assigned to subsection *Rotatae*) and *M. murex* is sister to *M. polymorpha* (currently assigned to subsection *Spirocarpos*). The molecular markers thus indicate that sect. *Spirocarpos* subsect. *Pachyspirae* appears to include at least two groups. The smaller group (including *M. murex*, *M. sphaerocarpos*, and *M. syriaca*) may represent a bridge to sect. *Spirocarpos* subsect. *Spirocarpos* (through the closely related pair of species *M. laxispira* and *M. polymorpha*).

5c. Subsection *Rotatae* (Urb.) Heyn, Annual Species *Medicago* 24–25 (1963)

Type: *M. rotata* Boiss.

Synonym:
Medicago sect. *Rotatae* Urb., Verh. Bot. Ver. Brandenb. 15: 36 (1873).

Diagnosis: Mature fruit walls relatively soft at maturity and lacking spongy tissue; fruit coil edge spineless or with irregular short spines (all other species in section *Spirocarpos*, except *M. italica* and *M. soleirolii*, are usually spiny-fruited); shoot often covered with gland-tipped trichomes (a much less common feature of species in other sections); young pods in a contracted spiral concealed within the calyx; seeds yellowish to brownish.

Species: *M. bonarotiana*, *M. noeana*, *M. rotata*, *M. rugosa*, *M. scutellata*, *M. shepardii*.

Phylogenetic relationships: Downie et al. (1998) observed a group comprising *M. bonarotiana* (as *M. blancheana*), *M. rotata*, and *M. shepardii* (the first two species hybridize and have long been known to be related, but the close relationship with *M. shepardii* was not suspected) that is quite separated from *M. rugosa*. Similarly, Bena (2001) observed three groups of the above six species: (1) *M. bonarotiana*, *M. rotata*, and *M. shepardii*; (2) *M. rugosa* and *M. scutellata* (the close relationship between these is well known); and (3) *M. noeana*. Maureira-Butler et al. (2008) did not examine *M. rugosa* and *M. scutellata*, but found exactly the results as Bena observed for (1) and (2) above. Steele et al. (2010) examined *M. bonarotiana* (as *M. blancheana*), *M. noeana*, *M. scutellata*, and *M. shepardii; M. bonarotiana* and *M. shepardii* grouped together, the other two were in a group with species of subsection *Pachyspirae*. These studies did not show a clear relationship of subsection *Rotatae* with other sections or subsections. It appears that subsection *Rotatae* is an artificial assemblage, but is maintained here in the absence of a clearer resolution of species relationships in section *Spirocarpos*.

5d. Subsection *Spirocarpos*

Synonyms:

Medicago sect. *Leptospirae* Urb., Verh. Bot. Ver. Brandenb. 15: 37 (1873).

Medicago subsect. *Leptospirae* (Urb.) Heyn, Annual Species *Medicago* 25 (1963).

Diagnosis: Mature fruit walls relatively soft at maturity and lacking spongy tissue (except sometimes *M. polymorpha*); dorsal suture usually with flexible, thin spines; young pods protruding sideways through the calyx; seeds yellowish or brownish.

Species: *M. arabica*, *M. coronata*, *M. disciformis*, *M. laciniata*, *M. laxispira*, *M. minima*, *M. polymorpha*, *M. praecox*, *M. sauvagei*, *M. tenoreana*.

Phylogenetic relationships: Downie et al. (1998) examined all ten species above except *M. coronata*. Bena (2001) examined all the above species except *M. laxispira*. Maureira-Butler et al. (2008) examined all the above species. Steele et al. (2010) examined four of the species. Some of the species grouped together but, on the whole, the subsection was not well supported. This is the least satisfactory of the four subsections recognized by Heyn (1963) and appears to be a "dust bin" taxon (i.e., a depository of disparate constituents that do not fit well into comparatively well defined, related taxa at the same rank).

5e. Subsection *Lupularia* (Ser. in DC.) E. Small, comb. nov.

Type: *M. lupulina* L.

Synonym:

Medicago sect. *Lupularia* Ser. in DC., Prodr. 2: 172 (1825).

Diagnosis: Annual herbs (*M. lupulina* is sometimes biennial or perennial); fruit a 1-seeded, non-dehiscent nutlet, the tip twisted in a small coil; seeds smooth; radicle half to two-thirds the length of seed; cotyledons epulvinate.

Species: *M. lupulina*, *M. secundiflora*.

Phylogenetic relationships: Simon (1979), on the basis of serological studies, concluded that *M. lupulina* and *M. secundiflora* were related, but that this pair of species was not closely related to other species of *Medicago*. Downie et al. (1998) observed that *M. lupulina* was not part of a well-supported group with other species of *Medicago* (*M. secundiflora* was not sequenced; it does have the *rpoC1* intron that *M. lupulina* lacks). Bena (2001) found *M. lupulina* in a strongly supported group with *M. tenoreana*, while *M. secundiflora* was not in such a group with any other species. Maureira-Butler et al. (2008) and Steele et al. (2010) also found *M. lupulina* occurred in a well-supported group with *M. tenoreana*, and there was no clear relationship for *M. secundiflora*. The two species are maintained together in sect. *Lupularia* on the basis of morphological similarity but, on balance, the evidence for recognition of this section is very weak. Based on results from phylogenetic analyses of molecular data, it appears that *M. lupulina* and *M. secundiflora* are allied to the species traditionally placed in section *Spirocarpos*, but the morphological differences between the former two species and the others are substantial. However, *M. lupulina* and *M. secundiflora*, by virtue of having single-seeded fruits, lack the numerous fruit characters that have traditionally defined section *Spirocarpos*, and it appears that convergent morphology may have misled taxonomists into the belief that the two species are closely related. In the phylogenetic studies reviewed here, both *M. lupulina* and *M. secundiflora* are each usually in a group with species traditionally placed in section *Spirocarpos*, but not with each other. Reduction in seed number to one, which characterizes *M. lupulina* and *M. secundiflora*, appears to have occurred independently (i.e., the character is not apomorphic). Therefore, the group *Lupularia* is artificial and is maintained here simply because a clear way of subdividing section *Spirocarpos* is not yet evident.

6. *Medicago* sect. *Lanigerae* E. Winkl. & Fedtsch., Bull. Jard. Bot. Petersbg. (Izv. Imp. S.-Petersburgsk. Bot. Sada) 5: 41 (1905)

Type: *M. lanigera* E. Winkl. & Fedtsch.

Diagnosis: Annual herbs; pods coiled, spineless, mostly one or two seeds per coil, densely covered with long white hairs (comparably dense pubescence only in *M. hypogaea* and *M. marina*); seeds smooth, radicle about two-thirds the seed length; cotyledons epulvinate.

Species: *M. lanigera*.

Phylogenetic relationships: *Medicago lanigera* was assigned by Heyn (1963) to section *Spirocarpos* subsection *Spirocarpos*, and this was accepted by Lesins and Lesins (1979). However, its relationships remain obscure. Downie et al. (1998) and Bena (2001) found it to be disassociated from all other species examined. Maureira-Butler et al. (2008) and Steele et al. (2010) observed it to have no close or clear relationship with other species.

7. *Medicago* sect. *Geocarpa* E. Small in Small and Brookes (1984*c*)

Type: *Trigonella aschersoniana* Urb, (*M. hypogaea* E. Small)

Synonym:

Factorovskya Eig, Zionist Org. Inst. Agric. & Nat. Hist. Bull. 6: 11 (1927). (A monotypic genus.)

Diagnosis: Annual, prostrate herbs; after flowering, peduncle, pedicel, and gynophore elongate to bury the ovary in the ground; fruit 2-seeded, deeply constricted between the seeds or one aborted and the fruit 1-seeded; seeds smooth; radicle about two-thirds the length of the seed; cotyledons epulvinate.

Species: *M. hypogaea*.

Phylogenetic relationships: Downie et al. (1998) found *M. hypogaea* was isolated from other species of *Medicago*. Steele and Wojciechowski (2003) found a group comprising *M. hypogaea* and species of sections *Platycarpae*, *Lunatae*, and *Lanigerae*, a quite varied assemblage indicative of a lack of a clearly defined relationship. In analyses of *trnK/matK* data, Steele et al. (2010) found *M. hypogaea* in a weakly supported group with sect. *Platycarpae* and sect *Lunatae*, but in analyses of *GA3ox1* data it was instead found to be a sister to a large group of species. Small and Brookes (1984*c*) speculated that *M. hypogaea* arose from *M. lanigera* (both species lack fruit spines and their fruits are densely pubescent).

8. *Medicago* sect. *Heynianae* Greuter, Candollea 25(2): 189 (1970)
TYPE: *M. heyniana* Greuter

DIAGNOSIS: Annual herbs; fruit coiled, spiny; seed rugose, transversely ridged; radicle about as long as seed; cotyledons epulvinate.

SPECIES: *M. heyniana*.

PHYLOGENETIC RELATIONSHIPS: Downie et al. (1998) found that *M. heyniana* was not closely related to other species. Bena (2001) found *M. heyniana* to be the sister species to *M. praecox*. Morphologically, *M. heyniana* is quite distinctive from *M. praecox*. Maureira-Butler et al. (2008), on the basis of one nuclear gene (β-*cop*), found exactly the same thing, but with a second nuclear gene (CNGC5), *M. heyniana* was a sister to a group formed by *M. murex*, *M. sphaerocarpos*, *M. polymorpha*, and *M. laxispira* (albeit the sister group relationship was weakly supported). Steele et al. (2010), on the basis of the plastid gene *trn*K/*mat*K, found *M. heyniana* to be a sister to *M. murex* and *M. polymorpha*; however, on the basis of the nuclear-encoded gene *GA3ox1*, it was not in any strongly supported group. Considering these results as a whole, there appears to be some proximity of *M. heyniana* to the related complex of *M. polymorpha*, *M. laxispira*, *M. murex*, and *M. sphaerocarpos*. However, as the relationships are still rather ambiguous, the traditional recognition of *M. heyniana* as a separate monotypic section is maintained here. *Medicago heyniana* is quite disparate from the species traditionally placed in section *Spirocarpos* in that is has rough-coated seeds (always smooth in the species currently placed in section *Spirocarpos*) and dehiscent fruit (*Spirocarpos* fruits are indehiscent), two characters that are common in *Trigonella*, the sister group to *Medicago*.

9. *Medicago* sect. *Orbiculares* Urb., Verh. Bot. Ver. Brand. 48 (1873)
TYPE: *M. orbicularis* (L.) Bartal.

DIAGNOSIS: Annual herbs; pods coiled, spineless, the coil edges paper-thin; seeds papillate; radicle as long as seed; cotyledons epulvinate.

SPECIES: *M. orbicularis*.

PHYLOGENETIC RELATIONSHIPS: The nucleotide-sequence data of Downie et al. (1998), Bena (2001), Maureira-Butler et al. (2008), and Steele et al. (2010) are quite inconsistent in indicating relationships for *M. orbicularis* (other than being a part of the group of all *Medicago* species). Given this ambiguity and the morphological distinctiveness of the species, the traditional recognition of *M. orbicularis* as a separate monotypic section is maintained here.

10. *Medicago* sect. *Hymenocarpos* Ser. in DC., Prodr. 2: 171 (1825)
TYPE: *M. radiata* L.

DIAGNOSIS: Annual herbs; fruit flat, curved through 0.5 to 1.5 coils, pectinate ciliate-spines (some forked) on the dorsal suture, shortly fimbriate-winged on the ventral suture; seeds rugose; radicle about as long as seed; cotyledons epulvinate.

SPECIES: *M. radiata*.

PHYLOGENETIC RELATIONSHIPS: The sequence data of Downie et al (1998) showed a very weak grouping of *M. radiata* with species of sect. *Lunatae*. The data of Bena (2001) showed *M. radiata* as a strongly supported sister group of some members of sections *Lunatae* and *Buceras*. The data for one of the two nuclear genes examined by Maureira-Butler et al. (2008) placed *M. radiata* within a group of species in those same sections (albeit with different species), but the data for the other nuclear gene examined indicated its placement in a weakly supported group with species of other *Medicago* sections. Steele et al. (2010), on the basis of *trn*K/*mat*K, recorded *M. radiata* as the sister group to species of sections *Buceras* and *Lunatae*, but on the basis of *GA3ox1*, *M. radiata* is the sister group to nearly all species of *Medicago*. As a whole, these phylogenetic analyses suggest a relationship with sections *Buceras* and *Lunatae*, some species of which share with *M. radiata* the feature of seed roughness. Nevertheless, sections *Buceras* and *Lunatae* are different from all other *Medicago* species in possessing pulvinate cotyledons. The traditional recognition of *M. radiata* as a separate monotypic section is maintained here.

11. *Medicago* sect. *Pectinatae* Boiss., Fl. Or. 2: 67 (1872)
TYPE: *Trigonella plicata* (Boiss. & Bal.) Boiss. (*Pocockia plicata* Boiss. & Bal., *Medicago plicata* (Boiss. & Bal.) Širj.) (lectotype here designated; Boissier included three quite disparate species in his section: *M. plicata*, *M. radiata*, and *Trigonella arabica* Delille).

DIAGNOSIS: Fruits flat, plicate (i.e., pleated, the only *Medicago* section with this character), both sutures denticulate with small, branched, pectinate spinules; seeds tuberculate; radicle about as long as seed; cotyledons epulvinate.

SPECIES: *M. plicata*.

PHYLOGENETIC RELATIONSHIPS: P. Lassen (personal communication) suggested *M. plicata* and *M. radiata* might constitute a group because of their similar sutural ornamentation (both species have short spines and/or short wings on both their sutures). Both species also have rough-surfaced seeds. The study of Maureira-Butler et al. (2008) did not indicate consistent support for any specific relationship of *M. plicata* with other species of *Medicago*. In particular, results of analyses of the two nuclear genes studied by Maureira-Butler et al. indicated that neither gene suggested a relationship of *M. plicata* and *M. radiata*. Given the lack of any clear relation with another species, *M. plicata* is placed here in its own section.

12. *Medicago* sect. *Ovales* E. Small, sect. nov.
TYPE: *Trigonella ovalis* Boiss. (*M. ovalis* (Boiss.) Urb.)

DIAGNOSIS: Annual herbs; flowers 4–8 in axillary capitula; fruits uncoiled, without spines, short, broad, flat, with a narrow wing (fringe) on the more convex (ventral) suture; seeds tuberculate, radicle almost as long as seed; cotyledons epulvinate. (Herbae annuae; capitulis 4–8-floris axillaribus sessilibus; legumina plana, nec circinata nec spinosos, sutura ventralis alata anguste; semina tuberculata, radicula cotyledones subaequilonga; cotyledones sine pulvino.)

SPECIES: *M. ovalis*.

PHYLOGENETIC RELATIONSHIPS: No studies based on analyses of molecular data have included this species. The reported chromosome counts of $2n = 32$ for this species (which should be verified) are very unusual for an annual species of *Medicago*, as indeed is its geographical isolation to Spain and North Africa. The flowers and pods grouped in the axils of the leaves is another unusual feature. The fruit morphology is somewhat reminiscent of section *Platycarpae*, but the four species of the latter are perennials of eastern Asia.

13. *Medicago* sect. *Buceras* (Ser.) E. Small
(Small 1987*c*, p. 180)
TYPE: *Trigonella polyceratia* L. (*Medicago polyceratia* (L.) Trautv.; lectotype designated by Small (1987*c*)).

Synonym:
Trigonella sect. *Buceras* Ser. in DC., Prodr. 2: 182 (1825).

DIAGNOSIS: Annual herbs; fruits linear, mostly subterete, conspicuously strangulate between seeds in a few species, never coiled or spiny; seeds mostly rugose, sometimes smooth; radicle somewhat shorter than cotyledons; cotyledons pulvinate.

PHYLOGENETIC RELATIONSHIPS: The three subsections are based more or less on the groupings of Širjaev (1928–1934), but with new combinations proposed by Small (in Small and Jomphe 1989*b*) and a reduction of subsection *Reflexae* (with *M. monspeliaca*) to subsection *Buceras*. Small (1987*c*) found the groups are well demarcated morphologically, although *M. monspeliaca* appears to be isolated (see Fig. 10-2).

Key to the subsections of *Medicago* section *Buceras*

1. Fruits not conspicuously strangulate 2
  2. Flowers straight and upright .... 13a. Subsection *Buceras* (in part)
  2. Flowers reflexed 3
    3. Fruits sharply reflexed (*M. monspeliaca*) .... 13a. Subsection *Buceras* (in part)
    3. Fruits horizontally deflexed .... 13b. Subsection *Deflexae*
1. Fruits conspicuously strangulate .... 13c. Subsection *Isthmocarpae*

13a. Subsection *Buceras*
Synonyms:
Trigonella sect. *Bucerates* subsect. *Erectae* Širj., Publ. Fac. Sci. Univ. Masaryk 128: 13 (1930) (part 3 of Širjaev (1928–1934)).
Medicago subsect. *Erectae* (Širj.) E. Small in Small and Jomphe (1989*b*, p. 3274).

DIAGNOSIS: Flowers subtended strictly (straight and upright); fruits strict or reflexed, not conspicuously strangulate.

SPECIES: *M. arenicola*, *M. astroites*, *M. carica*, *M. crassipes*, *M. fischeriana*, *M. halophila*, *M. heldreichii*, *M. medicaginoides*, *M. monantha*, *M. monspeliaca*, *M. orthoceras*, *M. pamphylica*, *M. persica*, *M. phrygia*, *M. polyceratia*, *M. rigida*.

PHYLOGENETIC RELATIONSHIPS: Bena (2001) recorded a group of *M. fischeriana*, *M. medicaginoides*, *M. monantha*, *M. orthoceras*, *M. phrygia*, *M. polyceratia*, and *M. monspeliaca*. Downie et al. (1998) observed that *M. fischeriana*, *M. medicaginoides*, *M. monantha*, *M. orthoceras*, *M. polyceratia*, and *M. monspeliaca* formed a group. In addition, all species tested lacked the *rpoC1* intron (Downie et al. 1998). Steele et al. (2010) examined the same species as Downie et al. (except for *M. orthoceras*) and found the same result. Maureira-Butler et al. (2008) observed for the nuclear gene CNGC5 that *M. astroites* and *M. crassipes* are in a group with *M. fischeriana* and for the nuclear gene β-*cop* that *M. crassipes* is in a group with *M. fischeriana* and *M. monspeliaca*, with *M. astroites* as the sister group to these three species. Accordingly, molecular data strongly support the hypothesis that the species examined constitute a monophyletic group.

Prior to the studies reviewed above, *M. monspeliaca* was considered to constitute a distinctive subsection of *Medicago* section *Buceras* (*Medicago* subsect. *Reflexae* (Širj.) E. Small in Small and Jomphe (1989*b*, p. 3275); *Trigonella* sect. *Buceras* subsect. *Reflexae* Širj., Publ. Fac. Sci. Univ. Masaryk 136: 20 (1931; part 4 of Širjaev 1928–1934)). Although morphologically distinctive, the studies reviewed here strongly support the placement of *M. monspeliaca* in subsection *Buceras*.

13b. Subsection *Deflexae* (Širj.) E. Small in Small and Jomphe (1989*b*, p. 3275)

Synonym:

Trigonella sect. *Bucerates* subsect. *Sessiles* series *Deflexae* Širj., Publ. Fac. Sci. Univ. Masaryk 136: 20 (1931) (part 4 of Širjaev 1928–1934).

Diagnosis: Flowers reflexed; fruits horizontally deflexed.

Species: *M. retrorsa*.

Phylogenetic relationships: No molecular data are available for this taxon.

13c. Subsection *Isthmocarpae* (Boiss.) E. Small in Small and Jomphe (1989*b*, p. 3275)

Type: *Trigonella isthmocarpa* Boiss. & Bal. (*M. isthmocarpa* (Boiss. & Bal.) E. Small).

Synonym:

Trigonella sect. *Isthmocarpae* Boiss., Fl. Orient. 2: 66 (1872).

Diagnosis: Flowers reflexed or erect; fruits reflexed or pendent, conspicuously strangulate (the only subsection with this feature).

Species: *M. isthmocarpa*, *M. rhytidiocarpa*.

Phylogenetic relationships: No molecular data are available for these two species. Morphologically, they seem very close.

14. *Medicago* sect. *Lunatae* (Boiss.) E. Small (Small 1987*c*, p. 180)

Type: *T. lunata* Boiss. (*Medicago biflora* (Griseb.) E. Small); lectotype designated by Small (1987*c*)).

Synonyms:

"*Trigonella Lunatae*" Boiss., Fl. Orient. 2: 67 (1872, undefined rank).

Trigonella sect. *Lunatae* (Boiss.) Širj., Publ. Fac. Univ. Masaryk 102: 12 (1928) (part 1 of Širjaev 1928–1934).

Diagnosis: Annual herbs; fruit flat or compressed, ovate or lunate, never coiled or spiny; seeds smooth or (more typically) rugose; radicle about as long as seed; cotyledons pulvinate.

An unusual feature of this section is shown by two of the species, *M. brachycarpa* and *M. edgeworthii*, both of which appear to show flexibility in longevity; sometimes growing as annuals, sometimes as perennials.

Species: *M. biflora*, *M. brachycarpa*, *M. edgeworthii*, *M. huberi*, *M. rostrata*.

Phylogenetic relationships: Bena (2001) observed that *M. brachycarpa* and *M. huberi* formed a strongly supported group that is part of a polytomy with *M. biflora* and another strongly supported group comprising seven species of section *Buceras*. Maureira-Butler et al. (2008) observed for both the nuclear gene CNGC5 and for β-*cop* that *M. brachycarpa*, *M. edgeworthii*, and *M. huberi* formed a well-supported group, but the placement of *M. biflora* relative to that group varied. Downie et al. (1998) also observed that *M. biflora*, *M. brachycarpa,* and *M. edgeworthii* grouped together, but very weakly. Steele et al. (2010) observed that *M. biflora* and *M. brachycarpa* formed a well-supported group based on *trn*K/*mat*K. On balance, it appears that the species examined do constitute a distinct section.

SUMMARY OF THE CONTRIBUTIONS OF MOLECULAR STUDIES TO THE CLASSIFICATION OF *MEDICAGO*

1. A dramatic accomplishment of the molecular studies conducted to date has been the clear demonstration that *Medicago*, as now understood in the large sense, is a monophyletic group clearly separated from related genera such as its sister group *Trigonella*.
2. The most important species of *Medicago* is *M. sativa*, (alfalfa, lucerne) which, as outlined in Chapter 2, is one of the world's chief crop plants, and may well contribute to products with an annual value as high as a trillion dollars. This polymorphic species (with several subspecies and varieties) is a member of section *Medicago*, treated here as composed of 11 perennial species. Gene transfer by conventional hybridization is known to be possible among many of these species, which are accordingly critical sources of germplasm for breeding improved cultivars. Molecular analysis has indicated degrees of relationships among the species, which should be helpful in the future breeding of alfalfa.
3. As presented in Chapter 2, other species of *Medicago* are currently being employed as crops, principally *M. arborea*, *M. italica*, *M. littoralis*, *M. lupulina*, *M. murex*, *M. polymorpha*, *M. rugosa*, *M. scutellata*, *M. sphaerocarpos*, and *M. truncatula*. Molecular studies have succeeded in pointing out the species of *Medicago* that are closely related to these and have shown in a number of cases that species previously thought to be closely related are in fact relatively distant. These studies should be helpful in the future in developing the agricultural potential of the species in question, and indeed in developing their relatives.
4. Classical morphological analysis of *Medicago* led to a number of very distinctive species being classified at

some rank by various taxonomists as monotypic taxa. Of the eight such species accepted as constituting monotypic sections of the genus, the molecular studies have demonstrated the genetic distinction of seven of these: *M. carstiensis*, *M. heyniana*, *M. hypogaea*, *M. lanigera*, *M. orbicularis*, *M. plicata*, and *M. radiata*. *Medicago ovalis*, which is accepted here as the eighth monotypic section, has not yet been examined using sequence data.

5. Section *Dendrotelis* is treated here as composed of three species of shrubs, a unique character for the genus. This section has not been examined in detail using sequence data, except for *M. arborea*.
6. Section *Platycarpae* is treated here as comprising four species of perennials with flat fruits. Molecular studies have indicated that three of the species group together.
7. The largest group of *Medicago* accepted here (with serious reservations) with 35 annual species, is section *Spirocarpos*. This has traditionally been divided into four groups, which are presented here as subsections. However, molecular studies have verified the validity only of subsection *Intertextae*, which includes four species that have long been known to be closely related. Analyses of nucleotide sequence data have shown that there are several small, well-supported groups of species, often including species from more than one traditional subsection, which are thus obviously artificial. As detailed in this chapter, some of the relationships among species were not suspected. *Medicago arabica*, traditionally accepted as a member of section *Spirocarpos* despite some unique characteristics, does not seem to fit well in the group.
8. *Medicago lupulina*, which is mostly annual, and the annual *M. secundiflora*, have traditionally been considered to be a group of two in *Medicago*, sometimes ranked as high as a subgenus. While there have been indications that the two species are not as closely related as they appear to be on the basis of external morphology, sequence analysis has dramatically demonstrated that they are most closely related to species of section *Spirocarpos*, not to each other. They are placed in this monograph together as a subsection of section *Spirocarpos*, albeit they do not deserve to be associated together in their own distinctive group. However, as noted above, the subsectional treatment of section *Spirocarpos* is already artificial, requiring additional data before a more satisfactory treatment is possible.
9. The remaining annuals are in section *Buceras* with 19 species and section *Lunatae* with five species. These species are distinctive within the genus *Medicago* in having jointed cotyledons, and all of them were, until recently, accepted as species of *Trigonella*. The data published to date suggest that the two subsections are natural groupings. The traditional subsectional classification of section *Buceras* presented here has not been checked using sequence data because experimental material of many of these species has not been available.

KEY PROBLEMS REMAINING IN THE INFRAGENERIC CLASSIFICATION OF *MEDICAGO*

1. The present morphological delimitation of section *Spirocarpos*, except for subsection *Intertextae*, is inconsistent with relationships indicated by molecular markers examined to date. The existence of strongly supported groups of various sizes has been demonstrated by analyses of nucleotide sequence data, but information is insufficient at present to provide a taxonomic structure that includes all or even most of the species.
2. As noted in Chapter 3, there are no, or virtually no, germplasm collections for 20 or so species of *Medicago*, and most of these have not been examined using molecular studies. There is a need to analyse these species.
3. While many of the sections of *Medicago* recognized above are reasonably supported, there remains uncertainty. The acquisition of additional evidence is necessary.
4. While a very high degree of certainty now is evident concerning many of the smaller groupings of species in *Medicago*, there is a progressively higher degree of uncertainty regarding species relationships of the larger groups and the eight monotypic sections of *Medicago* (the so-called "orphan taxa") recognized in this monograph. That is, the earliest branchings in the genus are unclear. Elucidation of a hierarchical structure (whether or not it deserves taxonomic delimitation) that shows how groups of *Medicago* are related (or how they have evolved) requires analysis based on more of the species and more molecular markers.
5. An amazing degree of incongruence between the relationships indicated by various molecular markers has been found, for which speculative hypotheses have been advanced. An evidence-based explanation for the discordance, which makes taxonomic classification uncertain, is required.

11. Identification of Species and Infraspecific Taxa

FRUIT CHARACTERS CRITICAL FOR IDENTIFICATION

Generally, vegetative plants of *Medicago* cannot be confidently identified to species, unless only a few species grow in a particular region and one is very familiar with them. As noted elsewhere, if flowers are present one can easily determine that a plant belongs to the genus *Medicago* as opposed to a related genus. However, most plants of *Medicago* cannot be identified to species with certainty unless mature fruits are present. Accordingly, it is critical to understand fruit structure and terminology in order to employ the identification key presented in this chapter.

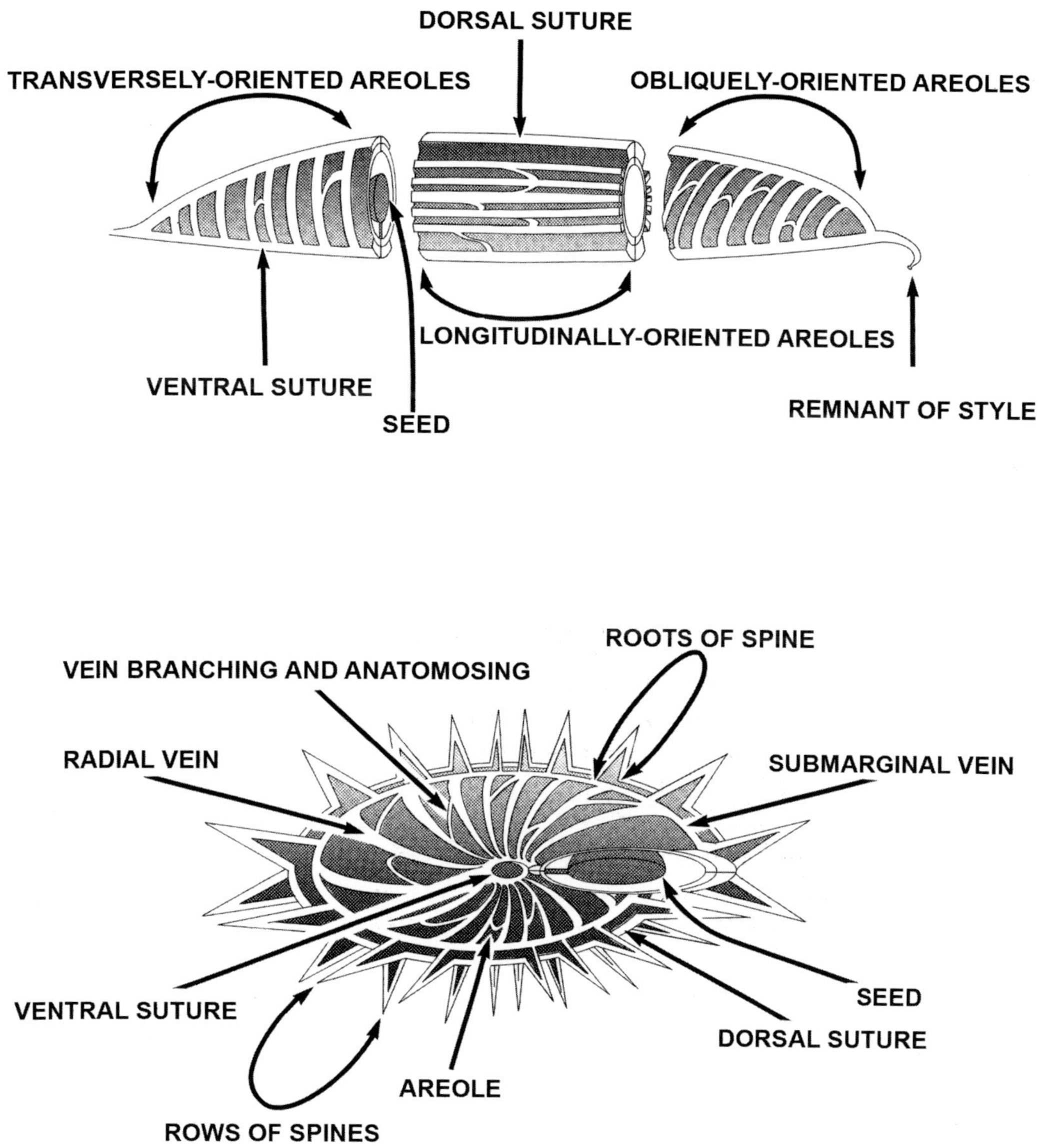

Fig. 11-1. Illustrations showing the structure and terminology of an uncoiled fruit (above) and a coiled fruit (below) (cf. Fig. 11-2). (Artist: M. Jomphe.)

About 40% of the species of *Medicago* have fruits that are straight or somewhat curved and are reminiscent of pea pods. Regardless of whether the pods are round or flat in cross section, there are two sides (valves) that are joined at two sutures (the dorsal suture and the ventral suture, the latter defined by the presence internally of attached ovules or seeds). Veins are more or less evident on the outer surfaces of the valves (the valve "faces") and the venation is often in patterns that are very important for identification. "Areoles" are the ultimate segments surrounded by veins, and these may be more or less isodiametrical or elongated. Principal veins typically run between the ventral and dorsal sutures, often anastomosing.

In more than half the species of *Medicago* the fruits are coiled, and coiled fruits present more characters than uncoiled fruits (indeed more complex characters), which are often essential for identification. Although there is usually a range in number of fruit coils for particular species (even on particular plants), the fruits of different species tend to differ in the number of coils. In coiled fruits, the ventral suture is positioned at the axis whereas the dorsal suture is on the outer side. The "valve faces" that are on uncoiled fruits are equivalent to the "coil faces" of coiled fruits. The coils are analogous to a stack of coins and the imprinted venation pattern (equivalent on both sides of the coils) frequently provides critical characters for identification. As a general rule in identifying species with coiled fruits, it is necessary to split a fruit between coils to observe the venation pattern. Veins are generally described as "running" from the axis of the fruit (ventral suture) to the outer rim of the coil (dorsal suture), and such veins are sometimes called "radial veins". In some species, the radial veins curve into an S-shape between the sutures, sometimes becoming almost concentric towards the dorsal suture. The degree of anastomosing between the veins is often important for identification. In a few species, the outer periphery of the coil face is devoid of veins; this is critical for identification of such species, but as a rule this character must be examined in young fruits because some of the species at

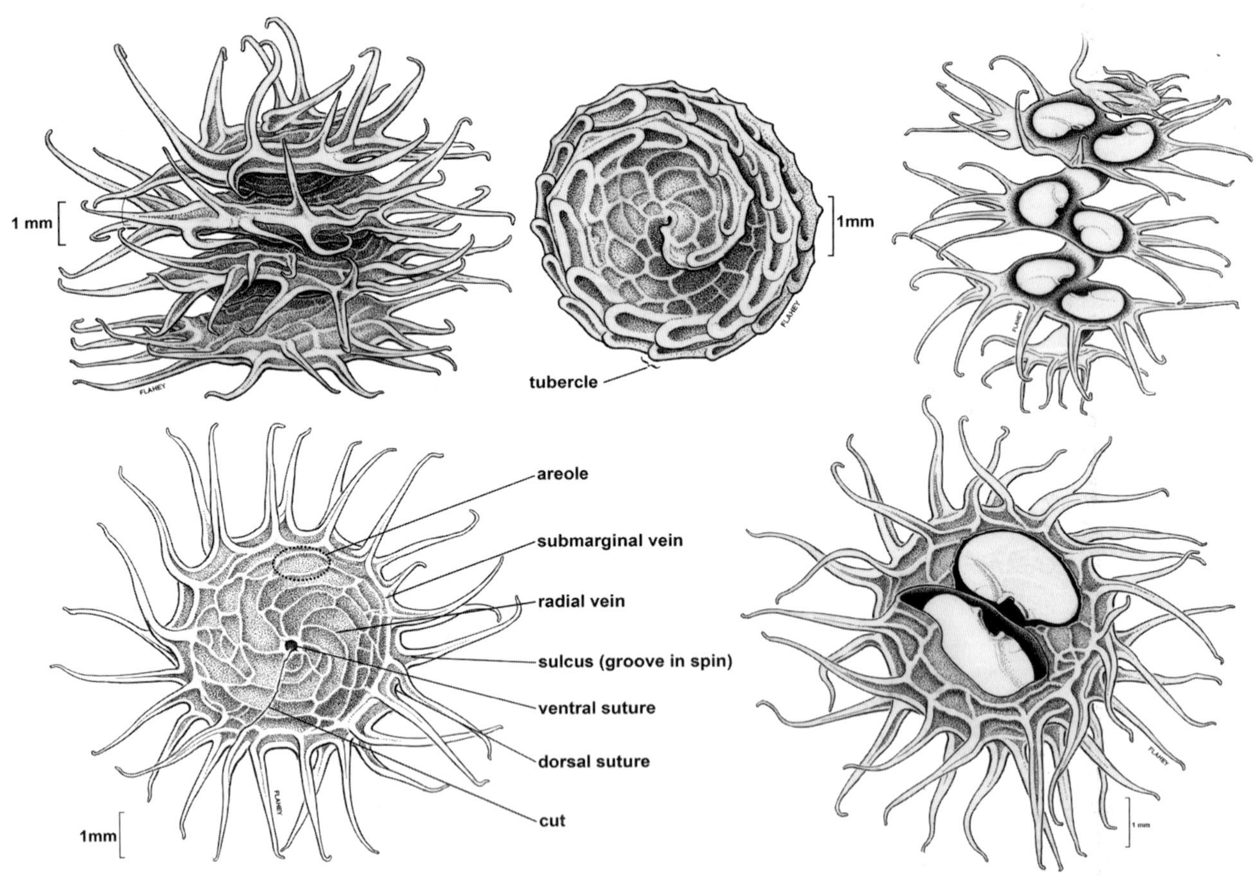

Fig. 11-2. *Medicago polymorpha* fruits. **Upper left:** A fruit with long spines. **Upper center:** A fruit with tubercles. **Lower left:** A fruit coil labeled with terminology used in this monograph. **Upper right:** A fruit that has been longitudinally sectioned and the coils stretched apart. Note the seeds attached at the axis. **Lower right:** A fruit that has been cross sectioned, showing the approximately two seeds that occur in a coil. (Artist: B. Flahey.)

maturity develop a tissue (called alveolar, spongy, or proliferant tissue) that often obscures the outer venation, thus mimicking a peripheral veinless area, which can lead to misidentification.

Very few of the species with uncoiled pods have sutures that are "ornamented" with short spinules (spicules) or short wings (sometimes fimbriated) on one or both fruit sutures, and no such species has well-developed spines. By contrast, most of the species with coiled pods have spines on the dorsal sutures (although forms within most of these species have shorter spines than normally present, or the spines have been reduced to small bumps called "tubercles"). The length, stiffness, stockiness, and orientation of the spines are often important for identification (in a few cases, the nature of tubercles is similarly important). During development, the spines are supplied by the venation from the coil faces, and sometimes in the mature fruit one sees a radial vein terminating at the base of a spine. In some species, the radial veins appear to attach to a vein that runs as a ring close to the edge of the coil; this is called a "submarginal vein" and often each spine will receive one branch from the submarginal vein (since the spines are on the dorsal suture, they receive one vein branch on either side of the coil). In a few species, just how close the submarginal vein is to the edge of the coil is important for identification. There are usually two rows of spines on a coil (one on either side of the coil edge), and sometimes the relative architecture of the spine bases and the coil edge are characteristic of given species. The coil edge may be arranged so that the tip of the dorsal suture is either elevated or at the same level as the spine bases, or there may be grooves between the tip of the dorsal suture and the adjacent spine bases.

DIFFICULTIES OF IDENTIFICATION DUE TO CONSIDERABLE VARIATION WITHIN SPECIES

Some of the very widespread species of *Medicago* are quite variable. These include *M. lupulina*, *M. medicaginoides*, *M. monantha*, *M. polymorpha*, *M. rigidula sensu lato* (including *M. rigiduloides*), and *M. sativa*.

DIFFICULTIES OF IDENTIFICATION DUE TO HYBRIDIZATION

The chief known "problems" with respect to hybridization in *Medicago* include the following:

1. The subspecies of *M. sativa*, which hybridize extensively in the wild (and are often hybridized in the production of cultivars), sometimes produce a confusing array of recombinants that defy identification.
2. *Medicago bonarotiana* hybridizes extensively with *M. rotata*. Introgression occurs primarily from *M. rotata* into *M. bonarotiana* so that *M. rotata* is usually distinctive, but *M. bonarotiana* is much less well defined. The widely employed name "*M. blancheana*" is based on *M. bonarotiana* plus the hybrids.
3. *Medicago truncatula* hybridizes and intergrades extensively with *M. littoralis*.
4. *Medicago littoralis* hybridizes extensively with *M. italica*, with gene flow occurring predominantly from *M. littoralis* to *M. italica*. Accordingly, not only are hybrid segregants common, but there is an especially confusing range of variation in *M. italica*.

DIFFICULTIES OF IDENTIFICATION DUE TO INCOMPLETE TAXONOMIC KNOWLEDGE

Some species or species complexes within *Medicago* are incompletely understood. At present, those seemingly most in need of clarification are as follows:

1. *Medicago polyceratia* is known from northern Africa and southern Europe, but some collections from India assigned to *M. orthoceras* appear to be *M. polyceratia*. The distinctiveness of *M. polyceratia* from *M. orthoceras* is unclear. Reported chromosome counts are variable but suggest that these species are polyploids and that studies of genome relationships would be helpful.
2. *Medicago suffruticosa* subsp. *suffruticosa* has been separated from *M. suffruticosa* subsp. *leiocarpa* primarily or exclusively on the basis of fruit pubescence, although various authors have alleged a range of additional differences. However, the nature of geographical, ecological, morphological, and genetic separation of the two subspecies requires clarification.
3. The distinctiveness and geography of *Medicago rigiduloides* and *M. rigidula* requires additional study.
4. Some "species" have been very poorly collected and studied and are in need of clarification. Most notable of these are *M. arenicola*, *M. laxispira*, and *M. sinskiae*.

MEDICAGO POLYMORPHA: THE MOST FREQUENTLY MISIDENTIFIED SPECIES

Medicago polymorpha is probably associated with more misidentifications of *Medicago* than any other species in the genus. Both with respect to herbarium collections and germbank accessions of *Medicago*, specimens that are identified as *M. polymorpha* are often not this species, and specimens that are not this species are often identified as

M. polymorpha. True to its specific epithet, the species is rather polymorphic, superficially mimicking dozens of other annual species of *Medicago* with coiled, spiny pods. However, *M. polymopha* is not especially difficult to identify. It is a very widespread weed and, among the *Medicago* species, it is second only to the easily identified *M. sativa* (alfalfa) with respect to spread around the planet. Its repeated misidentification appears to be the result mainly of how frequently it has been collected.

MEDICAGO SECTION *SPIROCARPOS* SUBSECTION *PACHYSPIRAE*: THE MOST DIFFICULT SPECIES TO IDENTIFY

In general in *Medicago*, the species of sect. *Spirocarpos* subsect. *Pachyspirae* are the most difficult to identify, even for those who have become quite expert on the genus (note in Chapter 10 that this group as presently defined is unsatisfactory, although many of the species do form a clade). This is because:

1. The diagnostic characteristics are quite variable and difficult to assess.
2. The coil face venation pattern and the presence of a peripheral veinless area on the coil are critical identification characteristics, but can usually be reliably assessed only in young fruits. In older fruits, a tissue proliferates in different areas of the fruits (a defining characteristic of the subsection), which not only obscures the critical identification features, but usually makes it appear (incorrectly) that there is a peripheral veinless area on the coils.
3. Both spiny and spineless forms (the spines reduced to tubercles in the latter) are common in most of the species. Non-spiny forms mimic a wide range of *Medicago* species making identification much more challenging.
4. The fruits become very hard at maturity (another defining characteristic of the subsection), and become quite troublesome (even dangerous when they are spiny) to separate into coils for assessment.

To assist in identification of this especially difficult group, key characters are tabulated in Table 11-1.

Table 11-1. Characters distinguishing species in *Medicago* section *Spirocarpos* subsection *Pachyspirae*.

Species	Chromosome number (2*n*)	Pubescence of shoot	Stipule dissection	Fruit shape	Three distinct ridges on fruit coil edge	Peripheral veinless area on fruit coil face	Curving of radial veins on fruit coil face	Anastomosing and branching veins on fruit coil face	Gaps between coils at fruit maturity	Simple hairs on mature fruit	Glandular hairs on mature fruit
M. constricta	14	Considerable	Limited to moderate	Spherical to ovoid	Absent	Absent	Strong	Weak	Absent	Absent	Absent
M. doliata	16	Considerable	Moderate	Spherical to ovoid	Absent	Absent	Weak	Strong	Usually absent	Often present	Often present
M. italica	16	Limited	Limited to moderate	Lenticular to cylindrical	Absent	Absent	Weak	Weak	Usually present	Absent	Usually absent
M. littoralis	16	Moderate	Moderate to strong	Discoid to barrel-like	Absent	Absent	Weak	Weak	Often present but absent in majority	Absent	Absent
M. murex	16	Very limited	Strong	Spherical to ovoid	Absent	Well developed	Weak	Weak	Absent	Absent	Absent
M. rigidula[1] and *M. rigiduloides*[1]	14 (16)	Considerable	Limited to moderate	Discoid, barrel-like, ovoid, or spherical	Absent	Absent	Strong	Moderate	Usually present	Usually absent	Usually present
"*M. sinskiae*"[1]	14,16	Considerable	Limited	Short-cylindrical	Absent	Absent	Moderate	Strong	Present	Absent	Present
M. soleirolii	16	Considerable	Moderate	Discoid to lenticular or short-cylindrical	Absent	Absent	Weak	Strong	Usually absent	Absent	Absent
M. sphaerocarpos	14	Very limited	Strong	Spherical to ovoid	Present	Well developed	Weak	Weak	Absent	Absent	Absent
M. syriaca	16	Very limited	Strong	Barrel-like	Absent	Poorly developed	Strong	Weak	Usually present	Absent	Absent
M. truncatula	16	Moderate	Moderate	Barrel-like	Absent	Absent	Weak	Weak	Usually absent	A few hairs present	Absent
M. turbinata	16	Considerable	Limited	Cylindrical or barrel-like	Absent	Well developed	Weak	Weak	Weak	Absent	Absent

[1] The status and differences among these very closely related "species" is discussed in detail in the chapters on them.

KEY TO SPECIES AND INFRASPECIFIC TAXA OF *MEDICAGO*

The dichotomous key has been constructed with extra leads to take account of the more likely misinterpretations of characters, and therefore particular alternatives do not always reflect the characters of the taxa. Narrowly distributed endemics (see Chapter 4, Biogeography) that are unlikely to be encountered except in their limited native areas are indicated in square brackets.

This key is in bracketed form, which economizes space and is relatively easier to use for keys that extend over many pages. All other (shorter) keys presented in this monograph are in indented form.

1A. Fruit produced underground (rarely observed dwarf ephemeral herb) .... *M. hypogaea*
1B. Fruit produced above ground 2

2A. Fruit (seeded portion) uncoiled or curving through no more than one coil 3
2B. Fruit (seeded portion) spiralling through at least one complete coil 6

3A. All fruit 1-seeded <u>and</u> the tips twisted in a small coil 4
3B. At least some fruits with two or more seeds, but if only 1-seeded fruits are present their tips are not twisted in a small coil 5

4A. Inflorescence a compact raceme of (5)14–24 flowers; on pod surface veins run obliquely from ventral to dorsal suture without reversing direction; mature pods ash-gray to black .... *M. lupulina*
4B. Inflorescence a one-sided (secund) raceme of 3–10 flowers; on pod surface veins run obliquely from ventral to dorsal suture, reversing direction towards dorsal suture; mature pods gray [endemic of North Africa, rarely introduced elsewhere] .... *M. secundiflora*

5A. Fruits with pectinate ciliate-spines (some forked) on the dorsal suture <u>and</u> shortly fimbriate-winged on the ventral suture; never plicate (pleated) .... *M. radiata*
5B. Fruits lacking the combination of <u>all three</u> characteristics of the above alternative 69

6A. Mature seeds smooth (viewed with at least 10× magnification) 9
6B. Mature seeds tuberculate, rugose, or ridged (viewed with at least 10× magnification) 7

7A. Fruit dorsal suture (at outer edge of coils) completely spineless .... *M. orbicularis*
7B. Fruit dorsal suture spiny or with short spinules 8

8A. Fruit dorsal suture edge with one row of short (ca. 1 mm or less), sometimes forked spines; ventral (axial) suture with fringed wing .... *M. radiata*
8B. Fruit dorsal suture edge with two rows of long (usually several millimeters), unforked spines; ventral suture (hidden at axis) without fringed wing [endemic of Turkey, Aegean Islands, Crete] .... *M. heyniana*

9A. <u>Mature</u> fruit glabrous, or with brownish, grayish, or gland-tipped hairs, or if white, not so thick as to obscure fruit 12
9B. <u>Mature</u> fruit covered with long white hairs (<u>not</u> gland-tipped hairs) 10

10A. Fruit spiny (plants confined to sandy seashores) .... *M. marina*
10B. Fruit spineless 11

11A. Flowers no longer than 4 mm, plants annual .... *M. lanigera*
11B. Flowers 4–8 mm long, plants perennial .... *M. prostrata* var. *pseudorupestris*

12A. Annual herbs, or perennial herbs with annual aboveground shoots not becoming particularly woody 15
12B. Woody shrubs (aboveground shoot perennial, developing very woody branches) 13

13A. Corolla lemon yellow [endemic of west Mediterranean islands] .... *M. citrina*
13B. Corolla orange yellow or yellow 14

14A. Fruit diameter 12–15 mm; fruits with 0.5–1.5 coils; seeds 3.5–4.5 mm long .... *M. arborea*
14B. Fruit diameter 5–10 mm; fruits with 1.5–2.5(3) coils; seeds 2–3.5 mm long [endemic of Crete] .... *M. strasseri*

15A. Roots annual (delicate at maturity; plants in native habitats usually dying before frost damage); many of the species with flowers shorter than 6 mm; flowers usually strongly selfing (depositing most of the pollen on the stigma within the flower); radicle of seed usually slightly shorter than or about equal to half length of seed; coil of fruit usually tight (lacking doughnut-like hole at axis) 33
15B. Roots perennial (woody and strongly branched at maturity, above-ground shoots in native habitats persisting until killed by frost); many of the species with flowers longer than 6 mm; flowers usually somewhat to strongly outcrossing (not depositing most of the pollen on the stigma within the flower); radicle of seed usually slightly longer than half length of seed; coil of fruit often lax (with apparent doughnut-like hole at axis) 16

16A. Spines or tubercles absent on coil edges of fruit 21
16B. Spines or tubercles present on coil edges of fruit 17

17A. Spines thin, flexible; 4–5 seeds per coil; extensive underground rhizomes developed .... *M. carstiensis*
17B. Spines thick, firm; 1–2 seeds per coil; extensive underground rhizomes not developed 18

18A. Corolla violet [endemic of Caucasus] .... *M. daghestanica*
18B. Corolla yellow 19

19A. Gland-tipped trichomes present on fruit [endemic of European Alps] .... *M. pironae*
19B. Gland-tipped trichomes absent on fruit 20

20A. Fruit diameter 2–4 mm [endemic of southern Bulgaria] .... *M. rhodopea*
20B. Fruit diameter 6–8 mm .... *M. saxatilis*

21A. Leaflets narrower (length:width usually narrower than 3:2) 23
21B. Leaflets broad (length:width usually wider than 3:2) 22

22A. Fruit with simple or gland-tipped trichomes; venation on coil face, if visible, somewhat net-like .... *M. suffruticosa* subsp. *suffruticosa*
22B. Fruit glabrous; venation on coil face prominently concentric .... *M. suffruticosa* subsp. *leiocarpa*

23A. Face of coil more or less obscurely veined 25
23B. Face of coil prominently reticulate-veined 24

24A. Fruit diameter 3–4 mm .... *M. rupestris*
24B. Fruit diameter 4–6 mm .... *M. cancellata*

25A. Seeds not separated or separated only by inconspicuous, thin partitions 27
25B. Seeds separated by conspicuous, thick, spongy partitions 26

26A. Fruit diameter less than 6.3 mm .... *M. papillosa* subsp. *papillosa*
26B. Fruit diameter greater than 6.3 mm .... *M. papillosa* subsp. *macrocarpa*

27A. Flowering pedicel distinctly longer than calyx, recurved .... *M. prostrata* subsp. *prostrata*
27B. Flowering pedicel shorter to somewhat longer than calyx, rarely recurved (*M. sativa*) 28

28A. Flowers yellow or variegated yellow-violet; or if uniformly violet or blue-violet, fruit with less than 1.5 coils 31
28B. Flowers uniformly violet or blue-violet and at least some of fruits with at least 1.5 coils (usually more) 29

29A. Calyx (base to tip of longest lobe) usually less than 4.5 mm long; fruit intersuture width usually less than 1.8 mm .... *M. sativa* subsp. *caerulea*
29B. Calyx usually more than 4.5 mm; midfruit intersuture width usually more than 1.8 mm 30

30A. Fruit lacking gland-tipped trichomes, or essentially so .... *M. sativa* subsp. *sativa*
30B. Fruit with gland-tipped trichomes .... *M. sativa* subsp. *sativa* × subsp. *glomerata*

31A. Fruit lacking gland-tipped trichomes .... *M. sativa* subsp. ×*varia*
31B. Fruit with gland-tipped trichomes 32

32A. Fruit densely covered with gland-tipped trichomes; flowers yellow .... *M. sativa* subsp. *glomerata*
32B. Fruit with a few gland-tipped trichomes; flowers variegated yellow-violet .... *M. sativa* subsp. *sativa* × subsp. *glomerata*

33A. If dorsal suture spines are present with a basal root vein connected to the venation on the coil face, that vein enters the spine directly (in a straight line from the coil axis to the spine) or not at all; mature seeds yellow or yellow-brown 37
33B. A vein from the coil face enters each dorsal suture spine very distinctly on one side of the spine base; mature seeds red-brown or black-brown 34

34A. Fruit short-cylindrical (ends truncate), diameter 8–12(15) mm, with 3–7 coils 36
34B. Fruit spherical to ovoid (ends rounded), diameter 10–15 mm, with (3)6–10(11) coils 35

35A. Fruit with gland-tipped trichomes .... *M. ciliaris*
35B. Fruit lacking gland-tipped trichomes .... *M. intertexta*

36A. Diameter of terminal coil of fruit less than 5.5 mm, fruit discoid (short-cylindrical, coils not greatly different in diameter); leaflets 3–8 mm long, distal stipule lobe longer than other lobes .... *M. muricoleptis*
36B. Diameter of terminal coil of fruit more than 5.5 mm, fruit barrel-shaped (coils in middle definitely larger than coils at ends); leaflets 10–24 mm long, distal stipule lobe not much longer than other lobes .... *M. granadensis*

37A. Fruit coils more or less flat, arranged as in a stack of coins 40
37B. Some fruit coils conspicuously imbricate (included in adjacent coils, as in a stack of bowls) 38

38A. Open part of "bowls" all point toward distal end .... *M. scutellata*
38B. Open part of "bowls" point toward middle 39

39A. Fruit without spines .... *M. bonarotiana*
39B. Fruit with small spines .... *M. bonarotiana* × *M. rotata*

40A. Fruit with numerous irregular (bent or hooked in various directions) short spines (0.2–0.5 mm) at right angles to pod face 41
40B. Fruit with more or less uniform spines, or spineless 42

41A. Leaflets not laciniate; or if laciniate, most inflorescences with no more than two flowers .... *M. rotata* var. *rotata*
41B. At least some leaflets laciniate <u>and</u> inflorescences often with more than two flowers .... *M. rotata* var. *eliezerii*

42A. Fruit with <u>oblique</u>, wing-like elevations on coil edge; calyx symmetrically appressed (star-like) at base of pod .... *M. noeana*
42B. Fruit with wing-like elevations <u>perpendicular</u> to coil edge, or without wing-like elevations; calyx appressed to one side of pod 43

43A. Fruit spiny and/or with more than 2.0 coils 45
43B. Fruit spineless (or merely with tubercles) and with 1–2.0 coils only 44

44A. Fruit pubescent (most apparent when young) [endemic of south-central Anatolia] .... *M. shepardii*
44B. Fruit glabrous .... *M. italica*

45A. Dorsal fruit suture spiny; or if spineless, radial veins on coil face extend on periphery to a submarginal veinless area or a submarginal vein close to dorsal suture, from which venation may extend to dorsal suture (in fruits with very hard walls, young fruits must be examined to avoid the misleading appearance of a peripheral veinless area) 47
45B. Dorsal fruit suture spineless; radial veins on coil face extend directly to dorsal suture 46

46A. Dorsal fruit suture with strongly thickened extensions of radial veins from coil faces; flowers 2.5–4 mm long .... *M. rugosa*
46B. Dorsal fruit suture smooth; flowers 7–10 mm long .... *M. soleirolii*

47A. Mature fruit extremely hard-walled (spiny and spineless forms of all species occur, and the coils of spiny forms cannot be separated by hand without probability of injury); the spines stocky and difficult to bend; base of spines often round, 2 roots often not being apparent at maturity; very young pod often in a contracted spiral concealed within calyx; venation pattern often obscure on coil face of mature fruit because of proliferation of alveolar tissue 57
47B. Mature fruit relatively soft-walled (sometimes hard in *M. polymorpha*); spines almost always present, these often relatively thin and flexible; base of spines 2-rooted, one root visibly arising in dorsal suture, the other in a submarginal vein; very young pod projecting sideways from calyx; venation often discernible on coil face of mature fruit 48

48A. Veins extend through entire face of coil 50
48B. Distal (outer) part of face of coil (from about half way from center to margin) veinless 49

49A. Spines all slanted towards base of fruit .... *M. disciformis*
49B. Spines not all slanting towards base of fruit .... *M. tenoreana*

50A. Fruit edge shows 4 ridges alternating with 3 grooves; leaflets often with a central anthocyanin blotch .... *M. arabica*
50B. Fruit edge shows fewer than 4 ridges, if any; leaflets usually lack blotches 51

51A. Stipules of larger leaves dissected 53
51B. Stipules of larger leaves entire or minutely dentate 52

52A. Fruit spines, when present, more or less parallel to coil face; one of the two venation roots of each spine originates on the coil face about one-third of coil radius away from the dorsal suture; raceme with (2)3–6(8) flowers .... *M. minima*
52B. Fruit spines, when present, often at 90° to coil face (thus giving a crown-like appearance); one of the two venation roots of each spine originates on the coil face very close to the dorsal suture; raceme with (2)6–12(17) flowers .... *M. coronata*

53A. Flowers (3)4–8 mm long; coils loosely to tightly appressed to each other; spines 90° to 180° to coil face 55
53B. Flowers 2–4 mm long; coils loosely appressed to each other (space evident between coils); spines 90° to 130° to coil face 54

54A. Fruit diameter 2.5–4.5 mm; adjacent coils loosely appressed .... *M. praecox*
54B. Fruit diameter 5–6 mm; adjacent coils rather separated [very rarely collected endemic of northern Iraq] .... *M. laxispira*

55A. Many anastomoses between radial veins on coil face .... *M. polymorpha*
55B. Very few if any anastomoses between S-shaped radial veins that run from center of coil face to submarginal vein 56

56A. Fruit diameter 3–5(7) mm; submarginal vein about one-fifth radius of coil away from dorsal suture; leaves often laciniate .... *M. laciniata*
56B. Fruit diameter 5–10 mm; submarginal vein about one-third to one-fourth radius of coil away from dorsal suture; leaves not laciniate [endemic of Morocco] .... *M. sauvagei*

57A. Peripheral part of coil face of young pods lacking a well-developed veinless area 61
57B. Peripheral part of coil face of young pods veinless or mainly veinless (mature pods must not be examined for this character because proliferation of alveolar tissue often misleadingly results in this appearance) 58

58A. Leaves pubescent on upper and lower surfaces .... *M. turbinata*
58B. Leaves glabrous, or glabrescent on lower surfaces 59

59A. Fruits more or less barrel-like; radial veins of coil face very strongly curved [endemic of Syria; perhaps Iran] .... *M. syriaca*
59B. Fruits ovoid to spheroid; radial veins of coil face weakly curved 60

60A. Coil edges of mature pods with 3 ridges; raceme with 1–4 flowers .... *M. sphaerocarpos*
60B. Coil edges of mature pods lack evident ridges; raceme with 1–2(3) flowers .... *M. murex*

61A. A "lateral vein" absent on coil face near edge of coil (i.e., near dorsal suture; this character must be examined in young fruit); dorsal suture spines or tubercles always absent .... *M. soleirolii*
61B. A "lateral vein" present on coil face near edge of coil; dorsal suture spines or tubercles often present 62

62A. Fruit coil faces with very strongly curving radial veins 63
62B. Fruit coil faces with weakly to moderately curving radial veins 66

63A. Shoots glabrescent; stipules strongly dissected; mature fruits lacking glandular hairs, usually with gaps between coils (endemic of Syria; perhaps Iran) .... *M. syriaca*
63B. Shoots pubescent; stipules slightly to moderately dissected; mature fruits with or without glandular hairs, with or without gaps between coils 64

64A. Some gaps evident between coils of mature pods; mature pods usually velvety-glandular (with short gland-tipped hairs); radial veins strongly curved or S-shaped, sometimes anastomosed peripherally 65
64B. No gaps evident between coils of mature pod (sometimes coils so strongly adpressed their limits are not distinguishable); mature pods glabrous; curving veins at periphery of pod face almost concentric, little anastomosed .... *M. constricta*

65A. Pollen grains predominantly 3-pored in polar view (predominantly pyramidal in side view); spine tips often strongly curved; pods often with no more than 5 coils [particularly present in Europe, northern Africa] .... *M. rigidula*
65B. Pollen grains predominantly 4-pored in polar view (predominantly cylindrical in side view); spine tips usually not strongly curved; pods often with more than 5 coils [particularly present in Asia] .... *M. rigiduloides*

66A. Pods lenticular, discoid, or barrel-like; veins weakly anastomosing in peripheral third (evident only in young pods) ... 67
66B. Pods spherical to ovoid; veins strongly anastomosing in peripheral third (evident only in young pods) ... *M. doliata*

67A. Fruit with at least a few simple trichomes; spines, when present, at 90° to plane of coil (inland plants; intergrades with *M. littoralis*) ... *M. truncatula*
67B. Fruit glabrous or at least lacking simple trichomes; spines, when present, at various angles to plane of coil (seaside plants; hybrids and introgressants frequent between the following two species) ... 68

68A. Flowers (1)2–17; edge of coil tapered; spines rare ... *M. italica*
68B. Flowers 1–5; edge of coil truncate; usually spiny ... *M. littoralis*

69A. Typical fruits wider (at widest point) than 3 mm; fruit very evidently flattened in plane connecting dorsal and ventral sutures (the maximum width between sutures is more than twice the thickest location between the sutures) ... 70
69B. Typical fruits no wider (at widest point) than 3 mm; fruit more or less terete or slightly flattened, rarely strongly flattened in plane connecting dorsal and ventral sutures (the maximum width between sutures is less than twice the thickest location between the sutures, except occasionally in *M. sativa* subsp. *falcata*) ... 82

70A. Fruit without gland-tipped trichomes ... 72
70B. Fruit with gland-tipped trichomes ... 71

71A. Leaflets denticulate; fruit 15–20 mm long, ventral suture concave ... *M. biflora*
71B. Leaflets entire; fruit 7–8 mm long, ventral suture straight [endemic of central and southern Anatolia] ... *M. rostrata*

72A. Fruit plicate [endemic of south-central Anatolia] ... *M. plicata*
72B. Fruit not plicate ... 73

73A. Fruit venation with prominent anastomoses between transverse veins toward dorsal suture; fruit glabrous to slightly pubescent; flowers yellow, but often tinged with purple, red or brown, especially when dried (perennials) ... 78
73B. Fruit venation prominently transverse, with few anastomoses; fruit glabrous to very pubescent; flowers yellow (annuals and perennials) ... 74

74A. Fruit with 4–15 seeds, with narrow fringed-hairy ventral wing ... 77
74B. Fruit with 1–2 seeds, lacking ventral fringe (sometimes narrowly winged) ... 75

75A. Fruit reniform; raceme with 4–6 flowers (perennial) ... *M. cretacea*
75B. Fruit lunate or lunate-ovate; raceme with 10–15 flowers (annual) ... 76

76A. Fruit more or less deflexed, 4–6 mm long (including beak); beak 1–1.3 mm long; leaflets oblong to obovate; corolla 2–5 mm long ... *M. brachycarpa*
76B. Fruit erect, 8–10 mm long (including beak); beak 3–3.5 mm long; leaflets obcordate; corolla 4–5 mm long [endemic of southwestern Anatolia] ... *M. huberi*

77A. Seeds smooth; flowers in prominently pedunculate racemes distributed over entire plant (perennial) ... *M. edgeworthii*
77B. Seeds rugose; flowers clumped in upper leaf axils (annual) ... *M. ovalis*

78A. Fruit shortly and broadly oval (both sutures strongly convex); ventral suture slightly winged ... *M. popovii*
78B. Fruit oblong-falcate (dorsal suture strongly convex, ventral suture weakly convex to straight); ventral suture not winged ... 79

79A. Seeds rugose .... *M. ruthenica*
79B. Seeds smooth 80

80A. Fruit 7–9 mm broad, 15–22 mm long; fruit with 4–6 seeds; fruit black at maturity .... *M. platycarpos*
80B. Fruit 3–5 mm broad, 7–14 mm long; fruit with 2–4 seeds; fruit brownish 81

81A. Venation on pod valve strikingly heterogeneous: veins near ventral suture are parallel, oblique, whereas veins near dorsal suture anastomose to form a prominent net [endemic of southern France] .... *M. hybrida*
81B. Venation on pod valve not strikingly heterogeneous: transverse veins anastomose irregularly towards dorsal suture .... *M. archiducis-nicolai*

82A. Radicle 75% or more of length of seed; plants (roots) annual 85
82B. Radicle 60% or less of length of seed; plants (roots) perennial (*M. sativa*) 83

83A. Flowers violet or variegated yellow-violet .... *M. sativa* subsp. ×*varia*
83B. Flowers yellow (*M. sativa* subsp. *falcata*) 84

84A. Fruit glabrous .... *M. sativa* subsp. *falcata* var. *falcata*
84B. Fruit with gland-tipped trichomes .... *M. sativa* subsp. *falcata* var. *viscosa*

85A. Fruit not conspicuously strangulate 87
85B. Fruit conspicuously strangulate 86

86A. Fruit straight to curved, with reticulately anastomosing veins; corolla 5–6 mm long; calyx teeth about equal to tube [endemic of central Anatolia] .... *M. isthmocarpa*
86B. Fruit falcate to semicircular, transversely veined (reticulation limited); corolla 7–9 mm long; calyx teeth longer than tube [endemic of southern Anatolia] .... *M. rhytidiocarpa*

87A. Fruit erect (or at least not bent back as much as 90° from flowering axis) 91
87B. Fruit reflexed, patent, or stellate-spreading (bent at least 90° back from flowering axis) 88

88A. Inflorescence of 1–2 flowers [endemic of Afghanistan] .... *M. retrorsa*
88B. Inflorescence of 6–20 flowers 89

89A. Dorsal fruit suture straight; infructescence non-pedunculate .... *M. monspeliaca*
89B. Dorsal fruit suture sinuate; infructescence pedunculate 90

90A. Fruit mainly straight, sometimes arcuate; inflorescence of 6–12 flowers; calyx teeth lanceolate-subulate, longer than or equal to tube .... *M. astroites*
90B. Fruit arcuate; inflorescence of 5–8 flowers; calyx teeth triangular-lanceolate, less than tube [endemic of southern Anatolia] .... *M. halophila*

91A. Fruit with 3–20 veins/cm at either suture 93
91B. Fruit with 30–40 veins/cm at either suture 92

92A. Fruit with more or less undulating sutures [endemic of southern Anatolia] .... *M. pamphylica*
92B. Fruit with straight sutures [endemic of southwestern Anatolia, East Aegean Islands] .... *M. carica*

93A. Fruit 6–10 mm long, 2–3 mm wide [endemic of southern Anatolia] .... *M. rigida*
93B. Fruit 10–100 mm long, 1–2 mm wide 94

94A. Peduncle 5–50 mm; inflorescence subumbellate or subracemose; calyx equals half of corolla length or less 99
94B. Peduncle 0–5(12) mm (excepting *M. medicaginoides*, with peduncles 0–50 mm); inflorescence umbellate; calyx equal or longer than half of corolla length 95

95A. Fruit (30)40–100 mm long, curved; inflorescence of 1–2(3) flowers; flowers (5)6–8 mm long; leaves sometimes incised .... *M. monantha*
95B. Fruit 10–50(60) mm long, straight or curved; inflorescence of (1)2–10 flowers; flowers 4–6 mm long 96

96A. Fruit falcate to semicircular; venation areoles transverse to oblique or random; peduncle 0–30 mm long; calyx teeth longer than or equal to tube .... *M. medicaginoides*
96B. Fruit straight to falcate; venation areoles substantially obliquely or randomly oriented; peduncle 0–5(12) mm; calyx teeth about equal to or shorter than tube 97

97A. Inflorescence of 4–10 flowers; fruit straight, clustered .... *M. crassipes*
97B. Inflorescence of 2–6 flowers; fruit straight or curved, erect-spreading 98

98A. Fruit straight, less frequently curved; (10)20–35(50) mm; venation areoles randomly arranged, but often longitudinal; inflorescence of 2–5 flowers; calyx tube less than or tending to equal teeth; flowers 4–5 mm long; leaves not incised .... *M. orthoceras*
98B. Fruit usually curved, (10)20–50 mm long; venation areoles randomly oriented, but often with sub-equal sides and diagonal orientation; inflorescence of 2–6(9) flowers; calyx teeth about equal to or often somewhat shorter than tube; flowers 4–6 mm; leaves sometimes incised .... *M. polyceratia*

99A. Fruit straight or slightly curved, not tortuous (flexuous), sutures not undulate 102
99B. Fruit distinctly curved (often especially terminal portion), more or less tortuous with more or less undulating sutures 100

100A. Flower 5–7(9) mm long .... *M. fischeriana*
100B. Flowers (6)7–10 mm long 101

101A. Inflorescence of 1–5(8) flowers; fruit (20)30–40 mm long; calyx teeth distinctly shorter than tube; flowers yellow [endemic of southwestern Iran, adjacent Iraq] .... *M. persica*
101B. Inflorescence of 8–14 flowers; fruit 10–20(30) mm long; calyx teeth about as long as tube; flowers yellow to orange .... *M. phrygia*

102A. Stem 10–30 cm; leaflets 4–8 mm; peduncle 10–50 mm; inflorescences of (8)10–20 flowers; fruit 15–30 mm long [endemic of southern Anatolia] .... *M. heldreichii*
102B. Stem 3–10 cm; leaflets 2–5 mm; peduncle 5–10 mm; inflorescence of 4–6 flowers; fruit 10–15 mm long [species of uncertain status, known only from type gathering from Anatolia] .... *M. arenicola*

12. *Medicago sativa* L.

M. sativa L., Sp. Pl. 778 (1753), (basionym).

The epithet *sativa* is Latin meaning sown or cultivated, descriptive of the fact that this is the most widely grown species in the genus, and indeed one of the most widely cultivated of all crops. The Latin epithets of the infraspecific taxa recognized below have the following derivations: *caerulea*, dark sky-blue, for the typical color of the flowers (the epithet was often rendered "*coerulea*" in the past, but the original spelling *caerulea* must be retained); *falcata*, falcate (like a crescent moon), in reference to the sickle-shaped fruits; *glomerata*, collected closely together into a head (a perception of the author of this name that is not helpful for identifying the group); *viscosa*, sticky (because of the secretion from the glandular hairs); *varia* (variable, referring specifically to the petal colors of this taxon).

Because the infraspecific taxonomy of *M. sativa* is much more complicated than that of any other species of the genus, and because it is by far the most important economically, the treatment in this chapter is more extensive than for any of the other species.

Overview of the Infraspecific Taxonomy of *Medicago sativa*

Medicago sativa is a polymorphic species with a variation pattern that has been complicated by hybridization, polyploidy, and domestication, a situation that is found in some other important forage crops (for example *Lotus*; see Small et al. 1984; Grant and Small 1996). Urban (1873), the first taxonomist to significantly monograph the genus *Medicago*, had little insight into the variation of *M. sativa*, and his rather elaborate classification system is quite confusing. Extremely extensive classifications of *M. sativa* were proposed in the first half of the 20th century by several Russian taxonomists who split *M. sativa* into dozens of species and innumerable infraspecific taxa. These classifications, as explained in a critique presented below, are based on outmoded taxonomic concepts and failure to understand hybridization, and have misled many alfalfa specialists into the acceptance of nonexistent "species".

As a result of extensive study of plants that I have collected during field trips to Eurasia and plants that I have cultivated, examination of thousands of herbarium collections, and review of the literature, I have concluded that a relatively simple infraspecific classification of *M. sativa* in a wide sense is appropriate, based on certain determinants: ploidy level and hybridization (Small 1983, 1984*b*, 1985*a*; Small and Bauchan 1984); flower color as a pollinator attractant (Small et al. 1981*b*; Small et al 1987; Brookes and Small 1988; Brookes et al. 1994; Small and Brookes 1984*b*; Small et al. 1997; Small et al. 1994); fruit coiling, a character adaptive against certain insects (Small and Brookes 1982, 1984*a*); and presence of glandular hairs, especially on the fruit, once again a character adaptive against insects (Small 1985*b*, 1986*a*; Small and Brookes 1986; Small et al. 1989).

In brief, the variation pattern in *M. sativa* is based on the following. Essentially all serious students of *M. sativa* in the broadest sense have recognized two fundamental groups, one with violet flowers and coiled pods (generally called *M. sativa*), the other with yellow flowers and sickle-shaped pods (often called *M. falcata*). Within both, there are diploids ($2n = 16$) and tetraploids ($2n = 32$). Moreover, also within both and at both chromosome levels, there are forms with glandular hairs, particularly on the pods. Finally, hybridization has occurred in nature between all the preceding, and many cultivars of alfalfa have been generated by hybridizing several of them.

Plants with violet flowers and coiled pods are much more common than any other kind. Because I (and others) have found that the diploids and tetraploids of this class of plant can be morphologically separated (largely because of the "gigas" effect that often accompanies polyploidization), they are recognized here as *M. sativa* subsp. *caerulea* (diploids) and *M. sativa* subsp. *sativa* (tetraploids). Most alfalfa cultivars are assignable to the tetraploid *M. sativa* subsp. *sativa* (albeit many have some parentage of subsp. *falcata*). The diploid subspecies *caerulea* is basically a wild taxon with a natural geographical range in Eurasia. Subspecies *sativa* is largely represented by cultivars, although it also occurs very widely as a ruderal (weedy) plant throughout the world. Determination of a natural range (assuming this existed before domestication occurred) is problematic because ruderal and presumptively wild plants generally cannot be distinguished with certainty. Moreover, there is considerable genetic exchange between cultivars and plants growing outside of cultivation (Jenczewski et al. 1998; Jenczewski et al. 1999*a*, 1999*b*; Muller et al. 2001*a*; Muller et al. 2001*b*; Muller et al. 2003; Muller et al. 2006; Greene et al. 2008). It cannot be determined now whether tetraploids originated from the diploids in nature or during domestication, or (more probably) under both circumstances. Indeed, because of genetic exchange between domesticated subsp. *sativa* and wild growing plants of this taxon, it is unlikely that wild races that existed 10 000 years ago have persisted in their original form.

Plants with yellow flowers and sickle-shaped pods are also widespread in Eurasia. However, unlike the situation described above, the diploids and tetraploids are not morphologically distinguishable (although the pollen does have some discriminating value). It seems likely that before domestication, the yellow-flowered, sickle-podded plants and the violet-flowered, coiled-podded plants were much more separated than they are today since the "pure" kinds of plants are dramatically physiologically and ecologically distinctive. If this ancestral situation existed today, recognizing them as separate species (*M. sativa* and *M. falcata*) would be justified. However, as a result of human activity, the two kinds have been very extensively intermingled and hybridized, and the original gene pools are combined to at least some extent probably in the majority of plants that now exist. Indeed, there is a long tradition of recognizing a hybrid group, often as a species. In this monograph, the hybrid group is delimited as *M. sativa* subsp. ×*varia*. Because it is not possible to distinguish morphologically plants of the yellow-flowered, sickle-podded group at the diploid and tetraploid level, both chromosome levels are included in subsp. *falcata*. Similarly, because diploids and tetraploids within the hybrid group are not distinguishable, both chromosome levels are included in subsp. ×*varia*.

Finally, there is the issue of glandular hairs, particularly on the fruits of only some plants of the species, a phenomenon that is very widespread in the genus *Medicago*. In *M. sativa*, glandular-haired plants tend very strongly to occur in given geographical areas within the range of given species and, as explained in Chapters 4 and 7, this seems to be an adaptive response to local herbivorous insects. In the *M. sativa* complex, glandular hairs are most prevalent in montane areas of the Caucasus and adjacent areas of the former USSR, and occur to a lesser extent in montane areas of southern Europe west of the former USSR and in North African mountainous areas. The character is most strongly developed in localized phases of both yellow-flowered, sickle-podded plants and yellow-flowered, coiled-podded plants. In the former, the plants are recognized as the taxon *M. sativa* subsp. *falcata* var. *viscosa* (which includes both chromosome levels). In the latter, the plants are recognized as the taxon *M. sativa* subsp. *glomerata* (which also includes the two chromosome levels). Hybrids between subsp. *glomerata* and subsp. *sativa* are fairly widespread, but the classification of *M. sativa* is best not complicated further by recognizing this hybrid group as an additional taxon.

Comparison of the Present Circumscriptions of Taxa of *Medicago sativa* with the Questionable Circumscriptions and/or Nomenclature of Some Key Recent Taxonomists

Tutin (1968) identified all European yellow-flowered plants of *M. sativa* with coiled pods as *M. sativa* subsp. *glomerata* (Balbis) Tutin (correctly Rouy, not Tutin). Although the critical taxonomic character pod glandularity is not mentioned, this treatment effectively identifies glandular *M. sativa*, except for hybrid forms, as I advocate. (Unfortunately, the widespread *M. sativa* subsp. ×*varia* is not taken into account in Tutin's key, although it is mentioned in his account; using Tutin's key, forms of this taxon will key out to his subsp. *ambigua* or to subsp. *glomerata*.)

Lubenets (1972) treated the names *M. glomerata* Balb. and *M. suffrutricosa* Ramond ex DC. as synonyms of *M. prostata* Jacq. However, these names clearly denote elements of three different species.

Gunn et al. (1978) treated *M. sativa* as comprising nine subspecies. As noted in the following table, there really are no equivalents of the taxa that they recognized by the subspecific names *glomerata*, *praefalcata*, and *ambigua* with those that I recognize in this volume, and I have concluded (Small 1986*a*) that their circumscriptions of these three taxa are without merit.

The name *M. sativa* subsp. *glomerata* (Balbis) Tutin was applied by Gunn et al. to a grouping with "nonglandular, appressed-pilose" pods. This, however, excluded the type specimen, which has glandular pods (Gunn et al. (1978) noted (p. 1) that their treatment of *M. sativa* was based on "the protolog of the type descriptions and not on a study of the types"). Balbis (1801) referred only to "M. leguminibus contortis villosis" and to "legumina...villosa", so that application of the name could not be determined without recourse to the type. In any event, the circumscription seems to stem from lack of familiarity with the fact that the epithet *glomerata* has traditionally been applied to glandular plants.

Gunn et al. (1978) keyed glandularly pubescent *M. sativa* to two of their subspecies, *praefalcata* and *viscosa*, the latter equivalent to var. *viscosa* recognized in this volume. They placed all glandularly pubescent forms with spiral legumes in "*M. sativa* subsp. *praefalcata* (Sinskaya) C.R. Gunn", based on *M. glutinosa* M. Bieb. subsp. *praefalcata* Sinskaya. In their circumscription of this taxon, Gunn et al. specifically excluded plants with yellow flowers, while referring to the legumes bearing numerous glandular hairs. However, Sinskaya's (1948) description of *M. glutinosa* subsp. *praefalcata* Sinskaya (Sinskaya 1950) is of a taxon of plants with yellow flowers and fruits completely lacking glandular hairs. I have not been able to determine whether a type was ever designated for this taxon, but in any event Sinskaya's (1948) description makes it clear that her subspecific epithet cannot be used for the taxon circumscribed by Gunn et al. The application of Sinskaya's name is erroneous, but nevertheless the combination *M. sativa* subsp. *praefalcata* (Sinskaya) Gunn was validly published and must apply to a nonglandular taxon lacking yellow flowers, a kind of *M. sativa* that exists only as occasional recombinants usually placed in *M. sativa* subsp. ×*varia*.

The name "*M. sativa* subsp. *ambigua* (Trautvetter) Tutin" adopted by Gunn et al. is based on the confused taxon adopted by Tutin, cited above; plants so recognized belong to *M. sativa* subsp. ×*varia*.

Table 12-1. Approximate equivalents of *Medicago sativa* subspecific names used by Gunn et al. (1978) and names used in this volume.

Gunn et al. (1978)	This volume
M. sativa subsp. *sativa*	*M. sativa* subsp. *sativa* var. *sativa* + part of var. *caerulea*
M. sativa subsp. *caerulea*	*M. sativa* subsp. *sativa* var. *caerulea*
M. sativa subsp. ×*varia*	*M. sativa* subsp. ×*varia*
M. sativa subsp. ×*hemicycla*	*M. sativa* subsp. ×*varia*
M. sativa subsp. *falcata*	*M. sativa* subsp. *falcata* var. *falcata*
M. sativa subsp. *viscosa*	*M. sativa* subsp. *falcata* var. *viscosa*
M. sativa subsp. *glomerata*	No equivalent (see text)
M. sativa subsp. *praefalcata*	No equivalent (see text)
M. sativa subsp. *ambigua*	*M. sativa* subsp. ×*varia*
No equivalent	*M. sativa* subsp. *glomerata*

Lesins and Lesins (1979) treated *M. sativa* as comprising four species, one of these with two subspecies (they also recognized hybrids among the groups, but did not apply formal names to these). On the whole, their treatment is quite conceptually consistent with the presentation given here, although some rankings are different. These authors found that *M. sativa* subsp. *glomerata* (as delimited here) of southern Europe is diploid ($2n = 16$) and referred this group to *M. glomerata* Balbis (s. str.). By contrast, these authors found that plants of *M. sativa* subsp. *glomerata* of Caucasia were tetraploid and they referred such plants to *M. glutinosa* M. Bieb. However, Lesins and Lesins suggested that both chromosome numbers might occur in both taxa, they did not indicate a clear geographical range for them, and they expressed uncertainty regarding related taxa that might be part of the same "species". Lesins and Lesins distinguished these two "species" by pod size, degree of coiling, and size of opening in the coil. However, they examined very few populations and I have been unable to separate diploid and tetraploid glandular *M. sativa* by using these characters — even after examining the Lesins' own herbarium vouchers at ALTA. Pollen diameter does tend to distinguish the ploidy levels (Small 1983), but only imperfectly. Pollen diameter suggests that *M. sativa* subsp. *glomerata* of North Africa is tetraploid, like the Caucasian plants. With our current knowledge, a taxonomic

Table 12-2. Approximate equivalents of names of taxa in the *Medicago sativa* complex used by Lesins and Lesins (1979) and names used in this volume.

Lesins and Lesins (1979)	This volume
M. sativa subsp. *sativa*	*M. sativa* subsp. *sativa*
M. sativa subsp. *caerulea*	*M. sativa* subsp. *caerulea*
M. falcata	*M. sativa* subsp. *falcata* var. *falcata*
Not recognized as a separate taxon	*M. sativa* subsp. *falcata* var. *viscosa*
Not recognized as a separate taxon	*M. sativa* subsp. ×*varia*
M. glomerata	*M. sativa* subsp. *glomerata* (diploid phase)
M. glutinosa	*M. sativa* subsp. *glomerata* (tetraploid phase)

partition of these glandular-podded plants on the basis of ploidy and/or morphology is not justified.

Brief Summary of Infraspecific Taxa of *Medicago sativa* Recognized in this Volume

1. Groups lacking glandular hairs on the fruits

Medicago sativa subsp. *sativa*

This tetraploid ($2n = 2x = 32$) taxon, defined by purple flowers and coiled fruits, includes most cultivated forms of alfalfa and escapes from cultivation. Many of the plants assignable here have been introgressed from subsp. *falcata*, but not sufficiently to warrant recognition as the hybrid taxon subsp. ×*varia*, discussed below.

Medicago sativa subsp. *caerulea* (Less. ex Ledeb.) Schmalh.

This diploid ($2n = 2x = 16$) taxon is similar to subsp. *sativa* in having decidedly coiled fruits and violet flowers, but is usually distinguishable from the latter by a stronger bluish tint to the flowers, smaller flowers, smaller fruits, smaller seeds, usually narrower leaflets, and a more delicate appearance. This is the very rarely cultivated wild ancestor that gave rise by autopolyploidy to the comparatively more vigorous subsp. *sativa*.

Medicago sativa subsp. *falcata* (L.) Arcang. var. *falcata*

This variety, defined by noncoiled fruits, yellow flowers, and a lack of glandular hairs, is very widespread in northern Eurasia. It has given rise to a few domesticated forms and is much used for breeding alfalfa resistant to cold, acid soils, and disease. It is both diploid ($2n = 16$) and tetraploid ($2n = 32$), but the ploidy classes cannot be reliably distinguished morphologically. This very variable taxon has been extensively and unjustifiably split into formal taxa.

Medicago sativa subsp. ×*varia* (T. Martyn) Arcang.

This taxon is widespread in and outside of cultivation. It is the result of hybridization of subsp. *falcata* with either subsp. *sativa* or subsp. *caerulea*. Most plants of this class are not F_1s, but the result of local backcrossing that has produced introgressants showing intermediacy in fruit characters (i.e., neither notably coiled nor more or less straight) and flower color (i.e., neither completely yellow nor completely violet, but usually a mixture or variegation of these colors). F_1 plants are generally recognizable by the presence of greenish flowers, the result of a blending of the yellow pigment of subsp. *falcata* and the violet color pigments of either subsp. *sativa* or subsp. *caerulea*. Stabilized wild introgressant populations, which are very common and widespread, have been commonly and unjustifiably recognized as taxa, and have greatly burdened the taxonomic literature. Since several of the diagnostic characters segregate and combine more or less independently, precise delimitation from the parental species can be difficult.

2. Groups with glandular hairs on the fruits

These groups occur in montane areas of the Caucasus and adjacent former USSR countries, and in mountains of Europe and Africa. This geographically restricted occurrence of fruits with glandular hairs in a species that is otherwise very widespread has been explained as the plants having evolved antiherbivorous hairs to discourage local insects (Small 1985*b*).

Medicago sativa subsp. *falcata* var. *viscosa* (Rchb.) Posp.

This yellow-flowered variety is defined by noncoiled fruit with glandular hairs. Gland-tipped trichomes are often also on the calyxes and sometimes on twigs and young leaves. As in subsp. *glomerata*, the glandular trichomes may have evolved to protect the fruits against local insects.

Grossheim (1919) described a form of *M. sativa* subsp. *falcata* that possesses high concentrations of glandular trichomes on the whole shoot, not just on the reproductive parts. This form, described as localized to the south coast of the Crimea, was given the name *M. falcata* var. *viscosissima* by Grossheim. Although it is not taxonomically distinguished here (and is assigned to the synonymy of *M. sativa* subsp. *falcata* var. *viscosa*), it is worth pointing out for its possible agronomic significance. In wild plants of the *M. sativa* complex, glandular hairs rarely occur in substantial concentrations on mature vegetative parts of the plant; by contrast, in several annual *Medicago* species, the leaves and stems are covered by glandular hairs and have proven to be insect resistant because of this (Small 1985*b*). This exceptionally glandular form of *M. sativa* could therefore be useful in alfalfa breeding.

Medicago sativa subsp. *glomerata* (Balbis) Rouy

This group includes yellow-flowered plants with coiled fruits. The fruits and sometimes also the calyxes are covered with gland-tipped trichomes. The glandular hairs may have evolved to repel a specialized insect predator (Small 1985*b*). As noted above, Lesins and Lesins (1979) alleged that diploid plants of Europe west of the former USSR constitute a species, *M. glomerata*, in contrast to a similar tetraploid species of the former USSR, *M. glutinosa*, but insufficient evidence for this was provided and the two are combined under the name subsp. *glomerata* in this chapter.

Medicago sativa subsp. *sativa* × subsp. *glomerata*

Hybrid swarms between subsp. *sativa* and subsp. *glomerata* are characterized by segregating flower colors (violet, yellow, and variegated) and the presence of a few gland-tipped trichomes on the fruits (but in much lower concentrations than found in subsp. *glomerata*). This group is not recognized taxonomically here (the most common scientific name applied to it is *M. polychroa*).

Table 12-3. Summary of some features of the taxa of *Medicago sativa*.

Taxon	Ploidy	Flower color	Pod coiling	Glandular hairs on pod
M. sativa subsp. *caerulea*	$2n = 16$	Violet (blue-violet)	Strongly coiled	No
M. sativa subsp. *sativa*	$2n = 32$	Violet	Strongly coiled	No
M. sativa subsp. ×*varia*	$2n = 16, 32$	Variegated (yellow-violet)	Weakly coiled	No
M. sativa subsp. *falcata* var. *falcata*	$2n = 16, 32$	Yellow	Not coiled	No
M. sativa subsp. *falcata* var. *viscosa*	$2n = 16, 32$	Yellow	Not coiled	Yes
M. sativa subsp. *glomerata*	$2n = 16, 32$	Yellow	Strongly coiled	Yes

Keys to Subordinate Taxa of *Medicago sativa*

A. Comprehensive key

1. Fruit without gland-tipped trichomes 2
   2. Flowers uniformly violet or blue-violet; some of fruits with at least 1.5 coils (usually more) 3
      3. Calyx (base to tip of longest lobe) usually less than 4.5 mm long;fruit intersuture width usually less than 1.8 mm........ subsp. *caerulea*
      3. Calyx length usually more than 4.5 mm; midfruit intersuture width usually more than 1.8 m........ subsp. *sativa*
   2. Flowers yellow or yellow-violet variegated; fruits often with less than 1.5 coils 4
      4. Flowers yellow; fruit with less than 1 coil (usually less than 0.5 coil) .... subsp. *falcata* var. *falcata*
      4. Flowers variegated yellow-violet and/or fruits with 0.8–1.4 coils .... subsp. ×*varia*
1. Fruit with at least some gland-tipped trichomes 5
   5. Flowers uniformly yellow; fruit densely covered with gland-tipped trichomes 6
      6. Fruit with less than 1 coil (usually less than 0.5 coil) .... subsp. *falcata* var. *viscosa*
      6. Fruit with more than 1 coil .... subsp. *glomerata*
   5. Flowers variegated yellow-violet; fruit lightly to moderately covered with gland-tipped trichomes .... *M. sativa* subsp. *sativa* × subsp. *glomerata*

B. Key to plants growing outside of Eurasia

Beyond the native Eurasian area of *M. sativa*, plants with glandular-haired fruit almost never grow outside of cultivation. Also, subsp. *caerulea* and hybrids of this and subsp. *falcata* are virtually absent, so the variation pattern of subsp. ×*varia* is simpler. Moreover, *M. sativa* subsp. *sativa* usually shows a syndrome of characteristics typical of domesticated plants, while *M. sativa* subsp. *falcata* tends to show characteristics more typical of wild plants, as illustrated in Fig. 12-1.

The following multiple-entry pictorial key is based on discriminant analysis of the best characters (flower color, growth habit, and degree of fruit coiling) that were found for separating subsp. *sativa*, subsp. *falcata*, and their hybrid subsp. ×*varia* (after Small and Brookes 1984*b*), based on North American material. This key is only appropriate for areas of the world outside of Eurasia. While flower color and degree of fruit coiling are equally valuable discriminating characters in and outside of Eurasia, plant growth habit is another matter. Outside of Eurasia, the discriminating value of the degree that the plant habit is erect lies in the fact that domestication has resulted in the dominance of erect cultivars in subsp. *sativa* whereas, by contrast, wild plants (most plants of subsp. *falcata* are wild) are much less erect.

Fig. 12-1. A comparison of flowering *Medicago sativa* subspecies *sativa* (left) and *falcata* (right). From Hallier (1885, vol. 23, plates 2334 and 2335, respectively). Notice the narrower leaflets in subsp. *falcata*. Generally in *M. sativa*, wild plants have narrower leaflets than domesticated plants. There are a few domesticated forms of subsp. *falcata* with wider leaflets, and there are wild forms of subsp. *sativa* with quite narrow leaflets. However, at least outside of Eurasia, the vast majority of plants of subsp. *sativa* have wider leaflets than the vast majority of plants of subsp. *falcata*.

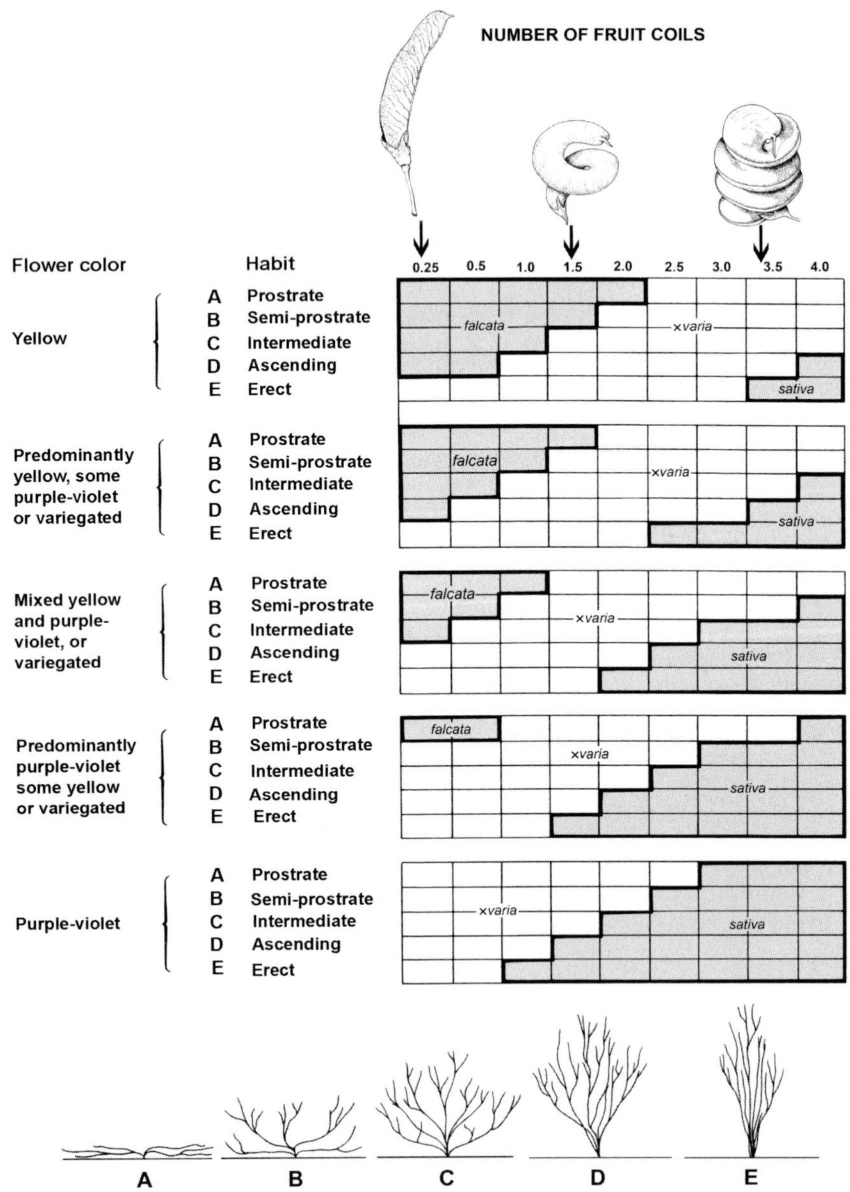

Fig. 12-2. Multiple-entry key to subspecies of the *Medicago sativa* – *M. falcata* continuum. These subspecies interbreed freely, generating recombinants with wide ranges of trait expression; clusters of traits are needed for correct subspecies assignment.

The groups discussed in the preceding are illustrated in Figures 12-3 to 12-16; in most cases vegetative and flowering organs were drawn from the same plants. *Medicago sativa* subsp. *falcata* var. *falcata* is notoriously variable, and three plates of illustrations are presented, each drawn from a different plant representative of this variability.

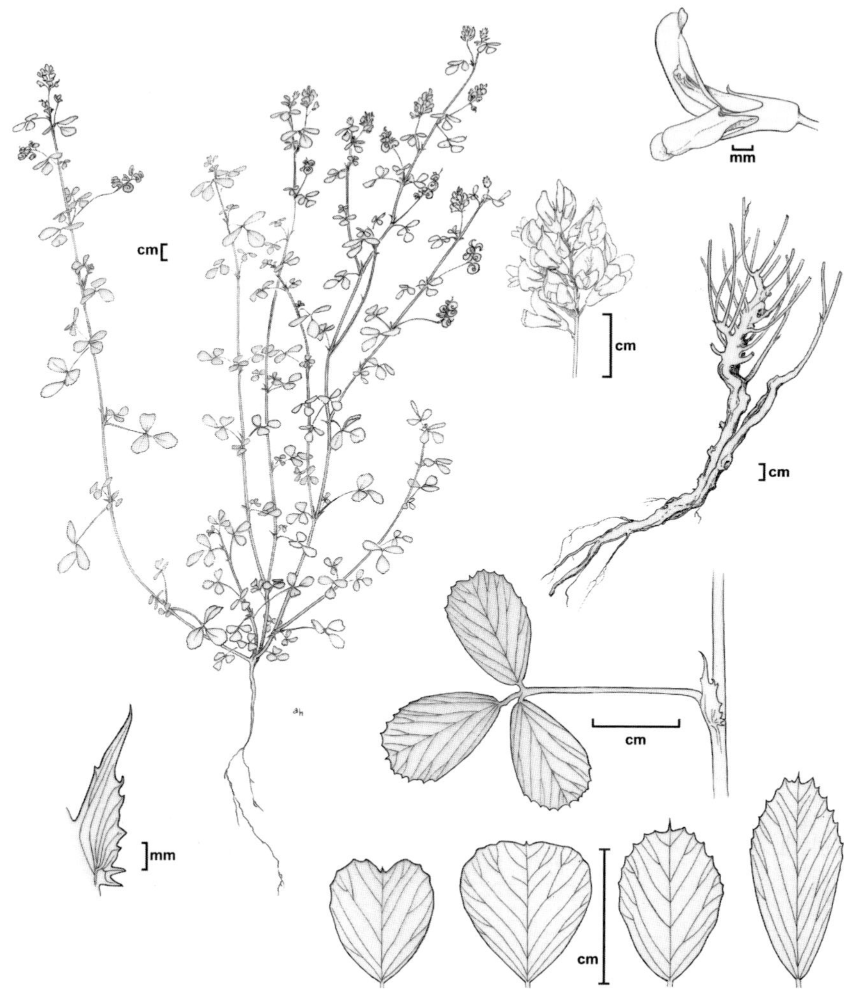

Fig. 12-3. *Medicago sativa* subsp. *sativa*. **Clockwise from top left:** Habit drawing of plant; inflorescence with peduncle; side view of flower; crown; leaf at stem node; leaflets; stipule. (Based on <u>E. Small S223</u>, DAO, except the crown, which is based on the University of Nanking 2488. Artist: A. Hanes.)

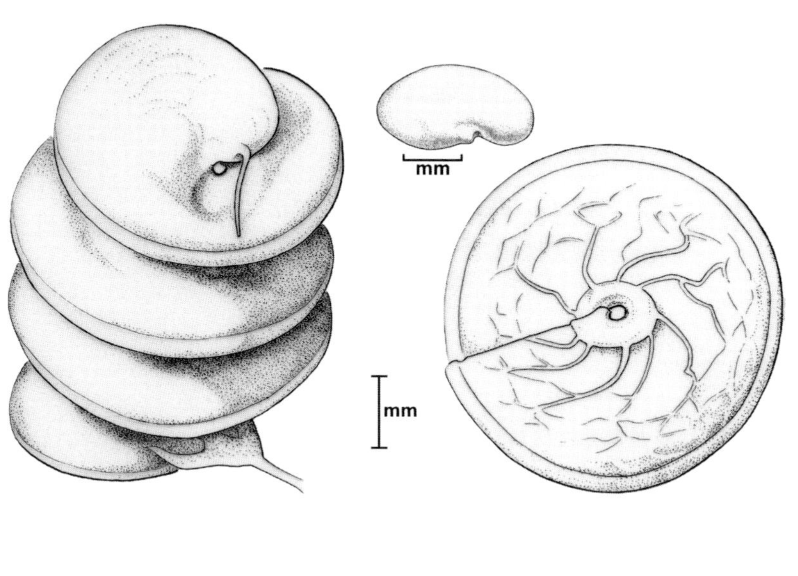

Fig. 12-4. *Medicago sativa* subsp. *sativa*. **Left:** Fruit. **Center:** Seed. **Right:** Fruit coil face showing venation. (Based on <u>E. Small M378</u>, DAO; artist: A. Hanes.)

Fig. 12-5. *Medicago sativa* subsp. *caerulea*. **Clockwise from top left**: Habit drawing of plant; stipule; infloresecence with peduncle; leaf and stipules at stem node. (Based on E. Small Turkey-106, DAO; artist: L. Yuzyk.)

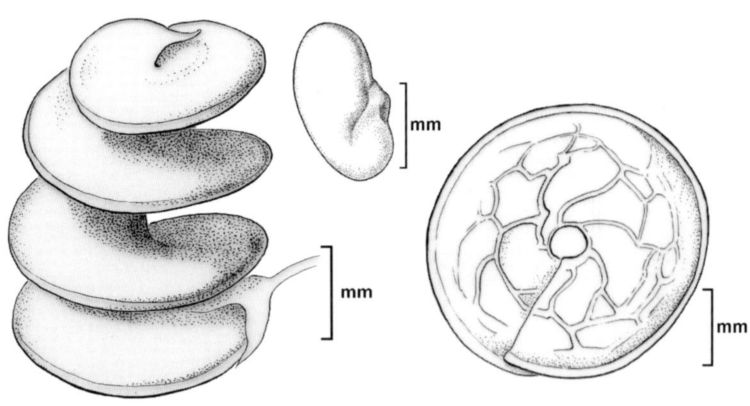

Fig. 12-6. *Medicago sativa* subsp. *caerulea*. **Left:** Fruit. **Center:** Seed. **Right:** Fruit coil face showing venation. (Based on E. Small S139, DAO; artist: A. Hanes.)

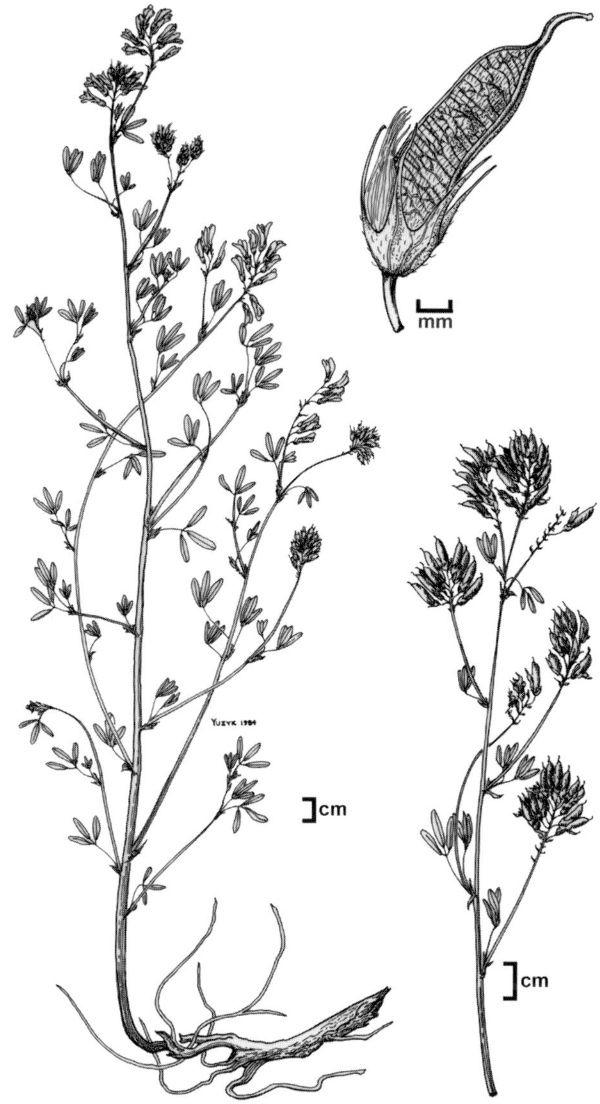

Fig. 12-7. *Medicago sativa* subsp. *falcata* var. *falcata*. Example #1. **Left:** Habit drawing of flowering plant. **Upper right:** Fruit. **Lower right:** Fruiting branch. (The flowering plant based on Anon. 595 (The Shuntuk Valley, 04 July 1931), WIR; fruiting branch and fruit based on Borkovskaya 1737, WIR; specimens identified as *Medicago quasifalcata*; artist: L. Yuzyk.)

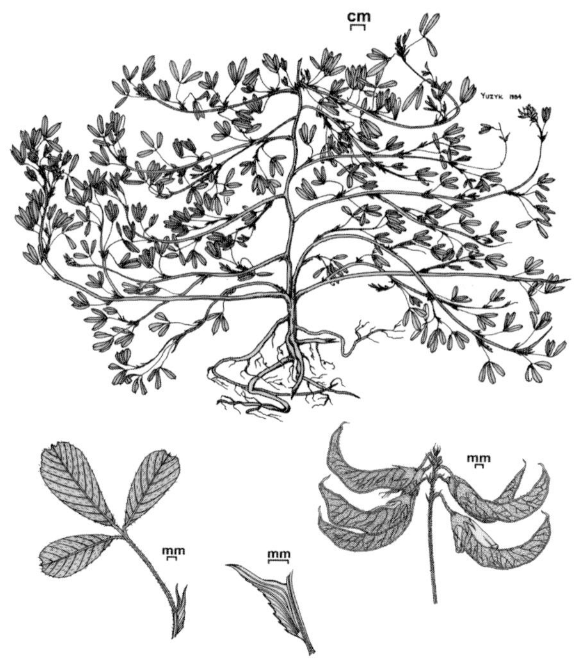

Fig. 12-8. *Medicago sativa* subsp. *falcata* var. *falcata*. Example #2. **Clockwise from top:** Habit drawing of flowering plant; infructescence; stipule; leaf with stipules at the base. (Based on Voskan'yan 21264a, WIR; specimens identified as *Medicago borealis*; artist: L. Yuzyk.)

Fig. 12-9. *Medicago sativa* subsp. *falcata* var. *falcata*. Example #3. **Left:** Flowering branch. **Upper right:** Fruit. **Lower right:** Fruiting branch. (Fruiting branch and fruit based on Ivanov 1364, WIR; flowering branch based on Anon 603 (Zonal Nursery of Lodi), WIR; specimens identified as *Medicago difalcata* Sinskaya; artist: L. Yuzyk).

Fig. 12-10. *Medicago sativa* subsp. *falcata* var. *viscosa*. **Clockwise from left:** Branch with flowers and some fruit; side view of flower; stipule; leaf attached to stem node. (Based on Berkovskaya 2572, WIR; artist: L. Yuzyk.)

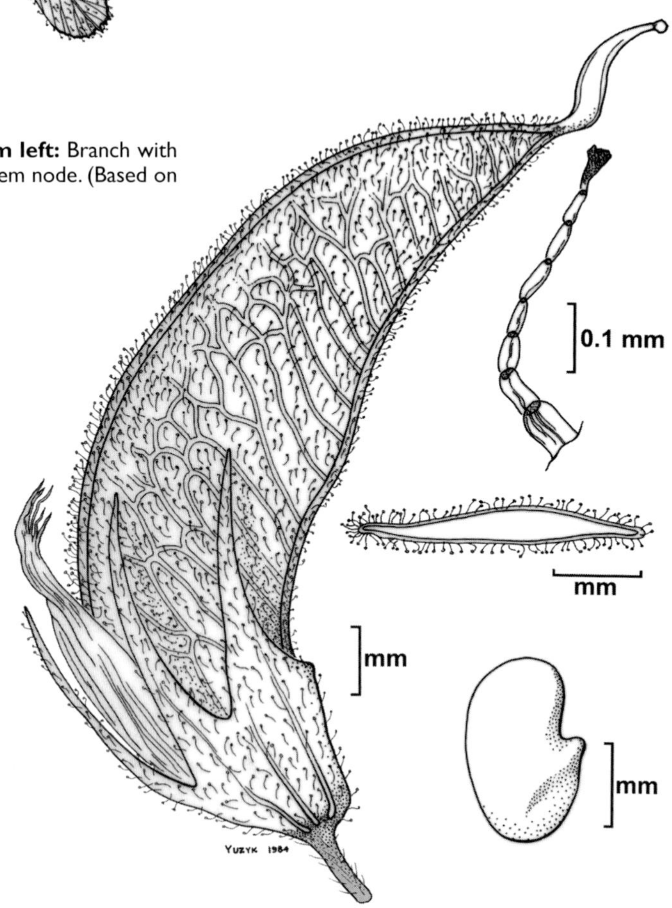

Fig. 12-11. *Medicago sativa* subsp. *falcata* var. *viscosa*. **Clockwise from left:** Fruit; gland-tipped hair; outline of cross section of fruit; seed. (Based on Berkovskaya 2572, WIR; artist: L. Yuzyk.)

Fig. 12-12. *Medicago sativa* subsp. *glomerata.* **Clockwise from left:** Habit drawing of branch; inflorescence; stipule; leaf attached to stem node. (Based on E. Small M275, DAO; artist: L. Yuzyk.)

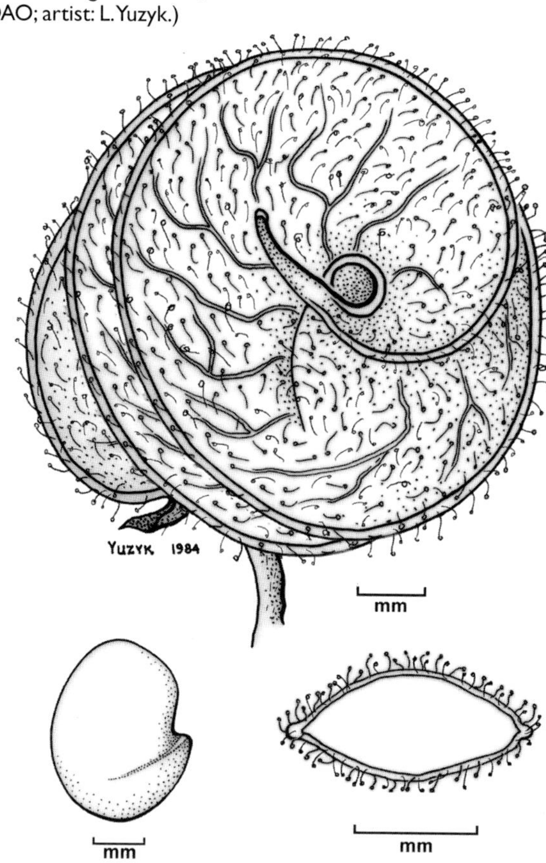

Fig. 12-13. *Medicago sativa* subsp. *glomerata.* **Clockwise from top:** Young fruit; outline of cross section of fruit coil; seed. (Based on E. Small M275, DAO; artist: L. Yuzyk.)

Fig. 12-14. *Medicago sativa* subsp *sativa* × *M. sativa* subsp. *glomerata.* **Clockwise from left:** Habit drawing of plant; inflorescence with peduncle; stipule; leaf with stipules at base of petiole. (Based on Yu. D. Soskov & A. Stankevitch 577, WIR; artist: L. Yuzyk.)

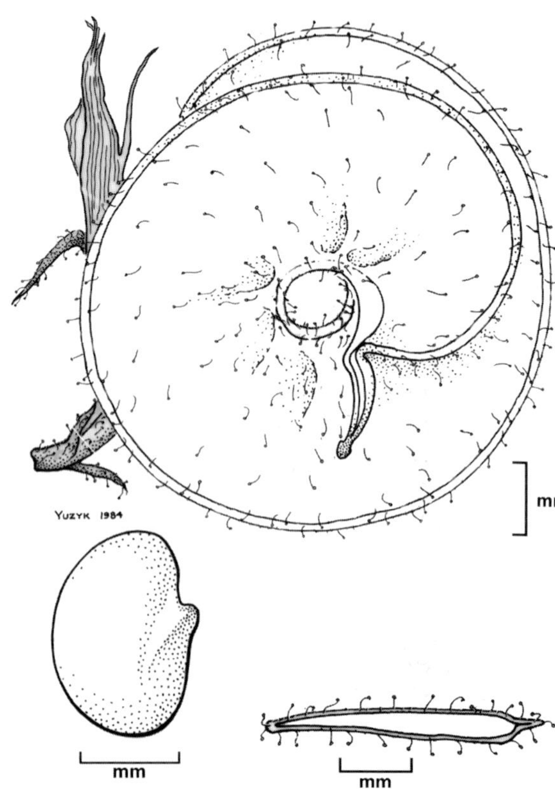

Fig. 12-15. *Medicago sativa* subsp *sativa* × *M. sativa* subsp. *glomerata.* **Clockwise from top:** Fruit; outline of cross section of fruit coil; seed. (Based on Yu. D. Soskov & A. Stankevitch 577, WIR; artist: L. Yuzyk.)

Excessive Taxonomic Splitting of *Medicago sativa* by some Russian Taxonomists

Medicago sativa has been treated by Russian taxonomists (and some alfalfa specialists who have adopted their taxonomies) as constituting dozens of species or infraspecific taxa (Grossheim 1919, 1945; Sumnevich 1932; Vassilczenko 1949; Sinskaya 1948, 1950; Lubenets 1972; Ivanov 1977, 1980). This is a reflection of the well known tendency of the classical Russian school of taxonomy to inflate taxonomic categories. It also reflects the tendency of the classical Russian school of phytosociology (primarily represented in this instance by E.N. Sinskaya) to utilize quasi-Latin nomenclature to label what are, at best, ecotypes. Regrettably, this has burdened the taxonomic literature with unjustified taxa and has severely misled alfalfa specialists. This rejection of the old Russian approach to the taxonomy of *M. sativa* has been made before (particularly see Lesins and Lesins 1979). However, I wish to make it clear that my criticism by no means invalidates much of the work referred to above as a guide to interesting variants; indeed, I have found the Russian publications cited to be very useful sources of information. Biological classification is in part an "art" that requires taxonomists to decide just how much formal classification is useful before users of the classification are misled or confused (see below and Figs. 12-16a to 12-16d for illustrations of the unacceptable extent to which "species" have been delimited.)

Extensive hybridization in the *M. sativa* complex has generated the localized formation of semi-stabilized introgressed or stabilized segregant populations in Eurasia, and the formal recognition of these (largely by the Russian school) has unnecessarily complicated the taxonomy of the species. However, such recognition is well outside the standards adopted by most modern taxonomists and has served to confuse rather than to help science and agriculture. Certainly, particular introgressed or segregant populations may harbor valuable combinations of genes, but as with the vast majority of cultivated plants today, this information is best simply recorded for given accessions, and the ecological and other information noted in scholarly articles.

Figures 12-16a to 12-16d illustrate type specimens of examples of "species" of *Medicago* recognized by Russian taxonomists (mainly I.T. Vassilczenko) that are referable to *M. sativa*. These do not merit taxonomic status as species (indeed, there is insufficient evidence of distinction at any taxonomic level), and they serve merely to illustrate the extensive confusion associated with attempting to impose taxonomic categories upon the patterns of hybridization, segregation, and introgression among the basic infraspecific taxa of *M. sativa* recognized in this monograph. (Photographs courtesy of LE.)

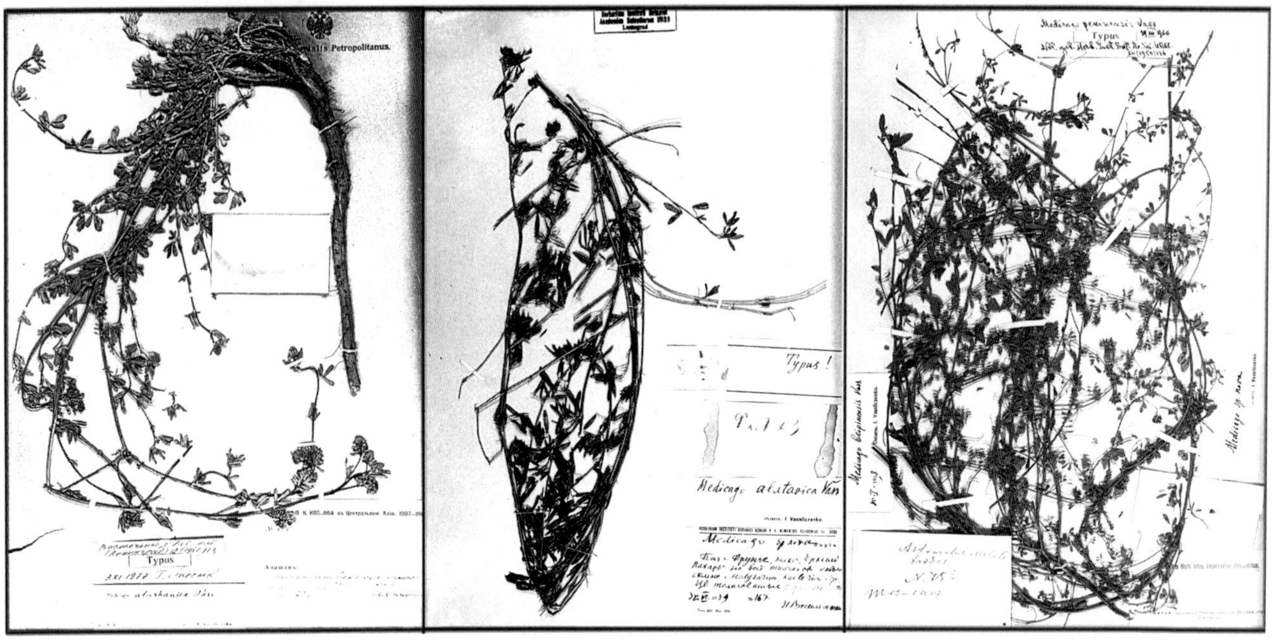

M. alaschanica Vassilcz *M. alatavica* Vassilcz. *M. beipinensis* Vassilcz.

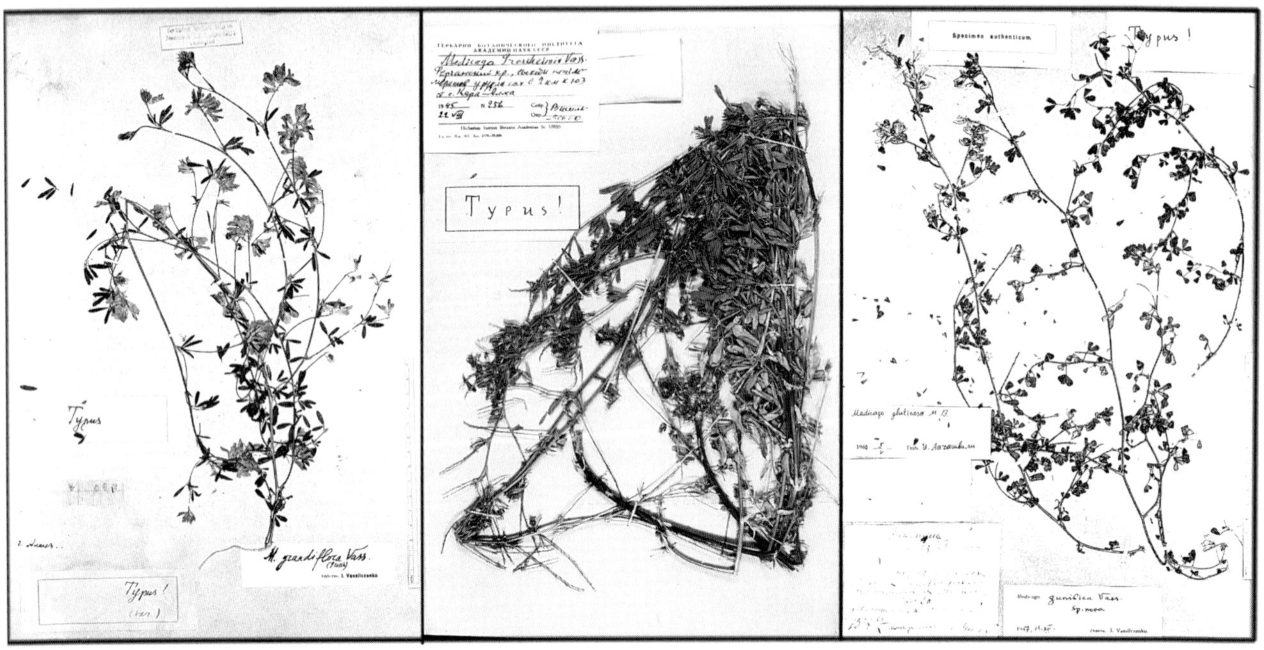

M. grandiflora (Grossh.) Vassilcz. *M. grossheimii* Vassilcz. *M. gunibica* Vassilcz.

Fig. 12-16a. Type specimens of "species" of *Medicago* recognized by the classical Russian school of taxonomy. None deserve recognition.

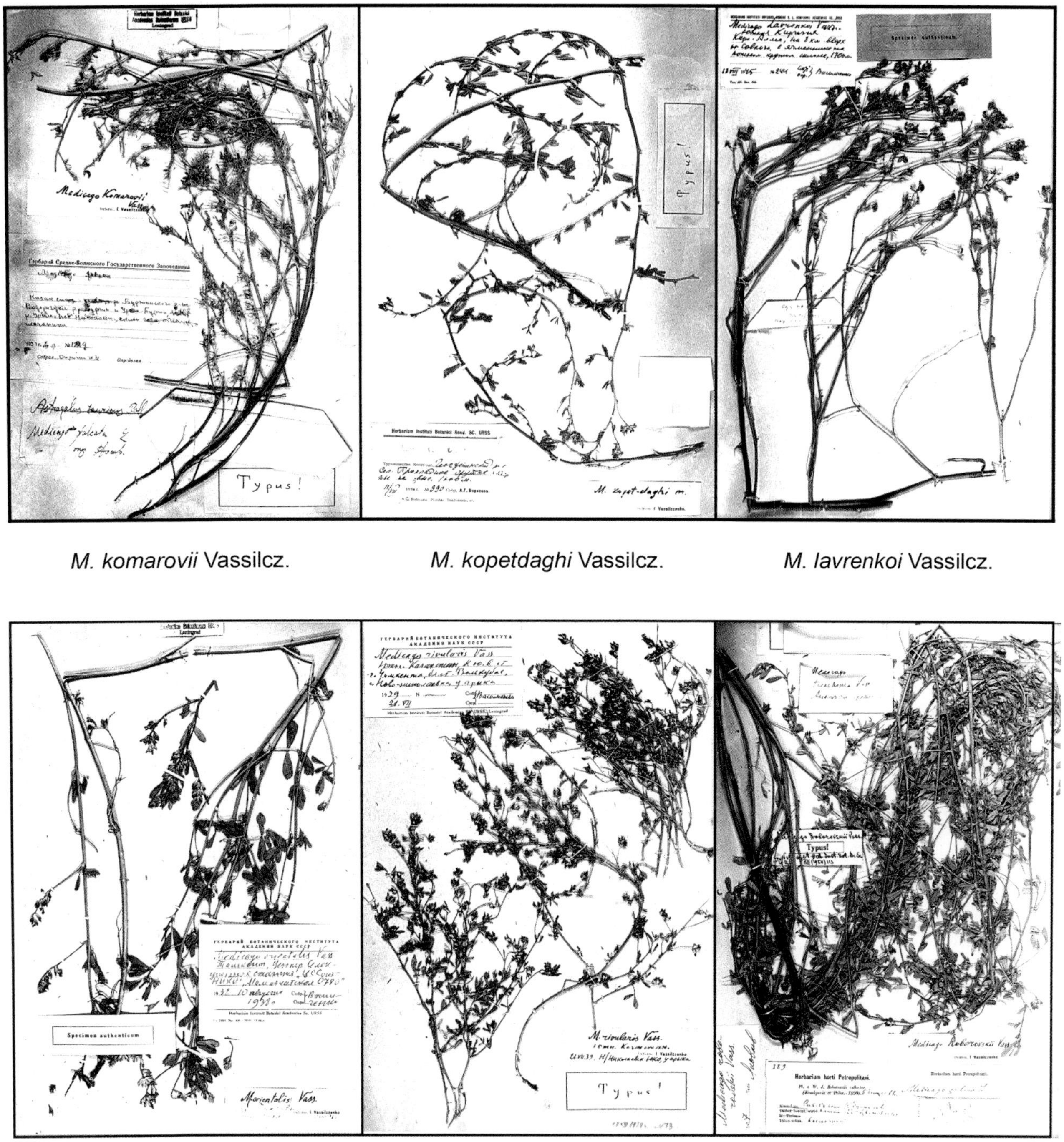

M. komarovii Vassilcz. *M. kopetdaghi* Vassilcz. *M. lavrenkoi* Vassilcz.

M. orientalis Vassilcz. *M. rivularis* Vassilcz. *M. roborovskii* Vassilcz.

Fig. 12-16b. Type specimens of "species" of *Medicago* recognized by the classical Russian school of taxonomy.

M. sogdiana Vassilcz. *M. tadzhicorum* Vassilcz. *M. tibetana* (Alef.) Vassilcz.

M. transcaucasica Vassilcz. in herb. *M. transoxana* Vassilcz. *M. vardanis* Vassilcz.

Fig. 12-16c. Type specimens of "species" of *Medicago* recognized by the classical Russian school of taxonomy.

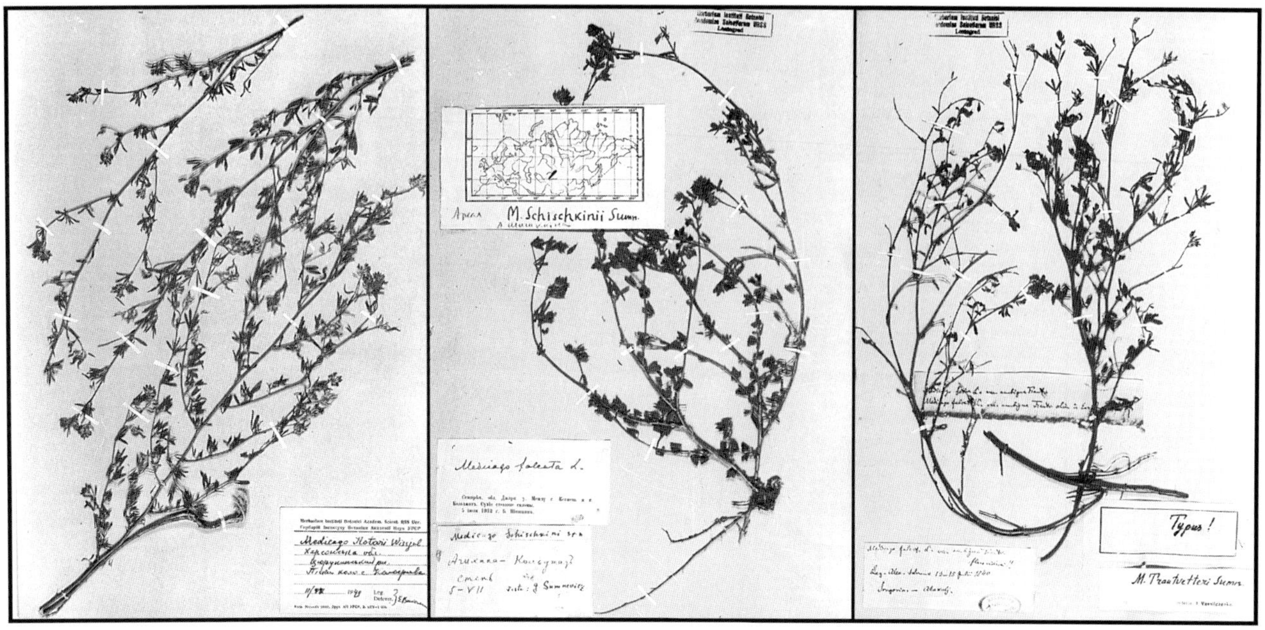

M. kotovii O.D.Wissjul. *M. schischkinii* Sumnev. *M. trautvetterii* Sumnev.

Fig. 12-16d. Type specimens of "species" of *Medicago* recognized by the classical Russian school of taxonomy.

Type Specimens (for taxa recognized in this monograph)

Medicago sativa L.

Heyn (1959: 162), in discussing the typification of *M. sativa*, stated that "the figure of Morison, tab. 16 f.2 (1680) cited by Linnaeus in the Sp. Pl. and bearing the same epithet is the most strongly connected with the species." This treatment is accepted in The Linnaean Plant Name Typification Project (http://www.nhm.ac.uk/research-curation/projects/linnaean-typification/; current June 2010; cf. Jarvis 2007): Type conserved. "Habitat in Hispaniae, Galliae apricis." Generitype of *Medicago* Linnaeus, nom. cons. Lectotype = "Medica falcata sativa siliqua cornuta magis tortili flore violaceo" in Morison, Pl. Hist. Univ., 2: 158, s. 2, t. 16, f. 2 (1680), designated by Heyn in Bull. Res. Council Israel, Sect. D, Bot. 7: 162 (1959).

Medicago sativa L. subsp. *caerulea* (Less. ex Ledeb.) Schmalh.

In his description of *M. caerulea*, Ledebour cited only "Lessing in herb. reg. berol. (Berlin)" and stated "Vidi tantum specimen unicum mancum in herb. supra indicato." Vassilczenko (1949) stated (translated from Russian) "Described from the plain adjoining the Caspian Sea, preserved in Berlin." However, R. Vogt, curator at B, informed me that "There is no material kept under the name *Medicago caerulea* at B. If the type material of this name has ever been part of our collections then it is very probable that it got lost in the fire of the Botanical Museum in WWII." Flora USSR states "Type in Leningrad". Heyn and Davis (1970) stated: "Holotype: (USSR) ad m. caspium, *Lessing* (LE)." I did not locate the type during a visit to LE and have not received a reply to an inquiry regarding the type. At this time, there is uncertainty regarding the type.

Medicago sativa L. subsp. *falcata* (L.) Arcang.

The Linnaean Plant Name Typification Project (cf. Jarvis 2007) (http://www.nhm.ac.uk/research-curation/projects/linnaean-typification/; current June 2010) provides the following information: "Provenance: 'Habitat in Europae pratis apricis, siccis.' Type: Lectotype : Herb. Linn. No. 933.8 (LINN) Designated by: Ali in *Taxon* 17: 541 (1968)."

Medicago sativa L. subsp. *xvaria* (Martyn) Arcang.

M. xvaria T. Martyn, Flora Rustica 3: 87 (1793), (basionym).

Martyn did not cite collections in his original description, but stated "Caspar Bauhin says that it is found in the south of France... Mr Ray observed it... between Norwich and Lynn." Vassilczenko (1949) stated (translated from Russian) that it is "Described from England...; preserved in London." However, the curators at BM, CGE, K, and OXF, the most likely places where a type might be located, informed me in 2007 that no collection that could be a candidate for the type is in their herbaria. In the apparent absence of a suitable specimen, the drawing in Martyn's publication (Fig. 12-17) is here designated as the lectotype.

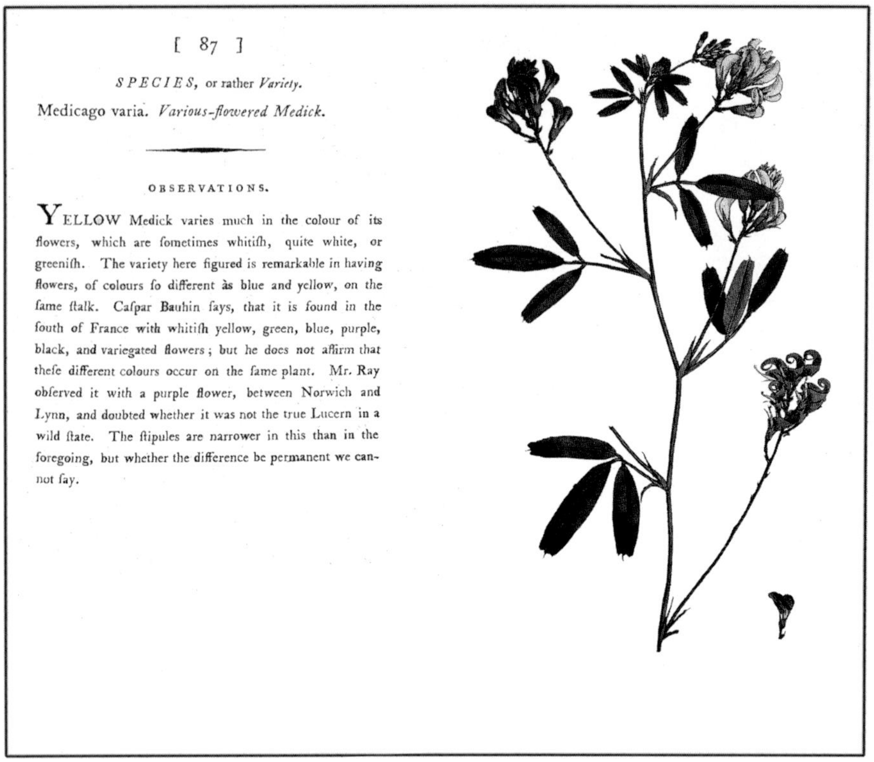

[87]

SPECIES, or rather *Variety*.

Medicago varia. *Various-flowered Medick.*

OBSERVATIONS.

YELLOW Medick varies much in the colour of its flowers, which are ſometimes whitiſh, quite white, or greeniſh. The variety here figured is remarkable in having flowers, of colours ſo different as blue and yellow, on the ſame ſtalk. Caſpar Bauhin ſays, that it is found in the ſouth of France with whitiſh yellow, green, blue, purple, black, and variegated flowers; but he does not affirm that theſe different colours occur on the ſame plant. Mr. Ray obſerved it with a purple flower, between Norwich and Lynn, and doubted whether it was not the true Lucern in a wild ſtate. The ſtipules are narrower in this than in the foregoing, but whether the difference be permanent we cannot ſay.

Fig. 12-17. An illustration of *M. sativa* subsp. ×*varia* and corresponding text from Thomas Martyn's *Flora Rustica*. The plate was hand-colored, with the inflorescence at the upper left in purple and the two inflorescences at the upper right in yellow. From the Missouri Botanical Garden copy of the book; scans courtesy of L. Fisher.

***Medicago sativa* L. subsp. *glomerata* (Balbis) Rouy**

The lectotype of *M. glomerata* Balb. is at TO (Dal Vesco et al. 1987; illustration on page 13 of this publication; the fruits have 25–30 glandular hairs/mm^2 (G. Forneris, personal communication)).

***Medicago sativa* L. subsp. *falcata* (L.) Arcang. var. *viscosa* (Rchb.) Posp.**

M. procumbens Besser (var.) β *viscosa* Rchb., Fl. Germ. Excurs. 504 (1832) (basionym). In this, Reichenbach states with regard to var. β (i.e., *viscosa*): "In Suchsen bei Dresden: R."

A complete citation for this taxon is *Medicago sativa* L. subsp. *falcata* (L.) Arcang. var. *viscosa* (Rchb.) Posp., Fl. Österr. Küsten. 2(2): 357 (1899). Here, Pospichal cites "Rchb. Fl. germ. exc. nr. 3214" and this could constitute the type. Reichenbach's exsiccatae are in many herbaria. His original collections at DR were substantially lost to a fire in 1849 and to the destruction of Dresden during World War II. Parts of his collections are at W, and a specimen there would appropriately be designated a lectotype.

Synonymy

Homotypic

Medicago sativa* L. subsp. *sativa

M. sativa L., Sp. Pl. 2: 778 (1753) (basionym).

Medica sativa (L.) Miller, Gard. Dict., ed. 8: no 1 (1768).

Medica sativa (L.) Lam., Fl. Franc. 2: 584 ("1778," 1779), nom. superfl.

***Medicago sativa* L. subsp. *caerulea* (Less. ex Ledeb.) Schmalh.**

M. caerulea Less. ex Ledeb., Fl. Ross. 1: 526 (1842), (basionym).

M. coerulea Less. ex Nyman, Consp. Fl. Eur. 166 (1878).

M. sativa L. subsp. *microcarpa* Urb. var. *caerulea* (Less. ex Ledeb.) Urb., Verh. Bot. Ver. Prov. Brandenb. 15: 57 (1873).

M. sativa L. subsp. *caerulea* (Less. ex Ledeb.) Schmalh., Fl. Sred. Juz. Ross. 1: 226 (1895).

M. sativa L. B. *microcarpa* II *caerulea* Aschers. & Graebn., Syn. VI(2): 403 (1906–1910).

Note: Urban's (1873) taxon *M. sativa* L. subsp. *microcarpa* Urb. is made up of three elements: var. *pauciflora* Urb. (= subsp. ×*varia*), var. *pilifera* Urb. (= subsp. *glomerata*), and var. *caerulea* (Less. ex Ledeb.) Urb. Urban's types were destroyed during World War II and his illustrations are hardly suitable as lectotypes. While one element of subsp. *microcarpa*, i.e., var. *caerulea*, could be made to represent subsp. *microcarpa*, and therefore the name subsp. *microcarpa* would take precedence over the name subsp. *caerulea*, this would be a pointless exercise.

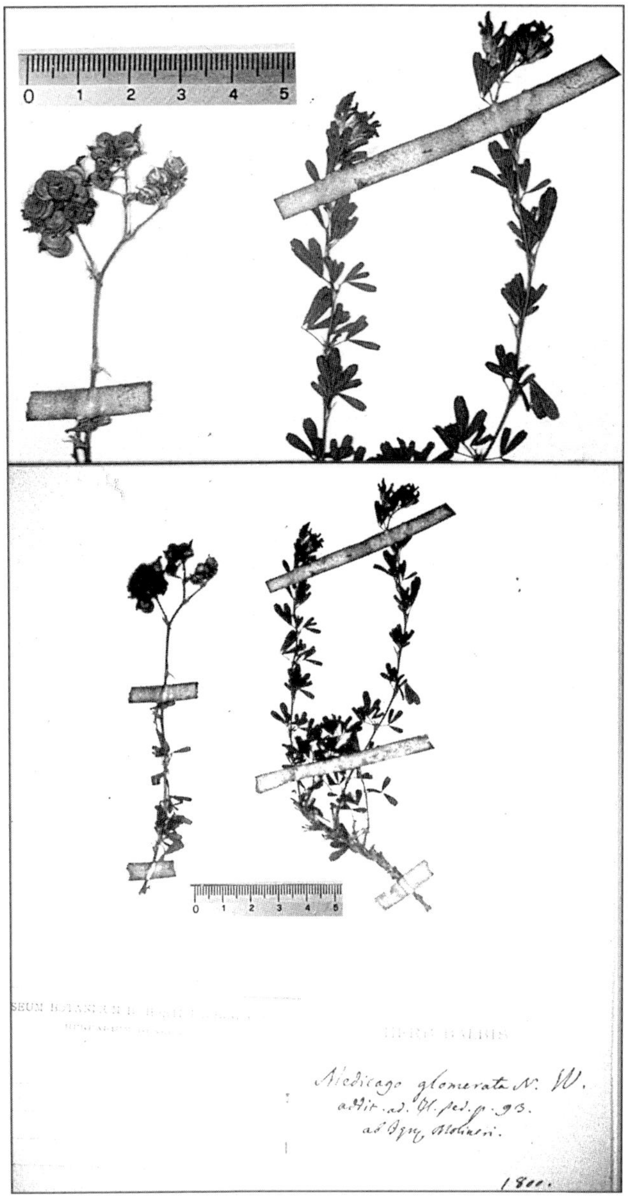

Fig. 12-18. Photograph of type specimen of *Medicago glomerata* Balbis (portion enlarged at top).

***Medicago sativa* L. subsp. *glomerata* (Balbis) Rouy**

M. glomerata Balb., Elenco 93 (1801), (basionym).

M. sativa L. subsp. *macrocarpa* Urb. var. *glandulosa* (Koch) Alef. f. *glomerata* (Balbis) Urb., Verh. Bot. Ver. Prov. Brandenb. 15: 56 (1873).

M. glutinosa M. Bieb. subsp. *glomerata* (Balb.) Nyman, Consp. Fl. Eur. 166 (1878).

(Nomenclaturally, the combination applies here; the name is also listed below in *M. sativa* subsp. *sativa* × subsp. *glomerata*, but only to point out that the circumscription has been applied more broadly. Similarly, *M. glutinosa* M. Bieb. var. *glomerata* Grossh. circumscriptionally has been applied to *M. sativa* subsp. *sativa* × subsp. *glomerata*.)

M. sativa L. var. *glomerata* (Balb.) Arcang., Comp. Fl. Ital. 160 (1882).

M. sativa L. *glandulosa* b. *glomerata* (Balb.) Arcang., Fl. Ital. 485 (1894).

M. sativa L. A. *macrocarpa* B. *glandulosa* II *glomerata* Aschers. & Graebn., Syn. 6(2): 400 (1906–1910).

M. sativa L. subsp. *M. glomerata* (Balb.) Rouy in Rouy & Fouc., Fl. France 5: 14 (1899). (Subspecific intention clear despite notation "subspec. – *M. glomerata* Balb.")

M. polychroa Grossh. var. *glomerata* (Balb.) Grossh., Novye dannye k poznaniyu kavkazskikh lyutsern (New data for studies on Caucasian alfalfa) 3 (1925).

M. falcata L. subsp. *glomerata* (Balbis) Soják, Cas. Nár. Muz., 150 (3-4): 138 (1982).

Not

M. glomerata Fisch. ex Moris, Mem. Acad. Sci. Torino 36: 190 (1833) (= *M. brachycarpa*).

Medicago sativa* L. subsp. *falcata* (L.) Arcang. var. *falcata

M. falcata L., Sp. Pl. 779 (1753), (basionym).

Medica falcata (L.) Miller, Gard. Dict., ed. 8 : no. 2 (1768).

M. falcata Lam., Encycl. 3(2): 639 (1792), nom. superfl.

M. sativa L. var. *falcata* (L.) Döll, Rhein. Fl. 802 (1843).

M. sativa L. subsp. *macrocarpa* Urb. var. *falcata* (L.) Döll, Verh. Bot. Ver. Prov. Brandenb. 14: 56 (1873).

M. sativa L. subsp. *falcata* (L.) Arcang., Comp. Fl. Ital. 160 (1882).

***Medicago sativa* L. subsp. *falcata* (L.) Arcang. var. *viscosa* (Rchb.) Posp.**

M. procumbens Besser (var.) β *viscosa* Rchb., Fl. Germ. Excurs. 504 (1832), (basionym).

M. sativa L. subsp. *macrocarpa* Urb. var. *falcata* (L.) Döll f. *viscosa* (Rchb.) Urb., Verh. Bot. Ver. Prov. Brandenb. 14: 56 (1873).

M. sativa L. subsp. falcata (L.) Arcang. var. *viscosa* (Rchb.) Posp., Fl. 2(2): 357 (1899).

M. falcata L. subvar. *viscosa* (Rchb.) Rouy in Rouy & Fouc., Fl. France 5: 11 (1899).

M. falcata L. var. *viscosa* (Rchb.) Grossh., Sci. Pap. Tiflis Bot. Gard. Part 1, 21 (1919).

M. romanica Prodán var. *viscosa* (Rchb.) Grossh. in Komarov (ed.), Fl. U.S.S.R. 11: 147 (1945).

M. sativa L. subsp. *viscosa* (Rchb.) C.R. Gunn, U.S. Depart. of Agriculture, Technical Bull. 1574: 18 (1978).

***Medicago sativa* L. subsp. ×*varia* (Martyn) Arcang.**

M. ×*varia* Martyn, Flora Rustica 3: 87 (1793), (basionym).

M. sativa L. subsp. *macrocarpa* Urb. var. *varia* (Martyn) Urb., Verh. Bot. Ver. Prov. Brandenb. 15: 57 (1873).

M. sativa subsp. ×*varia* (T. Martyn) Arcang., Comp. Fl. Ital., ed. 1, 160 (1882).

M. sativa subsp. ×*varia* (Martyn) O. Bolòs & Vigo, Butl. Inst. Catalan Hist. Nat 38: Bot. 1: 70 (1974). (Nom. superfl., as pointed out by Rabeler 1984).

Heterotypic

Medicago sativa* subsp. *sativa

Medica sativa f. *alba Benke*., Amer. Midl. Nat. 16: 424 (1935).

M. sativa L. gr. *afghanica* Bordzil. in Vavilov & Bukinich, Agricultural Afghanistan (Leningrad) 352 (1929).

M. afghanica (Bordzil.) Vassilcz., Bot. Journ. U.S.S.R. 31(3): 28 (1946).

M. sativa L. var. *anatolica* Bordakov, Bull. Appl. Bot. Pl. Breed. 19: 8 (1929).

M. orientalis Vassilcz. subsp. *anatolica* (Bordakov) Sinskaya, Fl. Cult. Pl. U.S.S.R. 13(1): 64 (1950). (Sinskaya recognized three forms based on petal color, none published validly.)

M. praesativa Sinskaya subsp. *armena* Sinskaya, Fl. Cult. Pl. U.S.S.R. 13(1): 52 (1950), nom. invalid.

M. asiatica Sinskaya, Fl. Cult. Pl. U.S.S.R 13(1): 71 (1950).

M. sativa L. var. *austro-europaea* Bordakov Bull. Appl. Bot. Pl. Breed. 19: 8 (1929).

M. sativa L. var. *erecta* Schweinf., Samml. arab.-aethiop. in Bull. Herb. Boiss. 4 (App. 2): 355 (1896).

M. sogdiana Vassilc. var. *chivensis* Vassilc., Acta Inst. Bot. Acad. Sci. URSS, Ser. 1, Fasc. 8, 92 (1949). (Type LE, cited by Vassilczenko 1949 p. 93.)

M. sativa L. gr. *Euasiaticae* Sinskaya, Phylogen. Taxon 411 (1940).

M. jemenensis Sinskaya subsp. *eujemenensis* Sinskaya, Fl. Cult. Pl. U.S.S.R. 13(1): 56 (1950), nom. invalid.

M. eusativa Sinskaya, Fl. Cult. Pl. U.S.S.R. 13(1): 65 (1950), nom. invalid. (Sinskaya recognized two color forms, but did not publish these validly.)

M. asiatica Sinskaya subsp. *geratica* Sinskaya, Fl. Cult. Pl. U.S.S.R. 13(1): 82 (1950). (Sinskaya described three forms based on petal color, none of the names published validly.)

M. sativa L. var. *germanica* Winkl., Cat. Hort. Vinob. (1877).

M. orientalis Vassilcz. subsp. *graeca* (Bordakov) Sinskaya, Fl. Cult. Pl. U.S.S.R. 13(1): 64 (1950).

M. jemenensis Sinskaya, Fl. Cult. Pl. U.S.S.R. 13(1): 55 (1950), nom. invalid. (Sinskaya recognized (but did not validly publish) two subspecies, noted here, and two forms based on petal color.)

M. sativa L. var. *afghanica* Bordakov subvar. *kandagharo-kabulica* Bordakov, Armenian Agr. 113 (1929).

M. asiatica Sinskaya subsp. *kandagharo-kabulica* (Bordakov) Sinskaya, Fl. Cult. Pl. U.S.S.R. 13(1): 79 (1950). (Sinskaya described two forms based on petal color, the names not published validly.)

M. karatschaica Latsch., Not. Syst. Geogr. Inst. Bot. Tibilissi 21: 29 (1959).

M. sogdiana Vassilc. var. *kaschgarica* Vassilc. (cited by Vassilczenko 1949 p. 93.)

M. asiatica Sinskaya subsp. *khivinica* Sinskaya, Fl. Cult. Pl. U.S.S.R. 13(1): 75 (1950). (Sinskaya recognized three forms based on petal color, none of the names published validly.)

M. kopetdaghi Vassilcz. in Vol. Sci. Works Leningrad 1941–1943, 101, ed. Schischk. (1946). (Vassilczenko (1984) stated that the type is "Turcomania: distr. Geok tepe vic. Prokhlaudnoje in declivibus abruptis, A. Borissova, LE; photo in this chapter.)

M. transoxana Vassilcz. var. *kopetdaghi* (Vassilc.) Khassan. (Khassanov 1972, p. 34).

Medica legitima Clus ex E.L. Greene, Man. Bot. San Francisco 101 (1894), nom. illeg.

M. sativa L. subsp. *macrocarpa* Urb., Verh. Bot. Ver. Prov. Brandenb. 15: 56 (1873), nom. invalid, nom. inadmiss. (Correctly, this would be *M. sativa* subsp. *sativa*. Urban's types at B were destroyed during World War II.)

M. macrocarpa (Urb.) Goldschmidt (1912), non Moris 1837, nom. illeg.

M. mesopotamica Vassilcz., Bot. Zhurn. SSSR 31(3): 27 (1946).

M. orientalis Vassilcz., Bot. Zurn. SSSR 31(3): 28 (1946). (Type LE, photo in this chapter.)

M. sativa L. var. *polia* C.J. Brand, U.S.D.A. Bur. Pl. Industr. Bull. 118: 23. (1907). (Type deposited in US according to author. Vassilczenko (1949) deposited a "type" of his species in LE, but of course Brand's type has precedence.)

M. polia (C.J. Brand) Vassilcz. in Journ. Bot. URSS, 31(3): 27 (1946).

M. sativa L. gr. *praecocissimae* Sinskaya, Phylogen. Taxon 411 (1940).

M. praesativa Sinskaya, Kul'turnaja Fl. SSSR (Fl. Cult. Pl. U.S.S.R.), 13(1): 51 (1950), nom. invalid. (Sinskaya recognized (but did not validly publish) two subspecies, noted here, and six forms within each subspecies based on petal color.)

M. sativa L. f. *prolifera* Dore, Can. Field-Nat. 73: 150 (1959) (holotype DAO!).

M. sativa L. var. *occidentali-europaea* Bordakov subvar. *provincialis* Bordakov, Bull. Appl. Bot. Pl. Breed. 19: 113 (1929).

M. sativa L. f. *salaverryensis* J.F. Macbr., Field. Mus. Nat. Hist. (Chicago), Bot. Ser. 8(2): 97 (1930). (In the publication, the type is given as Macbride & Featherstone 48 at F.)

M. asiatica Sinskaya, subsp. *sinensis* Sinskaya, Fl. Cult. Pl. U.S.S.R. 13(1): 82 (1950). (Sinskaya recognized two forms based particularly on petal color, none of the names published validly.)

M. sogdiana Vassilcz., Bot. Zhurn. SSSR (Journ. Bot. URSS) 31(3): 24 (1946). (Type LE, photo in this chapter.)

M. falcata L. var *stenocarpa* Rchb.

M. jemenensis Sinskaya subsp. *subarabica* Sinskaya, Fl. Cult. Pl. U.S.S.R. 13(1): 59 (1950), nom. invalid.

M. sativa L. var. *syriaco-palestinica* Bordakov, Bull. Appl. Bot. Pl. Breed. 19: 8 (1929).

M. syriaco-palestinica (Bordakov) Sinskaya, Fl. Cult. Pl. U.S.S.R. 13(1): 61 (1950).

M. tadzhicorum Vassilcz., Acta Inst. Bot. Acad. Sci. URSS, Ser. 1, Fasc. 8, 32 (1949), sine descr.-lat. (Type at LE, photo in this chapter.)

M. transoxana Vassilcz., Vol. Sci. Works Leningrad 1941–1943, 101, ed. Schischk., (1946). (photo of type in this chapter.)

M. sativa L. var. *tripolitanica* Bordakov, Bull. Appl. Bot. Pl. Breed. 19: 8 (1929).

M. tripolitanica (Bordakov) Sinskaya, Fl. Cult. Pl. U.S.S.R. 13(1): 59 (1950).

M. sativa L. var. *turcomanica* Bordakov, Armenian Agr. 113 (1929).

M. asiatica Sinskaya subsp. *turcomanica* (Bordakov) Sinskaya, Fl. Cult. Pl. U.S.S.R. 13(1): 74 (1950).

M. sativa L. var. *turkestanica* Vilbouchevich, Renseignements 8 (1895). (Agronomic reports on this are in Lamson-Scribner and Kennedy (1900))

M. sativa L. var. *turkestanica* (Vilbouchevich) Bordakov, Armenian Agr. 113 (1939).

M. asiatica Sinskaya subsp. *turkestanica* (Vilbouchevich) Sinskaya, Fl. Cult. Pl. U.S.S.R. 13(1): 71 (1950).

M. sativa L. var. *afghanica* Bordakov subvar. *vavilovi* Bordakov, Armenian Agr. 113 (1929).

M. vera Kirschleger, Fl. Vogés.-Rhén. 1: 122 (1870), nom. illeg.

M. sativa L. f. *vulgaris* Alef., Landw. Fl. 75 (1866), nom. inadmiss.

M. sativa L. subsp. *macrocarpa* Urb. var. *vulgaris* Urb., Verh. Bot. Ver. Prov. Brandenb. 15: 57 (1873), nom. inadmiss. (The autonym subsp. *sativa* would apply here.)

M. praesativa Sinskaya subsp. *sponanea* Sinskaya, Kul'turnaja Fl. SSSR (Fl. Cult. Pl. U.S.S.R.), 13(1): 51 (1950).

Medicago sativa subsp. *caerulea*

M. agropyretorum Vassilcz., Bot. Zurn. SSSR 26(1): 32 (1941). (Type LE, indicated in Vassilczenko 1949 p. 58).

M. tianschanica Vassilcz. var. *agropyretorum* (Vassilcz.) Vassilcz in Komarov, Fl. U.S.S.R. 11: 154 (1945).

M. varia Martyn f. *agropyretorum* (Vassilcz.) D.F. Cui in Adv. Pl. Taxon. Northwest China (eds. Z.Y. Yu, X.Y. Li and W.Z. Di): 47 (1992).

M. beipinensis Vassilcz. in Acta Inst. Bot. Acad. Sci. URSS, Ser. 1, Fasc. 8, 62 (1949), sine descr. lat. (Photo of type in this chapter. This taxon, described by Vassilczenko from a single collection from northern China, is provisionally placed in subsp. *caerulea*.)

M. caerulea Less. ex Ledeb. subsp. *caspica* Sinskaya, Fl. Cult. Pl. U.S.S.R. 12(1): 154 (1950).

M. hemicoerulea Sinskaya, Bot. Jour. U.S.S.R. 23(4): 327 (1938).

M. caerulea Less. ex Ledeb. var. *hemicoerulea* (Sinskaya) Grossh. in Komarov, Fl. U.S.S.R. 11: 151 (1945).

M. caerulea Less. ex Ledeb. subsp. *infradaghestanica* Sinskaya, Bull. Appl. Bot. Plant Breed 28(1) 22 (1948).

M. agropyretorum Vassilcz. var. *inundata* Vassilcz., Tr. Bot. Inst. Akad. Nauk. SSSR, Ser. 1, Flora Sist. Vyss. Rast. Fasc. 14: 60 (1949).

M. lessingii Fisch. & Mey. ex Kar., Bull. Soc. Nat. Mosc. 1: 150 (1839).

M. agropyretorum Vassilcz. var. *longiracemosa* Vassilcz. (mentioned by Khassanov 1972 p. 71)

M. caerulea Less. ex Ledeb. var. *melilotoides* Grossh., Sci. Pap. Appl. Sect. Tiflis. Bot. Gard. (Obz. Krym.-Kavk. *Medicago*) part 1, 32 (1919).

M. caerulea Less. ex Ledeb. var. *pallida* Grossh. in Komarov, Fl. URSS 11: 151 (1945).

M. pekinensis Vassilcz., Bot. Mater. Gerb. Bot. Inst. Komarova Acad. Nauk SSSR 12: 114 (1950). nom. inval.

M. agropyretorum Vassilcz. var. *prostrata* Vassilcz., Tr. Bot. Inst. Akad. Nauk. SSSR, Ser. 1, Flora Sist. Vyss. Rast. Fasc. 14: 60 (1949).

M. rupestris Eichv., Casp. Cauc. 6 (1825), non MB.

M. caerulea Less. ex. Ledeb. var. *sabulosa* Vassilcz., Tr. Bot. Inst. Akad. Nauk. SSSR, Ser. 1, Flora Sist. Vyss. Rast. Fasc. 14: 62 (1949).

M. caerulea Less. ex Ledeb. subsp. *semicoerulea* Sinskaya, Bull. Appl. Bot. Pl. Breed. 28(1): 19 (1948).

M. caerulea Less. var. *vulgaria* Grossh., Survey Crimean Caucasian representatives genus *Medicago* 32 (1919), nom. inadmiss.

M. caerulea Less. ex Ledeb. var. *vulgaris* Grossh., Sci. Pap. Appl. Sect. Tiflis. Bot. Gard. (Obz. Krym. -Kavk. *Medicago*) part 1, 33 (1919).

Medicago sativa subsp. *glomerata*

M. cirtensis Rouy in Rouy & Fouc., Fl. France 5: 14 (1899), in adnot. (Type: LY! lectotype designated by Small 1986*a*.)

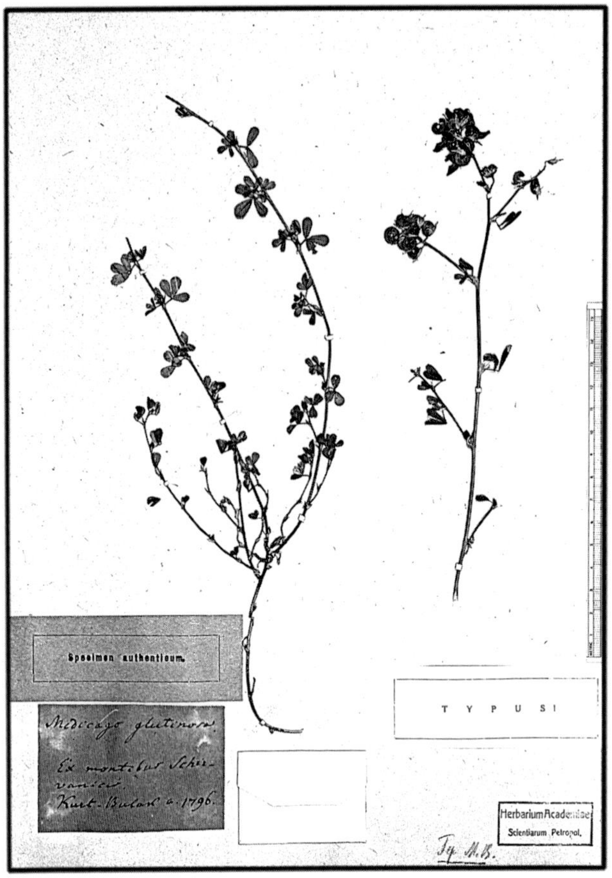

Fig. 12-19. Photograph of type specimen of *Medicago glutinosa* M. Bieb. at LE. This is assigned here to *M. sativa* subsp. *falcata* var. *glomerata*.

M. sativa L. subsp. *macrocarpa* Urb. var. *glandulosa* (Koch) Alef., Verh. Bot. Ver. Prov. Brandenb. 14: 56 (1873).

M. glutinosa M. Bieb., Tabl. Prov. Mer casp. 191 (1798); *M. glutinosa* M. Bieb., Fl. Taur. Cauc. 2: 224 (1808). (Type LE! photo DAO! and in this chapter.)

M. sativa L. subsp. *macrocarpa* Urb. var. *glandulosa* (Koch) Alef. f. *glutinosa* (M. Bieb.) Urb., Verh. Bot. Ver. Prov. Brandenb. 15: 56 (1873).

M. macrocarpa (Urb.) Aschers. & Graebn. B. *glandulosa* III *glutinosa* (M. Bieb.) Aschers. & Graebn., Syn. Mitteleur. Fl. 6(2): 400 (1907).

M. glutinosa M. Bieb. f. *grandiflora* Sinskaya, Bot. J. USSR 23(4): 393 (1938).

M. gunibica Vassilcz., Not. Syst. Herb. Inst. Bot. Acad. Sci. URSS (Bot. Mat. Germ. BIN), 11: 100 (1949). (Type LE! photo this chapter.)

M. glutinosa M. Bieb. var. *macrocarpa* (Urb.) Grossh., Sci. Pap. Appl. Sect. Tiflis Bot. Gard. Part 1, 23 (1919). (A confused application by Grossheim.)

M. glutinosa var. ß panzeriana DC. Prod. 2: 174. (In 1993 I studied the DC Herb at G, and observed a scrappy specimen with flowers ca. 10 mm long in a folder in Herb. DC. at G. This is minimal, but presumably is the type.)

M. sativa L. subsp. *microcarpa* Urb. var. *pilifera* Urb., Verh. Bot. Ver. Prov. Brandenb. 15: 57 (1873).

M. caerulea L. var./subsp. *pilifera* (Urb.) Vassilcz. (On page 61 of Vassilczenko (1949), he referred to the combination as var. *pilifera* (Urb.) Vassilcz.; on page 62 he called it subsp. *pilifera*.)

M. glutinosa M. Bieb. var. *prostrata* Vassilcz., Tr. Bot. Inst. Akad. Nauk. SSSR, Ser. 1, Flora Sist. Vyss. Rast. Fasc. 14: 68 (1949); nom. nud. (published without Latin diagnosis).

M. glutinosa M. Bieb. var. *typica* Grossh., Sci. Pap. Appl. Sect. Tiflis Bot. Gard. Part 4, 144 (1925), nom. illeg.

M. glutinosa M. Bieb. subsp. *typica* Sinskaya, Bull. Appl. Bot. Plant Breed. 28(1): 32 (1948), nom. illeg.

M. virescens Grossh., Sci. Pap. Appl. Sect. Tiflis Bot. Gard. Part 1, 26 (1919). (Type TBI.) (Based on description, this taxon appears to represent plants somewhat introgressed by *M. sativa* subsp. *sativa*.)

Medicago sativa subsp. *falcata* var. *falcata*

M. falcata L. var. *adscendens* Sumnevicz, Animadv. Syst. ex Herb. Univ. Tomsk. 1-2: 10 (1932).

M. falcata L. "S.-var." *albiflora* Babey ex Rouy in Rouy & Fouc., Fl. Fr. 5: 11 (1899).

M. falcata L. var. *alpigena* Schur, Enum. Pl. Trans. Exh. 151 (1866).

M. falcata L. var. *altaica* Sumnevicz, Animadv. Syst. ex Herb. Univ. Tomsk. 1-2: 13 (1932).

M. falcata L. var. *altissima* Grossh., Survey of the Crimean and Caucasian *Medicago*: 20 (1919).

M. angustifolia Chevallier, Fl. Gén. Env. Paris, 2 : 779 (1927).

M. romanica Prodán var. *angustifolia* Ptasch. ex Grossh. in Komarov (ed.) Fl. U.S.S.R. 11: 147 (1945).

M. aurantiaca Godr., Mém. Sect. Méd. Acad. Sci. Montpellier, Ser.1, 1: 425 (1853).

M. falcata L. var. *aurea* Schur, Verhandl. Naturf. Ver. Brunn 15(2): 171 (1876).

M. falcata L. var. *aureiflora* Rouy in Rouy & Fouc., Fl. Fr. 5: 11 (1899).

M. falcata L. var. *aureiflora* Retz. (cited by Vassilczenko 1949.)

M. falcata var. *aureiflora* Bab. & R. & F. (cited by Vassilczenko 1949.)

M. falcata L. var. *typica* Posp. subvar. *minor* Gaudin f. *aureiflora* (Hausmann) Hegi? (cited by Vassilczenko 1949)

M. romanica Prodán var. *australis* Vassilcz., Tr. Bot. Inst. Akad. Nauk. SSSR, Ser. 1, Flora Sist. Vyss. Rast. Fasc. 14: 48 (1949).

M. borealis Grossh. in Komarov (ed.), Flora U.S.S.R. 11: 291 (1945). (Type LE.)

M. falcata L. var. *brevipes* Vassilcz., Tr. Bot. Inst. Akad. Nauk. SSSR, Ser. 1, Flora Sist. Vyss. Rast. Fasc. 14: 42 (1949).

M. romanica Prodán var. *canescens* Vassilcz., Priroda 12: 72 (1940).

M. romanica Prodán subsp. *cinerea* Lindl. ex Vassilcz., Tr. Bot. Inst. Akad. Nauk. SSSR, Ser. 1, Flora Sist. Vyss. Rast. Fasc. 14: 48 (1949).

M. falcata L. var. *cretacea* Grossh. in Komarov (ed.), Flora U.S.S.R. 11: 145 (1945). (The illustration of a pod of this taxon in the publication appears to be of *M. cretacea*, although the description contrarily states that the pod is falcate. The description of the rest of the plant generally is not consistent with that for *M. cretacea*, and Grossheim is very unlikely to have confused the two, so perhaps the wrong illustration was inserted accidentally.)

M. difalcata Sinskaya, Trudy Prikl. Bot. 28(1): 29 (1948).

M. falcata L. var. *diffusa* Schur, Enum. Pl. Trans. Exh. 151 (1866).

M. falcata L. var. *dzumagalensis* Ubuk., Fl. Kirgizkoi SSR 7: 156 (1957).

M. falcata L. subsp. *erecta* Kotov, Priroda 1: 70 (1935). (Lectotypus at KW, indicated at http://www.nbuv.gov.ua/herbar/type_e.htm; current June 2007)

M. erecta Kotov, Journ. Bot. Acad. Sc. Ukraine, No. 2, 276 (1940), nom. illeg.

M. falcata L. var. *erecta* (Kotov) Grossh., Flora USSR 11: 146 (1941).

M. romanica Prodán var. *erecta* (Kotov) Grossh. in Komarov (ed.), Fl. U.S.S.R. 11: 146 (1945).

M. romanica Prodán var. *ergenensis* Vassilcz., Tr. Bot. Inst. Akad. Nauk. SSSR, Ser. 1, Flora Sist. Vyss. Rast. Fasc. 14: 48 (1949).

M. falcata L. var. *ferruginea* Sumnevicz, Animadv. Syst. ex Herb. Univ. Tomsk 1-2: 10 (1932).

M. romanica Prodán var. *ferruginea* (Sumnevicz) Gross. in Komarov (ed.), Fl. U.S.S.R. 11: 146 (1945).

M. falcata L. var. *foliolata* Vassilcz., Tr. Bot. Inst. Akad. Nauk. SSSR, Ser. 1, Flora Sist. Vyss. Rast. Fasc. 14: 42 (1949).

M. falcata L. var. *foliosa* Sumnevicz, Animadv. Syst. ex Herb. Univ. Tomsk. 1-2: 12 (1932).

M. falcata L. var. *typica* Posp. subvar. *minor* Gaudin f. *geisenheyneri* (Asch. & Graebn.) Hegi? (Vassilc. 1949 referred to mention by Hegi.)

M. falcata L. var. *genuina* Godron, Fl. Lorr. 169, pl. 1 (1843), nom. inadmiss.

M. romanica Prodán var. *steppe genuinae* Vassilcz., Tr. Bot. Inst. Akad. Nauk. SSSR, Ser. 1, Flora Sist. Vyss. Rast. Fasc. 14: 48 (1949), nom. inadmiss.

M. falcata L. var. *typica* Posp. subvar. *genuina* Duc. (mentioned by Vassilc. 1949)

M. falcata L. subsp. *glareosa* Vassilcz., Tr. Bot. Inst. Akad. Nauk. SSSR, Ser. 1, Flora Sist. Vyss. Rast. Fasc. 14: 40 (1949).

M. sativa L. subsp. *macrocarpa* Urb. var. *falcata* (L.) Döll f. *gracilis* Urb., Verh. Bot. Ver. Prov. Brandenb. 14: 56 (1873).

M. falcata L. var. *hirsuta* Trautv. ex [*M. sativa* var. *hirsuta* Trautv. in sched.] Sumnevicz, Animadv. Syst. ex Herb. Univ. Tomsk. 1-2: 11 (1932). (Cited by Sumnevicz as *M. falcata* L. var. *hirsuta* (Trautv.) Grossh.)

M. romanica Prodán var. *hirsuta* Trautv. ex Grossh. in Komarov (ed.) Fl. U.S.S.R. 11: 147 (1945).

M. intermedia Schultes, Observat. Bot. 160 (1809).

M. falcata L. var. *irghisica* Vassilcz., Tr. Bot. Inst. Akad. Nauk. SSSR, Ser. 1, Flora Sist. Vyss. Rast. Fasc. 14: 43 (1949).

M. falcata L. var. *jenisseica* Vassilcz., Tr. Bot. Inst. Akad. Nauk. SSSR, Ser. 1, Flora Sist. Vyss. Rast. Fasc. 14: 43 (1949).

M. difalcata Sinskaya subsp. *kazachstanica* Sinskaya, Fl. Cult. Pl. U.S.S.R. 13(1): 131 (1950).

M. kotovii O.D.Wissjul., Fl. URSR vi. 367 (1954). (Type LE, photo in this chapter.)

M. falcata L. var. *kuhartica* Vassilcz. in Komarov (ed.), Fl. U.S.S.R. 11: 144 (1945).

M. romanica Prodán var. *lanata* Grossh. in Komarov (ed.) Fl. U.S.S.R. 11: 147 (1945).

M. falcata L. var. *lasiagrostidetorum* Vassilcz. in Komarov (ed.), Fl. U.S.S.R. 11: 143 (1945).

M. falcata L. var. *latifolia* Grossh., Obz. Krym.-Kavk. *Medicago* 20 (1919).

M. laxiflora Schur, Verh. Naturf. Vereiins Brünn 15(2): 172 (1877).

M. falcata L. var. *lignescens* Sumnevicz, Animadv. Syst. ex Herb. Univ. Tomsk. 1-2: 9 (1932).

M. falcata L. var. *linearifolia* Grossh., Obz. Krym.-Kavk. *Medicago* 20 (1919).

M. romanica Prodán var. *linifolia* Kuschan ex Grossh. in Komarov (ed.) Fl. U.S.S.R. 11: 147 (1945).

M. falcata L. var. *macrantha* Vassilcz., Tr. Bot. Inst. Akad. Nauk. SSSR, Ser. 1, Flora Sist. Vyss. Rast. Fasc. 14: 39 (1949).

M. macrocarpa Moris, Fl. Sard. 1: 446, tab. 45 (1837).

M. falcata L. (var.) γ *major* W.D.J. Koch, Syn. Fl. Germ. Helv., ed. 1, 160 ("1835"; 1836).

M. borealis Grossh. var. *micrantha* Grossh. in Komarov (ed.), Fl. U.S.S.R. 11: 140 (1945).

M. falcata L. var. *microphylla* Cusin & Ansbergue, Herb. Fl. Franc., pt. 5/6, 1011 (1869).

M. falcata L. var. *typica* Posp. subvar. *minor* Gaudin

M. falcata L. var. *mongholica* Sumnevicz, Animadv. Syst. ex Herb. Univ. Tomsk. 1-2: 13 (1932).

M. falcata L. var. *montana* Boenninghausen, Prod. Fl. Monsst. Westph. 225 (1824).

M. difalcata Sinskaya subsp. *orientalcaucasica* Sinskaya, Fl. Cult. Pl. U.S.S.R. 13(1): 131 (1950).

M. falcata L. var. *orientalis* Sumnevicz, Animadv. Syst. ex Herb. Univ. Tomsk. 1-2: 11 (1932).

M. falcata L. var. *parviflora* Grossh. in Komarov (ed.), Fl. U.S.S.R. 11: 145 (1945).

M. falcata L. subsp. *pinetorum* Vassilcz., Tr. Bot. Inst. Akad. Nauk. SSSR, Ser. 1, Flora Sist. Vyss. Rast. Fasc. 14: 42 (1949).

M. falcata L. var. *pratensis* Boenninghausen, Prod. Fl. Monsst. Westph. 225 (1824).

M. procumbens Besser, Prim. Fl. Galiciae Austriac. 2: 127 (1809). (Neotypus at KW, indicated at http://www.nbuv.gov.ua/herbar/type_e.htm; current June 2007)

  M. falcata L. var. *procumbens* (Besser) Ledebour, Fl. Ross. 1: 526 (1842).

M. falcata L. var. *procumbens* Schmalh., Flora Central Southern Russia 1: 225 (1985), nom. superf.

M. falcata L. var. *major* W.D.J. Koch f. *prostrata* W.D.J. Koch in Rohlings, Deut. Fl. 5: 318 (1839).

M. falcata L. var. *typica* Posp. subvar. *minor* Gaudin f. *pseudoprostrata* Hausmann

M. falcata L. var. *pubescens* Rouy in Rouy & Fouc., Fl. Fr. 5: 11 (1899).

M. falcata L. var. *typica* Posp. subvar. *minor* Gaudin f. *pubescens* (Rouy) Hegi?

M. falcata L. var. *pygmaea* Sumnevicz, Animadv. Syst. ex Herb. Univ. Tomsk. 1-2: 12 (1932).

M. quasifalcata Sinskaya, Compt. Rend. Acad. Sci. URSS 48(4): 281 (1945); Proc. Lenin Acad. Agr. Sci. U.S.S.R. 18(4): 300 (1945).

M. falcata L. var. *ramosissima* Sumnevicz, Animadv. Syst. ex Herb. Univ. Tomsk. 1-2: 13 (1932).

M. falcata L. var. *revoluta* Sumnevicz, Animadv. Syst. ex Herb. Univ. Tomsk. 1-2: 12 (1932).

  M. romanica Prodán subsp. *revoluta* (Sumnevicz) Vassilcz., Tr. Bot. Inst. Akad. Nauk. SSSR, Ser. 1, Flora Sist. Vyss. Rast. Fasc. 14: 47 (1949).

M. falcata L. var *riparia* Boenninghausen, Prod. Fl. Monsst. Westph. 225? (1824).

M. falcata L. var. *robusta* Vassilcz., Tr. Bot. Inst. Akad. Nauk. SSSR, Ser. 1, Flora Sist. Vyss. Rast. Fasc. 14: 39 (1949)

M. romanica D. Brândză, Fl. Roman. Prodr. 1: 617 (1879).

M. romanica Prodán, Fl. Român. 1: 617 (1923).

  M. falcata L. subsp. *romanica* (Prodán) Hayek, Prodr. Fl. Penins. Balc. 1: 835 (1927).

  M. falcata L. var. *romanica* (Prodán) O. Schwarz & Klink., Verh. Bot. Ver. Prov. Brandenburg 74: 182 (1933).

M. sagarrae Sennen, Bull. Soc. Bot. France 68: 405 (1922). nom nud.

M. falcata L. var. *stenocarpa* Rchb., Fl. Germ. Excurs. 504? (1832).

M. falcata L. var. *suberecta* Lindemann, Bull. Soc. Imp. Nat. Moscow, pt. 1, 495, pl. 40 (1867).

M. falcata L. var. *sibirica* Sumnevicz, Animadv. Syst. ex Herb. Univ. Tomsk. 1-2: 11 (1932).

  M. romanica Prodán var. *sibirica* (Sumnevicz) Grossh. in Komarov (ed.), Fl. U.S.S.R. 11: 146 (1945).

M. falcata L. var. *sophiae* Bosse ex Grossh. in Komarov (ed.), Fl. U.S.S.R. 11: 143 (1945).

M. subfalcata Sinskaya, Bot. Jour. U.S.S.R. 23(4): 323 (1938), nom. invalid. (This kind of plant is slightly introgressed by subsp. *sativa*, but not enough to place it in subsp. ×*varia*; cf. *M. subfalcata* Schur., in the synonymy of subsp. ×*varia*.)

M. suffruticosa Grossh., Obz. Krym.-Kavk. *Medicago* 21 (1919), nom. illeg.

  M. falcata L. var. *suffruticosa* (Grossh.) Grossh. in Komarov (ed.), Fl. U.S.S.R. 11: 145 (1945).

M. sylvestris (*silvestris*) Fr., Novit. Fl. Suec., Mantissa 3: 92 (1842). (Commonly interpreted as *M. sativa* subsp. ×*varia*.)

  M. falcata L. proles *sylvestris* (Fr.) Rouy, Fl. Fr. 5: 12 (1899).

  M. falcata L. (var.) *sylvestris* (Fr.) Tourlet Cat. rais. d. plant. 115 (1903).

  M. sativa L (var.) *sylvestris* (Fr.) Asch. & Graebn., Syn. 6(2): 402 (1907).

Medica sylvestris Bubani, Fl. Pyren. 2: 471 (1900), nom. illeg.

M. falcata L. var. *tanaitica* Vassilcz., Tr. Bot. Inst. Akad. Nauk. SSSR, Ser. 1, Flora Sist. Vyss. Rast. Fasc. 14: 43 (1949).

M. falcata L. var. *tarbagatica* Sumnevicz, Animadv. Syst. ex Herb. Univ. Tomsk. 1-2: 9 (1932).

M. tenderiensis Opperman ex Klokov, Bot. Zhurn. (Kiev) 5(2): 45 (1948). (Paratype at KW (indicated at http://www.nbuv.gov.ua/herbar/type_e.htm; current June 2007).)

  M. falcata L. subsp. *tenderiensis* (Opperman ex Klokov) Vassilcz., Tr. Bot. Inst. Akad. Nauk. SSSR, Ser. 1, Flora Sist. Vyss. Rast. Fasc. 14: 43 (1949).

M. falcata L. var. *tenella* Sumnevicz, Animadv. Syst. ex Herb. Univ. Tomsk. 1-2: 10 (1932).

M. falcata L. var. *transbaicalensis* Vassilcz., Tr. Bot. Inst. Akad. Nauk. SSSR, Ser. 1, Flora Sist. Vyss. Rast. Fasc. 14: 43 (1949).

M. falcata L. var. *typica* Trautv., Bull. Soc. Imp. Nat. Moscow, pt. 1, 474, pl. 33 (1860), nom. inadmiss.

  M. falcata L. var. *typica* Trautv. ex Sumnevicz in Krylov, Fl. Zap. Sibir. 7: 1590 (1931?).

M. vardanis Vassilcz. in Journ. Bot. URSS (Bot. Zhurn.) 31(3): 24 (1946). (Type LE, photo in this chapter. This kind of plant is slightly introgressed by subsp. *sativa*, but not enough to place it in subsp. ×*varia*.)

M. falcata L. var. *vassilczenkoi* Ubuk., Fl. Kirgizkoi SSR 7: 156 (1957).

Medicago sativa subsp. falcata var. viscosa

M. falcata L. var. *glandulosa* W.D.J. Koch in Röhling, Deutschl. Fl., ed. 3, 5: 318 (1839). According to Chaudhri et al. (1972), Koch's German collections are at B, L, OXF, and WAG. An extant Koch collection of this variant is present only at WAG (Herb. Lugd. Bat. No. 910.43-317). This is a heterogeneous collection of eight branches, probably from a population segregating for taxonomically important fruit characteristics. The branch second from the right was designated as the lectotype by Small (1986*a*).

M. sativa L. (var.) *glandulosa* (W.D.J. Koch) Alef., Landw. Fl. 76 (1866).

M. sativa L. subsp. *glandulosa* (W.D.J. Koch) Arcang., Comp. Fl. Ital. 160 (1882).

M. falcata L. subsp. *glandulosa* (W.D.J. Koch) Greuter & Burdet, Willdenowia 19: 32 (1989).

M. glandulosa Davidov, Österr. Bot. Z., 52: 493 (1902). (Apparently described independently of Koch's var. *glandulosa*; also cited as *M. glandulosa* (Mert. & W.D.J. Koch) Davidov.

M. falcata L. subsp. *glandulosa* (Davidov) Kozhurarov, Izv. Bot. Inst. (Sofia) 15: 141 (1965).

M. falcata L. var. *tirnensis* Širj. (1928). (Lesins and Lesins 1979 referred to a specimen so identified as having glandular hairs on its peduncles, pedicels, and pods)

M. falcata L. subsp. *urumovii* Degen., Mag. Bot. Lap. 12: 214 (1914). (A type is not present at BP, where Degen's types are mainly deposited; however, the types were destroyed during the second world war (D. Konstantin, personal communication.) (Additional information on this grouping is given in Schwartz and Klinkowsi (1933).)

M. falcata L. var. *viscosissima* Grossh., Sci. Pap. Tiflis Bot. Gard. Part 1, 21 (1919).

Medicago sativa subsp. ×varia

M. tianschanica Vassilcz. var. *agrenica* Khassan. (Khassanov 1972, p. 62).

M. alaschanica Vassilcz., Not. Syst. Herb. Inst. Bot. Acad. Sci. URSS, xii: 113 (1949). (Type LE, photo in this chapter.)

M. alatavica Vassilcz. in Acta Inst. Bot. Acad. Sci. URSS, Ser. 1, Fasc. 8, 53 (1949), sine descr. lat. (Type at LE, photo in this chapter.)

M. alatavica Vassilcz. var. *lacunarum* Vassilc. (mentioned in Vassilczenko, cited above; status uncertain).

Fig. 12-20. Photograph of type specimen of *Medicago falcata* var. *glandulosa* W.D.J. Koch, at LD. This is assigned below to *M. sativa* subsp. *falcata* var. *viscosa*.

M. falcata L. var. *ambigua* Trautv., Bull. Soc. Imp. Nat. Moscou (Enum. Plant. soong.) 33(1): 474 (1860).

M. sativa L. subsp. *ambigua* (Trautv.) Tutin, Feddes Repert. 79: 53 (1968).

M. annularis Besser, Pri. Fl. Galiciae Austriac. 2: 127 (1809). (At G, I observed in a folder labelled var. ß *annularis* two sheets collected by Besser, possibly representing the type. These were assignable to subsp. ×*varia*.)

M. falcata L. var. *annularis* (Besser) DC., Prodr. 2: 172 (1825).

M. sativa L. (var.) *atriflora* Alef., Landw. Fl. 76 (1866).

M. caucasica Vassilcz., Trudy Bot. Inst. AN SSSR 1(8): 49 (1944). Also Not. Syst. Herb. Inst. Bot. Acad. Sci. URSS. (Bot. Mater. Gerb. Bot. Inst. Komarova Akad. Nauk SSSR) 13: 141 (1949).)

M. cibinensis Lerchenfeld ex Schur, Enum. Pl. Trans. Exh. 151 (1866), pro syn.

M. trautvetterii Sumnev. var. *coenosa* Vassilcz., Acta Inst. Bot. Acad. Sci. URSS, Ser. 1, Fasc. 8, 25 (1949).

M. komarovii Vassilcz. var. *cretacea* Vassilc., Acta Inst. Bot. Acad. Sci. URSS, Ser. 1, Fasc. 8, 56 (1949).

M. cyclocarpa Hy, Jour. Bot. (Paris) 9(23): 431 (1895).

M. falcata L. var. *cyclocarpa* (Hy) Rouy in Rouy & Fouc., Fl. Fr. 5: 12 (1899).

M. sativa L. var. *decolorans* Fries., Nov. Fl. Suec. Cont. 3: 91 (1842).

M. glutinosa M. Bieb. var. *denudata* Grossh., Novye dannye k poznaniyu Kavk. lyutsern 3 (1925).

M. falcata L. (var.?) *desertorum* Trautv., Enum. Pl. Soong. 474 (1860).

M. falcata L. var. *desertorum* (Trautv.) Rgl. & Herd., Enum. Pl. 33 (1866).

M. falcata L. var. *erectiuscula* Rouy in Rouy & Fouc., Fl. Fr., 5: 12 (1899).

M. falcata-sativa Rchb., Fl. Germ. Ex. 504 (1830–1833), nom. inadmiss. (hybrid formula).

M. tianschanica Vassilc. var. *fruticetorum* Vassilcz. in Grossh. (ed.), Fl. USSR 11: 153 (1945).

M. komarovii Vassilcz. var. *fructifera* Vassilc., Acta Inst. Bot. Acad. Sci. URSS, Ser. 1, Fasc. 8, 56 (1949).

M. sativa L. var. *gigantea* Grossh., Survey Crimean Caucasian representatives genus *Medicago* 26 (1919).

M. glutinosa var. *glabrata* Grossh., Sci. Pap. Appl. Sect. Tiflis Bot. Gard. Part 4, 144 (1925).

M. sativa L. var. *grandiflora* Grossh., Zap. Nauchno-Prokl. Otd. Tiflissk. Bot. Sada (Obs. Kr.-Kavk. *Medicago*) 1: 28 (1919). (Vassilczenko (1984) stated the type is "Mons Alaghez Transcaucasiae meridionalis prope Kosch-bulak, E. & N. Busch, LE"; photo in this chapter.)

M. grandiflora (Grossh.) Vassilcz., Vol. Sci. Works Leningrad 1941–1943, 101, ed. Schischk. (1945).

M. hemicycla Grossh., Papers Appl. Sect. Tiflis Bot. Gard. (Zap. Nauchno-Prikl. Otd. Tifl. Sada) 4: 147 (1925). (Type at TBI.)

M. sativa L. subsp. *hemicycla* (Grossh.) C.R. Gunn, U.S. Depart. Agriculture, Tech. Bull., 1574: 15 (1978).

M. falcata L. var. *hybrida* Gaudin, Fl. Helv. 4: 611 (1829).

M. komarovii Vassilcz. var. *incana* Vassilc., Acta Inst. Bot. Acad. Sci. URSS, Ser. 1, Fasc. 8, 56 (1949).

M. sativa L. (var.) *intermedia* Alef., Landw. Fl. 76 (1866).

M. sativa L. (var.) *kochiana* Alef., Landw. Fl. 73 (1866).

M. komarovii Vassilcz. in Vol. Sci. Works Leningrad 1941–1943, ed. Schischk., 101 (1946). (Type LE, photo in this chapter.)

M. kultiassovii Vassilcz., Acta Inst. Bot. Acad. Sci. URSS, Ser. 1, Fasc. 8, 53 (1949).

M. ladak Vassilcz., Bot. Zhurn. SSSR 31(3): 27 (1946).

M. lavrenkoi Vassilcz., Not. Syst. Herb. Inst. Bot. Acad. Sci. URSS (= Bot. Mater. Gerb. Bot. Inst. Komarova Akad. Nauk SSSR) 13: 141 (1950). (Photo of type in this chapter.)

M. tianschanica Vassilcz. var. *lavrenkoi* (Vassilc.) Khassan. (Khassanov 1972, p. 61).

M. hemicycla Grossh. var *leninacanica* (Sinsk.) Vasssilcz., Acta Inst. Bot. Acad. Sci. URSS, Ser. 1, Fasc. 8, 26 (1949).

M. lilacea Hy, Jour. Bot. (Paris) 9: 432 (1895).

M. komarovii Vassilcz. f. *longifolia* Vassilc., Acta Inst. Bot. Acad. Sci. URSS, Ser. 1, Fasc. 8, 56 (1949).

M. hemicycla Grossh. subsp. *loriensis* Sinsk., Tr. po prikl. bot., gen. i sel. (Bull. Appl. bot. Plant Breed.) 28(1): 14 (1948). (Sinskaya 1950 delimited 15 forms by Latin names, not validly published; eight of these forms were also recognized in her *M. hemicycla* Grossh. subsp. *medidaghestanica*.)

M. hemicycla Grossh. var. *maxima* Vassilcz., Acta Inst. Bot. Acad. Sci. URSS, Ser. 1, Fasc. 8, 26 (1949).

M. media Pers., Syn. Pl. 2(2): 356 (1807). (Vassilczenko (1949) interpreted the name as "*M. sativa media* Pers." and credited species recognition as follows: *M. media* (Pers.) Schult., Oester. Fl. 371 (1814).)

M. sativa L. subsp. *media* (Pers.) Schübler & Martens, Fl. Würtemb. 467 (1834).

Medica media (Pers.) Fourr., Ann. Soc. Linn. Lyon, sér. 2, 16: 359 (1868).

M. falcata L. subsp. *media* (Pers.) Bonnier & Layens, Fl. Fr. 74 (1894); ead. comb. Corb. 4, Nouv. Fl. Normandie 150 (1894).

M. falcata L. (var.) *media* (Pers.), Tourlet Cat. rais. d. plant. 115 (1903).

M. hemicycla Grossh. subsp. *medidaghestanica* Sinsk., Tr. po prikl. bot., gen. i sel. (Bull. Appl. bot. Plant Breed.) 28(1): 17 (1948).

M. mirandana Sennen & Elias, Bol. Soc. Iber. Ci. Nat. 26 : 89 (1927).

M. trautvetterii Sumnev. var. *mugodzharica* Vassilcz., Acta Inst. Bot. Acad. Sci. URSS, Ser. 1, Fasc. 8, 25 (1949).

M. tianschanica Vassilc. var. *oblonga* Vassilcz. in Grossh. (ed.), Flora USSR 11: 153 (1945).

M. tianschanica Vassilc. var. *obovata* Vassilcz. in Grossh. (ed.), Flora USSR 11: 153 (1945).

M. ochroleuca M. Kult. in Del., List Seeds Main Bot. Gard. USSR Acad. Sci. 1: 39 (1946). Type at LE according to Vassilczenko (1949).

M. tianschanica Vassilc. var. *ochroleuca* (M. Kult.) Khassan. (Khassanov 1972, p. 24).

M. media Pers. (var. β) *ochroleuca* Celak., Prodr. 659 (1867–1875).

M. sativa (var.) *pallida* Alef., Landw. Fl. 73 (1866).

M. sativa L. var. *parviflora* Grossh., Zap. Nauchno-Prokl. Otd. Tiflissk. Bot. Sada (Obs. Kr.-Kavk. *Medicago*) 1: 28 (1919).

M. pauciflora Ledeb., Fl. Ross. 1: 526 (1843). (Vassilczenko 1949. p. 50. illustrates "Ledebour's authentic specimen.")

M. sativa L. subsp. *microcarpa* Urb. var. *pauciflora* (Ledeb.) Urb., Verh. Bot. Ver. Prov. Brandenb. 15: 57 (1873).

M. falcata L. var. *pauciflora* (Ledeb.?) Lindemann, Fl. Chers. 1: 138 (1881). (Assignment provisional.)

M. caerulea Less. ex Ledeb. var. *pauciflora* (Ledeb.) Grossh. in Komarov, Fl. USSR 11: 151 (1945).

M. hemicycla Grossh. f. *pseudodzhavakhetica* Troitsk. (1926?), nom. illeg.

M. varia Martyn var. *pseudofalcata* Rouy in Rouy & Fouc, Fl. Fr. 5: 15 (1899), nom. illeg.

M. varia Martyn var. *pseudoglomerata* Rouy in Rouy & Fouc, Fl. Fr. 5: 15 (1899).

M. varia Martyn var. *pseudosativa* Rouy in Rouy & Fouc, Fl. Fr. 5: 14 (1899), nom. illeg.

M. tianschanica Vassilcz. var. *riparia* Vassilcz. in Grossh. (ed.), Fl. USSR 11: 154 (1945).

M. rivularis Vassilcz., Vol. Sci. Works Leningrad 1941–1943, ed. Schischk., 101 (1946). (Type LE; photo in this chapter.)

M. tianschanica Vassilcz. var. *rivularis* (Vassilc.) Khassan. (Khassanov 1972 p. 27, 71).

M. roborovskii Vassilcz. in Not. Syst. Herb. Inst. Bot. Acad. Sci. URSS, xii. 113 (1950). (Type LE; photo in this chapter.)

M. sativa L. var. *rosea* Alef., Landw. Flora 75 (1866).

M. sativa L. var. *rotundifolia* Alef., Landw. Flora 75 (1866).

M. sativa-falcata Ritschl., Fl. Posen 537 (1880), nom. inadmiss. (hybrid formula).

M. sativoides Léveillé, Bull. Géogr. Bot., sér. 4, 25: 50 (1915), nom. illeg.

M. tianschanica Vassilcz. var. *saxatilis* Vassilcz. in Grossh. (ed.), Flora USSR 11: 153 (1945).

M. schischkinii Sumnev., Animadvers. Syst. Herb. Univ. Tomsk No. 1-2: 5 (1932). (Type LE, photo in this chapter.)

M. spuria Hy, Jour. Bot. (Paris) 9(23): 431 (1895).

M. varia Martyn n.var. *spuria* (Hy) Rouy, Fl. Fr. 5: 14 (1899).

M. falcata L. var. *subdicycla* Trautv., Enum. Pl. Soong. 474 (1860).

M. subdicycla (Trautv.) Vassilcz., Bot. J. USSR 31(3): 23 (1946).

M. trautvetterii Sumnev. var. *subdicycla* (Vassilc.) Khassan. (Khassanov 1972, p. 28).

M. subfalcata Schur, Enum. Pl. Trans-silv. 151 (1866).

M. suffruticosa Grossh. (Grossheim 1919) , non *M. suffruticosa* DC.

M. trautvetterii Sumnev. var. *tenera* Vassilcz., Acta Inst. Bot. Acad. Sci. URSS, Ser. 1, Fasc. 8, 25 (1949).

M. tetrahemicycla Sinskaya, Fl. Cult. Pl. USSR 13(1): 83 (1950), nom. invalid. (Sinskaya delimited two forms on the basis of petal color, but did not validly publish the names.)

M. tianschanica Vassilcz., Bot. Zhurn. SSSR (Botanical Journal of the USSR) 25(3): 244 (1940). (Type at LE.)

M. sativa L. var. *tibetana* Alef., Landw. Flora 73 (1866). (Type at LE, photo in this chapter.)

M. tibetana (Alef.) Vassilcz., Vol. Sci. Works Leningrad 1941–1943, ed. Schischk., 101 (1946).

M. trautvetterii Sumnev. in Animadvers. Syst. Herb. Univ. Tomsk. Nos. 1-2: 3 (1932). (Type LE; photo in this chapter.)

M. sativa L. (var.) τ *tumida* Ser. ex DC. Prod. 2: 172 (1825). (I observed a collection of this at G in the general folder of *M. falcata*; this presumably is the type.)

M. subdicycla (Trautv.) Vassilc. var. *turgaiensis* (referred to in Vassilczenko 1949).

M. falcata L. var. *versicolor* Wallr., Sched. Crit. Pl. Fl. Hal. Sel. 1: 398 (1822). (In 1993 at G in DC's herbarium I observed in the *versicolor* file one specimen, that of "Mr. Wallr. 1822". By default this would seem to be the best choice of a type for the name.)

M. sativa L. (var. β) *versicolor* (Wallr.) Ser. ex DC., Prodr. 2: 173 (1825).

M. falcata L. (var.) β *versicolor* (Wallr.) W.D.J. Koch, Syn. Fl. Germ. Helv., ed. 1, 160 ("1835"; 1836).

M. sativa L. var. *versicolor* (Wallr.) W.D.J. Koch, Syn. Fl. Germ., ed. 2, 176 (1843), nom. superfl.

M. media Pers. (var. α) *versicolor* (Wallr.) Celak., Prodr. 659 (1867–1875).

Medicago sativa* subsp. *sativa* × subsp. *glomerata

M. polychroa Grossh. var. *denudata* Grossh., Novye dannye k poznaniyu kavkazskikh lyutsern (New data for studies on Caucasian alfalfa), Zap. Nauchno-Prikl. otd. Tifl. Sada (Sci. Pap. Appl. Sect. Tiflis Bot. Gard.) Part 4: 3 (1925).

M. sativa L. subsp. *faurei* Maire, Bull. Soc. Hist. Nat. Afrique Nord 24: 203 (1933). (Type MPU! Photo DAO; specimen cited in Representative Collections; also viewed online at: http://www.aluka.org; current March 2007.)

M. sativa L. subsp. *macrocarpa* (Urb.) d. (var.) *vulgaria* β (forma) *gaetula* Urb., Verh. Bot. Ver. Prov. Brandenb. 15: 57 (1873).

M. glutinosa M. Bieb. var. *glomerata* Grossh., Sci. Pap. Appl. Sect. Tiflis Bot. Gard. Part 1, 22 (1919) (i.e., var. *glomerata sensu* Grossheim; not Balbis's *M. glomerata*, although Grossheim thought it was).

M. glutinosa M. Bieb. var. *microcarpa* Grossh., Sci. Pap. Appl. Sect. Tiflis Bot. Gard. Part 1, 22 (1919).

M. grossheimii Vassilcz., Not. Syst. Herb. Inst. Bot. Acad. Sci. URSS (Bot. Mater. Gerb. Bot. Inst. Komarova Akad. Nauk SSSR) 12: 116 (1950). (Type LE, photo in this chapter.)

M. tianschanica Vassilcz. var. *grossheimii* (Vassilcz.) Ubukeeva.

M. polychroa Grossh., Novye dannye k poznaniyu kavkazskikh lyutsern (New data for studies on Caucasian alfalfa), Zap. Nauchno-Prikl. otd. Tifl. Sada (Sci. Pap. Appl. Sect. Tiflis Bot. Gard.) Part 4: 125 (1925). nom. superfl., illegit. (Type: TBI)

M. glutinosa M. Bieb. subsp. *polychroa* (Grossh.) Sinskaya, Bull. Appl. Bot. Plant Breed. 28(1): 34 (1948).

M. glutinosa M. Bieb. subsp. *praefalcata* Sinskaya, Bull. Appl. Bot. Pl. Breed. 28(1): 37 (1948).

M. sativa L. subsp. *praefalcata* (Sinskaya) C.R. Gunn in Gunn et al., U.S. Dept. Agric., Agric. Res. Serv. Tech. Bull. 1574: 10 (1978).

M. sativa L. subsp. *tunetana* Murb., Contrib. fl. Tunis., Act. Reg. Soc. Physiogr. Lund 33: 612 (1897). (Syntypes LD!, LE!; cited in Representative Collections.)

M. tunetana (Murb.) Vassilcz., Tr. Bot. Inst. Akad. Nauk. SSSR, Ser. 1, Fl. Sist. Vyss. Rast. Fasc. 14: 71 (1949).

M. polychroa Grossh. var. *typica* Grossh. nom. inadmiss.

Vernacular Names

Medicago sativa

ENGLISH: *Medicago sativa* is known as alfalfa in Iberia and North and South America, but more commonly as lucern(e) in South Africa, New Zealand, Australia, Oceania, and all European countries east of Spain.

FRENCH: La luzerne cultivée.

Medicago sativa subsp. *sativa*

ENGLISH: Alfalfa, lucerne (sometimes spelled lucern, luserne, and luzern), common lucerne, lucerne grass (in South Asia), purple lucerne, burgundy clover, Calvary clover (Plowden 1972), Chilean clover, snail clover, common alfalfa, purple alfalfa, purple medic (medic is also spelled medick). It has been claimed that the name "alfalfa" traces to the Persian *aspo-asti* (horse fodder), the Arabic *al-fasfasa* or the Kashmiri *ashwa-bal* (the latter two names meaning horse power; for additional information see Bolton 1972; Russelle 2001. Also see the discussion of the name *Medicago* in Chapter 9). It has been speculated that the name "lucerne" may derive from the Persian word läjwärd for lapis lazuli, the ultramarine blue mineral lazurite, an allusion to the blue flowers of *M. sativa* subsp. *sativa* and subsp. *caerulea* (Russelle 2001). It may also derive from the Lucerna river valley in Italy or the old Spanish *eruye* (Bolton et al. 1972).

FRENCH: La luzerne, lucerne, herbe aux bisons, herbe à vaches. Early French writers apparently referred to alfalfa as *sain foin* (healthy hay), although this is now the common name of a different legume species, *Onobrychis vicifolia*.

Medicago sativa subsp. *caerulea*

ENGLISH: Blue alfalfa.

FRENCH: No names found.

Medicago sativa subsp. *glomerata*

ENGLISH: Glandular alfalfa.

FRENCH: La luzerne en pelote, luzerne en forme de pelote.

Medicago sativa subsp. *falcata* var. *falcata*

ENGLISH: Crescent-shaped alfalfa, horned clover, Siberian alfalfa, sickle medic, sickle-podded alfalfa, sickle-podded medic, Swedish alfalfa, yellow-flowered alfalfa, yellow-flowered medic, yellow medic (medic is also spelled medick). The scientific synonym *M. romanica* has been associated with the common names Romanian alfalfa and steppe alfalfa.

FRENCH: Luzerne en faux, luzerne faucille, luzerne de Suède, luzerne jaune, luzerne sauvage.

Medicago sativa subsp. *falcata* var. *viscosa*

ENGLISH: No names found.

FRENCH: No names found.

Medicago sativa subsp. ×*varia*

ENGLISH: Bastard alfalfa, hybrid alfalfa, sand alfalfa, variegated alfalfa (also substitute lucerne for alfalfa in these names), semi-twisted alfalfa (a translation by Bolton (1962) of the Latin name *M. hemicycla*).

FRENCH: La luzerne bigarrée.

Medicago sativa subsp. *sativa* × subsp. *glomerata*

ENGLISH: No names found.

FRENCH: No names found.

Geographical Distribution

(N = native, I = introduced, U = uncertain status)

Medicago sativa is native to Eurasia and, because of its widespread cultivation in the temperate world and its small tendency to be weedy, it has become established in most countries. Subspecies *sativa*, the most widely cultivated form, is the most widespread. Domesticated forms tend to escape from cultivation to occupy open habitats such as roadsides and the borders of fields, but these forms are not aggressive weeds. In temperate and northern parts of Eurasia, the wild, diploid subspecies *caerulea* and some wild tetraploid forms of subspecies *sativa* grow well in wild habitats and appear to be wild, ancestral plants of cultivated alfalfa, although it is probably impossible to determine what the exact ancestral range was, or indeed whether apparently wild plants are at least partly the result of human activities. Subspecies ×*varia* is extensively cultivated (especially in more northern areas of

Fig. 12-21. Alfalfa growing wild in the courtyard of the Forbidden City (in the middle of Beijing), China. Alfalfa has become one of the world's most widespread weeds, albeit one that is benign. Photo courtesy of G.R. Bauchan.

the world) and has a greater tendency to weediness than the other taxa, and so it is also very widely distributed. Subspecies *falcata* is rarely cultivated; it is grown especially in cold areas. It has been less changed by domestication than the preceding two subspecies — it is able to thrive in many wild habitats and has become very widespread, especially in Eurasia. Subspecies *glomerata* and subsp. *falcata* var. *viscosa*, the two taxa distinguished by heavy concentrations of glandular hairs on the pods, have restricted natural distributions (due to apparent value of the glandular hairs as antiherbivorous adaptations against local insects, as explained in Chapters 4 and 7).

Today, alfalfa is the most widely grown of the temperate forage legumes, occupying a world area of more than 30 Mha. The United States (led by California, South Dakota, and Wisconsin), Argentina, and the countries of the former USSR account for about 70% of this area, followed by Canada with about 2.5 Mha and Italy and China with about 1 Mha each. Wherever alfalfa is cultivated, escapes are likely to be found in the vicinity.

Medicago sativa subsp. *sativa*

This tetraploid taxon appears to have originated in Eurasia, possibly in historical times in the western distribution range of its diploid ancestor, subsp. *caerulea*. The tetraploids grow in proximity to the diploids throughout their range, sometimes in mixed populations (Small and Bauchan 1984), but extend well beyond the diploid range. Subspecies *sativa* was adopted as a principal crop; consequently, it has been widely introduced and now is so extensively distributed as a ruderal plant that a native range, if such a range ever existed, cannot be discerned with confidence. Possibly in preagricultural times this taxon occupied most of eastern European and western Asia, but interpreting this as the native range is speculative.

Medicago sativa subsp. *caerulea*

This taxon is native to eastern Turkey, Iran, the European USSR region (including the trans-Volga region), the Crimea, Dagestan and the Caucasus (Grossheim 1945), and in the Asian USSR region, primarily in Kazakhastan in two areas — one north and northwest of the Caspian Sea as far east as the Aral Sea and the other in central-eastern Kazakhastan (Ivanov 1980).

Medicago sativa subsp. *glomerata*

Africa: Algeria (N), Tunisia (N).

Europe: Italy (N).

This group occurs in montane areas of the Caucasus and adjacent countries of the former USSR, and in mountains of southern Europe to Italy and North Africa (Tunisia, Algeria).

Medicago sativa subsp. *falcata* var. *falcata*

Africa: Morocco (I), South Africa (I).

Asia: Afghanistan (N), Armenia (N), Azerbaijan (N), China (N), East Aegean Islands (Greece; N), Gruziya (N), India (N), Iran (N), Japan (I), Kazakhstan (N), Kyrgyzstan (N), Lebanon (N), Mongolia (N), Nepal (N), Pakistan (N), Russia (N), Syria (N), Tadzhikistan (N), Turkey (N).

Australasia: Australia (I).

Europe: Albania (N), Austria (N), Balearic Islands (U), Belarus (N), Belgium (N), Bulgaria (N), Corsica (U), Crete (I), Czechoslovakia (N), Denmark (N), Estonia (N), France (N), Germany (N), Great Britain (N), Greece (N), Hungary (N), Ireland (N), Italy (N), Latvia (N), Liechtenstein (N), Lithuania (N), Luxembourg (N), Moldova (N), Netherlands (N), Norway (N), Poland (N), Portugal (N), Romania (N), Russia (N), Spain (N), Sweden (N), Switzerland (N), Turkey (N), Ukraine (N), Yugoslavia (N).

Northern America: Canada (I), USA (including Alaska (I); see Small (in press *a*)).

These plants are widespread in northern Eurasia (including the colder areas of Russia, Mongolia, Scandinavia, and China) and widely introduced elsewhere. Hansen (1909) stated that "It extends through a large part of Western Europe, cental and southern Russia, the Crimea, the Caucasus, and through ca. the western two-thirds of Siberia, at least as far as 69° North

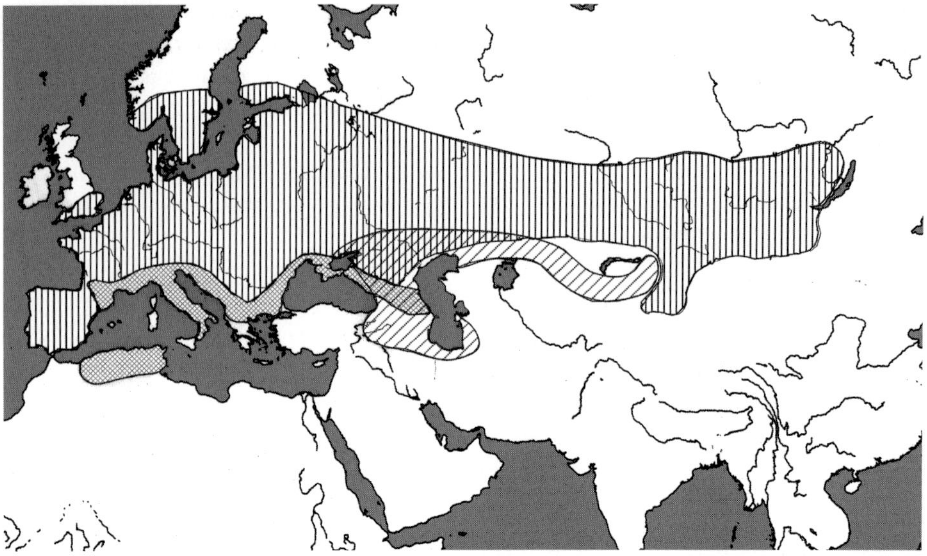

Fig. 12-22. The natural distributions of three of the subspecies of *Medicago sativa*. Vertical lines denote subsp. *falcata*; diagonal lines denote subsp. *caerulea* (note the substantial overlap with the other two subspecies between the Black and Caspian seas); and the double crosshatch pattern north of the Mediterranean, Black, and Caspian seas, as well as in northwestern Africa, denotes subsp. *glomerata*. The areas of these subspecies not only overlap with each other, but also with the very widespread ranges of subsp. *sativa* and subsp. ×*varia* — distributions that have been greatly expanded by humans.

latitude, through North China, the Trans-Caspian regions, including Turkestan and Persia, and through Afghanistan, Western India and Asia Minor." According to Oakley and Garver (1917) "There are no localities in which *Medicago falcata* is especially abundant throughout the wide region over which it occurs." Ecologically, the plants range from the colder portions of Siberia to hot semi-deserts. Depending on region and population, the plants are especially cold hardy, drought resistant, disease resistant, or able to tolerate relatively acidic soils. Some populations have creeping roots or rhizomes, which are naturally adapted to surviving frost heaving of the soil and seem to increase survival under heavy grazing.

Medicago sativa subsp. *falcata* var. *viscosa*

This taxon is more or less sympatric with the European range of subsp. *glomerata*. It has been collected from the Crimea, the north Caucasian shore of the Black Sea, the Caucasus, and southern Europe.

Medicago sativa subsp. ×*varia*

This heterogenous taxon is the result of both natural and artificial crossing between subspecies *sativa* (or *caerulea*) and *falcata*. Such hybrids (or hybrid swarms) are to be expected in the broad regions of geographical overlap of the two species, often exhibiting weedy tendencies. Within the past several hundred years, such hybrids allowed considerable expansion of cultivated alfalfa into northern regions of both North America and Europe, and onto relatively acidic soils. Today, most cultivars in cold climates have a parentage mainly of subsp. *sativa* that has been introgressed with small but important contributions from subsp. *falcata*, but are sufficiently similar to subsp. *sativa* to be identified as such. However, some cultivars in cold climates have substantial parentage of subsp. *falcata* and would be identified as subsp. ×*varia*.

Medicago sativa subsp. *sativa* × subsp. *glomerata*

These hybrids commonly occur within the range of subsp. *glomerata*, given above. They are frequent in the Caucasus and adjacent areas of the former USSR, and are found occasionally west of the former USSR in southern Europe, as well as in Algeria and Tunisia.

Description

Perennial herb (some annual domesticated forms), stems prostrate to erect, 10–100(150) cm long, branching, arising from a crown, rhizome, or roots. Vegetative parts of shoots more or less pubescent with simple appressed hairs, rarely also glandular hairs. Leaflets 5–30 mm long, 2–15 mm wide, obovate to linear or lanceolate, apical margin serrate, more or less pubescent below, more or less glabrescent above. Stipules lanceolate-acuminate, entire or basally toothed (especially at the base of lower, larger leaves). Inflorescence of 3–30(50) flowers, usually in a raceme, peduncle longer than subtending petiole. Flowers 5–15 mm long. Corolla violet, yellow, yellow-orange, variegated yellow-violet, or (rarely) pink, green or white. Calyx about half of length of flower, teeth about equal to length of tube, glabrous, or pubescent with simple hairs and/or glandular hairs. Pod spineless, brownish, glabrescent or pubescent with simple (and also sometimes glandular) hairs, straight, falcate, or with up to six coils, the coils loosely to tightly appressed, the pod face with veins running obliquely from ventral suture, branching somewhat, then anastomosing towards dorsal suture. Seeds 2–20 per pod, ovoid to deltoid, surface smooth, yellow, brownish, greenish-yellow, or violet-brown, 1–2.5 mm long, 1–1.5 mm wide, radicle slightly longer than half seed length.

Identification Hints

Subspecies *sativa*, ×*varia*, and *falcata* are not only widely cultivated but are also widespread as weeds, and are the most likely taxa of *M. sativa* to be encountered. Subspecies *caerulea* and *glomerata* are generally rarer and confined to Eurasia, but because they have been employed in alfalfa breeding programs, escapes could be encountered almost anywhere *M. sativa* is found.

Medicago plants with coiled, non-spiny fruits and purple or purple-yellow flowers are easily identified as *M. sativa* subsp. *sativa*, subsp. *caerulea*, or subsp. ×*varia*. Yellow-flowered plants are more problematic — if the fruit is coiled and non-spiny (with gland-tipped trichomes in subsp. *glomerata*, but lacking these in subsp. ×*varia*) there may be confusion with other perennials (some forms of which have fruit with gland-tipped trichomes), notably the much rarer and much more geographically restricted *M. rupestris* and *M. prostrata* (which have much more evident venation on the fruits), or with *M. papillosa* (which always has distinctive pads of tissue separating the seeds). Yellow-flowered plants with straight or falcate pods (*M. sativa* subsp. *falcata*) could also be confused with a variety of comparatively rarely encountered taxa, including *M. hybrida* (which normally has much wider leaflets) and some of the taxa of section *Platycarpae* (which mostly have much flatter pods, sometimes with rugose seeds).

Subspecies *glomerata* could be confused with *M. suffruticosa* subsp. *suffruticosa* of the Pyrenées region of Spain and France and the Atlas Mountains of Morocco: the latter usually has broader leaves, fewer flowers (three to 10 as opposed to four to 18), and somewhat longer radicles (about two-thirds the seed length as opposed to ca. half the length).

Habitat and Environmental Adaptation

Habitat

The species occurs in a variety of generally open habitats with moderately fertile soil.

M. sativa subsp. *sativa*: This taxon has been collected from a wide range of habitats, including steppes, rocky and grassy slopes, thickets, meadows, sand dunes, cultivated and fallow fields, and roadsides. Subspecies *sativa* includes wild, cultivated, and escaped forms that have become weeds. It includes an extensive range of ecotypes, differing in adaptations.

M. sativa subsp. *caerulea*: Some ecotypes of this group are adapted to semi-deserts and therefore appear to be drought-adapted; others, especially from the shores of the Caspian Sea, are thought to be tolerant of saline soils (Lubenets 1953).

M. sativa subsp. *glomerata*: This taxon occupies mesic, montane areas. Its gland-tipped hairs are protective against insects.

M. sativa subsp. *falcata* var. *falcata*: This group is adapted to the boreal, steppe conditions of Siberia. "It is on the dry steppes of Russia and Siberia that (it) attains its greatest importance, and over a large portion of that general region it is a fairly common constituent of the native vegetation... it seems to have a greater range of general adaptability than *Medicago sativa* (subsp. *sativa*), and is also less exacting in both its soil and climatic requirements" (Oakley 1917). Oakley also noted that it occurs from below sea level in Israel to about 2500 m in altitude in Afghanistan. Subspecies *falcata* occupies a huge and ecologically diverse range and its genes are very useful for extending the adaptive range of alfalfa, particularly with regard to cold resistance and the capacity to grow on relatively acidic soils.

M. sativa subsp. *falcata* var. *viscosa*: This taxon combines the general adaptive properties of subsp. *falcata* with the additional resistance to insects conferred by the presence of glandular hairs, as in subsp. *glomerata*. It occurs in lower parts of montane regions.

M. sativa subsp. ×*varia*: Having acquired tolerance to cold and acidic soils from subsp. *falcata*, this taxon has provided both cultivars and wild plants with considerable vigor. Subspecies ×*varia* includes wild, cultivated, and escaped forms that have become weeds, and contains an extensive range of ecotypes, differing in adaptations.

Phenology

Medicago sativa is a long-day plant, with the minimum photoperiod required to initiate flowering varying among cultivars (Major et al. 1991).

Soil

In cultivation, alfalfa achieves its best growth in deep, fertile, well-drained soils. Except for *M. sativa* subsp. *falcata* and cultivars with appreciable germplasm of this taxon, *M. sativa* is quite intolerant of significantly acidic soils, requiring a pH of 6.0–6.5 for optimal performance.

McCoy (1987) found that *M. sativa* has limited salt tolerance. Nevertheless alfalfa has a higher tolerance to saline soils than many other forage species and germplasm of ecotypes from saline habitats is available. Lesins and Lesins (1979) recorded the presence of *M. sativa* in sandy saline soils in eastern Anatolia.

With respect to subsp. *falcata*, Oakley (1917) commented that it is found in a great variety of soils ranging from stiff, heavy clay to almost pure sand. Lieth and Mochtchenko (2003) listed "*M. falcata*" in their halophyte database.

For *M. sativa*, a minimum pH of 4.9 is considered critical to allow effective nodulation (Sinskaya 1950), although soil acidity of 5.5 to 6.0 can inhibit nodule formation and suppress the growth of alfalfa (Ivanov 1980).

Temperature

There is great variation in *M. sativa* with respect to cold tolerance and the species can survive temperatures ranging from below –25 °C in winter to above 50 °C in summer. Alfalfa is grown over a wide range of environments, from Saudi Arabian deserts and subtropical areas of Australia through the much colder regions of Europe, western and Central Asia, and South and North America. Forms suited to cold regions are winter hardy and are adapted to a longer dormancy period and shorter growing season than less winter hardy forms suited to warm temperate regions.

Medicago sativa subsp. *falcata* is the most cold-adapted group within the species and hybrids with *M. sativa* (i.e., subsp. ×*varia*), including many cultivars, represent the most winter hardy kinds of alfalfa. Winter hardy cultivars harden relatively rapidly with deceasing temperatures and are slower to break dormancy when temperatures rise in the spring. Moreover, winter hardy kinds of alfalfa tend to be frost tolerant. Lesins and Lesins (1979) concluded that *M. sativa* subsp. *falcata* is not only cold tolerant, it is also heat tolerant, adapting it not only to the cold winters of steppes, but also to their hot summers. Oakley (1917) noted that in parts of India and southwest Asia, this taxon is exposed to extremely high summer temperatures coupled with a mild winter climate, whereas in Siberia winter temperatures are as low as –65 °C.

Water

Alfalfa is grown both under irrigated and dryland conditions. It is cultivated mainly under irrigation in the western United States and has been criticized as being wasteful of water. However, as pointed out by Putnam (1998), on a yield basis alfalfa is superior in water use to other major crops. Compared with many other forage species, alfalfa is an efficient user of water because of its deep taproot system. Water use efficiency is typically in the range of 5–9 cm evapotranspiration/hectare per tonne of forage dry matter produced (Sheaffer et al. 1988). Cultivars vary considerably in their ability to be productive in wet (irrigated) versus dry situations. The cold-tolerant alfalfas are generally less efficient water users than the less cold-hardy kinds. Plants become dormant during severe drought, resuming growth when moisture becomes available again (Hall et al. 1988). Such drought evasion unfortunately also reduces productivity. Alfalfa is a very productive crop, but only if provided with ample water, which it often uses extravagantly. In a world of shrinking water resources, there is a need to explore alfalfa and its relatives for the possibilities of finding or creating plant types that are both productive and water use efficient.

Alfalfa does not tolerate waterlogging. It has been suggested that the deep taproot of many variants of *M. sativa* is part of the problem (it may not be possible for oxygen to be transported down the taproots). Shallow-rooted cultivars withstand waterlogging better than deep-rooted cultivars (Cocks 2001). Intolerance of prolonged flooding is partly due to root and crown rots that may develop. When irrigation is provided or when alfalfa is grown in high-rainfall areas, well-drained soils are necessary to avoid soil saturation.

Subspecies *caerulea* is common in the dry, semi-desert region near the Caspian Sea where there is limited, if any, rainfall in the summer. Lesins and Lesins (1979) suggested its survival there is greatly aided by a deep taproot.

With respect to subsp. *falcata*, Oakley (1917) commented that "it is found in moist as well as in dry climates." Lesins and Lesins (1979) stated that subspecies *falcata* generally seems to prefer dry steppe conditions.

Sinskaya (1950) observed that alfalfa resistance to drought differs with respect to atmospheric (shoot) resistance and soil (root) resistance. Northern or mountainous forms of *M. sativa* seem to be relatively more resistant to soil drought than to atmospheric drought, a reflection of the relatively high moisture in the atmosphere but not in the soil. By contrast, southern forms that are irrigated are relatively more resistant to atmospheric drought than to soil drought. Since northern or mountainous climates are cooler than the hot oasis circumstances in which some alfalfas are cultivated, drought resistance and heat resistance are no doubt related. Further, atmospheric drought resistance may be related to disease resistance since high atmospheric humidity is conducive to the development of many fungal diseases. Sinskaya noted that Central Asian alfalfas, which are normally adapted to irrigation in low-humidity areas of Central Asia, have an unusually high susceptibility to fungal diseases when they are grown in humid areas.

Transpiration rates and water use efficiency vary among wild alfalfas, alfalfa landraces, and cultivars (Sinskaya 1950). This could represent adaptation to soil drought or to a combination of low atmospheric humidity and high temperature, or adaptation to desiccating winds. Numerous wild and cultivated forms of alfalfa simply shed leaves in response to strong drought so that, while survival is increased, productivity is strongly lowered. Of course, one wishes to optimize productivity in relation to local circumstances.

Biotic Relationships

Pollination

Small (1988) noted that the relatively large flowers and pollen to ovule ratio (374 for subsp. *sativa*, 334 for subsp. *falcata*) are consistent with *M. sativa* being at least partly cross-pollinated like other perennial species of *Medicago*.

Fig. 12-23. Special mechanisms of vegetative spreading in *Medicago sativa* subsp. *falcata*. **Left:** Reproduction by rhizomes (indicated by arrows; after Ivanov 1980, fig. 41). **Right:** Reproduction by suckering from "creeping roots" (indicated by arrows; after Sinskaya 1950).

Alfalfa on the whole is basically an outbreeder and selfing generally reduces seed yield. However, there is a large range of natural selfing and tolerance to inbreeding. Kovalenko et al. (1987) suggested that northern forms tended to develop a higher degree of selfing because the shorter seasons made setting at least some seed a greater priority than in southern climates.

According to Lesins and Lesins (1979), natural pollination of *M. sativa* is accomplished by ground-nesting bees, which are native to dry, low-precipitation regions and cannot survive in a rainy climate.

Dispersal

Lesins and Lesins (1979), in comparing forms of *M. sativa* with falcate pods (which they recognized as a separate species) and those with coiled pods (which they recognized as several species), stated: "The slightly bent, comparatively heavy pods of *M. falcata* are not adapted for distribution by rolling on the ground. Also, dehiscent pods scattering seeds indicate that *M. falcata* has originated under conditions where dispersal by whole pods was not suitable." They also stated that "The nondehiscent, roundish pods of *M. sativa* are adapted for dispersal by rolling with the wind, their spiral coils furthering an anchoring wherever loose drifting soil is encountered." Lesins and Lesins suggested that because *M. sativa* has some weedy tendencies and can establish in cultivated fields, and because its seeds are comparable in size and weight to some cereal grains, the species can be distributed along with grain, which is a mode of dispersal exhibited by many weeds.

Medicago sativa subsp. *falcata* has a greater ability to spread vegetatively than the other taxa of *M. sativa*. The plants often spread by rhizomes (i.e., underground stems), which are rarely present in subsp. *sativa* and more often present in subsp. ×*varia* because of some parentage of subsp. *falcata*. Also, subsp. *falcata* more often than the other taxa of *M. sativa* is "creeping-rooted", i.e., it has the capacity to produce adventitious stem buds at enlarged points along the length of its lateral

Fig. 12-24. Alfalfa is attacked by numerous insects, and resistant cultivars are a major breeding goal. **Clockwise from upper left:** Spotted alfalfa aphid; alfalfa seed chalcid with larva; tarnished plant bug with nymph; three-cornered alfalfa hopper with nymph; alfalfa butterfly with caterpillar; two spotted spider mite with nymph; alfalfa weevil and larva. (Artist: B. Flahey.)

underground roots (Grauman 1955; Daday 1962). The roots of creeping-rooted, plants "creep" or spread laterally, producing new shoots at some distance away from the crown. Creeping-rooted plants tend to have mat-like growth and nondistinct crowns and fibrous root systems rather than distinct, single taproots. See Chesneaux (1974) for a discussion of the relative ease of reproducing alfalfa vegetatively by root and stem cuttings of various ages.

Herbivory

More than 100 species of insects and nematodes are significant pests of alfalfa (Manglitz and Sorensen 1998). Breeding of alfalfa varieties for the past half century has been mainly concerned with creating cultivars that are resistant to pests and diseases. Multiple pest resistance is a key feature of modern alfalfa cultivars. Frame et al. (1998) list the following major pests of cultivated alfalfa: FOLIAGE: alfalfa weevils (*Hypera* spp.), caterpillars (*Colias eurythene*), cutworms/army worms (*Euxoa auxiliaris*), blister beetles (*Epicauta* spp.), spotted alfalfa aphid (*Therioaphis maculata*, *T. trifolii*), pea aphid (*Acyrthosiphon pisum*), blue alfalfa aphid (*Acyrthosiphon kondoi*), clover weevils (*Sitona* spp.), grasshoppers (*Melanophus* spp.), leafhoppers (*Agromyza frontella*), potato leafhoppers (*Empoacea fabae*), and spittlebugs (e.g., meadow: *Philaenus spumarius*); ROOTS/CROWNS: root knot nematode (*Meloidogyne hapla*), root lesion nematode (*Pratylenchus penetrans*), alfalfa stem nematode (*Ditylenchus dipsaci*), clover-root curculio (larvae, *Sitona hispidulus*), snout beetle (larvae, *Otiorhynchus ligustici*), and leatherjackets (*Tipula* spp.); SEED PODS: alfalfa seed chalcid (larvae, *Bruchophagus roddi*), and mirids (*Lygus* spp.).

Marble (1989) noted the principal insects of concern in the Near East.

Dawson et al. (1971) discussed the control of dodder (*Cuscuta* species) on alfalfa. These parasitic twining vines can seriously reduce alfalfa productivity.

Nodulation and mycorrhizae

USDA (2009) reported positive root nodulation for the various infraspecific taxa of *M. sativa*. The nodules of *M. sativa* are concentrated on the fibrous roots in the upper soil layer. Distinct strains of *Rhizobium meliloti* are required for different alfalfa genotypes. Seed inoculation is considered essential when introducing alfalfa to an area where the crop has not been grown recently or the soil pH is below 6.0. Below a soil pH of 6.0, *Rhizobium* activity and root nodulation may be markedly reduced.

It has been estimated that more than six million metric tons of nitrogen are fixed annually by leguminous crops in the United States, ca. one-third of which is because of alfalfa (Phillips and DeJong 1984). Moreover, the nitrogen yield from alfalfa is higher than from other legumes on a seasonal basis (Vance et al. 1988). The presence of rhizobia completely negates the need for nitrogen fertilizers for alfalfa and reduces the need for nitrogen fertilizers for crops rotated with alfalfa.

Fig. 12-25. Nodules on alfalfa roots (the arrow points to an area of many nodules). Photo courtesy of G.R. Bauchan.

Disease resistance

The principal diseases of cultivated alfalfa, along with their causal agents (after Frame et al. 1998), include: bacterial leaf spot (*Xanthomonas alfalfa*), common leaf spot (*Pseudopeziza medicaginis*), yellow leaf blotch (*Leptotrochila medicaginis*), Stemphylium (*Stemphylium botryosum*), Leptosphaerulina leaf spot (*Leptosphaerulina briosianna*, *L. trifolii*), downy mildew (*Peronospora trifoliorum*), spring black stem (*Phoma medicaginis* var. *medicaginis*), Alternaria (*Alternaria solani*), bacterial stem blight (*Pseudomonas medicaginis*, *P. syringae*), Stagonospora leaf spot (*Stagonospora meliloti*), rust (*Uromyces striatus*), summer black stem (*Cercospora medicaginis*), spring black stem (*Phoma medicaginis*), Fusarium wilt (*Fusarium oxysporum*), Verticillium wilt (*Verticillium albo-atrum*), bacterial wilt (*Clavibacter michiganense* subsp. *insidiosum*), Sclerotinia crown and stem rot (*Sclerotinia trifoliorum*), Rhizoctonia (*Rhizoctonia solani*), Phytophthora root rot (*Phytophthora megasperma*), anthracnose (*Colletotrichum trifolii*), Fusarium root rot (*Fusarium* spp.), and Aphanomyces root rot (*Aphanomyces euteiches*). Several viruses also cause significant losses, notably the alfalfa mosaic virus. For a more extensive list of alfalfa diseases, see Stuteville and Erwin (1990) and the American Phytopathological Society (2001). Marble (1989) examined the principal diseases of alfalfa in the Near East.

Economic History, Cultivars, and Uses

The origin of *M. sativa* traces to prehistoric times, with the selection of landraces having begun perhaps as long as 10 000 years ago, and probably in a variety of locations in Eurasia. As pointed out by Bolton (1962) "The common alfalfa (*Medicago sativa* L.) appears to be the only forage crop which was cultivated before recorded history — a distinction that limits the accuracy with which its center of origin can be deduced." Ivanov (1980), who suggested that alfalfa was cultivated 8000 to 9000 years ago, provides an extensive analysis of the numerous hypotheses that have been advanced regarding centers of domestication.

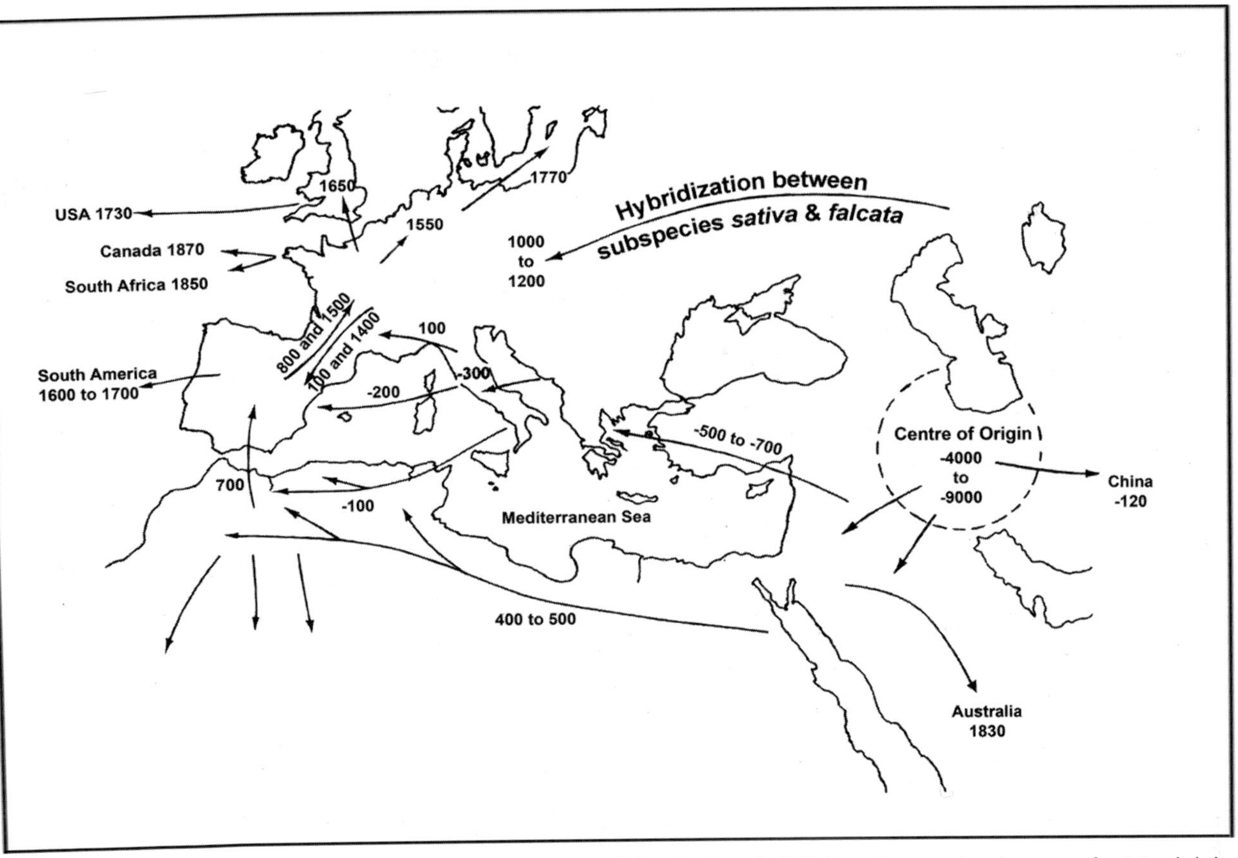

Fig. 12-26. An interpretation of the historical Eurasian routes of distribution of alfalfa from its postulated center of origin, slightly modified from Prosperi et al. (2001).

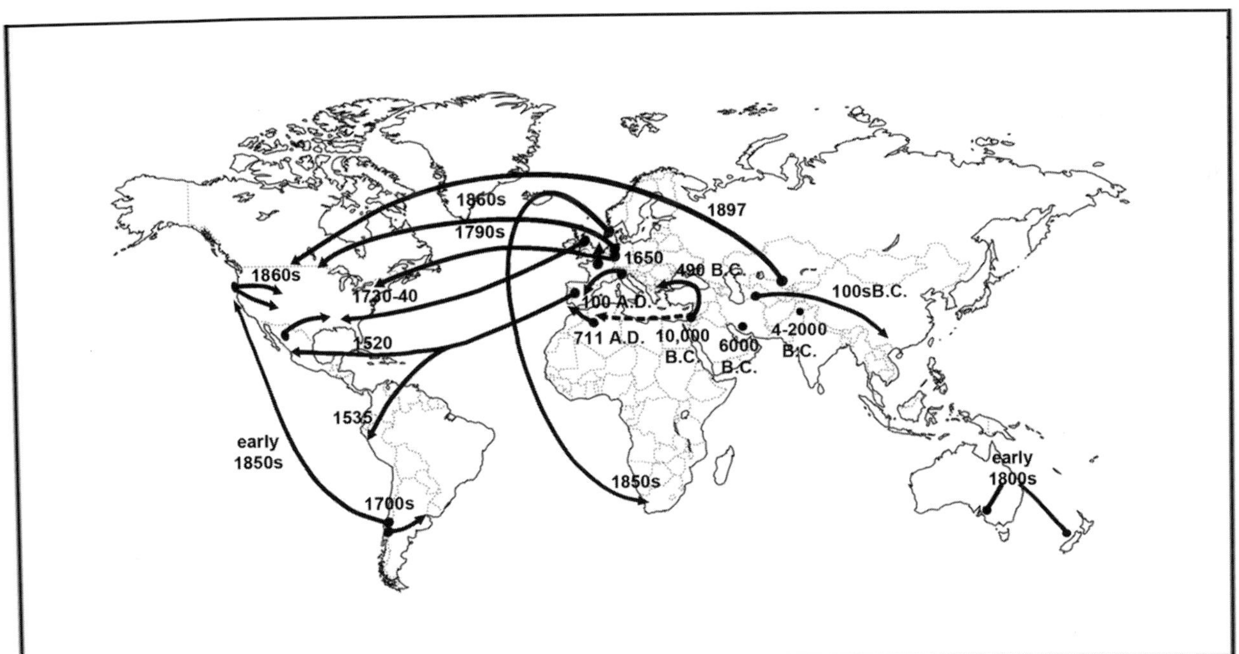

Fig. 12-27. An interpretation of world historical routes of alfalfa distribution, modified from Russelle (2001).

The postulated sites of alfalfa origin include the foothills and mountain valleys of Armenia, eastern Anatolia, Iran, Afghanistan, Central Asia, Jama, and Kashmir (Lesins 1976; Lesins and Lesins 1979). The highlands of Transcaucasia and Asia Minor and the adjoining areas of northwestern Iran are most frequently mentioned as the area of origin (Bolton et al. 1972). Asia Minor, Transcaucasia, Iran, and Turkistan constitute the "Near Eastern Center" of plant geographer N.I. Vavilov (Vavilov 1951), and there is a general consensus that alfalfa originated in this region. Perhaps alfalfa arose independently several times, rather than once. Sinskaya (1950) thought that there were two centers of origin, one in the mountainous region of Transcaucasia, Asia Minor, and adjoining areas of northwestern Iran, and the second in Central Asia, based on her observations that alfalfas in the latter area were in a less humid environment and were more susceptible to fungal diseases. Vavilov (1951) suggested that primary and secondary gene centers of cultivated plants are the best sources of germplasm resistant to numerous pests, which is not consistent with Sinskaya's deduction that fungus susceptibility in Central Asia reflects an origin there. However, Thies et al. (1985) found that resistance to bacterial wilt (*Corynebacterium insidiosum* (McCull.) H.L. Jens.) was highest in accessions from Central Asia. And, the work of A.A. Zarrabi (e.g., Zarrabi et al. 1985) indicated that resistance to the blue alfalfa aphid (*Acyrthosiphon kondoi* Shinji) is highest in south Central Asia (Afghanistan, India, Iran, Pakistan) and Transcaucasia. These observations suggest that Central Asia may well be a secondary center for alfalfa, where domestication subsequent to the original more westerly domestication may have occurred.

As noted below, the horse is likely the principal stimulus for the domestication and spread of alfalfa. Bolton (1962) concluded that "judging from all the old records, there can be little doubt that alfalfa was the principal fodder for the cavalry and chariot horses of the ancient Persians, Greeks, and Romans, and it was almost certainly carried along the old caravan routes by the travellers and invaders of those days." Invading armies and explorers would have brought alfalfa with them to feed their horses and other livestock. Alfalfa likely spread from its north-central Asian homeland both towards the East (to China and India) and towards the West (to Arabia, northern Africa, and Europe).

Postulated pathways of distribution of alfalfa are shown in Figs. 12-26 and 12-27.

Difficulties in tracing alfalfa domestication

The difficulty of distinguishing wild and domesticated alfalfa plants, coupled with a lack of fossil evidence, makes it hard to judge when domestication actually occurred. Also, the early literature on alfalfa is limited and is complicated by possible confusion by early writers of alfalfa with other plants, notably clover (*Trifolium*). For example, the word "sainfoin" (literally "true hay"), which used to denote alfalfa in France in the past (de Candolle 1885), is also used for other genera (e.g., *Hedysarum*, *Onobrychis*). Almost certainly the earliest recorded cultivation of alfalfa is predated by considerable domestication.

Early evolution of purple-flowered alfalfa

Alfalfa's early evolution begins primarily with the wild diploid *M. sativa* subsp. *caerulea*, which, as pointed out above, is native to a wide area of Asia and western Europe. Domesticated alfalfa belonging to subspecies *sativa* is entirely tetraploid, except for some experimental lines. As observed by Lesins and Lesins (1979): "the early agriculturists may have recognized the value of growing the more vigorous tetraploid *M. sativa* or, more likely, the spontaneously arisen tetraploids under cared-for growing-conditions ... naturally crowded out and replaced the less vigorous diploids." The general size increase associated with the tetraploid condition in the purple-flowered lineage is certainly preadaptive to cultivation and may explain the complete absence today of domesticated diploid purple-flowered alfalfa. Recently arisen tetraploids often have lower drought tolerance than their diploid progenitors (Stebbins 1986), but in cultivation alfalfa is usually provided reasonably moist circumstances. Stebbins (1947) believed that alfalfa is a segmental allopolyploid and, if so, the wide genetic range associated with this condition could account for the greater adaptability of the tetraploids. However, based on cytogenetic investigations and genetic studies that indicate tetrasomic inheritance, most students of alfalfa have concluded that the tetraploids arose by autopolyploidy (e.g., Lesins 1976). Segregation studies have not been conclusive, however. Some traits segregate according to an allopolyploid model, many others according to an autotetraploid model. This may be a result of differentiation in some sections of the genome (allopolyploid) and others that are not differentiated (autopolyploid). The rapid dissemination of *M. sativa* by humans in areas of the world outside its indigenous range makes it plausible that people were responsible for the expansion of the range of the tetraploids beyond that of the diploids.

The apparent center of domestication of purple-flowered alfalfa is in Vavilov's Near Eastern Center, mentioned above, which has moderately cold winters and hot dry summers, soils that are typically well drained, near-neutral pH, and subsoils with considerable lime (Michaud et al. 1988). The climatic and edaphic adaptation of domesticates that arose from this lineage limited the expansion of alfalfa cultivation to the adaptive limits of the wild plants.

Evolution of purple-flowered alfalfa in historical times

Charred seeds from archeological sites in Iran dating back 8000 years have been interpreted as alfalfa, speculatively consumed as a human crop (Russelle 2001). The earliest record of alfalfa use appears to be in ca. 3300-year-old Hittite brick tablets from central Turkey (Corum, Alacahoyük), dated from 1400 to 1200 BC (Bolton et al. 1972, based on a personal communication that the tablets "indicated that animals were fed on alfalfa all through the winter season and alfalfa was regarded as a highly nutritious animal feed"). The next earliest adequately documented evidence is a Babylonian text written at about 700 BC, in which alfalfa is recorded using a well-known Iranian name (Hendry 1923). The kind of alfalfa presently cultivated in much of Turkey, exemplified by the landrace Kayseri (Small 1981*b*, 1982; Small et al. 1990), has probably changed very little from the first domesticated alfalfa of western Asia. A closely related, less domesticated form called Yerli Yonca is cultivated in eastern Turkey, coincidentally where the wild diploids of *M. sativa* occur (Small 1981*b*, 1982). This class of cultivar is thought to have been acquired from Iran between 1400 and 1600 BC (Michaud et al. 1988).

Most authors accept that alfalfa of the purple-flowered kind was taken to Europe from southeastern Asia, about the fifth century BC, and migrated along with the Roman Empire (Hendry 1923). With the fall of the Roman Empire in the fifth century, alfalfa cultivation seems to have disappeared for the most part from Europe until the 16th century, when it was reintroduced. Purple-flowered *M. sativa* subsp. *sativa* appears to have been the principal kind of alfalfa cultivated in most of Europe until at least the 16th century. By the early 18th century, purple-flowered alfalfa was established in New Zealand and Australia, and was established in South Africa by the middle of the 18th century.

In the 16th century, European colonists introduced alfalfa to South America and introduced it to the eastern United States in 1736. By 1850, early cultivars, which were likely from South America, had reached California and from there the crop was progressively adopted in the western United States (Spanish missionaries are said to have introduced the crop from Mexico as early as the 1830s). These landraces were based on *M. sativa* subsp. *sativa* and lacked sufficient cold hardiness and tolerance to acidic soils to thrive in much of eastern North America and Canada.

Evolution of yellow-flowered alfalfa

The yellow-flowered alfalfa lineage (*M. sativa* subsp. *falcata*) exists mostly as wild plants, although there are a few cultivars (e.g., 'Anik' from Canada). 'Don' was selected for western United States rangelands rather than for pastures (Peel et al. 2009). Oakley and Garver (1917) observed that subsp. *falcata* has been utilized primarily as a wild forage plant since early times, and apparently was recently cultivated in India, southeastern Russia, and Chinese Turkestan. This Siberian-based subspecies is widespread across Eurasia (Oakley and Garver 1917), both as a diploid and as a tetraploid. Notwithstanding its cultivation in northern areas, yellow-flowered alfalfas have not developed much economic significance. Their importance lies in the expansion of the adaptive range of the domesticate of the purple-flowered lineage following hybridization.

Subspecies *sativa* and ×*varia* have provided the germplasm present in most cultivars. Subspecies *falcata* has contributed critical germplasm (much more to cultivars classified in subsp. ×*varia* than to cultivars classified in subsp. *sativa*). However, subsp. *falcata* is basically wild (i.e., undomesticated) and has undesirable characteristics, including slower regrowth, early onset of autumn dormancy, and decumbent growth habit. In northern areas of Russia, *M. sativa* subsp. *falcata* has been widely employed as a livestock species. As noted by Bolton (1962), this group has been widely distributed to farmers and researchers in North America and, consistent with its reputation, it has demonstrated some resistance to drought, cold, salinity, and pasturing. Nevertheless, due particularly to slow recovery after cutting and low seed set (there is a marked tendency for spontaneous shattering of the ripe seeds), agricultural use has been quite limited.

Evolution of variegated alfalfa

It is believed that when alfalfa was introduced to Germany and northern France in the 16th century, hybridization occurred between the purple-flowered and yellow-flowered lineages (Michaud et al. 1988). The resulting class of alfalfa is recognized as *M. sativa* subsp. ×*varia*, and includes not just the hybrids but also numerous hybrid derivatives. This hybridization resulted in the introduction of winter hardiness, drought resistance, disease resistance, procumbent habit, and tolerance to acidic soils into subspecies *sativa* from subspecies *falcata*. The resulting introgressed plants are commonly known as "variegated" varieties (for the variegated purple-yellow flowers), and include such informal classes as German "Franconia" and French "Provence" types of alfalfa. Although recorded history may not be exact as to the origin of such hybrid alfalfa, it is certain that such hybridization occurred. Wherever the purple-flowered and yellow-flowered alfalfa lineages grow in proximity in the wild, hybrids and stabilized introgressant populations commonly result. It is extremely probable that introgression has influenced the evolution of alfalfa in the northern limits of cultivation of the purple-flowered lineage wherever contact was made with the yellow-flowered lineage. The resulting plants would be superior adaptively to local conditions, and with selection for conditions of cultivation (unconscious or otherwise), improved alfalfa would have resulted.

Colonists from Europe imported to North America landraces based on *M. sativa* subsp. ×*varia*, and their natural resistance to cold and acidic soils, based on the ancestry of *M. sativa* subsp. *falcata*, led to the wide cultivation of alfalfa in the east and north of North America. Introduction of one particular variegated form, Grimm alfalfa (named for the German immigrant who imported it) led to the successful establishment of variegated types of alfalfa in the colder areas of North America (northern USA and Canada) in the latter part of the 19th century, where such forms now prevail. Today, more than half the world's alfalfa has parentage of both the purple-flowered and the yellow-flowered lineages (Lesins and Lesins 1979).

Cultivars

Thousands of cultivars of *M. sativa* have been bred (for detailed information see Hanson et al. 1960; Barnes et al. 1977; and Townsend et al. 1994). Most contemporary alfalfa cultivars are synthesized populations termed synthetics, developed by intermating six to 500 parents, often of diverse origin. In the past half century, breeding has produced numerous cultivars with resistance to pests and diseases. A particularly useful document, applicable to current cultivars grown in the United States, is the National Alfalfa & Forage Alliance's "Winter survival, fall dormancy, & pest resistance ratings for alfalfa varieties" (http://www.alfalfa.org/pdf/NAFA%2007-08%20Varieties%20Leaflet.pdf; current January 2010). The website of the North American Alfalfa Improvement Conference (http://www.naaic.org/; current January 2010) also provides a guide to cultivars.

Barnes et al. (1977) interpreted cultivated alfalfa introduced into the United States between 1850 and 1947 as traceable to nine basic germplasm sources: (1) subspecies *falcata* from Russia and Siberia; (2) Ladak, an introduction from Ladakh province of Kashmir in northern India, with considerable parentage from subsp. *falcata*; (3) subsp. ×*varia*, represented by natural hybrids from Europe; (4) African (probably from Arabia); (5) Chilean (from Spanish introductions to Mexico and Peru); and (6) Flemish (from northern France) — the last three types with apparent parentage of both subsp. *sativa* and subsp. *falcata*; as well as introductions from (7) India ("Indian" selected from two introductions from India); (8) Peru (Spanish introductions to Peru); and (9) Turkistan (from southern Russia, Iran, Afghanistan, and Turkey), composed primarily of subsp. *sativa*.

Subspecies *falcata* and ×*varia*, along with Ladak and Turkistan, are winter hardy; the Chilean and Flemish types have intermediate winter hardiness; and the African, Indian, and Peruvian types are nondormant, i.e., they are not winter hardy. Kidwell et al. (1994) examined DNA diversity among representatives of these nine alfalfa types and, except for subsp. *falcata* and the Peruvian group, were unable to distinguish the types genetically. Bauchan and Hossain (1998*b*, 2001*b*) and Bauchan et al. (2002) were able to identify some patterns among the historical groups using C- and N-banding of chromosomes.

Probably there are more cultivars of *M. sativa* subsp. *sativa* than of any other taxon of alfalfa. However, there is also an exceptional range of genotypes derived from crosses of subspecies *sativa* and *falcata* (many of these are of sufficiently obvious hybrid nature that they will be placed in subsp. ×*varia*). Such cultivars are widely employed in colder regions because the ancestry of subsp. *falcata* confers cold hardiness. 'Anik' is one of the few major cultivars of *M. sativa* subsp. *falcata* (Small 1992), although subsp. *falcata* has contributed considerable parentage to several alfalfa cultivars (Barnes et al. 1977). The glandular subsp. *glomerata* has been used for the genetic improvement of alfalfa, but only recently, and only to a limited extent (e.g., Sorensen et al. 1985). White et al. (1975) reported that subspecies *glomerata* (identified as *M. glutinosa*) has been used to improve the insect resistance of alfalfa, but their description of the plant suggests it was misidentified.

As noted below, "pasture-type" (for forage) and "hay-type" (for fodder) cultivars of alfalfa have been selected, the former relatively low yielding but capable of withstanding continuous grazing, the latter relatively high yielding but requiring a rest period between harvests. Some cultivars have been developed that combine the higher yield of the hay-type with the persistence under continuous grazing of the pasture type. The first cultivar of this type was released from the University of Georgia as 'Alfagraze' (Bouton et al. 1991).

Comparison of wild and cultivated alfalfa

A. Cryptic differences

Although alfalfa is the oldest of the cultivated forages, the cultivated phase has not evolved dramatically evident differences from the wild phase. Adaptations to disease organisms and invertebrate herbivores, and to local edaphic and climatic conditions, are not usually morphologically apparent, but are certainly among the most important characteristics distinguishing wild and cultivated alfalfa. Below, differences that have been selected between forage and fodder cultivars are discussed. Here, some general characteristics that have been selected under domestication are examined. Unlike many cultigens developed by humans, seed and fruit morphology and ecology are basically unchanged from those of wild plants.

Tolerance to repeated cutting or grazing is perhaps the most important agronomic and physiological change in cultivated versus wild plants. "Tolerance" to herbivory is a rather complex issue (Trumble et al. 1993; Rosenthal and Kotanen 1994; Verkaar 1986). Basically, tolerant plants maintain a high rate of recovery and growth in the face of loss of tissue. Coley et al. (1985; cf. reviews of Herms and Mattson 1992 and Mole 1994) pointed out that many species that grow very slowly put energy into such resistance features as deterrent toxins and mechanical structures; this seems adaptive to minimizing herbivory since recovery is necessarily very slow. On the other hand, many other species seem to put their energy into rapid growth, not defences, simply compensating for herbivory by quick growth. Obviously, under domestication plants that are grown as perennials for recurrent harvest and maximum production should belong to the latter class. Indeed, one would normally expect that under our continuous search for increased yield, such plants would be increasingly selected for tolerance to harvest. This analysis is consistent with what is known of alfalfa. The production of herbage and the longevity of perennial plants, including alfalfa, can be markedly reduced by too frequent grazing or harvesting. Nevertheless, few other forage species have been as selected to tolerate herbivory as alfalfa. Domesticated alfalfa sometimes tolerates (in long-season climates) 12 harvests annually for years (for example, this is possible with nondormant alfalfa in Hawaii). By contrast, wild plants tend not to withstand cutting very well.

B. Morphological differences

1. Stems

Except for certain types of low-growing, procumbent plants that mostly have some yellow-flowered background (Heinrichs 1963), domesticated alfalfas are erect, whereas wild plants show a range of variation but are often low to the ground, almost certainly to present a low profile to wild mammalian herbivores. Some of the changes under domestication are related to the growth of the crop in a dense monoculture or in competition with a few grasses and other legume species; this is in contrast with wild plants, which normally grow in competition with mixed vegetation. In a dense plantation, increased competition for light favors erect growth, hence selection of more robust stems. The stems of cultivated plants are often softer than those of wild plants.

2. Leaves

The leaflets of domesticated alfalfa are comparatively wide (the length to width ratio being in the order of 2 to 5), whereas often the leaflets of wild plants are narrower. Also, leaves of domesticated plants are usually larger than leaves of wild plants, and the leaf to stem ratio is often increased in cultivars, which may be achieved by profuse development of leaves. Selection for increased leaf to stem ratio has the desirable effect of increasing the protein content and tenderness of the crop, but generally makes plants more susceptible to herbivores. These leaf changes under domestication have been observed frequently in domesticated plants in addition to alfalfa, and likely reflect selection for photosynthetic productivity per unit area of land occupied.

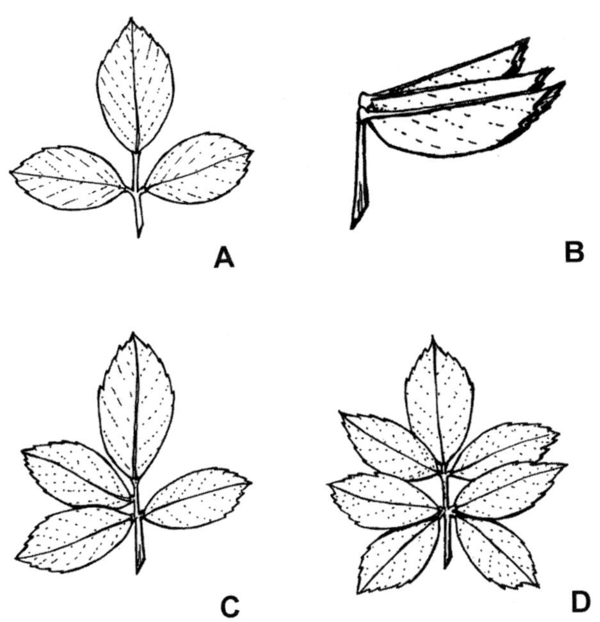

Fig. 12-28. Leaves of alfalfa with different numbers of leaflets. (A) Normal trifoliolate alfalfa leaf with leaflets in the open position (note that as is characteristic in *Medicago* species, the central leaflet is extended from the petiole via a petiolule that is longer than those of the lateral leaflets). (B) Normal trifoliolate leaf with leaflets folded in the protected position. (C) Multifoliolate alfalfa leaf with four leaflets. (D) Multifoliolate alfalfa leaf with seven leaflets. (Artist: B. Brookes.)

Normally, mature leaves of all species in the genus *Medicago* are trifoliolate. Selection of multifoliolate leaves (with more than three leaflets) is evident in Kayseri, the major landrace of Turkey, which quite likely is very similar to the earliest type of domesticated alfalfa (Small 1981*b*, 1982). Selection for multifoliolate leaves is, to this day, considered a means of improving alfalfa quality by plant breeders (e.g., Volenec and Cherney 1990; Juan et al. 1993), although this has been questioned. Multifoliolate leaves presumably increase the leaf to stem ratio, thereby improving quality. The multifoliolate trait per se does not necessarily confer greater herbage quality, which requires concurrent selection for quality measures such as digestibility. Trifoliolate leaves of *Medicago* species represent precision in evolutionary architecture — when young and vigorous, they fold elegantly at night to reduce their exposed surfaces by two-thirds (Small et al. 1981*c*; cf. Chapter 7). The leaflets are phototropic (they have a pulvinule), orienting to maximize photosynthesis or minimize stress from photodamage, extreme temperature, and desiccation (Travis and Reed 1990). A multifoliolate alfalfa leaf is an unnatural structure that sacrifices the natural orientation advantages of the trifoliolate condition, but does appear to represent a way of increasing the leaf to stem ratio, and thereby increasing quality. (As a matter of interest, a mutant phenotype of alfalfa with unifoliolate leaves (so-called "unifoliate" and "single-leaf") has been the subject of experimentation; see, for example, Brouwer and Osborn 1997.)

3. Roots

Root systems in cultivars may be adapted to occupation primarily of the arable soil layer (Ivanov 1980), and so are often shallower than the roots of wild plants, with less development of a deep taproot.

Forage

In the United States, dry matter yields of up to 20 t/ha have been achieved experimentally (Sheaffer et al. 1988). Yields of 9.4–17.6 t/ha have been reported in the United Kingdom (Aldrich 1984; Frame and Harkess 1987) and yields of 14.5–19.0 t/ha have been reported in France (Guy 1993). In the USA, seed yields average 750 kg/ha in the western states where the majority of crops are grown under irrigation, whereas in the Midwestern states, yields of 50 kg/ha are typical (Barnes and Sheaffer 1995). In Europe, an average seed yield is about 400 kg/ha.

Bloat may occur in ruminant animals when lush stands of alfalfa are grazed. Bloat does not occur in monogastric animals such as horses, but allowing horses to graze in luxuriant plantings of alfalfa can result in founder (a metabolic disorder resulting from stress, especially overeating, leading to a severe kind of laminitis, i.e., a hoof disorder that involves separation of live and horny portions of the hoof, which causes severe lameness). Estrogens that are produced in response to

Fig. 12-29. Photographs of the relatively primitive fodder harvest technology that has persisted in rural areas of eastern Turkey for centuries. **Top:** Sickle and its uses (a common injury for those using a sickle is the loss of one or more of the small digits of the left hand holding the plant material, especially when harvesting grain). **Bottom:** Scythe and its uses. Photos taken by E. Small near Ahlat village in 1981.

disease invasion can be present in *M. sativa* and are capable of reducing conception rates in cattle and sheep when they are grazed on or fed alfalfa prior to mating. Saponin content confers resistance to pests, but can affect some classes of livestock (see below). Estrogen content and saponin content differ among genotypes.

In some cases, alfalfa may be responsible for allergic reactions. Suspicion has been expressed that alfalfa has caused photosensitivity in animals following its ingestion (Mathews 1937; Hurst 1942; Gardner and Bennetts 1956; Watt and Breyer-Brandwijk 1962). In humans, alfalfa pollen is suspected of causing hayfever (Wodehouse 1971). Kaufman (1954) reported that injection of an infusion of alfalfa seed as a nostrum for people with arthritis produced edematous eruptions in two cases.

Forage versus fodder adaptation

Consistent with the definitions given in Chapter 2, throughout this monograph "forage" refers to fresh herbaceous vegetation that is consumed directly from the field by grazing livestock, whereas "fodder" refers to herbaceous vegetation that is harvested by humans, usually preserved (generally dried as hay or made into silage) and fed to livestock. Most alfalfa researchers refer to forage cultivars as defined here as "pasture-type" and fodder cultivars as defined here as "hay-type".

Under domestication, there has been disruptive selection of alfalfa for two kinds of harvest. To simplify, this has been for gradual but continuous harvest of the shoot versus occasional but massive harvest. "Pasture type" or forage cultivars have been selected for tolerance to continuous grazing. They are normally grown in mixtures along with grasses and other legumes, with which they must compete. Where animals are confined to an alfalfa pasture, the plants are continuously eaten for extended periods, trampled, and have animal waste dumped on them. Moreover, pastures are not usually irrigated, are less heavily fertilized, and often become weedy. By contrast, "hay type" or fodder cultivars have been selected for tolerance to intermittent harvest. These are allowed to grow undisturbed for an extended interval during which they can accumulate storage reserves. They must be adapted to withstanding and recovering quickly from periodic more or less complete harvests of the shoots. While frequently grown with grasses for mixed hay, fodder cultivars are often grown as a monoculture. By contrast with pastures, alfalfa monocultures are often irrigated, fertilized, weeded, and treated for insects. A little-appreciated benefit of clear-cutting alfalfa periodically is a drastic reduction of insect pests for several weeks — an effective means of controlling insects (Harper et al. 1990). In dryland, north-temperate regions, forage cultivars often have a procumbent growth habit, low-set crowns, drought tolerance, and marked fall dormancy. Forage (pasture) cultivars typically produce less herbage and less seed than fodder (hay) cultivars, and recovery is comparatively slow after cutting (Lowe et al. 1972). However, forage cultivars are capable of withstanding the greater stresses of their much less pampered environments.

Grazing tolerance in alfalfa has been associated with a number of features that contribute to generating and maintaining leaf area below the grazing height (see, for example, Brummer and Bouton 1991). These characteristics include decumbency or prostrateness, continuous bud production during the grazing season, high stem number, broad crowns, low (or underground) crowns, and creeping rootedness. The so-called creeping-rooted character is of special interest, having contributed to several grazing-tolerant cultivars and allowing persistence under heavy grazing (Heinrichs 1963). As pointed out above, creeping-rootedness is a much more common characteristic in *M. sativa* subsp. *falcata* than in other alfalfa lineages (see following paragraph). This should not be confused with rhizomatous forms of alfalfa, in which true rhizomes are initiated only from the original root axis of the parent plant, extend laterally only for a short distance, form a root, and emerge from the soil at the tip of the rhizome to

Fig. 12-30. Photographs of the relatively primitive alfalfa fodder technology that has persisted in rural areas of eastern Turkey for centuries. Photos taken by E. Small near Ahlat village in 1981 (Small 1981*b*). **Top left:** Wooden-wheeled oxcarts with alfalfa hay. **Top right:** Hay stacked on a house, a frequent practice to keep hay safe from pests. **Bottom left:** Weaving a baling rope out of alfalfa hay. **Bottom right:** Piles of dried manure clods used as fuel, especially for winter heat.

produce new stems (see Fig. 12-23). Because of meristematic activity in the phellogen, creeping-rooted alfalfa has the ability to develop new adventitious shoots and roots from buds on subsurface roots running horizontally 10–20 cm below the soil surface. Creeping-rooted alfalfa is normally very much more of a vegetatively spreading plant than rhizomatous alfalfa.

The high frequency of creeping-rootedness in *M. sativa* subsp. *falcata* may reflect the exceptional cold hardiness of this Siberian variant rather than adaptation to herbivory. Creeping roots provide protection against frost heaving, and the juvenile growth habit associated with creeping roots (which generate a continuous series of new plants) also promotes winter hardiness since juvenile growth is always more winter hardy than older growth (B.P. Goplen, personal communication). Deep-set crowns represent another characteristic that is useful both for winter hardiness (promoting thermal insulation) and inaccessibility to herbivores.

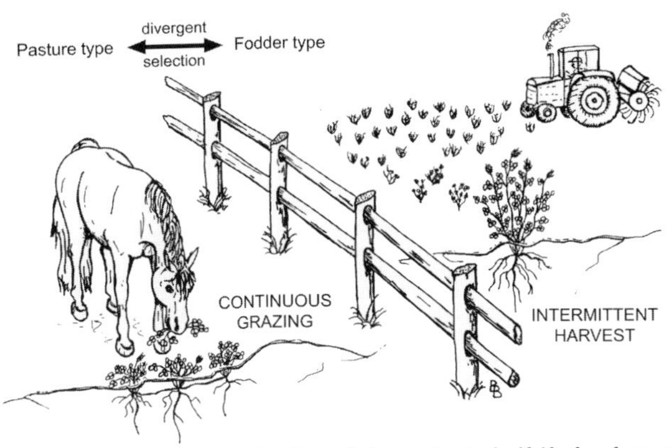

Fig. 12-31. Disruptive selection of domesticated alfalfa for forage (left) and fodder (right). A creeping-rooted form of alfalfa, adapted to continuous grazing, is shown on the left, while an erect form, selected to produce considerable biomass on a 4- to 6-week harvest cycle, is pictured on the right.

Fig. 12-32. Field choppers such as this machine can cut green alfalfa, chop it into smaller stem lengths, and blow it into a hopper. From the chopper the forage may be fed green (as "green chop") or put into a silo for long-term storage (as silage). Photo courtesy of J. Caddel, Oklahoma State University.

Fig. 12-33. Compressed alfalfa cubes. Photo courtesy of G.R. Bauchan.

Creeping-rooted alfalfa is obviously adaptive to grazing simply in that lateral vegetative spreading per se is a means of avoiding, albeit temporarily, being eaten by mammalian herbivores. However, there may be an additional adaptive explanation that is related to excrement deposition in pastures. As noted by Parish and Turkington (1990), urine and dung deposition cause considerable temporally and spatially varying pressures on plants in pastures. Urine contains nitrogen and potassium, which may stimulate the growth of grasses at the expense of some legumes such as alfalfa (which have a much lower need for supplemental nitrogen). Plants directly under dung pats may be killed or drastically inhibited. On the other hand, being near dung may be advantageous since cows rarely graze within 30 cm of fouled areas (references in Parish and Turkington 1990) and dung can be a useful soil amendment. Over time, microenvironments would wax and wane, and areas in which vegetation is killed would be opened to colonization as the effects of excrement become diluted. The ability of creeping-rooted alfalfa to spread laterally, and so find advantageous positions within its extremely heterogeneous environment, should therefore be of survival value. (Parish and Turkington (1989) did not notice interspecific differences in survival between spreading and nonspreading species in pastures, but such adaptive value could be much more evident intraspecifically, for example between spreading and nonspreading alfalfa cultivars.)

In addition to the morphological differences described above that distinguish forage-type and fodder-type alfalfa, physiological differences are also important. Avoiding exhaustion of root carbohydrate reserves appears to be a critical factor. The faster growth of fodder-type cultivars is associated with quicker use of the reserves than that used by the forage-type cultivars, and plants with more rapid use of the reserves tend to perish more quickly under grazing pressure (Berdahl et al. 1986). Basal leaf retention may also be important with respect to maintenance of root reserves by providing a source of new carbohydrates for replenishment (Brummer and Bouton 1992; Smith et al. 1992; Smith and Nelson 1967). Hodgkinson et al. (1972) observed that old leaves remaining on grazed alfalfa can increase their rate of photosynthesis to that of new leaves.

Fodder

Alfalfa is an excellent crop for production of hay, silage, and dehydrated material (cubes, pellets, and meal). Although alfalfa is currently used primarily to feed dairy cows, it is also employed as food for horses, beef cattle, sheep, chickens, turkeys, and other farm animals. Dehydrated alfalfa is used as a nitrogen supplement to feed a variety of livestock, including cows, rabbits, pigs, and poultry. In cold, northern areas, the aboveground portion can be harvested only once in a season, but in very warm climates as many as 12 cuts may be taken in a year. Alfalfa is the principal hay plant in numerous countries; for example, in the USA it is found in 40% of all hay (Hintz 1983).

Green manure, rotation crop, ground cover

Alfalfa is also known to considerably improve soil aeration and water infiltration into the soil because of the deeply penetrating, vigorous root system that creates openings. The deep roots enhance the soil by recycling nutrients from regions of the soil that are too deep for many other plants to reach. As a cover crop, alfalfa is particularly useful between trees in orchards and between widely spaced row crops. Because of its high nitrogen production, alfalfa is widely used as green manure. As a rotational crop, alfalfa not only improves soil nitrogen, but also serves to suppress weeds and break disease cycles. Curiously, some organically oriented horticulturists have found that alfalfa pellets are useful for fertilizing iris and rose beds, and even orchids. The alfalfa pellets are applied as a side dressing, or a tea is prepared and sprayed on the foliage.

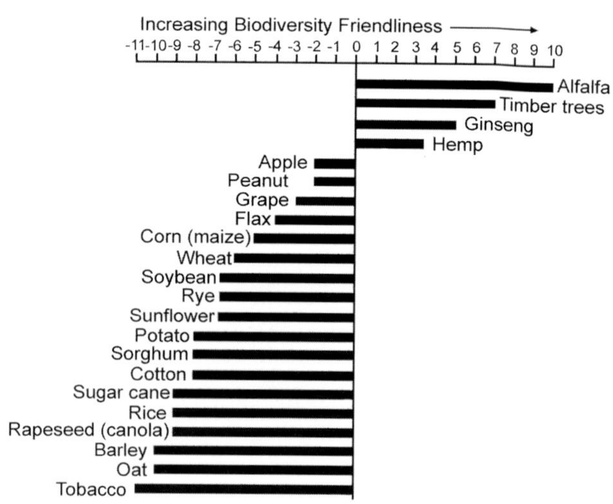

Fig. 12-34. A comparative evaluation of the biodiversity friendliness of major crops, in which alfalfa proved to be the best. See Montford and Small (1999*a*, *b*) for details of the methodology.

Some California cotton farmers have planted alfalfa as a trap crop for lygus bugs, an important cotton pest. The bugs prefer alfalfa to cotton. Rather than mowing all the alfalfa at once, the farmers alternately mow half the alfalfa habitat so that there are always fresh young alfalfa plants to retain the lygus populations. (Reported in The National Center for Appropriate Technology, archive for 2007; http://www.ncat.org/news/news2007.php; current September 2008.)

Putnam (1998) and Putnam et al. (undated) examined the contributions alfalfa makes to wildlife and the environment. Small and Catling (2007) reviewed the importance of alfalfa in terms of its biodiversity friendly values compared with other major crops. The friendliness of alfalfa for the welfare of biodiversity and the environment is remarkable. Alfalfa is extremely useful for reducing the ecological costs of agriculture, due in part to the fact that it is the most widely adapted agronomic major crop and the most energy-efficient crop to grow in temperate regions (Barnes et al. 1988). As a nitrogen-fixing plant, alfalfa is exceptionally efficient at reducing the amount of fertilizer nitrogen needed in agriculture and consequently of nitrate pollution (Peterson and Russelle 1991). Additionally, large amounts of manure are produced from livestock fed alfalfa, further contributing to organic agriculture.

The hydrological use of alfalfa in reduction of soil salinization in Australia

Parts of Australia are running out of drinkable water as well as natural groundwater that can support agricultural crops, despite the fact that there are usually substantial winter rains. Realizing that the country is faced with a serous environmental and resource management problem, the government has dedicated billions of dollars to the issue. The problem is based on the occurrence of salt at high levels in the subsoils of Australia. Before Europeans colonized Australia, deep-rooted perennials (trees, shrubs, and herbs) dominated the land, sucking water from considerable depth and consequently drying out the upper layers of soil and keeping the soil water level low. Europeans, however, replaced the native vegetation with shallow-rooted plants, often annual cereals for humans and grasses for sheep grazing, particularly in southern Australia, although Western Australia is also now at risk. This has led to the water table rising dramatically (in some places as much as 30 cm annually) and concentrating the salts from subsurface layers near the surface, which is contaminating drinking water and making it increasingly difficult for plants to grow because of soil salinization. Late in the 20th century, Australians realized that the situation was becoming critical and initiated remedial measures. One of the measures adopted has been the deliberate cultivation of alfalfa, the deep roots of which mimic Australia's primeval vegetation in drying out the upper levels of soil and thereby lowering the water table and the accompanying salt that has caused the problem. For a review of the role of alfalfa in salinity management in Australia, see Robertson (2006).

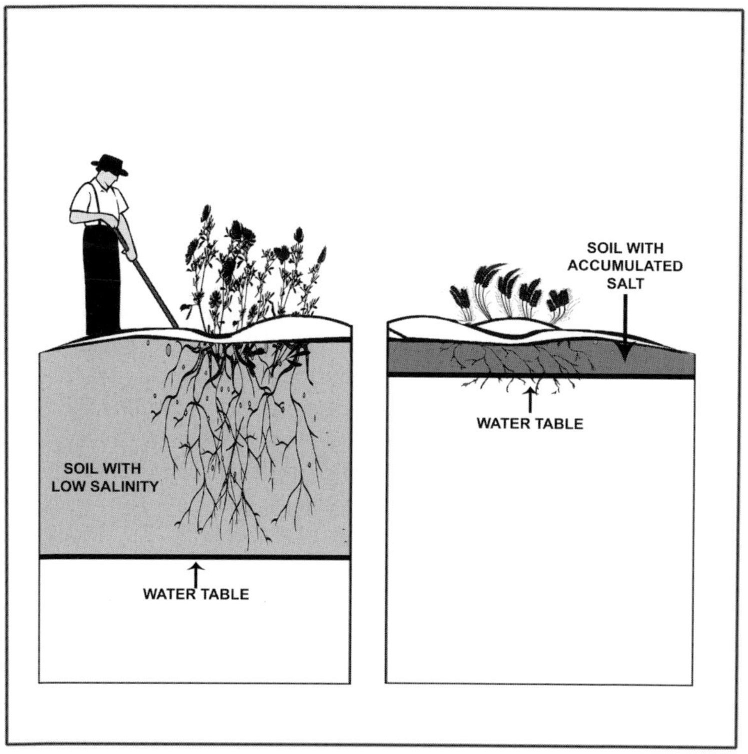

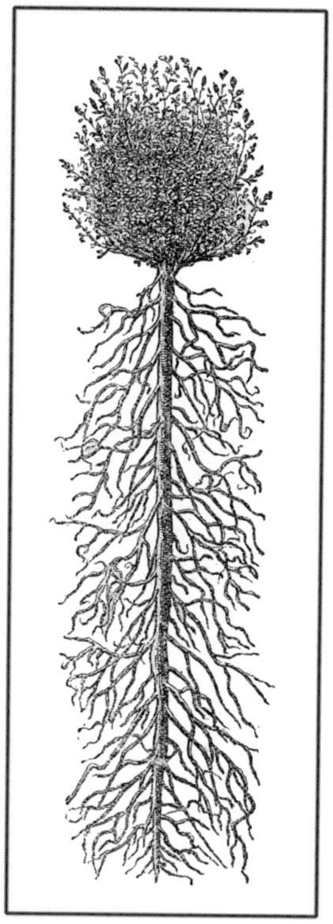

Fig. 12-35. The deep roots of alfalfa (shown in the left part of the left box) are useful in parts of Australia for drying out the upper soil layers and preventing salinization of the soil associated with cultivation of shallow-rooted crops (shown in the right part of the left box). The far right box shows an alfalfa plant with a long taproot, from Thompson (1895). (Figures prepared by B. Brookes.)

Other uses (salad sprouts, ornamental, plant products, etc.)

Alfalfa has long been used as human food, albeit in a relatively minor way. Curiously, the earliest documented use of the alfalfa genus was as human food. Flannery (1969) described evidence that, in a southwestern Iranian farming village dated from 7500 to 5600 BC, wild alfalfa seeds were consumed (likely from annual species of *Medicago*), probably in a mixed gruel, and possibly also wild alfalfa stems. He noted that the high protein wild legumes collected locally probably contributed to a diet better than that available in modern Iranian villages. The contemporary use of alfalfa as a vegetable is not extensive, although in recent times the young growth of established plants has been used as food in China, other parts of Asia, and South Africa (Fox and Wilson 1937). Dhar and Dhar (2000) recorded that in northern India the tender young leaves are consumed as a potherb. Bolton (1962) provided some recipes for using alfalfa foliage and stems as human food. Human consumption of alfalfa is now largely confined to the sprouted seeds. Fresh alfalfa sprouts are now commonly available in supermarkets and are widely thought to be an excellent, nutritious, gourmet food. They are also considered to be healthful, furnishing significant amounts of minerals, protein, and the vitamins thiamine, riboflavin, niacin, and ascorbic acid (Hesterman et al. 1981). Raw alfalfa sprouts are rapidly increasing in popularity for use in salads and sandwiches, and are also added just before they are served to cooked dishes such as stews, soups, omelets, vegetable stir-fries, and tacos.

Alfalfa saponins reduce levels of blood serum cholesterol in animals and this is considered good for people since elevated cholesterol levels clog arteries and can lead to heart attacks. However, it is not yet clear whether alfalfa saponins could be harmful to humans. The toxic, nonprotein amino acid canavanine in the seeds has caused anemia in monkeys (Bass et al. 1988) and has been alleged to be associated with the disease lupus (Foster and Duke 1990). Accordingly, some have contended that the consumption of alfalfa sprouts and alfalfa tea should be limited (e.g., Tyler 1993). Bacterial contamination is also a potential problem for alfalfa sprouts. Because of the surface features of the seed, surface sterilization is problematic and *Salmonella* and *Escherichia coli* O157:H7 outbreaks have occurred in these sprouts (see the Canadian Food Inspection Agency Bulletin "Sprouts Health Risk": http://www.inspection.gc.ca/english/fssa/concen/specif/sproutse.shtml; Also see the *Food Poison Journal*: http://www.foodpoisonjournal.com/2009/04/articles 26 April 2009, Raw alfalfa sprouts linked to salmonella contamination; current January 2010). Despite these problems, consumers should be confident that alfalfa sprouts are a healthful food.

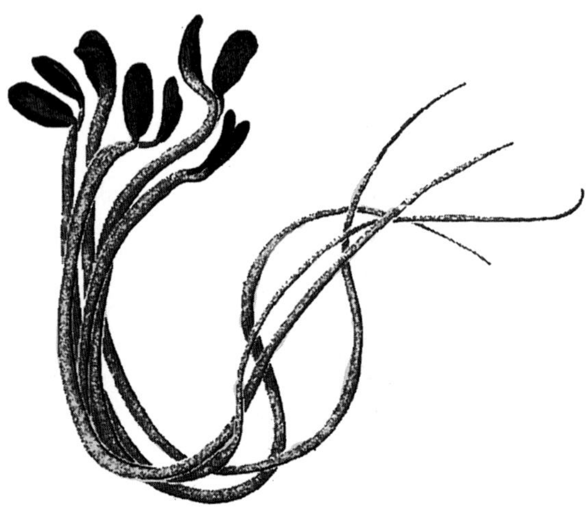

Fig. 12-36. Fresh alfalfa sprouts, which are popularly used in salads and sandwiches, and added to various cooked dishes. (Artist: M. Jomphe.)

Alfalfa is an important honey crop — in the United States, alfalfa is the primary honey crop, accounting for about one-third of the country's annual production (Putnam et al. undated).

Pablum was a bland, soft, precooked, dried cereal for infants marketed by the Mead Johnson Corporation, then by the Heinz Corporation, up until very recently. The cereal was invented in 1930 by scientists at the Hospital for Sick Children in Toronto. Pablum was the first fortified baby food, and its ease of preparation (water was simply added) greatly contributed to preventing infant malnutrition. Pablum was originally made from a mixture of wheat, oat, corn meal, bone meal, yeast, and alfalfa.

Alfalfa is the world's most efficient protein crop and the most important source of protein yield per hectare (Barnes et al. 1988), producing a remarkably high protein yield of more than 2 t/ha in temperate climates (e.g., Génier et al. 1992). It is therefore of major importance for humans, based on anticipated requirements for increased direct consumption of plant proteins. There is limited unexploited land available that can be converted to agricultural use, so efficiency of protein production will become critical in the near future. At present, alfalfa's food value to humans is chiefly as a precursor to meat, milk, and other animal products. It is becoming painfully clear that parts of the world can no longer afford the inefficiency of converting plants to meat. Although this inefficiency is less in arid regions, there is an increasing need to use plant protein more directly by using efficient crops. Commercial demand for proteins is growing rapidly. Alfalfa protein has extraordinary potential for alleviating world hunger and concomitantly reducing the ecological costs of agriculture. Methods have been developed to produce alfalfa leaf protein concentrate, involving the extraction of protein from the juice of green plant material (e.g., Olvera-Novoa et al. 1990; Huyghe et al. 2007). One day, alfalfa protein may be incorporated into various foods in the human diet. Other current research involves the insertion of specialized protein genes into alfalfa through conventional and genetic engineering techniques, which could lead to production of large quantities of desired speciality protein. As agricultural land is converted to urban uses, the efficiency of alfalfa protein production for feeding not just animals but humans as well could become extremely significant for addressing both the world's hunger problem and the technological development of new foods from basic plant constituents.

Perhaps the greatest new potential use of alfalfa lies in the possibility of using it as a biomass crop for the production of pharmaceutical and industrial products. Alfalfa is currently used to produce recombinant proteins commercially. The species has numerous benefits as a production host (D'Aoust et al. 2004). Its advantages include favorable agronomic characteristics (prolonged vegetative growth, ease of clonal propagation through stem cutting that is beneficial for batch-to-batch reproducibility), ability to fix nitrogen through symbiosis (thus reducing fertilizer inputs), and high leaf protein content (up to 30 mg total protein per gram of fresh weight). The Canadian biotechnology company Medicago Inc. (http:/www.medicago.com; current May 2010) has developed transformation methods and vectors for production of biopharmaceuticals in alfalfa leaves. The company claims to produce 10 g of recombinant protein for every square meter of greenhouse space. For alfalfa, yields of 1% of total soluble protein have been realized for the C5-1 antibody (Khoudi et al. 1999) as well as accumulations of 3.36% β-glucuronidase (D'Aoust et al. 2004). There is concern that the cultivation of outcrossing common food and livestock plants that have been genetically modified to produce pharmaceuticals could result in the transfer of the genes responsible for the medicinal compounds to varieties that are grown for consumption, with resulting toxicity. In the case of alfalfa, it should be possible to breed inbreeding varieties or make use of sterility genes to prevent unwanted genetic contamination. As of the preparation of this monograph, the subject of genetically engineered alfalfa is controversial. Alfalfa genetically engineered for herbicide resistance (Roundup, which contains glyphosphate, produced by the Monsanto Company) was marketed in the USA from 2005 to 2007, then prohibited from additional use pending examination of environmental concerns, particularly the transfer of the genes for herbicide resistance to weedy and cultivated forms of alfalfa. An appeal by Monsanto to the United States Supreme Court is under consideration. Van Deynze et al. (2008) reviewed concerns about gene flow in alfalfa.

Other materials that have been suggested as possible products that could be derived from alfalfa include biodegradable plastics; fermentation products such as alcohol and lactic acid; saponins with insecticidal properties; and carotene and xanthophyll in extracted juice (Frame et al. 1998).

Boe et al. (1998) noted that pheasants and ducks in the Midwestern states often use alfalfa fields for nesting sites in spring and early summer, and alfalfa mowing is very destructive to the nesting birds. They discussed the use of long-duration-flowering subsp. *falcata* as a late-harvest crop that would allow the game birds to employ the alfalfa fields as habitat during the early part of the season. Boutin et al. (1994) proposed that alfalfa be used as a set-aside cover crop for endangered birds, as it is one of the few crops that can produce high levels of biomass with minimum inputs that might affect ecosystems.

Alfalfa has considerable potential for production of biofuel. Although still at the experimental stage, several biofuel proposals are under investigation, including gasification of the less nutritious parts of the plant and conversion to electricity and, alternatively, fermentation of these parts to produce alcohol (Samac et al. 2006). A key to the choice of biomass plant species is the issue of lifetime energy costs, i.e., considering all factors, what is the relative amount of energy contained in the biofuel or biomass produced relative to the amount of energy invested? Studies have shown that energy costs associated with the production of alfalfa are low — indeed, they are far lower than for the production of corn (currently used to produce ethanol biofuel) and soybean (used to produce biodiesel).

Alfalfa has been experimentally examined for its potential as a fiber crop, especially for paper-making in Scandinavia, with the possibility of breeding cultivars with a higher content of stem fiber and cellulose, and fewer thin-walled cells, to improve the yield of pulp and reduce moisture retention (Berggren 1992).

The nature of polyploidy in *Medicago sativa* has been of particular interest to geneticists and alfalfa has been widely used as a model system in studies of autotetraploid inheritance (Barnes et al. 1988). Alfalfa has also been used to investigate alternate breeding strategies such as converting the tetraploid to the diploid level, breeding at that level, and then re-constituting the tetraploid (Bingham and McCoy 1979). Isogenic populations at the 2x, 4x, 6x, and 8x levels have also been employed for studying ploidy effects on fertility, and plant productivity and quality (Dunbier et al. 1975; Groose et al. 1988; Woodfield and Bingham 1995).

Chromosome Number

M. sativa subsp. *caerulea*

$2n = 16$: Bolton and Greenshields 1950; Lesins 1952, 1968 (as *M. sativa*); Fyfe 1957; Clement 1962 (as *M. caerulea* and 2 reports as *M. sativa*); Gillies 1968; Bolkhovskikh 1969 p. 307 (two citations), p. 308 (1 citation); Gillies 1972*a*, 1977 (as *M. sativa*); Kliphuis and Wieffering 1979; Magulaev 1980; Magulaev 1992; Agarwal and Gupta 1983*b*; Jahan et al. 1994 (as *M. sativa*); Small and Bauchan 1984; Khatoon and Ali 1993.

M. sativa subsp. *sativa*

$2n = 32$: Ghimpu 1928, 1929*a*, 1930; Kawakami 1930; Fryer 1930; Reeves 1930; Senn 1938; Skovsked 1939; Ledingham 1940; Julén 1944; Lesins 1952, 1968; Fyfe 1957; Lesins 1958; Clement 1962 (as *M. agropyretorum*, a taxon synonymized in this chapter with subsp.

caerulea; also as *M. sogdiana* and *M. transoxana*; and 4 reports as *M. sativa*); Bolkhovskikh et al. 1969, p. 307 (as *M. agropyretorum*), p. 308 (as *M. mesopotamica*), p. 309 (as *M. orientalis* and 3 reports as *M. praesativa*), p. 309 (4 reports as *M. sativa* and one report as *M. syriacopalastinica*); Gadella and Kliphuis 1970*a*; Gillies 1970*b*; Bijok et al. 1971; Fernandes and Santos 1971; Bhaumik 1976; Al-Mayah and Al-Shehbaz 1977; Kliphuis and Barkoudah 1977; Fernandes and Queirós 1978; Magulaev 1992 (as *M. caerulea* and as *M. karatschaica*); Majovsky et al. 1978; Uhrikova 1978; Kliphuis and Wieffering 1979; Magulaev 1980; Strid and Franzen 1981; Agarwal and Gupta 1983*b*; Mulligan 1984; Small and Bauchan 1984; Pogan et al. 1985; Zhu et al. 1985; Yang and Zhao 1986; Zary and Hickenbick 1986; Falistocco 1987; Schlarbaum 1988; Song et al. 1988; Yan et al. 1989; Khatoon and Ali 1991, 1993; Hollingsworth et al. 1992; Zhang et al. 1993; Jahan et al. 1994; Okumura et al. 1995; Savova et al. 1996.

Other numbers: Natural hexaploid ($2n = 48$) populations have not been documented. However, occasional hexaploid plants have been discovered and hexaploid alfalfa has been bred (Tomé 1947; Bingham 1970; Mariani 1975; Yen and Murphy 1979). Octoploids and other complements have been produced experimentally (Bingham and Saunders 1974; McCoy and Bingham 1988). Occasional aneuploid plants with $2n = 31, 33$, or 35 have been found (Bolton 1962).

M. sativa subsp. *falcata*

$2n = 16$: Ledingham 1940; Bolton and Greenshields 1950; Lesins 1952; Clement 1962; Lesins and Lesins 1964; Simon and Simon 1965; Kozuharov and Kuzmanov 1965; Bolkhovskikh et al. 1969, p. 308 (3 reports as *M. difalcata* and 2 reports as *M. falcata*), p. 309 (five reports as *M. quasifalcata* and one report as *M. romanica*); Gillies 1970*a*; Kozuharov et al. 1973, 1975 (as *M. falcata* var. *procumbens*); Majovsky et al. 1974; Kliphuis 1977; Strid and Franzen 1981; Small and Bauchan 1984; Yan et al. 1989; Měsiček 1992; Hojland and Poulsen 1994; Jahan et al. 1994; Savova et al. 1996.

$2n = 32$: Ghimpu 1929*b*, 1930; Cooper 1935; Julén 1944; Clement 1962; Skalińska et al. 1966; Bolkhovskikh et al. 1969, p. 308 (5 reports); Lesins and Lesins 1964; Fernandes et al. 1977; Fernandes and Queirós 1978; Kliphuis and Wieffering 1979; Loon and Kieft 1980; Magulaev 1980; Belaeva and Siplivinsky 1981; Agarwal and Gupta 1983; Small and Bauchan 1984; Baltisberger 1987; Lövkvist and Hultgård 1999.

$2n = 16, 18$: Bolkhovskikh et al. 1969, p. 307 (as *M. borealis*).

$2n = 16, 32$: Fryer 1930; Lesins 1952; Clement 1962 (2 reports); Mariani 1963; Lesins and Lesins 1979; Belaeva and Siplivinsky 1981; Magulaev 1992 (as *M. romanica*); Savova et al. 1996. Lesins and Lesins (1979) concluded that both chromosome races are widespread, the tetraploid occurring more frequently than the diploid. However, Small (1985*a*) concluded that the taxon is more frequently diploid than tetraploid.

M. sativa subsp. *falcata* var. *viscosa*

$2n = 16$: no reports found.

$2n = 32$: Fryer 1930 (as *M. media*); Bolkhovskikh et al. 1969, p. 308 (as *M. glandulosa*).

M. sativa subsp. *glomerata*

$2n = 16$: Ahuja 1955; Lesins and Lesins 1966; Lesins 1968; Gillies 1971*b*.

$2n = 32$: Cooper 1935, 1936 (as *M. glutinosa*); Clement 1962 (as *M. glutinosa*); Bolkhovskikh et al. 1969, p. 308 (3 reports, as *M. glutinosa*); Magulaev 1992 (as *M. glutinosa* and as *M. virescens*).

$2n = 16, 32$: Magulaev 1992 (as *M. glandulosa*).

M. sativa subsp. ×*varia*

$2n = 16$: Clement 1962 (as *M. lavrenkoi* and *M. pauciflora*); Bolkhovskikh et al. 1969, p. 308 (as *M. hemicylca*), p. 309 (2 reports as *M. trautvetteri*).

$2n = 16, 32$: Bolkhovskikh et al. 1969, p. 308 (3 reports as *M. hemicycla*).

$2n = 32$: Fryer 1930; Cooper 1935, 1936 (as *M. hemicycla*); Clement 1962 (as *M. hemicycla* and *M. tianschanica*); Bolkhovskikh et al. 1969, p. 309 (as *M. tetrahemicycla* and 3 reports as *M. tianschanica*); Majovsky et al. 1978; Murin 1978; Karshibaev 1992 (as *M. tianschanica*); Zhang et al. 1993 (as *M. media*); Savova et al. 1996.

$2n = 36$ (Bolkhovskikh et al. 1969, p. 309).

M. sativa subsp. *sativa* × subsp. *glomerata*

$2n = 32$: Clement 1962 (as *M. polychroa* and *M. tunetana*).

Genetic Relationships and Gene Transferability

Intraspecific

No barriers to hybridization within the species are known, except those associated with ploidy level (i.e., sterility results from hybridizing diploids with tetraploids), and with forms deliberately selected for sterility. Lesins (1952) found that artificially doubling the chromosome number of diploids generally resulted in excellent fertility of crosses with natural tetraploids, although he also observed that some plants naturally possessed sterility genes. Ledingham (1940) studied triploid hybrids between diploid subsp. *falcata* and tetraploid subsp. *sativa*. There has been very extensive artificial hybridization of different populations, and some synthetic varieties represent an intermingling of dozens, indeed sometimes hundreds, of other cultivars. Crosses between subspecies *sativa* and *falcata* generally produce agriculturally desirable heterosis (Riday et al. 2003; Riday and Brummer 2004). Armstrong and Gibson (1941) found there was no interbreeding barrier between subspecies

×*varia* and *glomerata*. Lesins (1968) found good seed set in crosses between subspecies *glomerata* and each of subsp. *sativa* and *falcata*, but based on a greater number of meiotic irregularities and relatively low seed set in F_1 hybrids, he concluded that subsp. *glomerata* should be considered to be a separate species (a conclusion not accepted in this monograph). McCoy and Smith (1983) found that certain clones of subsp. *glomerata* may be superior pollinators; accordingly, Lesins' observations may not indicate genetic divergence of subsp. *glomerata*. McCoy and Bingham (1988) nicely summed up the situation: "Species of the *M. sativa* complex belong to one biological species. All members freely intercross (especially when adjusted to the same ploidy level), have excellent chromosome pairing in F_1 hybrids, and have no hybrid sterility in the F_1 and later generations. Germplasm from all members of the complex can be readily incorporated into cultivated germplasm."

Interspecific

The following presents, in order, species that have proven to be progressively more difficult to hybridize with *M. sativa*.

Perennial Species

The species of *Medicago* that has proven easiest to hybridize with *M. sativa* is *M. prostrata*. Lesins (1962) hybridized the two at both the $2n = 16$ and $2n = 32$ levels and observed an interspecific barrier when *M. prostrata* was the female parent. Lesins (1968) hybridized $2n = 16$ *M. sativa* (as "*M. glomerata*") and $2n = 16$ *M. prostrata* and the hybrids were quite fertile (a hybridization barrier was again found when *M. prostrata* was the female parent). Sorensen et al. (1981) crossed diploid *M. prostrata* with tetraploid *M. sativa* and reported that the hybrid was tetraploid and fertile. See *M. prostrata* (Chapter 72) for additional information.

Lesins and Gillies (1972*b*) stated that unpublished work by Lesins found that *M. rhodopea* could be hybridized with *M. sativa*, but not at the diploid level. Gillies (1972*a*) studied a triploid hybrid between *M. rhodopea* and *M. sativa* (reported in Lesins (1972)). Lesins (1972) had produced the triploid by crossing diploid alfalfa with an artificially doubled tetraploid *M. rhodopea* (he also produce an aneuploid hybrid with $2n = 31$). Lesins (1970) reported a tri-species hybrid of *M. rhodopea*, *M. saxatilis*, and *M. sativa*.

Lesins (1961*b*) was the first to generate hybrids between *M. sativa* and *M. cancellata*. He produced several pentaploid ($2n = 40$) hybrids by crossing alfalfa ($2n = 32$) and *M. cancellata* ($2n = 48$), but produced just one hybrid by crossing hexaploid alfalfa and *M. cancellata*. Yen and Murphy (1979) crossed the hexaploid ($2n = 48$) *M. cancellata* with two hexaploid amphiploids of *M. sativa*, and found that while hybrids (both F_1 and F_2) could be established, there was very extensive sterility.

Lesins (1970) hybridized *M. saxatilis* (with $2n = 48$ chromosomes) with both tetraploid ($2n = 32$) and hexaploid ($2n = 48$) *M. sativa* and reported on the considerable chromosomal irregularities in the hybrids. Yen and Murphy (1979) crossed *M. saxatilis* with two hexaploid amphiploids ($2n = 6x = 48$) of *M. sativa* and found that, while hybrids (both F_1 and F_2) could be established, there was very extensive sterility. Lesins (1970) also reported on tri-species hybrids involving *M. rhodopea*, *M. sativa*, and *M. saxatilis*.

Medicago pironae and *M. daghestanica* are closely related, and their ability to hybridize with *M. sativa* is comparable. Lesins and Gillies (1968) were unable to generate hybrids of *M. sativa* with either species. McCoy and Echt (1993) were also unable to hybridize *M. daghestanica* with *M. sativa*. However, Lesins (1971) was able to produce tri-species hybrids between *M. pironae*, *M. daghestanica*, and *M. sativa* by crossing tetraploid *M. sativa* and an artificial tetraploid hybrid of the other two species. Lesins and Gillies (1972) stated that unpublished work by Lesins found that *M. daghestanica* could be hybridized with *M. sativa*, but not at the diploid level.

Lesins (1961*a*) found that he could not produce hybrids between *M. papillosa* and *M. sativa* when both were tetraploid ($2n = 32$), but that triploid hybrids resulted from crossing diploid *M. sativa* and tetraploid *M. papillosa*. McCoy and Smith (1984) repeated this hybrid (*M. papillosa* was identified as *M. dzawakhetica*), again finding uneven ploidy levels were required when one of the parents is tetraploid. They found the F_1s were male sterile although chromosome pairing was fair. They concluded that ovule–embryo culture could be used for gene transfer from *M. papillosa* to *M. sativa*. Clement (1963) hybridized diploid *M. papillosa* with diploid *M. sativa* and found excellent chromosome pairing and fair fertility in the F_1 hybrid and its backcrosses. McCoy and Quarisa (1989) studied diploid, triploid, and tetraploid hybrids of *M. papillosa* and *M. sativa*. McCoy et al. (1991) found that there was limited genomic affinity between the two species.

Lesins (1969) found that neither *M. hybrida* nor *M. suffruticosa* could be crossed with *M. sativa*.

Lesins (1952) generated artificial tetraploids ($2n = 32$) of *M. platycarpos* and *M. edgeworthii*, and was unable to hybridize these with tetraploid *M. sativa*. At the diploid ($2n = 16$) level, Lesins (1952) attempted crosses with *M. ruthenica*, and once again none of these crosses produced any viable seeds. Campbell (2003) cited personal unpublished information that neither classical hybridization nor protoplast fusion techniques had yielded viable hybrid embryos between *M. edgeworthii* and *M. sativa*.

Armour et al. (2008) succeeded in generating five "asymmetric hybrids" between *M. arborea* and *M. sativa*, but concluded that the hybrids contained only a single *M. arborea* chromosome or part of a chromosome.

Medicago lupulina

This is the only species in the genus that is often biennial (as well as annual and perennial). See Chapter 49, dealing with

M. lupulina, for information on alleged hybrids of *M. sativa* and this species. In fact, no one has documented that this cross can be made. Lesins (1952) generated an artificial tetraploid of *M. lupulina* and found that this could not be hybridized with tetraploid *M. sativa*.

Annual species

The only authenticated hybrid between *M. sativa* and an annual species of *Medicago* produced by conventional crossing was described by Sangduen et al. (1982). This hybrid was between *M. sativa* ($2n = 32$) and *M. scutellata* ($2n = 30$). The hybrid was a mixoploid, with shoots differing in chromosome number; it had very low pollen stainability and would not set seed (also see Fig. 6-1). Elgin et al. (1977) were unable to hybridize *M. scutellata* with *M. sativa*, and indeed the successful cross by Sangduen et al. was a very rare event.

Lesins (1952) conducted hybridization experiments among species at the same ploidy level. He attempted to cross diploid (2n = 16) *M. sativa* with *M. praecox* and *M. italica* (as *M. obscura*), and could not produce hybrids. He also attempted, without success, to hybridize an artificial tetraploid of *M. orbicularis* with tetraploid *M. sativa*.

Staszewski and Jakubowska (1992) reported that they pollinated *M. sativa* with pollen from the annual *M. doliata* and found one F_1 plant that appeared to be a hybrid; however, they were unable to generate hybrids with the following annual species: *M. arabica*, "*M. blancheana*", *M. coronata*, *M. disciformis*, *M. doliata*, *M. intertexta*, *M. italica*, *M. laciniata*, *M. littoralis*, *M. minima*, "*M. murex*," *M. orbicularis*, *M. polymorpha*, *M. radiata*, *M. rigidula*, *M. rotata*, *M. scutellata*, *M. truncatula*, and *M. turbinata*.

Other genera

Lesins (1952) generated artificial tetraploids of *Melilotus alba* Medik., *M. officinalis* (L.) Lam., *M. sulcatus* Desf., *Trigonella caerulea* Ser., and T. *foenum-graecum* L., and demonstrated by attempted hybridization that sterility between *M. sativa* and these species at the tetraploid level was not simply a matter of overcoming the ploidy difference (no viable seeds were set in any of these crosses). At the diploid ($2n = 16$) level, Lesins (1952) similarly attempted crosses between *M. sativa* and both *Melilotus macrorhiza* Bess. and *Trigonella calliceras* Fisch. ex M. Bieb., and once again none of these crosses produced any viable seeds.

Exchange by advanced biotechnology

Perennial species

McCoy (1985) used ovule–embryo culture to generate hybrids of *M. sativa* and *M. rupestris*. Using embryo rescue techniques, McCoy and Smith (1986) obtained hybrids between diploid *M. sativa* ($2n = 16$) and (at the same chromosome number) *M. hybrida*, *M. marina*, *M. papillosa*, *M. rhodopea*, and *M. rupestris*. The hybrids with both *M. rhodopea* and *M. rupestris* could be made with relative ease, while the hybrids with both *M. hybrida* and *M. marina* were obtained with considerable difficulty. As well, McCoy and Smith (1986) obtained tri-species hybrids between *M. sativa* × *M. papillosa* F_1 hybrids ($2n = 3x = 24$) and *M. cancellata* ($2n = 48$), and between *M. sativa* × *M. papillosa* F_1 hybrids ($2n = 3x = 24$) and *M. saxatilis* ($2n = 48$). McCoy and Bingham (1988) stated that (in unpublished work) interspecific hybrids had been produced between diploid *M. sativa* and diploid *M. daghestanica* and between diploid *M. sativa* and diploid *M. pironae*, using ovule–embryo culture. Genes can be transferred from *M. daghestanica* to alfalfa (*M. sativa*) by tri-species hybridization and embryo rescue (McCoy and Echt 1993, following work by Lesins and Gillies 1968). By using this technique, hybrids of *M. sativa* and *M. rupestris* were first produced, and these hybrids were then crossed with *M. daghestanica*.

Also using embryo rescue techniques, Piccirilli and Arcioni (1992*a*, *b*) crossed *M. sativa* ($2n = 32$) with *M. arborea* ($2n = 32$) and obtained hybrids. The same has been done by electrofusion of protoplasts (e.g., Nenz et al. 1996). A possible cross with *M. doliata* has also allegedly been made (Staszewski et al. 1992).

Annual species

Using embryo rescue techniques, Piccirilli and Arcioni (1992*a*, *b*) crossed *M. sativa* ($2n = 32$) with *M. noeana* ($2n = 32$), *M. rugosa* ($2n = 30$), and *M. scutellata* ($2n = 30$) and obtained hybrids with *M. rugosa*, but not with *M. scutellata* and *M. noeana*. Mizukami et al. (2006) produced somatic hybrids of *M. rugosa* and *M. sativa*. Mazin et al. (1994) produced fusion protoplasts of alfalfa and *M. lupulina*.

Conservation Status

An enormous number of accessions of *M. sativa* are in germplasm banks, so seemingly the species is not in need of conservation measures. However, the natural habitats of the diploid progenitors of *M. sativa* in Asia are rapidly being decimated, and thereby there is considerable danger that valuable genes are being lost. The majority of the endangered natural area of the diploids is in the former USSR, and for this area Ivanov (1980) provided the following analysis:

"Under the influence of civilization the local gene centers of rare species degenerate quickly; their naturalness is disturbed and they gradually disappear. In the context of major transformations in the national economy, especially large-scale construction of irrigation systems, dams, laying of gas and oil pipelines, and geological undertakings that involve extensive earthworks, the process of disruption of the natural plant canopy will be increasing significantly every year. To conserve centers of localization of rare species of alfalfa in regions of the active formation and differentiation, it is necessary to establish sanctuaries or protected areas and to enforce strict control over the work of these sanctuaries with the help of the society for the Conservation of Nature." Savova et al. (1996) contended that *M. sativa* subsp. *falcata* is native to parts of

Switzerland and the tetraploid phase is endangered there due to both introgression from local *M. sativa* subsp. *sativa* and the decline of habitats as a result of intensive land use. More generally, Savova et al. concluded that subsp. *falcata* as a whole (both the diploids and tetraploids) in Switzerland is in danger of genetic erosion and suggested that protective measures could include the prescription of a minimal distance between cultivated *M. sativa* and wild indigenous populations of *M. sativa* subsp. *falcata*.

Germplasm Sources

Bennett et al. (2006) recorded the following major genebank holdings identified as follows (cf. Chapter 3):

M. sativa (infraspecific taxa not identified): United States Department of Agriculture National Plant Germplasm System/Germplasm Resources Information Network (GRIN): 17; Systemwide Information Network for Genetic Resources/Consultative Group on International Agricultural Research (SINGER): 0; Australian Plant Genetic Resources Information System (SARDI, South Australian Research and Development Institute): 88; Vavilov Institute of Plant Industry (Leningrad) (VIR): 2014.

M. sativa subsp. *caerulea*: GRIN: 87; SINGER: 6; SARDI: 56; VIR: 0.

M. sativa subsp. *sativa*: GRIN: 2527; SINGER: 966; SARDI: 2202; VIR: 0.

M. sativa subsp. *glomerata*: GRIN: 7; SINGER: 3; SARDI: 36; VIR: 0.

M. sativa subsp. *falcata*: GRIN: 391; SINGER: 19; SARDI: 140; VIR: 0.

M. sativa subsp. ×*varia*: GRIN: 336; SINGER: 15; SARDI: 115; VIR: 0.

Sources of drought-resistant genes for alfalfa

The history of alfalfa may well explain why today it is rather wasteful of water. As explained in this chapter, wild, purple-flowered diploids (*M. sativa* subsp. *caerulea*), centered in dry regions near the Caspian Sea and therefore well adapted to drought, gave rise to tetraploids (*M. sativa* subsp. *sativa*) from which domesticated strains arose thousands of years ago, likely in Vavilov's Near Eastern Center — Asia Minor, Transcaucasia, Iran, and the highlands of Turkmenistan. The earliest cultivars were probably grown in moist environments, or were provided with irrigation, conditions in which alfalfa is highly productive. The lack of highly drought-resistant alfalfa today is likely a reflection of the lack of motivation historically for cultivating the crop in low-productivity circumstances associated with limited water supplies. As noted below, alfalfa conferred great advantages on societies that relied on riding horses because alfalfa is a concentrated, portable fodder with particular nutritive advantages for horses that are not provided to ruminant livestock. The considerable military advantages of war horses that could be fed quickly and kept working for long periods were a chief justification for providing irrigation to produce alfalfa. Natural grass and mixed grass pastures sufficed for livestock raised for meat, milk, and other products, especially with much lower grazing pressures historically.

Subspecies *caerulea*, the main ancestor of alfalfa, continues to occupy drought-prone areas and should be considered to be a source of genes for breeding drought resistance into alfalfa cultivars. Similarly, some populations of subsp. *falcata* are very drought resistant. The Russian works criticized in this monograph for their taxonomic treatments of the *M. sativa* complex are excellent sources of information on drought-resistant wild and cultivated materials that harbor genes that are likely invaluable for breeding. Grossheim (1945) wrote with regard to wild alfalfas of the USSR "as some of them are very drought resistant, their introduction into cultivation may be expected to play a considerable role in rehabilitation of arid and desert regions." Ivanov (1980) identified accessions of wild and cultivated alfalfas that are considered especially drought resistant. Wild accessions from very arid areas of course are likely to be especially well adapted to drought, but because of the absence of domestication, their general agronomic characteristics are inferior. Collections of the diploid subsp. *caerulea*, which is entirely wild, are of this nature, although the degree of drought resistance will depend on the aridity of the source site. Landraces from very arid areas are likely to offer a compromise between drought resistance and reasonable agronomic characteristics.

Research Problems and Priorities

Medicago sativa is the most intensively studied of all livestock crops, befitting its leading importance in much of the temperate world of forages and fodders. The priorities assigned to research on alfalfa differ according to country. Not surprisingly, the priorities of departments of agriculture are generally concentrated on agronomic management and on breeding improved cultivars for local circumstances, increased yield, and resistance to diseases, pests, and edaphic and climatic conditions.

Alfalfa cultivars have been bred especially for use in a variety of moist and dry habitats, and for soils of varying textures and fertility, but there remains immense potential for future improvement utilizing germplasm within the genus *Medicago*. Although there are significant limitations of the range of germplasm that is available, most of the material already at the disposition of breeders has not been adequately evaluated or exploited. Unfortunately, forages are often taken for granted, while more glamorous crops attract research and development support. This is extremely shortsighted because forages are generally less developed than other crop groups, and so are capable of providing especially great rewards for money invested. Private seed supply companies are sometimes reluctant to invest in developing crops such as alfalfa that do not require an annual purchase of seeds the supply of which they can closely control.

Moreover, there has been reluctance to develop new cultivars for dryland areas because of their comparatively low yield potential. There are, however, monumental advantages for society to increasing the water use efficiency of alfalfa, especially for semi-arid and arid regions, given the shrinking supply of this precious resource. Although dryland cultivars are available, alfalfa does not use water efficiently and requires irrigation in arid areas. Fortunately, there is a vast reservoir of diverse genetic resources of *M. sativa* available for the creation of new water-conserving cultivars to meet the needs of the future. This includes thousands of accessions in gene banks, including cultivars, landraces, and weedy forms that have been created as a result of thousands of years of selection and distribution by humans. Of the wild infraspecific taxa of *M. sativa*, some accessions of subspecies *falcata* and *caerulea* are particularly drought resistant. There are an additional 20 wild perennial species of *Medicago*, many of which are also very drought resistant, and gene transfer to alfalfa from the majority of these is possible through sexual crossing.

The following list suggests some general, long-term priorities.

1. The collection, preservation, and agronomic characterization of germplasm of wild species of *Medicago* for the purpose of alfalfa improvement. As discussed in Chapter 3, the rarer species of *Medicago* are in considerable danger of extermination and there is no or very limited representation in gene banks for dozens of species.
2. The development of *M. sativa* as a major source of protein to feed humans as an alternative in some circumstances to the rearing of livestock. As discussed above, as the world's most efficient protein crop, alfalfa is preeminently adapted to mankind's future need for increased direct consumption of plant proteins.
3. Genetic engineering of alfalfa for nontraditional products such as pharmaceuticals, industrial compounds, fiber, and biofuels. Of course, genetic engineering needs to be conducted responsibly, with due regard to possible gene escape to the wild and consequent effects on food and feed supplies, and on biodiversity.
4. Utilization of alfalfa for environmental preservation. As noted above, alfalfa is much less damaging to biodiversity and the environment than to other major crops, and has potential for increased use for maintenance of soils, landscapes, ecosystems, and wildlife.

Miller (2005) nicely summarized the status of alfalfa: "Plant breeding has improved the adaptation of alfalfa to a wide range of environments and has dramatically increased its seed and forage production potential. Genetic improvements have been made in the areas of improved pest resistance, salt tolerance, grazing tolerance, wheel traffic tolerance, and increased symbiotic nitrogen fixation. The quality of alfalfa as an animal feed has been improved in the areas of nutrient content and digestibility, resulting in increased animal performance. Future research with the aid of biotechnology holds the exciting promise of improvements in a wide range of hard to obtain traits, including biotic and abiotic stresses. The biotechnology-enhanced plant breeding techniques will further expand and diversify alfalfa's role in the twenty-first century."

General References on Alfalfa Agronomy and Technology

Bolton 1962; Hanson 1972; Hanson et al. 1988; Orloff et al. 1997; Summers and Putnam 2008; and Undersander 2004. For information on alfalfa (as "lucerne") in Australia see Bullen 2002; Lattimore 2008; McDonald et al. 1995; Ransom 2006; and Lolicato et al. 1998.

Representative Collections

M. sativa subsp. *sativa*

Afghanistan: Aitchison, J.E.T. 355, Kurrum Valley, V-VI-1879, in flower ("a very common field-weed, not cultivated now, nor could anyone tell me if it was formerly, it is certainly cultivated in the Punjab where it is also a weed, common from Shálizán to Alikhél") (BM, K). Grey-Wilson, C. & Hewer, T.F. 1410, Badakh shan-Wakhan, Deh Gulaman, east of Baba Tangi, alt. ca. 3150 m, sandy slopes and dry places, 24-VII-1971, in flower & fruit (K). **Australia:** Coveny R. 11403, Great Western Highway between Bullaburra and Wentworth Falls at base of Boddington Hill (33°44′S, 150°24′E), alt. 810 m, disturbed area by roadside, 17-XI-1982, in flower (K). **Azores:** Brooke, W.M.A. 11624, St. Maria, Vila Porto, growing on pastures near sea, 25-VI-1971, in flower (BM). **Balearic Islands:** Ferguson, L.F. & I.K. 2500, Mallorca, T.M. Estallenes, fields in sheltered and irrigated valley, alt. ca. 40 m, 17-VII-1969, in flower (BM). **China:** Steward, A.N. 1902, Kiangsu prov., Nanking, 3-V-1922, in flower & fruit (P). **England:** Hubbard, L.E., Berkshire, Kennington, in grassland, heavy soil, 4-VII-1942, in flower (K). Wade, A.E., Wales, Cardiff, Cathays Park, waste ground, VIII-1935, in flower (K). **Ethiopia:** Pappi, N. 1080, Eritrea-Amasen, Lungoil Mai-Belà presso Asmara, 5-V-1902, in fruit (BM). Schweinfurth, G. & Riva, D. 2105, Flora Abyssinae Septentrionalis, Colonia Eritrea, Amara, alt. 2300 m, 14-V-1892, in flower & fruit (P). **Fiji:** Koroiveibau, D 12306, Ra, Yagara, C.S.R. Estate, pasture, 20-I-1961, in flower (K). **France:** Vincent, prairies à Nancy (Meurthe), VII-1853, in flower & fruit, Flora Galliae et Germaniae Exsiccata de C. Billot No. 1115 (P). **Greece:** Lewis, G.P. 509, southwest Crete, 3 miles (1 mile = 1.6093 km) west of Paleohora, on roadside, 8-X-1979, in fruit (K). **Holland:** Berge, T.H. ten, Rotterdam, roadsides and debris in destroyed center, locally very common, 7-VII-1952, in flower & early fruit (K). **Iran:** Andersen, J.S. & Peterson, I.C. 336, 43 km east of Bojnurd, alt. 1100 m, semi-desert, 9-VII-1969, in flower (K). Hewer, T.W.

H3983, East Mazandaran Province, Golidagh, alt. 780 m, very common here as everywhere in wheat fields, 4-VII-1977, in flower (K). **Iraq:** Rawi 9156, Mergadraija near Haji Omran, alt. 1700–1800 m, mountainside, 21-VI-1947, in flower (K). **Israel:** Meyers, F. 2806, Jerusalem, cultivated ground, in flower, 10-VII-1903, in flower (K). **Japan:** Savatier 249, Yokoska, çà et là dans le voisinage des cultures, 1866–1874 (P). **Lebanon:** Makinson, D.J. 92, Qartaba, 20 miles northeast of Beirut, stony limestone bank, open sun, occasional, 27-VI-1964, in flower & fruit (K). **New Zealand:** Bangerter, E.B. 5118, North Island, Thames County, Tapu, by Tapu River (36°59′S, 175°30′E), waste ground densely overgrown with ruderal weeds and some garden outcasts, 29-XI-1973, in flower (K). **Pakistan:** Siddiqi, M.A. 1908, Baluchistan, Hanna, waste land near lake, escape?, 4-V-1963, in flower (K). **Saudi Arabia:** Boulos, L & Ads, A.S. 14130, Al Mahmoud, 35 km north of Abha, 10 km below Jabal Al-Sooda, wheat fields and canals by a spring, 21-V-1980, in flower (K). **Somaliland:** Beckett, J.J. 1396, Sheikh (9°57′N, 45°11′E), alt. 1530 m, weedy habitat, introduced?, 19-IX-1981, in flower & early fruit (K). **Spain:** Ball, P.W. et al. 407, Prov. Almeria, Sierra de Alhamilla, Peñon de Turrilas, alt. 900–1000 m, northwest facing stony slopes, 7-VI-1907, in flower (BM). Lange, Sierra de Randa, sur les terrains incultes et calcaires, rare, 28-VI-1889, in flower, E. Reverchon – Plantes de l'Andalousie 1889, No. 442 (K, P). ("*M. sativa* var. *orophila* Lange") Lewis, G.P. 458A, Andorra, road from Ordino to Llorts, alt. 1300–1400 m, on embankment by roadside, 19-VII-1979, in flower (K). **Sweden:** Hasslow, O.J., prov. Skane, paroecia Kviinge, ad pagum Kviinge, 3-VIII-1934, in flower & fruit, G. Samuelson Plantae Suecicae Exsiccatae No. 1050 (K). **Syria:** Dinsmore, J.E. 4806, Damascus, hillside, 17-V-1938, in flower (K). **Tibet:** Schlagintweit, herbarium of, prov. Bálti, Shigar to Skárdo, 31-VIII-1856, in flower & early fruit, Catalogue No. 6170 (P). (type? of *M. sativa tibetana* Alef.) **Turkey:** Davis & Hedge D30532, prov. Kars, Kisir dag above Susuz, alt. 2000 m, igneous slopes, 3-VII-1957, in flower (BM). Jenkins, C.R.F. 2337, Erzurum, west side of valley, about 3 km below turning to Ovaçik on the Ilica to Ispir road, north by northwest of Erzurum, steep rocky slope, 7-IX-1970, in flower & fruit (BM). **USSR:** Czernous, M., Ucraina, prope app. Umanj, seges Medicaginis, 20-VI-1904 (LE). Dybowski, W., Lithuania, Lubez, distr. Nowogródek Na suchych zdziczala, Zebr. 3-Sierp.-1900, Flora Polonica Exsiccata, E. Woloszezak (LE). **Yemen:** Deflers, A. 531, Vallée de Lana "a", alt. 2200 m, cultivé en grand comme fourrage, 7-VI-1887, in flower & fruit (P). Hepper, F.N. 6255, near Bukta village, ca. 40 km northeast of Dhamaq, ca. 14°44′N, 44°40′E, near limestone plateau, basalt over sandstone syncline, lucerne crop extensively cultivated in more or less irrigated plots, 3-XI-1925, in flower (K).

M. sativa subsp. *caerulea*

USSR: Čuba, J. & V. Vašák, Caucasus orientalis, Azerbaidzania, Apsheron peninsula, distr. Baku, in vicinitate pagi Khanlar, alt. 120 m, "6.8.1983", in flower & fruit (G). Grossheim, A., Transcaucasia, Azerbajdzhan, prov. Baku, distr. Salijany, dominium Kara-Tshala in steppa Shirvan; in pratis, in flower & fruit, 25-IX-25, A. Grossheim & B. Schischkin Plantae Orientales Exsiccatae 1928 No. 308 (BM, LE). Rubtzova, A.G., Transcaucasia, Azerbajdzhan, steppa Shirvan austro-orientalis, distr. Saliany, prope pag. Chaladzh, inter segetes Gossypii, 7-XI (sic)-1934, in flower (P).

M. sativa subsp. *glomerata*

Algeria: Cosson, E., prov. de Constantine, Oved Zenati, 7-VI-1880 (K). Gandoger, coteaux arides autour de Constantine, Galloprovincia loco dicto "le Luc" in arenosis ad margines agrorum, sylviumque, 4-VI-1850, in fruit (K). **France:** Faure, Le Luc (Var.), côteaux calcaires, 8-VI-1867, Soc. Volgéso-Rhénane (P). Faure, Le Luc (Var.), lieux incultes, 8-VI-1873, in flower & fruit, Société Helvétique (P). Hanry, H., Luc (Var.), Lieux incultes, bois, 16-VI-1853, in flower & early fruit, Flora Galliae et Germaniae Exsiccata de C. Billot, No. 1158 (BM, P). Maury, C., Luc, Var, 19-V-1853, in flower & fruit (BM). Müller, J. 107, au Luc, près de Toulon, 16-VI-1851, in flower & early fruit (K). Ounier, Provence, 6-VII-1846, in flower & fruit (K). **Italy:** Polini, L., prope Bussana nuova, Liguria, in graminosis Saxosis, 4-VII-1898, in flower & early fruit (K). **Tunisia:** Letourneux, M., Djebel osneor, in pinetis, 1-VII-1884, in flower & fruit (P). Letourneux, M., Oved Chereharas (?), in collibus arenosis, 26-VI-1884, in flower & fruit (P). Letourneux, M., prope Queter es sinan (?), in collibus calcareis (pinetis), 29-VI-1884, in flower & fruit (P). Letourneux, M., prope Ras El Airem (?), 30-VI-1884, in flower & fruit (P). **USSR:** Callier, A., Tauria, in declivibus montis "Pertschem" prope "Sudak", VI-1896, in flower & early fruit, Herbarium Normale Editum ab. I. Dörfler No. 5027 (BM). Golubkova, V. & Crautzova, E., Georgia, in declivibus ad vallen fl. Terek prope munimentum Darjaly, 25-VIII-1949, in flower & fruit (BM, K). Kozlovsky, V., Transcaucasia, Georgia, prov. Tiflis, distr. Dushet, prope St. Kobi, in Pratis, 3-VIII-1925, in flower & early fruit, A. Grossheim & B. Schischkin Plantae Orientales Exsiccatae 1928 No. 337 (K). **Yugoslavia:** Edmondson, J.R. 362, Macedonia, lower slopes of Krstvar, the hill 4 km south of Skopje, alt. 500–550 m, north-facing limestone slope, dryish, thin free cover, 27-VII-1970, in flower & very early fruit (BM).

M. sativa subsp. *falcata* var. *falcata*

Albania: Alston, A.H.G. & Sandwith, N.Y. 2331, South Albania, Cajup, Mali Lunxheriës, alt. 4000 feet (1 foot = 0.3048m), bare rocky sandstone slopes, shady spot, 6-VIII-1935, in flower & very early fruit (K). **Bulgaria:** Kojoucharov, S., Sozopol, sur les sables maritimes près de la ville, 9-VIII-1959, in flower & fruit (K). **Czechoslovakia:** Krist, V.L., Moravia Centr., in steppis collis "Kamenná horka"

(Sv. Michálek) ad Bohutice prope opp. Mon. Krumlov, alt. 278 m, 4-VII-1933, in flower, Flora Exsiccata Reipublicae Bohemicae Slovenicae No. 821 (K). Stearn, W.T., Mohelno, 7-VII-1937, in flower (K). **Denmark:** Young, D.P. 4508, Falster, Orohoved, by Storstrom br., rough waste ground by seashore, with *M. sativa* and hybrid (*M. sativa* subsp. ×*varia*), 31-VII-1952, in flower & very early fruit (K). **England:** Britton, C.E. 1455, near Teddington Lock, Surrey, waste land by cow path, 14-VII-1916, in flower (K). Gregory, E.S., southwest England, Somerset, Porlishead, waste ground, 30-VI-1904, in flower (K). **Finland:** Holmberg, K., Nylandia, par. Tenala, Lappvik, in saburra navali, in saburru etc., raro occurit, 23-VII-1908, in flower, Plantae Finlandiae Exsiccatae Mus. Univ. Helsingforsiensis No. 759 (K). Holmberg, K., Nylandia, par. Tenala, Lappvik, in Saburra navali, in flower, 23-VII-1908, Plantae Finlandicae Exsiccatae e Museo Botanico Universitatis Helsingforsiensis distributae No. 759 (LE, P). **France:** Le Grand, A., Cher, Saint-Germain-du-Puy-de-Bourges, le long de la voie ferrée, 20-IX-1901. Société Rochelaise 1901, No. 4708 (P) ("*M. falcata* var. *heterocarpa* Le Grand s.-var. *monospira* Le Grand"). **Greece:** Balls, E.K. & Gourlay, W.B. B3762, Metsovo, Epirus, alt. 4000 feet, dry sandy slopes and edges of cultivated land, 27-VII-1937, in flower & early fruit (K). Rechinger, K.H. & F. 10993, Macedonia orientalis, in monte Boz-Dagh prope Serrai (Seres), alt. 500 m, substr. calc., 15-VII-1936, in flower (K). Rechinger, K.H. & F. 10993, Macedonia orientalis, in monte Boz-Dagh prope Serrai (Seres), alt. 500 m, substr. calc., 15-VII-1936, in flower & early fruit (BM). Reverchon, G. 33, Crete, La Canée, moissons, 26-VI-1883, in flower (P). **Holland:** Reclaire, A., Rotterdam, along railway, on sand, VII-1898, in flower (K). Vermeulen, G. 266, Pays-Bas, Wageningen Lenkesveer, 3-VIII-1927, in flower (P). **Hungary:** Verdcourt 5442, Tata, roadside, 31-VIII-1983, in flower (K). **Latvia:** Meinertzhagen, R., Schwartzort, Kurtsher, Nehrung, sand soil, 24-VIII-1930, in flower & fruit (BM). **Poland:** Kulikowski, E., Rossia merid. ad Odessam, po Polach i obrywach na brzegach morza obficie, Zebr., 28-Sierpnia-1893, E. Woloszczak Flora Polonica Exsiccata, No. 421 (LE). Piekós, H., Carpati Occidentales, montes Piening, prope oppidum Krosienko (districtus Nowy Targ), supra locum dictum Gródek, in prato ad declive meridionale, alt. ca. 600 m, 15-VII-1970, in flower, Flora Poloniae Exsiccata No., 338 (BM). **Portugal:** da Cunha, A. Ricardo, Arredores de Lisboa serra de Monsanto VII-1888, in flower & early fruit, Flora Lusitanica Exsiccata No. 753 (P). **Romania:** Nyárády, E.I., Dobrogea, distr. Tulcea, in campestribus arenosis ad marginem silvae Letea, prope pagum Letea, alt. 4–5 m. 18-VII-1923. Museum Botanicum Universitatis cluj Flora Romaniae Exsiccata No. 1271 (K, P). ("*Medicago falcata* L. var. *filiformis* Nyár.") Wiśniewski, T. 1652, Dobroudja, Mamaia ad Constanta, 1-VII-1928, in flower & fruit (K). **Spain:** Reverchon, E., Province d'Andalousie, Sierra de Ronda, lieux incultes, sur le calcaire, 1200 mètres, VI-1889, in flower & early fruit, E. Reverchon-Plantes d'Espagne No. 553 (K, P). ("*M. falcata* var. *luteovittata* Willkomm?") Willkomm, Grazalema, lieux incultes sur le calcaire, 16-VI-1890, in flower & early fruit, east Reverchon-Plantes de l'Andalousie No. 553 (BM). **Sweden:** Alm, C.G. & Smith, H. 603, Uppland, Gryta Parish, Gryta, sunny grassy hillock, 11-VII-1951, in flower (BM). Fries, E.T., prov. Gotland, Visby, 19-VIII-1930, in flower & early fruit, G. Samuelsson Plantae Suecicae Exsiccatae No. 1048 (K). **Tibet:** Schlagintweit, herbarium of, prov. Ladák, Khárbu Kóma to Sháksi (southwest of Dah), 3-VII-1856, in flower, catalogue No. 5255 (P). **Turkey:** Davis & Hedge D30579, prov. Kars, Kars-Susuz, 8 km from Kars, alt. 1800 m, igneous slope, 5-VII-1957, in flower (BM). Davis & Hedge D32588, prov. Kars, Kisir Dag above Susuz, alt. 1800 m, rocky slopes (igneous), 20-VIII-1957, in early fruit (BM). **USSR:** Asarenkova & Parnassky, Ucraina, Prov. Cherson, insula Birjuczij, in declivibus arenosis, 1-VIII-1950, in flower & fruit (BM). ("*M. erecta* Kotov") Bobrov, E., prov. Krasnodar, distr. Gul, pag. Schuntuk, in praticulis, 1-VIII-1950, in flower & fruit (BM). ("*M. falcata* var. *altissima* Grossh.") Dowmar, N., Prov. Mohilew, praesertim circa urbem Mohilew ad Borysthenem, 1862 (LE). Frazer Jenkins, C.R. 2684, Ukrainskaya S.S.R., 1 mile west of river Nedra, west of Berezan, 50 miles east of Kiev, on Kharkov road, Kiyevskaya Oblast, beside track into fields, 19-VI-1971, in flower (BM). Gutik, J., Prov. Kiew, in pratis siccis pr. Uman, 6-18-VII-1899, in flower & early fruit, Herbarium Florae Rossicae, Museo Botanico Academ. Imper. Sc. Petropolitanicae Editum, Fasc. XV, No. 711 (LE). Kolakowvsky, A., prov. Krasnodar, distr. Tuapse, inter pag. Olginskoe et Dzhubga, in clivis lapidosis calcareis, 15-VI-1951, in flower (K). Mechmershausen, G. 5, Estonia, Virumaa, near Rakvere, in the quarry of Aluvere near the railroad of Kunda, abundant, 13-VIII-1930, in flower (K, LE). **Yugoslavia:** Edmondson, J.R. 369, Macedonia, Kriva Palanka, valley running north from town, between Popva Niva and Kibelica, dry slopes of valley mostly in openings in the woodland, alt. 670 m, 28-VII-1970, in flower & early fruit (BM). Gardner, M.F. & S.G. 2333, Slovenija, Juliske Alps, Mojstrana (46°27′N, 13°50′E), alt. 700 m, very rich alpine meadow, 25-VII-1984, in flower (BM).

M. sativa subsp. *falcata* var. *viscosa*

Bulgaria: Gilliat-Smith, B. 1006, Varna District, hills south of Varna, 9-X-1924, in flower & early fruit (K). **Greece:** Lemperg, F. 401, Macedonia, 8-VI-1937, in flower & very early fruit (K). Tedd, H.G. 387, Thrace, Xanthie-Macaclar, alt. 600 feet, dry rocky places, 4-VI-1930, in flower & early fruit (K). Tedd, H.G. 733, Thrace, Tchal Dagh, alt. 600 m, among stones, rocks, bushes, etc., 25-VI-1931, in flower & fruit (K). **Romania:** Malos, C., Olaru, M. & Paun, M., Oltenia, prope oppidum Craiova, in arenosis locis, loco nominato Fintîna Obedeanu, alt. ca. 100 m, 8-VII-1960, in flower & early fruit, Hortus Botanicus Inst. Agronom. T. Vladimiresc. (BM). **USSR:** Callier, A., Rossia, Tauria, in declinibus montis "Pertschem" prope "Sudak", VI-1896, in flower, Herbarium Normale editium ab I. Dörfler No. 5027 (P).

M. sativa* subsp. ×*varia

AUSTRIA: Woloszczak, Austria inferior, in pratis ditionis Vindobonensis inter parenter, solo argillaceo, alt. 250 m, (no date; received by P 20-XI-1889), Flora Exsiccata Austro-Hungaria No. 2015, (BM, P). CHINA: Licent, R.P., China Septentrionale, Siaotchoang Honann, 17-VI-1815, in flower & early fruit (P). Pelliot-Vaillant 809, Yong-Kou Ying, 6-VII-1908, in flower & early fruit (P). CZECHOSLOVAKIA: Rohlena, J, Behemia center, Praha in lapicidinis ad p. Bránik, 20-VIII-1927, Flora Exsiccata Republicae Bohemnicae Slovenicae 226 (P). FRANCE: Billot, C., aux environs de Nancy (Meurthe), prés secs et coleaux arides, Flora Gallicae et Germaniae Exsiccata de C. Billot (P). Hy, F., Ponts-de-Cê (Maine-et-Loire), sables des bords de la Loire, VIII-1895, in flower & fruit, Société Rochelaise 1895, No. 3751 (P) (Type ? – *M. cyclocarpa* Hy). ITALY: Davis & Sutton D65962, Arezzo, ca. 15 km southeast of Cortona Valecchie, dry grassy bank on noncalcareous soil, with *M. sativa* and *M. falcata*, 24-VI-1979, in flower & early fruit (BM). ROMANIA: Nyárády, E.I., Transsilvania, distr. Cluj, in declivibus apricis graminosis, ad "Santul Turcului" prope appid. Cluj, alt. 390 m, 9-VI-1923, Flora Romaniae Exsiccata No. 1270 (P). TIBET: Schlagintweit, Herbarium of, prov. Bálti, environs of Skárdo, alt. 6900 to 7500 feet, 6-VIII-4-IX-1856, Catalogue No. 877 (P). Schlagintweit, Herbarium of, prov. Ladák, 15-VII-5-VII-1856, in flower & early fruit, Catalogue No. 956 (P). YUGOSLAVIA: Justin, Carniola, ad viarum margines prope pagum Vreme in valle Reka, solo calcareo, alt. 360 m, VI, in flower, fruits in envelope, Flora Exsiccata Carniolica No. 487 (BM).

M. sativa* subsp. *sativa* × subsp. *glomerata

ALGERIA: Balansa, B. 976, plaine de Batna, mere spontanea, 24-VI-1853, in fruit (K, P). Faure, A., environ de Tlewcen (?), bord des champs, alt. 800 m, 19-VI-1933, in flower & fruit. Société Cènomane d'Exsiccata No. 2978 (P). (distributed as "*Medicago sativa* spp. *glandulosa* (Urb.) Maire var. *gaetula* Urb. f. *versicolor* Maire") Faure, A., Cerny près Tlemcen, champs cultivé, alt. 1200 m, 9-VIII-1932, in mature fruit (P). (Type! of *M. sativa* var. *faurei* Maire.) Maire, R., In dumetis prope Mausoura, in montibus Pomarientibus, solo calcareo, 800 m, 22-V-1933 (P). (Isotype(!) of *M. sativa* L. var. *gaetula* Urb. forma *versicolor* Maire; also viewed online at http://www.aluka.org, where the holotype (at MP) can also be seen; current March 2007. Note that the same website presents two herbarium collections, Maire, R. on 24-V-1933 (MP), as syntypes of the name.) AUSTRIA: Heller, L., Wiesen bei Gloggnitz, 1-IX-1886 (K). USSR: Grossheim, A, Transcaucasia, Georgia, prov. et distr. Tiflis, prope Kodzhori, 1200 m, in graminosis, 28-VI-25, A. Grossheim et B. Schischkin Plantae Orientales Exsiccatae 1928 No. 285 (K, LE). TUNISIA: Murbeck, S. Tunetia media, Maktar, in collib. calcar., 1-VI-1896, in flower (LD). (Syntype of *M. sativa* subsp. *tunetana* Murb.) Murbeck, Sv., Tunetia centralis, Dji Bargou, in collibus calcareis, 21 & 24-V-1903, Sv. Murbeck Plantae Selectae ex Africa Boreali No. 23 (P). (Possible type of *M. sativa* subsp. *tunetana* Murb.)

THE EVOLUTION OF *MEDICAGO SATIVA* IN RELATION TO THE HORSE

Medicago sativa is the only species in the genus that has been extensively domesticated. Humans have been altering it for thousands of years; by contrast, the dozen or so other species of *Medicago* that are cultivated have undergone intensive selection for, at most, less than half a century. Also, the economic importance of alfalfa dwarfs that of all other species of the genus combined, and accordingly the nature of its evolution under the hand of people is of exceptional importance. As detailed in the following, a remarkable and quite possibly unprecedented partnership among three species — humans, alfalfa, and the horse — has guided the course of the evolution of alfalfa for millennia. As noted below, historians generally appreciate that the domestication and use of the horse over the course of recorded history has been one of the most important factors in guiding the establishment and progress of civilizations in the world. What has been virtually ignored, however, is that without alfalfa, the horse could not have played its key roles nearly as well. Quite simply, the modern world owes an enormous debt to alfalfa.

Wild Horses and Their Relationships to Domesticated Horses

There is agreement that modern horses arose by domestication somewhere in Eurasia, no earlier than 8000 years ago. Domestication prior to this date could have occurred, but "there is...no conclusive evidence that the men of the Stone Ages ever domesticated the horse, although their use of horses as food and their keen interest and observation may properly be considered as the first step that was to lead to domestication" (Simpson 1936; cf. White 1989 regarding the possibility of paleolithic domestication).

All modern domesticated horses are generally assigned to *Equus caballus* L. One of two major camps contends that modern breeds are substantially the descendants of one wild progenitor, a view accepted by Darwin (1868). Proponents of the opposing camp hold that modern domesticated horses derive from several wild taxa (species or subspecies) and, moreover, that modern breeds can substantially be assigned to the distinct lineages. These authorities disagree extensively on both nomenclature and circumscriptions. "None of the many scholarly hypotheses as to the origin of the domestic horse has as yet been accepted universally" (Willoughby 1974).

Still another view was held by Simpson (1951), who suggested that probably horses were domesticated over a

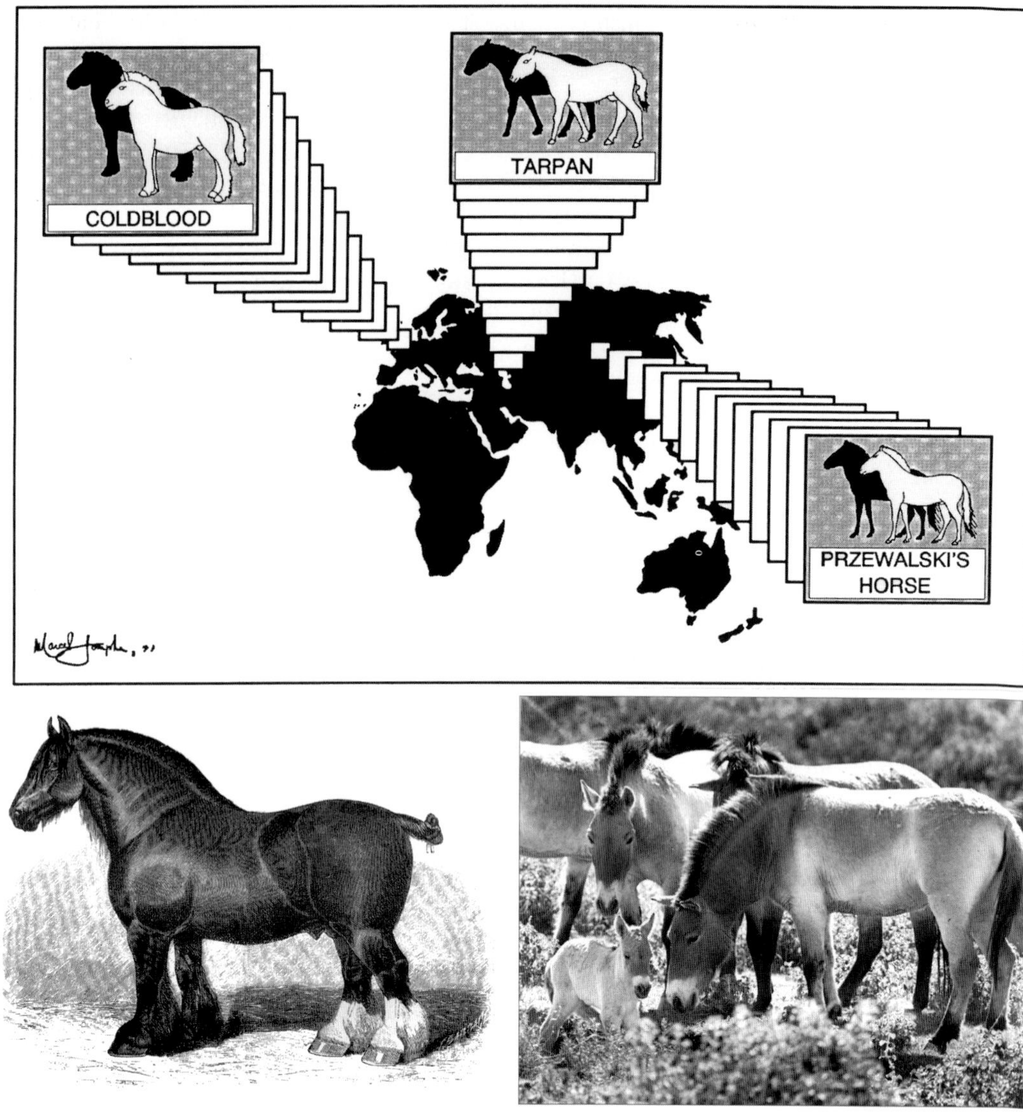

Fig. 12-37. Wild and domesticated horses. **Top left:** The postulated extinct "coldblood" (wild European "heavy horse"). **Bottom left:** The Clydesdale, thought to exemplify modern European heavy horses. **Top middle:** The extinct tarpan (south Russian wild horse) considered to be the ancestor of most modern horses. Note that its basic center of origin is thought to be sympatric with the basic center of origin of cultivated alfalfa. **Top right:** Przewalski's horse (Mongolian horse), the only extant genuinely wild horse. This horse, also shown at the bottom right, is distinguished from domesticated horses by its erect mane and lack of a forelock. Sources: The top picture was drawn by M. Jomphe; the bottom left figure was adapted from Thompson (1892); the photo on the bottom right is courtesy of the Zoological Society of San Diego and O.A. Ryder.

vast region with a variety of different wild races, all of which interbred and intergraded, and indeed that from very early times domesticated forms were interbred by people, so that it is not really possible to derive modern horse breeds from only a few original wild kinds. As Simpson noted, however, most students of horse evolution try to interpret the breeds in terms of a few original breeds or wild forms.

Based on cave paintings of extinct horses, and on Przewalski's horse discussed below, morphological characters of wild horses appear to include a relatively heavy head, thick neck, upright mane, short body, and graceful but relatively short legs (Zeuner 1963).

The most commonly postulated wild progenitors of domesticated horses are the extant Przewalski's horse, the extinct "tarpan" of southern Russia, and the extinct "coldblood" horse (so-named because of its apparent adaptation to cold European conditions) of western, northern and central Europe. Most students of horse domestication accept that horses similar to the tarpan were first and most substantially domesticated in the areas around the Black and Caspian Seas, and were exported to other areas of Eurasia where interbreeding with local horses (possibly independently domesticated) may have occurred. The local wild horses in western and northern Europe, and in eastern Asia, with which the tarpan's

Fig. 12-38. The so-called "pedigree seal", alleged to be the oldest pedigree in existence (5000 years of age, from Mesopotamia). The horse heads illustrated differ in mane type (erect, pendant, maneless) and profile (convex, straight, concave). The organization into horizontal rows and diagonal columns and the symbols (obscure, except for the upside-down female symbol at the lower left) are suggestive of selective breeding and hence domestication. (After Amschler 1935.)

domesticated descendants may have interbred, were probably the coldblood and Przewalski's horse, respectively. The role of Przewalski's horse is controversial, and the separateness of the coldblood horse has been widely argued.

As the only extant wild horse, Przewalski's horse has been examined extensively (Salensky 1907; Harper 1945; Mohr 1971). *Equus przewalskii* Poliakoff, or *E. caballus* subsp. *przewalskii*, described in 1881, was last seen in its native territory in the 1960s, in the Great Gobi Desert of Mongolia, and is believed to be extinct in the wild. No more than 1000 animals, bred in captivity, survive in zoos and preserves (Ryder 1988; Sattaur 1991). Cytogenetic studies have shown that domesticated horses examined to date have a somatic chromosome number of $2n = 64$, two fewer chromosomes than Przewalski's horse ($2n = 66$) (Ryder 1988; Sattaur 1991). Nevertheless, Przewalski's horse interbreeds readily with domestic horses, producing fertile offspring (Short et al. 1974). Interbreeding from feral escapes, as well as other "civilizing" influences, has resulted in the extermination of most of the wild forms. Chow Ben-Shun (1989) presented evidence that Przewalski's horse gave rise to early Chinese domesticated horses, so that very likely eastern Asian horses have some ancestry from Przewalski's horse.

The extinct tarpan or south Russian wild horse (*Equus caballus gmelini* Antonius) is, like Przewalski's horse, often considered as the sole ancestor of modern domesticated horses, although it has also been suggested that it is merely a feral escape from captivity. It apparently became extinct in the 19th or 20th centuries in the Ukraine; Groves (1986) stated that the last tarpan died in 1918. The only known illustration purportedly drawn from a living specimen is shown in fig. 79, Willoughby 1974 (cf. reconstructions in figs. 80 and 83 in Willoughby, and fig. 12:3 in Zeuner 1963). Many fossils from Siberia, Scandinavia, and France combine characteristics of the south Russian tarpan and Przewalski's horse, suggesting that the two alleged ancestors interbred freely over a wide area (Willoughby 1974). In fact, in much of the literature, the term tarpan includes most Eurasian horses, including Przewalski's horse.

A third type of horse has been held to be distinct by some authorities (Willoughby 1974), the so-called "coldblood" or "heavy horse" (*Equus caballus robustus* Pomel.). Willoughby proposed an alternate common name, the "Occidental horse". This is a pleistocene forest-dwelling wild horse of western, northern, and central Europe, thought to have become extinct about 1000 years BC (reconstructed illustrations are fig. 85, Willoughby 1974, and fig. 12:5, Zeuner 1963). Horses were found associated with humans in Europe towards the end of the fifth millennium BC, but not until about 2500 BC in appreciable numbers (Sherratt 1983). It has been claimed (e.g., Willoughby 1974) that the early European wild horses were large and of heavy build, pre-adapting them for draft usage. It is also widely believed that they gave rise (at least in part) to most modern large breeds of horses, such as the Belgian, Clydesdale, Lipizzaner, and Percheron. Zeuner (1963) noted that breeding for size is well known in domesticated animals, so size alone is a dubious basis for deducing that large European breeds trace to large ancestral European horses. Moreover, Holocene European horses were of limited stature (see size data in Bokonyi 1978), and the large modern European horse breeds may have originated no earlier than 1400 (D. Anthony, personal communication). The question of large size of European horses, accordingly, is problematic. Nevertheless, Zeuner (1963) concluded that there is sufficient evidence to indicate the reasonableness of the hypothesis of a western wild horse ancestor. Bokonyi (1978) stated that "although information is very vague about West European wild horses of the early Holocene...they easily could be the wild ancestors of the earliest domestic horses of western Europe." Goodall (1965) suggested that the Ardennes breed of France is the domesticated horse closest to the original primitive wild horse of the European forests, whereas Stock et al. (1944) suggested the Belgian breed is the least modified. Most authorities have emphasized that cultural diffusion from the cradle of horse domestication around the Caspian Sea undoubtedly witnessed considerable importation of domesticated horses to Europe so that horses of apparent European stock almost certainly trace much, if not all, of their ancestry to more easterly horses.

Predomestication Relationships of Alfalfa and Horses

The sympatry of wild horses and wild alfalfa over much of temperate Eurasia likely presented humans with two species preadapted for domestication in combination. Water is the principal limiting factor in the mutual wild habitats occupied by alfalfa and horses, and both require considerable amounts of water despite their adaptation to dry areas. In the case of alfalfa, deep roots (sometimes reaching down 9 m) allow access

Fig. 12-39. Threshing grain by use of a horse-drawn sled in eastern Turkey (photo by E. Small in 1981). Such labor-saving use of the horse has been carried out for thousands of years.

to ground water. While the plants occupy dry areas, periodic moisture is essential to replenish the ground water at root level, and of course for seed germination and seedling growth. Horses must also have access to water in their feeding ranges. Both the animal and the plant are preadapted to domestication in an oasis or irrigation situation, wherein water can be supplied, and in a sense both species benefitted under the hand of people in acquiring a regular supply of water. Both the animal and the plant are adapted to the same climate, being able to withstand cold winters and deep snow. (It should also be noted that humans selected unique forms of both alfalfa and the horse in the hotter region of the Middle East, i.e., nondormant alfalfas (Lehman and Marble 1986) and the Arabian horse.) In addition to common adaptation to their habitat, horses and alfalfa share other features that lend themselves to joint domestication. Both the animal and the plant are easily domesticated since horses bond quickly to people and alfalfa is undemanding in its requirements. Both the animal and the plant can tolerate limited care from humans, as reflected by the ease with which both species can return to ruderal existence.

There is good reason to believe that wild horses (like other domestic grazing animals), fed on wild species of the genus *Medicago* only incidentally and only as a minor dietary component. Collections of wild horses have been made primarily from areas with sparse, low quality vegetation, where wild alfalfas occur mostly in scattered patches. Alfalfa does not withstand grazing nearly as well as grasses. Horses are relatively unselective compared with most grazers with respect to the taxonomic composition of the plants they eat (Janis 1976) and are reported to prefer to feed on mixed as opposed to pure vegetation (Archer 1973).

The morphology of the horse is indicative of an animal adapted to open grassy savannahs and grasslands (Stebbins 1981) — the long slender legs terminating in hoofs reflect adaptation to rapid running or leaping in fairly open country and the high-crowned teeth (hypsodontic, i.e., growing continually and wearing down by contact) with complex patterns of enamel reflect a diet of hard siliceous leaves, particularly those of grasses. Indeed, it is widely believed that grasses and horses coevolved. The grasses eaten by the horses caused the evolution of the high-crowned teeth of modern horses. Conversely, the hard firm textures of the leaves of grasses that grow on open plains and steppes has resulted from selection by horses (Stebbins 1982). The digestive system of the horse provides excellent evidence that this animal is much more adapted to wild grasses than to alfalfa. As reviewed by Janis (1976), horses appear to have evolved digestive adaptations to a diet with a very high fiber content coupled with a very low protein content. Very few grazing animals of similar body size could survive on the high-fiber, low-protein diet that can sustain horses.

Place and Time of Horse Domestication and Their Bearing on Alfalfa Domestication

Early centers of alfalfa domestication likely occurred in areas where horse breeding became established. The status of agricultural development in a given region could be indicative of the probability that alfalfa was cultivated there at a given time. Nomadic peoples lacking a permanent home base are very unlikely to have played a role in the domestication of alfalfa. Clearly, the raising of domestic animals, in particular the horse, would have preceded the domestication of alfalfa and provides a critical context for historical analysis.

The history of irrigation farming is important in providing a probable framework for the ancient domestication of horses and alfalfa because of the desirability for water at least during part of the year for raising both. Irrigation appeared first in Khuzistan and resulted in the steppe of the Zagros region becoming a dominant growth center between 5500 and 4000 BC (Flannery 1969). A well-known site of horse breeding was around Lake Van (Lesins and Lesins 1979) in the northern part of Zagros (eastern Turkey). Since the new mode of enhanced production through irrigation did not spread rapidly out of Southwest Asia, there was ample opportunity for alfalfa cultivation to have arisen there. Lesins (1976) suggested that mountain-protected fertile valleys in the Zagros and the Elburz Mountains, and irrigated oases facing the Central Iranian Plateau, may have been the first places where alfalfa was cultivated.

Anthony (1991) provided comprehensive information on the history of horse domestication and its importance in the sequence of civilization. The earliest claims of domesticated horses are from the Ukraine north of the Black Sea, at about 4000 BC, where horses were ridden (Anthony and Brown 1991). The first literary mention of the horse is in a text of the Third Dynasty of Ur, at ca. 2100 BC (Drower 1969). Possibly the earliest record of horse breeding is discussed by Amschler (1935), who interpreted a 5000-year-old engraved stone seal from the highlands of southwestern Persia (east of Ur in Mesopotamia) as a primitive horse pedigree (illustrated here). The four horizontal rows of horse heads were interpreted as having different mane types (erect, pendant, and maneless) and head types (convex, straight, and concave profiles). The

Fig. 12-40. Aspects of the horse–human association. (A) Prehistoric man at the dawn of horse utilization, using the horse for food. (B) The use of the horse as a draft animal. (C, D) The use of the horse in warfare. (Artist: B. Flahey.)

concave profile of some of these figures was interpreted as belonging to early horses of the Arabian type, which has been accepted by some authors (e.g., Jones and Bogart 1973).

It seems that horse domestication first occurred in western Asia between 5000 and 6000 BC. Ivanov's (1980) suggestion that alfalfa domestication dates as far back as 8000 BP seems unlikely given the shorter period of horse domestication. A western Asian origin of alfalfa is consistent with the evidence reviewed in the previous section that the oldest records of alfalfa use (albeit much more recent than the records of horse use) trace to western Asia. The likelihood of a Central and eastern Asian origin of alfalfa is discussed below in relation to horse domestication.

According to Chow Ben-Shun (1989), horse remains in China dating to 5000 years ago closely resemble Przewalski's horse in stature. Archeological evidence of horse domestication in northwestern China has been found back to 2000 BC and, of particular significance, osteological evidence and Chinese Bronze Age illustrations (from 2000 to 300 BC) show that domesticated Chinese horses "retained throughout this time the characteristics of the Przewalski's horse, including relatively heavy heads, thick necks, upright manes, short bodies, and relatively short legs" (Chow Ben-Shun 1989). Only after the Western Han Dynasty (from 206 BC to AD 25) was there appreciable inflow of genes from horses of the tarpan class, with subsequent illustrations showing new larger breeds of graceful horses with smaller heads than the Przewalski horse. A Chinese expedition in 126 BC collected both horses and alfalfa seeds (Hendry 1923). Prior to this recorded spread of alfalfa eastwards, there is no evidence that alfalfa was fed to horses in China, so that while central Asia may well be a gene center for alfalfa and a center of origin of some alfalfa cultivars, it is probably not a center of domestication of wild alfalfa.

The hypothesis for the most easterly site of alfalfa origin is P.M. Zhukovsky's suggestion that it originated in Mongolia (for references and discussion of this and similar proposals, see Ivanov 1980). This is an interesting idea, given that the Mongolian horse represents the last surviving wild kind of horse and Mongolia is an ancient site of horse use. However, Mongolia is far removed from the center of distribution of the purple-flowered lineage, although the yellow-flowered lineage occurs there. As pointed out by Ivanov (1980), it is very unlikely that purple-flowered alfalfa was domesticated in Mongolia, first because it is poorly suited to overwintering there and second because agriculture in Mongolia was established in a substantial enough way for alfalfa cultivation no earlier than the 7th century AD.

India (Kashmir) is sometimes mentioned as a possible site of origin of alfalfa (cf. Ivanov 1980). De Candolle (1885), observing that the distribution of wild alfalfa extended west to Kashmir, thought this a possible place of origin, but noted the absence of a Sanskrit name, which seems curious if alfalfa was indeed domesticated there. An origin in "India" is improbable because the area is marginal to the principal areas of wild horse and wild alfalfa distribution, and evidence of early domestication of either is lacking. Thomas (1989) concluded that the horse was introduced in an already domesticated form in India, and horse breeding became prevalent only during the Iron Age (from 1000 to 400 BC).

The Co-Domestication of Alfalfa and Horses

In the following discussion the term "co-domestication", used to describe the relationship between alfalfa and horses, is employed in analogy with the term coevolution (the evolution of reciprocal adaptations of two taxa to each other) to denote our (conscious or not) genetic alteration of a pair of species guided by selection for the reciprocal species. In effect, co-domestication produces or improves a symbiosis between two domesticates, which complements the symbioses between humans and each of the domesticates (cf. Higgs and Jarman (1969) who interpreted domestication as a natural form of symbiosis).

Today, numerous livestock species, chiefly cattle, are fed alfalfa. Of the more than 4000 species of mammals, only about 16 have produced important domesticates (Wright 1954). However, the horse is the principal herbivore that needs to be examined in relation to adaptive properties of *Medicago sativa*. Alfalfa and horses possibly represent the longest history of human use of an interdependent pair of domesticated species, probably exceeding silkworm culture on mulberry, which seems to trace only to the first millenium BC (Tazima 1984). To a remarkable extent use of horses for transportation was a key innovation, arguably the most influential technology that has guided the course of civilization in the Old World for the past 6 millennia, and in the post-Columbian New World (Anthony 1986; Clutton-Brock 1992). Historical evidence of the intimate association between alfalfa and horses has been well documented, for example by Hendry (1923), Lesins (1976), and Lesins and Lesins (1979). The early use of both horses and alfalfa in various parts of Asia has led to various hypotheses concerning their primary and secondary centers of domestication. However, it seems clear that a primary domestication of both horses and alfalfa occurred in western Asia. Early centers of alfalfa domestication likely occurred in areas where horse breeding became established.

The earliest record of alfalfa use appears to be in ca. 3300-year-old Hittite brick tablets from central Turkey (Corum, Alacahoyük), dated from 1400 to 1200 BC (Bolton et al. 1972, based on a personal communication that the tablets "indicated that animals were fed on alfalfa all through the winter season and alfalfa was regarded as a highly nutritious animal feed"). Even today, alfalfa is "generally the most common and most accepted horse feed in most areas, the preferred horse feed in all circumstances" (Kerrigan 1986). Indeed, the word "alfalfa" originates from Arabic and Old Iranian for "horse fodder". Since domesticated horses have thrived according to the availability of alfalfa, this plant is really one of the select number of plants that have been critical to the shaping of the modern world. Indeed, livestock were still essential for warfare as late as the early 20th century. The British army used hundreds of thousands of mules and horses during the First World War, and it was noted that fodder constituted 25% of the weight of all materials shipped, exceeding the weight of armament shells. It is remarkable, therefore, that little consideration has been given to why alfalfa is so good for horses compared with other domesticated herbivores. It is hypothesized below that this is because of nutritional preadaptations of alfalfa that are unique to the horse.

A Dietary Explanation for Why Domesticated Horses Thrive on Alfalfa Better Than Other Domesticated Herbivores

Today, specialized mixtures that are high in alfalfa are used to feed a variety of domesticated mammals, especially those in high-input, high-production systems. The possibility arises that alfalfa was domesticated not just for the horse, but also for other animals. Cattle, sheep, goats, and pigs are the four major mammalian species domesticated as meat producers (Jewell 1969). These mammals, along with the wolf, were the earliest animals to be domesticated — this occurred in the cradle of prehistoric agriculture in the lower hills and adjacent plains of the area east of the Mediterranean extending to southwestern Iran (Reed 1969). Donkeys, camels, llamas, reindeer, and bovine species in eastern Asia serving the role of cattle are either far removed from sites of alfalfa domestication or are of limited importance in relation to alfalfa consumption. Cattle, sheep, and goats are ruminants (pigs, in a limited sense, function similarly, as will be explained) and, as argued below, this is important to understanding why alfalfa domestication developed principally in association with the horse rather than with the ruminants.

During the 7th millenium BC in the Fertile Crescent, sheep and goats were the first animals domesticated for meat production (Uerpmann 1987; see Hole 1989 for references to other proposed sites of domestication). Wild sheep (*Ovis orientalis* Gmelin) still live in the foothills and intermontane plains of the Zagros and Taurus ranges, where they feed primarily on wild grasses. Domestication starting as long as 9000 years ago continued in Iran and Anatolia (Flannery 1969). In these respects the history of domesticated sheep (*Ovis aries* L.) is quite similar to that of the horse. However, for the most part sheep have been shepherded on wild vegetation, limiting the possibility of co-domestication with a crop. In fact, overgrazing by sheep in the Old World has eliminated a substantial proportion of wild alfalfas, despite a tendency for wild alfalfa in sheep-grazed regions to evolve low-growing ecotypes (Small 1981*b*, 1982).

The domesticated goat (*Capra hircus* L.) is much like the sheep in its distribution, domestication, and ecology (Mason 1984). The wild ancestor (*C. aegagrus* Erxleben) is sympatric with wild *M. sativa* over much of the wild sheep's range. It is primarily a browser of mountainous regions, but is also adapted to semi-desert, and is capable of survival on very sparse forage. As with sheep, shepherding on natural vegetation has minimized the goat's need for artificial pastures and fodder.

The pig (*Sus scrofa* L.) eats a wide variety of plant and animal matter. It is therefore not a good candidate for co-domestication with any particular plant. Moreover, as noted below, the pig has a limited tolerance for alfalfa saponins. Wild pigs are very widespread in Eurasia, but while they are common in much of the range of wild alfalfa during the time of alfalfa's domestication (Vereshchagin 1959), the pig's relatively poor adaptation to low temperature (poor fur cover) and to deep snow cover (short legs) is indicative of an animal that is not suited to the natural range of alfalfa.

Domesticated cattle (*Bos taurus* L.) are thought to have originated in Greece or western Anatolia (Hole 1989). The wild ancestors of cattle (*Bos primigenius* Bojanus) were distributed throughout Europe, North Africa, and southwestern Asia, becoming extinct as late as 1627 (Reed 1959). Cattle were employed as draft animals before 5000 BC to pull plows (Reed 1969). According to Epstein and Mason (1984) "Cattle were the first farm animals used for work and have always been the most important draught animals the world over." Cattle have been, and continue to be, employed additionally as sources of meat and milk, and as milk sources they may be permitted to forage on wild pastures. They do seem adapted to spending most of their waking hours grazing, averaging 8 hours per day (Epstein and Mason 1984). However, the fact that cattle have been used so frequently as beasts of burden — much like horses — raises the question of why an early historical relationship with alfalfa as strong as with the horse did not develop.

Horses, cattle, sheep, and goats subsist on a fibrous diet with considerable cellulose, from which energy is obtained by a symbiotic association with cellulase-producing microbes (mostly bacteria, but also fungi and protozoans) in a fermentation chamber either in the hindgut or foregut. In horses this chamber is the cecum, positioned after the stomach in the first part of the large intestine, with absorption of the fermentation products taking place in the colon. In four-stomach ruminants such as cattle, sheep, and goats, the fermentation chamber occurs in the first two stomachs (reticulorumen, rumen). Fundamentally, the horse thrives on a diet similar to these ruminants (Frape and Boxall 1974). The pig merely has some enlargement of the forestomach and comparatively limited stomach micro-organisms (Janis 1976), reflecting its nonspecialization on plants. (Following Janis, the term *ruminant* designates all animals that ferment cellulose in a "forestomach".)

Bokonyi (1984) stated that the cecal fermentation of horses "is certainly inferior to the complex stomach and cud-chewing of the ruminants"; others have used this reasoning to explain the lack of success of the perissodactyl lineage of hoofed mammals, which includes horses and only a few other species, compared with the artiodactyl lineage, which includes cattle, sheep, goats, pigs, and many other species (Janis 1976). However, in an extensive review of the subject, Janis (1976) concluded that cecal grazers are not "inferior" to ruminants, their gut simply representing an adaptation to a different ecology than that of the ruminants. In ruminants such as the cow, the digestive tract and its contents comprise about 40% of their body weight whereas they comprise only about 15% of the body weight in horses. Accordingly, such ruminants digest cellulose more efficiently and need to eat less than horses on a per weight basis. Moreover, ruminants are able to utilize the bacterial protein created during the fermentation by simply digesting it in the digestive tract. Thus, they can use a variety of nitrogen sources in their feed supply. Nonruminants are not able to digest any of the bacterial protein produced, which is lost in the feces. However, the lesser weight of the digestive tract and its contents in horses must be advantageous to them in helping them flee from predators, and certainly for speed humans have always appreciated horses more than cattle. Many studies have compared the abilities of cows and horses to utilize grasses and legumes and it has consistently been found that, while horses are at a comparative disadvantage with grass forages since they digest fibrous components less efficiently than cows, they are equally efficient at extracting nutrients from alfalfa hay (Cymbaluk 1990). The differences between horse and cattle digestion suggest that cattle seem less in need of a supplement such as alfalfa (Janis 1976). In the rumen, protein and carbohydrates (including glucose) are fermented. Rumen bacteria synthesize all the amino acids needed by cattle from the fermented protein so that cattle are protected against dietary amino acid imbalance and are obviously less in need of a high-protein supplement such as alfalfa than horses are. Indeed, protein in alfalfa is so extensively degraded in the cow rumen that it is often wasted (Broderick and Buxton 1991). Although some amino acids are synthesized in the bodies of people and other monogastric animals, eight essential amino acids must be supplied in the diet so that a high-quality protein source is of much greater advantage to a horse than to a ruminant. Moreover, in cattle the fermentation of glucose prevents a rise in blood sugar level immediately after a meal, as occurs in the horse; so, for work purposes, horses may benefit more quickly than cattle from a high-energy supplement such as alfalfa. Additionally, there are unidentified growth and health factors for horses that occur with alfalfa consumption (Romic 1978, discussed below). The comparative advantages of an alfalfa diet for the horse and the cow are summarized in Fig. 12-41.

The horse is an animal that is naturally adapted to a low-quality diet by its ability to consume very large quantities and pass these through its gut quickly, albeit less efficiently than the ruminant (Ørskov and Kay 1987). It is curious that wild horses, adapted to a low-protein diet, took readily to the

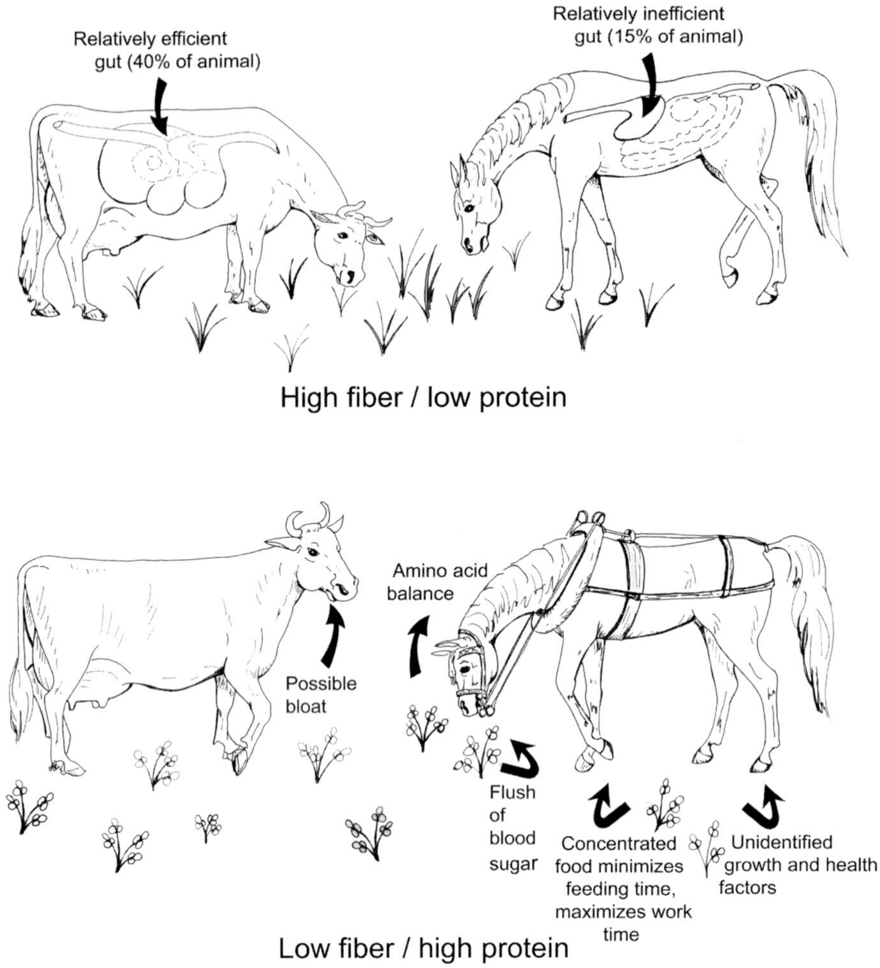

Fig. 12-41. A comparison of horse and cow nutrition on mixed grassland (above) and on alfalfa (below). **Above:** Both the horse and the cow do equally well on mixed grassland, but the horse must devote more energy and time to eating because of its smaller gut. **Below:** Alfalfa is of greater benefit to a working domesticated horse because of the dietary advantages discussed in this chapter. (Artist: B. Brookes.)

high-protein, concentrated alfalfa diet provided by people. Such specialized animal feeds that furnish nutrients in a concentrated form allow rapid animal growth and high productivity (Wright 1954). Rapid growth can benefit all livestock, but not especially the horse, except to the extent that it is raised for meat or secondary products. Because the horse is employed chiefly for work, a concentrated feed such as alfalfa allows more free time for work. However, there is strong evidence that the chief benefit of alfalfa to the horse is the special importance of alfalfa protein to the working horse. As reviewed above, the horse's cecal digestive system appears more suited to benefitting from a high-protein feed such as alfalfa than does the ruminant digestive system. Moreover, strongly exercising horses lose considerable nitrogen through sweat and use up protein stores quickly, and under such stress respond well to high levels of protein (Custalow 1991).

There are likely supplemental reasons why alfalfa was domesticated first for horses rather than for ruminants. Bloat is a digestive disturbance characterized by an abnormal distension of the abdomen due to accumulated gases in the rumen of cud-chewing animals. Bloat occurs when the animal fails to eructate (belch) the accumulated gas. Alfalfa is notorious for causing bloat in ruminant animals and it may cause death. Ruminants are subject to bloat from gorging on fresh alfalfa, whereas monogastric animals like horses are not. Another consideration is that sheep, goats, pigs, and cattle were domesticated initially to be used as living food reserves and were used only subsequently for draft power, for which nutritional considerations are particularly important. It is plausible that in addition to domesticating horses for their meat, humans may also have used them as working animals from the beginning of the domestication (Bokonyi 1984) since alternative, earlier-domesticated meat sources were already available.

Alfalfa Saponins and Their Possible Significance in the Co-Domestication of Alfalfa and the Horse

The most puzzling aspect of the relationship between alfalfa and livestock relates to the presence of hemolytic saponins in the plant. Curiously, some landraces of alfalfa in western Asia are extremely low in hemolytic saponin content, suggesting that there was selection against these chemicals, which are widely considered nutritionally undesirable. Nevertheless, there is some evidence suggesting that hemolytic saponins may actually be beneficial to horses. The relatively northern *M. sativa* subsp. *falcata* is notably higher in hemolytic saponin content than the more southern subsp. *sativa* and, as a consequence, horses in southern Europe and Asia may have been fed relatively low-saponin alfalfa for millennia, in contrast with the relatively high-saponin alfalfa that has been utilized in northern Europe and much of the New World for at least several centuries. These different dietary regimens raise the issue of whether, before the age of modern plant breeding, there was selection for a diet of hemolytic saponins in some horses.

Paradoxically for a livestock feed, alfalfa contains hemolytic saponins (capable of lysing red blood cells), which are widely believed to be antinutritional. Moreover, these saponins differ dramatically in their occurrence in alfalfa lineages and cultivars. The result has been that horses in some parts of the world have been (and continue to be) fed a steady diet of high-saponin alfalfa, whereas in other parts of the world horses have been maintained for millenia on low-saponin alfalfa. The adaptive inter-relationships of alfalfa and horses in light of these historical–geographical patterns are explored in this section.

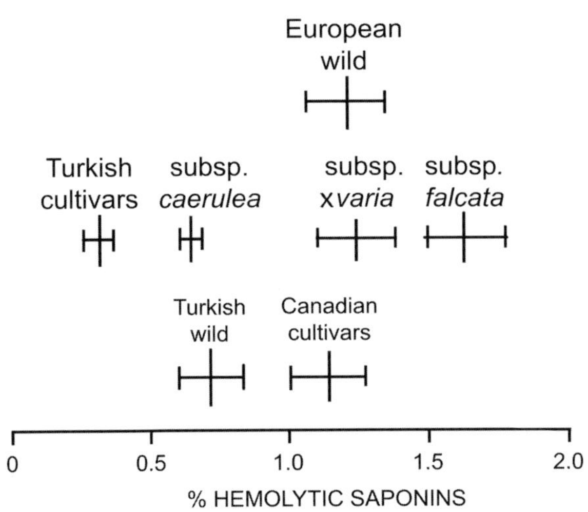

Fig. 12-42. The means and their 95% confidence limits for the hemolytic saponin content in the foliage of groups of *Medicago sativa*. The means of the groups in the vertical files are not significantly different. (After Small et al. 1990c.)

Saponins are of considerable economic interest, apparently imparting resistance to pests and diseases (see reviews by Birk 1969; Cheeke 1971; Bondi et al. 1973; Pederson 1978; Applebaum and Birk 1979; Oakenfull 1981; Price et al. 1987; and Oakenfull and Sidhu 1989). "More seems to be known about alfalfa saponins than those from any other source" (Oakenfull and Sidhu 1989). Major alfalfa saponins are hemolytic. As noted in Chapter 9, the most common alfalfa saponin, occurring as a glycoside of medicagenic acid, is found in leaves of all species of the genus *Medicago*, but not in related genera, and indeed is very rare in the plant kingdom (Jurzysta et al. 1988, 1992). The physiological effects induced by "alfalfa saponins" seem due primarily to medicagenic acid glycoside.

Alfalfa saponins were once believed to be responsible for contributing to bloat in ruminant animals (e.g., Oakenfull 1981), but are now generally not thought to be implicated (Majak et al. 1980; Bósca et al. 1986; Price et al. 1987; Hall and Majak 1989). While monogastric livestock are not susceptible to bloat, they are much more subject to the deleterious effects from alfalfa saponins than ruminants. Possibly, saponins are degraded by rumen microbial fermentation and so are not available in the remainder of the digestive tract. Alfalfa saponins have been reported not to affect growth of cattle or the performance of lactating dairy cows.

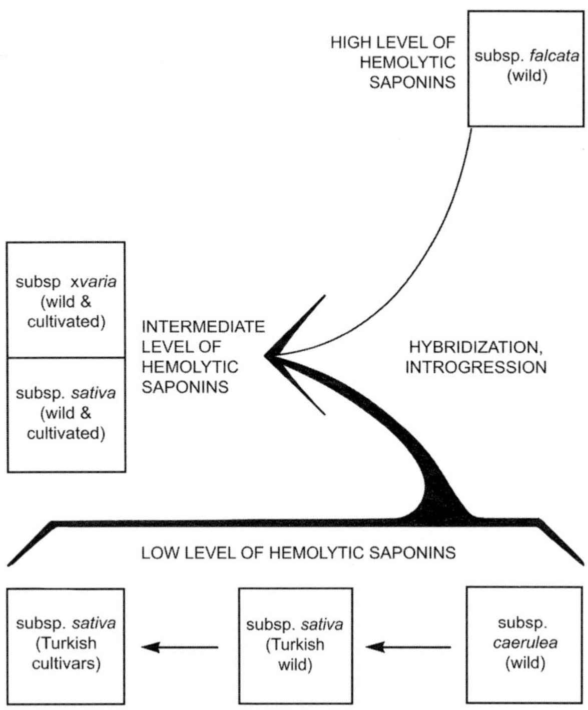

Fig. 12-43. An interpretation of the evolution of alfalfa with respect to the content of hemolytic saponins. The very low saponin content of many Turkish populations of subsp. *sativa* and of many Turkish cultivars is due to ancestry from *Medicago sativa* subsp. *caerulea* alone, which has low saponin levels, whereas the moderate saponin content in subsp. x*varia* and many populations of subsp. *sativa* appears to be due to the addition of germplasm from *M. sativa* subsp. *falcata*, which has high saponin levels.

Alfalfa saponins have been demonstrated to be very detrimental to poultry and somewhat detrimental to pigs. The harmful biological responses of animals consuming saponins have been attributed to erythrocyte hemolysis, reduced smooth muscle activity, enzyme inhibition, reduced blood and liver cholesterol levels, and reduced nutrient absorption. Deleterious effects have not been demonstrated in people, but the high level of hemolytic saponins in some alfalfa seedlings sold as salad sprouts (up to 8%) has raised suspicion, and has led to the attempt to chemically remove saponins from feed preparations made from sprouts (Livingston et al. 1985). Because of concern about the "antinutritional" and "antiquality" effects of alfalfa saponins, deliberate breeding to produce low-saponin alfalfa has been carried out (Pederson et al. 1973; Pederson 1978; Bósca et al. 1986; Lila et al. 1989).

Some saponins are phytoestrogenic, mimicking animal hormones (see reviews by Howarth 1988; Adams 1989). Such alfalfa saponins have been suspected of causing infertility and nymphomania in cattle. Detectable amounts from consuming alfalfa sprouts have been demonstrated in humans, but have not been associated with harmful effects. Increased growth and quality of lambs has been attributed to alfalfa estrogens, but beneficial effects have not been reliably established. Estrogenic alfalfa saponins are relatively limited and not thought to have nearly the importance of hemolytic saponins.

Lineages and cultivars of alfalfa differ notably in hemolytic saponin content (Small et al. 1990*c*). Subspecies *caerulea*, the major ancestor of most modern alfalfa cultivars, averages about 0.6% (dry leaf weight) and some forms of subsp. *sativa* (particularly in Turkey) also are similarly low in hemolytic saponin content. By contrast, subspecies *falcata* has a very high saponin content (averaging about 1.7% on a dry leaf weight basis). Cultivars of alfalfa have relatively high or low hemolytic saponin content depending on the relative parentage of subsp. *sativa* and *falcata*. Generally, only northern cultivars have considerable *falcata* parentage and therefore are relatively high in hemolytic saponins. As a result, horses in southern Europe and much of Asia appear to have been fed low-saponin alfalfa for millennia, whereas high-saponin alfalfa has been utilized in northern Europe and much of the New World since about the 16th century. Primitive Turkish landraces are almost devoid of hemolytic saponins (Small et al. 1990*c*). Since these likely reflect the kind of alfalfa present when horses were domesticated, it is possible that saponins are harmful to horses and selection against saponin content has occurred over the millennia. If so, the high saponin levels in northern European cultivars, acquired through introgression with subsp. *falcata*, could be deleterious to horses. On the other hand, as noted below, alfalfa saponins could be beneficial. If so, the extremely high saponin levels of a cultivar purely of the yellow-flowered lineage, such as 'Anik', might be especially beneficial (Small 1992). Since there has been an unintentional exposure of horses in northern Europe to high levels of alfalfa saponins for perhaps 500 years, and in northern North America (particularly Canada) for about 200 years, it is of interest to consider whether there have been selective consequences.

Surprisingly, given the well-known antinutritional effects of alfalfa saponins on some domesticated animals, the following study seems to be the only rigorous investigation examining the effects of alfalfa saponin levels on the health of horses (I.T. Johnson, personal communication). Fonnesbeck and Symons (1969) found that in horses "alfalfa hay significantly lowered the plasma cholesterol concentration both when fed alone and in combination with grains" and they specifically ascribed this effect to the high saponin content of the alfalfa they employed. This is in accord with many observations of alfalfa's hypocholesterimic property in other animals (Cheeke 1973), but leaves unanswered the question of whether it is good or bad for the horse.

The mechanism by which alfalfa saponins interfere with cholesterol absorption is uncertain, despite much study. The most commonly postulated mode of action is that the alfalfa saponins form insoluble complexes with cholesterol in the gut. An alternative explanation is that saponins induce absorption of bile acids by dietary fiber, possibly increasing conversion of cholesterol into bile acids in the liver (Oakenfull 1981; Price et al. 1987). Because serum cholesterol is widely assumed to be linked with atherosclerosis, a reduction may be desirable, and it has been suggested that consumption of saponins could be beneficial in reducing heart disease (references in Price et al. 1987). Alfalfa seed extract has been found to prevent the development of atherosclerosis under experimental conditions in rabbits (Dixit and Jain 1990). Alfalfa saponins have had the same effect in cynomolgus macaques, even serving to bring about regression of aortic and coronary atherosclerosis (references in Price et al. 1987).

There is one study that may demonstrate the extremely beneficial effects to horses (particularly foals) from consuming alfalfa saponins. Romic (1978) compared digestive disturbances and blood characteristics in horses fed alfalfa hay and "meadow hay" supposedly of equal dietary value, using fairly large samples of horses (768 horses in all). The study of blood features is particularly significant since the mode of physiological influence of hemolytic alfalfa saponins is specifically on the blood. Fewer cases of digestive difficulties and deaths were noted in the group fed alfalfa hay. More importantly, there were changes in blood composition, including increased erythrocyte number, increased hemoglobin level, alteration in relative types of leukocytes, and increased gamma globulins, proteins, hormonal and total iodine, and some minerals. Romic characterized the increased production of blood cells as "stimulative" and attributed the superior performance of the horses fed alfalfa hay to a strengthening of their defence system. Unfortunately, nowhere in his paper did Romic refer to alfalfa saponins and he did not specify the cultivar of alfalfa used, so the saponin levels of the experimental material is unknown (attempts at personal communication were unsuccessful). It is quite possible that the beneficial effects observed by Romic were due to the hemolytic saponins of alfalfa. Such a hypothesis is not simply unfounded speculation since saponins are well known as having anti-inflammatory properties and are used (although efficacy is debatable) in numerous human tonics, both Asian and Western (Oakenfull 1981). As the hardest worker among domesticates, the horse is the ideal candidate for a natural tonic. Romic's study needs to be conducted with a more appropriate experimental design to assess whether alfalfa saponins were beneficial, since the benefits that he observed could simply reflect superior nutrition of legume-based hay over grass hay.

The possibility of geographically based, differential, genetic, dietary adaptation in horses seems plausible in the context of what is known about dietary specialization in humans. For example, in people, lactose tolerance is associated with regions where milk has been produced historically (Newcomer 1978). Favism (a severe hemolytic anemia resulting from an inherited susceptibility to the broad bean, *Vicia faba* L.) varies considerably depending on racial group and geography, occurring in Mizrahi Jews, Mediterranean Europeans, Arabs, Asians, and Blacks (in 5% to 50% of the populations), but is virtually absent in Northern Europeans, European Jews, Native Americans, and Inuit (Cheeke and Shull 1985). Johns (1990) speculated that the Aymara of highland Bolivia, who have long used potato varieties containing potentially toxic levels of glycoalkaloids, can do so partly because of genetic selection for tolerance.

It is interesting to consider the possibility that northern European horses have adapted to high-saponin alfalfa in light of its postulated ancestor, the coldblood type of horse. The ancestral coldblood was a forest dweller, as opposed to the grassland dwellers to the east, and it is conceivable that disruptive dietary specialization evolved. Curiously, when such heavy northern horses have been imported to southern areas for draft purposes, the results have been poor (mules are usually more relied on), but when southern (Arabian-type) horses have been imported to the north (for speed) the results have been excellent (Wright 1954). These observations could reflect a tonic role of alfalfa saponins endemic in northern diets. However, it is possible that the results simply reflect climate effects versus feed and a proper controlled experiment is required to test the hypothesis.

A specialized microbiota of bacteria, protozoa, and fungi in the rumen or cecum of grazing mammals is largely responsible for the digestion of vegetation. In the horse, anaerobic Chytridiomycetes in the cecum are probably important in the digestion of low-quality fiber (Orpin 1981, 1988), the standard diet of the wild ancestors of modern horses, and it is unclear how this symbiosis is affected by providing high-quality vegetation such as alfalfa. The case of the tropical leguminous shrub *Leucaena leucocephala* (Lam.) de Wit., which has "nutritive value (discounting toxicities)... equal to or superior to alfalfa" (Duke 1981), is instructive. It is successfully used as ruminant forage only when the rumen is colonized by bacteria that degrade dihydroxypyridines, toxic intermediates in the metabolism of mimosine (Allison et al. 1990). These bacteria, however, have been found to be geographically limited, occurring in some areas but not others (Allison et al. 1990). Jones and Megarrity (1986) reported that *Leucaena* toxicity in Australian ruminants was overcome by transferring bacteria from Hawaiian goats. It is possible that the high levels of alfalfa saponins in northern Europe and North America could affect the cecal microbiota, and future research will need to consider this possibility.

Available information is insufficient to answer definitively whether hemolytic alfalfa saponins harm or benefit horses and, if so, whether differential adaptation to saponin levels has evolved. This review serves to point out the need to assess more carefully than has been done to date the effects of high- and low-saponin alfalfa on horses with a different ancestry and exposure to these respective diets.

13. *Medicago arabica* (L.) Huds.

M. arabica (L.) Huds., Fl. Angl. 288 (1762).

The epithet *arabica* is derived from Greek or Arabian. However, the species does not occur in Arabia. "Arabica" was used as part of the plant's name prior to 1753 when Linnaeus adopted the epithet.

Subordinate Taxa

None recognized here.

Type Specimen

Heyn in Bull. Res. Council Israel 7D: 167 (1959) lectotypified the basionym for *M. arabica* (*M. polymorpha* L. var. *arabica* L.) by an illustration of "Medica cochleata minor polycarpos annua capsula majore alba, folio cordato macula fusca", tab. 15 f. 12, in: Morison, R. 1680. Plantarum historia universalis Oxoniensis, vol 2(2). Oxford, UK. The Linnaean Plant Name Typification Project (http://www.nhm.ac.uk/research-curation/projects/linnaean-typification/; current April 2007; cf. Jarvis 2007) accepted this designation.

Synonymy

Homotypic

M. polymorpha L. var. *arabica* L., Sp. Pl. 780 (1753), (basionym).

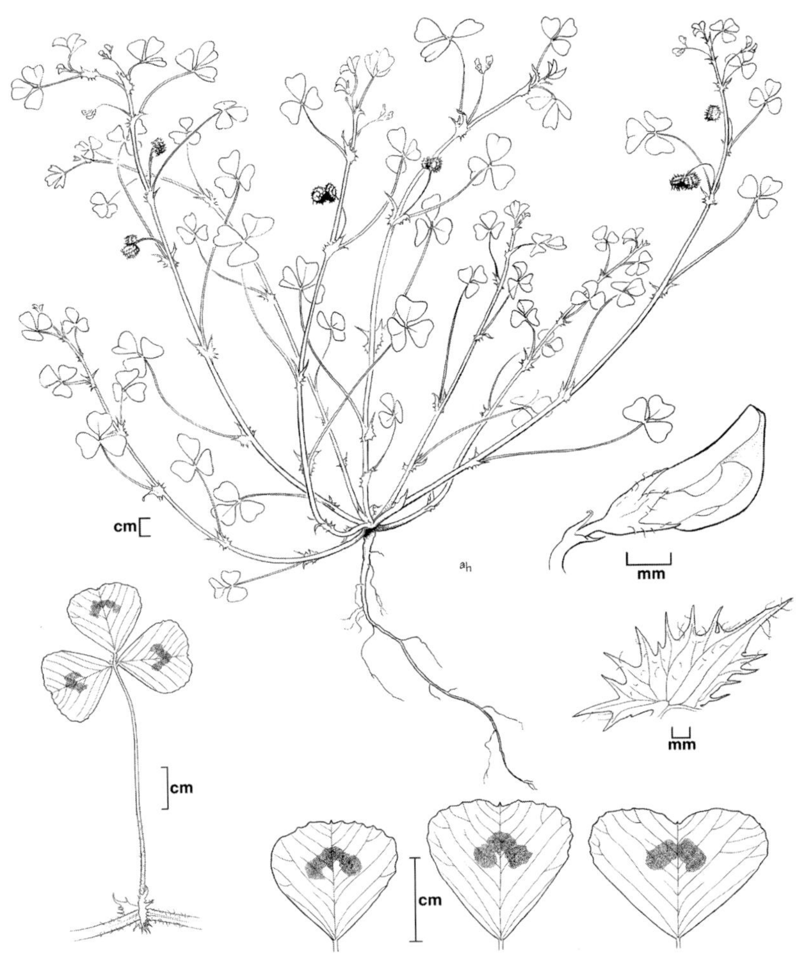

Fig. 13-1. *Medicago arabica.* **Clockwise from bottom left:** Leaf attached to stem; plant in flower and early fruit; flower; stipule; leaflets showing central spot. (The plant is based on Patten 194, US, the remainder on E. Small M888, DAO; artist: A. Hanes.)

M. arabica (L.) All., Fl. Pedem. 1: 315 (1785), isonym, nom. illeg. (The name *M. arabica* All., commonly encountered, is an incorrect attribution of authority to Allioni.)

Medica arabica (L.) Medik., Vorles. Churpfaelz. Phys.-Oecon. Ges. 2: 386 (1787).

Heterotypic

M. cordata Desr. in Lam., Encycl. Méthod. 3(2): 636 (1792), nom. illeg. (Burtt (1964) argued that the name *M. arabica* (L.) Huds. is illegitimate and that the correct name for the species is *M. cordata* Desr. Sales and Hedge (1994) pointed out that most authors have retained the name *M. arabica*.)

M. coronata Pall., Ind. Taur. ex Bieb. Fl. Taur.-Cauc. 2: 227 (1808).

Medica echinata Lam., Fl. Franç. 2: 587 (1779, "1778"), nom. illeg.

M. arabica (L.) Huds. f. *elegans* Kozuharov, Bulgarian Acad. Sci. 15: 152 (1965; type indicated).

M. arabica (L.) All. f. *heptacycla* Urb., Verh. Bot. Ver. Brandenb. 15: 74 (1873).

M. immaculata Schur, Verh. naturf. Ver. Brünn 15, 2: 174 (1877), pro syn.

M. maculata Willd. var. *immaculata* Schur, Verh. naturf. Ver. Brünn 15, 2: 174 (1877), pro syn.

Medica arabica (L.) Huds. subsp. *inermis* Ricker, U.S.D.A. Bur. Pl. Industr. Bull. 267: 33, plate 12, fig. 1 (1913). (Described from cultivated material, seeds from Chile.)

M. arabica All. var. *longispina* Rouy in Rouy & Fouc., Fl. France 5: 35 (1899).

M. maculata Sibth., Fl. Oxon. 232 (1794), nom. illeg.

M. maculata Willd., Sp. Pl. (ed. 4) 3(2): 1412 (1802), nom. illeg.

Medica maculata Bubani, Fl. Pyren., 2: 478 (1900), nom. illeg.

M. arabica (sine auct.) var. *minor* Loj. & Poj., Malphigia 20: 203 (1906).

M. oxalioides Schur, Verh. Naturf. Vereins Brünn 15(2): 174 (1877), pro syn.

M. arabica (L.) Huds. f. *pastoralis* Kozuharov, Bulgarian Acad. Sci. 15: 152 (1965; type indicated).

M. arabica f. *purpurea* Burkart in Fl. Prov. Buenos Aires 4(3a): 560 (1967).

M. salsuginosa Dumort. in sched. (in herb. B.R.)

M. talyschenisis Latsch. (Lachashvili), Georg. SSR Acad. Sci. Inst. Bot. Notes (Zametki) Plant. Syst. Geogr. 27: 76–77 (1969). (Holotype!: I. Latschaschvili, Talysh, Lerik district, near village of Lerik, grassy area, 30-V-1965, TBI.) (A form with tubercled fruits.)

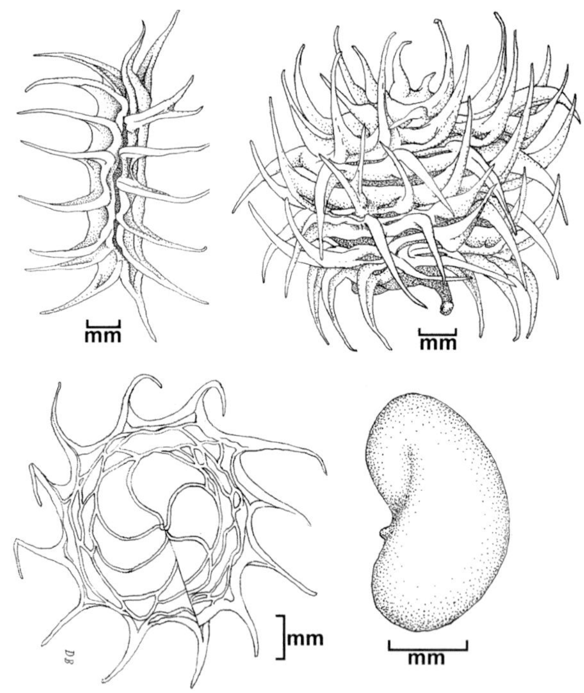

Fig. 13-2. *Medicago arabica.* **Clockwise from bottom left:** Fruit coil face showing venation; edge of coil showing central groove flanked by a lateral groove on each side; fruit; seed. (Based on E. Small M888, DAO; artist: D. Beddoes.)

Fig. 13-3. *Medicago arabica.* From Hallier (1885, vol. 23, plate 2348).

M. maculata Willd. var. *tetraphylla* Desportes, Fl. Sarthe Mayenne 51 (1838).

M. tinei Loj. & Poj., Fedde Rep. 5: 27 (1908), pro syn.

M. arabica All. var. *vulgaris* Rouy in Rouy & Fouc., Fl. France 5: 35 (1899), nom. inadmiss.

Vernacular Names

English: Arabian alfalfa, burclover (bur-clover), heart medic, heart trefoil, heartleaf clover, southern bur clover, spotted medic, spotted bur medic, spotted burclover (spotted bur-clover, spotted bur clover). (Bur is also spelled burr, medic is also spelled medick.)

French: la luzerne d'Arabie, luzerne hongroise, luzerne maculée, luzerne tachetée.

Geographical Distribution

(N = native, I = introduced, U = uncertain status)

Africa: Algeria (N), Azores (Portugal, N), Canary Islands (Spain, N), Egypt (N), Libya (N), Morocco (N), South Africa (I), Tunisia (N).

Asia: Azerbaijan (N), China (I), Cyprus (N), Dagestan (N), Gruziya (N), Iran (U), Israel (N), Japan (I), Lebanon (N), Syria (N), Turkey in Asia (N).

Australasia: Australia (I), Chatham Is. (New Zealand, I), Kermadec Is. (New Zealand, I), New Zealand (North, I), New Zealand (South, I), Tasmania (I).

Europe: Albania (N), Balearic Is. (Spain, N), Belgium (N), Bulgaria (N), Corsica (France, N), Crete (Greece, N), Czech Republic and Slovakia (I), East Aegean Islands (Greece, N), Estonia (N), France (N), Germany (N), Great Britain (N), Greece (N), Hungary (N), Ireland (I), Italy (N), Karpathos (Greece, N), Luxembourg (N), Malta (N), Netherlands (N), Portugal (N), Romania (N), Russia (southwestern, N), Sardinia (Italy, N), Sicily (Italy, N), Spain (N), Sweden (I), Switzerland (I), Turkey in Europe (N), Ukraine (Crimea, N), Yugoslavia (N).

Northern America: United States (I).

Southern America: Chile (I).

Along with *M. polymorpha*, *M. minima*, *M. sativa*, and *M. monspeliaca*, *M. arabica* is one of the likeliest *Medicago* species to be encountered. It is widely distributed in Eurasia and Africa, and widely adventitious elsewhere in the world. Prosperi et al. (2001) characterized *M. arabica* as a species that is common in some regions, such as Corsica, Sardinia, and Greece, but rare in Spain and North Africa.

Description

Annual herb 10–30(100) cm, procumbent to ascending, branching from base, young branches sparsely pubescent with simple and multicellular gland-tipped hairs. Leaflets 8–25 mm long, 6–25 mm wide, cuneate, obovate, or obcordate, 1–1.5 times as long as broad, upper third of margin serrate, often with a conspicuous dark (anthocyanin) blotch in center of upper surface, upper surface glabrous, lower with simple hairs, petiole with simple and multicellular gland-tipped hairs. Stipules ovate to sagittate-lanceolate, deeply dentate to lobed. Inflorescence of (1)2–5(8) flowers, peduncles about as long as subtending petiole. Flowers 4–5(6) mm, yellow (sometimes orange), pedicel 1–2 mm. Calyx 1.5–3 mm, with simple and sometimes also multicellular gland-tipped hairs, teeth equal to or longer than tube. Corolla about twice length of calyx or shorter. Young fruit protrudes sideways from calyx. Fruit glabrous, with 3–5(7) more or less adpressed coils, fairly flexible even at maturity, shortly ellipsoid to subglobose (i.e., ends rounded), or discoid to cylindriform (i.e., ends flattish), (4)5–7(8) mm in diameter, without an axial opening, with flexible, hooked, sulcate, dorsal suture spines up to 4 mm long (rarely lacking, in which case pod may be tubercled), often arched and recurved to intermesh with spines of adjoining coils, coil face with 4–8 curving radial veins that anastomose considerably on the outer third, reaching dorsal suture, somewhat thickened coil edge in end view shows a central groove flanked by lateral grooves to form a pattern of three grooves separating four ridges (not visible in side view of coil). Seeds 5–8(12) per pod, 2.0–3.5 mm long, 1.3–1.7 mm wide, yellow or yellow-brown, reniform, seed coat smooth, separated by a transverse, spongy fruit septum, radicle usually slightly more than half seed length.

The nomenclature in the synonymy presented above is suggestive of the range of variants encountered, albeit not deserving formal recognition: *immaculata* for forms with leaves lacking the characteristic spot; *heptacycla* for a form with numerous fruit coils; *longispina* for forms with long-spined fruits; and *minor* for small-fruited plants.

Nègre (1956) observed that leaflet blotching in *M. arabica* was somewhat unstable, depending often on seasonal environmental changes. Lesins and Lesins (1979) noted that the development of the leaf anthocyanin patches was influenced by light conditions and plant age.

Identification Hints

The fruit coil edge in *M. arabica*, showing four ridges alternating with three grooves, is unique in *Medicago* and an excellent character for its identification (see Fig. 13-2). Mature fruits of *M. sphaerocarpos* usually show three ridges on the coil edge, but this is a different pattern. The central purplish blotch usually found on the leaves is also a good character, although blotches are occasionally found on the leaves of other species. Large central leaf blotches are frequently found on *M. intertexta*, but these are usually rhomboid (see Color Photographs). Although the majority of species of *Medicago* cannot be identified to species with much certainty without fruits, in most cases vegetative plants with these leaf blotches can be reliably identified. If the above characters are not used,

M. arabica is likely to be confused with some similar annual species of *Medicago*. The venation pattern on the pod coil faces (inner two-thirds with a few curving radial veins, outer third with greatly anastomosed venation) is uncommon in *Medicago* and can be used for confirmation.

Habitat and Environmental Adaptation

Habitat

Medicago arabica occurs from 0 to ca. 1300 m, typically in grassy moist places, uncharacteristic of most annual *Medicago* species. It is found in edges of woods, shrub thickets, and meadows, and as a weed of cultivated and cleared or disturbed areas. Thanopoulos (2007) recorded its presence occasionally in shady areas.

Phenology

Like other annual species of *Medicago*, *M. arabica* is an annual or winter annual that flowers in the spring and completes its life cycle by setting fruit in early or mid-summer.

Soil

Medicago arabica typically occupies sandy soils, or at least soils with some sand. Piano and Francis (1992) described it as occupying sandy loam. In Algeria, Abdelguerfi et al. (1988*a*) noted that *M. arabica* tends to be in medium- and coarse-textured soils. The species is often in rocky soils. Grossheim (1945) observed that in the USSR, some sand in the soil seemed to be needed for good growth. Species of *Medicago* are generally intolerant of acidic soils, but compared with most other species of the genus, *M. arabica* is relatively acid-tolerant. Piano and Francis (1992) indicated that it occurs in soils ranging from acidic to slightly basic. In Sicily, Piano et al. (1991) observed the species growing in a pH range of 6.0–8.5. Rogers et al. (2005) ranked *M. arabica* as moderately sensitive to soil salinity.

Temperature

The species occurs from subtropical to temperate conditions. Bolton (1962) noted that it favors high temperatures. Krall et al. (1996*a*) and Krall et al. (1996*b*) found that four accessions of *M. arabica* tested for their ability to self-regenerate after overwintering in Wyoming failed to survive at the three locations tested. Derkaoui et al. (1991) found that fall-planted *M. arabica* in Oklahoma failed to survive the winter, indicating that the species does not have sufficient cold tolerance for successful self-replication in the southern Great Plains of the United States.

Water

This species typically occurs in humid steppes and fields. Piano and Francis (1992) recorded that in its native area it requires a minimum of 500 mm annual rainfall. Bolton (1962) noted that it does best in high rainfall areas, although it does well in parts of the USA where extended summer droughts occur. Grossheim (1945) observed that plants in the Caucasus tolerated the humid climate well. In Algeria, *M. arabica* occurs in the northeast, the wettest part of the country (Nègre 1959; Abdelguerfi et al. 1988*a*). Piano and Francis (1992) characterized the species as "relatively tolerant to temporary waterlogging occurring in poorly structured soils." Rogers et al. (2005) ranked *M. arabica* as moderately tolerant of waterlogging. However, Andrew and Hely (1960) noted that in Australia *M. arabica* tends to occur on well-drained soils.

Biotic Relationships

Pollination

Small (1988) found that the small flowers and pollen to ovule ratio (182) are consistent with *M. arabica* being largely self-pollinated.

Dispersal

The indehiscent spiny fruits are well adapted to dispersal in animal fur, accounting for the extremely widespread occurrence of this species.

Herbivory

Bialy et al. (2004) studied shoot saponins of *M. arabica*. These chemicals play an important role in palatability and nutrition (see Chapter 12).

Nodulation and mycorrhizae

USDA (2009) reported positive root nodulation for *M. arabica*. Yu et al. (1999) studied variation in nodulating ability of strains of *Rhizobium meliloti* and populations of *M. arabica* and found that there were marked differences depending on the geographical origin of the plants. Dear et al. (2003) found that *M. arabica* contributed significant quantities of nitrogen to agricultural areas in Australia.

Disease resistance

O'Neill et al. (2003) examined the comparative resistance of 36 annual species of *Medicago* to *Phoma medicaginis* (the cause of spring black stem and leaf spot, a major disease of alfalfa) and concluded that *M. arabica* is highly susceptible. Similarly, O'Neill and Bauchan (2000) found that two accessions of *M. arabica* were quite susceptible to anthracnose caused by *Colletotrichum trifolii*. Yaege and Stuteville (2002) tested the resistance of 33 annual species of *Medicago* to powdery mildew (*Erysiphe pisi*) and found that the two collections of *M. arabica* grown were susceptible (it was observed that often only one of a number of populations of other species tested was resistant).

Economic History, Cultivars, and Uses

Medicago arabica has attracted limited attention to date with respect to agricultural utility. Nevertheless, Bolton (1962) described it as "one of the more important annual or winter-annual species...rather widely cultivated in the United States and Australia" (a statement that exaggerates the importance of this species).

Cultivars

According to Bolton (1962), "Three varieties of spotted bur clover have been developed in the United States: 'Southern Spotted', 'Early Giant', and 'Manganese'. 'Manganese' is noted for its cold-resistance and 'Early Giant' for early growth and early seed production."

Forage

Medicago arabica is sown for forage only to a small extent, but likely contributes to forage utilized by livestock grazing on uncultivated lands, where it commonly occurs as a weed. Bolton (1962) stated that the ripe pods and seeds are nutritious and are a valuable supplement for livestock on pasture; and, that although the pods are spiny, the spines are soft and the pods are readily eaten. Bolton further stated that, although forms with spineless pods occur, these do not appear to have any particular advantage. Quinlavan et al. (1974) thought that *M. arabica* was unsuitable for agricultural use in Western Australia where its success as a weed is limited. Grossheim (1945) stated that the species' profuse growth and soft foliage justified its introduction as a forage plant, and that the soft fruits did not endanger livestock. Vitkus (1977) assessed *M. arabica* as suitable for cultivar selection for Lithuania.

Fodder

Grossheim (1945) observed that spontaneous *M. arabica* grew luxuriantly in hay meadows in the Lenkoran District of Azerbaidzhan, withstood trampling well, and greatly improved the quality of hay harvested. Mohammad and Qamar (1988) grew 14 annual species of *Medicago* in small test plots in Islamabad, Pakistan, located in a subtropical subhumid area that is notably hotter and wetter than the normal habitat of these species. The dry matter yield of *M. arabica* was about 5 t/ha.

Green manure, rotation crop, ground cover

Bolton (1962) stated that, although pasture is the principal use of the species, it is also grown for soil improvement through nitrogen fixation and green manuring.

Other uses (salad sprouts, ornamental, plant products, etc.)

Saniewska et al. (2005) demonstrated that saponin extracts from *M. arabica* shoots were active against 11 species of pathogenic fungi, and they suggested that *M. arabica* could become a source of natural fungicides.

Chromosome Number

$2n = 16$: Fryer 1930; Ghimpu 1928, 1929*a*, 1930 (as *M. maculata*); Heyn 1956; Clement 1962; Mariani 1963; Simon and Simon 1965; Fernandes and Santos 1971; Labadie 1976; Fernandes and Queirós 1978; Lesins and Lesins 1979; Abdelguerfi and Guittoneau 1979; Agarwal and Gupta 1983*b*; Abdelguerfi et al. 1988*b*; Buttler 1989; Wentworth et al. 1991; Magulaev 1992; Mariani et al. 1996.

Genetic Relationships and Gene Transferability

Intraspecific

No barriers to hybridization within the species are known, although artificial hybridization of different populations has not been attempted.

Interspecific

No information located.

Exchange by advanced biotechnology

No information located.

Conservation Status

Medicago arabica is an extremely widespread species, and so its gene pool is not endangered.

Germplasm Sources

Hundreds of accessions of *M. arabica* have been deposited in major germplasm collections (see Chapter 3).

Research Problems and Priorities

Medicago arabica is well adapted to moist temperate meadows, and agricultural experimentation to examine the possibility of developing cultivars is desirable.

Representative Collections

AUSTRALIA: Willis, J.H., Victoria, East Gippsland, Tambo river banks at camping ground south of Swift's Creek (ca. 65 km north of Bairnsdale), 20-VIII-1963, in flower & fruit (K). **BELGIUM:** Lambert, V., Ostende, voisinage de la mer, I-VI-1923, in fruit (K). Magnier, Ch., Visé, bord de la Mer, VII-1884, in fruit (P). Plantae Galliae et Belgii No. 463. **BULGARIA:** Kojoucharov, S. & Kouzmanov, B., les prairies près de la ville Ivailovgrad, 28-V-1959, in flower & fruit (K). Kozuharov, S., Bulgaria australis, prope urbem Petri…, 14-V-1963, in fruit (B). **CANARY ISLANDS (SPAIN):** Kunkel, G. 8982, Gran Canaria, San Mateo, alt. 850 m, 23-VI-1966, in flower (B). **CHILE:** Meyer, F.G. 9377, Juan Fernandez Islands, Mas Afuera, Quebrada Vacas, on talus slopes and shaded places near mouth of stream, 28-XI-1965, in fruit (K). **CYPRUS:** Casey, E.C. 1177, Kyrenia, alt. 120 feet., shady bank above little-used water channel among olive groves, 8-IV-1951, in flower & fruit (K). Meikle, R.D. 2718, Phlevas near Pyrgos, alt. 500 feet., on wall of old forest house, 27-IV-1902, in flower & fruit (K). **ENGLAND:** Craig-Christie, C., St. David's Fife, 4-VII-1903, in fruit (B). McCollum, M. Webster S918, Wickham, introduced with wool shoddy, 31-VIII-1961, in fruit (K). **FRANCE:** Jeanjean, Gironde: Villenave-d'Ornon, terrain nivelé par la Compagnie du Midi, V-1930, in flower & fruit, Société Française, 1932 Exsicc. Ch. Duffour No. 6595 (P). Legrand, Rec., près des environs de Montbrison (Loire), dans les champs, alt. 400 m, V-1874, in flower & fruit, F. Schultz Herbarium normale no. ser. Cent. 3 No. 245 (P).

Greece: Heldreich 833, Hagia Glykeria, near Athens, 30-IV-1897, in fruit (K). Orshan, G. 07503-5, central Crete, 8 km east of Ag. Barbara, Quercetum pubescentis, rendizna soil, alt. 480 m, 7-V-1964, in flower & fruit (HUJ). **Hungary:** Richter, Budapest, in herbidis humidis canalem subthermalem Aquinci, 13-V-1892, in fruit (P). **Iran:** Polunin, O. 11773, Banda Pahlair, V-1973, in flower & early fruit (K). **Italy:** Hiepko, P. 309, Apulien, Monte Gargano, Foresta Umbra, ca. 1 km south der Gant, alt. 700 m, 25-V-1972, in flower & young fruit (B). Levrier, E., Italia media, inter segetes et in herbosis incultis collis "la Sastra" ad septentrionen Florentiae, 22-V-1876, in flower & fruit (K). **Morocco:** Sennen & Mauricio, Gurugu à Cagmigriat, alt. 960 m, 16-VI-1931 (BM). **New Zealand:** Bangerter, E.B. 5260, North Island, Waitenata County, Mairangi Bay, Hastings Road (36°44′S, 174°45′E), on roadside bank, locally abundant, 14-X-1975, in flower & early fruit (K). **Spain:** Butcher, R., prov. Guipuzcoa, San Sebastian, sea cliff, 20-V-1952, in flower & fruit (K). Vicioso, B. & G., Calatayud, ad agrorum margines, 8-VI-1911, in flower & young fruit (B). **Sweden:** Tedin, H., prov. Skane, Paroecia Svalov, ad pagum Sualov, 16-VIII-1924, in flower, (G). Samuelsson Plantae Suecicae Exsiccatae No. 1011 (BM). **Syria:** Haradjian, M. 763, northern Syria, Mount Amanus, valées, alt. 2500 feet., VI-1906, in flower & fruit (K). **Turkey:** Kukkoken, I. 8369, Izmit, Sapancagölü Lake, 3 km west of Saracöglü Motel, alt. 40 m, gravelly roadside, 1-IX-1972, in fruit (G). Rechinger, K.H. 21811, Turcia europaea (Thracia orientalis), in nemoribus fluvii Maritza (Evros) prope Edirne (Adrianopel), alt. ca. 40 m, 2-III-1973, in early flower (B). **USA:** Bozeman, J.R. & Radford, A.E. 8841, South Carolina, Greenwood County, field at junction of highways 178 and 72 in Greenwood, 16-IV-1967, in fruit (BM). Pollard, H.M., California, Marin County, San Rafael, roadside, V-1936, in flower & fruit (B). Thomas, R.D. et al. 83038, Louisiana, beside US 167 about 1 mile north of Lafayette parish line south of Grand Coteau, 15-IV-1983, in flower & early fruit (BM). **Yugoslavia:** Richter, Dalamatia, ad margin. prat. humid. pr. Salona, VI-1870, in flower & fruit (K).

14. *Medicago arborea* L.

M. arborea L., Sp. Pl. 2: 778 (1753), (basionym).

The epithet *arborea*, from Latin, means of or pertaining to a tree, in allusion to the woody nature of the plant — really a shrub and not a tree.

Subordinate Taxa

None recognized here. *Medicago arborea* var. (or subsp.) *citrina* is treated as *M. citrina* in this monograph.

Type Specimen

Lectotypified by Heyn, Bull. Res. Council Israel 7D: 161 (1959) (Hb. Cliff., Hort. Sicc. Cliff. 377-6, BM!). Described by Linnaeus from the region of Naples and Rodhos (Rhodes Island). The Linnaean Plant Name Typification Project (cf. Jarvis 2007) (http://www.nhm.ac.uk/research-curation/projects/linnaean-typification/; current August 2010) provided the following additional information: "Provenance: 'Habitat in Rhodo, Neapoli.' Type: Lectotype: Herb. Clifford: 377, *Medicago* 6 (BM-000646774)." An image of the lectotype is available at this website.

Synonymy

Homotypic

Medica arborea (L.) Mill., Gard. Dict., ed. 8, no. 7 (1768).

M. arborea Aubl., Hist. Pl. Guiane ii. 779 (1775), nom. superfl.

Rhodusia arborea (L.) Vassilcz., Nov. Sist. Vyss. Rast. 9: 203 (1972).

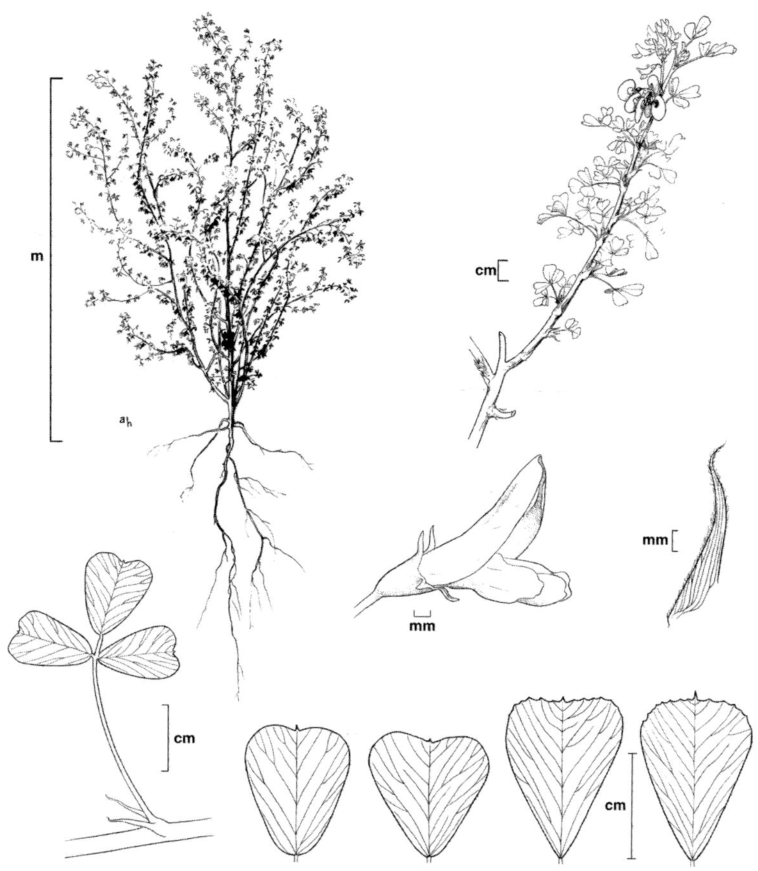

Fig. 14-1. *Medicago arborea.* **Clockwise from bottom left:** Leaf attached to stem; plant; branch with flowers and young fruits; stipule; leaflets. **Center:** Side view of flower. (Based on E. Small M220, DAO; artist: A. Hanes.)

Trigonella arborea (L.) Vassilcz., Bot. Mater. Gerb. Bot. Inst. Acad. Nauk SSSR (Leningrad) 14: 229 (1951).

Heterotypic

M. arborescens C. Presl., Fl. Sicul. 1: 20 (1826).

Vernacular Names

English: Shrub medic, shrubby medic (medic is also spelled medick), moon trefoil, tree alfalfa, tree lucerne, tree medic. The ancient Greeks and Romans referred to this species as "Cytisus" (Lesins and Lesins 1979), as did 18th century European botanists (Linnaeus 1753). (However, the genus presently recognized as *Cytisus* is an unrelated group of leguminous shrubs and small trees, some of which occur in the same region as *M. arborea*.) Some of these names could lead to confusion with another species, *Chamaecytisus palmensis* (Christ) F.A. Bisby & K.W. Nicholls (Fabaceae), also called tree lucerne (and false tree lucerne, lucerne tree, and tagasaste), an evergreen shrub native to the Canary Islands and now cultivated to feed livestock in warm countries.

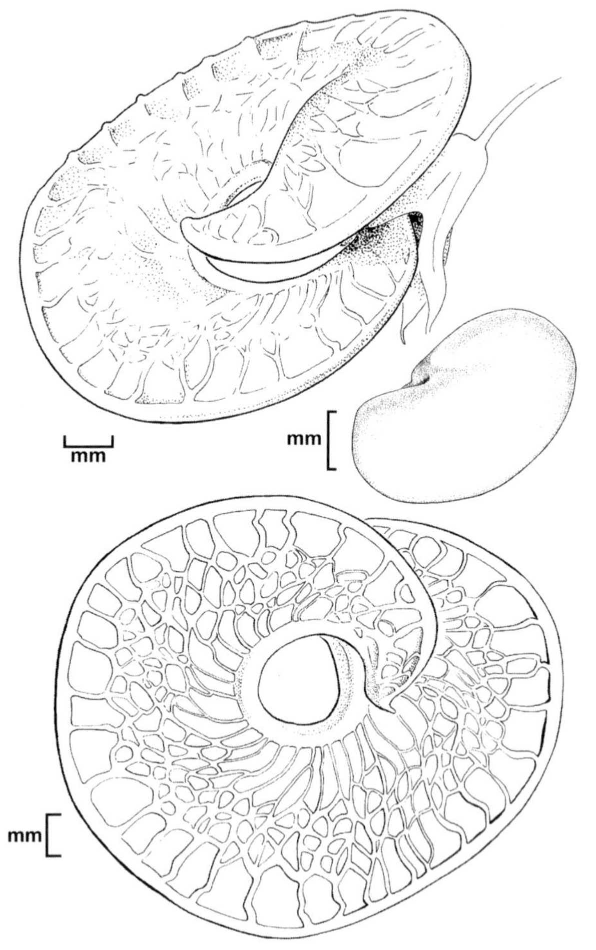

Fig. 14-2. *Medicago arborea.* **Top:** Fruit. **Center:** Seed. **Bottom:** Detail of venation on face of fruit. (Based on E. Small M220, DAO; artist: A. Hanes.)

French: amourette jaune, la luzerne arborescente (Laumont and L'Hermite 1953), luzerne en arbre.

Geographical Distribution

(N = native, I = introduced, U = uncertain status)

Africa: Canary Is. (U), Ethiopia (I), South Africa (I).

Asia: China (I), Israel (U), Jordan (U), Turkey in Asia (N).

Australasia: Australia (I), New Zealand (North, I), New Zealand (South, I), Tasmania (Australia, I).

Europe: Albania (N), Balearic Is. (Spain, N), Crete (N; cf. Thanopoulos 2007), Croatia (interpreted as I by Trinajstic and Pandza 2002), East Aegean Islands (Greece, N), France (N), Greece (N), Ireland (I), Italy (N), Malta (U), Monaco (N), Portugal (U), Sardinia (U), Sicily (U), Spain (N).

Southern America: Chile (I).

Medicago arborea is found on coasts and islands of the Mediterranean Sea, mostly in the northern area, and in the Canary Islands, generally in foothills along shores. Lesins and Lesins (1979) questioned the indigenous presence of the species in Algeria and indeed in northern Africa. González-Andrés et al. (1999) stated that the species was naturalized in Sicily, France, Spain, Portugal, Algeria, Tunisia, and Morocco. Much of the present range, especially in the West, may represent introductions for cultivation as an ornamental shrub. Greuter (in Greuter and Raus 1986) speculated that the species may have been confined to small islets of the Aegean Sea (for additional discussion see Thanopolous 2007). It may well be that humans are entirely responsible for spreading *M. arborea* to the West (once planted, the species may become naturalized in the area without further assistance from humans). However, the presence of the closely related endemic *M. citrina* on the Balearic Islands off the eastern mainland of Spain suggests that *M. arborea* was capable of occupying the western Mediterranean without the help of people.

Description

Woody perennial shrub 1–3(4) m, young twigs and leaves finely and densely pubescent with whitish simple hairs, aging twig surfaces cracking longitudinally to reveal developing bark. Leaflets 10–20 mm long, obovate to ovate, sometimes truncate or obcordate, generally not more than twice as long as broad, cuneate at base, more or less entirely or slightly denticulate near and at apex. Stipules lanceolate-cuspidate, entire. Inflorescence a very short, capitate raceme of 4–8(15) crowded flowers, peduncles 2–5 cm, often longer than the corresponding petiole. Flowers (10)12–15 mm, orange-yellow. Calyx teeth shorter than tube, calyx about as long as fruiting pedicel. Corolla more than twice length of calyx. Fruit with 0.5–1.5 coils, mostly in a flat plane, (5)10–15 mm in diameter with a central opening at least 1 mm wide, without dorsal suture spines, with short, appressed hairs, coil face with 20–30 radial veins anastomosing on the outer half, then reaching dorsal suture. Seeds 2–8 per pod, 3.5–4(4.5) mm long, brownish or yellow-brown, reniform, separated by a spongy transverse fruit septum, radicle equal to half seed length or slightly more.

Identification Hints

The three species of *Medicago* Section *Dendrotelis* are the only species of *Medicago* with very hard, woody stems, and the only species attaining the stature of shrubs (Chebbi et al. 1995). *Medicago arborea* is best distinguished from the related rare endemic *M. strasseri* by the fruits (for detailed comparison, see identification hints for *M. strasseri*). See *M. citrina* for hints about distinctions from it.

Habitat and Environmental Adaptation

Habitat

Medicago arborea occurs on hills, steep cliffs, seashore rocks, and rocky places, especially on limestone, from sea level to 800 m altitude. It frequently is present at low altitudes near the sea. The species is not generally considered to be significantly weedy, but has been described as invasive in the Mediterranean region (Bousquet-Mélou et al. 2005). In Greece, it is found in natural undisturbed habitats, but more often in anthropogenic localities, probably having been established by planting or escaping (Thanopolous 2007).

Phenology

Medicago arborea follows the typical Mediterranean pattern of starting vegetative growth in the autumn, after the first rains. Flowering occurs in the spring (February to June in its native area) following development of foliage. As noted by Stringi et al. (1994*a*), the flowering period can be very long, sometimes more than 6 months. The leaves and fruits tend to drop off in late season. Depending on climate, bud break and leaf production may occur more than once during the year (Laumont and L'Hermite 1953).

Soil

Medicago arborea mainly occupies limestone (calcareous) and volcanic soils. Its response to fertilization was studied by Bouzid and Papanastasis (1996).

Temperature

The characteristic proximity to the sea of this species reflects a limited capacity to withstand freezing. Plants grown outdoors failed to survive winters at Brandon, southern Manitoba (Lesins and Lesins 1979), where winter temperatures are very low. *Medicago arborea* also failed to survive at Aleppo in northern Syria, where winter frosts are rarely lower than –10 °C (Cocks 2001). Prosperi et al. (2001) stated that the distribution is limited in the north by frosts below –8 °C.

Water

The native habitats occupied are generally quite arid and *M. arborea* is considered to be an appropriate forage species for dry regions (De Koning et al. 2000*a*, *b*). Prosperi et al. (2001) noted that it occurs in areas with annual rainfall sometimes less than 250 mm. Stringi et al. (1994*a*) described it as "one of the most promising forage shrubs for Mediterranean environments with at least 400 mm of rainfall per year." The leaves are not particularly xeromorphic, and indeed the plant loses all its leaves during severe drought. Extensive root systems likely facilitate access to moisture. Lambert et al. (1989*a*) attributed the poor survival of an experimental planting of *M. arborea* to excessive soil moisture. Lefi et al. (2004*a*) found that irrigation could increase leaf production by 50%. Comparing the three woody species of section *Arborea*, Lefi et al. (2004*b*) and Lefi et al. (2004*a*) found water use efficiency increased in the following sequence: *M. citrina*, *M. arborea*, and *M. strasseri*. The chief production of biomass is normally in the spring with growth becoming quiescent with increasing drought.

Biotic Relationships

Pollination

Stringi et al. (1994*b*) found that *M. arborea* in Sicily was substantially outcrossing, whereas 6% of the flowers of plants protected from pollinators were able to set fruit. Lesins and Lesins (1979) observed that several collections of *M. arborea* were capable of self-pollination. *Medicago arborea* has the largest flowers in the genus except for those of *M. citrina*, and it may be naturally associated with different pollinating bees than most other *Medicago* species. Ricciardelli-D'Albore (1984) observed the following pollinators of *M. arborea* in Italy: *Anthophora acervorum* L., *Osmia rufa cornigera* Rossi, *Andrena ovatula* K., *Xylocopa violacea* L. and *Bombus* spp. He also noted that these bees are much less represented than honey bees on *M. arborea*, but are much more effective as pollinators since the honey bees are unable to trip the flowers.

Small (1988) reported that *M. arborea* has a pollen to ovule ratio of 328; Galloni et al. (2007) reported a ratio of 385. These ratios are consistent with the species being substantially outcrossed, but with some inbreeding.

Dispersal

Birds are the most likely agent of seed dispersal across seawater. There is no evidence of tolerance of the seeds to immersion in seawater, or indeed of their ability to survive the guts of birds. In fact, there is no evidence of adaptation to distribution by any particular class of animal. Once established on hillsides, dispersal is easily achieved, the seeds merely scattering downhill to suitable niches.

Herbivory

Insect herbivores do not appear significant, the plants generally not showing much evidence of attack. Mammalian herbivores readily browse on the young shoots, the often inaccessible locations doubtlessly providing protection. However, Cereti and Rossini (1994) observed that hares (*Lepus europaeus*) in Italy "neglected other species, such as alfalfa, and seemed to prefer gnawing the seedlings of *M. arborea* to the root collar, this way causing the death of the plants." Thanopolous (2007) concluded that grazing by small ruminants is the driving force restricting *M. arborea* to cliffs in Greece because the species spreads rapidly onto level ground that is protected.

Nodulation and mycorrhizae

Valdenegro et al. (2001) conducted an extensive study of the relationships of *M. arborea* with rhizobial bacteria and micro-organisms that stimulate rhizobacteria, and further studied relationships with arbuscular mycorrhizae. They found that *M. arborea* growth is stimulated by at least some representatives of all of these associates. However, the arbuscular mycorrhizal fungi were most consistent in stimulating growth, and indeed seemed to be the most significant microbial associate contributing to increased growth. The development of rhizobial nodules did not seem to be affected by the presence of any of the other microbial associates tested.

Disease resistance

The seeds of both *M. arborea* and the related *M. strasseri* have been described as rather susceptible to fungal attack (Perez-Garcia et al. 1992). Elgin and Ostazeski (1992) found that *M. arborea* was more than 40% resistant to races 1 and 2 of anthracnose caused by *Colleotrichum trifolii* Bain.

Economic History, Cultivars, and Uses

As noted below, the species has some usefulness as natural browse in the Old World and has been exploited to a limited extent for forage and fodder. *Medicago arborea* has been the subject of considerable agronomic research in recent decades and there has been some breeding of the species for forage purposes. Breeding based on Greek germplasm was reported by Vaitsis and Konstantakis (1994). *Medicago arborea* has been used as an ornamental for hundreds of years at least, and possibly in classical Greek and Roman times. Selection for tolerance to cultivation as an ornamental may have occurred, but there is no evident difference between wild plants and plants grown in cultivation.

Cultivars

Vaitsis (1993) recorded a forage cultivar named 'Naxos', registered in Greece.

Forage

Although Prosperi et al. (2001) characterized *Medicago arborea* as intolerant of grazing, it is considered to be one of the most potentially valuable forage shrubs in the Mediterranean environment because of its palatability and nutritive value and its high preference by small ruminants (Amato et al. 2004). It has been esteemed as a browse plant for goats and cattle for centuries. Alegre et al. (1993*a*) measured its protein content as more than 18%. Pistoia et al. (2003) studied the principal chemical–nutritional parameters of *M. arborea* and concluded that although forage production was notably lower than that of alfalfa, the protein content and digestibility were high, whereas crude fiber content was very low. Saponin levels were found to be relatively low compared with some alfalfa varieties (see Chapter 12). *Medicago arborea* provides natural forage for goats in some regions. In the dry season in several marginal and dry areas of Tunisia, when herbaceous vegetation is limited, it is an important range plant (Saadani et al. 1989). It has been explored as a browse plant in New Zealand, but proved unsuited to the environment (Lambert et al. 1989*a*, 1989*d*). A study in Italy by Mastrorilli and Colucci (1997) found that 5-year-old plants of *M. arborea* grown at experimental spacings of 2 × 1, 1.5, or 2 m could withstand sheep grazing at intervals of 60 days during the growing season and 90 days during the dormant season. Based on experiments in Italy, Corleto et al. (1993) recorded a dry matter production of 18 t/ha when plants were cut at 180 days. Watered plants produced, on average, 21.1 t/ha in contrast with 9.7 t/ha for unirrigated plots. Stringi et al. (1993) noted that in Sicily it was possible to obtain a production of 1.6 t/ha of dry matter, which is much lower than the 9 t/ha that has been recorded in New Zealand (Radcliffe 1983).

Fodder

The young shoots are occasionally harvested for fodder before becoming woody, and indeed the ancient Greeks and Romans praised *M. arborea* as a fodder plant (Lesins and Lesins 1979). In experiments with sheep and goats, harvested *M. arborea* was found to have good preference ratings (Lambert et al. 1989*b*) and have excellent digestibility (Lambert et al. 1989*c*).

Green manure, rotation crop, and ground cover

This species is useful for binding soil on steep slopes. Perhaps the most notable situation where this is evident is at the base of the Acropolis in Athens where protection from grazing animals has allowed the shrubs to persist, probably for millennia. Indeed, in Greece *M. arborea* is extensively planted for range improvement on dry and infertile sites (Droukales 1976). Experimental studies have shown that *M. arborea* can be used to control erosion on steep slopes and for stabilizing land following fire (Andreu et al. 1994; Andreu et al. 1998; Mastrorilli et al. 1999). The species is also considered to have potential for alley farming and plantation systems intended to reduce wind and water erosion (De Koning et al. 2000*a*, *b*).

Other uses (salad sprouts, ornamental, plant products, etc.)

Medicago arborea is cultivated mostly as an ornamental and is the principal *Medicago* cultivated for this purpose. (Several of the annual species, such as *M. scutellata*, have also been grown as ornamentals.) The wood of old plants of *M. arborea* is hard and as dark as ebony, and has been used for saber handles, canes, and beads.

Chromosome Number

$2n = 32$: Ghimpu 1929*b*, 1930; Clement 1962; Bolkhovskikh et al. 1969 p. 307; Lesins and Lesins 1979; Falistocco 1987; Baltisberger and Voelger 2006.

Lesins and Lesins (1979) noted that a diploid ($2n = 16$) report appears to have been based on a misinterpretation.

Genetic Relationships and Gene Transferability

Intraspecific

No information located.

Interspecific

Armour et al. (2008) made 2000 pollinations between *M. arborea* and *M. sativa* and succeeded in generating five "asymmetric hybrids". In the most carefully studied of these, they concluded that "only a single *M. arborea* chromosome or chromosome part had been transferred." Nevertheless they wrote that "The work demonstrates that agronomically useful traits can be introgressed into *M. sativa* from *M. arborea*." Research by E.T. Bingham has involved the use of such hybrids to transfer large seed size to alfalfa (*Medicago* Genetic Reports, University of Wisconsin-Madison: http://www.medicago-reports.org/; current August 2010). Natural hybridization between *M. arborea* and the closely related *M. citrina* has not been observed, and artificial hybridization has not been attempted. Because they differ in ploidy level, intersterility is to be expected.

Exchange by advanced biotechnology

Damiani and Arcioni (1991) transformed *M. arborea* with *Agrobacterium* rhizogenes. Pupilli et al. (1991) electrofused protoplasts of *M. arborea* and *M. sativa* to produce heterokaryons, but were unable to establish replicating heterokaryon or hybrid cultures. Nenz et al. (1996) and Monteiro (2000) studied protoplast fusion between *M. sativa* and *M. arborea* and the consequent production of somatic hybrids. Piccirilli and Arcioni (1992*a*, *b*) crossed *M. arborea* ($2n = 32$) with *M. sativa* ($2n = 32$) and obtained 56 viable embryos from which they were able, by using embryo rescue techniques, to produce three hybrid plants.

Conservation Status

The natural range of *M. arborea* is somewhat uncertain, so it is not possible to determine whether there are unique natural populations in danger of extirpation. It is likely that some island and/or mountain populations have evolved distinctive genotypes, but at present this has not been documented. Throughout much of its range the species is not threatened, and indeed its long-lived perennial nature serves to perpetuate it quite reliably.

Germplasm Sources

More than 100 accessions of *M. arborea* are available from major germplasm collections (Chapter 3).

Research Problems and Priorities

The systematic relationships of *M. arborea* and the related narrow endemics *M. citrina* and *M. strasseri*, as well as other odd, clearly related endemic populations (Greuter et al. 1982), remain to be examined in depth. Genetic studies, particularly isozyme studies, may clarify the biogeographical relationships of various populations, and the role of humans in spreading the plant. From an economic viewpoint, *M. arborea* has proven to be useful in agriculture principally as browse and fodder in dry regions, and its future value may lie principally as an experimental source of genes for the improvement of economic species and as a host of perhaps unique strains of *Rhizobium*.

Representative Collections

France: Brenan, J.P.M. 394, Alpes-Maritimes, Cap Martin, by roadside in maquis-pine wood, 19-III-1939, in flower & fruit (K). Leredde, C., Alpes-Maritimes, Cap Martin, naturalisa au bord de la mer, 11-IV-1950 in flower & fruit, Société Française pour l'Echange des Plantes Vasculaires - Exsicc. B. De Retz, Fasc. 4 (1950) No. 304 (K). **Greece:** Brummitt, R.K. 6305, prov. Fokís, ancient site of Delphi, outside museum, I-V-1969, in flower & fruit (K). Danin, A. & Heyn, C. 75.53.8, Delphi, limestone cliffs near Castalian fountain, alt. 600 m, 2-X-1975, vegetative (HUJ). Heldreich, Attica, prope Athenas, in rupibus excelsis collis Lycabetti, alt. 800–900 feet., 16-V-1896, in flower & fruit, Dr. Heldreich Herbarium Graecum Narmale 1317 (P). Lewis, G.P. 659, Athens, growing around and on the Acropolis, 16-V-1980, in flower & fruit (K). Rechinger, K.H. & F. 4815, Cycladuminsula, parva calcarea ad occidentem insulae Heraklia, 2-V-1934, vegetative (BM). **Israel:** Meyers, S. & Dinsmore, J.E. 1885, Jerusalem, rocky hills, 2 m tall, alt. 800 m, 4-V-1921, in flower & fruit (G). **Italy:** Chabert, E., Montboron, près Nice (Alpes-Marit), coteaux secs, 25-III-1882, Société dauphinoise, 1884 No. 4072 (P). Ross, H., Palermo, in fructicetis maritimis, 3-V-1891, Dr. C. Baenitz Herbarium European (P). Strobl, P.G, Sicily, ad oras maris Tyrriheni prope Mondel lo solo calcareo, quasi indigena, 4-VII-1973, in fruit (K). **Spain:** Sennen, F. 1173, Catalogne, Massif du Tibidabo, Vedado du Marquis de Sentmenat, V-1931, in flower & fruit (BM). Verdcourt, B. 4284A, Catalonia, S' Agsro, 7-VI-1966 (K).

Notes

The classification of *M. arborea* depends on where taxonomists stand on the lumper-splitter continuum. Vassilczenko (1972) raised the species to a separate genus, *Rhodusia*, reflecting the traditional old-school Russian tendency to inflate taxonomic categories. *Medicago citrina* and *M. strasseri* have been treated as subspecies of *M. arborea*.

Based largely on the common assumption that woody plants (at least plants with perennial aboveground woody stems) are primitive, *M. arborea* is widely stated to be the most primitive species in the genus *Medicago*. A contrary viewpoint can be based on the observation that a "primitive" trait does not necessarily mean that the species is primitive. Conceivably, woodiness in *M. arborea* represents a reactivated previously dormant section of the *Medicago* genome, which is still dormant in most other species of the genus.

15. *Medicago archiducis-nicolai* Širj.

M. archiducis-nicolai Širj., Kew Bull. Misc. Inform. 7: 270 (1928), (basionym).

Širjaev gave the epithet *archiducis-nicolai* to this species as a dedication to Archduke Nicholas, i.e., Russian Grand Duke Nikolay Nikolayevich (1856–1929), whose father was also known as Grand Duke Nikolay Nikolayevich (1831–1891). The Russian Revolution took place in 1917 and the Czar of Russia (also named Nicholas) abdicated. The Czar and his family were imprisoned and executed by the new Bolshevik rulers in 1918, ending the Romanov dynasty. In 1919 Nikolayevich wisely emigrated to France, where he was chosen to be the leader of the Russian monarchists in exile in France.

Subordinate Taxa

None recognized here.

Type Specimen

Learner, Frank D., China, Kansu, and northeast Tibet, Received by Kew 6-IX-1921, in flower & fruit (K, holotype!).

Synonymy

Homotypic

Medicago sirjeaewii ("*schirjaewii*" or "*schirjaëwii*") Hand.-Mazz., Oësterr. Bot. Zeitschr. 87: 123 (1938).

Trigonella archiducis-nicolai (Širj.) Vassilcz., Acta Inst. Bot. Acad. Sci. URSS, Ser. 1, Fasc. 10, 172 (1953).

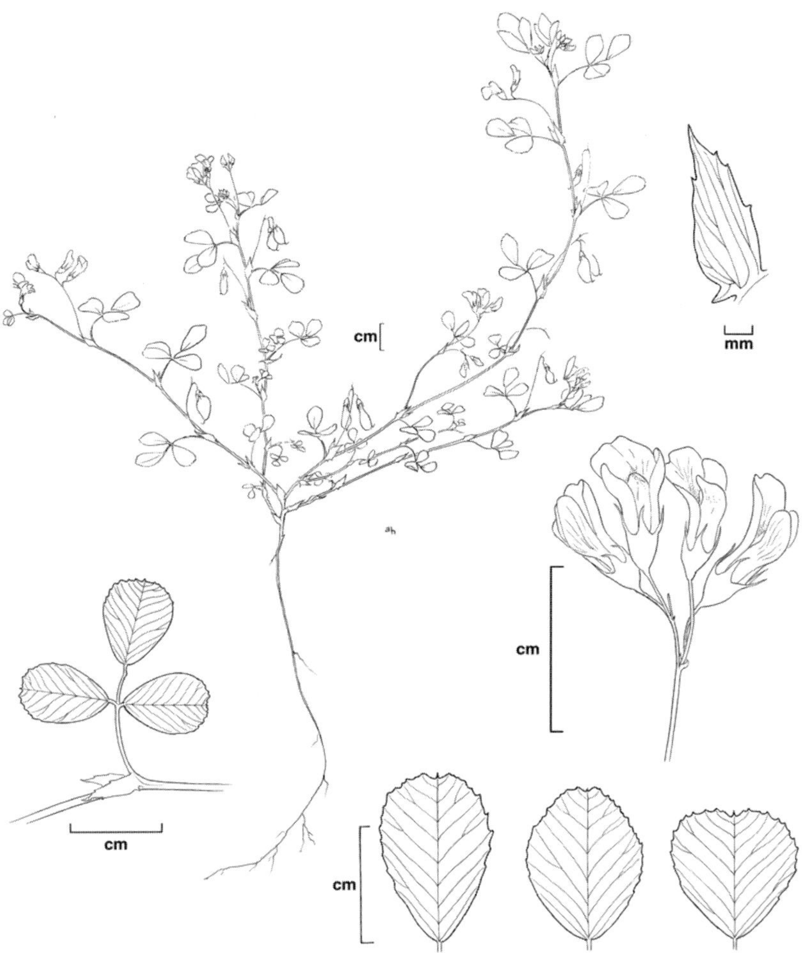

Fig. 15-1. *Medicago archiducis-nicolai.* **Clockwise from bottom left:** Leaf attached to stem; plant; stipule; group of flowers; leaflets. (Based on #5664, W; artist: A. Hanes.)

Melissitus archiducis-nicolai (Širj.) I.Y. Latsashvili, Zam. Sist. Geogr. Rast., 32: 18 (1976).

Kamiella archiducis-nicolai (Širj.) Vassilcz. Novosti Sist. Vyssh. Rast. 16: 134 (1979).

Melilotoides archiducis-nicolai (Širj.) Yakovlev, Rast. Tsentral. Azii, Mater. Bot. Inst. Komarov 8a: 86 (1988).

Heterotypic

Vassilczenko (1953) assigned *M. ruthenica* var. *alpina* Maxim., an unpublished name, recorded on a herbarium specimen, to this species.

Vernacular Names

English: Nicholas's medic (medic is also spelled medick), Nicholas's fenugreek.

French: No names found.

Geographical Distribution

(N = native, I = introduced, U =uncertain status).

Asia: China (reported from the Chinese provinces of Gansu, Ningxia, Quinghai, Shaanxi, Sichuan, and Xizang Zizhiqu (northeast Tibet), N). A distribution map by Yakovlev (1988; cf. Yakovlev et al. 1996) shows the sites of 11 collections.

Description

Perennial herb, 8–20 cm long, procumbent to ascending, numerous, slender, angular stems, twigs weakly pilose with simple hairs. Leaflets 5–20 mm long, 5–14 mm wide, oval or obovate to round, very finely denticulate with almost entire margins, apically slightly notched and with a cuspidate tip, glabrous. Stipules of upper leaves 7–13 mm long, semi-sagittate at the base, dentate, pointed, glabrous. Inflorescence capitate, with ca. 4–5 flowers, peduncles twice as long as subtending petiole, pilose, 1–2 cm in flower, elongating up to 3.5 cm in fruit. Pedicels about 2 mm long, pilose, initially erect, later declined. Flowers 7–9 mm, yellow or reddish on the outside and yellow on the inside. Calyx 3–4 mm, campanulate, with simple hairs, teeth triangularly lanceolate, equal in length or shorter than the tube. Fruit about 13 mm long, 4 mm wide, flat, oblong to linear, weakly falcately curved, slightly thickened along both sutures, reticulately veined, glabrous. Seeds two to three per pod, ca. 2 mm long.

Identification Hints

Medicago archiducis-nicolai is rather similar to *M. platycarpa* and *M. ruthenica*. Unlike *M. ruthenica* (which has rugose seeds) it has smooth seeds, whereas the fruits are smaller than those of *M. platycarpos*. It has longer pods than does *M. popovii*.

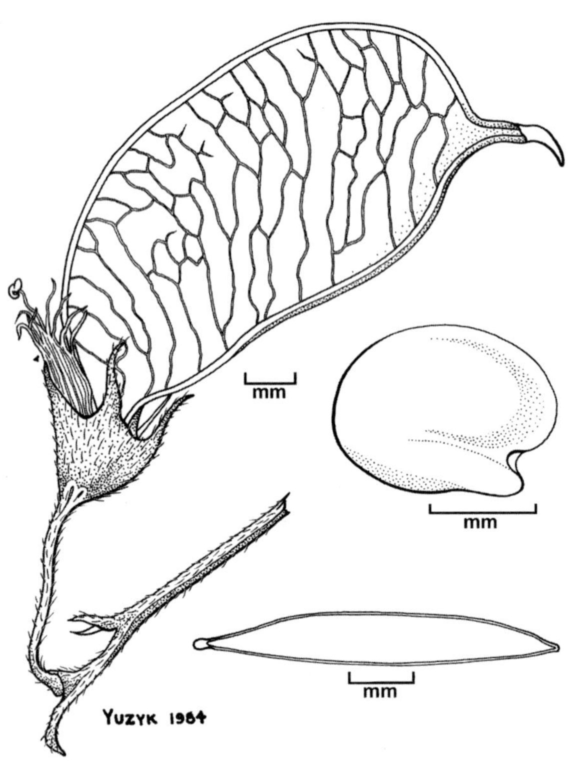

Fig. 15-2. *Medicago archiducis-nicolai.* **Top and left:** Fruit attached to stem. **Center right:** Seed. **Bottom right:** Cross section of fruit. (Based on H. Smith 10595, W; artist: L. Yuzyk.)

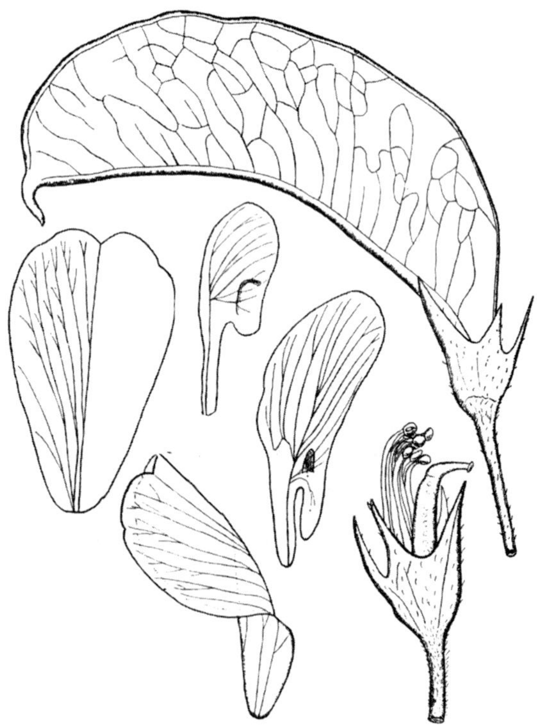

Fig. 15-3. *Medicago archiducis-nicolai.* Fruit and a dissected flower. **Clockwise from upper right:** Fruit; flower with petals removed; standard petal (folded); standard petal (venation shown on one side). **Top center:** Keel petal. **Bottom center:** Wing petal. Source: Širjaev (1928).

Fig. 15-4. Photograph of holotype of *Medicago archiducis-nicolai.*

Habitat and Environmental Adaptation

Habitat

Medicago archiducis-nicolai is found in valleys, on the banks of lakes and rivers, in steppes, and on the borders of coniferous forests, at altitudes of 2400–4000 m. Du and Wang (1990) observed that in Gansu province, after cultivated fields were abandoned, they were first occupied by weeds, then forbs and grasses, including *M. archiducis-nicolai*, and finally by species of permanent grasses and sedges.

Phenology

The species flowers June to August and starts to form fruit in August.

Soil

The substrates occupied include sandy and rubbly slopes, and often soil with considerable humus.

Temperature

No information located. Based on its habitats, this appears to be a plant of mesic climates.

Water

The species typically occurs in moist environments.

Biotic Relationships

Pollination

No information located.

Dispersal

No information located.

Herbivory

No information found.

Nodulation and mycorrhizae

USDA (2009) reported positive root nodulation for *M. archiducis-nicolai*.

Disease resistance

No information located.

Economic History, Cultivars, and Uses

This species has not attracted agricultural interest to date.

Cultivars

No cultivars have been bred.

Forage

Medicago archiducis-nicolai is not sown for forage.

Fodder

This species is not harvested for fodder.

Green manure, rotation crop, ground cover

No information located.

Other uses (salad sprouts, ornamental, plant products, etc.)

No information located.

Chromosome Number

$2n = 16$: Wu et al. 1986; Qi 1997.

Genetic Relationships and Gene Transferability

Intraspecific

No information located.

Interspecific

No information located.

Exchange by advanced biotechnology

No information located.

Conservation Status

This is a rare species in great need of conservation evaluation. Because there are no genebank holdings of it, collections are urgently required.

Germplasm Sources

No source of seeds has been located.

Research Problems and Priorities

Medicago archiducis-nicolai is very poorly collected and examined and there is an immediate need to obtain germplasm for study.

Representative collections

CHINA: Hummel, D. 3859, prov. Kansu, Ma-tze-kuo, in graminoso sicco, alt. ca. 2300 m, 28-VI-1930, in flower (S). Przewalski, N.M., China occidentalis: prov. Kansu, Terra Tangetorum, 21-VI-1872 (LE). Smith, H. 2620, prov. Sze-ch'uan: Chen-chiang-kuan, in prato subhumido, alt. ca. 2500 m, 5-VII-1922, in flower & very early fruit (S). Smith, H. 3399, prov. Sze-ch'uan: Hsioeh-shan, in prato, 11-VIII-1922, in flower (S). Wu yuhu 1763, Xining, Quinghai, alt. 2300 m, in gully, 23-VII-1985, in late flower & early fruit (B).

16. *Medicago arenicola* (Huber-Mor.) E. Small

(a species of uncertain status)

M. arenicola (Huber-Mor.) E. Small in Small, Lassen & Brookes, Willdenowia 16(2): 433 (1987).

The epithet *arenicola* is derived from the Latin *arena*, a sandy place, plus *cola*, dweller, an allusion to the sand dune habitat.

Subordinate Taxa

None recognized here.

Type Specimen

Holotype: Huber-Morath, A. 9859, Turkey, Province C3 Antalya, District Antalya, Sandstrand bei Lara 7–8 km east of von Antalya, 23-V-1950, in late flower & late fruit (fruits dehiscing) (G!).

Synonymy

Homotypic

Trigonella arenicola Huber-Mor., Notes Royal Bot Gard. Edinb. 29(3): 322 (1969), (basionym).

T. halophila subsp. *arenicola* (Huber-Mor.) Ponert, Feddes Repert. 83: 639 ("1972" = 1973).

Heterotypic

No names found.

Vernacular Names

ENGLISH: No names found.

FRENCH: No names found.

Geographical Distribution

(N = native, I = introduced, U = uncertain status).

ASIA: Turkey (N).

This species is known only from the type gathering in Antalya, Turkey.

Description

Annual herb, branched from base with several stems, stems adpressed-pubescent, prostrate, 3–10 cm. Leaflets small,

Fig. 16-1. *Medicago arenicola.* (Based on Huber-Morath 9859, G; artist: B. Flahey.)

2–5 mm long, 2–4 mm wide, obovate to obcordate-cuneate, denticulate towards tip, stipules semi-sagittate, entire or dentate. Inflorescence a spherical, umbellate head with 4–6 flowers, peduncle 5–10 mm. Flowers yellow, 6–7 mm, calyx campanulate, ca. 3 mm, with lanceolate teeth of more or less equal length, shorter than tube. Legumes spreading, linear, somewhat compressed, slightly curved, adpressed-puberulent, reticulate-netted, 10–15 mm long, 1.5–2 mm wide, sutures distinct, straight, with 4–6 seeds. Seeds oblong, wrinkled, about 2 mm long, 1 mm wide, radicle about 80% of seed length.

Identification Hints

This taxon was described by Huber-Morath (in Huber-Morath and Zohary 1969) as "related to *Trigonella halophila* Boiss. [i.e., *Medicago halophila*], but clearly differing in the legumes curved upwards, not stellate-spreading, with regularly curved sutures, and distinctly wrinkled seeds. It also differs from *T. halophila* in the larger calyx and corolla."

Habitat and Environmental Adaptation

Habitat

Sand dunes.

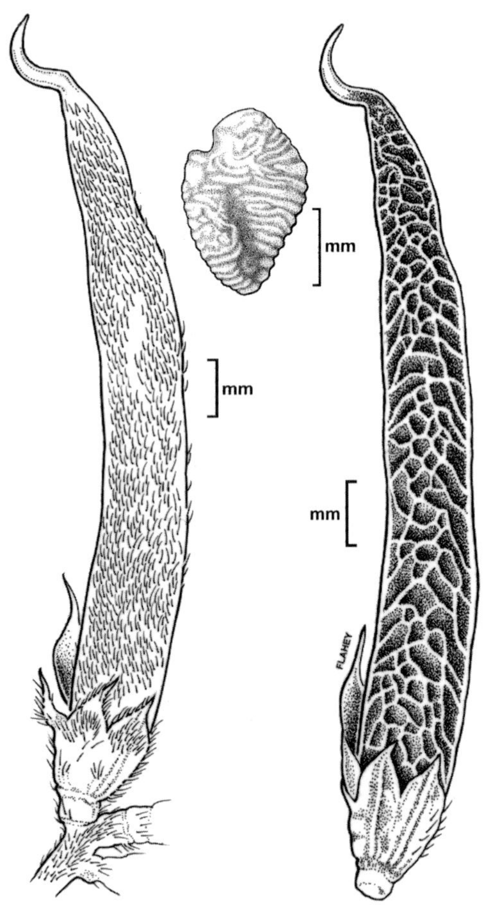

Fig. 16-2. *Medicago arenicola.* **Left:** Fruit. **Top center:** Seed. **Right:** Fruit drawn without hairs to show venation. (Based on Huber-Morath 9859, G; artist: B. Flahey.)

Phenology

The type collection was in late flower and early fruit on 23 May 1950.

Soil

Sand dunes.

Temperature

No information located.

Water

No information located. The sand dune habitat suggests that the plant is adapted to dry soils.

Biotic Relationships

Pollination

No information located.

Dispersal

No information located.

Herbivory

No information located.

Nodulation and mycorrhizae

No information located.

Disease resistance

No information located.

Economic History, Cultivars, and Uses

This species is barely known and has not attracted agricultural interest.

Cultivars

No cultivars have been bred.

Forage

This taxon has not been used for forage.

Fodder

This taxon has not been used for fodder.

Green manure, rotation crop, ground cover

No information located.

Other uses (salad sprouts, ornamental, plant products, etc.)

This taxon has not been employed for any recorded purpose.

Chromosome Number

Not determined.

Genetic Relationships and Gene Transferability

Intraspecific

No information available.

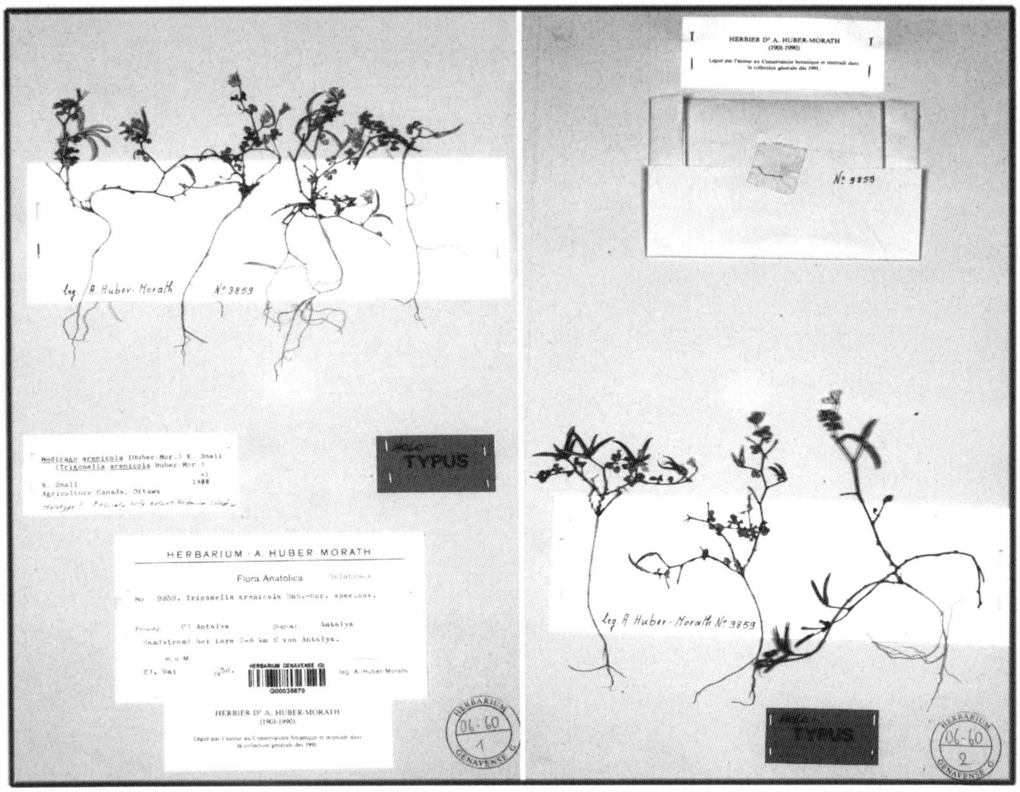

Fig. 16-3. Holotype of *Trigonella arenicola* Huber-Mor., at G.

Interspecific

No information available.

Exchange by advanced biotechnology

No information available.

Conservation Status

The status of this species is extremely poorly understood. Whatever taxonomic status it is eventually afforded, the collection on which this taxon is based likely is unique and its re-collection is urgently required. The Bern Convention on the Conservation of European Wildlife and Natural Habitats lists *M. arenicola* (under the name *Trigonella arenicola*) as a "strictly protected species" (the species is Asian, not European).

Germplasm Sources

No germplasm available.

Research Problems and Priorities

Medicago arenicola is very poorly understood and it is possible that further evaluation could result in it being reassigned as merely a variant of *Medicago halophila*. The immediate priority is to collect a germplasm sample so that the status of this taxon can be properly evaluated.

Representative Collections

This taxon is known only from the type collection.

Notes

In a personal communication I received from A. Huber-Morath (1901–1990), discoverer and describer of *M. arenicola*, dated 4 May 1987 (3 years before his death), he related the following: "The Holotype, discovered in 1950, exists only in very few examples. The new species has never been collected again. *Trigonella arenicola* is a good species near to *T. halophila*, but easy to distinguish by legumes with two straight sutures (one subundulate suture in *T. halophila*) with more numerous and very close transversal nerves."

In a personal communication to me dated 16 October 1999, D. Pearman, Frome St. Quintin, Dorchester, England, wrote: "I have concentrated on legumes in general and *Medicago* and *Trigonella* (s.l.) in particular. In 1998 we stayed at Lara (east of Antalya) for 5 days, trying to collect *Trigonella polycarpa* and *T. arenicola* (your *Medicago heldreichii* and *M. arenicola*). We found abundant *T. polycarpa* and no *arenicola*. In fact, we were not certain *T. arenicola* exists — we looked very thoroughly indeed over 15 miles of coast."

17. *Medicago astroites* (Fisch. & C.A. Mey.) Trautv.

M. astroites (Fisch. & C.A. Mey.) Trautv., Bull. Sci. Acad. Imp. Sci. St. Petersb. 8: 271 (1841).

The epithet *astroites* in Greek signifies star-like, an allusion to the stellately spreading arrangement of the fruits.

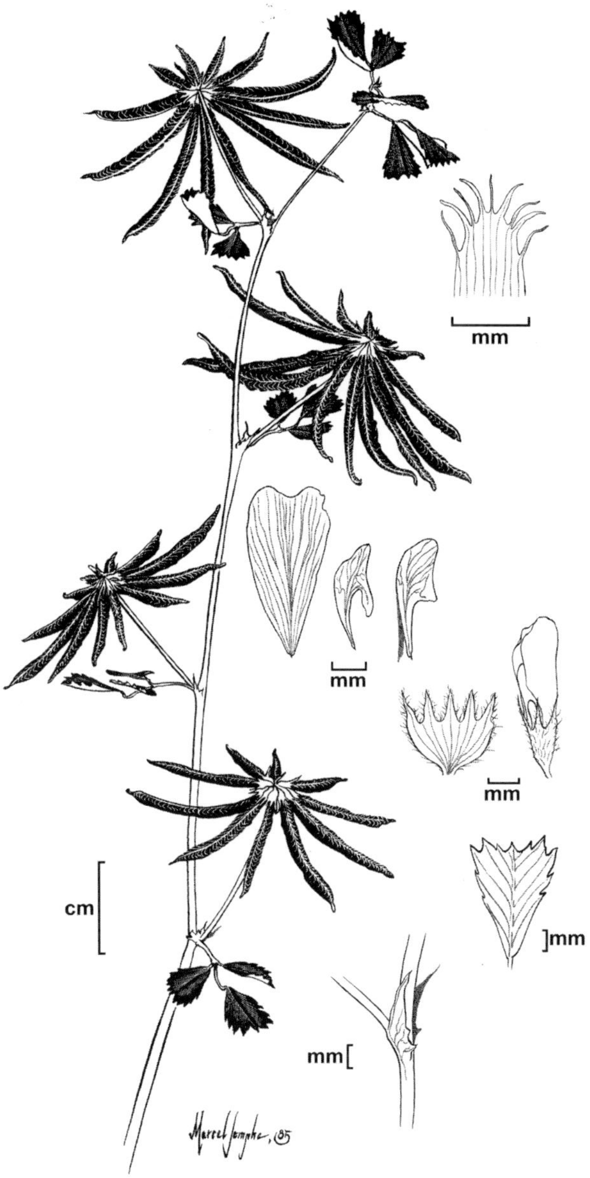

Fig. 17-1. *Medicago astroites.* **Left:** Plant. **Top right:** Spread out androecial column (free stamen not shown). **Right center:** (left to right) standard petal; wing petal; keel petal; calyx; side view of flower. **Lower right:** Leaflet and stem node showing stipules. (Based on isotype, Szovits 232, K; artist: M. Jomphe.)

Subordinate Taxa

None recognized here.

Type Specimen

Type: Szovits 232, Transcaucasia, prov. Karabagh orientalis, ad rivum Inçay, in collibus siccis, 27-V-1829, in fruit. (Holotype LE!, isotypes G! (Herb. Boiss.) and K!)

Synonymy

Homotypic

Trigonella astroites Fisch. & C.A. Mey., Index Sem. Horti Petrop. (St. Petersb.) 1: 40 (1835), (basionym).

Heterotypic

T. astroites f. *amasiaca* Širj., Generis *Trigonella* revisio critica, Spisy Pfir. Fac. Masarykovy Univ., 3: 27 (1930). (This was characterized as having strongly curved pods and leaflets that are wider and more oval than typical of the species.)

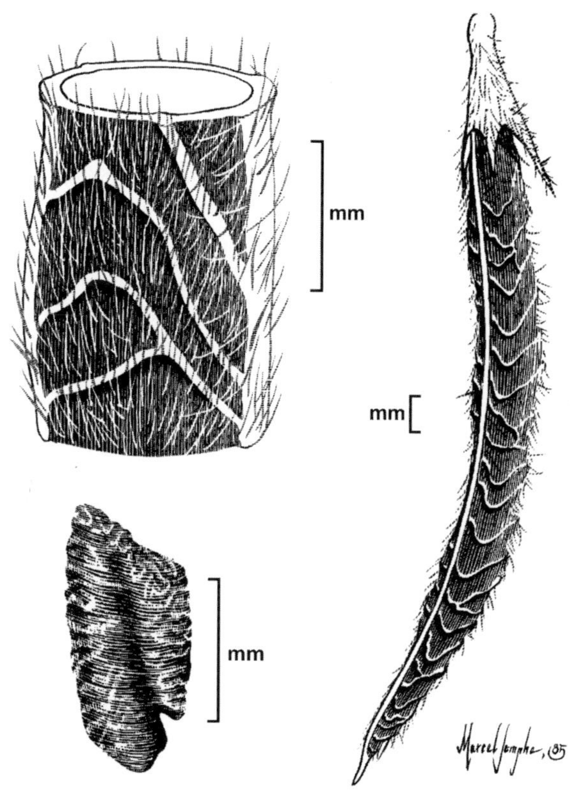

Fig. 17-2. *Medicago astroites.* **Clockwise from bottom left:** Seed; section of fruit showing venation; entire fruit. (Based on Davis & Hedge D. 28098, K; artist: M. Jomphe.)

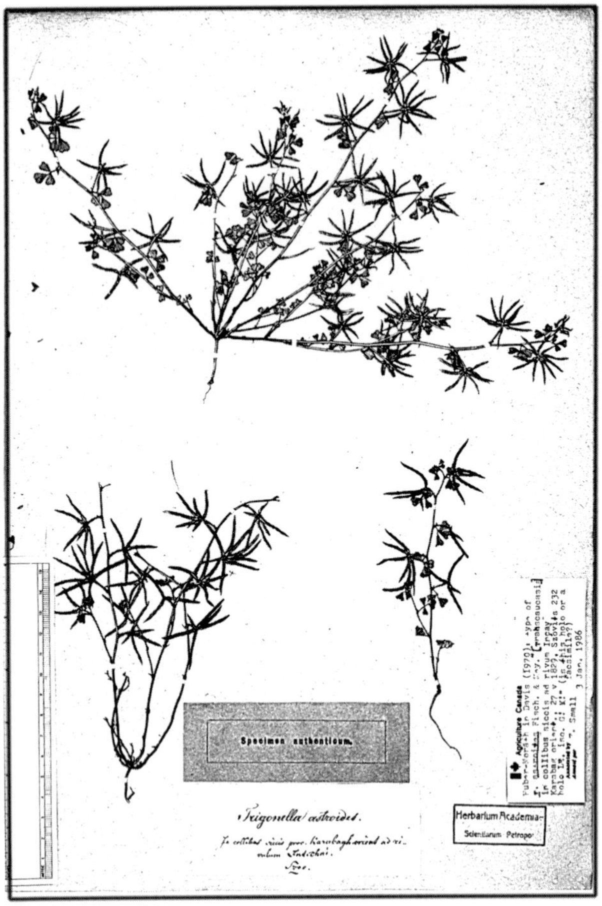

Fig. 17-3. Apparent holotype of *Trigonella astroites* Fisch. & Mey.

T. sinuata Boiss., Diagn. Pl. Orient., ser. 1, 2(9): 18 (1849). (B. Balansa 447, sables maritimes entre Mersina et les ruines des Pompeïopolis (Cilicie), 1855; this was annotated as "Typus" by Širjaev in 1928.)

T. astroites var. *subsessilis* Zoh., Flora Palaestina 2: 132 (1972). (Zohary (1972, p. 132) recognized this by "peduncles very short or missing," but on page 461 described this as an erratum.)

Not

M. astroites Bertol. Fl. Ital. 8: 272 (1850) (= *M. italica*).

Vernacular Names

English: No names found.

French: No names found.

Geographical Distribution

(N = native, I = introduced, U = uncertain status):

Asia: Armenia (N; Sagatelyan 1980), Iran (N), Iraq (N), Israel (N), Jordan (N), Lebanon (N), Syria (N), Turkey (in Asia, N).

This species is native to the eastern Mediterranean and eastward.

Description

Annual herb, stems diffusely branched from base, prostrate to ascending, slightly adpressed-pubescent, 5–25 cm long. Leaflets 3–8 mm long, 2.5–8 mm wide, oblanceolate to obovate, truncate to obtuse at apex, markedly denticulate toward tip, terminal half to two-thirds of leaflet finely serrate-dentate, sparingly adpressed-pilose on lower side, mostly glabrous or sometimes also pubescent on upper surfaces. Stipules semi-sagittate, lanceolate-acuminate, dentate from broadened base or incised or entire. Inflorescence an umbellate-capitate raceme of (2)6–10(13) flowers, peduncles (0)1–2 cm, thickened when fruit develops. Flowers 4–5 mm, yellow, sessile or pedicels very short. Calyx about half as long as corolla, 1.5–3.5 mm, almost glabrous to moderately hairy, campanulate or tubular-campanulate, teeth lanceolate-subulate, equal or slightly longer than tube. Pods stellately divaricate or erect, linear, straight, or slightly curved, somewhat compressed-cylindrical, tapering to an acute or sharpened apex, reticulately transverse-veined, 10–20(30) mm long, 0.8–1.5 mm wide, finely appressed-pilose to glabrescent, with 4–6 seeds, seed-bearing (ventral) suture straight, the other (dorsal) subundulate. Seeds finely tuberculate, oblong-cylindrical, brownish, about 2 mm long, 0.75–1 mm wide, radicle about 80% of seed length.

Identification Hints

The most conspicuous feature of this species is the stellately spreading arrangement of the fruits. It is closest to *M. halophila* (both species have a star-like spreading of fruits and a somewhat subundulate dorsal fruit suture). *Medicago stellata* has calyx teeth that are lanceolate-subulate and longer than or equal to the tube, whereas *M. halophila* has calyx teeth that are triangular-lanceolate and shorter than the tube. *M. halophila* often has more than seven flowers in the inflorescence whereas *M. stellata* rarely has this many, and the pods of *M. halophila* are curved whereas those of *M. stellata* are mainly straight.

Habitat and Environmental Adaptation

Habitat

Medicago astroites occurs on dry rocky slopes, sands, dry gravelly riverbeds, steppes, among shrubs and in pine forests, and in cultivated land, at altitudes of 0–1800 m. It often inhabits shrub and steppe associations.

Phenology

This species flowers from April to June in its native area.

Soil

Medicago astroites typically occupies rocky and gravelly limestone-based substrates and sandy soils.

Temperature

Medicago astroites appears to be adapted to the spring and early summer conditions of the Middle East.

Water

This species is a drought-avoiding annual ephemeral, sometimes occurring on sand dunes and in subdeserts.

Biotic Relationships

Pollination

The small flowers of this species indicate that it is largely an inbreeder.

Dispersal

No information located.

Herbivory

No information located.

Nodulation and mycorrhizae

No information located.

Disease resistance

No information located.

Economic History, Cultivars, and Uses

This uncommon species has not attracted agricultural interest to date.

Cultivars

No cultivars have been bred.

Forage

Medicago astroites is not sown for forage.

Fodder

No information located.

Green manure, rotation crop, ground cover

No information located.

Other uses (salad sprouts, ornamental, plant products, etc.)

No information located.

Chromosome Number

$2n = 16$: Reasat et al. 2003.

Genetic Relationships and Gene Transferability

Intraspecific

No information located.

Interspecific

No information located.

Exchange by advanced biotechnology

No information located.

Conservation Status

This species is in need of conservation evaluation and of collection of germplasm for evaluation of variation patterns for conservation purposes.

Germplasm Sources

At least 50 accessions have been deposited in major germplasm institutions (Chapter 3).

Research Problems and Priorities

Medicago astroites is very poorly studied, and its botanical and agricultural potential require evaluation.

Representative Collections

Iran: Anders, O.2535, south of Mosul, alt. 250 m, 10-IV-1969, in flower & fruit (HUJ). Koelz, Walter 15717a, Durud, Luristan, alt. 5000 feet., in sand, 22-V-1940, in flower & fruit (E). Zohary, M. & Orshan, G., northern Iran, env. of Teheran on the Karuj road, Artemisia steppe on foothills, alt. 1200 m, 7-V-1961, in fruit (HUJ). **Iraq:** Al-Rawi, A. 20481, 50 km east of Samarra, Asia village (10 km north of Shaikh Mohammed), alt. ca. 50 m, sandy soil, 5-V-1957, in fruit (K). Gillett, J.B. 10982, Kursi, Jebel Sinjere, alt. ca. 500 m, limestone mt., bare slopes, occasional, 23-V-1948, in fruit (K). Guest, E. 13401, Jazira (south of Sinjar), alt. ca. 350 m, open Artemisia herba-alba, 10-V-1984, in fruit (K). Haines, R. Wheeler 858, Shahraban, sandy area with small dunes, 11-IV-1957, in flower & early fruit (E). Poore, M.E.D. 518, northeast of Sar Qal'a in the valley of the Nasin Chai, alt. 1200 feet, steppe of Artemisia herba-alba, 8-IV-1958, in fruit (K). **Israel:** Heyn, C., Danin, A. & Plitmann, U., C. Negev, Mischer Haruhot, 10 km west of Mitzpeh-Ramon, loess plain, 18-IV-1967, in late flower & fruit (HUJ). Danin, A., Judean Mts., Jerusalem near Ora, marly chalk, 20-V-1987, in flower & fruit (HUJ). Danin, A, C. Negev Highlands, 1 km south of Rosh Horsha, in Nahal Horsha, cork crevices near rock, alt. 950 m, 16-IV-1986, in flower & fruit (HUJ). Gilboa, M., Shmida, A., & Lur, M., C. Negev, Har HaNegev, Nahal Loz, junction of Ma'ale 'Arod, wide loessial bed, 27-V-1972, in fruit (HUJ). Danin, A., C. Negev, west of Mizpe Ramon, broad loess wadi, 19-IV-1967, in flower & fruit (HUJ). **Jordan:** Eig, A. & Zohary, M., Ammon, 2–3 km southeast of Ein Suela, 14-IV-1929, in flower & early fruit (HUJ). Simpson, N.D. S3372, near Zarga River bridge, Deir Abba, alt. 250 m, 24-III-1953, in flower & fruit (K). Western, A.C. WH79a, opposite camp at Beidha, mixed annuals, legumes and grasses from sandy slope, facing northeast to east, III-IV-1967, in flower & very early fruit (K). Western, A.C. WH82J, east side of Wadi Allumti near cliff and bordering barley fields, III-IV-1967, in flower & early fruit (K). Western, A.C. WH99c, upper terrace of cornfields on slopes of Jebel Shahir above heads of wadis, limestone soil, III-IV-1967, in flower & very early fruit (K).

Syria: Barkoudah, Y.L. 1256, Jabal Druze, north of Shahba Tell Shihan, lava soil, fallow fields, 18-V-1964, in flower & fruit (E). Dinsmore, J.E. 7745, south of Jarud, alt. 670 m, fields, 16-IV-1940, in flower & fruit (K). Samuelsson, G. 3540, desertum palmyrense, ad radices avstr. montis Djebel el Abiad, in valliculo lapidoso ("ouadi") ca. 15 km ad bor.-occid. a fonte Ain el Beida, alt. ca. 500 m, 18-IV-1933, in flower & fruit (K). Samuelsson, G. 4418, Antilibanon, Ouadi Barada, solo calcareo, Louk Ouadi-Barada, ca. 1025 m, in rupestribus, 6-V-1933, in fruit (K). Turkey: Coode, M.J.E. & Jones, B.M.G. 332, Adana, near Tuzle, near sea level, sand dunes, open sand, 2-V-1965, in flower & fruit (E, K) (Note: this specimen was identified as *T. halophila* Boiss. and is so much like it that a case could be made for combining *M. astroites* and *M. halophila*.) Balansa, B. 1216, base occidentale du Boulgas-Dagh (Phrygie), dans les forêts de Quercus agilops, 14-VI-1857, in fruit (BM, P). Balansa, B. 909, Kamechly-Tchaï près de Béréketly (Cappadoce), 1300 m d'alt., 15-VI-1856, in fruit (K, P). Balls, E.K. 2277, Malatya, alt. 3900 feet, Limestone skree, 14-V-1935, in flower & fruit (E). Balls, E.K. 861, Gaziantep, alt. 2700 feet, cultivated ground, 27 -IV-1934, in flower & early fruit (E). Balls, E.K. B.2277, Malatya, limestone scree, alt. 3900 feet, 14-V-1935, in flower & fruit (K, P). Bornmüller 2805, Pontus-Galaticus, in vinetis, alt. 500 m, 16-V-1890, in fruit (BM). Campbell, F. 41 & 44, prov. Malatya, Euphrates Valley between Malatya and Elazig, alt. 900 m stony and grassy hillside, 19-V-1967 (K) (flowers atypically purple). Coode, M.J.E. & Jones, B.M.G., 2144, Ankara, 15 km north of Ankara on Çankiri Road, stony chalky slopes, 8-V-1965, in fruit (E). Coode, M.J.E. & Jones, B.M.G. 840 (mixed collection), Icel, Mut, road, Mut to Büyük Egri Dag, alt. 1000 m, Pinus brutia forest, 12-V-1965, in flower & early fruit (E). Davis 43337, Bitlis, 9 km from Ahlat to Adilcevaz, alt. 1750 m, light gravelly hills, 21-V-1966, in flower & fruit (E). Mera, Alinoglu, N. 187, Universtesi Merasi, Pasture, 7-VI-1963, in flower & fruit (E). Mera, N. Alinoglu 184, prov. Ankara, Ankara, Orta dogu, Universitesi Merasi, pasture, 20-V-1963, in flower and early fruit (E). Polunin, O. 159222, above Peristrama Valley, near Aksarey, cornfields, V-1981 (E). Stapf 728, Kharpuk, in campis ad Korakasch supra pekenik, 12-VI-1889, in flower & fruit (K). Tobey 2575, Sinop, before Nacikamza Köy, alt. 600 m, limestone, 11-IV-1969, in flower & early fruit (E). Tobey, 2696, Kargi, Sinop, under Kôs Dag, alt. 400 m, sandy, gravelly hills, 1-V-1969, in fruit (E).

18. *Medicago biflora* (Griseb.) E. Small

M. biflora (Griseb.) E. Small in Small, Lassen & Brookes, Willdenowia 16(2): 433 (1987).

The epithet *biflora* is from the Latin *bi*, two, plus *flor*, flowered, an allusion to the frequent occurrence of two flowers in the inflorescence.

Subordinate Taxa

None recognized here.

Type Specimen

Grisebach in Spicelegio (1843) cited the original collection site of this taxon as "In Bithynia pr. Modurlu (Pestalozza)". Strid (2000) noted that the type is at GOET and an isotype is at G-Bois.

Synonymy

Homotypic

Trigonella biflora Griseb., Spicil. Fl. Rumel. 1(1): 41 (1843), (basionym). The names *T. biflora* Griseb. and *T. lunata* Boiss. were both proposed for the previously undescribed species in 1843. When considered to be a species of *Trigonella*, the name *T. lunata* Boiss. has been traditionally given priority (e.g., see discussion in Širjaev (1928–34)). Strid (2000) stated that "Boissier's name was apparently published 1–2 months earlier."

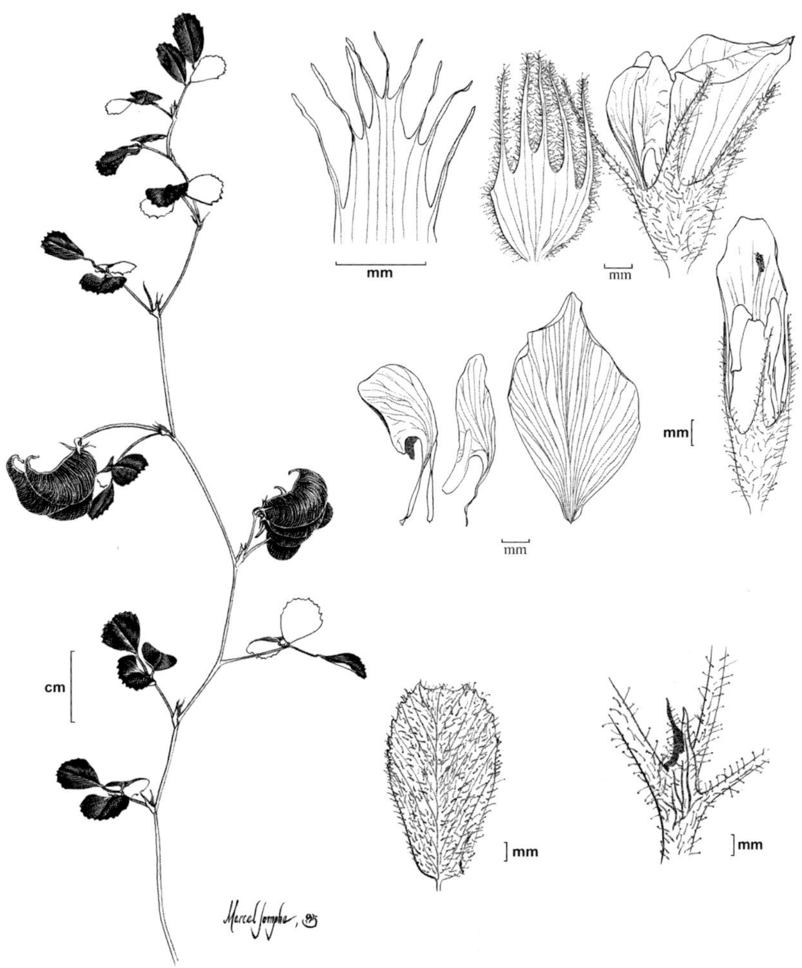

Fig. 18-1. *Medicago biflora.* **Left:** Fruiting branch. **Upper right:** Floral parts (clockwise from upper left: spread out androecial column (free stamen not shown); calyx; two views of flowers; standard petal; wing petal; keel). **Lower right:** Stem node showing stipules. **Center bottom:** Leaflet. (Based on E. Small Turkey-189, DAO; artist: M. Jomphe.)

Pocockia biflora (Griseb.) Walp., Ann. Bot. Syst. 2(2): 347 (1852).

Melissitus biflorus (Griseb.) Latsch., Trudy Tiflissk. Bot. Inst. 19: 22 (1958).

Melilotoides biflora (Griseb.) Czerep., Vasc. Pl. Rus. Adjac. States: 251 (1995).

Heterotypic

Trigonella lunata f. *coerulans* Bornm. ex Širj., Generis *Trigonella* L. Revisio Critica, Spisy Pfir. Fac. Masarykovy Univ. 1: 39 (1928). (Refers to a blue-flowered plant.)

T. lunata Boiss., Diagn. Pl. Orient ser. 1(2): 19 (1843). (Type: Boissier, E., Turkey, in umbrosis regionis montanae montis Cadmi supra Colossam, VI-1842, in late flower & fruit, Herb. Boiss. G!; syntype: Aucher 1175 K; photo DAO.) Not *M. lunata* Reichenb. Fl. Germ. Excurs. 501 (1832), nom. illeg.

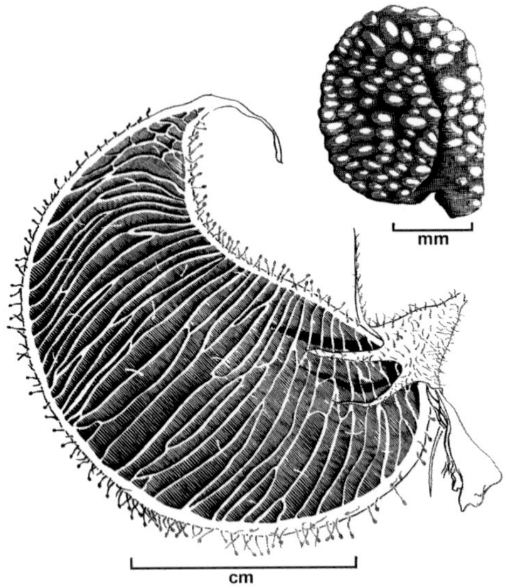

Fig. 18-2. *Medicago biflora.* **Bottom:** Fruit showing venation on flat face. **Upper right:** Seed. (Based on E. Small Turkey-157, DAO; artist: M. Jomphe.)

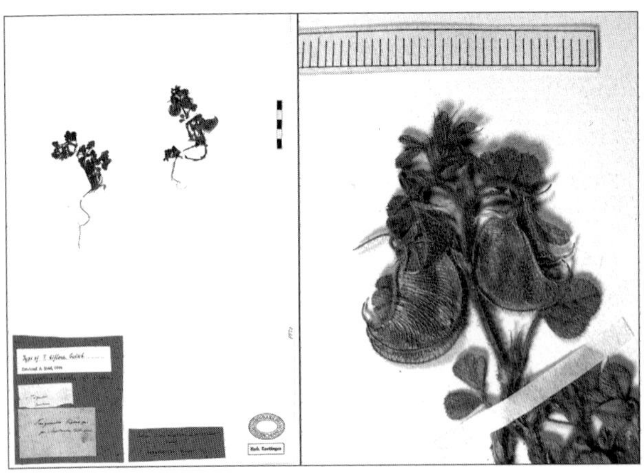

Fig. 18-3. Photos of holotype of *Trigonella biflora* Griseb. (Courtesy of J. Heinrichs, GOET; Id: 8961; registration number: 8958.)

M. lunata (Boiss.) Boiss., Diagn. Pl. Orient. 1(7): 94 (1846).

Pocockia lunata (Boiss.) Boiss., Diagn. Pl. Orient. ser. 1, 2(9): 12 (1849).

Telis lunata (Boiss.) O. Kuntze, Rev. 1: 209 (1891).

T. lunata f. *flaviflora* Bornm. ex Širj., Generis *Trigonella* L. Revisio Critica, Spisy Pfir. Fac. Masarykovy Univ. 1: 39 (1928). (Refers to a yellow-flowered plant.)

Vernacular Names

English: Lunate medic (medic is also spelled medick), lunate fenugreek.

French: No names found.

Geographical Distribution

(N = native, I = introduced, U = uncertain status):

Asia: Armenia (N), Azerbaijan (N), Iran (N), Jordan (N; Boulos and Al-Eisawi 1977), Syria (N), Turkey in Asia (N).

Medicago biflora is native to the Middle East and southern Trans-Caucasus.

Description

Annual herb, (5)10–20(35) cm, branched from base, prostrate to ascending or erect, twigs and leaves pubescent with simple and glandular hairs. Leaflets obovate-cuneate, apically notched, distal margins denticulate, (4)5–10 mm long, 3–5(7) mm wide, tomentose on both sides. Stipules ovate-lanceolate to lanceolate, acuminate, slightly dentate or occasionally entire. Inflorescence an umbellate raceme of 1–3 flowers, peduncle 3–10 mm, terminal awn 3–8 mm. Flowers yellow (rarely bluish-yellow), 5–9 mm, with short pedicels (ca. 1 mm), calyx 5–6 mm, campanulate, teeth linear-subulate or linear-lanceolate, teeth somewhat longer to twice as long as tube, pubescent with glandular and simple hairs. Pods lunately or semi-lunately plano-convex, weakly falcately curved, tip bent to side and tapering to a beak, (10)15–22 mm long, (7.5)9–10(12) mm wide, densely covered with glandular hairs, with numerous fine transverse and interconnected veins, pedicels bent to side, each pod with 7–10 seeds. Seeds about 2–2.5 mm long, 1–1.5 mm wide, finely tuberculate, radicle almost as long as length of seed.

Identification hints

The only other species of *Medicago* with lunate pods covered with gland-tipped hairs is the closely related *M. rostrata*. *Medicago biflora* has denticulate leaflets and pods that are generally 15–20 mm long with a concave ventral suture, whereas *M. rostrata* has entire leaflets and fruits that are usually 7–8 mm long with a straight ventral suture.

Habitat and Environmental Adaptation

Habitat

Medicago biflora occurs on dry, stony slopes, among scattered shrubs and trees, in dry river beds, along river banks, in valleys, meadows, and as a weed of fallow and cultivated land, in low and middle montane belts, with collections from 400 to 1300 m in altitude.

Phenology

The species flowers from March to May and fruits from May onward in the indigenous area.

Soil

Medicago biflora has been collected in basic (typically limestone-based), stony substrates.

Temperature

The species is adapted to the spring and early summer conditions (cool to increasingly hot) of the Middle East.

Water

Based on its occupation of well-drained soils, *M. biflora* appears to be moderately drought-resistant.

Biotic Relationships

Pollination

Small (1988) noted that the flowers are small and the pollen to ovule ratio is low (316), observations that are consistent with *M. biflora* being largely self-pollinated and substantially an inbreeder.

Dispersal

The flattened fruit may be an adaptation to wind dispersion.

Herbivory

No information located.

Nodulation and mycorrhizae

No information located.

Disease resistance

No information located.

Economic History, Cultivars, and Uses

This species has not attracted agricultural attention to date.

Cultivars

No cultivars have been bred.

Forage

Medicago biflora is not sown for forage.

Fodder

This species is not considered to be a fodder plant.

Green manure, rotation crop, ground cover

No information located.

Other uses (salad sprouts, ornamental, plant products, etc.)

No information located.

Chromosome Number

No information located.

Genetic Relationships and Gene Transferability

Intraspecific

No information located.

Interspecific

No information located.

Exchange by advanced biotechnology

No information located.

Germplasm Sources

Only a few accessions have been deposited in germplasm institutions (Chapter 3).

Research Problems and Priorities

Medicago biflora has received little study to date and its botanical and agricultural characteristics remain to be evaluated.

Representative Collections

Syria: Haradjian, M. 4450, Syriae borealis, environs Killis, 2000–25 000 feet, 26-V-3-VI-1911, in fruit (G). **Turkey:** Alinoglu, N./Mera 187, Prov. Ankara, Ankara, Ortadagu Universtesi Merasi, pasture, 7-VI-1963, in flower & fruit (E). Balansa, B. 110, petite collines pierreuses bordant la rive droite du Kamechly-Tchaï, près de Béréketly, alt. 1300 m, 17-VI-1856, in flower & fruit (G); also from same location Balansa 909 on 15-VI-1856 (G, K, P). Balansa 275, Base occidentale du Boulgas-Dagh près de Boulgas-Keui, alt. 1000 m, fôrets de *Quercus aegilops*, 14-VI-1857, in fruit (G) (also Balansa 1216 from same location, 14-VI-1857, in fruit (G, BM, P). Balls, E.K. B2277, Malatya, limestone scree, alt. 3900 feet, 14-V-1935, in flower & fruit (K, P); also Balls 2277 at E. Bornmüller 2805, Pontus-Galaticus, in vinetis, alt. 500 m, 16-V-1890, in fruit (G, BM). Bourgeau, E. 195, Elmalu, dans les ravines, 17-VI-1860, in fruit (G). Campbell, F. 41 (& 44), prov. Malatya, Euphrates Valley between Malatya and Elazig, alt. 900 m, stony and grassy hillside, 19-V-1967 (K) (unusual in having purple flowers). Coode, M.J.E. & Jones, B.M.G. 2144, Ankara, 15 km north of Ankara on Çankiri road, stony chalky slopes, 8-V-1965, in fruit (E). Stapf, Herb. O., 728, Kharpuk, in campis ad Korakasch supra Pekenik, 12-VI-1889, in flower & fruit (K; probably = Sintensis, P. 728 (G)). Tobey, 2696, Kargi, Sinop, under Kôs Dag, alt. 400 m, sandy, gravelly hills, 1-V-1969, in fruit (E). Polunin, O. 159222, above Peristrama Valley near Aksarey, cornfields, V-1981 (E).

19. *Medicago bonarotiana* Arc. sensu stricto[1]

M. bonarotiana Arcang., Nuovo Giorn. Bot. Ital. 8: 6 (1876), (basionym).

The epithet *bonarotiana* commemorates Michelangelo, i.e., Michelangelo Buonarroti (1475–1564), Italian sculptor (most notably *Statue of David*), painter (most notably the ceiling of the Sistine Chapel), architect (Chief Architect to St. Peter's, charged with the completion of the new church, the most prestigious architectural commission in Christendom) and poet. Arcangeli wrote in 1876 (translated from Italian): "In the search for an appropriate name for this new species, I considered the fact that it was discovered in an area very close to the villa of the illustrious Michelangelo Buonarroti, and that this very year, we will be celebrating his 400th birthday."

[1]As discussed in the following, this taxon is part of a confusing complex that has been interpreted differently by various authors. Almost all the literature today includes *M. bonarotiana* as part of "*M. blancheana* Boiss.", a more comprehensive taxon that includes numerous hybrids between *M. bonarotiana* and *M. rotata*. The interpretation of *M. bonarotiana* adopted here excludes these hybrids, as discussed below.

Subordinate Taxa

None recognized here.

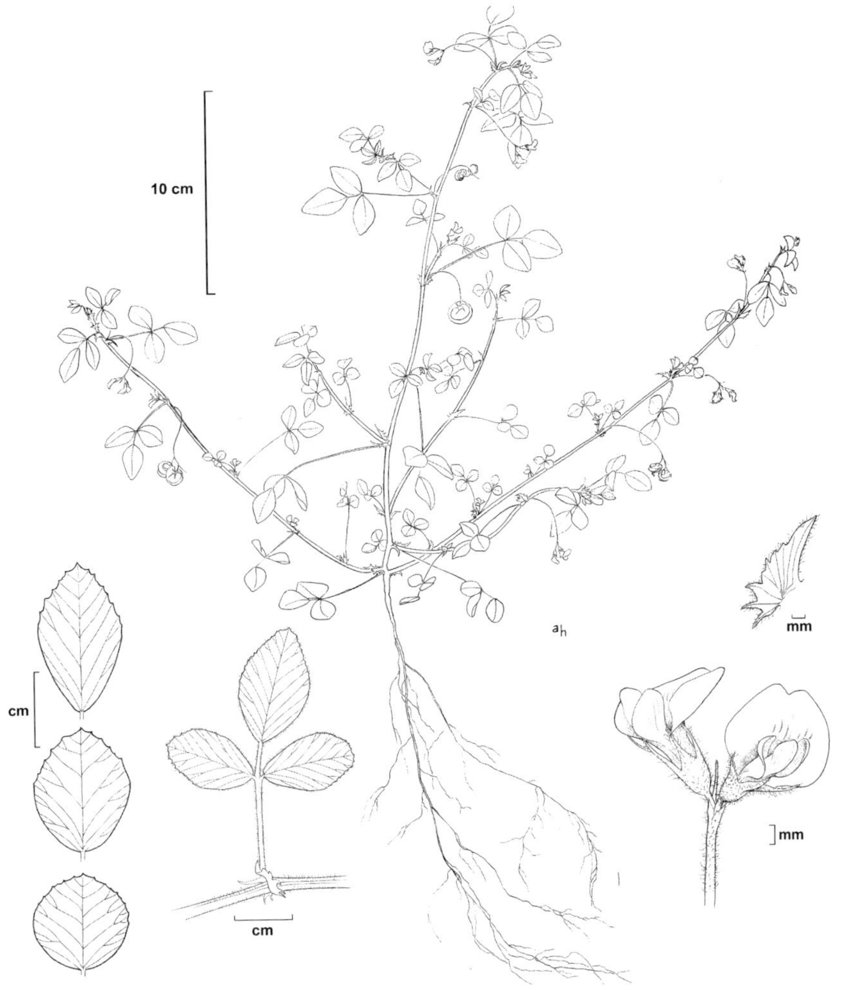

Fig. 19-1. *Medicago bonarotiana.* **Clockwise from top:** Plant; stipule; group of flowers; leaf attached to stem; leaflets. (Based on E. Small M353, DAO; artist: A. Hanes.)

Type Specimen

Heyn (1963) gave the following details in lectotypifying *M. bonarotiana* Arc.: "...In agris ad aggeret prope Septiniam Florentia, Maio, Arcangeli 1875 (G)... Arcangeli distributed a great number of specimens of '*Medicago bonarotiana*'... the author chose the specimens collected in 1875, deposited at the Conservatoire Botanique in Geneva, as the lectotype." However, I was unable to find a specimen bearing an indication that it was the lectotype chosen by Heyn at G. Of the four Arcangeli 1875 specimens at G, three bore the same reproduced label. I (Small 1994) designated the fourth as the lectotype in the absence of other information. It bore the inscription: "In agris et ad aggeres prope Septinianum ...Florentia Maio. ? Arcangeli. 1875."

Synonymy

Homotypic

M. blancheana var. *bonarotiana* (Arcang.) Arcang., Comp. Fl. Ital. 61 (1882).

M. rotata Boiss. subsp. *bonarotiana* (Arcang.) Ponert, Feddes Repert. 83(9-10): 639 ("1972" = 1973).

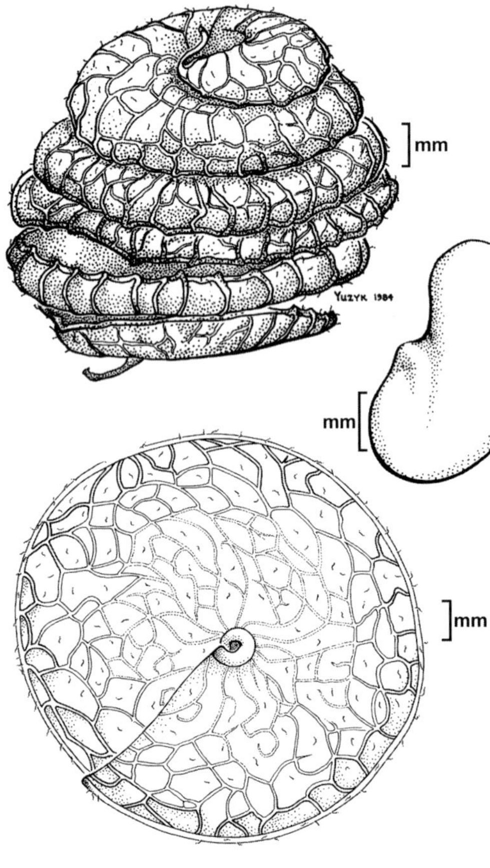

Fig. 19-2. *Medicago bonarotiana*. **Top:** Fruit. **Center right:** Seed. **Bottom:** Fruit coil face showing venation. (Based on E. Small M353, DAO; artist: L. Yuzyk.)

Heterotypic

M. blancheana var. *inermis* Post Fl. Syr. Pal. Sin. 227 (1883–1896).

M. recurvata Post (unpublished name on a herbarium specimen, noted by Heyn (1963)).

Vernacular Names

English: Wheel medic (medic is also spelled medick). This name has also been applied to *M. rotata* × *M. bonarotiana*

French: No names found.

Geographical Distribution

(N = native, I = introduced, U = uncertain status)

Asia: Cyprus (N), Israel (N), Lebanon (N), Syria (N), Turkey in Asia (N).

Europe: Italy (I), Malta (I), Netherlands (I) (to be expected as an introduction elsewhere in Europe).

Medicago bonarotiana is basically an eastern Mediterranean species.

Description

Annual herb 10–30(40) cm, procumbent or more typically ascending, branching from near base, young branches pubescent with simple and gland-tipped hairs. Leaflets 10–18 mm long, 5–8 mm wide, obovate or sometimes ovate or oblanceolate, with a small terminal tooth, upper third of margin serrate, both surfaces with simple hairs, gland-tipped hairs on lower surfaces, very rarely laciniate. Stipules lanceolate-subulate, dentate. Inflorescence of 1–3(5) flowers, peduncles about twice as long as subtending petiole, curved downward. Flowers 6–9 (10) mm, pale to dark yellow. Calyx 3–4 mm, with simple and gland-tipped hairs, teeth equal to tube. Corolla up to twice length of calyx. Young fruit contracted and concealed within calyx. Fruit 8–12 mm wide, with glandular hairs (rarely lacking), 4–6 coils, middle coils broadest, coils except middle coils like bowls pointing towards the middle of the fruit, coils with a dense net of veins forming a strong subquadrate network and connecting to the dorsal suture, often blackish at maturity. Fruit spines absent. Seeds 4–5 mm long, 1–2 in each coil, sometimes separated by very thin membranous partitions, seed coat smooth, yellow to light brown, radicle about half length of seed.

Identification Hints

Medicago bonarotiana is the only species of *Medicago* with imbricate (bowl-like) coils arranged so that the open part of the bowl always points towards the middle, i.e., the bowls at the fruit tip are stacked differently than the bowls at the fruit base. This taxon hybridizes extensively with *M. rotata*. As discussed below, introgression occurs primarily from *M. rotata* into *M. bonarotiana* so that *M. rotata* is usually distinctive, but *M. bonarotiana* is much less well defined.

Habitat and Environmental Adaptation

Habitat

Medicago bonarotiana is most often observed as a weed of fields. Presumably in presettlement times it occupied recently disturbed, open habitats and became associated with anthropogenic habitats as civilization progressed.

Phenology

This species is in flower from March to May in much of its native area.

Soil

Medicago bonarotiana typically occupies heavy clay and will also do well in loam soils. Lesins and Lesins (1979) noted that during their collecting expedition to acquire *Medicago* seeds, they found *M. bonarotiana* characteristically in heavy, reddish clay soils.

Temperature

Like most annual species of *Medicago*, *M. bonarotiana* is a winter annual, germinating in the cool conditions of the Mediterranean winter, but tolerating at least some hot temperatures that occur in the early summer, before the plant sets seed and dies.

Water

Piano and Francis (1992) observed that "*M. blancheana*" in its native area requires a minimum of 400 mm annual precipitation. *Medicago bonarotiana* is a drought evader like most other annual Mediterranean region species of *Medicago*. It tolerates some lack of water, but completes its life cycle before water deficits become extreme in the middle of the summer. Like the majority of weeds, when faced with environmental stress, it has the capacity to complete its life cycle while still a very small plant.

Biotic Relationships

Pollination

Small (1988) noted that the small flowers and low pollen to ovule ratio (151) are consistent with *M. bonarotiana* being largely self-pollinated.

Dispersal

The indehiscent fruits lack spines and so are not well adapted to dispersal in animal fur. No obvious adaptations for dispersal are present, although the fruits are quite round and could be expected to roll when blown by the wind in open terrain.

Herbivory

Shade and Kitch (1983) noted that the glandular hairs of "*M. blancheana*" resulted in resistance to the pea aphid (*Acyrthosiphon pisum* (Harris)). MacLean and Byers (1983) observed that the glandular hairs of "*M. blancheana*" tended to deter the alfalfa blotch leafminer (*Agromyza frontella* (Rondani)) from ovipositing. Shade et al. (1979) found that "*M. blancheana*" was very resistant to the potato leafhopper (*Empoasca fabae* (Harris)). Johnson et al. (1980*a*, 1980*b*) demonstrated resistance to the alfalfa weevil (*Hypera postica* (Gyllenhal)). Ferguson et al. (1982) found that "*M. blancheana*" was quite resistant to the spotted alfalfa aphid (*Therioaphis maculata* (Buckton)).

Nodulation and mycorrhizae

USDA (2009) reported positive root nodulation for *M. blancheana*.

Disease resistance

Yaege and Stuteville (2002) tested the resistance of 33 annual species of *Medicago* to powdery mildew (*Erysiphe pisi*) and found that the six collections of "*M. blancheana*" grown were susceptible, whereas one showed some resistance. O'Neill and Bauchan (2000) in a comparative test of 36 annual species of *Medicago* found that seven accessions of "*M. blancheana*" were quite susceptible to anthracnose caused by *Colletotrichum trifolii*.

Economic History, Cultivars, and Uses

Medicago bonarotiana has not attracted agricultural interest to date.

Cultivars

No cultivars have been bred.

Forage

Medicago bonarotiana is not sown for forage. Mohammad and Qamar (1988) grew 14 annual species of *Medicago* in small test plots in Islamabad, Pakistan, located in a subtropical subhumid area that is notably hotter and wetter than the normal habitat of these species. The dry matter yield of "*M. blancheana*" was 3.5 t/ha. St. John-Sweeting et al. (1996) compared several species in South Australia and found that "*M. blancheana*" was comparable in both herbage and seed production to cultivar 'Caliph' of *M. truncatula*, and suggested that further study of commercial development was warranted.

Fodder

This species is not considered to be a fodder plant.

Green manure, rotation crop, ground cover

No information located.

Other uses (salad sprouts, ornamental, plant products, etc.)

Medicago bonarotiana has been occasionally grown as an ornamental in Europe (Heyn 1963).

Chromosome Number

$2n = 16$

As *M. bonarotiana*: Lesins and Lesins 1979.

As *M. blancheana*: Heyn 1956; Clement 1962; Simon and Simon 1965; Fernandes and Santos 1971; Fernandes and Queirós 1978; Mariani et al. 1996.

Genetic Relationships and Gene Transferability

Intraspecific

No information located.

Interspecific

M. bonarotiana hybridizes naturally with *M. rotata* and there is extensive natural introgression between the two (as discussed in the chapter on *M. rotata*). Simon and Millington (1967) were unable to produce hybrid seed by crossing "*M. blancheana*" with *M. rotata*, although the two should be easily crossable (presumably if *M. bonarotiana* is the female parent).

Exchange by advanced biotechnology

No information located.

Conservation Status

Medicago bonarotiana is a fairly common and somewhat weedy species, not in need of special conservation measures.

Germplasm Sources

Germplasm institutions usually list accessions under the name "*M. blancheana*", a taxon (as noted below) I split into *M. bonarotiana* and hybrids between *M. bonarotiana* and *M. rotata*. Close to 300 accessions are available (see Chapter 3).

Research Problems and Priorities

The agricultural potential of *M. bonarotiana* has been studied only to a minor extent to date. The species might have potential as a pasture legume and deserves additional investigation in this regard.

Representative Collections

Cyprus: Casey, E.C. 1144, Kyrenia, alt. 150 feet, uncultivated olive grove, 22-III-1951, in flower & fruit (K). **Israel:** Feldman, M. 9042, Upper Galilee, Alma to Dishon, field borders and roadside, 12-V-1956, in fruit (HUJ). Liston, A. & Monias, H., Judean mountains, Jerusalem east of Givat Ram, field, 13-IV-1982, in flower & fruit (HUJ). Natolsky, N., Mt. Carmel, Shefeyia, rocks, 12-V-1924, in flower & fruit (HUJ). Shrinkar, C., Hula Valley, Lahauot Habashan, basalt soil, west slope, 8-IV-1954, in flower & fruit (HUJ). Zohary, M. M/9170, Upper Galilee, between Kabii and Hanita Wadi Qarn, 23-III-1955, in flower & fruit (HUJ) (leaves somewhat laciniate and designated "f. *lacinicata*"). Zohary, M. & Heller, D., Lower Galilee, Algamor, basalt soil, 12-IV-1971, in fruit (HUJ). **Israel/Syria (Golan):** Zohary, M., Golan, near the village Jarba, basalt soil, wet field, 17-IV-1968, in fruit (HUJ). **Italy:** Arcangeli, F., Italia media, prope pagum "Settigirano" ad orientem Florentiae, in agris, abunde, 21-V-1877, in fruit (BM). Fiori, A., Etruria, Florentia (Firenze), a Mensola inter segetes ad Villa Strozzi, alt. 100 m, solo vario, 24-V-1908, in flower & fruit, Flora Italia Exsiccata series II Fiori et Béguinot No. 1070 (BM, K). Levrier, E., Italia media, in vincis ad agris pagi Settignano, prope Florentiam (Etruria), 22-V-1875, in fruit (K). Levrier, E., supra pagum Settignano ad orientem Florentiae, in agris cultis, 23 & 30-V-1886, F. Schultz herbarium normale nov. ser. cent. 22 No. 2155 (P). **Lebanon:** Meinertzhagen, R., Beyrouth, Syria, in crops, 21-III-1933, in flower & fruit (BM). Mouterde, P. 2031, Hadeth, 12-IV-1933, in fruit (P). **Malta:** Borg, J., Ins. Melitae, inter segetes ad Ghain il Ghira, sp. rarissima, 4-IV-1931, in flower & fruit (K). **Syria:** Samuelsson, G. 4591, Djebel El Drouz, supra Toweida, alt. ca. 1250 m, in agro lapidoso, solo basaltico, 9-V-1933, in flower & fruit (K). Zohary, M. M/9093, Suada, Jebel Druz, 15-V-1933, in flower & fruit (HUJ).

Note

Lesins and Lesins (1979) observed that, in cultivated plants, the first trifoliolate leaf developed tended to have comparatively small lateral leaflets, a characteristic they stated is shared by the closely related *M. rotata*. Other characteristics they observed to indicate a close relationship to *M. rotata* included block-shaped pollen grains with four or five germinal apertures and a long after-ripening period of the seeds.

MEDICAGO BONAROTIANA ARC. × *M. ROTATA* BOISS. (= *M. BLANCHEANA* BOISS. PRO PARTE)

There is substantial hybridization between the two annual eastern Mediterranean species *M. bonarotiana* and *M. rotata* (Small 1984*a*, 1994). The type of the name *M. blancheana* Boiss. is a hybrid of the two species and, therefore, is inappropriately designated as the type of *M. bonarotiana*, which as circumscribed above excludes the hybrids.

Type Specimen of *Medicago blancheana* Boiss.

In his description of this species, Boissier (1856) noted the information "Hab. in segetibus prope Saida cl. Blanche Maio 1853." Boissier (1856) mentioned two specimens, "Hab. inter segetes Syriae prope Sidonem (BL!)" and "ad Marasch Syriae bor. (Haussknl)." The two specimens that Boissier indicated that he had observed (as signified by !) were present in the Boissier herbarium at Geneva. Both collections have fruits that are spiny and biconvex, and are therefore unquestionably the hybrid between *M. rotata* and *M. bonarotiana*. Heyn (in Heyn and Davis 1970) specified the first specimen (M. Blanche 587, collected 19-V-1853) as the type, thereby lectotypifying the name *M. blancheana* Boiss.

Synonymy for *Medicago blancheana* (= *M. bonarotiana* + *M. bonarotiana* × *M. rotata*)

M. blancheana Boiss., Diagn. Pl. Orient. Nov. Ser. 2(5): 75 (1856), (basionym).

The epithet *blancheana* commemorates M. Blanche, the collector of the type specimen of *M. blancheana*.

Homotypic

M. rotata subsp. *blancheana* (Boiss.) Ponert, Repert. 83(9-10): 639 ("1972" = 1973).

Heterotypic

No names found.

Should the hybrid be afforded species status?

The species name *M. blancheana* (or *M.* ×*blancheana*) could be used to designate the hybrids between *M. bonarotiana* and *M. rotata*, but this would be misleading in suggesting that there is a stabilized taxon of hybrid origin. Almost certainly there is continuing hybridization and introgression and I believe this is more clearly indicated by the designation *M. bonarotiana* × *M. rotata*. There appears to be predominantly unidirectional introgression from *M. rotata* into *M. bonarotiana*. Based on the many collections I have examined at the HUJ herbarium, it appeared that, in Israel, for every two specimens assignable to *M. bonarotiana*, one was assignable to *M. bonarotiana* × *M. rotata*, suggesting that the amount of gene transfer is very substantial.

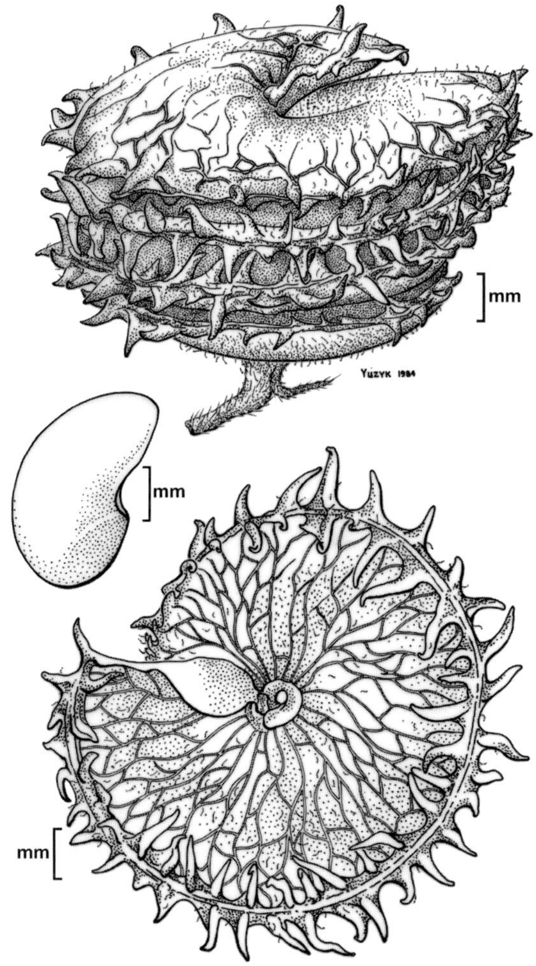

Fig. 19-3. Fruit of *Medicago bonarotiana* × *M. rotata* (part of "*M. blancheana*"). **Top:** Fruit: **Center left:** Seed. **Bottom:** Coil face showing venation. (Based on C. Heyn M/9062, HUJ; artist: L. Yuzyk.)

KEY TO DISTINGUISHING *MEDICAGO BONAROTIANA*, *M. ROTATA*, AND THEIR HYBRID

1. Fruit without spines (fruit always with oppositely imbricate coils, i.e., like facing stacks of bowls, open part of bowls pointed toward middle) .... *M. bonarotiana*
1. Fruit with discernible spines (fruit with or without oppositely imbricate coils) 2
   2. Fruit with oppositely imbricate coils .... *M. rotata* × *M. bonarotiana*
   2. Fruit lacking imbricate coils .... *M. rotata*

Comparison of circumscriptions and nomenclature for *Medicago bonarotiana*, *M. rotata*, and their hybrid of Heyn (1963), Lesins and Lesins (1979), and the classification presented here

Heyn (1963) and Lesins and Lesins (1979), principal students of *Medicago* (all deceased), had rather different interpretations of the *Medicago bonarotiana* – *Medicago rotata* complex than that given here. What Lesins and Lesins would have actually included under the various names is not entirely clear since they did not annotate specimens in herbaria. An interpretation of species and hybrid concepts, based on the keys in Heyn (1963) and Lesins and Lesins (1979), is given in the following table. Heyn annotated numerous specimens as *M. blancheana* var. *blancheana*, some of which I would include in *M. bonarotiana*, others in *M. bonarotiana* × *M. rotata*. Heyn appears to have had a narrower conception of the hybrid than I do. I identify the hybrid group as plants that have the biconvex fruit (with oppositely imbricate coils) of *M. bonarotiana* and at least some of the fruit spines of *M. rotata*. Based on the large collection of the complex I studied at the HUJ herbarium in 1992, it appears that, while *M. rotata* introgresses into *M. bonarotiana*, the former remains fairly distinct. Indeed, Heyn's inclusion in the same species of the hybrid (as *M. blancheana* var. *blancheana*) and *M. bonarotiana* (as *M. blancheana* var. *bonarotiana*) is tacit recognition of the predominantly one-way flow of germplasm into *M. bonarotiana* from *M. rotata*.

Comparison of circumscriptions and nomenclature of principal previous students of the *Medicago blancheana–bonarotiana–rotata* complex with those in this monograph.

Heyn (1963)	Lesins and Lesins (1979)	This monograph
M. blancheana var. *bonarotiana*	*M. bonarotiana*	*M. bonarotiana*
M. blancheana var. *blancheana*	*M. bonarotiana* × *M. rotata*	*M. bonarotiana* × *M. rotata*
M. rotata var. *rotata*	*M. rotata*	*M. rotata* var. *rotata*
M. rotata var. *eliezeri*	*M. bonarotiana* × *M. rotata*	*M. rotata* var. *eliezeri*

Representative Collections of *M. bonarotiana* Arc. × *M. rotata* Boiss.

Cyprus: Chapman, E. 168, Kyrethrea, Ayios Demetrianos, alt. 600 feet, common weed in cornfield, 6-IV-1950, in flower & early fruit (K). Chapman, E. 505, Mia Milia, alt. 500 feet, in cornfield, not very common, 18-IV-1950, in flower & fruit (K). Chapman, E. 95, Kythrea, Ayios Demetrianos, alt. 600 feet, in wheat field on edge of Kythrean hummocks, "3/4" – 1950, in flower (K). **Israel:** Davis, P.H. 4533, Wadi Beidan (Nablus), 23-IV-1942, in fruit (K). Feinbrun, N. & Jacobovitch, E., Jerusalem, Mt. Scopus, crop field, 13-V-1938, in fruit, Universitatis Herbraicae Hierosylmitanae Flora Palestinae Exsiccata No. 335 (BM, P) (a rather variable population was collected in this exsiccata set; collections at G, K = *M. bonarotiana*). Gordon, B., Mt. Scopus, Jerusalem, 13-V-1938, in fruit (HUJ). Grizi, A., Upper Galilee near Wadi Hindaj, field, terra rossa, 12-V-1956, in fruit (HUJ). Heyn, C. M/9062, Esdraelon Plain, Sarid, abandoned field, 13-VI -1954, in fruit (HUJ). **Jordan:** Meyers, F.S. & Dinsmore, J.E. B1701, Ain-el-Tabigha, alt. 200 m, fields, 31-III-1911, in flower & fruit (K). **Lebanon:** Herb. B.V.D. Post, Beirut, College grounds, IV-1900, in fruit (G). **Syria:** Haradjian, M. 2043, Djebel seman, alt. 1200 feet, 15-20-V-1908, in flower & fruit (K).

20. *Medicago brachycarpa* M. Bieb.

M. brachycarpa Fischer ex M. Bieb., Fl. Taur.-Caucas. 3: 517 (1819), (basionym).

The epithet *brachycarpa* is from the Greek *brachy*, which means short, plus *carpus*, which means fruited, descriptive of the fruit.

Subordinate Taxa

None recognized here.

Type Specimen

Iberia prope Kadshari, Wilhelm (Caucasia) (LE). Vassilczenko (2002*a*) provided the following information: "Type: Transcaucasia ('in montosis prope Tiflin circa Kadschari')."

Synonymy

Homotypic

"*M. brachycarpa* Fischer" (i.e., *M. brachycarpa* Fischer in M. Bieb., a commonly encountered miscitation ascribing authority to Fischer, not M. Bieberstein).

Trigonella brachycarpa (Fischer ex M. Bieb.) Mem. Reale Accad. Sci. Torino 36: 190 (1833).

Melissitus brachycarpus (Fischer ex M. Bieb.) Latsch., Trudy Tiflissk. Bot. Inst. xix: 19 (1958).

Melilotoides brachycarpa (Fischer ex M. Bieb.) Soják, Sborn. Nár. Muz. Praze (Acta Mus. Nat. Pragae 38 B), 1982(1–2): 103 (1982).

Heterotypic

T. aucheri Boiss., Diagn. Pl. Orient. 1, 1(2): 20 (1843). (Boissier 1872 (p. 89) cited Auch. exs. 1342, which is present in Herb. Boiss. and may qualify as type material.)

T. brachycarpa (Fischer ex M. Bieb.) Moris f. *aucheri* (Boiss.) Širj., Generis *Trigonella* L. Revisio Critica, Spisy Pfir. Fac. Masarykovy Univ. 1: 36 (1928).

Pocockia cilicica Boiss. & Bal., Boiss. Diagn. Pl. Orient. ser 2, 3(5): 76 (1856). (Boissier 1872 cited a Boiss. specimen and a Balansa specimen, both of which are in Herb. Boiss. G; one of the following may be an appropriate type: Boissier, G., in regione media montis Casii Syriae borealis, V-VI-1846, in fruit; Balansa, M 256, in Tauro Cilicico prope Gulek, 21-VI-, in fruit.)

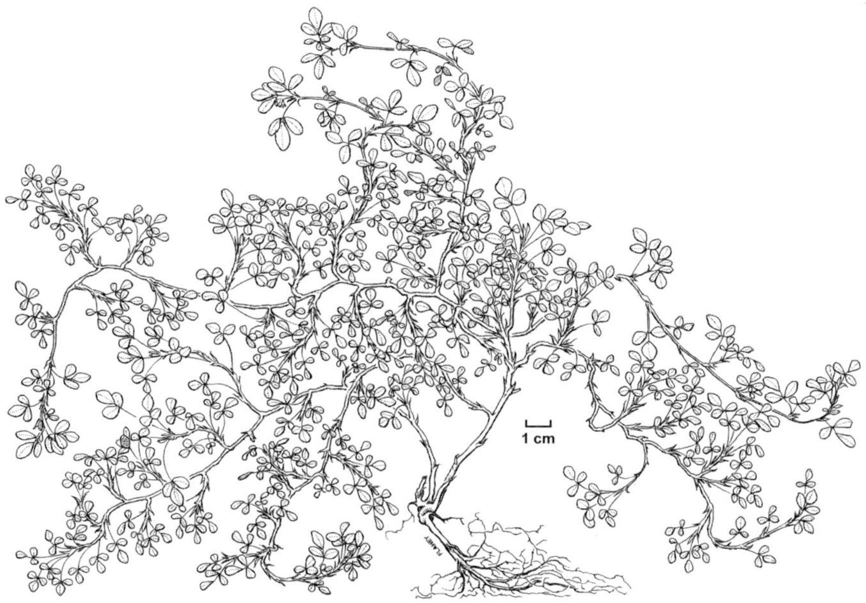

Fig. 20-1. *Medicago brachycarpa.* A greenhouse-grown, vegetative plant. (Based on a plant grown from seed from the United States Department of Agriculture (accession PI 352705, from Turkey); artist: B. Flahey.)

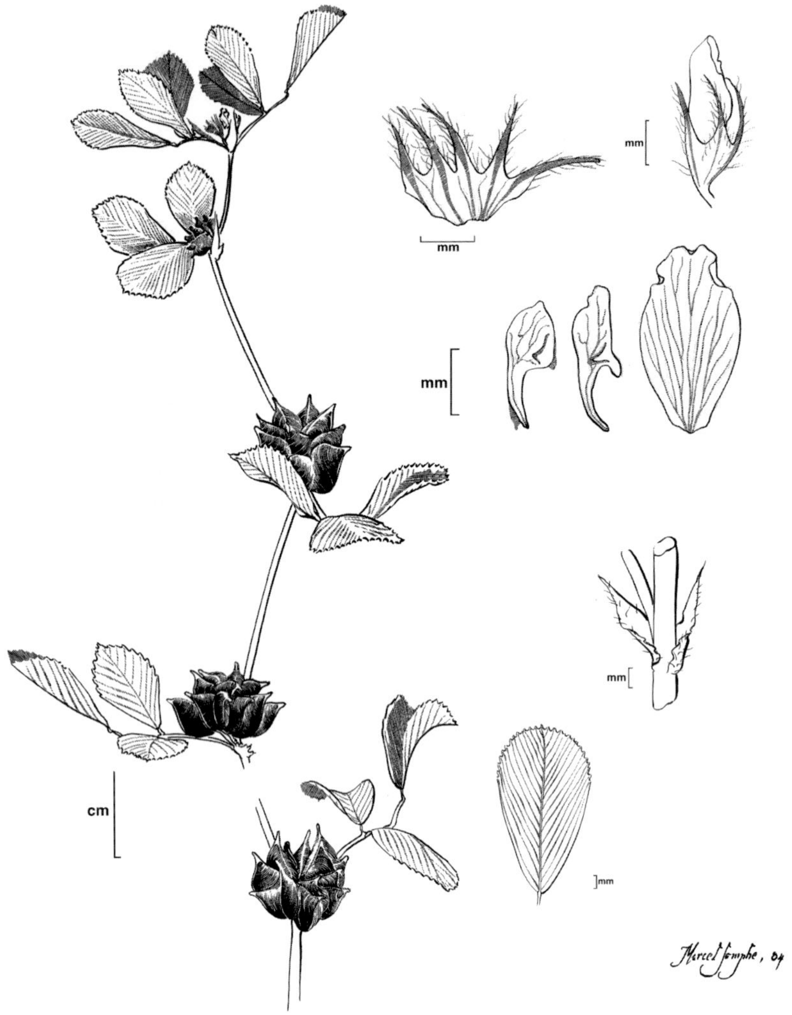

Fig. 20-2. *Medicago brachycarpa.* **Left:** Fruiting branch. **Upper right:** Floral parts (clockwise from upper left: calyx; side view of flower; standard petal; wing petal; keel). **Middle right:** Stem node showing stipules. **Lower right:** Leaflet. (Based on specimen annotated "specimen utilisé par Marcel Jomphe pour illustration scientifique", GH; artist: M. Jomphe.)

T. brachycarpa (Fischer ex M. Bieb.) Moris f. *cilicica* (Boiss.) Širj., Generis *Trigonella* L. Revisio Critica, Spisy Pfir. Fac. Masarykovy Univ. 1: 36 (1928).

M. brachyloba Fischer ex Moris, Mem. Reale Accad. Sci. Torino, ser 2, 36: 190 (1819).

M. glomerata Fischer in lit. in DC., Prodr. 2: 172 (1825).

M. glomerata Fischer ex Moris, Mem. Reale Acad. Sci. Torino ser 2, 36: 190 (1833).

T. glomerata (Fischer) Hort. Paris ex Fisch. & Mey., Bull. Soc. Imp. Nat. Moscou 11: 340 (1838). (*T. glomerata* auct.), nom. illeg.

Pocockia glomerata (Fischer) Boiss., Diagn. Pl. Orient. ser. 1 2(9): 12 (1849).

Telis glomerata (Fischer) O. Kuntze, Rev. 1: 209 (1891).

M. pilosa Fischer ex Moris, Mem. Reale Accad. Sci. Torino, ser 2, 36: 190 (1833).

T. brachycarpa (Fischer ex M. Bieb.) Moris f. *typica* Širj., Generis *Trigonella* L. Revisio Critica, Spisy Pfir. Fac. Masarykovy Univ. 1: 36 (1928), nom. inadmiss.

Vernacular Names

English: Short-fruited fenugreek (a concocted name not in common use; "fenugreek" is based on considering this species as a member of *Trigonella*).

French: No names found.

Geographical Distribution

(N = native, I = introduced, U = uncertain status)

Asia: Armenia (N), Azerbaijan (N), Gruziya (N), Iraq (N), Iran (N), Lebanon (N), Syria (N), Turkey in Asia (N).

Europe: Ukraine (Crimea; U).

Medicago brachycarpa occurs in the Middle East and Transcaucasia.

Description

Annual or reportedly sometimes a biennial herb 3–25(30) cm, prostrate to ascending or erect, branching from base, densely puberulent, the hairs sometimes reddish when dry. Leaflets 5–15 mm, oblong to obovate-cuneate, rounded at tip, denticulate, upper half to one-third of margin serrate. Stipules ovate-lanceolate or ovate-subulate, entire or denticulate. Inflorescence densely capitate with 10–16 flowers. Peduncles 0(10) mm. Flowers 2.5–3 mm, yellow or pale yellow, pedicels short (flowers subsessile). Calyx ca. 3 mm, teeth linear-subulate, shorter or about equal to tube. Calyx about as long as corolla. Pods flat, more or less deflexed (horizontally spreading), half-ovate or ovate-lunate, pubescent, transversely nerved (the veins not coalescing), 4–6 mm long (including beak), 3–5 mm wide, the short straight beak 1–1.25 mm. Seeds 1–2 per pod, ovate, 2–2.5 mm, tuberculate, radicle about three-fourths the seed length to as long as seed.

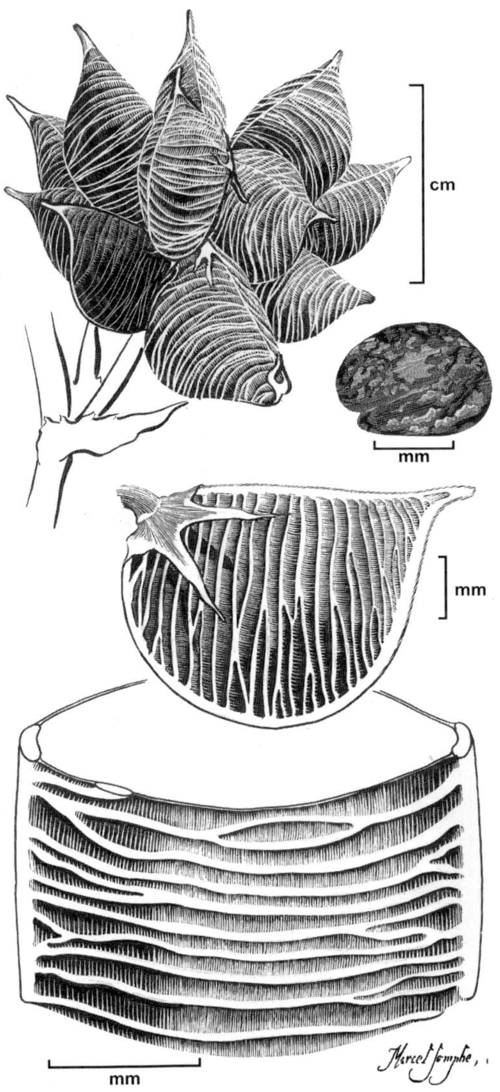

Fig. 20-3. *Medicago brachycarpa.* **Top to bottom:** Cluster of fruits; seed; fruit; portion of fruit. (Based on Herbarium G.C. Joad, specimen collected 3.07.1856 (B. Balansa?), GH; artist: M. Jomphe.)

Identification Hints

This species is most similar to *M. huberi*, differing most evidently from the latter by its shorter, more deflexed fruit. Both of these species could be confused with some of the other flat-fruited species of *Medicago*, perhaps most likely *M. cretacea*, which, however, has fewer flowers in the raceme and a reniform (not a lunate) fruit shape.

Habitat and Environmental Adaptation

Habitat

Medicago brachycarpa occurs in limestone rocks, steppes, among xerophytic vegetation including pine and juniper woods, oak scrub, grassy and gravelly slopes, cultivated and fallow fields, ruderal places, and in a low montane belt from 350 to 2100 m.

Phenology

The species flowers from April to June in Turkey.

Soil

The species occupies calcareous, coarse-textured substrates, as well as a range of cultivated agricultural soils.

Temperature

Based on its habitat, *M. brachycarpa* tolerates early summer, Mid-Eastern conditions.

Water

Medicago brachycarpa is somewhat drought resistant.

Biotic Relationships

Pollination

Small (1988) recorded a pollen to ovule ratio of 627, which is very unusual for annual species of *Medicago* and is indicative of an outcrossing species. However, the extremely small flowers suggest that *M. brachycarpa* is largely self-pollinated and substantially an inbreeder.

Dispersal

The wide pods of this species may be adapted to wind distribution.

Herbivory

No information located.

Nodulation and mycorrhizae

No information located.

Disease resistance

No information located.

Economic History, Cultivars, and Uses

This species has not attracted agricultural interest to date.

Cultivars

No cultivars have been bred.

Forage

Medicago brachycarpa is not sown for forage.

Fodder

This species is not harvested for fodder.

Green manure, rotation crop, ground cover

No information located.

Other uses (salad sprouts, ornamental, plant products, etc.)

No information located.

Chromosome number

2*n* = 16: Tschechow 1932 (as *Trigonella glomerata*); also Tschechow 1936 cited in Bolkhovskikh et al. 1969.

Genetic Relationships and Gene Transferability

Intraspecific

No barriers to hybridization within the species are known, although artificial hybridization of different populations has not been attempted.

Interspecific

No information located.

Exchange by advanced biotechnology

No information located.

Conservation Status

This species is not rare, but occurs in areas of the world where *Medicago* has not been well collected. Therefore, basic evaluation of its present conservation status is desirable.

Germplasm Sources

Little more than a dozen accessions are available from major *Medicago* germplasm collections (see Chapter 3).

Research Problems and Priorities

Medicago brachycarpa is a poorly understood species and its agricultural potential remains to be evaluated.

Representative Collections

Azerbaijan: Grossheim, A. 132, Transcaucasia, Azerbajdzhan, distr. Zangelan, prope p. Pirtshevan, in lapidosis, 29-V-1937, in fruit (E). **Georgia:** Hohenaker, R.F., Georgiae caucasicae prope Helendorf, in collibus aridis, V-VI-1835, in flower & early fruit (G, K, P). Holmberg, O.R. 1715, Atskhur, Prov. Tiflis, Distr. Akhaltsikh, Georgia, 19-VI & 2-VII-1912, in flower & fruit (LD). Holmberg, O.R. 1767, Benara, Prov. Tiflis, Distr. Akhaltsikh, Georgia, alt. 1200 m, 20-VI & 3-VII-1912, in late flower & fruit (LD). **Iran:** Persia: Szovits 234, Persia, Karabagh, in fruit (G, K). **Israel/Syria (Hermon):** Liston, A. & Schmida, A. 497/5, Hermon, 6.5 km north of Najdal Shams, alt. 1900 m, 4-VI-1986, in fruit (HUJ). Schmida, A., S. Hermon, 4 km north-northwest of Majdal Shams, hard Jurassic limestone, alt. 1650 m, 10-VI-1976, in fruit (HUJ). Zohary, D., Mt. Hermon, ascent from Shib'ah, alt. 2360–2400 m, 11-X-1943, in fruit (HUJ). **Syria:** Samuelsson, G. S442, Attik, alt. ca. 1000 m, in rupestribus calcareis machiarum, 27-V-1933, in fruit (BM). Trotsiluy, circa Swediem Syrie, Taurus No. 3, 4, in fruit (K). Wall, E. 226, Ammanus, Attik, alt. 1000 m, 27-V-1933, in flower & fruit (LD). **Turkey:** Alingoglu, N.OR 165, Prov. Ankara, Ankara, Atatürk Orman Çiftligi, near Dolapdere, 3-VII-1962, in fruit (E). Balansa, B. 102, Plaine de Césarie (Cappodoce), 1107 m d'alt., 25-VI-1856, in fruit (G). Balansa, B. 274, in arenosis et cultis derelictis regionis montanae Anatoliae in Phrygia prope Ouchak, 18-VI-1857, in fruit (Herb. Boiss. G, two sheets). Balansa, B. 465, (double collection) flowers — forêts de pin situées au-dessus du village d'Alla-Dagh, à 7 lieues au NO. de Mersina (Cilicia), 16-V-1855; fruits — village de Gulek-Boghas (Cilicie) dans les vignes, 21-VI-1855 (BM, P). Balansa, B. 466, le village de Sédichig, à 2 lieues au No. de Mersina, région chaude, coteaux calcaires, 20-V-1855, in fruit (P). Balls, E.K. & Gourlay W.B. B1136a, Erujik, Cilician Taurus, alt. 3500 feet, cultivated land, 7-VI-1934, in fruit (K). Balls, E.K. 1136A Buriyik, Cilician Taurus, alt. 3500 feet, cultivated land, 7-VI-1934, in fruit (E). Balls, E.K. B1136, Marash, waste ground, alt. 2300 feet, 23-V-1934, in flower & fruit (BM). Boissier, E., Caria in monte Cadmo supra Denisleh, VI-1942, in fruit (Herb. Boiss. G, three sheets in folder). Bourqeau, E., Elmalu, in collibus, 3-VI-1860, in fruit (G, K, P). Coode, M.J.E. & Jones, B.M.G. 2574, Kütahya, 5 miles west of Abide, alt. 800 m, chalk slopes, 17-VI-1965, in flower and fruit (A). Davis 4429, Van, 8 km from Van to Erçek, alt. 2100 m, sloping limestone rocks, 6-VI-1966 in flower & fruit (K). Helbaek 2586, Prov. Konya, Konya to Cumra, Küçük Köy, alt. 980 m, cultivated land, 16-VI-1962, in flower & early fruit (E). Huber-Morath, A. 5082, prov. Denizli (region C2, Davis 1970), district Tavas, Tavas — Denizli, 18 km nach Tavas, alt. 1300 m, 4-VI-1938, in flower & fruit (G). Huber-Morath, A. 5807, prov. Burdur (region C3, Davis 1970), district Tefenni, 7 km south ob Dirmil (Altinyayla), alt. 1500 m, 9-VI-1938, in flower & fruit (G). Huber-Morath, A. 5084, Kleinasien: Carien (region C2, Davis 1970), Kale Tavas — Muğla, Pinuswald 35 km nach Kale Tavas, 5-VI-1938, in fruit (G). Huber-Morath, A. 16549, prov. Denizli (region C2, Davis 1970), district Denizli, Pinus brutia — west foot of Honaz mountain 27 km south Denizli, alt. 1100 m, 1-VI-1962, in fruit (G). Tobey, C. 676, Vilayet, Amasya, Erbaa, Catalan, alt. 700 m, with pines, hot, stony, south slope, 2-VI-1964, in flower (E).

21. *Medicago cancellata* M. Bieb.

M. cancellata M. Bieb., Fl. Taur.-Caucas. 2: 226 (1808), (basionym).

The epithet *cancellata* is Latin for latticed, an allusion to the lattice-like raised venation of the fruit coil faces.

Subordinate Taxa

None recognized here.

Type Specimen

In regard to the type, Vassilczenko (1949) noted that it was "described from the southeast (from the region of Krasnoyarsk, former Sarepta)." Vassilczenko (2002*c*) provided the following information: "Type: Lower Volga ('in deserto Caucasico-caspio rarius necnon in campestribus wolgicis circa coloniani Sareptanam')."

Synonymy

Homotypic

No names found.

Heterotypic

M. ciscaucasica B. Fedtsch. in Not. Syst. Herb. Inst. Bot. Acad. Sci. URSS, 8: 176 (1940). (Vassilczenko (1949)

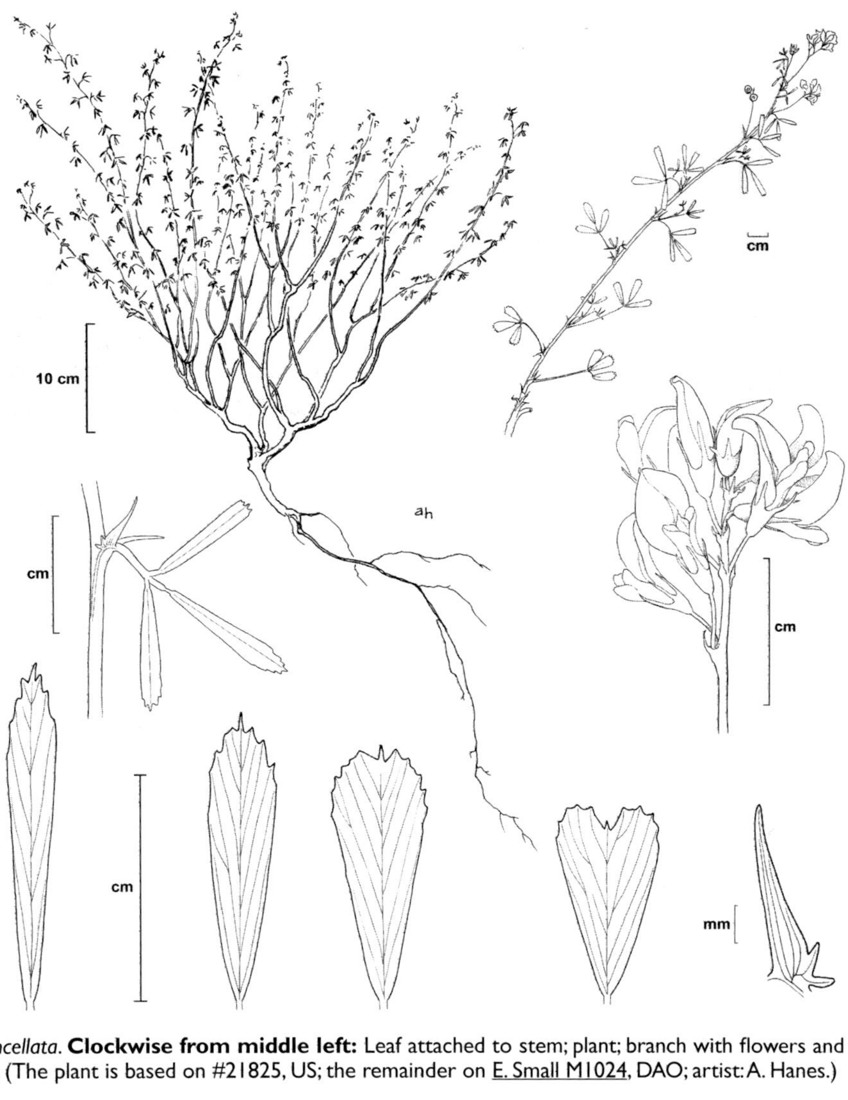

Fig. 21-1. *Medicago cancellata.* **Clockwise from middle left:** Leaf attached to stem; plant; branch with flowers and young fruits; group of flowers; stipule; leaflets. (The plant is based on #21825, US; the remainder on E. Small M1024, DAO; artist: A. Hanes.)

referred to "var. *ciscausica* (B. Fedtsch.)" from the Stavropol region as having taller stems (up to 40 cm) and generally larger measurements of its constituent parts. However, there is no indication of taxonomic delimitation. Vassilczenko also noted a plant collected by Lessing in 1933 near Orsk on the Ural river with very small pods (2–2.5 mm in diameter) and noted that Lessing's label bore the note "*Medicago prostrata* var. *tatarica* Less., Guberlinsk. Leg. Lessing, 1833)". Vassilczenko further noted "It should probably be regarded as a distinctive variety of *M. cancellata* M.B. (*M. c.* var. *tatarica* (Less.) Vass." These varietal names have no status.)

Not

M. cancellata Ten. (= *M. tenoreana*).

M. cancellata Kryl. (= *M. medicaginoides*).

Vernacular Names

English: Checkered alfalfa (Bolton 1962).

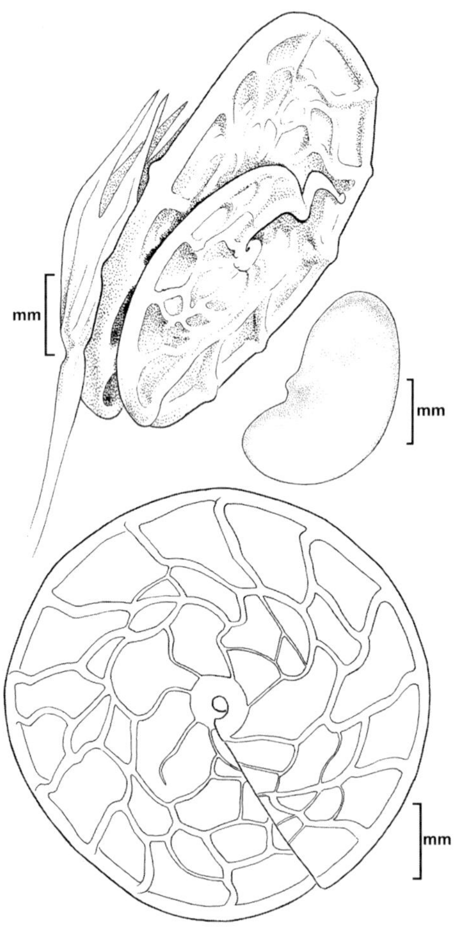

Fig. 21-2. *Medicago cancellata.* **Top:** Fruit. **Center right:** Seed. **Bottom:** Fruit coil face showing venation. (Based on E. Small M150, DAO; artist: A. Hanes.)

Fig. 21-3. Type of *Medicago cancellata* at LE.

French: No names found.

Geographical Distribution

(N = native, I = introduced, U = uncertain status)

Asia: Kazakhstan (N), Russia in Asia (North Caucasus, N).

Europe: Russia in Europe (west of Caspian Sea: Don Basin, Lower Volga; N).

This species is native to the Caspian region and the North Caucasus.

Description

Perennial forb producing numerous, ascending to erect stems 15–25(40) cm, from a stout rhizome. Twigs and leaves pubescent with simple hairs. Leaflets 5–12 mm long, obovate to narrowly wedge-shaped (generally about twice as long as broad), dentate at apex, glabrous or slightly appressed-pubescent (upper surface glabrous or essentially so, lower surface with some simple trichomes). Stipules subulate-linear, entire or occasionally the lower ones with one or a few teeth at their base. Inflorescence capitate, of 5–12 flowers, peduncle slender, equalling or exceeding the subtending leaf. Flowers 5.5–7 mm long, yellow, calyx shorter than half length of flower. Calyx tube shorter to slightly longer than teeth. Stipules subulate-linear, entire or occasionally the lower ones one- or two-toothed at their base. Young fruit protrudes sideways from calyx. Fruit glabrous at maturity, with 1.5–3 loosely adpressed coils 3–4.5 mm in diameter, without (or essentially without) a central (axial) opening, the outer coil margin strongly thickened (lacking spines), coil face with prominent, thick, spreading radial veins that anastomose on the outer part of the coil and then reach the thickened dorsal suture. Seeds 2–6 per pod (1–2 per coil), 2.5–3 mm long, light brownish-yellow or rusty-brown, smooth, reniform, not evidently separated by a transverse spongy fruit septum, radicle equal to half seed length or slightly longer.

Identification Hints

This hexaploid ($2n = 48$) is very similar in appearance to one of its putative ancestors, *M. rupestris* ($2n = 16$), in having narrow leaflets and fruit venation with strong relief; it is distinguishable by its larger fruits.

Habitat and Environmental Adaptation

Habitat

Medicago cancellata occurs in infertile, dry soils of steppe and semi-desert areas and on rocky slopes, notably with open exposures (south- or east-facing). In steppes it is often in mixed grass associations.

Phenology

Flowering occurs in June and July in its native area.

Soil

The species has been collected growing in sand, particularly in sandstones and weathered soils. Rogers et al. (2005) ranked *M. cancellata* as moderately tolerant of soil salinity.

Temperature

Vassilczenko (1949) described *M. cancellata* as having considerable resistance to cold and frost.

Water

Medicago cancellata occupies semi-deserts and is obviously quite drought resistant. Its deeply penetrating roots serve to obtain water only available deep in the soil.

Biotic Relationships

Pollination

Small (1988) noted that the relatively large flowers and pollen to ovule ratio (434) are consistent with *M. cancellata* being at least partly cross-pollinated, like other perennial species of *Medicago*.

Dispersal

The indehiscent fruits lack spines and are therefore not adapted to dispersal by attaching to fur, as in some other perennial *Medicago* species. On slopes, the fruits presumably are dispersed by rolling downhill. The fruits are probably consumed by animals with a few of the seeds surviving after passing through their digestive systems.

Herbivory

No information located.

Nodulation and mycorrhizae

USDA (2009) reported positive root nodulation for *M. cancellata*.

Disease resistance

Borges et al. (1976) found that *M. cancellata* is a potentially important source of resistance to *Stemphylium botryosum* leafspot of alfalfa.

Economic history, cultivars, and uses

This species has attracted only limited agricultural interest to date.

Cultivars

No cultivars have been bred.

Forage

Grossheim (1945) suggested that *M. cancellata* has promise as a drought-resistant forage plant. It appears to be course, with limited productivity compared with alfalfa cultivars. Lavrent'-ev and Chernenko (1975) compared the productivity of an alfalfa cultivar with that of 10 wild species, including *M. cancellata*, in Rostov province, Russia and, while the first cut harvest was notably larger in the alfalfa cultivar, subsequent stem production was highest in *M. cancellata*.

Fodder

This species is not considered to be a source of fodder at present.

Green manure, rotation crop, ground cover

No information located.

Other uses (salad sprouts, ornamental, plant products, etc.)

Medicago cancellata has occasionally been grown as a rock garden ornamental.

Chromosome Number

$2n = 6x = 48$: Lesins 1959; Lesins and Lesins 1979; Yen and Murphy 1979; Magulaev 1980, 1982.

It has been proposed that *M. cancellata* arose by the combination of two genomes of *M. sativa* and four genomes of *M. rupestris*. As noted by McCoy and Bingham (1988), cytogenetic studies are necessary to support this hypothesis.

Genetic Relationships and Gene Transferability

Intraspecific

No barriers to hybridization within the species are known, although artificial hybridization of different populations has not been attempted.

Interspecific

Lesins (1961*b*) was the first to generate hybrids between *M. sativa* and *M. cancellata*. He produced several pentaploid ($2n = 40$) hybrids by crossing alfalfa ($2n = 32$) and *M. cancellata* ($2n = 48$), but produced just one hybrid by crossing hexaploid alfalfa and *M. cancellata*. Yen and Murphy (1979) crossed *M. cancellata* with two hexaploid amphiploids ($2n = 6x = 48$) of *M. sativa* and found that while hybrids (both F_1 and F_2) could be established, there was very extensive sterility. Lesins (1970) reported a tri-species hybrid of *M. cancellata*, *M. saxatilis*, and *M. sativa*. Lesins (1970) found that *M. saxatilis* could be easily crossed with *M. cancellata* and that the F_1 was quite fertile.

Exchange by advanced biotechnology

Using embryo rescue techniques, McCoy and Smith (1986) obtained tri-species hybrids between *M. cancellata* ($2n = 48$) and *M. sativa* × *M. papillosa* F_1 hybrids ($2n = 24$).

Conservation Status

This species is poorly understood and, in view of its potential for alfalfa improvement, should be evaluated for its need for special conservation measures.

Germplasm Sources

Slightly more than a dozen accessions are available in major germplasm institutions (Chapter 3).

Research Problems and Priorities

Medicago cancellata has notable winter hardiness and adaptation to poor soils, and may be useful for transferring these characteristics to alfalfa. There is an urgent need to enlarge the limited number of available germplasm collections.

Representative Collections

RUSSIA: Becker, A., Flora Rossiae merid-orient. Gouvern. Saratow, Sarepta, VII-1896, in flower & very early fruit (K). Becker, A., Rossica Sarepta, 1894, in flower & fruit (K). Becker, ca. Sarepta, ad fl. river Wolgam, in campestribus, 1853, in flower & early fruit (BM). Becker, Sarepta, 3-VI, in flower (K). Fisher, South Wolga, 1828, in flower & fruit (K).

22. *Medicago carica* (Huber-Mor.) E. Small

M. carica (Hub.-Mor.) E. Small in Small, Lassen & Brookes, Willdenowia 16(2): 433 (1987).

The specific epithet *carica* denotes Karia, an ancient division of Turkey (southwest Anatolia) where this species is confined.

Subordinate Taxa

None recognized here.

Type Specimen

Huber-Morath, A. 16553, Prov. Muğla, District Muğla, Muğla-Kale, *Pinus brutia* - Wald, Kalkgeröll, 16 km nordöstlich ob Muğla, 1250 m, 28-V-1962, in fruit (holotype G, isotype and topotype also at E (see below)).

Synonymy

Homotypic

Trigonella carica Huber-Mor., Bauhinia 2(3): 299 (1965), (basionym).

Heterotypic

Trigonella alindana Huber-Mor., in herb. (name not published; see information in Representative Collections).

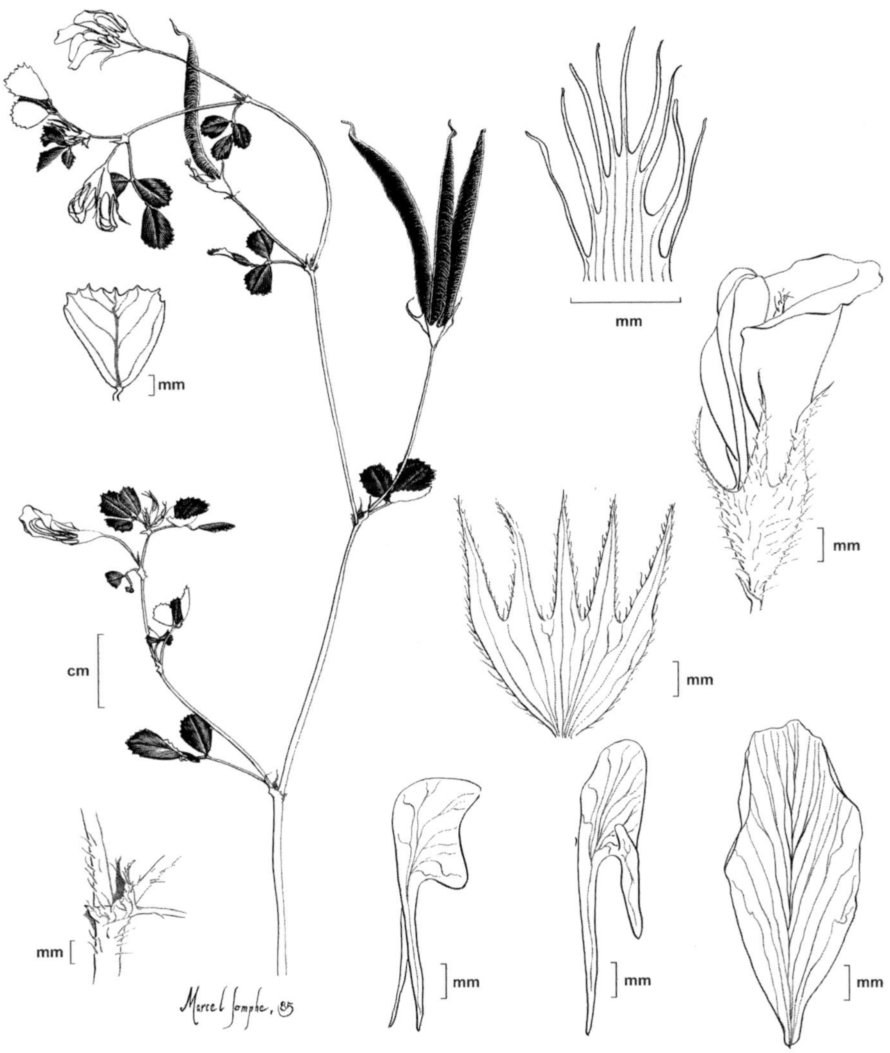

Fig. 22-1. *Medicago carica.* **Bottom left:** Stem node showing stipules; **Left:** Branch with fruits and flowers; leaflet. **Right:** Floral parts (clockwise from top: spread out androecial column (free stamen not shown); side view of flower; standard petal; wing petal; keel; calyx). (Based on Davis 41037, K; artist: M. Jomphe.)

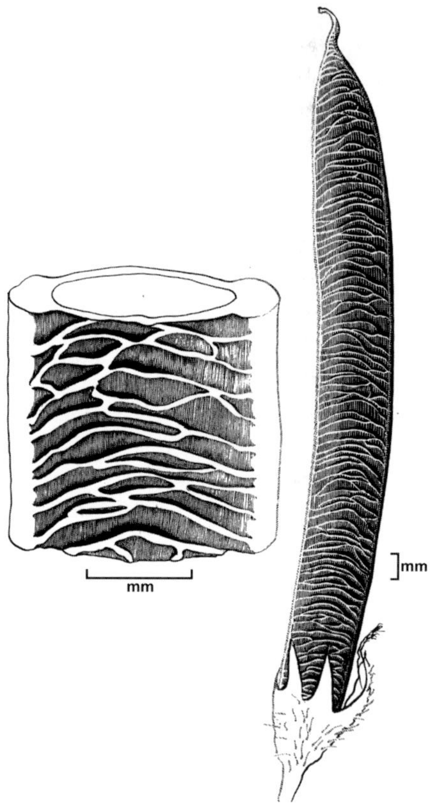

Fig. 22-2. *Medicago carica.* **Left:** Portion of fruit. **Right:** Fruit. (Based on Dudley D.35081, K; artist: M. Jomphe.)

Vernacular Names

ENGLISH: No names found.

FRENCH: No names found.

Geographical Distribution

(N = native, I = introduced, U = uncertain status)

ASIA: Turkey (N).

EUROPE: East Aegean Islands (Greece; N).

This species is a local endemic of southwestern Anatolia and the nearby East Aegean Islands. In Greece, it grows only in the most southeastern part, on the islands of Kastellorizo and Strongili (Greuter 1979).

Description

Annual herb stems 10–20 cm long, ascending to suberect, branched from base, twigs slim, pubescent with adpressed simple hairs. Leaflets 5–9 mm long, obovate-cuneate or triangular, generally about two-thirds as broad as they are long, upper third of margin serrate. Stipules lanceolate, with or without basal teeth. Inflorescence capitate, of 2–3 flowers, peduncles 1–3 cm, about half as long to as long as subtending petiole. Flowers 8–10 mm long, yellow. Calyx campanulate, 5–6 mm long, with simple hairs, lanceolate-linear teeth as long as tube. Corolla twice

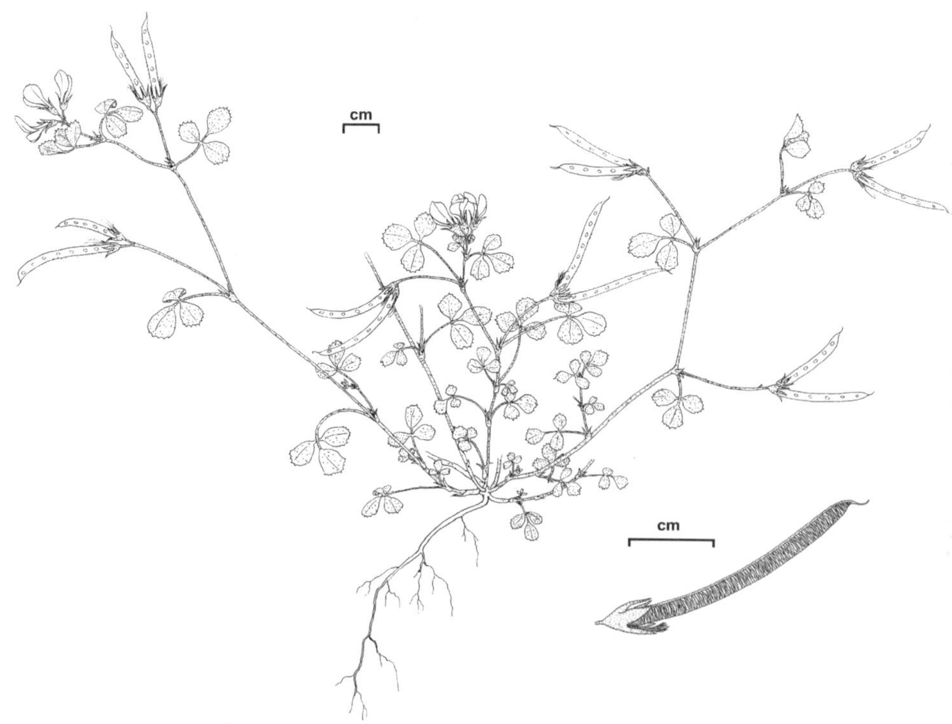

Fig. 22-3. *Trigonella carica* from Huber-Morath (1965). Reproduced with the permission of the editor of Bauhinia.

Fig. 22-4. A topotype at E of *Trigonella carica* Huber-Mor. (collected at the type locality 3 days prior to the collection of the holotype; on the herbarium sheet, Huber-Morath tentatively labelled the specimen as "*Trigonella alindana* Huber-Mor.", but published his new species as *T. carica*).

length of calyx. Fruit linear, erect, compressed, 20–35 mm long, 2–3 mm wide, with adpressed hairs, with more than 100 veins running from one suture to the other, the veins anastomosing. Seeds 8–12 per pod, about 1.5 mm long, surface smooth, radicle three-fourths length of seed or almost as long as seed.

Identification hints

Medicago carica resembles *M. pamphylica* and is most easily distinguished by the latter species having one of its two sutures undulating.

Habitat and Environmental Adaptation

Habitat

Medicago carica is found in rocky limestone slopes and scrub at altitudes of 0–1250 m. In Greece, it has been recorded on calcareous ground, karstic rocks, stony and rocky slopes, semi-natural garrique, amidst low evergreen shrubs, and in ruderal places (Greuter 1979; Thanopoulos 2007).

Phenology

The species flowers from March to May.

Soil

Medicago carica typically occupies limestone soils.

Temperature

The species germinates in the cool late Mediterranean winter and completes its life cycle in the increasingly hot, drying summer.

Water

No information located.

Biotic Relationships

Pollination

Small (1988) noted that the flowers are small and the pollen to ovule ratio is extremely low (92), observations that are consistent with *M. carica* being largely self-pollinated and substantially an inbreeder.

Dispersal

No information found. No adaptations specialized for dispersal are present.

Herbivory

No information located

Nodulation and mycorrhizae

No information located.

Disease resistance

No information located.

Economic History, Cultivars, and Uses

This species has not attracted agricultural attention to date.

Cultivars

No cultivars have been bred.

Forage

Medicago carica is not sown for forage.

Fodder

This species is not considered to have fodder usage at present.

Green manure, rotation crop, ground cover

No information located.

Other uses (salad sprouts, ornamental, plant products, etc.)

No information located.

Chromosome Number

The chromosome number has not been reported.

Genetic Relationships and Gene Transferability

Intraspecific

No barriers to hybridization within the species are known, although artificial hybridization of different populations has not been attempted.

Interspecific

No information located.

Exchange by advanced biotechnology

No information located.

Conservation Status

Medicago carica is a poorly studied, poorly collected, rare, narrowly endemic species and is therefore in need of conservation measures. *Medicago carica* is listed in the Red Data Book of Rare and Threatened Plants of Greece (Phitos et al. 2009).

Germplasm Sources

No source of seeds has been located.

Research Problems and Priorities

Medicago carica is very poorly understood. There is an urgent need to clarify its basic botanical characteristics and agricultural potential and to collect seeds for deposition in germplasm institutions.

Representative Collections

Turkey: Davis 25447 & Polunin, O., Prov. Muğla, dist. Fethiye, Kalkan, alt. 10 m, rocky limestone slopes, facing south, 29-III-1956, in flower (BM, E). Davis 41037, Muğla, d. Marmaris, above Inisdibi, alt. 300 m., maquis on limestone, 13-IV-1965, in flower & fruit (E). Dudley, D.35081, prov. Muğla, Muğla to Kale, 16 km from Muğla, alt. 1250 m, *Phlomis* - *Quercus* macchie, in scree, alt. 1250 m., 25-V-1962, in fruit (K, E). (Labelled "*Trigonella alindana* sp. nov. Huber," a name apparently Huber-Morath had in mind but did not validate; this specimen was annotated by Huber-Morath in 1962 as *Trigonella carica* Hub-Mor., and bears an "Isotypus" label, although it is not an isotype.)

23. *Medicago carstiensis* Jacq.

M. carstiensis Jacq., Collecteana 1: 86 ("1786" published 1787), (basionym). Associated figure: Jacq. Icon. Pl. Rar. Vol. 1, tab. 156 (1787), (illustration reproduced below).

The epithet "carstiensis", Latin for "of the Karst region", reflects the discovery of this species in mountains of the Karst region of Carniola (northwest Yugoslavia, now the Republic of Slovenia).

Note: The citations "*Medicago carstiensis* Wulfen in Jacq." and *Medicago carstiensis* Wulfen" are widely employed. Franz Xavier Wulfen (1728–1805) and Nikolas Joseph Jacquin (1727–1817) were both residents of Austria, and Wulfen is considered to have contributed to the published treatments in Jacquin's publications.

Subordinate Taxa

None recognized here.

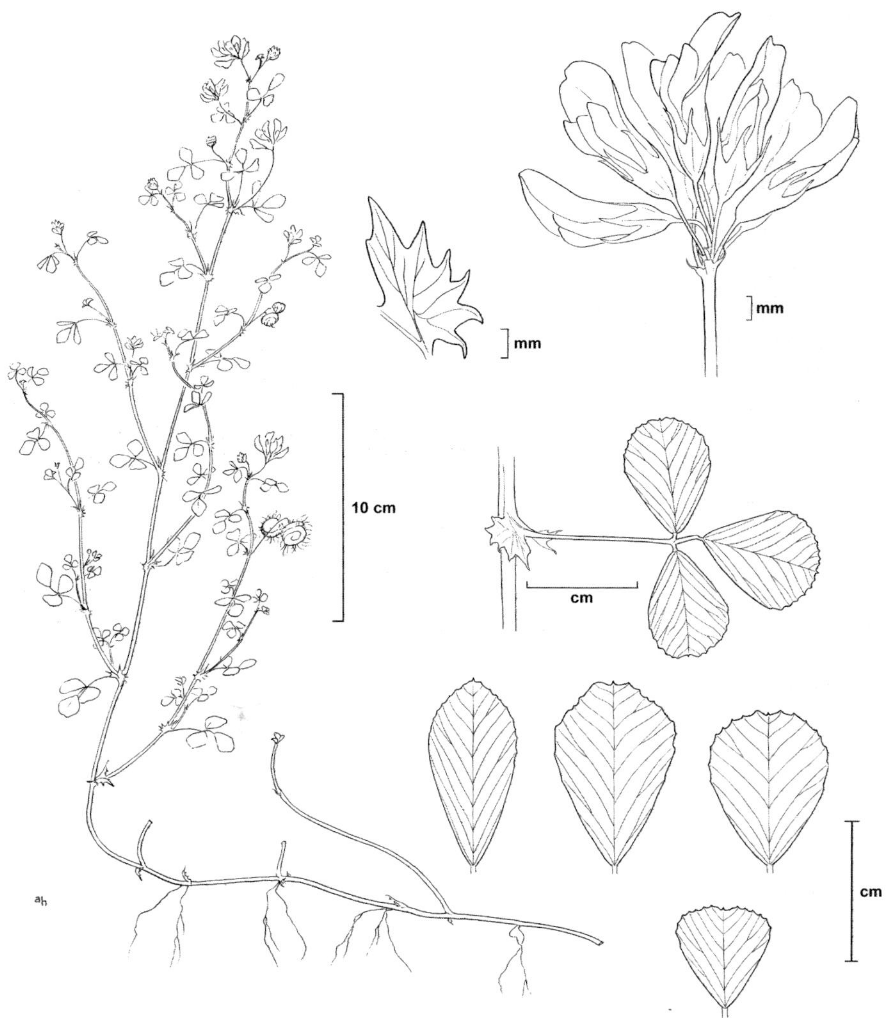

Fig. 23-1. *Medicago carstiensis.* **Clockwise from left:** Plant; stipule; group of flowers; leaf attached to stem; leaflets. (Based on Fiori et al. 297, Flor. Ital. Exsic., in many herbaria; artist: A. Hanes.)

Type Specimen

Jacquin's types are notoriously poor, scattered, and difficult to determine (see for example D'Arcy 1970). F.X. von Wulfen's herbarium is mainly at W. In April 2007, E. Vitek, Director of the Department of Botany at W, informed me that they had searched for a possible type specimen without success, but were in possession of the original drawing published for the new species (Fig. 23-3). The original drawing may be considered as the type, pending location of a more suitable specimen.

Synonymy

Homotypic

No names found.

Heterotypic

M. carstiensis subsp. *belasicae* Kožuharov, Bulgarian Acad. Sci. Reports (Izv. Bot. Inst. (Sofija)) 15: 150 (1965) (type indicated in this publication; said to have silky hairs on the undersides of the leaves and larger flowers and pods).

M. umbellata hort. ex DC. Prod. II: 178 (1825) (name merely mentioned incidentally).

Vernacular Names

English: Creeping-rooted medic (medic is also spelled medick). This name could be confused with creeping-rooted forms of alfalfa (*M. sativa*).

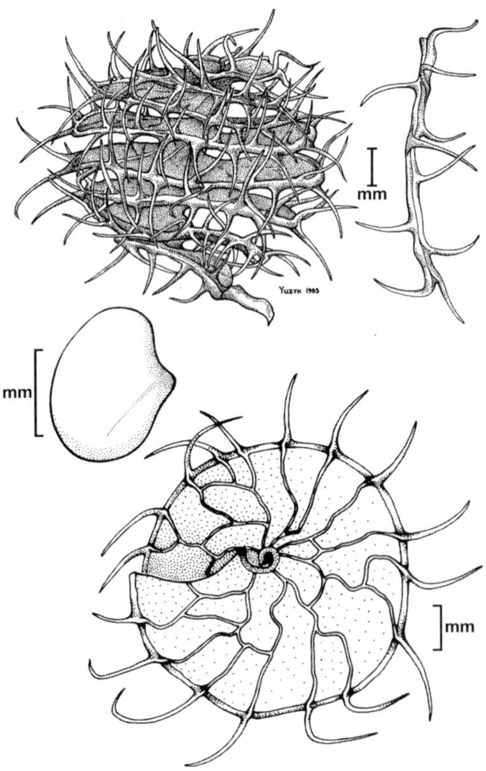

Fig. 23-2. *Medicago carstiensis.* **Clockwise from top left:** Fruit; edge of coil; detail of venation on fruit coil face; seed. (Based on E. Small F14, DAO; artist: L. Yuzyk.)

Fig. 23-3. Illustration of *Medicago carstiensis* from N.J. Jacquin, 1787, Icones plantarum rariorum, plate 156, various publishers (e.g., B. White et filium, London, England).

Fig. 23-4. *Medicago carstiensis.* From Hallier (1885, vol. 23, plate 2353).

French: Luzerne du Karst.

Geographical Distribution

(N = native, I = introduced, U = uncertain status)

Asia: Japan (I, according to "Alien species recognized to be established in Japan or found in the Japanese wild as of Oct. 27 2004" http://www.env.go.jp/en/nature/as/041110.pdf; current June 2010).

Europe: Albania (N), Austria (N), Bulgaria (N), Greece (N), Italy (N), Yugoslavia (N).

This is an endemic species of the coastal regions of the eastern part of the Adriatic Sea, distributed in northeast Italy, southeast Austria, the former Yugoslavia, Albania, and Bulgaria. Thanopoulos (2007) noted its occurrence in only one locality in Greece (Mount Pageon), apparently the southernmost location in Europe. The species is uncommon and not abundant where it is found.

Description

Perennial herb, stems 30–60 cm long, arising from a rhizome, stem branching, ascending to erect, nearly or quite glabrous, hairs simple if present. Leaflets 11–22 mm long, elliptical to ovate, generally about 1.5 times as long as broad, upper two-thirds of margin serrate. Stipules ovate-lanceolate or lanceolate, entire or dentate at base. Inflorescence of (1–)5–12(–20) flowers, peduncles usually longer than subtending petiole. Flowers 6–8 mm, yellow (standard petal with reddish-brown veins acting as nectar guides). Calyx teeth shorter than tube. Calyx about half as long as corolla. Fruit glabrous, blackish at maturity, shortly cylindrical, 6–8 mm wide, with 5–8 coils, end coils flattish with slender, flexible, nearly straight dorsal suture spines half as long to as long as the coil diameter, coil face with a few, strong, scarcely anastomosing veins that extend into the spines, dorsal suture in a shallow groove between the elevated margins of the coil edge. Seeds 4–5 per coil, 1.7–2.5 mm long, smooth, light yellow to brown, not evidently separated by transverse spongy fruit septa, radicle about two-thirds length of the seed.

Identification hints

This is the only *Medicago* species exclusively with rhizomes (which are found sporadically in a few other species, especially *M. sativa*).

Habitat and environmental adaptation

Habitat

Medicago carstiensis occurs in rocky and wooded areas, at altitudes of 1000–1500 m, often in association with shrubs. Unlike most other *Medicago* species it grows well in shade.

Phenology

This species flowers from June to August in its native area. The plants require a cold period for flowering to occur (Lesins and Lesins 1979). Vitkus et al. (1977) studied plant development in cultivation for 3 years, observing a progressive increase in height.

Soil

This species typically occupies calcareous soil. Rogers et al. (2005) ranked *M. carstiensis* as moderately tolerant of soil salinity.

Temperature

Lesins and Lesins (1979) observed that *M. carstiensis* underwent some winter killing on the Canadian prairies when snowfall was light, but that it generally survived cold winters well. Lubenets and Vitkus (1978) noted that the species was tolerant of winters in Lithuania.

Water

No information located.

Biotic Relationships

Pollination

Lesins and Lesins (1979) observed that the species set seed when pollinators were in short supply, indicating that considerable selfing can take place.

Dispersal

No information located. The fruit spines are weak, but may nevertheless serve to anchor the fruit to passing animals.

Herbivory

No information located.

Nodulation and mycorrhizae

USDA (2009) reported positive root nodulation for *M. carstiensis*.

Disease resistance

The species is moderately resistant to *Pseudopeziza medicaginis* f. sp. *sativa* (Schmiedeknecht and Lesins 1968).

Economic History, Cultivars, and Uses

This species has not attracted agricultural interest to date.

Cultivars

No cultivars have been bred.

Forage

Medicago carstiensis is not sown for forage.

Fodder

Lesins and Lesins (1979) reported that "forage" production of *M. carstiensis* was about 60% that of alfalfa.

Green manure, rotation crop, ground cover

Medicago carstiensis spreads by rhizomes, which binds the soil and therefore offers the potential for use as a soil stabilizer.

Other uses (salad sprouts, ornamental, plant products, etc.)

No information located.

Chromosome Number

$2n = 16$: Fryer 1930; Clement 1962; Lesins and Lesins 1979.

Genetic Relationships and Gene Transferability

Intraspecific

No barriers to hybridization within the species are known, although artificial hybridization of different populations has not been attempted.

Interspecific

Lesins and Lesins (1979) reported that they were unsuccessful in attempting to hybridize *M. carstiensis* with either diploid ($2n = 16$) or tetraploid ($2n = 32$) *M. sativa*.

Exchange by advanced biotechnology

No information located.

Conservation Status

The extent to which *M. carstiensis* requires conservation measures is not clear. The species is considered to be a threatened plant in Bulgaria (Peev et al. "Biodiversity of Vascular Plants in Bulgaria", http://www.worldwildlife.org/bsp/publications/europe/bulgaria/bulgaria3.html; current June 2010). In Bulgaria, *M. carstiensis* has been given the status of "Protected" (Republic of Bulgaria Biological Diversity Act, 2002). *Medicago carstiensis* is listed in the Red Data Book of Rare and Threatened Plants of Greece (Phitos et al. 2009). Given the species' limited distribution, unusual nature, and limited representation in gene banks, it seems desirable that steps be taken to promote its conservation in nature.

Germplasm Sources

Only a dozen or so accessions are available from major germplasm institutions (Chapter 3).

Research Problems and Priorities

The creeping rootstock of *M. carstiensis* is very unusual in *Medicago*, although it is present in some strains of *M. sativa*. The trait may be useful for pasturage (since it confers resistance to grazing) and for erosion control.

The phyletic relationships of *M. carstiensis* and *M. orbicularis* require study. Urban (1873) placed both species in section *Orbiculares* of the genus, although they differ considerably. It has been speculated that *M. carstiensis* is a relic species that is ancestral to the much more widespread *M. orbicularis* (Bennett et al. 2006).

Given the very limited number of germplasm collections currently available, there is a need to acquire additional material.

Representative Collections

Albania: Grande, Albania, nei querceti di Rabiha, 29-VIII-1918 (BM). **Austria:** De Tommasini, M., Karst près d'Orlic aux environs de Trieste (Illyrie), bois des montagnes calcaires, 25-V-1860, en fleur en mai en fruits en juin, F. Schultz Herbarium Normale Cent. 7. No. 636 (P). Jabornegg, Carthinia, in fruticetis ad St. Primus et Wallendorf prope Klagenfurt, alt. 500 m, in flower & early fruit; (on same sheet:) Marchesetti, in agro Tergestino, in dumetis valleculae prope Oleg, solo calcareo, alt. 350 m, in flower & fruit (BM, K). Rechinger, K., Carinthia (Kärnten), Jerolitsch bei Krumpendorf a. Würthersee, 12-VII-1939, in flower & fruit (K). Satintz, south of Klagenfurt, south Carinthia, 24-VII-1863, in flower & fruit (K). **Bulgaria:** Kozuharov, S., in silvaticis prope pag. Kolarovo, distr. Petric, Bulgaria australis, 5-V-1961, in fruit (B) (one fruit with seven coils!). Kuzmanov, B. 80792, Belasica Mt., above village Samuilova, distr. Petric, grassy open habitats in a mixed forest (*Fagus sylvatica, Castanea sativa*), 10.VI-1980, in flower (B). **Italy:** Daemen, Lombardie, pasturo, VIII-1853, in fruit (BM). Marchesetti, C., Ager Tergestinus: Orleg, in graminosis valleculae, alt. 550 m, solo calcareo, Flora Italica Exsiccata Series III No. 2693 (BM, P). Porta, Venetia, ditio Petavina, in lucis Euga neorum collium, alt. 1–200 feet, sol. vulcan, 10-VI-1871, in flower & fruit (K). Rigo, Italia septentr., provintia Veronae, in memorib, montanis Baldi montis, sol. calcareo, alt. 1000–12 000 m, VII-1882, in flower & fruit (K). Rigo, Venetia, in nemoribus Baldi montis, alt. 400–1000 m, sol. calcar., VI-VII-1886, in flower & fruit (K). **Yugoslavia:** Justin, Carniola, in dumetis ad pagum Vreme in valle Reka, solo calcareo, alt. 360 m, mense junio, in flower & fruit, Flora Exsiccata Carniolica No. 490 (BM). Krenol, E. & F., Kosovol Metohija, west von Pec, Rugova-Schlucht, wegang zur lumbardska planina, alt. ca. 1550 m, 23-VII-1975 (BM). Krenol, F., Makedonija, Stogova Pl., ob von Koçista, alt. 1500–2000 m, 12-VII-1976, in flower & fruit (BM).

24. *Medicago ciliaris* (L.) All.

M. ciliaris (L.) All., Fl. Pedem. 1: 315 (1785).

The epithet *ciliaris*, constructed from the Latin *cilium*, eyelid, has been applied to mean fringed (e.g., *Habenaria ciliaris* (L.) R. Br. is the yellow fringed orchis). With respect to *M. ciliaris*, this is a reference to the ciliate (not really fringed) appearance of the gland-tipped hairs, especially on the fruit. The epithet *ciliata* would have been a preferable choice.

Subordinate Taxa

None recognized here. *Medicago ciliaris* is very closely related genetically to *M. intertexta* (e.g., Schlarbaum et al. 1984; Rose et al. 1988; Abdelkefi et al. 2001) and is often considered to be a subspecies or variety of *M. intertexta*.

Type Specimen

Heyn (1963) stated that the only possible choice for a lectotype is the Linnaean herbarium (LINN) specimen 933-24, designated as "*arabica*". However, The Linnaean Plant Name Typification Project (http://www.nhm.ac.uk/research-curation/projects/linnaean-typification/; current June 2010; cf. Jarvis 2007) provides the following information: "Linnaean Name: *Medicago polymorpha* Linnaeus var. *ciliaris* Linnaeus. Reference: *Species Plantarum* 2: 780 (1753). Type: Neotype : France. Aude: Craboules, prés salés et bord des chemins, Jun 1889, Abbé Pons s.n. (Magnier Fl. Select. Exsicc. 2353) (LD). Designated by: Lassen in Turland & Jarvis (ed.), Taxon 46: 475 (1997)... Heyn (in Bull. Res. Council Israel, Sect. D, Bot.

Fig. 24-1. *Medicago ciliaris*. Plant with mature fruits. (Based on J. Gillett 10638B, DAO; artist: B. Flahey.)

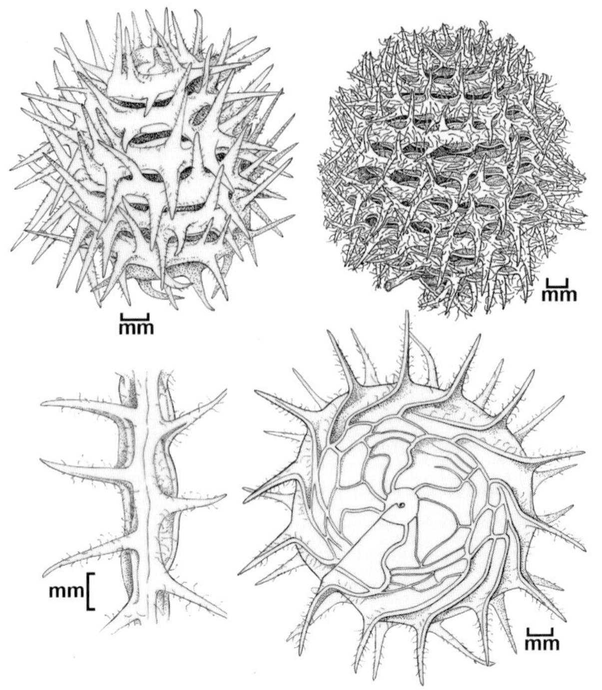

Fig. 24-2. *Medicago ciliaris.* **Top:** Fruits (lightly pubescent at left, heavily pubescent at right). **Bottom left:** Edge of coil. **Bottom right:** Fruit coil face showing venation. (Fruit at upper right based on E. Small M1037; artist: B. Flahey; remainder based on E. Small M583, DAO; artist: A. Hanes.)

7: 167. 1959) discussed the name but was unable to locate a suitable type. Lassen subsequently designated a neotype."

Synonymy

Homotypic

M. polymorpha L. var. *ciliaris* L., Sp. Pl. 780 (1753) (basionym).

Medica ciliaris Medik., Vorles. Churpfälz Phys.-Öcon. Ges. 2: 386 (1787).

M. ciliaris (L.) Krock., Flora Silesiaca 2(2): 244 (1790).

Medica ciliaris (L.) Bubani, Fl. Pyr. 2: 479 (1899).

M. intertexta L. var. *ciliaris* (L.) Heyn, Scripta Hierosolymitana 12 (1963).

M. intertexta (L.) Mill. subsp. *ciliaris* (L.) Ponert in Feddes Repert. 83(9-10): 640 (1972, publ. 1973).

M. intertexta All. subsp. *ciliaris* (L.) O. Bolòs & Vigo in But. Inst. Catalana Hist. Nat., 38 Bot., 1: 69 (1974), comb. superfl.

Heterotypic

The International Plant Names Index (http://www.ipni.org/index.html; current June 2010) reported that *M. intertexta* Willd., Sp. Pl. ed. 4 (Willdenow) 3(2): 1411 (1802) = *M. ciliaris*.

Not

M. ciliaris Luce, Topogr. Nachr. Oesel 253 (1823) (= *M. lupulina*).

M. ciliaris Balb. ex DC., Fl. Franç. (ed. 3, Suppl.) 548 (1815) (= "*M. tuberculata*" (presumably *M. turbinata*)).

M. ciliaris Savi, Due Cent. Fl. Etrusc. 148 (1804); (ex Ser. in DC., Prodr. 2: 176 (1825), pro. syn.). (= *M. polymorpha*).

M. intertexta (L.) Mill. var. *decandollei* (Tin.) Urb., Verh. bot. Ver. Brandenb. 15: 62 (1873) (= *M. muricoleptis*).

Vernacular Names

English: Ciliate medic (medic is also spelled medick).

French: Luzerne ciliée.

Fig. 24-3. Comparison of fruits of *Medicago ciliaris* and *M. intertexta*. (A) *M. intertexta* (based on E. Small 1985-211, DAO). (B) *M. ciliaris* (based on E. Small M9, DAO). Note gland-tipped hairs of the latter. (Scanning electron micrographs prepared by B. Brookes.)

Geographical Distribution

(N = native, I = introduced, U = uncertain status)

Africa: Algeria (N), Canary Islands (N), Egypt (N), Madeira (N), Morocco (N), Tunisia (N).

Asia: Cyprus (N), Iran (N; note Mozaffarian 1988), Iraq (N), Israel (N), Jordan (N), Japan (I), Lebanon (N), Syria (N), Turkey in Asia (N).

Europe: Balearic Islands (N), Corsica (N), Crete (N), East Aegean Islands (Greece; N), France (N), Greece (N), Italy (N), Malta (N), Portugal (N), Sardinia (N), Sicily (N), Spain (N), Turkey in Europe (N).

Medicago ciliaris is omni-Mediterranean. The western range is sympatric with the very closely related *M. intertexta*, but according to Heyn (1963), *M. intertexta* does not extend to the eastern range of *M. ciliaris*.

Description

See *M. intertexta* and the following identification hints.

Identification Hints

Medicago ciliaris is much like *M. intertexta*, but the fruits are always covered with gland-tipped trichomes (unlike those of *M. intertexta*) and purple leaf blotches are rare (they are often present in *M. intertexta*). Less reliable differences are as follows: *M. ciliaris* tends to have somewhat smaller and relatively spherical pods (the pods are usually ovoid in *M. intertexta*) and shorter fruit spines (2–4 mm long, up to 8 mm in *M. intertexta*), and the spines of *M. ciliaris* tend not to interweave as they often do in *M. intertexta*.

Habitat and Environmental Adaptation

Habitat

Medicago ciliaris occurs in damp soils, swamps, grasslands, and other moist habitats. In Greece, Thanopoulos (2007) recorded the species in macchie and other shrub associations, open grasslands, abandoned terraced slopes, in olive groves, and beside roads and fields. Abdelguerfi and Laouar (2000) observed ecotypic differentiation among populations of three groups: those collected from semi-arid, subhumid, and humid regions (plants adapted to the driest regions tended to flower earliest). Abdelguerfi-Laouar et al. (2003) studied the comparative ecology of *M. intertexta* and the closely related *M. ciliaris* in Algeria and observed that *M. intertexta* was much more widely distributed, occurring in soils that were relatively rich in sodium. Another indication of the difference in habitat preferences of these two related species is the observation by Thanopoulos (2007) that in Greece *M. ciliaris* appears to be more widespread than *M. intertexta*.

Phenology

This species flowers in late spring and early summer. Yahiaoui and Abdelguerfi (1999) noted that populations from high rainfall sites in Algeria flowered later and for a shorter period.

Soil

Piano and Francis (1992) observed that in its native area the species occurs in clay soils. *Medicago ciliaris* typically occupies heavy soils (Heyn 1963). In Algeria, Abdelguerfi et al. (1988*a*) noted that *M. intertexta* occurs in fine- and very fine-textured soils. The species is found in soils that range from neutral to moderately basic. In Sicily, Piano et al. (1991) recorded the species growing in a pH range of 7.0 to 9.0. Rogers et al. (2005) ranked *M. ciliaris* as moderately tolerant of soil salinity, but Merabet et al. (2006) found this species had only limited tolerance to saline conditions. Abdelly et al. (1996) observed that although *M. ciliaris* was regularly found on the edges of sebkhas (closed depressions with hydromorphic, saline soil, periodically submersed) in Tunisia, the species had limited tolerance of soil salinity.

Temperature

Medicago ciliaris is a typical Mediterranean climate winter annual, completing its life cycle in early summer before increasing temperatures become severe.

Water

Piano and Francis (1992) observed that in its native area the species requires a minimum of 350 mm annual precipitation. *Medicago ciliaris* typically occurs in moist soils. In Tunisia, it occupies a humid to subhumid zone with 600–1200 mm of precipitation annually (Zoghlami and Hassen 2004). In Algeria, it occupies areas with a wide range of precipitation, including the relatively moist northeast, but is present in water-retaining, fine-textured soil (Abdelgeurfi et al. 1988*a*).

Biotic Relationships

Pollination

The species is probably largely an inbreeder, but its relatively large flowers (for an annual species of *Medicago*) suggest that at least some outcrossing occurs. In a study of sympatric *M. ciliaris* and *M. intertexta* in Algeria, Abdelkefi et al. (2001) observed that the former was more of an inbreeder than the latter.

Dispersal

The indehiscent spiny fruits are adapted to dispersal in animal fur.

Herbivory

Medicago ciliaris is most notable for the presence of glandular trichomes, especially on the fruit. Glandular trichomes tend to be associated in *Medicago* with resistance to insects (see Chapter 7). The close relative *M. intertexta*, which lacks glandular trichomes, is comparatively rarely present in the eastern Mediterranean range of *M. ciliaris*, and it is possible that certain insect species in this eastern range stimulated the evolution of protective glandular hairs in *M. ciliaris*.

Nodulation and mycorrhizae

USDA (2009) reported positive root nodulation for *M. ciliaris*.

Disease resistance

Yaege and Stuteville (2002) tested the resistance of 33 annual species of *Medicago* to powdery mildew (*Erysiphe pisi*) and reported that the five collections of *M. ciliaris* grown were susceptible to the disease whereas one showed partial resistance. In a comparative test of 36 annual species of *Medicago*, O'Neill and Bauchan (2000) found that seven accessions of *M. ciliaris* were quite susceptible to anthracnose caused by *Colletotrichum trifolii*.

Economic History, Cultivars, and Uses

This species has not attracted significant agricultural attention to date.

Cultivars

No cultivars have been bred.

Forage

Medicago ciliaris is not sown for forage, although it is consumed by livestock grazing in natural, unmanaged lands in the Mediterranean area. Tadmor et al. (1971) conducted studies of ecotypes selected for their ability to survive and persist in the Negev Desert of Israel in an area where rainfall averaged only 90 mm annually, but an additional 263 mm of water was received annually in the form of run-off from erratic winter storms. The dry matter production of an ecotype of *M. ciliaris* was 4.7 t/ha (averaged over 6 years), considered rather good given the arid environment.

Fodder

This species is not harvested for fodder.

Green manure, rotation crop, ground cover

No information located.

Other uses (salad sprouts, ornamental, plant products, etc.)

No information located.

Chromosome Number

$2n = 16$: Fryer 1930; Ghimpu 1929*b*, 1930; Clement 1962; Bolkhovskikh et al. 1969 p. 307; Simon and Simon 1965; Scrugli and Bocchieri 1977; Fernandes et al. 1977; Fernandes and Queirós 1978; Lesins and Lesins 1979; Agarwal and Gupta 1983*b*; Schlarbaum et al. 1984; Abdelguerfi et al. 1988*b*.

$2n = 16 + 1B$: Agarwal and Gupta 1983*b*.

$2n = 18$: Heyn 1956; Abdelguerfi and Guittoneau 1979; Fiorini and Raffaelli (1996; "$2n = 18$ or $2n = 16 + 2B$").

Genetic Relationships and Gene Transferability

Intraspecific

No barriers to hybridization within the species are known, although artificial hybridization of different populations has not been attempted.

Interspecific

Lesins et al. (1971) found that *M. ciliaris* could be intercrossed readily with *M. intertexta*, but the hybrids had reduced pollen fertility. They noted that the cross of *M. ciliaris* and *M. muricoleptis* produced hybrids with only about 35% pollen fertility. Lesins and Lesins (1979) recorded that they were unable to hybridize *M. ciliaris* with *M. granadensis*.

Exchange by advanced biotechnology

No information located.

Conservation Status

Medicago ciliaris is a fairly common species, not in need of conservation.

Germplasm Sources

Hundreds of accessions of *M. intertexta* are available in germplasm institutions (see Chapter 3).

Research Problems and Priorities

Medicago ciliaris is a vigorous species that has potential as an annual pasture legume of comparatively moist areas. It requires agronomic evaluation.

Representative Collections

Algeria: Balansa, B. 928, Biskra, dans les moissons, 20-IV-1853, in fruit (P). Chevrallier, L. 42, Biskra, in arvis incultis sat frequens, 24-IV-1896, in fruit (P). Gandogher, M., Mustapha, IV-1879, in fruit, M. Gandogher Flora Algeriensis Exsiccata 302 (P). Chapman, E., Mile 3, Larnaca-F'gusta Road, sea level, in Juncau marsh near sea, 13-IV-1950, in flower & fruit (K). Genneou, A. 1461, Yialousa, alt. 100 m, in wet places near a marsh, 16-IV-1920, in flower & fruit (K). Simpson, N.D. 716, Gemmeiza, frequent, 5-III-1922, in flower & early fruit (K). **England:** Lousley, J.E. et al., Flitwick, Bedfordshire, 25-X-1958, in flower & fruit, Wool Adventive Flora of Britain – Lousley No. W850 (K). **France:** Gauthier, G., Craboules, près de Narbonne (Aude), terrains salés, 10-VI-1879, in fruit, Société dauphinoise, 1880 No. 747 (K, P). Joad, G.C., Mentone, Alpes Maritimes, 5-IV-1876, in flower & fruit (K). Maillard, A., Narbonne, 15-VI-1858, in fruit (K). **Germany:** Holm-Nielsen, L.I. Nielsen, and S.P. Pinnerup 211, Dry sandy hill near

Tralau, northwest of Bad Oldesloe, in open vegetation, 8-VIII-1974, in flower & fruit (L). **Greece:** De Heldreich, in vinetis Atticae rara, versus Phalerum, 26-IV-1856, in fruit, De Heldreich Herbarium Graecum Normale No. 475 (K). Guiol, F. N1409, Attica pr. Vraona, 1-VI-1930, in fruit (BM, HUJ). **Iraq:** Gillet & Rawi 7309, Shahreban, alt. 60 m, ground flooded from flow canal, occasional, 11-IV-1947 in flower & fruit (K). Polunin, O. 5185, Dilthawa near Baghdad, alt. 300 feet, bank by irrigation channel, I-IV-1988, in flower & fruit (K). **Israel:** Bornmüller, J., Palaestina australis, Jaffa, in arenosis humidis ad Wadachnin, 9-IV-1897, in fruit, Bornmüller Iter Syriacum 1897 No. 395 (K, P). **Italy:** Levrier, E., Italia med., Etruria (Joscana) juxta villam Arcangeli pagi Settignano ad orientem Florentiae, in herbosis incultis, 23-V-1886, in fruit (K). Levrier, E., prope Florentiam (podere de Fattojo), in herbosis incultis ad margines agrorum pagi Settignano, 22-V-1875, in flower & fruit (BM). **Italy (Sicily):** Davis, D. & Sutton, S. 63263, Caltanisetta, ca. 20 km south of Piazza Armerina, banks over limestone rocks, alt. 100 m, 17-V-1979, in flower & fruit (BM). Davis, D. & Sutton S. 63378, Agrigento, ca. 25 km northeast of Sciaco, east of Caltabellotta, alt. 200 m, damp areas near road drains (no date, but about 1979), in flower & fruit (BM). Rigo, G., Sicilia Prov. Siracusa, Noto, in herbidis collium, solo tuffaceo, 15-IV-1898, in flower & fruit, G. Rigo Iter Italicum quartum anni 1898 (P). **Malta:** Duthie, J.F., insula Gaulos, Guirdan, in argillosis, 19-IV-1874, in flower & fruit (K). **Morocco:** Jahandiez, E. 3, Rabat, lieux herbeux, 11-IV-1924, in fruit (BM). Sennen & Mauricio 8762, Beni-Said à Dar-Kebdani, champs, 13-V-1934, in fruit (BM). **Spain:** Ball, P.W. 1507, prov. Sevilla, Camas, dry arable land and trackside, 21-V-1967, in fruit (BM). (Plants only 5–15 cm high, atypically small, no doubt because of dry habitat unusual for this taxon). Bornmüller, J., Madeira, Funchal, in campis maritimis prope rupem Gurgulho, 26-III-1900, in fruit, J. Bornmüller Planta Exsiccatae Maderenses 1900 No. 472 (P). Bourgeau, E. 160, Medina Sidonia, dans les champs, 21-III-1849, in fruit (BM, P). **Syria:** Maitland, T.D. 424, Shemban, alt. 715 m, early-VI-1959, in fruit (K). **Tunisia:** Cosson, M.M.E. et al., Nebeul, presqu'île du Cap Bon, 13-V-1883, in fruit (BM). Rouy, M., Béja, (received by P in 1881) (P).

25. *Medicago citrina* (Font Quer) Greuter

M. citrina (Font Quer) Greuter in Greuter & Raus, Willdenowia 16(1): 112 (1986).

The epithet *citrina* is modern Latin meaning citron-yellow, an allusion to the lemon-yellow petals.

Subordinate Taxa

None recognized here.

Type Specimen

Font Quer's Latin citation of his new taxon *M. arborea* var. *citrina* includes "Hab. in rupestribus calcareis parva insula "Espartar" dicta, pr. Ebuso, ubi Gros, d. 23-V-1918, legit." Bolòs & Vigo (1974) stated (in Spanish) "Loc. class.: Illa de l'Espartar, pr. Eivissa." Type presumably at BC.

Synonymy

Homotypic

M. arborea var. *citrina* Font Quer, Mem. Mus. Ci. Nat. Barcelona Sér. Bot. 1(2): 7 (1924), (basionym).

M. arborea subsp. *citrina* (Font Quer) O. Bolòs & Vigo, Bull. Inst. Catalana Hist. Nat. 38 Bot, 1: 69 (1974).

Heterotypic

No names found.

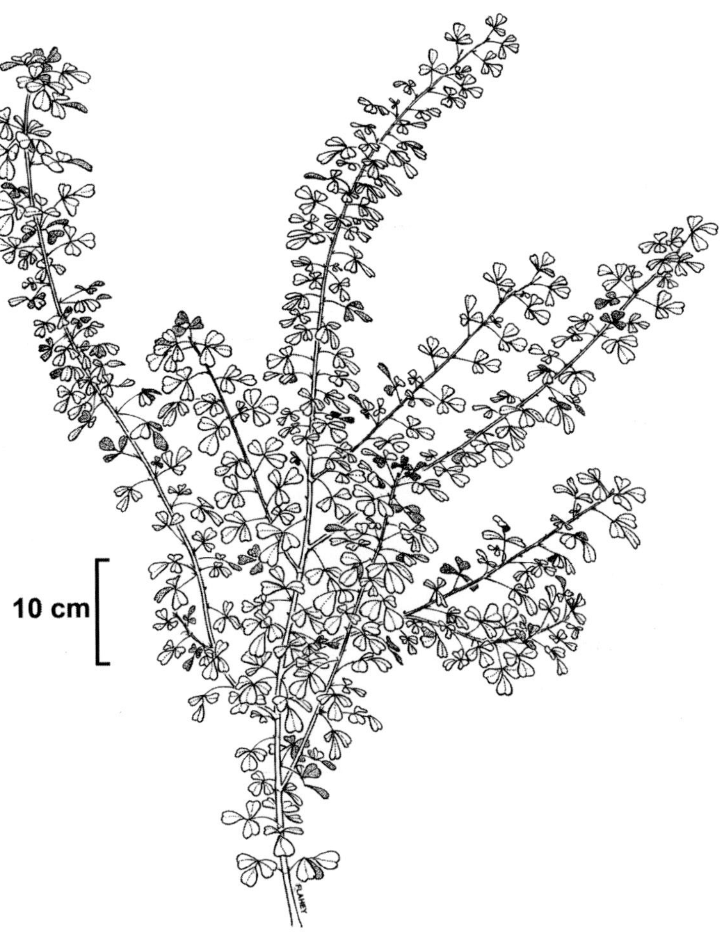

Fig. 25-1. *Medicago citrina*. A greenhouse-grown, vegetative plant. (Voucher: E. Small M2071, DAO. Grown from seeds supplied by R. Vogt, Botanical Garden Berlin-Dahlem; original collection — Spanien: Balearen, Ibiza, Insel Espartar, VI-1980, leg. H. Kuhbier; artist: B. Flahey.)

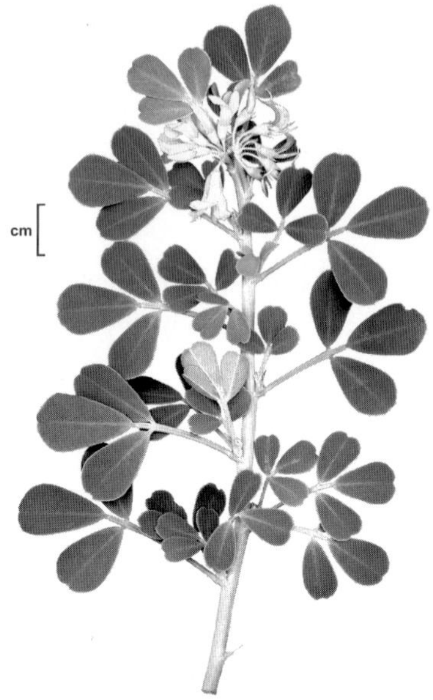

Fig. 25-2. *Medicago citrina.* Flowering branch. (Source: Universitat de les Illes Balears, http://herbarivirtual.uib.es; current August 2010, reproduced with the permission of J.R. Larrucea, Dept. Biologia, Universitat de les Illes Balears, Spain.)

Vernacular Names

ENGLISH: No names found.

FRENCH: No names found.

Geographical Distribution

This taxon is confined to a few Mediterranean islands off the eastern coast of Spain, including the Columbretes archipelago (recorded on three small islets: Ferrera, Grossa, and Foradada), some islets around Ibiza, around Cabrera within the Balearic Islands, and the small islet Escull del Cap near Alicante. It is the dominant woody species on Ferrera and Foradada in the Columbretes archipelago, found mainly on northern slopes, and was reintroduced to Grossa by humans after extinction from rabbit overgrazing and fires (Pérez-Bañon et al. 2003). Serra Laliga et al. (2001) discovered the species on the Spanish mainland. *Medicago citrina* occurs in the western Mediterranean, whereas the other taxa of section *Dendrotelis* are found in the eastern Mediterranean (Robledo et al. 1993).

Fig. 25-3. *Medicago citrina.* Fruiting branch. (Source: Universitat de les Illes Balears, http://herbarivirtual.uib.es; current August 2010, reproduced with the permission of J.R. Larrucea, Dept. Biologia, Universitat de les Illes Balears, Spain.)

Fig. 25-4. *Medicago citrina.* Two views of a very old plant growing on S'Espartar Island, Ibiza, Spain. Note the very large woody base of this plant, which was found growing on the northern rocky slopes of the island where there is very limited habitat that can be occupied. Photographs by A.I.J. Gallardo in March 2000, and used with permission. Dr. Gallardo, a specialist on *M. citrina*, noted that these plants are the oldest ones on the island and have the largest stems she had ever seen.

Description

Medicago citrina closely resembles *M. arborea*, and the description of the latter may be examined. Differences from its close relatives *M. arborea* and *M. strasseri* are noted below.

Chebbi et al. (1995) compared a small number of populations of *M. arborea*, *M. citrina*, and *M. strasseri*, all grown under similar conditions, and observed significant differences between *M. citrina* and the other taxa with respect to numerous characteristics. Compared with the other taxa, *M. citrina* had more pronounced branching, with short, feathery branches developed more laterally, resulting in a wider plant. In addition to the usual trifoliolate leaves of *Medicago*, *M. citrina* leaves sometimes had four or five leaflets. Compared with the other two taxa, *M. citrina* had the most pubescent leaves (on the undersurface averaging 35.6 hairs/mm^2 compared with 28.3 hairs/mm^2 for *M. arborea* and 26.0 hairs/mm^2 for *M. strasseri*). Moreover, in addition to the normal simple hairs, *M. citrina* often had hairs branched into two at the base. Compared with the other two taxa, *M. citrina* had the longest stipules and apparently a thicker wax layer on the leaves. However, the observations of Chebbi et al. (1995) were based on only a small number of populations.

Identification Hints

The most reliable character difference for *M. citrina* compared with *M. arborea* and *M. strasseri* is the possession of lemon-yellow flowers (versus yellow or orange-yellow in the other two species). *Medicago citrina* also has larger pods, seeds, and leaves (Robledo et al. 1993), but there is overlap in the ranges of these characteristics.

Habitat and Environmental Adaptation

Habitat

This species grows mostly on rocky slopes. Font Quer (1924) found the plants growing on the island of Espartar in inaccessible gorges, collection requiring that they be dislodged by throwing stones at them from the top of the gorge. Detailed studies of the habitat were conducted by Juan and Crespo (1999), Juan et al. (1999), and Serra Laliga et al. (2001).

Phenology

Pérez-Bañon et al. (2003) observed that flowering of *M. citrina* in the Columbretes archipelago occurred over a long period, from the end of January until the end of May, with peak flowering from the middle of February to the middle of April. Such an extended period of flowering is unusual for Mediterranean shrubs. The seasonal productivity dynamics of *M. citrina* and *M. arborea* have been studied under natural conditions and under irrigation by Lefi et al. (2004*a*), Lefi et al. (2004*b*), and Lefi et al. (2004*c*). Both species showed maximum leaf growth in the autumn, followed by a moderate growth in the winter. *Medicago citrina* was more productive of leaf biomass, but the reductions due to drought were comparable in both species.

I have grown all three species of section *Dendrotelis* (*M. citrina*, *M. arborea*, and *M. strasseri*) for many years in a greenhouse in Ottawa, Canada, and the plants have almost always remained vegetative, very rarely flowering and never setting seed, although increasing in size over the years. By contrast, every other species of *Medicago* that I have grown in the same greenhouse has flowered and set seed. The reasons for the failure to reproduce under this circumstance are unclear, but may be related to a requirement for higher light and a different photoperiod than I have provided.

Soil

"The populations occur on a range of different substrates including calcareous and volcanic soils, although these are always strongly nitrified because of guano deposits" (Juan et al. 2004). *Medicago citrina* grows better than its close relative *M. arborea* under saline conditions (Sibole et al. 1994).

Temperature

This species occurs in a Mediterranean climate, much like the closely related *M. arborea*. In the Columbretes archipelago the mean annual temperature is 18.6 °C (Pérez-Bañon et al. 2003).

Water

Medicago citrina occupies a "semi-arid-dry, thermo-Mediterranean bioclimate" (Juan et al. 2004). In the Columbretes archipelago the mean annual rainfall is 315 mm (Pérez-Bañon et al. 2003). Correal (1997) noted that the species is resistant to drought stress, apparently in part because it has a deeper root system than the related *M. arborea*. Although resistant to drought, fruit set is considerably reduced under very dry conditions. González-Andrés et al. (1999) characterized *M. citrina* as more tolerant to summer drought than its close relatives *M. arborea* and *M. strasseri*.

Biotic Relationships

Pollination

Large bees are the major pollinators of *M. citrina* in the Balearic Isles (Pérez-Bañon et al. 2003) and the plants appear to be largely outbreeding. In contrast, Pérez-Bañon et al. (2003) studied the reproductive ecology of *M. citrina* in the Columbretes archipelago of the Mediterranean Sea where bees are absent, and the plants appeared to be largely inbreeding. They observed that the flowers are obviously structurally adapted to outcrossing. However, the flowers are fully self-compatible, but require pollinators for most tripping, with spontaneous self-pollination (auto-tripping) occurring at a rate of only about 20%. Wind and high temperatures appear to be responsible for some of the tripping and self-pollination that occurs. In the absence of bees, flies act as the main agent causing flower tripping. Chebbi et al. (1995) found that the standard petal of *M. citrina* of the Columbretes was as long as the other petals, whereas the standards of *M. arborea* and *M. strasseri* were usually shorter than the carina, and they suggested that this limited the outbreeding capacity of the flowers of *M. citrina* of the Columbretes.

Dispersal

Juan et al. (2004) considered wind, water, and animals as potential seed distributors and suggested that most likely seabirds were responsible for the distribution of *M. citrina* among the islands inhabited. It appears that seed germination improves after passing through an animal's digestive tract.

Herbivory

In the Columbretes archipelago "the plants suffer from outbreaks of herbivores" such as rabbits and insects, and this can be serious given the low populations of the plants (Pérez-Bañon et al. 2003). See below for information on depredation by a scale insect.

Nodulation and mycorrhizae

No information located.

Disease resistance

No information located.

Economic History, Cultivars, and Uses

This species has not attracted agricultural development to date.

Cultivars

No cultivars have been bred and there is no evidence of selection by humans. This narrowly distributed endemic has the same general appearance and properties of the closely related *M. arborea*, and may have been used similarly as an ornamental and browse plant within its area of distribution, although documentation of this is generally lacking.

Forage

González-Andrés et al. (1999) suggested that because *M. citrina* retains its foliage under dry conditions that cause its close relatives *M. arborea* and *M. strasseri* to shed their leaves, it is a better choice for summer forage in dry areas. González-Andrés and Ceresuela (1998) suggested that the species would be good winter forage in winter-cold, semi-arid environments.

Fodder

Font Quer (1924) stated that a farmer in Eivissa grew *M. citrina* in his fields for cattle feed, the seed originating from Espartar gorge.

Green manure, rotation crop, ground cover

No information located.

Other uses (salad sprouts, ornamental, ground cover, etc.)

No information located.

Chromosome Numbers

$2n = 48$: Lesins and Lesins 1979; Boscaiu et al. 1997 (as *M. arborea*); Fernandes and Santos 1971 (as *M. arborea*); Fernandes and Queirós 1978 (as *M. arborea*).

This taxon is a hexaploid variant of the closely related tetraploid *M. arborea*.

Genetic Relationships and Gene Transferability

Intraspecific

Based on DNA affinities, Juan et al. (2004) studied genetic relationships among the 10 known isolated subpopulations of *M. citrina*, including four from Ibiza, three from Cabrera, two from Columbretes, and one from an offshore islet in northern Alicante province. They concluded that the overall genetic system was severely fragmented and that the 10 populations fell into three groups.

Interspecific

Natural hybridization among the three related woody species of section *Dendrotelis* has not been recorded, nor has artificial hybridization.

Exchange by advanced biotechnology

No information recorded.

Conservation Status

Medicago citrina was listed in the International Union for Conservation of Nature (IUCN) Red List of Threatened Species (http://www.iucn.org/about/work/programmes/species/red_list/index.cfm; current June 2010), with the status of CR (critically endangered). It has a very narrow and fragmented distribution range, with relatively few populations, all of which deserve to be protected. There has been a decline in the number of mature individuals and in the quality of the habitat. In 1997, the population on the Columbretes was decreased 40% by a scale insect (*Icerya purchasi*), which is native to Australia; this insect was also detected on the Balearic Islands in 2001. Other threats include introduced rabbits, alien species such as *Opuntia maxima*, and periodic, severe attacks by the parasitic plant *Cuscuta*.

More so than for any other species of *Medicago*, protective conservation measures have been enacted for *M. citrina*. It is listed in Annex I of the Spanish Royal Decree 439/1990 as "sensitive to the disturbance of its habitat," the legislation protecting the natural habitat. Since 1985, it has been strictly protected within the region of Valencia by a regional government decree. On the Columbretes, the species occurs within a federally owned nature reserve. Except for scientific study, access to populations on the islands of Ferrera and Foradada is prohibited and these islands have been designated as microreserves, with a management plan developed for them in 1993. *Medicago citrina* was reintroduced to the island of Grossa after being eradicated there by rabbits brought to the island in the 18th and 19th centuries, and the rabbits were eliminated by 1987. An action plan protects a microreserve of about two dozen plants on "Illot de la Mona", within the boundaries of El Montgó Nature Park. All the small Balearic Islands are protected as "Natural Areas of Special Interest" under the Parliament of the Balearic Islands Law 1/1991. On these islands, most of the populations are protected because they occur within the National Park of Cabrera. The Botanical

Garden of Sóller is active in monitoring the islets around Cabrera and conducting reintroduction measures that are also intended for the islets of Ibiza.

Recommended conservation actions are as follows (IUCN website http://www.iucn.org/themes/ssc/our_work/plants/Top50/English/Medicago_citrina_EN.htm; current May 2007). "The priority is to control attacks by the scale insect. It has been noted that the scale insect is a problem in citrus farms on the Spanish mainland, since farmers spray their citrus trees to kill a leaf miner, and in doing so kill a ladybird [beetle], which is the main predator of the scale insect. If correct, measures are needed to maintain ladybirds in the areas where *Medicago* grows. A method of controlling *Cuscuta* also needs to be identified, and invasive alien species (e.g., *Opuntia maxima*) need to be managed. In general, conservation efforts require more information about the population trends of *Medicago citrina* over a prolonged period. Reintroduction work needs to be continued."

Germplasm Sources

Medicago citrina is cultivated and seeds are stored in the Botanical Garden of Valencia. The species is also cultivated in the Botanical Garden of Sóller and Instituto Murciano de Investigación y Desarrollo Agrario y Alimentario(IMIDA) of Murcia.

Research Problems and Priorities

Inasmuch as this hexaploid species is rare and relatively little-studied, there is need to collect and preserve germplasm, to analyse genetic relationships with the related, more common tetraploid *M. arborea*, and to examine the agricultural potential of its characteristics. Since it has larger flowers and a different flower color than the tetraploid form that is widely used as an ornamental, it may be particularly useful as an ornamental.

Representative Collections

Spain: Beltrán, M. et al. 6204SC, Castellon, islas Columbretes, Isla Ferrera, suelo volcánico muy nitrificados, 5-IV-1982, in flower & fruit (K). Bowden, L.F. & Sims, P.A. 781, Menorca, T.M. Mahon, about 3 km from Mahon, Balearic Islands, alt. ca. 30 m, in river bed and on banks, 2-IV-1967, in flower & fruit (BM). Cannon, J.F.M. & M.J. 3218, Ibiza, Es Cana near Santa Gulalia, at side of track to Punta Arabi, 28-V-1966, in flower & fruit (BM).

26. *Medicago constricta* Durieu

M. constricta Durieu, Actes Soc. Linn. Bordeaux 29(2): 15 (1873), (basionym).

The epithet *constricta* is Latin for constricted. The fruit has the appearance of being constricted between the coils, but this interpretation could be placed on the fruits of many of the species of *Medicago*, so fruit constriction is not helpful as a diagnostic character.

Subordinate Taxa

None recognized here. Mehregan et al. (2002), based on material collected in Iran, treated *M. sinskiae*, *M. constricta*, *M. rigidula*, and *M. rigiduloides* as one species — *M. rigidula* — recognizing the first two as subspecies of *M. rigidula*.

Type Specimen

Bourgeau, Greece, Isle of Rhodes, champs incultés près Triandra, 12-V-1870, in fruit (Herb. Boiss., G!).

Synonymy

Homotypic

M. rigidula (L.) All. subsp. *constricta* (Durieu) Ponert, Feddes Repert., 83(9-10): 640 ("1972" publ. 1973).

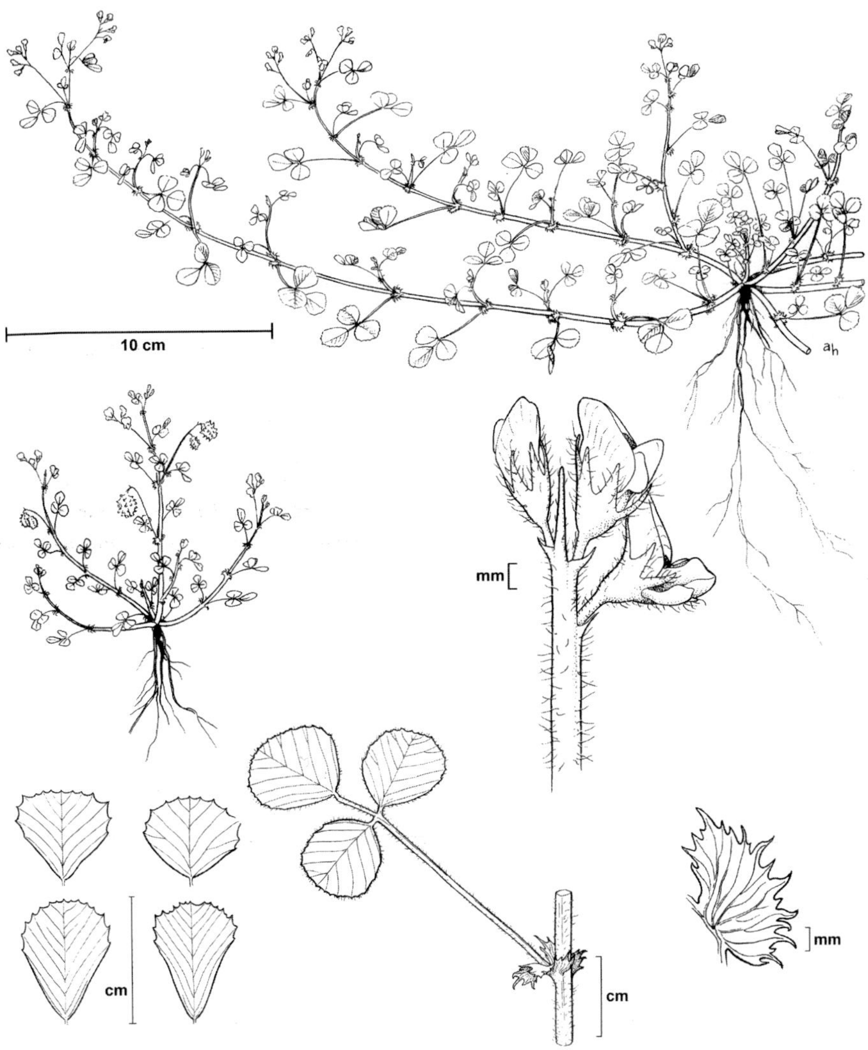

Fig. 26-1. *Medicago constricta.* **Clockwise from bottom left:** Leaflets; small plant with fruit; large plant with flowers and immature fruits; group of flowers; stipule; leaf attached to stem. (Artist: A. Hanes.)

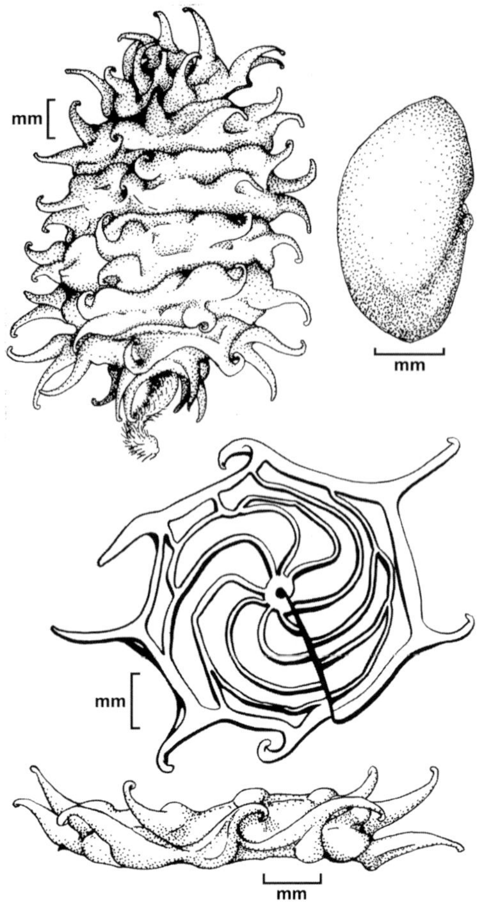

Fig. 26-2. *Medicago constricta.* **Top left:** Fruit. **Top right:** Seed. **Center:** Fruit coil face showing venation. **Bottom:** Edge of coil (Artist: L. Yuzyk.)

Heterotypic

M. globosa Presl, Delic. Prag. 45 (1822). (Heyn (1963) did not accept Presl's description and name as applicable to *M. constricta*, but did interpret other authors' (including Urban 1873) use of the name *M. globosa* as applicable to *M. constricta*. Some authors have accepted *M. globosa* as a synonym of *M. rigidula*.)

Vernacular Names

English: Globe medic (name encountered in Bolton (1962) and Radwan et al. (1980) and obviously based on the old name *M. globosa*). (Medic is also spelled medick.)

French: No names found.

Geographical Distribution

(N = native, I = introduced, U = uncertain status)

Asia: Cyprus (N), Gruziya (N), Iran (N), Iraq (N), Israel (N), Jordan (N), Lebanon (N), Syria (N), Turkey in Asia (N).

Europe: Bulgaria (rare; N), Greece (including Crete and East Aegean islands, N), Turkey in Europe (N).

Medicago constricta is native to the Eastern Mediterranean from Greece to the Middle East.

Description

Annual herb, stems (5)10–35 cm long, procumbent (usually) to ascending, branched from base, twigs and leaves moderately to densely covered with simple hairs. Leaflets 3–7 mm long, 2–4 mm wide, obovate to cuneate, upper third of margin serrate. Stipules dentate at base. Inflorescence of 1–2(5) flowers, peduncles longer than subtending petiole. Flowers 4–5.5 mm long, yellow, corolla about 1.5 times length of calyx. Calyx 2–3 mm long, with simple and glandular hairs, teeth subulate, about as long as tube. Young fruit contracted and concealed within calyx, then turning sideways out of calyx. Very young fruit with simple and glandular hairs, becoming glabrescent and glabrous when mature, with 6–8 strongly adpressed coils, mature pods very hard, spherical to ovoid, 6–12 mm long, middle coils 5–7 mm in diameter with or without a small central (axial) opening, with dorsal suture spines inserted at about 180° to coil face, coil face with 6–10 strongly arched radial veins, becoming nearly concentric in distal region with a few anastomoses. Seeds 1–2 per coil, 3.5–4.5 mm long, 1.5–2 mm wide, smooth, yellow or yellow-brown, reniform, separated by a transverse, spongy fruit septum, radicle slightly less than half seed length.

Identification Hints

This species is much like *M. rigidula* and *M. rigiduloides*, but distinguishable by the following: in mature pods of *M. constricta* there are no gaps evident between the coils (some gaps occur in *M. rigidula* and *M. rigiduloides*), and indeed the coils may become so strongly adpressed in *M. constricta* that their limits may be difficult to distinguish; the mature pods of *M. constricta* are glabrous whereas those of *M. rigidula* and *M. rigiduloides* are usually velvety-glandular due to the presence of gland-tipped hairs; and the curving veins on the coil face (viewable in immature pods) may be almost concentric and do not anastomose much, whereas those of *M. rigidula* and *M. rigiduloides* may also be very strongly curved or S-shaped, but they sometimes anastomose peripherally. *Medicago constricta* pods may also be reminiscent of glabrous-fruited forms of *M. doliata*, but the immature pods of the latter have coil faces showing radial veins that are only weakly to moderately curved.

Habitat and Environmental Adaptation

Habitat

Medicago constricta has been recorded from several habitats: sandy and stony ground, slopes, hillsides, and grassy areas. In Greece, Thanopoulos (2007) observed that the species occurs in macchie, garrique, and various other shrublands, dry meadows, woodlands, forests, and waste places and fields. Although fairly widespread geographically, it is not common.

Phenology

Flowering typically occurs from April to June in the native area.

Soil

This species typically occupies well-drained, rocky, calcareous soils. In Greece, Thanopoulos (2007) recorded it on rocky limestone, in sandy areas near the sea, and on saline soils. Piano and Francis (1992) indicated that *M. constricta* occupies soils of variable texture that are neutral to moderately basic.

Temperature

The species is a typical drought- and heat-evading winter annual, germinating in cool, late-winter conditions and completing its life cycle as heat and drought are increasing in early summer. Radwan et al. (1977) found that winter temperature was a critical factor limiting the distribution of *M. constricta* in northern Iraq.

Water

Piano and Francis recorded that in its native area *M. constricta* requires a minimum annual rainfall of 350 mm.

Biotic Relationships

Pollination

Small (1988) noted that the small flowers and pollen to ovule ratio (138) are consistent with *M. constricta* being largely self-pollinated.

Dispersal

The indehiscent spiny fruits are well adapted to dispersal in animal fur.

Herbivory

No information located.

Nodulation and mycorrhizae

USDA (2009) reported positive root nodulation for *M. constricta*.

Disease resistance

O'Neill et al. (2003) examined the comparative resistance of 36 annual species of *Medicago* to *Phoma medicaginis* (the cause of spring black stem and leaf spot, a major disease of alfalfa), and concluded that *M. constricta* had high resistance. Similarly, in a comparative test of 36 annual species of *Medicago*, O'Neill and Bauchan (2000) found that three accessions of *M. constricta* were quite susceptible to anthracnose caused by *Colletotrichum trifolii*. Yaege and Stuteville (2002) tested the resistance of 33 annual species of *Medicago* to powdery mildew (*Erysiphe pisi*) and found that three collections of *M. constricta* grown were susceptible (it was observed that often only one of a number of populations of other species tested was resistant).

Economic History, Cultivars, and Uses

Medicago constricta has not attracted agricultural interest to date.

Cultivars

No cultivars have been bred.

Forage

Medicago constricta is not sown for forage. St. John-Sweeting et al. (1996) compared several *Medicago* species in South Australia and suggested that further study of commercial development of *M. constricta* was warranted.

Fodder

This species is not considered to be a fodder plant.

Green manure, rotation crop, ground cover

No information located.

Other uses (salad sprouts, ornamental, plant products, etc.)

No information located.

Chromosome Number

$2n = 14$: Simon 1965; Gillies 1971*a*; Lesins and Lesins 1963*a*, 1979.

Genetic Relationships and Gene Transferability

Intraspecific

No information located.

Interspecific

Lesins and Lesins (1979) reported that they were unable to hybridize *M. constricta* with *M. sphaerocarpos* (identified as "*M. murex*") at the same chromosome number ($2n = 14$). Simon and Millington (1967) were unable to produce hybrid seed by crossing *M. constricta* with each of *M. truncatula*, *M. littoralis*, and *M. italica*.

Exchange by advanced biotechnology

No information located.

Conservation Status

Medicago constricta is sufficiently widespread that there is no need for special conservation measures.

Germplasm Sources

Several hundred accessions have been deposited in germplasm institutions (see Chapter 3).

Research Problems and Priorities

Medicago constricta is one of more than a dozen annual species of *Medicago* that has received only cursory analysis of its agricultural potential to date. It may be useful, like the annual species that have been developed, as a special-purpose forage for stressful environments and may additionally harbor genes that are valuable for improvement of the major agricultural species of *Medicago*.

Representative Collections

Cyprus: Casey, E.C. 1021, Kyrenia, uncultivated olive grove, alt. 100 feet, 23-IV-1950, in fruit (K). Kennedy, E.W. 1271, Platnes, alt. 3400 feet, dry mountainside, 19-V-1938, in fruit (K). **Greece:** Archley, S.C. 2437, Mt. Pentilihin, alt. 2000 feet., 31-IV-1931, in fruit (K). Archley, S.C. 755, M. Kimka (Attica), alt. 800 feet, fields, waysides, waste ground, V-1931, in fruit (K). De Heldreich, Attica, in vinetis prope Hagia Glykeria, ad radices collis Turcovuni, 30-IV-1884, in fruit (K). Shay, J.M. 80-292, Crete, Kommos, southwest coast Crete, Pitsidia, flat limestone hill south of village 7-IV-1980, in fruit (K). Sledge, W.A., Rhodos, Kamiros, 17-IV-1981, in fruit (K). **Iran:** Jacobs, M. 6380, S. Lorestan – Sheshom (33°06′N, 47°43′E), alt. 700–750 m, dry pastured and trampled soil with short herbaceous vegetation on disturbed limestone rock, 24-IV-1963, in fruit (K). **Iraq:** Gillett, J.B. 10912, Kursi Jebel Sinjar, alt. 650 m, cereal fields, occasional, 23-V-1948, in fruit (K). Gillett, J.B. 7967, Garadagh Village alt. 850 m, overgrazed stony fallow field, locally common, 26-IV-1947, in flower & fruit (K). Rawi, A. 22129, ca. 10 km west of Tawela (on road between Halabja and Tawela), alt. ca. 1400 m, on hillside, 15-VI-1957, in fruit (K). **Israel:** Berliner, B. 18799, Samaria, Umm el Fahm, herbaceous batha, Tuff and chalk, 5-IV-1976, in late flower & fruit (HUJ). Eig, A., Philistean Plain, env. of Nahalt-Yehuda, lime-sandy-loam hills, 17-III-1924, in flower & fruit (HUJ). Eig, A., Zohary, M. & Grizi, A., Philistean Plain, env. of Mrar on way to Rehovot Hills, gray compact soil covered with gravel, 7-IV-1935, in fruit (HUJ). Jolles, P., Acre Plain, Acre, sand-dunes, 15-IV-1926, in fruit (HUJ). Weinstein, Philistean Plain, Shehunat Boruchov, sandy-loam soil, 11-III-1925, in flower & fruit (HUJ). **Syria:** Haradjian, M. 2832, Monts Misairy, Bahamra, 15 miles east du Ladikie, alt. 1000 feet., iv-1909, in flower & fruit (G) (Haradjian 2835 from same location). **Turkey:** Alava, R. 6675, Icel, Silifke, rocky places among small cultivated fields on the hill near the fortress, 17-V-1967, in fruit (HUJ). Davis & Polunin, O. 25705B, prov. Antalya, Kumkoy, between Antalya and Serik, alt. 5 m, fixed dunes in open *Pinus pinea – Myrtus* forest, 6-IV-1956, in flower & young fruit (K). Davis & Polunin, O. 25,839, prov. Antalya, Manavgat-Kara Point, alt. 3 m, fixed dunes, 10-IV-1956, in flower & fruit (K). Sintensis, P. 1076, Iter Trojanum 1883, Renkoei, in graminosis, 1883, in flower & fruit (BM). Townsend, C.C. 67/61, Aspendos, west of Alanya, on bare stony ground near the roadside just west of the theatre, common, 20-IV-1967, in flower & fruit (K).

27. *Medicago coronata* (L.) Bart.

M. coronata (L.) Bartal., Catal. Piante Siena: 61 (1776).

The epithet *coronata* is Latin for crowned, a reference to the crown-like appearance of both ends of the fruits.

Subordinate Taxa

None recognized here.

Type Specimen

Heyn in Bull. Res. Council Israel 7D: 167 (1959) lectotypified the basionym for *M. coronata* (*M. polymorpha* L. var. η *coronata* L.) by the illustration accompanying the phrase "Medica coronata cherleri parva" on page 386 in: Bauhin, J., Cherler, J.H., et al. 1650–1651. Historia Plantarum Universalis, vol. 2. Typographia Caldoriana (printer), Yverdon, Switzerland. This was accepted in The Linnaean Plant Name Typification Project (http://www.nhm.ac.uk/research-curation/projects/linnaean-typification/; current June 2010; cf. Jarvis 2007).

Synonymy

Homotypic

M. polymorpha L. var. η *coronata* L., Sp. Pl. 780 (1753), (basionym).

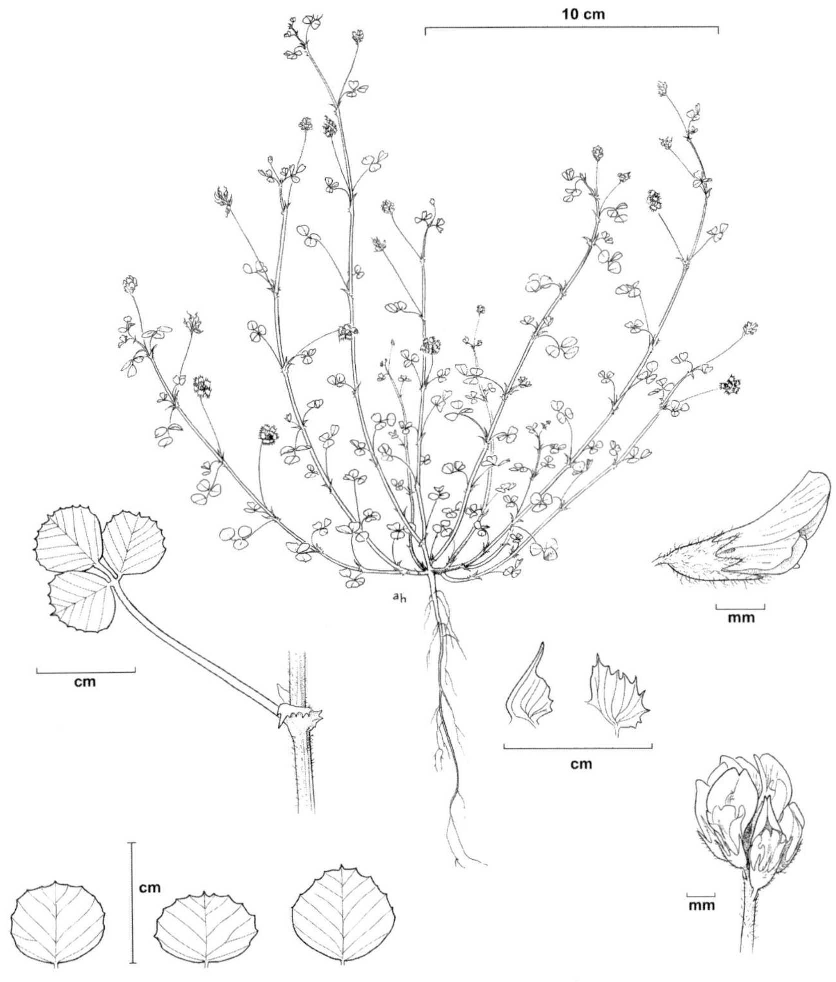

Fig. 27-1. *Medicago coronata.* **Clockwise from bottom left:** Leaflets; leaf attached to stem; plant; side view of flower; group of flowers; stipules. (Plant based on US 137240, remainder on E. Small M101, DAO; artist: A. Hanes.)

Medica coronata (L.) Medik. in Vorles. Churpfälz. Phys.-Öcon. Ges. 2: 386 (1787). (Presumptively synonymous.)

M. coronata (L.) Desr. in Lam. Encycl. Méthod. 3: 634 (1792), nom. superf.

Medica coronata (L.) Bubani, Fl. Pyr. 2: 476 (1899), nom. superf.

Heterotypic

M. coronata (L.) Desr. var. *brevipedunculata* Eig., Contr. Fl. Pal. 2: 22 (1927).

M. coronata (L.) Desr. f. *glandulosa* Koč. (1961). (Cited p. 163, Kozuharov 1965.)

M. coronata (L.) Desr. var. *glandulosa* (Koč.) Kozuharov, Bulgarian Acad. Sci. 15: 163 (1965).

Diploprion medicaginoides Viv., Fl. Lib. Spec. Enum.: 48 (1924) (as discussed by Heyn (1963), Viviani separated a form of *M. coronata* as the monotypic genus *Diploprion*).

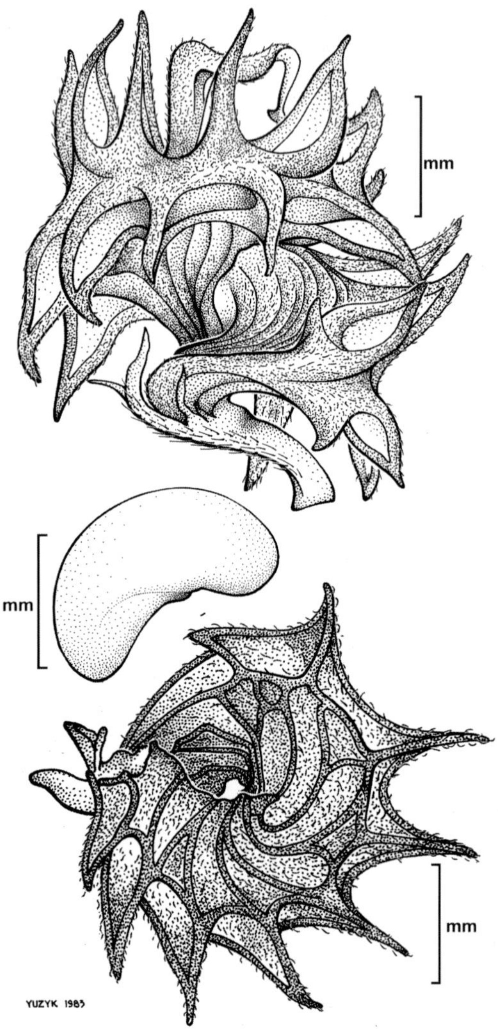

Fig. 27-2. *Medicago coronata*. **Clockwise from bottom left:** Seed; fruit; fruit coil face showing venation. (Based on E. Small M12, DAO; artist: L. Yuzyk.)

M. coronata (L.) Desr. var. *minor* Heldr. in Cat. Herb. Orph. 18 (1877), nom. nud.

M coronata (L.) Desr. var. *minor* Heldr. ex Hal. Consp. Fl. Graec. 1: 367 (1901), pro syn.

M. coronata (L.) Desr. var. *multiflora* Eig., Contr. Fl. Pal. 2: 22 (1927).

M. coronata (L.) Desr. var. *uncata* Boiss., Fl. Or. Suppl. 165 (1888). (Type: De Heldreich Plantae Exsicc. Flora Hellenicae, in collibus...1875, in fruit. Herb. Boiss. G!)

M. urbani Heldr. in herb. ex Boiss., Fl. Or. Suppl. 165 (1888), pro syn.

M. vermicularis Cesati in Friedrichsthal, Bot. Anh. Ces. & Fenzl (Reise Neu-Griechenl.) 274 (1838).

M. coronata (L.) Desr. var. *vermicularis* (Ces.) Hal., Consp. Fl. Graec. 1: 367 (1901).

(At HUJ, I observed a folder labelled "*M. coronata* var. *uncata* Boiss. = var. *gileadense* Eig. unpublished" and a folder of *M. coronata* var. *brevipedunculata* Eig. At G, I noticed material labelled *M. coronata* var. *parviflora* Boiss.)

Not

M. coronata (L.) Gaertn., De Fruct. Sem. Pl. 2: 349 (1791). Heyn 1963 listed this in synonymy with *M. polymorpha* and noted that this name was synonymous according to the combination, but not according to the description and illustration (quoad descr. et illustr. non quoad syn.). The International Plant Names Index listed this (in 2006) as *M. coronata* Gaertn., Fruct. ii t. 155, f. 7 = *M. tuberculata* (an ambiguous name, usually interpreted as *M. turbinata*).

M. coronata Pall., Ind. Taur. ex Bieb. Fl. Taur.-Cauc. 2: 227 (1808) (= *M. arabica*).

Unresolved

Heyn (1963) included *Medicago libyca* Spreng., Syst. Veg. Lin. 3: 289 (1826) as a synonym of *M. coronata* "with some reservation."

Vernacular Names

English: Crown medic, crowned medic (medic is also spelled medick).

French: La luzerne couronnée, luzerne en couronne.

Geographical Distribution

(N = native, I = introduced, U = uncertain status)

Africa: Egypt (N), Libya (N).

Asia: Cyprus (N), Iran (N), Iraq (N), Israel (N), Jordan (N), Lebanon (N), Syria (N), Turkey (N), Turkmenistan (N).

Europe: Albania (N), Bulgaria (N), Corsica (France, N), Crete (Greece, N), East Aegean Islands (Greece, N),

France (N), Greece (N), Italy (N), Portugal (I?), Sicily (Italy, N), Turkey (N), Spain (N), Yugoslavia (N).

Medicago coronata is omni-Mediterranean (mainly north-Mediterranean) but rare in Africa.

Description

Annual herb, stems (5)7–25(30) cm long, ascending to erect, branched from base, twigs and leaves pubescent with simple hairs, glandular hairs also present in some plants. Leaflets 4–8(12) mm long, 3–6 mm wide, usually obovate, typically somewhat longer than broad, ca. upper third of margin serrate. Stipules entire or dentate with a few teeth at base. Inflorescence of (3)6–12(15) flowers in a dense, rounded head, peduncles usually longer than subtending petiole. Flowers 2–3 mm long, pale yellow. Calyx teeth subulate, shorter than or equalling tube (typically one tooth longer than others). Corolla usually less than twice length of calyx. Young fruit protrudes sideways from calyx. Fruit with simple hairs, glandular hairs also sometimes present, short-cylindric, 1–4 mm long, with 1.5–2.5(3.5) loosely adpressed coils 1.5–4.5 mm in diameter, coil face with 5–8 strongly curving radial veins (sometimes appearing concentric) that anastomose somewhat before reaching a submarginal vein, then reach the dorsal suture, with a furrow between the submarginal vein and the dorsal suture (the suture thus often in a shallow depression). Spines 0.5–1.5 mm long present on dorsal suture, sometimes hooked, often at 90° to coil face (thus giving a crown-like appearance). Seeds 2–3 mm long, 1–2 per coil, , 1–3 per pod, smooth, yellow or yellow-brown, reniform or oblong-ovoid (irregular seed shapes often recorded), not separated by a transverse spongy fruit septum, radicle more than half as long as the cotyledons.

Identification Hints

Medicago coronata is not likely to be misidentified, except as *M. praecox*, which has much more dissected stipules and far fewer flowers and fruits on the peduncles.

Habitat and Environmental Adaptation

Habitat

Medicago coronata occurs largely on stony plains, hillsides, and dry grassy places, typically in herbaceous plant associations of the Mediterranean region, and as a weed in cultivated and fallow fields.

Phenology

The species flowers in winter in Mediterranean areas, often from March to May.

Soil

Piano and Francis (1992) recorded that in its native area *M. coronata* occurs on loam-clay soils that are slightly to moderately basic. The species is often found on limestone substrates, chalky rock soils, and stony places. In Bulgaria it has been collected from leached maroon forest soils. Bolton (1962) stated that the species grows on very poor soils.

Temperature

Medicago coronata is a typical winter annual, germinating in the cool late winter of the Mediterranean area and completing its life cycle in the increasing temperatures of mid-summer.

Water

Piano and Francis (1992) recorded that in its native area *M. coronata* requires a minimum annual rainfall of 250 mm. Bolton (1962) claimed that *M. coronata* shows a high degree of drought resistance.

Biotic Relationships

Pollination

The small flowers (among the smallest in the genus *Medicago*) are consistent with *M. coronata* being largely self-pollinated. Small (1988) reported a pollen to ovule ratio of 506, which is rather similar to the ratios in the outcrossing perennial species of the genus. Nevertheless, it is clear that the species is a strong inbreeder.

Dispersal

Medicago coronata fruit spines are often hooked (at least in long-spined forms), and fruits with such spines are well adapted to dispersal in animal fur.

Herbivory

No information located.

Nodulation and mycorrhizae

No information located.

Disease resistance

Yaege and Stuteville (2002) tested the resistance of 33 annual species of *Medicago* to powdery mildew (*Erysiphe pisi*) and found that the two collections of *M. coronata* grown were susceptible (it was observed that often only one of a number of populations of other species tested was resistant). In a comparative test of 36 annual species of *Medicago*, O'Neill and Bauchan (2000) found that two accessions of *M. coronata* grown were quite susceptible to anthracnose caused by *Colletotrichum trifolii*.

Economic History, Cultivars, and Uses

This species has not attracted agricultural interest to date.

Cultivars

No cultivars have been bred.

Forage

Medicago coronata is not sown for forage. Metawi et al. (1999) mentioned that the species is a significant natural forage for sheep in the Sinai region of Egypt.

Fodder

This species has not been considered as a source of fodder.

Green manure, rotation crop, ground cover

No information located.

Other uses (salad sprouts, ornamental, plant products, etc.)

No information located.

Chromosome Number

$2n = 14$: Clement 1962; however, Simon and Simon (1965) concluded that Clement had in fact examined *M. polymorpha*.

$2n = 16$: Fryer 1930; Heyn 1956; Simon and Simon 1965; Bolkhovskikh et al. 1969, p. 308; Lesins and Lesins 1962, 1979; Natarajan 1978; Agarwal and Gupta 1983*b*.

Genetic Relationships and Gene Transferability

Intraspecific

No barriers to hybridization within the species are known, although artificial hybridization of different populations has not been attempted.

Interspecific

Lesins and Lesins (1979) recorded that their attempt to hybridize *M. coronata* with the very similar *M. praecox* was unsuccessful.

Exchange by advanced biotechnology

No information located.

Conservation Status

This species is sufficiently widespread that there is no need for conservation measures.

Germplasm Sources

More than 150 accessions are available from major germplasm institutions (see Chapter 3).

Research Problems and Priorities

Medicago coronata is one of more than a dozen annual species of *Medicago* that has received only cursory analysis of its agricultural potential to date. It may be useful, like the annual species that have been developed, as a special-purpose forage for stressful environments and may additionally harbor genes that are useful for improvement of the major agricultural species of *Medicago*.

Representative Collections

Albania: Alston, A.H.G. & Sandwich, N.Y. 1234, South Albania, district of Sarandë (santi Quaranta), Sarandë, rocky limestone slope above the town, alt. ca. 200 feet, 30-V-1933, in flower & fruit (BM, K). **Bulgaria:** Kojoucharov, S. & Kouzmanov, B., près de la ville revailograd, les collines calcaires, 28-V-1959, in flower & fruit (K). **Cyprus:** Meikle, R.D. 2797, Perapedhi, Kafkalla Hill, alt. 2200 feet, on chalky slope, 2-V-1962, in flower & fruit (K). **Egypt:** Elliot, G.F.S. 3758, (sine data) (BM). Letourneux, Alexandria, 1877, in flower & fruit (K). **France:** Chambeiron, près de Toulon, bords des bois au pied des rochers de la montagne de Faron, 25-V-1860, in flower & fruit, Flora Galliae et Germaniae Exsiccata de C. Billot No. 3037 (P). Jousset, Touffles de Kermès au Prégnon, près Aix (Bouche-du-Rhône), V-1892, in flower & fruit, Société Rochelaise 1893 No. 3444 (P). **Greece:** Bourgeau, E., Lycia, Adalia, in graminosis, 5-V-1860, in fruit (P). Casey, E.C. 1150, Crete, Kyrenia, alt. 160 feet, rocky land, 22-III-1951, in flower & fruit (K). Griffith, H. 1797, near Yassikevy, alt. 500 feet, in *Quercus coccifera* scrub, sides of path, etc., 14-VI-1936, in fruit (K). Heldreich, Attica, prope Piraeum, in collibus apricis, 12-IV-1884, in flower & fruit, Dr. Heldreich Herbarium Graecum Normale No. 883 (P). Lewis, G.P. 575, Spetsai, above Spetsai town, alt. ca. 800 feet, in open area of pine forest on limestone hill, 6-V-1980, in flower & fruit (K). Lewis, G.P. 620, small island with caves just opposite Kuilas on mainland, north of island of Spetsai, alt. ca. 10 feet, on rough rocky limestone close to the shore, 8-V-1980, in fruit (K). Rechinger, K.H. & F. 9585, Thracia occidentalis, Montes Rhodope supra pagum Iasmos (Jasi-Köj), substr. silic., 14-VI-1936, in fruit (BM). Townsend, C.C. 63/18, Delphi, path to the stadium, alt. ca. 2000 feet, on stony ground, 17-V-1963, in flower & fruit (K). Shay, J.M. 80-276, Crete, Kommos, southwest coast, Crete, Pitsidia, valley above spring near small chapel, steep hillside with sparse, patchy vegetation, 7-IV-1980, in flower & fruit (K). Shay, J.M. 80-468, Crete, Kommos, hill overlooking dig site, low wind-trained maquis, 15-IV-1980, in flower & fruit (K). **Iran:** Davis & Bokhari D55805, D5 (Times Atlas), 48 km from Bebehan to Omidge, near spring and café by palms, alt. 200 m, bare calcareous hills of escarpment, 21-III-1974, in flower & fruit (K). Jacobs, M. 6373, S. Lorestan-Sheshom (33°06′N, 47°43′E), alt. 700–750 m, dry pastured and trampled soil with short herbaceous vegetation on disturbed limestone rock, 24-IV-1963, in flower & fruit (K). **Iraq:** Eig, A. & Zohary, M., foot of Jabal Bashiquah (northeast of Mosul) between Bashiquah and Fad'hiliyah, alt. ca. 450 m, ruderal grazing place, gray stony soil, 21-IV-1933, in flower & fruit (HUJ). Low, A. 89, Tigris Plain and Khurdish Hills, alt. 800–2000 feet, waste ground, poor soil and stony, 12-III-1936, in flower & early fruit (BM). Poore, M.E.D. 472, Kani Kirmaj, alt. 1600 feet, steppic grassland, 19-III-1958, in flower & fruit (K). **Israel:** Berliner, R., C. Samaria, Har Eyval, terra rossa, maquis and batha, Eocene limestone rocks, IV-1977, in flower & fruit (HUJ) Bornmüller, J., Palaestina, Galilaea, in saxosis ad Nazareth (solo calc.), 26-IV-1897, in fruit, Bormüller Iter Syriacum 1897 No. 393 (P). Eig, A., Zohary, M. & Feinbrun, N., Gildead, Wadi Warran, 10-V-1927, in late flower & fruit (HUJ) (identified as var. *uncata*

Boiss.). Feinbrun, N., Philistean Plain, env. of Benei-Beraq, sand–sandy loam soil, 18-III-1926, in flower & fruit (HUJ) (identified as var. *pedunculata* Eig). Jaffa, D. & Sheinkar, C., Hula Valley, Lahavot-Habashah, basalt soil, west slope, 6-IV-1954, in late flower & fruit (HUJ). Leinkram, S. & Parmet, H., Judaean Mountains, Jerusalem, Givat Ram, fallow field, 20-IV-1956, in fruit, Universitatis Hebraicae Hierosolymitanae Flora Terrae Israëlis Exsiccata No. 648 (BM, P). Melville, R. 70/26, Northern outskirts of Tel Aviv, coastal dunes, 25-III-1970, in fruit (K). Zohary, M., Mt. Carmel, weed in lawn, alt. 350 m, 16-IV-1968, in flower & fruit (HUJ). Zohary, M. & Feinbrun, N., coastal Galilee opposite Nahariya, west slope, loam on Kurkar rock, 15-III-1940, in flower & early fruit (HUJ). **Israel/Syria (Golan):** Lack, W., Heyn, C., Herrnstadt, I., Pazii, B., & Liston, A. 7-84-180/1, south Golan, hill south of Susita, Susita chalk, alt. 100 m, 27-III-1984, in flower & fruit (HUJ). Shmida & Noy-Meir, South Hermon, Qalaat Nemrod, the road Banias-Majdel Chams, among the fortress ruins, Jurassic hard limestone, alt. 750 m, 22-IV-1975, in flower & early fruit (HUJ). **Jordan:** Eig, A., Zohary, M. & Feinbrun, N., Ammon, Amman, 7-V-1927, in fruit (HUJ). **Lebanon:** Meinertzhagen, R., Beyrouth Syria, in crops, 31-III-1933, in flower & fruit (BM). Trench, C.P.C., Bechmezzine, waste places, alt. 1000 m, 16-IV-1945, in fruit (BM). **Spain:** Huter, Porta & Rigo 775, Legnium Granatense, pr. Malaga, loc. arenas ad Cerro Coronado, 1-V-1879, in fruit (K). **Syria:** Norris, F.H., Mt. Cassius, alt. 4000 feet, 12-III-1945, in flower (BM). **Turkey:** Alava, R. 66892, Adana, Toprakkale, crevices of basaltic rock near the fortress, 17-V-1967, in flower & fruit (HUJ). Davis & Hedge D27135, prov. Hatay, dist. Antakya, Iskenderun – Antakya, west of Amik Göl, alt. 150 m, rocky limestone slope, 25-IV-1957, in flower & fruit (BM). Balansa, B. 208, environs de Koukouloudja, près de Smyrne, 9-V-1854, in flower & fruit (P). Barclay, C. 1468, Ismir province, Bergama (Pergamon), 109 km north of Ismir, alt. ca. 1000 feet, among antique ruins, 17-V-1969, in flower & fruit (K). **Yugoslavia:** Pichler, T., Dalmatia, pr. Dizman ad Clissa, V-1880, in flower & fruit (K).

28. *Medicago crassipes* (Boiss.) E. Small

M. crassipes (Boiss.) E. Small in Small, Lassen & Brookes, Willdenowia 16(2): 433 (1987).

The epithet *crassipes* is derived from the Latin *crassus*, which means thick or heavy, plus *pes*, base or foot, adopted by the original author Boissier for what he described as extremely thickened peduncles.

Subordinate Taxa

None recognized here.

Type Specimen

Boissier, E., in cultis planitiei Cariensis ad septentrionem pagi Davas sitae, in Cadmi vallibus supra Denisleh, VI-1842. (Holotype G, syntypes BM, G, K; isotype at W.)

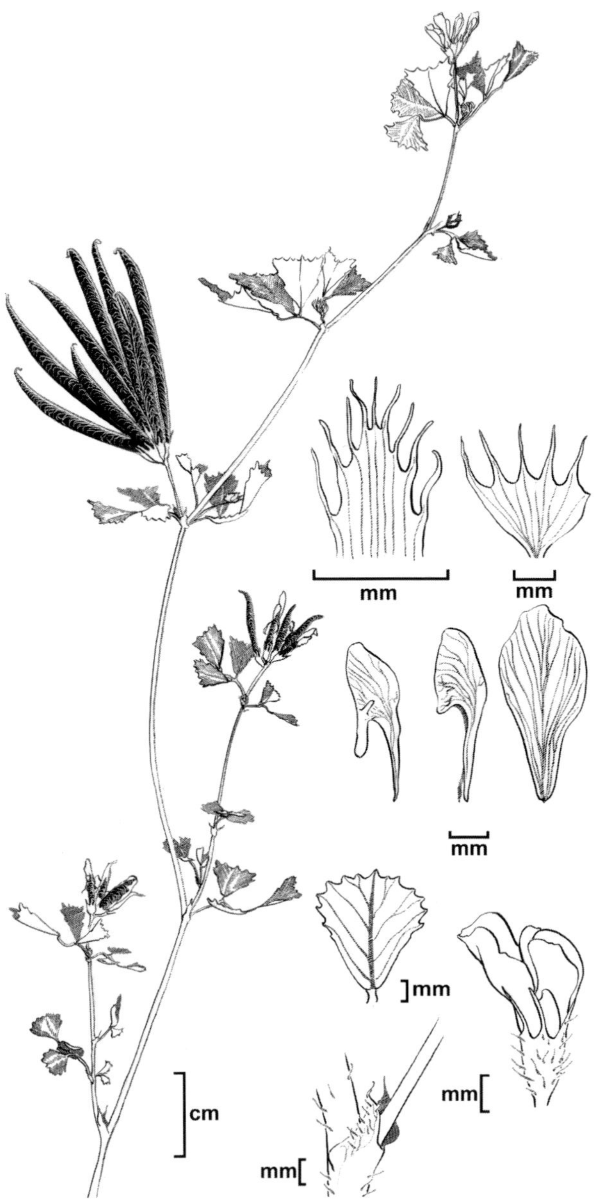

Fig. 28-1. *Medicago crassipes.* **Left:** Branch with fruit and flowers. Remaining figures clockwise from upper left: spread out androecial column (free stamen not shown); calyx; standard petal; side view of flower; stem node showing stipules; leaflet; wing petal; a keel petal is between the wing and standard petals. (Based on Pl. R. Hissar 16 May 1915, G; artist: M. Jomphe.)

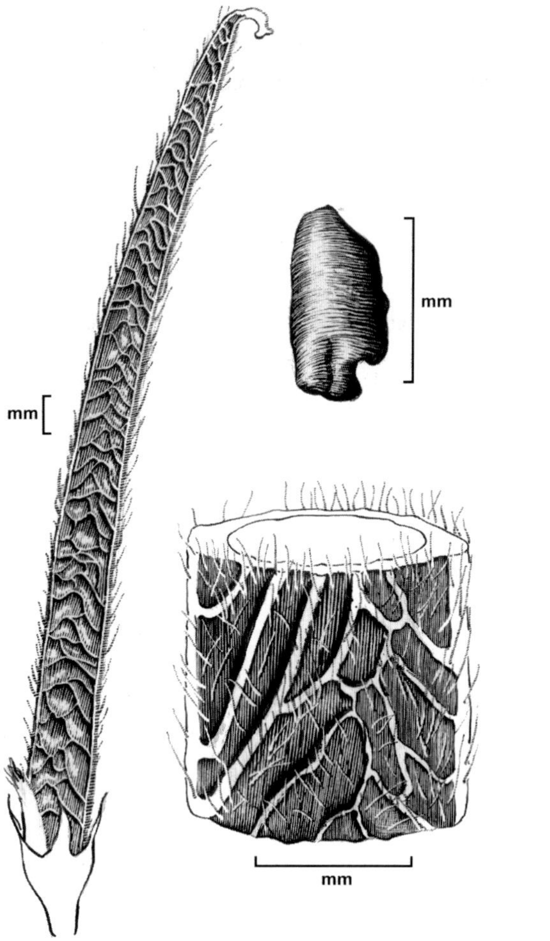

Fig. 28-2. *Medicago crassipes.* **Left:** Fruit. **Upper right:** Seed. **Lower right:** Portion of fruit. (Based on Pl. R. Hissar 16 May 1915, G; artist: M. Jomphe.)

Synonymy

Homotypic

Trigonella crassipes Boiss., Diagn. Pl. Orient. ser. 1, 1(2): 23 (1843), (basionym).

Heterotypic

No names found.

Vernacular Names

ENGLISH: No names found.

FRENCH: No names found.

Geographical Distribution

(N = native, I = introduced, U = uncertain status):

ASIA: Lebanon (N), Iran (N), Iraq (N), Syria (N), Turkey (N).

This species occurs in the eastern Mediterranean, including Anatolia, Anti-Lebanon (Syria-Lebanon), northern Iraq, and western Iran.

Description

Annual herb, branches (2)5–20(25) cm long, prostrate, branched from base, twigs and leaves sparsely adpresssed-pubescent with simple hairs. Leaflets 5–8 mm long, obovate or obcordate, somewhat longer than broad, upper portion of leaflets serrate. Stipules lanceolate-acuminate, dentate. Inflorescence of 4–10 flowers in an umbel, peduncles 2–3(18) mm long, strongly thickening as fruits develop. Flowers 4–6 mm, yellow. Calyx 3–4 mm, with simple hairs,

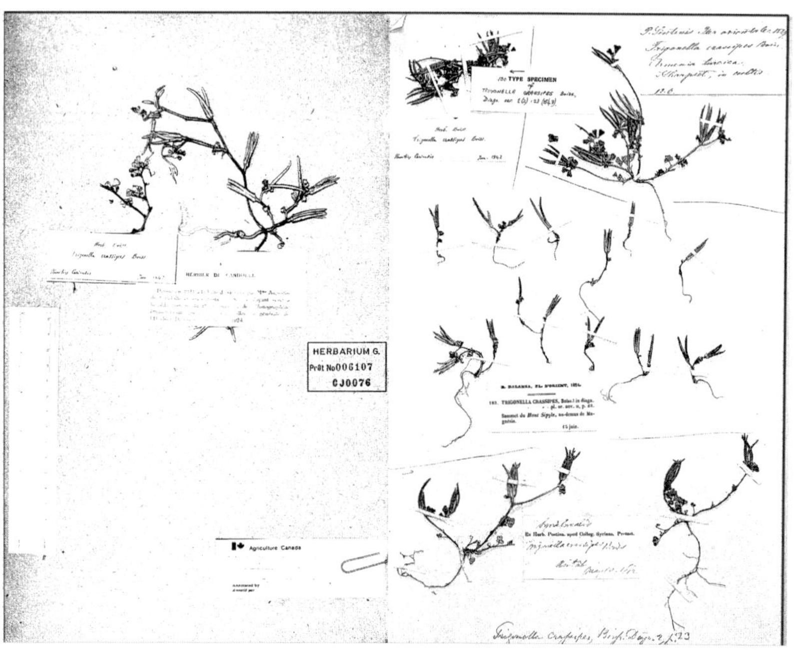

Fig. 28-3. Some type folder collections of *Trigonella crassipes* Boiss. in the Boissier herbarium at G. The holotype is at the upper left corner of the sheet at right.

teeth lanceolate-subulate, about equal to tube. Corolla about twice length of calyx. Fruits 12–35 mm long, 1–2 mm wide, linear, straight, tapering towards the tip, with a slightly hooked beak, somewhat compressed, appressed-pilose, with obliquely transverse reticulating veins and thickened sutures. Seeds oblong or rhomboid, about 2 mm long, surface finely pebbled. Seeds 8–10 per pod, radicle about 80% of length of seed.

Identification Hints

Medicago crassipes is similar to *M. rigida*, but the latter has smaller fruits (12–35 mm long in *M. crassipes*, 6–10 mm long in *M. rigida*). Both species occur in Anatolia.

Habitat and Environmental Adaptation

Habitat

Medicago crassipes occurs in scrub, pine forests, steppe, dry slopes, fallow and cultivated land, typically at altitudes of 600–1700 m.

Phenology

The species flowers from April to July in its native eastern Mediterranean area.

Soil

Medicago crassipes has generally been collected from calcareous soils of various textures.

Temperature

No information located.

Water

This species appears to be a drought-evading winter annual.

Biotic Relationships

Pollination

Small (1988) noted that the flowers are small and the pollen to ovule ratio is low (183), observations that are consistent with *M. crassipes* being largely self-pollinated and substantially an inbreeder.

Dispersal

There are no evident morphological adaptations for seed dispersal in this species.

Herbivory

No information located.

Nodulation and mycorrhizae

No information located.

Disease resistance

No information located.

Economic History, Cultivars, and Uses

This species has not attracted agricultural interest to date.

Cultivars

No cultivars have been bred.

Forage

Medicago crassipes is not sown for forage.

Fodder

This species is not used as a source of fodder.

Green manure, rotation crop, ground cover

Evans et al. (2004) examined the potential of numerous fodder species to survive winter in North Korea and produce nitrogen, organic matter, and soil stabilization in maize and rice fields. *Medicago crassipes* was judged to have good winter survival and excellent green biomass production, and to produce excellent ground cover.

Other uses (salad sprouts, ornamental, plant products, etc.)

No information located.

Chromosome Number

No information located.

Genetic Relationships and Gene Transferability

Intraspecific

Artificial hybridization of different populations has not been reported.

Interspecific

No information located.

Exchange by advanced biotechnology

No information located.

Conservation Status

This species has a relatively restricted distribution range and germplasm collections are limited. It requires an analysis of its conservation requirements and collections of seeds need to be made and preserved.

Germplasm Sources

Only two accessions of *M. crassipes* were located (see Chapter 3).

Research Problems and Priorities

Medicago crassipes has received little attention to date, in part because until recently it was considered to be a species of *Trigonella* and so its potential as an agricultural species, like the important species of *Medicago*, was not considered. Like many of the other annual species of *Medicago*, it has received little botanical and agricultural analysis. It may have potential as a special legume crop for certain marginal agricultural conditions, and it may harbor genes useful for improvement of other species of *Medicago*.

Representative Collections

Lebanon: Boissier, E., Antilibanus, pr. Rashey (pr. Rasheya in a second collection, presumably Rasheiya), IV-1846, in fruit (G) (plants dwarfed, 3–4 cm high). Mouterde, P. 2243, Baalbeck, 14-V-1933, in fruit (G). Peyron, E., Antilibanus, Racheyn, 6-VI-1884, in fruit (G). **Turkey:** Alava, Reino 6707, Gaziantep, along road to Urta, 2 km east of town, limestone hill on left side of road, 23-IV-1967, in flower & fruit (E). Balansa, B. 106, partie supérieure de la vallée de Kamechly-Tchaï (Cappodoce), vers 1700 m, 18-VI-1856, in fruit (G). Balansa, B. 183, sommet du Mont Sipyle, au-dessus de Magnésie, 15-VI-1854, in fruit (G) (dwarf plants). Balansa, B. 247, alluvious du Guzel-Déré à Moersina, 23-V-1855, in fruit (G). Balansa, B. 1224 & Balansa 866 , Ouchak (Phrygie), alt. vers 910 m, champs en friche, VI-1857, in fruit (G, K). Balls, E.K. 859, Gaziantep, cultivated land, alt. 2700 feet, 27-IV-1934, in fruit (K, E). Balls, K., B2223, Gaziantep, Kara Tash, vineyards and fallow land, nonlime, alt. 3300 feet, 3-V-1935, in fruit (BM, K, E). Bornmüller, J., Lydia, Bunarbaschi, in monte Takhtali-dagh, alt. 900 m, 26-V-1906, in fruit, J. Bornmüller Lydiae et Cariae Pl. Exsicc. 1906 No. 9256 (K) (identified as *Trigonella orthoceras*). Bourgeau, E., Elmalu, in collibus, 3-VI-1860, in flower & fruit (G). Davis & Hedge D28051, prov. Gaziantep, Dülük Baba, 7 km north of Gaziantep, alt. 1100 m, disturbed calcareous steppe, 13-V-1957, in flower & fruit (E). Davis & Hedge D28376, prov. Mardin, Mardin, alt. 1100 m, chalky vineyards, 21-V-1957, in fruit (BM, K, E). Davis 42907, Mardin, 11 km from Midyat to Gerçüs, alt. 1100 m, limestone steppe, 13-IV-1966, in fruit (E). Sintensis, P. 724, Armenia Turcica, Kharput, in cult. derelict, supra Pekenik, 12-VI-1889, in fruit (E, G, K, LD). **Syria:** Haradjian, M. 3, & Haradjian 50, environs d'Aintab, 3000 feet, V-1906, in flower & fruit (G). Haradjian, M. 1367, Mt. Dübük, au nord d'Aintab, 3–4000 feet, VI-1907, in fruit (G). Haradjian, M. 3963, Entre Alexandretti (?) et Aintab, 1–300 feet, 15-20-III-1910, in flower & very early fruit (G). Haussknecht, C., in graminibus, Aleppo & Aintab, Syria borealis, 22-IV-1865, in fruit (G). Mouterde, P. 5118, Alep (= Aleppo), 18-IV-1936, in late fruit (G). Peyron, E. 1380, Libano: Hammana (?), 11-VI-1882, in fruit (G).

29. *Medicago cretacea* M. Bieb.

M. cretacea M. Bieb., Fl. Taur.-Caucas. 2: 223 (1808), (basionym).

The epithet *cretacea* is Latin for chalky, of chalk, inhabiting chalky soils, a reference to the calcareous soils occupied.

Subordinate Taxa

None recognized here.

Type Specimen

Vassilczenko (1953) identified the type as Steven 1816, with the information (translated from the Russian "from the vicinity of Karasubazar [in the Crimea], type preserved in Leningrad." Vassilczenko (2002*b*) provided the following information for the type: "Crimea, vicinity of Belogorsk ('in Tauriae abruptis sterilissimis cretaceis circa Krassubasur')."

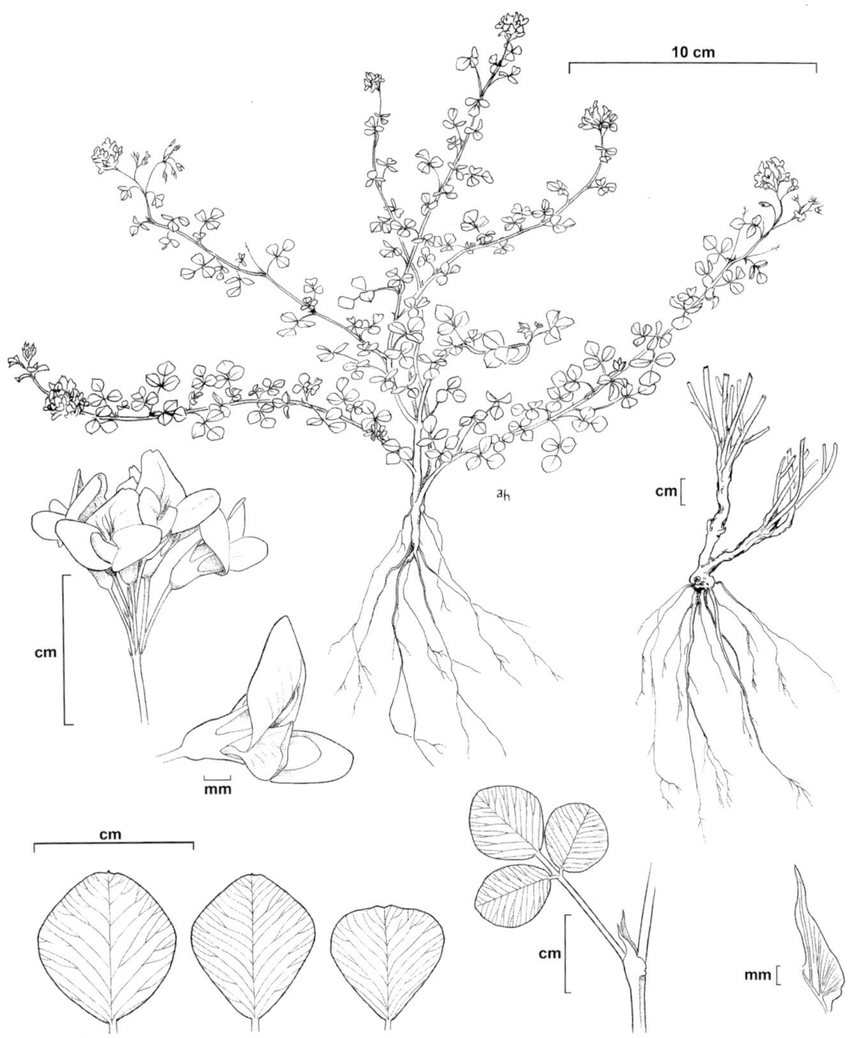

Fig. 29-1. *Medicago cretacea.* **Clockwise from bottom left:** Leaflets; side view of flower; group of flowers; plant; root and crown; stipule; leaf attached to stem. (The root and crown are based on Kolakovsky 3684, A; the remainder on E. Small M348, DAO; artist: A. Hanes.)

Synonymy

Homotypic

Trigonella cretacea (M. Bieb.) Taliev in Rast. mel. obnazh. Yuzhn. Ross. 2: 69 (1905).

Melissitus cretaceus (M. Bieb.) Latsch., Trudy Tiflissk. Bot. Inst. 19: 13 (1958).

Crimaea cretacea (M. Bieb.) Vassilcz., Novosti Sist. Vyssh. Rast. 16: 131 (1979).

Melilotoides cretacea (M. Bieb.) Soják, Sborn. Nár. Muz. Praze, 1982(1–2): 103 (1982).

Heterotypic

M. nunnularia Steven in Bess., Cat. Hort. Cremeneci 85 (1816). (Not *M. nummularia* DC., Cat. Pl. Horti Monsp. 124 (1813) = *Hymenocarpos circinatus* (L.) Savi var. *nummularius* (DC.) Bornm.)

Vernacular Names

ENGLISH: No common English name is available. Vassilczenko (1953) termed the species "cretaceous fenugreek" (translated from the Russian; he considered this to be a species of *Trigonella*, hence he used the common name of the genus "fenugreek").

FRENCH: No names found.

Geographical Distribution

(N = native, I = introduced, U = uncertain status):

ASIA: Russia (Krasnodar; N).

EUROPE: Ukraine (Crimea; N).

This narrowly distributed species occurs north of the Black Sea, including the southern coast of the Crimea and the northern Black Sea coast of the Caucasus. It is endemic to the southern Crimea and the northern part of western Transcaucasia.

Description

Perennial forb, numerous branching stems 10–30 cm long arising from woody base, ascending to erect, twigs and leaves almost glabrous or sparsely pubescent with simple, short, appressed hairs. Leaflets (4)5–8(10) mm long, slightly less broad than long, obovate to suborbicular, lacking serrations, glaucous, somewhat rigid, glabrous above, glabrous or somewhat pilose below. Stipules slender, lanceolate,

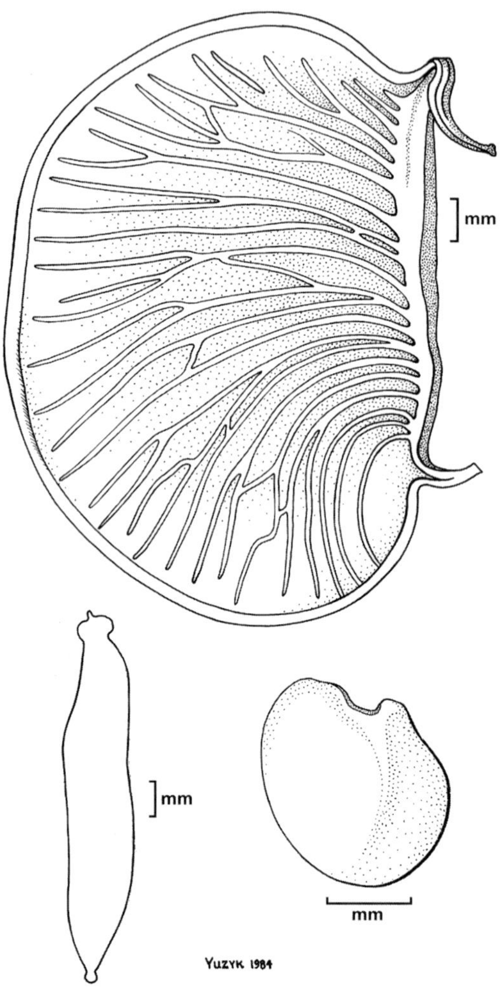

Fig. 29-2. *Medicago cretacea.* **Top:** Flat face of fruit. **Bottom left:** Outline of cross section of fruit. **Bottom right:** Seed. (Based on Kolakovsky 3684, A; artist: L. Yuzyk.)

Fig. 29-3. Type of *Medicago cretacea* at LE.

entire, 4–8 mm long. Inflorescence capitate, of 3–8 flowers. Peduncles 2–8 mm long, longer, as long as, or (most typically) shorter than subtending petiole. Flowers mostly 4–5 mm long, yellow. Calyx about half as long as corolla, calyx teeth longer than tube. Fruit flat, elliptic or reniform, ventral suture almost straight, dorsal suture semi-circularly curved, 5–12 mm long, 4–8 mm broad, initially diffusely tomentose and later glabrous, lacking spines, the sutures thickened, the radial veins prominent, almost parallel and scarcely anastomosing, the pod with a short, slightly curved or straight beak. Pods usually one-seeded (rarely with two seeds), the seeds smooth-surfaced, yellow-brown, 2–3.5 mm long, radicle almost as long as the seed.

Identification Hints

The fruit of this species (reniform, 1(2)-seeded, with little-anastomosed transverse veins) is distinctive and usually serves to identify the species.

Habitat and Environmental Adaptation

Habitat

Medicago cretacea occurs in dry, calcareous soils, often in lower ranges of mountains.

Phenology

The species flowers from May to July, with fruiting commencing in July in its native area.

Soil

Medicago cretacea typically occupies dry limestone mountain slopes, chalky outcrops, and calcareous, stony steppes.

Temperature

The semi-montane habitat of this perennial species indicates that it is tolerant of winter temperatures.

Water

The species typically occurs in dry conditions.

Biotic Relationships

Pollination

Small (1988) recorded a pollen to ovule ratio of 1338, which is exceptionally high for a species of *Medicago* and suggests that the plant is outcrossed. However, the flowers are relatively small for a perennial species, suggesting that the plant is an inbreeder. Probably *M. cretacea* is at least partly cross-pollinated, like other perennial species of *Medicago*.

Dispersal

The fruit is quite flat and therefore could be adapted in part for wind dispersal.

Herbivory

No information located.

Nodulation and mycorrhizae

No information located.

Disease resistance

Schmiedeknecht and Lesins (1968) recorded that of a number of species of *Medicago* examined this was the only one susceptible to *Pseudopezia meililoti*, although it was resistant to *Pseudopezia medicaginis*.

Economic History, Cultivars, and Uses

This species has not attracted agricultural development to date.

Cultivars

No cultivars have been bred.

Forage

Medicago cretacea is not sown for forage.

Fodder

This species has not been utilized as a fodder plant, although grazing animals will consume it.

Green manure, rotation crop, ground cover

No information located.

Other uses (salad sprouts, ornamental, plant products, etc.)

This species is occasionally grown in gardens.

Chromosome Number

$2n = 16$: Lesins and Lesins 1979.

$2n = 32$: Gardé and Malheiros-Gardé 1953.

Genetic Relationships and Gene Transferability

Intraspecific

Artificial hybridization of different populations has not been reported.

Interspecific

No information located.

Exchange by advanced biotechnology

No information located.

Conservation Status

Given that this species is a perennial, has a very limited distribution range, and has rarely been collected, it should be considered to have very high priority for a conservation assessment and associated protection measures.

Germplasm Sources

Only seven accessions of this species have been located in major germplasm institutions (see Chapter 3).

Research Problems and Priorities

Medicago cretacea is an endemic species that has received limited study to date. Very few germplasm collections have been made and additional collections should be acquired.

Very limited information is available on the characteristics of the species that may be of potential agronomic use, and such studies need to be undertaken.

Representative Collections

Ukraine: Callier, A., in declivibus montis Pertschem prope Sudak, 1896, in flower (G). Callier, A., same location, 25-V-1896, in fruit (G). Callier, A. 60, same location, 25-V-1896, in flower & fruit (S). Callier, A., In declivibus montis "Pertschem" prope "Sudak" V-VI-1900, in flower & fruit, Herbarium Normale editium ab I Dörfler No. 4226 (B, BM, G, LE, P). **Russia:** Kolakovsky, A., prov. Krasnodar, jugum Markotch, prope pag. Kabardinka, in clivis rupestribus calcareis, 19-VI-1951, in fruit (BM, K, S).

Note

Podol'naia and Efimova (1986) studied fruit development in *M. cretacea*.

30. *Medicago daghestanica* Rupr. ex Boiss.

M. daghestanica Rupr. ex Boiss., Fl. Orient. 2: 95 (1872), (basionym).

The epithet *daghestanica* alludes to the geographical restriction of this species to the Republic of Dagestan and its adjacent regions.

Subordinate Taxa

None recognized here.

Type Specimen

Ruprecht on 09 June 1861, Caucasus orientalis, Dagestania borealis (from Gimry, about 1000 m altitude), Museum botan. Acad. Petrop. (LE) (Fig. 30-3).

Synonymy

Homotypic

No names found.

Heterotypic

M. daghestanica Rupr. ex Boiss. var. *coerulescens* Trautv., Plantarum Messses Anno 1874, in Armenia a Dre. G. Radde et U. Dagiiestania ab A. Becher, p. 10 (1875; http://www.archive.org/stream/miscellaneouspap02trau/miscellaneouspap02trau_djvu.tx; current January 2010).

M. daghestanica Rupr. ex Boiss. var. *pallida* Trautv., op. cit. supra, p. 45.

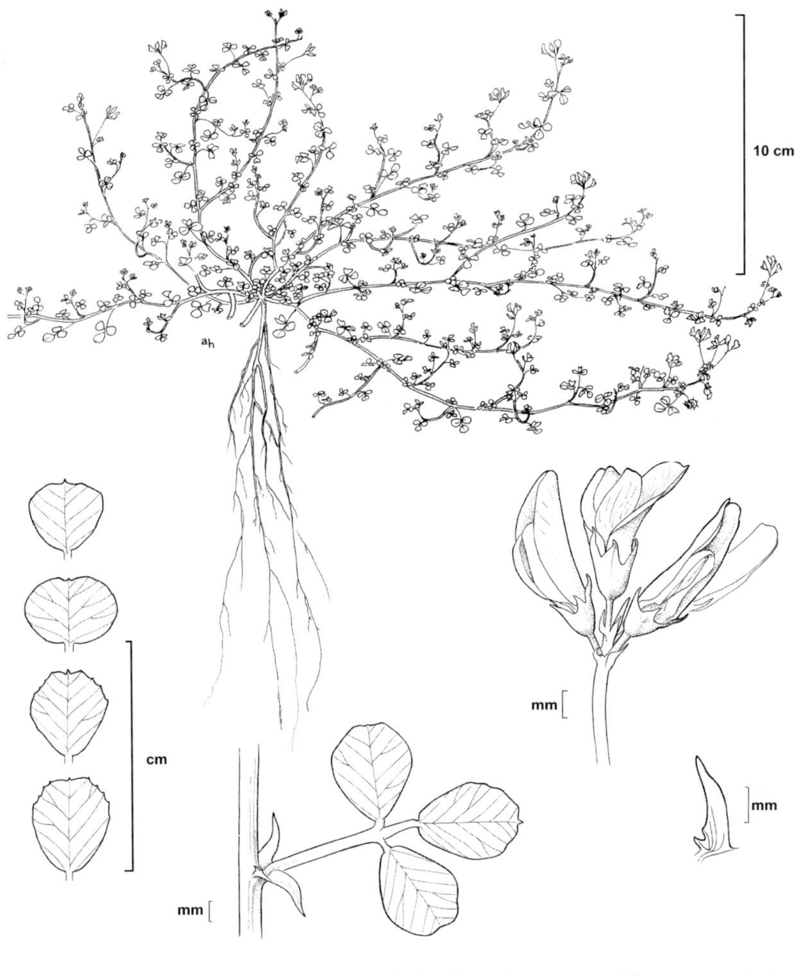

Fig. 30-1. *Medicago daghestanica.* **Clockwise from bottom left:** Leaflets; plant; group of flowers; stipule; leaf attached to stem. (Based on E. Small M1023, DAO; artist: A. Hanes.)

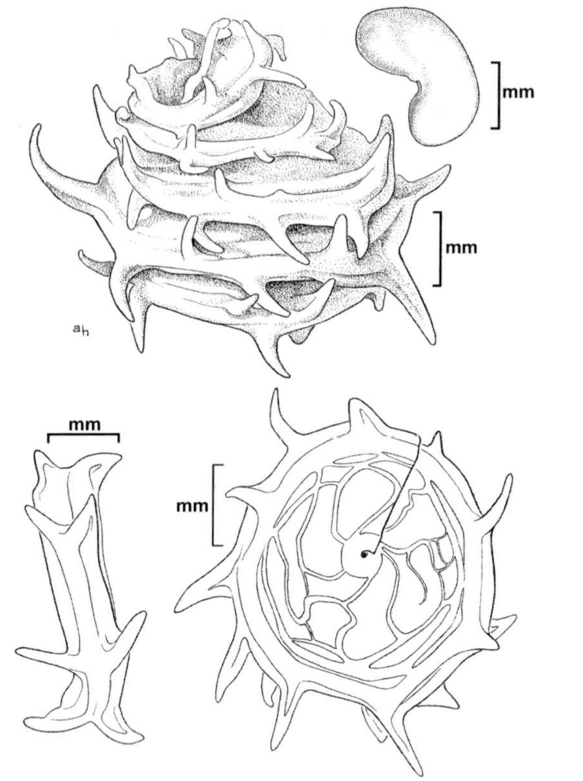

Fig. 30-2. *Medicago daghestanica.* **Clockwise from top left:** Fruit; seed; fruit coil face showing venation; edge of coil. (Based on E. Small M1023, DAO; artist: A. Hanes.)

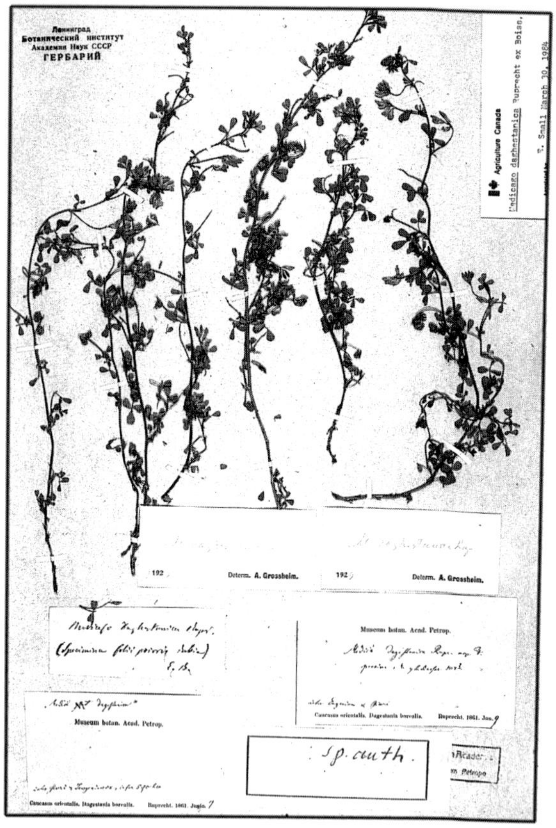

Fig. 30-3. Type specimen of *Medicago daghestanica* at LE.

Vernacular Names

ENGLISH: Dagestan alfalfa (Bolton 1962), Dagestanian medic (medic is also spelled medick).

FRENCH: No names found.

Geographical Distribution

(N = native, I = introduced, U = uncertain status):

ASIA: Azerbaijan (N), Russia (Checheno-Ingushetia, Dagestan; N).

Medicago daghestanica is endemic to subalpine regions of the Caucasus.

Description

Perennial forb with a root deeply penetrating the soil, numerous stems arising from the woody crown, stems 20–40 cm long, slender, decumbent or ascending. Twigs and lower surface of leaves pubescent with simple, adpressed hairs. Leaflets (3)5–8 mm long, triangular-obcordate, obovoid, or obcuneate, dentate in upper third (the serrations rather small and sometimes absent), somewhat stiff, veins prominent. Stipules of lower leaves semi-sagittate, ovoid-lanceolate, with acute apex and dentate base, prominently veined (stipules of upper leaves narrower and often entire). Inflorescence capitate, with 4–11 flowers, peduncles equalling or somewhat longer than the length of the subtending petiole. Flowers 6–8 mm long, white, light blue, or light violet, corolla twice length of calyx, the calyx campanulate, pubescent, the slender teeth slightly longer than the tube. Young fruit protrudes sideways from calyx. Fruit cylindrical, covered with glandular hairs or subglabrous, with 3–6 weakly adpressed coils, the coils 3–5 mm in diameter, the coil faces with slender, inconspicuous veins that anastomose in the outer third before entering a lateral vein, the outer margin of the coil with two deep grooves and short, slender, often slightly curved spines (occasionally reduced to tubercles). Seeds ovate or weakly reniform, 1.8–2.5 mm long, yellowish or light to dark brown. Seeds one or two in each coil, somewhat separated by a transverse, spongy septum, radicle half or less than half the length of the seed.

Identification Hints

Medicago daghestanica is the only species of *Medicago* with the combination of spiny, coiled fruits and (white or light blue to) violet flowers.

Habitat and Environmental Adaptation

Habitat

Medicago daghestanica occurs in mountainous areas, on calcareous taluses and rocks, and in dry meadows, at altitudes of 500–1500 m.

Phenology

The species flowers in June and July in its native area.

Soil

Medicago daghestanica typically occupies weathered, calcareous rocky soils.

Temperature

Lesins and Lesins (1979) observed that *M. daghestanica* survived the cold winters of Brandon, Manitoba when protected by snow or straw mulch.

Water

Medicago daghestanica often occupies dry sites.

Biotic Relationships

Pollination

Small (1988) noted that the relatively large flowers and large pollen to ovule ratio (795) of the species are consistent with *M. daghestanica* being at least partly cross-pollinated, like other perennial species of *Medicago*.

Dispersal

The short spines on the indehiscent pods may assist in distribution by becoming attached to the fur of animals, although the much more developed spines of other *Medicago* species suggest that this is not of major significance. The species occupies sloping hillsides and the sides of mountains, and the spines may be important in anchoring the pods to the substrate.

Herbivory

No information located.

Nodulation and mycorrhizae

No information located.

Disease resistance

No information located.

Economic History, Cultivars, and Uses

This species has attracted very little agricultural interest to date.

Cultivars

No cultivars have been bred.

Forage

Grossheim (1945) described *M. daghestanica* as "of no interest for forage use." By contrast, Vassilczenko (1949) stated the species "may be of value as a pasture-land plant."

Fodder

No information located. This species would seem to have some potential as a fodder plant.

Green manure, rotation crop, ground cover

No information located.

Other uses (salad sprouts, ornamental, plant products, etc.)

Medicago daghestanica is occasionally used as a rock garden plant, an appropriate usage given its small, tufted appearance and adaptation to rocky soils.

Chromosome Number

$2n = 16$: Lesins and Lesins 1961, 1979; Lesins and Gillies 1968; Gillies 1972*a*; Karshibaev 1992.

Genetic Relationships and Gene Transferability

Intraspecific

Hybridization among different populations of *M. daghestanica* has apparently not been attempted.

Interspecific

Medicago daghestanica has been crossed with *M. pironae*, although the hybrids were completely sterile (Lesins and Lesins 1961; Lesins and Gillies 1968; Lesins 1971). Lesins and Lesins (1979) reported that they were unable to hybridize *M. daghestanica* with *M. marina*. Lesins (1971) produced tri-species hybrids between *M. pironae*, *M. daghestanica*, and *M. sativa* by crossing tetraploid *M. sativa* and an artificial tetraploid hybrid of the other two species. McCoy and Echt (1993) were unable to hybridize *M. daghestanica* with *M. sativa* by conventional hybridization (but see information below on tri-species crosses and embryo rescue). Lesins and Gillies (1972) stated that unpublished work by Lesins showed that *M. daghestanica* could be hybridized with *M. sativa*, but not at the diploid level.

Exchange by advanced biotechnology

McCoy and Bingham (1988) stated that (in unpublished work) interspecific hybrids had been produced between diploid *M. sativa* and diploid *M. daghestanica* using ovule–embryo culture. Genes can be transferred from *M. daghestanica* to alfalfa (*M. sativa*) by tri-species hybridization and embryo rescue (McCoy and Echt 1993, following work by Lesins and Gillies 1968). By using this technique, hybrids of *M. sativa* and *M. rupestris* are first produced, and these hybrids are then crossed with *M. daghestanica*.

Conservation Status

The narrow distribution range and few germplasm collections of this species indicate that a high priority should be given to protecting its habitat and acquiring additional seed.

Germplasm Sources

Only one accession of this species has been located in germplasm institutions (see Chapter 3).

Research Problems and Priorities

Medicago daghestanica is a rather rare species that is poorly understood. Germplasm collection, detailed ecological analysis of the plants in their native area, and studies of the agronomic characteristics and potential for improving alfalfa are required.

Representative Collections

Russia: Alexeenko 6146, prov. Dagestan, distr. Dargni, Alp. Devaschi, in herbidis lapidosis, alt. 3800, 30-V-1901 (LE). Alexeenko 6147, prov. Dagestan, distr. Awarsk, in decliv. meridionale montis Gimri, in saxosis calcareis, alt. 3600 feet, 22-V-1901 (LE). Alexeenko, Th. 6143, prov. Dagestan, distr. Gaunib, Ad viam inter pag. Gunibet Czech, alt. 3100–5300 feet, 6-VII-1897 (LE). Alexeenko, Th. 6145, Dagestan, distr. Dargi, pr. pag. Akoscha, in abruptis argillosis, alt. 4500 feet, 13-VII-1898 (LE).

31. *Medicago disciformis* DC.

M. disciformis DC., Cat. Pl. Horti Monsp. 124 (1813), (basionym).

The epithet *disciformis* is derived from the Greek *disciformis*, *disci*, disc, plus the Latin *formis*, which means "shaped" or "resembling", an allusion to the disk-shaped pods.

Subordinate Taxa

None recognized here.

Type Specimen

The holotype, cited by Heyn and Davis (1970), is "In aridis apricis rupestribus medio vere prope Castelnau in agro Monspessulano" (P). As noted by Heyn (1963), this collection from France was used by De Candolle in recognizing his new species, *M. disciformis*.

Synonymy

Homotypic

Medica disciformis (DC.) Bubani, Fl. Pyren. 2: 477 (1899).

Heterotypic

M. disciformis DC. var. *strumensis* Velč. & Bond., Bulg. Acad. Sci. Reports 8: 218 (1961). (Described as a Balkan endemic.)

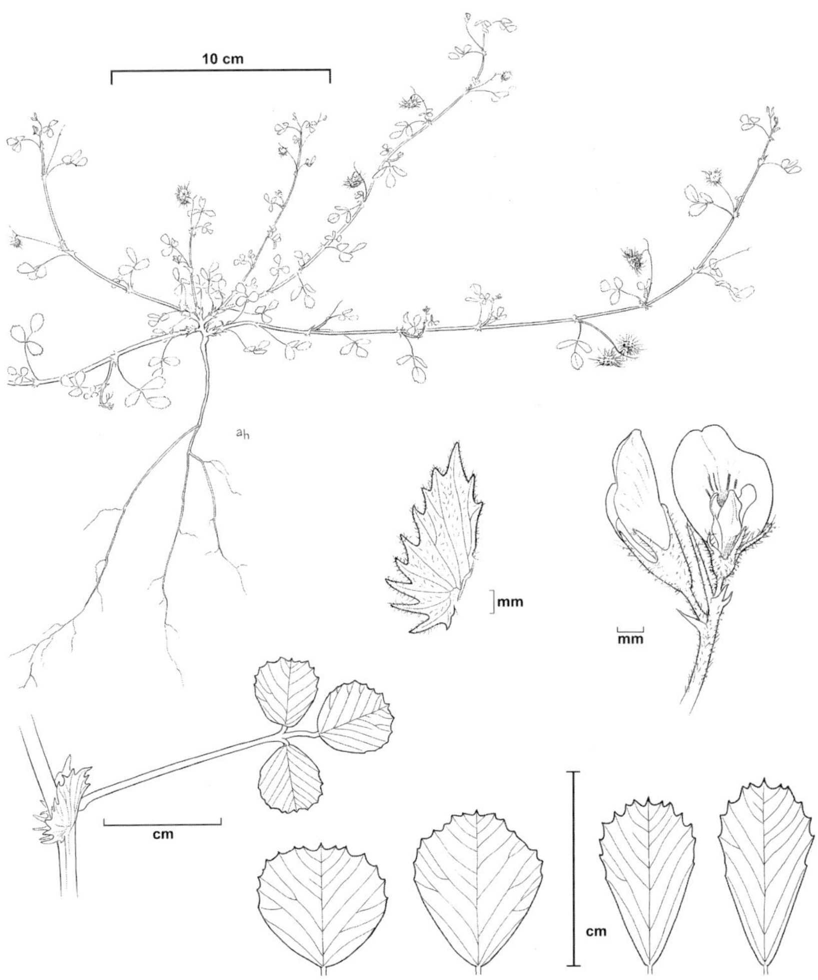

Fig. 31-1. *Medicago disciformis.* **Clockwise from bottom left:** Leaf attached to stem; plant; group of flowers; leaflets. **Center:** Stipule. (Plant based on Chapman et al. 86, US, remainder based on E. Small M1013, DAO; artist: A. Hanes.)

M. disciformis var. *strumensis f. parvicarpa* Kozuharov, Bulgarian Acad. Sci. 15: 163 (1965).

Kozuharov (1965) cited the name *M. disciformis* DC. var. *apiculata* Urb., referring to Verh. Bot. Ver. Brandenb. 15: 79 (1873), but Urban did not publish this combination there. Heyn (1963) cited "forma *apiculata*" without reference to a publication.

Vernacular Names

English: No common English name is in usage. A literal translation of the Latin name would be "disc medic(k)". However, this name is applied to *M. italica* (usually called "*M. tornata*") and its cultivars, whereas "large disc medic(k)" and "black disc medic(k)" refer to *M. orbicularis*.

French: Luzerne en disque, luzerne à fruits en disque (contrived names, rarely encountered, based on literal translation from the Latin name).

Geographical Distribution

(N = native, I = introduced, U = uncertain status):

Africa: Algeria (N), Libya (N).

Asia: Cyprus (N), Turkey (N).

Europe: Bulgaria (N), Czech Republic (I), France (including Corsica; N), Greece (including Crete and East Aegean Islands; N), Italy (including Sicily; N), Spain (islands; N), Turkey (N), Yugoslavia (N).

Medicago disciformis occurs mostly in the northwest Mediterranean area, but extends to the eastern Mediterranean.

Description

Annual herb (10)15–25(30) cm high, usually procumbent, branched from base. Vegetative parts covered with simple and gland-tipped hairs. Leaflets 4–6(12) mm long, about one-third longer than broad, obovate to obcordate, upper third with serrate margin. Stipules 4–6 mm long, entire or dentate. Inflorescence of 1–2(4) flowers, peduncle longer than subtending petiole. Flowers 4–6(7) mm long, yellow. Corolla less than twice length of calyx. Calyx with simple and glandular hairs, teeth slightly shorter than tube. Young fruit protrudes sideways from calyx. Fruit discoid, 3–5 mm long, glabrous at maturity (gland-tipped hairs often present on young fruit), with 5–7(8) coils, the coils flexible and not tightly compressed, the largest coils 5–8 mm wide, the coils decreasing gradually towards the ends. Coil face with S-shaped (nonanastomosing) radial veins that enter into a veinless margin in outer half of

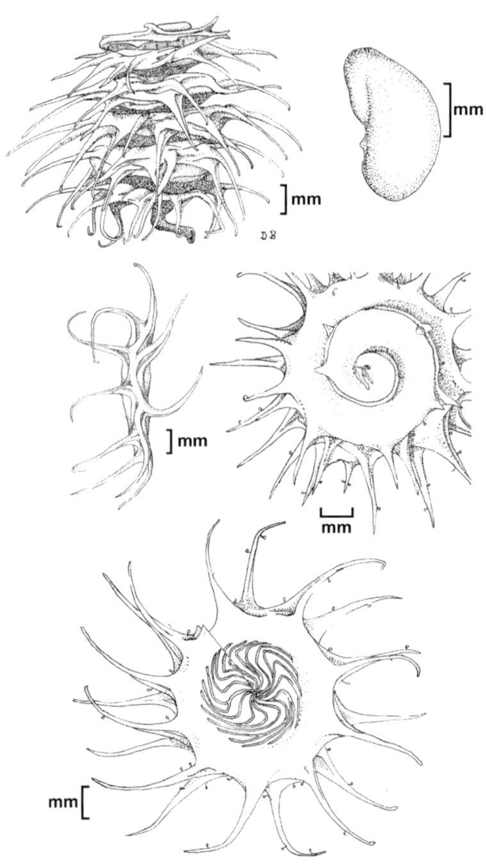

Fig. 31-2. *Medicago disciformis.* **Clockwise from top left:** Side view of fruit; seed; view of stigma end of fruit; fruit coil face showing venation; edge of coil. (Based on E. Small M17, DAO; artist: D. Beddoes.)

Fig. 31-3. *Medicago disciformis.* From Hallier (1885, vol. 23, plate 2347).

coil. Spines slender, about 4 mm long (or rarely very short), distal coils spineless, spines grooved, inserted at angle of 180° to outer fruit surface (or at a narrower angle towards ends of fruit). Seeds 2.5–4.5 mm long, 1.3–1.8 mm wide, oblong or reniform, the surface smooth and yellow to light brown, 1–2 per coil, not separated by transverse spongy septa, radicle about half length of seed or up to two-thirds of the length.

Identification Hints

Medicago disciformis is usually distinguishable by the peripheral veinless area of the fruit coil face, spines slanted towards the base, and the (usually) spineless (or with very reduced spines) apical coil. The most likely species to be confused with *M. disciformis* is the similar *M. tenoreana*. *Medicago disciformis* has spines slanted towards the fruit base, whereas in *M. tenoreana* the spines point towards the base and the tip. *Medicago tenoreana* also has thicker coil edges.

Habitat and Environmental Adaptation

Habitat

Medicago disciformis typically occurs in abandoned fields, rocky hillsides, and woodlands. Thanopoulos (2007) recorded its presence in Greece on gravelly beaches, ravines, littoral places, garrique, macchie, other shrub and woodland associations, open and stony places, grasslands, and as a weed of cereal crops, orchards, fallow fields, and ruderal places, from sea level to 1500 m in altitude.

Phenology

This species flowers from March to May in its native area.

Soil

Piano and Francis (1992) observed that in its native area *M. disciformis* occurs on loam-clay soils that are neutral to moderately basic. The species often occupies rocky soils.

Temperature

The species occurs in the Mediterranean climate typical of annual *Medicago* species. Krall et al. (1996*a*) found that an accession of *M. disciformis* tested for its ability to self-regenerate in Wyoming failed to survive over the winter at three different locations.

Water

Medicago disciformis typically occurs in moderately dry areas. Piano and Francis (1992) recorded that in its native area it requires a minimum annual rainfall of 350 mm.

Biotic Relationships

Pollination

Small (1988) noted that the small flowers and small pollen to ovule ratio (176) of the species are consistent with *M. disciformis* being largely self-pollinated. This species is basically an inbreeder.

Dispersal

The spiny fruits of *M. disciformis* are adapted to dispersal in animal fur. Spineless forms, sometimes recognized as *M. disciformis* f. *apiculata*, have been found in Cyprus and Turkey, i.e., the eastern part of the range of the species, which suggests there is a different mode of seed distribution there.

Herbivory

Shade et al. (1975) examined the resistance of 21 species of *Medicago* to attack by larvae of the alfalfa weevil (*Hypera postica* (Gylh.)) and found that *M. disciformis* was 100% resistant. The exudate from the secretory trichomes was found to be responsible for larval antibiosis, and newly hatched larvae did not survive when placed on the plant or when the exudate was applied to them topically. Johnson et al. (1980*a*, 1980*b*) similarly demonstrated resistance to the alfalfa weevil. Shade et al. (1979) documented the strong resistance of *M. disciformis* to the potato leafhopper (*Empoasca fabae* Harris). Shade and Kitch (1983) noted that the glandular hairs of *M. disciformis* resulted in resistance to the pea aphid (*Acyrthosiphon pisum* (Harris)). Ferguson et al. (1982) found that *M. disciformis* was quite resistant to the spotted alfalfa aphid *Therioaphis maculata* (Buckton). MacLean and Byers (1983) observed that the glandular hairs of *M. disciformis* tended to deter the alfalfa blotch leafminer (*Agromyza frontella* (Rondani)) from ovipositing.

Nodulation and mycorrhizae

USDA (2009) reported positive root nodulation for *M. disciformis*.

Disease resistance

Yaege and Stuteville (2002) tested the resistance of 33 annual species of *Medicago* to powdery mildew (*Erysiphe pisi*) and found that three of the four collections of *M. disciformis* grown were resistant to the disease (it was observed that often only one of a number of populations of other species tested was resistant). O'Neill and Bauchan (2000) in a comparative test of 36 annual species of *Medicago* found that four accessions of *M. disciformis* were quite susceptible to anthracnose caused by *Colletotrichum trifolii*.

Economic History, Cultivars, and Uses

Medicago disciformis has attracted limited agricultural interest to date.

Cultivars

No cultivars have been bred.

Forage

Medicago disciformis is not sown for forage, although it is cropped by grazing animals in areas where it occurs. Mohammad and Qamar (1988) grew 14 annual species of *Medicago* in small test plots in Islamabad, Pakistan, located in a subtropical, subhumid area that is notably hotter and wetter than the normal habitat of these species. The dry matter yield of *M. disciformis* in this environment was 8.5 t/ha, higher than all the other species examined.

Fodder

This species is not grown as a fodder crop.

Green manure, rotation crop, ground cover

No information located.

Other uses (salad sprouts, ornamental, plant products, etc.)

No information found (but see note below).

Chromosome Number

$2n = 16$: Ghimpu 1928, 1929*b*, 1930; Simon and Simon 1965; Lesins and Lesins 1979; Song et al. 1988.

Genetic Relationships and Gene Transferability

Intraspecific

Barriers to hybridization within the species have not been examined.

Interspecific

Medicago disciformis is morphologically closest to *M. tenoreana*, although Lesins and Lesins (1979) were unable to obtain hybrids between these species. G.R. Bauchan (personal communication) attempted without success to hybridize *M. disciformis* with both diploid and tetraploid *M. sativa*, with the goal of transferring insect resistance. In addition, colchicine was used to produce a tetraploid *M. disciformis* that was utilized in hybridization studies with *M. sativa* without success (Bauchan 1984).

Exchange by advanced biotechnology

No information located.

Conservation Status

Medicago disciformis is a moderately widespread species, not in danger of reduction because of habitat loss, and so not requiring conservation measures.

Germplasm Sources

More than 200 accessions of *M. disciformis* are available from germplasm institutions (see Chapter 3).

Research Problems and Priorities

Medicago disciformis is a typical annual *Medicago*, but one that has not been exploited economically. It has not received much research to date and requires study of its economic potential.

Representative Collections

ALGERIA: Azeudt 118, Alger, in fruit (BM). **BULGARIA:** Kozuharov, S. & Kuzmanov, B. 673a, prope pag. Lebnica, distr. Blagoevgrad, 6-V-1960 (BM). Kozuharov, S. & Kuzmanov, B., Bulgaria australis, prope pag. Marik ostino, distr. Blagoeugrad, in herbidis siccis, II-V-1961, in fruit, Inst. Bot. Acad. Scientiarum Bulgarine Plantae Bulgaricae Exsiccatae Cent. VII No. 673b (B, BM). **CROATIA:** Pichler, T., Dalmatia, ad S. Girolomo prope Spalato, V-1885, in fruit (K). Pichler, T., Dalmatia, in Mth. Mariano pr. Spalato, V-1880, in fruit (K). Pichler, T., Dalmatien Ragusa, 9-VI-1868, in fruit (K). **FRANCE:** Chambeiron, Toulon (Var.), bords des champs calcaires incultes au pied du Mont Faron, 28-V-1860, Flora Galliae et Germaniae exsiccata de C. Billot No. 3035 (P). Humbert, H., Cassis, Bouches-du-Rhône, lieux herbeux, "fin avril"-1912, in fruit (P). Messeray, R., Montpellier, garigue calcaire, 15-IV-1936 (also 20-IV-1936), in flower & fruit (K). **GREECE:** Bourgeau, E., Lycica, Adalia, in graminosis, 13-IV-1860, in fruit (P). De Heldreich, Attica prope Athenas, in locis herbidis, in colle Lycabetto, 8-IV-1891, in fruit, De Heldreich Herbarium Graecum Normale No. 1129 (K, P). Brummitt, R.K. 6578, Kirkladhes, island of Delos (Dhilos), west side, 10-V-1969, in fruit (K). Griffith, H. 1798, Thrace, above Yassikevy, alt. 500 feet, in scrub of *Quercus coccifera*, 14-VI-1936, in fruit (K). Lewis, G.P. 585, Saronic Islands, Spetsai, Spetsai town, alt. ca. 800 feet, meadowland near summit of limestone hill, 6-V-1980, in fruit (K). Rechinger, K.H. & F. 9361, Thracia occidentalis, in fauce fluvii, Nestos (Mesta) prope Toxotai (Oktschilar), alt. ca. 60–100 m, substr. calc., 12-VI-1936, in fruit (K). Townsend, C.C. 640417/10, Cyclade 5, Delos, several plants, behind the archaic lions, 17-IV-1964, in fruit (K). **ITALY:** Porta & Rigo, Italia Austral., Apulia in collibus graminosis ad Manfredamum, alt. 100–400 feet, 25-IV-1875, in flower & fruit, Porta & Rigo ex. itinere II Aalico No. 110 (K, P). **SPAIN:** Bourgeau, E. 697, Pyrenées Espagnoles, Martourel, près de Mont-Serrat, 6-V-1892, in fruit (K). Bourgeau, E., province de Murcia, Cabo de Palos, champs incultes, 10-IV-1952, in fruit, E. Bourgeau Pl. d'Espagne 1852 No. 1585 (P). **TURKEY:** Balansa, B., Smyrne, 9-V-1954, in fruit (P). Barclay, C. 1466, Ismir Province, Bergama (Pergamon), 109 km north of Ismir, alt. ca. 1000 feet, among antique ruins, 17-V-1969, in fruit (K). Bornmüller, J. 4595, Blithynia prope Mudania, in siccis collium, 14-V-1899, in fruit (P).

Note

Bird and Wingham (1983) studied the presence of certain lectins in a dozen species of *Medicago* (lectins are proteins that selectively bind carbohydrates and carry out several important metabolic functions in cells), and observed that these were only detectable in *M. disciformis* and *M. turbinata*. The lectins in *M. disciformis* are used in the classification of cryptantigens and for red blood cell polyagglutinability as a means of blood typing (Bird and Wingham 1983).

32. *Medicago doliata* Carmign.

M. doliata Carmign., Giorn. Sci. Lett. Accad. Ital. Sci. 1: 48 (1810), (basionym).

The epithet *doliata* is from the Latin *dolium*, cask, descriptive of the typical barrel shape of the pods of this species.

Subordinate Taxa

None recognized here.

Type Specimen

Sales and Hedge (1994) lectotypified the name *M. doliata* Carmign. by "a specimen [at Pisa] bearing the name '*Medicago doliata* Carmignani' in the handwriting of Gaetano Savi, Herb. Hort. Pisani p. 422."

Synonymy

Homotypic

No names found.

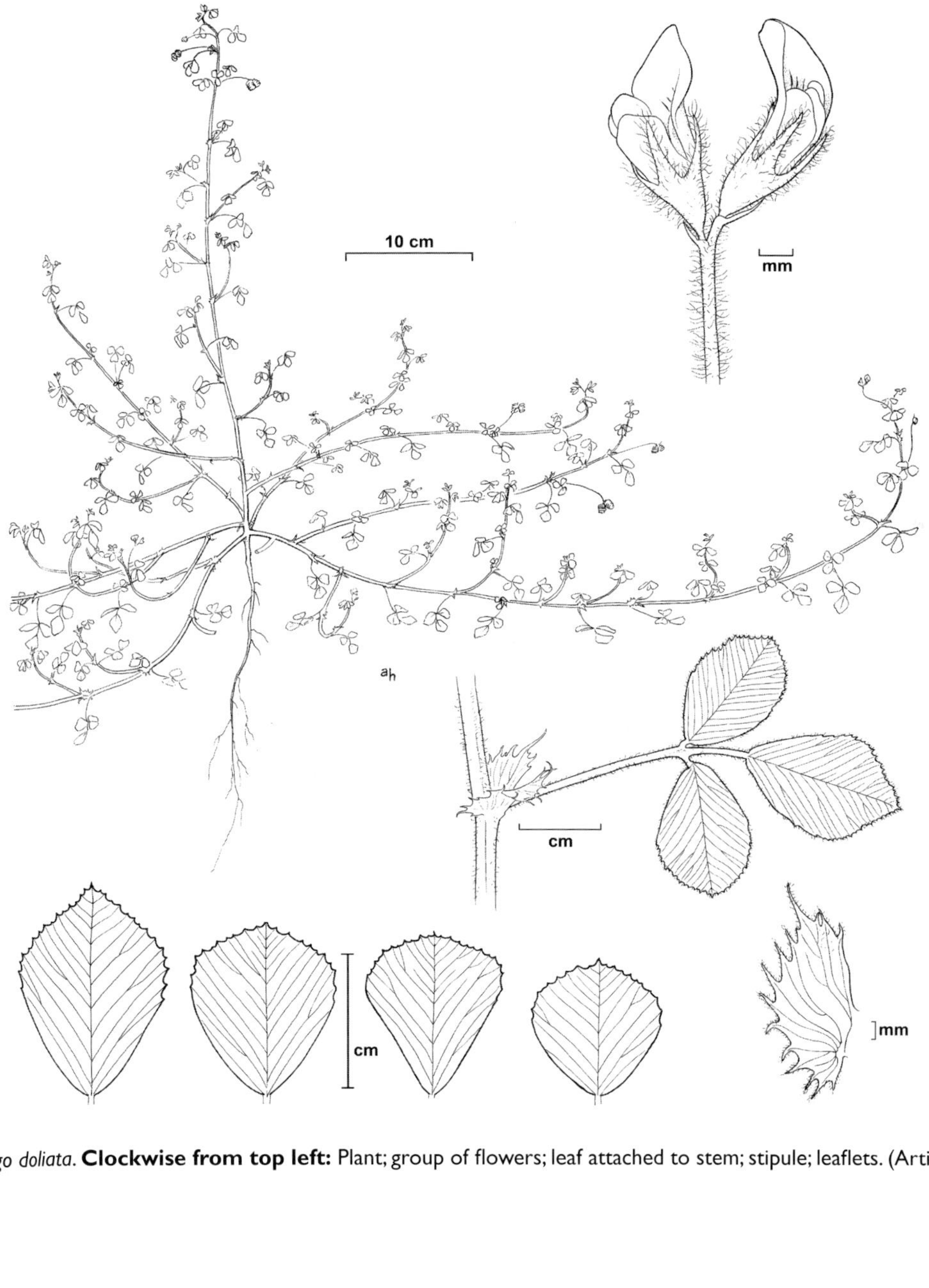

Fig. 32-1. *Medicago doliata.* **Clockwise from top left:** Plant; group of flowers; leaf attached to stem; stipule; leaflets. (Artist: A. Hanes.)

Heterotypic

M. aculeata Willd. Sp. Pl. 3(2): 1410 (1802), non Gaertner, Fruct. Sem. Pl. 2: 349 (1791). (Heyn (1963, p. 113) noted that the type of *M. aculeata* Willd. is the specimen Herb. Willdenow 1317 at B; I have observed this specimen, which bears the handwriting in the bottom right hand corner "Bouché. W." and the annotation slip Photo-Nr. 875 as well as IDC cat. nr. 14317. Heyn (1963) concluded that *M. aculeata* Gaertn., Fruct. Sem. Pl. 2: 349 (1791) is a nomen confusum, sometimes applied in the past to what is here termed *M. doliata*).

"*M. turbinata* Willd." var. *aculeata* Moris, Fl. Sardoa 1: 445, tab. 45, fig. B (1837).

"*M. turbinata* Willd." subsp. *aculeata* (Willd.) Nyman, Consp. Fl. Eur. 168 (1878).

"*M. turbinata* Willd." var. *sphaerocarpos* Nègre subvar. *aculeata* (Gaertn.) Nègre, Luzern. Maroc. 58 (1956).

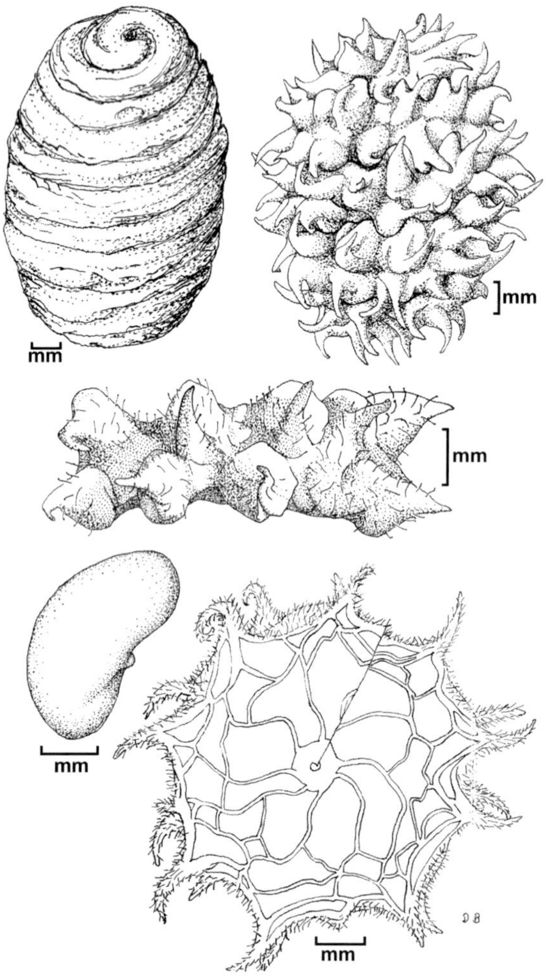

Fig. 32-2. *Medicago doliata.* **Clockwise from top left:** Spineless fruit; spiny fruit; edge of coil; fruit coil face showing venation; seed. (Spineless fruit based on E. Small M333, DAO, remainder based on E. Small M816, DAO; artist: D. Beddoes.)

M. oliviformis ("*olivaeformis*") Guss. var. *crassa* Porta et al., Nuov. Giorn. Bot. Ital. 11: 249 (1879).

"*M. turbinata* Willd." var. *inermis* Asch. f. *dextrorsa* Asch., Cat. Hort. Berol. App. (1871).

"*M. turbinata* Willd." var. *aculeata* (Moris) Urb. f. *dextrorsa* Asch., Cat. Hort. Berol. App. (1871).

M. globulosa Desv., J. Bot. Agric. Appl. 3: 77 (1814).

"*M. turbinata* Willd." var. *inermis* Asch., Cat. Hort. Berol. App. (1871). (Heyn (1963, p. 114) noted that the type of "*M. turbinata* Willd. var *inermis* Aschers." is the specimen labelled by Willdenow "Monspel. Herb. Willdenow 14315/1" at B; I observed this. It is labelled IDC microfiche reproduction 14315 (loose annotation slip), with Photo-Nr. 873 stamped in blue on the specimen.)

"*M. turbinata* Willd." var. *laevis* Boiss., Voy. Bot. Esp. 165 (1839–1845).

M. muricata Benth., Cat. Pl. Pyr. Bas-Languedoc 102 (1826). (Heyn and Davis (1970) cited as the type: Bentham, collected in Montpellier, France (K)). (The name *M. muricata* is applied to *M. doliata* by some authors, but not all.)

M. doliata var. *muricata* (Benth.) Heyn, in Davis, P.H. (ed.), Flora of Turkey 3: 507 (1970).

M. neglecta Guss., Fl. Sicul. Prodr. 2: 575 (1828, i.e., 1828–1832).

"*M. turbinata* Willd." var. *neglecta* (Guss.) Batt., Batt. & Trab. Fl. Algér. 2: 230 (1888).

"*M. turbinata* Willd." var. *sphaerocarpos* Nègre subvar. *neglecta* (Guss.) Nègre, Luzern. Maroc. 58 (1956).

M. oliviformis ("*olivaeformis*") Guss., Pl. Rar. 390 (1826).

"*M. turbinata* Willd." subsp. *oliviformis* (Guss.) Nyman, Consp. Fl. Eur. 168 (1878).

"*M. turbinata* Willd." var. *oliviformis* (Guss.) Batt., Batt. & Trab. Fl. Algér. 2: 230 (1888).

"*M. turbinata* Willd." var. *cylindracea* Nègre subvar. *oliviformis* (Guss.) Nègre, Luzern. Maroc. 58 (1956).

"*M. turbinata* Willd." var. *inermis* Asch. f. *sinistrorsa* Asch., Cat. Hort. Berol. App. (1871).

"*M. turbinata* Willd." var. *aculeata* (Moris) Urb. f. *sinistrorsa* Asch., Cat. Hort. Berol. App. (1871).

"*M. turbinata* Willd." var. *turbinata* (L.) Thell., Fl. Adv. Montp. 312 (1912).

"*M. turbinata* Willd." var. *typica* Hal., Consp. Fl. Graec. 1: 363 (1901), nom. inval.

Not

M. aculeata Gaert., Fruct. et Sem. Pl. 2: 349 (1791). Heyn (1963) interpreted the name *M. aculeata* Gaertn. as a nomen confusum, but assigned it to *M. aculeata* Willd., i.e., *M. doliata*. However, Heyn and Davis (1970) wrote "Here we accept the opinion of Greuter (Greuter & Rechinger, Chl. Kyther. (Boissiera 18): 81 (1967)) that

M. doliata Carmign. is the only name for this species not based on doubtful identification. *Medicago aculeata* Gaertn. is based on an ambiguous illustration, and *M. aculeata* Willd. (1902) is a later homonym."

Numerous authors have (mistakenly) used the name *M. turbinata* Willd. to refer to *M. doliata*, hence "*M. turbinata* Willd." in the above synonymy.

Vernacular Names

English: Keg medic (medic is also spelled medick).

French: Luzerne à fruits épineux.

Geographical Distribution

(N = native, I = introduced, U = uncertain status):

Africa: Algeria (N), Libya (presence uncertain), Morocco (N), Tunisia (N).

Asia: Israel (N), Jordan (N), Iran (N; Mehregan et al. 2002), Lebanon (N), Syria (N), Turkey (N).

Europe: Albania (N), Bulgaria ("reported in error" according to Greuter et al. 1989), Greece (including Crete; N), France (including Corsica; N), Italy (including Sardinia and Sicily; N), Malta (N), Portugal (N), Spain (N), Yugoslavia (N).

Medicago doliata is an omni-Mediterranean species. Prosperi et al. (2001) noted that the species is abundant in North Africa and southern Spain, but is rarer in the northern Mediterranean countries.

Description

Annual herb, 20–50 cm high, ascending or procumbent, branched from base, vegetative parts covered with simple hairs. Leaflets 10–15(20) mm long, obovate to oblanceolate, upper third of margin serrate. Stipules dentate. Inflorescence of 1–2(5) flowers, peduncles longer than subtending petiole. Flowers 4–6 mm long, corolla yellow, slightly longer than calyx, calyx with simple hairs, teeth about equal to length of tube. Young pod contracted and concealed within calyx. Fruit spherical to ovoid (proximal coil truncate), 7–15(20) mm long, 6–10 mm at widest coil, covered with glandular and/or simple hairs or rarely glabrescent, spiny or spineless, strongly hardened at maturity. Coils 5–7(9), strongly adpressed at maturity, boundary between adjacent coils often wavy, coil surface with slightly bent radial veins anastomosing in distal third, the most distal veins entering into a lateral vein that arches between adjacent spines (vein features not visible in mature fruit). Ridge between lateral vein and dorsal vein not apparent at maturity. Spines (when present) inserted at angle of 180° to coil surface, conical, the tips sometimes hooked, the base often covered by alveolar tissue in mature fruit, spines sometimes absent or reduced to tubercles. Seeds (3)4–6 mm long, 1–2 per coil, separated by transverse, spongy septa, smooth, yellow to yellow-brown, radicle less than half of length of seed.

Identification Hints

Medicago doliata is very similar to most of the other hard-fruited annual species of the genus. If only mature fruits are available, on which alveolar tissue obscures the venation, one might mistakenly conclude that a veinless zone on the pod face is present, and the material could be misidentified as *M. turbinata, M. murex*, or *M. sphaerocarpos*. *Medicago doliata* is particularly likely to be confused with *M. rigidula* and *M. rigiduloides* (the three species frequently have fruits covered with gland-tipped trichomes; the venation of the coil face of *M. doliata* is less anastomosed peripherally than is the case in the other two species); *M. doliata* also resembles *M. constricta*, which has much less evident, if any, gaps between the fruit coils at maturity.

Habitat and Environmental Adaptation

Habitat

Medicago doliata occurs in open woodlands and arable land from sea level to 900 m in altitude. Illustrative of the range of habitats occupied, Thanopoulos (2007) recorded that in Greece the species grows in colluvial loam, on gravel beach, near coasts, in grasslands, by roadsides, and as a weed in open olive groves.

Phenology

The species flowers in the spring in its native area, usually between March and June depending on location.

Soil

Medicago doliata often occupies clay soils, as observed by Nègre (1956) and Bounejmate et al. (1992) in Morocco and Heyn (1963) in Israel. In Algeria, Abdelguerfi et al. (1988*a*) noted that it occurs in medium- to very fine-textured soils. Piano and Francis (1992) recorded the species generally in clay and loam soils that are slightly to moderately basic. In Sicily, Piano et al. (1991) observed the species growing in a pH range of 6.5–9.0.

Temperature

Medicago doliata is a winter annual that passes its life cycle during the progressively warming winter to summer Mediterranean climate. *Medicago doliata* occupies a range of altitudes, but is often found at higher altitudes than most other medics. Most commercial medic (i.e., annual *Medicago*) cultivars have been bred for the Mediterranean climates of southern Australia where they experience mild winters (Crawford 1983*a*), so there has been a desire to find cold-tolerant genotypes. Piano and Francis (1992) observed that the species is often one of the dominant medics in the cold environments of western Asia and North Africa. Bounejmate et al. (1994) looked for genotypes of *M. doliata* in Morocco that might survive in the low temperatures of inland Morocco and found that the most frost-tolerant populations came from the highest altitudes of the distribution range in the country. However, genotypes from lower altitudes had larger seeds and produced more shoot dry matter. Krall et al. (1996*a*) found that an accession of *M. doliata* tested for its ability to self-regenerate in Wyoming failed to survive over winter at three different locations.

Water

The species is often in moist soils, although also not uncommonly in dry habitats. In Tunisia, it prefers relatively moist areas, occupying a humid to subhumid zone with 600–1200 mm of precipitation annually (Zoghlami and Hassen 1999, 2004). Lesins and Lesins (1979) recorded that it did not tolerate heavy watering in their greenhouse. Abdelguerfi et al. (1988*a*) noted that *M. doliata* appears to prefer well-drained soils, in areas receiving more than 450 mm of precipitation. Piano and Francis (1992) observed that in its native area the species requires a minimum of 350 mm annual precipitation.

Biotic Relationships

Pollination

Small (1988) noted that the small flowers and pollen to ovule ratio (257) of the species are consistent with *M. doliata* being largely self-pollinated. This species is basically an inbreeder.

Dispersal

The spiny fruits of *M. doliata* are adapted to dispersal in animal fur. Forms of the species with reduced spines or entirely lacking spines may have different dispersal methods, or may simply be adapted to growing in localized areas.

Herbivory

Bolton (1962) stated that the species (as *M. aculeata*) has been reported in California to be little affected by the clover seed chalcid (*Bruchophagus platyptera* Walker).

Nodulation and mycorrhizae

USDA (2009) reported positive root nodulation for *M. doliata*.

Chebouti and Abdelguerfi (2003) compared the root weight, number of nodules, and nodule weight responses of *M. doliata* when it was grown under either drought or irrigation and found that the three variables decreased more or less proportionately to each other under drought conditions.

Disease resistance

O'Neill et al. (2003) examined the comparative resistance of 36 annual species of *Medicago* to *Phoma medicaginis* (the cause of spring black stem and leaf spot, a major disease of alfalfa) and concluded that *M. doliata* had high resistance. Similarly, in a comparative test of 36 annual species of *Medicago*, O'Neill and Bauchan (2000) found that three accessions of *M. doliata* were quite susceptible to anthracnose caused by *Colletotrichum trifolii*, but one accession showed 33% resistance to the disease. Yaege and Stuteville (2002) tested the resistance of 33 annual species of *Medicago* to powdery mildew (*Erysiphe pisi*) and found that the four collections of *M. doliata* grown were susceptible to it (it was observed that often only one of a number of populations of other species tested was resistant).

Economic History, Cultivars, and Uses

Medicago doliata has attracted only limited agricultural interest to date.

Cultivars

Carneiro et al. (2004) described 'Atalaia', a Portuguese-selected cultivar of *M. doliata*, as having good adaptation to grazing because of its prostrate to semi-prostrate growth habit, and having at least as good levels of dry matter production as available as Australian annual *Medicago* cultivars for purposes of stocking long-duration, rain-fed pastures under Mediterranean conditions (no such Australian commercial cultivars of *M. doliata* are available at this time). 'Manesty' is an Australian experimental frost-tolerant line from SeedCo. Ltd. (Walsh et al. 2001).

Forage

Medicago doliata is rarely sown for forage. Bolton (1962) reported that the species (as *M. aculeata*), on the basis of experiments in Sicily, is palatable and of good nutritive value. St. John-Sweeting et al. (1996) noted that one accession studied in South Australia had poor dry matter production and so was not of potential commercial interest. Mohammad and Qamar (1988) grew 14 annual species of *Medicago* in small test plots in Islamabad, Pakistan, located in a subtropical, sub-humid area that is notably hotter and wetter than the normal habitat of these species. The dry matter yield of *M. doliata* was about 2.8 t/ha.

Fodder

This species is not used as a fodder plant.

Green manure, rotation crop, ground cover

No information located.

Other uses (salad sprouts, ornamental, plant products)

No information located.

Chromosome Number

2*n* = 16: Ghimpu 1929*b*, 1930 (as *M. olivaeformis*); Tschechow 1932 (as "*M. oliviformis*"); Simon and Simon 1965 (as *M. aculeata*); Fernandes and Santos 1971 (as *M. aculeata*); Fernandes and Queirós 1978 (as *M. aculeata*); Lesins and Lesins 1979; Abdelguerfi and Guittoneau 1979 (as *M. aculeata*); Abdelguerfi et al. 1988*b*; Mariani et al. 1996; Oberprieler and Vogt 1996*a*.

Genetic Relationships and Gene Transferability

Intraspecific

Barriers to hybridization within the species have not been examined.

Interspecific

Simon and Millington (1967) were unable to produce hybrid seed by crossing *M. doliata* and *M. truncatula*. Staszewski and Jakubowska (1992) reported that they pollinated *M. sativa* with pollen from *M. doliata* and found one F_1 plant that appeared to be a hybrid.

Exchange by advanced biotechnology

No information located.

Conservation Status

This species is fairly widespread and has some weedy tendencies. As its survival is not threatened, it is not in need of conservation measures.

Germplasm Sources

Numerous seed accessions have been deposited in germplasm banks, including more than 1000 at SARDI (see Chapter 3).

Research Problems and Priorities

Medicago doliata has not received much development to date, but has potential as a special legume for agriculturally stressful circumstances. Its agricultural potential needs to be assessed.

Representative Collections

Algeria: Bandel, Prov. de Constantine, Vallée de L'Oued Berda (prairie Drouin), calcaire argileux, arrosé, defendement fortement au soleil, alt. 630 m, 22-V-1858, in fruit (P). **France:** Albert, A., près Solliès-Pont (Var.), champs au pied de la colline, 27-V-1889, in fruit, Société Rochelaise 1889 No. 2612 (P). Albert, A., près Toulon, dans les champs à la Garde, 25-V-1885, in flower & fruit, Flora Selecta Exsiccata publié par. Ch. Magnier No. 1125 (P). De Hartsen, F.A., Alpes Maritimes, 3-V-1869, in flower & fruit (K). **Greece:** De Heldreich, Attica, Hagia Glykeria prope Athenas, in locis herbosis, ad vias, 30-IV-1884, in fruit, De Heldreich Herbarium Graecum Normale No. 833 (P). **Iran:** Mehregan, I. 650 m (alt.), 5 km, Pol-e Dokhtar to Andimeshk, Luristaan, southwestern Iran, in fruit (DAO). **Israel:** Eig, A., Zohary, M. & Feinbrun, N., Upper Galilee, Ein Zeitim, 11-V-1927, in fruit (HUJ). Eig, A., Zohary, M. & Feinbrun, N., Samaria, Wadi 'Ara, 12-V-1927, in fruit (HUJ). **Israel/Syria (Golan):** Peri, D., Mount Hermon, south slope, Jura, chalky soil, alt. 1250–1500 m, "4.7.67", in fruit (HUJ). **Italy:** Huter, Porta & Rigo 200, Calabria I. occid., ad montem Cataria pr. Reggio, alt. 4–5000 m, sol argillos-granit., 28-IV-1877, in fruit (K). Huter, Porta & Rigo 200b, Callabrio I. orient. ad agrorum margines, ad pedem mt. St. Andrea pr. Locella, alt. 4–500 m, loc. argillos, 11-V-1877, in fruit (K). Levrier, E., ex agro Florentino ad vias campestres juxta villam Arcangelli, page Settignano, ad orientem urbis; 13-V-1886, in flower & fruit (K). Porta & Rigo 358, Italia Austral., Tapygia inter segetes pr. Ostrante, sol. calcareo, 15-V-1875, in flower & fruit, (P). Rigo, G., Sicilia, Prov. Siracusa, in collibus circa "Avola", solo tuffaceo, IV-1898, in flower & fruit, Herbarium Normale editum ab I Dörfler No. 4230 (P). **Italy (Sicily):** Davis, D. & Sutton, S. 63262, Caltanissetta, ca. 20 km south of Piazza Armerina, alt. 100 m, banks over limestone rocks, 17-V-1979, in flower & fruit (BM). Davis, D. & Sutton, S. 64185, Messina, ca. 3 km southeast of Cesarò, alt. ca. 100 m, eroded bank at edge of arable field, 31-V-1979, in fruit (BM). De Heldreich, in Monte Muarte prope Panurmum, 20-V-1840, in flower & fruit (K). Rigo G. 53, Prov. Siracusa, Avola, in collibus solo tuffaceo, 13-IV-1898, in flower & fruit (K). **Morocco:** Mauricie, H., Tafersit, Beni-Hedien, 8-V-1932, in fruit (BM). **Portugal:** Daveau, J., près du Cap d'Espichel, collines argilo-calcaires, IV-1883, in flower & fruit. Daveau, J., Valée d'Alcantara, environs de Lisbonne, champs & collines, IV-1879, in fruit (P). Matos, Y. & A., Coimbra, Loages, terreno calcáreo, cultivado, 20-V-1953, in fruit (BM). Scarlett 239 (mixed collection with *M. polymorpha*), Lisbon area, sea level upwards, cultivated places, 1935, in fruit (K). **Spain:** Bourgeau, E. 706, Barcelone, route de Garcia, 24-IV-1844, in flower & fruit (K). Bourqeau, E., env. de Chiclana, commun, 29-IV-1849, in fruit (P). Ellman, E. 668, north of Algeciras near Los Barrios, in sandy railway bank, 18-V-1924, in flower & fruit (K). **Syria:** Zohary, M., Syria bor., Muslemiie, 6-V-1931, in fruit (HUJ). Zohary, M., Syria bor., about Aleppo, 3-V-1931, in flower & fruit (HUJ). **Turkey:** Townsend, C.C. 640422/3, Mersim district, frequent on arable (areas) near the ruins of Pompeiopolis, 22-IV-1964, in flower & fruit (K).

33. *Medicago edgeworthii* Širj.

M. edgeworthii Širj. Oesterr. Bot. Zeitschr. 87: 123, in obs. (1938). (Nomen novum for *M. pubescens* (Edgew. ex Baker) Širj. The combination *M. edgeworthii* Širj. was mentioned in the text of this one-page publication (cited above), entitled *Medicago schirjaëwi* Hand.-Mzt., sp. nova" authored by H.R.E. Handel-Mazzetti. Handel-Mazzetti's new species *M. schirjaëwi* was dedicated in the publication to G. Širjaev, and presumably was published with the approval of Širjaev. Accordingly, the suggestion has been made that the name should be *Medicago edgeworthii* Širj. ex Hand.-Mazz.)

The epithet *edgeworthii* was adopted in honor of botanist Michael Pakenham Edgeworth (1812–1881), a British civil servant who conducted botanical studies in Bengal.

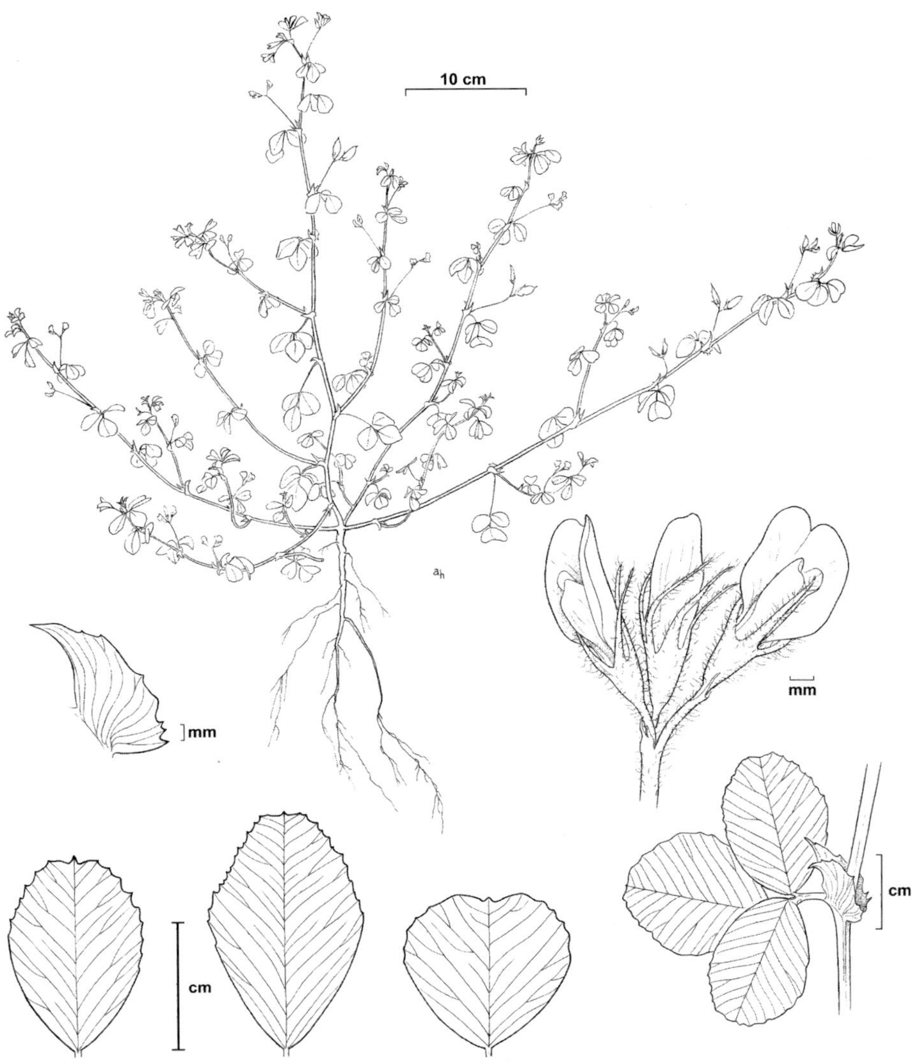

Fig. 33-1. *Medicago edgeworthii.* **Clockwise from top:** Plant; group of flowers; leaf attached to stem; leaflets; stipule. (Based on E. Small M957, DAO (= United States Department of Agriculture accession PI245002); artist: A. Hanes.)

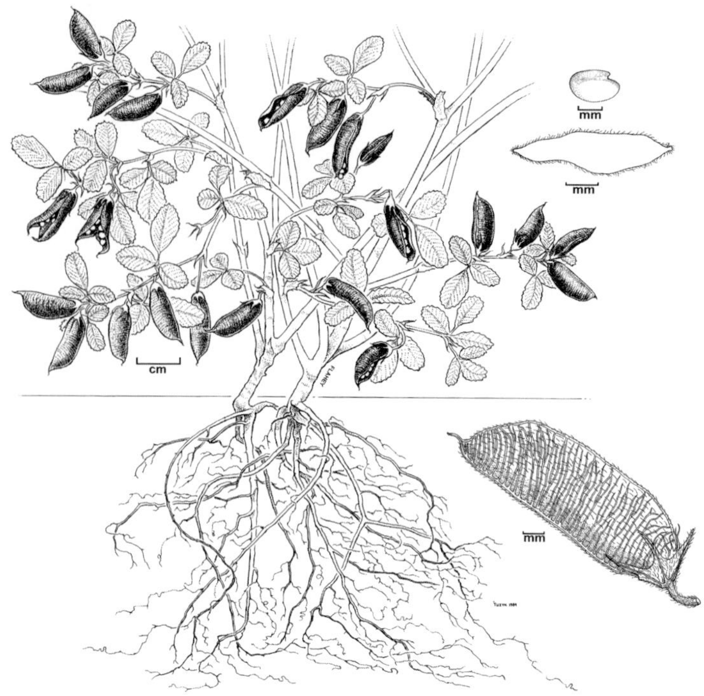

Fig. 33-2. *Medicago edgeworthii.* Two plants in fruit. (Based on greenhouse plants grown from seed from the United States Department of Agriculture, accession code USDA ACS19960, ACNO1; artist: B. Flahey.) **Inserts at right, top to bottom:** Seed; outline of cross section of fruit; fruit. (Based on W. Koelz 9970, GH; artist: L. Yuzyk.)

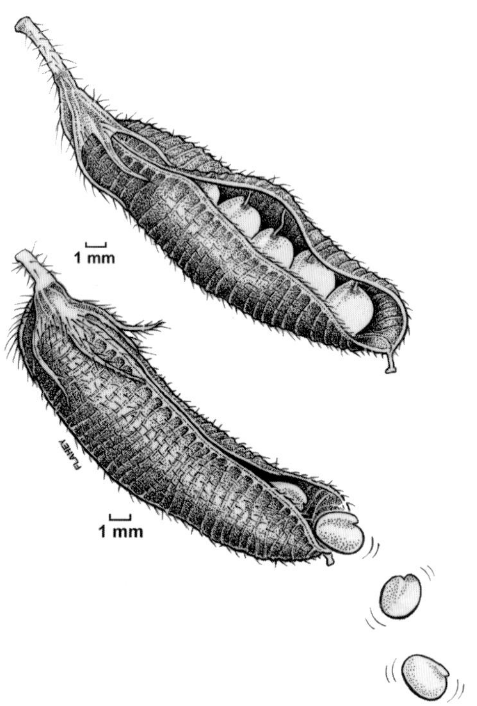

Fig. 33-3. Dehiscing fruits. (Based on greenhouse plants grown from seed from the United States Department of Agriculture, accession code USDA, ACS19960, ACNO1; artist: B. Flahey.)

Subordinate Taxa

None recognized here.

Type Specimen

Lectotype (of the replaced synonym *Trigonella pubescens* Edgew. ex. Baker): M.R. Edgeworth 30, Garhwhal, Mana (northwestern India), 10 000 feet, X-1838 (K). (Lectotype cited by Ali (1977, p. 297) and Rechinger (1984, p. 230).)

Synonymy

Homotypic

Trigonella pubescens Edgew. ex Baker in Hook. f., Fl. Brit. India 2(4): 88 (1876) (basionym).

M. pubescens (Edgew. ex Baker) Širj., Kew Bull. Misc. Inform. 1: 271 (1928). (The names *M. pubescens* Guss., Pl. Rar. 313 (1826) and *M. pubescens* Hornem., Hort. Hafn. 2: 726 (1815) prevent the use of Širjaev's name *M. pubescens* for *M. edgeworthii*. The name *M. pubescens* DC. & Lam., Fl. France 5: 569 (1815) was interpreted by Heyn (1963, p. 118) as a synonym of *M. turbinata*, so presumably *M. pubescens* DC., Cat. Pl. Horti Monsp. 1234 (1813) is also a synonym of *M. turbinata*.)

Melilotoides pubescens (Edgew. ex Baker) Yakovlev, Rast. Tsentral. Azii, Mater. Bot. Inst. Komarov 8a: 88 (1988).

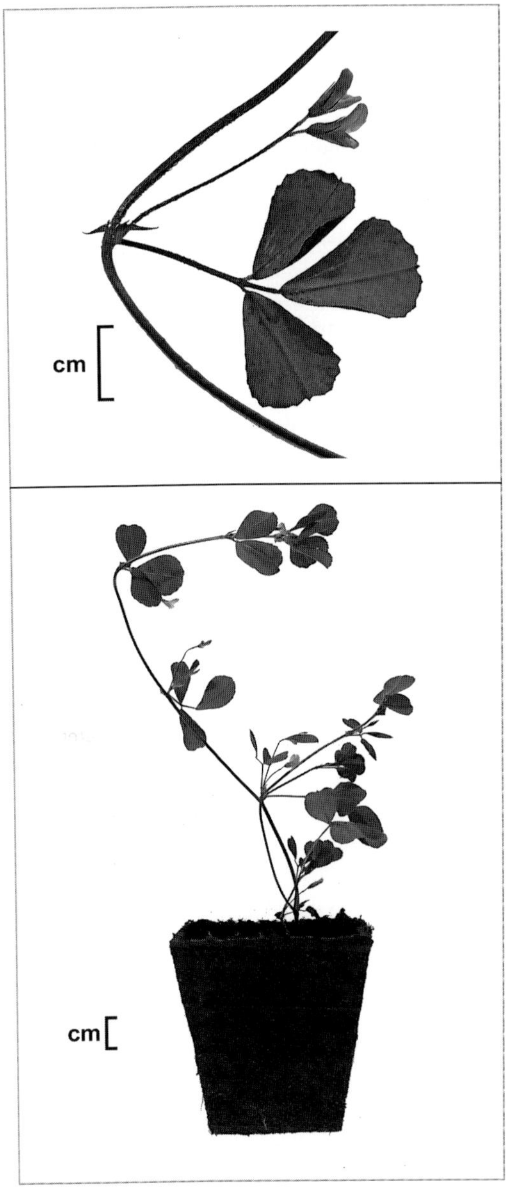

Fig. 33-4. An 18-day-old seedling of *M. edgeworthii* in flower (grown from seeds from United States Department of Agriculture accession ACS19960, ACNO9).

Heterotypic

M. pubescens Širj. var *aitchisonii* Širj. (see comment for Afghanistan collection, cited below).

T. pubescens Edgew. ex Baker var. *aitchisonii* (Širj.) Vassilcz. (cited by Vassilczenko 1953, p. 174).

Trigonella tibetica Vassilcz., Not. Syst. Leningrad 14: 234 (1951).

Melilotoides tibetica (Vassilcz.) Yakovlev, Rast. Tsentral. Azii, Mater. Bot. Inst. Komarov 8a: 89 (1988).

T. lasia Blatt., J. Bombay Nat. Hist. Soc. 36: 479 (1933).

Vernacular Names

English: No names found.

French: No names found.

Geographical Distribution

(N = native, I = introduced, U = uncertain status):

Asia: Afghanistan (U), China (N, including Quinghai, Sichuan, Xizang Zizhiqu (Tibet), and Yunnan), India (N; northwestern India; Himachal Pradesh, Jammu-Kashmir, Punjab, and Uttar Pradesh), Nepal (N), Pakistan (N; northern).

Medicago edgeworthii is native to the Himalayas and alpine areas west to the mountains of Afghanistan. The range includes tropical India and temperate Asia.

Description

Perennial forb, numerous, slim, prostrate, procumbent, ascending or erect stems 30–40(50) cm long, branched at base, twigs and leaves pubescent with simple hairs (considerable variation in degree of pubescence has been recorded; Vassilczenko (1953) noted that the pubescence may be shed to a considerable extent as the pods mature; rarely, glandular hairs are present (Shah and Kothari (1975)). Leaflets (4)6–15(20) mm long, generally about 1.5 times as long as broad, obovate-oblong, obovate, or oval, cuneately narrowed towards base, the distal half (or less) serrate, both sides covered with appressed simple hairs (or upper side glabrescent), the petioles usually shorter than the blades. Stipules 5–8(10) mm long, linear-lanceolate to ovate-lanceolate, semi-sagittate, the margin entire or inconspicuously dentate. Inflorescence with 1–3(6) flowers, the peduncle 1–2 cm long, longer than the subtending petiole, the inflorescences both axillary and terminating branches. Flowers 4–6 mm long, yellow, the corolla slightly exserted (only slightly longer than calyx). Calyx 3–5 mm long, pubescent, the subulate teeth two or more times the length of the tube. Fruit (10)12–16 mm long, 4–5 mm wide, ovally-oblong, flat, ventral suture straight or weakly convex, dorsal suture convex (and indistinctly undulate), spineless, with a slender, apical, straight beak. The fruit surface with numerous fine transverse veins that branch towards the dorsal suture, the surface profusely covered with appressed hairs (less so at maturity, and the fruits sometimes more or less glabrous). Fruits with 4–15 seeds, not separated by transverse, spongy fruit septa. Seeds ovate-elliptic, brown, smooth, radicle longer than half seed length.

Identification Hints

A combination of fruit characters serves to distinguish this species from other species of *Medicago* with flat fruits: almost straight ventral suture, curved dorsal suture, and pubescence (less marked at maturity). A distinctive feature of this species is that the flowers and fruits are distributed over the whole plant, beginning from the very base of the stems (Vassilczenko 1953).

Habitat and Environmental Adaptation

Habitat

Medicago edgeworthii occurs in mountains at altitudes sometimes higher than 3000 m, typically on steep rocky banks and dry hill slopes, as well as in grasslands, stony pastures, and open fields, and at roadsides.

Phenology

The species flowers from April to August in much of its native area, the exact time of flowering and fruiting depending on the local climate. Singh (1989) noted that in northern Rajasthan state, India (in the Great Indian Desert), *M. edgeworthii* flowers and fruits from January to April. *Medicago edgeworthii* grown in a greenhouse in Beltsville, Maryland continued to grow after seeds were harvested, but when it was cut back like alfalfa it failed to grow back (G. Bauchan, personal communication). In Ottawa, I observed that some first-year plants grown in a greenhouse produced fruits and seeds extremely copiously, but then died, whereas others either remained vegetative or continued to live. Lesins and Lesins (1979, page 55) referred to an annual population (likely grown in a greenhouse in Edmonton) that was receives as *M. edgeworthii*, but that they thought could not be that taxon, which they believed was exclusively perennial. In a greenhouse in Ottawa, five of 108 seedlings came into flower by 18 days of age under a 16-hour photoperiod at 20 °C, with good moisture and light, indicating a capacity for precocious flowering. It appears that, at least in greenhouses, *M. edgeworthii* can function both as an annual and a perennial and possesses greater phenological plasticity than all other species of *Medicago* examined to date.

Soil

The species has been collected primarily from shallow soils and substrates that are sandy or rocky.

Temperature

Medicago edgeworthii is adapted to extremely cold temperatures (Campbell and Bauchan 2002).

Water

Medicago edgeworthii has been collected for the most part on dry mountain ridges and dry hill slopes, mostly on well-drained soils. However, Campbell and Bauchan (2002) characterized the species as being adapted to moist locations.

Biotic Relationships

Pollination

Small (1988) recorded a pollen to ovule ratio of 469, which is consistent with *M. edgeworthii* being at least partly cross-pollinated like other perennial species of *Medicago*. The small flowers with corollas scarcely projecting out of the calyxes are suggestive of inbreeding. I observed that this species grown in a greenhouse protected from insects nevertheless set pods and seeds in very large quantities, so at least some populations are completely self-fertile. However, Wang and Li (2005) observed bumblebees foraging on *M. edgeworthii* in Yunnan Province, China. Campbell and Bauchan (2002) referred to *M. edgeworthii* as "allogamous".

Dispersal

There are no obvious morphological adaptations for seed dispersal in this species, but the seeds are produced copiously. The pods are dehiscent, scattering many seeds often while still attached to the plants. The flatness of the pods would tend to prevent the pods from rolling down hillsides, which would tend to keep the seeds near their mother plants. However, the flat pods could also be an adaptation for wind distribution, which would scatter the seeds more widely.

Herbivory

No information located.

Nodulation and mycorrhizae

USDA (2009) reported positive root nodulation for *M. edgeworthii*. Feng (2000) examined 201 rhizobial isolates from root nodules of *M. edgeworthii* from Yunnan Province, China and found that these were genetically very variable, but were related to *Rhizobium mongolense*.

Disease resistance

No information located.

Economic History, Cultivars, and Uses

This species has attracted little agricultural interest to date.

Cultivars

No cultivars have been bred.

Forage

Medicago edgeworthii is not sown for forage.

Fodder

This species has not been employed for fodder.

Green manure, rotation crop, ground cover

No information located.

Other uses (salad sprouts, ornamental, plant products, etc.)

No information located.

Chromosome Number

$2n = 16$: Yan et al. 2000; Campbell and Bauchan 2002.

Lesins (1952) has been mistakenly cited as a $2n = 32$ count; however, tetraploidy was artificially induced.

Genetic Relationships and Gene Transferability

Intraspecific

No barriers to hybridization within the species are known, although artificial hybridization of different populations has not been attempted.

Interspecific

Lesins (1952) was unable to hybridize *M. edgeworthi*, that was artificially doubled in chromosome number to $2n = 32$ and *M. sativa* with the same number of chromosomes. Campbell (2003) cited personal unpublished information that neither classical hybridization nor protoplast fusion techniques had yielded viable hybrid embryos between *M. edgeworthii* and *M. sativa*.

Exchange by advanced biotechnology

Campbell (2003) studied genes in *M. edgeworthii* that he considered potentially transferable to *M. sativa* for improvement of the latter.

Conservation Status

This species occurs in remote areas of Asia that have received very little conservation attention to date and, accordingly, a conservation assessment of *M. edgeworthii* is required. There is very little *M. edgeworthii* germplasm available and therefore there is a need for collection.

Germplasm Sources

The principal source of seeds of *M. edgeworthii* is the United States Department of Agriculture National Plant Germplasm System, the material collected by T.A. Campbell.

Research Problems and Priorities

For purposes of improving *M. sativa*, *M. edgeworthii* appears to offer genes that confer frost-resistance and tolerance to poor soils. There is a need for collection of germplasm, and of evaluation of agronomic potential for the possibility of gene transfer to alfalfa.

Representative Collections

Afghanistan: Aitchison, J.E.T. 712, Kurrum Valley, Kaiwás, not common, 25-VII-1879, in early flower (K, LE; the LE specimen was annotated in 1927 as *M. pubescens* Širjaev v. *Aitchisonii* Širjaev, and may represent a type for Širjaev's var. *aitchisonii*). **India:** Drummond, J.R. 1970, Kunáwar, Sohha Rám, in. glareosis etc., alt. ca. 8000 feet, IV-1884, in flower & fruit (K). Drummond, J.R. 21772, Punjab, Krol, 11-IV-1885, in flower & fruit (K, P). Drummond, J.R. 21775, Punjab, Urni, Kunawar, 17-VII-1885, in flower & fruit (K). Drummond, J.R. 22281, Punjab, Kunawar, 1885, in flower & fruit (K). Drummond, J.R. 24766, Punjab, Limla-Kalha, alt. 4000 feet, 1908, in flower & early fruit (K). Parker, R.N. 3420, Punjab, Rawalpindi District, Banni, Murree Hills, alt. 1740 m, on dry ridges, in shallow soil, 5-IV-1938, in flower & fruit (K). Rich, H.H. 530, (Punjab) below Jutojk, 14-V-1917, in flower & fruit (K). Stewart, H.M., Gyantri Plains, "Flora of Tibet", 7-VII-1818, in flower & early fruit (P). Stewart, R.R. & Rahman, A. 25388, Swat State, N.W.F.P. (Punjab), Ushu, alt. ca. 8000 feet, 26-VII-1953, in flower & fruit (K). Stewart, R.R., Kashmir, Sonamarq, 4-VII-1921, in flower & early fruit (K). Stewart, R.R. 9270, Pahalgam, Kashmir, 7500 feet, 3-VIII-1927, in flower (S). Thompson, J., Himal. Bor. Occ., alt. 7–1000 feet, in flower & fruit, Herb. Ind. Or. Hook fil & Thompson (K, P). **Nepal:** Dobremez, J.F. & Rausundhai 3076, Narpha (28°45′N, 83°40′E), dans champs d'orge, alt. 2650 m, 20-05-74, in fruit (G). **Tibet:** Kingdon-Ward, F. 20133, Rima, alt. 5000 feet, a weed of cultivated ground and stony pastures, 15-VII-1950. Ludlow & Sherriff 9742, hills south of Lhasa, ca. 91°15′E, 29°30′N, on steep, rocky banks, alt. 12 500 feet, 11-VII-1943, in flower & fruit (BM). Ludlow, Sherriff & Elliot 12213, Sangyü, Yigrong Chu, Pome, ca. 95°5′E, 30°10′N, alt. 7000 feet, on edge of cultivation, 10-II-1942, in flower & fruit (BM). Soulié, J.A., Thibet Oriental, Tongolo, Principauté de kiala, 1893, in flower & fruit (K). Walton, H.J. 46, (Eastern Himalayas-Sikkim), Gyangtse, VII-IX-1904, in late flower & fruit, another sheet in flower & fruit (K) (according to Vassilczenko (1953): *Trigonella tibetica* Vassilcz. (1951)).

Note

Until 2009, when I observed that the cotyledons of *M. edgeworthii* are pulvinate (see Fig. 4-1), I did not appreciate that this species was a member of section *Lunatae* and I assigned it to section *Platycarpae*. It is sometimes perennial, like other species in section *Platycarpae*, but it is also (flexibly) annual, like *M. brachycarpa* of section *Lunatae*. The phylogenetic studies reviewed (Chapter 10) make it clear that *M. edgeworthii* is very closely related to other members of section *Lunatae*.

34. *Medicago fischeriana* (Ser.) Trautv.

M. fischeriana (Ser.) Trautv., Bull. Sci. Acad. Imp. Sci. St. Petersb. 8: 271 (1841).

The epithet *fischeriana* commemorates Friedrich Ernst Ludwig von Fischer (1782–1854), a Russian botanist of German origin who was director of the St. Petersburg Botanical Garden.

Subordinate Taxa

None recognized here.

Type Specimen

Wilhelm, Tiflis (Caucasia), holotype LE!

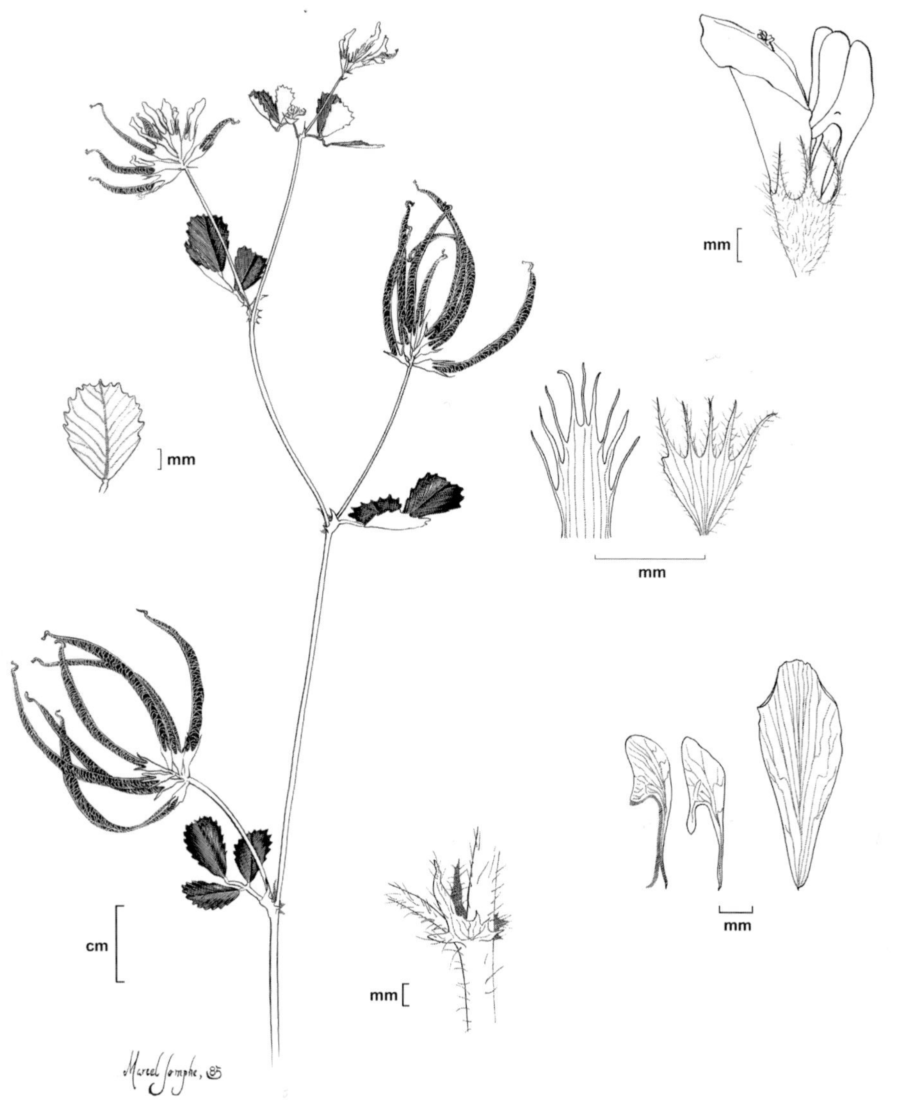

Fig. 34-1. *Medicago fischeriana.* **Left:** Branch with fruits and flowers; leaflet. **Upper right:** Side view of flower. **Center and lower right:** Floral parts (clockwise from upper left: spread out androecial column (free stamen not shown); calyx; standard petal; wing petal; keel). **Lower center:** Node showing stipules. (Based on C. Tobey 669, E; artist: M. Jomphe.)

Synonymy

Homotypic

Trigonella fischeriana Ser. in DC., Prodr. 2: 183 (1825), (basionym).

Heterotypic

T. flexuosa Fisch. ex M. Bieb. Fl. Taur.-Cauc. 3: 515 (1819). (non *T. flexuosa* Delil.)

T. fischeriana Ser. var. *perennans* Bornm. (A name mentioned in passing in Širjaev, Generis *Trigonella* revisio critica, Spisy Pfir. Fac. Masarykovy Univ., 3: 23 (1930). Of interest because it implies that there is a perennial form of the species. However, this was based on a herbarium collection without fruits, from Anatolia, so the identification of this is suspect, as is the possibility of a perennial form of the species.)

T. subracemosa Boiss. Fl. Orient. 2: 73 (1872). (Type is Balansa 912!, cited below.)

T. fischeriana var. *subracemosa* (Boiss.) Širj. Generis *Trigonella* revisio critica, Spisy Pfir. Fac. Masarykovy Univ., 3: 23 (1930). (This taxon was alleged to differ by having cuneiformly obovate leaflets, shorter peduncles of inflorescences, and a smaller calyx and a smaller corolla.)

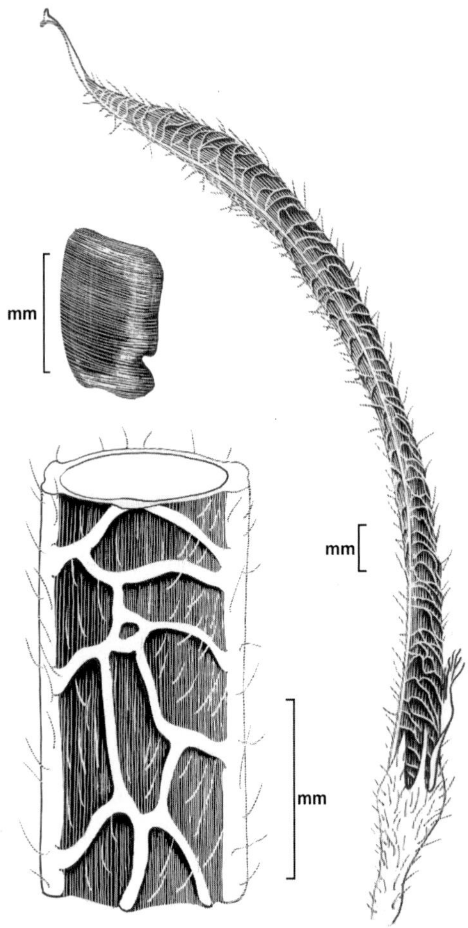

Fig. 34-2. *Medicago fischeriana.* **Clockwise from lower left:** Portion of fruit showing venation; seed; fruit. (Based on C. Tobey 669, E; artist: M. Jomphe.)

Vernacular Names

English: No names found.

French: No names found.

Geographical Distribution

(N = native, I = introduced, U = uncertain status):

Africa: Algeria (I).

Asia: Armenia (N), Azerbaijan (N), Gruziya (Georgia, N), Iran (western, N), Iraq (N), Turkey (Anatolia, N).

Europe: Greece (Samos Island, just west of Anatolia; found only near the summit of Mount Kerkis at 1400 m: see Burton (1999)), Russian Federation (southwestern, N), Ukraine (Crimea, N).

This species is native to southeastern Europe and southwestern Asia.

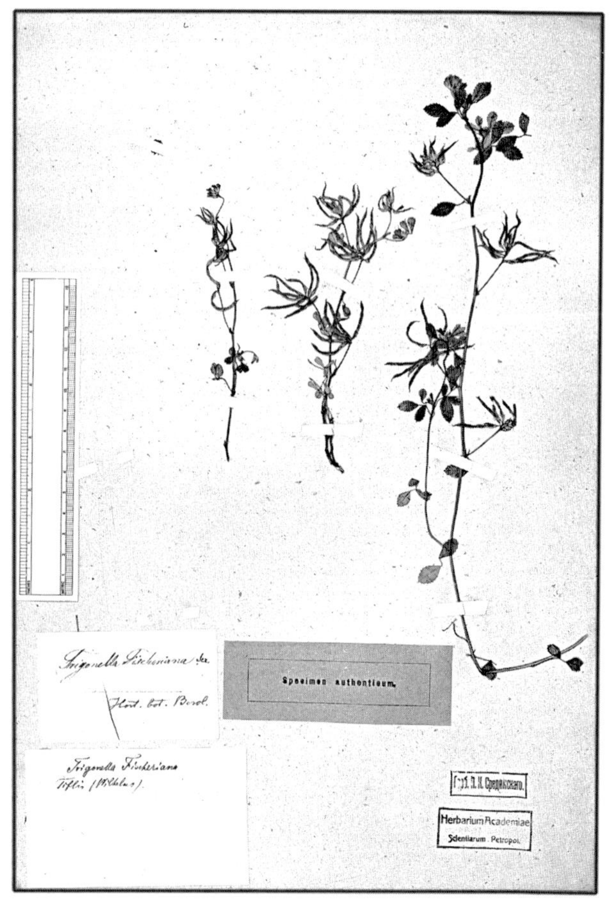

Fig. 34-3. Type specimen of *Trigonella fischeriana* Ser.

Description

Annual herb, stems branched from base, procumbent or ascending, (5)10–30(40) cm long, slightly to moderately adpressed-pilose. Leaflets obovate to oblong, 3–10(15) mm long, 3–7 mm wide, distal half to two-thirds of margin serrate, pilose with simple trichomes on lower surface, slightly pubescent on upper surface. Stipules semi-sagittate or lanceolate-subulate, dentate. Inflorescence of 4–10 flowers in a short raceme or subumbellate head, peduncle 5–40 mm, slender, longer than subtending petiole. Flowers 5–8(9) mm, yellow, corolla longer than calyx, calyx 3–5.5 mm, campanulate, faintly appressed-pilose, teeth lanceolate-subulate, about equal to tube. Fruit linear, straight to arcuate, subcompressed, sutures slightly thickened, convex on dorsal suture, more or less sinuous (tortuous, more or less constricted between the seeds), transversely reticulate-veined, puberulent, (10)15–30 mm long, 1–1.5 mm wide. Seeds 5–8 per pod, elongate-cylindrical, 2–2.5 mm long, 0.7–1 mm wide, finely tuberculate, greenish-yellow or brown, radicle about 80% of length of seed.

Identification Hints

Medicago fischeriana is similar to *M. phrygia*, but differs in having somewhat tortuous fruits with more or less undulate sutures and smaller flowers (5–7 mm versus (6)7–10 in *M. phrygia*).

Habitat and Environmental Adaptation

Habitat

Medicago fischeriana occurs in the subalpine and alpine zone, and is found in steppe, on rocky slopes, in open places among shrub associations, at the edges of woods, and in fallow fields, at altitudes of 600 to 2000 m.

Phenology

Flowering occurs from April to June in the native area.

Soil

This species has been collected from a variety of substrates, including limestone rocky soils and clays.

Temperature

Medicago fischeriana occurs in countries with Mediterranean and/or southern Eurasian continental climates. It appears to be adapted to warm to hot summers.

Water

This species appears to be a winter annual, relying particularly on water accumulated in the soil during the wet season.

Biotic Relationships

Pollination

Small (1988) noted that the flowers are small and the pollen to ovule ratio is low (262), observations that are consistent with *M. fischeriana* being largely self-pollinated and substantially an inbreeder.

Dispersal

No information available. Presumably, like other annual species of the genus with no evident adaptations for seed distribution, the seeds are distributed by a combination of wind and animal vectors, and occasionally by gravity and rain.

Herbivory

No information located.

Nodulation and mycorrhizae

No information located.

Disease resistance

No information located.

Economic History, Cultivars, and Uses

This species has not been employed as an agricultural crop.

Cultivars

No cultivars have been bred.

Forage

Medicago fischeriana has not been sown for forage.

Fodder

This species has not been used as a fodder source.

Green manure, rotation crop, ground cover

Evans et al. (2004) examined the potential of numerous fodder species to survive winter in North Korea and produce nitrogen, organic matter, and soil stabilization in maize and rice fields. *Medicago fischeriana* was judged to have good winter survival and good green biomass production, and to produce fair ground cover.

Other uses (salad sprouts, ornamental, plant products, etc.)

There is no record of economic use of *M. fischeriana*.

Chromosome Number

No information located.

Genetic Relationships and Gene Transferability

Intraspecific

No information found.

Interspecific

No information found.

Exchange by advanced biotechnology

No information located.

Conservation Status

This species occurs in areas of Asia where plants are subject to genetic erosion, and it therefore requires a conservation assessment.

Germplasm Sources

A small number (less than 20) of accessions are available from major germplasm institutions (see Chapter 3).

Research Problems and Priorities

Medicago fischeriana has not been the subject of much research (in part because until recently it has been considered to be a species of *Trigonella*) and its potential economic usefulness is unknown. Germplasm collection and basic botanical and agronomic evaluation are needed.

Representative Collections

Gruziya (Georgia): Hohenacker, R.F., Unio itiner, Georgia, Caucasus, 1838, in fruit & flower (BM, G, P). **Iran:** Archibald, J.C. 1089A, Luristan, 30 km west of Khorramabad, dryish stony clay on hill top ridge, alt. 1340 m, 1-V-1966, in flower & fruit (E). Furse, P. 1477, Golpaigan, Kuh Alwand, alt. 6500 feet, limestone rocky mountain with pockets of light earth and earthy scree, 17-IV-1962, in flower & early fruit (E). **Turkey:** Alava, Reina 7096, Kayseri, Ürgüp, along road to Nevsehir, up to 1 km from the city limits, on waste places, edges of road, and cultivated fields, 14-VI-1968, in flower & fruit (E). Alava, Reina 7147, Kayseri, Ürgüp, highway between Ürgüp and road fork to Göreme, road banks and hills, 15-VI-1968, in flower & fruit (E). Balansa, B. 912, près de Béréketly (Cappadoce), collines bordant la rive droite du Kamechly-Tchaï, 1300 m d'alt., 15-VI-1856, in fruit (G, P). Bourgeau, E. Gumusch-khané, champs incultes, 8-VI-1862, in flower & fruit (G). Coode, M.J.E. & Jones, B.M.G. 1962, Amasya, 17 km from Merzifon to Hausa, alt. 700 m, roadside, 2-VI-1965, in flower & fruit (E). Coode, M.J.E. & Jones, B.M.G. 1990, Tokat, Tokat to Yildizeli, north side of Camlibel, Dag., 1500 m, open woodland, 4-VI-1965, in flower & fruit (E). Davis 42190, Nevsehir, 3 km east of Ürgüp, 1100 m, fallow fields, 30-IV-1966, in flower & fruit (E). Davis & Hedge D. 29281, Prov. Tunceli, Pülümür, alt. 1550 m, igneous south slope, 8-VI-1957, in flower & fruit (E). Dudley 36129 (also 36134), prov. Kütahya, Kütahya to Tavsanli, 30 km from Tavsanli, alt. 800–900 m, cultivated fields and road verges, 23-VI-1962, in flower & fruit (K). Dudley, D. 36132, Prov. Kütahya to Tavsanli, 30 km from T.; 800–900 m, cultivated fields and road verges, 23-VI-1962, in fruit (E). Kotschy, T. 253 Tschomakli ad argaeum montem, alt. 6200 feet, 30-V-1859, in flower & young fruit (G). Sintenis, P. 5881, Armenia turcica, Szandschak Gümüchkhane, 12-VI-1894, in flower & early fruit (E). Tobey, C. 669, Vilayet, Amasya, Erbaa, Çatalan, alt. 900 m, with Cedrus, shale rock and rock chips, southern exposure, 2-VI-1964, in flower & fruit (E).

35. *Medicago granadensis* Willd.

M. granadensis Willd., Enum. Pl. Hort. Berol. 803 (1809), (basionym).

The epithet *granadensis* is the Latin form for "of Granada", Spain, where the type specimen was collected. As noted below, the species is not native to Spain.

Subordinate Taxa

None recognized here.

Type Specimen

"*M. granadensis*. Leguminibus cochleatis...Habitat in Hispania" Herb. Willdenow, 14321/1 (B! Observations recorded: mature pods 8–9.5 mm wide, ca. 6.5 coils, spines somewhat intertwined, middle leaflets ca.15 mm long, stipules with about seven lobes, and middle coil clearly broadest).

Synonymy

Homotypic

M. granatensis Willd. ex. Ser. in DC. Prodr. 2: 180 (1825). (Merely a change in spelling of the specific epithet by Seringe.)

Heterotypic

Medicago aegagropila Desv., J. Bot. Appl. Agr. 3: 76 (1814). Type: Desvaux, Loire inf., France (G), according to Heyn (1963).

M. galilaea Boiss., Diagn. Pl. Or., Ser. 1, 2(9): 10 (1849). (Type: Boissier, E., Palaestina, IV-V-1846, in fruit (G! Observations recorded: One of three specimens at G bears the notation "Typus". This has fruits 9 mm wide, without glands or hairs, ca. six coils, and interlocking

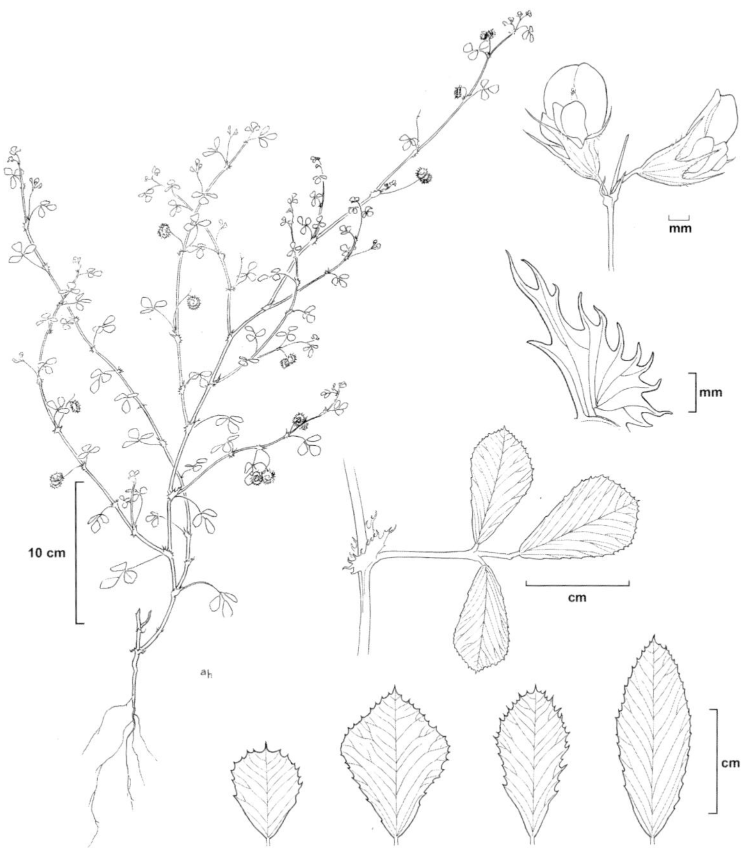

Fig. 35-1. *Medicago granadensis*. **Clockwise from left:** Plant; group of flowers; stipule; leaf attached to stem; leaflets. (Based on Feinburn 65, A; artist: A. Hanes.)

right-angle spines, but not as pronounced as in typical *M. intertexta*.)

M. intertexta Mill. var. *glandulosa* Eig, Bull. Inst. Agr. Nat. Hist. 6: 23 (1927).

M. galilaea Boiss. var. *incisa* Post, Fl. Syr. Pal. Sin. 229 (1883–1896).

Vernacular Names

English: Galilee medic (Bolton 1962; medic is also spelled medick).

French: Luzerne de Grenade.

Geographical Distribution

(N = native, I = introduced, U = uncertain status):

Africa: Egypt (N).

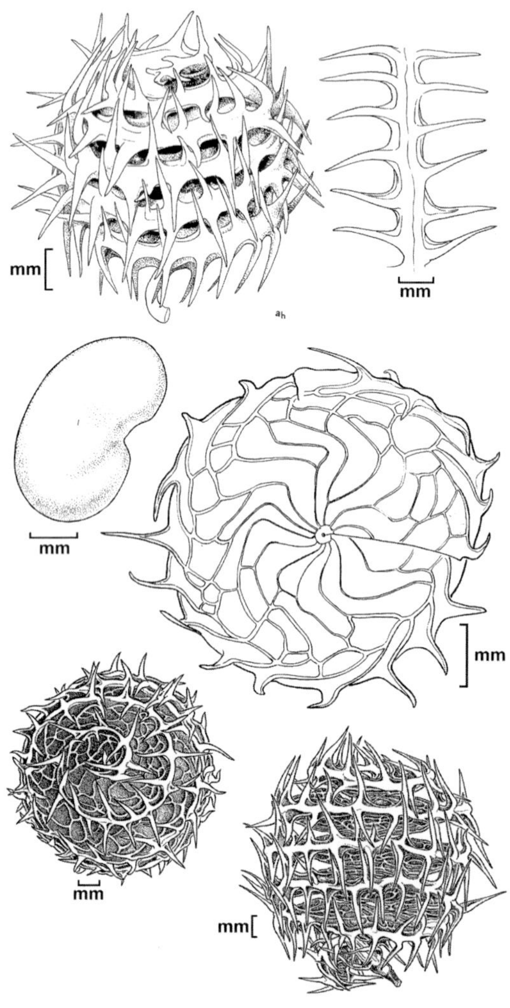

Fig. 35-2. *Medicago granadensis.* **Clockwise from center left:** Seed; fruit; edge of coil; fruit coil face showing venation. (Based on E. Small M345, DAO; artist: A. Hanes.) **Bottom left:** View of fruit from stigma end. **Bottom right:** Side view of fruit (based on E. Small M1980, DAO; artist: B. Flahey.)

Asia: Israel (N), Jordan (N), Lebanon (N), Syria (N), Turkey (N).

Europe: France (I; casual alien), Spain (I; Heyn (1963) noted that the type specimen, collected near Malaga, is an adventitious plant), Turkey (I).

Southern America: Chile (I).

Medicago granadensis is native to the eastern Mediterranean and is introduced elsewhere.

Description

Annual herb, procumbent or ascending, 15–40 cm high, branches covered with multicellular hairs, rarely glabrous, branched from lower part of plant. Leaflets 10–20 mm long, 5–8 mm wide, obovate to oblanceolate or elliptical, sometimes blotched, glabrous or with a few simple hairs (rarely with multicellular hairs), margin of distal third serrate. Stipules ovate to oblong, dentate-laciniate at base. Inflorescence of (1)2–4(6) flowers, peduncle longer than subtending petiole. Flowers 4–5 mm long, yellow to orange-yellow, corolla more than twice as long as calyx, calyx with simple hairs (rarely multicellular hairs), teeth as long as tube. Young pod protrudes sideways from calyx. Pod 5–8 mm long, glabrous or rarely with a few multi-celled hairs, with 5–7 coils that are pergamentaceous and loose (not adpressed), diameter of middle (largest) coil 7–10(–12) mm, the other coils decreasing towards the ends, coil faces with about 12 radial veins that anastomose in the distal half, from the anastomosed veins diagonal veins emerge and each enters one dorsal suture spine, the spines 1–2 mm long, grooved, the rows of dorsal spines on the coil faces inserted at about 90° to the coil surface (more or less appressed to pod), spines of adjacent coil faces interlocking. Seeds ca. 5 mm long, smooth, dark red-brown to black-brown, 1–2 per coil, not separated by transverse, spongy fruit septa, radicle often less than half the seed length.

Identification Hints

Medicago granadensis is very similar to *M. muricoleptis*, but is reliably separable from it by the diameter of the terminal coil of the fruit (more than 5.5 mm in *M. granadensis*, less than 5.5 mm in *M. muricoleptis*).

Habitat and Environmental Adaptation

Habitat

Medicago granadensis occurs in cultivated and fallow fields, waste places, and orchards. As all of these are anthropogenic habitats, presumably in preagricultural times the species was a generalist pioneer, occupying open or recently cleared places.

Phenology

The species flowers from April to May in its native area.

Soil

Medicago granadensis typically occupies heavy soils. Piano and Francis (1992) observed that in its native area it occurs on clays that range in pH from neutral to moderately basic. In Israel it has been collected from terra rossa and basalt.

Temperature

Based on the geographical distribution, this species is adapted to warm to hot summers.

Water

This species is a typical Mediterranean drought-avoiding winter annual. Piano and Francis (1992) recorded that in its native area it requires a minimum annual rainfall of 400 mm.

Biotic Relationships

Pollination

The small flowers of this species indicate that it is largely an inbreeder.

Dispersal

The indehiscent spiny fruits of *M. granadensis* are adapted to dispersal in animal fur, accounting for the occasional adventitious presence of this species outside of its native area.

Herbivory

No information located.

Nodulation and mycorrhizae

USDA (2009) reported positive root nodulation for *M. granadensis*.

Disease resistance

Yaege and Stuteville (2002) tested the resistance of 33 annual species of *Medicago* to powdery mildew (*Erysiphe pisi*) and found that the three collections of *M. granadensis* grown were susceptible to the disease (it was observed that often only one of a number of populations of other species tested was resistant). In a comparative test of 36 annual species of *Medicago*, O'Neill and Bauchan (2000) found that four accessions of *M. granadensis* were quite susceptible to anthracnose caused by *Colletotrichum trifolii*.

Economic History, Cultivars, and Uses

Medicago granadensis has not been employed for any particular purpose.

Cultivars

No cultivars have been bred.

Forage

Medicago granadensis is not sown for forage. Noy-Meir et al. (1989) studied the effects of protecting natural Mediterranean grazing lands by fencing and observed no consistent effects on *M. granadensis* populations. Díaz et al. (2001) characterized *M. granadensis* as "grazing resistant".

Fodder

The species has not been employed for fodder purposes.

Green manure, rotation crop, ground cover

No information located.

Other uses (salad sprouts, ornamental, plant products, etc.)

No information located.

Chromosome Number

$2n = 16$: Simon and Simon 1965; Lesins and Lesins 1979; Schlarbaum et al. 1984.

$2n = 16, 17, 18$: Heyn 1963.

$2n = 18$: Heyn 1956.

The counts of 17 and 18 require confirmation.

Genetic Relationships and Gene Transferability

Intraspecific

Lesins et al. (1971) crossed a form of *M. granadensis* that had black seeds with a form that had brown seeds and noted that subsequent segregation was in a 3 to 1 ratio, black being dominant over brown. No barriers to hybridization were observed.

Interspecific

Lesins and Lesins (1979) reported that *M. granadensis* could not be hybridized with its nearest relatives — *M. muricoleptis*, *M. intertexta*, and *M. ciliaris*. Schlarbaum et al. (1984) studied morphological and karyological relationships of the related species *M. granadensis*, *M. muricoleptis*, *M. intertexta*, and *M. ciliaris*. The analysis suggested that *M. granadensis* is relatively isolated from the other species. Rose et al. (1988) studied chloroplast DNA relationships of these species. Small et al. (1999) studied isozyme relationships of the same species and concluded that the presence of several unique alleles in *M. granadensis* was indicative of a complete absence of natural hybridization with the other three species.

Exchange by advanced biotechnology

Skinner (1992) cloned a DNA fragment that occurred in a high number of copies in *M. granadensis* but in a low number of copies in 17 other *Medicago* species (this was not used to transfer the DNA).

Conservation Status

This species is poorly understood and, given its relative rarity, a conservation assessment is desirable. There is a need to acquire germplasm, both for conservation and development purposes.

Germplasm Sources

Fewer than 100 accessions are available from major germplasm institutions with large *Medicago* collections (see Chapter 3).

Research Problems and Priorities

Medicago granadensis is an uncommon species that has not received much study. Its potential in agriculture as a crop or as a contributor to the genetic improvement of other species requires examination.

Representative Collections

France: Bourgeau, E. 108, Castignaux, dans les prés, 8-VI-1848, in fruit (P). Gandoger, M. 302, Arnas (Rhône), 18-VI-1863, in fruit (K) (an annotation by C. Heyn suggests that this is an unusual specimen). **Israel:** Boissier, Palaestina, Jerusalem (presumably from the Plain of Esdraelon), IV-1846, in fruit (P). Feinbrun, N., Hertzlia, near Tel-Aviv, borders of ditches, in fruit, Universitatis Hebraicae Hierosolymitanae Flora Palaestina Exsiccata No. 65 (B, K, P). Samuelsson, G. 703, Judaea: Kiryath Anavim pr. Jerusalem, inter segetes, alt. ca. 600 m, 29-IV-1932, in fruit (K). Samuelsson, G. 936, Judaea: Bet Hakerem prope Jerusalem, in hortulo, 8-V-1932, in fruit (K). **Jordan:** Meyers, F.S. & J.E. Dinsmore 3768, Huleh, fields, alt. 2 m, 12-V-1911, in fruit (K). **Spain:** Reverchon, Spanien (B). Herbarium Splitgerberianum: Herb. Lugd. Bat. 908.118-105, in Hispania prope Malagam, July, in fruit (L). **Turkey:** Balansa, B., au nord de Mersina (cilicie), coteaux calcaires, 23-V-1855, in fruit (P).

36. *Medicago halophila* (Boiss.) E. Small

M. halophila (Boiss.) E. Small in Small, Lassen & Brookes, Willdenowia 16(2): 433 (1987).

The epithet *halophila* is Latin for salt-loving, an allusion to the maritime habitat.

Subordinate Taxa

None recognized here.

Type Specimen

Balansa, B. 251, sables maritimes entre Mersina et Pompeiopolis (Pompeiopolim; in flower 3-IV-1855) (holotype G!). Huber-Morath (1970) indicated an isotype was at W (referred to as a "holotype" by Huber-Morath on his 1967 label).

Synonymy

Homotypic

Trigonella halophila Boiss., Diagn. Pl. Orient. 2(5): 78 (1856), (basionym).

Heterotypic

Trigonella sinuata Boiss. var. *brachycarpa* Boiss., Diagn. 2, 5: 78 (1856).

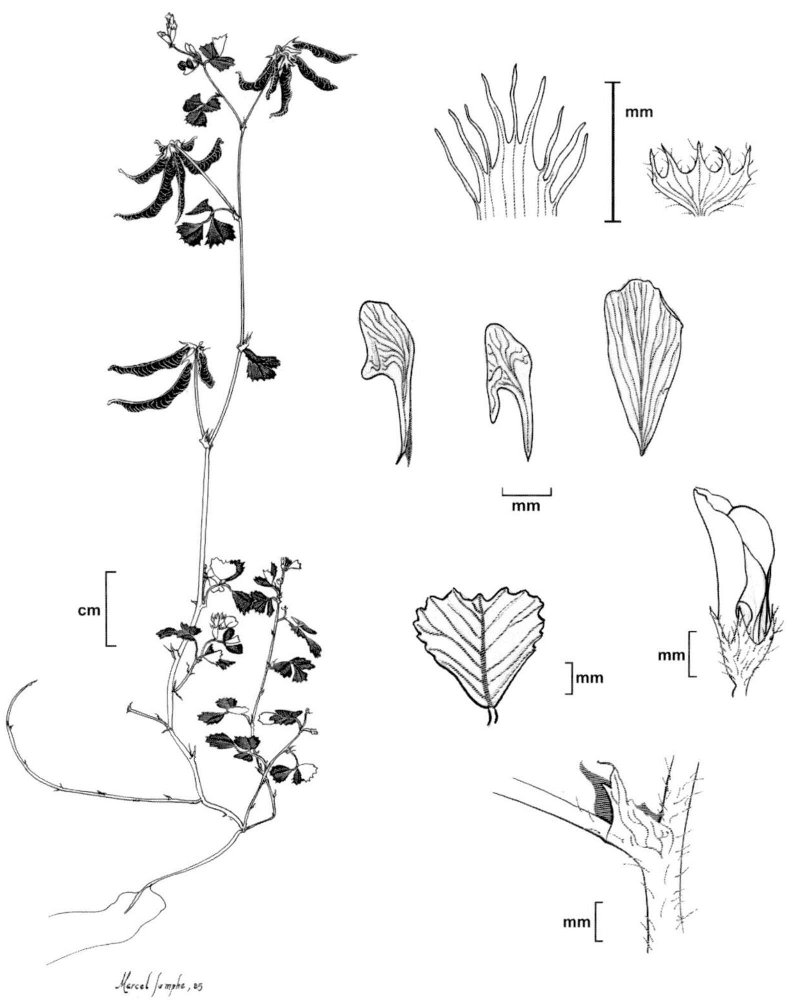

Fig. 36-1. *Medicago halophila*. **Left:** Plant. **Upper right:** Floral parts (clockwise from upper left: spread out androecial column (free stamen not shown); calyx; standard petal; wing petal; keel). **Center right:** Side view of flower. **Lower right:** Node showing stipules. **Center:** Leaflet. (Based on holotype, B. Balansa 251, G; artist: M. Jomphe.)

Vernacular Names

ENGLISH: Salt trefoil (Çakan et al. 2005).

FRENCH: No names found.

Geographical Distribution

(N = native, I = introduced, U = uncertain status):

ASIA: Turkey (N).

Medicago halophila is endemic to southern Anatolia (Cilicia).

Description

Annual herb, branched from base, stems prostrate, procumbent or ascending, 3–15 cm long. Twigs and leaves sparingly appressed-pilose with simple hairs. Leaflets very small (2–4 mm long), longer than wide, obovate-cuneate to obcordate-cuneate, obtusely denticulate towards tip. Stipules small, subulate from a broadened base. Inflorescence umbellate-capitate, of 5–8 flowers, peduncles 5–10 mm long, longer than subtending petioles. Flowers ca. 5 mm long, yellow, calyx campanulate, 2–2.5 mm long, teeth triangular-lanceolate, somewhat shorter than tube, faintly appressed-pilose. Pods stellately divaricate, linear, 10–13 mm long, 1–1.5 mm wide, slightly compressed, ventral (seed-bearing) suture slightly curved (arcuate), dorsal suture slightly subundulate, venation transversely reticulately netted, with 3–4 seeds. Seeds about 2 mm long, oblong, finely tubercled, radicle approximately 80% of length of seed.

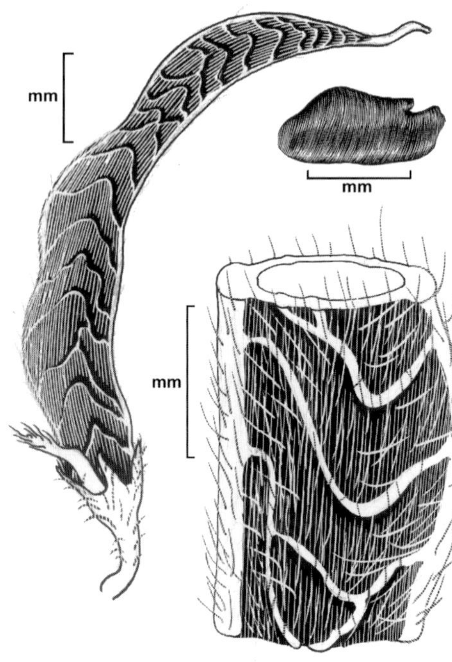

Fig. 36-2. *Medicago halophila.* **Clockwise from left:** Fruit; seed; portion of fruit. (Based on holotype, B. Balansa 251, G; artist: M. Jomphe.)

Identification Hints

Medicago halophila most closely resembles *M. astroites*, differing in its arcuate-curved fruit (rarely curved in *M. astroites*) having fewer flowers in the inflorescence (5–8 in *M. halophila*, 6–12 in *M. astroites*), and in calyx features (teeth lanceolate-subulate and longer than or equal to the tube in *M. astroites*, triangular-lanceolate and less than the tube in *M. halophila*).

Habitat and Environmental Adaptation

Habitat

Medicago halophila has been collected on sand dunes near the sea.

Phenology

Flowering occurs in the native Turkish distribution range from April to June.

Soil

The species typically occupies marine sands.

Temperature

Medicago halophila is found near Mediterranean seashores, and therefore presumably the prevailing temperatures are moderated by the sea.

Water

Based on its sand dune habitat, the species is ephemeral and drought-avoiding, presumably taking advantage of winter rains.

Biotic Relationships

Pollination

Small (1988) noted that the flowers are small and the pollen to ovule ratio is low (316), observations that are consistent with *M. halophila* being largely self-pollinated and substantially an inbreeder.

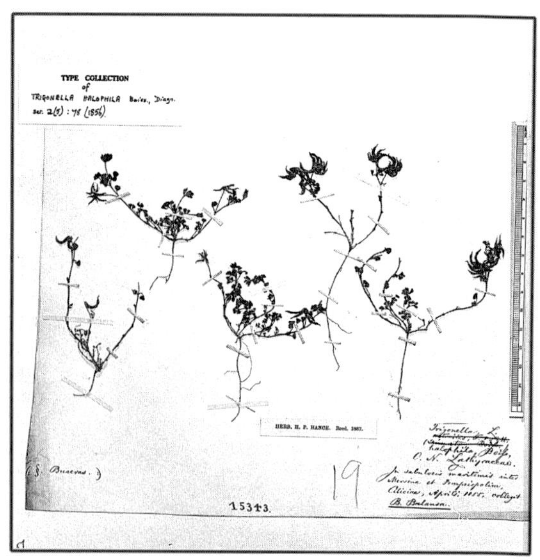

Fig. 36-3. Photograph of type collection of *Trigonella halophila* Boiss. at G.

Dispersal

No specializations for seed distribution are apparent. The seeds presumably simply fall away from the capsules.

Herbivory

No information located.

Nodulation and mycorrhizae

No information located.

Disease resistance

No information located.

Economic History, Cultivars, and Uses

This species has not been employed for any purpose.

Cultivars

No cultivars have been bred.

Forage

The species has not been used for forage.

Fodder

No fodder usage has been recorded.

Green manure, rotation crop, ground cover

No information located.

Other uses (salad sprouts, ornamental, plant products, etc.)

No information located.

Chromosome Number

No information located.

Genetic Relationships and Gene Transferability

Intraspecific

No information located.

Interspecific

No information located.

Exchange by advanced biotechnology

No information located.

Conservation Status

Çakan et al. (2005) predicted that unless provided with protective measures, *M. halophila* was "likely to become extinct within the next 10 years" in the Çukurova Deltas of southern Turkey. The Bern Convention on the Conservation of European Wildlife and Natural Habitats lists *M. halophila* (under the name *Trigonella halophila*) as a "strictly protected species" (the species is Asian, not European). There is a great need to acquire germplasm of this species, both for conservation purposes as well as for agricultural assessment.

Germplasm Sources

No accessions are available in the major germplasm institutions.

Research Problems and Priorities

This species is one of the least studied in the genus *Medicago*, in part because it was considered to be a member of *Trigonella* until recently. It is in need of basic botanical studies as well as assessment of its potential in agriculture, both as a special-purpose legume crop and as a source of genes for improving the major agricultural species of *Medicago*.

Medicago halophila is apparently related to *M. astroites*. *Medicago arenicola* (Chapter 16) is based on a single collection and this may be simply a marginal population of *M. halophila*. The relationships among these species requires clarification.

Representative Collections

TURKEY: Balansa, B. 447, entre Mersina et les ruines de Pompeiopolis (Cilicie), in flower 3-IV, in fruit 19-IV-1855 (BM, P). Balansa, B. 845, sables maritimes entre Mersina et Pompeiopolis, Cilicie, IV-1855, in fruit (HUJ). Balls, E.K. 704, Mersine, growing on seashore, alt. 10 feet, 9-IV-1934, in flower & fruit (BM, E). Bilger, K. 5334, suf feuchless, shelle, zwischen, Dunung, bei Taursus, 26-IV-1955, in fruit, (G). Reese, H, Prov. C5 Içel, district Mersin, 51 km east of Silifke, 10-VI-1950, in flower & early fruit (G).

37. *Medicago heldreichii* E. Small

M. heldreichii E. Small in Small, Lassen & Brookes, Willdenowia 16(2): 434 (1987).

The epithet *heldreichii* commemorates Theodor von Heldreich (1822–1902), a German-born botanist and director of the Athens Botanical Garden and Natural History Museum. He is the collector of the type specimen and co-author of the former name of this species, *Trigonella polycarpa*.

Subordinate Taxa

None recognized here.

Type Specimen

Boissier & Heldreich 590, Asie Mineure, Adalia, in campis sterilibus incultis, 28-V-1845, in flower & fruit. (Holotype & isotype Herb. Boiss. G!; isotypes at E!, K!, P!; isotype also at W teste Huber-Morath 1970.)

Synonymy

Homotypic

Trigonella polycarpa Boiss. & Heldr. in Boiss., Diagn. Pl. Orient. Ser. 1 2(9): 18 (1849); non *M. polycarpa* Willd. ex Godron in Gren. & Godron, Fl. France 1: 389 (1848).

Heterotypic

No names found.

Vernacular Names

ENGLISH: Heldreich's alfalfa (a contrived name at the United States Department of Agriculture, Natural Resources

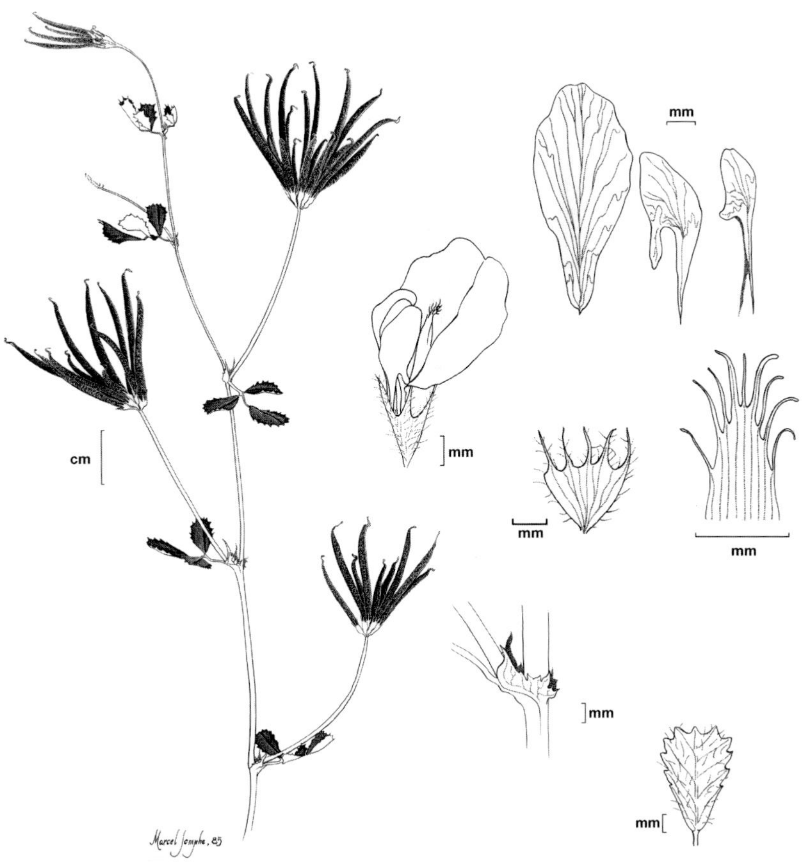

Fig. 37-1. *Medicago heldreichii.* **Left:** Fruiting branch. **Upper right:** Floral parts (clockwise from left: side view of flower; standard petal; wing petal; keel; spread out androecial column (free stamen not shown); calyx. **Lower center:** Node showing stipules. **Lower right:** Leaflet. (Based on holotype, Boissier & Heldreich 590, G; artist: M. Jomphe.)

Conservation Service, Introduced Plants of the United States website, PLANTS Database; http://plants.usda.gov/java/introduced?startChar; current June 2010).

French: No names found.

Geographical Distribution

(N = native, I = introduced, U = uncertain status):

Asia: Turkey (N).

Northern America: USA (I, not established).

This species is endemic to southwestern Anatolia.

Description

Annual herb, branched from base, stems slender, procumbent to ascending, 10–30 cm long. Leaves (upper and lower surfaces) and twigs pilose with adpressed hairs. Leaflets (3)4–6(8) mm long, (3)4–6 mm wide, obovate, oval, or almost round, distal half of margin denticulate. Stipules ovate-lanceolate to lanceolate, dentate to incised. Inflorescence of (8)–10–20 flowers, peduncles 1–5 cm, much longer than subtending petiole. Flowers yellow to orange, 5.5–7 mm long, calyx about 4 mm long, campanulate, teeth triangular-lanceolate or lanceolate-acuminate, about as long as or shorter than tube. Legumes closely clustered, linear, 15–30 mm long, 1–2 mm broad, slightly compressed, more or less erect, straight or tapered towards apex, transversely reticulately veined, appressed-pilose, thickened along sutures, with many seeds. Seeds oblong-cylindrical, ca. 2 mm long, finely tubercled, radicle about 80% of length of seed.

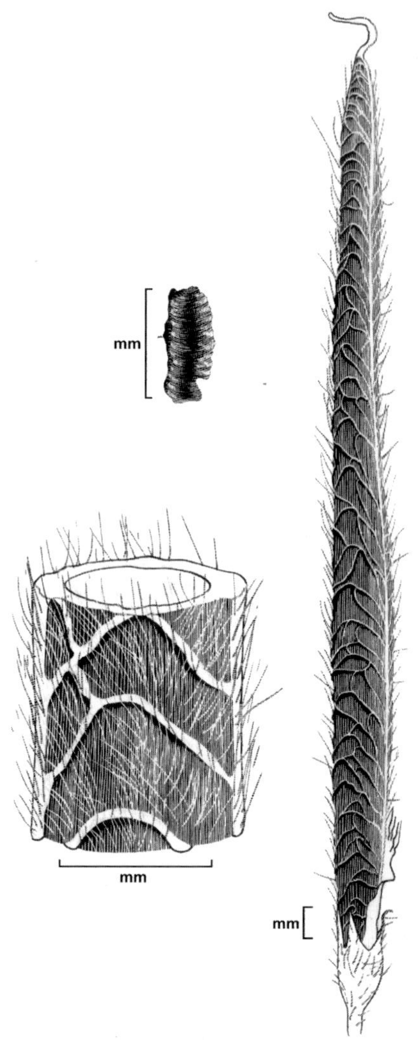

Fig. 37-2. *Medicago heldreichii.* **Upper left:** Seed. **Right:** Fruit. **Lower left:** Portion of fruit. (Based on holotype, Boissier & Heldreich 590, G; artist: M. Jomphe.)

Identification Hints

Medicago heldreichii is most likely to be confused with *M. phrygia* (they are probably closely related), although the latter has curved pods. It could also be confused with *M. arenicola*, but the latter is known from only the type collection.

Habitat and Environmental Adaptation

Habitat

Medicago heldreichii occurs on sandy coasts, fixed sand dunes, grassy slopes, pastures, and in open pine forests, near sea level.

Phenology

Medicago heldreichii flowers from March to May in its native area.

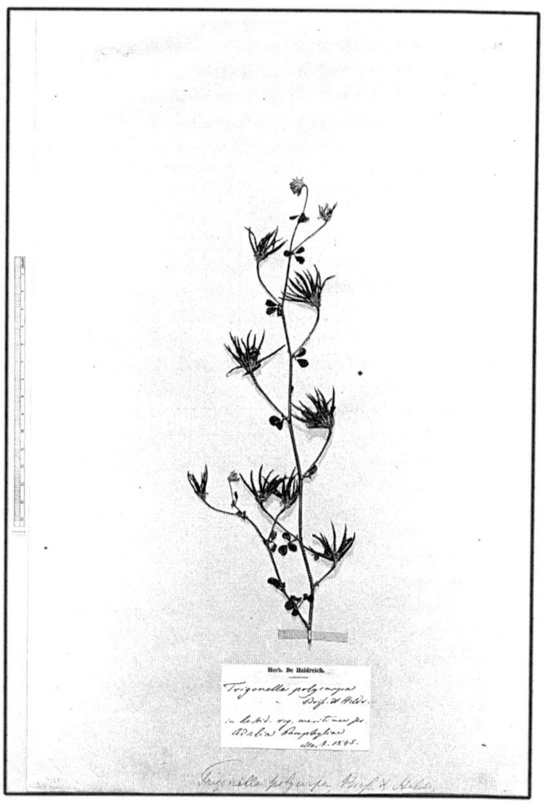

Fig. 37-3. Photograph of holotype of *Trigonella polycarpa* Boiss. & Heldr., at G.

Soil

This species typically occupies sandy and gravelly soils.

Temperature

The proximity to the sea of the natural habitats of this species suggests that although it is adapted to hot areas, the extreme temperatures are moderated by the sea.

Water

The habitats occupied suggest that this species is adapted to low rainfall areas.

Biotic Relationships

Pollination

Small (1988) noted that the flowers are small and the pollen to ovule ratio is low (183) in this species, observations that are consistent with *M. heldreichii* being largely self-pollinated and substantially an inbreeder.

Dispersal

No information is available. Based on the lack of specific adaptations for distribution, the seeds are likely dispersed by a combination of gravity occasional rain, and by accidental adherence to feathers or fur.

Herbivory

No information located.

Nodulation and mycorrhizae

No information located.

Disease resistance

No information located.

Economic History, Cultivars, and Uses

This species has apparently not been used by people.

Cultivars

No cultivars have been bred.

Forage

Medicago heldreichii is not sown for forage.

Fodder

This species is not employed for fodder.

Green manure, rotation crop, ground cover

No information located.

Other uses (salad sprouts, ornamental, plant products, etc.)

No information located.

Chromosome Number

No information located.

Genetic Relationships and Gene Transferability

Intraspecific

Artificial hybridization of different populations has not been recorded.

Interspecific

No information located.

Exchange by advanced biotechnology

No information located.

Conservation Status

The Bern Convention on the Conservation of European Wildlife and Natural Habitats lists *M. heldreichii* (under the name *Trigonella polycarpa*) as a "strictly protected species" (the species is Asian, not European). The species requires protection and, towards this goal, germplasm collections need to be made and conserved.

Germplasm Sources

Accessions of this species are not available.

Research Problems and Priorities

Medicago heldreichii is a rare, little studied species in great need of germplasm acquisition and characterization.

Representative Collections

TURKEY: Davis & Polunin, O. 25704, prov. Antalya, Kumköy, between Antalya & Serik, alt. 5 m, fixed dunes in open pinus-Myrtus forest, 6-IV-1956, in flower & early fruit (BM, K, E). Heldreich, Turkey, Prov. Antalya (zone C3), district Antalya, in campis sterilibus incultis à l'Est d'Adalia, près de la mer, 28-III-1845, in fruit (isotype, labelled as holotypus by Huber-Morath, who recently contributed this collected to G from his personal herbarium). Simon 76-389, Prov. Konya (region C4), district Ermenek, Ermenek-Mut, 13 km nach Ermenek, bei Kayaönü, alt. 1210 m, 12-VII-1976, in fruit (G). Tengwall, T.A. & Huber-Morath, A. Kleinasien: Pamphylien: Mellan Lara bei Antalya, alt. 30 m, 3-V-1936, in flower & early fruit (G) (this specimen was recently contributed to G from the personal herbarium of Huber-Morath; it is labelled as *T. polycarpa* var. *Tengwallii* Širjaev var. nov., although there is no evidence that this name was published.)

38. *Medicago heyniana* Greuter

M. heyniana Greuter, Candollea 25: 190 (1970), (basionym).

The epithet *heyniana* commemorates Israeli taxonomist Chaia Clara Heyn (1924–1998), a professor at the Hebrew University of Jerusalem who specialized on annual species of *Medicago* (see Fig. 9-8).

Subordinate Taxa

None recognized here.

Type Specimen

Greuter 5346, W. Greuter Iter graecum VI, regio aegaea austr. in latere austro-orientali et ad cacumen montis Kollas insulae Karpathos perrara, alt. 750 et 900 m.s.m., 17.5.1963 (herb. Greuter! (W. Greuter's personal herbarium; Greuter is at B); Fig. 38-3).

Synonymy

Homotypic

Medicago sect. *Heynianae* Greuter, op. cit. p. 189.

Heterotypic

No names found.

Vernacular Names

English: None found. An appropriate name would be Heyn's medic (medic is also spelled medick).

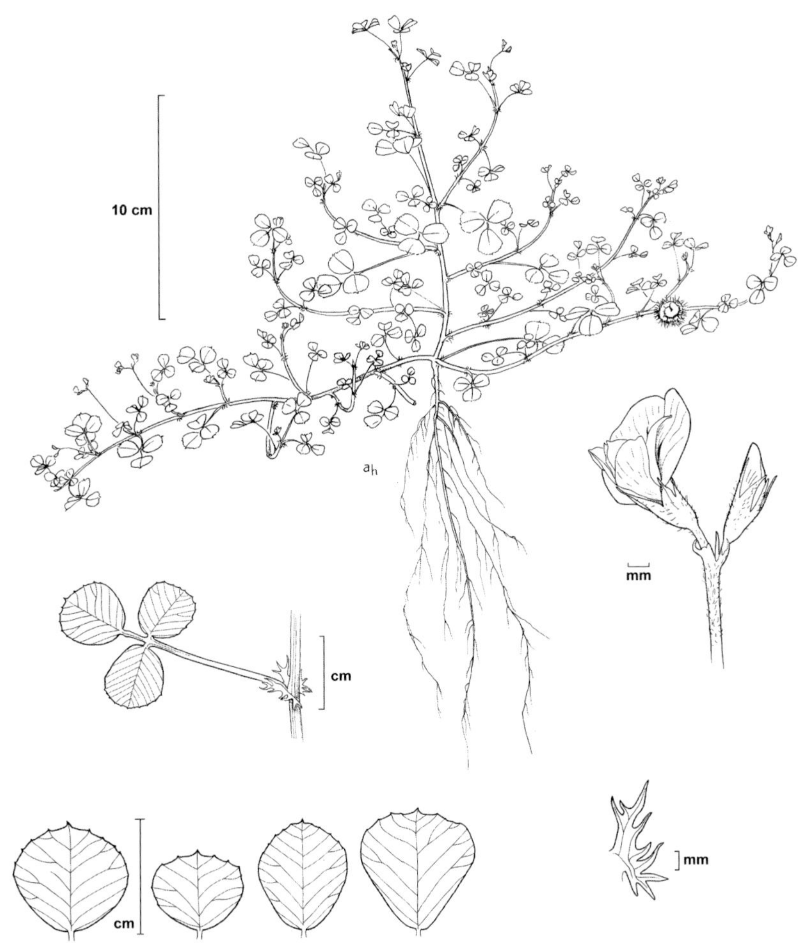

Fig. 38-1. *Medicago heyniana.* **Clockwise from top:** Plant; group of flowers; stipule; leaflets; leaf attached to stem. (Based on E. Small M342, DAO; artist: A. Hanes.)

French: No names found.

Geographical Distribution

(N = native, I = introduced, U = uncertain status):

Asia: Turkey (Marmaris Peninsula of the southwest; N).

Europe: Greece (including Crete, East Aegean Islands; N).

Medicago heyniana has been collected only from Turkey and the following Aegean (Dodecanese) Islands: Karpathos, Tilos, Amorgos, Rhodes (= Rodos; Greuter et al. 1989). Lassen (1983) pointed out that future research might reveal other localities beyond those already known.

Description

Annual herb, branched from base, branches 20–45 cm long, prostrate to ascending, sparsely pubescent with subadpressed simple hairs. Leaflets 3–10 mm long, 3–8 mm wide, glabrous above, hairy on midrib below, obovate to obcordate, upper third of margin with small serrations (less evident in smallest leaves). Stipules toothed at base, triangular. Inflorescence of one or two flowers, peduncles 9–15 mm long, about as long as subtending petiole. Flowers 5–9 mm long, yellow. Calyx 3–3.5 mm long, scarcely pubescent, teeth triangular, about equal to tube length. Young spiralled fruit emerging sideways from calyx. Mature fruit glabrous, yellowish-gray (often with a violet hue), with 2.5–4.5 coils (disc-shaped when few coils are present, short-cylindrical when more coils are present), widest coil 8–12 mm in diameter, laxly spiralled (coils not tightly adpresssed), coil face with 10–12 veins arising from ventral suture and curving almost concentrically (with weak anastomoses) toward dorsal suture where they end in sclerenchymatous tissue. Dorsal suture spines thin, 2–4 mm long, more or less parallel to the coil faces, grooved at base, one basal prong arising from the sclerenchymatous tissue and a second basal prong from the dorsal suture, two rows of spines on each dorsal suture. Seeds yellow-brown, 3–5 per coil, about 3 mm long, verrucose (longitudinally raised ridges are transverse to seed axis), not separated by spongy partitions, radicle as long as the seed.

Identification Hints

This species is superficially similar to many annual *Medicago* with pergamentaceous (i.e., relatively soft-walled) coiled fruits, but is easily distinguished by its rough (verrucose) seeds.

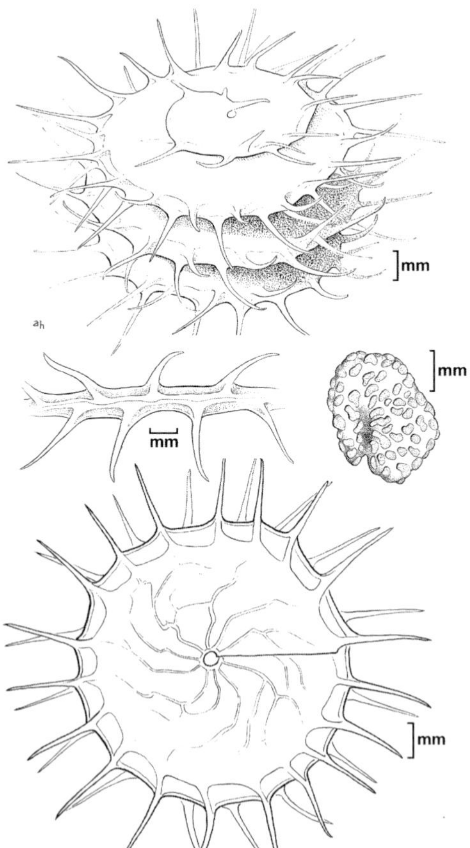

Fig. 38-2. *Medicago heyniana.* **Clockwise from top:** Fruit; seed; fruit coil face showing venation; edge of coil. (Based on E. Small M342, DAO; artist: A. Hanes.)

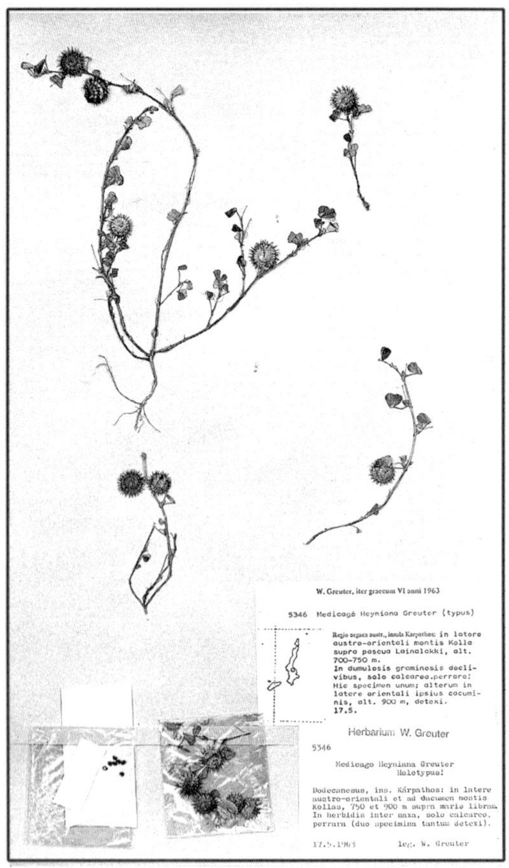

Fig. 38-3. Holotype of *Medicago heyniana*. Photograph courtesy of W. Greuter.

Habitat and Environmental Adaptation

Habitat

Medicago heyniana occurs in open areas occupied by shrubs or herbs, particularly on dry rocky hillsides of eastern Mediterranean islands, at altitudes of 700–900 m. It has been collected on cliffs, burnt pinewood, garrique, and cultivated ground (Thanopoulos 2007).

Phenology

The species is a winter annual, germinating in late winter and setting seed by early summer.

Soil

Medicago heyniana has been collected from rocky, calcareous soils, clay soil patches, and cultivated grounds (Turland et al. 1993).

Temperature

The species occurs in hot Mediterranean locations that are moderated by the sea.

Water

Medicago heyniana is found in very dry situations and is a winter annual, depending on stored winter rainfall and occasional rains during the season of growth.

Biotic Relationships

Pollination

Small (1988) noted that the relatively low pollen to ovule ratio (232) of *M. heyniana* is consistent with it being largely self-pollinated. However, the flowers are larger than in most other annual species of *Medicago*, suggesting that the species may be relatively attractive to pollinators.

Dispersal

The spiny fruits seem to be an adaptation to dispersal by animal fur, but the spines are not hooked so they probably are not efficient in this regard.

Herbivory

No information located.

Nodulation and mycorrhizae

No information located.

Disease resistance

O'Neill et al. (2003) examined the comparative resistance of 36 annual species of *Medicago* to *Phoma medicaginis* (the cause of spring black stem and leaf spot, a major disease of alfalfa) and concluded that *M. heyniana* had high resistance. In a similar comparative test of 36 annual species of *Medicago*, O'Neill and Bauchan (2000) found that one accession of *M. heyniana* was quite susceptible to anthracnose caused by *Colletotrichum trifolii*. Yaege and Stuteville (2002) tested the resistance of 33 annual species of *Medicago* to powdery mildew (*Erysiphe pisi*) and found that the single collection of *M. heyniana* grown was resistant to it (it was observed that often only one of a number of populations of other species tested was resistant). Yaege and Stuteville commented that of all species examined, *M. heyniana* seemed exceptionally resistant to the disease (although only one population was examined).

Economic History, Cultivars, and Uses

This very rare species has not been employed for any purpose.

Cultivars

No cultivars have been bred.

Forage

Medicago heyniana is not sown for forage, and indeed it is so rare that it cannot contribute as a wild source of forage.

Fodder

This species has not been used as a fodder source.

Green manure, rotation crop, ground cover

No information located.

Other uses (salad sprouts, ornamental, plant products, etc.)

No information located.

Chromosome Number

2*n* = 16: Lesins and Lesins 1979; Kamari and Matthäs 1986.

Genetic Relationships and Gene Transferability

Intraspecific

Artificial hybridization of different populations has not been attempted.

Interspecific

Lesins and Lesins (1979) were unable to hybridize *M. heyniana* with *M. orbicularis* or *M. radiata*, species sometimes hypothesized to be its close relatives or at least share characteristics considered primitive in *Medicago*.

Exchange by advanced biotechnology

No information located.

Conservation Status

Medicago heyniana was listed in the International Union for Conservation of Nature Red Data List (1993) under the category of (rare) threatened plants, and it is also noted as a rare and threatened plant of Greece (Turland 1995, Phitos et al. 2009). Accordingly, there is a need to preserve its seeds in germplasm banks.

Thanopoulos (2007) noted that "It is the only Greek *Medicago* species legally protected by the Presidential Decree no. 67/1981, falling into the category of the nonendemic, rare, and endangered species. Under this legislation its collection needs an official permission." The species has been considered to be "Vulnerable" in Greece, (Turland 1995). However, the conservation status of the species on Tilos Island is considered good (Panitsa and Tzanoudakis 2000).

Turland (1995) recorded that "Overgrazing by sheep and goats is likely to be a threat in that it causes damage to plants, lowers seed production and inhibits the establishment of seedlings. Changes in land use could also deplete populations." Turland specified measures to conserve the plant, including monitoring, conservation of the habitat by fences, and collection of seeds for deposit in seed banks.

Germplasm Sources

Five accessions (probably including duplicates) are listed in the major germplasm collections of *Medicago* of the world (see Chapter 3).

Research Problems and Priorities

Because it has been little studied to date, this species is in need of basic botanical examination and agronomic evaluation.

Representative Collections

See type specimen, cited above. Also note E. Small M342, DAO, used as the model for the illustrations.

Note

Lesins and Lesins (1979) recorded large, central purple blotches on the first true (unifoliate) leaf of some seedlings of this species.

39. *Medicago huberi* E. Small

M. huberi E. Small in Small, Lassen & Brookes, Willdenowia 16(2): 434 (1987).

The epithet *huberi* commemorates Swiss botanist Arthur Huber-Morath (1901–1990), who first named and described this species (as *Trigonella sirjaevii*).

Subordinate Taxa

None recognized here.

Type Specimen

Huber-Morath (1970) provided the following details under "Type:" Huber-Morath, A. 5083 & Reese, Region C2 (of Davis 2008), Denizli: Kale-Muğla, Quercetum 7 km nach Kale, 1100 m, 4-VI-1938 (holo Hb. Hub.-Mor., iso. Hb. Basler Bot. Ges.) (BASBG lists type, presumably isotype material of *Trigonella sirjaevii* Huber-Mor. (http://pages.unibas.ch/botges/herbar/typen.htm; current June 2010)). In 1970, Huber-Morath's collections were in his private herbarium (Hb.

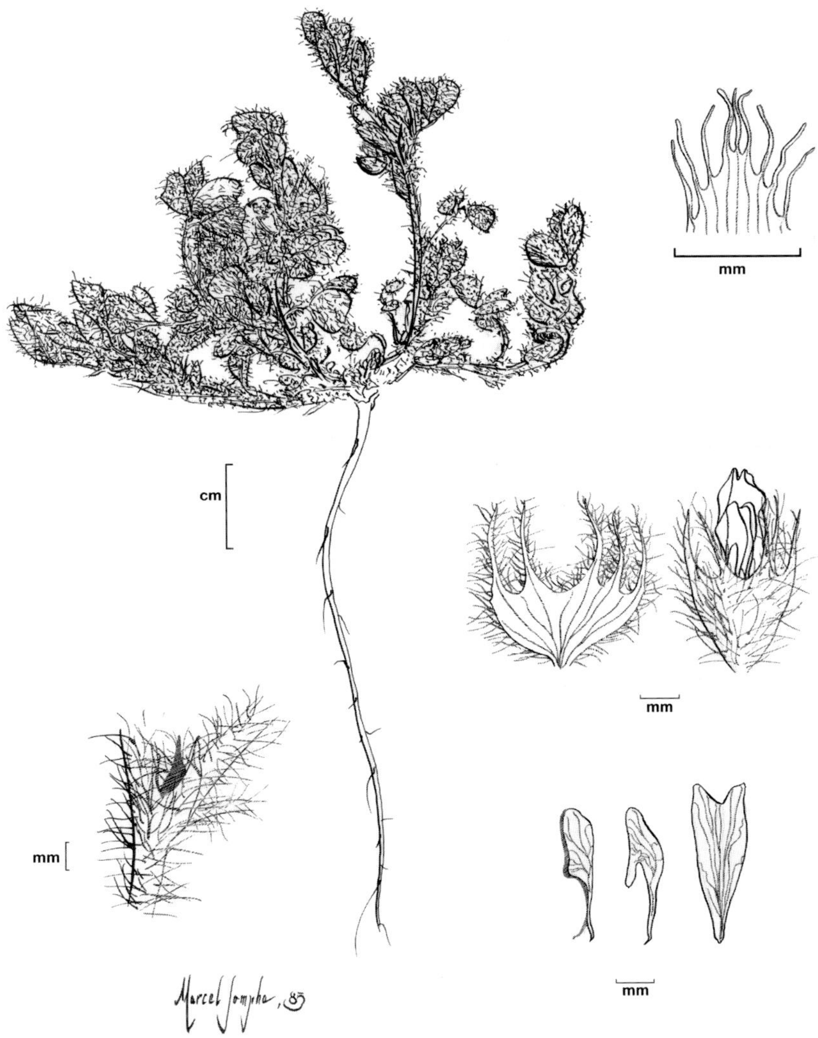

Fig. 39-1. *Medicago huberi.* **Top left:** Plant. **Right:** Floral parts (**Top:** Spread out androecial column (free stamen not shown). **Center from left:** Calyx; side view of flower. **Bottom from left:** Keel; wing petal; standard petal). **Bottom left:** Stem node showing stipules. (Based on Dudley D.35326, K; artist: M. Jomphe.)

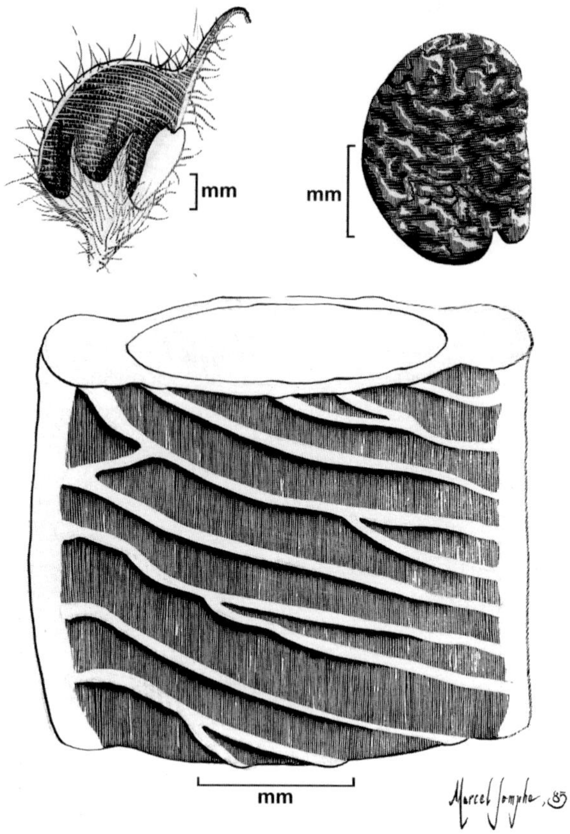

Fig. 39-2. *Medicago huberi.* **Upper left:** Fruit. **Upper right:** Seed. **Bottom:** Portion of fruit showing venation. (Based on Dudley D. 35326, K; artist: M. Jomphe.)

Hub.-Mor.). Subsequently, he contributed his collections to G and I observed the following collection labeled "holotypus" by him: Huber-Morath, A. 5083, Kleinasien: Carien (region C2 of Davis 2008), Kale Tavas – Muğla, Pinetum 7 km nach Kale Tavas, 5-VI-1938 (G). I have observed instances of Huber-Morath imprinting "Holotypus" on two collections of the same species so that his indication of type status on a sheet needs to be considered circumspectly. His designation of the type in his 1970 publication is accepted here.

Synonymy

Homotypic

Trigonella sirjaevii Huber-Mor., Repert. Spec. Nov. Regni. Veg. (Fedde Repert.) 46: 135-136 (1939). (The combination *Medicago sirjaevii* is precluded by the name *Medicago schirjaëwii* (*sirjeaewii shirjaewii*); Hand.-Mazz., Öesterr. Bot. Zeitschr. 87: 123 (1938), syn. subst.) (= *M. archiducis-nicolai*)).

Heterotypic

No names found.

Vernacular Names

English: Širjaev's fenugreek (a concocted name not in common use; "fenugreek" is based on considering this species as a member of *Trigonella*).

French: No names found.

Geographical Distribution

(N = native, I = introduced, U = uncertain status):

Asia: Turkey (N).

This species is endemic to southwestern Anatolia.

Description

Annual dwarf herb, 3–6 cm tall, prostrate, stems 20–60 cm long, branched from base, densely villous. Leaflets 6–10 mm long, 5–8 mm wide, obcordate-cuneate, truncate or emarginate, denticulate from tip to as far as middle, densely pubescent on both sides. Stipules ovate-lanceolate, entire. Inflorescences umbellate, with 10–16 flowers, peduncles almost absent (rarely short-pedunculate). Flowers yellow, 4–5 mm, pedicels ca. 1 mm. Calyx campanulate, villous, 4–5 mm, teeth lanceolate-linear, as long as or somewhat longer than tube. Fruits flat, half-ovate, erect, 8–10 mm (including beak), 3–3.5 mm wide, transversely nerved veins not anastomosing. Seeds one or two per pod, ovate, ca. 3 mm long, tuberculate, radicle about two-thirds length of seed to almost its length.

Identification Hints

Medicago huberi is closely related to *M. brachycarpa*; see the latter for hints on identification.

Habitat and Environmental Adaptation

Habitat

Medicago huberi has been recorded on stony slopes, in open scrub, and in pine forest, at elevations of 1100–1500 m.

Phenology

This species flowers in May and June in its native area.

Soil

No information located.

Temperature

No information located.

Water

No information located.

Biotic Relationships

Pollination

Small (1988) recorded a pollen to ovule ratio of 415, indicative that the species is an outcrosser. However, the flowers are relatively small, suggesting that *M. huberi* is largely self-pollinated and substantially an inbreeder, like essentially all other annual species of *Medicago*.

Dispersal

No specializations for seed distribution are present.

Herbivory

No information located.

Nodulation and mycorrhizae

No information located.

Disease resistance

No information located.

Economic History, Cultivars, and Uses

This species has not been employed for any purpose.

Cultivars

No cultivars have been bred.

Forage

Medicago huberi is not sown for forage and is so rare it cannot be significant as wild forage.

Fodder

Medicago huberi is not harvested for fodder.

Green manure, rotation crop, ground cover

No information located.

Other uses (salad sprouts, ornamental, plant products, etc.)

No information located.

Chromosome Number

Not determined.

Genetic Relationships and Gene Transferability

Intraspecific

No information located.

Interspecific

No information located.

Exchange by advanced biotechnology

No information located.

Conservation Status

This extremely rare species is in desperate need of conservation measures, particularly the collection and preservation of germplasm.

Germplasm Sources

In 1983 in response to my request for seeds, the discoverer of *M. huberi*, Huber-Morath, replied that there were no seeds available. This situation does not seem to have changed to the present.

Research Problems and Priorities

Medicago huberi is a rare, little-studied endemic that requires collection and evaluation. Its small size probably precludes its development as a crop, but it may possess genes of value for improvement of other species.

Representative Collections

Turkey: Dudley D. 35326, prov. Denizli, Yesilova-Denizli, 44 miles from Y, alt. 1019 m, denuded Pinus, 1-VI-1962, in fruit (E). Dudley D. 35357, prov. Denizli, Kizilacabölük-Denizli, 10 km from K., alt. 1000 m, open denuded pine woods, 2-VI-1962, in flower & fruit (E).

40. *Medicago hybrida* (Pourr.) Trautv.

M. hybrida (Pourr.) Trautv., Bull. Sci. Acad. Imp. Sci. St. Petersb., 8: 271 (1841).

The epithet *hybrida* is Latin for hybrid, reflecting a misperception that the species originated by hybridization.

Subordinate Taxa

None recognized here.

Type Specimen

Type preserved in Toulouse according to Vassilczenko (1949).

Synonymy

Homotypic

Trigonella hybrida Pourr., Hist. & Mém. Acad. Roy. Sci. Toulouse, Sér. 1, 3: 331 (1788), (basionym).

Heterotypic

M. pourretii Noulet, Fl. Bass. Sous-Pyren. 151 (1837), nom. illeg.

Medica pourretii (Noulet) Bubani, Fl. Pyren., 2: 471 (1899–1900), nom. illeg.

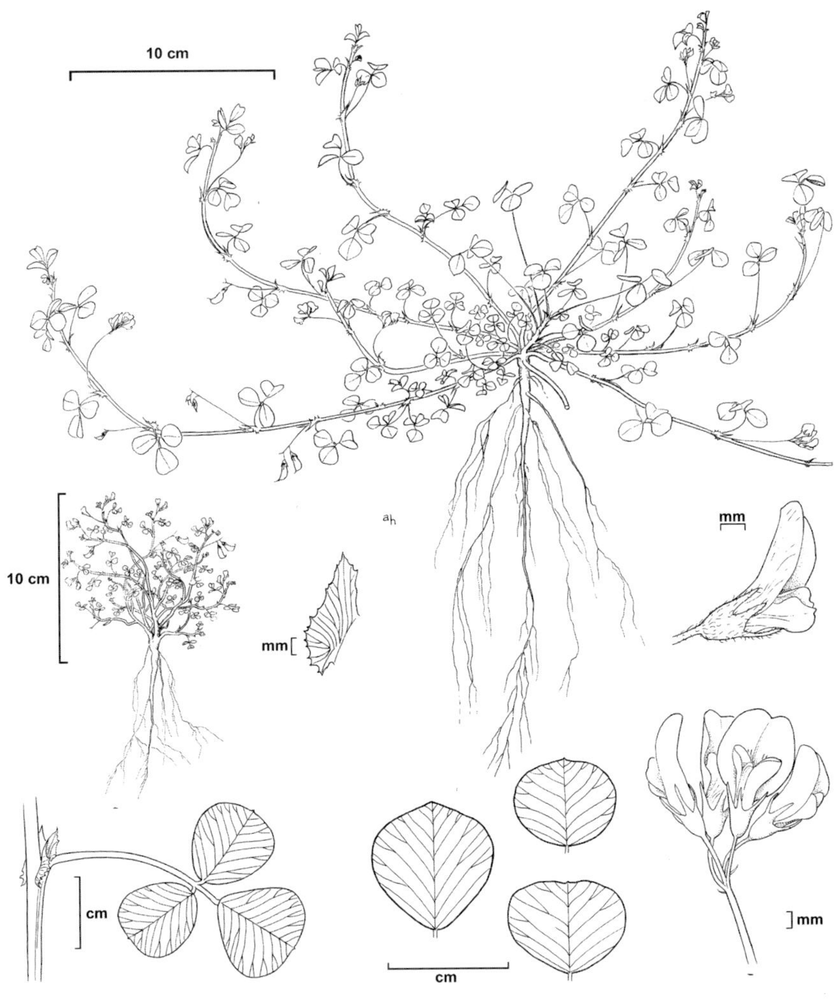

Fig. 40-1. *Medicago hybrida.* **Clockwise from top:** Plant; side view of flower; group of flowers; leaflets; leaf attached to stem; young plant. **Center:** Stipule. (Young plant based on Théveneu, 12 May 1866, GH; remainder based on E. Small M350, DAO; artist: A. Hanes.)

Vernacular Names

ENGLISH: Hybrid alfalfa (a contrived name, based on the Latin name; this name could be confused with "hybrid medic", a name used for the *M. italica* cultivar 'Toreador'), Pyrenean medick (medick is also spelled medic).

FRENCH: Luzerne hybride.

Geographical Distribution

(N = native, I = introduced, U = uncertain status):

EUROPE: France (N).

NORTHERN AMERICA: USA (Indiana; I (not established)).

This species is native to southern France, including Languedoc and Corbière, and from the Eastern Pyrenées almost to the Rhône.

Description

Perennial herb, stems prostrate, procumbent or ascending, arising from the crown, (10)20–40(50) cm long, weakly pilose. Leaflets obovate, broadly oval, or almost round, (7)10–15(18) mm long, (5)7–12(14) mm wide, denticulate along upper half or third of margin, glabrous on upper side, pubescent with simple hairs on lower midrib. Stipules entire or dentate at base, semi-sagittate. Inflorescence of (2)4–5 flowers, peduncles more or less equal to subtending petiole (sometimes shorter or longer). Flowers (5)6–8 mm long, yellow, calyx 3–4 mm, teeth about equal to tube, pubescent. Young pods emerge straight from calyx, then bend somewhat sideways, mature pods flat, convex on one side and concave on the other, weakly falcate, glabrous, (7)10–12 mm long, 3.5–4.5 mm wide, the flat sides with oblique, transverse veins that arise from the concave (ventral) suture, branch, and anastomose toward the convex (dorsal) suture, apex of pod with a short, curved beak. Seeds two to four per pod, not separated by spongy partitions, smooth, oval, 3–3.5 mm long, radicle about two-thirds of length of seed.

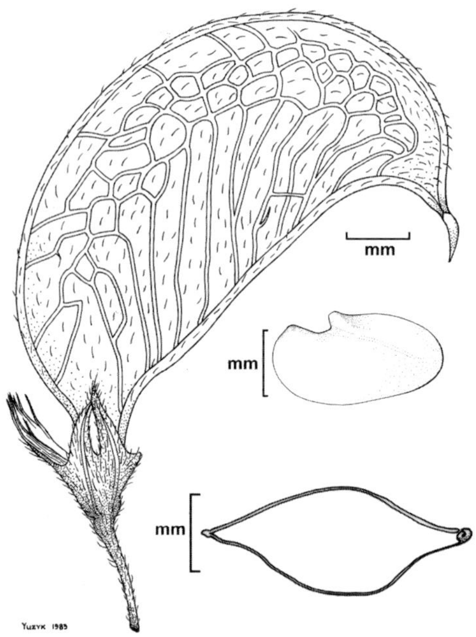

Fig. 40-2. *Medicago hybrida.* **Clockwise from top left:** Fruit; seed; outline of cross section of fruit. (Based on E. Small M57, DAO; artist: A. Hanes.)

Identification Hints

The fruits of *M. hybrida* are rather like those of the species of section *Platycarpae* (*M. archiducis-nicolai*, *M. platycarpos*, *M. popovii*, and *M. ruthenica*), as well as *M. cretacea* and *M. edgeworthii*, all farther to the east of *M. hybrida* (in India, Pakistan, former USSR, and/or in China), but lacking as distinctive a heterogeneity of venation of the pod (veins parallel, straight, oblique near ventral (straight-concave) suture, branching towards dorsal (concave) suture to form numerous roundly polygonal alveoli).

Habitat and Environmental Adaptation

Habitat

Medicago hybrida occurs in grassy glades, on dry rocky hillsides, and on limestone slopes, often among pine trees and scrubby shrubs, at altitudes of 500–1500 m.

Phenology

The species flowers in June and July in its native area.

Soil

Medicago hybrida typically occupies calcareous substrates.

Temperature

Lesins and Lesins (1979) found that *M. hybrida* could not survive the very cold winters at Brandon, Manitoba.

Water

Lesins and Lesins (1979) indicated that *M. hybrida* prefers areas with high humidity. However, the species occurs on dry hillsides, suggesting it can withstand some drought.

Biotic Relationships

Pollination

Lesins (1969) and Lesins and Lesins (1979) noted that *M. hybrida* is partially an inbreeder, more so than other *Medicago* perennials, setting seed in the absence of pollinators. These authors suggested that pollinators were limited because of the high humidity in the native habitat of this species, and that inbreeding is an adaptation to this condition. Small (1988) noted that the relatively large flowers and pollen to ovule ratio (576) are consistent with *M. hybrida* being at least partly cross-pollinated, like other perennial species of *Medicago*.

Dispersal

No evident adaptations for seed dispersal are present. Given the mountainous habitat, gravity likely plays a role. Dispersal of seeds may also be accomplished by seeds clinging to animals, although there is no specific adaptation for this. Some seeds may also be capable of passing through the guts of herbivores, but this has not been examined.

Herbivory

No information located.

Nodulation and mycorrhizae

USDA (2009) reported positive root nodulation for *M. hybrida*.

Disease resistance

No information located.

Economic History, Cultivars, and Uses

This species has not been utilized in agriculture.

Cultivars

No cultivars have been bred.

Forage

Medicago hybrida is not sown for forage, although in its native area it has limited use as natural forage.

Fodder

Medicago hybrida is not harvested for fodder.

Green manure, rotation crop, ground cover

No information located.

Other uses (salad sprouts, ornamental, plant products, etc.)

No information located.

Chromosome Number

$2n = 16$: Lesins and Lesins 1966; Lesins 1969.

Gillies (1972*b*) noted that the karyotype of *M. hybrida* was very similar to that of *M. suffruticosa*, indicating the close relationship of these two species.

Genetic Relationships and Gene Transferability

Intraspecific

Artificial hybridization of different populations has not been attempted.

Interspecific

Lesins (1969) found that *M. hybrida* could not be crossed with *M. sativa* (but see below). Lesins (1969) was able to hybridize *M. hybrida* with both subspecies of *M. suffruticosa*. The hybrids produced some seeds, indicating a close relationship of the two species.

Exchange by advanced biotechnology

Using embryo rescue techniques, McCoy and Smith (1986) obtained hybrids between *M. hybrida* and (at the same chromosome number) diploid *M. sativa* ($2n = 16$), although this hybrid combination was obtained with considerable difficulty.

Conservation Status

This species has a restricted geographical range and may be subject to genetic erosion. It is in need of a conservation assessment and germplasm collection.

Germplasm Sources

Only three accessions appear to be available from germplasm institutions (see Chapter 3).

Research Problems and Priorities

Medicago hybrida has not received much attention to date, although it may have some value as feed for livestock under certain pasture conditions, and may also possess genes useful for the improvement of alfalfa.

Representative Collections

France: Boullu, l'Abbé, forêt des Fanges (Aude), clairières, 16-VI-1888, in flower & early fruit (Société dauphinoise 1889 No. 5568) (BM, G, P). Delpont, Laroque-de-Fa (Aude), Le Bernadié, bois, 31-V-1931, in flower & early fruit (BM). Dromer, E. 6673, Château de Quiribus (France, Pyrenées-Orientales), chemi de ce château sur sol calcaire (specimen cultivated from seeds from this location) (G). Endrels, (ex herb. H. Van Heurck), ad sacelium St. Antoine de Calamus prope St. Paul de Fenouilhèdes, Pyr. or., Unio itiner, 26-VII-1830, in fruit (P). Endrels, prope sacellum St. Antoine de Calamus près St. Paul de Fenouilhèdes (Montes Corbarii), VI-1829, in flower & fruit (K). Fages, Ondes (Haute-Garonne), 6-VII-1886, in early fruit, F. Schultz Herbarium Normale Nov. Ser. Cent. 22 No. 2157 (BM, K, P). Humbert, H., près Axat (Aude), rochers des gorges de pierre lys, vers 350 m d'alt., 2-VI-1910, in flower & very early fruit (P). Luizet, D., Narbonne (Aude), Mount Alaric, 4-VI-1887, in flower & fruit (P). Maille, A., près Perpignan, montagnes des Corbières, VI-1834, in flower & fruit (K). Marty, L., Aude, Palaja, bords du ruisseau de Villemaury, 20-VI-1904, in flower & early fruit, Société pour l'étude de la flore franco-helvétique No. 2209 (P). Timbal-Lagrave, bords de l'Ariège près de son embouchure (Haute-Garonne), sables, 1-VII-1853, in flower & early fruit, Flora Galliae et Germaniae, Exsiccata de C. Billot No. 1645 (BM, P).

41. *Medicago hypogaea* E. Small

M. hypogaea E. Small, Taxon 33: 633 (1984), (basionym).

The epithet *hypogaea* is based on the Greek *hypo*, below, plus *gaia*, earth, reflecting the underground development of the fruits.

Medicago hypogaea was first assigned to the genus *Trigonella*, by Urban (1882), then reassigned by Eig (1926, 1927) to the monotypic genus *Factorovskya*, and finally Small and Brookes (1984*c*) recognized it as a species of *Medicago*. The species is noteworthy for its geocarpic fruit, the fertilized aerial flowers becoming imbedded in the soil by the elongation of the peduncle, pedicel, and especially the gynophore. The geocarpic nature of *M. hypogaea* so impressed earlier taxonomists that they ignored the fundamental similarities

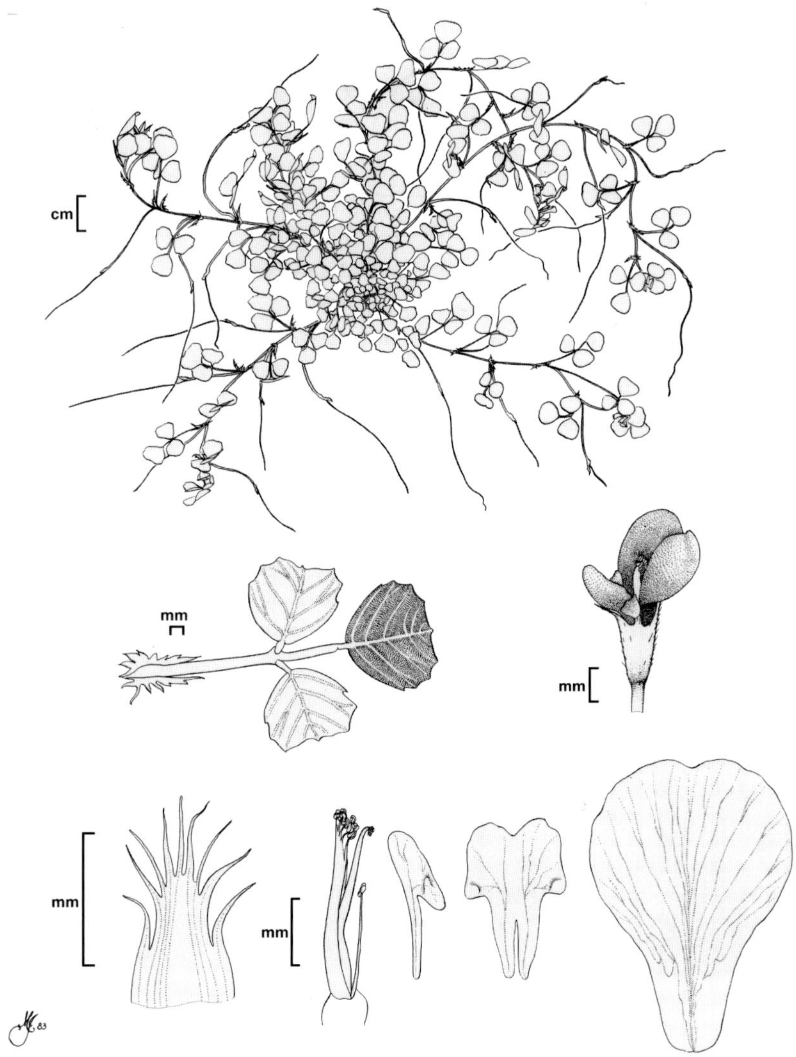

Fig. 41-1. *Medicago hypogaea.* **Top:** Plant (root not shown; the thin, long structures extending from the branches are the gynophores). **Center left:** Leaf. **Center right:** Side view of flower. **Bottom, from left:** Androecial column spread out (free stamen not shown); side view of androecial column (note 9+1 arrangement of stamens); wing petal; keel; standard petal. (Based on E. Small F177, DAO; artist: M. Jomphe.)

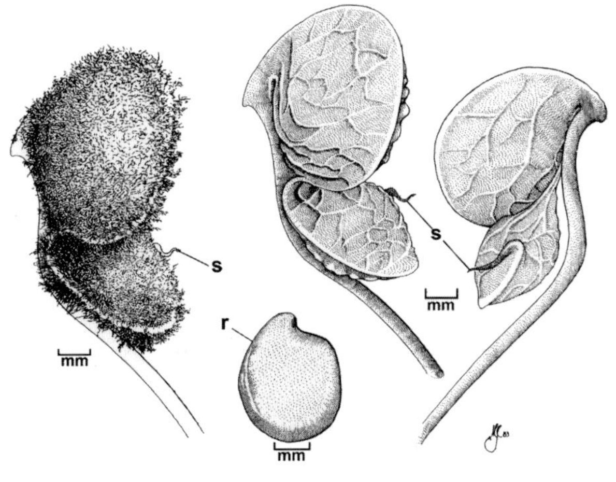

Fig. 41-2. *Medicago hypogaea.* **From left:** A very pubescent fruit; seed; two views of a glabrous fruit; seed. (s = style remnant, r = radicle; artist: M. Jomphe.)

in numerous characters that place the species squarely in *Medicago* (see Chapter 9). The placement of this species in *Medicago* has been confirmed by molecular studies (Downie et al. 1998; Steele and Wojciechowski 2003).

Subordinate Taxa

None recognized here.

Type Specimen

Small, E. F177, seed obtained from collection 124, Royal Botanic Gardens, Kew; Krukoff Collection 124 (from Y. Tankus, Tel-Aviv, Israel, 1977), seeds planted in Ottawa greenhouse May 1986, plant collected 21 January 1987, in fruit (holotype DAO!; isotype B!).

Synonymy

Homotypic

No names found.

Heterotypic

Trigonella aschersoniana Urb., Sitz. Ber. Bot. Ver. Brandenb. 33: 67 ("1881", printed 1882).

Factorovskya aschersoniana (Urb.) Eig, Contr. Fl. Pal. (Bull. Inst. Agric. Nat. Hist., Tel-Aviv) 6: 18 (1927).

The combination *Medicago aschersonia* (Urban) E. Small was precluded by the name *M. aschersoniana* Urb. in Verh. Bot. Vereins Prov. Brandenburg 15: 77 (1873) (syn. subst. = *M. laciniata* (L.) Mill.). Urban gave the following information for "15 specimens collected by Ascherson at various stages of age, from the bud to the most mature stages" (translated from German), and noted "habitat in Aegypto inferiore prope Mariut in collibus calcareis m. Febr. florif: leg Ascherson a 1880." As noted in Small and Brookes (1984*c*), since the holotype at Berlin was destroyed during World War II, a search was made for additional potential type material in numerous herbaria, based on possible collections in Egypt by Ascherson during his third voyage in 1889, and in the absence of locating any collections, the type of *M. hypogaea* was proposed as the neotype of *T. aschersoniana*.

Vernacular Names

English: Sand fenugreek (Çakan et al. 2005; a name that considers the species to belong to *Trigonella*). Suggestion: "subterranean medic" (sub medic for short), by analogy with subterranean (sub) clover, mentioned below.

French: No names found.

Geographical Distribution

(N = native, I = introduced, U = uncertain status):

Africa: Egypt (N), Libya (N).

Asia: Cyprus (N), Greece (Rhodos, N; see Conservation Status), Iraq (U), Israel (N), Lebanon (N), Syria (N), Turkey (N).

Medicago hypogaea is native to the southeastern Mediterranean area.

Description

Annual, dwarf herb, branched from base, stems prostrate, 2–15 cm long, glabrous or nearly so. Leaflets triangular-semiorbicular, cuneate at base, 2–4 teeth on distal margins of sides, 2–3.5(6) mm long, 2.5–4.5(6.5) mm wide, glabrous above, slightly pilose below (especially on midrib), stipules ovate to ovate-lanceolate, laciniate, or pinnatifid-dentate, petioles 4–12 cm long. Inflorescence one-flowered, axillary, peduncle 2–8 mm long at anthesis, about half as long as subtending petiole. Flowers 2–4 mm, corolla yellow, nearly twice length of calyx and often remaining closed, calyx pilose, lobes triangular to lanceolate-triangular, half as long as tube. Floral pedicel and peduncle bend downward and elongate (about three times their original length at initial flower maturity: peduncle to 5–20 mm, pedicel to 1.5–2 mm), and a gynophore develops at base of ovary and also elongates (1.5–10 cm, generally 1.5–6 cm), pushing the young ovary into the ground to a depth of 2–4 cm, where it develops into a fruit (failure to enter the ground results in the ovary drying up and not developing into a fruit). Fruit deeply two-parted (often constricted between the seeds), the upper one often abortive, each part one-seeded (if not abortive), lenticular or irregular in shape, 5–6(10) mm long, 3–5 mm wide, young fruit glabrous, mature fruit covered with appressed wooly or fleecy tomentum (occasionally glabrous). Seed compressed-ovoid, yellowish, smooth, ca. 4 mm long, 3 mm wide, radicle about as long as seed.

Identification Hints

Medicago hypogaea is rarely collected because it is an ephemeral, very small, drought-evading plant, often completing its life cycle in a month during the wet spring, and reproducing more or less in the same place. It is very distinctive in that it produces its fruits underground like the peanut (*Arachis hypogaea* L.). The fruits are covered with cottony hairs, a feature likely to be confused only with *M. lanigera*. The leaves are perhaps the smallest in the genus *Medicago*.

Habitat and Environmental Adaptation

Habitat

Medicago hypogaea occurs on mountain slopes, hills, grasslands, sand dunes, and other open sites of the Mediterranean and Sahara region, near sea level. Eig (1926) noted that in Israel "As the localities inhabited by *Trig. Aschersoniana* Urb. are always old untilled places, this plant is also of a certain practical value in agricultural life, when land purchases are made, and in similar cases, as it indicates, that the land it grows upon, had been left untilled for scores of years." For a discussion of the special adaptations of *M. hypogaea*, see Chapter 7.

Phenology

The species flowers in early winter, from January to February (to a lesser extent, to April), in its native area. On 24 July 2007 I planted seeds of 200 accessions representing 62 species of *Medicago* (every taxon of the genus for which seeds were available) in a greenhouse in Ottawa, Canada; in early September, the first plants to develop flowers were *M. hypogaea* and *M. truncatula* var. *laxicycla*, both extremely short-lived taxa.

Soil

Medicago hypogaea typically occupies calcareous soils.

Temperature

The species occurs in the seasonally cool to warm winter and early spring of the Eastern Mediterranean.

Water

Medicago hypogaea is a winter annual and an ephemeral, relying on Mediterranean seasonal winter rains to complete its life cycle and evade drought. Townsend (1974) commented "By its leaves it is definitely a mesophytic plant but, being ephemeral and geocarpic, it accommodates itself well to some pronounced steppe associations."

Biotic Relationships

Pollination

This species is a strict, obligative inbreeder. Small (1988) reported that the small flowers and its small pollen to ovule ratio (338) are consistent with *M. hypogaea* being self-pollinated. Eig (1926) noticed the occasional domestic bee visiting the flowers in Israel, but generally insects were absent.

Dispersal

No obvious mechanisms for seed dispersal are evident. Many of the seeds likely simply germinate in the soil near the maternal plant, where they were pushed underground by the mechanism described above.

Herbivory

No information located.

Nodulation and mycorrhizae

USDA (2009) reported positive root nodulation for *M. hypogaea*.

Disease resistance

No information located

Economic History, Cultivars, and Uses

This species has not been employed for any economic purpose. After I published the name "*Medicago hypogaea*", there was some interest in Australia in examining the pasture merits of the species, because subterranean clover (*Trifolium subterraneum* L., also known as sub clover), another of the very few species that produces all of its seeds underground, is well known as a pasture species in Australia. As noted below, however, apart from sharing the very unusual feature of geocarpy, *M. hypogaea* is much less suitable as a pasture plant than is *T. subterraneum*.

Cultivars

No cultivars have been bred.

Forage

Medicago hypogaea is not sown for forage and is such a small plant that it is unlikely to be of much feeding significance for livestock. Nevertheless, Eig (1926) observed a herd of sheep feeding on the plant in Israel.

Fodder

No information located.

Green manure, rotation crop, ground cover

No information located.

Other uses (salad sprouts, ornamental, plant products, etc.)

No information located.

Chromosome Number

$2n = 14$: Goldblatt 1981; Small and Brookes 1984*c*.

Genetic Relationships and Gene Transferability

Intraspecific

No information located.

Interspecific

No information located.

Exchange by advanced biotechnology

No information located.

Conservation Status

Medicago hypogaea is fairly widespread, but is inconspicuous and often overlooked. The species is considered to be threatened in the Çukurova Deltas of southern Turkey

(Çakan et al. 2005, who noted that *M. hypogaea* had been given the status of "regionally extinct" under old International Union for the Conservation of Nature and Natural Resources (IUCN) Red List categories). *Medicago hypogaea*, recently discovered on Rhodos Island (R. Thanopoulos, personal communication), has been added to the new edition of the Red Data Book of Rare and Threatened Plants (Phitos et al. 2009).

Germplasm Sources

Very few sources of seeds are available, almost none from the world's largest gene banks.

Research Problems and Priorities

Medicago hypogaea is a peculiar, little-studied species. Its small size and ephemeral nature suggest that it does not have a role to play in directly supplying forage, but this may be misleading, and it is possible that there are natural pasture applications. The species may possess useful genes that can be transferred to other species of *Medicago*.

Representative Collections

Cyprus: Casey, E.C. 958, Kyrenia, grassy place on rocks, alt. 200 feet, 14-III-1950, in flower (K). Casey, E.C. 971, Kormakiti (district of Kyrenia), alt. 4000 feet, in short uncultivated turf, 18-II-1950, in flower (K). Casey, E.C. 995, Kyrenia, alt. 200 feet, short turf on thin soil over rock, 26-III-1950, in flower & fruit (K). Chapman, E. 35, mile 5. Nicosia-F'gusta Rd., alt. 500 feet, frequent in wastelands in Asphodel association, 30-III-1950, in flower & early fruit (K). Chapman, E. 390, mile 3, Nicosia-F'gusta Rd., alt. 500 feet, in waste lands in Poterium-Asphodel association, covering fairly large patches, 12-IV-1950, in flower (K). Chapman, E. 480, Mia Milia, alt. 600 feet, frequent in dry trampled grassland, 17-IV-1950, in fruit (K). Oswald, P.H. 120, south of Xeri, Nicosia, alt. 900 feet, spaces between stones of threshing floors, 20-IV-1959, in flower & fruit (K). **Israel:** Danin, A., south Judean Mountains, Har Amassa, shallow soil at the foot of limestone rock, alt. 750 m, 11-I-1973, in flower & fruit (HUJ). Danin, A., N. Negev, 10 km west of Dimuna, loessial soil, 14-III-1973, in fruit (HUJ). Eig, A., Esdraelon Plain, Ein Taba'un, slopes of wadi, among rocks, 21-II-1924, in flower & early fruit (HUJ). Eig A., Feinbrun, N. & Grizi, A., Shefela, env. of Sarafand, sandy-clay soil, uncultivated places, 31-I-1936, in flower & early fruit, Univ. Hebraicae Hierosolymitanae Flora Palaestinae Exiccata No. 332 (BM, G, K). Feinbrun, N., Tel-Aviv, Philistean Plain, calcareous sandstone, 16-I-1926, in flower & fruit (HUJ). Hayne, W.A., "Plants of the Holy Land", Wady Zerka Ma'in, III-1872, in flower (K). Mosheiyov, G., env. of Dead Sea, III-1935, in flower & fruit (HUJ). Naftolsky, N., Lower Galilee, Wadi Malik, 22-II-1927, in flower & fruit (HUJ) (an unusually large plant, with branches 20 cm long, leaflets mostly 8 mm long, 6 mm wide). Naftolsky, N., Philistaean Plain, Ben Shemen, hill, rocks, 13-I-1923, in flower & fruit (HUJ). Naftolsky, N., Sharon Plain Petah-Tigra, 14-III-1928, in flower & fruit (HUJ). Snogerup, S. & B. 3087, ca. 6 km east of Jerusalem, near the road to Jericho, very dry Mediterranean shrub vegetation, 16-II-1984, in flower & fruit (BM). Weichselfish, M., Mt. Carmel, Wadi Falah opps. 'Atlit, 22-II-1961, in flower (HUJ). Zohary, M., Judaean Desert, near km 10 on Jerusalem-Jericho road, soil among rocks, 10-III-1947, in flower & fruit (HUJ). Zohary, M. & Weisel, Y., Shefela, 1 km east of Negba, sandy-loam hill, 21-II-1958, in flower & fruit (HUJ). **Israel/Syria (Golan):** Kushnir, T., Syria, Mt. Hermon, slopes, 8-IV-1942, in fruit (HUJ). **Libya:** Pampanini, R. N4130, Libia-Cirenaica, Bengasi, Giocchi el-Chebir (Lete), 9-III-1933, in fruit (K). **Syria:** Zohary, M., Syria bor., Kerkhan – Alexandretta, 9-V-1931, in fruit (HUJ). **Turkey:** Sintensis, P., Mersina, 21-II-1888, in flower (K).

42. *Medicago intertexta* (L.) Mill.

M. intertexta (L.) Mill., Gardeners Dictionary 8th edition, *Medicago* no. 4 (1768).

The epithet *intertexta*, from the Latin *inter-*, between, plus *textus*, tissue, means interwoven, an allusion to the interwoven spines of the fruits. The epithet was originally spelled *intortexta*, a misprint that was emended by Heyn (1963).

Subordinate Taxa

None recognized here.

Type Specimen

Heyn (1959: 164) lectotypified the basionym *M. intertexta* (*M. polymorpha* L. var. *intertexta* L.) by two illustrations of the same pod in Morison Pl. Hist. Oxon. 2: tab. 15, figs. 8 & 9 (1680). The accompanying text: "Medica cochleata spinosa major dicarpos capsula seu spinis longioribus sursum & deorsum tendentibus". This was accepted in The Linnaean Plant Name Typification Project (http://www.nhm.ac.uk/research-curation/projects/linnaean-typification/; current June 2010; cf. Jarvis 2007).

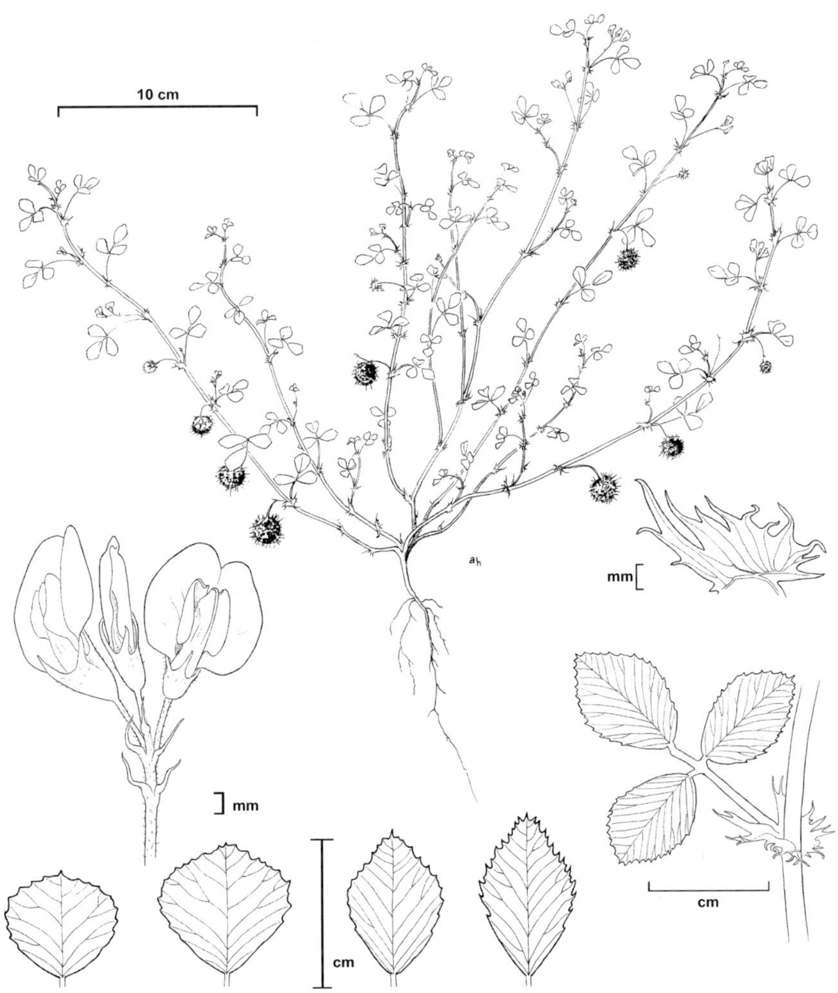

Fig. 42-1. *Medicago intertexta.* **Clockwise from top:** Plant; stipule; leaf attached to stem; leaflets; group of flowers. (The plant is based on H. Graves & J. Ball, 1890, GH, the remainder on E. Small M321, DAO; artist: A. Hanes.)

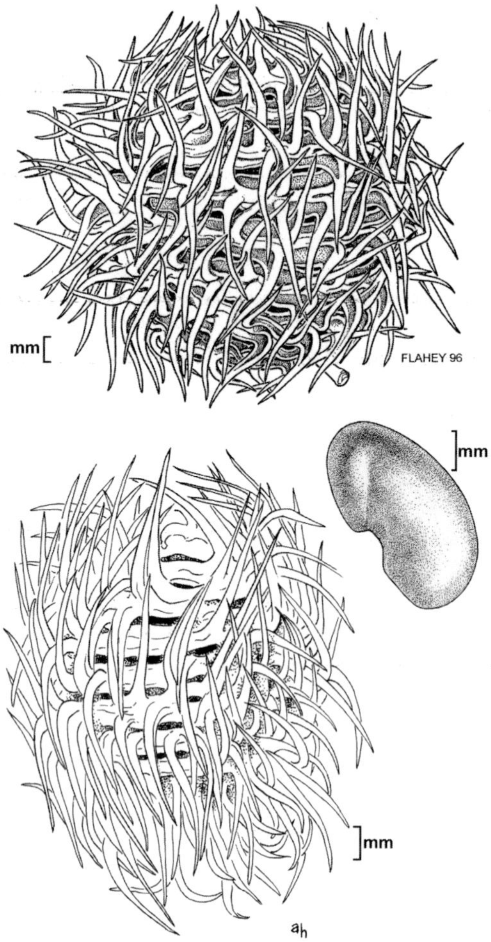

Fig. 42-2. *Medicago intertexta*. **Top:** Fruit (artist: B. Flahey). **Center right:** Seed. **Bottom:** Fruit (artist: A. Hanes). (All based on E. Small M321, DAO.)

Synonymy

Homotypic

M. polymorpha L. var. *intertexta* L., Sp. Pl. 780 (1753), (basionym).

M. intertexta (L.) All., Fl. Pedem. 1: 315 (1785). (Some taxonomists have interpreted the fact that Miller did not provide a reference to Linnaeus's name, above, as sufficient to justify Allioni, rather than Miller, as the author of the name.)

Medica intertexta (L.) Medik. in Vorles. Churpfälz. Phys.-Öcon. Ges. 2: 386 (1787).

M. intertexta (L.) Gaertn., De Fruct. 2: 350 (1791).

Medica intertexta (L.) Moench, Meth. 117 (1794).

Heterotypic

M. intertexta (L.) Mill. subsp. *aculeata* Urb., Verh. Bot. Ver. Brandenb. 15: 62 (1873). (Urban (p. 44) stated that in his monograph varieties were designated with Latin letters and forms by Greek letters. Despite some inconsistencies in his monograph, I have interpreted Urban's taxa accordingly throughout this monograph. Where he presented a hierarchy of two Latin letters, the first is interpreted as a subspecies, the second as a variety.)

M. crinita C. Presl, Fl. Sicul. (Presl) 1: xx (1826).

M. echinus DC. ("*M. echinus* Lam.") in Lam. & DC., Flore Française ed. 3, 4: 546 (1805), nom. illeg. (As interpreted by Heyn (1963), who noted that she had been unable to find the type specimen; de Candolle appears to have had *M. ciliaris* in mind when he recognized his taxon. However, Heyn also noted that numerous herbarium specimens that correspond to *M. intertexta* have been annotated as *M. echinus*. That the name has been interpreted as *M. intertexta* and not *M. ciliaris* is suggested by the following "Type Information" in The International Plant Names Index (in 2006): "Elle croît dans les provinces méridionales; aux environs de Nice (All.) ... nom. inval. as de Candolle cited *Medicago intertexta* var. α Lam. [*M. intertexta*] in synonymy, and as this is the type sensu stricto of *M. intertexta* de Candolle's name is illegitimate.")

- *M. intertexta* (L.) Mill. subsp. *aculeata* Urb. var. *echinus* (DC.) Urb., Verh. Bot. Ver. Brandenb. 15: 62 (1873).
- *M. ciliaris* (L.) All. subsp. *echinus* (DC.) Bonnier & Layens, Fl. Fr. 75 (1894).
- *M. intertexta* (L.) Mill. var. *echinus* (DC.) Burn., Fl. Alp. Marit. 2: 97 (1896), comb. superf.

M. echinata St. Lag., Ann. Soc. Bot. Lyon 7: 130 (1880).

M. panormitana Tineo in Tod. Fl. Sic. Exsicc. No. 651, ex Urb., Verh. Bot. Ver. Brandenb. 15: 62 (1873) pro syn.

- *M. intertexta* (L.) Gaertn. var. *panormitana* (Tin. ex Urb.) Urb., Verh. Bot. Ver. Brandenb. 15: 62 (1873).

M. intertexta (L.) Mill. subsp. *aculeata* Urb. var. *echinus* (DC.) Urb. f. *pilifera* Urb., Verh. Bot. Ver. Brandenb. 15: 62 (1873).

M. polycycla Moris, Mem. Accad. R. Sci. Torino 38: xxxi (1835).

M. echinus DC. var. *tuberculata* Moris, Fl. Sard. 1: 453 (1837).

M. intertexta (L.) Mill. subsp. *aculeata* Urb. var. *echinus* (DC.) Urb. f. *variegata* Urb., Verh. Bot. Ver. Brandenb. 15: 62 (1873).

- *M. echinus* DC. var. *variegata* (Urb.) Ricker in McKee & Ricker, USDA Bur. Pl. Ind. Bull. 267: 32 (1913).

Vernacular Names

English: Calvary medic, Calvary clover (Medic is also spelled medick; "Calvary" is based on the dark red (or purple) blotches often present on the leaves, said to be reminiscent of Christ's blood spilled during the crucifixion on Calvary Hill in Jerusalem. The spiny seed pods have been said to resemble Christ's crown of thorns. Vickery (1995) refers to a custom of sowing the seeds on Good Friday), hedge-hogs (for the spiny appearance).

French: La luzerne hérisson, luzerne à épines enchevêtrées, luzerne enchevêtrée, luzerne entremêlée.

Geographical Distribution

(N = native, I = introduced, U = uncertain status):

AFRICA: Algeria (N), Canary Islands (N), Egypt (U), Madeira (N), Morocco (N), South Africa (I), Tunisia (N).

AUSTRALASIA: Australia (I).

EUROPE: Balearic Islands (N), Corsica (N), Crete (N), France (N), Greece (N), Italy (N), Malta (N), Portugal (N), Sardinia (Italy; N), Sicily (Italy; N), Spain (N), Turkey in Europe (N), Yugoslavia (I).

This species is native to the western Mediterranean and adventitious in Central and Western Europe. Numerous floras indicate that *M. intertexta* is not only native to countries of the western Mediterranean (such as Cyprus and the East Aegean Islands of Greece), but also occurs in Asia (including Iran and Turkey) and the Middle East (including Israel, Iraq, Jordan, Lebanon, and Syria). However, these records appear to be based on considering *M. ciliaris* as a variety of *M. intertexta* (i.e., *M. intertexta* var. *ciliaris*).

Description

Annual herb, stems 20–50(70) cm long, procumbent to ascending, branched from the base, twigs and leaves sparsely covered with simple hairs or glabrous. Leaflets (6)15–22 mm long, 8–14 mm wide, 1.5 to twice as long as broad, obovate to oblanceolate, distal half or third of leaf margin serrate, leaflets sometimes with a large central anthocyanin blotch. Stipules dissected or dentate. Inflorescence of 1–4(10) flowers, peduncles about as long as subtending petiole (elongating in fruit). Flowers 5–9 mm long, yellow or light yellow. Calyx with simple hairs, teeth equal to tube or shorter. Corolla about twice length of calyx or somewhat less. Young fruit protrudes sideways from calyx. Fruit ovoid or rarely spherical, glabrous or glabrescent, with (3)6–10 loosely adpressed coils, the largest 9–15(17) mm in diameter, coils with or without a central (axial) opening 1–3 mm wide, coil faces with a distinct net of veins mostly anastomosing in distal half of coil face, each dorsal suture spine receiving a diagonal vein emerging from the coil face net of veins, spines 1–5(8) mm long, each row of spines usually inserted about 90° to 100° to surface of coil, the spines of adjacent coil faces usually interlocking, rarely tubercles present instead of spines. Seeds 5–20 per pod, 4–5 mm long, 2.3–2.8 mm wide, the surface smooth, dark red-brown to black-brown, reniform, not separated by transverse, spongy fruit septa, radicle less than half as long as the cotyledons.

Heyn (1963) noted that populations with tubercles instead of spines on the fruits (sometimes recognized as f. *tuberculata* Moris) seem to be endemic to Sardinia.

Identification Hints

Medicago intertexta is usually a much taller and more vigorous plant than other annual species of *Medicago*. It often has a large purple blotch at the center of each leaflet (a feature sometimes found in other *Medicago*, notably *M. arabica*) and its large fruits lack gland-tipped trichomes. These features serve to distinguish it from its nearest relatives, *M. ciliaris*, *M. granadensis*, and *M. muricoleptis*.

Laouar and Abdelguerfi (2000*a*, 2000*b*, 2003) compared populations of *M. intertexta* with *M. ciliaris* in Algeria, and concluded that *M. ciliaris* develops precociously and produces more pods per plant, but the number of seeds per pod is reduced compared with *M. intertexta*. However, the differences are too small to be of diagnostic value.

Habitat and Environmental Adaptation

Habitat

Medicago intertexta occurs in relatively more mesic, fertile habitats than most other annual species of *Medicago*. Abdelguerfi-Laouar et al. (2003) studied the comparative ecology in Algeria of *M. intertexta* and the closely related *M. ciliaris*, and observed that *M. intertexta* was mostly restricted to relatively humid regions of the northeast of the country, generally at altitudes of less than 250 m, and in soils that were relatively deficient in calcium. Laouar and Abdelguerfi (2000*a*, 2000*b*, 2003) found that *M. ciliaris* occupied a much wider range of habitats in Algeria than did *M. intertexta*.

Phenology

Flowering occurs in the spring, generally from March to July depending on location in the native area.

Soil

Piano and Francis (1992) observed that in its native area the species usually occupies clay soils that are slightly to moderately basic. In Sicily, Piano et al. (1991) observed the species growing in a pH range of 8.0–9.0. In Israel, *M. intertexta* typically is in heavy soils (Heyn 1963). In Algeria, Abdelguerfi et al. (1988*a*) observed that it occurs in fine- and very fine-textured soils. Rogers et al. (2005) ranked *M. intertexta* as moderately tolerant of soil salinity and Prosperi et al. (2001) noted that the species sometimes occurs in salty soils, but Zoghlami and Hassen (1999) found that in Tunisia *M. intertexta* was associated with soils that were low in salt.

Temperature

Medicago intertexta is a typical Mediterranean winter annual, germinating in relatively cool late winter conditions and maturing seed by early summer as the climate becomes hot.

Water

Piano and Francis (1992) observed that in its native area the species requires a minimum of 400 mm annual precipitation. *Medicago intertexta* typically occupies moist soils. In Algeria, it usually occurs in areas receiving more than 700 mm of precipitation annually (Abdelgeurfi et al. 1988*a*). In Tunisia, it occupies a humid to subhumid zone with 600–1200 mm of precipitation (Zoghlami and Hassen 2004). Based on trials in New South Wales, Australia, Desborough (1976) commented that "This species has the characteristic of nodal rooting on wet soils and should be investigated on waterlogged soils." Piano and Francis (1992) characterized the species as "relatively tolerant to temporary waterlogging occurring in poorly structured soils."

Biotic Relationships

Pollination

Small (1988) noted that the relatively low pollen to ovule ratio (176) is consistent with *M. intertexta* being largely self-pollinated. However, the flowers are larger than in most other annual species of *Medicago*, suggesting that the species may be relatively attractive to pollinators. See information for the very closely related *M. ciliaris* (Chapter 24).

Dispersal

The indehiscent spiny fruits of this species are well adapted to dispersal by attachment to animal fur.

Herbivory

In tests of the alfalfa weevil (*Hypera postica*) on 13 annual species of *Medicago*, Pandey et al. (1984) found that *M. intertexta* showed resistance.

Nodulation and mycorrhizae

USDA (2009) reported positive root nodulation for *M. intertexta*.

Disease resistance

Yaege and Stuteville (2002) tested the resistance of 33 annual species of *Medicago* to powdery mildew (*Erysiphe pisi*) and found that the five collections of *M. intertexta* grown were susceptible to it (it was observed that often only one of a number of populations of other species tested was resistant). In a comparative test of 36 annual species of *Medicago*, O'Neill and Bauchan (2000) found that six accessions of *M. intertexta* were quite susceptible to anthracnose caused by *Colletotrichum trifolii*.

Economic History, Cultivars, and Uses

This species has not been developed to date for agricultural purposes.

Cultivars

No cultivars have been bred.

Forage

Medicago intertexta is not sown for forage, although it is grazed where it occurs. Photosensitization known as trefoil dermatitis has been associated with the grazing of this species by cattle, sheep, and horses (Gardner and Bennetts 1956). Mohammad and Qamar (1988) grew 14 annual species of *Medicago* in small test plots in Islamabad, Pakistan, located in a subtropical, subhumid area that is notably hotter and wetter than the normal habitat of these species. The dry matter yield of *M. intertexta* was about 4.3 t/ha.

Fodder

This species is not used as a source of fodder.

Green manure, rotation crop, ground cover

The species is occasionally used as cover for dry, sunny banks.

Other uses (salad sprouts, ornamental, plant products, etc.)

Medicago intertexta was once grown as an ornamental potted plant, but is rarely cultivated for this purpose today.

Chromosome Number

$2n = 16$: Fryer 1930 (as *M. echinus* and *M. intertexta*); Ghimpu 1928, 1929*a*, 1930 (as *M. echimus*); Heyn 1956; Clement 1962; Bolkhovskikh et al. 1969; Simon and Simon 1965; Fernandes et al. 1977; Lesins and Lesins 1979; Abdelguerfi and Guittoneau 1979; Agarwal and Gupta 1983*b*; Schlarbaum et al. 1984; Abdelguerfi et al. 1988*b*; Diosdado et al. 1993; Mariani et al. 1996; Oberprieler and Vogt 1996*b*; Mohamed 1997; Parra et al. 1999.

$2n = 16 + 1B$: Heyn 1963; Agarwal and Gupta 1983*b*.

$2n = 16, 32$: Fernandes and Queirós 1978.

$2n = 32$: Fernandes and Santos 1971. Lesins and Lesins (1979) noted that the reported count of $2n = 32$ could not be confirmed.

Genetic Relationships and Gene Transferability

Intraspecific

Artificial hybridization of different populations has not been attempted.

Interspecific

Medicago intertexta is very closely related to *M. ciliaris* and the wisdom of separating it as a species has been debated. The two species can be hybridized readily (see information for *M. ciliaris*). Lesins et al. (1971) reported that hybridization of *M. intertexta* with either *M. ciliaris* or *M. muricoleptis* produced hybrids with reduced pollen fertility (40% to 50%). Hybridization with *M. granadensis* was unsuccessful (Lesins and Lesins 1979).

Exchange by advanced biotechnology

Thomas et al. (1990) fused protoplasts of *M. intertexta* and *M. sativa*, but were unable to generate plants from resulting calli.

Conservation Status

This species is fairly widespread and has some tendency to weediness. Accordingly, it is not in need of conservation.

Germplasm Sources

About 500 accessions of *M. intertexta* are available from germplasm institutions (see Chapter 3).

Research Problems and Priorities

Medicago intertexta is a vigorous species adapted to relatively moist situations. It would appear to have good productivity, and deserves to be examined for its agricultural potential.

Representative Collections

Algeria: Davis 52850, Les Falaises to Mansouriah (Bejaï-Djidjelli), alt. 30 m, shaley hills, one clump at edge of orchard, 26-V-1971, in flower & fruit (BM). **France:** Bourgeau, E. 108, env. de Toulon, dans les prés à Castignaux, 8-VI-1848, in flower & fruit (BM). Mosseray, R., près de Narbonne, prairie humide, 12-XI-1936, in flower (fruits presumably shed) (K). **Italy:** Fiori, Lucania, Matera, in herbosis humidis secundum torrentem prope oppidum, alt., ca. 320 m, solo calcareo, 7-VI-1913, in flower & fruit, Flora Italica Exsiccata Series III Fiori & Béguinot No. 2099 (BM). Groves, H., La Corocia, près Sienne (Toscane), bords des champs, 5-X-1882, in fruit, Société dauphinoise No. 3670 (P). Riggo, G. 172, Calabria, Prov. Reggio Brancaleone, in locis herbosis, 15-V-1898, in fruit (P). Todaro, Sicily, Ficuzza, in arvis argillosis, Todaro Flora Sicula Exiccata No. 1136 (K). **Italy (Sicily):** Davis & Sutton, S. 63371, Agrigento, ca. 5 km northeast of Sciacca, below Rocca Nadore, alt. 200 m, 19-V-1979, in fruit (BM). Davis & Sutton, S. 63540, Trápani, ca. 25 km northeast of Trápani, Monte Cofano, alt. ca. 300 m, 21-V-1979, in flower & fruit (BM). Stace, C.A. & Cotton, R., lower slopes of Madonie mountains above Isnello, south of Caphalu, alt. ca. 700–800 m, 26-V-1972, in flower & fruit. **Malta:** Borg, J., in S. Melitae, ad Ghain-il-Gbira, in herbosis, 25-IV-1931, in flower & fruit (K). **Morocco:** Guiol, F., juxta El Araix, Imp. Maroc, in herbosis, alt. 20 m, 23-V-1930, in fruit (BM). Maire, R., prope urbem Kenitra in arvis argillaceis, 27-IV-1925, in flower & fruit (P). Roffey, J., Tangier, 5-V-1916, in flower & fruit (BM). **Portugal:** Scarlett 237, Lisbon area, sea level upwards, growing with crops, 1935, in fruit (K). **Spain:** Bourgeau, E., Medina Sidonia, dans les champs, 21-III-1849, in fruit (P). Brinton-Lee, D.1550, near Zahena, meadow, alt. 0 feet, 26-VI-1975, in fruit (BM). Ellman, E. & Hubbard, C. 665, north of Algericas, near Los Barrios, in sandy railway bank, 18-V-1924, in fruit (K). Sandwith, N.Y. 6223, prov. Cádiz, near Madina Sidonia, finca de los Alburejos, in arable fallow, 11-VI-1963, in flower & fruit (K). Stace, A. 393, Cadiz, Algar, near Arios, in rich grassy meadows, alt. ca. 300 feet, 2-V-1961, in flower with young fruit in envelope (BM). **Tunisia:** Davis & Lamond D57832a, Plain of Oued Kebir, south of Tabarka, alt. 0–10 m, in fields, 13-V-1975 (BM). **Yugoslavia:** Krieger, Dalmatia, Rag., in campis, V-1910, in flower & fruit (K).

43. *Medicago isthmocarpa* (Boiss. & Bal.) E. Small

M. isthmocarpa (Boiss. & Bal.) E. Small in Small, Lassen & Brookes, Willdenowia 16(2): 433 (1987).

The epithet *isthmocarpa*, from the Greek *isthmos*, meaning a narrow passage such as a neck of land between two seas, plus *karpos*, fruit, is an allusion to the succeeding wide and narrow portions of the fruit.

Subordinate Taxa

None recognized here.

Type Specimen

Balansa, B. 115, Turkey, Pleine de Césarie (Cappodoce) (Kayseri), alt. 1107 m, VII-1856 (holotype G!; isotype LE). (Širjaev (1928–1934) presented a photo of a fruit that he termed the type.)

Synonymy

Homotypic

Trigonella isthmocarpa Boiss. & Bal. in Boiss., Diagn. Pl. Orient. Ser. 2, 3(6): 44 (1859), (basionym).

Heterotypic

No names found.

Vernacular Names

ENGLISH: No names found.

Fig. 43-1. *Medicago isthmocarpa.* **Left:** Fruiting branch. **Upper right:** Spread out androecial column (free stamen not shown). **Center from left:** Keel; wing petal. **Lower right:** Stem node showing stipules and leaflet. (Based on holotype, B. Balansa 115, G; artist: M. Jomphe.)

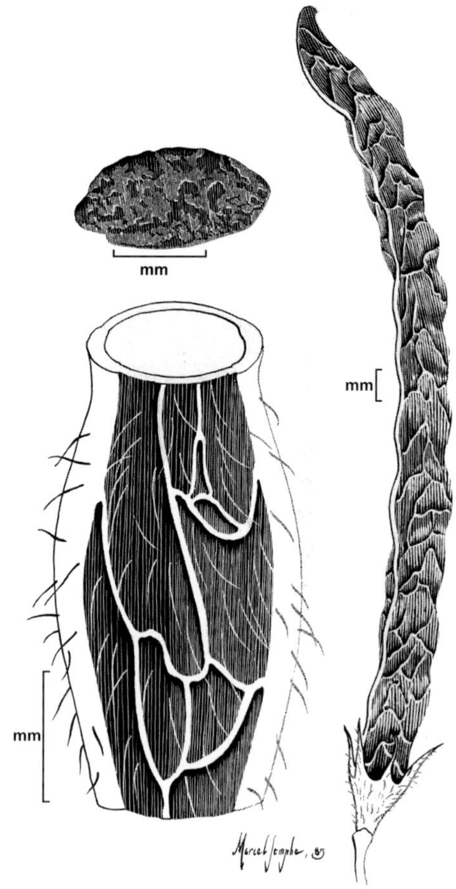

Fig. 43-2. *Medicago isthmocarpa.* **Upper left:** Seed. **Right:** Fruit. **Lower left:** Portion of fruit. (Based on holotype, B. Balansa 115, G; artist: M. Jomphe.)

French: No names found.

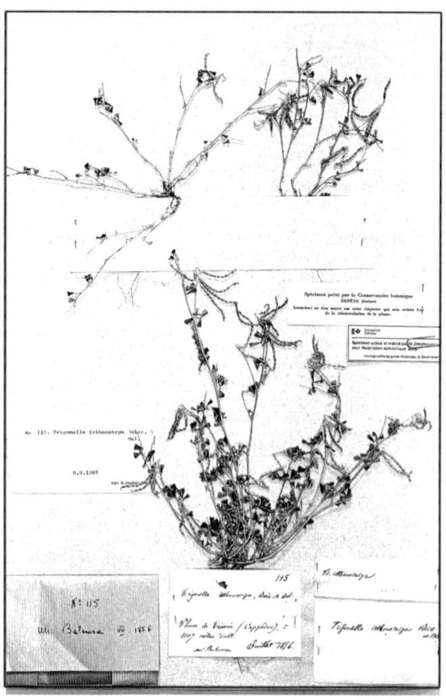

Fig. 43-3. Photograph of holotype of *Trigonella isthmocarpa* Boiss. & Bal.

Geographical Distribution

(N = native, I = introduced, U = uncertain status):

Asia: Turkey (N).

This species is a local endemic of central Anatolia.

Description

Annual herb, stems prostrate to ascending-erect, densely appressed-pubescent, 4–20 cm long. Leaflets oblong-triangular, a few small marginal serrations in distal part, 3–7 mm long, 2–4 mm wide, pubescent on both surfaces (especially beneath) with simple hairs, stipules lanceolate-acuminate, dentate or incised. Inflorescence of 2–6 flowers in a loose raceme, peduncle 10–25 mm, longer than subtending petiole. Flowers yellow, ca. 5 mm, calyx campanulate, 3–4 mm, the lanceolate teeth as long as or shorter than the tube. Fruits linear, cylindrical, straight to slightly curved, pedicels becoming deflexed (initially straight in flowers), 10–30 mm long, 1.5–2 mm wide, constricted between seeds, reticulate-netted, hooked at tip (beak), with 2–10 seeds, thickened and pilose on sutures. Seeds oblong, brownish, about 2–2.5 mm long and 1–1.3 mm wide, tubercled, radicle approximately 80% of length of seed.

Identification Hints

Medicago isthmocarpa is most similar to (and most obviously related to) *M. rhytidiocarpa*, the only other *Medicago* with fruits so constricted between the seeds that the pods are more or less moniliform. By contrast with the latter, *M. isthmocarpa* has less-curved fruits (straight to curved in contrast with the falcate to semi-circular fruits of *M. rhytidiocarpa*), a smaller corolla (5–6 mm versus 7–9 mm in *M. rhytidiocarpa*), and calyx teeth about equal to the length of the tube (longer than the tube in *M. rhytidiocarpa*).

Habitat and Environmental Adaptation

Habitat

Medicago isthmocarpa occurs on rocky igneous slopes and steppes at altitudes of 1000–1600 m.

Phenology

The species flowers from May to July in its native area.

Soil

No information located.

Temperature

No information located.

Water

No information located.

Biotic Relationships

Pollination

Based on the small flowers, this species appears to be an inbreeder.

Dispersal

No obvious adaptations for seed dispersal are apparent.

Herbivory

No information located.

Nodulation and mycorrhizae

No information located.

Disease resistance

No information located.

Economic History, Cultivars, and Uses

This rare endemic has not been used in agriculture.

Cultivars

No cultivars have been bred.

Forage

Medicago isthmocarpa is not sown for forage and is so uncommon it is unlikely to be of any significance as natural forage in its native area.

Fodder

The species is not employed for fodder.

Green manure, rotation crop, ground cover

No information located.

Other uses (salad sprouts, ornamental, plant products, etc.)

No information located.

Chromosome Number

No information located.

Genetic Relationships and Gene Transferability

Intraspecific

No information located.

Interspecific

No information located.

Exchange by advanced biotechnology

No information located.

Conservation Status

Given the rarity of this species, it deserves high conservation protection. In particular, seeds need to be collected and preserved.

Germplasm Sources

No germplasm accessions of this species have been located.

Research Problems and Priorities

Medicago isthmocarpa is a very poorly researched species. There is an urgent need to acquire seeds of this rare endemic and to conduct basic botanical and agricultural studies of it.

Representative Collections

TURKEY: Huber-Morath, A. 16176, Prov. Nigde, Distr. Aksaray, Eruptivgestein, Steppe am Fuss des Hasan Dagi, 28 km südlich Aksaray, alt. 1100 m, 27-VI-1959, in flower & early fruit (G). Reese, H., Prov. B5 Kayseri, district Kayseri, 3 km south Kayseri, 10-VI-1939, in flower & fruit (G).

44. *Medicago italica* (Mill.) Grande

M. italica (Mill.) Grande, Bull. Orto Bot. Regia Univ. Napoli 5: 230 (1916); ead. comb. Fiori in Fiori & Paoletti, Iconogr. Fl. Ital., ed. 2: 237 (1921), emend. Fiori, Nuov. Fl. Ital. 1: 832-833 (1925).

The epithet *italica*, Latin for Italian, reflects the collection area from which the species was first described. In fact, it is quite widely distributed outside of Italy.

Subordinate Taxa

No subdivisions of *M. italica* are recognized here. By contrast, previous students of this species have formally split it extensively. Lesins and Lesins (1979) recognized 14 infraspecific taxa in a complex hierarchy and stated that it "is the most polymorphic species in the genus." Heyn (1963), who specialized on *Medicago* section *Spirocarpos* of which *M. italica* is a constituent, recognized five varieties in it (one of which has since been assigned to *M. shepardii*). Nègre (1956, 1959) analyzed *M. italica* of Morocco, where the species is especially abundant; his publication of 1959 recognized three subspecies with a total of six varieties and some formae. Oostroom and Reichgelt (1958) recognized two subspecies.

As detailed in Small and Brookes (1990*b*), it appears that past students of *M. italica* have failed to recognize the main generator of variation in the species — introgression from the allied species *M. littoralis*. Heyn (1963) appreciated that hybridization between the two species produced plants that are difficult to identify. However, prior to 1990 it was not understood that gene flow appears to be predominantly from *M. littoralis* to *M. italica*. Both species are mostly found on sandy seashores bordering the Mediterranean Sea. Two considerations favor unidirectional gene transfer. First, *M. littoralis* is by far the more abundant species, often forming

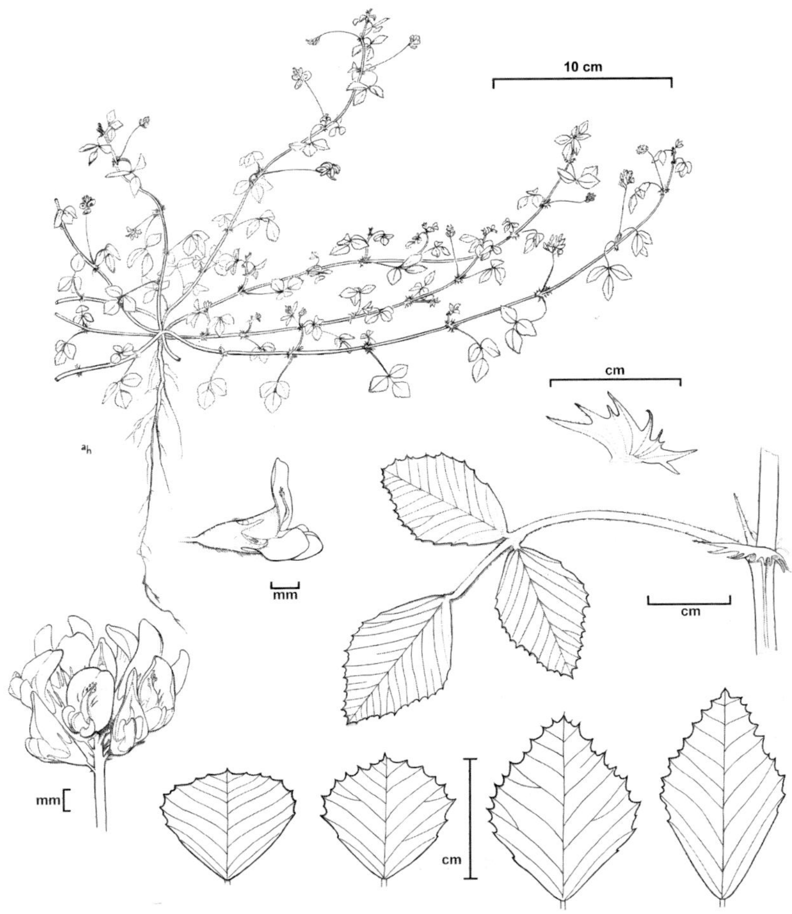

Fig. 44-1. *Medicago italica.* **Clockwise from top:** Plant; stipule; leaf attached to stem; leaflets; group of flowers; side view of flower. (Based on E. Small M970, DAO; artist: A. Hanes.)

more or less continuous stands that stretch for kilometers, whereas *M italica* is much less common; therefore, of the pollen transferred between the two species, *M. italica* is likely to receive proportionately more "foreign" pollen than vice-versa. Second, a unidirectional interbreeding barrier was found by Lesins and Erac (1968*a*), expressed as chlorophyll deficiency in F_1 plants and poor survival in F_2 plants when *M. littoralis* served as the maternal parent, but not when pollen transfer was from *M. littoralis* to *M. italica*. Collections of "*M. italica*" from a given locale often show what seem to be genetically segregating characters suggestive of recent hybridization with *M. littoralis* (e.g., segregation for number of fruit coils and for fruit spininess, two characters that strongly tend to differentiate the species). As noted by Small and Brookes (1990*b*), the hybrid with *M. littoralis* is rare in Israel (a region largely disjunctive from the main western Mediterranean area of distribution of *M. italica*), for reasons that are not clear (Small and Brookes suggested that it is related to the fact that *M. littoralis* of Israel is distinctive from the remainder of its species).

The following key distinguishes *M. italica* from *M. italica* × *M. littoralis*.

1. Dorsal sutures of pods completely lack spines .... *M. italica*
1. Dorsal sutures of pods spiny 2
   2. Pods with fewer than 3.5 coils .... *M. italica*
   2. Pods with more than 3.5 coils .... *M. italica* × *M. littoralis*

Type Specimen

Sales and Hedge (1994) lectotypified the basionym *Medica italica* Mill. as follows: "the illustration of *Medicago maritima*, trifol. πολνκαρποζ [polýkarpos], fructu toroso, no spinoso in Tilli, Cat. Pl. Horti Pisani, tab. 42, fig 2 (1723)."

Synonymy

Note: The following includes the pure form of the species (*M. italica*) as well as the hybrid (*M. italica* × *M. littoralis*), as I have not examined most of the types and the descriptions are usually insufficient for assignment.

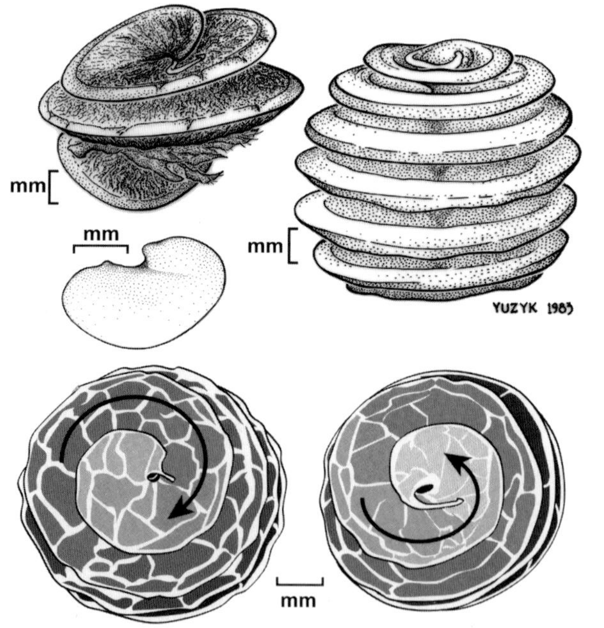

Fig. 44-2. *Medicago italica.* **Clockwise from center left:** Seed (based on E. Small M223, DAO; artist: L. Yuzyk); typical kind of fruit (based on E. Small M223, DAO; artist: L. Yuzyk); fruit with numerous coils (probably the result of introgression from *M. littoralis*; based on E. Small M758, DAO; artist: L. Yuzyk); stigma-end views of fruits showing opposite directions of coiling (artist: M. Jomphe.)

Homotypic

Medica italica Mill., Gard. Dict., ed. 8: no. 5 (1768), (basionym).

M. italica (Mill.) Steud., Nomencl. Bot., ed. 2, 2(9): 109 (1841), pro syn.

M. italica (Mill.) Steud. ex Fiori, Fl. Ital 1: 832 (1925).

Heterotypic

M. obscura Retz. var. *aculeata* Guss., Pl. Rar. 315 (1826).

- *M. obscura* Retz. subsp. *lenticularis* (Desr.) Urb. var. *aculeata* (Guss.) Urb., Verh. Bot. Ver. Brandenb. 15: 66 (1873).
- *M. obscura* Retz. subsp. *helix* (Willd.) Urb. var. *aculeata* (Guss.) Urb., Verh. Bot. Ver. Brandenb. 15: 66 (1873). (Note for the above two names the extraordinary use by Urban of an epithet from one taxon to designate two different taxa.)
- *M. tornata* (L.) Mill. var. *aculeata* (Guss.) Heyn (1963, p. 89).

M. astroites Bertol. Fl. Ital. 8: 272 (1850).

M. italica (Mill.) Grande subsp. *belasicae* Kožuharov, Izv. Bot. Inst. (Sofija) 15: 150 (1965).

M. calcar Lowe in Hook. Journ. Bot. 8: 291 (1856). (= *M. italica* × *M. littoralis*)

M. helix Willd. var. *calcarata* Lowe, Man. Fl. Mad. 1: 163 (1868). (= *M. italica* × *M. littoralis*)

M. canariensis Benth. ex Webb in Webb & Benth., Phytogr. Canar. 3, 2: tab 56 (1837). (= *M. italica* × *M. littoralis*)

M. commutata Tod., Ind. Sem. Hort. Bot. Panorm. 23 (1867). (= *M. italica* × *M. littoralis*).

M. corrugata Durieu in Duchartre, Rev. de Bot., Recueil Mens., 1: 365 (1845).

- *M. italica* (Mill.) Grande subsp. *corrugata* (Durieu) Nègre, Bull. Soc. Hist. Nat. Afr. N. 50: 291 (1959).

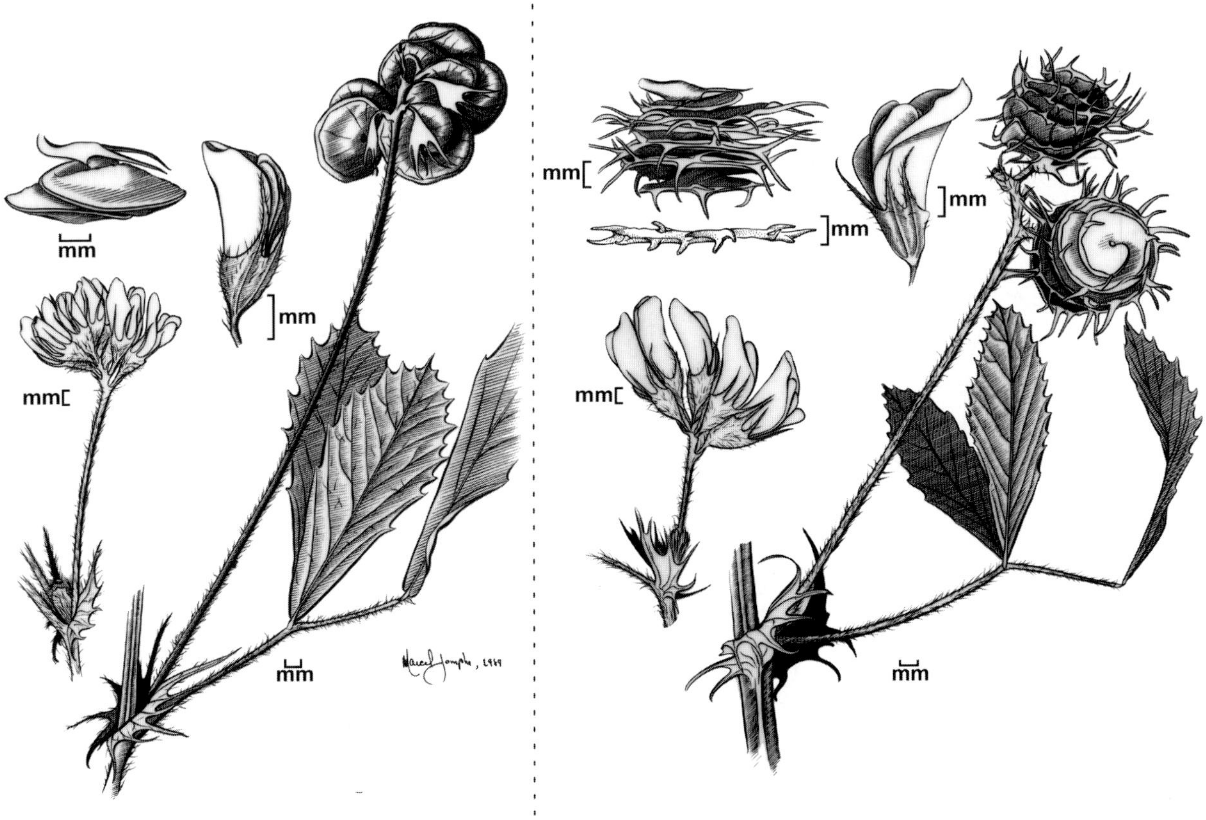

Fig. 44-3. A contrast of the pure form of *Medicago italica* and the hybrid between *M. italica* and *M. littoralis*. **Left side:** Pure form of *Medicago italica*. **Clockwise from center left:** Group of flowers; fruit; side view of flower; fruiting branch. (Based on E. Small M1478, DAO; artist: M. Jomphe.) **Right side:** *Medicago italica* × *M. littoralis*. **Clockwise from top center:** Side view of flower; fruiting branch; group of flowers (based on E. Small M1364, DAO; artist: M. Jomphe); fruit coil edge; fruit (based on Eig, Zohary, and Feinbrun 28 August 1927, HUJ; artist: L. Yuzyk).

M. obscura Retz. subsp. *lenticularis* (Desr.) Urb. var. *inermis* Urb. f. *dextrorsa* Urb., Verh. Bot. Ver. Brandenb. 15: 66 (1873).

M. obscura Retz. subsp. *helix* (Willd.) Urb. var *inermis* (Lowe) Urb. f. *dextrorsa* Urb., Verh. Bot. Ver. Brandenb. 15: 66 (1873).

M. obscura Retz. subsp. *helix* (Willd.) Urb. var. *aculeata* (Guss.) Urb. f. *dextrorsa* Urb., Verh. Bot. Ver. Brandenb. 15: 66 (1873).

M. obscura Retz. subsp. *tornata* (L.) Urb. var. *inermis* Urb. f. *dextrorsa* Urb., Verh. Bot. Ver. Brandenb. 15: 66 (1873).

M. obscura Retz. subsp. *tornata* (L.) Urb. var. *muricata* (L.) Urb. f. *dextrorsa* Urb., Verh. Bot. Ver. Brandenb. 15: 67 (1873).

M. laevis Desf. var. *dicycla* Maire (holotype: Maire, R., A. Lodi, Djebel Nador, 1000 m, Algeria (MP); viewed online at http://www.aluka.org; current June 2010)

M. helix Willd., Sp. Pl. 3(2): 1409 (1802).

- *M. obscura* Retz. subsp. *helix* (Willd.) Urb., Verh. Bot. Ver. Brandenb. 15: 66 (1873).
- *M. obscura* Retz. subsp. *helix* (Willd.) Arcangeli, Comp. Fl. Ital. 162 (1882), nom. superfl.
- *M. obscura* Retz. subsp. *helix* (Willd.) Batt. in Batt. & Trabut, Fl. Algérie 1: 228 (1889).
- *M. obscura* Retz. var. *helix* (Willd.) Fiori & Béguinot in Fiori Fl. Anal. Ital. 2: 35 (1900).
- *M. italica* (Mill.) Grande var. *helix* (Willd.) Fiori, Fl. Ital. 1: 832 (1925).
- *M. italica* (Mill.) Grande subsp. *helix* (Willd.) Emberger & Maire in Jahandiez & Maire, Catal. Pl. Maroc 4: 1038 (1941).
    - *M. tornata* (L.) Mill. subsp. *helix* (Willd.) Ooststrom & Reichgelt, Bull. Jard. Bot. État 27: 671 (1957).

M. helix Willd. var. *inermis* Lowe, Man. F1. Mad. 164 (1868).

- *M. obscura* Retz. subsp. *helix* (Willd.) Urb. var *inermis* (Lowe) Urb., Verh. Bot. Ver. Brandenb. 15: 66 (1873).
- *M. italica* (Mill.) Grande subsp. *helix* (Willd.) Urb. var. *inermis* (Urb.) Emberger & Maire in Jahandiez & Maire, Catal. Pl. Maroc 4: 1038 (1941).

M. obscura Retz. subsp. *lenticularis* (Desr.) Urb. var *inermis* Urb., Verh. Bot. Ver. Brandenb. 15: 66 (1875).

M. obscura Retz. subsp. *tornata* (L.) Urb. var *inermis* Urb., Verh. Bot. Ver. Brandenb. 15: 66 (1873). (Urban used the epithet *inermis* to designate two different infraspecific taxa of "*M. obscura*".)

M. obscura Retz. subsp. *tornata* (L.) Urb. var. *inermis* Urb., Verh. Bot. Ver. Brandenb. 15: 66 (1873).

Fig. 44-4. Lectotype of *Medicago obscura* Retz. This is a quite representative plant of *M. italica*.

M. intermedia DC. ex Ser. in DC., Prodr. 2: 174 (1825).

M. laevis Desf., Fl. Atlant. 2: 213 (1799).

M. italica subsp. *obscura* var. *laevis* (Desf.) Maire & Weiller, Bull. Soc. Hist. Nat. Afr. Nord 272 (1939).

M. italica subsp. *tornata* var. *laevis* (Desf.) Nègre, Trav. Inst. Sci. Chérifien, Sér. Bot. 5: 44 (1956).

M. lenticularis Desr. in Lam., Encycl. Méthod., Bot. 3(2): 630 (1792), nom. illeg.

M. italica (Mill.) Grande subsp. *maroccana* Nègre, Trav. Inst. Sci. Chérifien, Sér. Bot. 5: 44 (1956). (Nègre proposed a var. *inermis* (i.e., without spines or tubercles on the fruits) and a var. *spinosa* (i.e., with spines on the fruits) within his subspecies, apparently ignoring the issue of an autonym.)

M. muricata (L.) Willd. Sp. Pl. 3: 1414 (1802). (In the *Medicago muricata* folder at Herb. Willd. in B, I observed three sheets, all of *M. italica* × *M. littoralis*.)

M. obscura Retz. subsp. *tornata* (L.) Urb. var. *muricata* (L.) Urb., Verh. Bot. Ver. Brandenb. 15: 67 (1873).

M. italica (Mill.) Grande var. *muricata* (L.) Fiori, Fl. Ital. 1: 833 (1925).

M. italica subsp. *tornata* var. *muricata* Nègre, Trav. Inst. Sci. Chérifien, Sér. Bot. 5: 44 (1956).

M. obscura Retz., Observ. Bot. 1: 24, tab. 1 (1779) (non *M. obscura* Visiani). (Sales and Hedge (1994) noted that "the type is the figure in this work or (?) a specimen in the Retzius herbarium at Lund." Lassen lectotypified the name in 1977 by a collection at Lund (LD; 84/76.5111; Fig. 44-5)).

M. italica (Mill.) Grande var. *obscura* (Retz.) Fiori, Fl. Ital. 1: 832 (1925).

M. tornata (L.) Miller subsp. *obscura* (Retz.) Kerguélen, Bull. Assoc. d'Informatique Appl. Bot. (Marseille), 1: 183 (1994).

M. plumbea Bertol., Fl. Ital. 8: 267 (1850).

M. obscura Retz. var. *rugulosa* Ser. in DC., Prodr. 2: 174 (1825).

M. tornata (L.) Mill. var. *rugulosa* (Ser.) Heyn (1963, p. 88), nom. illeg.

M. obscura Retz. subsp. *lenticularis* (Desr.) Urb. var. *inermis* Urb. f. *sinistrorsa* Urb., Verh. Bot. Ver. Brandenb. 15: 66 (1873).

M. obscura Retz. subsp. *helix* (Willd.) Urb. var. *inermis* (Lowe) Urb. f. *sinistrorsa* Urb., Verh. Bot. Ver. Brandenb. 15: 66 (1873).

M. obscura Retz. subsp. *helix* (Willd.) Urb. var. *aculeata* (Guss.) Urb. f. *sinistrorsa* Urb., Verh. Bot. Ver. Brandenb. 15: 66 (1873).

M. obscura Retz. subsp. *tornata* (L.) Urb. var. *inermis* Urb. f. *sinistrorsa* Urb., Verh. Bot. Ver. Brandenb. 15: 67 (1873).

M. obscura Retz. subsp. *tornata* (L.) Urb. var. *muricata* (L.) Urb. f. *sinistrorsa* Urb., Verh. Bot. Ver. Brandenb. 15: 67 (1873).

M. helix Willd. var. *spinosa* Guss., Pr. Fl. Sic. 2: 557 (1882).

M. helix Willd. var. *spinosa* Webb in Webb. & Benth., Phytogr. Canar. 3, 2: 56 (1845). (= *M. italica* × *M. littoralis*)

M. obscura Retz. var. *spinosa* Guss., Fl. Sic. Syn. 2: 364 (1844).

M. italica (Mill.) Grande subsp. *helix* (Willd.) Emberger & Maire var. *spinosa* (Guss.) Fiori (combination cited by Emberger & Maire in Jahandiez & Maire, Catal. Pl. Maroc 4: 1038 (1941)).

M. helix Willd. var. *spinulosa* Moris, Fl. Sard. 1: 438 (1837). (= *M. italica* × *M. littoralis*)

M. tornata (L.) Mill. var. *spinulosa* (Moris) Heyn (1963, p. 87).

M. striata Bast. in Desvaux's J. Bot. Appl. Agric. 3(1-2): 19 (1814).

M. littoralis Rohde ex Loisel. subsp. *striata* (Bast.) Douin in Bonnier, Fl. Compl. Fr. 3: 21 (1914).

M. tornata var. *striata* (Bast.) K.A. Lesins & I. Lesins (1979, p. 159).

M. tornata (L.) Mill. subsp. *striata* (Bast.) Kerguélen, Lejeunia, Nouv. Sér. 120: 127 (1987).

M. italica (Mill.) Grande subsp. *striata* (Bast.) Greuter & Burdet in Greuter & Raus, Willdenowia 19: 32 (1989).

M. polymorpha L. var. γ *tornata* L., Sp. Pl. 780 (1753). (Lectotype = "Medica tornata minoris lenis" in Parkinson, Theatr. Bot., 1116, 1116, 1640, designated by Heyn in Bull. Res. Council Israel, Sect. D, Bot. 7: 165 (1959). This was accepted in The Linnaean Plant Name Typification Project (http://www.nhm.ac.uk/research-curation/projects/linnaean-typification/; current June 2010; cf. Jarvis 2007). Sales and Hedge (1994) also accepted this as the type. Heyn (1963) contended that the similar name *M. tornata* Willd. Sp. Pl. 3: 1409 (1802) belonged in synonymy if the name, description, and synonymy presented by Willdenow was considered, but not if the type was considered; she pointed out that Willdenow's type specimen of *M. tornata* (Herb. Willd. 14314-1) was identified by Urban as a specimen of *M. littoralis*.)

M. tornata (L.) Mill., Gard. Dict., ed. 8 (no. 3), (1768). (Sales and Hedge (1994) discuss the interpretation that the name *M. italica* Mill. is preferably chosen for the species.)

Medica tornata (L.) Medik., Vorles. Churpfälz. Phys.-Öcon. Ges., 2: 386 (1787).

M. obscura Retz. subsp. *tornata* (L.) Urb., Verh. Bot. Ver. Brandenb. 15: 66 (1873).

M. obscura Retz. subsp. *tornata* (L.) Batt. in Batt. & Trabut, Fl. Algér., Dicotyl. 228 (1889), comb. illeg. & superfl.

M. obscura Retz. proles *tornata* (L.) Rouy in Rouy & Fouc., Fl. Fr. 5: 24 (1899), comb. illeg.

M. italica (Mill.) Grande var. *tornata* (L.) Fiori, Fl. Ital. 1: 833 (1925).

M. obscura Retz.var. *tornata* (L.) P. Fourn., Quatre Fl. Fr. 547 (1936), comb. illeg.

M. italica (Mill.) Grande subsp. *tornata* (L.) Emberger & Maire in Jahandiez & Maire, Catal. Pl. Maroc. 4: 1038 (1941).

M. tornata Willd. ex Ces., Passer, & Gibelli, Comp. Fl. Ital. 2: 723 (1883). (= *M. italica* × *M. littoralis*)

M. tricycla DC., Adnot. Catal. Pl. Horti Monsp. 125 (1813).

M. littoralis Rohde ex Loisel. subsp. *inermis* (Moris) Urb. var. *tricycla* (DC.) Urb., Verh. Bot. Ver. Brandenb. 15: 69 (1873).

M. littoralis Rohde ex Loisel. subsp. *tricycla* (DC.) J.M. Laínz, Anales Inst. Forest. Invest. (Madrid), 12: 26 (1968).

M. obscura Retz. var. *triflora* Ser. in DC., Prodr. 2: 174 (1825).

M. obscura Retz. subsp. *lenticularis* (Desr.) Fiori var. *typica* Fiori in Fiori & Paol., Fl. Ital. 2: 35 (1900), nom. illeg.

Note: As documented above, for *M. obscura* Urban (1873) coined var. *inermis* (for non-spiny-podded) and var. *aculeata* or var. *muricata* (for spiny-podded) for three infraspecific categories; and f. *dextrosa* (pods coiling to the right) and f. *sinistrorsa* (pods rotated to the left) for five infraspecific categories.

Not

Medicago italica E.H.L. Krause in Sturm, Fl. Deutschl., ed. 2, 9: 125 (1901) ≡ *Melilotus italica*.

Vernacular Names

English: Disc medic (medic is also spelled medick). Quinlivan et al. (1974) noted that "The common name disc medic describes the pods of the cultivar 'Tornafield', which looks like a series of discs, the coils being pressed together only in the center. However, cultivar 'Murrayland'... has barrel-shaped pods with the coils pressed together right to the edges." The name disc medic could lead to confusion with *M. disciformis*. Also called hairy medick (name proposed in Haslam et al. (1977)).

French: La luzerne d'Italie, luzerne italienne, luzerne obscure, luzerne limaçon, luzerne ronde.

Geographical Distribution

(N = native, I = introduced, U = uncertain status):

Africa: Algeria (N), Libya (N), Morocco (N), Tunisia (N).

Asia: Cyprus (N), Israel (N), Jordan (N).

Europe: Belgium (I), France (including Corsica; N), Italy (including Sardinia and Sicily; N), Malta (N), Netherlands (I), Portugal (N), Spain (N).

Medicago italica is native principally to the Western Mediterranean area, but has also been collected occasionally from Israel. It is adventitious in parts of Europe.

Description

Annual herb, vegetative parts with simple hairs, usually procumbent, branches 15–40 (100 or more) cm long. Leaflets 8–16 mm long, 5–8 mm wide, obovate to oblanceolate, distal third of margins serrate, stipules broadly ovate, margins vary from nearly entire to laciniate. Inflorescence with (3)8–15(25) flowers, peduncle usually much longer than subtending petiole. Flowers 4–7(10) mm long, corolla yellow to bright yellow, about twice as long as calyx, calyx with simple hairs, teeth longer than tube. Young fruit contracted and concealed within calyx. Pod glabrous, discoid to cylindrical, 5–8 mm wide, with 1.2–8 more or less adpressed coils, coils with ca. 10 radial veins that scarcely anastomose but branch slightly just before entering a lateral vein, that vein close to the dorsal suture, a furrow not evident between the lateral vein and the dorsal suture, dorsal suture lacks spines or has mostly slender small spines. Seeds 1–2 per coil, 1.9–3.5 mm long, surface smooth, yellow, yellow-brown or brown, more or less reniform, separated by relatively thick, transverse, spongy fruit septa, radicle equal to or less than half seed length.

Identification Hints

The fruit walls of *M. italica* often become quite hard and the coil surfaces may develop some proliferant spongy tissue, like the other species placed in *Medicago* sect. *Spirocarpos* subsect. *Pachyspirae*, and *M. italica* is most likely to be confused with certain other species of this section. *Medicago italica* is most likely to be confused with *M. littoralis*. The best character for distinguishing the two is the number of flowers in the inflorescence (i.e., on the peduncle) — usually one to three in *M. littoralis*, usually five or more in *M. italica*. Sometimes *M. italica* fruits resemble those of *M. rigidula*; however, the veins on the coil faces of the latter (observed in young pods, in which case the veins are not obscured by proliferant tissue) are strongly bent, unlike those of *M. italica*. *Medicago italica* could be confused with *M. soleirolii*, but the latter can be distinguished by the presence of some gland-tipped hairs on the leaves.

Habitat and Environmental Adaptation

Habitat

Medicago italica almost always occurs near the sea, predominantly in sand. It has occasionally been collected in Mediterranean grasslands.

Phenology

In its native range, *M. italica* flowers in the spring (typically February to April). Clarkson and Russell (1975) found that flowering was accelerated by vernalization (long vernalization slightly delayed flowering), long photoperiods (18 and 24 hours), and high temperature.

Soil

Wild forms of *M. italica* grow mainly on sandy, calcareous soils close to the coast. Piano and Francis (1992) noted that in its native area the species grows on neutral to slightly basic soils. In Algeria, Abdelguerfi et al. (1988*a*) observed that *M. italica* occurs in medium- and coarse-textured soils. Available Australian cultivars are suited to light-textured, neutral to alkaline soils. Cultivar 'Tornafield' is suited to neutral or slightly acidic sand or sand over clay soils. Cultivar 'Rivoli' has been found to grow well in coastal calcareous sands of Australia with a pH of 7.0–8.0, as well as on neutral sands with a pH of 5.8.

Temperature

Medicago italica occurs naturally mostly on Mediterranean seasides, and so is adapted to a warm climate, with a consequent lack of tolerance to cold (Piano and Francis 1992). Krall et al. (1996) found that three accessions of *M. italica* examined for ability to self-regenerate in Wyoming failed to survive the winter at any of the three locations tested. Christiansen and Svejcar (1989) sowed *M. italica* in Oklahoma in March 1985 and observed that all the plants were killed by frost and cold in the following winter. Derkaoui et al. (1991) also found that fall-planted *M. italica* in Oklahoma failed to survive the winter, indicating that the species does not have sufficient cold tolerance for successful self-replication in the southern Great Plains of the United States.

Water

As with most Mediterranean annual species, *M. italica* is native to well-drained soils. It occurs frequently in dry environments, with a minimum annual rainfall of 200 mm (Piano and Francis 1992). Francis and Poole (1973) observed that it is very susceptible to waterlogging. Cultivar 'Rivoli' has been described as adapted to medium and higher rainfall areas above 400 mm in Australia. Cultivar 'Murrayland' is a deeply rooting plant that is very tolerant of dry conditions.

Biotic Relationships

Pollination

Small (1988) noted that the small flowers and pollen to ovule ratio (172) are consistent with *M. italica* being largely self-pollinated. This species is basically an inbreeder, although introgression with *M. littoralis* indicates that some natural outcrossing occurs.

Dispersal

The fruits are indehiscent, like those of most coiled *Medicago* fruits, and constitute the basic dispersal unit. Spiny-fruited forms of *M. italica* can be distributed by becoming tangled in animal fur. Cultivars of this species are basically spineless, which is advantageous in cultivation in keeping the seeds localized in the area where the plant has been established. The mechanisms of distribution of the non-spiny-fruited forms are unclear. Given that the plants normally occur on sea beach sand, it seems plausible that wind is important; pods with very few coils present a flat face that could act as a sail in the wind, whereas pods with many coils could be rolled along the sand.

Herbivory

Cultivars of *M. italica* are susceptible to the blue-green aphid (*Acyrthosiphon kondoi*) and, to a lesser extent, the spotted alfalfa aphid (*Therioaphis trifolii* fm. *maculata*); Lake (1989) reported that surveys of the large South Australian Department of Agriculture's National Annual *Medicago* collection failed to find resistant accessions. However, cultivar 'Toreador' (a hybrid with *M. littoralis*) has resistance to blue-green aphids, but (like most other medics) is attacked by the cowpea aphid (*Aphis craccivora* Koch) and the lucerne flea (*Sminthurus viridis* (L.)), and occasionally by sitona weevils (*Sitona* species) and brown pasture loopers (*Ciampa arietaria* (Guenée)). It may also be a host for root lesion nematode (*Pratylenchus neglectus*). Shade et al. (1975) examined the resistance of 21 species of *Medicago* to attack by larvae of the alfalfa weevil (*Hypera postica* (Gylh.)) and found that *M. italica* was very resistant.

Nodulation and mycorrhizae

USDA (2009) reported positive root nodulation for *M. italica*. Brockwell et al. (1976) evaluated *R. meliloti* strains suitable for inoculating cultivars 'Murrayland' and 'Tornafield'. They found that *M. italica* formed effective nodules with most groups of *R. meliloti* strains, but was slightly more strain specific than *M. sativa*.

Disease resistance

Yaege and Stuteville (2002) tested the resistance of 33 annual species of *Medicago* to powdery mildew (*Erysiphe pisi*) and found that five of the collections of *M. italica* grown were susceptible to the disease whereas two showed partial resistance. In a comparative test of 36 annual species of *Medicago*, O'Neill and Bauchan (2000) found that seven accessions of *M. italica* were quite susceptible to anthracnose caused by *Colletotrichum trifolii*, but one accession showed 32% resistance. Diseases recorded on cultivars include root fungus (*Rhizoctonia* spp.) and black stem fungus (*Phoma* spp.).

Economic History, Cultivars, and Uses

Medicago italica has been utilized in agriculture primarily in Australian natural "pastures" (i.e., rangelands).

Cultivars

Information on the following four Australian cultivars is available in the *Register of Australian Herbage Plant Cultivars*. The cultivar 'Tornafield' (with spineless pods), the first cultivar of *M. italica*, was developed at the Institute of Agriculture, University of Western Australia and released in 1970 (see Poole (1970) for an agronomic evaluation). 'Murrayland' (with spineless pods) was selected by the Commonwealth Scientific and Industrial Research Organisation (CSIRO), 'Deniliquin' (Australia) and registered in 1972. 'Rivoli' (with predominantly spineless pods) was originated by R.S. Martyn, South Australian Department of Agriculture and registered by R.N. Oram (see Martyn and Oram 1990). 'Swani' was registered in 1979.

'Toreador', a hybrid between *M. italica* and *M. littoralis*, was bred for southwestern New South Wales, Australia and is particularly suited to deep sandy soils. The varietal rights belong to Seedmark/PlantTech (Adelaide, Australia). This variety has been termed "hybrid medic", a name that could be confused with *M. hybrida*.

Forage

Cultivars of *M. italica* are sown, mostly in Australia, primarily for semi-managed, dryland, grazing. Mohammad and Qamar (1988) grew 14 annual species of *Medicago* in small test plots in Islamabad, Pakistan, located in a subtropical, subhumid area that is notably hotter and wetter than the normal habitat of these species. The dry matter yield of *M. italica* was about 2.80 t/ha.

Fodder

Medicago italica is only occasionally harvested as hay.

Green manure, rotation crop, ground cover

Cultivars of *M. italica* are sown, mostly in Australia, for crop rotation, particularly for increasing soil nitrogen levels for subsequent cereal crops.

Other uses (salad sprouts, ornamental, plant products, etc.)

These uses have not been recorded.

Chromosome Number

$2n = 16$: Ghimpu 1928, 1929*a*, 1929*b*, 1930 (as *M. helix* and *M. tornata*); Ahuja 1955; Heyn 1956 (as *M. tornata*); Clement 1962 (as *M. obscura* and *M. tornata*); Lesins and Lesins 1963*a* (as *M. obscura*), 1979; Simon and Simon 1965; Lesins and Erac 1968*a* (as *M. striata*); Fernandes and Queirós 1978 (as *M. obscura*); Abdelguerfi and Guittoneau 1979 (as *M. tornata*); Agarwal and Gupta 1983*b*; Abdelguerfi et al. 1988*b*; Oberprieler and Vogt 1996*a*.

"$2n = 16, 17$, or 18": Fryer 1930 (as *M. obscura*).

Genetic Relationships and Gene Transferability

Intraspecific

A lack of interbreeding barriers among populations of *M. italica* was observed by Lesins and Erac (1968a) and Lesins and Singh (1973).

Interspecific

Lesins (1952) was unable to hybridize *M. italica* (as *M. obscura*) with *M. sativa*, both of which were at the diploid ($2n = 16$) level. Lesins and Singh (1973) found that *M. shepardii*, considered to be a variety of *M. italica* by Heyn (1963), could not be hybridized with *M. italica*. Simon and Millington (1967) were unable to produce hybrid seed by crossing *M. italica* with *M. turbinata*, *M. rigidula* (*M. rigiduloides*?), *M. murex* (*M. sphaerocarpos*?), or *M. constricta*. Lesins and Singh (1973) observed a strong breeding barrier between *M. italica* and *M. littoralis*. However, as noted above, Lesins and Erac (1968*a*) found that unidirectional interbreeding was possible, with *M. littoralis* pollen fertilizing *M. italica*. Additionally, Simon (1965) and Lake (1989) produced hybrids between *M. littoralis* and *M. italica*. Also as noted above, 'Toreador' is a cultivar that originated as a hybrid of the two species. Lake (1989) attempted hundreds of crosses of *M. italica* and *M. truncatula*, with only one successful cross, and the resulting hybrid was inviable; Jenczewski et al. (1997) crossed these two species and noted very odd segregation in the hybrids. Lesins and Lesins (1979) reported that they crossed *M. soleirolii* with *M. italica* and generated several F_1 plants that were chlorophyll deficient and died within 2 months.

Exchange by advanced biotechnology

Lake (1989) discussed the transfer of genes for aphid resistance from *M. truncatula* via *M. littoralis* to *M. italica*.

Conservation Status

M. italica is a widespread species that is not in need of conservation protection.

Germplasm Sources

More than a thousand accessions identified as *M. italica* are available from germplasm institutions (see Chapter 3).

Research Problems and Priorities

Medicago italica is a very variable species that has proven of value in agriculture. However, only a small part of its genetic pool has been adequately examined and exploited, so additional study and development are required. Piano and Francis (1992) described the species as having "regional potential".

Representative Collections

Algeria: Allard, C., Ravins des Beni-Atia près de l'Arba (Algérie), pentes, alt. 400–1000 m, V-VI-1883, in fruit, Société dauphinoise 1884 No. 4073 (P). Maire, R. & Weiller M. 435, Tripolitania, Leptis Magna, solo arenoso, 15-IV-1938, in fruit (P). **Australia:** Copley, B. 402, South Australia, Northern Yorke Peninsula, Hundred of Wiltunga (ca. 140 km north-northwest of Adelaide), 26-VIII-1967, in flower & young fruit (BRI). **France:** Blanchet, Basses-Pyrenées, de Bayonne au Boucau, sables de la zone salée, flowers 25-IV, fruits 23-V-1882, Flora Selecta Exsiccata publié par Ch. Magnier No. 231 (P). Bouby, H. h.038, Le Bois en Ré, Ile de Ré. (Charente-maritime), sables maritimes, 25-IV-1957, in flower & early fruit (P). Neyraut, E.J., Basses-Pyrenées, Le long de l'Adour, de l'extrémité des Allées marines au lazaret de Bayonne en suivant le pied des dùnes, 21-VI-1891, in flower & fruit (US). **Israel:** Eig, A., Zohary, M. & Feinbrun, N., Philistean Plain, Sarafand, 18-III-1927, in fruit (HUJ). Orshan, G., Acco Plain, Kirjath Haim, light soil, 31-III-1964, in flower & fruit (UC). Zohary, M., coastal Negev, Rafiah, IV-1925, in fruit (UC). Zohary, M. & Grizi, A., Sharon, Gan-Shemuol, sandy soil, 25-III-1953, in flower & fruit, Universitatis Herbraicae Hierosolymitanae Flora Terrae Israëlis Exsiccata No. 531 (GH, K, NY, P, UC, US). **Italy:** Bucknall, C. & White, J.W., Sicily, Leasaudo south of Girgenti, 3-V-1906, in fruit (K). Davis, D. & Sutton, S. 63281, Sicily, Catánia, 15 km south of Piazza Armerina, alt. 400 m, Eucalyptus woodland on sandy soil, 17-V-1979, in fruit (BM). Todaro 568, Palermo a catalfano, in collibus et campis arenosis, IV (received by P, in 1882) (P). **Morocco:** Balansa, B., Ile de Magador, 25-IV-1867, in fruit (P). Ibrahim, environs d'Agadir, Maroc méridional, VI-1877, in flower & fruit (GH). Roffey, J., Tangier, 29-IV-1916, in fruit (BM). **Portugal:** Ball, J., ex arenosis Lusitaniae Transtaganae prope Olyssiponem, IV-1879, in fruit & flower (GH). Matos, J., Figueira da Foz-Gala, in arenosis maritimes, 11-V-1950, in flower & fruit (BM). Buchtien, O., Porto, in fruit, 1891 (US). **Spain:** Porta & Rigo 636, prov. Gaditana, ad Puerto de Santa Maria, in arenos. maritimis, 5-V-1895, in flower & fruit (GH, K). Sennen, F., Cataluñ, dunes, 20-V-1918 (P). Bourgeau, E. 163, Puerto Santa Maria, sables maritimes, 14-IV-1849, in flower & fruit (K, P).

Representative collections of *M. italica* × *M. littoralis*

Algeria: Reveillère, Maison Carée à Alger, 15-V-1862, in fruit. **Australia:** Green, R., New South Wales, Old homestead area, Wongablock, Pooncarie, open sandy rises, in soft red sand, X-1974, in flower & fruit (BRI). **France:** Lloyd, J., Gallia occidentalis, South Brévin, Loire inférieure, in flower & fruit (US). **Portugal:** Barbey, W. 150, Madeira, cliffs to the eastward, 26-IV-1855, in flower & fruit (US). **Spain:** Cook, A.C. 530, Gran Canaria, Guia, 1897, in flower & early fruit (US). Leresche, L., Santander, VII-VIII-1878, in flower & fruit (US).

(Note: No such collections were present in the herbarium in Jerusalem (HUJ), indicating the relative purity of *M. italica* in Israel.)

45. *Medicago laciniata* (L.) Mill.

M. laciniata (L.) Mill., Gard. Dict., ed. 8: no. 5 (1768).

The epithet *laciniata*, Latin for slashed into narrow divisions with taper-pointed incisions, is an allusion to the frequent occurrence of incised leaves in the species.

Subordinate Taxa

None recognized here. Much of the literature recognizes *M. laciniata* (L.) Mill. var. *brachyacantha* Boiss., characterized by pods that are cylindrical (not spherical or ovoid), with fewer pod coils (less than five, not five or more), spines often relatively short, stipules deeply dentate rather than laciniate, leaflets less often dissected, and a peduncle that is shorter than the subtending petiole (not longer), and said to occupy relatively arid habitats compared with the remainder of the species. However, documented studies indicating that such a separation is warranted are lacking (a conclusion also reached by Stirton (1982)).

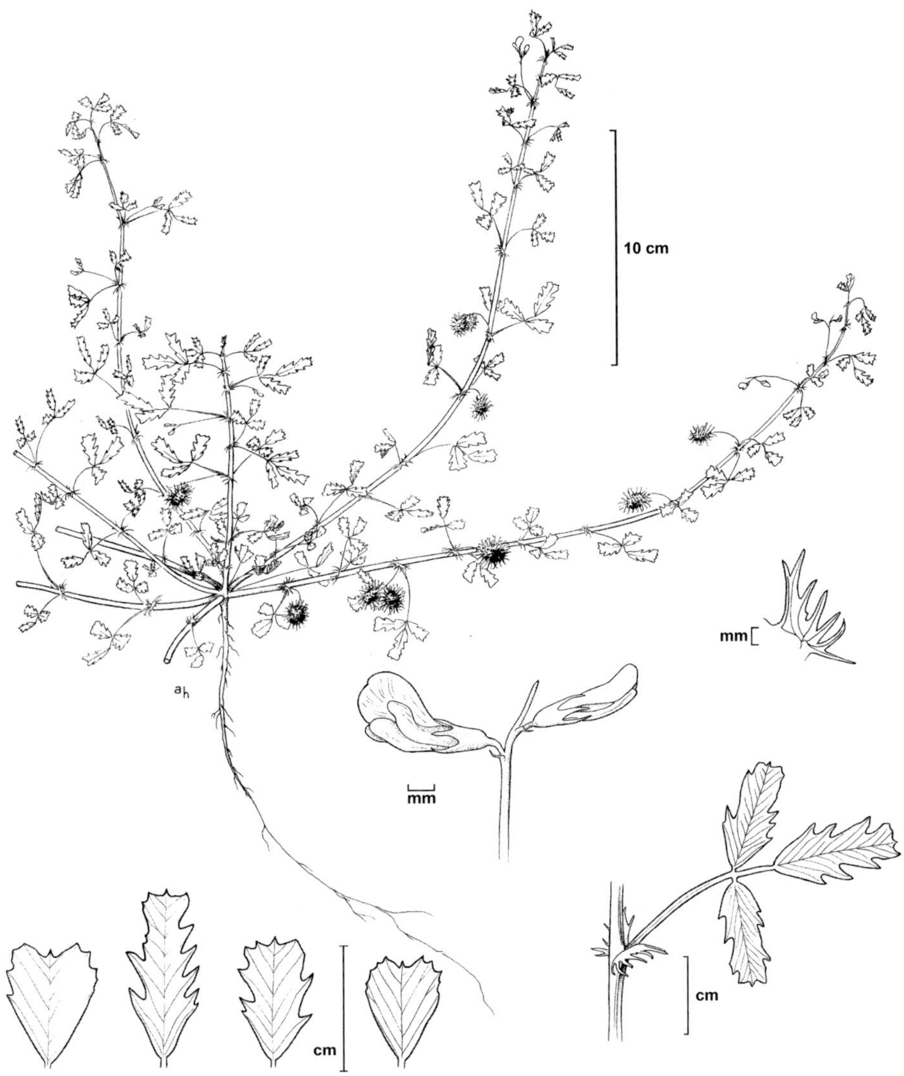

Fig. 45-1. *Medicago laciniata.* **Clockwise from top:** Plant; stipule; leaf attached to stem; leaflets. **Center:** Group of flowers. (Based on E. Small M594, DAO; artist: A. Hanes.)

Type Specimen

Heyn (1959: 169) lectotypified the basionym for *M. laciniata* (*M. polymorpha* L. var. ν *laciniata* L.) by an illustration ("iconotype"), t. 34, in: Breyne, J. 1678. Exoticarum aliarumque minus cognitarum plantarum, centuria prima, etc. David-Fridericus Rhetius, Danzig (Gdansk, modern Poland) (Breyn, Exot. Pl. Cent., 81). The Linnaean Plant Name Typification Project (http://www.nhm.ac.uk/research-curation/projects/linnaean-typification/; current June 2010; cf. Jarvis 2007) provided the accompanying phrase "Trifolium cochleatum spinosum Syriacum, laciniatis foliis."

Synonymy

Homotypic

M. polymorpha L. var. ν *laciniata* L., Sp. Pl. 781 (1753), (basionym).

Heterotypic

M. aschersoniana Urb., Verh. Bot. Ver. Prov. Brandenb. 15: 77 (1873).

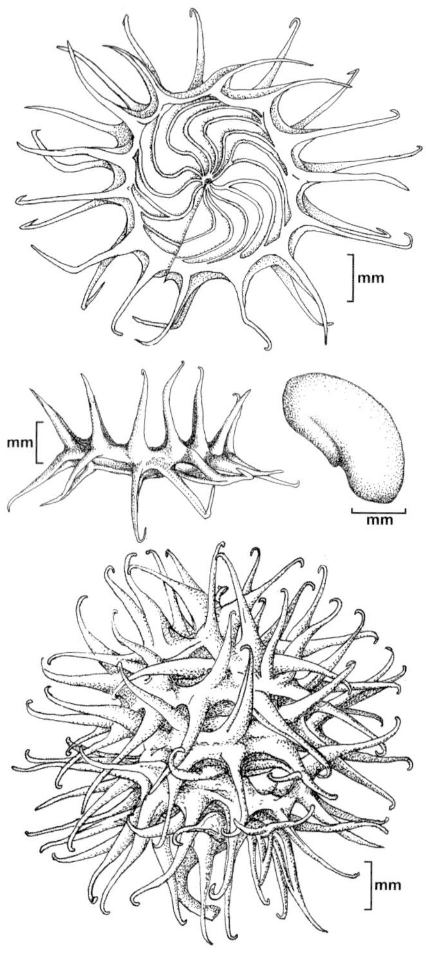

Fig. 45-2. *Medicago laciniata*. **Clockwise from top left:** Fruit coil face showing venation; edge of coil; seed; fruit. (Based on E. Small M597, DAO; artist: A. Hanes.)

M. laciniata (L.) Mill. subsp. *aschersoniana* (Urb.) Thell., Vierteljahrsschr. Zürich 64: 755 (1919).

M. laciniata (L.) Mill. var. *brachyacantha* Boiss., Diagn. Pl. Nov. Ser. 1, 9: 10 (1849). (Type: Boissier, E., Arabia Petraea, III-1846, in fruit Herb. Boiss. G!; two sheets, one of which I labelled as the lectotype.) (Not *M. brachyacantha* Kern., Oest. bot. Zeitschr. 18: 386 (1868) = *M. minima*.)

M. aschersoniana Urb. var. *brachyacantha* (Boiss.) Urb., Verh. Bot. Ver. Brandenb. 15: 77 (1873).

M. laciniata (L.) Mill. var. *brevidenticulata* Cand., Bull. Soc. Bot. France 45: 185 (1898).

M. laciniata (L.) Mill. var. *brevispina* Benth. (Cat. Pyr. 104 (1826), nom. nud.) ex Urb., Verh. Bot. Ver. Brandenb. 15: 77 (1873).

M. laciniata (L.) Mill. subsp. *brevispina* Arcangeli, Comp. Fl. Ital.: 165 (1882).

M. diffusa Poiret in Lam., Encycl. Méth., Bot., Suppl. 3: 524 (1814).

M. laciniata (L.) Mill. var. *integrifolia* Godr., Fl. Juven. 74 (1853).

M. laciniata (L.) Mill. var. *integriuscula* Cuisin, Ansb. Herb. Fl. France 6, t. 1043 (1870).

M. laciniata (L.) Mill. var. *intermedia* Nègre, Luzern. Marcoc. 74 (1956).

M. laciniata (L.) Mill. var. *longispina* Benth. (Cat. Pyr. 104 (1826), nom. nud.) ex Urb., Verh. Bot. Ver. Brandenb. 15: 77 (1973).

M. laciniata (L.) Mill. subsp. *longispina* Arcangeli, Comp. Fl. Ital. 165 (1882).

M. aschersoniana Urb. var. *macrantha* Urb., Verh. Bot. Ver. Brandenb. 15: 78 (1873). (Type, indicated in Heyn (1963): Boissier, Arabia, Petraea, 3, 1846 (G).)

M. laciniata (L.) Mill. var. *micrechinus* Maire (holotype: Maire, R. 16-IV-1932, in rupestribus calcareis: Teniet-el-Aouedj ad orientem oasis Beni-Ounif, 900 m, Algeria; (type at MP, viewed online at http://www.aluka.org; current June 2010). (Status?)

M. monantha Blatt., J. Bombay Nat. Hist. Soc. 36: 969 (1933), non *M. monantha* (C.A. Meyer) Trautv. (Chapter 53).

M. laciniata (L.) Mill. var. *pilosa* Nab., Publ. Fac. Sci. Univ. Masaryk 35: 68 (1923), p.p.

M. aschersoniana Urb. var. *brachyacantha* (Boiss.) Urb. f. *pilosula* Urb., Verh. Bot. Ver. Brandenb. 15: 78 (1873).

M. pseudogranatensis Blatt., J. Bombay Nat . Hist. Soc. 36: 969 (1933).

M. laciniata (L.) Mill. var. *saharae* Nègre, C.R. Soc. Sci. Nat. Maroc. 90 (1951).

M. schimperiana Hochst. ex Boiss., Fl. Orient. 2: 104 (1872), pro syn.

M. laciniata (L.) Mill. subsp. *schimperiana* P. Fourn., Quatre Fl. France 545 (1936), nom. invalid.

The name *Medicago leonis* Sennen & Mauricio, Cat. Fl. Rif Or. 146 (1933) was assigned to synonymy under *M. laciniata* in Flore d'Afrique du Nord (http://www.tela-botanica.org/eflore/BDAFN/1.00/nn/147168/synonymie; current June 2010).

Vernacular Names

ENGLISH: Cutleaf medic, cut-leaf medic, tattered medic (medic is also spelled medick).

FRENCH: La luzerne laciniée, luzerne découpée, luzerne dentelée.

Geographical Distribution

(N = native, I = introduced, U = uncertain status):

AFRICA: Algeria (N), Botswana (I), Canary Islands (Spain, N), Djibouti (U), Egypt (N), Ethiopia (N), Kenya (N), Lesotho (I), Libya (N), Madeira Islands (Portugal, N), Morocco (N), Namibia (I), Socotra (N), Somalia (N), South Africa (I), Tanzania (N), Tunisia (N).

ASIA: Afghanistan (N), Bahrain (N), Egypt (Sinai, N), India (N), Iran (N), Iraq (N), Israel (N), Japan (I), Jordan (N), Kuwait (N), Lebanon (N), Oman (N), Pakistan (N), Qatar (N), Saudi Arabia (N), Syria (N).

AUSTRALASIA: Australia (I).

EUROPE: Corsica (France, I), Estonia (I), France (I), Italy (I), Spain (I), Sweden (I), Ukraine (I)

NORTHERN AMERICA: Canada (I), USA (I).

This species is probably indigenous to countries of the southern Mediterranean Circle and those east of the Mediterranean (where it seems adapted to soils of steppes and deserts), but it has been introduced worldwide as a ruderal weed, making determination of the primeval range difficult.

Description

Annual herb, procumbent to ascending, sparingly branched from base, branches 5–30(50) cm long, pubescent with simple hairs. Leaflets 3–12 mm long, 2–6 mm wide, obovate to oblong-cuneate, laciniate, incised-dentate, incised-pinnatifid, or margins of distal half serrate to deeply serrate (laciniate and nonlaciniate leaves often on same plant), pubescent on abaxial surface, stipules ovate to lanceolate, deeply dentate to laciniate. Peduncle with 1–2(3) flowers, shorter to longer than subtending petiole. Flowers 3–6(8) mm, corolla pale to dark yellow, less than twice length of calyx, calyx 2–4 mm, pubescent with simple hairs, teeth shorter, longer, or about equal to tube. Young pod protrudes sideways from calyx, mature pod glabrous, glabrescent or (rarely) pubescent with simple hairs (very rarely glandular hairs), short-cylindrical, spherical, or ovoid, 3–8(10) mm long, with 3–7(9) coils in a lax spiral, broadest coil 2.5–6 mm in diameter, coil face with 6–16 radial S-shaped (sigmoid) veins, some branched, veins enter a broad lateral vein near dorsal suture, furrow between lateral vein and dorsal suture, spines on dorsal suture 1–6 mm long, grooved (especially sulcate at base), longer spines hooked, spines at angle of 90° to 180° to coil faces. Seeds 1–2 per coil, 2–3 mm long, 1–1.3 mm wide, reniform, not separated by spongy transverse septa, surface smooth, yellow to yellow-brown, radicle more than half and up to two-thirds length of seed.

Identification Hints

Medicago laciniata is closely related to the rare *M. sauvagei*, but is more likely to be misidentified as (1) *M. minima*, which has much hairier foliage, much less-dissected stipules, and a submarginal vein more remote from the coil margin; or (2) *M. polymorpha*, the coil face of which has quite reticulate venation, whereas that of *M. laciniata*, has notably S-shaped radial veins that anastomose little.

Habitat and Environmental Adaptation

Habitat

In its natural habitat, *M. laciniata* occurs particularly in dry, stony deserts, less commonly in woodlands and grasslands. Where it has been introduced, the species tends to be a weed in ruderal situations. Heyn (1963) and Abdelguerfi et al. (1988*a*) noted that *M. laciniata* appears to tolerate an exceptionally wide range of edaphic and climatic conditions.

Phenology

Medicago laciniata flowers in the spring, from February to June (in the Northern Hemisphere) depending on local climate. Young et al. (1992) examined relative flowering times (in a common garden) of 32 ruderal populations collected from New South Wales, Australia and found no significant differences among them (they suggested a common origin may explain the uniformity).

Soil

Piano and Francis (1992) recorded that in its native area the species occupies a range of soil textures, with pH varying from slightly to moderately basic. Bounejmate et al. (1992*b*) found that in Morocco, *M. laciniata* tended to occupy sites with low clay content. Abdelguerfi et al. (1988*a*) noted that in Algeria *M. laciniata* tends to be in medium- and coarse-textured soils. Zoghlami et al. (1996), in reference to Tunisia, commented that *M. laciniata* is abundant in coarse-textured soils poor in calcium and organic matter. In deserts, the species typically occurs in sand or stony sand.

Temperature

This species is a winter annual, but occurs under very hot desert conditions.

Water

Medicago laciniata is often the only *Medicago* species found in dry deserts and appears to be especially drought-resistant. Indeed, the species seems to be restricted to very low rainfall regions (Piano and Francis 1992). In North Africa, it occurs in areas with 25–50 mm annual rainfall (Gintzburger and Blesing 1979). Zoghlami et al. (1996) observed that *M. laciniata* tended to develop best in areas of Tunisia receiving less than 300 mm of rainfall annually. Francis (1981) found that *M. lacinata* occurred in areas of Libya with 50 mm or less rainfall annually. Abdelguerfi et al. (1988*a*) found that in Algeria this species occupied drier areas than any other native species of *Medicago*. Friedman and Elberse (1976) concluded that productivity was directly related to the amount of water present, with *M. laciniata* simply growing more as more water was available, but quite able to survive as a dwarf herb under drought conditions. Brahim and Smith (1993) observed that *M. laciniata* in a field plot in Arizona produced seeds, with only 125 mm winter precipitation. The species normally occurs in well-drained situations that dry out rapidly in the spring (Andrew 1964).

Ghorbani Javid et al. (2006) examined a drought-tolerant ecotype of *M. laciniata* that was able to sustain high production of dry matter during drought in Iran, comparing it with a drought-intolerant ecotype. They suggested that the tolerant ecotype's superiority was due to leaf proline and soluble sugars that accumulated in the tolerant ecotype during drought conditions, producing tolerance perhaps by osmotic adjustment and high relative water content. Ghorbani Javid et al. (2005) suggested that laciniate leaves were adaptive in reducing transpiration area while minimizing reduction of photosynthesis. Moradi et al. (2005) suggested that drought tolerance in the species seems to involve "a plethora of adaptive strategies including morphological and physiological characteristics like less leaf area, plant height, and greater root-shoot ratio, higher RWC due to better OA, and greater partitioning of K, Mg, Ca, and Zn in leaves along with higher production of organic solutes, proline, sugars, and proteins and the up-regulation antioxidant system."

Biotic Relationships

Pollination

Small (1988) noted that the relatively low pollen to ovule ratio (91) and small flowers are consistent with *M. laciniata* being largely self-pollinated and the species being mainly an inbreeder.

Dispersal

Friedman and Orshan (1974) attributed seed distribution of *M. laciniata* in the Negev Desert largely to rodents and ruminants that carry the spiny pods tangled in their fur. They pointed out that seeds are also dispersed by these animals through their feces, since at least some seeds remain viable after passing through the ruminant digestive system. As with some other spiny-fruited annual *Medicago* species that become entangled in sheep wool, *M. laciniata* sometimes appears spontaneously in waste areas around wool-combing mills.

Herbivory

In its natural desert habitat, *M. laciniata* is grazed both by domestic and wild mammals, including goats, sheep, gazelles (*Gazella gazella*), rodents, and hares (Friedman and Orshan 1974). Shade et al. (1975) examined the resistance of 21 species of *Medicago* to attack by larvae of the alfalfa weevil (*Hypera postica* (Gylh.)) and found that *M. laciniata* was very resistant to it.

Nodulation and mycorrhizae

USDA (2009) reported positive root nodulation for *M. laciniata*. Jordan (1981) noted that *M. laciniata* is particularly resistant to root nodulation by most strains of *Rhizobium meliloti*. Similarly, of 11 species of annual medics, *M. laciniata* was the only one tested that was not consistently nodulated by the naturalized soil rhizobia of 28 South Australian soils (Ballard and Charman 2000).

Villegas et al. (2006) found that the nitrogen-fixing rhizobia on *M. laciniata* from Mediterranean soils in Tunisia and France differed markedly from other *Sinorhizobium meliloti* or *S. medicae* isolates found on three other annual *Medicago* species (*M. truncatula*, *M. polymorpha*, and *M. sauvagei*). They proposed two infrasubspecific (biovar) divisions within *S. meliloti*: bv. *medicaginis* for *Sinorhizobium* efficient on *M. laciniata* and bv. *meliloti* for the classically known *S. meliloti* group.

Disease resistance

O'Neill et al. (2003) examined the comparative resistance of 36 annual species of *Medicago* to *Phoma medicaginis* (the cause of spring black stem and leaf spot, a major disease of alfalfa) and concluded that *M. laciniata* had high resistance to it. In another comparative test of 36 annual species of *Medicago*, O'Neill and Bauchan (2000) found that nine accessions of *M. laciniata* were quite susceptible to anthracnose caused by *Colletotrichum trifolii*. Yaege and Stuteville (2002) tested the resistance of 33 annual species of *Medicago* to powdery mildew (*Erysiphe pisi*) and found that the ten collections of *M. laciniata* grown were susceptible to the disease (it was observed that often only one of a number of populations of other species tested was resistant).

Economic History, Cultivars, and Uses

This species has attracted only limited agricultural interest to date.

Cultivars

No cultivars have been bred.

Forage

Medicago laciniata is not sown for forage, but it is a natural component of Mediterranean grazing lands. Townsend (1974) noted that "it is considered a valuable grazing plant in Kuwait."

Fodder

This species is not treated as a hay plant.

Green manure, rotation crop, ground cover

No information located.

Other uses (salad sprouts, ornamental, plant products, etc.)

Phillips (1917) recorded that finely ground *M. laciniata* in an infusion is used as a love philtre (potion). Al-Yahya et al. (1987) examined chemical constituents of the species with a view to medicinal potential.

Chromosome Number

2*n* = 16: Fryer 1930; Ghimpu 1929*b*, 1930; Heyn 1956; Reese 1957; Lesins and Lesins 1961, 1979; Clement 1962; Simon and Simon 1965; Lesins and Erac 1968*b* (as *M. aschersoniana* and *M. laciniata*); Bolkhovskikh et al. 1969; Van Loon 1974; Abdelguerfi and Guittoneau 1979; Abdelguerfi et al. 1988*b*; Oberprieler and Vogt 1996*a*; Malallah and Brown 1999; Baltisberger and Widmer 2006.

Genetic Relationships and Gene Transferability

Intraspecific

Lesins and Erac (1968*b*) hybridized a laciniate-leaved form with a nonlaciniate-leaved form and observed that the hybrids were fertile.

Interspecific

Singh and Lesins (1972) were able to hybridize *M. laciniata* with *M. sauvagei*. When *M. sauvagei* was the female parent, a chlorophyll deficiency occurred in the F_1 plants, but the F_1 plants of the reverse direction cross was more viable. Although the F_1 plants had near-normal meiosis (indicating a fairly close relationship of the two species), their pollen set was low and the F_2 plants exhibited a high mortality (indicating that potential gene exchange between the two is difficult).

Exchange by advanced biotechnology

No information located.

Conservation Status

Medicago laciniata is a very widespread species that is expanding its range. It is not in special need of conservation.

Germplasm Sources

About a thousand accessions of this species have been deposited in major germplasm institutions (see Chapter 3).

Research Problems and Priorities

Among annual species of *Medicago*, none is naturally adapted to drought and heat more than *M. laciniata*. These features suggest that it deserves to be studied and developed as a crop for arid, hot situations, and that additionally its genes may be useful for improving other species of *Medicago* for occupation of extreme conditions.

Representative Collections

Algeria: Balansa, B. 929, Biskra, graviers de l'Oued-Biskra, 20-III-1853 in flower & fruit (P). Chevallier, L. 419, Planta Saharae algeriensis, Bouhoura prope Ghardaïa, sub palmis, 7-III-1950, in fruit (P). Chevallier, L. 43, Ghardaïa, in arenosis et glareosis, 5-IV-1899, in flower & fruit (P). Jamain, M., Biskra, sables, 21-I-1853, in flower & fruit (P). Maire, R., M. Djebel Arusitten, gravicus calcaires, alt. 300 m, 29-III-1922, in flower & fruit (P). **Australia:** Ererist, S.L. 7937, Augathella-Tambo Road about 30 miles north of Augathella, in gray clay, common in disturbed ground on roadside, 15-IX-1966, in flower & fruit (K). **Bahrain Islands:** Good, R. 89, 7-III-1950, in fruit (BM). Phillips, A.C. 56, Dumistan, alt. ca. 200 feet sandy plain, 13-III-1986, in fruit (K). **Belgium:** Lousley, J.E., Vesdre, Wool alien, IX-1958, in fruit (K). **Britain:** Lousley, J.E., Flitton, Bedfordshire, 8-IX-1957, in fruit, Wool Adventive Flora of Britain – J.E. Lousley No. 500 (K). **Canary Islands:** Bornmüller, J., La Palma, Santa Cruz, in dedivitatibus ad mare, 27-IV-1901, in flower & fruit, J. Bornmüller Planta Exsicc. Canariensis 1901 No. 2233 (P). Bourgeau, E. 318, Teneriffa, Guimar, in petrosis basalticos, 18-II-1855, in flower & fruit (P). Bourgeau, E., Teneriffa, in arvis vulgaris, 1845, in flower & fruit, Bourgeau Plantae Canarienses No. 505 (P). Pitard, C.J. 120, Tenerife, Barranco de Bufadero, in incultis, 2-I-1905, in flower & fruit (P). **Egypt:** Davis, P.H. 10283, km 24, Cairo-Suez road, 2-IV-1945, in fruit (K). Palmly, C.L. 65, Solum, IV-1933, in flower & fruit (K). Simpson, N.D. 4740, near Amria Causeway, 22-IV-1927, in fruit (K). Simpson, N.D. 5873, wadi near West Sennûr, 26-III-1928, in fruit (K). **Ethiopia:** Fries, I., Tadesse, M. & Vollesen, K. 3172, Sidamo, 40 km east of Neghelle along the road to Filtu, alt. 1450 m, Acacia-Commiphora bushland, 20-V-1982, in flower & fruit (K). **India:** Drummond, J.R. & Kabir 14855, in feratis ca. 1275 m ad Mt. Talinos dit Lhalipur, 19-IV-1904, in fruit (P). Hooker, Inde, vegetative, Herb. Ind. Or. Hook. fil. & Thompson (P). Schlagintweit, Herbarium of, prov. Panjáb, Musakél (south of Kalabághh on the Indus), along the Salt range via Varcha and Chóia to Gujrat, alt. 1400–2500 feet, 17-II-5-III-1857, vegetative, Catalogue No. 11151 (P). **Iran:** Davis & Bokhari D56287, F7 (Times Atlas), 70 km from Lar to Jahrom, alt. 900 m, rocky north limestone slope, 1-IV-1974, in flower & fruit (K). Davis & Bokhari D5748, DS (Times Atlas), near Omidye, 74 km from Bebehan, alt. 100 m, fixed sand dunes at foot of escarpment, 21-III-1974, in flower & fruit (K). Kunkel 16424, southwest corner Hormoz Island, alt. 40 m, rocky slope, 1-IV-1974, in fruit (K). Léonard J., 8660, bordure montagneuse ouest de la dépression du Jaz Murian, 57°45′E, 27°24′N, alt. 520 m, végétation suffrutescente, 27-IV-1972, in fruit (K). **Iraq:** Guest, Rawi & Rechinger 17535A, Jabal al-Muwaila', near Kuwait (ca. 70 km east of Jabal Hamrin), north of Amara, alt. ca. 80 m, barren, much-denuded conglomerate hills, 27-III-1957, in flower & fruit (K).

Guest, Rawi & Rechinger 18986, 12 km west-northwest of Ansab (ca. 143 km south of Salman, just over Saudi border), alt. ca. 340 m, flat silty depression on rocky limestone desert plateau, 24-IV-1957, in fruit (K). **Israel:** D'Angelis, J. & Amdursky, I., Negev, near Bir Asluj, stony hill, 27-III-1952, in fruit, Universitatis Hebraicae Hierosolymitanae Florae Terrae Israëlis Exsiccata No. 532 (BM, K, P). Naftulsky, N. 57, upper Jordan Valley, Gesher Nehalim, 25-II-1926, in flower & fruit (K). **Jordan:** Al-Eisawi, D. 2291, South Ghor, Ghor As Safi, Um Al-Hashiem, 8-II-1978, in flower (BM). Gillett, J.B. 15781, southeast coast Dead Sea, 31°3′ – 13′N, alt. ca. 350 m, stony alluvial slopes with sandy soil, locally common, 23-IV-1963 in flower & fruit (K). Meyers, F.S. & Dinsmore, J.E. 4941, Gamala, fields, 3-IV-1911, in flower & fruit (K). **Kenya:** Verdcourt 3232, Naivash District, 8 miles from Naivasha on the Nakuru Road, alt. 6400 feet, scattered in grass with many weedy ruderals, 3-4-XII-1981, in flower & fruit (K). **Kuwait:** Boulos, L. & Al-Hasan, R. 15322, run-off along the coastal highway to Al-Nuwaisib, facing Al-Ahmadi Port, 27°05′N, 48°05′E, alt. 20 m, loose sandy soil, 7-III-1985, in fruit (BM). Dickson, V. 724A, Adaliyeh, 4 miles south of Kuwait Town, sea level, on edge of corn crops, 31-III-1951, in fruit (K). **Libya:** Guinard, K.M. KG/LIB/478, Mendil, 26-II-1952, in fruit (BM). Turner, D. DT6/82, Fezzan, Brack, Wadi Zig Zah, XII-1981-IV-1982, in fruit (BM). **Morocco:** Hooker, South Morocco, Greater Atlas, V-1871, in fruit (mixed collection with *M. italica* × *M. littoralis)* (P). Miller, Russel & Sutton 562, 6 km north of Tiznit, near El-Mader-el-Kebir (29°8′N, 9°7′W), alt. 200 m, sandy roadside verge, 3-IV-1974, in flower & fruit (BM). **Pakistan:** Lamond, J. 549, West Pakistan, Baluchistan, Makran, Hoshab to Panjgur, north of Balgather Plain, sandy roadside, alt. ca. 850 m, 19-IV-1965, in fruit (G). **Oman:** Maxwell Darling, R.C. 442, Sharga (Trucial coast), 1937, in flower & fruit (K). Radcliffe-Smith, A. 3840, Tanuf, alt. 2000 feet, in dry wadi bed, in shade of Acacia, and on banks of wadi bed, 8-III-1926, in flower & fruit (K). **Qatar:** Boulos, L. 11050, 8 km west of Kar'anah, 2-IV-1977, in fruit (K). Boulos, L. 11073, 7 km southwest Al Wukair (Eocene), along the road to Al Kharanah, 2-IV-1977, in fruit (G). Wilcox, C. 117, near Bir Uwainat Ali, alt. 20 m, desert scrub, sand, stony, 18-IV-1971, in fruit (K). **Saudi Arabia:** Leyland, M. A32, 15 km west of Badarah pumping station extremely stony katti wadi, 23-IV-1969, in fruit (K). Mandaville, J.P., Jr. 2351, 65 km northeast by East Riyadh, sandy wadi bank, 13-III-1969, in flower & early fruit (BM). Mandaville, J.P., Jr. 2754, 25°43′N, 48°7′E, dry water course, sand on silt, 10-III-1970, in flower (BM). **Socotra:** Balfour, B. 668, II-III-1880, in flower & fruit (BM). **Somaliland:** Bally, P.R.O. & Melville, R. 15742, Gallgallo, 64 km south-southwest of Bosaso, at east end of Al Mado Range, on grassy slope at foot of cliffs of limestone escarpment, 10-I-1973, in flower & early fruit (K). **South Africa:** Brierley, E.M. 194, Sepani, G.F.S., alt. 4900 feet, veld, III-1932, in flower & fruit (BM). Dieterlen, A. 147, Afrique australe, Léribé, Basutoland, pentes, ravines, talus, in flower & fruit (received by P in 1914) (P). Leistner, G.A. 1507, Kuruman, 7 miles northeast of Van Zylsrus, alt. ca. 3100 feet, white silty sand in dry river bed of Kuruman River, locally abundant, 24-IX-1959, in flower & fruit (K). **Sweden:** Blom, C., Göteborg, Ringön, in ruderatis, 11-IX-1956, in flower & fruit (BM). **Tanganiyaka:** Greenway, P.J. & Talbot, L. 10150, Musoma District, northwest Beacon on Bounday from Lake Lagarja towards the Oldohyo Labai Hills, alt. ca. 5000 feet, grassland on a gray-clay loam, rare and local, 7-V-1961, in flower & fruit (K). **Tunisia:** Cosson, M.M.E. et al., Zeram-el-Din, nord l'El-Djem, 11-IV-1883, in fruit (P) (and another collection on 13-V-1883) (P). Ducouret, environs de Sfax, "donnée par le Ministère de l'Instruction publique, 1856" (P).

Note

Young and Lagudah (1994) used DNA genotype analysis to compare accessions from eastern Australia, Africa, and Israel. On this basis they were able to distinguish (introduced) Australian and (also introduced) South African accessions, although no consistent differences were found among accessions from native areas of distribution.

46. *Medicago lanigera* Winkl. & B. Fedtsch.

M. lanigera Winkl. & B. Fedtsch., Bull. Jard. Bot. Petersbg. (Izv. Imp. S.-Petersburgsk. Bot. Sada) 5: 41 (1905), (basionym). Eduard Winkler (1799–?) was deceased at the time of publication, but nevertheless was a full co-author of the name *M. lanigera*. Lesins and Lesins (1979) cited the name as "*Medicago lanigera* Winkler et Fedtschenko ex Fedtschenko," which would be correct if Fedtschenko were to be considered as the sole authority.

One encounters various abbreviations of the names of Winkler and Fedtschenko; the versions presented here are those of The International Plant Names Index (http://www.ipni.org/; current June 2010).

The epithet *lanigera* is Latin for wooly, a reference to the dense whitish hairs of the plant, especially those covering the fruits.

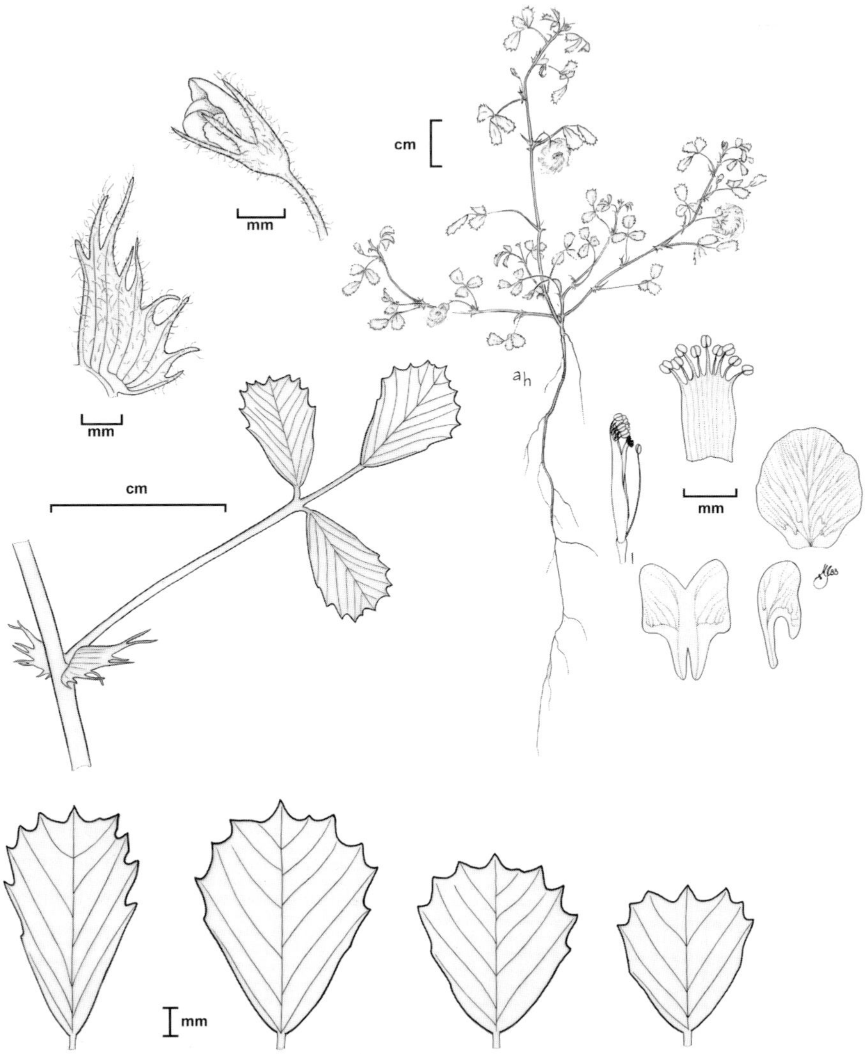

Fig. 46-1. *Medicago lanigera.* **Center right:** Floral parts (clockwise from top: flattened androecial column (free stamen not shown); standard petal; wing petal; keel; side view of androecial column showing free stamen). (Based on E. Small M346, DAO; artist: M. Jomphe.) **Remaining figures, clockwise from top left:** Stipule; side view of flower; plant; leaflets; leaf attached to stem. (Based on E. Small M346, DAO; artist: A. Hanes.)

Subordinate Taxa

None recognized here.

Type Specimen

Bukhara in Herb. Hort. Bot. Leningrad (LE).

In their description of their new species, Winkler and Fedtschenko provided the following information (translated from Russian): "Bukhara: the western slope of mount Hodja-Kadian near Kabadian, on the left bank of the Kabadian river, 20 April [Julian calendar used at the time; = Gregorian 2 May] 1883 (A. Regel!) and between the Vahsh and the Kabadian rivers, 2–3000', 17 [Julian calendar used at the time; = Gregorian 29] April 1883 (A. Regel!). A rather small plantlet, whose downy fruits make it extremely distinct from all the known species of *Medicago*. The late K. Yu. Winkler had acknowledged this species as a new one but, unfortunately, he only went as far as to assign it (in a herbarium) a name, and this is the one that I have maintained for this plant."

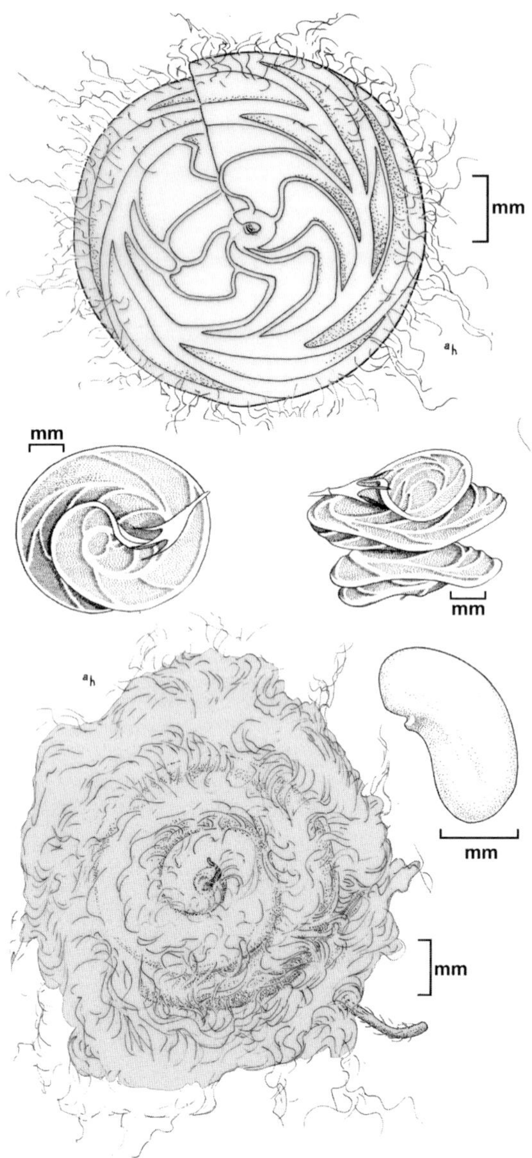

Fig. 46-2. *Medicago lanigera.* **Center left:** View of fruit (without pubescence). **Center right:** Side view of fruit (without pubescence). (Based on E. Small M346, DAO; artist: M. Jomphe.) **Top:** Fruit coil face showing venation. **Bottom left:** Fruit. **Bottom right:** Seed. (Based on E. Small M346, DAO; artist: A. Hanes.)

Synonymy

Homotypic

No names found.

Heterotypic

No names found.

Vernacular Names

English: Wooly (or woolly) alfalfa (Bolton 1962).

French: No names found.

Geographical Distribution

(N = native, I = introduced, U = uncertain status):

Asia: Afghanistan (N), Tajikistan (N), Turkmenistan (N), Uzbekistan (N).

Medicago lanigera is endemic to temperate Asia, including the mountain ranges of Turkmenia and Pamir-Alai.

Description

Annual herb, branched from base, decumbent to ascending, branches 4–20 cm long, vegetative parts more or less densely covered with simple hairs, leaflets 6–12 mm long, 3–7 mm wide, cuneate, obovate, or lanceolate, distal one- to two-thirds of margin serrate, stipules basally lobed. Peduncle bearing a single flower, shorter than or equal to corresponding petiole. Flowers 2–4 mm long, corolla pale yellow, calyx more than half length of corolla, densely pubescent with simple hairs, teeth slender, longer than or equal to tube. Young pod protrudes sideways from calyx. Mature pod spineless and covered with many-celled hairs, discoid, 2–3 mm long, with 3–4 coils, coils 5–8 mm wide, pergamentaceous (soft-walled), venation on coil surface (visible only upon removal of hairs) presents about 10 'S-shaped', unbranched radial veins that intersect a lateral vein about one-fifth of the radius from the dorsal suture, veins then emerge from the lateral vein and enter the dorsal suture. Seeds 1–2 per coil, surface smooth, yellow, not separated by a spongy transverse membrane, 3–4 mm long, radicle more than half as long as the seed length.

Identification Hints

Medicago lanigera is a distinctive species in having the (spineless) pods densely covered with long white hairs. The radicle is longer (about two-thirds of the seed length) than in most annual, spiral-fruited *Medicago* species.

Habitat and Environmental Adaptation

Habitat

Medicago lanigera occurs in dry, stony, or rocky (sometimes grassy) slopes of the middle altitudes of mountains.

Phenology

The species flowers in June in its native area.

Soil

Medicago lanigera typically occupies rocky soils.

Temperature

The habitat includes relatively cool, low montane areas of central Asia.

Water

Medicago lanigera apparently occurs in dry and/or well-drained substrates.

Biotic Relationships

Pollination

Small (1988) noted that the relatively low pollen to ovule ratio (235) and small flowers of the species are consistent with *M. lanigera* being largely self-pollinated. The species appears to be an inbreeder.

Dispersal

No information located. The curiously cottony fruits may play a role in seed distribution, but the nature of this is uncertain. The pubescence may simply be protective against herbivores.

Herbivory

No information located.

Nodulation and mycorrhizae

No information located.

Disease resistance

O'Neill et al. (2003) examined the comparative resistance of 36 annual species of *Medicago* to *Phoma medicaginis* (the cause of spring black stem and leaf spot, a major disease of alfalfa) and concluded that *M. lanigera* is highly susceptible to it. In a similar comparative test of 36 annual species of *Medicago*, O'Neill and Bauchan (2000) found that one accession of *M. lanigera* was quite susceptible to anthracnose caused by *Colletotrichum trifolii*. Yaege and Stuteville (2002) tested the resistance of 33 annual species of *Medicago* to powdery mildew (*Erysiphe pisi*) and found that the single collection of *M. lanigera* grown was susceptible to the disease (it was observed that often only one of a number of populations of other species tested was resistant).

Economic History, Cultivars, and Uses

The species has no record of being utilized for any purpose.

Cultivars

No cultivars have been bred.

Forage

Medicago lanigera is not sown for forage and Grossheim (1945) described it as "of no value for forage purposes."

Fodder

No information located.

Green manure, rotation crop, ground cover

No information located.

Other uses (salad sprouts, ornamental, plant products, etc.)

No information located.

Chromosome Number

$2n = 16$: Lesins and Lesins 1961; Simon and Simon 1965; Astanova 1981.

Genetic Relationships and Gene Transferability

Intraspecific

No information located.

Interspecific

No information located.

Exchange by advanced biotechnology

No information located.

Conservation Status

Little information is available about this endemic, which is largely inaccessible because of its remote mountainous location. No measures appear to have been adopted to date to assess the need for conservation, and the extent to which it is threatened has not been examined. Because it is so poorly known and collected, the species deserves to be the subject of immediate seed acquisition for the purpose of germplasm conservation.

Germplasm Sources

Only four accessions were located in germplasm institutions (see Chapter 3).

Research Problems and Priorities

Medicago lanigera is a very poorly understood species that is in immediate need of study with respect to its conservation status, basic botany, and agricultural potential.

Representative Collections

Afghanistan: Anders, O. & Podlech, D. 6113, prov. Qunduz, Surkhab Tal bei Qezel Say, an der Strasse von

Raghlan nach Qunduz, alt 500 m, Lössboden, 2-V-1971, in fruit (G). Rechinger, K.H. 33829, Northeast Afghanistan, Kataghan, inter Pul-i Khumri (35°55′N, 68°45′E) et Ribat (36°09′N, 68°25′E), alt. 700–1200 m, 5-V-1967, in fruit (G, K, S). Rechinger, K.H. 34075, North Afghanistan, Mazar-i-Sharif, Tashkurghan, (36°44′N, 67°41′E), alt. 500 m, in collibus argillosis, 7-V-1967 (G). Carter, J.E. 83 between Khulm and Mazar-i-Sharj, alt. 420 m, in roadside ditch, very dry in summer following flood water, 19-IV-1969, in flower & young fruit (K). **Tajikistan:** Koczkareva, T.F. & Kinzikaeva G.K., Tadzhikistania Australis (USSR), Jugum Aruktau in vicinitae trajectus Darbaza-Kum, ad declive australe angustiarum, in gypsaceis, alt. 600 m, 7-IV-1976, in flower & fruit (K). **Uzbekistan:** Vvedensky, Pamir-Alaj, ad declivia argillosa secus fl. Surchan- darja contra pag. Jangi-Cazar, 28-IV-1928, in fruit (G). **Turkmenistan:** See http://www.mobot.org/MOBOT/Research/LEguide/collections/35/761.html (current June 2010) for details and photograph of Botschantzev V.P. & Mikhailova M.A., 17-IV-1978.

Note

Indicative of the inapparent relationships of *M. lanigera* with other species, in describing their new species *M. lanigera*, Winkler and Fedtschenko placed it in a new, unique section, *Medicago* sect. *Lanigerae* (this treatment is adopted in Chapter 10). Small and Brookes (1984c) suggested that *M. lanigera* was related to *M. hypogaea*, the latter apparently also having enigmatic relationships to other species of the genus *Medicago*.

47. *Medicago laxispira* Heyn

M. laxispira Heyn in Rechinger, Flora Iranica 157 (text): 266 (1984) + 157 (tab.): tab. 166 (1984), (basionym).

The epithet *laxispira* is based on the Latin *laxus* (loose) plus *spira* (from the Greek *speira*), meaning loosely spiralled, a reference to the very unusual wide spacing of the coils of the fruit.

Subordinate Taxa

None recognized here.

Type Specimen

Anders 2771, Kurd.: Mosul: South Mosul, in collibus, 250 m, (date uncertain: 28-IV-1969?), Iraq (W!).

Synonymy

Homotypic

No names found.

Heterotypic

No names found.

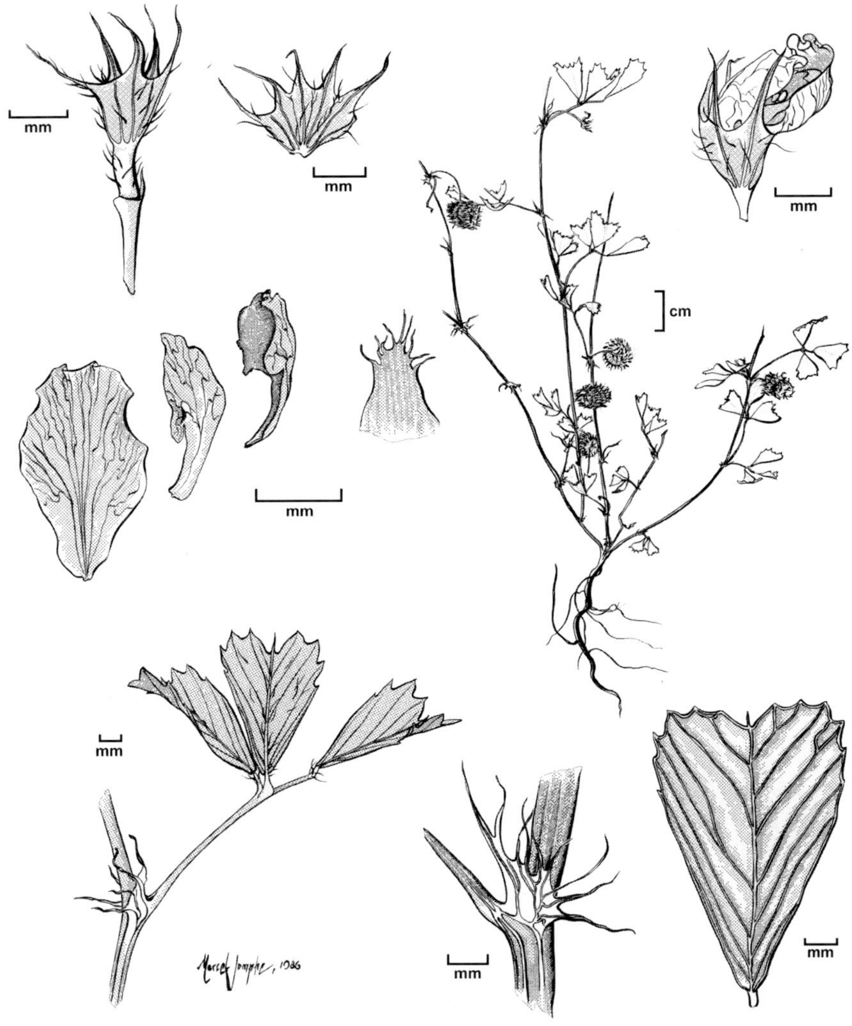

Fig. 47-1. *Medicago laxispira.* **Clockwise from top left:** Calyx and pedicel; calyx; plant; side view of flower; leaflet; stem node with stipules; leaf attached to stem; floral parts (from left: standard petal; wing petal; keel; flattened androecial column with free stamen not shown). (Based on E. Small H51 (DAO); artist: M. Jomphe.)

Vernacular Names

ENGLISH: No names found.

FRENCH: No names found.

Geographical Distribution

(N = native, I = introduced, U = uncertain status):

ASIA: Iraq (N).

This taxon was described on the basis of three collections from Iraq. Mosul (= Al Mawşil), the site of two of the collections, is located in northern Iraq at 36°18′N, 43°05′E.

The third collection is from Athba, presumably Al Athba south of Mosul.

Description

Annual herb, branches 20–40 cm long, vegetative parts sparsely pubescent. Leaflets 10–15 mm long, 5–10 mm wide, the distal margins serrate. Stipules laciniate. Inflorescence with one or two flowers, the peduncle longer than the subtending petiole. Flowers 2–3 mm long, the corolla yellow, slightly longer than the calyx. Fruit cylindrical, about 6 mm long, 5–6 mm wide, lacking pubescence, with 3–4.5 coils, dorsal suture of coils spiny, the spines at 90° to the coil faces, the coil faces with 6–10 radial veins that branch and anastomose. Seeds about 3 mm long, surface smooth, yellow to brown, reniform, radicle shorter than half of seed length, 1–2 per coil.

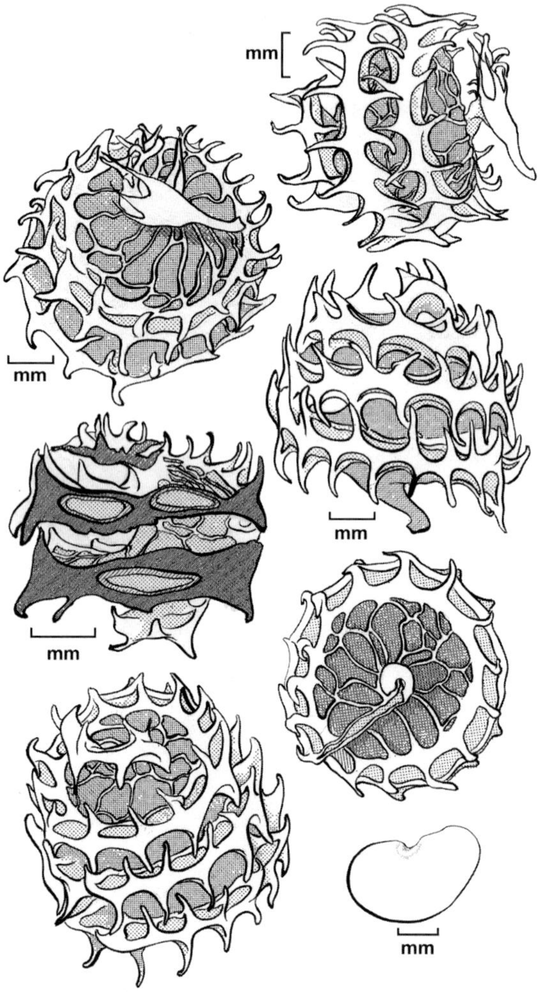

Fig. 47-2. *Medicago laxispira.* Views of fruit; seed (based on E. Small H51 (DAO); artist: M. Jomphe.)

Identification Hints

In her description, Heyn (1984) stated that "*M. laxispira* resembles in its vegetative parts some forms of *M. polymorpha* from which it differs in fruit characters: remote thick coils with spines at 90° to their surface as compared with much thinner (most often soft) not distant coils with spines at 180° at least in the middle coils; the number of veins on the surface of the coils in *M. laxispira* is smaller, they are less strongly bent and far less anastomosing in their distal part.... The fruit resembles superficially from side view the East Mediterranean *M. rotata* Boiss., but differs in venation of coils and shape and indumentum of leaves. In surface view the fruit resembles *M. sauvagei* Nègre (a Moroccan endemic species) but is easily distinguished by the different coil venation and spine insertion."

The combination of very loosely appressed fruit coils, spines perpendicular to fruit coil face, and reticulate coil venation should serve to identify this taxon which, in any case, is so rare that it will not normally be encountered.

Habitat and Environmental Adaptation

Habitat

Medicago laxispira has been collected from hillsides and as a weed of grain fields.

Phenology

The species flowers in the spring in its native Iraq.

Soil

No information located.

Temperature

No information located. Given the location, this species presumably germinates in the cool conditions of late winter and spring and matures seeds before summers become very hot.

Water

No information located.

Small et al. (1991) examined the germination characteristics of a form of *M. truncatula* (*M. truncatula* var. *laxicycla* E. Small) with very laxly coiled fruits, and suggested that the loose coiling facilitated germination of many of the seeds in response to rare rains. (Normally the fruit of *M. truncatula* is indehiscent, locking all the seeds together.) In support of this, they demonstrated that indeed most of the seeds germinate immediately while still within their pod, when provided with water. The same experiment was conducted with *M. laxispira*, but only about 1 seed/pod germinated, suggesting the adaptive explanation for lax fruit coiling in *M. truncatula* var. *laxicycla* does not apply to *M. laxispira*.

Biotic Relationships

Pollination

Based on the small flowers, this species is clearly a strong inbreeder.

Dispersal

Spiny fruits of *Medicago* normally are an adaptation to dispersal in animal fur but the short, appressed spines of this species doubtfully adhere well to fur.

Herbivory

No information located.

Nodulation and mycorrhizae

No information located.

Disease resistance

No information located.

Economic History, Cultivars, and Uses

There is no record of use of this species for any purpose.

Cultivars

No cultivars have been bred.

Forage

This species has not been employed for forage.

Fodder

Medicago laxispira has not been used for fodder.

Green manure, rotation crop, ground cover

No information located.

Other uses (salad sprouts, ornamental, plant products, etc.)

No information located.

Chromosome Number

$2n = 16$: Small et al. 1991.

Genetic Relationships and Gene Transferability

Intraspecific

No information located.

Interspecific

No information located.

Exchange by advanced biotechnology

No information located.

Conservation Status

In view of the rarity of this species, known populations deserve to be protected.

Germplasm Sources

Seeds of Anders 2641 (corresponding to Small, E. H51) and Anders 2771 (corresponding to Small, E. H52) were used by Small for study purposes. Seeds generated from H51 and H52 were distributed to other *Medicago* researchers and are the basis for all experimental work (particularly DNA genetic relationship studies) reported in the literature to date.

Research Problems and Priorities

Medicago laxispira is a taxon of very uncertain status, having been described on the basis of only three collections from Iraq. It resembles *M. polymorpha* and study is required to confirm its distinctness. There is a need to acquire collections and germplasm of this species and to study its botanical and agricultural characteristics.

Representative Collections

Iraq: Anders 2641, west of Athba, Iraq, hillsides and grain fields, 250 m, 10-V-1969 (W). Barkley, F.A. & Barklay, E.D., Digala near Mosul, 7-V-1963 (W). Small, E. H51 (seven herbarium sheets, designated a–g), grown in greenhouse, Ottawa, Canada, from seed obtained from Anders 2641: H51a harvested 01 May 1987; H51b harvested 03 October 1988; H51c, H51d, and H51e harvested 19 October 1988; H51f harvested 19 November 1988 (voucher for chromosome count of $n = 8$); H51g harvested 02 March 1989 (all collections at DAO). Small, E. H52a, grown in greenhouse, Ottawa, Canada, from seed obtained from the holotype, Anders 2771 cited above; harvested 17 July 1986 (DAO). Small, E. H52b, grown in greenhouse, Ottawa, Canada, from seed obtained from the holotype, Anders 2771 cited above; harvested 15 November 1988 (DAO).

48. *Medicago littoralis* Rohde ex Loisel.

M. littoralis Rohde ex Loisel., Not. Pl. Fl. France 118 (1810), (basionym).

The epithet *littoralis*, from the Latin *littoralis*, also spelled *litoralis* in Latin, means of the seashore, a reference to the seashore habitat. Although the correct current English word is littoral, one finds seashore species with the epithet spelled with either one letter 't' or two; the author's original spelling must be retained since both spellings are acceptably derived. Although litoralis is not a spelling error, the spelling *M. litoralis*, which is frequently encountered, is an error since it is not the original spelling.

Subordinate Taxa

None recognized here.

Type Specimen

Sales and Hedge (1994) lectotypified the name *M. littoralis* Rohde ex Loisel. as follows: "the specimen *Medicago littoralis* Rohde à Nice dans les sables du bord de la mer, par M. Rohde" in the Loiseleur-Deslongchamps herbarium at Avignon (AV). (Heyn (1963, p. 93) wrote that "Rohde's specimen cited by Loiseleur-Deslongchamps could not be located. However, from the description it is evident that it is

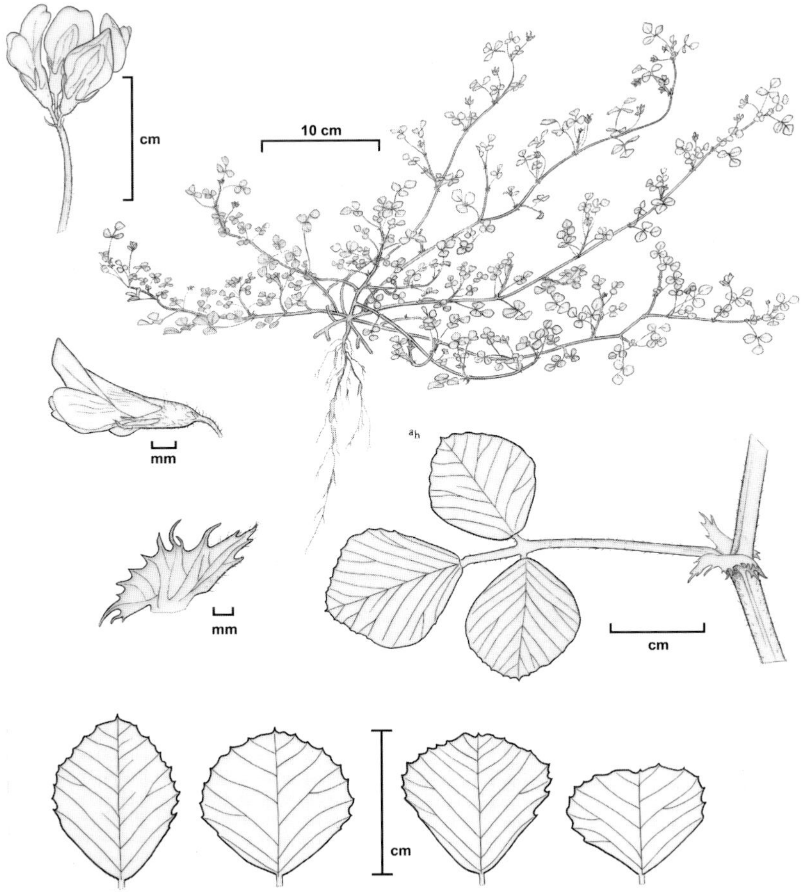

Fig. 48-1. *Medicago littoralis.* **Clockwise from top left:** Group of flowers; plant; leaf attached to stem; leaflets; stipule; side view of flower. (Based on E. Small M972, DAO; artist: A. Hanes.)

a specimen of the most widespread form of *M. littoralis* with spiny pods and more than three coils.")

Synonymy

Homotypic

Medica littoralis (Loisel.) Bubani, Fl. Pyren. 2: 476 (1899/1900).

M. truncatula Gaertn. subsp. *littoralis* (Rohde ex Loisel.) Ponert, Feddes Repert. 83 (9-10): 639 (1973).

Heterotypic

M. arenaria Ten., Cat. Pl. Hort. Neapol., App. 1: 59 (1819).

M. braunii Godron in Gren. & Godron, Fl. France Prosp. 1: 393 (1849).

M. littoralis Rohde ex Loisel. subsp. *braunii* (Godron) Bonnier & Layens, Fl. Fr. 75 (1894).

M. littoralis Rohde ex Loisel. var. *breviseta* DC. in Lam. & DC., Fl. France, ed. 3, 5: 568 (1815). (Four sheets were observed in Herb. Lam. (P), with more than one collection on two sheets. The name is hereby lectotypified by the specimen bearing the label "*Medicago* trouvé par ...1808.")

M. littoralis Rohde ex Loisel. subsp. *breviseta* (DC.) Urb., Verh. Bot. Ver. Brandenb. 15: 70 (1873).

M. cylindracea DC., Adnot. Cat. Pl. Horti Monsp. 123 (1813). (A single specimen is present in Herb. DC. (P), bearing the inscription "Medicago tornata minima" and is here taken as the type).

M. littoralis Rohde ex Loisel. subsp. *breviseta* (DC.) Urb. var. *cylindracea* (DC.) Urb., Verh. Bot. Ver. Brandenb. 15: 70 (1873).

M. littoralis Rohde ex Loisel. subsp. *cylindracea* (DC.) Nyman, Consp. Fl. Eur. 167 (1878).

M. littoralis Rohde ex Loisel. proles *cylindracea* (DC.) Rouy, Fl. Fr. 5: 30 (1899).

M. littoralis Rohde ex Loisel. subsp. *breviseta* (DC.) Urb. var. *depressa* Urb., Verh. Bot. Ver. Brandenb. 15: 70 (1873).

M. littoralis Rhode ex Loisel var. *dextrorsa* Dur., Expl. Sci. Algérie, Atlas, 39 (1855).

M. littoralis Rohde ex Loisel. subsp. *inermis* (Moris) Urb. var. *tricycla* (DC.) Urb. f. *dextrorsa* Urb., Verh. Bot. Ver. Brandenb. 15: 69 (1873).

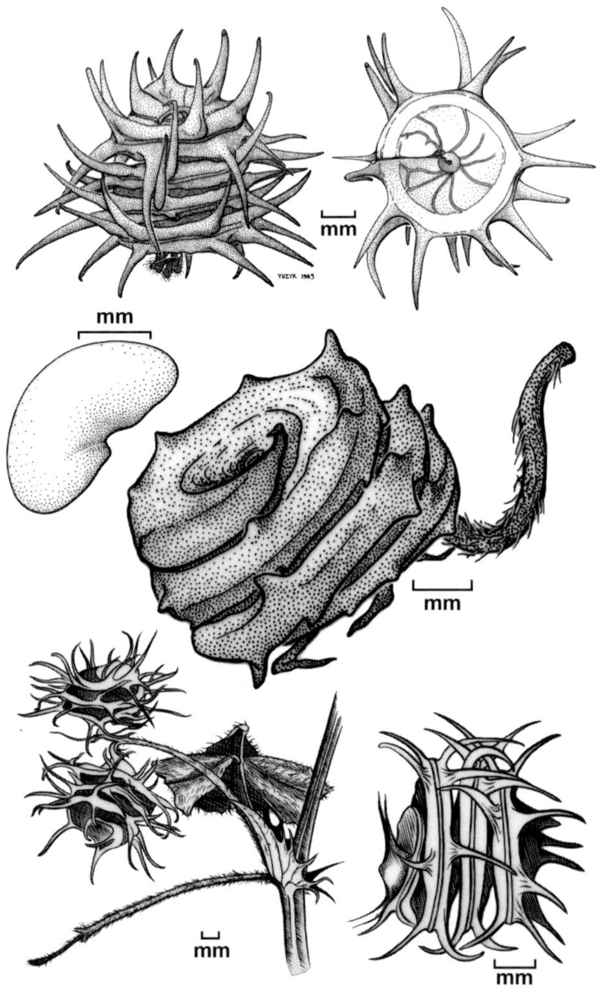

Fig. 48-2. *Medicago littoralis.* **Top left:** Spiny fruit. **Top right:** Fruit coil face showing venation. **Center left:** Seed (based on Hereault 21662, USA; artist: L. Yuzyk). **Center right:** Short-spined fruit (based on L. Kralik 21-3-1854, GH; artist: L. Yuzyk.). **Bottom left:** Portion of fruiting branch. **Bottom right:** Fruit (based on E. Small M231, DAO; artist: M. Jomphe).

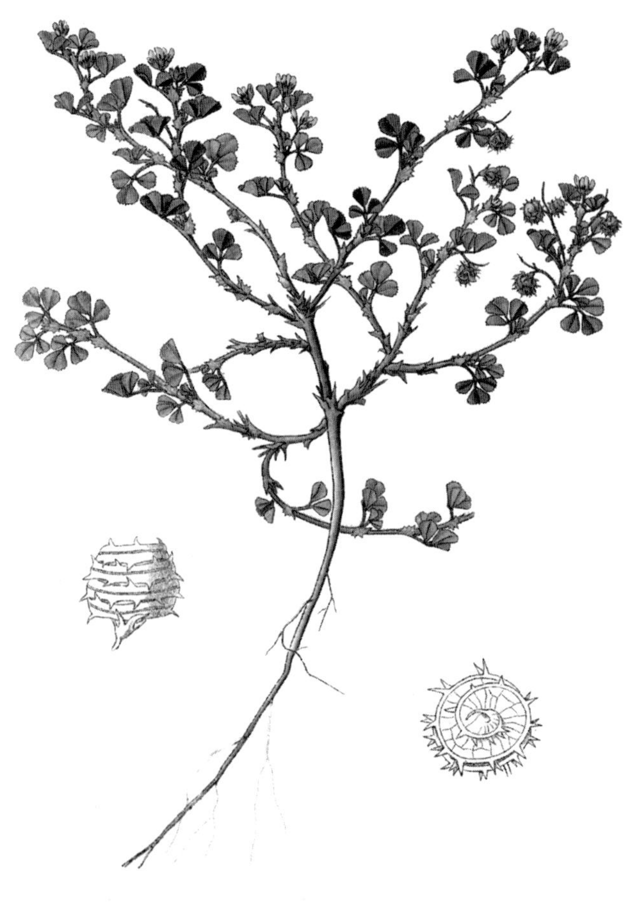

Fig. 48-3. *Medicago littoralis.* From Hallier (1885, vol. 23, plate 2345).

M. littoralis Rohde ex Loisel. subsp. *inermis* (Moris) Urb. var. *pentacycla* Urb. f. *dextrorsa* Urb., Verh. Bot. Ver. Brandenb. 15: 70 (1873).

M. littoralis Rohde ex Loisel. subsp. *longiseta* (DC.) Urb. f. *dextrorsa* Urb., Verh. Bot. Ver. Brandenb. 15: 70 (1873).

M. littoralis Rohde ex Loisel. subsp. *breviseta* (DC.) Urb. var. *cylindracea* (DC.) Urb. f. *dextrorsa* Urb., Verh. Bot. Ver. Brandenb. 15: 70 (1873).

M. gracilis Biv. in Bivona, Nuov. Plant. Ined. 10, Tod. Ind. Sem. Hort. Bot. Panorm. 26 (1858).

M. littoralis Rohde ex Loisel. var. *inermis* Moris, Fl. Sard. 1: 439 (1837).

M. littoralis Rohde ex Loisel. subsp. *inermis* (Moris) Urb., Verh. Bot. Ver. Brandenb. 15: 69 (1873).

M. littoralis Rohde ex Loisel. var. *inermis* Rouy in Rouy & Fouc., Fl. France 5: 30 (1899). (Relation to above taxa not determined.)

M. littoralis Rohde ex Loisel. var. *longeaculeata* Moris, Fl. Sard. 1: 440 (1837).

M. littoralis Rohde ex Loisel. var. *longiseta* DC. in Lam. & DC., Fl. France, ed. 3, 5: 568 (1815).

M. littoralis Rohde ex Loisel. subsp. *longiseta* (DC.) Urb., Verh. Bot. Ver. Brandenb. 15: 70 (1873).

M. littoralis Rohde f. *microsolen* Maire & Sennen (a name viewed at the website at http://www.aluka.org; current June 2010; this shows two collections at MP, both by Sennen on 16 December 1934, and states (unacceptably since a holotype cannot be two specimens) that each is both the holotype of f. *microsolen* and the type of "*Medicago littoralis* Rohde subvar. *cylindraea* (DI.) Urb.")

M. littoralis Rohde ex Loisel. subsp. *inermis* (Moris) Urb. var. *pentacycla* Urb., Verh. Bot. Ver. Brandenb. 15: 70 (1873).

M. pusilla Viv., Pl. Aegypt. Arab. Dec., 4: 14 (1830).

M. littoralis Rohde ex Loisel. var. *rouyana* Fiori in Fiori & Paol., Fl. Ital. 2: 37 (1900).

M. rugulosa Batt. in Batt. & Trab., Fl. de l'Alger. (Dicot.) 239 (1889). (Type!: *Battandier, J.A.*, V-1883, Nador de Médéa, sables, Algeria (P), a mixed collection with *M. italica*; viewed online at http://www.aluka.org; current June 2010.)

M. littoralis Rohde ex Loisel. subsp. *inermis* (Moris) Urb. var. *tricycla* (DC.) Urb. f. *sinistrorsa* Urb., Verh. Bot. Ver. Brandenb. 15: 69 (1873).

M. littoralis Rohde ex Loisel. subsp. *inermis* (Moris) Urb. var. *pentacycla* f. *sinistrorsa* Urb., Verh. Bot. Ver. Brandenb. 15: 70 (1873).

M. littoralis Rohde ex Loisel. subsp. *longiseta* (DC.) Urb. f. *sinistrorsa* Urb., Verh. Bot. Ver. Brandenb. 15: 70 (1873).

M. littoralis Rohde ex Loisel. subsp. *breviseta* (DC.) Urb. var. *cylindracea* (DC.) Urb. f. *sinistrorsa* Urb., Verh. Bot. Ver. Brandenb. 15: 70 (1873).

M. subinermis Bertol., Fl. Ital. 8: 290 (1852).

M. littoralis Rohde ex Loisel. var. *subinermis* (Bertol.) Boiss., Fl. Or. 2: 99 (1872).

M. tetracycla Presl, Fl. Sicul. (Presl) 1: 20 (1826).

M. tricycla Senn. non DC. in Bull. Soc. Bot. France 68: 405 (1921), nom. nud. (merely placed in a list by Sennen, without any indication of status).

M. trigyra Senn., Bol. Soc. Iber. Ci. Nat. 26: 88 (1927).

Not

M. littoralis Rohde ex Loisel. subsp. *striata* (Bast.) Douin in Bonnier, Fl. Compl. Fr. 3: 21 (1914). = *M. italica*.

Medicago ×casellasii Montserrat, Collect. Bot., Barcinone 5: 61 (1956) is an alleged hybrid of *M. littoralis* and *M. praecox*.

Vernacular Names

English: Coast medic, coastal medic, Harbinger medic, littoral medic, shore medic, strand medic, water medic (medic is also spelled medick).

French: La luzerne du litoral, luzerne du littoral, luzerne littorale, luzerne des rivages.

Geographical Distribution

(N = native, I = introduced, U = uncertain status)

Africa: Algeria (N), Egypt (N), Libya (N), Madeira Islands (Portugal; N), Morocco (N), Tunisia (N).

Asia: Azerbaijan (N), Cyprus (N), Egypt (N), Gruziya (Georgia; N), Iran (U), Israel (N), Jordan (N), Lebanon (N), Russia (Krasnodar; N), Syria (N), Turkey (N), Turkmenistan (N).

Australasia: Australia (I).

Europe: Albania (N), Belgium (I), Bulgaria (N), France (including Corsica; N), Greece (including Crete and Karpathos and East Aegean Islands; N), Italy (including Sardinia and Sicily; N), Malta (N), Netherlands (I), Portugal (N), Spain (including Balearic Islands, Canary Islands; N), Turkey (N), Yugoslavia (N).

Northern America: United States (New Jersey; I).

Medicago littoralis is the likeliest *Medicago* species to be encountered in Old World maritime habitats. This Ponto-Mediterranean species is widely distributed around the Mediterranean Sea and adjacent regions, and is adventitious on the southern coast of the North Sea.

Description

Annual herb, procumbent, ascending or (infrequently) ascending, branched from base, branches 7–20(40) cm long, vegetative parts pubescent with simple hairs. Leaflets 3–8 mm long, 2–7 mm wide, pubescent on both surfaces, distal margins serrate, stipules dentate to laciniate. Inflorescence of 1–3(5) flowers, peduncle about equal to subtending petiole. Flowers 3–6 mm, corolla yellow, about twice as long as calyx, calyx with simple hairs, teeth shorter than tube. Young pod contracted and concealed within calyx. Mature pod discoid to cylindrical, glabrous, hard at maturity, 3–10 mm long, with 2–6(8) adpressed coils, 3–7 mm wide. Coil surface with about 10 somewhat curving radial veins arising from ventral suture (at axis of pod) that branch slightly and enter lateral vein (near dorsal suture). Pod usually spiny (sometimes tubercular or spineless), the spines 90° to 180° to pod surface. Seeds 1–2 per coil, 2.5–3.7 mm long, yellow to brownish-yellow, reniform, separated by transverse spongy membranes, the surface smooth, radicle less than half length of seed.

Identification Hints

As with other annual species of *Medicago* that develop very hard fruit at maturity, the face of the coils and indeed the bases of the spines may be obscured by the development of spongy tissue.

Extremely variable populations exist, likely the result of introgression with *M. truncatula* and *M. italica*. The following two paragraphs discuss discrimination of *M. littoralis* from the other two species.

Medicago littoralis is most likely to be confused with the similar but inland *M. truncatula* with which it hybridizes and intergrades. Lesins and Lesins (1963*b*) observed that in *M. littoralis* the pollen grains are cylindrical and spindle-shaped, with three germinal apertures, whereas in *M. truncatula* shapes are mixed (such as spindle, bisphenoid, and pentahedral), with three to four (up to six) germinal apertures. However, this is a difficult character to employ for identification purposes and has not been extensively examined for consistency. A detailed study of interspecific variation between the two species was conducted by Small and Brookes (1990*a*). Heyn (1963) had earlier differentiated the species in part by longer branches and leaves of *M. truncatula*. However, Small and Brookes (1990*a*) demonstrated, by growing plants in a common garden, that these size differences were due to growth in comparatively fertile soils by *M. truncatula* (it grows on heavier inland soils than the mainly sandy seashores on which *M. littoralis* grows). Heyn (1963) observed that, in Israel, intermediate forms are common where the two species meet on soils disturbed by humans. Small and Brookes (1990*a*) found that of the 26 traits they examined, density of hairs on the fruits was the best discriminating character. The best additional discriminating character was the presence of a groove on either side of the fruit dorsal suture (more pronounced in *M. truncatula* than in *M. littoralis*), but this character is best viewed in relatively young fruit. Fruits were often larger in *M. truncatula*, as claimed in the literature, but this is an unreliable character for identification. As claimed in the literature, spine angle proved to have some value for separating the two species: spines between adjacent coils in *M. truncatula* tend to overlap, but this rarely occurs in *M. littoralis*. For practical purposes, the species can usually be identified by their mature fruit, which is glabrous or essentially so in *M. littoralis*, but usually has at least a few trichomes in *M. truncatula*.

Medicago littoralis also hybridizes and intergrades with *M. italica* (Small and Brookes 1990*b*). See *M. italica* (Chapter 44) for information and a key to distinguishing *M. littoralis* from the widespread hybrid, *M. italica* × *M. littoralis*.

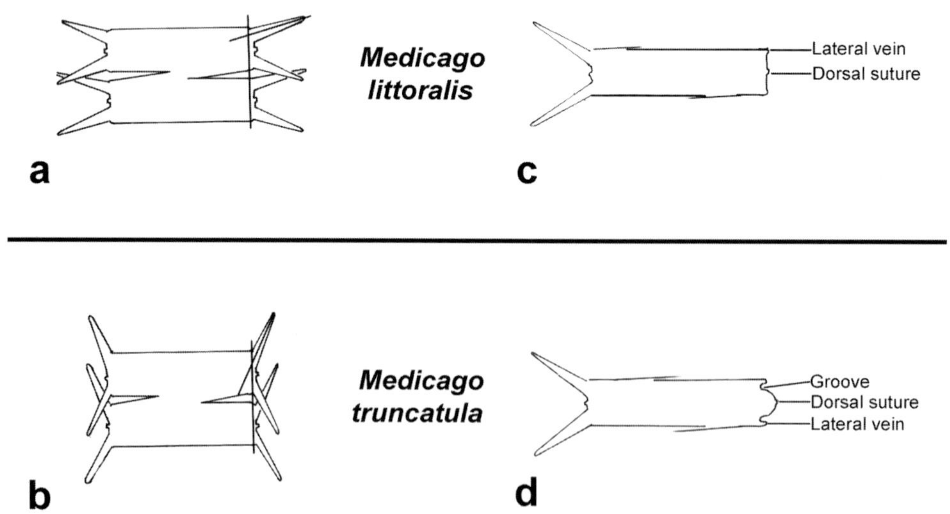

Fig. 48-4. Some characters that aid in distinguishing *M. littoralis* and *M. truncatula*. **Left:** Diagrammatic long section of two adjacent fruit coils of a, *M. littoralis* and b, *M. truncatula*. Notice the spines often tend to lie closer to the fruit axis in *M. truncatula*. **Right:** Diagrammatic long sections of a coil of c, *M. littoralis* and d, *M. truncatula* (spines not shown on right hand side for clarity). Notice that in *M. littoralis* the dorsal suture and lateral veins tend to be at the same level, whereas in *M. truncatula* the dorsal suture projects beyond the lateral veins, and also there tends to be a groove on either side of the dorsal suture.

Habitat and Environmental Adaptation

Habitat

Medicago littoralis occurs primarily on sandy seashores, where (together with *M. marina* and sometimes *M. italica*) it is often the main if not the only *Medicago*, sometimes forming extensive beach populations. It is also found to a smaller extent in Mediterranean grasslands and arid, sandy inland areas. Thanopoulos (2007) noted that in Greece it occurs mainly in coastal environments, but also on steep cliffs and sometimes as a weed in olive groves, fruit orchards, and ruderal sites.

Phenology

This winter annual flowers February to May, at the end of winter and in the spring, in its native Mediterranean area. Clarkson and Russell (1975) found that flowering was accelerated by vernalization (but not long vernalization, which delayed flowering), long photoperiods (18 and 24 hours), and high temperature.

Soil

Medicago littoralis typically occupies maritime rocks, gravel, and sand, although it is capable of growing well in loam (clay–sand mixtures). Piano and Francis (1992) recorded that in its native area it occurs on sandy soils that are slightly to moderately basic. The Register of Australian Herbage Plant Cultivars for Legumes (http://www.pi.csiro.au/ahpc/legumes/legumes.htm; current September 2010) provided the following information on cultivars of *M. littoralis*: 'Harbinger', 'Harbinger AR', and 'Herald' are well adapted to alkaline sands and loams. Zoghlami et al. (1996), in reference to Tunisia, commented that *M. littoralis* is often in coarse-textured soils poor in calcium and organic matter. Abdelguerfi et al. (1988*a*) observed that in Algeria *M. littoralis* occurs in medium- and coarse-textured soils. Rogers et al. (2005) ranked *M. littoralis* as moderately tolerant of soil salinity. Lieth and Mochtchenko (2003) listed the species in their halophyte database.

Temperature

The species is found mostly in the warm, sea-moderated temperatures of Mediterranean seashores and has a reputation for susceptibility to cold (Piano and Francis 1992). Cocks and Ehrman (1987) found that *M. littoralis* in Syria was almost totally intolerant of frost. Krall et al. (1996*a*) found four accessions of *M. littoralis* examined for their ability to self-regenerate in Wyoming failed to survive over the winter at the three locations tested. Christiansen and Svejcar (1989) sowed *M. littoralis* in Oklahoma in March 1985 and observed that all the plants were killed by frost and cold in the following winter period. Derkaoui et al. (1991) also found that fall-planted *M. littoralis* in Oklahoma failed to survive the winter, indicating that the species does not have sufficient cold tolerance for successful self-replication in the southern Great Plains of the United States.

Water

Medicago littoralis predominantly occurs in well-drained sand, sometimes in more water-retentive soils. It occurs frequently in dry environments. Piano and Francis (1992) recorded that in most areas of the Mediterranean it requires a minimum annual rainfall of 175 mm. Zoghlami et al. (1996) observed that *M. littoralis* tended to develop best in areas of Tunisia receiving less than 300 mm of rainfall annually. Francis (1981) found that *M. littoralis* was well adapted to areas of Libya with less than 100 mm rainfall annually. The Australian cultivars 'Harbinger', 'Harbinger AR', and 'Herald' are well adapted to regions of Australia with 275–400 mm annual rainfall. Cultivar 'Harbinger' is extremely sensitive to waterlogging (Clarkson 1977). Brahim and Smith (1993) observed that *M. littoralis* in a field plot in Arizona produced seeds with only 125 mm winter precipitation.

Biotic Relationships

Pollination

Small (1988) noted that the relatively low pollen to ovule ratio (257) and small flowers are consistent with *M. lttoralis* being largely self-pollinated. The species appears to be an inbreeder.

Dispersal

The indehiscent spiny fruits of *M. littoralis* are well adapted to dispersal in animal fur, accounting in part for the extremely widespread occurrence of this species. Kotzé et al. (1995) showed that intact seeds will pass through the gut of sheep and remain viable.

Herbivory

Small et al. (1989) found that less than 3% of wild plants of *M. littoralis* have fruits infested by the bean weevil *Bruchidius bimaculatus* (Olivier) (see discussion of fruit size in Chapter 7). Pérez-Mellado et al. (2000) noted that the lizard *Podarcis lilfordi* consumed *M. littoralis* in the Balearic Islands. Hensen (2002) found evidence that in Spain fruits of *M. littoralis* were collected by the harvester ant *Messor bouvieri*.

The Register of Australian Herbage Plant Cultivars for Legumes (http://www.pi.csiro.au/ahpc/legumes/legumes.htm; current June 2010) provided the following information on cultivars of *M. littoralis*: 'Harbinger AR' was selected from 'Harbinger' for resistance to spotted alfalfa aphid (*Therioaphis trifolii*) and blue-green aphid (*Acyrthosiphon kondoi*). 'Herald' is described as equally resistant to the aforementioned aphids as the Harbingers and, in common with the Harbingers, is moderately resistant to cowpea aphid (*Aphis craccivora*), susceptible to sitona weevil (*Sitona discoideus*), redlegged earth mite (*Halotydeus destructor*), and lucerne flea (*Sminthurus viridis*).

Nodulation and mycorrhizae

USDA (2009) listed positive root nodulation for *M. littoralis*. Maremmani et al. (2003) reported that arbuscular mycorrhizae were present in *M. littoralis* growing in sand dunes on the coast of Italy.

Disease resistance

Yaege and Stuteville (2002) tested the resistance of 33

annual species of *Medicago* to powdery mildew (*Erysiphe pisi*) and found that the seven collections of *M. littoralis* grown were susceptible to the disease (it was observed that often only one of a number of populations of other species tested was resistant). Elgin and Ostazeski (1992) found that *M. littoralis* was more than 90% resistant to Races 1 and 2 of anthracnose caused by *Colletotrichum trifolii* Bain. However, in a comparative test of 36 annual species of *Medicago*, O'Neill and Bauchan (2000) found that seven accessions of *M. littoralis* were all quite susceptible to anthracnose caused by *Colletotrichum trifolii*.

Economic History, Cultivars, and Uses

Medicago littoralis provides the principal source of fixed nitrogen in many neutral to alkaline sandy-loam soils, which are frequent in South Australia; the species is often grown in near-monoculture conditions in farming systems employed there (Nair et al. 2004). The first cultivar, 'Harbinger', was introduced to Australia (released in 1959) from California plants, but is thought to have originated from plants from Iran (Clarkson 1977).

Cultivars

The Register of Australian Herbage Plant Cultivars for Legumes (http://www.pi.csiro.au/ahpc/legumes/legumes.htm; current June 2010, the document updated January 2007) contains registration statements that were contained in the *Register of Australian Herbage Plant Cultivars* (3rd ed., 1990) and those published subsequently in the *Australian Journal of Experimental Agriculture*, and provides information on the origin and development of each cultivar, a full morphological description, and referenced information on the agronomic uses and merits of the cultivar. The following cultivars of *M. littoralis* are included: 'Harbinger' ("registered prior to December 1971", described as possibly a hybrid with *M. truncatula*), 'Harbinger AR' (a successor to 'Harbinger', registered 1988), and 'Herald' (very similar to the Harbingers, but distinguishable by the fact that its pods coil in a clockwise direction, opposite to 'Harbinger' and 'Harbinger AR'. Also, 'Herald' has a prominent dark brown blotch on each leaflet that fades with the onset of warmer weather, as well as purple flecking on the underside of the leaflets — a feature that is rare or absent in the Harbingers. Other varieties include 'Angel' (Seedmark/PlantTech; a herbicide-tolerant form generated by mutagenesis, described by Oldach et al. 2008) and 'Jaguar' (Keith Seeds/Auswest; registered in 2004). The cultivar 'Toreador' is a hybrid between *M. italica* and *M. littoralis* (see Chapter 44 on *M. italica* for additional information).

Forage

With regard to forage yield, *M. littoralis* is one of the annual species of *Medicago*, as noted above, that is widely planted in Australia for forage. It is also one of the species planted in South Africa (Kotzé et al.1995). It has potential for providing good pasturage in sand dune areas (Abdelguerfi et al. 1988*a*). Sanadgol et al. (2006) found that the unirrigated yield of *M. littoralis* over a 3-year period in the Gorgan Plain of Iran was 979 kg/ha. In a nursery plot on alluvial soil in the steppic zone (220 m elevation) of northern Iraq, during a season with rainfall of 475 mm (November to May), forage yields for 'Harbinger' were recorded as follows:

Forage yield of *Medicago littoralis* 'Harbinger' in northern Iraq (kg/ha), after Kernick 1978.

Fertilized with 60 kg/ha P_2O_5		Unfertilized	
Green matter	Dry matter	Green matter	Dry matter
5000	2600	1200	400

Zhu and Sheaffer (1997) recorded aboveground dry matter yields (in kg/ha) of cultivar 'Harbinger' at two sites in Minnesota (compared with the alfalfa cultivar Nitro, yields in parentheses): 5676 (3815) and 3909 (4596).

Mohammad and Qamar (1988) grew 14 annual species of *Medicago* in small test plots in Islamabad, Pakistan, located in a subtropical, subhumid area that is notably hotter and wetter than the normal habitat of these species. The dry matter yield of *M. littoralis* was about 3.5 t/ha.

With regard to seeds, Kotzé et al. (1995) examined the nutritional value of *M. littoralis* seeds for feeding sheep in South Africa and found that cultivar 'Santiago' is suited for utilization by sheep during the summer months when medic pods are a major component of their diet. Seed yields of up to 1250 kg/ha for 'Harbinger' have been realized in the littoral zone of Algeria (Kernick 1978).

Fodder

As with other annual *Medicago* species for which cultivars have been selected, *M. littoralis* is occasionally used as a fodder source.

Green manure, rotation crop, ground cover

Ramert (1993) found that intercropping carrots with *M. littoralis* reduced damage by the carrot fly (*Psila rosae* (F.)), although this led to a reduction in carrot yield.

Other uses (salad sprouts, ornamental, plant products, etc.)

No information located.

Chromosome Number

2*n* = 14: Agarwal and Gupta 1983*b*.

2*n* = 16: Fryer 1930; Ghimpu 1929*b*, 1930; Heyn 1956; Larsen 1956 (as *M. longiseta*); Simon and Simon 1965; Lesins and Erac 1968*a*; Bjorkqvist et al. 1969; Dahlgren et al. 1971; Fernandes and Santos 1971; Nilsson and Lassen 1971; Fernandes et al. 1977; Fernandes and Queirós 1978 (2 reports); Oliva et al. 1978; Abdelguerfi and Guittoneau 1979; Abdelguerfi et al. 1988*b*; Slavik et al. 1993; Oberprieler and Vogt 1996*a*.

Genetic Relationships and Gene Transferability

Intraspecific

Simon and Millington (1967) observed considerable pollen sterility and some inviability in some intraspecific crosses of *M. littoralis*.

Interspecific

Simon and Millington (1967) found that *M. littoralis* and *M. truncatula* could be hybridized easily and that they generally produced fully fertile hybrids (although there was some F_1 sterility due to chromosomal rearrangements between the two species, and F_2 breakdown was observed). Lake (1989) found that a cross between *M. truncatula* and *M. littoralis* produced partially sterile plants, some of which had reduced viability. Despite interspecific incompatibility, Lake (1989) demonstrated that, for breeding improved cultivars, gene exchange is possible among *M. truncatula*, *M. littoralis*, and *M. italica*. Attempts (more or less successful) to hybridize *M. littoralis* and *M. italica* are discussed in Chapter 50. Simon and Millington (1967) were unable to produce hybrid seed by crossing *M. littoralis* with *M. rigidula* (*M. rigiduloides*?), *M. murex* (*M. sphaerocarpos*?), *M. turbinata*, or *M. constricta*. See Chapter 44 for an alleged hybrid (*M.* ×*sabulensis* H. Lév.) of this species with *M. littoralis*. A putative hybrid between *M. littoralis* and *M. praecox* was named *M.* ×*casellasii* P. Montserrat, in Collect. Bot. (Barcelona) 5: 61 (1956).

Exchange by advanced biotechnology

No information located.

Conservation Status

This is one of the most well-represented species in *Medicago* gene banks and is extremely common in the Old World. It is not in need of special conservation measures.

Germplasm Sources

More than 3000 accessions have been deposited in germplasm collections (see Chapter 3).

Research Problems and Priorities

Medicago littoralis has been studied much more intensely than all other annual species of *Medicago* except its close relative *M. truncatula*. Several cultivars have been developed and it is a continuing subject of agronomic studies. Its value is now well recognized and the species continues to attract research activities and agricultural development.

Representative Collections

Algeria: Durieu, M., Bone, sables maritimes, 17-III-1841, in fruit (P). **Azerbaijan:** Holmberg, O.R. 144, Transcaucasia, prov. Baku, distr. Baku, Tschorny Gorod prope Baku, 27-IV in flower & fruit, 10-V-1912 in fruit (K). **Canary Islands (Spain):** Bourgeau, E., Teneriffa, Barranco de Martianez ad Tortum Orotavensem, (no date; ex Herb John Ball 1890), in fruit (P), in flower & fruit (GH). **Cyprus:** Casey, E.C. 1103, 6 miles east of Kyrenia, alt. 60 feet, sand-hills by sea, 25-II, 6-IV-1951, in flower & fruit (K). Casey, E.C. 1173, Kyrenia, rocky land by seashore, alt. 15 feet, 6-IV, 28-IV-1951, in fruit (K). **Egypt:** Ascherson, P., Alexandria, Ramleh, 15-III-1887, in flower & fruit (P). Davis, P.H., Mariut, Abusir, 3-III-1944, in flower & fruit (K). Hurst, H.A., near Alexandria, IV-1871, in fruit (K). Letourneux, A., Râs Kenâis, 11-IV-1879, in flower & fruit (K). **France:** Cartier, A., Saint Raphaël (Var), sables maritimes, 16-VI-1879, in fruit, Société Dauphinoise 1880 No. 2423 (K). Mabille, P., Corsica, sables de l'Arenella à Biguglia, 20-30-IV-1868, in flower & fruit. Mabille Herbarium Corsicum No. 362 (P). Mabille, P., Corsica, Bonifacio, 17-IV-1868, in flower & fruit, Mabille Herbarium Corsicum No. 363 (GH) (identified as *M. braunii* Gren. & Godr. var. *inermis*). **Greece:** Cannon, J.F.M. & M.J. 4105, Attiki, Nea Makri, roadside above sandy beach, 1-IV-1973, in flower & early fruit (BM). Heldreich, Attica, Hagia Glykeria, ad radices collis Turcovuni, in locis herbosis, vinetis et ad vias, 30-IV-1884, in flower & fruit, Dr. Heldreich Herbarium Graecum Normale No. 834b (P). Heldreich, Attica, in arenosis maritimis Phaleri, 8-V-1884 in fruit, Dr. Heldreich Herbarium Graecum Normale No. 834a (P). Reverchon, E., Kissamos, lieux arides, 13-IV-1884, in fruit (P). **Israel:** Davis, P.H. 4305, Binyamina-Caesarea, common in the sand dunes, 8-IV-1842, in fruit (K). Feinbrun, N., Schechnowitz, L. & Spindel, M., Ramath-Gan, near Tel-Aviv, roadsides, sandy soil, 12-III-1928, in fruit, Univ. Hebraicae Hierosolymitanae Flora Palaestina Exsiccata No. 64 (BM, K). Zohary, D. & Segal, T., Sharon Plain, Pardes Hanna, 17-III-1951, in flower & fruit (K). **Italy:** Fiori, A., Etruria, prov. di Lucca, Viareggio, in arenosis maritimis obvia, loco pineta dicto, 5-VI-1915, in flower & fruit, Flora Italica Exsiccata Series III Fiori & Beguinot No. 2284 (K). Pease, A.S. 8889, Sicily, Siracusa, dry soil, 24-III-1906, in flower & fruit (GH). Rigo, G. 201, Sicilia, Prov. Messina, inter Galati et Tremestieri, in locis herbosis, 1-V-1898, in fruit (P). **Lebanon:** Meinertzhagen, R., Beyrouth, Syria, in sand dunes, 21-III-1933, in flower & very early fruit (BM). Townsend, C.C. 74/54, between Tyre and Sidon, Khaizaran Springs, on loose sand by the sea shore, 7-V-1974, in fruit (K). **Libya:** Davis 49777, Sabratha, west of Tripoli, alt. 2–5 m, sandy/rocky slopes near sea, 20-III-1970, in flower & fruit (K). Galli, M. 24, Gout Forest 28 km east of Tripoli, at bottom of stabilized dune area, 30-V-1954, in flower & fruit (US). Maitland, D. 79, 10 miles south of Tripoli, sea level, open sunny bank by road, 24-III-1970, in fruit (K). **Portugal:** Bourgeau, E., prés Faro, Algarve, 4-IV-1853, in fruit (GH). Daveau, J. 1018, Berlenga, granits désagrégés de l'île, V-1883, in flower & fruit (P). **Sardinia (France):** Reverchon, E., Poutzo, sables de la plage, 30-IV-1882, in flower & fruit (K). **Spain:** Gardner, M.F. & S.G. 1822, prov. Valencia, Cullera, Torre del Cabo (39°9′N, 0°13′W), alt. 30 m, rocky coastal slope, 24-IV-1982, in flower & fruit (BM). Loscos, Castelserás, in herbidis frequens, 11-V-1875, in fruit, Series Exiccata Florae Aragonensis Centuria Prima No. 33 (P). Perris, J.B. & Stubing, G., Alicante, Guardamar del Segura, Pastizales anuales sobre dunas, Malcolmietalia Rivas Goday 1957, 26-III-1984, in fruit, Herbario de la facultad de Farmacia V.F. Universidad de Valencia No. 57 (P). Raynal, J.A. 8849, Miami Playa (Espagne, Terragona), sables littoreaux à env. 2 km NE de la station, 22-IV-1962 (P). Sennen, Fre. 3007, Casteloefels, sables maritimes, VI-1918, in flower & fruit (US). **Tunisia:** Pitard, C.J. 91, Gabès (Menira), in incultis desertis, III-1907, in flower & fruit (K). **Turkey:** Barclay, S.C. 1563, Izmir Province, Çeome, 90 km west of Izmir, sea level, beach, 21-V-1969, in fruit (K). Davis 40779, C1, Aydin, Didyma, alt. 10 m, sand dunes, 9-IV-1965, in fruit (K). Davis 41285, BI, Izmir, Çesme to Signe, alt. 30 m, fallow field, 26-IV-1965, in flower & fruit (K). **Yemen:** Bisset, E. 171, Telhán (44°26′E, 14°22′N), alt. 2540 m, stony slopes, 8-IX-1982, in flower & fruit (K). **Yugoslavia:** Jackson, A.K. & Turrill, W.B. 115, La pad Peninsula, 19-IV-1935, in fruit (K). Pichler, T., Dalmatien, Almissa, 16-V-1868, in fruit (K).

49. *Medicago lupulina* L.

M. lupulina L., Sp. Pl. 779 (1753), (basionym).

The epithet *lupulina* is Latin for hop-like, in allusion to the hop-like fruit clusters, i.e., comparable with the "cones" of *Humulus lupulus* L. that are used to brew beer.

Subordinate Taxa

None recognized here. *Medicago lupulina* is an extremely variable species (Blaser and Stokes 1946; Goertzen and Small 1993), but the variation is not structured in ways that can reasonably be classified formally. Of the many criteria that have been used to delimit infraspecific groups, the presence of gland-tipped hairs and whether annual and/or biennial or perennial have been most frequently employed. The presence of gland-tipped hairs is extremely variable (Goertzen and Small 1993) and does not constitute a reasonable taxonomic character. As noted below, all forms of *M. lupulina* appear to be potentially perennial in a greenhouse, and perennial populations appear sporadically throughout the Old World, especially where the climate is relatively mild or near the equator.

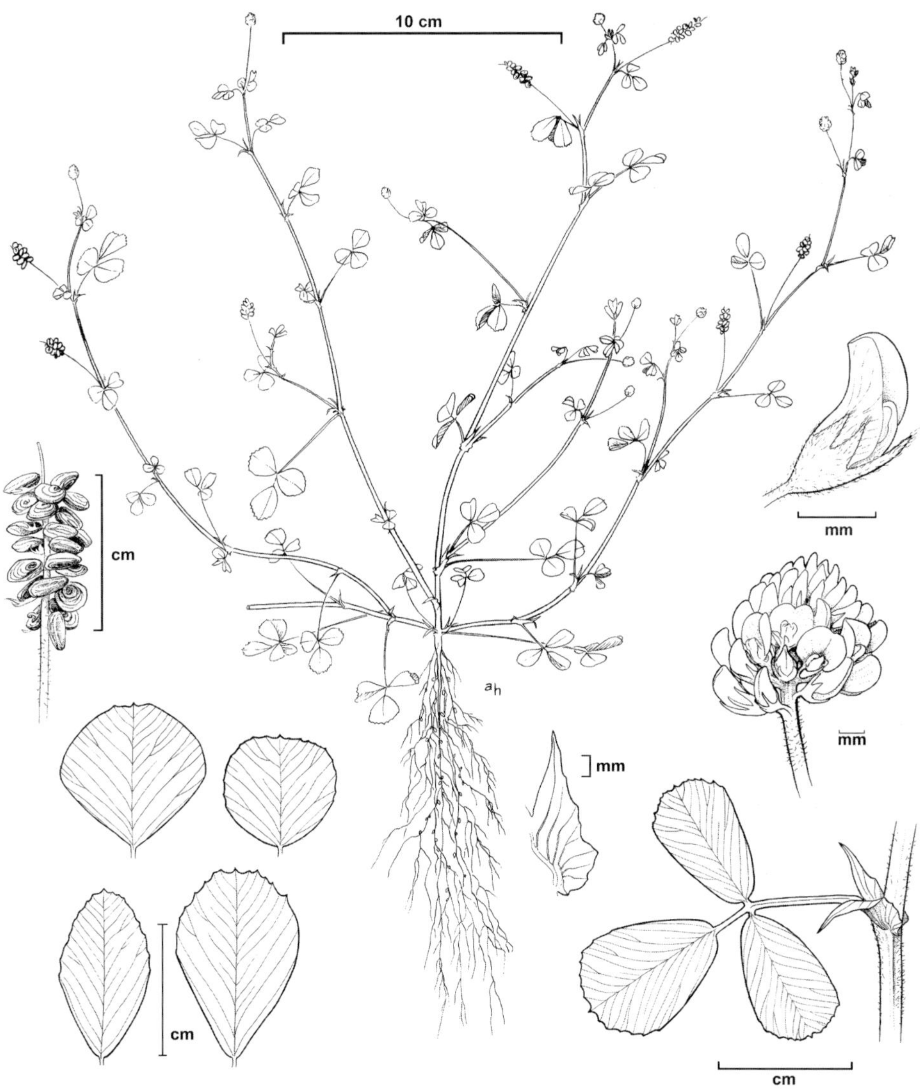

Fig. 49-1. *Medicago lupulina.* **Clockwise from top:** Plant; side view of flower; group of flowers; leaf attached to stem; stipule; leaflets; cluster of fruit. (Based on E. Small M20, DAO; artist: A. Hanes.)

Type Specimen

Heyn (1959) provided the following analysis: "*Medicago lupulina*. The phrase-name is cited from the Fl. Suec. In its outer appearance this species is quite unlike any other species of the genus *Medicago* and was included in the Hort. Cliff. within the genus *Trifolium*. On transferring it to the genus *Medicago*, Linnaeus adhered to the phrase of the Hort. Cliff., changing only the name of the genus. Nevertheless, we cannot accept the specimen in the Hort. Sicc. Cliff. bearing the phase as being the type, because the transfer of the species to another genus was doubtlessly based on additional specimens examined by Linnaeus after the completion of the Hort. Cliff. (as stated above, the specimens of the Hort. Sicc. Cliff. were not seen by Linnaeus after 1738). The only specimen in the Linn. Herb. "933–10" bearing the name *M. lupulina* on the herbarium sheet, does not give any indication of being connected with the Sp. Pl. As in the case of *M. falcata* and for the same reasons, we decided not to choose any lectotype at present. ... If no type specimen should be found, we could choose as the most plausible lectotype the figure of J. Bauhin and Cherler (1650–1651), p. 389 accompanying the description of 'Trifolium pratense luteum mas, flore minore...' This source is cited by Tournefort as synonym to the phrase-name which Linnaeus cites as synonym in the Fl. Suec. for his later *M. lupulina*."

Fig. 49-3. *Medicago lupulina.* From Hallier (1885, vol. 23, plate 2340).

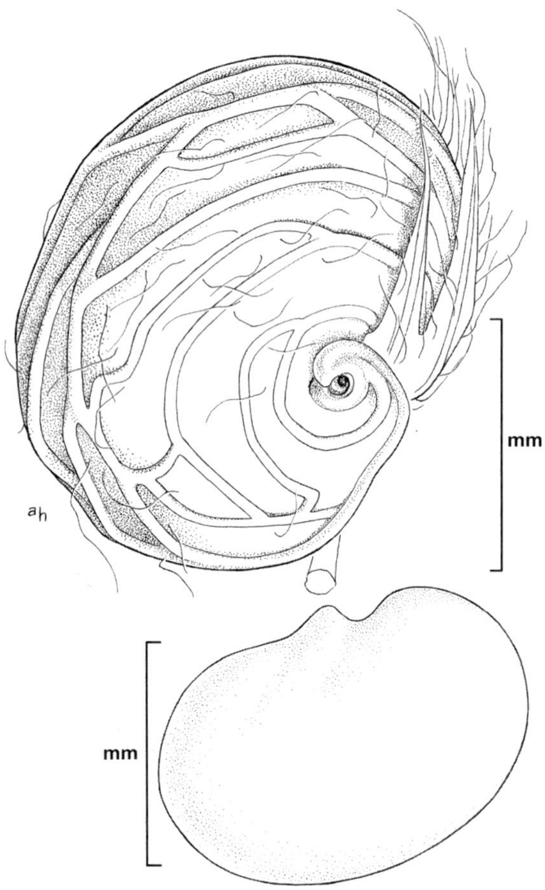

Fig. 49-2. *Medicago lupulina.* **Top:** Fruit. **Bottom:** Seed. (Based on E. Small M20, DAO; artist: A. Hanes.)

Fig. 49-4. A perennial form ("f. *perennans*" of authors) of *Medicago lupulina*. Notice the well-developed root (arrow).

However, The Linnaean Plant Name Typification Project (cf. Jarvis 2007) (http://www.nhm.ac.uk/research-curation/projects/linnaean-typification/; current June 2010) provided the following information: "Reference: *Species Plantarum* 2: 779 (1753). Provenance: "Habitat in Europae pratis." Type: Lectotype : Herb. Linn. No. 933.10 (LINN) Designated by: Ali in Taxon 17: 540 (1968)... Comments: Although Heyn (in Bull. Res. Council Israel, Sect. D, Bot. 7: 162. 1959) discussed some of the original material, she did not designate a lectotype. Despite her comments, 933.10 (LINN) is linked with the *Species Plantarum* account by the number ("7") of this species written on the sheet by Linnaeus. It is original material for the name and was designated as the type by Ali."

Synonymy

Homotypic

Medica lupulina (L.) Scop., Fl. Carniol., ed. 2., 2: 88 (1772).

Medicula lupulina (L.) Medik., Vorles. Churpfaelz. Phys.-Öcon. Ges. 2: 386 (1787).

Trifolium lupulinum Savi, Fl. Pis. 2: 171 (1798).

Lupulina aurata Noulet, Fl. Bass. Sous-Pyren. 157 (1837).

Melilotus lupulinus (L.) Trautv., Bull. Sci. Acad. Imp. Sci. St. Petersb. 8: 271 (1841), non Lamarck 1779.

Lupularia parviflora Opiz, Seznam 61 (1852), nom. illegit.

Heterotypic

M. apennina J. Woods, Tourist's Flora : 84 (1850).

M. breviflora Gilib., Fl. Lit. Inch 2: 95 ("1781", 1782).

M. canescens Menyharth, Kal. Vid. Noven. 61 (1877).

M. corymbifera W.L.E. Schmidt ex Schltdl., Linnaea 4: 74 (1829).

M. lupulina L. var. *corymbosa* Ser. in DC., Prodr. 2: 172 (1825).

M. lupulina L. f. *corymbosa* (Ser.) Urb., Verh. Bot. Vereins Prov. Brandenburg 15: 52 (1873).

M. cupaniana Guss., Fl. Sicul. Syn. 2: 362 (1844).

M. lupulina L. var. *cupaniana* (Guss.) Boiss., Fl. Orient. 2: 105 (1872).

M. lupulina L. subsp. *cupaniana* (Guss.) Nyman, Consp. Fl. Eur. 170 (1878).

M. lupulina L. proles *cupaniana* (Guss.) Rouy in Rouy & Fouc., Fl. Fr. 5: 9 (1899).

M. lupulina L. var. *dzumgalensis* Ubuk., Fl. Kirgizkoi SSR 7: 156 (1957).

M. lupulina L. subsp. *eurasiatica* Braun-Blanq. in Bull. Soc. Bot. France 77: 288 (1930), nom. inval.

M. lupulina L. var. *glandulosa* W.D.J. Koch in Sturm, Deutschl. Fl. 5: 324 (1806).

M. lupulina L. var. *glanduligera* Neum. in Neum. & Ahlfv., Sveriges Fl. 331 (1901).

M. lupulina L. var. *integristipula* Rouy & Foucaux (recognized in Kozuharov 1965).

M. lupulina L. subsp. *jalasii* Rothm., Feddes Reprt. Spec. Nov. Regni Veg. 67: 6 (1963).

M. leiocarpa Guss., Syn. Fl. Sic. 2: 362 (1844).

M. lupulina L. var. *cupiana* Guss. f. *leiocarpa* Urb., Verh. Bot. Vereins Prov. Brandenburg 15: 52 (1873).

M. mniocarpa Wallr. ex Ser. in DC., Prodr. 2: 172 (1825). (Mentioned in synonymy.)

Trigonella mniocarpa Wallr. ex Ser. in DC., Prodr. 2: 172 (1825). (Mentioned in synonymy.)

M. lupulina L. f. *pedunculata* Font-Quer (Type: Font-Quer, P. 25-V-1928, in herbosis montis Kalaa, alt. 1000 m., Morocco (MP), type viewed online at http://www.aluka.org; current June 2010). (Status?)

M. lupulina L. var. *perennans* Hal., Consp. Fl. Gr. 1: 362 (1901).

M. lupulina L. var. *perennans* Grossheim, Fl. Kavk. 2: 260 (1930), nom. superf.

M. lupulina L. var. *polystachya* Ser. in DC., Prodr. 2: 172 (1825).

M. lupulina L. f. *polystachya* (Ser.) Urb., Verh. Bot. Vereins Prov. Brandenburg 15: 52 (1873).

M. reniformis Dulac, Fl. Hautes-Pyrenées 278 (1867), nom. illeg.

M. stipularis Wallr., Erst. Beitr. Fl. Hercyn. 260 (1840).

M. lupulina L. var. *stipularis* (Wallr.) Hallier et al. in Koch, Syn. Fl. Germ. Helv., ed. 3. 534 (1891).

M. lupulina L. var. *typica* Urb. f. *stipularis* (Wallr.) Urb., Verh. Bot. Vereins Prov. Brandenburg 15: 52 (1873).

M. lupulina L. var. *typica* Urb., Verh. Bot. Vereins Prov. Brandenburg 15: 52 (1873), nom. inadmiss.

M. lupulina L. var. *unguiculata* Ser. in DC., Prodr. 2: 172 (1825).

M. lupulina L. f. *unguiculata* (Ser.) Urb., Verh. Bot. Vereins Prov. Brandenburg 15: 52 (1873).

M. lupulina L. var. *vassilczennkoi* Ubuk., Fl. Kirgizkoi SSR 7: 156 (1957).

M. lupulina L. var. *vulgaris* Koch, Syn. Fl. Germ. Helv. 161 (1835), nom. inadmiss.

M. willdenowii Boenninghausen, Prodr. Fl. Monast. Westphal. 161 (1824), non Mérat 1812.

M. lupulina L. var. *willdenowiana* (sic) (Boenninghausen) Koch, Syn. Fl. Germ. Helv. 161 (1835).

M. lupulina L. var. *willdenowii* (Boenninghausen) Ascherson, Fl. Brandenburg 1: 139 (1867), nom. illegit.

M. lupulina L. var. *typica* Urb. f. *willdenowii* (Boenninghausen) Urb., Verh. Bot. Vereins Prov. Brandenburg 15: 52 (1873).

M. lupulina L. subsp. *willdenowiana* (sic) (Boenninghausen) Soják, Sborn. Nár. Mus. v Prace, Rada B, Prír. Vedy 19 (1963: 155 (1963).

Medicula lupulina (L.) Medik. subsp. *willdenowiana* (sic) (Boenninghausen) Holub, Folia Geobot. Phytotax. 18(2): 204 (1983).

M. willdenowii Mérat, Nouv. Fl. Env. Paris, ed 1: 296 (1812).

M. lupulina L. subsp. *willdenowii* (Mérat) Arcangeli, Comp. Fl. Ital. 159 (1882).

In the general collection at G in 1993, I observed "*Medicago othrysiensis* sp. nov. aff. lupulinae. sommet de l'Othrys", a collection from Herbiers Boissier et Barbey-Boissier, "series n'ayant pas servi à la redaction du Flora Orientalis." This was merely a very scrappy, dwarfed specimen of *M. lupulina*. The name does not seem to have been published.

Vernacular Names

English: Black medic (medic is also spelled medick), black medic clover, black seed (blackseed), hop clover, hop medic, non(e)such, trefoil, yellow trefoil. Most of these names are discussed by Spedding and Diekmahns (1972). According to Pieters (1927), "this plant is called yellow clover or hop lucerne in Germany and trefoil, yellow trefoil, medic and nonesuch in England." Quinlivan et al. (1974) noted that "black medic (*Medicago lupulina*), under the name English trefoil, was the first medic offered for sale in the colony of Western Australia." Clarkson (1977) noted that the name English trefoil persists in Australia.

French: Bujoline, la luzerne lupuline, lupine, lupin noir, lupuline, luzerne houblon, luzerne-houblon, luzerne houblonnée, luzerne lupuline, mignonnette, minette, minette dorée, trèfle jaune, trèfle noir, triolet.

Geographical Distribution

(N = native, I = introduced, U = uncertain status)

Africa: Algeria (N), Azores (Portugal; N), Canary Islands (Spain; N), Egypt (N), Ethiopia (N), Kenya (U), Libya (N), Madeira Islands (Portugal; N), Mauritius (I), Morocco (N), Reunion (France; I), Somalia (U), South Africa (U), Tanganyika (N?), Tanzania (U), Tunisia (N).

Asia: Afghanistan (N), Arabia (N), Armenia (N), Azerbaijan (N), China (N), Cyprus (N), Dagestan (N), Gruziya (N), India (N?, I?), Iran (N), Iraq (N), Israel (N), Jordan (N), Korea (?), Japan (I), Kazakhstan (N?), Kyrgyzstan (N?), Lebanon (N), Mongolia (N), Nepal (N), Pakistan (N), Philippines (I), Russia (N), Saudi Arabia (N), Syria (N), Taiwan (I), Tajikistan (N?), Tibet (N), Turkey (N), Turkmenistan (N?), Uzbekistan (N?).

Australasia: Australia (I), New Zealand (I), Tasmania (I).

Europe: Albania (N), Austria (N), Belarus (N), Balearic Islands (Spain; N), Belarus (N), Belgium (N), Bulgaria (N), Czechoslovakia (N), Denmark (N), Estonia (N), Finland (N), France (N), Germany (N), Greece (including Crete, East Aegean Islands, and Karpathos; N), Hungary (N), Ireland (N), Italy (including Sardinia and Sicily; N), Latvia (N), Liechtenstein (N), Lithuania (N), Luxembourg (N), Malta (N), Moldova (N), Netherlands (N), Norway (N), Poland (N), Portugal (N), Romania (N), Russia (N), Spain (N), Sweden (N), Switzerland (N), Turkey (N), Ukraine (N), United Kingdom (N), Yugoslavia (N).

Northern America: Canada (I), Mexico (I), USA (including Alaska; I).

Southern America: Argentina (I), Bahamas (I), Bermuda (Britain; I), Bolivia (I), Brazil (I), Chile (I), Columbia (I), Dominican Republic (I), Guatemala (I), Haiti (I), Peru (I), Puerto Rico (I).

Pacific: Hawaii (I).

Medicago lupulina is a native of Europe, temperate Asia, North Africa, and perhaps tropical Africa, but is a widely established weed throughout temperate and subtropical regions.

Description

Annual, biennial, or short-lived perennial herb, prostrate, decumbent, ascending, or erect, stems branched (or often not when crowded by other plants), (5)20–100 cm long, twigs and leaves glabrescent to very pubescent, simple appressed hairs present, sometimes also gland-tipped hairs. Leaflets 6–20 mm long, 4–18 mm wide, elliptical, ovate, rhomboidal, suborbicular, or obovate, distal half with serrate margins. Stipules lanceolate, entire or irregularly toothed. Inflorescence of (5)15–50 flowers in a short, cylindrical head up to 15 mm long, peduncles much longer than subtending petiole (and elongating in fruit). Corolla yellow, 2–4 mm long, about twice length of calyx or shorter. Calyx 1.5–2.5 mm long, campanulate, pubescent with simple hairs (gland-tipped hairs sometimes also present), teeth unequal, subulate, about equal to or longer than tube. Young pod contracted within calyx, covered with simple (and sometimes also gland-tipped) hairs. Fruits one-seeded, black (or occasionally ash-gray) at maturity, up to 3 mm long, 1.5 mm wide, rough-ridged, slightly curved (sometimes forming almost a single coil), the tip slightly coiled, the pod face with 3–6 strongly curved veins, venation arcuate-reticulate (somewhat anastomosing veins starting at ventral suture and running obliquely to dorsal suture). Seeds oval to reniform, about 1.5–2 mm long, 1–1.5 mm wide, yellow to olive-green or brown, radicle longer than half length of seed.

Identification Hints

The only species similar to *M. lupulina* is the rare *M. secundiflora*, which is easily distinguished by its secund inflorescence or infructescence (flowers or fruits on one side of the subtending axis).

Habitat and Environmental Adaptation

Habitat

Medicago lupulina is naturally adapted to pastures. It is a common weed, often found in lawns, on riverbanks, and in disturbed areas such as roadsides, railway embankments, and wastelands. In Greece, Thanopoulos (2007) recorded its presence in shrub associations, open deciduous woodlands, both wet and dry grasslands, in fallow fields, and as a weed of orchards and ruderal sites, from sea level to 2300 m, occasionally growing in shade.

Phenology

There are annual, biennial, and short-lived perennial forms of *M. lupulina* (Turkington and Cavers 1979). The species appears frequently to be perennial in Mediterranean mountain habitats and in much of Greece (P. Lassen, personal communication), but is predominantly annual in ruderal habitats. Under greenhouse conditions, the plants can be maintained indefinitely, unlike obligate annuals that normally die after setting seed (Lesins and Lesins 1979). Flowering date can vary considerably (Lammerink 1968). I cultivated 100 collections of *M. lupulina* in a greenhouse in Ottawa and found a very wide range of flowering times. Vitkus (1987) observed that plants sown in early spring in Lithuania all produced seeds and died the same year; plants sown in late summer remained vegetative or reached the early flowering stage by fall, overwintered, and produced seeds in the second year; and plants sown in late summer produced more vegetative growth in the second year before producing seeds, compared with the plants sown in early spring in the first year.

Soil

Medicago lupulina grows on a wide variety of soils, with some preference for calcareous substrates (Spedding and Diekmahns 1972). The reported pH of soils occupied ranges from 4.8 to 7.8 (Turkington and Cavers 1979). Samuel Roberts Noble Foundation (1959) for Oklahoma recommended that *M. lupulina* be grown in medium- and fine-textured soils, but not sandy soils, at a pH of 6.2 or above.

Temperature

Medicago lupulina thrives in temperate conditions and withstands subtropical climates. Samuel Roberts Noble Foundation (1955) noted that the species was not severely damaged by a hard freeze. Based on growth chamber experiments, Brandsaeter et al. (2002) observed that *M. lupulina* cultivar 'Virgo Pajberg' showed medium freezing resistance. Krall et al. (1996) found that cultivar 'George', examined for ability to self-regenerate in Wyoming, failed to survive the winter at the three locations tested. Stepanova et al. (1996) reported that the self-regeneration ability of some accessions of *M. lupulina* after overwintering in Russia was 80% or more, although other accessions performed much more poorly.

Water

This species is common in moist to dry pastures. It appears to tolerate periodic drought and has been reported to prefer dry soils in England (Spedding and Diekmahns 1972).

Biotic Relationships

Pollination

This species is largely an inbreeder, although some populations exhibit outcrossing (Lammerink 1968). Olivieri et al. (1991) characterized *M. lupulina* as facultatively selfing. While the flowers are visited by bees and other pollinating insects, in most cases this makes no difference for seed set (Turkington and Cavers 1979). Small (1988) noted that the relatively low pollen to ovule ratio (320) and small flowers of the species are consistent with *M. lupulina* being largely self-pollinated and mainly an inbreeder.

Dispersal

The single-seeded fruits of *M. lupulina* lack obvious features that facilitate dispersal and tend to fall in the vicinity of the maternal plant (Sidhu and Cavers 1977). However, the fact that the species is one of the world's common weeds indicates that it has the ability to travel considerable distances. Birds and grazing animals are thought to be the principal natural distributors, with distribution also occurring as contaminants in crops and seeds distributed by people. The seeds can pass through the digestive system of grazing animals (Dunbier 1972), including sheep (Lammerink 1968), which can distribute the seeds long distances and, since the seeds can float, they may occasionally be water-distributed (Turkington and Cavers 1979).

Herbivory

Barney (1987) observed that deer and rabbits grazed cultivated *M. lupulina*. Goertzen and Small (1993) found experimentally that the density of glandular hairs on the leaves was negatively correlated with damage to the leaves by insects, demonstrating the protective function of the hairs.

Nodulation and mycorrhizae

USDA (2009) reported positive root nodulation for *M. lupulina*.

Disease resistance

Yaege and Stuteville (2002) tested the resistance of 33 annual species of *Medicago* to powdery mildew (*Erysiphe pisi*) and found that 13 of the collections of *M. lupulina* grown were susceptible to the disease whereas two showed slight resistance to it. In a comparative test of 36 annual species of *Medicago*, O'Neill and Bauchan (2000) found that 15 accessions of *M. lupulina* were all extremely susceptible to anthracnose caused by *Colletotrichum trifolii*. Stepanova et al. (1996), based on study of numerous strains of *M. lupulina* in Russia, recorded that the main disease affecting black medic is root rot caused by the *Fusarium* species *F. culmorum*, *F. oxysporum*, *F. moniliforme F. solani*, and *F. sambicium*.

Economic History, Cultivars, and Uses

Grossheim (1945) noted that "The common annual *M. lupulina* has long been cultivated in Europe, but it has not attained wide distribution... In England, *M. lupulina* was introduced into cultivation in 1659, in France about 1785; in Germany it did not gain currency before the second half of the

19th century." Bolton (1962) stated that at one time seed from *M. lupulina* was used to adulterate alfalfa and clover seed, and as a result the species acquired an unfavorable reputation.

Cultivars

'George' black medic was released by the Montana State Agricultural Experiment Station (Sims et al. 1985) and is reputed to be frost-tolerant. 'Virgo Pajberg' is a cultivar produced in Denmark. 'Arka' is a Lithuanian cultivar.

Forage

Medicago lupulina is a valued constituent of spring and early summer pastures. According to Lesins and Lesins (1979), "*M. lupulina* is sometimes added in seed mixtures for pastures. It reseeds readily and persists even under severe grazing conditions. Its yielding capacity, however, is low." Grossheim (1945), in reference to conditions in the USSR, commented that "The crude protein is very high, exceeding even that of wild forms of *M. sativa* and related species. The perennial habit of var. *perennans* also constitutes an important attribute of this form as regards feed value... This variety can be recommended for undersowing in exhausted grass stands even without previous soil cultivation... and its cultivation can be recommended without hesitation for those places in the USSR where the cultivation of *M. sativa* and other legumes is impossible for climatic and other reasons." Grossheim also noted that since *M. lupulina* tends to lodge in pure stands, it is often used in mixtures with more erect legumes. For England, Spedding and Diekmahns (1972) stated that "Although black medick has been regarded by some as a valuable herbage plant...it has not been generally recommended except for inclusion in small quantities in seed mixtures... Mixtures containing a high proportion of black medick have been grown, but have not been particularly favoured... Black medick is reported to be of no value in permanent pastures, but is useful for establishing short-duration swards on poor, shallow, sandy soils where the better clovers will not grow... However, it has been stated that it should not be sown alone on account of its weak stems... It is sometime sown in mixture with cereals for green manure... As black medick is less susceptible than red clover to clover rot (*Sclerotinia*), it can replace the latter on clover-sick land... It can be grazed by sheep early in the year, but is not very acceptable to cattle." Vitkus (1977) assessed *M. lupulina* as suitable for cultivar selection for Lithuania.

Fodder

Although rarely cut for hay, yields of 4.5–6.7 t/ha have been recorded (Wheeler 1950).

Green manure, rotation crop, ground cover

Medicago lupulina is often employed as a green manure, typically plowed under in the fall as part of a crop rotation. Pieters (1927) noted that "In Germany and France (it) has given good results when used as a companion crop to be hoed under for a crop in spring or in fall for a following grain crop." The cultivar 'George' has been recommended for use as a green manure crop in Montana (Sims et al. 1985).

Other uses (salad sprouts, ornamental, plant products, etc.)

The species is considered useful as a bee plant for production of honey. It is occasionally employed to attract butterflies.

Chromosome Number

$2n$ = 16: Ghimpu 1928, 1929*a*, 1930; Fryer 1930; Tishler 1934; Cooper 1935; Heiser and Whitaker 1948; Pólya 1948; Heyn 1956; Löve and Löve 1956; Mulligan 1957; Clement 1962; Gadella and Kliphuis 1966, 1968, 1970*b*; Bolkhovskikh et al. 1969, p. 308 (2 reports); Podlech and Dieterle 1969; Majovsky et al. 1970*a*; Fernandes and Santo 1971; Kozuharov et al. 1972; Löve and Kjellqvist 1974; Kozuharov et al. 1975; Labadie 1976; Sareen and Trehan 1976, 1977; Fernandes et al. 1977; Fernandes and Queirós 1978; Natarajan 1978; Lesins and Lesins 1979; Abdelguerfi and Guittoneau 1979; Kliphuis and Wieffering 1979; Loon and Kieft 1980; Loone and Setten 1982; Löve and Löve 1982; Pogan et al. 1982; Agarwal and Gupta 1983*b*; Mulligan 1984; Queirós and Ormonde 1984; Šopova et al. 1983; Zhu et al. 1985; Montmollin 1986; Abdelguerfi et al. 1988*b*; Galland 1988; Hrušovská-Osuská 1988; Baltisberger and Charpin 1989; Schlarbaum et al. 1989; Sokolovskaya et al. 1989; Yan et al. 1989; Baltisberger 1990; Karshibaev 1992; Kumari and Bir 1990; Luque and Díaz Lifante 1991; Měsiček and Soják 1992; Vitek et al. 1992; Al-Bermani et al. 1993; Khatoon and Ali 1993; Dempsey et al. 1994; Jahan et al. 1994; Gadnidze et al. 1998; Lövkvist and Hultgård 1999; Khatoon and Ali 2006.

$2n$ = 18: Xu et al. 1992; Zhang et al. 1993 ($2n = 16 + 2B$).

$2n$ = 16, 32: Heyn 1956; Mariani 1963; Bolkhovskikh et al. 1969, p. 308 (3 reports); Mizianty et al. 1981; Agarwal and Gupta 1983*b*; Ward 1983; Semerenko 1989; Semerenko and Shvets 1989; Bir and Kumari 1979.

$2n$ = 30: Colombo et al. 1987 (as *M. cupianiana*).

$2n$ = 32: Tschechow 1932; Heyn 1956; Kodama 1989.

Chromosome counts of $2n = 32$ are quite rare and are confined to Europe, whereas counts of $2n = 16$ are very common and have been found in Europe and elsewhere. The $2n$ = 30 count (based on an Italian collection) is unique and would benefit from additional study. Bilis (2003) discussed doubling the chromosome number of a diploid ($2n = 16$) cultivar for breeding purposes.

Genetic Relationships and Gene Transferability

Intraspecific

No barriers to hybridization within the species are known, although presumably crossing diploids and tetraploids will produce sterile triploids. Artificial hybridization of different populations has not been reported.

Interspecific

There are several reports of hybrids between *M. lupulina* and *M. sativa* (Southworth 1914, 1928; Fryer 1930; Schröck 1943; Atwood and Grunn 1952) — all of these alleged successful crosses are doubtful (cf. Lesins and Lesins 1979). Lesins (1952) was unable to hybridize *M. lupulina* that was artificially doubled in chromosome number to $2n = 32$ with *M. sativa* carrying the same number of chromosomes. *Medicago* ×*revolii* H.J. Coste & Soulié (in Le Monde des Plantes 22(129): 6–7) is a putative natural hybrid between *M. lupulina* and *M. sativa* subsp. *falcata*. However, the description of the fruits as essentially identical to those of the latter make it very unlikely that the specimen described was really as alleged. Lesins and Lesins (1979) reported that they were unable to cross *M. lupulina* with *M. platycarpos* or with *M. secundiflora*, its apparent closest relative (see Chapter 10 for evidence that the two species are not closely related).

Exchange by advanced biotechnology

Mazin et al. (1994) produced somatic hybrids between *M. lupulina* and *M. sativa*.

Conservation Status

This species is extremely widespread and is not in need of conservation measures.

Germplasm Sources

At least 800 accessions are available in germplasm institutions (see Chapter 3).

Research Problems and Priorities

Medicago lupulina has been considered to have minor agricultural value for centuries and has been cultivated to a limited extent in many areas of the world. Its values lie in its usefulness for ground cover and as a special forage in areas that are too stressful for alfalfa — the species definitely has potential for expanded development for these applications. Although this cosmopolitan weed is nutritious, productive, and resistant to grazing, it is not known to be particularly well suited to dryland agriculture. However, *M. lupulina* possesses an enormous range of adaptive variation throughout the world and a search for perennial drought-resistant forms would be worthwhile.

Representative Collections

Afghanistan: Furse, P. 7949, Paghman, 15 miles northwest of Kabul, alt. 8000 feet, irrigated highly shaded wood of willows and poplars, 21-VI-1966, in flower & fruit (K). Griffith, Herbarium of, the late East India company No. 1486 (distributed 1862-63) (P, L). **Algeria:** Bové, N., Alger, bords de l'Arach, V-1837, in flower & fruit (K, P). Maire, A. HG1921, M. Timhadit, rochers humides de calcaires marneux, alt. 1800–1900 m, 26-VII-1921, in flower (P). **Argentina:** Fischer, W. 76, vicinty of General Roca, Rio Negro, alt. 250–360 m, IX-1914-II-1915, in flower (BM). Tirel 24, prov. de Cordoba, Barrio Alto Palermo, 10-XII-1973, in flower & early fruit (P). Tirel 279, prov. de Cordoba, ciudad de America, le long d'un ruisseau, 28-IX-1974, in flower & early fruit (P). **Australia:** Hubbard, C.E. 8245, New South Wales, Glen Innes, alt. 3520 feet, weed on roadside, "8.4"-1931, in flower & fruit (K). Tilden, J.E. 792, Victoria, Point Lonsdale, X-XI-1912, in flower & fruit (BM). **Azores (Portugal):** Goncalves, B. 2939, Graciosa, Santa Cruz, 7-VII-1971, in flower & fruit (BM). **Bermuda Islands (British territory):** Collins, F.S. 207, Flatts, roadside 15-VIII-1913, in flower & early fruit (P). Collins, F.S. 417, South Road, common, 29-XI-1915, in flower & fruit (P). **Bolivia:** Bang, A.M. 189, Vic, La Paz, alt. 10 000 feet, 1890, in flower (BM). **Britain:** McCollun Webster, M. 1367, South. Devon, Bradley Wool Mill, Newton Abbot, introduced with wool, 19-XI-1957, in flower & fruit (K). **Bulgaria:** Kojoucharov, S., Bachkovo, Mt. Rodopi-Centrales, 21-VI-1959, in flower & early fruit (K). **Canada:** Michaux, André, Herbier de l'Amérique septentrionale, Malbaye et Baye St. Paul (P). **Canary Islands (Spain):** Bourgeau, E. 765, Insul Palma-la Caldera, 3-VIII-1845, in flower & young fruit (P). **China:** Bartholomew, B. & Boufford, D.E. 1929, Yunnan prov., grounds of the Kunming Institute of Botany, Heilongtan, Kunming municipality, growing in a disturbed area, alt. ca. 2000 m, 16-VIII-1984, in flower & very early fruit (BM). Fang, W.P. 4254, Szechuan, Sungpan Hsien, on grassy slope, 17-VIII-1928, in flower (P). Tsang, W.T. 23650, Hunan, Yi Chang district, P'ing T'ou Shan T'ang Wan village, meadow, sandy soil, fairly common, 17-30-IV-193(?), in flower & very young fruit (P). **Cyprus:** Casey, E.C. 864, Prodromos, alt. 4500 feet, hillside, 2-VII-1949, in flower & fruit (K). Merton, L.F.H. 2343, Kakopetria, alt. 24 000 feet, streamsides and by irrigation ditches, 24-V-1955, in flower & fruit (K). **Czechoslovakia:** Nitka, J., Bohemia centralis, distr. Praha-vychod, in agro prope vicum Dobrejovice haud procul ab oppido Ricany, alt. 300 m, 8-IX-1960, in flower & early fruit, Plantae Cechoslovacae Exsiccatae Bot. Mus. Nat. Prague Centuria II No. 168 (BM). **Denmark:** Dahl, H.S. D.45b, Nordsjaelland, Lynaes, 30-VI-1942, in flower (BM). Nielsen, I. & Pedersen, P., Gjesing, between Grenaa and Randers, south-exposed slopes, 30-V-1968, Flora Jutlandica Exsiccata No. 418 (BM, L). **Dominican Republic:** Liogier A.H. 12528, Sierra de Neiba, cloud forest, near La Doscientos army post, roadside, alt. 1750–1850 m, 5-7-IX-1968, in fruit (P). Liogier A.H. 16754, Los Arroyos, along the international highway from Pedernales to Duvergé, wet woods on roadsides, fairly common, alt. 1600 m, 8-XI-1969, in flower (P). **Egypt:** Simpson, N.D. 2387, north of Kharge Oasis, bank on field margin, 24-I-1924, in flower & fruit (K). **Ethiopia:** Hall, J.P. 50, Wollo province, Dessie, alt. 2600 m, 15-VIII-1946, in flower (BM). Robertson, S.A. 1186, Eritrea, near Ad Teclesan (15°34′N, 38°52′E), alt. 2200 m, Eucalyptus plantation with stream, lush vegetation for area, "2.10"-1969, in flower & fruit (K). Schweinfurth, G. & Riva,

D. 796, Flora Abyssiniae septentrionalis (colonia Eritrea), environs de Saganeïti, alt. 2200 m, 7-9-III-1892, in flower & fruit (P). **Finland:** Nordström, A. 892, Åland. Soltunga, Centre, Eriksson's Manor, by road to Skage, field margin, I-VII-1981, in flower (BM). Strom, Inga, Nylandia, par Tenada, Lappvik, in arenosa litore, 12-VII-1907, Planta Finlandiae Exsiccatae e Museo botanico Universitatis Helsingforsiensis distributae No. 758 (LE, P). **France:** Brachet, F., Hautes-Alpes Remollon, prairies fraîches à la Gravière, 10-VI-1901, in flower & fruit, Société Rochelaise 1901 No. 4707 (P). Duffour, Ch., Côte-d'Or, Beaune, champs, 13-VIII-1917, in flower & fruit, Société Française 197 Exsicc. Ch. Duffour No. 2350 (P). **Greece:** Guiol, F. 1703, Euboae – Mt. Dirphys, 8-VI-1930, in flower & fruit (BM). **Haiti:** Ekman, E.L. 1154, Massif de la Salle, Horve Tranchant, alt. ca. 1800 m, on limestone slopes, 28-VII-1924, in flower & early fruit (K). **India:** Drummond, J.R. 21795, Punjab (P). Polunin, O. 56/357, Kashmir, Tang marg, alt. 6000 feet, waste ground among cultivation, 17-VIII-1956, in flower & fruit (BM). **Iran:** Jardine, N. 793, Arak, Darun, (50°24′E, 32°57′N), streamside, 25-VII-1963, in flower & fruit (E). Lindsay, N. 342, Sultanabad near Kasvin, alt. 5000 feet, grass fields, V-1935, in flower & early fruit (BM). Lindsay, N. 925, prov. Guilan, Lahijan, dryish places, very common, V-VI-1936, in flower & very young fruit (BM). **Iraq:** Gillett 8036, Shaqlawa, alt. 900 m, irrigated fields and gardens, 8-V-1919, in flower & early fruit (K). Rawi, A. & Serhang, I. 26555, Pushtashan, northeast of Rania, lower slopes of Quandil Range, alt. ca. 1050 m, shade in Quercus forest near stream, 17-VIII-1958, in flower & fruit (K). **Israel:** Eig., A. & Zohary, M., Upper Galilee, between wadi Abu Ali & Gebel el 'Arus, 4-VI-1926, in flower & fruit (BM). **Israel/Syria (Golan):** Eig, A., "Flora syriaca", Mt. Hermon, Wadi Shibbah, near river, alt. ca. 1200 m, 20-VII-1924, in flower & fruit (HUJ). **Italy:** Jury, S.L. et al. 6503, Reg. Abruzzi, Prov. L'Aquila, Gran Sasso d'Italia, campo Imperatore, near top of ski lift (42°28′N, 13°35′E), alt. 2000 m, dry mountain pasture, 18-VII-1985, in flower (BM). **Japan:** Faurie 512, Tokyo, 25-VI-1887, fruit (P). Ichikawa, K., Hakozaki near Fukuoka City, Fukuokaken, Kyusho Island, alt. 3 m, 10-V-1928, in flower & early fruit, Tanaka Distribution of Japanese Plants No. 13 (P, L). **Jordan:** Oran, S. 404, Tafila, 10-XI-1974, in flower & fruit (BM). **Kenya:** Roberts, S.A.L. H181/55, New Nasha, Koromgo Farm (0°44′S, 36°16′E), alt. 6200 feet, 26-IX-1955, in flower & early fruit (K). **Latvia:** Syreysczikow, S. 3337, ad viam ferream prope Stationem Assari, 6-VI-1910 (LE). **Lebanon:** Bertschinger, A., Djebel Makmal, Westhang, alt. 1900 m., 30-VII-1937, in flower & very early fruit (G). **Libya:** Sandwith, N.Y. 2554, Cyrenaica, Bsiara, near Giovanni Berta, alt. ca. 600 m, on flat limestone rock of altiplano, 11-IV-1939, in flower & fruit (K). **Lithuania:** Dybowski, W., Niañkow, distr. Nowogródek, 1895, Flora potonica exsiccata Dr. E. Wotoszcak No. 422 (LE). **Mauritius:** Johnston, H.H., Plaines Wilhelms, alt. 1390 feet, roadside, flowering branch 10-IX-1888, fruiting branch 17-IX-1888 (K). **Mexico:** Basurto, P. and Durán, G. 301, Puebla, 6 km al NE de Huauchinango, 17-V-1979, in flower & early fruit (BM). **Mongolia:** Potanin, G.N., Mongolia borealis, circa lacus Ubsa, 1879, in flower (P). **Morocco:** Cosson, M., SE de la prov. de Demrat, Ibrahim, 3-VII-1882. **Nepal:** Schlaginweit, Herbarium of, environs of Kathmándu, alt. 5000–7000 feet, 4-8-III-1857 in flower, Catalogue No. 13042 (P). **Netherlands:** Maas Geesteranus, R.A. 4296, prov. Zuid-Holland, between Leiden and Rijnsburg, on fallow land, 4-VIII-1948, in flower & fruit (K). **New Zealand:** Bangerter, E.B. 5145, North Island, Waitemata County, Mairangi Bay, Hastings Road (36°44′S, 174°45′E), on waste ground with other ruderal species, occasional to locally frequent, 24-I-1974, in fruit (K). **Peru:** Hutchinson, P.C. & Wright, J.K. 5260, prov. Celendin, Dept. Cajamarca Marañon above Balsas, 3–4 km below summit of the road to celendin, alt. 2950 m, roadsides, common, 22-V-1964, in flower & fruit (P). **Philippines:** Merrill, E.D. 7762, Benguet subprovince, Luzon, V-1911, in flower & fruit (BM). **Poland:** Jasiewicz, A. & Kucowa, I., Polonia meridionalis districtus Miechów, prope pagum Kalina Wielka, in declivi stepposo, solo cretacei-margaceo, vegetatione ad associationem Inuletum ensifoliae accedente, 15-VI-1968, in flower & early fruit (BM). **Saudi Arabia:** Hilleoat, D. 61, ASIR, Dalaghan National Park, 30 km southeast Abha, "1.3.1981", in flower (BM). **Somaliland:** Gillett, J.B. & Watson, R.M. 23838, North Somalia, north of Erigaviol (10°48′N, 47°22′E), alt. 1900–2060 m, sunny grassy spot with shallow soil at base of cliff, locally frequent, 8-VII-1981, in flower & fruit (K). Glover, P.E. & Gilliland, H.B. 1138, Erigavo District, British Somaliland, shade in Dalo Forest glade, 7-VII-1945, in flower & early fruit (BM). **South Africa:** Repton, J.E. 4619, prov. Transvaal, District Pretoria, Eastern Sports Grounds, Loftus, alt. 4500 feet, weed in lawns, very common, 18-X-1957, in flower (K). **Spain:** Grimes, J. 145, Piscos de Europa (1°7′50″ long., 43°9′50″ lat.), 27-VII-1969, in flower & early fruit (BM). **Sweden:** Norrman, C.M., Öland, Vickleby par., St. Vickleby, roadside, 7-VI-1964, in flower (BM). **Turkey:** Huber-Morath, A., Prov. erzincan, district Kelkit, Kelkit, Bachbett, alt. 1200 m, 1-VII-1966, in flower & early fruit (G). Kukkonen, I. 8126, Trabzon, ca. 1 km north of Hamsiköy, ca. 60 km southwest of Trabzon on the road to GümüŞhane, wet meadow on a steep slope, alt. 1200 m, 23-VIII-1972, in flower & fruit (G). **Switzerland:** Ayasse, E., Genèveve, prairies artificielles, bord des champs, 1-VII-1885, in fruit, Société dauphinoise 1886 No. 4858 (P). Favrat, L. & Barbey, W., Vernayaz, Valais, 3-X-1870, in flower & fruit (K). **Taiwan:** Tanaka, T. & Shinada, W. 10971, Tomida-cho, Taihoku-shi, 25-IV-1932, in fruit (P, L). **Tanganyika:** Greenway, P.J. & Kovnuri 12527, Monduli District, Ngorongoro Crater Rim, south side, alt. 7400 feet, very local on the margins of a lawn and a foot path in a garden, 4-VII-1966, in flower & fruit (K). **Tibet:** Schlagintweit, Herbarium of, prov. Ladák, right shore of the Indus near Leh, 4-19-VII-1856, in flower & early fruit, Catalogue No. 1130 (P). Soulié, J.A., Thibet

Oriental, Ta-tsien-lou (principauté de Kiala), 23-VI-1893, in flower & fruit (P). Walton, H.J. 72, Gyangtse, VII-IX-1904, in flower & fruit (P). **Tonga:** Ilolahia, V. 70, Kukw'alofa (Tongatapu), alt. 15 feet, by Tonga Club in field, 7-VIII-1958, in flower & fruit (K). **Turkey:** Balls, E.K. B1425, Ergias Dagh, Kayserie, hard beaten waste ground on edges of cultivated field, alt. 6500 feet, 18-VI-1934, in flower & fruit (BM). Davis & Hedge D29841, prov. Coru (Artvin), Borçka-Hopa, alt. 450 m, banks above Borçka, 21-VI-1957, in flower & fruit (BM). **USA:** Abrams, L. 2738, southern California, San Bernardino Co., Lytle Creek Canyon, alt. 3000 feet, 14-VII-1902, in flower (P). Wooton, E.O., Montana, Lewiston, 12-VI-1921, in flower & early fruit (P). **USSR (Europe):** Androssow, N., prov. Charkow, distr. Sumy, Grebenniki, 2-VII-1906 (LE). Niemann, R., Distr. Novo-Ladoga, ad ripas fluv. Pascha, 14-VI-1898 (LE). Saposhinikow, J., prov. Kiew, in pratis pr. u. Uman, 22-VI-2-VII-1898, Herbarium Florae Rossicae, à Musco Botanico Academ. Imper. Sc. Petropolitanae editum Fasc. IX No. 411 (LE). **Yemen:** Wood, J.E.I. Y/75/143, Wadi Dhar near Sanáa, alt. ca. 2400 m, in an orchard, 7-V-1975, in flower & early fruit (BM). Wood, J.E.I. Y/75/681, Jebel Shibam, Menacha, alt. ca. 2500 m, frequent on bare grassy banks, by rocks, etc., 1-IX-1975, in fruit (BM). **Yugoslavia:** Duyfjes, B.E.E., Hengeveld, R. & van der Voet, C.W. 36, Dalmatia, island Lokrum near Dubrovnik, roadside, 24-V-1962, in flower (BM).

Fig. 49-5. *Medicago lupulina* painted on dinnerware. **Top:** Cup, by Royal Copenhagen, based on an illustration from Flora Danica (Oeder et al. 1761–1883). **Bottom:** "Fairy plate" by Heinrich Villeroy and Bock, based on an illustration of Cicely Mary Barker (1895–1973).

Note

Medicago lupulina has sometimes been considered to be the shamrock of Ireland, although this honor is usually given to *Trifolium dubium* Sibth. (Humphries 1992).

50. *Medicago marina* L.

M. marina L., Sp. Pl. 779 (1753), (basionym).

The epithet *marina* is based on the Latin *marinus*, of the sea, a reference to the seashore habitat of this species.

Subordinate Taxa

None recognized here.

Type Specimen

Heyn (1959: 163) stated that "the specimen 'Hort. Sicc. Cliff. 377[sic]-10' accompanied by the phrase from the Hort. Cliff. on the herbarium sheet is the holotype." The Linnaean Plant Name Typification Project (cf. Jarvis 2007) (http://www.nhm.ac.uk/research-curation/projects/linnaean-typification/; current June 2010) provided the following information:

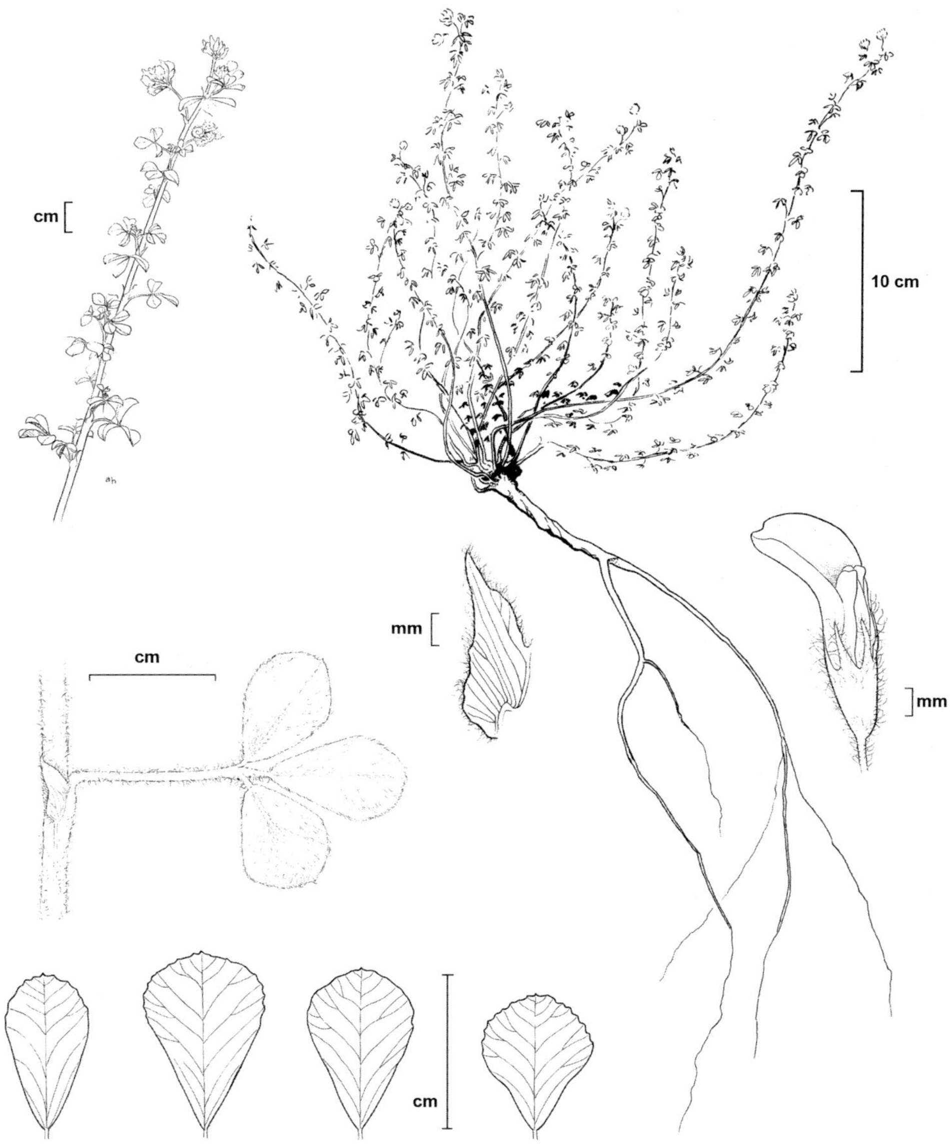

Fig. 50-1. *Medicago marina.* **Clockwise from top left:** Branch with flowers and young fruit; plant; side view of flower; leaflets; leaf attached to stem. **Center:** Stipule. (Plant based on Borza 429, US, remainder based on E. Small M639, DAO; artist: A. Hanes.)

"Provenance: 'Habitat in maris Mediterranei littoribus.' Type: Lectotype : Herb. Clifford: 378, *Medicago* 10 (BM-000646780) Designated by:... Heyn (1959)." An image of the lectotype is available at this website.

Synonymy

Homotypic

Medica marina (L.) Lam., Fl. Fr. 2: 584 (1779), nom. illeg.

Heterotypic

M. marina L. f. *genuina* Asch. & Graebn. Syn. 6(2): 404 (1906), nom. illeg.

M. marina L. f. *genuina* Grossh., Fl. Kavk. 2: 264 (1930), nom. illeg.

M. marina L. var. *inermis* Rouy in Rouy & Fouc., Fl. Fr. 5: 17 (1899).

M. marina f. *longispina* (Sen.) Maire (cited by Nègre 1956, p. 46 as a taxon not worthy of recognition)

Medicago maritima Pall. ex Georgi, Beschr. Russ. Reich. 3(4): 1197 (1800). (An early version of Index Kewensis equated this to *M. rigidula*, but this is certainly erroneous. Georgi used Linnaeus' numerical coding system, and his *M. maritima* corresponds to Linnaeus' *M. marina*.)

Medica maritima Bubani, Fl. Pyren. 2: 475 (1900), nom. illeg.

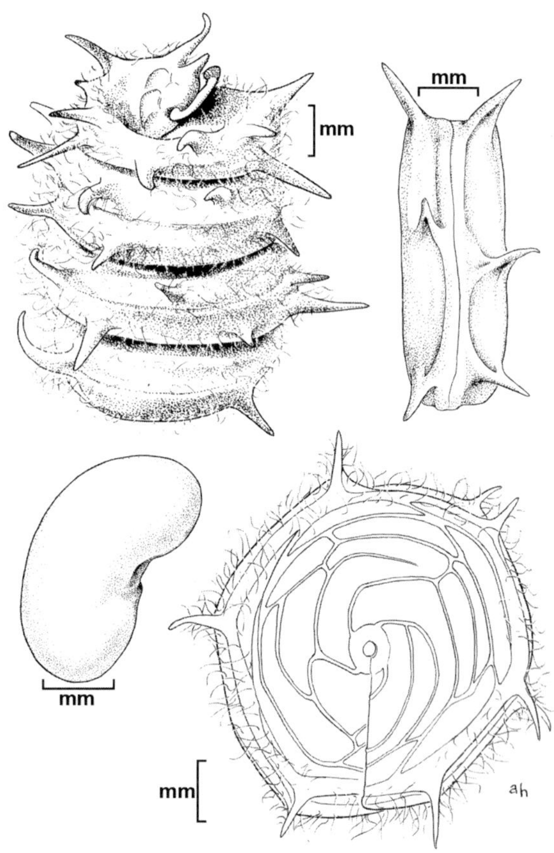

Fig. 50-2. *Medicago marina.* **Clockwise from top left:** Fruit; edge of coil; fruit coil face showing venation; seed. (Based on E. Small M110, DAO; artist: A. Hanes.)

Medica tomentosa Moench, Meth. Suppl. 38 (1802), nom. illeg.

M. marina L. var. *tuberculata* Rouy in Rouy & Fouc., Fl. Fr. 5: 17 (1899).

Medicago marina L. f. *tuberculata* (Rouy) Asch. & Graebn., Syn. 6(2): 404 (1906).

The following is an alleged hybrid (*M. littoralis* × *M. marina*) recognized by Léveillé, supposedly growing on a beach between the parents:

M. ×*sabulensis* H. Lév., Repert. Spec. Nov. Regni Veg. 7: 200 (1909), and (cited as "*M. sabolensis* H. Lév.") in Bull. Acad. Int. Géogr. Bot. 19: 212 (1909).

Vernacular Names

English: Coastal medic, sea alfalfa, sea medic (medic is also spelled medick), marine alfalfa (Bolton 1962).

French: La luzerne marine, luzerne maritime, luzerne des dunes, luzerne des sables, minette des sables.

Geographical Distribution

(N = native, I = introduced, U = uncertain status)

Africa: Algeria (N), Canary Islands (Spain; N), Egypt (N), Libya (N), Morocco (N), Tunisia (N).

Asia: Cyprus (N), Israel (N), Jordan (N), Lebanon (N), Russia (N), Sinai (Egypt; N), Syria (N), Turkey (N).

Europe: Albania (N), Bulgaria (N), France (including

Fig. 50-3. *Medicago marina.* From Hallier (1885, vol. 23, plate 2337).

Corsica; N), Gruziya (N), Italy (including Sardinia, Sicily; N), Greece (including Crete, Karpathos, East Aegean Islands; N), Malta (N), Portugal (N), Romania (N), Spain (including Balearic Islands; N), Turkey (N), Ukraine (N), Yugoslavia (N).

Medicago marina occurs on Mediterranean and Black Sea shores and on the Atlantic coasts of Iberia and France.

Description

Perennial herb, stems arising from a crown ("woody rootstock") or from a spreading rhizome, prostrate, decumbent, or ascending, 10–60 cm long, root stout, often taprooted. Stems and leaves densely covered with simple, whitish hairs, producing a gray appearance. Leaflets thick, 8–16 mm long, 5–11 mm wide, obovate, the distal margin serrate (but not visible because of dense hairs). Stipules ovate-lanceolate, entire or obscurely toothed at base. Inflorescence of 5–16 flowers, flowering peduncle shorter or about equal to subtending petiole, fruiting peduncle often longer than subtending petiole. Flowers 6–10 mm long, corolla lemon yellow or bright yellow, about twice as long as calyx, calyx 2–4 mm, densely appressed-pubescent, the teeth shorter, equal, or longer than length of tube. Young fruit protrudes from opening in calyx tube, then turns sideways. Mature pod yellowish-gray to dark gray, covered with matted hairs, cylindrical (truncate at both ends) or occasionally discoid, spiny or tubercled (rarely almost spineless), the 2–4 coils not tightly appressed, 4–6 mm wide, lacking or with a small central (axial) opening. Coil face with approximately six veins arising from ventral suture, bending slightly, then anastomosing in outer portion of coil face and running almost parallel to lateral vein. Shallow furrow between lateral vein and dorsal suture. Dorsal suture spines conical, grooved at base, slightly oblique to coil face. Seeds 1–2 per coil, yellowish or brownish, 3–4 mm long, ca. 2 mm wide, smooth, separated by thin, transverse, spongy fruit septa, radicle less than half seed length.

Identification Hints

This common Old World seashore plant is easily recognized by the dense white tomentum over most parts of the plant and is unlikely to be confused with any other species of *Medicago*.

Habitat and Environmental Adaptation

Habitat

Medicago marina is almost always in loose sand very close to the sea (rarely, it is found on river banks or fields near the sea). Thanopoulos (2007) noted that in Greece it also occurs in dune systems, often playing a pioneer role. The only other species of *Medicago* likely to be found in similar circumstances are *M. littoralis* and *M. italica*, although both tend to be in somewhat less sandy soils than *M. marina*. The species has occasionally been collected on shingle or sea cliffs. *Medicago marina* has been characterized as a "psammo-xero-halophyte" since it grows in sand and has resistance both to salt and desiccation, the chief stresses of its habitats.

Phenology

The species flowers in the spring in its native area, often from February to June (depending on local climate).

Soil

Medicago marina virtually always occupies loose, seaside sand. Given this habitat, it is not surprising that it is salt-tolerant (Baburina and Shevyakova 1987, 1995; McCoy 1987). Lieth and Mochtchenko (2003) listed it in their halophyte database.

Temperature

This species occurs on Old World seashores and so temperatures are moderated by the sea. Lesins and Lesins (1979) found that *M. marina* overwintered in Brandon, Manitoba, which has very cold winters, but showed little growth during the following summer.

Water

Medicago marina occurs in well-drained sand. Its very hairy shoots suggest that the plant restricts transpiration and so is adapted to dry circumstances. Indeed, Jeanrenaud (1975) found that *M. marina* leaves were quite resistant to water loss. Winter et al. (1976) found that a small closure of the stomata at midday decreased transpiration in *M. marina*. Jeanrenaud (1974) noted that when growing in sand dunes, *M. marina* produces extensive root systems to reach water supplies.

Biotic Relationships

Pollination

Olivieri et al. (1991) characterized *M. marina* as selfing (referring to unpublished work by Olivieri), while characterizing most of the other perennial species of the genus as outcrossing.

Small (1988) noted that the relatively large flowers and pollen to ovule ratio (465) are consistent with *M. marina* being at least partly cross-pollinated, like other perennial species of *Medicago*. Pías and Guitián (2001) also examined the pollen to ovule ratio for this species, characterizing it as exemplifying "facultative xenogamy". Silici and Gökceoglu (2007) did not report pollen of *M. marina* in honey samples examined from Turkey, although pollen of the species was available in some sampled localities, suggesting that pollinators were not visiting the flowers.

Dispersal

The spiny fruits could be spread in animal fur, although the spines are generally not hooked. Given that the species is almost completely confined to sandy seashores, it is more likely that the indehiscent fruits are spread by wind and the spines act to anchor the fruits in the sand.

Herbivory

Shade et al. (1975) examined the resistance of 21 species of *Medicago* to attack by larvae of the alfalfa weevil (*Hypera postica* (Gylh.)) and found that *M. marina* was very resistant to it.

Nodulation and mycorrhizae

USDA (2009) reported positive root nodulation for *M. marina*. Maremmani et al. (2003) reported that arbuscular mycorrhizae were present in *M. marina* growing in sand dunes on the coast of Italy.

Disease resistance

No information located.

Economic History, Cultivars, and Uses

This species has not attracted agricultural interest.

Cultivars

No cultivars have been bred.

Forage

Medicago marina is unsuitable for forage given its ecological restriction to sandy seashores, spiny pods, and densely hairy shoots (the spines and hair are likely to deter consumption). Grossheim (1945) commented that "the morphological properties of *M. marina* are very undesirable and the species can hardly be considered as a potential subject for cultivation." Nevertheless, *M. marina* was reported to have good value as natural forage for camels, sheep, and goats in seashore habitats of Egypt after more palatable species had been severely grazed (Egyptian National Action Program to Combat Desertification: http://www.unccd.int/actionprogrammes/africa/national/2005/egypt-eng.pdf; current June 2010).

Fodder

No record of this species being harvested as hay was located.

Green manure, rotation crop, ground cover

Its natural sea-beach habitat suggests that *M. marina* could be valuable as a soil stabilizer on sandy seashores and sand dunes, and indeed Elci (1983) proposed that *M. marina* is a useful plant for coastal soil stabilization. Bennett (1999) suggested it could be used to assist in regenerating the vegetation of salt pans.

Other uses (salad sprouts, ornamental, plant products, etc.)

Medicago marina is occasionally (but rarely) used as a garden ornamental, useful for producing large mats of strikingly grayish foliage, along with bright yellow flowers.

Chromosome Number

$2n = 16$: Ghimpu 1928, 1929*b*, 1930; Heyn 1956; Bolton 1962; Clement 1962; Delay 1969; Fernandes and Santos 1971; Colombo et al. 1978; Oliva et al. 1978; Lesins and Lesins 1979; Baltisberger 1990; Baltisberger et al. 1993; Ruíz de Clavijo 1993; Slavik et al. 1993.

Genetic Relationships and Gene Transferability

Intraspecific

Artificial hybridization of different populations has not been attempted, so no crossing barriers within the species are known.

Interspecific

Lesins and Lesins (1979) reported that they were unable to hybridize *M. marina* with *M. sativa*, *M. daghestanica*, and *M. pironae*.

Exchange by advanced biotechnology

Using embryo rescue techniques, McCoy and Smith (1986) obtained hybrids between *M. marina* and diploid *M. sativa* ($2n = 16$, i.e., at the same chromosome number); this hybrid combination was obtained with considerable difficulty and the hybrids were extremely weak (McCoy and Bingham 1988).

Conservation Status

Maxted et al. (1998) provided an ecogeographic analysis of *M. marina* for conservation purposes. This widespread species is not in need of conservation protection except in the Black Sea region (the regional Red Data List includes it as a threatened species). The International Union for Conservation of Nature status is NE (not endangered), except in the Ukraine. The only localities with *M. marina* in the Crimea (Sevastopol and Evpatoria areas and Cape Tarkhankut) are considered in strong need of protection because of the conversion of the coast into a resort zone.

Germplasm Sources

Fifty accessions have been reported in the major germplasm institutions examined (see Chapter 3).

Research Problems and Priorities

Medicago marina is a quite divergent species in its genus, with an unusual habitat. Although it does not appear to have potential directly as an agricultural plant, it may possess genes, particularly for salt tolerance, that could contribute to the improvement of cultivars of other species of *Medicago* and deserves to be investigated in this regard.

Representative Collections

Albania: Pennington, R.V. 16, Durazzo, near sea level, on sandy dunes, in thin layer of vegetation mould near Durazzo, 13-V-1934, in flower (K). Pennington, R.V. 192, Durazzo, alt. 100 feet, below the old walls and in the meadows, clay soil, 10-V-1936, in flower (K). **Algeria:** Davis 52161, Seraidi (Bugeand) Plage, near Annaba (Bone), sea level, sandy shore, 10-V-1971, in young fruit (BM). Reverchon, E. 112, Kabylie, Bougie, les sables, sur la plage, V-1896, in flower (P). **Spain (Balearic Islands):** Bowden, L.F. & Sims, P.A. 2046, Ibiza, Cala Portinatx, northernmost part of island, sand dunes behind main beach, 14-V-1968, in flower & fruit (BM). Cannon, J.F.M. & M.J. 3224, Ibiza, Es Cana near Santa Eulalia, in loose dry sand at top of beach, 29-V-1966, in flower (BM). **Bulgaria:** Kojoucharov, S., près de la ville Bourgas, sur les sables maritimes, 10-VI-1959, in flower (K). Turrill, W.B. 1167, Burqas, sandy seashore, 14-VIII-1926, vegetative (K). **Cyprus:** Ingram, C. 43A, Polis, sandy shore, sea level, 20-II-1948, in flower (BM). Laukkonen, P. 287, distr. Famagusta, Apostolos Andreas Monastery, on the beach, 22-IV-1973, in flower & fruit (K). **Egypt:** Bornmüller, J. 10599, Alexandria, in arenosis ad Sidi-Gaber, 8-IV-1908, in flower & fruit (BM, P). Danin, A., northern Sinae, Hof H'asara, near Nahal Dikla, 15 km southwest of Rafah, sands on loessial ground, near seashore, 13-V-1973, vegetative (HUJ). Hurst, H.A., near Alexandria, IV-1871, in flower (BM). **France:** Reau, N., Dunes de Plaissance en Oleron (Ch.-Inf.) 20-IX-1882, in fruit, Société Rochelaise 1892 (P). Reverchon, E. 235, Corsica, Porto, sables des plages, 17-V-1885, in flower & fruit (K). Stagg, F.N., Corsica, Porto Plage, seashore, IV-1938, in flower (K). Théveneau, A., Agde (Hérault), sables maritimes, VI-1859, in flower & fruit, Flora Galliae et Germaniae Exsiccata de C. Billot No. 2650 (P). **Greece:** Barclay, C. 179, Crete, Isle of Elaphonisi, south coast, west of Paleochorou, Prov. of Selino, in sand, 11-V-1967, in flower & fruit (K). Cannon, J.F.M. & M.J. 4301, Crete, Nomos Lasithloi, 2 km east of Sitia, loose sand at top of beach, 9-IV-1974, in flower & fruit (BM). Rechinger, K.H. & F. 8338, inter Triandra et Kromasto, in arenosis maritimis, 22-VI-1935, vegetative (BM). **Israel:** Leinkram, S., coastal Galilee, Akhgiv, coastal rocks, 23-III-1955, in flower & fruit (BM). Zohary, M., Tel-Aviv., dunes, 27-III-1928, in flower & fruit, Univ. Hebraicae Hierosolymitanae Flora Palaestinae Exsiccata No. 63 (BM, K, P). **Italy:** Davis, D. & Sutton, S. 64739, Calabria (Aspromonte), ca. 42 km southwest of Reggio di Calabria, Bova Marina, sea level, pebble beach, 9-VI-1979, in flower & early fruit (BM). Humphries, C.J. & Richardson, I.B.K. 320, Sardinia, Maddalena, near Abbatogia northeast of Sardinia, coastal sand, alt. 0–5 m, 16-IV-1973, in flower & fruit (BM). Rigo, G. 231, Calabria, Reggio, in arenosis maritimis, 1-V-1898, in flower (P). Stace, C.A. & Coton, R., Sicily, Spadafora – San Martino, just east of Milazzo, alt. to 5 m, sandy seashore with small dunes, 23-V-1972, in fruit (BM). **Lebanon:** Meinertzhagen, R., Beyrouth, dunes, 21-III-1933, in flower (BM). **Libya:** Guichard, K.M., Leptis Magna, 11-III-1952, in flower & young fruit (BM). Johnson, G.F. 47, Keren, Tripoli, alt. 500 feet, sandy ground, 16-II-1949, in flower (BM). **Morocco:** Davis 50992, Tangier to Asitah, sea level, dunes and sandy shore, 6-IV-1971, in flower & fruit (BM). Davis 54250, 10 km north of Ovalidia, near sea level, dunes, 8-IV-1972, in flower & fruit (BM). **Portugal:** Brummitt, R.K. & Ernst, W.R. 5846, prov. Estremadura, Nazaré, on sandy shore, 20-IV-1967, in flower & fruit (K). Carvalho, A. Goltz de, Buarcos, Lusitania, 4-V-1885, F. Schultz herbarium normale nov. ser. cent. 20 No. 1957 (P). Goltz de Carvalho, A., Lusitania, Buarcos, 4-V-1885, in flower, F. Schultz Herbarium Normale Nov. Ser. Cent. 20 no. 1957 (K). Matos, J., Matos, A. & Marques, A., Aveiro, Barra de S. Jacinto, nasarelas maritimas, 21-V-1954, in flower & fruit, Univ. Conimbrigencis Flora Lusitanica Exsiccata No. 4871 (BM). **Romania:** Pantu, Z.C. & Solacolu, Th., Dobrogea, Distr. Constanta, in arenosis Ponti Euxini (Marea, Neagra) prope pagum Agigea, in ditione stationis biologicae, alt. 3 m, VI-1923, in flower & fruit, Museum Botanicum Universitatis Cluj, Flora Romaniae exsiccata No. 429 (K, P). Toma, M., distr. Constanta, Agigea, dune maritime, 6-VII-1963, in fruit (BM). **Spain:** Bourgeau, E. 162, Plage de Santa Cathalina, près Puerto Santo Maria, 13-V-1849, in flower (P). Gardner, M.F. & S.G. 1815, prov. Valencia, El. Saler, (39° 15′N, 0° 18′W), alt. 15 m, coastal woodlands, shrub community beneath coniferous trees, 23-IV-1982, in flower & fruit (BM). Stace, C.A. 66, Tarragona, Torredembarra near Tarragona, on yellow sand dunes, colonising bare sand a few yards from the Mediterranean, alt. 6 feet, 19-IV-1961, in flower & fruit (BM). **Tunisia:** Cosson, M.M.E. et al., Hammam-El-Lif, 8-V-1883, in flower & fruit (P). Simpson, N.D. 38.076, Zargis, white sand dunes, 31-III-1938, in flower (BM). **Turkey:** Davis & Polunin 25825, prov. Antalya, Manavgat-Kara point, alt. 3 m, white dunes, 10-IV-1956, in flower (BM). Péronin, A. 18, près Anamour, sables maritimes, V-1872, in flower & fruit (P). **USSR:** Kolakovsky, A., Abchazia, distr. Gulripsch, prope pag. Pschapy, in arenosis maritimis 6-V-1951, in fruit (BM, K, LE). Vassilczenko, I.T., Tauria, in littoribus maris pontici (Nigri), in arenis contra hortum "Sinop", 13-14-IV-1916, in flower & fruit (BM, K, LE). **Yugoslavia:** Krieger, Dalmatia, Furule Mezzo, saxosis maritimis, V-1910, in flower & fruit (K).

51. *Medicago medicaginoides* (Retz.) E. Small

M. medicaginoides (Retz.) E. Small in Small, Lassen & Brookes, Willdenowia 16(2): 434 (1987).

The epithet *medicaginoides* is based on *Medicago* plus *oides* (which means "resembling" or "like", based on the Greek *eides*), pointing out the resemblance to the genus *Medicago* when this species was described originally as a member of *Lotus*.

Subordinate Taxa

None recognized here. Peduncle length, which has been widely used to split *M. medicaginoides* into species, is a very variable and unreliable feature for taxonomic delimitation.

Type Specimen

Medicago medicaginoides is based on *Lotus medicaginoides* Retz. Lectotype (!) of *Lotus medicaginoides*

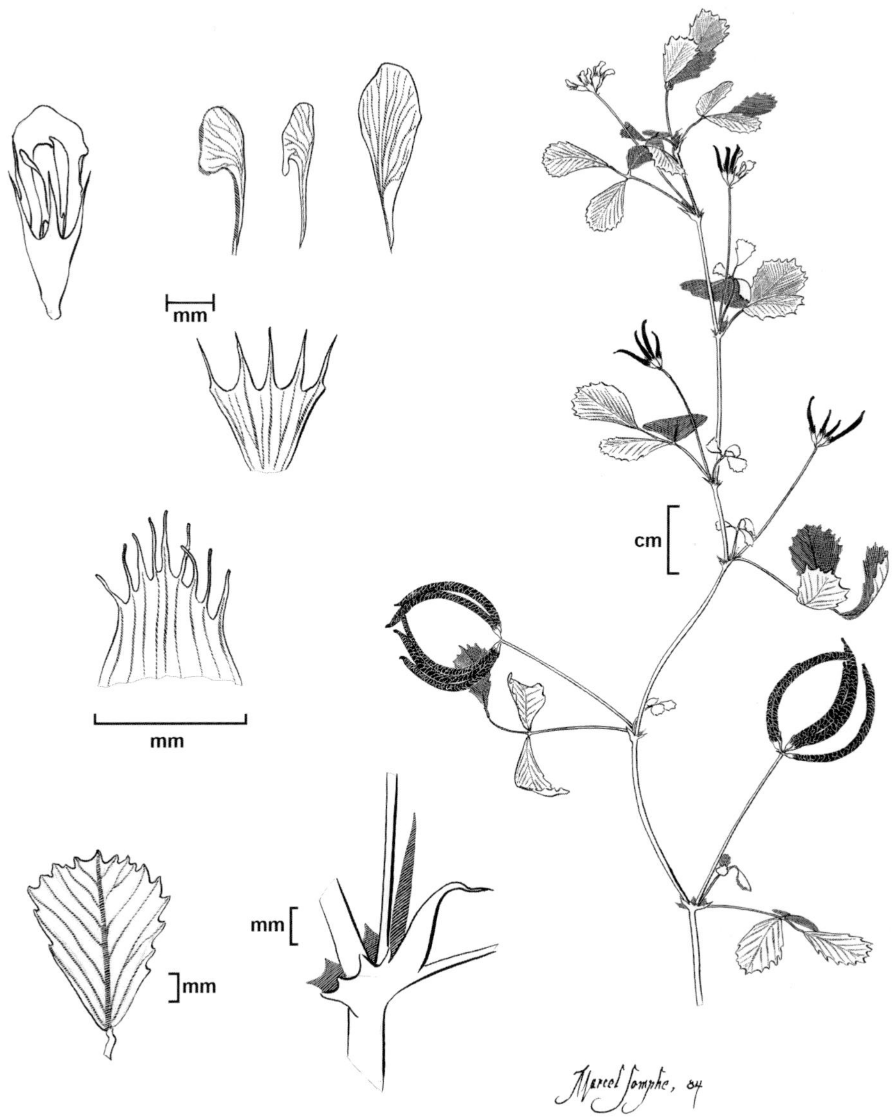

Fig. 51-1. *Medicago medicaginoides.* **Upper left:** Floral parts (clockwise from left: view of flower; keel; wing petal; standard petal; calyx; flattened androecial column (free stamen not shown)). **Right:** Fruiting branch. **Lower center:** Stem node showing stipules. **Lower left:** Leaflet. (Based on lectotype of *Lotus medicaginoides* Retz. at LD; artist: M. Jomphe.)

Retz. at LD (annotation slip states "Lectotypus! Per Lassen 1975").

Synonymy

Homotypic

Lotus medicaginoides Retz. Observ. 2: 23 (1781), (basionym).

Heterotypic

Trigonella arcuata C.A. Mey., Ver. Pfl. Casp. Meer. 136 (1831). (Type: Meyer, C.A., Azerbaijan, Talysch, in collibus lapidosia prope pagum Svant, 20-VI-1831, LE, lectotype by Huber-Morath (1970), isolectotype Herb. Boiss. G!)

M. arcuata (C.A. Mey.) Trautv. Bull. Acad. St. Petersb. 8: 272 (1841).

T. cancellata Desf. ex Pers. var. *arcuata* (C.A. Mey.) Širj., Publ. Fac. Sci. Univ. Masaryk (Brno) 128: 17 (1930).

T. cancellata Pers. Syn. 2: 355 (1807). (Type P, isotype LE, according to Vassilczenko 1953.)

M. cancellata (Pers.) Kryl., Fl. Zap. Sib. 8: 1587 (1933).

M. connivens Trautv., Acta. Horti Petrop. 4: 125 (1876).

T. cancellata Pers. var. *genuina* Širj. f. *daghestanica* Širj., Publ. Fac. Sci. Univ. Masaryk (Brno) 128: 17 (1930).

T. filiformis Fisch. ex Ledeb., Fl. Ross. 1: 532 (1842).

M. connivens Trautv. var. *genuina* Trautv., Acta Hort. Petrop. 4 125 (1876), nom. inadmiss.

T. cancellata Pers. var. *genuina* Širj., Publ. Fac. Sci. Univ. Masaryk (Brno) 128: 17 (1930), nom. inadmiss.

T. cancellata Pers. subsp. *issyk-kulensis* Vassilcz. (cited in Vassilczenko (1953, p. 216–217, type stated to be at LE).

T. striata L. f. var. *monantha* Panč., Verh. Zool.-Bot. Ges. Wien. 6: 481 (1856).

T. retzii Spreng., Syst. Veg. III: 283 (1826).

T. striata L. fil., Suppl. Pl. 340 (1781), non *M. striata* Bast. 1914.

T. tenuis Fisch. ex M. Bieb., Fl. Taur.-Cauc. 3: 514 (1819) (type: D. Wilhelms, Tiflis, circa Kadschari, LE).

M. tenuis (Fisch. ex M. Bieb.) Trautv., Bull. Acad. St. Petersb. 8: 271 (1841).

Buceras torosa Moench, Suppl. 51 (1802).

B. tortuosa Steud., Nom., Ed. 2, I: 233 (1840).

T. cancellata Pers. f. *uniflora* Širj., Publ. Fac. Sci. Univ. Masaryk (Brno) 128: 17 (1930).

T. arcuata C.A. Mey. var. *uniflora* (Širj.) Vassilcz. (1953, p. 221).

T. cancellata Pers. var. *genuina* Širj. f. *vulgaria* Širj., Publ. Fac. Sci. Univ. Masaryk (Brno) 128: 17 (1930).

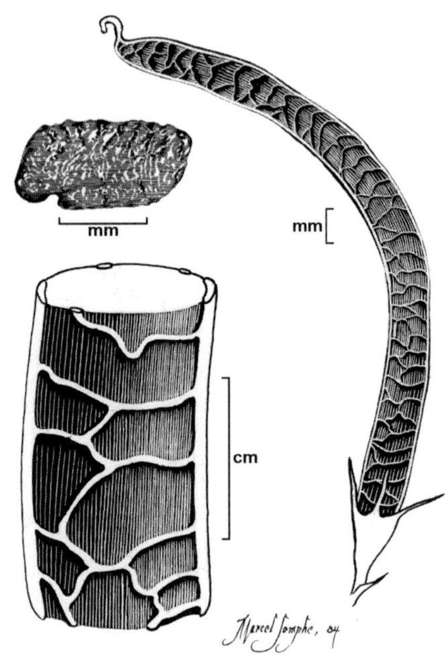

Fig. 51-2. *Medicago medicaginoides.* **Clockwise from bottom left:** Portion of fruit; seed; fruit. (Based on lectotype of *Lotus medicaginoides* Retz. at LD; artist: M. Jomphe.)

Vernacular Names

English: No names found.

French: No names found.

Geographical Distribution

(N = native, I = introduced, U = uncertain status)

Asia: Armenia (N), Azerbaijan (N), China (N), Gruziya (N), Iran (N), Kazakhstan (N), Kyrgyzstan (N), Mongolia (N), Russia (N), Tadzhikistan (N), Turkey (N). Turkmenistan (N), Uzbekistan (N).

Europe: Bulgaria (N), Greece (N), Russia (Dagestan; N), Ukraine (N), Yugoslavia (N).

Medicago medicaginoides occurs in Asia and Europe. The southernmost European distribution is in Greece, recorded at three disjunct localities ranging in altitude from 1100 to 1800 m (Thanopoulos 2007).

Description

Annual herb, stems prostrate to ascending, (5)10–40 cm long, branching from base, twigs and leaves sparingly adpressed-pubescent with simple hairs. Leaflets triangular, obtriangular oblong, obovate, or elliptical, 5–12 mm long, 3–9 mm wide, distal half with serrate margins. Stipules lanceolate, semi-sagittate, margin dentate at base or smooth (rarely incised). Inflorescence of (1)2–8 flowers in umbellate head, peduncle 0–30 mm long. Corolla yellow, 4.5–6 mm long, calyx 3.5–6 mm long, campanulate, teeth linear-subulate or lanceolate, about as long as or longer than tube. Pod linear, cylindrical to somewhat compressed, with adpressed simple hairs, 15–35 mm long, 1–2 mm wide, almost straight or curved (falcate to semi-circular), veins transversely reticulate,

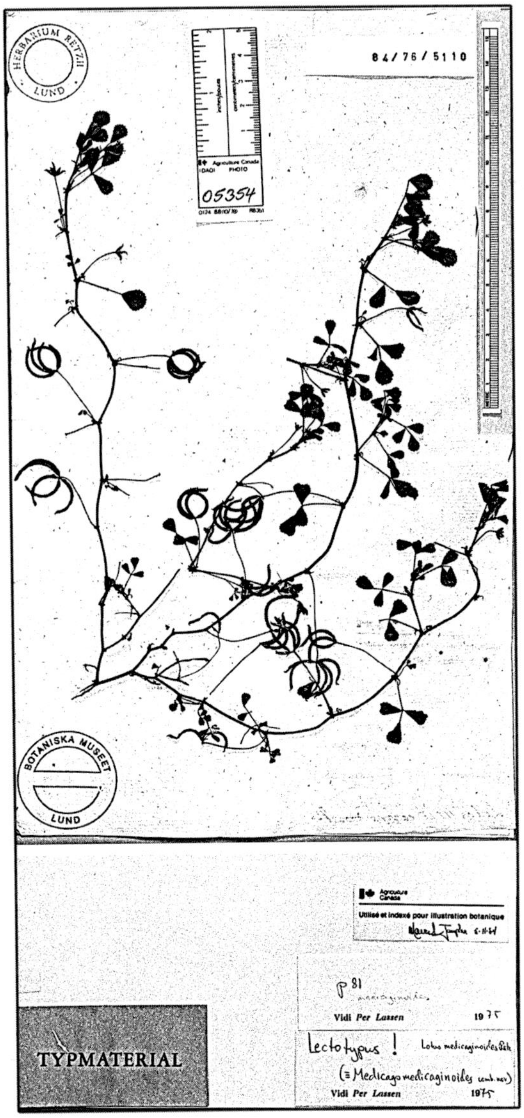

Fig. 51-3. Photograph of lectotype of *Lotus medicaginoides* Retz at LD.

with short pointed beak, with 7–14 seeds. Seeds oblong, 1.5–2.5 mm long, 1–1.2 mm wide, oblong-cylindrical, tuberculate, brown, radicle about 80% of length of seed.

Identification Hints

The characters of this species are rather generalized and it may easily be misidentified as several other species.

Habitat and Environmental Adaptation

Habitat

Medicago medicaginoides occupies stony slopes, hillsides, ravines, steppes, gravelly dry river beds and banks, rocky pastures, semi-deserts, and fallow and cultivated fields, reaching to the alpine zone. Occasionally it is a weed.

Phenology

This species flowers from March to August, depending on local climate.

Soil

Medicago medicaginoides is found on a variety of soils, mostly alkaline, such as pebbly or stony limestone. It has been collected from sand, clay, and solonetzic soils.

Temperature

Medicago medicaginoides is a widespread species, found occasionally in deserts and subalpine areas, indicating a range of temperature tolerance. However, like most other annual species of *Medicago*, it matures its seeds before the hottest, driest times of summer arrive.

Water

This species is noted for often occupying dry conditions.

Biotic Relationships

Pollination

Small (1988) noted that the flowers are small and the pollen to ovule ratio is low (286), observations that are consistent with *M. medicaginoides* being largely self-pollinated and substantially an inbreeder.

Dispersal

The pods and seeds have no evident adaptations for distribution and presumably are dispersed by some combination of adherence to animals, passing through the guts of animals and/or by being carried by water, wind, or gravity.

Herbivory

No information located.

Nodulation and mycorrhizae

USDA (2009) reported positive root nodulation for *M. medicaginoides*. Gan et al. (2007) studied the nodules of *M. medicaginoides*. Shi et al. (2007) observed arbuscular mycorrhizae in *M. medicaginoides*.

Disease resistance

No information located.

Economic History, Cultivars, and Uses

This species has not attracted agricultural interest.

Cultivars

No cultivars have been bred.

Forage

Medicago medicaginoides is not sown for forage.

Fodder

This species has not been reported to be of use as forage. However, Popenko and Pershina (1979) studied ensilage of ephemeral vegetation, including *M. medicaginoides*, and attributed the poor results to an almost total absence of lactic acid bacteria and great amounts of putrefactive and saprophytic bacteria.

Green manure, rotation crop, ground cover

Evans et al. (2004) examined the potential of numerous fodder species to survive winter in North Korea and produce nitrogen, organic matter, and soil stabilization in maize and rice fields. *Medicago medicaginoides* (identified as *M. arcuata*) was judged to have good winter survival and fair green biomass production, but to produce poor ground cover.

Other uses (salad sprouts, ornamental, plant products, etc.)

No information located.

Chromosome Number

$2n = 16$: Gardé 1948; Bolkhovskikh et al. 1969; p. 322 (as *Trigonella striata*); Astanova 1981 (as *T. striata*); Magulaev 2001 (as *T. arcuata*).

According to Vassilczenko (1953, p. 216, translated from Russian) "The number of chromosomes in this species (identified as *Trigonella tenuis*) is evidently 16 (see p. 133 in V.I. Chekhov and the note on the possibility of an error in the determination)." Also according to Vassilczenko (1953, p. 217, translated from Russian), and referring to *Trigonella cancellata*, "The number of chromosomes in this species is evidently 16."

Genetic Relationships and Gene Transferability

Intraspecific

Artificial hybridization of different populations has not been attempted, so information regarding barriers to hybridization within the species is unavailable.

Interspecific

No information located.

Exchange by advanced biotechnology

No information located.

Conservation Status

Medicago medicaginoides is a quite widespread species, which would not seem to be in evident need of conservation measures. However, it has not been studied to any appreciable extent, and it is not known whether there are variants of special interest that might require conservation. *Medicago medicaginoides* is listed in the Red Data Book of Rare and Threatened Plants of Greece (Phitos et al. 2009).

Germplasm Sources

About a dozen accessions have been deposited in major germplasm institutions (see Chapter 3).

Research Problems and Priorities

Medicago medicaginoides is closely related to *M. polyceratia* and additional study of the relationships between these two species is required. *Medicago medicaginoides* has not been examined for its economic potential. There is a need to assess its basic botanical characteristics and its possible agricultural utility.

Representative Collections

Iran: Archibald, J.C. 1508, Bekhtiari Country, between Azna and Dorud, alt. ca. 1740 m, southwest facing slope, stony clay, 24-IV-1966, in flower & early fruit (K). Bowles Scholarship Bot. Exp. 1262, Mianeh, Tehran-Tabriz Rd., alt. 5000 feet, roadsides, 3-V-1963, in fruit (BM). Rechinger, K.H. 41681, Persia, prov. Azerbaijan occid., in valle fluvii Qotur West Khvoy versus fines Turcicas, alt. 1800–2000 m, 10-VI-1971, in fruit (BM). Rechinger, K.H. 52838, Persia, Gorgan (Mohammad Reza Shah National Park), in planitie ("dasht") ad viam versus Almeh ducentem, North Robat-e Qareh Bil, (37° 19′N, 56° 26′E), alt. 1200 m, 5-VI-1975, in fruit (G). Rechinger, K.H. 55872, Persia, prov. Khorasan, in montibus serpentinicis prope Robat-e Safid inter Mashhad et Torbat-e Heydariyeh, alt. 1600–1700 m, 29-V-1977, in flower & fruit (G). **Turkey:** Davis 43563A, Kars, 16 km from Igdir to Tazluca, Aras Valley, alt. 920 m, sandy gravel plain, 25-V-1966, in flower & fruit (K). Davis 43601, Kars, 15 km from Tuzluca to Kagizman, Aras Valley, alt. 950 m, gravel terrace of Aras River, 25-V-1966, in flower & fruit (K, E). Davis 43713, Kars, Aras valley, between Igdir D.Ü.C. & Sedar Bulak, alt. 900 m, steppe, sandy soil, 27-V-1966, in flower & fruit (E). Karamanoglu 66-34, Prov. Kars, district Horasan, Horasan-Karakurt, alt. 1600 m, 29-VI-1966, in fruit (G). Rechinger, K.H. 57344, prov. Agri, in agris derelictis prope Karabulagh, 25 km a Dogubayazit septentriones versus, alt. 1700 m, 3-VII-1977, in flower & fruit (G). Tobey 2652, Çerum, after Gumushaciköy, alt. 800 m, rocky, scrub, 31-IV-1969, in fruit (E). **USSR:** Callier, A. 574, Karasubasar (Caucasus), In declivibus montis cretacei Akkaja 7-VI-1900, in fruit (K, P). Grossheim, A., Transcaucasia, prope Erivan, in lapidosis, 10-V-1922, in flower & fruit, Grossheim & Schischkin Plantae orientalis exsiccatae 115 (G, LD). Hohenacker, T.F., Georgiae caucasicae, Elisabethpol, in aridis ditionis, 1834, in flower & fruit (K). Sintensis, P., Regio transcaspia, Aschabad, in cultis, V-1900, in flower & fruit (P). Sintensis Iter transcaspico-persicum 1900–1901 No. 155 (P). Szovits, caucasus, Somchetiae prope Catharinenfeld, in collibus lapidosis, in flower & fruit (K). Holmberg 1766, Benara, prov. Tiflis, district Akhaltsikh, Georgia, alt. 1200 m, 20-VI & 8-VII-1912, in late flower & maturing fruit (LD).

52. *Medicago minima* (L.) Bartal.

M. minima (L.) Bartal., Cat. Piante Siena 61 (1776).

The epithet *minima* is based on the Latin *minimus* (least, smallest), and was applied by Linnaeus to this taxon because of the generally small size of the plant compared with other species of *Medicago*.

Subordinate Taxa

None recognized here.

Type Specimen

Heyn (1959: 168) lectotypified the basionym for *M. minima* (*M. polymorpha* L. var. λ *minima* L.) by an illustration, plate 1651 on page 386, accompanying the phrase "Medica echinata minima" in: Bauhin, J., Cherler, J.H., et al. 1650–1651. *Historia Plantarum Universalis*, vol. 2. Typographia Caldoriana (printer), Yverdon, Switzerland. The Linnaean Plant Name Typification Project (cf. Jarvis 2007) (http://www.nhm.ac.uk/research-curation/projects/linnaean-typification/;

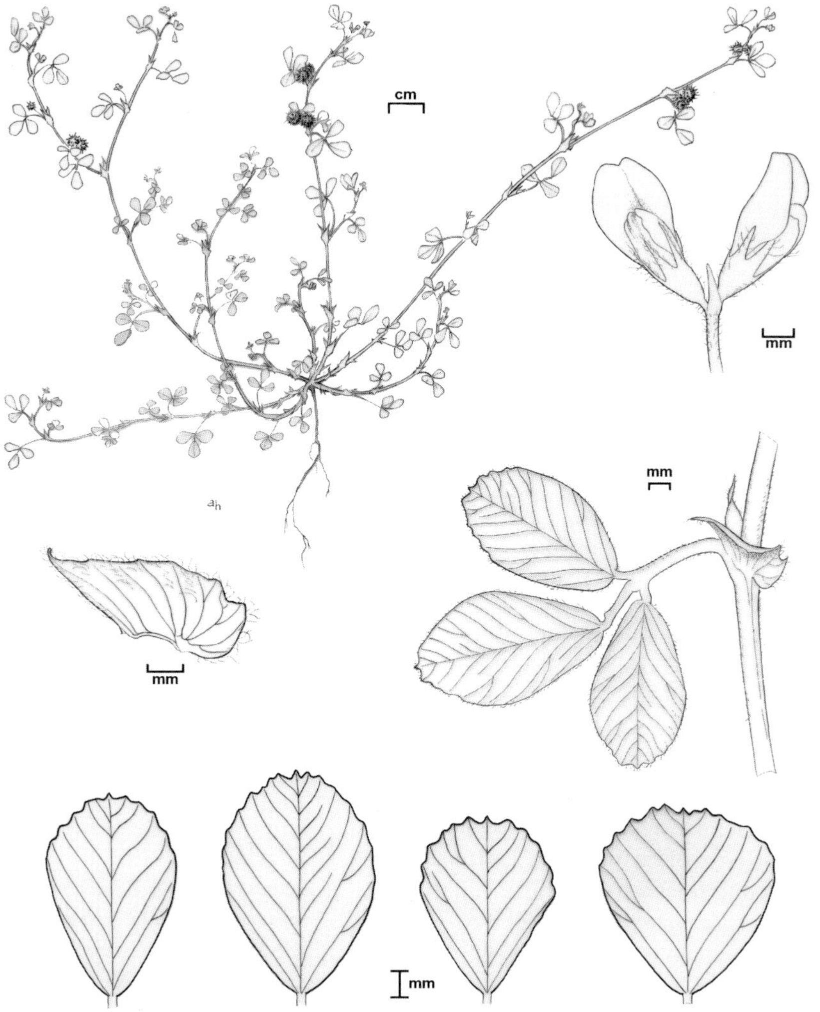

Fig. 52-1. *Medicago minima.* **Clockwise from top left:** Plant; flowers; leaf attached to stem; leaflets; stipule. (Plant based on Haradjian 2833, collection 99/36, HUJ, remainder based on E. Small M332, DAO; artist: A. Hanes.)

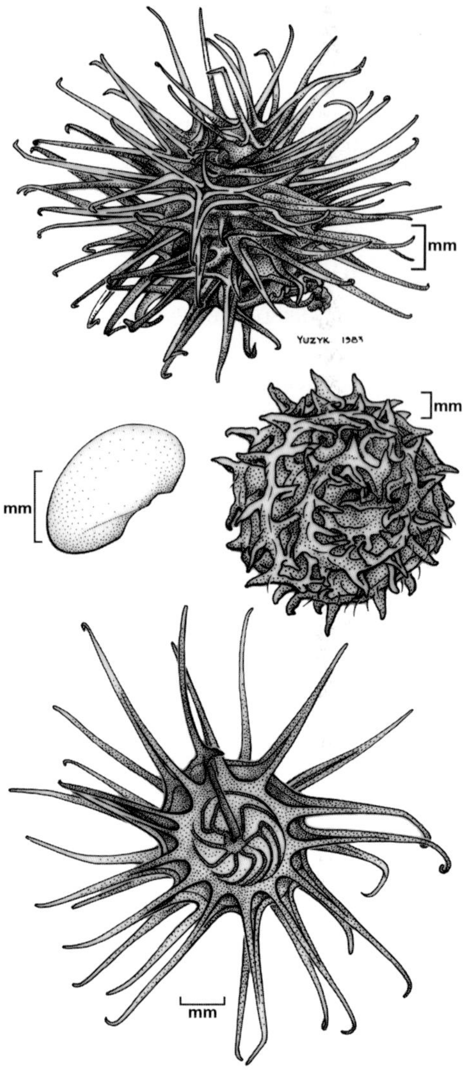

Fig. 52-2. *Medicago minima.* **Clockwise from top:** Fruit (side view); short-spined fruit (end view); sectioned fruit showing venation on face of coil; seed. (Short-spined fruit based on "var. *brevispina* Benth.", A. Eig & M. Zohary M/214, HUJ; remainder based on E. Small M22, DAO; artist: L. Yuzyk.)

current June 2010) provides the following information: "Linnaean Name: *Medicago minima* (Linnaeus) Linnaeus, nom. inval. Reference: Flora Anglica: 21 (1754) Type: Type not relevant. Current Name: *Medicago minima* (L.) Bartal.... Comments: Shinners (in Rhodora 58: 3. 1956) questioned whether the recombination of var. *minima* at species rank was made by Linnaeus (in Fl. Anglica: 21. 1954). Shinners suggested that this is an error of interpretation, as Linnaeus subsequently continued to treat the taxon as a variety of *M. polymorpha*. Shinners accepted Bartalini (1776) as the first to make the combination at species rank, as did Stearn (Introd. Ray's Syn. Meth. Stirp. Brit. (Ray Soc. ed.): 57. 1973). However, Greuter (in Candollea 44: 563. 1989) disagreed, arguing that the combination dates from 1754. Brummitt & Meikle (in Watsonia 19: 181–183. 1993) argued, however, that there are typographical variants of the thesis which make

Fig. 52-3. *Medicago minima.* From Hallier (1885, vol. 23, plate 2349).

it unreasonable to interpret names such as this (though they focused particularly on two names in *Primula*) as a new combination."

Synonymy

Homotypic

M. polymorpha L. var. λ *minima* L. Sp. Pl. 780 (1753), (basionym).

M. minima (L.) Grufb. in L., Fl. Angl. 21 (1754), nom. inval. (GRIN Taxonomy for Plants, http://www.ars-grin.gov/cgi-bin/npgs/html/taxdump.pl?Medicago%20minima%20(L.)%20Grufb; current July 2010) noted that "Grufberg probably did not intend to make this combination.")

M. minima (L.) L. Invalidly published by C. Linnaeus in Flora Anglica, 21 (14 April 1754); accordingly, the correct name may be cited as *M. minima* (L.) L. ex Bartal.

Spirocarpus minimus (L.) Opiz, Seznam 93 (1852).

Heterotypic

M. brachyacantha A. Kerner, Öesterr. Bot. Zeitschr. 18: 386 (1868). (The following exsiccata of the name is at B, and therefore would seem to be a candidate for its type: Kerner, J., Austria inferior, in campis sterilibus ad oppidum Wiener Neustadt; solo argillaceo; alt 300 m, in flower & fruit, Flora Exsiccata Austro-Hungarica 2014.)

M. minima (L.) Bartal. var. *brachyodon* Reichenbach, Fl. Germ. Excurs. 2: 502 (1832).

M. minima (L.) Bartal. var. *brevispina* Benth. in Smith, (Sowerby's) Engl. Bot., Suppl. 1: tab. 2635, Pl. 2E, a, b (1831). (Heyn (1963) stated the type is Sowerby, Engl. Bot. tab. 2635.)

M. minima (L.) Bartal. subsp. *brevispina* (Benth.) Ponert, Feddes Repert. 83 (9-10): 639 (1973).

M. minima var. *canescens* Ser. in DC. Prodr. 2: 178 (1825). (Herb. DC. at P has a single sheet with two collections, of which Bellay in 1822 is a suitable candidate as type.)

M. minima (L.) Bartal. var. *vulgaris* Urb. subvar. *canescens* (Ser.) Rouy in Rouy & Fouc., Fl. France 5: 41 (1899).

M. minima (L.) Bartal. var. *compacta* Neyr. Bull. Soc. Linn. Bord. 108 (1897).

M. arabica L. var. *echinata* Weston, Bot. Univ. 3: 481 (1772).

M. minima (L.) Bartal. var. *elongata* Roch., Pl. Banat. Rar. 51 (1828).

M. elongata (Roch.) Simonk. ex A. Kerner, Sched. Fl. Exsicc. Austr.-Hung., 6: 4 (1893).

M. minima (L.) Bartal. var. *vulgaris* Urb. subvar. *elongata* (Roch.) Rouy in Rouy & Fouc., Fl. France 5: 41 (1899).

M. minima (L.) Bartal. f. *elongata* (Roch.) Asch. & Graebn., Syn. 6(2): 438 (1906).

M. minima (L.) Bartal. var. *exilis* Lange, Pug. 362 (1860–1865).

M. graeca Hornem., Hort. Bot. Hafn. 2: 728 (1815).

M. minima (L.) Bartal. var. *graeca* (Hornem.) Cand., Bull. Soc. Bot. France 45: 192 (1898).

M. polymorpha L. var. *hirsuta* L., Sp. Pl. 2: 780 (1753). (Lectotype = "Medica echinata hirsuta" in Bauhin & Cherler, Hist. Pl. Univ., 2: 386, 386 (1651), designated by: Heyn (1959: 168). Accepted by The Linnaean Plant Name Typification Project (http://www.nhm.ac.uk/research-curation/projects/linnaean-typification/; current June 2010)

M. hirsuta (L.) Bartal., Cat. Piante Siena 61 (1776).

M. hirsuta (L.) All., Fl. Pedem. 1: 315 (1785).

M. minima (L.) Bartal. var. *hirsuta* (L.) Desportes, Fl. Sarthe Mayenne 50 (1838).

Medica hirsute (L.) Bubani, Fl. Pyren. 2: 477 (1899).

M. hirsutissima Cyrillo ex Steud. Nomencl. Bot. ed 2, 2(9): 109 (1841), pro syn.

M. inconspicua Nevski, Trudy Bot. Inst. Akad. Nauk SSSR, Ser. 1, Fl. Sist. Vyss. Rast. 4: 250 (1937).

M. minima (L.) Bartal. var. *intermedia* Kozuharov, Bulgarian Acad. Sci. 15: 161 (1965) (type referred to in publication).

M. lineata Tausch, Flora (Regensb.) 14: 210 (1831).

M. minima (L.) Bartal. var. *longiseta* DC. ex Seringe in DC., Prodr. 2: 178 (1825). (In Herb. DC. at P, I observed two collections on one sheet in a folder (and a second folder with numerous collections); from the sheet in a single folder, the collection from Teneriffe in 1807 is taken as the type.)

M. minima (L.) Bartal. var. *longispina* Lowe, Man. Fl. Mad. 136 (1868).

M. minima (L.) Bartal. var. *maior* Moris, Fl. Sard. 1: 451 (1837).

M. meyeri Gruner, Bull. Soc. Imp. Naturalistes Moscou 40(2): 416 (1867).

M. mollissima Roth, Bot. Abh. Beobacht. 49 (1787).

M. minima (L.) Bartal. var. *mollissima* (Roth) Koch, Syn. Fl. Germ. Helv. 164 (1836 ("1835")).

M. minima (L.) Bartal., f. *mollissima* (Roth) Urban, Verh. Bot. Vereins Prov. Brandenburg 15: 78 (1873).

M. minima (L.) Bartal., f. *mollissima* (Roth) Kozuharov, Bulgarian Acad. Sci. 15: 160 (1965), nom. superfl.

M. minima (L.) Bartal. subsp. *ononidea* de Coincy ex Rouy in Rouy & Fouc, Fl. Fr. 5: 42 (1899).

M. ononidea (Rouy) A.W. Hill, Ind. Kew., Suppl. 6: 125 (1926).

M. minima (L.) Bartal. var. *pubescens* Webb. Hist. Canar. 3(2): 65 (1842).

M. minina (L.) Bartal. f. *pubescens* (Webb) Urb., Vereins Bot. Ver. Brandenberg 15: 78 (1873).

M. minima (L.) Bartal. var. *vulgaris* Urb. subvar. *pubescens* (Webb) Rouy in Rouy & Fouc., Fl. France 5: 41 (1899).

M. pulchella Lowe, Trans. Cambridge Philos. Soc. 6 (reimpt.): 547 (1858, originally 1838).

M. minima (L.) Bartal. var. *pulchella* (Lowe) Lowe, Man. Fl. Madeira 1: 166 (1868).

M. minima (L.) Bartal. subsp. *pulchella* (Lowe) Arcangeli, Comp. Fl. Ital. 165 (1882).

M. pulchella Tod. non Lowe, Ind. Sem. Hort. Panorm: 36 (1868).

M. polymorpha L. var. *recta* Desf., Fl. Atl. 2: 212 (1800).

M. recta (Desf.) Willd. Sp. Pl. 3: 1415 (1802). (A folder of *M. recta* at Herb. Willd. in B has 2 specimens (1 and 2 in upper right hand corner). Heyn (1963: 64) cited 1 as the type.!)

M. minima (L.) Bartal. var. *recta* (Desf.) Burnat, Fl. Alp. Marit. 2: 109 (1896).

M. rigidula Hoffm. ex Steud., Nomencl. Bot. (ed. 2) 2(9) : 109 (1841), pro syn.

M. sessilis Peyr. ex Post, Fl. Syr. Palest. & Sinai 10 (1896).

M. minima (L.) Bartal. var. *stenophylla* Clav., Fl. Gir. 263 (1884).

M. minima (L.) Bartal. var. *vulgaris* Urb. subvar. *stenophylla* (Clav.) Rouy in Rouy & Fouc., Fl. France 5: 41 (1899).

M. minima (L.) Bartal. var. *viscida* Koch, Syn. Fl. Germ. Helv. 1: 164 (1835).

M. minima (L.) Bartal. f. *viscida* (Koch) Urban, Verh. Bot. Vereins Prov. Brandenburg 15: 78 (1873).

M. minima (L.) Bartal. var. *vulgaris* Urb. subvar. *viscida* (Koch) Rouy in Rouy & Fouc., Fl. France 5: 41 (1899).

M. viscida (Koch) Schur, Verh. Bot. Ver. Brünn 15(2): 173 (1877).

M. minima (L.) Bartal. var. *vulgaris* Urb., Verh. Bot. Vereins Prov. Brandenburg 15: 78 (1873), nom. inadmiss.

M. minima (L.) Bartal. f. *vulgaris* (Urb.) Morariu, Fl. Rom. 5: 136 (1957).

Vernacular Names

English: Goldfields medic (so-named for the Eastern Goldfields in Western Australia), Kaalgoorlie medic (Australia), small bur clover (burclover), least medic, lesser bur medic, little medic, little bur clover, small alfalfa, small medic, small bur medic, small wool(l)y bur medic, sticky-weed, wool(l)y bur medic. Occasionally, simply the name bur medic is applied, although this is more usually reserved for *M. polymorpha*. (Bur is also spelled burr, medic is also spelled medick.)

French: La luzerne naine (luzerne nâne encountered once), petite luzerne.

Geographical Distribution

(N = native, I = introduced, U = uncertain status)

Africa: Algeria (N), Canary Islands (Spain; N), Djibouti (N), Egypt (N), Ethiopia (N), Libya (N), Madeira Islands (Portugal; N), Morocco (N), Socotra (Yemen; U), Somalia (N), South Africa (I), Sudan (N), Tunisia (N).

Asia: Afghanistan (N), Armenia (N), Azerbaijan (N), China (N), Cyprus (N), Egypt (N), Gruziya (N), India (I), Iran (N), Iraq (N), Israel (N), Japan (I), Jordan (N), Kazakhstan (N), Kyrgyzstan (N), Lebanon (N), Pakistan (N), Russia (N), Saudi Arabia (N), Syria (N), Tajikistan (N), Turkey (N), Turkmenistan (N), Uzbekistan (N).

Australasia: Australia (including Tasmania; I), New Zealand (I).

Europe: Albania (N), Austria (N), Balearic Islands (Spain; N), Belarus (I), Belgium (N), Bulgaria (N), Corsica (France; N), Denmark (N), Crete (Greece; N), Czech Republic and Slovakia (N), Denmark (N), East Aegean Islands (Greece, N), Estonia (I), France (N), Germany (N), Great Britain (N), Greece (N), Hungary (N), Italy (N), Liechtenstein (N), Luxembourg (N), Malta (N), Moldova (N), Netherlands (N), Poland (N), Portugal (N), Romania (N), Russia (N), Sardinia (Italy; N), Sicily (Italy; N), Spain (N), Sweden (N), Switzerland (N), Turkey (N), Ukraine (N), Yugoslavia (N).

Northern America: Argentina (I), Chile (I), Mexico (I), USA (I).

Pacific Ocean: Hawaii (I), New Caledonia (France; I).

Southern America: Argentina (I), Chile (I).

Medicago minima is one of the likeliest *Medicago* species to be encountered, rivaling *M. sativa*, *M. polymorpha*, and *M. arabica* in terms of probability of being observed. It is widely distributed in Eurasia and Africa, and widely adventitious elsewhere in the world. The species is naturalized in Argentina and Australia where it is an extremely important component of rangelands used for livestock grazing.

Description

Annual herb, ascending or procumbent, branched from base, branches 5–40 cm long. Vegetative parts, calyx, and pod covered with simple hairs, gland-tipped hairs, or both. Stipules lanceolate to ovate-lanceolate, entire or minutely dentate at base. Leaflets 4–8(12) mm long, 2–7 mm wide, obovate or rarely oblanceolate, distal third of margin serrate. Inflorescence capitate, with (1)2–5(8) flowers, peduncle shorter to longer than subtending petiole. Flowers 2–4.5(6) mm long, calyx appressed-hairy, teeth unequal, about as long as tube, corolla yellow, less than twice length of calyx. Young pod protrudes sideways from calyx. Mature pod discoid, cylindrical, suborbicular, or ovoid, more or less pubescent, 3–5 mm long, 2.5–6 mm in diameter, with 3–5 loosely adpressed pergamentaceous coils. Coil face with 6–8 strongly curved unbranched radial veins entering a broad, veinless margin. Dorsal suture spines varying in different plants from short tubercles to twice as long as coil diameter and hooked, more or less parallel to coil face. Spines grooved. Seeds 1–2 per coil, somewhat reniform or oblong-ovoid, 1.5–2.5 mm long, ca. 1.3 mm wide, separated by a thin spongy fruit septum, seedcoat smooth and yellow to light brown, radicle about half as long as seed.

Identification Hints

The fruit coil face of *M. minima* has a submarginal vein more remote from the dorsal suture coil margin (about a third of the coil radius away) than any other species of annual *Medicago*. This species is most frequently confused with *M. laciniata* but, in addition to the (less developed) submarginal vein characteristic, *M. minima* has much hairier foliage and much less-dissected stipules.

Habitat and Environmental Adaptation

Habitat

Medicago minima occurs in diverse habitats, especially in semi-arid temperate climates. It is common on hillsides, steppes, sand dunes, fields, and forest edges, and as a weed of waste ground, roadsides, cultivated fields, and orchards. Compared with other annual species of *Medicago*, *M. minima* is exceptionally tolerant of a wide range of edaphic and climatic conditions (Heyn 1963; Abdelguerfi et al. 1988*a*).

Phenology

As with other annual species of *Medicago*, *M. minima* flowers in the spring and early summer, the exact time (March to May in the Mediterranean area) depending on local climate.

Soil

Medicago minima occurs naturally in a variety of mostly alkaline soils, including sand. Zoghlami et al. (1996) observed that in Tunisia *M. minima* did not show any apparent preference for soil texture. However, Abdelguerfi et al. (1988*a*) observed that in Algeria *M. minima* occurs in medium- and coarse-textured soils, and Piano and Francis (1992) described it as occupying loam-clay ranging from slightly to moderately basic. In Sicily, Piano et al. (1991) observed the species growing in soil with a pH range of 6.0–9.0. Fresnillo Fedorenko (2001) summarized soil adaptation of *M. minima* in Australia and Argentina. He suggested that the species has, compared with other medics, a greater ability to utilise less soluble calcium phosphates or absorb phosphate at high pH or higher calcium concentrations. *Medicago minima* has a low tolerance to soil salinity, although Lieth and Mochtchenko (2003) listed it in their halophyte database. Abdelly et al. (1996) observed that although *M. minima* was regularly found on the edges of sebkhas (closed depressions with hydromorphic, saline soil, periodically submersed) in Tunisia, the species had limited tolerance of soil salinity. Loader (1979) found that application of sulphur to a naturalized meadow of *M. minima* increased yield.

Temperature

The species is considered to be relatively cold tolerant (Piano and Francis 1992). Cocks and Ehrman (1987) found that, of 22 species of *Medicago* examined in Syria, *M. minima* was among the most frost tolerant. Nevertheless, subfreezing temperatures in December 1998 in West Central Texas caused considerable seedling mortality of cultivar 'Devine' (Ueckert et al. 2003). Derkaoui et al. (1991) found that fall-planted *M. minima* in Oklahoma failed to survive the winter, indicating that the species does not have sufficient cold tolerance for successful self-replication in the southern Great Plains of the United States.

Water

As with most Mediterranean annual species, *M. minima* is native to well-drained soils. It appears to withstand periods of drought, but Francis and Poole (1973) observed that it is extremely susceptible to waterlogging. Piano and Francis (1992) indicated that it requires a minimum annual rainfall of 225 mm. In Tunisia, the species occupies a semi-arid zone with 200–450 mm of precipitation annually (Zoghlami et al. 1996; Zoghlami and Hassen 2004). Abdelguerfi et al. (1988*a*) observed that in Algeria *M. minima* occurs in the driest, best-drained soils in areas receiving less than 520 mm of precipitation. Fresnillo Fedorenko et al. (1995) found that *M. minima* was able to survive experimental drought, albeit with lower productivity. Fresnillo Fedorenko (2001) made the following observations. He suggested that there is ecotypic differentiation within *M. minima*, with strains adapted to low rainfall producing plants that are smaller, flower earlier, mature faster, produce less seed, have a higher harvest index, and have a shorter growth cycle compared with plants adapted to wetter environments. Water stress has been observed to reduce shoot more than root growth and to stimulate a larger allocation of dry matter to reproductive structures. The shallow root system likely allows *M. minima* to take advantage of small rainfalls and compete against native perennial grasses.

Biotic Relationships

Pollination

Small (1988) noted that the relatively low pollen to ovule ratio (194) and small flowers are consistent with *M. minima* being largely self-pollinated. The species appears to be an inbreeder.

Dispersal

The numerous indehiscent spiny fruits of *M. minima* have relatively long, soft spines that are well adapted to dispersal in animal fur, accounting for the extremely widespread occurrence of this species. It has been widely distributed in wool and often is spontaneous around wool factories.

Herbivory

Larval alfalfa weevils (*Hypera postica*) have been observed to attack cultivar 'Devine', but the insect was less of a problem than for other annual medics (Muir et al. 2005).

Nodulation and mycorrhizae

USDA (2009) reported positive root nodulation for *M. minima*. O'Connor et al. (2002) found that *M. minima* was very responsive to arbuscular mycorrhizas established in Australian soils.

Disease resistance

O'Neill et al. (2003) examined the comparative resistance of 36 annual species of *Medicago* to *Phoma medicaginis* (the cause of spring black stem and leaf spot, a major disease of alfalfa) and concluded that *M. minima* is highly susceptible to it. In a similar comparative test of 36 annual species of *Medicago*, O'Neill and Bauchan (2000) found that four accessions of *M. minima* were all quite susceptible to anthracnose caused by *Colletotrichum trifolii*. Yaege and Stuteville (2002) tested the resistance of 33 annual species of *Medicago* to powdery mildew (*Erysiphe pisi*) and found that the three collections of *M. minima* grown were susceptible to the disease (it was observed that often only one of a number of populations of other species tested was resistant). Ocumpaugh et al. (2007) noted that powdery mildew (*Erysiphe polygoni*) was a minor problem on cultivar 'Devine'.

Economic History, Cultivars, and Uses

Medicago minima has been only slightly developed to date for agricultural purposes, although in areas where it grows it is considered to be good forage. Vitkus (1977) assessed *M. minima* as suitable for cultivar selection for Lithuania.

Cultivars

'Devine', described in detail by Ocumpaugh et al. (2007), is the first cultivar of *M. minima* to be registered. It was developed by the Texas Agricultural Experiment Station and released in 2005. 'Devine' originated from a naturalized stand in a pasture west of Devine, Texas, southwest of San Antonio.

Forage

Medicago minima is very palatable, with a high nutritive value (Fresnillo Fedorenko 2001). It is reported to be a valuable forage species in semi-arid regions of Argentina (Busso et al. 1998) and can be equally useful in similar areas of Australia (Robertson and Smith 2006). In Central Argentina, it makes a significant contribution to cattle production on rangelands during late winter and spring, and has been observed to produce nearly 3 t/ha and represent more than 60% of aboveground herbaceous material (Fresnillo Fedorenko 2001). Ocumpaugh et al. (2007) indicated that yields in Texas were about 4.5 t/ha/year, with seed yields of 200–250 kg/ha. While *M. minima* dry matter production has been demonstrated to be substantial in stressful, semi-arid conditions, other commercial legumes are notably more productive under good growing conditions (Robertson and Smith 2006). St. John-Sweeting et al. (1996) noted that one accession studied in South Australia had poor dry matter production and so was not of potential commercial interest.

Gardner and Bennetts (1956) recorded that trefoil dermatitis (a type of photosensitization) has been associated with the grazing of *M. minima* by cattle, sheep, and horses.

Fodder

No information located.

Green manure, rotation crop, ground cover

No information located.

Other uses (salad sprouts, ornamental, plant products, etc.)

Representing the oddest use ever recorded for a species of *Medicago*, *M. minima* was chosen as a standard test material for examination of the effects of Unidentified Flying Objects (UFOs) on local vegetation (In: Bounias, M.C.L. "Biochemical traumatology as a potent tool for identifying actual stresses elicited by unidentified sources: evidence for plant metabolic disorders in correlation with a UFO landing." http://www.scientificexploration.org/journal/jse_04_1_bounias.pdf; current June 2010.)

Chromosome Number

$2n = 16$: Ghimpu 1928, 1929*a*, 1929*b*, 1930; Tischler 1934; Larsen 1956; Clement 1962 (as *M. minima* and *M. minima* vars. *sorentinii* and *brachyodon*); Simon and Simon 1965; Bolkhovsikih et al. 1969, pp. 308, 309 (2 reports); Majovsky et al. 1970*b*; Dahlgren et al. 1971; Fernandes and Santos 1971; Nilsson and Lassen 1971; Kozuharov et al. 1972; Kozuharov et al. 1973; Löve and Kjellqvist 1974; Kozuharov et al. 1975; Fernandes et al. 1977; Fernandes and Quieros 1978; Lesins and Lesins 1979; Abdelguerfi and Guittoneau 1979; Loon 1980; Magulaev 1980; Astanova 1981; Abdelguerfi et al. 1988*b*; Šopova and Sekovski 1989; Baltisberger 1991*a*, *b*; Karshibaev 1992; Slavik et al. 1993; Oberprieler and Vogt 1996*b*; Dobeš and Hahn 1997.

Genetic Relationships and Gene Transferability

Intraspecific

No barriers to hybridization within the species are known. Lesins and Lesins (1979) conducted some intraspecific hybridization and did not comment on any barriers to breeding.

Interspecific

Murr (1986) alleged a herbarium collection was a hybrid between *M. minima* and *M. falcata* (i.e., *M. sativa* subsp. *falcata*), this under the name *M. heterocarpa* Dürrnberger 1892 in sched.

Exchange by advanced biotechnology

No information located.

Conservation Status

Medicago minima is very widely distributed and indeed is quite weedy in many areas. Accordingly, it is not considered in need of conservation. One International Union for the Conservation of Nature and Natural Resources (IUCN) Red List of Threatened Species for the United Kingdom (http://www.bsbi.org.uk/Plant_Status_List_2005.pdf; current May 2007) stated that the species is "vulnerable" but this assessment has only very local significance.

Germplasm Sources

More than 1500 accessions are available from major germplasm institutions (see Chapter 3).

Research Problems and Priorities

Medicago minima has attracted limited agricultural development to date, but is so successful as a weed it would seem to merit close attention for its potential.

Representative Collections

Afghanistan: Archibald, J.C. 1260, Fars, Kuh-i-Saidum, alt. ca. 2230 m, southwest- and northwest-facing limestone cliffs, 10-IV-1966, in flower & fruit (K). Purse, P. 8156, Katagan province, Farqar upper valley, alt. 5000 feet, edge of ajube, 28-VI-1966, in flower & fruit (K). **Algeria:** Reboud, V., Constantine, 1874, in fruit, Société Dauphinoise No. 347 (P). **Argentina:** Burkart, A., Gob. de la Pampa, Lihuel-Calel, 30-XI-1959, in flower & fruit (P). **Australia:** Everist, S.I. 5999, Queensland, Belmont near Brisbane, in light brown sandy loam along drains, 11-IX-1956, in fruit (K). **Balearic Islands:** Bowden, L.F. & Sims, P.A. 1733, Ibiza

San Antonia Abad, north of town, limestone hills, frequent, 7-V-1968, in fruit (BM). Bowden, L.F. & Sims, P.A. 2108, Ibiza, hill between harbour and Talamanca Bay, shaley limestone, 17-V-1968, in flower (BM). **Bulgaria:** Kojoucharov, S. & Kouzmanov, B., près de la ville Asenovgrad, sur les sables de la rivière Arda, 29-V-1959, in flower & fruit (K). **Canary Islands:** Bourgeau, E., Canaria, in arvis vulgaris, 18-III-1896, in flower & fruit, E. Bourgeau Plantae Canarienses No. 760 (P). Pitard, C.J. 116, Palma, Playa de Bajamar, in arenosis maritimis, 22-III-1905, in flower & fruit (P). **China:** Delavay, Yunnan prov., Meao ki tse, 24-III-1882, in late flower & fruit (P). Licent, R.P. 1762, Chine septentrionale, Chansi meridional vers Kon tchong tcherm, 4-V-1916, in flower & fruit (P). **Cyprus:** Casey, E.C. 1142, Kyrenia, alt. 150 feet, uncultivated olive grove, in flower & fruit 22-III-1956, in flower 22-VI-1951 (K). Chapman, E. 57, Lacovounera Forest, alt. 600 feet, common on southern aspects and unstable soils in Kythrean hummocks, 31-III-1950, in flower & fruit (K). **Ethiopia:** Schimper, prope Axum, in agris requietis, 17-IX-1842, in fruit (P). Thulin, M. 1576, Arussi Prov., Chilalo awrajo, at Wabe Shebelle River near Asasa, alt. ca. 2250 m, in rocky slope to the river, 28-X-1971, in fruit (K). **France:** Gillot, Montagne de Chassey (Saône-et-Loire), pelouses dans l'enceinte même de la station prehistorique dite le Camp de Chassey, sur le calcaire jurassique, alt. 570 m, 25-VI-1879, in flower & early fruit, Société dauphinoise 1880 No. 347 bis. (P). Lassimónne, S.E., Allier, Yzeure, Robé, 11-VI-1930, in flower & early fruit, Société Linnéenne de la Seine Maritime No. 938 (P). **Germany:** Koch, Schlossberg prope Wachenheim, in clivibus incultis collis, F. Schultz, herbarium normale, nov. ser. cent. 8, No. 768 (P). Stearn, W.T. 69, about 8 km south of Heidelberg, Baden, sand dunes near Sandhausen, 1-VIII-1935, in fruit (K). **Greece:** Reverchon, E., Crête, Kissamos, les rochers, rare, 9-V-1884 (P). Orshan, G. 06501-10, central Crete, southern foothills of the Psiloritis mountain ridge, 8 km on way to Kemari, Thymus thymbra batha, light rendzina soil, alt 530 m, 6-VI-1964, in fruit (HUJ). **Hungary:** Simonkai, L., ad oppidum Arad, in graminosis, 31-V-1890, in flower & fruit, F. Schultz Herbarium Normale No. Ser. Cent. 27 nov. 2625 (K). Simonkai, L., ad oppidum Arad. Hungaria, in graminosis, 31-V-1890, in fruit, F. Schultz herbarium normale nov. ser. cent. No. 2625 (P). **Iran:** Bornmüller, J. & A. 6588, Persia borealis, Enseli, in arenosis, 22-IV-1902, in flower & fruit (P). Bunge, M. 1868-13, pr. Teheran, V-1859, in flower & fruit (P). Kotschy, Th., prope pagum Gere inter Buschir et Schiras, 1842, in flower & fruit (P). Lambinon, J. 76/250, Prov. du Mâzandérân, Bâbolsar, alt. ca. 20 m, petite digue empierrée au contact des dunes, rudéralisées du bord de la Mer Caspienne, 16-IV-1976, in flower & early fruit (HUJ). **Iraq:** Polunin, N. et al., Sudur, Mansuriyat al-Jabal, Diyala Liwa, stony waste, 9-III-1956, in flower, Univ. Baghdadensis Flora of Iraq Exsiccatae No. 60 (BM). Thesiger, W., Iraqui Kurdistan, Handiyan Village at the foot of the Baradort Mtns., 14-V-1951, in fruit (BM). Wahab, A. et al. 6794, Liyala Liwa, Sud ur Mansuria, course gravel, 27-III-1964, in fruit (K). **Israel:** Zohary, M., Upper Galilee, Tsuriel (Sukhmata) marl soil, 10-V-1955, in fruit, Univ. Hebraicae Hierosolymitanae Israëlis Exsiccata 649 (K). Zohary, M., Upper Galilee, Tsuriel (Sukhmata), marl soil, 10-V-1955, in fruit, Universitatis Hebraicae Hierosolymitanae Flora Israëlis Exsiccata No. 649 (BM, P). Zohary, M., Mt. Carmel, weed in lawn, alt. 350 m, west exposure, 16-IV-1968, in fruit (HUJ). **Israel/Syria (Golan):** Liston, A. & Lev-Ari, Y. 7-85-377/3, S. Hermon, Har Kata, basalt, moist ground, nitrogen rich (sheep stop), alt. 1100 m, 26-V-1985, in fruit (HUJ). **Italy:** Rogers, J.A. 730, Fiesoli, IV-1920, in flower & fruit (K). **Japan:** Dickins, M., Yokahama, 1877, in flower & early fruit (P). **Lebanon:** Davis, P.H. 5956, Beirut, 26-IV-1943, in flower & fruit (K). Maitland, T.D. 332, Shemlan, alt. 715 m, 19-IV-1958, in fruit (K). Norris, F.H., Beirut, sand hills, alt. 100 feet, 2-IV-1945, in fruit (BM). **Libya:** Guichard, K.M. KG/LIB/248, Leptis Magna, 11-III-1952, in flower & very early fruit (BM). Maire, A. & Weiller, M. 434, Tripolitania: in glareosis Amis Kam, 15-IV-1938, in fruit (P). (Type! of *M. minima* (L.) Grufl., var. *recta* (Desf.) Burnat, subv. *polyadenia* Maire; viewable online at http://www.aluka.org; current June 2010). **Morocco:** Davis, P.J. D48889, ca. 16 km from Irherm to Taroudannt, alt. 1350 m, rocky slope, 29-III-1969, in flower & fruit (BM). Maire, A., infra Igherur, in rupestribus siliceis Auti-Atlantis, alt. 1400–1500 m, 5-IV-1934, in flower & fruit (P). **New Caledonia:** McKee, H.S. 5195, Anse Vata, approx. sea level, weed in grass, 2-IX-1956, in flower & fruit (K). **Poland:** Kulikowski, E., Rossia austr., Grzbówko prope Ovidiopolim, 15-V-1900, Flora Polonicae Exsiccata Dr. F. Wutoszcak No. 813 (LE). **Romania:** Baenitz, C., Auf Felsen am "eisernen Thor" (Vercirova) alt. 70 m, 17-V-1895, in flower (P). Borza, Al. & Nyárády, E., Dobrogea, distr. Caliacra ad littorem Ponti Euxini (Marea neagra) prope "Farul de la Sabla", alt. 3–5 m, 14-VI-1925, in fruit, Museum Botanicum Universitatis Cluj, Flora Romaniae Exsiccata No. 1268 (K, P). **Socotra:** Balfour, B. 667, II-III-1880, in fruit (K). **Spain:** Brummitt, R.K. & Ernst, W.R. 5813, prov. Salamanca, Ciudad Rodrigo, on walls of town, 24-IV-1967, in flower & early fruit (K). **Sudan:** Aylmer, G. 600, Red Sea district in vicinity of Erkowit (18°50′N, 37°4′E), alt. ca. 4000 feet, 2-7-III-1936, in fruit (BM). **Sweden:** Sterner, R., prov. Öland, paroecia Räpplinge, juxta castellum oppidi Borgholm, 3-VII-1925, in flower & fruit, G. Samuelsson Plantae Suecicae Exsiccatae No. 1054 (K). **Switzerland:** Aellen, P., Basel, Badische Lagerhäuser, Grasplatz, 6-VI-1931, in flower & fruit (K). **Syria:** Haradjian, M. 2833 & 27888, Monts Nusairy, Bahama, 15 miles à l'E. de Ladikie, alt. ca. 1000 feet, IV-1909, in flower & fruit (K). Zohary, M., Antiochia, 8-V-1931, in fruit (HUJ). **Tibet:** David, Chine Thibet Oriental, Prov. de Moupin, IV-1870, in flower & fruit (P). **Turkey:** Balls, E.K. B2093, Adana, alt. 500 feet, grass slopes, stony clay and lime, 9-IV-1935, in fruit (BM). Bourgeau, E., Elmalu, in collibus, 3-VI-1860, in flower & fruit (P). Davis & Dodds 18764, prov. Ankara, Hacikadun Valley near Kecieren, fallow fields,

11-VI-1952, in flower & fruit (BM). **USA:** Bicknell, E.P., New York, Yonkers Wool Mill, Yonkers, in fruit, 1-IX-1894 (NY; collection mixed with *M. rigidula*). McCart, W.L. 1683, Texas, Collin Co., south of Wylie, roadside, 5-V-1939, in flower & fruit (P). Tharp, B.C. 44092, Texas, Travis County, Austin, 31-III-1944 (K). **USSR:** Goloskokov, V., Kazachstania, rami austro-occidentales jugi Alatau Transiliensis, in fluxumedio fl. Zapadnyj Dzhenyschke, in declivibus australibus lapidosis, 18-VI-1963 (BM, LE). Kulikowsky, E., prov. Cherson, distr. Odessa, 28-V-1900, (LE). Meyer, F.N. 503, prov. Daghestan, Caucasus, on dry steppes near Petrovsk, 15-V-1910, in flower & fruit (K). Sintensis, P., Regio transcaspica, Aschabad, in montibus supra Nephton, 4-V-1900, in flower & fruit, Iter Transcaspico-Persicum 1900–1901 No. 246a (P). Sintensis, P., Regio transcaspica, Stara Karakala in steppis, 26-V-1901, in fruit, Iter Transcaspico-Persicum 1900–1901 No. 1820 (LE). Voronov, G., Georgia, Caucasus, in rupestribus meridionalibus in Horto Tbilisiensi, 23-IV-1900, in flower & fruit (K). Voronov, G., Georgia, in rupestribus meridionalibus in Horto Tbilisiensi, 23-IV-1900, in flower & fruit (BM). **Yemen:** Wood, J.R.I., y/75/677, Jebel Shibam, Menacha, alt. ca. 2600 m, on rocky banks near a stream, I-IX-1975, in flower & fruit (BM). **Yugoslavia:** Gilliat-Smith, B. 3244, outside Sarajevo, along the Miljacka, roadside, 4-VII-1933, in flower & fruit (K).

53. *Medicago monantha* (C.A. Mey.) Trautv.

M. monantha (C.A. Mey.) Trautv., Bull. Sci. Acad. Imp. Sci. St. Petersb. 8: 272 (1841).

The epithet *monantha* is Greek for one-flowered, which is descriptive of some variants of the species.

Subordinate Taxa

Infraspecific taxa are not recognized here. The treatment follows Small and Fawzy (1992) in informally recognizing an "eastern" and a "western" group. This classification differs from most of the literature and, accordingly, the following historical background is presented for orientation.

In most floristic literature, the plants included here in *M. monantha* are placed in one of the following: *Trigonella monantha* C.A. Meyer, *T. geminiflora* Bunge (*T. monantha* subsp. *geminiflora* (Bunge) Rech. f.), *T. incisa* Benth. (*T. monantha* subsp. *incisa* (Benth.) Ali), or *T. noeana* Boiss. (*T. monantha* subsp. *noeana* (Boiss.) Huber-Morath). Critical details of previous taxonomies are summarized in the following.

The most authoritative comprehensive taxonomic work available on the genus *Trigonella* is that of Širjaev (1928–1934). He accepted four species in the *M. monantha* complex, keyed as follows (translated from Latin):

Fig. 53-1. *Medicago monantha* (western phase). (Based on R. Görz 547, DAO; artist: B. Flahey.)

1. Racemes scarcely pedunculate (up to 4 mm) or not pedunculate 2
1. Racemes shortly pedunculate (5–9 mm) 3
  2a. Flowers mostly paired (3); calyx teeth linear, glabrous; areoles of legume oblong with subcontiguous sides (ratio 3–4:1), sometimes strongly thickened; keel slightly shorter than wings; lateral leaflets obovate or elliptical, occasionally incised-dentate .... *T. noeana*
  2b. Lower flowers solitary, upper ones paired; calyx teeth lanceolate, glabrous; areoles of legume subquadrate, the sides not subcontiguous; keel slightly shorter than wings; lateral leaflets triangular, never incised (Turkestan) .... *T. geminiflora*
  2c. Flowers mostly solitary; calyx teeth subulate, pubescent; areoles of legume very narrow, with subcontiguous sides; keel notably shorter than wings; lateral leaflets obovate or elliptical, sometimes incised .... *T. monantha*
    3. Areoles of terete legume more or less elongately curved (ratio 4:1) (Mesopotamia) .... *T. noeana* f. *pedunculata*
    3. Areoles of compressed legume rather straight, pentagonal, or quadrate (ratio 1:1) (northern India) .... *T. incisa*

Vassilczenko (1953) also studied *Trigonella* extensively. He distinguished taxa in the *M. monantha* complex much as did Širjaev (translated from Russian):

1. Pedicels with fruit 5–10 mm long; mature fruits to 3 mm wide; areoles of fruit square or pentagonal, their length equal to their width (India) .... *T. incisa*
1. Combination of features different 2
  2. Areoles of fruits almost square; lateral leaflets inversely triangular .... *T. geminiflora*
  2. Areoles of fruits elongated, drawn-out along the fruit, 3–4 times longer than wide; lateral leaflets ovate or obovate 3
    3. Areoles of fruits with a distinct lumen medially, walls of areolae comparatively thin; teeth of calyx linear, almost glabrous .... *T. noeana*
    3. Areoles of fruits with an almost imperceptible lumen medially, the walls thickened and almost contiguous; teeth of calyx subulate, pubescent .... *T. monantha*

For the treatment of *Trigonella* in regional floras, Huber-Morath (1970) reduced *M. noeana* to a subspecies of *T. monantha*, Ali (1977) reduced *M. incisa* to a subspecies of *T. monantha*, and Rechinger (1984) considered all four taxa alleged to be species as subspecies of *T. monantha*.

The most recent and thorough investigation of the *M. monantha* complex is by Rechinger (1984), who separated the taxa as follows (translated from Latin):

1. Inflorescences pedunculate .... subsp. *incisa*
1. Inflorescences sessile 2
  2. Areoles of legumes isodiametric; legume veins thickened; marginal suture of legume thick .... subsp. *geminiflora*
  2. Areoles of legumes about as long as wide; legume veins thin; marginal suture of legume thin 3
    3. Flowers mostly paired, sometimes 3, rarely solitary; areoles of fruit not much longer than wide .... subsp. *noeana*
    3. Flowers mostly solitary, rarely paired; areoles of fruit clearly longer than wide .... subsp. *monantha*

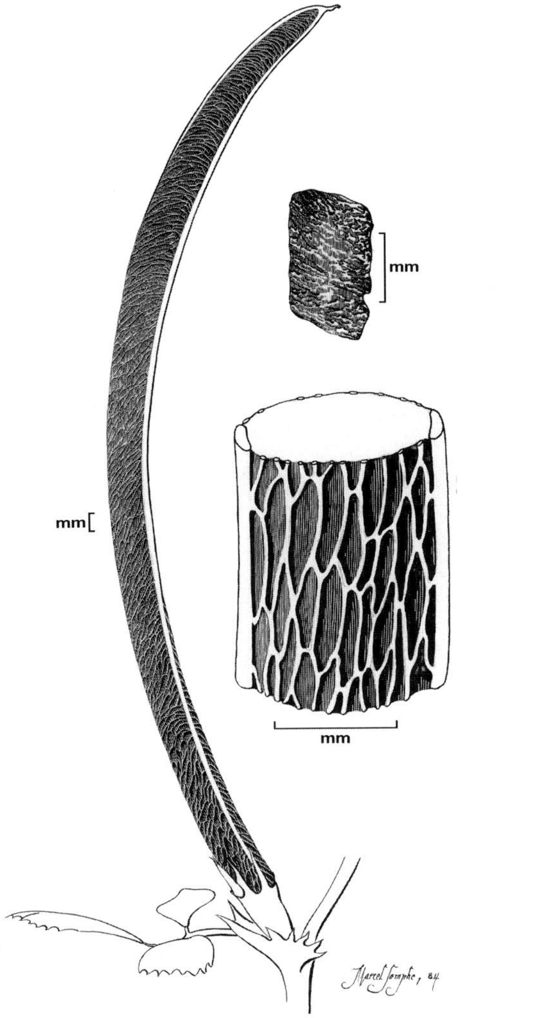

Fig. 53-2. *Medicago monantha* (western phase). **Clockwise from left:** Fruit; seed; portion of fruit showing venation. (Based on E. Small Turkey-E38, DAO; artist: M. Jomphe.)

Fig. 53-3. *Medicago monantha* (eastern phase). Clump of vegetative plants. (Based on greenhouse-grown plants, E. Small M2051, DAO; seeds obtained from The United States Department of Agriculture, PI 222273 (misidentified in the GRIN (Germplasm Resources Information Network) database as *Medicago medicaginoides* (Retz.) E. Small = *T. arcuata* C.A. Mey.), seeds originally from Shah's Reserve, Tehran, Iran; artist: B. Flahey.)

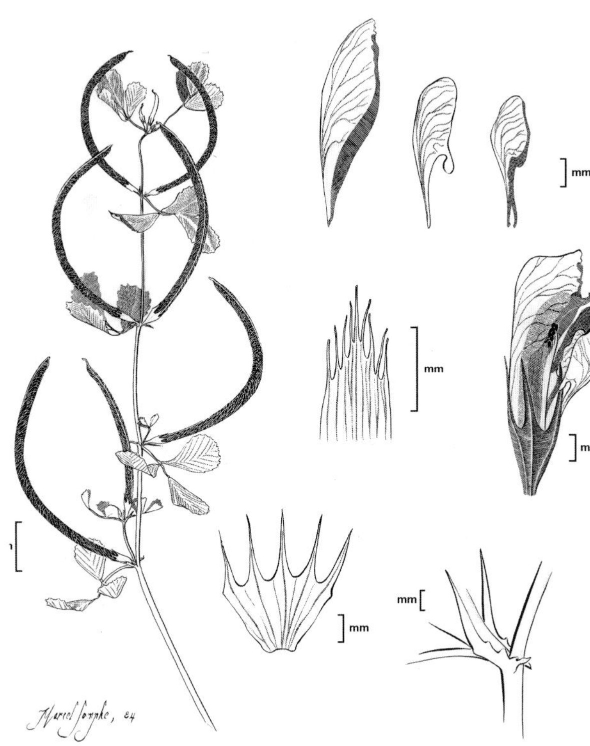

Fig. 53-4. *Medicago monantha* (eastern phase). **Left:** Plant. **Top right:** Floral parts (clockwise from left: standard petal; wing petal; keel; side view of flower; flattened androecial column (free stamen not shown)). **Lower right:** Stem node showing stipules. **Lower center:** Calyx. (Based on V. Goloskokov 1963, A; artist: M. Jomphe.)

Rechinger (1984) gave the following distribution areas for the subspecies he recognized (basically much like the ranges given in Širjaev 1928–1934): *monantha* — western area of the range; *noeana* — western and central area of the range (but extending to Afghanistan); *geminiflora* — southern Persia, Afghanistan, Pakistan, Central Asia; *incisa* — Persia, Afghanistan, Pakistan, India. It is worthwhile noting that Rechinger pointed out that his four subspecies were distributed from west to east in the sequence *monantha*, *noeana*, *geminiflora*, and *incisa*. Vassilczenko (1953) gave the following rather comparable ranges: *monantha* — Asia minor, Transcaucasus, Talysh, Eastern Mediterranean, northwestern Iran; *noeana* — Eastern Mediterranean, Iraq, Turkey, Caucasus, Transcaucasus, Talysh, Iran, Afghanistan; *geminiflora* — Afghanistan, Pakistan, Mongolia, China (Tientsin); *incisa* — northern India, Pakistan, Afghanistan.

The fruits of *M. monantha* (Figs. 53-2, 53-5) are normally curved, but sometimes the curving at the end of the pod is almost hook-like. The taxonomic interpretation of certain specimens of the *M. monantha* complex with hook-shaped pods is problematical. These are often placed in *Trigonella macroglochin* Dur. and *T. uncinata* Banks & Soland. Small et al. (1987) pointed out that the latter name is not applicable to the *M. monantha* complex. Rechinger (1984) remarked with respect to *T. macroglochin* (translated from

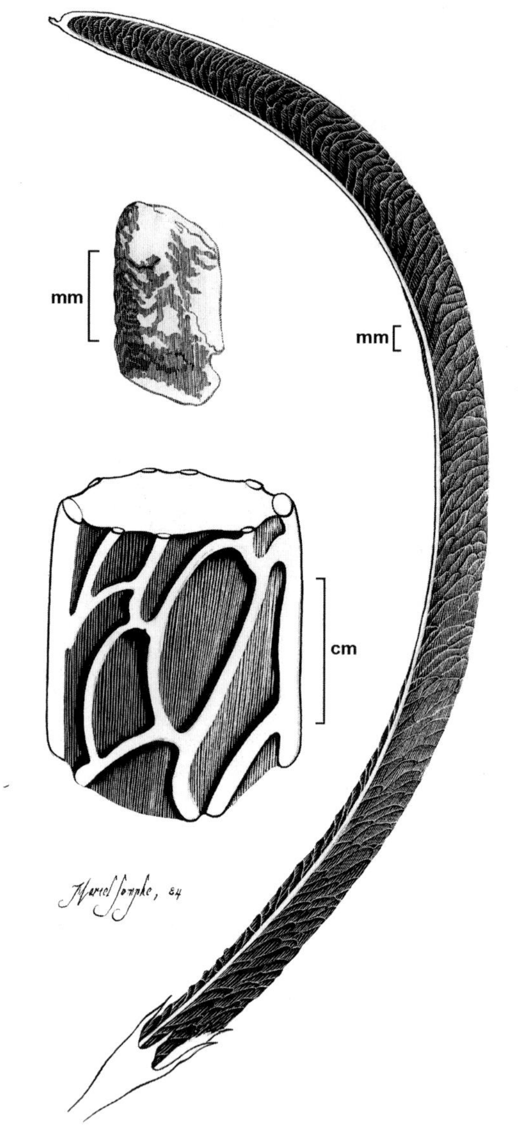

Fig. 53-5. *Medicago monantha* (eastern phase). **Right:** Fruit. **Top left:** Seed. **Center left:** Portion of fruit showing venation. (Based on V. Goloskokov 1963, A; artist: M. Jomphe.)

German): "I have given this taxon specific rank not so much out of the conviction that it is indeed a separate species but rather for practical reasons, namely to avoid complicating the differentiation of the extremely polymorphic *T. monantha* even further." Širjaev (1928–1934) and Townsend (1974) recognized this as a variety, *T. monantha* var. *macroglochin* (Durieu) Širj. This very poorly collected group does not appear to merit separate distinction at any rank from *M. monantha*.

As discussed in Small and Fawzy (1992), previous taxonomists did not perceive that the four commonly recognized taxa making up the *M. monantha* complex replace each other in an east-west direction in a smooth, clinal fashion with respect to morphology and geography. Since a smooth (as opposed to a stepped cline) can only be divided into taxa arbitrarily, the splitting of the variation pattern into taxa is questionable.

The most evident difference is with respect to the areas delimited on the fruit surface by the intersecting veins: the fruit areoles are more elongated in the western phase than in the eastern. As a rough guide, most collections west of 45° E longitude are assignable to the western phase and most collections east of 65° E longitude are assignable to the eastern phase.

Type Specimen

C.A. Meyer 148, Talysch, in campis et collibus siccis lapidosis prope pagum Svant (Azerbaijan), 20-vi-1830, LE!

Synonymy

Homotypic

Trigonella monantha C.A. Mey., Verz. Pfl. Casp. Meer. 137 (1831), (basionym).

Heterotypic

T. noeana Boiss. var. *aleppica* Zohary, Beih. Bot. Centralbl. LI: 280 (1934).

T. monantha (C.A. Meyer) Trautv. var. ß *brachycarpa* Boiss., Fl. Or. 2: 78 (1872) (type: Ky. exs. 470: Kotschy, Th., rara in arvis prope Marra pagum, alt. 4000 feet, 25-V-1855, in fruit (Herb. Boiss. G!).

T. noeana Boiss. f. ε *brachycarpa* (Boiss.) Širj., Generis

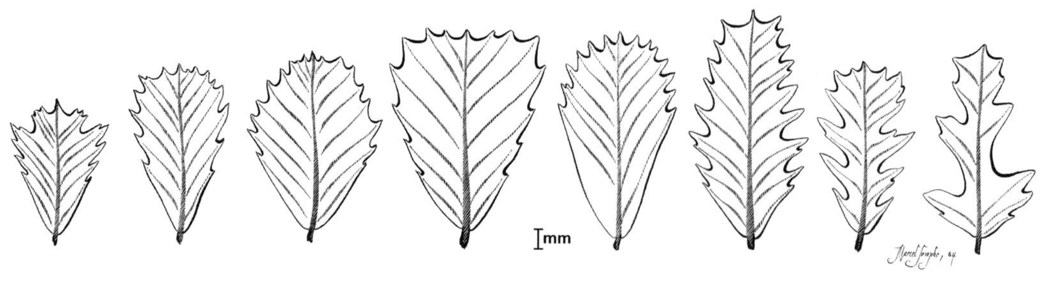

Fig. 53-6. *Medicago monantha* (eastern phase). Leaflets from various collections, illustrating the range of leaflet incision in the species. (Artist: M. Jomphe.)

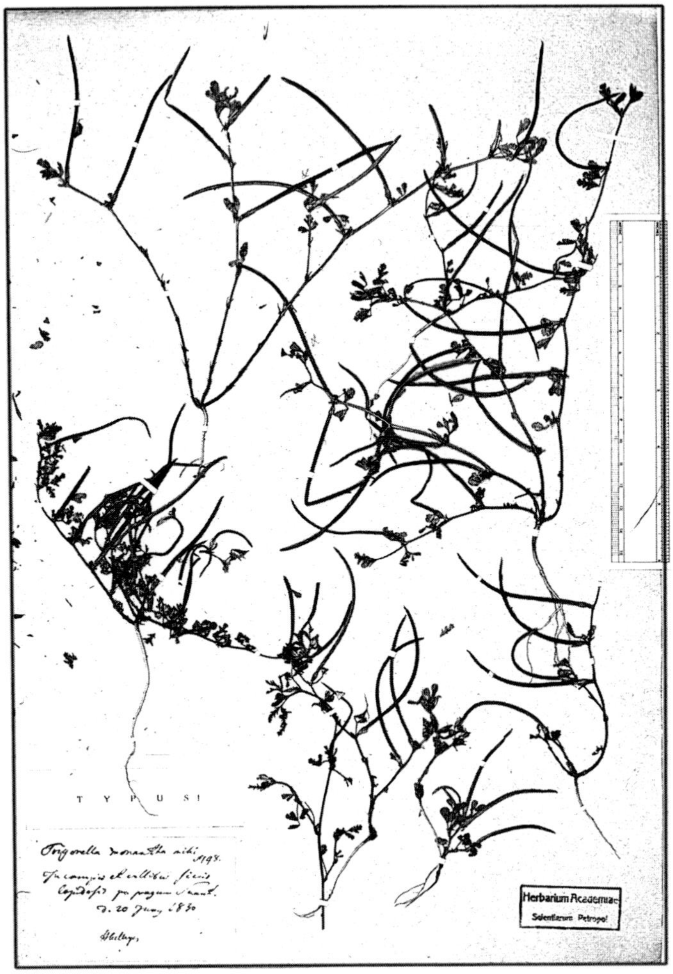

Fig. 53-7. Holotype of *Trigonella monantha* C.A. Meyer at LE.

Trigonella L. revisio critica, Spisy Pfir. Fac. Masarykovy Univ., part 4: 8 (1931).

T. brahuica Boiss., Fl. Orient. 2: 73 (1872). (At Herb. Boiss. at G, I observed three specimens labelled *T. brahuica* Boiss., which were annotated by K.H. Rechinger as *T. monantha* subsp. *incisa* and by A. Huber-Morath as *T. incisa*; one of these specimens may be appropriate as the type.)

T. geminiflora Bunge var. *curvata* Širj. & Rech. f., Dan. Biol. Skr. 9, 3: 12 (1957).

M. dicarpa Trautv., Trudy Glavn. Bot. Sada 1: 24: (1870).

T. geminiflora Bunge, Arbeiten Naturf. Vereins Riga 1: 219 (1847). (Reliq. Lehm. 105 (1847) cited in older literature.) (The type is: Lehmann, Kizyl Kum: Tjumenbaj N Bachara. Holotype LE!; isotypes at K! & W! At G, I observed a scrappy specimen in an envelope with Bunge inscribed, which was annotated isotypus (21.9.1967) by A. Huber-Morath.)

T. monantha subsp. *geminiflora* (Bunge) Rech. f., Flora Iranica 2: 244 (1984).

T. incisa var. *geminiflora* (Bunge) Boiss., Fl. Or. 2: 76 (1872).

T. monantha (C.A. Meyer) Trautv. var. *genuina* Širj., Generis *Trigonella* L. revisio critica, Spisy Pfir. Fac. Masarykovy Univ., part 4: 4 (1931), nom. inadmiss.

T. incisa Benth. in Royle, Illustr. Bot. Himal. Mount. 197 (1835). (Type at K; appears to be a collection by Royle on a sheet along with two other collections: Royle 1035 Herbarium Benth., in India Orientale, in flower & fruit. Širjaev (1931 p. 34) presented a photo of a fruit of what he termed a cotype.)

T. monantha (C.A. Meyer) Trautv. subsp. *incisa* (Benth.) A. Jelen. & G. Kupat., Biologischeskii Zhurnal Armenii 21(6): 69 (1968). (Ali (1977) made the same combination in Fl. W. Pakistan 100: 294 (1977), but it is precluded by Jelenevskii and Kupatadze (1968).)

M. monantha (C.A. Meyer) Trautv. subsp. *incisa* (Benth.) Verloove & Lambinon, Syst. Geog. Plants 76(2): 220 (2006).

T. incisa Fr., Bull. Herb. Boiss. 3(4): 41 (1904), nom. illeg. (non *T. incisa* Benth.).

T. noeana Boiss. f. α *integrata* Širj., Generis *Trigonella* L. revisio critica, Spisy Pfir. Fac. Masarykovy Univ., part 4: 7 (1931). (ex Godr. in sched., Sec. Thell. Fl. Adv. Montp. 301 (1912).)

T. macroglochin Dur., Bull. Soc. Bot. France. 6: 593 (1859).

T. monantha var. *macroglochin* (Durieu) Širj., Generis *Trigonella* L. revisio critica, Spisy Pfir. Fac. Masarykovy Univ., part 4: 4 (1931).

T. noeana Boiss. var. *minor* Post, J. Linn. Soc. XXIV, 425 (1888).

T. incisa Benth. f. *multiflora* Širj., Generis *Trigonella* L. revisio critica, Spisy Pfir. Fac. Masarykovy Univ., part 4: 11 (1931).

T. noeana Boiss., Diagn. P. Orient., Ser. 2, 3(2): 11 (1856). (The type of *T. noeana* Boiss. is given by Huber-Morath (1970) as: Noë, Iraq, in herbidis Babyloniae ad Kutt, V-1851 (Herb. Boiss., G). Neither A. Huber-Morath nor K.H. Rechinger, who indicated in their publications that they saw the type, actually labelled a specimen as the type. Two specimens could be chosen as the holotype, Herb. Noëanum 1020 and Herb. Noëanum 54. I chose the former and attached a label in 1993. The specimen bears the information: "Herbarium Noëanum No. 1020, Kutt in herbidis legi, Majo 1951" (Babylonia ad Kutt collection of Noë is one of the specimens cited by Boissier 1972 p. 77).)

T. monantha C.A. Meyer subsp. *noeana* (Boiss.) Huber-Morath in P.H. Davis, Fl. Turkey 3: 475 (1969).

M. monantha (C.A. Meyer) Trautv. subsp. *noeana* (Boiss.) Greuter & Burdet, Willdenowia 19: 32 (1989).

T. monantha C.A. Meyer f. *pinnatifida* Thell., Fl. Adv. Montp. 301 (1912).

T. noeana Boiss. f. β *pinnatifida* Beck. ex Širj., Generis *Trigonella* L. revisio critica, Spisy Pfir. Fac. Masarykovy Univ., part 4: 8 (1931).

T. noeana Boiss. f. ζ *pedunculata* Širj., Generis *Trigonella* L. revisio critica, Spisy Pfir. Fac. Masarykovy Univ., part 4: 8 (1931).

M. psilocarpa Trautv., Linn. 15: 97 (1841).

T. noeana Boiss. f. δ *reticulata* Širj., Generis *Trigonella* L. revisio critica, Spisy Pfir. Fac. Masarykovy Univ., part 4: 8 (1931).

T. noeana Boiss. f. *tricarpa* Beck in Stapf, Polak Exp. 2: 63 (1886).

"*T. uncinata*" Eig, Journ. Bot. 75: 188 (1937). Not *T. uncinata* Banks & Soland., in Russ., Nat. Hist. Aleppo, ed. 2, 2: 261 (1794).

Not

M. monantha Blatter, J. Bombay Nat. Hist. Soc. 36: 969 (1933) (= *M. laciniata*).

Trigonella polyceratia L. (but *M. monantha* has been identified as "*T. polyceratia*" by some authors, e.g., *T. polycerata* (sic) Bak. in Hook. (frequently known by this name in India, according to Cristurean 1972), Fl. Br. Ind. 2: 87 (1879); *T. polyceratia* Anth., Notes Roy. Bot. Gard. Edinb. 18: 285 (1939)).

Vernacular Names

English: No names found.

French: No names found.

Geographical Distribution

(N = native, I = introduced, U = uncertain status)

Asia: Afghanistan (N), Armenia (N), Azerbaijan (N), Bhutan (N), China (N), India (N), Iran (N), Iraq (N), Israel (N), Jordan (N), Lebanon (N), Mongolia (N), Pakistan (N), Qatar (N), Saudi Arabia (N), Syria (N), Tadzhikistan (N), Turkey (N), Turkmenistan (N), Uzbekistan (N).

Europe: Belgium (I; for discussion, see Verloove and Lambinon 2006).

Northern America: USA (Massachusetts; I; not established).

Medicago monantha is native to a wide area of Asia.

Description

Annual herb, prostrate to ascending, branching from base, stems (5)15–30(40) cm long, vegetative parts, calyx and pod diffusely pubescent with adpressed simple hairs. Leaflets 5–12(20) mm long, 3–8(10) mm wide, obovate, oblong or elliptical, the distal margin denticulate, sometimes laciniate, pinnatifid or incised-dentate. Stipules semi-sagittate, the margin of the base denticulate or not. Inflorescence of 1–3(7) flowers, peduncle 0–5 mm long. Calyx campanulate, (3)5–6 mm long, teeth lanceolate to subulate and about as long as tube. Corolla yellow, 6–8 mm long. Fruit linear, erect or spreading, 4–10 cm long, 1–2 mm wide, subcompressed, straight to semi-circular, the distal end sometimes uncinnately curled, with anastomosing veins producing a reticulate venation. Seeds numerous, oblong-cylindrical, 2–3.5 mm long, 1 mm in diameter, brown to yellowish, minutely tubercled, radicle about 80% of length of seed.

Identification Hints

This species is fairly easily distinguished from other *Medicago* by its long, curved, single or paired fruits.

Habitat and Environmental Adaptation

Habitat

Medicago monantha occurs in a variety of habitats, including steppes, dry rocky slopes, semi-deserts, and fallow and cultivated lands.

Phenology

Flowering occurs in the spring, from March to July depending on location. Halloran and Pennell (1981), based on experiments in Australia, concluded that because *M. monantha* tended to flower quite early in its development, it had potential as an early-flowering forage for growing in the lower rainfall agricultural areas of southern Australia.

Soil

This widespread species occurs in a variety of soils, including clay-silt and gravelly or stony substrates.

Temperature

Medicago monantha tolerates the temperatures associated with a range of continental, Mediterranean, and semi-desert conditions in spring and early summer.

Water

This species typically occurs in moderately dry soils.

Biotic Relationships

Pollination

Small (1988) found a mean pollen to ovule ratio of (238), an observation that is consistent with *M.monantha* being largely self-pollinated and substantially an inbreeder. The flowers are slightly larger than most annual species of *Medicago*, suggestive that at least some outcrossing occurs.

Dispersal

No specific dispersal mechanisms have been identified, and the generalized fruit likely distributes its seeds through a combination of animals, wind, water, and gravity.

Herbivory

Leaf laciniation, as discussed in Chapter 7, may be an adaptation to deterring insect herbivorves, since it is suggestive that insect attack has already occurred, and the fact that laciniation in species of *Medicago* is often much more common in some particular area of the entire distribution range may indicate that the feature is there to repel a local insect. With respect to *M. monantha*, Halloran and Pennell (1981) noted that "one character exhibiting wide variation is the degree of indentation of the leaf, which varied from entire to deeply indented. The cut-leaf character occurred in this species at 550 m and above, indicating the possibility that the variation in this character is of adaptive significance or is closely linked to, or a pleiotropic expression of, genes conferring some measure of adaptability."

Nodulation and mycorrhizae

USDA (2009) reported both positive and negative root nodulation for *M. monantha* (the only species of the genus for which a negative root nodulation report was available). Athar and Mahmood (1978) and Balodi and Rao (1991) reported the presence of nodules in *M. monantha*.

Disease resistance

No information located.

Economic History, Cultivars, and Uses

Medicago monantha has not been developed as an agricultural crop.

Cultivars

No cultivars have been bred.

Forage

Medicago monantha is not sown for forage.

Fodder

According to Bentham in his original description of his new species *M. incisa*, it is "used as fodder for cattle" in northern India, presumably as material simply harvested from rangelands.

Green manure, rotation crop, ground cover

No information located.

Other uses (salad sprouts, ornamental, plant products, etc.)

The species (as *Trigonella incisa*) has been used in folk medicine in India (http://envis.frlht.org.in/botanical_search.php?gesp=1645%7CPhyllanthus+amarus+SCHUM.+%26+THENN; current June 2010).

Chromosome Number

$2n = 30$: Singh and Singh 1976.

$2n = 44$: Gardé 1948 (as *Trigonella geminiflora*); Murin and Sheikh 1971 (as *T. noeana*); Kliphuis and Barkoudah 1977 (as *T. noeana*); Bir and Kumari 1979 (as *T. incisa*); Kumari and Bir 1990 (as *T. incisa*).

$2n = 48$: Khatoon and Ali 1991 (as *T. monantha* subsp. *incisa* of Pakistan).

The numbers reported here are exceptional. They require confirmation and exploration of their significance.

Genetic Relationships and Gene Transferability

Intraspecific

Artificial hybridization of different populations has not been attempted, so it is not known whether there are barriers to hybridization within the species.

Interspecific

No information located.

Exchange by advanced biotechnology

No information located.

Conservation Status

This species is widespread and so is not in danger of extinction. However, it is very likely that there are regional variants of interest not yet discovered, so there is a need for botanical research towards the end of ascertaining the necessity of conservation measures. It has been listed (as *Trigonella incisa*) as a "traded plant of conservation concern" in India http://envis.frlht.org.in/botanical_search.php?gesp=1645%7CPhyllanthus+amarus+SCHUM.+%26+THENN; current June 2010).

Germplasm Sources

More than 500 accessions of *M. monantha* are available in major germplasm centers (see Chapter 3).

Research Problems and Priorities

Medicago monantha has received limited research to date. From a basic botanical perspective, this very widespread and variable species requires taxonomic clarification. One of the chromosome counts recorded, $n = 24$, is higher than that of any other annual species of *Medicago*, and is indicative of high polyploidy and indicates that it is necessary to examine chromosomal number variation throughout its geographical range. From an agricultural perspective, this species is exceptionally adapted to stressful climates (that have limited water, high heat, and poor soils) and so it likely harbors genes valuable for the improvement of alfalfa.

Representative Collections

Afghanistan: Ekberg, L. 9021, Prov. Kunduz, 25 km north of Kunduz, ca. 5 km east of the road, grassy slopes, 8-VI-1969, in fruit. Hedge, I. & Wendelbo, P. 3066, prov. Kabul, Kabul, upper part of Tang-i-Gharu, stony slopes, alt. 1800 m, 8-V-1962, in flower & fruit (E). Hedge, I., Wendelba, P. & Ekberg, L. 8201, Prov. Badghis, Darrah-i-Bum, alt. ca. 800 m, stony slopes, 19-V-1969, in fruit (E). **Bhutan:** Grierson, A.J.C. & Long, D.G. 2932, Phuntsholing district, Phuntsholing, 26°52′, 89°23′, alt. ca. 250 m, cultivated ground in park, 15-II-1982, in flower & fruit (E). **Iran:** Archibald, J.C. 1768, Arak area, east of Arak, alt. 1645 m, low volcanic outcrops, lava debris and lava sand in full sun, 3-V-1966, in flower and fruit (E). Godmanso, E. & E., Persepolis, alt. ca. 4000 feet, stony clay limestone, 9-IV-1960, in flower & fruit (BM). Koelz, W. 16475, Shahkuh, Mazenderan, in bare torrent bed, 19-VII-1940, in flower and fruit (E). Lamond, J. & Iranshahr, M. 3391, Prov. Azerbaijan, Mianeh to Zanjan, Kaflan Kuh, electricity station road opposite old Turkish bridge, amongst hillside cornfields, 29-V-1971, in flower & fruit (E). Parris, B.S. 75.306, Kerman prov., nr. Khabr village, west side of Kuh-e-Khabr, ca. 28° 51′N, 56° 23′N, alt. 2300–2400 m, under shrubs amongst limestone rocks on valley slopes, 2-V-1975, in flower & fruit (E). Rechinger, K.H. & F. 4158, prov. Khorasan, Shoukatabad prope Birdjand, in incultis, alt. ca. 1400 m, 24-V-1948, in fruit (E). Rechinger, K.H. & F. 4397, Prov. Khorasan, in collibus ad Turbat-e Haidari, alt. ca. 1300 m, 27-V-1948, in flower & fruit (E). Rechinger, K.H. & F. 5831, Prov. Shahrud-Bustam, in declivibus australibus montium Shahvar ad Nekarman (Nigarman), alt. 2000 m, 20-26-VII-1948, in flower & fruit (E). **Iraq:** Eog. A. & Zohary, M., northern Iraq, 79 km northeast of Deltawah (north of Baghdad), foothill of Jabol Hamrin, alt. 100 m, light sandy soil with projecting rocks, 12-IV-1933, in flower & early fruit (HUJ). Low, A. 107, Tigris plain, stony hillside, alt. 800–1000 feet, 20-III-1936, in flower (BM). Wheeler Haines, R. 909, Jebel Hamrin, plentiful in disturbed stream bottoms, 5-IV-1957, in flower & fruit (E). Wheeler Haines, R. 633, Shaglawah, alt. 2800 feet, rich grass of irrigated orchard, pasture, 9-IV-1956, in flower & fruit (E). Wheeler Haines, R., Daoudia, near Baghdad, river bank in mud from Adheim flood, 25-III-1961, in flower & fruit (E). **Israel/Syria (Golan):** Liston, A. 7-84-220/12, Hermon, Nemneman ridge, above Mann Valley, rocky slope, alt. 1500 m, 28-V-1984, in fruit (HUJ). Liston, A. & Lev-Ari, Y. 7-85-371/2, S. Hermon, upper Nahal Arar, scree, alt. 1500–1550 m, 25-V-1985, in fruit (HUJ). Shmida, A., S. Hermon, 4 km north-northwest of Majdel Chams, hard Jurassic limestone, alt. 1650 m, 10-VI-1976, in fruit (HUJ). **Jordan:** Eig, A. & Zohary, M., Edom, Wadi Musa, mountains, 17-IV-1929, in fruit (HUJ). Eig, A. & Zohary, M., Edom, 26 km south of Ma'an, 18-IV-1929, in fruit (HUJ). Eig, A. & Zohary, M. Ammon, Ein Suela, Nubian sandstone, 14-IV-1929, in fruit (HUJ). **Pakistan:** Burtt, B.L. B.504, District Peshawar, Near Akora, ca. 4 miles west of Altock bridge, Nullah leading down to Kabul R, 1-IV-1958, in flower & fruit (E). Lamond, J. 1017, Baluchistan, Quetta, Quetta to Chaman, Yaro, near Bostan, ca. 40 km from Quetta, alt. ca. 1450 m, sandy flat, just after branch of Pishin road, 8-V-1965, in flower & fruit (E). Lamond, J. 1425, Quetta, Mekhtar to Murgha Kibzai, alt. ca. 1500–1600 m, field-side weed, 18-V-1965, in flower & fruit (E). Lamond, J. 885, Baluchistan, Quetta, Bolan Pass, south of Quetta, alt. 1500 m, rocky roadside slopes, 6-V-1965, in fruit (E). **Syria:** Eig, A. & Zohary, M., top of Jebel Abd el Aziz, alt. 620 m, Cenomanian stones, gray steppic soil, 29-IV-1933, in fruit (HUJ). Kotschy, Th. 470, prope Marra pagum alt. 4000 feet, rara in arvis, 25-V-1855, in fruit, Iter Syriacum 1855, a Damasco versus Palmyrum (P). Zohary, M., northern Syria, Telejin to Abudhur, 5-V-1931, in flower & fruit (HUJ). Zohary, M., northern Syria, Muslemie, 6-V-1931, in fruit (HUJ). **Turkey:** Balansa, B. 1223, Ouchak (Phrygie), vers 910 m d'alt., champs en friche, 8-VI-1957, in fruit (P). Coode, M.J.E. & Jones, B.M.G. 1720, Çorum, Iskilip, 26 miles from Çorum to Iskilip, alt. 600 m, saline marshy ground, 29-IV-1965, in fruit (E). Coode, M.J.E. & Jones, B.M.G. 1999, Amasya 5 km east of Merzifon, alt. 600 m, wheatfield weed, 2-VI-1965, in flower and fruit (E). Davis & Hedge D.28096, Prov. Urfa, Akçakale, alt. 400 m, waste ground at edge of field in plain, 17-V-1957, in flower & fruit (E). Davis & Hedge D.28918, Prov. Elazig, Elazig-Kale, on Euphrates, 16 miles from Elazig, alt. 1300 m, 4-VI-1957, in flower & fruit (E). Davis 42405, Urfa, 40 km from Viransehir to Urfa, alt. 700 m, fallow calcareous fields, 2-V-1966, in fruit (E). Davis 42888, Mardin, 11 km from Midyat to Gerçüs, alt. 1100 m, limestone steppe, 13-V-1966, in flower & fruit, (E). Davis 42997, Mardin, 2 km north of Gerçüs, alt. 900 m fallow-fields, 14-IV-1966, in flower & fruit (E). Davis 44299, Van, 7 km from Van to Erçek, alt. 1850 m, igneous slopes, in steppe, 5-VI-1966, in flower & fruit (E). Helbaek 2468, Prov. Konya, Konya to Çumra, Kücük Köy, alt. 980 m, saline areas, 18-V-1962, in flower & fruit (E). **USSR:** Goloskakov, V. 1847, Kazachstania, montes Karatau Syrdariensis, planities in viciniis pag. Aral-TJjube, in declivibus australibus lapidosis, 26-V-1963, in flower & fruit (BM, E). Grossheim, A., distr. Nachitshevan, in montibus, Darry-Dagh, 21-V-1928, in flower & early fruit (HUJ). Grossheim, A. & Gurvitsh, C., Transcaucasia, respublica Nachitschevan, prope Shach-bux, Nachischevan-tschaj, 19-V-1934, in flower & fruit (HUJ).

54. *Medicago monspeliaca* (L.) Trautv.

M. monspeliaca (L.) Trautv. in Bull. Sci. Acad. Imp. Sci. St. Petersb. 8: 272 (1841).

The epithet *monspeliaca* is based on the Montpellier region (near the mouth of the Rhône in southern France), where specimens were collected and used for the original description of this species.

Subordinate Taxa

Infraspecific taxa are not recognized here. *Medicago monspeliaca* is extremely polymorphic and has been extensively subdivided, as indicated below in Synonymy, by petiole length (often notably long in the species), pubescence and fruit length, and whether the leaflets are incised. None of

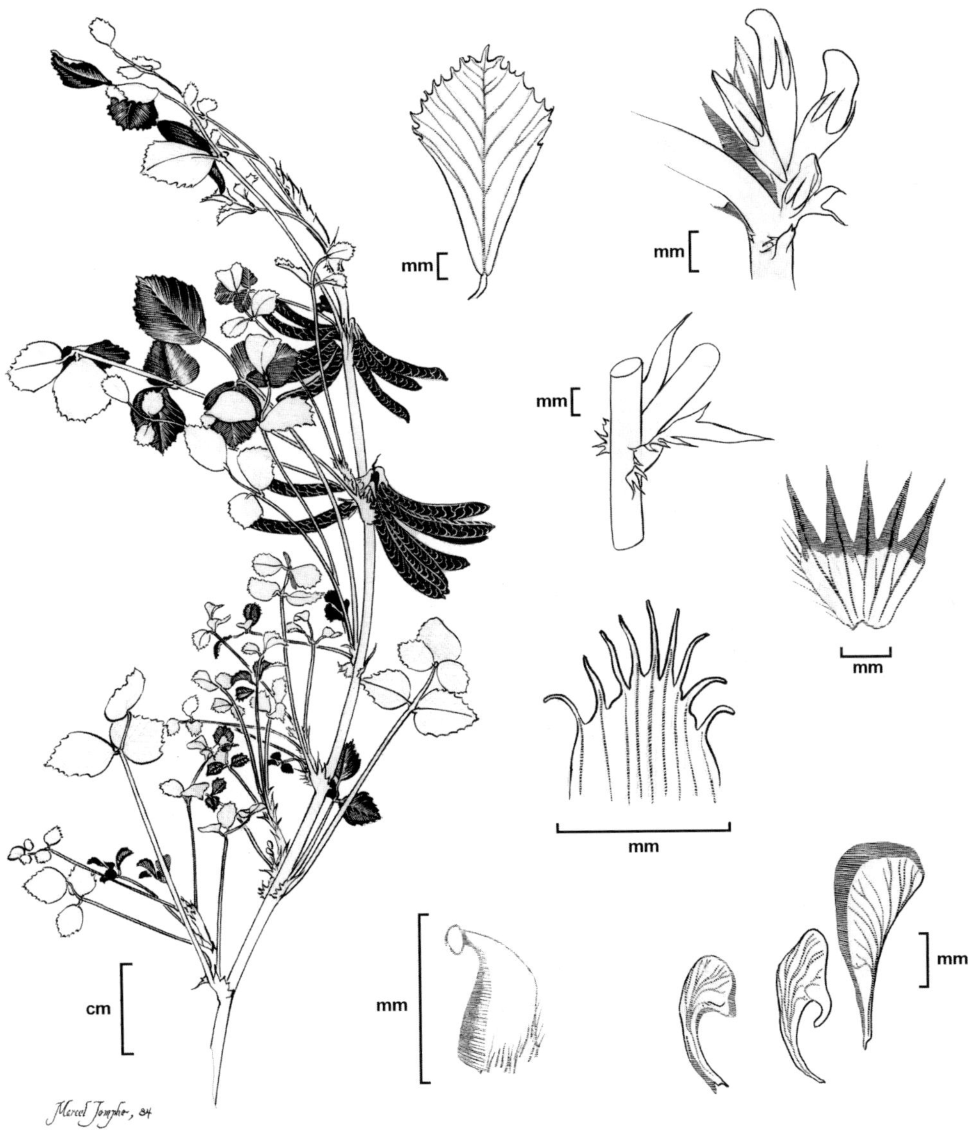

Fig. 54-1. *Medicago monspeliaca.* **Clockwise from left:** Fruiting branch; leaflet; group of flowers; stem node showing stipules; calyx; flattened androecial column with the free stamen not shown); petals (left to right: keel; wing petal; standard petal); pistil. (Based on E. Small T252, DAO; artist: M. Jomphe.)

these characters has been shown to be geographically based or correlated with other characters, and accordingly worthy of taxonomic recognition.

Type Specimen

The Linnaean Plant Name Typification Project (http://www.nhm.ac.uk/research-curation/projects/linnaean-typification/; current June 2010) provides a photograph of the type and the following information: "Linnaean Name: *Trigonella monspeliaca* Linnaeus. Reference: *Species Plantarum* 2: 777 (1753). Provenance: 'Habitat Monspelii.' Type: Lectotype: Herb. Linn. No. 932.13, right specimen (LINN). Designated by: Lassen in Turland & Jarvis (ed.), Taxon 46: 482 (1997)... Comments: Ali (in Nasir & Ali, Fl. W. Pakistan 100: 292. 1977) and others have treated Herb. Linn. No. 932.12 (LINN) as the lectotype, but this cannot be regarded as relevant original material because it lacks Linnaeus' annotation with the *Species Plantarum* species number (in this case '5') and is a post-1753 addition to the collection and not original material for the name."

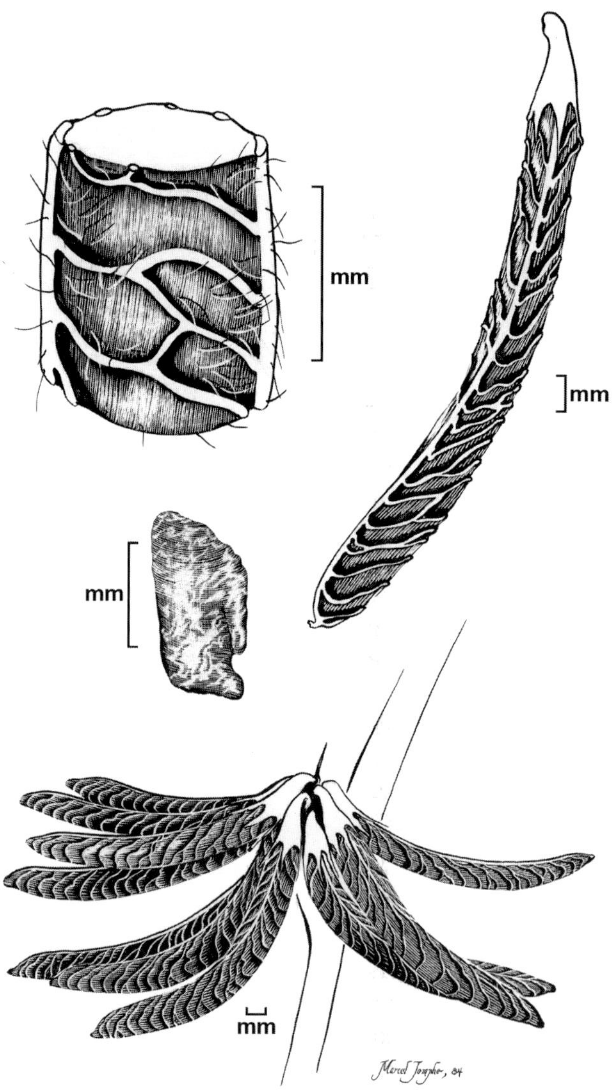

Fig. 54-2. *Medicago monspeliaca.* **Clockwise from bottom:** Group of fruits; seed; portion of fruit showing venation; fruit. (Detail of venation based on E. Bourgeau 04 April 1951, G, remainder based on E. Small T252, DAO; artist: M. Jomphe.)

Synonymy

Homotypic

T. monspeliaca L., Sp. Pl. 777 (1753), (basionym).

Buceras monspeliacum (L.) All., Fl. Pedem. 1: 313 (1785).

Falcatula monspeliaca (L.) Fourr., Ann. Soc. Linn. Lyon sér. 2, 16: 360 (1868).

Telis monspeliaca (L.) Kuntze, Revis. Gen. Pl. 209 (1891).

Heterotypic

T. monspeliaca L. subsp. *caulescens* Feinbr., Beih. Bot. Centralb. Bd. 51(2): 396 (1934).

T. divaricata Clairv., Man. Herb. 132 (1811), nom. illeg.

T. monspeliaca L. f. *eigii* Širj., Generis *Trigonella* revisio critica, Spisy Pfir. Fac. Masarykovy Univ. 4: 24 (1931). (Treated by Vassilczenko (1953) and Zohary (1972) as varieties, neither making clear that they were authoring new combinations, likely because of uncertainty as to whether Širjaev was recognizing his forma as a variety.)

Buceras elliptica Moench, Methodus (Moench) 143 (1794), nom. illeg.

T. monspeliaca L. var. *glabrata* Eig, Bull. Inst. Agr. Nat. Hist. 6: 10 (1927).

T. monspeliaca L. f. *incisa* Širj., Generis *Trigonella* revisio critica, Spisy Pfir. Fac. Masarykovy Univ. 4: 24 (1931). (Listed merely as a synonym by Širjaev.)

Fig. 54-3. *Medicago monspeliaca.* From Hallier (1885, vol. 23, plate 2356).

T. monspeliaca L. var. *leiocarpa* Koch, Syn. ed. 3, I: 551 (1837).

T. monspeliaca L. f. *leiocarpa* (Koch) Širj., Generis *Trigonella* revisio critica, Spisy Pfir. Fac. Masarykovy Univ. 4: 24 (1931).

T. monspeliaca L. var. *macrocarpa* Bornm., Verh. Zool.-Bot. Ges., Wien 48: 577 (1898).

T. monspeliaca L. f. *macrocarpa* (Bornm.) Širj., Generis *Trigonella* revisio critica, Spisy Pfir. Fac. Masarykovy Univ. 4: 24 (1931).

T. monspeliaca L. f. *moabitica* Širj. & Eig, Moab Palaestinae 7: V (1927).

T. monspeliaca L. var. *naftolskyi* Eig, Bull. Inst. Agr. Nat. Hist. 6: 10 (1927).

T. monspeliaca L. f. *naftolskyi* (Eig) Širj., Generis *Trigonella* revisio critica, Spisy Pfir. Fac. Masarykovy Univ. 4: 24 (1931).

T. monspeliaca L. f. *nuda* Širj., Generis *Trigonella* revisio critica, Spisy Pfir. Fac. Masarykovy Univ. 4: 24 (1931).

T. monspeliaca L. var. *petiolata* Post., Fl. Syr. Pal. Sin. 222 (1883–1896). (Musselman and Saoud (2004) found no type at BEI or G.)

T. monspeliaca L. f. *petiolata* (Post.) Širj., Generis *Trigonella* revisio critica, Spisy Pfir. Fac. Masarykovy Univ. 4: 24 (1931).

T. monspeliaca L. subsp. *subacaulis* Feinbrun, Beih. Bot. Centralb. 51(2): 396 (1934).

T. monspeliaca L. f. *typica* Eig, Bull. Inst. Agr. Nat. Hist. 6: 9 (1927). (Širjaev's forma was interpreted as a variety by Vassilczenko (1953). In any event it is a nom. illeg.)

Note: *Medicago ovalis* was mistakenly treated as a synonym of *T. monspeliaca* L. in *Flora Europaea* (Tutin 1968).

Vernacular Names

English: Hairy medic (a contrived name that could apply to some other species; medic is also spelled medick), star-fruited fenugreek (considers the species to belong to *Trigonella*). Suggested name: Montpellier medic.

French: Luzerne de Montpellier (trigonelle de Montpellier when considered to be a species of *Trigonella*).

Geographical Distribution

(N = native, I = introduced, U = uncertain status)

Africa: Algeria (N), Egypt (N), Libya (N), Morocco (N), Tunisia (N).

Asia: Armenia (N), Azerbaijan (N), Cyprus (N), Gruziya (N), Iran (N), Iraq (N), Israel (N), Jordan (N), Lebanon (N), Pakistan (N), Russia (N), Syria (N), Turkey (N), Turkmenistan (N).

Australasia: Australia (I).

Europe: Albania (N), Austria (N), Balearic Islands (Spain; U), Belgium (N), Bulgaria (N), Czech Republic (N), Estonia (I), France (including Corsica; N), Greece (including Crete and Karpathos; N), Hungary (N), Italy (including Sardinia and Sicily; N), Malta (N), Moldova (N), Portugal (N), Romania (N), Slovakia (N), Spain (N), Switzerland (N), Turkey (N), Ukraine (N), Yugoslavia (N).

Northern America: USA (Alabama, Delaware, New York; I).

Southern America: Chile (I).

This species is native to Western Europe, North Africa, the Eastern Mediterranean Circle, the Middle East, and Central Asia. It has been collected as a waif in the northeastern United States and (allegedly) in adjacent Canada (Turkington and Cavers 1979).

Description

Annual herb, stems prostrate to ascending, branched from the base, 3–40 cm long. Vegetative parts more or less pubescent with simple appressed hairs. Leaflets obovate, obovate-cuneate, oval, ovate, or globose, (3)4–10(13) mm long, (2.5)3–7(10) mm wide, the distal half to three-fourths serrate, leaflets sometimes laciniate or incised. Stipules semi-sagittate, often denticulate at base. Peduncle extremely short, bearing 4–18 flowers in a capitate or subumbellate inflorescence in an axil. Corolla yellow, ca. 4 mm long, slightly longer than calyx. Calyx 2.5–3 mm long, teeth lanceolate-subulate, longer than the tube, tube pubescent. Legumes declinate (deflexed), stellately arranged, linear, 7–15(25) mm long, 1–2.0 mm wide, slightly compressed, somewhat curved, pubescent or glabrous, with obliquely transverse, anastomosing, prominent veins and a short, somewhat pointed beak at the apex. Seeds numerous, rhomboid-ovate, finely tuberculate, ca. 1.5 mm long, brownish, radicle about 80% of length of seed.

Identification Hints

This is a widespread species of Eurasia and so relatively likely to be encountered there. It is easily distinguished from other species of *Medicago* by the many-fruited sessile or subsessile racemes with stellately spreading, usually deflexed fruits. However, several species of *Trigonella* are remarkably similar in appearance and the key to genera of Fabaceae, tribe Trifolieae (Chapter 9) should be consulted for distinction of *Medicago* and *Trigonella*.

Habitat and Environmental Adaptation

Habitat

Medicago monspeliaca occurs on dry rocky slopes, sand dunes, and in scrub and forest. It is also a very widespread weed of fallow and cultivated fields and roadsides. In Greece, Thanopoulos (2007) noted its occurrence at altitudes up to 1700 m.

Phenology

Flowering occurs from February to June, depending on local climate in the native distribution area.

Soil

Medicago monspeliaca is found in a variety of soils, including gravel and sand.

Temperature

The species occurs as a spring and summer annual, mostly in continental and Mediterranean climates (i.e., cool to warm springs and hot summers). It extends to altitudes sometimes exceeding 1500 m in mountains, tolerating cool conditions. Cocks and Ehrman (1987) found that of 22 species of *Medicago* examined in Syria, *M. monspeliaca* was among the most frost tolerant.

Water

The species is capable of growing in sand dunes, suggestive of considerable drought tolerance. In Tunisia, it occupies a semi-arid zone with 250–450 mm of precipitation annually (Zoghlami and Hassen 2004).

Biotic Relationships

Pollination

Small (1988) recorded a pollen to ovule ratio of 385, rather high for an annual species of *Medicago* and suggestive of outcrossing. However, the flowers are quite small, indicating that *M. monspeliaca* is largely self-pollinated and substantially an inbreeder, like other annual *Medicago*.

Dispersal

No fruit specializations are apparent for seed distribution, which presumably occurs by a combination of animal, wind, water, and gravity vectors.

Herbivory

No information located.

Nodulation and mycorrhizae

USDA (2009) reported positive root nodulation for *M. monspeliaca*.

Disease resistance

No information located.

Economic History, Cultivars, and Uses

Robertson and Smith (2006) noted "there seems to be nothing known of the agronomy of *M. monspeliaca*."

Cultivars

No cultivars have been bred.

Forage

Medicago monspeliaca is not sown for forage, but is a component of natural pastures in the Old World. Robertson and Smith (2006), on the basis of studies in Australia, concluded that "the low production of *M. monspeliaca* limits its value as a component of pastures."

Fodder

The species is not grown or collected as fodder.

Green manure, rotation crop, ground cover

No information located.

Other uses (salad sprouts, ornamental, plant products, etc.)

No information located.

Chromosome Number

$2n = 16$: Gardé 1948; Bolkhovskikh et al. 1969, p. 321; Fernandes and Santos 1971; Fernandes et al. 1977; Fernandes and Queirós 1978; Murin and Ferakova 1981; Dvorak and Dadakova 1984; Luque and Díaz Lifante 1991; Kirschner and Štepánek 1992; Oberprieler and Vogt 1996*a*; Slavik et al. 1993; Pavlova 1996; Magulaev 2001; Reasat et al. 2003.

Genetic Relationships and Gene Transferability

Intraspecific

Artificial hybridization of different populations has not been reported, so it is not known whether there are barriers to hybridization within the species.

Interspecific

Širjaev (Publ. Fac. Sc. Univ. Masaryk, Brno No. 136, 31, (1931)) recognized a hybrid, *T.* ×*elmaluensis* = *T. monspeliaca* × *T. crassipes* (*M. monspeliaca* × *M. crassipes*), an interpretation that requires confirmation.

Exchange by advanced biotechnology

No information located.

Conservation Status

Because this species is extremely widespread and shows weedy tendencies, the natural populations are not endangered. It has not yet been demonstrated that there are local populations that are genetically distinctive, but since the species is polymorphic and occurs over a wide range of habitats, there may be such populations and so it is necessary to be sensitive to the possibility that local rare populations deserve conservation protection.

In France, *M. monspeliaca* has been protected locally: "Relatif à la liste des espèces végétales protégées en région Poitou-Charentes complétant la liste nationale, NOR : ENVN8800085A (Journal officiel du 10 mai 1988)...Vu la loi nE76-629 du 10 juillet 1976 relative à la protection de la nature, notamment ses articles 3 et 4; Vu le décret nE77-1295 du 25 novembre 1977, pris pour son application et concernant la protection de la flore et de la faune sauvages du patrimoine naturel français; Vu l'arrêté du 20 janvier 1982 modifié relatif à la liste des espèces végétales protégées sur l'ensemble du territoire national; Vu l'avis du Conseil national de la protection de la nature" (information from expired website; cf. http://www.tela-botanica.org/eflore/BDNFF/4.02/nn/41359; current July 2010). The species is similarly protected in the Ile-de-France region (http://www.mnhn.fr/mnhn/cbnbp/flore/textes/detail/iledefrance.htm; current June 2010).

Germplasm Sources

More than 600 accessions are available in major germplasm institutions (see Chapter 3).

Research Problems and Priorities

Medicago monspeliaca does not seem to have evident agriculturally desirable properties and has received relatively little study to date. It may have stress-resistant genes that can be usefully transferred to alfalfa.

Representative Collections

Albania: Alston, A.H.G. & Sandwith, L.H. 1782, above Leskovik, alt. ca. 3500 feet, rocky limestone slopes of Melesin, 19-VI-1933, in flower & fruit (K). **Algeria:** Davis, P. 51666, Ol, Oran to Misserghin, alt. 100–300 m, limestone hills (terra rossa), 27-IV-1971, in fruit (HUJ). Faure, A., Oued-Imbert, vers les Trembles (Oran), pelouses sèches, 22-IV-1917, in flower & fruit (K). **Austria:** Kerner, J., Austria inferior, in agris prope oppidum wiener Neustadt, solo argillaceo, alt. 300 m, (no date), Flor Exsiccata Austro-Hungarica No. 2010 (P). **Chile:** Phillipé, in alveo fluminis Ancagus, VII-1861, in flower & fruit (G). **Cyprus:** Chapman, E. 625, Kyrenia-Myrtou Rd., mile 14, alt. 50 feet, abundant in old fallow, 4-V-1950, in fruit (K). **Czechoslovakia:** Podpera, J., Moravia meridionalis, Hodonin, in arenosis inter vicos Ceje et Mutenice, alt. 180–200 m, VII-1923, in flower & fruit, Flora Exsiccata Republicae Bohemicae Slovenicae No. 38 (K, P). **France:** Aymonin, G.G. & Keraudren, A. 28091 & 28113, Montelimar, defilé de Trente-Pas, 1-VI-1973 (P). Ellman, E. & Sandwith, N.Y. 267, Pyrenées Orientales, Vermet-les-bains, alt. 2100 feet, roadside, 7-VI-1925, in flower & fruit (K). Héribaud, Fr., Puy-de-Dôme, chanturgues, près de Clermont, champs incultes, 4-VI-1900 (P). Foucaud, J., près Fouras (Ch.-Inf.), sables maritimes, 12-VI-1885 (P). **Greece:** Lewis, G.P. 587, Saronic Islands, Spetsai, above spetsai town, in loose rubble, alt. 200–800 feet, 6-V-1980, in fruit (K). Rechinger, K.H. fil. 1250, Crete, distr. Kissamos, ins. Grabusa Dimitraki, in saxosis calc., 20-IV-1942, in fruit (K). Sintensis, P. & Bornmüller, J. 295, Insula Thasos, Limenas, 1891 (P). **Gruziya:** Hohenacker, T.F., Georgiae caucasicae, V-1854, in flower & fruit (K). **Iraq:** Gillett, J.B. 10898, Kursi jebel sinjar, alt. ca. 650 m, limestone, lower limit of forest, denuded dry slopes, occasional, 23-V-1948, in fruit (K). Rawi, A. 22153, ca. 10 km west of Tawela, on road between Halabja & Tawela, alt. ca. 1400 m, on hillside, 15-VI-1957, in fruit (K). Wheeler Haines, R. 865, Shahraban, frequent in sandy area near dunes, 11-IV-1957, in flower & fruit (E). **Iran:** Danin, A. & Plitmann, U. 65-1447, Azerbaidjan, 20 km south of Jolfa on the road to Tabriz, alt. 1350 m, gravelly slopes near wadi, west and east exposures, 3-VII-1965, in fruit (HUJ). Rechinger, K.H. & F. 7202, Prov. Khorasan, Montes Kopet-Dogh, Inter Kucan et jugam Alamli, alt. ca. 1600 m, 3-VI-1948, in flower & fruit (E). Zohary, M. & Orshan, G., northern Iran, Shimran (env. of Teheran), gravelly slopes on the foothills of the Elburs, alt. 1350 m, 5-V-1961, in flower & fruit (HUJ). **Israel:** Meyers, F. 24456, Jerusalem, fields, 17-III-1912, in flower & fruit (K). Samuelsson, G. 2407, Judea, Jerusalem, in ruderatis, 23-III-1933, in flower & fruit (K). Zohary, M., Judean Mts., Bar-Giora bridge, bathah and garigue, terra-rossa, 20-IV-1963, in fruit (HUJ). Zohary, M. & Plitmann, U., Upper Galilee, env. of Metula, roadsides and abandoned yards, 11-III-1963, in fruit (HUJ). Zohary, M. & Plitmann, U., Mt. Gilboa, Zarayin turn, calceareous hill, gray soil, 29-III-1963, in fruit (HUJ). Zohary, M. & Plitmann, U., Upper Jordan Valley, env. of Naharayim, submediterranean batha, marly basalt soil, 30-III-1963, in fruit (HUJ). **Israel/Syria (Golan):** Danin, A., Golan, near Ortal, basalt soil, 6-V-1988, in fruit (HUJ). **Jordan:** Naftolsky, N., Ammon, Ein Karpet, 14-IV-1929, in fruit (HUJ). Naftolsky, N., Ammon, fields, 20-IV-1929, in fruit (HUJ). **Italy:** Béguinot, A., Venetia, Patavium (Padova), supra urbis moenia locis dictis Polveriera et Grate di S. Massimo, ubi copoiosa, solo siliceo et calcareo, 13-V-1906, Flora Italica Exsiccata Fiori, Béguinot & Pampanini No. 460 (K). De Heldreich, Sicily, in collibus herbidis Sicilae centralis, ad Caltanisotta, 17-IV-1840, in flower & early fruit (K). Todaro 196, Palermo, in collibus aridis, V (received 1882 by P), (P). Todaro, Sicily, Palermo, in collibus aridis, V, in flower & fruit (K). **Malta:** Duthie, J.F., Insula Gaulos, prope Rabbat, in herbosis siccis, 10-IV-1874, in flower & fruit (K). **Morocco:** Balls, E.K. B2459, Amizmiz, riverbed, alt. 4300 feet, 30-VI-1936, in fruit (K). Pittard, C.J. 2346, Settat ad Guïcu, in seqetibus, 31-IV-1912, in fruit (P). **Spain:** Bourgeau, E. 720, Pyrenées Espagnoles, Barcelone, Mont-Jovy, 21-IV-1847, in flower & fruit. Bourgeau, E., Cartagena, champs, 4-IV-1851 (P). **Syria:** Eig, A. & Zohary, M., 16 km south of Homs, beyond the village Chounliar, uncultivated hill, gray-reddish soil, 2-V-1933, in fruit (HUJ). Eig, A. & Zohary, M., 28 km west of Selimié, alt. 475 m, fields fallow 2–4 years, with somewhat gravelly soil, 2-V-1933, in fruit (HUJ). Zohary, M., about Damascus, 14-V-1931, in fruit (HUJ). **Switzerland:** Faurat, L. & Barbey, W., Sion, Valais, IV-V-1879, in flower & fruit (K). Thomas, près Lion, Vallais, alt. 2000 pieds, VII-1866, in fruit (K). **Turkey:** Balls, E.K. 863, Gaziantep, 27-IV-1934, in fruit (E). Coode, M.J.E. & Jones, B.M.G. 1728, Çorum, Iskilip, 26 miles from Çorum to Iskilip, Kizilirmak valley, alt. 600 m, saline marshy ground, 29-IV-1965, in flower and fruit (E). Davis & Hedge D28469, prov. Mardin, Mardin-Nusaybin, 12 km from Mardin, alt. 750 m, fallow fields at foot of encampment, 22-V-1957, in flower & fruit (BM). Davis & Polunin 25995, prov. Mersin, distr. Anamur, Anamur – Gilindire, alt. 20 m, rocky limestone slopes, 14-IV-1956, in flower & fruit (BM, E). Davis & Polunin, O. 25338, Prov. Muğla, Marmaris-Emecik, alt. 200 m, roadside on serpentine, 25-IV-1956, in flower & fruit (E). Davis 41332, Muğla, d. Marmaris, Bay east of Datça Port, sea level, sand dunes, 18-IV-1965, in flower & fruit (E). Tobey, C. 931, Vilayet, Samsun, Havza, Sekizgöz, alt. 1000 m, roadside, sandy clay, 1-IV-1965, in flower & fruit (E).

55. *Medicago murex* Willd.

M. murex Willd., Sp. Pl. 3(2): 1410 (1802), (basionym).

The epithet *murex* is based on a mollusc called murex (*Murex trunculus*, more commonly classified as *Hexaplex trunculus*), from which a celebrated purple dye was obtained in antiquity. Its spiny, coiled shell is reminiscent of the fruit of *M. murex* (but also resembles the fruit of numerous other species of *Medicago*).

Warning Regarding Possible Ambiguity of Names

Lesins et al. (1970) discovered that two completely intersterile kinds of plants, one with $2n = 16$ chromosomes and the other with $2n = 14$, occurred in what was then known as *Medicago murex* Willd. Subsequently, the type specimen of this name, housed in the Willdenow Herbarium at Berlin

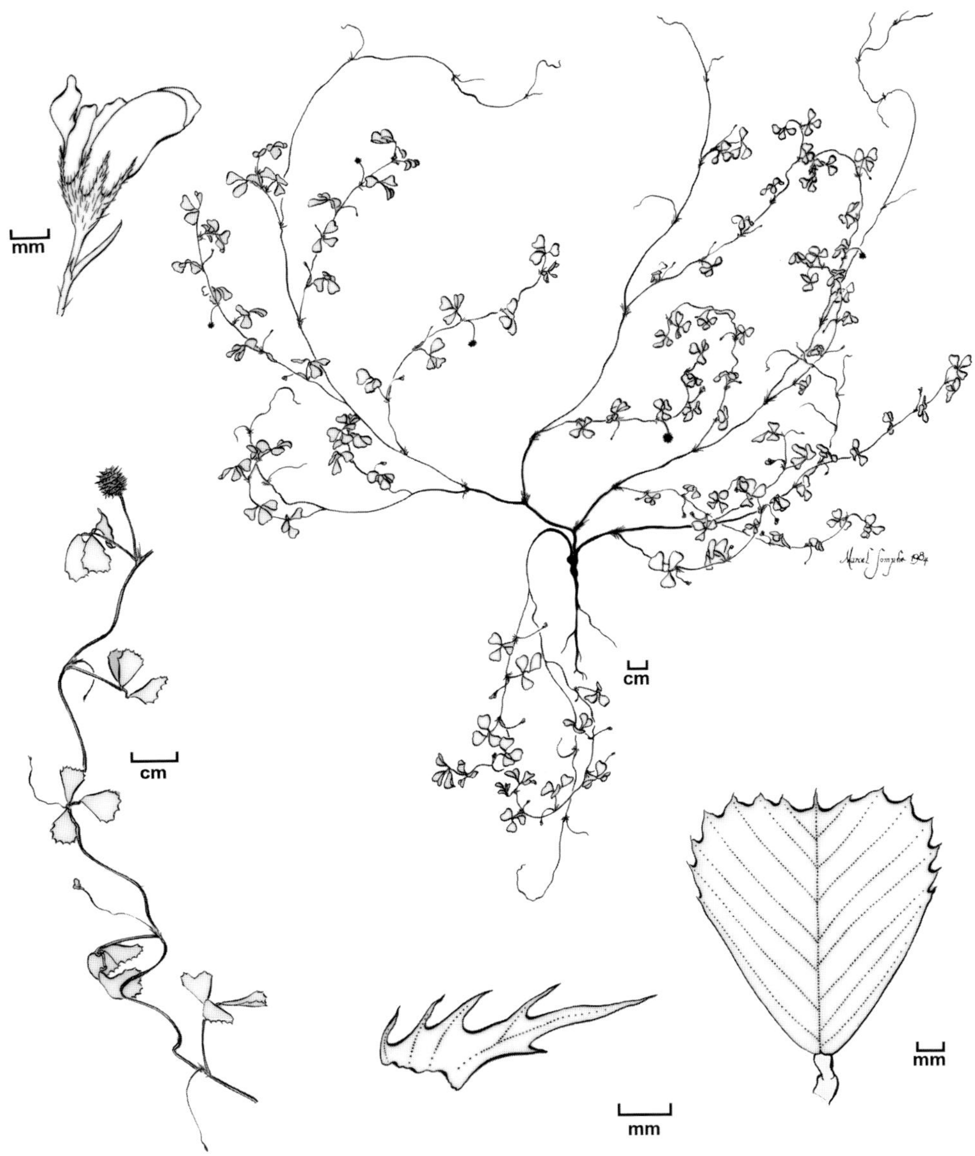

Fig. 55-1. *Medicago murex*. **Clockwise from top left:** Side view of flower; plant; leaflet; stipule; branch with fruit. (Based on E. Small M1149, DAO (type collection of *M. lesinsii*; greenhouse-grown plant); artist: M. Jomphe.)

(Botanischer Garten und Botanisches Museum Berlin-Dahlem), became critical to which of the chromosome races would bear this name when the two classes of plant were recognized as separate species. As noted in this chapter, several morphological characters distinguish the two species. However, the type of *M. murex* Willd. has very little material and information, and only the fruits can be used to predict its probable chromosome number. K. Lesins sent some fruit samples of the two kinds to the German plant taxonomist H. Scholz in Berlin, and on 24 October 1969 Scholz wrote that "with some reservation it is my opinion the Willdenow's specimens of *Medicago murex* likes in the best way your sample of the $2n = 16$ type. The ridges on the coils are very indistinctly expressed..." However, my colleague G. Baillargeon, while studying for his Ph.D. in Berlin in 1983, advised me that the fruits on the type specimen could be interpreted as the $2n = 14$ kind of plant. Additionally at my request, W. Greuter and R. Vogt, who were in charge of the Willdenow Herbarium, examined the Willdenow type and concluded (personal communication) that with present evidence "the question cannot in our opinion be definitely resolved." I subsequently described the $2n = 16$ kind of plant as a new species, *M. lesinsii* E. Small (Small and Brookes 1985). However, Gillespie and McComb (1991), based on examination of the type by the Australian botanist G. Perry, concluded that Willdenow's type corresponds to the $2n = 16$ species.

In 1993 I personally visited B and examined Herb. Willdenow. There are two Willdenow collections, one merely with a piece of a plant with an immature pod, which does not show the diagnostic characters clearly enough to decide whether the collection is of the $2n = 14$ or $2n = 16$ kind. (In correspondence to Lesins in 1969, Scholz referred to this (Willdenow No. 14318/2) as an isotype.) The main collection (Willdenow No. 14318/1) has a mounted plant and three pods in an envelope (it is possible that the pods in the envelope do not come from the collected plant, but this is unlikely and there is no reason to believe this). The two maturest pods do not bear ridges that are more than subtly evident, and are accordingly unlikely to belong to the $2n = 14$ form. Therefore, I am in agreement with the conclusions of H. Scholz and G. Perry (mentioned above) that the name *M. murex* Willd. must refer to the $2n = 16$ class of plant that is described in this chapter. The $2n = 14$ class of plant is correctly named *M. sphaerocarpos*.

Unfortunately, my published name *M. lesinsii* E. Small for the $2n = 16$ class, and my previous use of the name *M. murex* for the $2n = 14$ class, were adopted by several authors for the purpose of distinguishing the two species. No harm has been done by the use of the name *M. lesinsii* (which is merely an unambiguous junior synonym). However, some authors (including me) have specifically used the name *M. murex* to indicate the $2n = 14$ class of plant, which could lead to confusion if the context of use is not well understood. Moreover, since both classes of plant have largely been included in the same species, it is often unclear when the name *M. murex* has been used just what species really is meant.

Subordinate Taxa

None recognized here.

Type Specimen

Herb, Willdenow (B! photo HUJ!). Note above discussion.

Synonymy

Homotypic

M. ciliaris (L.) All. subsp. *murex* (Willd.) Bonnier & Layens, Fl. Fr. 75 (1894).

Heterotypic

M. heterocarpa Spach ex Cosson & Durieu, Explor. Sci. Algér., Atl. tabl. 89 (1889). (Placement in *M. murex* rather than in *M. sphaerocarpos* follows Abdelguerfi and Guittoneau (1989). These authors gave the following details for the type: Mauritania prope Mostaganem, Cl. Delestre 1844, Herb. Spach (P!).)

M. murex Willd. subsp. *heterocarpa* (Cosson & Durieu) Jauzein, Fl. Champs cult. 863 (1995).

M. murex Willd. subsp. *inermis* (Guss.) Arcangeli, Comp. Fl. Ital. 164 (1882).

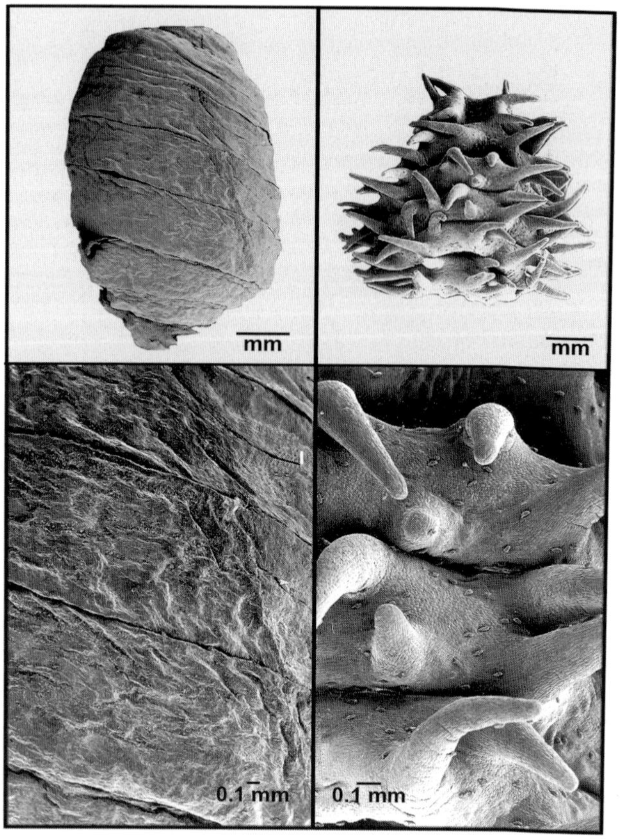

Fig. 55-2. *Medicago murex*. **Top left:** Spineless fruit. **Bottom left:** Close-up of spineless fruit. (Based on E. Small M1104, DAO). **Top right:** Spiny-fruit. **Bottom right:** Close-up of spiny fruit. (Based on E. Small M1102, DAO). Scanning electron photomicrographs by B. Brookes.

M. murex Willd. var. *inermis* (Guss.) Urban, Verh. Bot. Ver. Prov. Brandenb. 15: 73 (1873).). (Placement in *M. murex* rather than in *M. sphaerocarpos* follows Abdelguerfi and Guittoneau (1989).)

M. lesinsii E. Small in E. Small & Brookes, Canad. J. Bot. 63(4): 733 (1985), (type: E. Small M1149, DAO).

M. muricula St. Lag., Ann. Soc. Bot. Lyon 7: 130 (1880). (This binomial seems to have been proposed by St. Lager merely as a preferred etymological rendition for *M. murex*, and no type specimen is at GE or indeed seemingly even designated.)

M. sphaerocarpos Bertol. var. *ovalis* Moris, Fl. Sardoa 1: 446 (1837). (I follow Abdelguerfi and Guittoneau (1989) in placing this name in *M. murex* rather than under *M. sphaerocarpos*.)

M. sorrentini Tineo in Todaro, Linnaea, 30: 759 (1860). nom. nud. (I follow Abdelguerfi and Guittoneau (1989) in placing this name in *M. murex* rather than under *M. sphaerocarpos*. These authors (1989, p. 426) give "type" details as follows: Todaro Fl. Sic. Exs. 755, Alcano, Sicily (BM, U); however, as a nom. nud. there is no type. This collection has spineless fruits.)

M. murex Willd. var. *sorrentinii* (Tin.) Fiori & Béguinot in Fiori, Fl. Anal. Ital. 2: 38 (1900).

M. murex Willd. var. *inermis* (Guss.) Urb. f. *sorrentini* (sic) (Tin.) Urb., Verh. Bot. Ver. Prov. Brandenb. 15: 73 (1873).

Note: See Chapter 92 on *M. sphaerocarpos* for names that have not been determined whether they are synonymous with *M. murex* or *M. sphaerocarpos*.

Vernacular Names

ENGLISH: Murex medic (medic is also spelled medick). "Spiny medic" is also sometimes encountered, but because there are numerous species that have spiny fruit, this name should be avoided. (Note: because *M. murex* and *M. sphaerocarpos* have only recently been separated, it is often not possible to distinguish which common name an author is applying to these species.)

FRENCH: Luzerne hérissée, luzerne murex.

Geographical Distribution

(N = native, I = introduced, U = uncertain status)

AFRICA: Algeria (N), Morocco (N), Tunisia (N).

ASIA: Israel (N), Turkey (presumably N)

EUROPE: France (N), Germany (U), Greece (N), Italy (N), Sardinia (N), Sicily (N), Spain (N), Portugal (N).

Medicago murex is native to Mediterranean Circle countries.

Description

Annual herb 10–30(50) cm, branched from the base, stems usually procumbent, vegetative parts glabrous or weakly pubescent with simple hairs. Leaflets 7–15 mm long, generally somewhat longer than broad, obovate to cuneate, upper third of margin serrate. Stipules laciniate or dentate-laciniate. Peduncles short, about as long as subtending petiole. Flowers 3–5 mm long, yellow. Calyx pubescent with simple hairs, teeth as long as tube, corolla less than twice length of calyx. Young fruit initially concealed within calyx, then turning sideways through the calyx teeth. Fruit glabrous, spherical, or ovoid, with 5–9 strongly adpressed coils, 4–10 mm wide in middle (widest) coils, usually with dorsal suture spines (rarely with tubercles or spineless), coil face with 4–8 slightly curved, unbranched radial veins terminating in a veinless submarginal border extending about one-third of radius of coil. Seeds 1–2 per coil, 3–5 mm long, smooth, yellow to yellowish-brown or reddish-brown, radicle half or less length of seed.

Small and Brookes (1985) observed distortion and twisting of the stem in *M. murex* (note Fig. 55-1). However, Gillespie and McComb (1991) failed to observe this, and suggested that the observations of Small and Brookes could have been due to the glasshouse conditions in which they were grown or by the presence of some disease or virus.

Identification Hints

The above description applies both to *M. sphaerocarpos* and the very similar *M. murex*. See "Identification Hints" for *M. sphaerocarpos* for distinction of the two species.

Habitat and Environmental Adaptation

Habitat

Medicago murex commonly occurs in heavily grazed, lower-altitude parts of the Mediterranean region. In Greece, Thanopoulos (2007) recorded it in sandy shores, open places in garrique, in other shrubland associations, grasslands, and as a weed in cereal crops, olive groves, and fallow fields, from sea level to 900 m in altitude.

Phenology

The species flowers in early summer (April to June) in the Mediterranean area. In Australia, cultivar 'Zodiac' is considered to be a middle- to late-season maturity medic.

Soil

Abdelguerfi and Guittoneau (1989) recorded that *M. murex* occupies a variety of sandy soils in North Africa. Abdelguerfi et al. (1988*a*) observed that in Algeria *M. murex* occurs in soils of all textures, from coarse to fine. In Australia, cultivar 'Zodiac' is well suited to a variety of loam soils, as well as gravelly sands and sand over clay. It has been grown on clay soils.

Species of *Medicago* are generally intolerant of acidic soils, but compared with most other species of the genus, *M. murex* is relatively acid tolerant (Howieson and Ewing 1986; Howieson et al. 1988; Howieson et al. 1992; Bounejmate and Robson 1992; Piano and Francis 1992; Cheng et al. 2004) and tolerant of the manganese toxicity associated with acidic soils (Carneiro et al. 2001). Francis and Gillespie (1981) found that *M. murex* was common on the island of Sardinia, where acidic soils are preponderant. Pinkerton and Randall (1993) investigated the need for potassium. The species is considered unsuitable for soils where exchangeable aluminium is more than 15%. In Algeria, where *M. murex* is native, Abdelguerfi and Guittoneau (1989) collected it from five sites, the pH ranging from 5.8 to 7.8. In Sicily, Piano et al. (1991) observed the species growing in soil with a pH range of 6.0–9.0. In Australia, the pH tolerance range recommended for planting is 4.5–8.0 (Gillespie 1994). A pH of 4.8–6.5 is preferred.

Temperature

Medicago murex is a typical Mediterranean winter annual, germinating in the cool winter and setting seed by early summer when temperatures may become high. Cultivar 'Zodiac' tolerates mild frosts.

Water

Medicago murex is adapted to low rainfall, Mediterranean areas that receive most of their precipitation in the winter growing season. However, by the standards of medic species, it is adapted to relatively wet conditions. In Algeria, where *M. murex* is native, Abdelguerfi et al. (1988*a*) and Abdelguerfi and Guittoneau (1989) collected it from sites where the annual rainfall ranges from 700 to 1100 mm. Nègre (1956, 1959) pointed out that the species occurs in the relatively wet Atlantic plains of Morocco and in the relatively wet region of northeastern Algeria. Piano and Francis (1992) noted that in its native area the species requires a minimum annual rainfall of 400 mm. In Australia, cultivar 'Zodiac' has been recommended for areas receiving more than 450 mm rainfall, most of it falling during the winter growth season. The species will tolerate waterlogging for short periods, but like other medics it is not suitable for soils that become inundated or waterlogged for long periods. Blumenthal and Ison (1993*a*, *b*) examined the relative productivity of *M. murex* in relation to water use.

Biotic Relationships

Pollination

Small (1988) noted that the relatively low pollen to ovule ratio (345) and small flowers of the species are consistent with *M. murex* being largely self-pollinated. The pronounced isozyme uniformity of its populations and rarity of heterozygotes (Small et al. 1992) also showed that outbreeding is very rare.

Dispersal

The indehiscent spiny fruits are well adapted to dispersal in animal fur, accounting for the widespread occurrence of this species. According to Gillespie (1989) "pods of [cultivar] Zodiac are unlikely to adhere to wool in large quantities because the spines are short" (as a rule, Australian cultivars have been selected to be spineless or have short spines to prevent contamination of wool).

Herbivory

Cultivar 'Zodiac' is susceptible to the spotted alfalfa aphid (*Therioaphis trifolii*) and the lucerne flea (*Sminthurus viridis*), but has some tolerance to the blue-green aphid (*Acyrthosiphon kondoi*) and the pea aphid (*Acyrthosiphon pisum*), as well as the red-legged earth mite (*Halotydeus destructor*). *Medicago murex* remains prostrate when heavily grazed by livestock (this feature makes it resistant to grazing), but does tend to grow more erect when not grazed, making it suitable for harvest for hay or silage. As with other Australian medics sown for pasture, its high level of hardseededness results in germination over several years, considered very desirable for persistence (Dear and Jenkins 1992).

Nodulation and mycorrhizae

USDA (2009) reported positive root nodulation for *M. murex*. Howieson and Ewing (1989) observed that *M. murex* was outstanding in its ability to achieve nodulation in acidic soils. Cheng et al. (2002) reported that although *M. murex* is adapted to relatively acidic soils, when it is nodulated by *Sinorhizobium meliloti* in soil with a pH of 4.3, fewer nodules were produced, although these grew at the same rate as nodules formed at a pH of 7.0.

Disease resistance

O'Neill et al. (2003) examined the comparative resistance of 36 annual species of *Medicago* to *Phoma medicaginis* (the cause of spring black stem and leaf spot, a major disease of alfalfa) and concluded that *M. murex* (identified as *M. lesinsii*) had high resistance to it. Barbetti (2007) identified three accessions of *M. murex* with a very high level of resistance to *P. medicaginis*. In a similar comparative test of 36 annual species of *Medicago*, O'Neill and Bauchan (2000) found that two accessions of *M. murex* (identified as *M. lesinsii*) were quite susceptible to anthracnose caused by *Colletotrichum trifolii*. Cultivar 'Zodiac' is tolerant of alfalfa mosaic virus. Yaege and Stuteville (2002) tested the resistance of 33 annual species of *Medicago* to powdery mildew (*Erysiphe pisi*) and found that the two collections of *M. murex* (identified as *M. lesinsii*) grown were susceptible to the disease (it was observed that often only one of a number of populations of other species tested was resistant).

Economic History and Uses

Cultivars

'Zodiac', the only available cultivar of the species, was developed by the Western Australian Department of Agriculture and registered in 1988. It is based on collections from Sardinia and has been described by Gillespie (1987, 1988, 1989; the 1989 publication is available online at The

Register of Australian Herbage Plant Cultivars for Legumes http://www.pi.csiro.au/ahpc/legumes/legumes.htm; current June 2010). Its pods are generally short-spined.

Forage

Medicago murex is sown for forage in Australia. Livestock disorders from grazing that have been noted with respect to *M. murex* include photosensitization in horses, occasional red gut in sheep, and frequent bloat in cattle. Conlan et al. (1994) studied the species' tolerance to grazing.

Fodder

As noted under "Herbivory", *M. murex* can be harvested for fodder.

Green manure, rotation crop, ground cover

No information located.

Other uses (salad sprouts, ornamental, plant products, etc.)

No information located.

Chromosome Number

$2n = 16$: Ghimpu 1928, 1929*a* (as *M. sphaerocarpos*), 1929*b*, 1930 (as both *M. sphaerocarpos* and *M. murex*); Fryer 1930; Heyn 1956, 1963; Simon and Simon 1965; Bolkhovskikh et al. 1969, p. 309; Lesins et al. 1970; Abdelguerfi and Guittoneau 1979; Lesins and Lesins 1979; Small and Brookes 1985 (as *M. lesinsii*); Abdelguerfi et al. 1988*b*; Mariani and Falistocco 1990; Gillespie and McComb 1991; Mariani et al. 1996 (as *M. lesinsii*).

Genetic Relationships and Gene Transferability

Intraspecific

No barriers to crossing within the species are known, although artificial hybridization of different populations has not been attempted.

Interspecific

Attempts to hybridize *M. murex* with *M. sphaerocarpos* have failed (see information for the latter). Lesins et al. (1970) were unable to hybridize *M. murex* with another annual medic, *M. doliata*, with the same chromosome number ($2n = 16$).

Exchange by advanced biotechnology

No information located.

Conservation Status

Medicago murex is a fairly common species, well represented by germplasm collections, and so is not in need of conservation measures.

Germplasm Sources

Several hundred accessions of this species are present in major germplasm collections of *Medicago* (see Chapter 3).

Research Problems and Priorities

Medicago murex has demonstrated its agricultural value and deserves additional economic development. Piano and Francis (1992) described it as having "regional potential".

Representative Collections

Algeria: Bové, N., champs, V-1988 (P). Durieu, M., atterissements du Bon Merzoug à Constantine, 10-V-1840, in fruit (P). Reboud, V., Djebel-Ouach, près de Constantine, prairies, 12-V-1881, in flower & fruit, Société Dauphinoise No. 3592 (P). Small, E. M1101, grown in an Ottawa greenhouse (1983–1984) from K. Lesins' seed collection (University of Alberta seed stock 353), originally from Institut Scientifique Cherifien (seed stock 1586), (DAO). Small, E. M1148, grown in Ottawa greenhouse (1983–1984) from seed collected by A. Abdelguerfi (174d), Dra el Alaag, Djebel Ouahche (Constantine), fruits spineless (DAO). Small, E. M1149, grown in an Ottawa greenhouse (1983–1984) from seeds collected by A. Abdelguerfi (174f), from Dra el Alag. Djebel (Ouahche (Constantine) (DAO; also note Abdelguerfi and Guittoneau 174f at AL). **Israel:** Lesins, K. 1983, grown from seed from C.C. Heyn collection 3172, Sharon Beit Lid (ALTA). Small, E. M1104, grown in an Ottawa greenhouse (1983–1984) from K. Lesins' seed collection, University of Alberta seed stock 1982, originally from Sharon Beit Lid, fruits spineless (DAO). **Italy:** Small, E. M1102, grown in an Ottawa greenhouse (1983–1984) from K. Lesins' seed collection, University of Alberta seed stock 1075, originally collected by Lesins from Capoterra, Sardinia (DAO).

56. *Medicago muricoleptis* Tineo

M. muricoleptis Tineo, Pl. Rar. Sicul. Pugill. 1: 18 (1817), (basionym).

The epithet *muricoleptis* is based on the Latin *murico* — spiny, like the mollusc known as murex, plus the Greek word *leptos* — thin, slender, an allusion to the slender spines of the spiraled fruit.

Subordinate Taxa

None recognized here.

Type Specimen

Heyn (1963) stated: "In the Pl. Sic. Rar., Tineo named as type locality 'Prope Panormum a Bucca di Falcu.' The

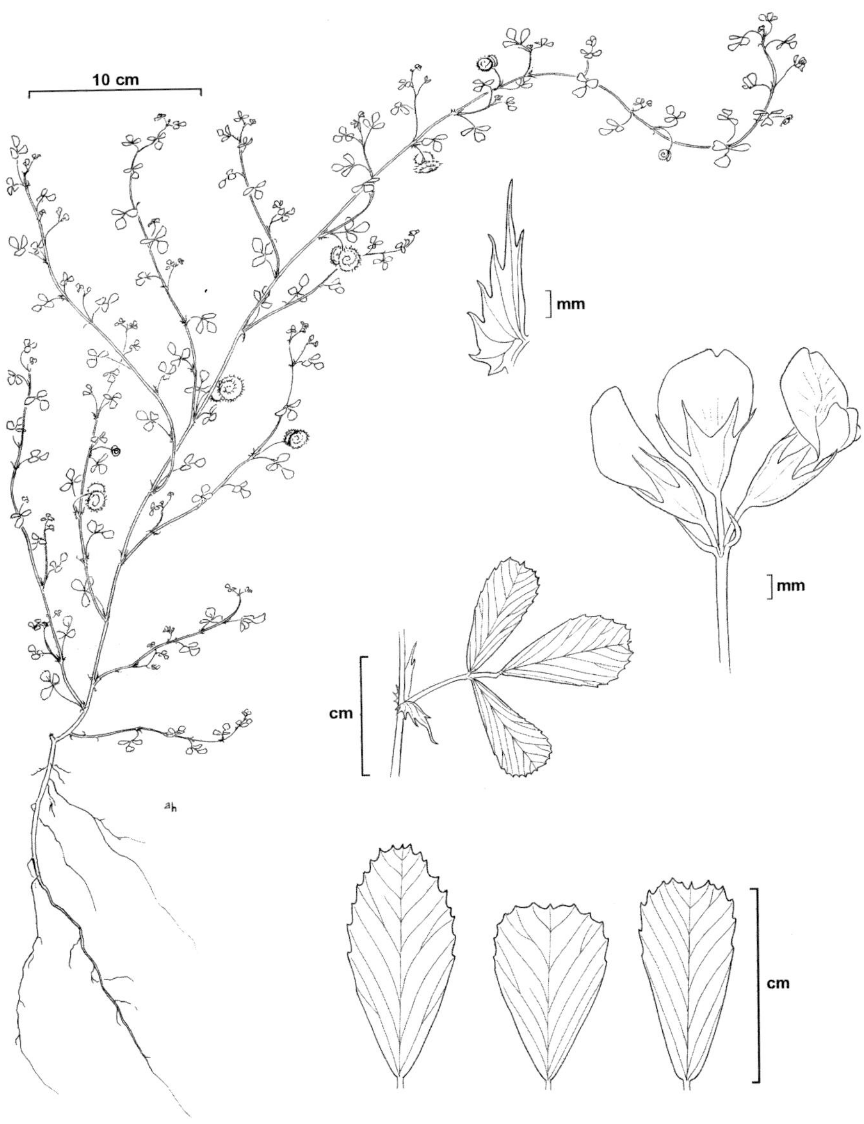

Fig. 56-1. *Medicago muricoleptis.* **Clockwise from left:** Plant; stipule; group of flowers; leaflets. **Center:** Leaf attached to stem. (Group of flowers based on G.A. Stevenson 3607, DAO, remainder based on Fioro 2099 (Flor. Ital. Exs.), K; artist: A. Hanes.)

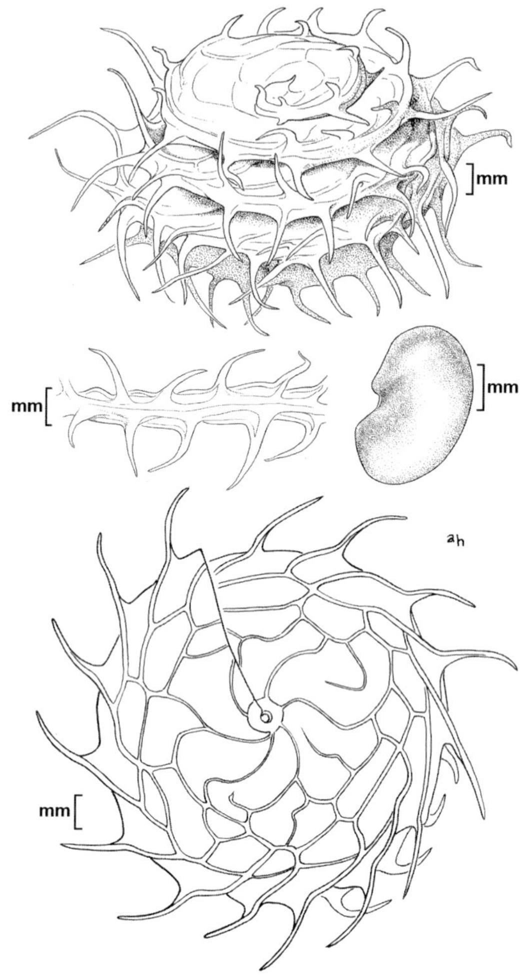

Fig. 56-2. *Medicago muricoleptis.* **Clockwise from top:** Fruit; seed; fruit coil face showing venation; edge of fruit coil. (Based on E. Small M347, DAO; artist: A. Hanes.)

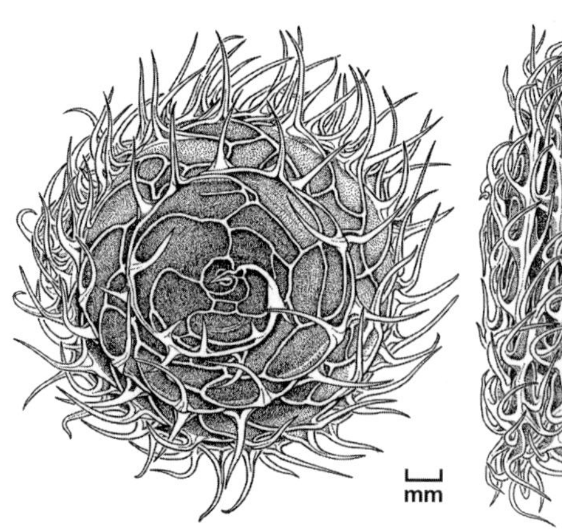

Fig. 56-3. *Medicago muricoleptis.* **Left:** View of fruit from stigma end. **Right:** Side view of fruit (based on E. Small M2034, DAO; artist: B. Flahey.)

only specimen of *M. muricoleptis* collected by Tineo, seen by us, is from the conservatoire Botanique in Geneva and is labelled '*Medicago muricoleptis* prope Panornum Tineo'. This specimen might be taken as lectotype."

Synonymy

Homotypic

M. intertexta Mill. var. *muricoleptis* (Tineo) Fiori in Fiori & Paol., Fl. It. 2: 41 (1900).

Heterotypic

M. decandollei Tineo ex Guss., Fl. Sicul. Syn. 2: 369 (1844).

M. intertexta (L.) Mill. var. *decandollei* (Guss.) Urb., Verh. Bot. Ver. Brandenb. 15: 62 (1873).

M. muricoleptis Tineo var. *minor* Tineo ex Loj. & Poj., Fl. Sic. 1(2):68 (1891).

(*M. muricoleptis* Ser. in DC., Prodr. 2: 179 (1825) was recognized as *M. interxtexta* var. *decandollei* by Heyn (1963), but Seringe's name is merely a later homonym of *M. muricoleptis* Tineo.)

Vernacular Names

ENGLISH: No names found.

FRENCH: No names found.

Geographical Distribution

(N = native, I = introduced, U = uncertain status)

EUROPE: France (southeastern; U), Greece (N), Italy (including Sicily; N).

This rare plant occurs in the western Mediterranean area, primarily in southern Italy.

Description

Annual herb, decumbent to ascending, branched from base, branches 15–30 cm long. Vegetative parts sparsely covered with simple hairs or glabrous. Leaflets 3–8(10) mm long, 2–4(5) mm wide, obovate or oblanceolate, distal two-thirds (or less) with serrate margins. Stipules with pointed lobes. Peduncle with (1)2–4(6) flowers, shorter to longer than subtending petiole. Flowers 5–7.5 mm long, calyx 2.5–4 mm long, with simple hairs, teeth about as long as tube, corolla yellow, more than twice length of calyx. Young pod protrudes sideways from calyx in a loose spiral. Mature pod discoid to cylindrical with smaller end coils, 4–5 mm long, glabrous, spiny, with 3–6 loosely appressed coils of diameter 8–12 mm. Coil face with a distinct net of 7–10 branching radial veins, one diagonal vein entering into each spine. Dorsal suture spines emerge from both faces of coil at an angle of about 90° and interlock with spines from facing coil, spines grooved, 1–2 mm long. Seeds 4–5 mm long, 1–2(3) per coil, not separated (or barely separated) by spongy fruit septa, radicle less than half as long as seed, seed smooth, dark red-brown to black-brown.

Identification Hints

Medicago muricoleptis is easily mistaken for *M. granadensis*, but can be reliably separated from it because *M. muricoleptis* has smaller fruit (the diameter of the terminal coil of the fruit is less than 5.5 mm in *M. muricoleptis*, more than 5.5 mm in *M. granadensis*).

Habitat and Environmental Adaptation

Habitat

Medicago muricoleptis has been collected from abandoned fields, wasteland, olive groves, and pastures. Presumably in presettlement times it was a generalist pioneer of disturbed habitats.

Phenology

Medicago muricoleptis is typical of annual Mediterranean species of *Medicago*, germinating during winter rains and flowering in the spring.

Soil

This species has generally been recorded in heavy soils. However, in Greece Thanopoulos (2007) observed that it had been collected on sandy substrates. Piano and Francis (1992) noted that in its native area *M. muricoleptis* occupies clays that are slightly to moderately basic.

Temperature

Medicago muricoleptis occurs in a typical Mediterranean climate: cool at germination time in late winter and increasingly hot by the time the seeds have matured and the pods are distributed in early summer.

Water

The species often occupies moist soils and appears to avoid summer drought characteristic of Mediterranean areas by completing its life cycle and setting seeds in the summer. Piano and Francis (1992) recorded that in its native area *M. muricoleptis* requires a minimum annual rainfall of 400 mm.

Biotic Relationships

Pollination

Small (1988) noted that the relatively low pollen to ovule ratio (295) is consistent with *M. muricoleptis* being largely self-pollinated. However, the flowers are larger than those of most annual species of *Medicago*, which are inbreeders, suggesting that this species is relatively attractive to pollinators. Nevertheless, *M. muricoleptis* is likely an inbreeder for the most part.

Dispersal

Because the spines of the fruits of this species tend to be appressed to the fruit surface, they are much less adapted to aiding in the dispersal of the pods via animal fur than most other spiny-fruited annual *Medicago*.

Herbivory

No information located. The loose fruit coiling would seem to make the fruits susceptible to insect attack (small insects could simply enter the fruit between the coils), but the appressed spines seem to compensate by acting as a barrier.

Nodulation and mycorrhizae

No information located.

Disease resistance

Yaege and Stuteville (2002) tested the resistance of 33 annual species of *Medicago* to powdery mildew (*Erysiphe pisi*) and found that the single collection of *M. muricoleptis* grown had slight resistance to the disease (it was observed that often only one of a number of populations of other species tested was resistant). In a comparative test of 36 annual species of *Medicago*, O'Neill and Bauchan (2000) found that one accession of *M. muricoleptis* was very resistant to anthracnose caused by *Colletotrichum trifolii*, whereas most accessions of annual *Medicago* were quite susceptible to it.

Economic History, Cultivars, and Uses

This species has not attracted significant agricultural attention to date.

Cultivars

No cultivars have been bred.

Forage

Medicago muricoleptis is not sown for forage, although it is consumed by livestock that graze in natural, unmanaged lands in the Mediterranean area.

Fodder

This species is not harvested for fodder.

Green manure, rotation crop, ground cover

No information located.

Other uses (salad sprouts, ornamental, plant products, etc.)

No information located.

Chromosome Number

$2n = 16$: Lesins and Lesins 1979; Schlarbaum et al. 1984; Mariani et al. 1996.

Genetic Relationships and Gene Transferability

Intraspecific

No barriers to hybridization within the species are known, although artificial hybridization of different populations has not been attempted.

Interspecific

Lesins et al. (1971) were able to cross *M. muricoleptis* with its close relatives *M. intertexta* and *M. ciliaris*, but the F_1 plants exhibited considerable sterility. The cross of *M. ciliaris* and *M. muricoleptis* produced hybrids with only about 35% pollen fertility. Lesins and Lesins (1979) reported that *M. muricoleptis* could not be hybridized with *M. granadensis*.

Exchange by advanced biotechnology

No information located.

Conservation Status

Medicago muricoleptis is listed in the Red Data Book of Rare and Threatened Plants of Greece (Phitos et al. 2009). This species is poorly understood and, given its rarity, a conservation assessment is desirable. There is a need to acquire germplasm, both for conservation and development purposes.

Germplasm Sources

Thirty accessions are listed among the holdings of the major germplasm institutions with large collections of *Medicago* seeds (see Chapter 3).

Research Problems and Priorities

Medicago muricoleptis is a rare species that has not received much study. Its potential in agriculture as a crop or as a contributor to the genetic improvement of other species requires examination.

Representative Collections

France: Bourgeau, E. 108, dans les prés à Castignaux, 8-VI-1848, in late flower & fruit (K). **Italy:** Groves, H., inter segetes et in incultis, prope Sienam (Coroncina) "6/76", in late flower & fruit (K; numerous collections). Fiori, A. et A. Béguinot 2099, Flora Italica Exsiccata Series III, in herbosis humidis secundum torrentem prope oppidum, Lucania, Matera, alt. 320 m, solo calcareo, 7-VI-1913, in late flower & fruit (K). Levier, E., San Casciano de Bagni, inter segetes, 12-"Ju"-1881, in fruit (L). **Sicily:** Splitgerber, F.L. 2.279, 12-V-1833, in fruit (L).

Note

Some southern Italian and Sicilian collections assigned to *M. intertexta* var. *decandollei* (Tin.) Urb. by C. Heyn (1963) were placed in *M. muricoleptis* by Lesins and Lesins (1979). These appear intergradient between the two species and may be the result of introgression. They may be identified as *M. intertexta* (L.) Miller × *M. muricoleptis* Tin. Representative collections follow:

Italy: Huet du Pavillon, E. & A., Sicily, prope Villafrati, in incultis, 20-V-1856, in fruit (BM; specimen with limited information apparently from this site is in a type folder (57/7241) at K. A similar specimen at K is E. & A. Pavillon 81). Todaro, Sicily, Capo di Orlando, in argillosis, Todaro Flora Sicile Exiccata No. 467, V, in flower & fruit (listed by Heyn, p. 132, as *M. muricoleptis*) (K; Todaro Exiccata 1136 and Exiccata 1136 at K are similar).

57. *Medicago noeana* Boiss.

M. noeana Boiss., Diagn. Pl. Orient. Nov. Ser. 2(2): 10 (1856), (basionym).

The epithet *noeana* commemorates Friedrich Wilhelm Noë (1798?–1858), a German botanist, pharmacist, and director of the Botanic Garden of the École Impériale de Médicine de Galata Serai in Constantinople, who collected the type specimen of this species.

Subordinate Taxa

None recognized here.

Type Specimen

Noë, Herb. Noëanum no. 242, Kurdistan in culti, V-1849 (G!).

Synonymy

Homotypic

No names found.

Heterotypic

No names found.

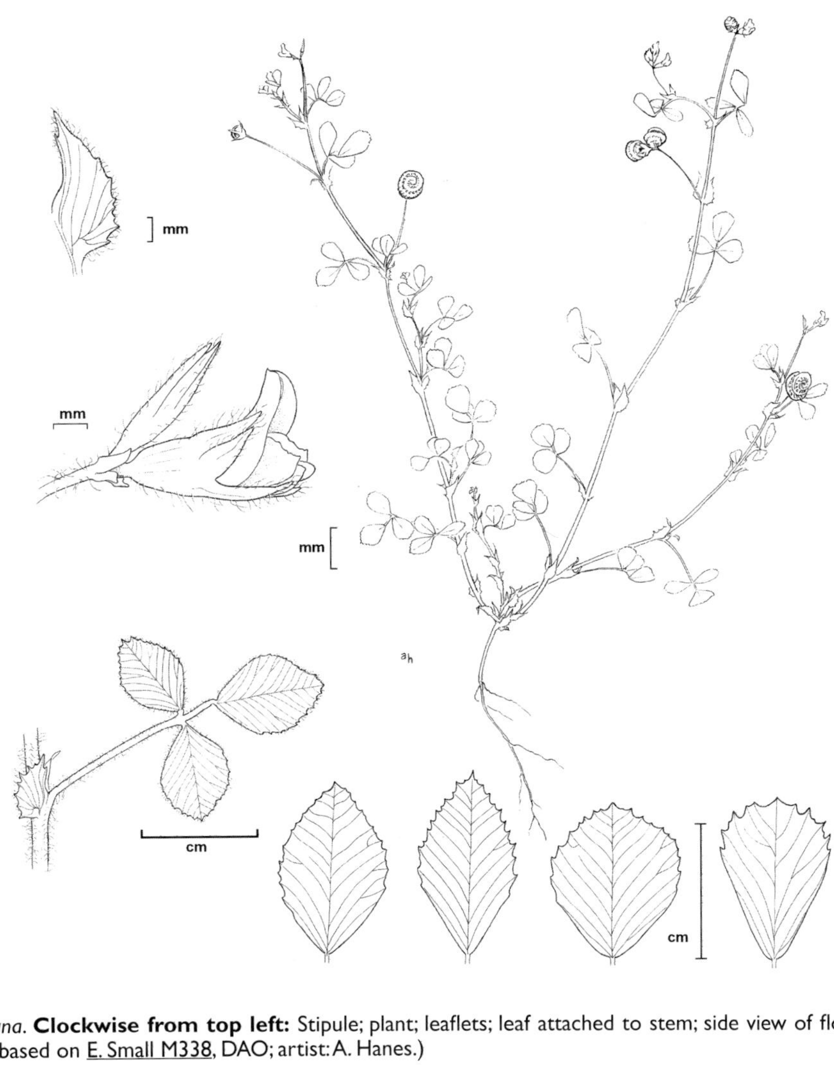

Fig. 57-1. *Medicago noeana.* **Clockwise from top left:** Stipule; plant; leaflets; leaf attached to stem; side view of flower. (Plant based on Poore 340, K, remainder based on E. Small M338, DAO; artist: A. Hanes.)

Vernacular Names

ENGLISH: No names found.

FRENCH: No names found.

Geographical Distribution

(N = native, I = introduced, U = uncertain status)

ASIA: Iraq (N), Syria (N), Turkey (N).

Medicago noeana is a local endemic of northern Iraq, Anatolia, and Syria.

Description

Annual herb, branched from near base, branches procumbent to ascending, 10–30(40) cm long. Vegetative parts more or less pubescent with simple hairs. Leaflets 6–15 mm long, 3–8 mm wide, obovate, ovate, elliptical, rhomboid, or occasionally cuneate, distal half with serrate margins. Stipules entire or serrate at base. Inflorescence of 1–2(7) flowers, peduncles longer than subtending petiole. Flowers 5–7 mm long, calyx more or less pilose with simple hairs, teeth subulate, as long as tube, corolla yellow, less than twice length of calyx. Young pod contracted and concealed inside calyx. Calyx distinctly membranous and symmetrically accrescent around base of mature fruit. Mature pod short-cylindrical, ca. 5 mm long, spineless, glabrous, with (2.5)3–5 loosely appressed coils, becoming thick and hard at maturity, widest coil 4–7 mm in diameter, coil face with ca. 15 thick radial veins, slightly branched near lateral vein near dorsal suture, thick oblique folds joining lateral vein to dorsal suture (a herring-bone pattern appears on the coil edges). Seeds 1–2 per coil, 2.7–4.5 mm long, ca. 1.5 mm wide, oblong, separated by a thin, spongy, transverse fruit septum, radicle equal to or less than half length of seed, seed coat smooth, yellow to light brown.

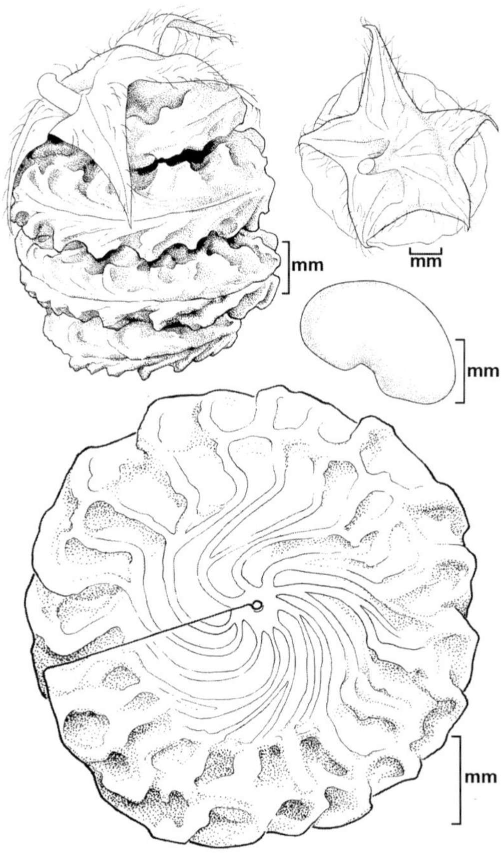

Fig. 57-2. *Medicago noeana.* **Clockwise from top left:** Fruit; bottom of fruit showing calyx; seed; fruit coil face showing venation. (Based on E. Small M338, DAO; artist: A. Hanes.)

Identification Hints

The oblique ridges on the coil edge and the appressed, symmetrically disposed calyx (not displaced to one side as in other annual *Medicago* species) at the base of the pod uniquely characterize this species.

Habitat and Environmental Adaptation

Habitat

Medicago noeana has been recorded in open forests and deserts.

Phenology

This species flowers in April to June in its native area.

Soil

Medicago noeana has been collected from heavy, calcareous soils. Piano and Francis (1992) indicated that in its native area it occurs on loam and clay soils that are slightly to moderately basic.

Temperature

The desert habitat occupied by some populations of this species indicates it has a high tolerance for summer temperatures. Most cultivars of annual *Medicago* have been bred in Australia and lack tolerance to frost (Cocks and Ehrman 1987). In this respect, *M. noeana* is considered to be a better possibility for West Asia than the available Australian cultivars (Abd El Moneim and Cocks 1986); compared with most other medics, the species grows in relatively cool areas (Piano and Francis 1992).

Water

Medicago noeana occurs in dry areas. Piano and Francis (1992) noted that in its native area it requires a minimum annual rainfall of 400 mm.

Biotic Relationships

Pollination

Small (1988) noted that the relatively low pollen to ovule ratio (224) and small flowers are consistent with *M. noeana* being largely self-pollinated. The species appears to be an inbreeder.

Dispersal

The spineless, indehiscent fruits are not adapted to dispersal in animal fur, as in most other annual *Medicago* species. Ghassali et al. (1998) studied the ability of sheep to disperse seeds that they had consumed and found that less than 2% of the seeds of *M. noeana* successfully survived travel through the sheep's gut.

Herbivory

M. noeana appears to be highly resistant to some insects, particularly the potato leafhopper *Empoasca fabae* (Harris) (Shade et al. 1975; Shade et al. 1979). Shade and Kitch (1983) noted that the glandular hairs of *M. noeana* resulted in resistance to the pea aphid (*Acyrthosiphon pisum* (Harris)). MacLean and Byers (1983) observed that the glandular hairs of *M. noeana* tended to deter the alfalfa blotch leafminer (*Agromyza frontella* (Rondani)) from ovipositing.

Nodulation and mycorrhizae

USDA (2009) reported positive root nodulation for *M. noeana*. Lorenzetti et al. (1987) tested the symbiotic response of *Rhizobium meliloti* strains on *M. noeana* and observed a high level of specificity, with some strains greatly increasing productivity, others having little effect. Materon and Zibilske (2001) observed similar results from their studies of the ability of various rhizobial strains to innoculate *M. noeana*. Vázquez et al. (2001) examined the performance of *M. noeana* inoculated concurrently with *Sinorhizobium meliloti* and arbuscular mycorrhizal fungi of the genus *Glomus* and found that occasional enhancement resulted from the combination of micro-organisms, depending on the conditions of growth.

Disease resistance

Yaege and Stuteville (2002) tested the resistance of 33 annual species of *Medicago* to powdery mildew (*Erysiphe pisi*) and found that one collection of *M. noeana* was susceptible to the disease whereas another was resistant to it (it was observed that often only one of a number of populations of other species tested was resistant). In a comparative test of 36 annual species of *Medicago*, O'Neill and Bauchan (2000) found that two accessions of *M. noeana* exhibited strong susceptibility to anthracnose caused by *Colletotrichum trifolii*.

Economic History, Cultivars, and Uses

There is no record of this species having been employed for any economic purpose.

Cultivars

No cultivars have been bred.

Forage

Medicago noeana is not sown for forage and is so rare that it contributes little to natural wildland forage.

Fodder

This species is not harvested for fodder.

Green manure, rotation crop, ground cover

No information located.

Other uses (salad sprouts, ornamental, plant products, etc.)

No information located.

Chromosome Number

$2n = 16$: Lesins and Lesins 1963*a*; Falistocco and Falcinelli 1993.

Genetic Relationships and Gene Transferability

Intraspecific

Artificial hybridization of different populations has not been attempted and, accordingly, barriers to hybridization within the species are not known.

Interspecific

Lesins and Lesins (1979) doubled the chromosome number of *M. noeana* to $2n = 32$ and crossed this polyploid with *M. rugosa* (which they thought had $2n = 32$, but as noted in Chapter 82 probably has $2n = 30$), but were unable to obtain hybrids.

Exchange by advanced biotechnology

Piccirilli and Arcioni (1992*a*, *b*) attempted to produce hybrids between *M. noeana* ($2n = 16$) and *M. sativa* ($2n = 32$) using embryo rescue techniques; they obtained viable embryos but were unable to mature these into viable plants using these techniques.

Conservation Status

A conservation assessment of *M. noeana* is required. There is a great need to acquire germplasm of this rare, endemic species, both for conservation purposes as well as for agricultural assessment.

Germplasm Sources

More than 80 accessions are present in major germplasm institutions (see Chapter 3).

Research Problems and Priorities

Medicago noeana is one of the least studied species in the genus *Medicago* because of its rarity and inaccessibility. Piano and Francis (1992) described it as having "regional potential". It is in need of basic botanical studies, as well as assessment of its potential in agriculture, both as a special-purpose legume crop and as a source of genes for improving the major agricultural species of *Medicago*.

Representative Collections

IRAQ: Gillett & Rawi 7517, Qaranjir, between Kirkuk and Chem-chemal, alt. 700 m, grassy slopes, 14-IV-1947, in flower & fruit (K). Poore, M.E.D. 340, Chem-chemal, protected area of steppic grassland, 16-IV-1958, in flower & fruit (K). **Syria:** Haradjian, M. 70, Plantae Syriae borealis, environs d'Aintab, alt. 3000 feet, VI-1906, in flower & fruit (G). Haradjian, M. 142, Plantae Syriae borealis, environs Aintab, alt. 3000 feet, V-1906, in flower & fruit (G). Haradjian, M. 2057, Plantae Syriae borealis, région inf. on Djebel Siman, alt. 1200 feet, 15-20-V-1908, in late flower & fruit (G). **Turkey:** Balansa, B., bords de la route coudaisant d'Ouchak à Boulgas – Keü (Phrygie), 12-VI-1857, in flower & fruit (G). Davis 42442, C8, Mardin, 10–12 km west of Savur, rocky limestone slopes in Quercus aegilops forest, 5-V-1966, in flower & fruit (K). Davis 43290, C8, Sürt, 3–5 km south of Sürt, alt. 700 m, rocky limestone slopes, 17-V-1966, in fruit (K). Sintensis P. 783, Mesopotamia, in flower & fruit (K). Townsend, C.C. 640422/19, 40 km west of Mersin, frequent amid the ruins of Elalussa Sebasto, 22-IV-1964, in fruit (K).

Note

Boissier (1872) segregated *M. noeana* and *M. rugosa* in a unique section, *Elegantes*. Heyn (1963) also suggested that the two species are closely related. Mehregan et al. (2002, p. 215) recognized *M. noeana* as the sole species in a new taxon: *Medicago* L. section *Spirocarpos* Seringe in DC. subsection *Astrocalyx* I. Mehregan.

58. *Medicago orbicularis* (L.) Bartal.

M. orbicularis (L.) Bartal., Cat. Piante Siena 60 (1776).

The Latin epithet *orbicularis*, circular, is an allusion to the round shape of the fruits.

Subordinate Taxa

None recognized here. Heyn (1963), who was not at all reluctant to subdivide annual species of *Medicago* into infraspecific taxa, nevertheless concluded that "It proved impossible to subdivide the great complex of variable forms of *M. orbicularis* into subspecies or varieties, for examination of herbarium material showed that no discontinuity of characters could be found."

Type Specimen

Heyn (1959: 164) lectotypified the basionym *M. polymorpha* L. var. α *orbicularis* L. by an illustration, tab. 15, fig. 1, in: Morison, R. 1680. Plantarum historia universalis Oxoniensis, vol. 2, 152, s.2. Oxford, UK. The accompanying

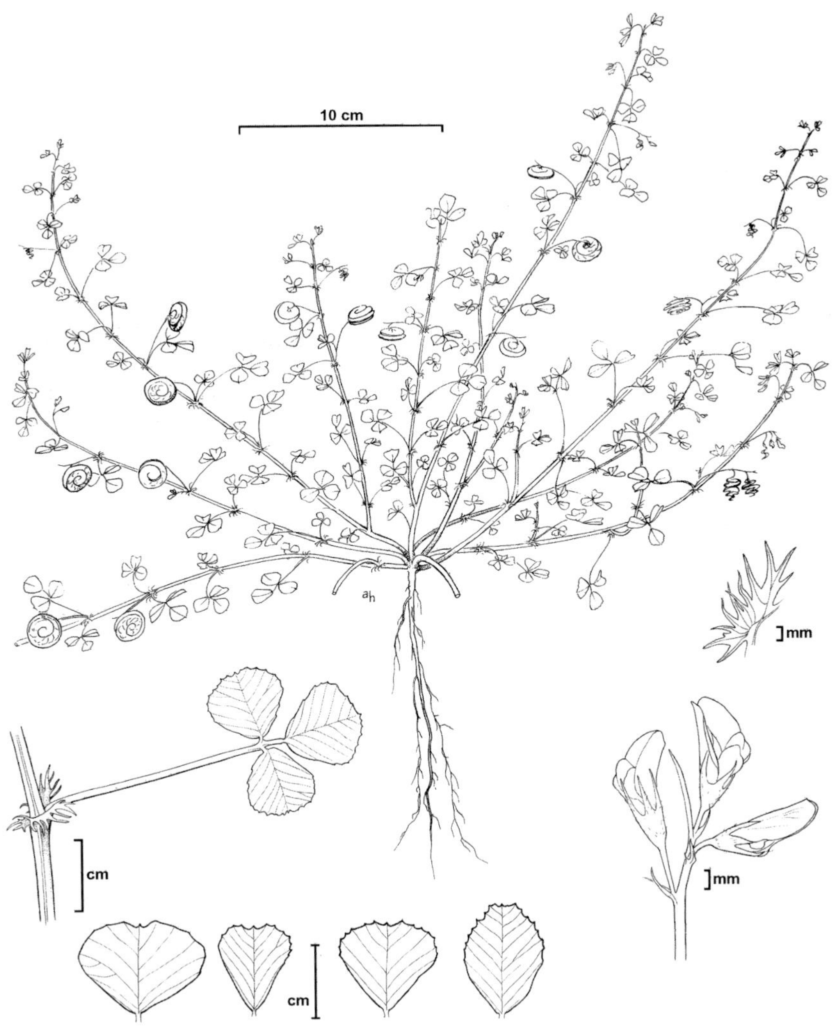

Fig. 58-1. *Medicago orbicularis.* **Clockwise from top:** Plant; stipule; group of flowers; leaflets; leaf attached to stem. (Plant based on accession 6538, HUJ, remainder based on E. Small M877, DAO; artist: A. Hanes.)

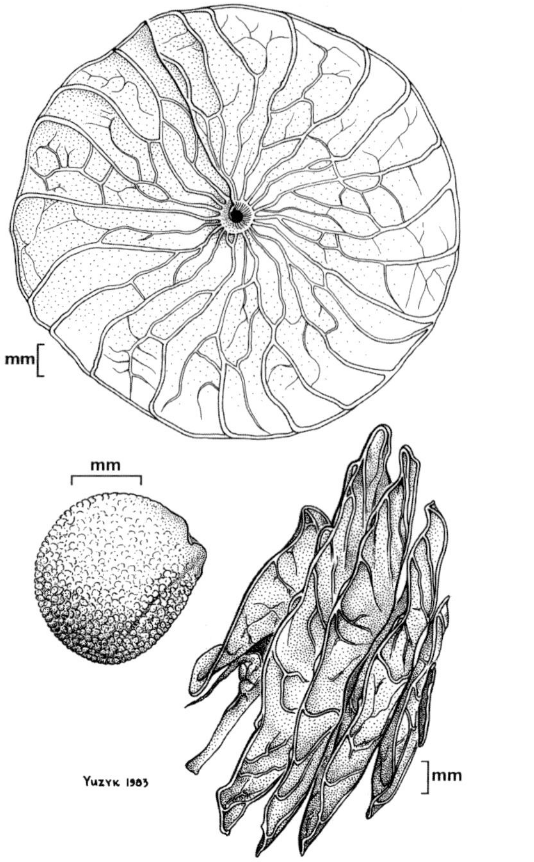

Fig. 58-2. *Medicago orbicularis.* **Clockwise from top:** Fruit coil face showing venation; side view of fruit; seed. (Based on E. Small M316, DAO; artist: L. Yuzyk.)

Fig. 58-3. *Medicago orbicularis*, from Hallier (1885, vol. 23, plate 2342).

text is: "Medica cochleata major dicarpos, fructus capsula compressa orbiculata, nigra, plana, oris crispis." This lectotypification was accepted in The Linnaean Plant Name Typification Project (http://www.nhm.ac.uk/research-curation/projects/linnaean-typification/; current June 2010; cf. Jarvis 2007).

Synonymy

Homotypic

M. polymorpha L. var. α *orbicularis* L., Sp. Pl. 779 (1753) (basionym).

Medica inermis Lam., Fl. Franç. 2: 586 (1778, publ. 1779), nom. illeg.

M. orbicularis (L.) All., Fl. Pedem. 1: 314 (1785).

Medica orbicularis (L.) Medikus, Vorles. Churpfaelz. Phys.-Oecon. Ges. 2: 386 (1787).

Medica orbiculata Bubani, Fl. Pyren., 2 : 475 (1900), nom. illeg.

Heterotypic

M. ambigua Jordan ex Boreau, Fl. Centre France ed. 3, 2: 147 (1857).

M. applanata Willd. ex Schlecht. Enum. Pl. Hort. Reg. Bot. Reg. Berol., Suppl. 52 (1813), nom. nud.

M. applanata Willd. ex Hornem., Hort. Reg. Bot. Hafn. Suppl. 85 (1819).

M. orbicularis (L.) Bartal. f. *biancae* Urb., Verh. Bot. Ver. Brandenb. 15: 60 (1873). (Type specimens at K, BM, U, and G according to Heyn (1963).)

M. biancae (Urb.) Pinto da Silva, Agron. Lusit., 35: 298 (1974). (Silva stated "*Medicago biancae* Todaro ined. = *M. orbicularis* Bianca, non Mill.")

M. orbicularis (L.) Bartal. f. *canescens* Urb., Verh. Bot. Ver. Brandenb. 15: 60 (1873).

M. orbicularis Bartal. var. *castellana* J. Casellas, Coll. Bot. (Barcelona) 6: 224–225 (1962). (Vivant (2000) described this as having smaller pods with rounder and less rough seeds.)

M. orbicularis Bartal. subsp. *castellana* (J. Casellas) P. Monts., Bull. Soc. Échange Pl. Vasc. Eur. Occid. Bassin Médit. 15: 76. 1974 (1972–1973 publ. 1974).

M. clypeata Lindl., Edward's Bot. Reg., Misc. 60 (1839).

M. cuneata J. Woods, Tourist's Fl. 84 (1850). Lectotype at K, according to Heyn (1963).

M. echinata Steud., Nomencl. Bot. (ed. 2) 2(9): 109 (1841). nom. inval., in syn.

M. orbicularis (L.) Bartal. f. *glabella* Abdull. in Khassanov Dikorastuschnie lyutserny Srednei Assii 86 (1972), nom. nud.

M. orbicularis (L.) Bartal. f. *glandulosa* Urb. Verh. Bot. Ver. Brandenb. 15: 60 (1873).

M. orbicularis (L.) Bartal. f. *glanduligera* Kozuharov, Bulgarian Acad. Sci. Reports 15: 149 (1965) (type indicated in this publication).

M. orbicularis (L.) All. var. *hebecarpa* Panč., Verh. Zool.-Bot. Ges. Wien. 6: 479 (1856).

M. marginata Willd., Enum. Pl. Hort. Reg. Bot. Berol. 2: 802 (1809). (Holotype Herb. Willd., B!)

M. orbicularis (L.) Bartal. var. *marginata* (Willd.) Benth., Cat. Pl. Pyr. Bas-Languedoc 100 (1826).

M. orbicularis (L.) Bartal. subsp. *marginata* (Willd.) Nyman, Consp. Fl. Eur. 167 (1878).

M. orbicularis (L.) Bartal. proles *marginata* (Willd.) Rouy in Rouy & Fouc., Fl. Fr. 5: 18 (1899).

M. orbicularis (L.) Bartal. var. *microcarpa* Rouy in Rouy & Fouc., Fl. France 5: 18 (1899), non Seringe in DC. Prodr. 2: 174 (1825) (illegitimate later homonym).

M. orbicularis (L.) Bartal. var. *macrocarpa* Rouy in Rouy & Fouc., Fl. France 5: 18 (1899).

M. orbicularis (L.) Bartal. f. *macrocarpa* (Rouy) Casellas (Casellas 1962, p. 223).

M. orbicularis (L.) Bartal. var. *pilosa* Benth. Cat. Pyr. 101 (1826).

M. orbicularis (L.) Bartal. f. *pilosa* (Benth.) Urb., Verh. Bot. Ver. Brandenb. 15: 60 (1873).

M. orbicularis (L.) Bartal. f. *praecox* Sinsk., Fl. Cult. Pl. U.S.S.R. 13(1): 176 (1950), nom. illegit.

M. orbicularis (L.) Bartal. f. *rubra* Sinsk., Fl. Cult. Pl. U.S.S.R. 13(1): 177 (1950), nom. illegit.

M. scutellata Pall., Ind. Taur. ex Bieb. Fl. Taur.-Cauc. 2: 226 (1808). (Listed as a synonym of *M. orbicularis* by Heyn (1963.))

M. orbicularis (L.) Bartal. f. *tardiflora* Sinsk., Fl. Cult. Pl. U.S.S.R. 13(1): 176 (1950), nom. illegit.

M. orbicularis (L.) Bartal. var. *typica* Asch. & Graebn., Syn. 4(2): 405 (1907), nom. inadmiss.

M. orbicularis (L.) Bartal. var. *undulata* Rouy in Rouy & Fouc., Fl. France 5: 18 (1899).

M. orbicularis (L.) Bartal. f. *viridis* Sinsk., Fl. Cult. Pl. U.S.S.R. 13(1): 177 (1950), nom. illegit.

M. orbicularis (L.) Bartal. var. *marginata* (Willd.) Benth. f. *vulgaris* Abdull. in Khassanov, Dikorastuschnie lyutserny Srednei Assii 86 (1972), nom. nud.

Vernacular Names

English: Black disc medic (blackdisk medic), button clover, button medic, flat-podded medic, large disc medic, orbicular alfalfa (Bolton 1962), round-fruited medic (medic is also spelled medick; disc is sometimes spelled disk).

French: Herbe au limaçon, limaçon, la luzerne marginée, luzerne orbiculaire, luzerne ronde.

Geographical Distribution

(N = native, I = introduced, U = uncertain status)

Africa: Algeria (N), Canary Islands (Spain; N), Egypt (N), Ethiopia (U), Libya (N), Madeira (Portugal; N), Morocco (N), South Africa (I), Tunisia (N).

Asia: Afghanistan (N), Armenia (N), Azerbaijan (N), Cyprus (N), Gruziya (N), India (N), Iran (N), Iraq (N), Israel (N), Japan (I, according to "Alien species recognized to be established in Japan or found in the Japanese wild as of Oct. 27, 2004"; http://www.env.go.jp/en/nature/as/041110.pdf; current June 2010), Jordan (N), Kazakhstan (N), Kyrgyzstan (N), Lebanon (N), Oman (I), Pakistan (N), Russia (N), Saudi Arabia (I), Syria (N), Tajikistan (N), Turkey (N), Turkmenistan (N), Uzbekistan (N).

Australasia: Australia (I).

Europe: Albania (N), Bulgaria (N), Czech Republic (I), France (including Corsica; N), Greece (including Crete, Karpathos, East Aegean Islands; N), Hungary (I), Italy (including Sardinia and Sicily; N), Malta (N), Portugal (N), Romania (N), Spain (including Balearic Islands; N), Turkey (N), Ukraine (N), United Kingdom (I), Yugoslavia (N).

Northern America: USA (I).

Southern America: Chile (I).

This species is certainly omni-Mediterranean and West Asian. Its status in Europe and Central Asia is debatable, and it is widely adventive elsewhere.

Description

Annual herb, branched from base, branches 10–40(100) cm long, usually procumbent (sometimes ascending). Vegetative parts usually glabrous or glabrescent, very rarely with gland-tipped hairs. Leaflets 7–13 mm long, 5–10 mm wide, obovate to cuneate, distal third to two-thirds with serrate margins. Stipules 3–6 mm long, laciniate. Inflorescence of 1–2(5) flowers, peduncles shorter or longer than subtending petiole. Flowers 3–4(6) mm long, calyx glabrous or sparsely hairy, teeth about as long as tube, corolla yellow, up to twice as long as calyx. Young pod protruding sideways from calyx in a loose spiral that gradually contracts. Mature pod lenticular or discoid, spineless, 5–10 mm long, glabrous, glabrescent, or covered with gland-tipped hairs. Mature pod soft, brown-yellow or blackish-purple, with (2)3–7 coils that are papery at margins, diameter of middle coils 8–20(24) mm, usually just one pod developing on each peduncle. Coil face with anastomosing radial veins, often somewhat thickened at coil margin, lacking a lateral vein. Seeds triangular or deltoid-orbicular, compressed, 3–6 per coil, usually separated by thin partitions, 2.5–3 mm long, seedcoat tubercular, yellow, brownish-yellow or reddish-brown, radicle about equal to length of seed.

Identification Hints

The rather papery coiled pods of *M. orbicularis* are likely to be confused only with those of *M. bonarotiana* and *M. bonarotiana* × *M. rotata,* or possibly with *M. scutellata*, which can be distinguished from *M. orbicularis* by their imbricate (bowl-like) coils and smooth seeds.

Habitat and Environmental Adaptation

Habitat

Medicago orbicularis commonly occurs on hilly slopes and in cultivated and ruderal habitats. Zoghlami et al. (1996), in reference to Tunisia, observed that *M. orbicularis* is found principally on the margins of cultivated lands, fallow areas, and foothills. Thanopoulos (2007) recorded its presence in Greece at altitudes from sea level up to 1800 m, as a constituent of natural shrub and woodland associations, meadows (sometimes wet), and grasslands, and as a weed of crop fields, orchards, and wasteland.

Phenology

As with other annual Mediterranean region species of *Medicago*, *M. orbicularis* is a winter annual flowering in the spring and early summer.

Soil

Bennett (1999) commented that "The soils of all but a few of the sites in southwest Turkey were assessed to be freely drained." This is in contrast to remarks made by Lesins and Lesins (1979) and Heyn (1963) who observed that *M. orbicularis* occurs in heavy, waterlogged soils, possibly a reflection of the time of year at which the sites were visited. However, Bounejmate et al. (1992*b*) also found that in Morocco *M. orbicularis* tended to occupy sites with a high clay content. *Medicago orbicularis* also appears to prefer highly alkaline soils, with a large proportion of the sites in Turkey, Algeria, and Syria having a pH of 9.0 or above. Abdelguerfi et al. (1988*a*) observed that in Algeria *M. orbicularis* occurs in soils of all textures, from coarse to fine. Piano and Francis (1992) described the species as occupying loam and clay soils ranging in pH from slightly to moderately basic. In Sicily, Piano et al. (1991) observed the species growing in a pH range of 6.5–9.0. McKee (1916) noted that in cultivation in the United States, *M. orbicularis* grows best on rich loam, including heavy lowlands. Ewing (1956) remarked that the species has a wide soil adaptation and can be grown on a range of soil types. Bolton (1962) stated that the species has shown adaptation to saline soils in the southern USA. However, Rizk et al. (1978) found that of four annual species of *Medicago* tested, *M. orbicularis* was the least tolerant of sodium chloride. This is consistent with Zoghlami and Hassen (1999), who observed that in Tunisia *M. orbicularis* was associated with soils that were low in salt.

Temperature

Medicago orbicularis occurs in areas that may become very hot during the early summer. Piano and Francis (1992) described it as more or less cold tolerant, by the standards of other medics. Krall et al. (1996) found that two of the ten accessions of *M. orbicularis* examined for their ability to self-regenerate after overwintering in the continental climate of Wyoming survived at one of the three locations tested. Subfreezing temperatures in December 1998 in West Central Texas caused considerable seedling mortality of cultivar 'Estes' (Ueckert et al. 2003).

Water

Piano and Francis (1992) observed that in its native area the species requires a minimum of 250 mm annual precipitation. In Tunisia, the species occupies a semi-arid zone with 250–450 mm of precipitation annually (Zoghlami and Hassen 2004). Note the remarks above with respective to the water-holding capacity of the soils occupied. Conway et al. (2001) studied the performance of *M. orbicularis* in Central Queensland, an area with summer-dominant rainfall that differs from the rain-fed winter Mediterranean climate to which the species is adapted (winter rainfall in Central Queensland is typically 80–100 mm). They concluded that *M. orbicularis* performed among the best of several *Medicago* species studied. Townsend (1974) noted that in Iraq the species is often found in relatively moist situations. McKee (1916) observed that in cultivation in the USA, *M. orbicularis* grows most productively when provided with adequate moisture, although it will survive and make fair growth under rather arid conditions.

Biotic Relationships

Pollination

Small (1988) noted that the small flowers and pollen to ovule ratio (154) of the species are consistent with *M. orbicularis* being largely self-pollinated. This species is basically an inbreeder.

Dispersal

By comparison with other annual species of *Medicago* lacking spiny pods, *M. orbicularis* is exceptionally adventive and widespread outside of its indigenous area. Heyn (1963) suggested that the pods are distributed by wind. Olivieri et al. (1991) stated "We do not agree with Heyn (1963) that it might be dispersed by wind. Rather, the main reason for its success could be the high number of seeds per pod (about 20, instead of the four or five for most other species), and moreover, the high percentage of hard (dormant) seeds." Heyn's viewpoint is supported by the presence of the large wind-sail surface of the pods. The viewpoint of Olivieri et al. is contradicted by the fact that most annual species of *Medicago* have hard seed and there is no apparent reason why many seeds per pod is more favorable for wide dispersal than many seeds per plant. As is well known, numerous plant species are extraordinarily invasive despite the fact that they lack apparent specializations for seed dispersal.

Herbivory

No information located.

Table 58-1. Forage yield of native *M. orbicularis* in northern Iraq (kg/ha), after Kernick (1978).

Fertilized with 60 kg/ha P_2O_5		Unfertilized	
Green matter	Dry matter	Green matter	Dry matter
4200	1600	3200	1100

Nodulation and mycorrhizae

USDA (2009) reported positive root nodulation for *M. orbicularis*. In Central and Western Spain it has been shown that, when infected by an effective strain of *Rhizobium*, *M. orbicularis* resists freezing temperatures and long periods of winter drought, whereas noninfected plants die (Feuntes 1981). Chebouti and Abdelguerfi (2003) compared the root weight, number of nodules, and nodule weight responses of *M. orbicularis* when it was grown under either drought or irrigation and found that the three variables decreased more or less proportionately to each other during drought.

Disease resistance

O'Neill et al. (2003) examined the comparative resistance of 36 annual species of *Medicago* to *Phoma medicaginis* (the cause of spring black stem and leaf spot, a major disease of alfalfa) and concluded that *M. orbicularis* had high resistance to it. In a similar comparative test of 36 annual species of *Medicago*, O'Neill and Bauchan (2000) found that eight accessions of *M. orbicularis* were very susceptible to anthracnose caused by *Colletotrichum trifolii*. Yaege and Stuteville (2002) tested the resistance of 33 annual species of *Medicago* to powdery mildew (*Erysiphe pisi*) and found that the eight collections of *M. orbicularis* grown were susceptible to the disease (it was observed that often only one of a number of populations of other species tested was resistant). Pathipanawat et al. (1996) tested the resistance of 75 accessions of *M. orbicularis* to alfalfa mosaic virus and found a wide range of resistance.

Economic History, Cultivars, and Uses

Sinskaya (1950) considered *M. orbicularis* to be a promising species for winter culture in the subtropical regions of the USSR. Townsend (1974) stated with respect to Iraq that *M. orbicularis* is "another species which is locally considered to be of certain value for grazing in natural pastures." *Medicago orbicularis* was sown for pasture in New South Wales, Australia as early as the 1890s (Brownlee 1985). Saunders et al. (1993) considered it to be a potential medic for cultivar development in southern Australia. The species has been grown to a small extent in the southern United States, especially in Texas and Tennessee. To date, however, it has attracted relatively limited interest as a crop plant.

Cultivars

In 1957 the Texas Research Foundation at Renner released an unnamed selection of *M. orbicularis* (Davis et al. 1957). According to Ueckert et al. (2003), cultivar 'Estes' "is believed to be a rediscovery of the material originally released as button clover by the Texas Research Foundation." 'Estes' is used in North Central Texas.

Forage

This species produces high-quality forage for livestock (Derkaoui et al. 1993; Zhu et al. 1996) and has high seed productivity (Conway et al. 2001). St. John-Sweeting et al. (1996), in reference to South Australia, stated that "*M. orbicularis* has been shown to be a prolific seed setter... however it exhibits erratic breakdown of hard seed making it a nuisance in crops and reducing the likelihood of a dense pasture." Nevertheless, on the basis of a trial conducted by these authors, they found that the species was comparable both in herbage and seed production with cultivar 'Caliph' of *M. truncatula*, and noted that the pods are spineless, a characteristic of considerable importance to the wool industry; they suggested that further study of it for commercial development was warranted. Belov (1971) examined 37 annual species of *Medicago* for suitability as new crops in the Tashkent region of Uzbekistan and *M. orbicularis* was one of the three species recommended for development. McKee (1916) recommended *M. orbicularis* as pasturage in areas of the USA with a mild winter (including the western and southern coasts). He warned that, like some other annual medics, it is readily eaten by livestock in the green state, but sometimes causes bloat in cattle and sheep, so caution should be exercised when turning stock into fields for the first time, especially in the early spring when succulent growth occurs. The numerous pods produced at maturity are also eaten readily (Bolton 1962). *Medicago orbicularis* has been seeded sporadically in northeastern Texas (Wheeler 1950; Davis et al. 1957). Forage yields and protein concentrations were increased when *M. orbicularis* was added to grass or cereal pastures (Rumbaugh and Johnson 1986). Ibragimova and Smolenskaya (2001) examined 24 annual species of *Medicago* for their agronomic performance in the Novosibirsk region of Russia and concluded that only *M. scutellata* and *M. orbicularis* showed promise for the area. Vitkus (1977) assessed *M. orbicularis* as suitable for cultivar selection for Lithuania. Nasirzadeh et al. (2006) evaluated *M. orbicularis* as suitable for the dryland or range regions of Iran. Sanadgol et al. (2006) found the unirrigated yield of *M. orbicularis* over a 3-year period in the Gorgan plain of Iran was 1076 kg/ha. Tadmor et al. (1971) conducted studies of ecotypes selected for their ability to survive and persist in the Negev Desert of Israel in an area where rainfall averaged only 90 mm annually, but an additional 263 mm of water was received annually in the form of run-off from erratic winter storms. Dry matter production of an ecotype of *M. orbicularis* grown in this area was 3.3 t/ha (averaged over 6 years). Table 58-1 lists the forage yields of *M. orbicularis* in a nursery plot on alluvial soil in the steppic zone (220 m elevation) of northern Iraq during a season with rainfall of 475 mm (November to May).

Mohammad and Qamar (1988) grew 14 annual species of *Medicago* in small test plots in Islamabad, Pakistan located in a subtropical, subhumid area that is notably hotter and wetter than the normal habitat of these species. The dry matter yield of *M. orbicularis* was about 4 t/ha.

Fodder

McKee (1916), with respect to the areas of the United States where *M. orbicularis* can be grown, recommended it as hay, noting that it is readily eaten by all kinds of stock. He wrote: "For hay of the best quality, it should be cut when most of the pods are still green but well developed and just before the leaves begin to fall. When allowed to become more mature, many of the pods as well as the leaves are lost in handling and the value of the hay is greatly reduced. In growing for hay, a supporting crop of oats or barley or some other grain should be sown with the button clover, to facilitate cutting. The decumbent habit of the plant makes this desirable (when grown with grasses, the plants become much more erect), and aside from this, the mixture, consisting of grain and button clover, makes excellent feed. The use of button clover for hay production will no doubt be very limited, as there are other legume crops that are more desirable for this purpose. However, its habit of growth and not its feeding value as hay is the factor limiting its use." Additional information on growing the species for hay or ensilage is provided in Davis et al. (1957), who noted that "Button clover has been found to yield an average of one and one-half tons of air dry hay per acre over a five year period."

Green manure, rotation crop, ground cover

McKee (1916) and Sinskaya (1950) observed that *M. orbicularis* can be used as green manure.

Other uses (salad sprouts, ornamental, plant products, etc.)

Naturalized *M. orbicularis* has been described as an important wild food for white-tailed deer in the Cross Timbers area of Oklahoma and Texas. "Button clover seeds are utilized by wildlife and the plant attracts insects which can be consumed by wild turkey and other birds." (ftp://ftp-fc.sc.egov.usda.gov/GA/PMC/TDT/PlantWildlifeRev1.pdf; current June 2010).

Chromosome Number

$2n = 16$: Ghimpu 1928, 1929*a*, 1929*b*, 1930; Fryer 1930; Larsen 1955; Heyn 1956; Clement 1962; Mariani 1963; Simon and Simon 1965; Bolkhovskikh et al. 1969, p. 309; Dahlgren et al. 1971; Fernandes and Santos 1971; Strid 1971; Löve and Kjellqvist 1974; Horjales 1976; Labadie 1976; Sareen and Trehan 1976; Fernandes and Queirós 1978; Lesins and Lesins 1979; Abdelguerfi and Guittoneau 1979; Loon 1980; Loon and Kieft 1980; Magulaev 1980; Agarwal and Gupta 1983*b*; Abdelguerfi et al. 1988*b*; Colombo and Trapani 1990; Baltisberger 1991*a;* Díaz Lifante et al. 1992; Karshibaev 1992; Magulaev 1992; Oberprieler and Vogt 1996*a*; Karadağ and Gülcan 1997; Ibragimova 1999; Baltisberger et al. 2006.

Lesins (1952) has been mistakenly reported as representing a natural $2n = 32$ count (e.g., Bolkhovskikh et al. 1969 p. 309, as "Larsen 1952"); however, tetraploidy was artificially induced.

Genetic Relationships and Gene Transferability

Intraspecific

No barriers to hybridization within the species are known. Lesins and Lesins (1979) reported that they hybridized rather different forms of the species to each other.

Interspecific

Lesins (1952) was unable to hybridize *M. orbicularis* that was artificially doubled in chromosome number to $2n = 32$ with *M. sativa* that had the same number of chromosomes. Lesins and Lesins (1979) were unable to hybridize *M. orbicularis* with *M. heyniana*.

Exchange by advanced biotechnology

No information located.

Conservation Status

Medicago orbicularis is a widespread, somewhat weedy species and therefore apparently not in need of conservation protection. Nevertheless, it has been listed as a protected species in the Limousin region of France (http://www.mnhn.fr/mnhn/cbnbp/flore/textes/detail/limousin.htm; current June 2010). It has also been listed as a protected species in Hungary (http://econets.iucn-ce.org/files/1/060125204354_Species%20Hungary; current July 2010).

Germplasm Sources

Approximately 3000 accessions of *M. orbicularis* are available in major germplasm institutions (see Chapter 3).

Research Problems and Priorities

Medicago orbicularis has good potential for development as an annual legume fodder crop and it is surprising that it has not received more attention than it has. It deserves to be examined more thoroughly for its agronomic properties in dryland situations, and cultivar development should be encouraged.

Representative Collections

Algeria: Bové, N., Alger, V-1883, in fruit (P). Durieu, M., Constantine, 28-V-1860, in fruit (P). **Australia:** Kelso, R.P. 152, New South Wales, East Tamworth, roadside, 17-XI-1954, in flower & fruit (BM). **Britain:** Smith, R.L., near Pengam, Cardiff, rubbish heap, VII-1937, in flower & fruit (K). **Bulgaria:** Kojoucharov, S. & Kouzmanov B., les prairies près du village Mezek, la vallée de la rivière Moritza, 1-V-1959, in flower & fruit (K). **Canary Islands:** Bourgeau, E., Teneriffa, in Convalle bco, del Agua, prope oppidum Joad de los Yivos, IV-1845, in flower & fruit, E. Bourgeau Plantae Canarienses No. 506 (P). Pitard, C.J. 119, Teneriffe, Barranco de Bufadero, in incultis regionis maritimae, 18-IV-1905, in fruit (P). **Cyprus:** Casey, E.C. 1168, Kyrenia, alt. 100 feet, uncultivated olive grove, 4-IV-1951, in fruit (K). Meikle, R.O. 2496, Elisis near Galinoporni, alt. 100 feet, in wet ground by stream, 6-IV-1962, in fruit (K). **Ethiopia:** Parker, C. E265, Ambalage Pass, alt. 2750 m, weed in flax, "5.10"-1969; in flower & fruit (K). Schimper 1896, prope Axum in agris requietis, 17-X-1842, in fruit (BM). Schimper, prope Axum, in agris requietis, 17-IX-1842, in fruit (P). **France:** Déséglise, A., coteaux calcaires de l'Yèvre, Fleurs le 4, fruit le 2 juillet-1860, Flora Galliae et Germaniae Exsiccata de C. Billot No. 2823 (P). Mailho, bord des champs Arignac (Ariège), 10-VI-1889, in flower & fruit, Société Rochelaise 1889 No. 1002 (P). **Greece:** Cannon, J.F.M. & M.J. 4206, Crete, Nomos Irakliov, near Agii Deka, stony roadside, 7-IV-1974, in flower & fruit (BM). Guiol, F. N507, Laconia pr. Masclina, 3-V-1929, in fruit (BM). Guiol. F., Attica, pr. Athenas, V-1929, in fruit (BM). **Iran:** Lindsay, N. 1012, prov. Guitan, Lahijan, cliff in the hills, IV-1936, in fruit (BM). Zohary, M. & Orshan, G. A320/24, northern Iran, 17 km south of Zirub, Zelkova- Lonicera forest, alt. 860 m, 9-V-1961, in fruit (HUJ). **Iraq:** Gillett 8293, near Jindian above Rawanduz, by irrigation channel, 15-V-1947, in fruit (K). H.F. Mooney 4307, northern Iraq, Mosul Liwa, Chaluki, alt. 3300 feet, in fallow field, 30-IV-1951, in flower & fruit (K). **Israel:** Davis, P.H. 4222, near Waldheim (d. Haifa), dry bushy places, 28-III-1942, in fruit (K). Maitland, T.D. 25, Sheikh Jarrah, local, IV-1963, in flower & fruit (K). **Italy:** Davis & Sutton D62814, Calabria (Aspromonte), ca. 30 km south of Reggio, south of Leonardo, alt. 100 m, riverbed on gravel, 10-V-1979, in fruit (BM). Davis & Sutton D65772, Campobasso, ca. 15 km west of Isérnia towards Colli a Volturni, alt. ca. 400 m, edge of cornfield on calcareous soil, 22-VI-1979, in flower & fruit (BM). **Jordan:** Western, A.C. WH132d, El-Gib – Latroun Road, rocky hilltop, limestone and terra rossa, facing northeast, III-IV-1967, in flower & fruit (K). **Lebanon:** Graham, J.A., Coast road about 12 miles north of Tyre (towards Sidon, near Jemjam), dry wayside ditch, 5-V-1945, in flower & fruit (BM). Meinertzhogen, R., Beyrouth, Syria, crops, 31-III-1933, in fruit (BM). Meinertzhogen, R., Beyrouth, Syria, crops, 31-III-1933, in fruit (BM). **Libya:** Keith 515, Benghazi, Cyrenaica, Attag, Fueihat, alt. 10 m, 17-III-1959, in fruit (K). **Malta:** Bricknell, C.A. 161, on dry soil, alt. 50 feet, "5.4"-1964, in fruit (K). **Morocco:** De Wilde J.J.F.E. & P.A.W.J. & Dorgelo, J. 2535, Forêt between Al-Hajeb and Ifrane, 18 km northwest Ifrane, arable field in forest, 25-V-1961, in fruit (BM). Roffey, J., Tangier, 28-IV-1916, in fruit (BM). **Oman:** Mandaville, J.P. Jr. 6407, vicinity of an-Nid (23°13′N, 57°19′E), alt. 2500–2700 feet, weed at edge of garden, 9-IV-1975, in flower & fruit (BM). **Portugal:** Moller, A., Coimbra, Quinta das Monicas VI-1889, in fruit, Flora Lusitanica Exsiccata No. 757 (P). Perris, J.B. & Stübing, G. 53, Alicate, Moraira, 15-V-1984, in fruit (K). Scarlett 236, Lisbon area, sea level upwards, cultivated places, 1935, in fruit (K). **Romania:** Borza, Al. & Nyárády, E.I., Dobrogea, distr. Caliacra, ad littorem ponti euxini (Marea neagra) prope "Farul de la Sabla", alt. 3–5 m, 14-VI-1925, in fruit, Museum Botanicum Universitatis Cluj Flora Romaniae Exsiccata No. 1269 (K, P). Toma, M., distr. Adamclisi, Carpinis, 3-VI-1967, in flower & fruit (BM). **Sicily:** Stace, C.A. & Cotton R., near South Nicola, halfway from Messina to Villa-franca, south-facing grassy, cut and burnt slopes, 23-V-1972, in fruit (BM). **Spain:** Peris, J.B. & Stübing G., Alicate, Moraira, 15-V-1984, in fruit, Universidad de Valencia Exsiccata II No. 53 (BM). Stace, C.A. 78, Tarragona campsite at Torrembarra, near Tarragona, alt. less than 50 feet, in grassland under Cupressus and Eucalyptus trees, 19-IV-1961, in flower & fruit (BM). **Syria:** Berton, A. 126, 1 km S.S.O. de Rachaya, champs pente du ravin, alt. 1250 (sic), in flower & fruit (P). Zohary, M. M/723, Syria bor., about Aleppo, 3-V-1931, in fruit (HUJ). **Tunisia:** Barratte, G., Baja-gare, 10-IV-1888, in flower & fruit (P). Barratte, G., La Manouba, 14-V-1888, in fruit (P). Davis & Lamond D57503, Kasserine to El kef, ca. 6 km south of Thala, alt. 1030 m, calcareous plateau, edge of fields, 9-V-1975, in flower & fruit (BM). **Turkey:** Balls, E.K. B832, Nurgana, Gaziantep, alt. 2700 feet, cultivated ground, 25-IV-1934, in flower & fruit (BM). Bornmüller, J. 9274, Lydia, Smyrna in collibus prope "Kokarialü" et "Giöstepe", 11-V-1906, in fruit (P). **USSR:** Davis 33182, Crimea, distr. Yalta, Nikita, alt. 5 m, fallow fields, 30-V-1959, in fruit (K). Puring, N., Flora Taurica, In declinibus hetbosis horto Massandra prope Jalta, 14-27-IV-1900 (LE). Yunge, A., Tauria, prope opp. Jalta et Kuktebel, 6-V-1905, in fruit (K). **Yugoslavia:** Jackson, A.K. & Turrill 124, Lapad Peninsula, in half-shade, 19-IV-1935, in flower & fruit (K). Pichler T., Dalmatia, in agris prope Joretto pr. Spalato, VI-1880, in fruit (K).

59. *Medicago orthoceras* (Kar. & Kir.) Trautv.

M. orthoceras (Kar. & Kir.) Trautv., Trudy Imp. St.-Petersburgsk. Bot. Sada (Acta Horti Petrop.) 5: 421 (1877).

The epithet *orthoceras* is based on the Greek *ortho*, meaning straight, plus *keras*, meaning horn, indicating the straight (uncurved), horn-like shape of the pods.

Subordinate Taxa

None recognized here.

Occasional plants with deeply incised or partite leaflets occur, as they do in several other species of *Medicago*, but because such plants grow in the company of typical forms of the species and are linked with them by intermediate forms, taxonomic recognition of these is unwarranted.

Type Specimen

Karelin & Kirilov, in arenosis ad flumen Lepsa (Turkestan), 1840, MW (Fig. 59-3). (In disagreement,

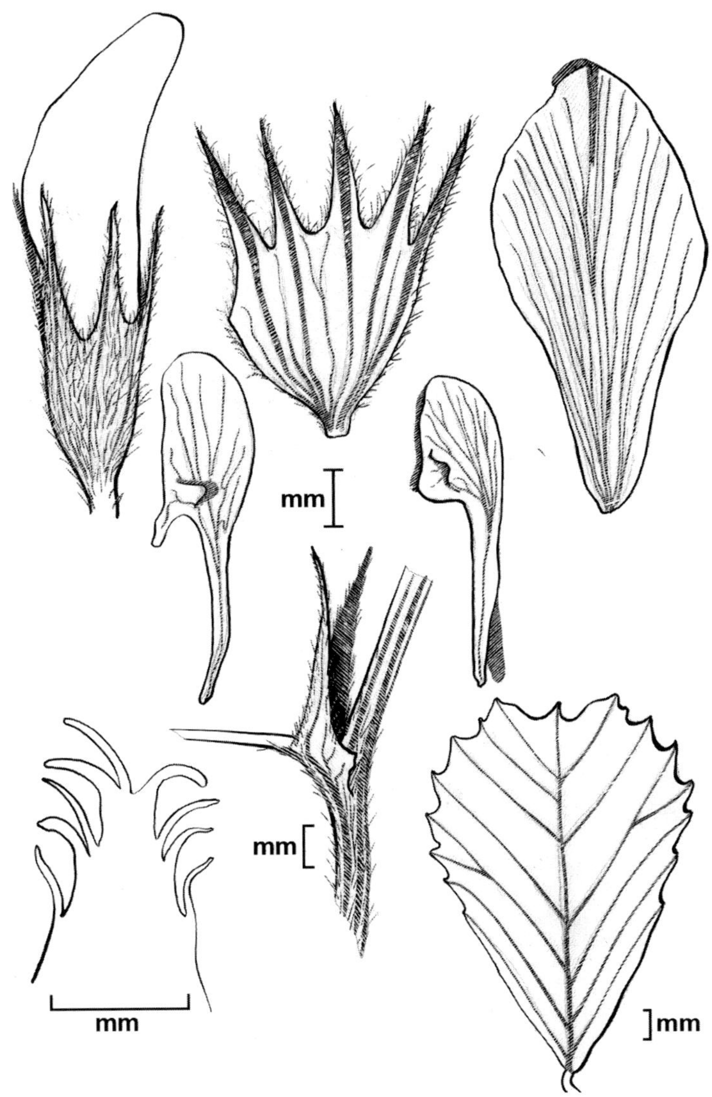

Fig. 59-1. *Medicago orthoceras.* **Top (left to right):** Side view of flower; calyx; standard petal. **Center (left to right):** Wing petal; keel petal. **Bottom (left to right):** Flattened androecial column (free stamen not shown); stem node showing stipules; leaflet. (Based on L. Polijako, 26-V-1953, LD; artist: M. Jomphe.)

Vassilczenko (2002*a*) provided the following information: "Type: Kazakhstan, Tarbagatai ('Circa montes Tarbagatai sub finem maji deflorata')".)

Synonymy

Homotypic

Trigonella orthoceras Kar. & Kir., Bull. Soc. Imp. Naturalistes Moscou 14: 399 (1841), (basionym).

Heterotypic

Trigonella anatolica Boiss. & Bal. in Boiss., Diagn. Pl. Orient. 2, 3(6): 46 (1859). (Type, Herb. Boiss., G!)

T. orthoceras Kar. & Kir. var. *anatolica* (Boiss. & Bal.) Boiss., Fl. Or. 2: 77 (1872).

T. orthoceras Kar. & Kir. var. *baylissi* Blakelock, Kew Bull. 3: 423 (1948). (Type: Bayliss, B.E. 134A, cited below, K.)

"*M. polycerata*" var. *chlorosperma* Trautv., Bull. Soc. Imp. Naturalistes Moscou, 33: 475 (1860).

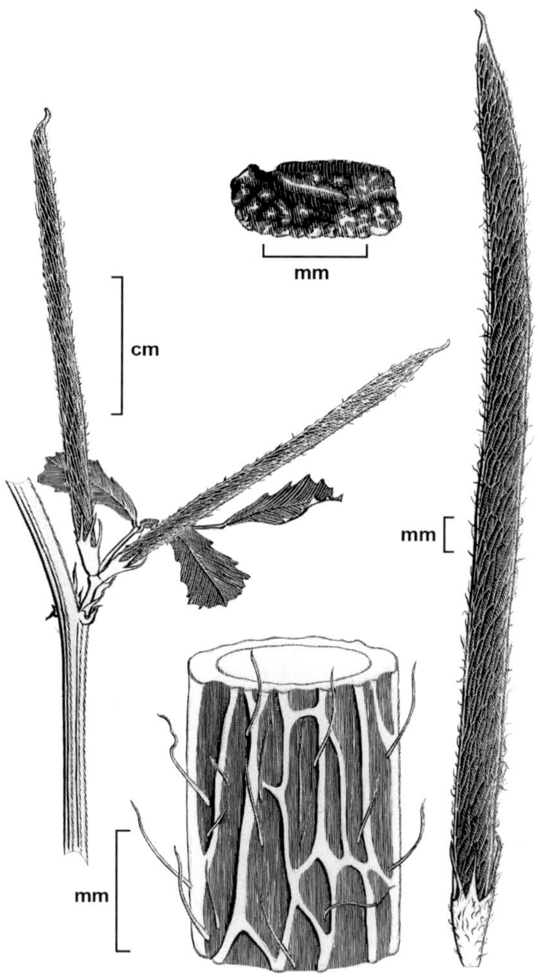

Fig. 59-2. *Medicago orthoceras.* **Clockwise from left:** Branch with two fruits; seed; fruit; portion of fruit showing venation. (Based on L. Polijako, 26-V-1953, LD; artist: M. Jomphe.)

T. orthoceras Kar. & Kir. f. *pedunculata* Širj., Generis *Trigonella* revisio critica, Spisy Pfir. Fac. Masarykovy Univ., 4: 17 (1931).

"*T. polycerata*" of various authors (e.g., *T. polycerata* M. Bieb., Fl. Taur.-Cauc. 2: 222 (1808)).

T. orthoceras Kar. & Kir. f. *typica* Širj., Generis *Trigonella* revisio critica, Spisy Pfir. Fac. Masarykovy Univ., 4: 17 (1931), nom. illeg.

Vernacular Names

English: No names found.

French: No names found.

Geographical Distribution

(N = native, I = introduced, U = uncertain status)

Asia: Armenia (N), Azerbaijan (N), China (N), Gruziya (N), Kazakhstan (N), Kyrgyzstan (N), Iran (N), Iraq (N), Lebanon (I), Pakistan (N), Russia (N), Syria (I), Tadzhikistan (N), Turkey (N), Turkmenistan (N), Uzbekistan (N).

Europe: Russia (N). According to Thanopoulos (2007): "There is only one record of *Medicago orthoceras* known in Greece, based on a specimen kept in LD (!). The specimen has been found in the southern parts of Evros prefecture, close to the city of Alexandroupoli. This place probably forms the westernmost boundary of this Anatolian element."

This species occurs in Asia and southwestern Russia and has been introduced into Western Europe.

Description

Annual herb, branched from base, stems ascending or rarely erect, (5)20–40(50) cm long. Vegetative parts more or less pubescent with simple, appressed hairs, or glabrescent. Leaflets (4)6–15 mm long, 3–8(9) mm wide, obovate, cuneate, obovate-triangular, or oblong-obovate, distal half with serrate margins, abaxial surface slightly pubescent, leaflets sometimes incised. Stipules semi-sagittate, dentate at base, 2–4 mm long. Inflorescence of 1–5 flowers, peduncle (0)3–12 mm long. Flowers ca. 5 mm long, calyx shortly tubular-campanulate, 3–4(5) mm long, appressed-pilose, slightly shorter or about as long as corolla, the teeth lanceolate-acuminate, about as long as tube, corolla yellow. Fruit erect, linear, or slightly curved, somewhat compressed, 15–30(45) mm long, 1–1.5(2) mm wide, more or less appressed-hairy, with a very short beak, reticulate-netted, areoles (inter-vein sections) oblong lengthwise, 2–3 times as long as broad. Pod with numerous seeds, seeds oblong-cylindrical or oblong-rhomboid, about 2 mm long, 1 mm wide, brownish, finely tuberculate, radicle about 80% of length of seed.

Fig. 59-3. Holotype of *Trigonella orthoceras* Kar. & Kir. at MW.

Identification Hints

Medicago orthoceras is very difficult to separate from *M. polyceratia*, and the taxonomy of these two species requires additional study (see information for *M. polyceratia*, Chapter 68). Both species occur in similar habitats, but are geographically separated. *Medicago polyceratia* is native to Morocco, Algeria, Tunisia, northern Portugal, Spain, and southern France, whereas *M. orthoceras* is native to Asia and southwestern Russia. Previously published keys are often not reliable for morphological separation.

Širjaev (1928–1934) distinguished the species as follows (translated from Latin):

Legumes straight (not curved), 2–3 cm long, 1–3(4) (per inflorescence), areoles (areas delimited by veins) longitudinally elongate (diameter 2–3:1), calyx teeth as long as calyx tube, leaflets serrate .... *M. orthoceras*

Legumes more or less curved (up to semi-circular), much rarer straight, 2–5 cm long, 1–6(8) (per inflorescence), areoles equal-sided (1:1), calyx teeth shorter than calyx tube, leaflets occasionally incised-dentate .... *M. polyceratia*

Vassilczenko (1953) distinguished them much as did Širjaev (translated from Russian):

Fruit 20–30 mm long, 1–1.5 mm wide, straight (not curved), 1–3(4) in a raceme, teeth of calyx linearly-lanceolate, areoles on valves more elongated .... *M. orthoceras*

Fruit (20)30–50 mm long, 2 mm wide, curved (to semi-circular), very rarely straight .... *M. polyceratia*

Small and Fawzy (1991) found seven characters that differ significantly between the two species. Širjaev's (1928–1934) observations of differences were verified only with respect to fruit curvature, length, and areole shape, but these differences proved too weak for identification. Vassilczenko's (1953) key did not improve significantly over that of Širjaev; although his perception that there is a difference in fruit width was verified, once again the character proved to be weakly discriminating. The most useful character discovered by Small and Fawzy (1991) is that the valves of the fruits of *M. polyceratia* rarely have fewer than 35 veins crossing the longitudinal axis of the fruit valve, whereas those of *M. orthoceras* rarely have more than 35.

Habitat and Environmental Adaptation

Habitat

Medicago orthoceras occurs on dry rocky slopes, sands, steppes, deserts, semi-deserts, and pastures, and is sometimes a weed of cultivated and fallow fields. It has been recorded at altitudes ranging from 1200 to 1900 m.

Phenology

Medicago orthoceras flowers from April to June in much of its native range. Its life cycle in the Zhungaer Desert of China was documented by Yao and Tan (2005). These authors observed that seed germination occurred in the spring, summer, and autumn, but was most prevalent in the spring and autumn. Germination was very closely related to environmental factors such as temperature and precipitation. The plants matured quickly, with the fruits maturing before the arrival of the arid summer season.

Soil

This species occupies soils varying in texture, which may be stony, rubbly, gravelly, sandy, or fine textured.

Temperature

This species occurs in deserts and has the capacity to withstand considerable heat.

Water

Medicago orthoceras is often found in semi-deserts, deserts, and dry soils, indicating tolerance to drought.

Biotic Relationships

Pollination

Small (1988) noted that the flowers are small and the pollen to ovule ratio is low (209), observations that are consistent with *M. orthoceras* being largely self-pollinated and substantially an inbreeder.

Dispersal

No specializations for seed distribution are evident.

Herbivory

No information located.

Nodulation and mycorrhizae

USDA (2009) reported positive root nodulation for *M. orthoceras*.

Disease resistance

No information located.

Economic History, Cultivars, and Uses

This species has not attracted agricultural interest to date.

Cultivars

No cultivars have been bred.

Forage

Medicago orthoceras is not sown for forage. Beresnev et al. (1980) recorded that wild stands of *M. orthoceras* were a preferred food of sheep in Kazakhstan.

Fodder

No information located.

Green manure, rotation crop, ground cover

No information located.

Other uses (salad sprouts, ornamental, plant products, etc.)

No information located.

Chromosome Number

$2n = 44$: Astonova 1981.

Genetic Relationships and Gene Transferability

Intraspecific

Artificial hybridization of different populations has not been attempted, so it is not known whether there are barriers to hybridization within the species.

Interspecific

No information on hybridization studies was located. Given the difficulty of separating *M. orthoceras* from *M. polyceratia*, there is certainly a need to ascertain the genetic relationships between these species. Širjaev (1928–1934) commented that there were specimens suggestive of intermediacy with *T. geminiflora* and *T. cancellata* (both referable to *M. medicaginoides*). Vassilczenko (1953) also interpreted a herbarium collection as a hybrid between *T. orthoceras* (i.e., *M. orthoceras*) and *T. cancellata* (i.e., *M. medicaginoides*). These observations suggest that these two species may hybridize naturally.

Exchange by advanced biotechnology

No information located.

Conservation Status

Medicago orthoceras has received limited study to date and its conservation status is not well appreciated. Although the species is widespread, there may well be local ecotypes that are unique and deserve protection.

Germplasm Sources

A dozen accessions are listed in major germplasm institutions with large collections of *Medicago* seeds (see Chapter 3).

Research Problems and Priorities

Medicago orthoceras is closely related to *M. polyceratia* and additional study of the relationships between these two species is required to determine whether they are only geographical phases of the same species. The agricultural potential of *M. orthoceras* has not been examined to date. It clearly has considerable adaptation to drought stress, it may be directly usable as a managed wild pasture legume, and it may also have genes that could be usefully transferred to *Medicago* species currently utilized in agriculture.

Representative Collections

Iraq: Bayliss, B.E. 134A, Qaiyarah, Mosul, alt. 640 feet, common on fairly good soils, 1-5-39, in flower & fruit (K). **Pakistan:** Lamond, J. 171A, Baluchistan, Quetta, Nushki to Quetta, alt. 700 m, steep hillsides, in rocks at edge of road, 30-III-1965, in flower & fruit (E). Lamond, J. 610, Baluchistan, Makran, Panjgur Irrigation Scheme, alt. 1000 m, sandy river bed below dams, 20-IV-1965, in flower & fruit (E). Popov, G. 251, Baluchistan, near Nushki, alt. 1500 m, silty plain, 3-VI-1963, in flower & fruit (BM). **Turkey:** Bornmüller, J., Lydia, Bunarbaschi, in monte Takhtali-dagh, alt. 900 m, 26-V-1906, in fruit, J. Bornmüller Lydiae et Cariae Pl. Exsicc. No. 9256 (P). Calvert, H.H. & Zohrab, J. 1290, Armenia, 1867(?), in flower & fruit (K). Karamanoglu 66-33, Prov. Kars, district Horasan, Horasan-Karakurt, alt. 1600 m, 28-VI-1966, in fruit (G). Reese, H., Prov. B6 Sivas, district Sivas, 17 km east von Sivas, 16-VI-1939, in fruit (G). **USSR:** Becker, A., Russia merid., in deserto nec non in faucibu ad flumen wolga prope Sarepta, V-1877, in flower & fruit, F. Schultz Herbarium Normale Nov. Ser. Cent. 11 No. 1064 (P). Goloskokov, V., Kazachstania, in montes Karatau Syrdariensis, planites in viciniis pag. Aral-Tjube, in declivibus australibus lapidosis, 26-V-1963, in flower & fruit (BM, K, E). Karelin & Kiriloff 175, in arenosis ad rivulum Burgan circa montes Targatbai, 1840, in fruit (P). Townsend, C.C. 85/848, Armenia, Caucasus, among the ruins of Garni, hills above Yevevan, alt. ca. 1500 m, 20-VI-1985, in fruit (K). Vasák, V., Asia Centralis, Tian-shan, montes "Kingizski Khrebet", 10 km ad boreo-orientem versus ab oppido Frunze (Pishpek), in vicinitate pagi chop-aryk, alt. 1000–1200 m, 29-V-1974, in fruit (K).

60. *Medicago ovalis* (Boiss.) Urb.

M. ovalis (Boiss.) Urb., Verh. Bot. Vereins Prov. Brandenburg 15: 81 (1873). (Commonly cited as *M. ovalis* (Boiss.) Širj.)

The Latin epithet *ovalis*, meaning oval, points out the shape of the fruits.

Subordinate Taxa

None recognized here.

Type Specimen

Hedges and Sales (2001) noted "Ind. loc. 'Reperi in arenosis secús fluvium Guadalhorce unâ leucâ ad urbe Malaga remotis'." A specimen I observed at G in 1993, from Herb. Boissier (but in the general collection), was labelled "Typus or isotypus?" by A. Hansen; it bore the information: Malaga versus ostriam fl. Guadalfeo, mai. Ex. unicum! 1837.

Synonymy

(**Note:** Clement and Smythies (1976) reviewed the nomenclature of *M. ovalis*.)

Homotypic

Trigonella ovalis Boiss., Elenchus Pl. Nov. 36 (1838), (basionym).

Pocockia ovalis (Boiss.) Boiss., Diagn. Pl. Orient., Ser. 1, 2(9): 12 (1849).

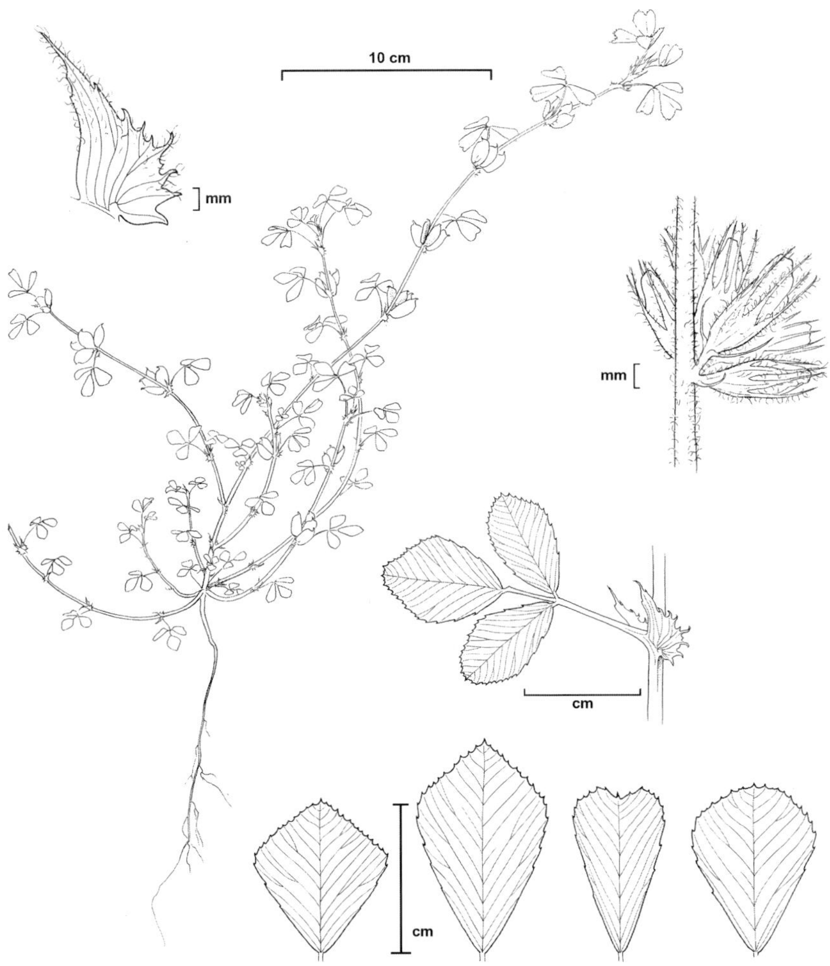

Fig. 60-1. *Medicago ovalis.* **Clockwise from left:** Plant; stipule; group of flowers; leaf attached to stem; leaflets. (Based on accession H727/79, K; artist: A. Hanes.)

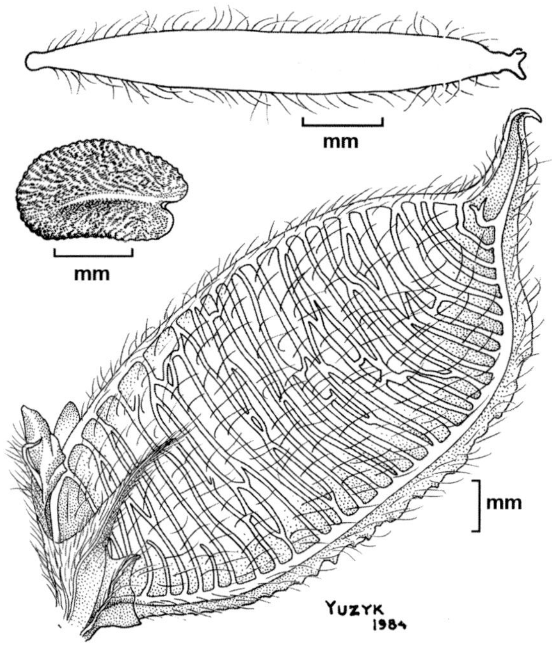

Fig. 60-2. *Medicago ovalis*. **Clockwise from top:** Outline of cross section of fruit; side view of fruit; seed. (Based on Illegible 21 April 1873, GH; artist: L. Yuzyk.)

Melissitus ovalis (Boiss.) Latsch., Trudy Tiflissk. Bot. Inst. 19: 7 (1970).

T. monspeliaca var. *ovalis* (Boiss.) Malag., Sin. Fl. Ibér. 37: 589 (1976).

Heterotypic

No names found. *Medicago ovalis* was mistakenly treated as a synonym of *T. monspeliaca* L. in *Flora Europaea* (Tutin 1968).

Not

M. ovata hort. ex Steud., Nomencl. Bot. (Steudel) 514 (= *M. murex* or *M. sphaerocarpos*).

Vernacular Names

English: No names found.

French: No names found.

Geographical Distribution

(N = native, I = introduced, U = uncertain status)

Africa: Algeria (N), Mauritania (N), Morocco (N).

Europe: Spain (N).

This narrowly distributed species occurs in southern Spain (where it is rare) and northwestern Africa.

Description

Annual herb, branched from base, stems prostrate to ascending, (7)10–20(30) cm long. Vegetative parts more or less pubescent with simple hairs. Leaflets obovate or obtriangular, 5–10(13) mm long, 4–8 mm wide, distal half to two-thirds of margin serrate (occasionally weakly incised-dentate). Stipules ovate or broadly lanceolate, dentate at base, upper stipules more or less entire. Flowers (and fruits) in groups of 4–8 in leaf axils, almost sessile. Peduncle absent or essentially so. Calyx 4–5 mm long, pubescent, teeth linearly setiform, ca. three times longer than tube. Corolla about equal in length to calyx, yellowish. Fruit broadly oval, somewhat asymmetrical, pubescent, (4.5)5–6(7) mm wide, 8–12(15) mm long (excluding beak, which is 2–3 mm long and bent obliquely upwards and to the side), with a narrow wing (fringe) along the more convex (ventral suture) side, finely and densely transversely veined. Fruit with ca. four seeds, ca. 2.5 mm long and 1 mm wide, rotund-compressed, oblong, finely tuberculate, radicle almost as long as seed.

Identification Hints

Medicago ovalis is unlikely to be mistaken for other species of the genus as within its range it is the only one with short, broad pods. Hedges and Sales (2001) commented that *M. ovalis* appears to be similar to *M. brachycarpa*, but the two species are isolated geographically. Additional distinguishing characters (none of which is exclusive to this species) are the fringed ventral wing, rugose seeds, and flowers and pods grouped in the axils of the leaves.

Habitat and Environmental Adaptation

Habitat

Medicago ovalis occurs on dry limestone slopes, sands, lower mountains, and fields, and has been collected along roadsides and railways.

Phenology

Medicago ovalis flowers in March.

Soil

This species typically occupies limestone-based substrates and sand.

Temperature

Medicago ovalis is adapted to survive from winter to midsummer under Mediterranean climate conditions.

Water

The geographical distribution and annual habit of the species suggest that it is a drought-evading winter annual. Its common occurrence in sandy soil suggests that it is adapted to well-drained soil.

Biotic Relationships

Pollination

The small flowers indicate that this species is largely an inbreeder.

Dispersal

The flat pods may represent an adaptation to wind distribution.

Herbivory

No information located.

Nodulation and mycorrhizae

No information located.

Disease resistance

No information located.

Economic History, Cultivars, and Uses

Medicago ovalis has not attracted agricultural interest to date.

Cultivars

No cultivars have been bred.

Forage

Medicago ovalis is not sown for forage.

Fodder

This species is not harvested for fodder.

Green manure, rotation crop, ground cover

No information located.

Other uses (salad sprouts, ornamental, plant products, etc.)

No information located.

Chromosome Number

$2n = 32$: Tschechow 1932; Bolkhovskikh et al. 1969, p. 309. Vassilczenko (1953) and Hedges and Sales (2001) also reported $2n = 32$ but did not cite the original research publications.

Genetic Relationships and Gene Transferability

Intraspecific

Artificial hybridization of different populations has not been attempted, so it is not known whether there are barriers to hybridization within the species.

Fig. 60-3. *Medicago ovalis*. **From:** Boissier, E.P. 1840. Voyage botanique dans le midi de l'Espagne pendant l'année 1837. Vol. 1, Fig. 51. **Left:** Plant. **Top right:** Side view of flower (above); calyx (below). **Center right:** Petals (clockwise from top: standard; wing; keel; wing). **Below petals:** Androecial column (above); pistil (below). **Bottom right (from left to right):** Fruit; section of fruit showing ovules; leaflet.

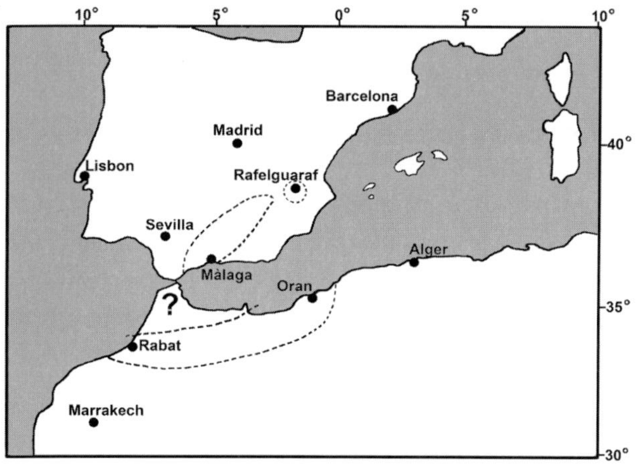

Fig. 60-4. The main distribution areas of *Medicago ovalis* (shown by broken lines). Adapted from Hansen (1977). Carretero (1990) observed that while the species was thought to be restricted in southern Spain from the province of Malaga through the Andalucia region to the province of Albacete, he collected a specimen in the region of Rafelguaraf in the province of Valencia.

Interspecific

No information located.

Exchange by advanced biotechnology

No information located.

Conservation Status

Since this is a very rare species it appears to require a conservation assessment.

Germplasm Sources

No accessions were located.

Research Problems and Priorities

Medicago ovalis is an enigmatic species that has been little studied to date. It requires germplasm collection, basic botanical study, and agricultural analysis of its potential.

Representative Collections

(**Note:** Hansen (1977) cites many collections of this rare species from Algeria and Morocco.)

Algeria: Balansa, B., bords de la route qui va du pont de la Macta à Saint-Leu, 11-V-1851, in fruit (K, LD, P). Faure, A., Les Trembles, près Sidi-bel-Abbès, bords de la voie ferrée, 10-V-1917, in fruit (LD). Warion, A., in pascuis inter Chamoercopem ad Sidi-bel-Abbès, 1-V-1876, in flower & fruit, A. Warion Pl. Atlanticae Selectae 1878 No. 119 (K, P). **Morocco:** Jahandiez, E. 162, Meknès, vers Dkhissa, bord des champs, 9-V-1929, in fruit (BM). Jahandiez, E. 275b, Faza, berges argileuses de l'oued Termas, 24-V-1929, in fruit (BM). Samuelsson, G. 7208, Distr. boreali-centralis: Dar Kaid Medboh, ad marg. vrae, ca. 750 m, 29-IV-1936, in flower & fruit (S). **Spain:** Bourgeau, E., Andalusie, champs incultes à Riopar, 29-VI-1850, in fruit (G, P). Cosson, E., cultivés à Thurilles de graines d'espagne, 1851, in flower & fruit (K). Huter, Porta & Rigo 766, regnum Grenatense, lot. arenos, ad Cerro Coro, nado et Jarro S. Umo, prope Malaga, arenas aridis, 1-V-1879, in fruit (G, K). Huteri, ex seminibus in hispania prope Malaga, 1879 lectis, 1880, in flower (K).

61. *Medicago pamphylica* (Huber-Mor. & Širj.) E. Small

M. pamphylica (Huber-Mor. & Širj.) E. Small in Small, Lassen & Brookes, Willdenowia 16(2): 434 (1987).

The epithet *pamphylica* is based on the Greek *pam* (all) plus *phyllikos* (leaves), a reference to the trifoliolate leaves, with each leaflet in turn often trifid, a characteristic that, in the opinion of Huber-Morath and Širjaev who first named this species, serves to distinguish it from its nearest relatives.

Subordinate Taxa

None recognized here.

Type Specimen

Tengwall, T.A., Turkey, Kleinasien, Pamphylien, Hafis Passa, ca. 50 km nördl. von Antalya, alt. 800 m, 31-V-1936 (S). (The collection is in flower and early fruit and has somewhat dissected leaflets.)

In the original description of their new species, Huber-Morath wrote: "Hab.: Asia-Minor; Pamphylia, prope Hafis PaŞa inter Antalya (Adalia) et Buçac, 800 m; leg. T.A. Tengwall 31.V.1936. Spec. orig. in herb. E. Wall, Stockholm et in herb. A. Huber-Morath, Basel."

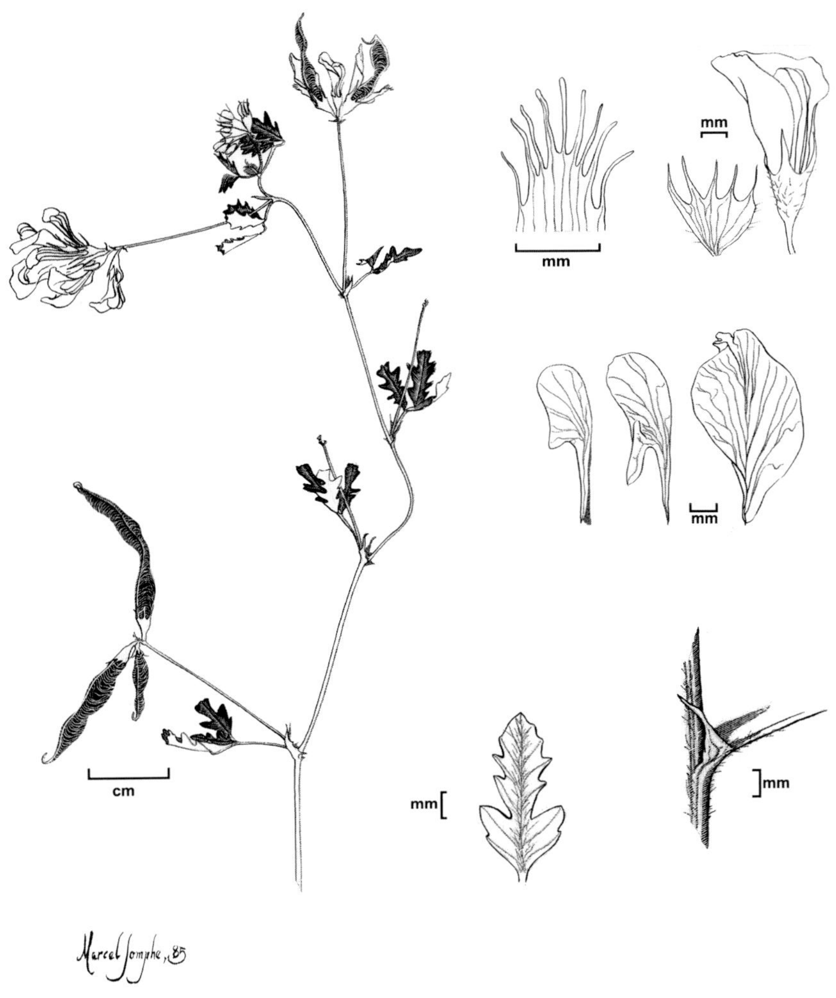

Fig. 61-1. *Medicago pamphylica.* **Left:** Branch with flowers and fruits. **Top right:** Floral parts (clockwise from left: flattened androecial column (free stamen not shown); calyx; side view of flower; standard petal; wing petal; keel). **Lower right:** Stem node showing stipules. **Lower center:** Leaflet. (Based on isotype, T.A. Tengwall 716, K; artist: M. Jomphe.)

Huber-Morath (1970) stated "holo. Hb. E. Wall, Stockholm! iso. Hb. Hub.-Mor.!" Huber-Morath contributed the specimen from his personal collection (Herb. A. Huber-Morath) to G. The G specimen, which I have seen, is an isotype and there is also an isotype at K (Figs. 61-1 and 61-3). BASBG also lists type material of *Trigonella pamphylica* Huber-Mor. & Širj. (http://pages.unibas.ch/botges/herbar/typen.htm; current June 2010).

Synonymy

Homotypic

Trigonella pamphylica Huber-Mor. & Širj., in Huber-Mor., Repert. Spec. Nov. Regni Veg. 44: 322–323 (1938), (basionym).

Heterotypic

No names found.

Vernacular Names

ENGLISH: No names found.

FRENCH: No names found.

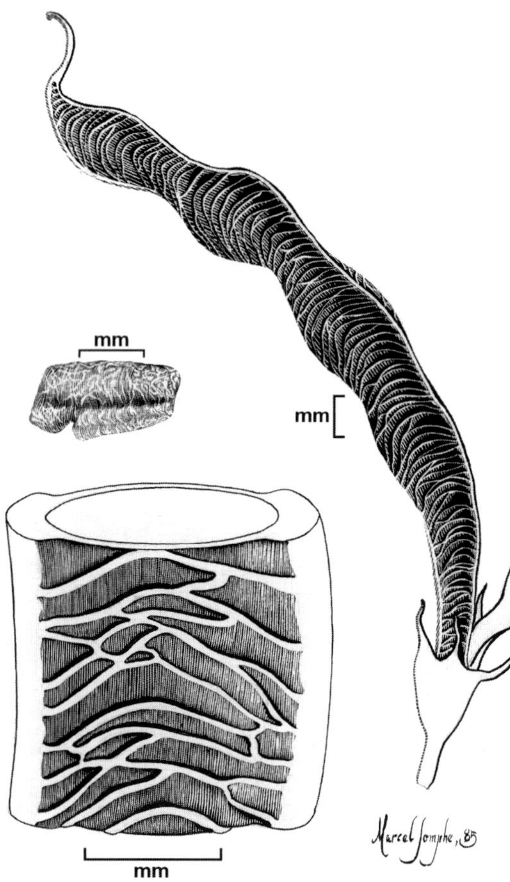

Fig. 61-2. *Medicago pamphylica.* **Clockwise from center left:** Seed; fruit; portion of fruit showing venation. (Based on A. Huber-Morath 9854, K; artist: M. Jomphe.)

Geographical Distribution

(N = native, I = introduced, U = uncertain status)

ASIA: Turkey (N).

Medicago pamphylica is a very localized endemic, with its native distribution confined to southern Anatolia.

Description

Annual herb, stem branched from base, branches ascendent to erect, 10–30 cm long. Vegetative parts diffusely pubescent with appressed simple hairs. Leaflets obovate or obcordate-cuneate, 3–10 mm long, 3–8 mm wide, distal margins serrate, leaflets sometimes incised or pinnatifid (often trifid or less frequently quinquepartite). Stipules semi-sagittate, entire or dentate. Inflorescence of 3–6 flowers in a short raceme forming a spherical head, peduncle 1–5 cm long, longer than subtending petiole. Calyx campanulate, 3.5–4 mm long, teeth lanceolate-linear, about as long as tube, sparsely pilose. Corolla yellow, 6–9 mm long. Fruits linear, straight, or somewhat curved, erect or spreading, with more or less undulate sutures and numerous close transverse anastomosing veins, glabrous or slightly pilose, 17–30 mm long, 1.5–2 mm wide. Seeds oblong-rhomboid, "wrinkled" (according to original description), ca. 2 mm long and 1 mm wide, radicle about 80% of length of seed.

Identification Hints

In the original description (cited above) of *T. pamphylica* (i.e., *M. pamphylica*), Huber-Morath distinguished it from what he presumably regarded as the nearest or at least most similar Anatolian species, as follows. Differences from *T. crassipes* (i.e., *M. crassipes*) include: frequent occurrence of trifid leaflets (not present in *M. crassipes*); longer peduncles (1–5 cm in *M. pamphylica*; 2–3(18) mm in *M. crassipes*); inflorescence of 3–6 flowers (4–10 in *M. crassipes*); larger flowers (6–9 mm long in *M. pamphylica*, 4–6 mm in *M. crassipes*); and fruit characters (undulate sutures and numerous close, transverse anastomosing veins in *M. pamphylica*, obliquely transverse reticulating veins and thickened sutures in *M. crassipes*). Differences of *T. pamphylica* (i.e., *M. pamphylica*) from *T. rigida* (i.e., *M. rigida*) include: frequent occurrence of trifid leaflets (not present in *M. rigida*); longer peduncles (only 3–4 mm in *M. rigida*); inflorescence of 3–6 flowers (7–12 in *M. rigida*); larger flowers (usually only about 5–6 mm long in *M. rigida*); and fruit characters (obliquely anastomosing reticulating veins in *M. rigida*).

Habitat and Environmental Adaptation

Habitat

Medicago pamphylica occurs in scrub associations and forests, at altitudes of 800–1400 m.

Phenology

The species flowers from May to July.

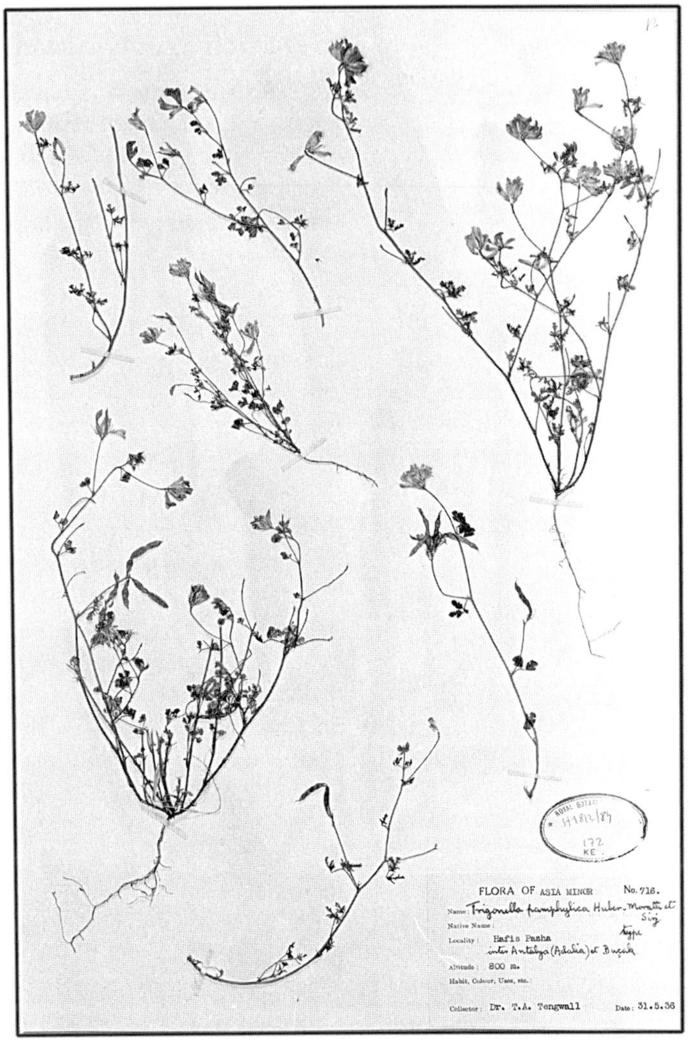

Fig. 61-3. Photograph of isotype of *Trigonella pamphylica* Huber-Mor. & Širj. at K.

Soil

No information located.

Temperature

This species occurs in the Mediterranean climate conditions of southern Anatolia.

Water

No information reported. The climate of southern Anatolia is Mediterranean, with drought increasing in the summer.

Biotic Relationships

Pollination

Small (1988) noted that the flowers are small and the pollen to ovule ratio is low (142), observations that are consistent with *M. pamphylica* being largely self-pollinated and substantially an inbreeder.

Dispersal

Adaptations for seed dispersal are not apparent.

Herbivory

No information located.

Nodulation and mycorrhizae

No information located.

Disease resistance

No information located.

Economic History, Cultivars, and Uses

This rare species has not attracted agricultural interest to date.

Cultivars

No cultivars have been bred.

Forage

Medicago pamphylica is not sown for forage.

Fodder

This species is not collected for livestock feed.

Green manure, rotation crop, ground cover

No information located.

Other uses (salad sprouts, ornamental, plant products, etc.)

No information located.

Chromosome Number

Not reported.

Genetic Relationships and Gene Transferability

Intraspecific

Artificial hybridization of different populations has not been attempted, so it is not known whether there are barriers to hybridization within the species.

Interspecific

No information located.

Exchange by advanced biotechnology

No information located.

Conservation Status

This rare endemic species is in need of a conservation assessment.

Germplasm Sources

No accessions have been reported.

Research Problems and Priorities

Medicago pamphylica is a rare species that has been little studied. It is in need of germplasm collection, basic botanical studies, and agricultural assessment with regard to its potential as a crop for dryland regions and as a source of genes for improving cultivars of the important agricultural species of *Medicago*.

Representative Collections

TURKEY: Akuran, Y. & Quézel 6083, Cu-Mersia, Anamur, Abanoz, Abies cilicica, 14-VI-1976, in flower & fruit (E). Davis, P.H. 15769, Prov. Antalya, distr. Gebiz (P1), Bozburun Da., between Tasli Y. and Kozlu De. 27-VII-1949, in flower & fruit (E). Huber-Morath, A. 5068, Pamphylien, Vilayet Antalya, Quercus Cerris – Macchie, Passhöhe 3 km ob Hafis Paşa an der Strasse Burdur-Antalya, alt. 800 m, 11-VI-1938, in fruit (B, G, HUJ) (leaflets somewhat dissected). Huber-Morath, A. 9854, Kleinassien, Pamphylien, Vilayet Antalya, Passhöhe 3 km ob Hafis Paşa, *Quercus coccifera* macchie, alt. 890 m, 21-V-1950, in flower (G, HUJ) (leaflets laciniate). Huber-Morath, A. 13377, Vilayet Antalya, Passhöhe 3 km südlich ob Hafis Paşa, Quercetum cocciferae, alt. 860 m, 13-V-1956, in flower (G). Huber-Morath, A. 17173, Prov. Antalya, district Akseki, Akseki-Beyşehir, Abies cilicica – Wald 10 km North Akseki, alt. 1380 m, 4-VII-1964, in flower & fruit (G). Tengwall, T.A. 716, Hafis Passa, ca. 800 m, 31-V-1936, in flower & early fruit (B). Wängsjö, E. & G. 2675, Termessos, Vilayet Antalya, Güllük Dagi, alt. 700–800 m, in silva frondosa montana, 22-VI-1968, in fruit (LD).

62. *Medicago papillosa* Boiss.

M. papillosa Boiss., Diagn. Pl. Orient. Ser. 1, 1(2): 23 (1843), (basionym).

The epithet *papillosa* is based on the Latin *papillosus*, meaning with papillae (i.e., with tiny, often nipple-like protuberances). This adjective has been used in descriptive botany to refer to a variety of plant organs, but the phrase "pilis papillaribus" in Boissier's description of the species makes it clear that the epithet is a reference to the glandular trichomes on the fruits of many of the plants.

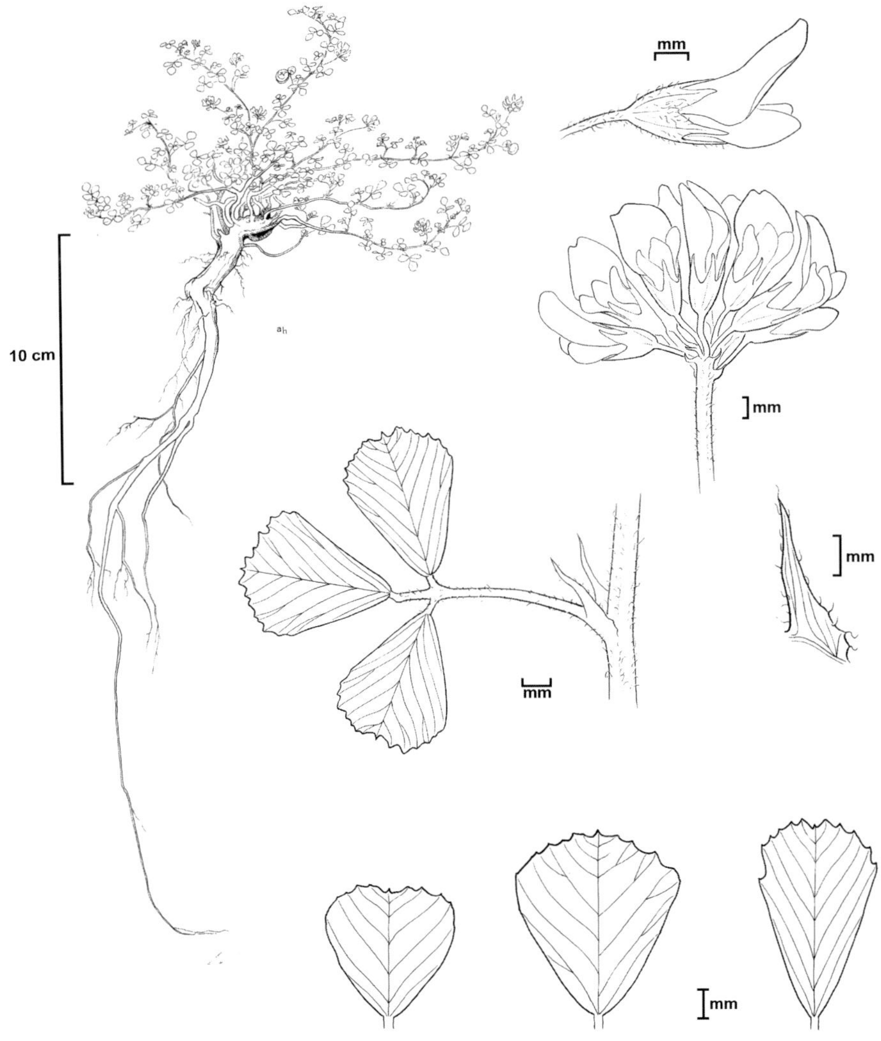

Fig. 62-1. *Medicago papillosa* subsp. *papillosa*. **Clockwise from left:** Plant; side view of flower; group of flowers; stipule; leaflets. **Center:** Leaf attached to stem. (Based on a field-collected plant, A. Grossheim 16 June 1928, LE; artist: A. Hanes). Note limited lobing of stipule and how it attenuates gradually toward the tip; also note rather obovate shape of leaflets (cf. *M. papillosa* subsp. *macrocarpa*, Fig. 62-3).

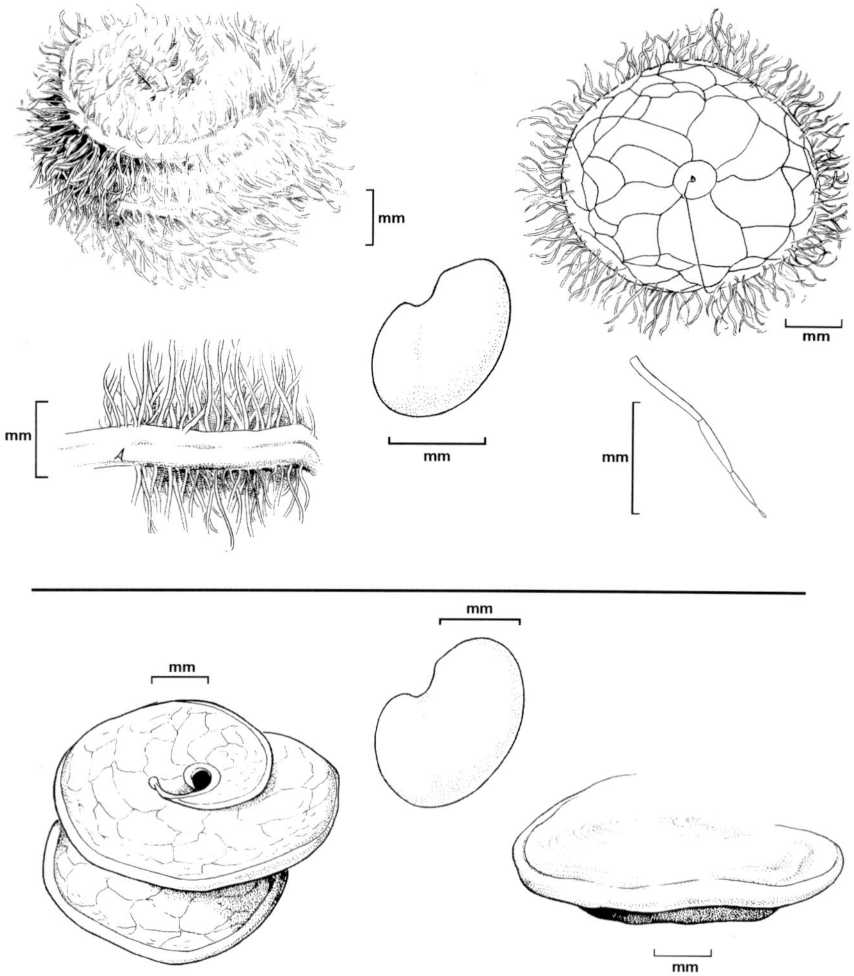

Fig. 62-2. *Medicago papillosa* subsp. *papillosa*. **Above line:** Glandular-pubescent fruit. **Clockwise from upper left:** Fruit; face of fruit coil showing venation; glandular trichome; seed; coil edge (hairs removed on outer edge). (Based on V. Sapozhnikov 03 July 1916, LE; artist: A. Hanes.). **Below line:** Glabrous fruit. **From left:** Fruit; seed; coil. (Based on G.A. Stevenson 3919, DAO; artist: A. Hanes.)

Key to Subordinate Taxa

The following key discriminates the *M. papillosa* taxa and the similar *M. sativa sensu lato*.

Key discriminating the subspecies of *M. papillosa* from *M. sativa*

1. Plants fruiting 2
  2. Seeds not separated in fruit by spongy, membranous partitions .... *M. sativa*
  2. Seeds separated in fruit by spongy, membranous partitions .... *M. papillosa*, 3
    3. Fruit diameter 6.3 mm or larger .... *M. papillosa* subsp. *macrocarpa*
    3. Fruit diameter less than 6.3 mm .... *M. papillosa* subsp. *papillosa*
1. Plants flowering 4
  4. Plant lacking purple flush on calyx tube .... *M. sativa*
  4. Native-grown[1] plants with purple flush on calyx tube .... *M. papillosa*, 5
    5. Flowers usually longer than 8 mm .... *M. papillosa* subsp. *macrocarpa*
    5. Flowers usually less than 8 mm .... *M. papillosa* subsp. *papillosa*

[1]As reported in Small (1986*b*), plants of *M. papillosa* grown in Canada did not develop a purple (anthocyanin) flush on the calyx tube, an almost invariable feature of plants collected from the wild in Asia. It may be that the appearance of anthocyanin in plants grown in their native high altitudes is related to the greater intensity of ultraviolet radiation at these heights. This feature is very rarely present in other species of *Medicago*, most of which characteristically have habitats at lower altitudes than *M. papillosa*.

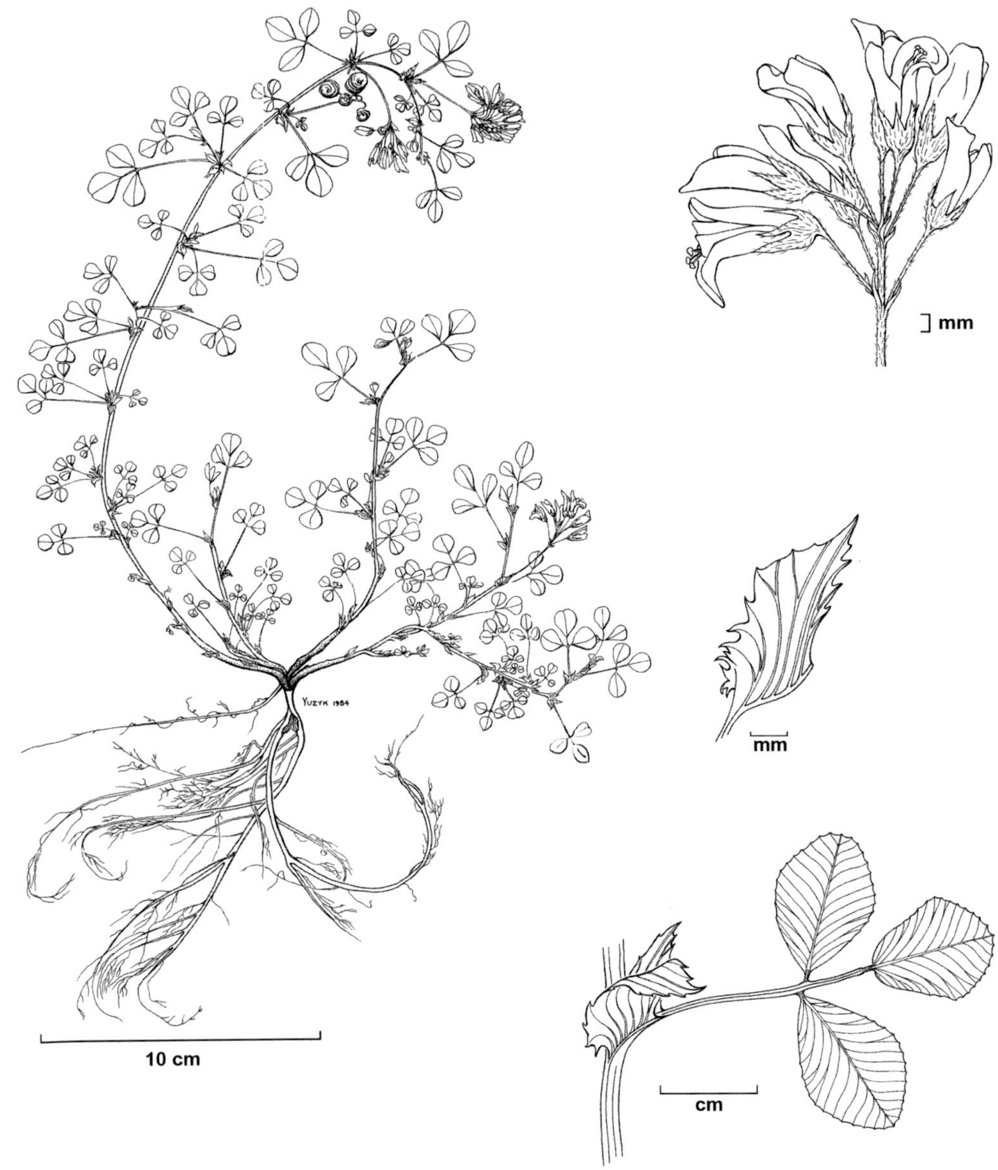

Fig. 62-3. *Medicago papillosa* subsp. *macrocarpa*. **Clockwise from left:** Plant; group of flowers; stipule; leaf attached to stem. (Based on a greenhouse-grown plant, K.A. Lesins 890, the "type" of "*M. ignatzii*", ALTA; artist: L. Yuzyk.) Note extensive lobing of stipule, how it is "leafy", and how its terminal portion attenuates relatively abruptly (cf. *Medicago papillosa* subsp. *papillosa*, Fig. 62-1). Also note the rather elliptical shape of the leaflets.

The taxonomy of the *M. papillosa* complex was treated in detail in Small (1986*b*). In the past, "*M. papillosa*" was commonly interpreted as a species of the Pontus Mountains of northeastern Turkey and "*M. dzhawakhetica* Bordz." as a closely related species of the mountains of Transcaucasia (i.e., the Trans-Caucasus, the Caucasus Minor). Contributing to the confusion is the apparent (but superficial) similarity of *M. papillosa* with some wild forms of the *M. sativa* complex. In both complexes, some populations have fruits with a covering of glandular trichomes and the taxonomic use of this character has compounded the confusion. Ponert (1973) relegated *M. papillosa* to *M. glutinosa* subsp. *papillosa* (Boiss.) Ponert, i.e., he considered it part of the *M. sativa* complex.

Medicago papillosa subsp. *papillosa* is distinguished from *M. papillosa* subsp. *macrocarpa* by having a fruit diameter of no more than 6.3 mm, flowers usually less than 8 mm, and stipules of mature leaves entire or with a few basal serrations (as opposed to having a few serrations distally as well as proximally). Fruits with and without gland-tipped hairs occur in both subspecies of *M. papillosa*. Above, a key is presented distinguishing the two subspecies as well as the similar *M. sativa*. The following table compares the two subspecies.

Table 62-1. A comparison of the subspecies of *Medicago papillosa* (after Small 1986*b*).

	subsp. *papillosa*	subsp. *macrocarpa*
Mature stipules	Distal portion narrows gradually or abruptly; 0–7 marginal teeth, frequently located at base of outer margin	Distal portion narrows gradually; 3–12 marginal teeth, distributed more or less evenly over outer margin
Mature leaflet shape	Elliptical to obovate, but tending to obovate	Elliptical to obovate, but tending to elliptical
Anthesic flower (corolla) length (mm)	5.5–9.0	7.5–12.0
Mature pod diameter (mm)	3.5–6.0	6.5–8.5
Glandular hairs on pod	Present in about half of collections with pods	Present in almost all collections with pods
Chromosome number	$2n = 16, 32$	$2n = 16$
Geography	Mostly east of longitude 41° E; Turkey, Armenia, Azerbaijan, Gruziya	West of longitude 41° E; Turkey

Geographical Distribution

(N = native, I = introduced, U = uncertain status)

ASIA: Armenia (N), Azerbaijan (N), Gruziya (N), Turkey (N).

Medicago papillosa is native to northeastern Turkey and the adjacent Transcaucasus, including Armenia, Azerbaijan, and Gruziya (Georgia). *Medicago papillosa* subsp. *papillosa* occurs in Transcaucasia and northeastern Turkey east of longitude 40°. *Medicago papillosa* subsp. *macrocarpa* is native to northeastern Turkey west of longitude 40° (see Fig. 62-6). The species occurs at altitudes of 1000–3000 m, higher than for most other species of *Medicago*.

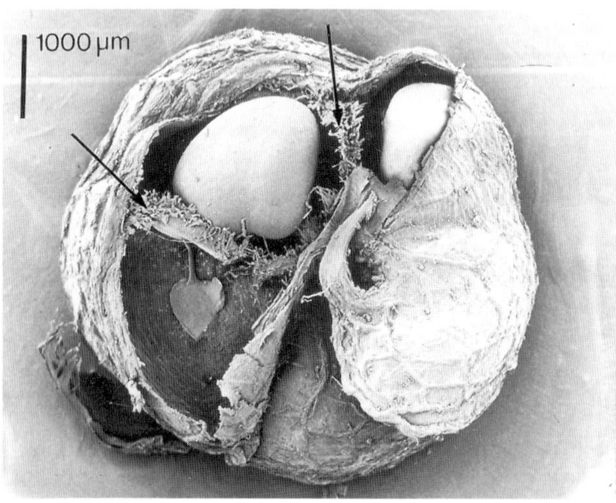

Fig. 62-4. Spongy layer (arrows) separating seeds in fruit of *Medicago papillosa* complex, the critical feature separating the complex from the very closely related *M. sativa* complex (based on K.A. Lesins 98, DAO; scanning electron micrograph by B. Brookes).

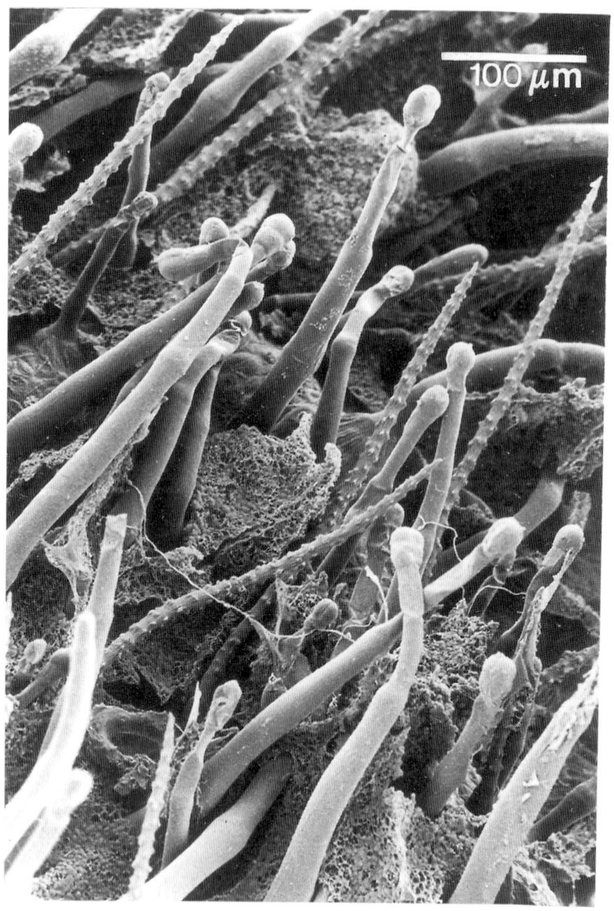

Fig. 62-5. Glandular hairs on fruit of *Medicago papillosa* subsp. *papillosa* (based on E. Small-Turkey-135, DAO; scanning electron micrograph by B. Brookes). Interspersed with the gland-tipped hairs, which are multicellular, are a few nonglandular hairs, which are slimmer, rough-surfaced, and unicellular.

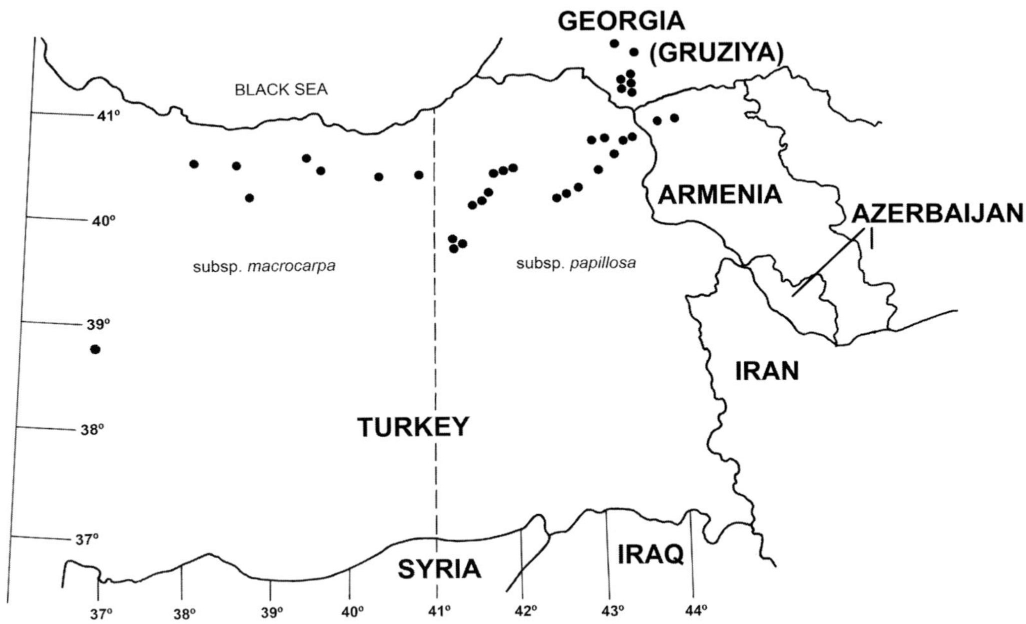

Fig. 62-6. Distribution of the two subspecies of *Medicago papillosa* (based on Small 1986*b*). A few plants of subspecies *papillosa* have been found in the eastern range of subsp. *macrocarpa*.

NOMENCLATURE OF *MEDICAGO PAPILLOSA* SUBSP. *PAPILLOSA*

Type Specimen (of *M. papillosa* and *M. papillosa* subsp. *papillosa*)

M. Aucher-Eloy 928, in subalpestribus Armeniae, (Heyn and Davis (1970) stated this is in "NE Anatolia"), 1837, specimen in flower & early fruit, Herb. Boiss., G!

Synonymy

Homotypic

M. glutinosa M. Bieb. subsp. *papillosa* (Boiss.) Ponert, Feddes Repert. 83: 639 (1973).

Heterotypic

M. papillosa Boiss. var. *citrina* Lesins & Lesins, Genus *Medicago* 129 (1979), nom. illegit. This variant was based merely on lemon-yellow petals. "Type!": K.A. Lesins 1862, at ALTA, grown in Edmonton from seeds received from F. Tosun, University Farm, Erzurum, and presumably collected near Erzurum, Turkey.

M. dzhawakhetica Bordz., Prot. Zased. Kievsk. Obschch. Estestvoispyt. 24 (1907). Type!: E. Bordzilowski 30.7.1907, "Transcaucasia Armenia. Prope Akhalkalaki (prov. Tiflis) in locis herbosis, KW (Herbarium Eug. Bordzilowski, Flora Rossica exsiccata).

M. papillosa Boiss. var. *dzhawakhetica* (Bordz.) Bordz., Zap. Kiev. Obsch. Estestvoispyt. 25: 86 (1915).

M. papillosa Boiss. subsp. *microcarpa* Urb., Verh. Bot. Ver. Prov. Brandenb. 15: 58 (1873). (Urban's types at B were destroyed during the Second World War.)

M. dzhawakhetica Bordz. var. *timofeewi* Troitz., Rep. (Vestn.) Bot. Gard. Tiflis, New Ser. 1: 90 (1923). The type, said by Vassilczenko (1949) to be from Borchalo in the Trans-Caucasus, is presumably in Tiflis. Troitzky's (1923) variety is based on a lowland collection and, as he pointed out, its diagnostic features may well simply be the result of ameliorated lowland habitat in comparison with the alpine environment where the species normally occurs.

NOMENCLATURE OF *MEDICAGO PAPILLOSA* SUBSP. *MACROCARPA*

M. papillosa Boiss. subsp. *macrocarpa* (Boiss.) Urb., Verh. Bot. Vereins Prov. Brandenburg 15: 58 (1873).

Type Specimen

Balansa, B. 1414, Vallée de Dzimil (Zjimil) (Lasistan (Lazistan)), vers 2000 mètres d'altitude, collines incultes, VII-1866, in flower and fruit (E, G, K, P) (type (Herb. Boiss. G!) and syntypes!).

Synonymy

Homotypic

M. papillosa Boiss. var. *macrocarpa* Boiss. Fl. Orient. 2: 96 (1872–1873), (basionym).

Heterotypic

M. ignatzii Lesins and Lesins, genus *Medicago* 131 (1979), nom. nud. ("Type": K.A. Lesins 89, plants grown in Edmonton using seeds from Istavris, Turkey (=Istavros, 40°15′N, 39°45′E); ALTA.)

Vernacular Names

ENGLISH: Alpine alfalfa (Bolton 1962, as *M. dzhawakhetica*).

FRENCH: No names found.

Description

Perennial herb arising from a stout, woody rhizome, the root system penetrating deep into the ground. Stems numerous from rhizome, 5–30(40) cm long, procumbent to ascending, appressed-hairy. Leaflets 3–12 mm long, 2–12 mm wide, obovate, oboval, or cuneate-obovate, glabrous or essentially so above, appressed-hairy beneath, denticulately serrate on distal portion. Stipules triangular-lanceolate, appressed-pubescent, outer margin dentate or incised (mostly at base). Corolla 5–8 mm long, yellow, calyx appressed-hairy, turbinate-campanulate, with unequal triangular-lanceolate teeth more or less as long as the tube. Inflorescence almost capitate with 5–15 flowers. Peduncle as long as or longer than subtending petiole. Pod glabrous or with a few hairs (rarely densely pubescent) or with pellucid articulate hairs, with 2–4(5) loosely adpressed coils, 3.5–8 mm wide, the coil face with 8–12 curving radial veins that anastomose. Seeds 2.3–3 mm long, 1.3–1.5 mm wide, 1–2 seeds per coil, separated by thick, spongy septa, yellow to brownish-yellow, radicle about half length of seed.

Identification Hints

Medicago papillosa can be confused with other yellow-flowered, montane perennials with coiled, nonspiny fruits, especially *M. sativa*; however, the prominent interseed fruit partitions make it quite distinctive.

Habitat and Environmental Adaptation

Habitat

Medicago papillosa occurs on dry, rocky slopes, cliffs, talus slopes, and mountain pastures, and infrequently in grassy places, at altitudes ranging up to 2600 m. It is an infrequent species. The western subsp. *macrocarpa* seems to occupy somewhat more mesic circumstances than the eastern subsp. *papillosa* (based on information in Vassilczenko 1949).

Phenology

The species flowers from June to August.

Soil

Medicago papillosa has been collected from igneous and calcareous soils. McCoy (1987) found that *M. papillosa* (identified as *M. dzhawakhetica*) had limited salt tolerance.

Temperature

The species occurs under the often-cool alpine conditions of a continental climate.

Water

Medicago papillosa typically occupies dry, well-drained sites.

Biotic Relationships

Pollination

Small (1988) noted that the relatively large flowers and pollen to ovule ratio (408) of *M. papillosa* are consistent with it being at least partly cross-pollinated, like other perennial species of *Medicago*.

Dispersal

No specializations for seed distribution are evident. Given the alpine habitat, gravity probably plays a major role.

Herbivory

No information located.

Nodulation and mycorrhizae

USDA (2009) reported positive root nodulation for *M. papillosa*.

Disease resistance

No information located.

Economic History, Cultivars, and Uses

Medicago papillosa has attracted the interest of alfalfa (*M. sativa*) breeders because of the closeness of the two species, offering the possibility of transferring genes to the latter.

Cultivars

No cultivars have been bred.

Forage

Medicago papillosa is not sown for forage, although Koç and Tan (1997) considered it to have several agronomic characteristics. According to Grossheim (1945), the forage properties of "*M. dzhawakhetica*" are reasonable and the plant has considerable protein and is of high feed value, but "*M. papillosa*" "displays no morphological characteristics that might be considered valuable for forage use, although the plant deserves attention on account of its drought resistance and should be considered for trials, if not for its own sake then with a view to hybridization." Small et al. (1990) found that the content of hemolytic saponin was quite high in *M. papillosa*, and because saponins are antinutritional to some classes of livestock (see Chapters 9 and 12), this needs to be considered for feeding purposes.

Fodder

No information located.

Green manure, rotation crop, ground cover

No information located.

Other uses (salad sprouts, ornamental, plant products, etc.)

No information located.

Chromosome Number

Subspecies not determined

$2n = 16$: Cooper 1935, 1936 (probably subsp. *papillosa*); Lesins 1952, 1956; Lesins and Lesins 1963*a*; Clement

1962 (as *M. dzhawakhetica*); Bolkhovskikh et al. 1969, p. 308 (2 reports).

$2n = 32$: Lesins 1958.

Subsp. *papillosa*

$2n = 32$: Lesins and Lesins 1966.

$2n = 16, 32$: Lesins and Lesins 1979; McCoy and Smith 1986; Small 1986*b*.

Subsp. *macrocarpa*

$2n = 16$: Small 1986*b*.

Genetic Relationships and Gene Transferability

Intraspecific

Lesins and Lesins (1979) referred to unpublished work they had conducted attempting hybridization at the diploid level of "*M. ignatizii*" (*M. papillosa* subsp. *macrocarpa*) with "*M. papillosa*" (*M. papillosa* subsp. *papillosa*), the results suggesting that the two taxa were intersterile. As noted in Small (1986*b*), this requires confirmation. As noted above, diploids and polyploids occur within *M. papillosa*, so hybrids between these can be expected to be generated with difficulty (if at all) and to exhibit sterility.

Interspecific

Lesins (1961*a*) found that he could not produce hybrids between *M. papillosa* and *M. sativa* when both were tetraploid ($2n = 32$), but that triploid hybrids resulted from crossing diploid *M. sativa* with tetraploid *M. papillosa*. McCoy and Smith (1984) repeated this hybrid (*M. papillosa* was identified as *M. dzhawakhetica*), again finding that uneven ploidy levels were required when one of the parents is tetraploid. They found that the F_1 plants were male sterile, although chromosome pairing was fair. They concluded that ovule–embryo culture could be used for gene transfer from *M. papillosa* to *M. sativa*. Clement (1963) hybridized diploid *M. papillosa* with diploid *M. sativa* and found excellent chromosome pairing and fair fertility in the F_1 hybrid and its backcrosses. McCoy and Quarisa (1989) studied diploid, triploid, and tetraploid hybrids of *M. papillosa* and *M. sativa*. McCoy et al. (1991) found that there was limited genomic affinity between the two species.

Exchange by advanced biotechnology

Using embryo rescue techniques, McCoy and Smith (1986) obtained hybrids between *M. papillosa* and diploid *M. sativa* ($2n = 16$, i.e., at the same chromosome number). They also obtained tri-species hybrids between *M. sativa* × *M. papillosa* F_1 hybrids ($2n = 3x = 24$) and *M. cancellata* ($2n = 6x = 48$), as well as between *M. sativa* × *M. papillosa* F_1 hybrids ($2n = 3x= 24$) and *M. saxatilis* ($2n = 6x = 48$).

Conservation Status

Medicago papillosa is an uncommon species of considerable agricultural interest. Because it is potentially in peril from grazing pressure, a conservation assessment is required, and indeed it would be prudent to increase holdings of its seeds in germplasm banks.

Germplasm Sources

A dozen accessions are listed in major *Medicago* germplasm collections (see Chapter 3).

Research Problems and Priorities

Medicago papillosa seems to have agricultural potential as germplasm for disease resistance and for improving the adaptation of alfalfa to mountain meadows (McCoy and Smith 1984). Its potential in this regard deserves attention.

Representative Collections of *M. papillosa* subsp. *papillosa*

(Additional collections cited in Small 1986*b*).

TURKEY: Alpay 2420, Meryenena (40°41′N, 38°33′E), Trabzon, 1960 (E). Alpay 2458, Oltu (40°33′N, 41°59′E), alt. 2000 m, 1958 (E). Bordzilowski, E., Transcaucasia, Armenia, prov. Tiflis, prope Akhalkalaki (41°26′N, 42°29′E), in locis herbosis, 30-VII-1907 (KW) (Herbarium Eug. Bordzilowski, Flora Rossica exsiccata, type of name *M. dzhawakhetica* Bordz.). Davis, P.H. & Hedge, D.29472, Karaugan-Sarikamis (40°18′N, 42°26′E), igneous slopes, alt. 2100 m (K, E). Davis, P.H. & Hedge, D.29561, Kars, prov. Kars (40°37′N, 43°05′E) 1800 m, fallow fields on plain, 15-VI-1957 (E, K). Davis, P.H. & Hedge, D.30528, Kisir dag above Susuz (40°58′N, 43°04′E) 2000 m, rocky igneous bank, 3-VII-1957 (E, K). Huet du Pavillon, circa Tortum (39°55′N, 41°19′E), VI-1853, in flower & young fruit (G). Lesins, K.A. 1862, seeds received from F. Tosun, University Farm, Erzurum (presumably collected near Erzurum, 39°55′N, 41°19′E) (ALTA (two sheets of plants grown in Edmonton, one of these designated the "type" of *M. papillosa* var. *citrina* Lesins & Lesins), DAO). Lesins, K.A. 1864, seeds from same source as Lesins 1862, (ALTA (plants grown in Edmonton), DAO (Small, E. M1098) three sheets of plants grown in Ottawa from Lesins' seeds). Lesins, K.A. 96, seeds from B. Wise, Botanical Garden, Tbilisi, Georgia, USSR, Tbilisi Bot. Gard. No. 56-146, (ALTA) (two sheets of plants grown in Edmonton). Lesins, K.A. 97, seed from B. Wise, Botanical Garden, Tbilisi, Georgia, USSR (ALTA) (plants grown in Edmonton); Small, E. 1094 (DAO) (grown in Ottawa from Lesins' seeds). Lesins, K.A. 98, Akhalkalaki, Georgia, USSR (41°26′N, 42°29′E), (ALTA, DAO). Small, E. Turkey 118, Gonulalan, 11 km northwest of Arpacay (40°56′N, 43°25′E), (DAO) (2 sheets of plants from Turkey plus 4 sheets of plants grown in Ottawa). Small, E. Turkey 121, 18 km northwest of Arpacay (40°55′N, 43°20′E), (DAO). Small, E. Turkey 123, 22 km northwest of Susuz (40°56′N, 42°56′E) (DAO). Small, E. Turkey 124, 14 km southwest of Selim (40°23′N, 42°40′E), (DAO) (2 sheets of plants from Turkey plus

4 sheets of plants grown in Ottawa). Small, E. Turkey 131, 10 km southwest of Tortum (40°15′N, 41°35′E), (DAO). Small, E. Turkey 134, 14 km northeast of Tortum (40°0′N, 41°38′E) (DAO). Small, E. Turkey E-120, 11 km southwest of Tortum (40°17′N, 41°37′E) (DAO). Stapf, O. Herbarium Orientale 1652, Istavros (G, K, P). **USSR:** Lachashvili, I. & Khintibidze, L., Bakuriani, Georgia, village of Bakuriani, the Tskhratskaro mountain pass (41°45′N, 43°31′E), in dry rubbly areas, 17-VII-1959 (E). Kozlowksy, W., Borzhom (=Borzhomi, 41°49′N, 42°23′E), 28-VI-1923, in flower (Schischkin, B. & Grossheim Planta Orientalis Exsic. 187) (G, K). Mulandzhanyan, Ya. & Austisyan, V., Pokr Sepasar, Akhuryan River, Gukasyanskii District (Gukasyan = 41°01′N, 43°50′E; Akhuryan "Stream" = 40°06′N, 43°39′E), 18-VII-1957 (E). R.K. & Sh. A., Blagodarnoe (Blagodarnoye), Kalininskii District (Kalininskii = 41°05′N, 44°11′E) Armenia, Lorplemsoukhoz State Farm, 15-VI-1953 (E).

Representative Collections of *M. papillosa* subsp. *macrocarpa*

(Additional collections cited in Small 1986*b*)

Turkey: Anon. 206, Vallée de Djimil (Lazistan) vers 2000 m (G). Balls & Gourlay 258, Trabzon (41°00′N, 39°43′E) (this is a seaport and, as *M. papillosa* is a plant of mountains, this collection was presumably made well inland), (K). Balls, E.K. 1612a, Haldizan, southeast Trebizond vilayet, alt. 8500 feet, in nonlime soil in turf at edges of dry field, 11-VIII-1934, in flower & fruit (K). Balls, E.K. B1612a, Haldizan 8500 feet, growing in nonlime soil in turf at edges of hay fields, (40°37′N, 40°23′E), 11-VIII-34 (E, K). Davis 21962, prov. Kayseri/Malata, east side of pass between Pinarbasi and Gurun, (38°43′N, 36°51′E), 6000 feet, rocky calc. slopes, 18-VI-1954 (K, E) (most southern and western collection known; Davis (1970) described the plant as "depauperate"). Dorfler, J. 7154, Szandschak, province Gumuschkane (G, P). Du Parquet R., Dzimil, Lazistan, alt. 2000 m, VII-1886, in fruit (BM). Kotschy, T., 26-VII-1859, "Iter Cilicico-Kurdicum" (cited by Boisser (1872) in describing his new var. *macrocarpa*: "in regione alpina Ponti alt. 6000 feet ad Alischeri Khan"). Lesins, K.A. 894, Alisher (= Alicheri = Aliseri, 40°18′N, 38°36′E) (DAO) (Small M1097, grown in Ottawa from Lesins' seeds). Sintensis, P. 1652, Armenia turcica Gumuschkane, Istavros, in herbidis, 9-VIII-1889, in flower & fruit (G, P). Sintensis, P. 7154, Armenia turica, Szandschak Gümüschkane, Kanagoelldagh, in pascuis alpin, 6-VIII-1894, in flower & early fruit (G, P). Tobey 1247a, A6, Ordu, Cambasi, 2000 m (40°39′N, 37°59′E), open alpine pasture, 14-VII-1965 (E).

63. *Medicago persica* (Boiss.) E. Small

M. persica (Boiss.) E. Small in Small, Lassen & Brookes, Willdenowia 16(2): 434 (1987).

The epithet *persica* is based on the Latin *persicus*, belonging to Persia. The species does, in fact, occur in Iran (i.e., ancient Persia).

Subordinate Taxa

None recognized here.

Type Specimen

Aucher 4325 (Aucher-Eloy herb. 4329), Persia australis, in fruit. (Type: Herb. Boiss. G!; Isotype K!; cited by Širjaev (1928–1934) at B and MF.)

Synonymy

Homotypic

Trigonella persica Boiss., Diagn. Pl. Orient. 1, 1(2): 22 (1843), (basionym).

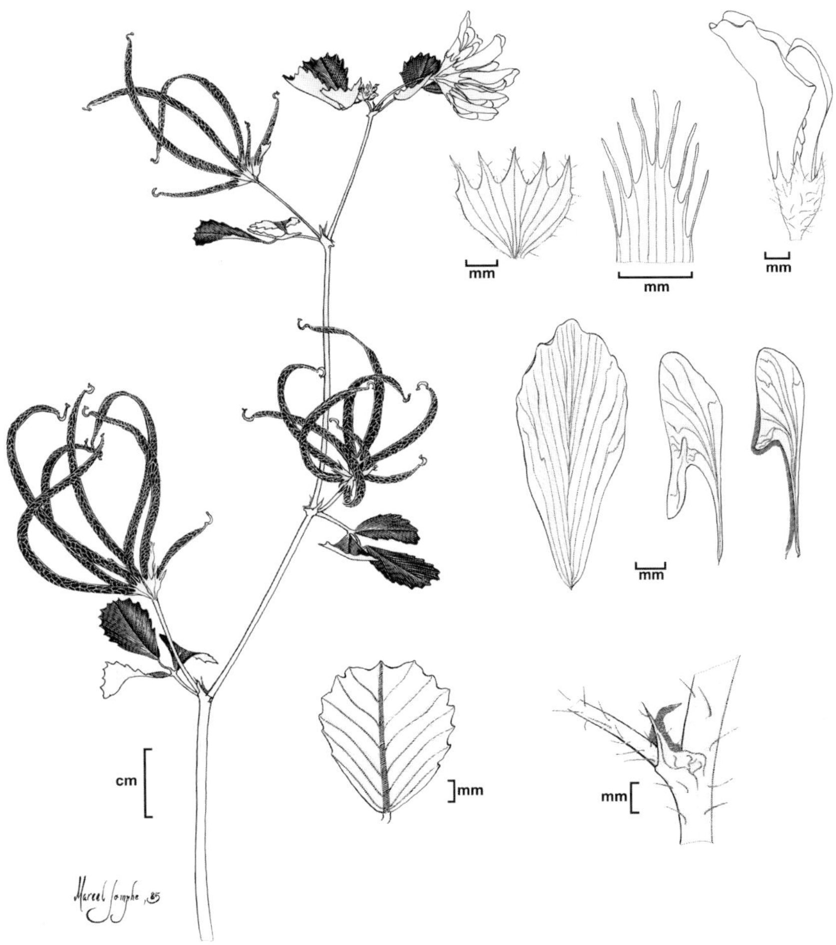

Fig. 63-1. *Medicago persica.* **Left:** Branch with fruits and flowers. **Upper right:** Floral parts (clockwise from top left: calyx; flattened androecial column (free stamen not shown); side view of flower; keel; wing petal; standard petal). **Lower right:** Stem node showing stipules; **Lower center:** Leaflet. (Based on specimen annotated "Bowles Scholarship Bot. Exp. 1959, May 24, 1963", K; artist: M. Jomphe.)

Heterotypic

T. persica Boiss. f. *nabelekii* Širj., Generis *Trigonella* revisio critica, Spisy Pfir. Fac. Masarykovy Univ., 3: 24 (1930).

T. persica Boiss. f. *typica* Širj., Generis *Trigonella* revisio critica, Spisy Pfir. Fac. Masarykovy Univ., 3: 24 (1930), nom. inadmiss.

Not

M. persica Stev. ex Fisch., Cat. Hort. Gorenk. ed 2: 69 (1812), nom. nud. (ex Ledeb., Fl. Ross. 1: 530 (1842)) = *M. polymorpha*.

T. persica Stapf, Verh. Zool.-Bot. Ges. wien 39: 208 (1889) = *T. elliptica* Boiss.

Vernacular Names

English: No names found.

French: No names found.

Geographical Distribution

(N = native, I = introduced, U = uncertain status)

Asia: Iran (N), Iraq (N).

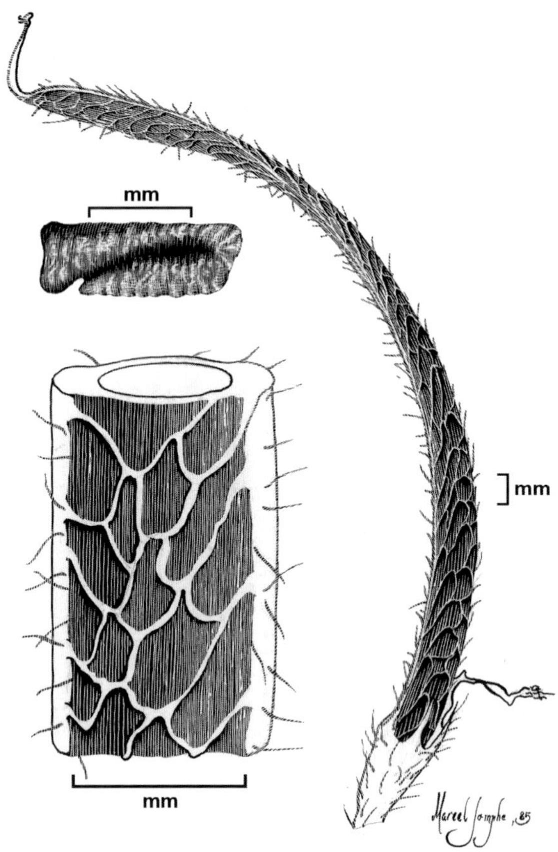

Fig. 63-2. *Medicago persica.* **Clockwise from center left:** Seed; fruit; portion of fruit showing venation. (Based on holotype; artist: M. Jomphe.)

Medicago persica occurs in southwestern Iran and adjacent Iraq.

Description

Annual herb, single-stemmed or branched from base, stems slender, ascending, 4–15 cm long, more or less pubescent with appressed hairs. Leaflets obovate or obtriangular, 3–9 mm long, 2–5(6) mm wide, two-thirds of apical margin serrate, appressed-pilose on both surfaces or just on lower surface. Stipules semi-sagittate, linear-subulate, pubescent. Inflorescence subumbellate, of 1–10 flowers, peduncle 1–2 cm long, longer than the subtending petiole. Calyx ca. 3 mm long, sparsely pubescent, the teeth lanceolate, shorter than the tube. Corolla yellow, 8.5–9 mm long. Fruit 2–4 cm long, ca. 1.5 mm wide, pubescent with appressed hairs, slightly compressed, linear, curved semi-circularly, veins running obliquely on the pod face, prominently anastomosed, the ends of the fruits converging at maturity. Seeds ca. 1.5 mm long, 0.4 mm wide, cylindrical, surface rugulose, radicle about 80% of length of seed.

Identification Hints

Medicago persica is similar to *M. phrygia* and *M. fischeriana*. It differs from *M. phrygia* in usually having an inflorescence of less than eight flowers (usually more in *M. phrygia*), fruits often longer than 30 mm (rarely as long as 30 mm in *M. phrygia*), calyx teeth distinctly shorter than the corresponding tube (about as long as the tube in *M. phrygia*), and yellow flowers (yellow or orange in *M. phrygia*). *Medicago persica* differs from *M. fischeriana* in having flowers that are often 7 mm or more in length (normally no longer than 7 mm in *M. fischeriana*) and nonsinuous fruits (sinuous in *M. fischeriana*).

Habitat and Environmental Adaptation

Habitat

Medicago persica has been collected in pebbly river beds and stony places.

Phenology

This species flowers in March and April in its native area.

Soil

Medicago persica typically occupies gravelly and alluvial areas, occasionally clay. The substrate tends to be derived from limestone.

Temperature

The species occurs in areas receiving cool early spring conditions that are subsequently subjected to very hot continental summers.

Water

This species, like other annuals in the genus, is a drought evader, occupying dry areas but completing its life cycle before a killing drought. It tends to be located in moist depressions.

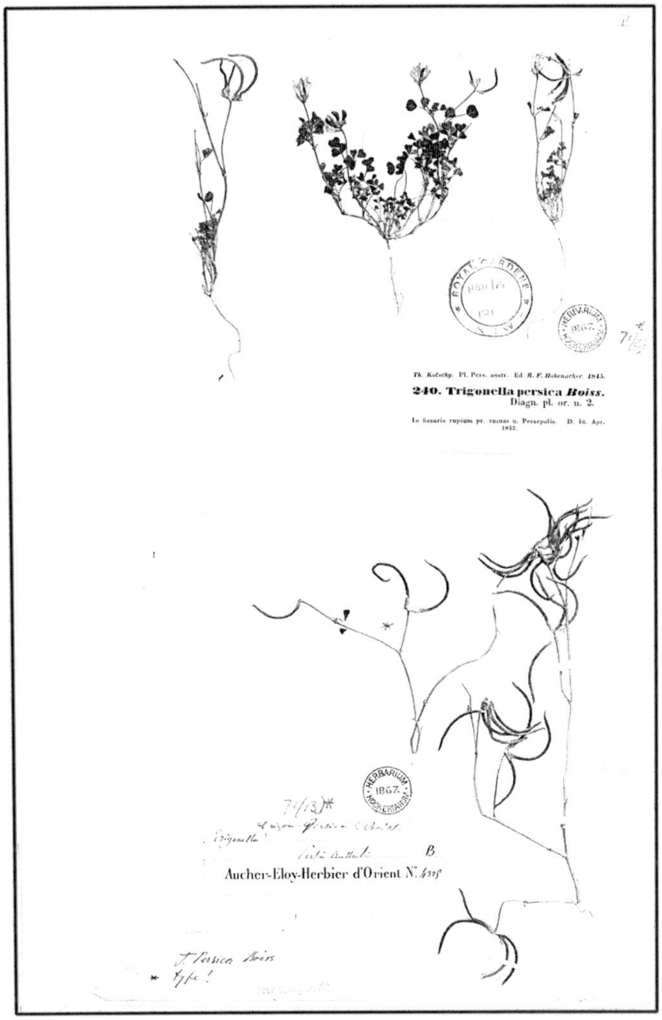

Fig. 63-3. Photograph of isotype of *Trigonella persica* Boiss., at K.

Biotic Relationships

Pollination

The small flowers indicate that this species is an inbreeder.

Dispersal

No obvious adaptations for seed distribution are evident.

Herbivory

No information located.

Nodulation and mycorrhizae

No information located.

Disease resistance

No information located.

Economic History, Cultivars, and Uses

This species has not attracted agricultural interest to date.

Cultivars

No cultivars have been bred.

Forage

Medicago persica is not sown for forage.

Fodder

This species is not harvested for livestock feed.

Green manure, rotation crop, ground cover

No information located.

Other uses (salad sprouts, ornamental, plant products, etc.)

No information located.

Chromosome Number

No information located.

Genetic Relationships and Gene Transferability

Intraspecific

Artificial hybridization of different populations has not been attempted, so it is not known whether there are barriers to hybridization within the species.

Interspecific

No information located.

Exchange by advanced biotechnology

No information located.

Conservation Status

Because this species has a very limited distribution range, it is in need of a conservation assessment and very likely requires the implementation of protective measures.

Germplasm Sources

No accessions were located.

Research Problems and Priorities

Medicago persica is a narrowly distributed endemic with no obvious characteristics that would suit it to being developed as a crop, except for the fact that it grows under stressful conditions. Its adaptation to a hostile climate and substrate suggests that it may harbor genes useful for improvement of the more economic *Medicago* species. Very little is known about the species and it requires basic botanical study. There is an urgent need to collect and preserve its seeds.

Representative Collections

Iran: Anon., West Persia, Abadan, sea level, IV-1927, Flora of West Persia 93, IV-1927, in flower & fruit (K). Archibald, J.C. 1239, Fars, Kuh-i-Saidun, alt. ca. 2230 m, ledges on west-, southwest-, and northwest-facing limestone cliffs, 10-IV-1966, in flower & early fruit (E). Archibald, J.C. 1508, Bakhtiari Country, between Azna and Dorud, southwest facing slope, alt. ca. 1740 m, stony clay in Euphorbia–Acantholimon association, 24-IV-1966, in flower & early fruit (E). Danin, A., Baum, B. & Plitmann, U. 65-534, West Iran, west of Toyserkan, alt. 1720 m, small wadi, alluvial wet soil, 27-VI-1965, in flower & fruit (HUJ). Davis & Bokhari D55711, ca. 10 km from Bebehan to Omidiye, alt. 350 m, gravelly calcareous hills, 21-III-1974, in flower (E). Davis & Bokhari D56405, 14 km from Babameydan to Yasuj, alt. 1600 m, rocky limestone slopes facing south, Quercus zone, 7-IV-1974, in flower (E). Furse, P. 1765, Shuhturun-kuh foothills, Azna, alt. 6000 feet, rocky slopes above river, almost ungrazed, 6-V-1962, in flower & early fruit (E). Jacobs, M. 6267, northern and western Iran, 33°17′N, 46°15′E, West Lorestan, Ilam, alt. 400–450 m, alluvial riverine plain, calcareous but not sulphatic, largely bare, recently disturbed, in moister depression, 18-IV-1963, in flower & fruit (E). Kotschy Th. 240, pr. ruinas u. Persepolis, in fissuris rupium, 16-IV-1842, in flower & early fruit, Th. Kotschy Pl. Pers. austr. Ed. R.F. Hohenacker 1845 No. 240 (BM, P, G).

64. *Medicago phrygia* (Boiss. & Balansa) E. Small

M. phrygia (Boiss. & Balansa) E. Small in Small, Lassen & Brookes, Willdenowia 16(2): 433 (1987).

The epithet *phrygia* is based on Greek and means Phrygian. Phrygia was an ancient country of west-central Asia Minor where specimens used to describe the plant were collected.

Subordinate Taxa

None recognized here.

Type Specimen

Balansa, B. 1220, Ouchak (Phrygie), vers 950 m d'alt., vignes, fl. 24-V & fr. 1-VII-1887(the collection has five plants

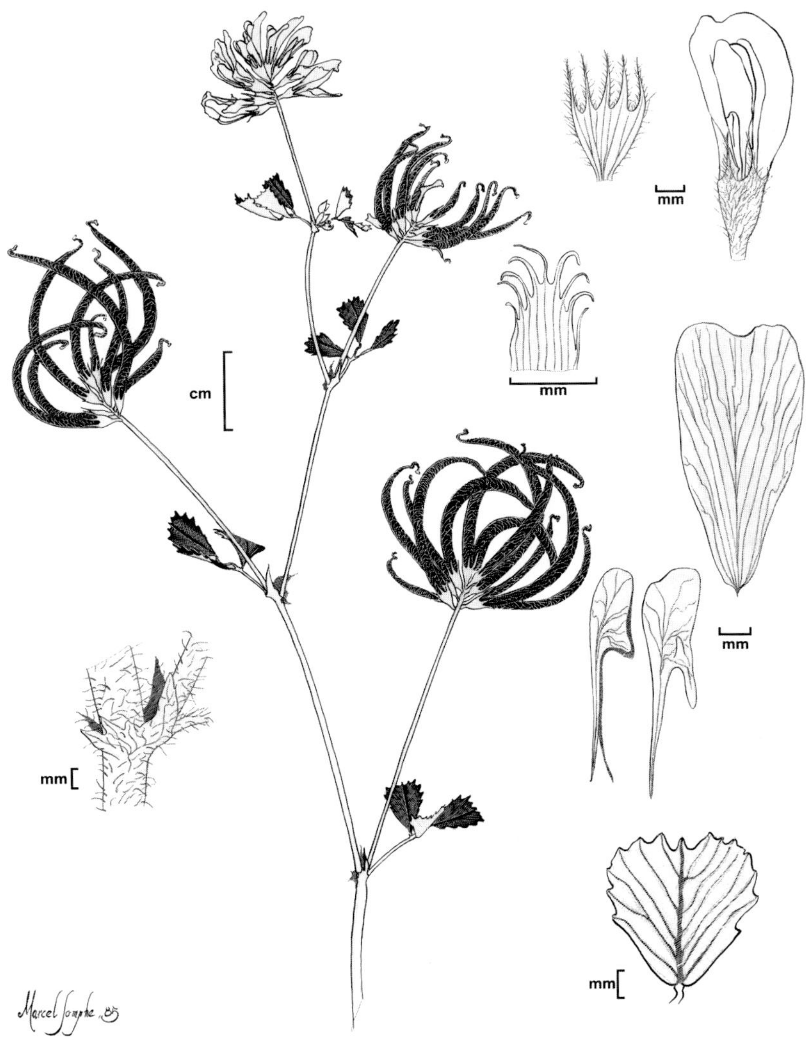

Fig. 64-1. *Medicago phrygia.* **Lower left:** Stem node showing stipules. **Left:** Branch with fruits and flowers. **Upper and center right:** Floral parts (clockwise from top left: calyx; side view of flower; standard petal; wing petal; keel; flattened androecial column (free stamen not shown)). **Lower right:** Leaflet. (Based on Coode & Jones 2418, E; artist: M. Jomphe.)

in flower, one in fruit) (G, P) (As noted below, the G collection has been chosen as the lectotype of the name *T. aurantiaca* var. ß *phrygia* Boiss.).

Synonymy

Homotypic

Trigonella phrygia Boiss. & Balansa in Boiss. Diagn. Pl. Orient. Ser. 2, 3(6): 45 (1859), (basionym).

T. aurantiaca var. ß *phrygia* (Boiss.) Boiss., Flora Orient. 2: 74 (1872). (Boissier noted "Legumina subteretia magis incurva apice conniventia" and cited Balansa exs. 1220 & Bourg. exs. 82 as examples of the variety. The Herb. Boiss. (G) var. ß *phrygia* folder contains Bourgeau 82, Balansa 1220, and Kotschy 216; I annotated Balansa 1220 as lectotype in 1993.)

Heterotypic

T. aurantiaca Boiss., Diagn. Pl. Orient. 1, 1(2): 22 (1843). (Type: Boissier, E., Turkey (Anatolia) Herb. Boiss. G! (plural syntypes; four sheets in a folder, collected VI-1842, the plants in flower and in fruit; specimens also at W, E.) Not *M. aurantiaca* Godr., Mém. Sect. Méd. Acad. Sci. Montpellier 1: 425 (1853) = *M. sativa* subsp. *falcata* var. *falcata*.

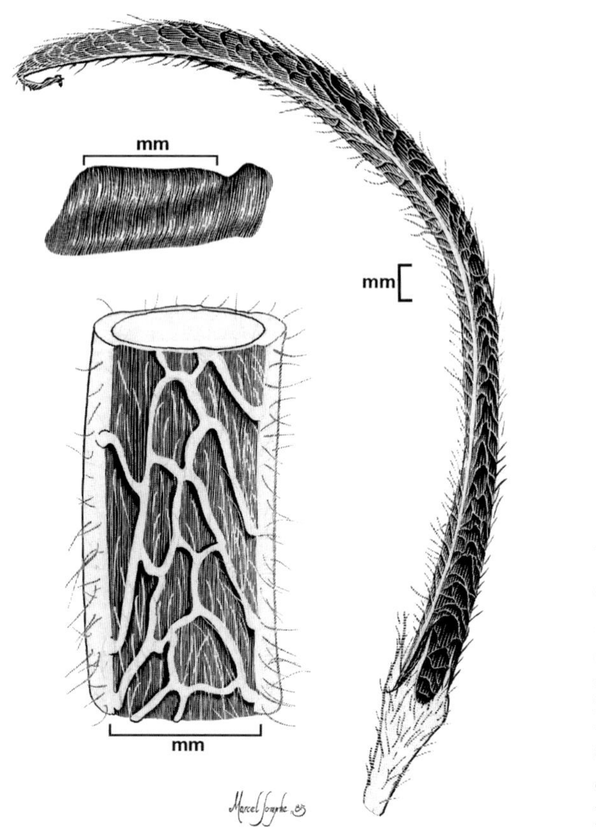

Fig. 64-2. *Medicago phrygia.* **Clockwise from center left:** Seed; fruit; portion of fruit showing venation. (Based on Coode & Jones 2418; artist: M. Jomphe.)

T. aurantiaca Boiss. var. *cariensis* Širj., Generis *Trigonella* revisio critica, Spisy Pfir. Fac. Masarykovy Univ., 3: 20 (1930).

T. kotschyi (non Fenzl.) Rawi, Dep. Agr. Iraq Tech. Bull. 14: 78 (1964) (according to Townsend (1974)).

T. mareschiana Hand.-Mazz., Ann. Nat. Mus. Wien 27: 81 (1913). (Vassilczenko (1953), who recognized this as a separate species, stated that the type is in Vienna.)

T. aurantiaca Boiss. var. *pallida* Bornm., Mitth. Thür. Bot. Ver., N.F. 6: 53 (1894) (Cotype cited in Iraq, representative specimens below).

Vernacular Names

English: No names found.

French: No names found.

Geographical Distribution

(N = native, I = introduced, U = uncertain status)

Asia: Iran (N), Iraq (N), Syria (N), Turkey (N), United Arab Emirates (N; reported from the northeast coast in El-Amry 1995).

Medicago phrygia is indigenous to western Asia. Thanopoulos (2007) provided the following information: "The only two Greek localities of this species are found on the islands of Samos and Ikaria. This is an Anatolian element, and the East Aegean Islands form the westernmost known boundary in species distribution. In Greece, it is found on marble and schist substrates, on sandy shores and in open dry-rocky places of grazed slopes with patches of evergreen shrubs and scree. Unger (1862) cited this species in Kefallinia (Ionian Islands) a locality which would constitute a remarkable extension of its geographic distribution. Unfortunately, no voucher was found in GJO... Unger's report may be a misidentification."

Description

Annual herb, branched from base, stems procumbent to ascending, sparsely appressed-pubescent, 5–30 cm long. Leaflets oblong-cuneate to broadly obovate-cuneate, upper margin serrate, (3)5–12 mm long, 3–8 mm wide, slightly to moderately hairy abaxially, more or less glabrescent above. Stipules lanceolate-acuminate, lower ones dentate. Inflorescence of 8–14 flowers in a short raceme (or subumbellate), forming a spherical to ovate head, peduncle (15)30–40 mm long, longer than subtending petiole. Corolla yellow to orange, 7–10 mm long. Calyx campanulate, half or less of corolla length, 3.5–4 mm long, teeth linear-lanceolate, about as long as tube. Fruits linear, usually arcuate to semicircular, slightly compressed, somewhat pubescent, 10–30 mm long, 1–1.5 mm wide, transversely reticulate netted, with a short, bent beak and 6–8 seeds. Seeds oblong-cylindrical, brownish, tubercled, 1.5–2 mm long, 0.7–1 mm wide, radicle about three-fourths of seed length to almost as long as seed.

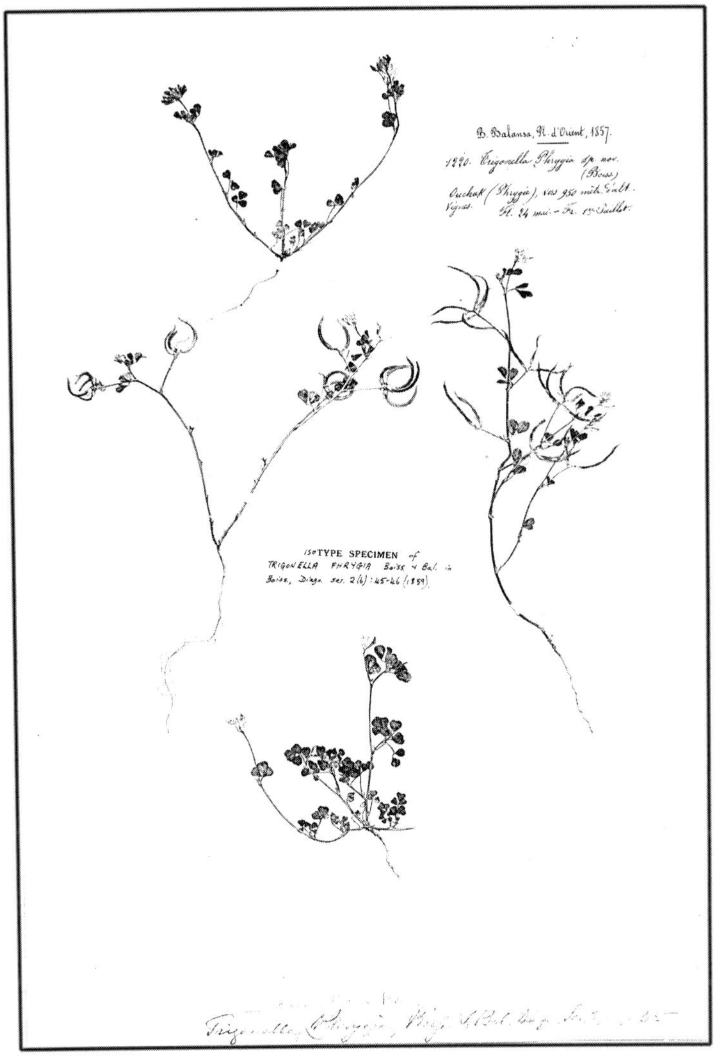

Fig. 64-3. Photograph of type of *Trigonella phrygia* Boiss. & Balansa, at G.

Identification Hints

Medicago phrygia is very similar to *M. persica* and *M. fischeriana*. See *M. persica* (Chapter 63) for key distinctions.

Habitat and Environmental Adaptation

Habitat

Medicago phrygia occurs from sea level to the subalpine zone (0–1600 m) in oak scrub, pine forest, rocky slopes, and fallow and cultivated land, sometimes in subdeserts.

Phenology

This species flowers from March to July in its native area. In experiments in Australia, Halloran and Pennell (1981) found that *M. phrygia* was exceptionally early flowering.

Soil

Medicago phrygia typically occupies basic soils of a variety of textures.

Temperature

This species occurs in a fairly wide range of habitats, requiring it to tolerate cool conditions during germination to increasingly hot temperatures during seed maturation.

Water

Medicago phrygia appears to be moderately drought tolerant.

Biotic Relationships

Pollination

Small (1988) noted that, compared with other annual species of *Medicago*, the flowers are relatively large and the pollen to ovule ratio is comparatively high (466). These observations suggest that *M. phrygia* is unlike almost all other annual *Medicago* species, which are largely self-pollinated and substantially inbreeding.

Dispersal

Adaptations for seed distribution are not apparent.

Herbivory

No information located.

Nodulation and mycorrhizae

USDA (2009) reported positive root nodulation for *M. phrygia*.

Disease resistance

No information located.

Economic History, Cultivars, and Uses

This species has not attracted agricultural interest to date.

Cultivars

No cultivars have been bred.

Forage

Medicago phrygia is not sown for forage.

Fodder

This species is not harvested for animal feed.

Green manure, rotation crop, ground cover

Evans et al. (2004) examined the potential of numerous fodder species to survive winter in North Korea and produce nitrogen, organic matter, and soil stabilization in maize and rice fields. *Medicago phrygia* was judged to have good winter survival and good green biomass production, and to produce excellent ground cover.

Other uses (salad sprouts, ornamental, plant products, etc.)

No information located.

Chromosome Number

No information located.

Genetic Relationships and Gene Transferability

Intraspecific

Artificial hybridization of different populations has not been attempted, so it is not known whether there are barriers to hybridization within the species.

Interspecific

No information located.

Exchange by advanced biotechnology

No information located.

Conservation Status

This species is uncommon, with almost no germplasm collections. There is a need for a conservation assessment and collection of seeds for gene bank preservation.

Germplasm Sources

Among germplasm institutions, SARDI has 28 accessions (see Chapter 3).

Research Problems and Priorities

Medicago phrygia is a poorly researched species in great need of basic botanical and agricultural evaluation, as well as germplasm collection.

Representative Collections

Iran: Davis & Bokhari D56258, 50 km from Lar to Jahrom, alt. 1200 m, wadi on limestone conglomerate under Paliurus, 1-IV-1974, in flower & early fruit (K, E). Davis & Bokhari D56405, 14 km from Babameydan to Yasuf, alt. 1600 m, rocky limestone slopes facing south, 7-IV-1974, in flower & very early fruit (K). Lee, A.F. 138, 8 miles north-east of Masjid-l-sulaiman, alt. 1500 feet, steep, well-drained, south-facing slope of deep marly soil, 8-IV-1949, in flower & fruit (K). Macmillan, H.F. 94, Abadan, sea level, IV-1927, in flower & early fruit (K). Merton, L.F.H. 3814, Bundasht, almost due north of Lar, in Yazd-e-khast, alt. ca. 1250 m, in grazed pastures receiving winter run-off, 18-IV-1959, in flower & fruit (K). Merton, L.F.H. 3862, Eg. Klan-e-Zinium (southwest of Shiraz), alt. 1930 m, in open grazed steppe on low limestone hills, 27-IV-1959, in flower & fruit (K). **Iraq:** Agnew, A.D.Q., Agnew, S., Chiad, F., & Butty, S., Erbil, Forestry nursery weed, 8-IV-1962, in flower & fruit (E). Al-Rawi, A. 20372, 13 km east of Samarra, alt. ca. 65 m, irrigated field, 5-V-1957, in flower & fruit (K). Alizzi & Husain 33736, 10 km west of Tikrit, gravelly soil 2-5-1965, in flower & fruit (K). Bayliss, B. 75, Quiydrah, Mosul, common in wadis, channels, and depressions, 16-IV-1939, in flower & very early fruit (K). Börnmuller, J. 309, Dsinibel, Assyria, 23-IV-1893, in fruit (K; identified as *T. aurantiaca* var. *pallida* Börn. and annotated as cotype of *T. mareschiana* Hand-Mzt). Gillett & Rawi 7233, Jebel Makhul near Ain Dibs, alt. 220 m, dry steppe by wadi, occasional, 2-IV-1947, in flower & very early fruit (K). Gillett, J.B. 10684, Hawija Experimental Farm, alt. ca. 200 m, irrigated lawn, dry steppe formation, occasional, 2-IV-1948, in flower (K). Omar, S. & Hamid, H. 36304, Bet-Shargat & Tikrit, sandy clay soil, 7-IV-1969, in flower & fruit (K). Wheeler Haines, R., Jadriyah, Tigris bank after Adheim flood, 3-IV-1961, in flower & early fruit (E). **Turkey:** Balls, E.K. 944, Akhev Dagh. Marash, alt. 5000 feet, limestone screes and open slopes between scrub, 1-V-1934, in flower (E). Bourgeau, E. 82, Lycia, Elmalu, in collibus, 18-V-1860, in flower & fruit (P). Coode, M.J.E. & Jones, B.M.G. 2418, Usak, 6 miles south of Uzak towards Sivasli, alt. 900 m, wheatfield, light soil over chalk, 14-VI-1965, in flower & fruit (E). Coode, M.J.E. & Jones, B.M.G. 2607, Kütahya, 2 miles west of Abide on road to Simav, alt. 700 m, weeds of thin poor wheat, 17-VI-1965, in flower & fruit (E). Davis & Dodds 18692, prov. Konya, Yavsan Memlehasi near Tuz Golu, in *Artemisia* steppe, 8-VI-1952, in flower & fruit (K, E). Davis 18863, Dodds, Cetik, Prov. Nigde, between Taspinar and foot of Hasan Dag, 15-VI-1952, in flower & early fruit (E). Davis 41464, Muğla, Yatagan to Gökbel, alt. 300 m, Open *Quercus coccifera* marquis on granite substrate, 20-IV-1965, in flower (E). Davis 42436, Mardin, Midyat, alt. 950 m, fallowfield, 5-V-1966, in flower & fruit (E). Dudley D35250, Prov. Antalya, Korkuteli to Kizilcadag, 4 miles from Kork, alt. 1100 m, neglected and cultivated fields, 31-V-1962, in flower & early fruit (E). Dudley D35652, Prov. Afyonkarah isar, Denizli-Çardak, 10 km from Dinar, alt. 1000–1200 m, *Quercus coccifera*, clay, 10-VI-1962, in flower & fruit (E). Helbaek 2507, Prov. Konya, Konya to Çumra, Kuçük Köy, alt. 980 m, cultivated land, 23-V-1962, in flower & fruit (E). Kotschy, T., prope Everek, in collinis vulcanicis, alt. 4100 feet, 1-VI-1859, in fruit, Iter cilicico-kurdicum 1859 No. 216 (P). Manissadjian 555, Um Merzifoun 1891/92, Sandstein hügel, in flower (P). Sintensis, P., Armenia turcica, Szandschak Gümüschkhane, Godonands in chromdere, 12-VI-1894, in flower & very early fruit, Iter orientale 1894 No. 5881 (P). Sintensis, P., Paphlagonia, Wilajet, Kastambuli, Tossia, Beschetscham, "3/6"-1892, in flower & early fruit, Iter orientale No. 4108 (P).

65. *Medicago pironae* Vis.

M. pironae Vis., Cat. Index Sem. Hort. Bot. Patavia 1855: 7 (1856); reprinted in Linnaea 28: 365 (1857).

The epithet *pironae* commemorates Italian botanist and physician Giulio Andrea Pirona (1822–1895).

Subordinate Taxa

None recognized here.

Type Specimen

Kammerer, in ditione Illyrico-litorali, in rupium fiss uris ad Modrea prope Tolmein in valle fluvii Isonzom, solo calc, alt. 200–300 m, no date, comm. Mus. Bot. Univ. Vindob. Flora Exsiccata Austro–Hungarica No. 803 (BM!, DAO!, P!). The type collection is from Mount Matajura on the Italian–Slovenian border, ca. 18 km west of Tolmin and ca. 60 km north of Trieste, near the Isonzo River. (Holotype not determined. An inquiry to PAV regarding the possibility that a collection there should be considered to be the holotype did not result in a reply.)

Synonymy

Homotypic

No names found.

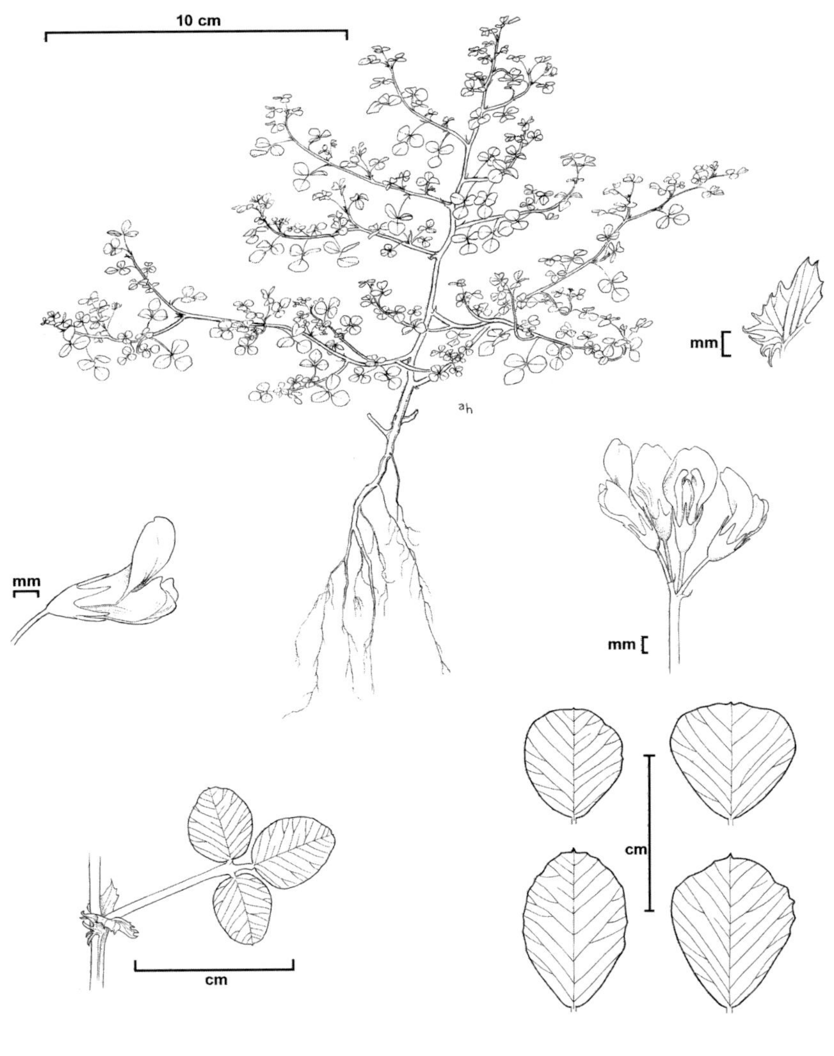

Fig. 65-1. *Medicago pironae.* **Clockwise from top:** Plant; stipule; group of flowers; leaflets; leaf attached to stem; side view of flower. (Based on E. Small M672, DAO; artist: A. Hanes.)

Heterotypic

M. rupestris Pirona, Fl. Forojul. Syll. 41, non *M. rupestris* MB.

Vernacular Names

ENGLISH: Piron's medic (medic is also spelled medick).

FRENCH: No names found.

Geographical Distribution

(N = native, I = introduced, U = uncertain status)

EUROPE: Austria (N), Italy (N), Yugoslavia (N; recorded in Slovenia by Dakskobler 2005).

This is a local endemic of the Alps of northeastern Italy and areas of Austria, as well as the former Yugoslavia.

Description

Perennial herb, branches 20–40 cm long, branches arising from crown, glabrescent. Leaflets obovate to obcordate, apex serrate, 4–8(11) mm long, 4–8(10) mm wide. Stipules sagittate, ovate-lanceolate, upper stipules entire, lower ones incised-dentate at base. Inflorescence of (1)2–6(10) flowers, the peduncle about equalling the length of the subtending petiole. Corolla 6–8.5 mm long, yellow. Calyx 3–5 mm long, shorter than half length of flower, sparsely pubescent with simple hairs, teeth longer than tube. Young pod with sparse pubescence, protrudes sideways through the calyx teeth. Fruit 3.5–7 mm wide, shortly-cylindrical, with 1.5–4 loosely adpressed coils, glandular-pubescent, spiny, the spines grooved due to two-rooted insertion on dorsal suture, spines typically inserted at ca. 90° to coil face, coil face with veins anastomosing in outer half. Seeds 1–2 per coil, yellowish, ca. 2.5 mm long, 1.2 mm wide, not separated by transverse, spongy septa, radicle more or less half seed length.

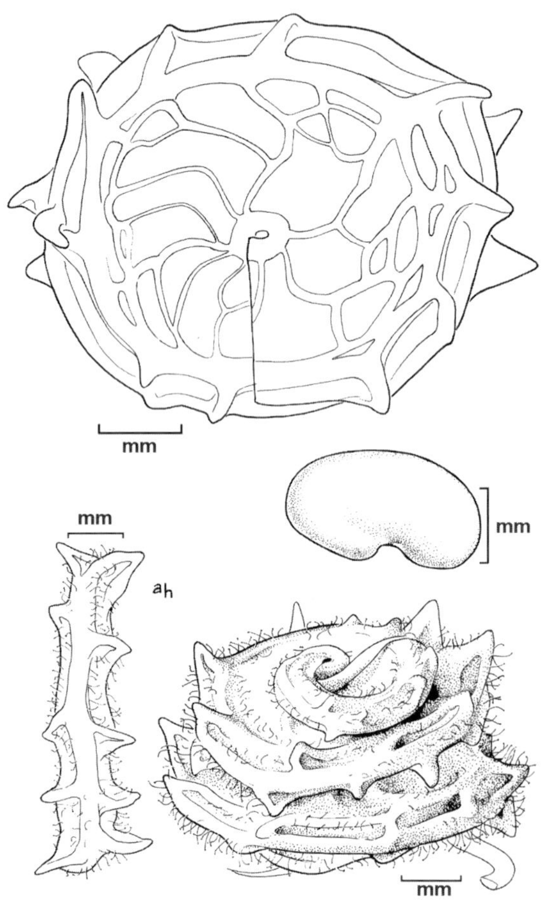

Fig. 65-2. *Medicago pironae*. **Clockwise from top:** Fruit coil face showing venation; seed; fruit; edge of coil. (Based on E. Small M159, DAO; artist: A. Hanes.)

Identification Hints

This is the only perennial *Medicago* with spiny, coiled fruit covered with gland-tipped hairs. However, the fruits are remarkably similar to those of the annual species *M. rigidula*.

Habitat and Environmental Adaptation

Habitat

Medicago pironae occurs on rocky hillsides.

Phenology

Peak flowering occurs from late April to June in its native area.

Soil

This species characteristically occupies calcareous, rocky soils.

Temperature

Medicago pironae occurs in alpine conditions. Winter killing was observed in plants established at Brandon, Manitoba (Lesins and Lesins 1979).

Water

This species is often found in dry, well-drained locations and these considerations, as well as its deep roots, suggest adaptation to a limited water supply.

Biotic Relationships

Pollination

Small (1988) noted that the relatively large flowers and pollen to ovule ratio (553) are consistent with *M. pironae* being at least partly cross-pollinated, like other perennial species of *Medicago*.

Dispersal

The indehiscent spiny fruits are adapted to dispersal in animal fur, but not ideally so because the spines are not hooked.

Herbivory

Fiori and Dolan (1981) found in an experimental field that *M. pironae* received only half as much damage from the potato leafhopper *Empoasca fabae* (Harris) as did most alfalfa cultivars.

Nodulation and mycorrhizae

No information located.

Disease resistance

No information located.

Economic History, Cultivars, and Uses

This species has not attracted agricultural development to date.

Cultivars

No cultivars have been bred.

Forage

Medicago pironae is not sown for forage. In experimental cultivation, Lesins and Lesins (1979) noted that it produced about 30% of the herbage of a commercial alfalfa cultivar and regrowth after cutting was poor.

Fodder

The species is not grown as fodder.

Green manure, rotation crop, ground cover

No information located.

Other uses (salad sprouts, ornamental, plant products, etc.)

No information located.

Chromosome Number

2*n* = 16: Lesins and Lesins 1961; Lesins and Gillies 1968; Gillies 1972*a*; Lovka et al. 1972.

Genetic Relationships and Gene Transferability

Intraspecific

Artificial hybridization of different populations has not been attempted, so it is not known whether there are barriers to hybridization within the species.

Interspecific

Medicago pironae has been crossed with *M. daghestanica*, although the hybrids were completely sterile (Lesins and Lesins 1961; Lesins and Gillies 1968; Lesins 1971). However, Gillies (1972*a*) noted that the pachytene chromosomes were very similar, indicating that these two species are related. McCoy and Echt (1993) were also unable to hybridize *M. pironae* with *M. sativa*. Lesins (1971) was able to produce tri-species hybrids between *M. pironae*, *M. daghestanica*, and *M. sativa* by crossing tetraploid *M. sativa* and an artificial tetraploid hybrid of the other two species. Lesins and Gillies (1972) stated that unpublished work by Lesins showed that *M. pironae* could be hybridized with *M. sativa*, but not at the diploid level.

Exchange by advanced biotechnology

McCoy and Echt (1993) were unable to hybridize *M. pironae* with *M. sativa*, but were able to hybridize *M. pironae* with the hybrid of *M. sativa* and *M. rupestris* by using embryo rescue to obtain viable seeds.

Conservation Status

The 1997 International Union for Conservation of Nature Red List of Threatened Plants included *M. pironae* (http://www.ceu.hu/envsci/soe/ref/nature/threpl11.htm; current July 2007). A conservation analysis for Croatia (http://croatia.usaid.gov/pdf-stories/biodiversity-report.pdf; current June 2007) listed *M. pironae* as "DD" (data deficient, i.e., a lack of sufficient data available to assess the risk of extinction).

Germplasm Sources

Seven accessions of *M. pironae* are listed among the major germplasm institutions with large collections of *Medicago* (see Chapter 3).

Research Problems and Priorities

Medicago pironae is a poorly known species in great need of basic botanical and agricultural study. It is very poorly represented in gene banks and collection of its seeds should be a priority.

Representative Collections

Austria: de Tommasini, M., Province de Gorice, entre Saint Lucien et Modrea, rochers calcaires, 10-VI in flower, 13-VII in fruit, F. Schultz herbarium normale, Cent. 11 No. 1043 (BM, K,P). Kammerer, In ditione Illyrico-litorali, in rupium fiss uris ad Modrea prope Tolmein in valle fluvii Isonzom, solo calc, alt. 200–300 m, no date, comm. Mus. Bot. Univ. Vindob. Flora Exsiccata Austro-Hungarica No. 803 (BM, P). **Italy:** Ball, J., ex alpium Carniacarum, covvalle Isonzo Thal, Inter Canale e Volzano, 21-V-1862, in flower & early fruit (K). Bampanini (Pampanini), B., Venetia, Prov. de Treviso, Vittorio, abundans in nemorosis montanis, praecipue aridis, prope pagum Tovena, in latere meridionali montium Cimone et Torresella (Passo di S. Uboldo), Praealpium Bellunensium, alt. 600 m, solo calcareo, flor. 11, fruct. 29-VI-1905, Flora Italica Exsiccata Fiori, Béguinot & Pampanini No. 297 (distributed as *M. carstiensis* with later correction) (BM, GH, K). Erichiutti G., Venetia, prov. di Udine, in rupibus calcareis et locis aridis montis Glemina (Gemona), in glareosis calcareis secundun viam Gemona-Maniaglia, nec non in montibus Quarnan et Bernadia circa Torlano et Picat in valle rivi Scelsò di Prestento, V-VI-1904, in flower & early fruit, Flora Italica Exsiccata Fiori, Beguinot & Pampanini No. 100 (BM, K). Mall, ex saxosis juxta Gemona, V-1862, in flower & fruit (K). Marchesetti, Alpi Giulie, Modrea pr. Tólmino, VIII-1888, in flower & fruit (K). Pichler, T., Venetia in glareos saxosis pr. Romana, V-1879, in flower & fruit (K). Pichler, Venetia orient. ad rupes p. Gemonam VII-1868, in flower & fruit (K). **Yugoslavia:** Poldini, L. 6676, M. Santo prope Gorica (Yugoslavia, Slovenia), alt. 600 m, sol calcareo, in rupestribus apud Seslerio-Ostryetum, 15-VII-1970, in flower & early fruit (G).

66. *Medicago platycarpos* (L.) Trautv.

M. platycarpa (sic) (L.) Trautv., Bull. Sci. Acad. Imp. Sci. St. Petersb. 8: 271 (1841).

The epithet *platycarpos* is based on the Greek *platys*, meaning flat, plus *karpos*, meaning fruit, in allusion to the flat fruits. Trautvetter's rendition *platycarpa* would be correct if used as an adjective, and indeed the spelling *Medicago platycarpa* is widely used. However, *platycarpos* is employed here as an appositive noun, and Linnaeus's original rendition (as *Trigonella platycarpos*) is correct.

Subordinate Taxa

None recognized here.

Type Specimen

The Linnaean Plant Name Typification Project (http://www.nhm.ac.uk/research-curation/projects/linnaean-typification/; current June 2010) provided the following information: "*Trigonella platycarpos* Linnaeus. Reference: *Species Plantarum* 2: 776 (1753). Provenance: 'Habitat in

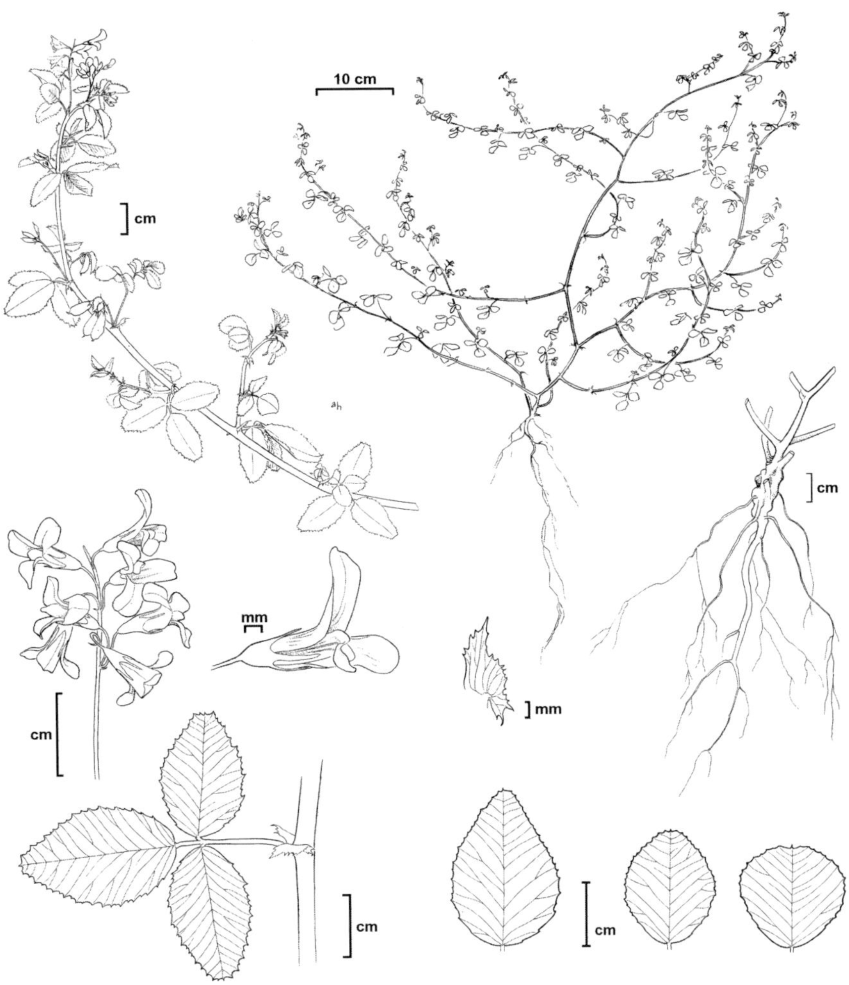

Fig. 66-1. *Medicago platycarpos.* **Clockwise from top left:** Branch with flowers; plant; root and crown; leaflets; leaf attached to stem; group of flowers. **Center:** Side view of flower; stipule. (Based on E. Small M164, DAO; artist: A. Hanes.)

Sibiria.' Type: Lectotype: Herb. Linn. No. 932.2 (LINN) Designated by: Lassen in Turland & Jarvis (ed.), Taxon 46: 483 (1997). Current Name: *Medicago platycarpos* (L.) Trautv."

Synonymy

Homotypic

Trigonella platycarpos L., Sp. Pl. 2: 776 (1753), (basionym).

Medicago platycarpos (L.) Ledeb., Fl. Ross. 1(3): 523 (1843), nom. superf.

Melissitus platycarpus (sic) (L.) Golosk. Fl. Kazakhst. 5: 29 (1961).

Turukhania platycarpos (L.) Vassilcz., Novosti Sist. Vyssh. Rast. 16: 132 (1979).

Melilotoides platycarpos (L.) Soják, Sborn. Nár. Muz. Praze, 1982(1-2): 104 (1982).

Heterotypic

Trigonella karkarensis Semen. ex Vassilcz., Not. Syst. Herb. Inst. Bot. Acad. Sci. URSS, xiv: 232 (1951). (Vassilczenko (1953) noted that the type ("Karkara fl., 7000 feet, leg. Semenov") is in LE.)

Melissitus karkarensis (Semen. ex Vassilcz.) Golosk., Fl. Kazakhst. 5: 29 (1961).

Melilotoides karkarensis (Semen. ex Vassilcz.) Soják, Sborn. Nár. Muz. Praze, 1982(1-2): 104 (1982).

Melilotoides platycarpos (L.) Soják var. *karkarensis* (Semen. ex Vass.) G.P. Yakovlev, Rast. Tsentral. Azii, 8a: 88 (1988).

T. platycarpos var. *potanini* Sumnev., Animadvers. Syst. Herb. Univ. Tomsk. Nos. 1-2: 14 (1932). (Exceptional to the description below, this taxon was described as having glandular pubescence in the upper part.)

T. ruthenica Patrin ex Ledeb., Fl. Altaic. 3: 253 (1831), non L. (according to Grossheim 1945).

Vernacular Names

ENGLISH: No English common name is available, although several Russian names are in use in Russia, including Siberian lucerne, yellow lucerne, flat-fruited fenugreek, and flat-fruited lucerne (Balabaev 1934). The name flat-fruited fenugreek was based on the interpretation that this species belonged to the genus *Trigonella* (of which the most well-known species is known as fenugreek). Flat-fruited lucerne (or alfalfa) is more appropriate, although the same name is equally descriptive of the flat-fruited species allied to *M. platycarpos*. Siberian lucerne and yellow lucerne are names that are more typically applied to *M. sativa* subsp. *falcata*.

FRENCH: No names found.

Geographical Distribution

(N = native, I = introduced, U = uncertain status)

ASIA: China (western Manchuria and in the south to the Peking-P'o Hua Shan mountains; N), Kazakhstan (north-eastern mountain regions; N), Kyrgyzstan (N), Mongolia (northern; N), Russia (N).

EUROPE: Russia (Bashkiria, Karelia, Siberia; N).

The known range is principally in mountainous regions, their river valleys, and their meadows, chiefly in Siberia, the

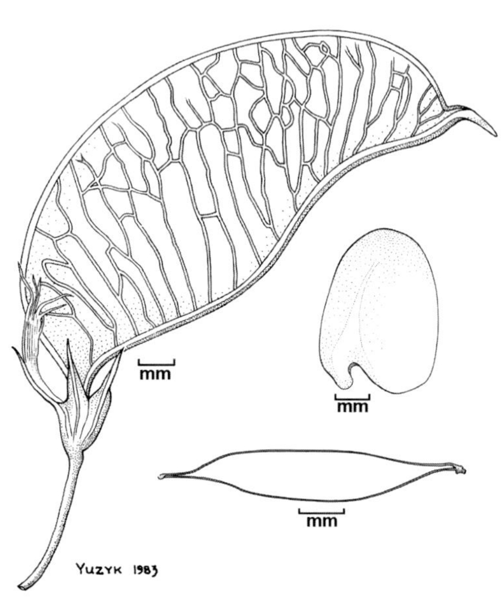

Fig. 66-2. *Medicago platycarpos.* **Clockwise from top:** Fruit; seed; outline of cross section of fruit. (Based on E. Small M164, DAO; artist: L. Yuzyk.)

Fig. 66-3. *Medicago platycarpos.* From Gmelin (1769, vol. 4, tab. ix).

Transbaikal, and the mountains of Kazakhstan, Mongolia, and the Far East. The species seems most at home along the northern fringe of the steppe regions of the former USSR. *Medicago platycarpos* extends from Bashkiria and western Siberia eastwards to Lake Baikal and the Amur River. Towards the south, higher latitudes are occupied (up to 3000 m in eastern Tien Shan and Narym), with occurrence in the unglaciated mountain Ussuri region between Russia and China. The northernmost locations occur in the Yenisei Valley of Siberia, ranging to 69°N.

Balabaev (1934) provides extensive information and biogeographical discussion of the areas of the former USSR occupied by *M. platycarpos*. The area of occupation is widespread but discontinuous, probably reflecting a retreat from a more continuous preglaciation distribution.

Description

Perennial herb, branches (15)25–80(100) cm long, reproducing from perennial rhizome growing near soil surface, stems erect or ascending, branching from crown, branching usually limited, branches sometimes violet and glabrate or sparsely pubescent with simple hairs. Leaflets (8)20–30(40) mm long, broadly oval, rounded-obovate, or oblong-rhombic, generally (1)1.3–1.8 times as long as broad, serrate with uneven mostly sharp teeth from tip almost to base, upper surface glabrous, lower surface sparsely tomentose (mostly on veins) to glabrescent, on drying leaves sometimes purplish. Stipules triangular-lanceolate, 5–8 mm long, toothed. Inflorescence a short raceme of (4)8–12(15) flowers, peduncles slender, axillary, more or less as long as subtending petiole. Flowers 7–9(13) mm long, yellow with violet or violet-red streaks (sometimes yellow-blue; transitions from yellow to violet encountered; petals often drying to blue). Calyx broadly campanulate, (2.5)3–4(5) mm long, with some simple hairs or almost glabrous, teeth lanceolate, approximately equal to tube. Corolla about twice length of calyx. Fruit glabrous, pendulous, flat, oblong-oval or elliptical, weakly convex to straight or concave on ventral suture, convex on dorsal suture, (15)18–22(25) mm long, 7–9(10) mm broad, black at maturity, fruit face with slightly curving veins arising from ventral suture that anastomose slightly, then reach dorsal suture, the beak short and curved. Seeds 4–6(7) per pod, 2.5–3.0 mm long, yellow or yellow-brown, ovoid, smooth, radicle as long as, or almost as long as, seed.

Identification Hints

Medicago platycarpos is rather similar to the closely related *M. ruthenica*. The two species are best distinguished by roughness of the seed coat (smooth in *M. platycarpos*, rugose in *M. ruthenica*) and by pod length (usually 18–22 mm long in *M. platycarpos* and usually 8–12 mm long in *M. ruthenica*). Other notable differences are larger leaflets and less pubescence in *M. platycarpos*.

Habitat and Environmental Adaptation

Habitat

Medicago platycarpos occurs in moist soils in river valleys and in meadows bordered by temperate forest and alpine trees, and within partially open regions of forest, along the slopes of ravines, on river banks, and in coppices, wood margins, and dense grass. Dorogostaiskaya (1978) noted that *M. platycarpos* is a weedy plant in northeastern Siberia, but generally the species is not noted for weediness. In its southern range it is found at elevations up to 3000 m.

In general the species is rather rare, occurring in restricted numbers in dispersed sites. Further, it is not dominant in the sites to which it is best adapted and is almost always found growing among other plant species rather than forming pure stands.

Phenology

Mikhailova (1979) described the species' developmental life cycle and characterized it as similar to that of *M. sativa*. *Medicago platycarpos* flowers from May to July and produces fruits beginning in July in much of its range. Vassilczenko (1953) observed that some northern plants (from the Turukhansk region on the Yenisei River) remained vegetative, and he suggested that they may have been introduced naturally along the river system from southern plants in the Altai not adapted to maturing seed in the northern part of the range. Balabaev (1934) also commented that in the northernmost part of the range the species is frequently sterile and produces a low seed yield.

Soil

Medicago platycarpos typically occupies podzolic soils bordering forested regions and soils associated with river valleys. It is found in steppe soil, rocky soil, and sand.

Temperature

Medicago platycarpos occurs in northern areas of Eurasia, including Siberia, indicating adaptation to cold, continental climates. It extends farther north (to 69° in the Yenisei Valley of Siberia) than any other species of *Medicago*.

Water

As reflected by the habitats occupied, the species is adapted to quite moist conditions such as meadows, although sometimes it occurs in seasonally dry habitats such as dry rocky outcrops and gravelly soils. The shallow root system and mesophytic, large glabrous leaves are morphological indices of a plant adapted to moist conditions. However, Mikhailova (1984), on the basis of the anatomy of the leaves, described *M. platycarpos* as having intermediate adaptation between mesic and xeric conditions.

Biotic Relationships

Pollination

Small (1988) recorded a pollen to ovule ratio of only 295, suggestive of inbreeding by comparison with other perennial

Medicago. However, the flowers are relatively large and the prominent honey guides on the petals are likely attractive to northern indigenous bees, suggesting the species is at least partly cross-pollinated, like other perennial species of the genus.

Dispersal

Medicago platycarpos does not seem to have any specializations for seed distribution. Its characteristic association with river systems suggests water dispersal or dispersal by animals using river valleys.

Herbivory

No information located.

Nodulation and mycorrhizae

USDA (2009) reported positive root nodulation for *M. platycarpos*.

Disease resistance

No information located.

Economic History, Cultivars, and Uses

No domestication has occurred and *M. platycarpos* has not been grown for agricultural purposes. As noted below the species provides natural forage and has considerable economic potential.

Cultivars

No cultivars have been bred.

Forage

Bolton (1962) wrote, "Russian authors state that *M. platycarpos* has low productivity, poor recovery, and is often sterile in the north." Nevertheless, some Russian authors have been more favorable. Vassilczenko (1953) characterized *M. platycarpos* as a valuable forage plant, providing considerable biomass. Kucherov and Mikhailova (1978) described it as rich in protein, carotenoids, and vitamin C. Drozdov (1974) and Ryakhova and Kucherov (1975) suggested that *M. platycarpos* may be useful as a cultivated crop in Russia. Experimentally cultivated plants survived well in Edmonton, Alberta, but less so in Brandon, Manitoba (Lesins and Lesins 1979).

Fodder

No information located.

Green manure, rotation crop, ground cover

No information located.

Other uses (salad sprouts, ornamental, plant products, etc.)

No information located.

Chromosome Number

$2n = 16$: Fryer 1930; Cooper 1935; Oldenmeyer 1956*a*; Clement 1962; Simon and Simon 1965; Bolkhovskikh et al. 1969, p. 309 (2 reports); Rostovtseva 1977; Lesins and Lesins 1979; Stepanov 1994; Chepinoga et al. 2009 (2 reports).

Lesins (1952) has been mistakenly reported as a $2n = 32$ count; however, tetraploidy was artificially induced.

Genetic Relationships and Gene Transferability

Intraspecific

Artificial hybridization between populations of the species has not been reported.

Interspecific

Attempts to hybridize *M. platycarpos* with *M. lupulina*, *M. orbicularis*, and *M. rigidula* failed (Oldenmeyer 1956*a*). Oldenmeyer (1956*b*) claimed that he hybridized *M. platycarpos* with *M. ruthenica* when the latter was the pistillate parent, but the reciprocal cross was unsuccessful. Most of the seed that was produced in the "successful" hybridization aborted during development, and the few flowering plants produced mostly showed marked meiotic irregularities. However, Campbell et al. (1997) suggested that in fact Oldenmeyer actually hybridized two *M. ruthenica* accessions. Lesins (1952) was unable to hybridize *M. platycarpos* that was artificially doubled in chromosome number to $2n = 32$ with *M. sativa* that had the same number of chromosomes. Various crosses between *M. platycarpos* and *M. sativa* at different chromosome levels proved unsuccessful (Oldenmeyer 1956*a*; Lesins and Lesins 1979).

Exchange by advanced biotechnology

No information available.

Conservation Status

Medicago platycarpos has rarely been collected or studied, and since it occurs in areas vulnerable to human activity, it is in need of a conservation assessment.

Germplasm Sources

Twenty-five accessions of *M. platycarpos* are recorded among the world's major *Medicago* germplasm institutions (see Chapter 3).

Research Problems and Priorities

Medicago platycarpos is a perennial plant adapted to cold climates, two ways in which this species is rather similar to *M. sativa*. However, since the former extends farther north than any other *Medicago* species, it must be considered to be of exceptional interest for growth in cold climates. Since it appears to provide excellent forage, there is a good possibility that it can be domesticated, especially for use in cold climates. Balabaev (1934) discussed *M. platycarpos* as perhaps the most promising wild legume for domestication for forage in northern Russia, since it is a species with sufficient resistance to the cold climate and sufficiently high productivity that it might provide a stable feed base for the animal-raising economy of the settled Polar North. However, as also noted by Balabaev, the wild plant is low in productivity, regrows poorly after cutting, and is not very competitive. A program of domestication would require extensive collection and analysis of the range of variation of this rather elusive species, and subsequent selection. *Medicago platycarpos* also offers germplasm that may be useful for the improvement of alfalfa but, in view of the intersterility with alfalfa, studies of means of transferring genes are required. In addition to the possibility of exploiting *M. platycarpos* for its adaptation to cold conditions, it should also be kept in mind that the species is well adapted to rocky and sandy soils. Seed collections in germplasm banks are presently limited and, in view of the uncommonness of the species and the possibility of its extermination in given regions, it is highly desirable to increase these germplasm holdings as soon as possible.

Representative Collections

China: Chiao, C.Y., 3078, Shatung Prov., Lung Tung, Tsinanfu, alt. 300 m, along rocky slope, 7-IX-1930, in fruit (E). **USSR:** Brenner, M., Siberia, Jenisei, Karasina, 10-VII-1876 (LE). Karelin & Kirilof 218, in rupestribus montium Tarbagatai ad torrentem Terektly rario. In montosis sylvatis inter Ustkamenogorsk et Buchtarminsk, 1840, in flower & early fruit (G). Karelin & Kiriloff 1351, in glareosis ad fl. Aksu, regionis montane Alatan, 1841, in fruit (BM, P). Ledebour, Altai, Herbarium Hookerianum 1867, in fruit (K). Malzew, I., prov. Irkutzk (Siberia), distr. Balagansk, in fruiticetis umbrosis secus fl. Iret pr. P. Baschejewsky 29-VII-1901, in flower & fruit, (G, K, LE). Meyer 904, F.N., Southwest Siberia, Sminogorsk, 21-IX-1911, in fruit (K). Roldugin, I., Kazachstania, rami septentrionali-orientales, jugi Alatau Dshungarici, montes Kendyrsakal, in angustiis fl. Kendyrsakal, prope stat. viae ferreae Druzhba, in silva betulino-piceetina collucata, 11-VII-1960, in flower (BM), in flower & early fruit (G). Schrenk, Songarei, Herbarium Hookerianum 1867, in flower (K).

Notes

Medicago platycarpos is usually found in shade or partial shade, rarely in settings receiving full insolation, and in this respect it is very unusual in the genus *Medicago* (*M. carstiensis* also is adapted to shade). The purplish anthocyanin coloration usually observed in the flowers and sometimes seen in the leaves on drying is of interest. Anthocyanins are often developed in leaves of mountainous plants that receive full sunlight, possibly because they are protective against ultraviolet radiation. Perhaps anthocyanins play a related role in *M. platycarpos* since it is both a mountain-inhabiting plant and one that does not seem adapted to full sunlight.

Ignasiak and Lesins (1973) noted some similarities in the petal carotenoids of *M. platycarpos* and *M. ruthenica*, seemingly more evidence of the close relationship of these two species.

67. *Medicago plicata* (Boiss. & Balansa) Širj.

M. plicata (Boiss. & Balansa) Širj., Spisy přír. Fak. Masarykovy Univ. 102: 33 (1928).

The epithet *plicata* refers to the plicate or plaited (i.e., with folds alternating in direction) appearance of the fruits.

Subordinate Taxa

None recognized here.

Type Specimen

Balansa, B. 246, alluvions du Guzel-Déré à Mersina, 23-V-?, in flower & fruit (G!).

Synonymy

Homotypic

Pocockia plicata Boiss. & Balansa in Boissier, Diagn. P. Orient. Ser. 2, 3(5): 76 (1856), (basionym).

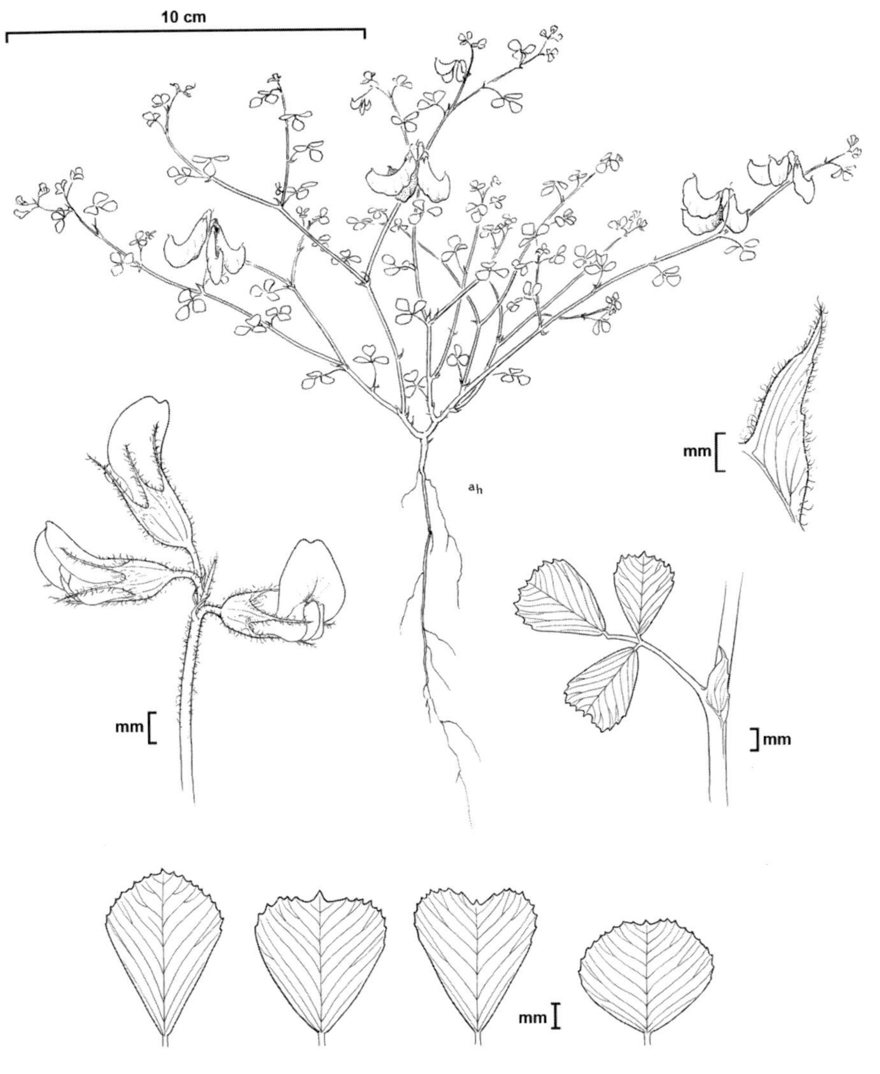

Fig. 67-1. *Medicago plicata.* **Clockwise from top:** Plant; stipule; leaf attached to stem; leaflets; group of flowers. (Based on B. Balansa in 1856, GH; artist: A. Hanes.)

Trigonella plicata (Boiss. & Balansa) Boiss., Fl. Orient. 2: 90 (1872).

Heterotypic

No names found.

Vernacular Names

ENGLISH: No names found.

FRENCH: No names found.

Geographical Distribution

(N = native, I = introduced, U = uncertain status)

ASIA: Turkey (N).

Medicago plicata is endemic to south-central Turkey.

Description

Annual herb, branched from base, stems ascending to erect, 5–20(25) cm long, appressed-pubescent. Leaflets obovate to obcordate, pubescent (especially on lower surface), distal margins serrate, 4–9 mm long, 3–9 mm wide. Stipules lanceolate-acuminate, pilose, the base entire or denticulate. Inflorescence of 2–6 flowers, umbellate, the peduncle 0.5–1.5 cm long, slightly longer than subtending petiole. Calyx campanulate, tomentose, ca. 3.5–4 mm long, teeth lanceolate-subulate, somewhat longer than tube. Corolla yellow, ca. 4.5–5 mm long. Fruits 10–15 mm long, 5–6 mm wide, flat, pubescent, parchment-like, linear-oblong, falcate, transversely nerved, with 3–5 pleats, both sutures denticulate (with small, white, branched, short pectinate spinules), with 4–6 seeds. Seeds ovate-oblong, compressed, 2–2.5 mm long, ca. 1.5 mm wide, finely tuberculate, radicle about equal to seed in length.

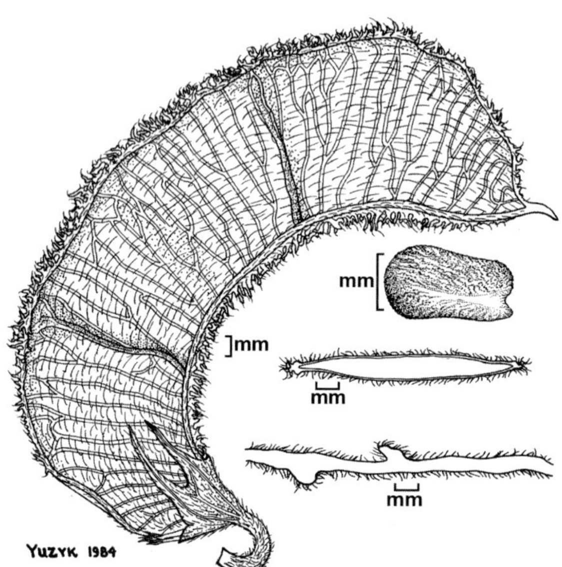

Fig. 67-2. *Medicago plicata.* **Clockwise from top left:** Fruit; seed; outline of cross section of fruit through area lacking folds; outline of portion of longitudinal (sagittal) section of fruit through area with folds. (Based on B. Balansa in 1856, GH; artist: L. Yuzyk.)

Identification Hints

Medicago plicata is the only species of *Medicago* with plicate pods. The short, branched spines on the sutures are reminiscent of *M. radiata*, and it has been suggested that it be grouped with the latter (P. Lassen, personal communication).

Habitat and Environmental Adaptation

Habitat

Medicago plicata has been collected especially from river valleys.

Phenology

In its native area the species flowers in May and June.

Soil

Medicago plicata has been collected from gravelly and rock soils.

Temperature

Like most other annual species of *Medicago*, *M. plicata* is a winter annual, completing its life cycle in a Mediterranean climate before the onset of extreme summer drought and heat.

Water

No information located.

Biotic Relationships

Pollination

The very small flowers indicate that this species is an inbreeder.

Dispersal

The peculiar pods of this species, with their tiny sutural spines, may be adapted to attachment to fur for the purpose of seed distribution.

Herbivory

No information located.

Nodulation and mycorrhizae

No information located.

Disease resistance

No information located.

Economic History, Cultivars, and Uses

This species has not attracted agricultural interest to date.

Cultivars

No cultivars have been bred.

Forage

Medicago plicata is not sown for forage.

Fodder

This species is not employed as fodder.

Green manure, rotation crop, ground cover

No information located.

Other uses (salad sprouts, ornamental, plant products, etc.)

No information located.

Chromosome Number

No information located.

Genetic Relationships and Gene Transferability

Intraspecific

Artificial hybridization of different populations has not been attempted, so it is not known whether there are barriers to hybridization within the species.

Interspecific

No information located.

Exchange by advanced biotechnology

No information located.

Conservation Status

This is a very rare species and, as such, it needs a conservation assessment and collection of its seeds for deposition in gene banks.

Germplasm Sources

Among germplasm institutions, the only accession of *M. plicata* located is in SARDI (see Chapter 3).

Research Problems and Priorities

Medicago plicata is one of the least understood species of the genus. Its botanical and agricultural properties and potential are essentially unknown. Seeds are almost unavailable and their collection is an immediate priority.

Representative Collections

Turkey: Balansa, B., Kamechly-Tchaï au dessous de Béréketly (Cappadoce), collines pierreuses longeant la rive droite, vers 1300 m d'alt., 15-VI-1856, in fruit (G, GH, P). Coode, M.J.E. & Jones, B.M.G., 841B, ICEL, Mut, Mut to Büyük Egri Dag, alt. 1000 m, rodd, *Pinus brutia* forest, 12-V-1965, in flower & fruit (E). Huber-Morath, A. 9871, Kleinasien: Içel (Cilicien), Vilayet Mersin, *Quercus* macchie 1–2 km nordöstl und nördl. Gülnar, alt. 950–980 m, 7-VI-1950, in late flower & fruit (G, HUJ). Huber-Morath, A. 9872, Kleinasien: Cilicien, Vilayet Mersin, Mut – Kirobaşi, Pinus brutia Wald 11 km ob Mut., alt. 800 m, 14-VI-1950, in flower & fruit (G). Huber-Morath, A. 13381, Içel, Distr. Tarsus, Pinetum brutiae bei Gülek Boĝas, alt. 820 m, 29-V-1956, in fruit (G).

68. *Medicago polyceratia* (L.) Trautv.

M. polyceratia (L.) Trautv., Bull. Sci. Acad. Imp. Sci. St. Petersb. 8: 271 (1841). "*M. polyceratia* (L.) Sauvages ex Trautv." according to Hedge and Sales (2001).

The epithet *polyceratia* (frequently misspelled "*polycerata*") is based on the Greek *poly,* meaning many or very, plus *keration*, meaning little horn, an allusion to the many, horn-like pods in the infructescence.

Subordinate Taxa

None recognized here. This species is very variable in peduncle length and degree of pinnatisection of the leaflets (which are often incised or partite), and these characters have led to extensive recognition of infraspecific taxa (for a recent example see Amich and Sanchez 1983).

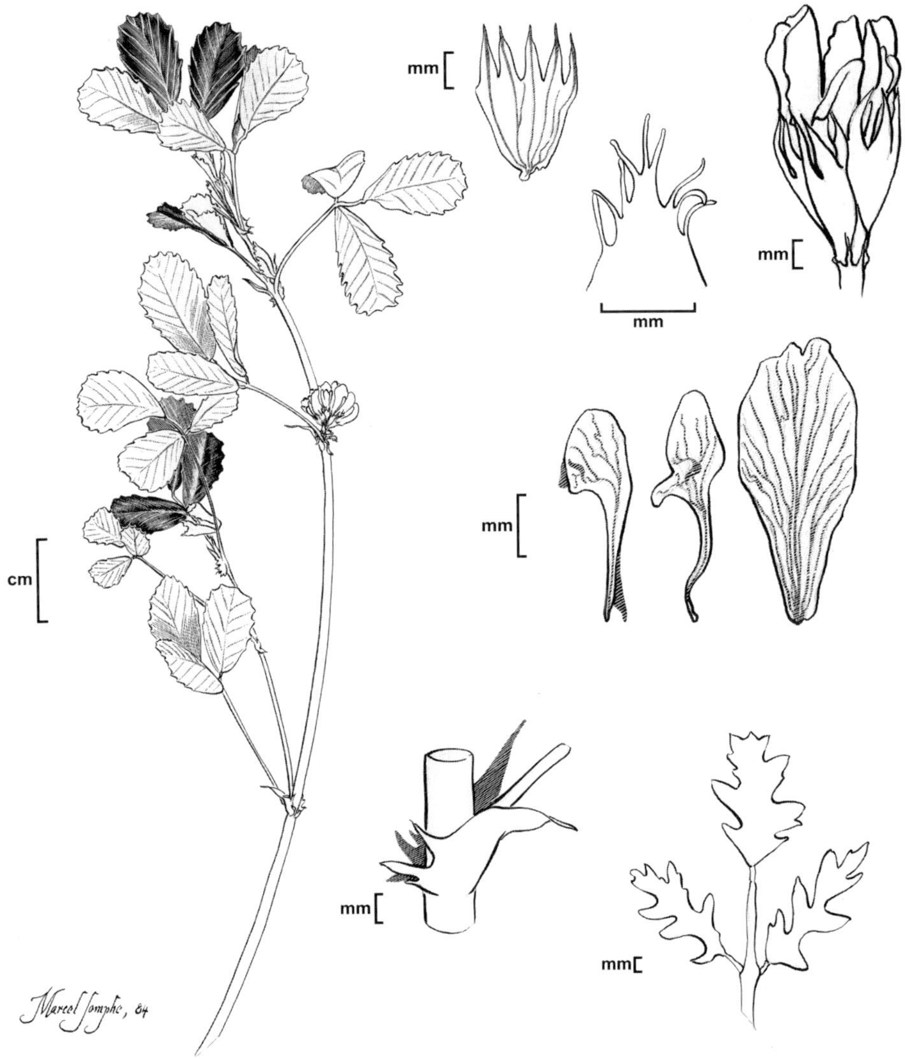

Fig. 68-1. *Medicago polyceratia.* **Left:** Branch with fruits and flowers. **Top right:** Floral parts (clockwise from top left: calyx; flattened androecial column (free stamen not shown); group of flowers; standard petal; wing petal; keel); incised leaf; stem node showing stipules. (Incised leaf based on E. Small T150, DAO; remainder on E. Small T152, DAO; artist: M. Jomphe.)

Type Specimen

The Linnaean Plant Name Typification Project (http://www.nhm.ac.uk/research-curation/projects/linnaean-typification/; current June 2010) provided the following information: "Linnaean Name: *Trigonella polyceratia* Linnaeus. Reference: *Species Plantarum* 2: 777 (1753). Provenance: 'Habitat in Hispania.' Type: Lectotype: Herb. Linn. No. 932.4 (LINN). Designated by: Lassen in Turland & Jarvis (ed.), Taxon 46: 483 (1997). Current Name: *Medicago polyceratia* (L.) Trautv."

Synonymy

Homotypic

Trigonella polyceratia L., Sp. Pl. 777 (1753), (basionym).

Heterotypic

Trigonella amandiana Samp., Ann. Sci. Nat. (Oporto) 6: 143 (1900). (Vassilczenko (1953) stated "Type in Lisbon?")

T. polyceratia L. var. *amandiana* (Samp.) Samp., Manual Fl. Port. 235 (1911).

T. polyceratia L. subsp. *amandiana* (Samp.) Amich & J. Sánchez, Stud. Bot. Univ. Salamanca 2: 129 (1983).

M. polyceratia (L.) Trautv. subsp. *amandiana* (Samp.) Greuter & Burdet, Willdenowia 19: 32 (1989).

T. ×ambigua Samp., Ann. Sci. Nat. (Oporto) 6: 144 (1900).

T. polyceratia L. f. *ambigua* (Samp.) Širj., Generis *Trigonella* revisio critica, Spisy Pfir. Fac. Masarykovy Univ. 4: 14 (1931).

T. polyceratia L. var. *atlantica* Ball., J. Linn. Soc. 305 (1873).

T. atlantica (Ball.) Vassilcz., Tr. Bot. Inst. Akad. Nauk SSSR, Ser. 1, Flora Sist. Vyss. Rast. Fasc. 14, p. 226 (1953).

M. fissa Trautv., Bull. Sci. Acad. Imp. Sci. St. Petersb. 8: 272 (1841).

T. polyceratia L. f. *glabrata* Širj., Generis *Trigonella* revisio critica, Spisy Pfir. Fac. Masarykovy Univ. 4: 14 (1931).

Telis incisa Kuntze, Revis. Gen. Pl. 209 (1891).

T. pinnatifida Cav. var. *intermedia* Amo, Mem. Real Acad. Ci. Exact. Madrid 5: 291 (1861), nom. nud.

T. polyceratia L. var. *longipes* Samp., Manual Fl. Portug. 235 (1911).

T. polyceratia L. var. *major* Lange, Vidensk. Meddel. Dansk Naturhist. Foren. Kjøbenhavn 1865: 367 (1866).

T. polyceratia L. var. *melanosperma* Maire, Bull. Soc. Nat. Hist. Afr. Nord 4:14: 139 (1923). (Type cited in "Representative Collections" below.)

T. polyceratia L. f. *melanosperma* (Maire) Širj., Generis *Trigonella* revisio critica, Spisy Pfir. Fac. Masarykovy Univ. 4: 14 (1931).

T. polyceratia L. var. *minor* Rouy, Bull. Soc. Bot. France 31: 74 (1884).

Buceras mutica Moench, Methodus 142 (1794).

T. pinnatifida Cav., Icon. 1: 26, plate 38 (1791).

T. polyceratia L. var. *pinnatifida* (Cav.) Lange, Vidensk. Meddel. Dansk Naturhist. Foren. Kjøbenhavn 1865: 172 (1866).

T. polyceratia L. f. *pinnatifida*, Willk. & Lge. Prodr. Fl. Hisp. 3: 390 (1880).

T. amandiana Samp. (b) *pinnatifida* Samp., in sched., nom. nud.

T. amandiana Samp. var. *pinnatifolia* Samp, Ann. Sci. Nat. (Oporto) 6: 144 (1900).

T. polyceratia L. subsp. *amandiana* (Samp.) Amich & J. Sánchez var. *pinnatifolia* (Samp.) Amich & J. Sánchez, Stud. Bot. Univ. Salamanca 2: 130 (1983).

Buceras polyceration All., Fl. Pedem. 1: 313 (1785).

T. polyceratoides Lange, Vidensk. Meddel. Dansk Naturhist. Foren. Kjøbenhavn 1865: 172 (1866).

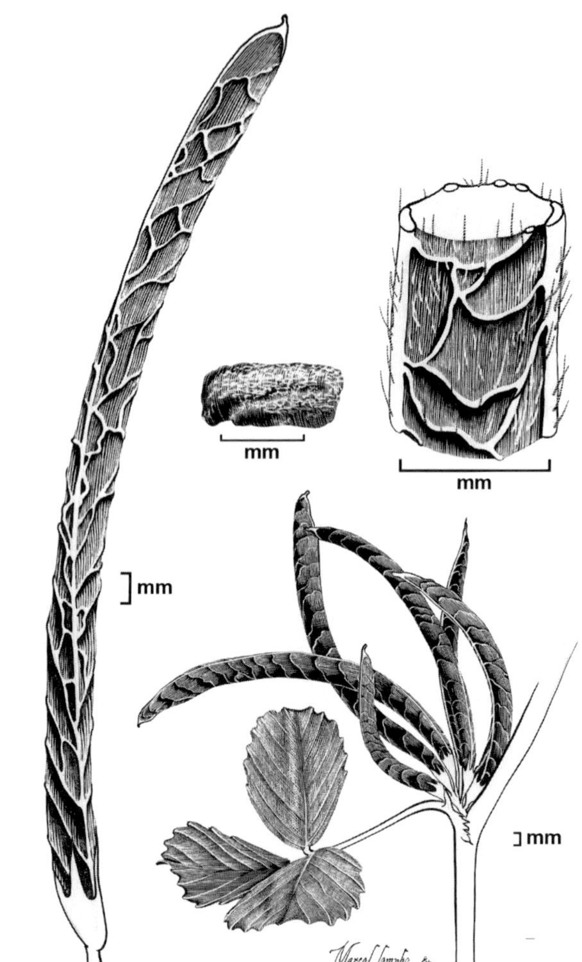

Fig. 68-2. *Medicago polyceratia.* **Clockwise from left:** Fruit; seed; portion of fruit showing venation; group of fruits. (Based on E. Small T152, DAO; artist: M. Jomphe.)

T. polyceratia L. var. *polyceratoides* (Lange) Malag., Sin. Fl. Ibér. 37: 589 (1976).

T. polyceratia L. var. *prostrata* Lange, Vidensk. Meddel. Dansk Naturhist. Foren. Kjøbenhavn 1865: 367 (1866). (Vassilczenko (1953, p. 226) interpreted Lange's taxon as a forma and proposed: *T. polyceratia* L. var. *prostrata* (Lange) Vassilc.)

T. amandiana Samp. var. *serratifolia* Samp., Ann. Sci. Nat. (Oporto) 6: 144 (1900).

T. polyceratia L. subsp. *amandiana* (Samp.) Amich & J. Sánchez var. *serratifolia* (Samp.) Amich & J. Sánchez, Stud. Bot. Univ. Salamanca 2: 130 (1983).

T. polyceratia L. var. *sericea* Emberger (Type: Emberger, L., 10-VII-1934, Plateau des lacs, autour du grand lac, Morocco (MP) ; information viewed online at http://www.aluka.org; current June 2010.)

T. polyceratia L. var. *subpinnatifida* Willk., Ser. Inconf. Pl. Aragon. 28 (1863).

T. polyceratia L. var. *vulgaris* Lange, Vidensk. Meddel. Dansk Naturhist. Foren. Kjøbenhavn 1865: 367 (1866).

Not

T. polycerata (sic) Bak. in Hook., Fl. Br. Ind. 2: 87 (1879) = *M. monantha*.

T. polyceratia Anth., Notes Roy. Bot. Gard. Edinb. 18: 285 (1939) = *M. monantha*.

Vernacular Names

English: No names found.

French: Luzerne à nombreuses cornes, sieures cornes (name used by the classical French botanist A.P. de Candolle) (Trigonelle à nombreuses cornes, when considered to be a species of *Trigonella*).

Geographical Distribution

(N = native, I = introduced, U = uncertain status)

Africa: Algeria (N), Morocco (N), Tunisia (N).

Asia: Iran (U), Saudi Arabia (I).

Europe: France (southern; N; however Širjaev (1928–1934) indicated the French collections are adventive), Ireland (I; rare casual; http://www.biochange.ie/alienplants/result_extinct.php?extinct=449; current July 2010), Portugal (northern; N), Spain (N).

Medicago polyceratia is native to northern Africa and southern Europe. Some collections from India assigned to *M. orthoceras* (which see) appear to be *M. polyceratia*, but I have not identified them as the latter species, as additional study is necessary. Examples of such specimens are cited below in "Representative Collections".

Description

Annual herb, branched from base, stems prostrate to ascending (occasionally erect), (15)25–45(60) cm long. Vegetative parts diffusely pubescent with appressed simple hairs. Leaflets 4–12(17) mm long, (3)5–9(11) mm wide, obovate, elongately obovate, or suborbicular, margin dentate to pinnatipartite. Stipules semi-sagittate, dentate at base. Inflorescence of (1)2–6(8) flowers, subumbellate, peduncle typically very short (less than 3 mm long), but sometimes to 4 cm long. Calyx 2.5–5 mm long, the subequal teeth subulate and pilose, shorter than the tube. Corolla yellow, 3–7(9) mm long. Pods usually in groups, (10)20–50 mm long, 1.4–2.3 mm wide, linear, slightly compressed, slightly arcuate or rarely straight, appressed-pilose, the sutures thickened. Transversely veined, veins anastomosing, venation areoles with length about equal to width. Seeds 1.4–2 mm long, 0.8–2 mm wide, elongately cylindrical or oblong-rhomboidal, pale yellow or brownish, finely tuberculate, radicle about 80% of length of seed.

Identification Hints

Medicago polyceratia is difficult to distinguish from *M. orthoceras*. See Chapter 59 on *M. orthoceras* for information on how best to separate the two species.

Habitat and Environmental Adaptation

Habitat

Medicago polyceratia occurs in fields, forests, and on mountainside slopes. It has been collected from altitudes ranging from 100 to 1600 m.

Phenology

This species flowers from March to May in its native area.

Soil

Medicago polyceratia typically occupies stony soils.

Temperature

No information located.

Water

Medicago polyceratia typically occupies rather dry soils.

Biotic Relationships

Pollination

Small (1988) observed a pollen to ovule ratio of 389, rather high for an annual species of *Medicago* and suggestive of outbreeding. However, the flowers are small, suggesting that *M. polyceratia* is largely self-pollinated and substantially an inbreeder.

Dispersal

Ramos et al. (2006) showed that about a third of *M. polyceratia* seeds consumed by sheep remain viable in the dung, indicating that sheep may be a major distribution agent.

Herbivory

No information located.

Nodulation and mycorrhizae

USDA (2009) reported positive root nodulation for *M. polyceratia*. Jain (2003) provided a detailed description of the anatomical development of nodules in the species.

Disease resistance

No information located.

Economic History, Cultivars, and Uses

Medicago polyceratia has not attracted agricultural development to date.

Cultivars

No cultivars have been bred.

Forage

Medicago polyceratia is not sown for forage. However, Robledo et al. (1991) found evidence that it is a significant wild resource for feeding sheep on fallow land in Spain.

Fodder

This species is not collected for animal feed.

Green manure, rotation crop, ground cover

Hernandez et al. (2001) employed *M. polyceratia* as an experimental self-seeding cover crop in olive orchards to prevent soil erosion and nitrify the soil. Kapoor and Ramakrishnan (1975) found that weedy *M. polyceratia* in wheat fields, except at the very highest density of wheat plants, actually benefited the wheat by increasing nitrogen in the soil.

Other uses (salad sprouts, ornamental, plant products, etc.)

Khan and Ahmad (1993) noted that in Bhopal the seeds of *M. polyceratia* are used medicinally (as a diuretic and carminative). Kamal et al. (1997) suggested that the species is a potential source of rotenoids, a group of natural insecticides. Kamal and Sharma (1981) examined steroidal sapogenin content in *M. polyceratia* seeds, but found these important precursors of sex hormones, corticosteroids, and oral contraceptives were present in relatively limited amounts.

Chromosome Number

$2n = 24$: Sareen and Singh 1976.

$2n = 28$: Coutinho and Santos 1943; Bir and Sidhu 1967; Bolkhovskikh et al. 1969, p. 321 (4 reports); Fernandes et al. 1977; Fernandes and Queirós 1978; Luque and Díaz Lifante 1991.

$2n = 28, 30, 32$: Gardé 1948.

$2n = 44$: Singh and Roy 1970; Agarwal and Gupta 1983*a*.

The numbers reported here are exceptional. They require confirmation and exploration of their significance.

Genetic Relationships and Gene Transferability

Intraspecific

Artificial hybridization of different populations has not been attempted, and therefore the existence of barriers to hybridization within the species is not known.

Interspecific

No information located.

Exchange by advanced biotechnology

No information located.

Conservation Status

Medicago polyceratia is a protected species in the (political) region of Paris, France (http://cbnbp.mnhn.fr/cbnbp/especeAction.do?action=taxo&cdNom=141935; current July 2010), and in the region of Provence-Alpes-Côte d'Azur, France (http://www.indrenature.net/expert/PPACA.html; current July 2010). It is also protected in the province of Catalonia, Spain (http://www.gencat.net/mediamb/eng/lleis/alaw013.htm; current June 2007). It has been listed as a "traded plant of conservation concern" in India (http://envis.frlht.org.in/botanical_search.php?gesp=1645%7CPhyllanthus+amarus+SCHUM.+%26+THENN; current July 2010).

Germplasm Sources

Several dozen accessions are present among the germplasm institutions with large *Medicago* collections (see Chapter 3).

Research Problems and Priorities

Medicago polyceratia is closely related to *M. medicaginoides*, and additional study of the relationships of these two species is required. *Medicago polyceratia* and *M. orthoceras* are even more closely related, and additional study could demonstrate that these are perhaps only geographical phases of the same species, or that they differ basically by ploidy level. *Medicago polyceratia* has received limited study to date. This species is in need of basic botanical and agricultural studies.

Representative Collections

Algeria: Davis 58579, Tiaret to Aflou, between Ain Deheb and Hassiane-Ed-Dih, alt. 1100 m, sandy hills on plateau, *Stipa, Artemisia, Lygeum,* 1-VI-1975, in fruit (E). Davis 58731, Dj. Amour, Aflou to El Ghicha (El Richa), 2 km on piste from Aflou – El Bayadh road, alt. 1400 m, sandy field margins, 4-VI-1975, in fruit (BM, E). Faure, A., Hauts-plateaux oranais, bedeau, pelouses sablonneauses, alt. 1100 m, 15-V-1921, in fruit (BM). **Morocco:** Davis 53830A, 5–10 km southwest of Tefraoute, alt. 1000 m, sandy gravel of dry oued bed, 29-III-1972, in flower & early fruit (E). Davis 53997, Nr. Irherm, alt. 1700 m, rocky east slope, 2-IV-1972, in flower & fruit (E). Davis, P. & J. D.49006, south

Sous plain, 32 km from Taroudannt to Aoulouz, alt. 400 m, sandy plain, protected from grazing, 31-III-1969, in flower & fruit (BM, E). Lindberg, H. 3306, Atlas magnum, in convalle fl. Reraïa, prope pag., Asni, alt. ca. 1200 m, 29-V-1926, in fruit (K). Maire, R. 1927, Zerhoun prope Moulay Idris, in dumetis montis solo calcareo, alt. 700 m, 21-IV-1926 (P). Maire, R., A monts de Djalja, Aïn-Bahrara, forêts de *Pinus halepensis*, 16-5-1921 (MPU) (Information at http://www.aluka.org (current 2010) states that this is the holotype of "*T. polycerata* var. *melanosperma* Maire".). **Saudi Arabia:** Mandaville, J.P. Jr. 8086, Eastern Province, northern Bath Al-Faruq (25°43′N, 48°53′E), sandy flat, former Bedouin camp site, many Mediterranean weeds here, probably from spilled grain, 1-IV-1983, in late flower & fruit (K). **Spain:** Borja, J., Campos de la Moncloa (Madrid), V-1960, in fruit (BM). Bourgeau 2130, Cerro de Alcala de Heneares, près Madrid, 30-V-1854, (E, P). Bourgeau, E. 625, Cartagena, champs, 28-IV-1849, in fruit (K). Elisée Reverchon 783, Prov. de Teruel, Camarena, alt. 1500 m, lieux arides et sablonneux, sur le calcaire, V-1892, in flower & fruit (E). Goday, R. & Pérez, C., Madrid, campos de los alrededores de la ciudad, Lugarclásico, IV-1947, in flower & fruit, Flora Hispánica Herbaria Normal Centuria III no. 238 (K). Graells, M.P., environs de Madrid, dans les champs incultes, V-1855, in flower & fruit (K, P). Porta & Rigo 190, Albacate, prope Almansa et Chinchilla, in percuis et arvis incultis, alt. 4–5000 m, 7-16-VI-1891, in fruit (K). Reverchon, E., Prov. de Jaen, Barrancon, Vallentina, lieux arrides, sur le calcaire, 1600 m, rare, VI-1904 (P). Reverchon, E., Prov. de Tervel, Camarena, lieux arides et sablonneux, sur le calcaire, alt. 1500 m, VI-1892 (P). Sennen, F. 55, Castille, Valladoliel, champs caillouteux, 23-VI-1906, in fruit (BM). **Tunisia:** Davis & Lamond D57200, north of Sbeitla, 16 km from Sbiba, alt. 730 m, fallow fields on sandy soil, 4-V-1975, in flower & fruit (E). Davis & Lamond D57463, north slope of Djebel Chambi, west of Kasserine, alt. 750–800 m, open sandy slopes, 9-V-1975, in flower & fruit (E).

Representative Collections from India that are confusingly like *M. polyceratia*, although assigned to *M. orthoceras*: Drummond, J.R. 13780, Punjab, in arvis etc. Gujrat, alt. ca. 800 feet, in Mart. 1899, in flower & fruit (E). Drummond, J.R. 4375, Punjab, Samrala in dit Ludhiana, III-1885, in flower & fruit (E). Schlagintweit, herbarium of, northwest India, prov. Panjab, from the Bari Duab in a direction of 60 miles northeast of Lahor, 18-23-III-1857, in flower & fruit, catalogue no. 10653 (P). Thomson, Punjab, Herb. Ind. Or. Hook fil. & Thomson (P). Voyage de V. Jacquemont aux Indes orient, sheets bearing numbers 33, 54, 520 (P).

69. *Medicago polymorpha* L.

M. polymorpha L., Sp. Pl. 779 (1753), (basionym).

The epithet *polymorpha* is based on the Greek *polys*, which means many, plus *morphe*, which means form or shape, an allusion to the wide range of appearance of plants of this species.

Subordinate Taxa

None recognized here.

Type Specimen

Heyn (1959: 163) lectotypified *M. polymorpha* by the specimen Hort. Sicc. Cliff. 378-11 δ (Herb. Linn). The Linnaean Plant Name Typification Project (http://www.nhm.ac.uk/research-curation/projects/linnaean-typification/; current June 2010; cf. Jarvis 2007) provided the following information (along with a photograph of the type): "Linnaean Name: *Medicago polymorpha* Linnaeus var. *polymorpha* Reference:

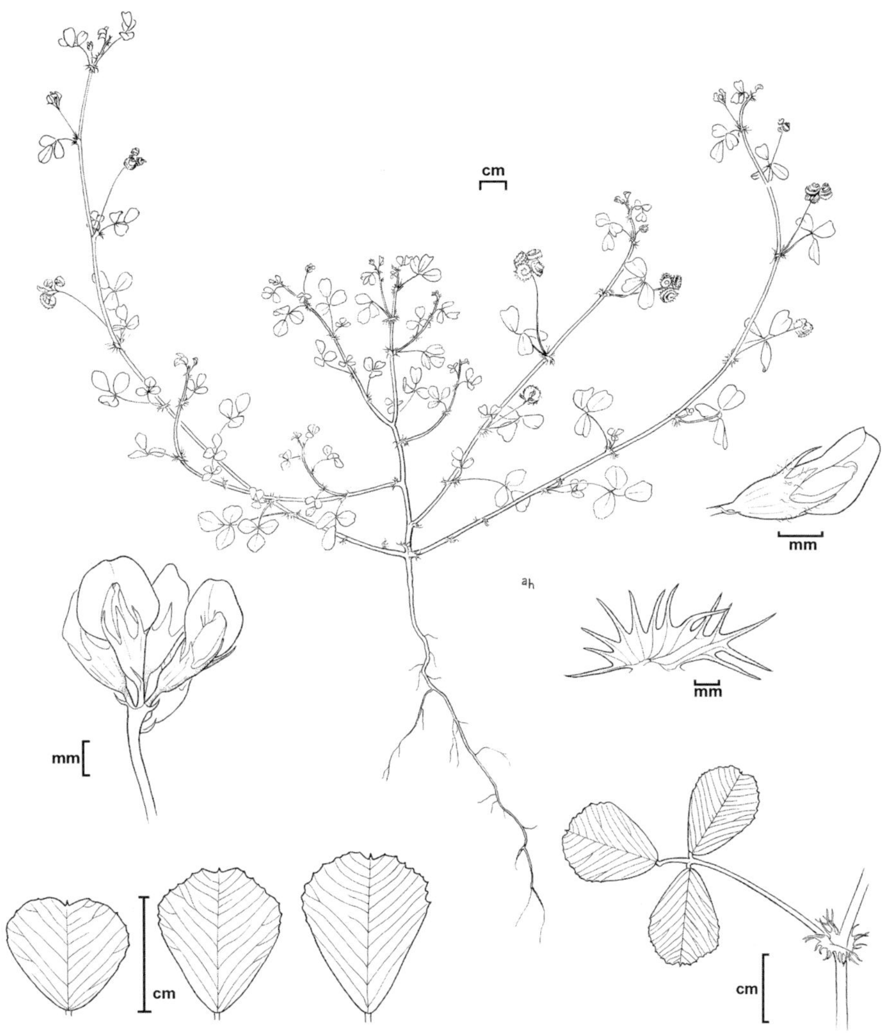

Fig. 69-1. *Medicago polymorpha*. **Clockwise from top:** Plant; side view of flower; stipule; leaf attached to stem; leaflets; group of flowers. (Plant based on Eig, collection 99/177, HUJ; remainder based on E. Small M978, DAO; artist: A. Hanes.)

Species Plantarum 2: 779, 1200 (1753) Provenance: 'Habitat (p. 1200:) Europa australi.' Type: Lectotype: Herb. Clifford: 378, *Medicago* 11 (delta) (BM-000646786) Designated by: Heyn in Bull. Res. Council Israel, Sect. D, Bot. 7: 163 (1959).... Comments: Shinners (in Rhodora 58: 5–7. 1956) provided an extensive discussion, concluding '(the name) is here typified by the element later named var. *nigra*', but he cited no specimen or illustration to serve as the type. Heyn (1959) subsequently designated Clifford material as the type. However, Greuter (in Boissiera 13: 80. 1967) argued that an implicit typification by Krocker in 1790 meant that the name must be regarded as a nomen ambiguum. Lambinon (in Taxon 30: 363. 1981) subsequently proposed the rejection of the name but this was not approved by the Committee for Spermatophyta (in Taxon 35: 560. 1986)."

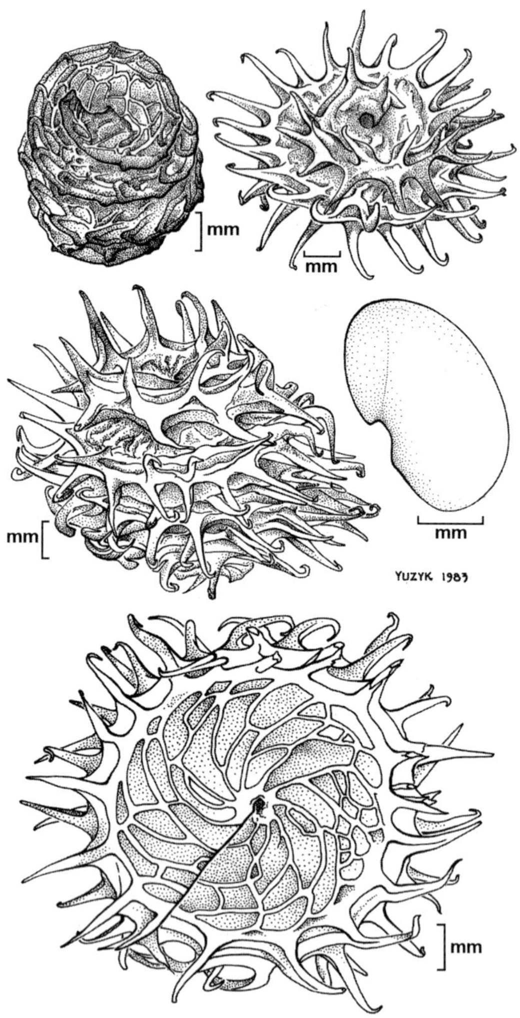

Fig. 69-2. *Medicago polymorpha.* **Clockwise from top left:** Tuberculate fruit (based on Heyn, L.S. Smith, S.D. Everest 802; GH); spiny fruit (based on Eig, Zohary, Feinbrun 5-4-1927, HUJ); seed and fruit coil face showing venation (based on Eig, Zohary, Feinbrun 5-4-1927, HUJ); another spiny fruit (based on R. Alaya 6662, HUJ). (Artist: L. Yuzyk.)

Synonymy

Homotypic

Medica polymorpha (L.) Scop., Fl. Carniol., ed. 2, 2: 89 (1772).

M. hispida Gaertn. subsp. *polymorpha* (L.) Rouy, Fl. Fr. 5: 36 (1899); ead. comb. Sudre, Florule Toulousaine 59 (1907), comb. illeg.

M. polycarpa Willd. ex Schlecht. subsp. *polymorpha* (L.) Cadevall & Sallent, Fl. Catalunya, 2: 85 (1913), comb. illeg.

Heterotypic

M. hispida Gaertn. subsp. *microcarpa* Urb. var. *polygyra* Urb. f. *aculeata* Urb., App. Observ. Bot. Berol. 4 (1872). (Note that Heyn (1963) assigned *M. aculeata* Gaertn. to "*M. aculeata* Willd." (i.e., *M. doliata*), a treatment followed here, although others have assigned it to *M. polymorpha*.)

M. hispida Gaertn. subsp. *polymorpha* Willd. (race) *polygyra* (Urb.) Rouy var. *aculeata* Rouy in Rouy & Fouc., Fl. France 5: 37 (1899).

M. apiculata Willd., Sp. Pl. 3(2): 1414 (1802).

M. hispida Gaertn. subsp. *microcarpa* Urb. var. *oligogyra* Urb. f. *apiculata* (Willd.) Urb., App. Sem. Hort. Berol. 3 (1872).

Medica apiculata (Willd.) E.L. Greene, Man. Bot. San Francisco Bay 102 (1894).

M. hispida Gaertn. subsp. *apiculata* (Willd.) Corb., Nouv. Fl. Normandie 151 (1894).

Fig. 69-3. *Medicago polymorpha.* From Hallier (1885, vol. 23, plate 2350).

M. hispida Gaertn. var. *apiculata* (Willd.) Burnat, Fl. Alpes Marit. 2: 106 (1896).

M. hispida Gaertn. subsp. *polymorpha* Willd. var. *apiculata* (Willd.) Rouy in Rouy & Fouc., Fl. France 5: 36 (1899).

M. denticulata Willd. var. *apiculata* (Willd.) Posp., Fl. Oest. Kuestenl. 2: 362 (1899).

M. polymorpha L. f. *apiculata* (Willd.) Shinners, Rhodora 58: 11 (1956).

M. polymorpha L. var. *apiculata* (Willd.) Ooststr. & Reichg., Acta Bot. Neerl. 7: 115 (1958).

M. nigra (L.) Krocker var. *apiculata* (Willd.) O. Bolòs & Vigo, Buttl. Inst. Catal. Hist. Nat., Sect. Bot. 1, 38: 70 (1974).

M. denticulata Willd. var. *apiculata* Moris in Moris & De Not., Fl. Caprar. 35 (1839).

M. polycarpa Willd. var. *apiculata* Godr. in Gren. & Godr., Fl. France 1: 390 (1848).

M. denticulata Willd. var. *apiculata* Boiss., Fl. Or. 2: 103 (1872).

M. berteroana Moris, Ann. Storia Nat. 4: 59 (1830).

Medica bimuricata Moench, Methodus 117 (1794).

M. lappacea Desr. var. *brachyacantha* Lowe, Man. Fl. Mad. 158 (1868).

M. hispida Gaertn. subsp. *macrocarpa* Urb. var. *pentagyra* Urb. f. *breviaculeata* Urb., App. Ind. Sem. Hort. Bot. Berol. 4 (1872).

M. hispida Gaertn. subsp. *macrocarpa* Urb. var. *pentacycla* (DC.) Urb. f. *breviaculeata* Urb., Verh. Bot. Ver. Brandenb. 15: 75 (1873).

M. denticulata Willd. var. *brevispina* Benth., Cat. Pl. Pyr. Bas-Languedoc 103 (1826).

M. pentacycla DC. var. *brevispina* Benth., Cat. Pl. Pyr. Bas-Languedoc 104 (1826).

M. polymorpha L. var. *brevispina* (Benth.) Heyn, Scripta Hierosolymitana 12: 77 (1963).

M. hispida Gaertn. var. *macracantha* (Webb & Berth.) Briq. subvar. *brevispina* Nègre, Luzern. Maroc. 69 (1956).

M. caspia Grunes ex Boiss., Fl. Orient. Suppl. 167 (1888).

M. caspica Jacq. ex Spreng., Nov. Prov. Hort. Berol. 27 (1818).

M. denticulata Willd. var. *caspica* (Jacq.) Grossh. in Schischkin (ed.), Fl. USSR 11: 171 (1945).

M. catalonica hort ex Urb., Ind. Sem. Hort. Berol. 4 (1872), name merely listed in synonymy.

M. ciliaris Savi, Due Cent. Fl. Etrusc. 148 (1804), non *M. ciliaris* (L.) All. (A specimen collected in 1808 in Herb. DC, bearing the information *Medicago ciliaris* Savi, could be taken as the type of the name, were it necessary.)

M. denticulata Willd. var. *ciliaris* Savi ex Ser. in DC., Prodr. 2: 176 (1825), pro syn.

M. polymorpha L. var. *ciliaris* (Ser.) Shinners, Rhodora 58: 9 (1956).

M. apiculata Willd. var. *confinis* W.D.J. Koch, Syn. Fl. Germ. Helv. 1: 164 (1836). ("*M. confinis* Koch" presented merely as a synonym.)

M. hispida Gaertn. var. *confinis* (W.D.J. Koch) Burnat, Fl. Alpes Marit. 2: 106 (1896).

M. polymorpha L. var. *confinis* (W.D.J. Koch) N. Diklic, Flora Socialist Republic Serbia 10 (Suppl. 2): 115 (1986).

M. confinis (Koch) J.M. Black, Flora South Australia 2 (1924).

M. denticulata Willd. var. *confinis* Batt. in Batt. & Trab., Fl. Algér. 2: 232 (1888).

M. conica Schkuhr, Handb. 2: 420 (1808).

M. cyrenaica Maire & Weiller, Bull. Soc. Hist. Nat. Afr. Nord. 30: 273 (1939) (isotype P! holotype MP; types online at http://www.aluka.org; current June 2010.)

M. denticulata Willd., Sp. Pl. 3: 1414 (1802). (A folder at Herb. Willd., B, had four collections, of which one bearing the label "Photo-Nr. 885" and "IDC Cat. Nr. 14327" appears suitable as a type.)

M. hispida Gaertn. var. *denticulata* (Willd.) Godron, Fl. Lorr. 1: 393 (1849).

M. polycarpa Willd. ex Schlecht. var. *denticulata* (Willd.) Godr. in Gren. & Godron, Fl. Fr., 1: 390 (1849), comb. illeg.

M. denticulata (Willd.) Boiss., Fl. Orient. 2: 102 (1872).

M. hispida Gaertn. f. *denticulata* (Willd.) Urb., App. Ind. Sem. Hort. Bot. Berol. 3 (1872).

M. hispida Gaert. subsp. *microcarpa* Urb. var. *oligogyra* Urb. f. *denticulata* (Willd.) Urb., Verh. Bot. Ver. Prov. Brand. 15: 74 (1873).

M. lappacea Desr. in Lam. subsp. *denticulata* (Willd.) Nyman, Consp. Fl. Eur. 169 (1878).

Medica denticulata (Willd.) Greene, Man. Bot. San Francisco 102 (1894).

M. polymorpha L. var. *denticulata* (Willd.) Kerguélen, Ind. Syn. Fl. Fr. XIV (1993).

M. nigra (L.) Krocker var. *denticulata* (Willd.) O. Bolòs & Vigo, Buttl. Inst. Catal. Hist. Nat., Sect. Bot. 1, 38: 70 (1974).

M. distans Poir. in Lam., Encycl. Méthod. Suppl. 3: 526 (1813).

M. denticulata Willd. var. *genuina* Boiss., Fl. Or. 2: 102 (1872), nom. inadmiss. (Boissier, E., Arabia Petraea, III-46 (G. Boiss.!) would be a suitable choice for a type specimen.)

M. gracillima Tineo ex. Urb., App. Ind. Sem. Hort. Berol. 3 (1872), pro syn. (Based on Tineo exsic. 850.)

M. hispida Gaeartn. subsp. *polymorpha* (Willd.) Rouy var. *gracillima* (Tin.) Rouy in Rouy & Fouc., Fl. France 5: 37 (1899).

M. hamata Presl, Delic. Prag. 45 (1822).

M. heterocarpa hort ex Urb., Ind. Sem. Hort. Berol. 3 (1872) (merely a listed name).

M. hispida Gaertn., De Fruct. Sem. Pl. 2: 349 (1791), nom. illeg.

M. polymorpha L. subsp. *hispida* (Gaertn.) Ponert, Feddes Repert., 83(9-10): 639 (1973).

M. hispida Gaertn. subsp. *microcapa* Urb. var. *polygyra* Urb. f. *inermis* Urb., App. Ind. Sem. Hort. Bot. Berol. 4 (1872).

M. hispida Gaertn. subsp. *polymorpha* Willd. (race) *polygyra* var. *inermis* Rouy in Rouy & Fouc., Fl. France 5: 37 (1899).

M. intermedia DC. ex G. Don, Gen. Hist. 2: 169 (1832). (A later homonym.)

M. hystrix Ten., Prod. Fl. Nap. 45 (1811). (Heyn (1963) commented that although considered to be a synonym of *M. polymorpha* by Urban "the description does not seem to describe any variety of *M. polymorpha*.")

M. lappacea Desr. in Lamarck, Encycl. Méthod. 3: 637 (1792), nom. illegit.

M. denticulata Willd. var. *lappacea* (Desr.) Benth. Cat. Pl. Pyr. Bas-Languedoc 103 (1826), nom. illegit.

M. hispida Gaertn. var. *lappacea* (Desr.) Burnat, Fl. Alp. Marit. 2: 107 (1892).

M. hispida Gaertn. subsp. *macrocarpa* Urb. 'race' *lappacea* (Desr.) Urb. in Aschers. & Graebner, Syn. Mitteleur. Fl. 6, 2: 432 (1907).

M. hispida Gaertn. subsp. *macrocarpa* Urb. var. *pentagyra* Urb. f. *longeaculeata* Urb., App. Ind. Sem. Hort. Bot. Berol. 4 (1872).

M. hispida Gaertn. subsp. *macrocarpa* Urb. var. *pentacycla* (DC.) Urb. f. *longeaculeata* Urb., Verh. Bot. Ver. Brandenb. 15: 75 (1873).

M. pentacycla DC. var. *longeaculeata* Rouy in Rouy & Fouc., Fl. France 5: 38 (1899).

M. pentacycla DC. var. *longiseta* Benth., Cat. Pl. Pyr. Bas-Languedoc 104 (1826).

M. hispida Gaertn. subsp. *macrocarpa* Urb. var. *trigyra* Urb. f. *longispina* Urb., App. Ind. Sem. Hort. Bot. Berol. 4 (1872).

M. hispida Gaertn. subsp. *macrocarpa* Urb. var. *tricycla* (Godr.) Urb. f. *longispina* Urb., Verh. Bot. Ver. Brandenb. 15: 75 (1873).

M. loreti Albert, Bull. Soc. Et. France 1 (1892). (= *M. loretii* Albert, Bull. Herb. Boissier 1, App. I: 13 (1893).)

M. denticulata Willd. var. *macracantha* Webb & Berth., Hist. Nat. Iles Canaries 3(2(2)): 64 (1842).

M. lappacea Desr. var. *macracantha* (Webb & Berth.) Lowe, Man. Fl. Madeira 1: 158 (1862).

M. hispida Gaertn. var. *macracantha* (Webb & Berth.) Briq., Prodr. Fl. Cors. 2, 1: 271 (1913).

M. denticulata Willd. var. *macrocarpa* Moris & De Not., Fl. Caprar. 35 (1839).

M. hispida Gaertn. subsp. *macrocarpa* Urb., App. Ind. Sem. Hort. Bot. Berol. 4 (1872).

M. denticulata Willd. f. *maculata* Maire (Isotype! at P, viewed online at http://www.aluka.org; current June 2010; holotype (2 herbarium sheets so labelled) at MPU, also viewed online at http://www.aluka.org; current June 2010)). (Status?)

M. hispida Gaertn. var. *lappacea* (Desr.) Burnat f. *maculifolia* (type? at P, information (of undetermined status) found online at http://www.aluka.org; current June 2010).

M. hispida Gaertn. subsp. *microcarpa* Urb., App. Ind. Sem. Hort. Bot. Berol. 3 (1872).

M. nigra (L.) Krocker subsp. *microcarpa* (Urb.) O. Bolòs & Vigo, Buttl. Inst. Catal. Hist. Nat., Sect. Bot. 1, 38: 69 (1974).

M. microdon Ehrenb., Cat. Sem. Hort. Berol. (1827) (characterized by Heyn (1963) as "nomen nudum?").

M. obscura Visiani var. *microdon* (Ehrenb.) Vis., Fl. Dalm. 3: 283 (1852).

M. hispida Gaertn. subsp. *macrocarpa* Urb. var. *trigyra* Urb. f. *microdon* (Ehrenb.) Urb., App. Ind. Sem. Hort. Bot. Berol. 4 (1872).

M. hispida Gaertn. subsp. *macrocarpa* Urb. var. *tricycla* (Godr.) Urb. f. *microdon* (Ehrenb.) Urb., Verh. Bot. Ver. Brandenb. 15: 75 (1873).

M. lappacea Desr. var. *microdon* Batt. in Batt. & Trab., Fl. Algér. 2: 231 (1888).

M. hispida Gaertn. var. *microdon* (Batt.) Nègre, Luzern. Maroc. 69 (1956).

M. mitra Willd. ex Urb., App. Ind. Sem. Hort. Bot. Berol. 4 (1872), pro syn.

Medica muricata (Willd.) Bubani, Fl. Pyren., 2: 479 (1900).

M. neurocarpos Kunze (Syn. Pl. Amer. Austr. Msc. Diar. 224,) ex. Urb., App. Ind. Sem. Hort. Berol. 3 (1872), pro syn.

M. obscura Vis., Fl. Dalmat. 3: 283 (1852), non Retz. (1779).

M. polymorpha L. var. *nigra* L., Mant. Pl. 454 (1771). (Lectotype: Morison, Pl. Hist. Univ. 2(2): 154, s. 2, t. 15, f. 13 (1680); lectotypified by Lassen in Turland and Jarvis (1997). Epitype: France. Var, dans les champs argilo-calcaires ... à Solliès-Toucas, 10 Juin 1892, A. Albert s.n. (Magnier Fl. Sel. Exsicc. 2961) (LD). The Linnaean Plant Name Typification Project (http://www.nhm.ac.uk/research-curation/projects/linnaean-typification/; current June 2010; cf. Jarvis 2007) provides the following information: "Comments: The lectotype is Morison's f. 13, not f. 19 as cited by Linnaeus or f. 18 as given by Morison (cf. Heyn (in Bull. Res. Council Israel, Sect. D, Bot. 7: 168, 173. 1959).")

M. nigra (L.) Krock., Fl. Siles. 2(2): 244 (1790).

M. hispida Gaertn. var. *nigra* (L.) Burnat, Fl. Alp. Marit. 2: 108 (1896).

M. obscura Visiani, Sp. Pl. 3: 1418 (1802). (At Herb. Willd. in B the folder for this had only one collection, bearing Bouché. W. in the lower right hand corner and an affixed label Photo-Nr. 899, as well as a loose label IDC microfiche Cat. Nr. 14331.)

M. oligocarpa Corb. Nouv. Fl. Normandie150 (1893). (Status unclear according to Heyn (1963); listed as a hybrid in The International Plant Names Index, viewed online at http://www.ipni.org/; current June 2010.)

M. hispida Gaertn. subsp. *microcarpa* Urb., var. *oligogyra* Urb., App. Ind. Sem. Hort. Bot. Berol. 3 (1872).

M. hispida Gaertn. subsp. *oligocarpa* Corb., Nouv. Fl. Normandie 150 (1893).

M. hispida Gaertn. subsp. *polymorpha* (Willd.) Rouy var. *oligocarpa* (Corb.) Rouy in Rouy & Fouc., Fl. Fr. 5: 36 (1899).

M. lappacea Lam. var. *paucigyrosa* Guss., Fl. Sic. Syn. 2: 366 (1844).

M. pentacycla DC., Cat. Pl. Horti Monsp. 124 (1813). (Herb. DC. in P has one sheet with several collections; the specimen labelled 18 June 1807 can be taken as the type.)

M. lappacea Desr. var. *pentacycla* (DC.) Godr. in Gren. & Godr., Fl. France 1: 390 (1847).

M. denticulata Willd. var. *pentacycla* (DC.) Boiss., Fl. Or. 2: 102 (1872).

M. hispida Gaertn. subsp. *macrocarpa* Urb. var. *pentacycla* (DC.) Urb., Verh. Bot. Vereins Prov. Brandenburg 15: 75 (1873).

M. hispida Gaertn. subsp. *pentacycla* (DC.) Arcangeli, Comp. Fl. Ital. 164 (1882).

M. hispida Gaertn. proles *pentacycla* (DC.) Rouy, Fl. Fr. 5: 38 (1899).

Medica pentacycla (DC.) Bubani, Fl. Pyren., 2: 475 (1900).

M. nigra (L.) Krocker var. *pentacycla* (DC.) O. Bolòs & Vigo, Buttl. Inst. Catal. Hist. Nat., Sect. Bot. 1, 38: 70 (1974).

M. hispida Gaertn. subsp. *macrocarpa* Urb. var. *pentagyra* Urb., App. Ind. Sem. Hort. Bot. Berol. 4 (1872).

M. persica Stev. ex. Fisch., Cat. Hort. Gorenk. ed 2, 69 (1812), nom. nud., ex. Ledeb. Fl. Ross. 1: 530 (1842), non *M. persica* (Boiss.) E. Small.

M. polyantha Dulac, Fl. Hautes-Pyr. 278 (1867), nom. illeg.

M. polycarpa Willd., Enum. Pl. Horti Berol. Suppl. 52 ("1813" =1814), nom. nud., ex Benth. Cat. Pl. Pyr. Bas-Languedoc 103 (1826), pro syn. (*M. polycarpa* Willd. ex Schlecht) (A folder of *M. polycarpa* in Herb. Willd. at B had 1 collection with an affixed label "Photo-Nr. 897" and a loose label "IDC Cat. Nr. 14339.")

M. polycarpa Willd. ex Godr. in Gren. & Godr., Fl. France 1: 389 (1848).

M. hispida Gaertn. subsp. *polycarpa* (Willd.) Bonnier & Layens, Fl. Fr. 75 (1894).

M. polimorpha (sic) subsp. *polycarpa* (Willd.) Romero Zarco, Lagascalia 14(1): 146 (1986).

M. hispida Gaertn. var. *polygyra* Urb., App. Ind. Sem. Hort. Bot. Berol. 4 (1872).

M. hispida Gaertn. subsp. *polygyra* (Urb.) Arcangeli, Comp. Fl. Ital. 164 (1882).

M. hispida Gaertn. proles *polygyra* (Urb.) Rouy, Fl. Fr. 5: 37 (1899).

M. hispida Gaertn. var. *polygyra* (Urb.) P. Fourn., Quatre Fl. Fr. 546 (1936).

M. nigra (L.) Krocker var. *polygyra* (Urb.) O. Bolòs & Vigo, Buttl. Inst. Catal. Hist. Nat., Sect. Bot. 1, 38: 70 (1974).

M. polymorpha L. var. *polygyra* (Urb.) Shinners, Rhodora 58: 9 (1956).

M. polyphorma Larrañaga, Escritos de Don Dámaso Antonio Larrañaga 2: 233 (1923).

M. reticulata Benth., Cat. Pl. Pyr. Bas-Languedoc 101 (1826).

Medica reticulata (Benth.) Bubani, Fl. Pyren. 2: 480 (1900).

M. polycarpa Willd. ex Schlecht. subsp. *reticulata* (Benth.) Coste, Fl. Fr. 3: 719 (1906).

M. hispida Gaertn. var. *reticulata* (Benth.) Urb. in Asch. & Graebn., Syn. Mitteleur. Fl. 6(2): 431 (1907).

M. reynieri Albert, Bull. Herb. Boissier 1, Append. 1: 14 (1893).

M. hispida Gaertn. subsp. *polymorpha* (Willd.) Rouy var. *reynieri* (Alb.) Rouy in Rouy & Fouc., Fl. France 5: 37 (1899).

M. sardoa Moris ex G. Don, Gen. Hist. 2: 170 (1832) (*M. sardoa* Moris, Stirp. Sard. Elench 1: 15 (1827)).

M. hispida Gaertn. f. *sardoa* (Moris ex G. Don) Asch. & Graebn., Syn. Mitteleur. Fl. 6(2): 433 (1907).

M. hispida Gaertn. f. *serratifolia* Blakelock, Kew Bull. 3: 416 (1949).

M. serratifolia Rawi, Dep. Agr. Iraq Tech. Bull. 14: 3 (1965). nom. illegit.

M. stricta Wallr. ex Urb., App. Ind. Sem. Hort. Berol. 3 (1872), pro syn.

M. terebella St.-Lag., Ann. Soc. Bot. Lyon 7: 130 (1880).

M. terebellum Willd., Sp. Pl. 3(2): 1416 (1802), nom. illeg. (At Herb. Willd. in B there are two specimens.)

M. hispida Gaertn. (race) *terebellum* Urb. in Aschers. & Graebn. Syn. Mitteleur. Fl. 6, 2: 432 (1907).

M. lappacea Desr. var. *tricycla* Godr. in Grenier & Godron, Fl. France 1: 390 (1849).

M. hispida Gaertn. var. *tricycla* (Godr.) Urb., App. Observ. Bot. 4 (1872).

M. hispida Gaertn. subsp. *macrocarpa* Urb. var. *tricycla* (Godr.) Urb., Verh. Bot. Ver. Brandenb. 15: 75 (1873).

M. polymorpha L. var. *tricycla* (Godr.) Shinners, Rhodora 58: 8 (1956).

M. nigra (L.) Krocker var. *tricycla* (Godr.) O. Bolòs & Vigo, Buttl. Inst. Catal. Hist. Nat., Sect. Bot. 1, 38: 70 (1974).

M. hispida Gaertn. subsp. *macrocarpa* Urb. var. *trigyra* Urb., App. Ind. Sem. Hort. Bot. Berol. 4 (1872).

M. polycarpa Willd. var. *tuberculata* Godr. in Gren. & Godron, Fl. France 1: 389 (1848).

M. lappacea Desr. var. *tuberculata* Godr., in Gren. & Godron, Fl. France 1: 390 (1848), nom. illegit.

M. hispida Gaertn. subsp. *microcarpa* Urb. var. *oligogyra* Urb. f. *tuberculata* (Godr.) Urb., App. Ind. Sem. Hort. Bot. Berol. 3 (1872).

M. hispida Gaertn. subsp. *polymorpha* Willd. var. *tuberculata* (Godr.) Rouy in Rouy & Fouc., Fl. France 5: 36 (1899).

M. polymorpha L. var. *vulgaris* (Benth.) Shin. f. *tuberculata* (Godr.) Shin., Rhodora 58: 11 (1956).

M. denticulata Willd. var. *tuberculata* Moris, Fl. Sard. 1: 164 (1837).

M. ttuberculata Urb., Ind. Sem. Hort. Berol. 4 (1872), name merely listed in synonymy.

M. denticulata Willd. var. *typica* Pos., Fl. Oest. Kuestenl. 2: 362 (1899), nom. inadmiss.

Medicago uncinata Urb., Ind. Sem. Hort. Berol. 4 (1872), name merely listed in synonymy.

M. uncinata Nym., Consp. 169 (1878), pro syn.

M. denticulata Willd. var. *vulgaris* Benth., Cat. Pl. Pyr. Bas-Languedoc 103 (1826), nom. inadmiss., nom. invalid.

M. polymorpha L. var. *vulgaris* (Benth.) Shinners, Rhodora 58: 310 (1956), nom. invalid.

M. pentacycla DC. var. *vulgaris* Benth., Cat. Pl. Pyr. Bas-Languedoc 104 (1826), nom. inadmiss.

Not

M. flexuosa Ten., Fl. Nap., 1: Prodr. p. xiv (1811–1815). This name was assigned to the synonymy of *M. polymorpha* by Urban (1873) and Bentham (1826). However, Heyn (1963: 73) noted that Tenore's diagnosis is consistent with a different species, and a specimen she observed labelled as *M. flexuosa* was in fact *M. littoralis*.

Vernacular Names

English: Burr medic, bur clover (burrclover, burclover), bur trefoil, California bur clover, denticulate medic, fimbriate medic, southern bur clover, toothed bur clover, toothed medic. (Bur is also spelled burr, medic is also spelled medick.)

French: La luzerne hérissée, luzerne polymorphe.

Geographical Distribution

(N = native, I = introduced, U = uncertain status)

Africa: Algeria (N), Azores (Portugal; N), Canary Islands (Spain; N), Djibouti (N), Egypt (N), Ethiopia (U), Kenya (U), Libya (N), Madeira Islands (Portugal; N), Mauritius (I), Morocco (N), Reunion (I), Socotra (N), South Africa (I), Tanzania (U) Tunisia (N).

Asia: Afghanistan (N), Azerbaijan (N), China (N), Cyprus (N), Gruziya (N), India (U), Iran (N), Iraq (N), Israel (N), Japan (I), Jordan (N), Kazakhstan (N), Kuwait (U), Kyrgystan (N), Lebanon (N), Nepal (U), Pakistan (U), Philippines (I), Russia (N), Saudi Arabia (N), Syria (N), Taiwan (I), Tadzhikistan (N), Tajikistan (N), Turkey (N), Turkmenistan (N), Uzbekistan (N), Vietnam (I), Yemen (U).

Australasia: Australia (I), Chatham Is. (I), Kermadec Is. (I), New Zealand (I).

Europe: Albania (N), Austria (N), Belarus (I), Belgium (I), Bulgaria (N), Czech Republic (I), England (U), Estonia (I), France (including Corsica; N), Germany (N), Greece (including Crete, East Aegean Islands; N), Hungary (I), Ireland (U), Italy (including Sardinia, Sicily; N), Latvia (I), Malta (N), Netherlands (I), Portugal (N), Romania (N), Russia (I), Slovakia (I), Spain (including Baleares; N), Switzerland (I), Turkey (N), Ukraine (N), Yugoslavia (U).

Northern America: Canada (I), Mexico (I), USA (I).

Pacific Ocean: Hawaii (I), Marianas (I).

Southern America: Argentina (I), Bolivia (I), Brazil (I), Chile (I), Costa Rica (I), Dominican Republic (I), Ecuador (I), Peru (I), Uruguay (I), Venezuela (I).

Medicago polymorpha is indigenous to much of Eurasia and Africa, and widely introduced and adventive elsewhere where it is not limited by cold, drought, and waterlogging of soil.

Description

Annual herb, stems 15–40(70) cm long, branched from base, glabrescent, procumbent or decumbent to ascending. Leaflets obovate, obcordate, or cuneate-obovate, 6–20 mm long, 5–18(20) mm wide, upper third to half with serrate margin (leaflets rarely laciniate), glabrous above, glabrous or sparsely hairy underneath. Stipules lanceolate to ovate-lanceolate, laciniate. Inflorescence of (1)2–6(10) flowers in a dense head, the peduncle longer or shorter than the corresponding petiole. Flowers 3.5–6 mm long, corolla yellow, usually less than twice length of calyx, calyx equal to or slightly more than half the flower length, glabrous or sparsely covered with simple hairs, teeth lanceolate-subulate, about equal to length of tube. Young pod protrudes sideways from the calyx, pubescent, mature pod discoid, short- to long-cylindrical or conical-truncate, gray to black, 2–12 mm long, usually glabrous, spiny, tuberculate, or lacking appendages, coils 1.5–7, in a lax spiral 4–10 mm wide, sometimes quite hard at maturity, coil face with transverse veins anastomosing in outer part of coil before entering the lateral vein, spines if present grooved and 2-rooted, one prong from the dorsal vein and the other from the marginal vein. Seeds 2–4 mm long, 1.5–2.2 mm wide, light-yellow to brownish, 1–2 per coil, separated by spongy transverse fruit septa, radicle about half as long as seed, seedcoat smooth, yellow to yellow-brown.

Identification Hints

Medicago polymorpha is the most likely species of the genus to be collected in numerous countries (Piano et al. 1982) and the most likely to be confused with other species. The fruits are very generalized for an annual, spirally coiled species of *Medicago*. The fruits occasionally approach in hardness and proliferation of alveolar tissue the hard-fruited pods of species of section *Spirocarpos* subsect. *Pachyspirae*. However, they are usually similar to the pods of certain mostly soft-fruited species of section *Spirocarpos* subsect. *Spirocarpos* (= subsect. *Leptospirae*), particularly *M. laciniata and M. minima*, two species that are also quite common as weeds. The fruit coil faces of both of the latter species have submarginal veins that are much more remote from the coil margin than is the case in *M. polymorpha*. The coil face of *M. polymorpha* has quite reticulate venation, whereas that of *M. laciniata* has notably S-shaped radial veins that anastomose little. *Medicago minima* is a quite hairy plant, whereas *M. polymorpha* almost always essentially lacks hairs.

Habitat and Environmental Adaptation

Habitat

Medicago polymorpha occurs in a large range of soils and climates, but not in true deserts or high mountains. Ehrman and Cocks (1990) found that in Syria *M. polymorpha* occurs in a very wide range of ecological zones compared with other medics. Carneiro and Serrao (1996) observed the same in Portugal. Francis and Katznelson (1977) described it as "ubiquitous" in Crete and Greece. This is the most common ruderal *Medicago* (with the possible exception of escaped alfalfa), occurring both in weedy habitats and among cultivated crops and, indeed, it is listed as a noxious weed in many jurisdictions.

Phenology

In its native areas, *M. polymorpha* usually flowers from March to May. Clarkson and Russell (1975) found that flowering was accelerated by vernalization (a long vernalization slightly delayed flowering), long photoperiods (18 and 24 hours), and high temperature. The Register of Australian Herbage Plant Cultivars for Legumes (http://www.pi.csiro.au/ahpc/legumes/legumes.htm; current June 2010) has information on the relative flowering times of cultivars 'Serena', 'Circle Valley', and 'Santiago', and notes on the effects of temperature, late planting (which speeds up flowering), and moisture limitation in the spring (which prolongs the vegetative period) on flowering. Bolland (1986) found that 'Serena' was especially early flowering compared with other medics in Australia. Del Pozo et al. (2000) and Del Pozo et al. (2001) found that in Chile wild collections of *M. polymorpha* from arid and semi-arid zones were programmed to come into flower earlier than populations from more humid zones.

Soil

Abdelguerfi et al. (1988*a*) concluded that *M. polymorpha* is adapted to essentially all soil zones of Algeria and is indifferent to soil texture. Zoghlami et al. (1996), in reference to Tunisia, found that *M. polymorpha* is often in fine-textured soils (30% to 35% clay) that are rich in calcium and organic matter. Similarly, Bounejmate and Beale (1992) found that in Morocco, *M. polymorpha* tended to occupy sites with high clay content. In Australia, naturalized *M. polymorpha* grows on a variety of soils, but does particularly well on alkaline clay soils (Young et al. 1996). Cultivars 'Serena', 'Circle Valley', and 'Santiago' have been described as well adapted to a range of Australian soils, including alkaline sandy clay loams and lighter-textured sandy loams, and shallow sands over gray clays (Ewing 1986). Species of *Medicago* are generally intolerant of acidic soils, although compared with most other species of the genus, *M. polymorpha* is relatively acid tolerant (Piano and Francis 1992). However, Piano and Francis noted that in its native habitats normally it is found in soils ranging from slightly acidic to slightly basic. In Sicily, Piano et al. (1991) observed the species growing in soil with a pH range of 6.5–9.0. Strains of *M. polymorpha* have been selected that are relatively acid tolerant (Howieson and Ewing 1986; Howieson et al. 1988; Howieson et al. 1992). In Australia, the pH tolerance range recommended for planting *M. polymorpha* is 4.5–8.0 (Gillespie 1994).

Medicago polymorpha performed very well in an experimental study of cover crops suitable for saline soils of the Central Valley of California (Mitchell 1996). Ibrar and Hussein (2003) examined the response of *M. polymorpha* to salinity and concluded that although the plant is susceptible to high levels of salinity, it can be cultivated on marginal saline lands for fodder and as a protective cover for reducing soil erosion. Rogers et al. (2005) considered *M. polymorpha* to be a moderately suitable choice as a pasture fodder plant for salt-affected soils in Australia. Dunbabin and Slavich (1990) and Merabet et al. (2006) also found that *M. polymorpha* is moderately salt tolerant. Lieth and Mochtchenko (2003) listed the species in their halophyte database. Abdelly et al. (1996) observed that, although *M. polymorpha* was regularly found on the edges of sebkhas (closed depressions with hydromorphic, saline soil, periodically submersed) in Tunisia, the species had limited tolerance of soil salinity.

Temperature

Medicago polymorpha is very tolerant of a range of temperatures. Piano and Francis (1992) described it as more or less cold tolerant. It grows better in cooler climates than most of the annual species of *Medicago* (Lesins and Lesins 1979), but also tolerates very hot climates. Azizi et al. (2004) recommended *M. polymorpha* for cultivar development for cool regions of Asia and North Africa. Hekneby et al. (2001) found that *M. polymorpha* 'Santiago' and 'Anglona' exhibited good cold resistance compared with several other species. Krall et al. (1996) found that eight accessions of *M. polymorpha* examined for their ability to self-regenerate in Wyoming failed to survive over the winter at any of the three locations tested. Christiansen and Svejcar (1989) sowed two accessions of *M. polymorpha* in Oklahoma in March 1985 and observed that they were all killed by frost and cold in the following winter period. Subfreezing temperatures in December 1998 in West Central Texas caused considerable seedling mortality of cultivar 'Armadillo' (Ueckert et al. 2003).

Water

Rogers et al. (2005) ranked *M. polymorpha* as moderately tolerant to waterlogging. Piano and Francis (1992) characterized the species as "relatively tolerant to temporary waterlogging occurring in poorly structure soils." In Sardinia (where the species is native) it occupies sites ranging in annual rainfall from 50 to 1100 mm, whereas in Chile (where it is introduced), rainfall varies from 100 to 1300 mm annually (Del Pozo et al. 2001). In Tunisia, the species occupies a semi-arid zone with 250–450 mm of precipitation annually (Zoghlami and Hassen 2004). Brahim and Smith (1993) observed that *M. polymorpha* in a field plot in Arizona produced seeds with only 125 mm of winter precipitation. Lesins and Lesins (1979) reported that *M. polymorpha* is more tolerant than other annual species of *Medicago* to conditions of elevated humidity, and the species competes more successfully with grasses in moister soils than most annual species of the genus. Tadmor et al. (1971) conducted studies of *M. polymorpha* ecotypes selected for their ability to survive and persist in the Negev Desert of Israel in an area where rainfall averaged only 90 mm annually, but received an additional 263 mm of water in the form of run-off from erratic winter storms. The dry matter production of nine ecotypes varied from 2.1 to 4.5 t/ha (averaged over 6 years), which is considered rather good given the dry environment.

Biotic Relationships

Pollination

Small (1988) noted that the small flowers and pollen to ovule ratio (281) are consistent with *M. polymorpha* being largely self-pollinated and basically an inbreeder. Hannachi et al. (1998) studied isoenzyme variability of wild populations of *M. polymorpha* in Tunisia and concluded that the level of genetic variation was in accordance with a strictly autogamous system of reproduction. Based on genetic analyses, Vitale et al. (1998) found that crossing rates were very low, but suggested that rare outcrossing events may contribute to the high genetic variability that has been observed in the species.

Dispersal

The spiny fruits of *M. polymorpha* are well adapted to dispersal in animal fur, accounting for the extremely widespread occurrence of this species. Forms with spiny fruits are much more widely distributed than forms with spineless or tubercled fruits. Del Pozo et al. (2001) noted that in Sardinia and Chile, populations with spineless pods or very short, stiff spines were more common in drier environments, and they suggested that sheep were likely responsible for distributing the type of plant with spiny pods. Kotzé et al. (1995) showed that some intact seeds will pass through the gut of sheep and remain viable.

Herbivory

Information on cultivar susceptibility is available in the Register of Australian Herbage Plant Cultivars for Legumes. Cultivar 'Santiago' has intermediate resistance to the bluegreen aphid (*Acyrthosiphon kondoi*). Cultivars 'Santiago', 'Serena', and 'Circle Valley' are considered to be tolerant to the cowpea aphid (*Aphis craccivora*). The three cultivars are only slightly resistant to the red-legged earth mite (*Halotydeus destructor*). 'Santiago' is susceptible to the lucerne flea (*Sminthurus viridis*).

Nodulation and mycorrhizae

USDA (2009) reported positive root nodulation for *M. polymorpha*. Jain (2003) provided a detailed description of the anatomical development of nodules in *M. polymorpha*. Materon and Zibilske (2001) studied the ability of various rhizobial strains to inoculate the species. Ewing and Howieson (1989) and Howieson and Ewing (1989) found that *M. polymorpha* was outstanding in its ability to achieve nodulation in Australian acid soils. Charman and Ballard (2004) noted that *M. polymorpha* is nodulated by rhizobia (*Sinorhizobium meliloti* and *S. medicae*) that reside in many Australian soils, but many of the rhizobial strains do not effectively nodulate *M. polymorpha*, and none of the plant accessions studied was able to be nodulated by a wide range of the rhizobial strains. These authors also studied selections of *M. polymorpha* with superior N_2-fixing symbiotic ability with naturalized soil rhizobia. Loi et al. (1993) found that exposure to considerable acidity (pH 5.4) resulted in differences between genotypes in reduction of nodulation. Denton et al. (2007) studied the ecology of nodulation of *M. polymorpha* in relation both to rhizobial strain and presence of naturalized rhizobia in the soil, and found that there was an interaction such that some strains nodulated well in *M. polymorpha* when there were other rhizobial strains present.

Disease resistance

Cultivars 'Santiago', 'Serena', and 'Circle Valley' have intermediate resistance to *Phoma medicaginis* (see the Register of Australian Herbage Plant Cultivars for Legumes). Powdery mildew (*Erysiphe pisi*) is a concern for cultivated *M. polymorpha* (Jones and Nicholas 1992). Yaege and Stuteville (2002) tested the resistance of 33 annual species of *Medicago* to powdery mildew and found that the 33 collections of *M. polymorpha* grown were all susceptible to the disease (it was observed that often only one of a number of populations of other species tested was resistant). In a comparative test of 36 annual species of *Medicago*, O'Neill and Bauchan (2000) found that 34 accessions of *M. polymorpha* exhibited a range of susceptibility to anthracnose caused by *Colletotrichum trifolii* (most accessions were quite susceptible to it, a few were quite resistant). De Haan et al. (2002) found two of 16 accessions of *M. polymorpha* that were resistant to Phytophthora root rot (caused by *Phytophthora medicaginis*).

Economic History, Cultivars, and Uses

Medicago polymorpha is one of the more important annual medics that have been developed for use as pasture forage for dry, hot environments.

Cultivars

The Register of Australian Herbage Plant Cultivars for Legumes (http://www.pi.csiro.au/ahpc/legumes/legumes.htm; current June 2010, the document updated to January 2004) contains registration statements that were published in the *Register of Australian Herbage Plant Cultivars* (3rd ed., 1990) and those published subsequently in the *Australian Journal of Experimental Agriculture*, both of which provide information on the origin and development of each cultivar, a full morphological description, and referenced information on the agronomic uses and merits of the cultivar. The following cultivars of *M. polymorpha* are included: 'Circle Valley' (registered 1976, a short-spined form of the species), 'Santiago' (registered 1988, selected for having spineless pods, occasionally with white spots on the upper leaves that fade at higher temperatures), and 'Serena' (registered 1976, selected for having spineless pods).

Cultivar 'Armadillo' from the United States was selected from a naturalized ecotype from Pasture 18 at the Texas Agricultural Research Station, Beeville, South Texas, released in 1998 (Ocumpaugh 1999), and registered in 2004 (Ocumpaugh et al. 2004). It is employed in Central and West Texas. 'Armadillo' performs poorly on sandy acidic soils.

Carneiro et al. (2004) described 'Lentisca', a Portuguese-selected cultivar, as having good adaptation to grazing because of its prostrate to semi-prostrate growth habit, and at least as good levels of dry matter production as the available Australian annual *Medicago* cultivars for purposes of stocking long-duration, rain-fed pastures under Mediterranean conditions.

The French cultivar 'Mauguio' is recommended for use in rangelands for sheep rearing (additional information below; also see Prosperi et al. 1999).

Experimental frost-tolerant cultivars from SeedCo Ltd. (Australia) that are mentioned in Walsh et al. (2001) include 'Carpet', 'Mauguio', and 'Polygraze'. 'Scimitar', from Seedmark (Australia), is described as a spineless burr medic intended to replace 'Santiago' (http://www.seedmark.com.au/files/medics_scimitar.pdf; current June 2010).

'Cavalier', another "spineless-podded" Australian cultivar developed at SARDI, was released in 2003 and sold by Seed Distributors.

'Combarbala' is a Chilean early flowering cultivar described as suitable for Mediterranean dryland areas (Del Pozo et al. 2001). 'Cauquenes' is another Chilean cultivar suitable for Mediterranean dryland areas (Ovalle et al. 2001).

'Anglona' is the first annual *Medicago* cultivar from Italy (Porqueddu et al. 1998).

Abdelguerfi et al. (2000) refer to cultivars of *M. polymorpha* selected in Morocco.

Forage

In the littoral zone of Tunisia, green forage yields in the order of 20 000 kg/ ha in dryland cultivation have been reported for *M. polymorpha* (Kernick 1978). Sanadgol et al. (2006) found that the unirrigated yield of *M. polymorpha* over a 3-year period in the Gorgan plain of Iran was 1487 kg/ ha. Sitzia and Fois (1999), based on a study in Sardinia, found that *M. polymorpha* herbage and seeds provided excellent feed for sheep. Fois et al. (2000), also based on studies in Sardinia, observed productivities of 3.1 t/ha for pods and 1.4 t/ha for seeds, with a crude protein content of 21% in pods and 11% in stubble. Ewes preferred to consume the pods over the stubble and showed no effects on live weight or body condition, indicating that *M. polymorpha* is an important feed resource for sheep grazing during the summer. Loi et al. (1995) reported that herbage production of dry matter in Sardinia varied between 2 and 8 t/ha, with up to 1.5 t/ha of seed produced. Cultivar 'Armadillo' has produced annual yields of about 4.5 t/ha (Ocumpaugh et al. 2007). Radwan et al. (1978) observed a dry matter (including pods) productivity of a local strain of *M. polymorpha* on unirrigated land in northern Iraq of 4.38 t/ha, and noted that selection of suitable cultivars and rhizobia would probably result in a higher yield (Australian cultivars performed much more poorly). Shrestha et al. (1998) concluded from experimental growth of *M. polymorpha* in Michigan that its productivity for the year of planting was comparable with that of alfalfa.

Zhu and Sheaffer (1997) recorded the following above-ground dry matter yields (in kg/ha) of cultivar 'Santiago' at two sites in Minnesota (compared with the alfalfa cultivar 'Nitro', yields in parentheses): 5390 (3815) and 3505 (4596). Mohammad and Qamar (1988) grew 14 annual species of *Medicago* in small test plots in Islamabad, Pakistan located

Table 69-1. Forage yields of native *M. polymorpha* in northern Iraq (kg/ha), after Kernick (1978).

Fertilized with 60 kg/ha P_2O_5		Unfertilized	
Green matter	Dry matter	Green matter	Dry matter
4500	1900	3400	1400

in a subtropical, subhumid area that is notably hotter and wetter than the normal habitat of these species. The dry matter yield of *M. polymorpha* was about 4 t/ha. Sitzia et al. (2000) concluded that burr medic provides high-quality forage for ewes during the winter–spring period in the Mediterranean environment and that the pods can be subsequently grazed during the summer. Kotzé et al. (1995) examined the nutritional value of *M. polymorpha* seeds for feeding sheep.

In a nursery plot on alluvial soil in the steppic zone (220 m elevation) of northern Iraq, during a season with rainfall of 475 mm (November to May), forage yields were recorded as follows:

Medicago polymorpha seed yield under irrigated conditions in South Africa was 62 kg/ha (Wasserman 1974).

Some problems have been associated with grazing on *M. polymorpha*. Mechanical injury can result from grazing by some animal species on the spiny pods (cultivars rarely have spines). Dodd (1916) documented that the so-called aphis disease of sheep, horses, and cattle in Australia is a photosensitization response caused by ingestion of *M. polymorpha*. This disease occurs in the spring. It is characterized by swelling of unpigmented and unprotected parts of the skin, reddening and exudation, scab formation, and accompanying itching, biting, scratching, or rubbing of the affected areas. Photosensitization due to consumption of this species has also been cited by Steyn (1934), Hurst (1942), Gardner and Bennetts (1956), and Watt and Breyer-Brandwijk (1962). Mathews (1937) referred to a record of a similar effect in humans, but this has not been substantiated. Eyal et al. (1975) found that a combination of seeded *M. truncatula* and *M. polymorpha* provided considerable palatable pasture for sheep in a semi-arid area of Israel, but a fertility disorder apparently due to estrogenic activity was noted.

Fodder

Medicago polymorpha is mostly cultivated as forage, but can be harvested and used like alfalfa hay when desired.

Green manure, rotation crop, ground cover

Nieddu et al. (2000) and Porqueddu et al. (2000) found that *M. polymorpha* was a good cover crop for vineyards in Sardinia, but did not suppress weeds as effectively as subterranean clover. Veronesi and Rosellini (2000) evaluated a population of *M. polymorpha* for use as a cover crop in Italy. Nair et al. (2006*b*) noted that the French cultivar 'Mauguio' was bred in part "to maintain soils around infrastructures: roadsides, railway lines embankments, and disturbed areas; to avoid soil erosion from rainfall and winds, reduce the use of herbicides, limit infiltrations of pesticides in soils and enhance underground life in sloping vineyards."

Other uses (salad sprouts, ornamental, plant products, etc.)

Townsend (1974) stated that the species is "sometimes eaten with potatoes as a leafy vegetable in India." Similarly Dhar and Dhar (2000) recorded that in northern India the tender young leaves are consumed as a pot herb. The leaves and flowers of *M. polymorpha* are eaten raw in the Near East according to a Food and Agriculture Organization (of the United Nations) document (http://www.fao.org/DOCREP/003/X6879E/X6879E24.htm; current June 2010). According to Larkcom (2007), the plant's leaf tips, leaf stalks, and tender leaves are cooked in various ways in soups, and boiled, steamed, and stir-fried in parts of China and in Taiwan. "Burclover attracts insects which can be consumed by wild turkey and other birds" (ftp://ftp-fc.sc.egov.usda.gov/GA/PMC/TDT/PlantWildlifeRev1.pdf; current June 2010).

Chromosome Number

$2n = 14$: Fryer 1930 (4 accessions, as *M. hispida*); Larsen 1956; Clement 1962; Simon and Simon 1965 (3 accessions); Bolkhovskikh et al.1969, p. 308 (as *M. hispida*); Fernandes and Santos 1971; Gillies 1971*a*; Lesins and Lesins 1962, 1979; Mariani 1963; Dahlgren et al. 1971; Kliphuis and Barkoudah 1977; Fernandes and Queirós 1978; Labdie 1976; Loon and Jong 1978; Agarwal and Gupta 1983*b*; Luque et al. 1988; Falistocco and Falcinelli 1991; Mariani et al. 1996; Oberprieler and Vogt 1996*a*; Karadağ and Gülcan 1997.

$2n = 14, 16$: Heiser and Whitaker 1948; Heyn 1956; Bir and Kumari 1979; Kumari and Bir 1990.

$2n = 16$: Ghimpu 1929*b* (as *M. apiculata*); Ghimpu 1930 (as *M. apiculata*, *M. denticulata*, *M. lappacea*, *M. nigra*, and *M. pentacycla*); Bolkhovskikh et al.1969, p. 308 (as *M. denticulata*); Fernandes et al. 1977; Fernandes and Queirós 1978; Labadie 1979; Sidhu 1979; Bir and Sidhu 1980; Astanova 1981 (as *M. denticulata*); Sidhu and Bir 1983; Díaz Lifante et al. 1992; Karshibaev 1992 (as *M. denticulata*); Magulaev 1992 (as *M. denticulata*); Mohamed 1997.

Genetic Relationships and Gene Transferability

Intraspecific

De Hann and Barnes (1998) successfully hybridized forms of *M. polymorpha* that had different pod types, stem color, and growth habit. Lesins and Lesins (1979) reported that they hybridized a spiny-fruited form with a form lacking fruit spines. These studies did not indicate any intra-specific barriers to hybridization.

Interspecific

Simon and Millington (1967) were unable to produce hybrid seed by crossing *M. polymorpha* with *M. praceox* and *M. rigidula* (*M. rigiduloides*?).

Exchange by advanced biotechnology

No information located.

Conservation Status

Medicago polymorpha is a very widespread weed, and given this status it does not seem in need of conservation.

Germplasm Sources

Approximately 7500 accessions are listed among the holdings of germplasm institutions with large collections of *Medicago* (see Chapter 3).

Research Problems and Priorities

Medicago polymorpha is extensively used in agriculture and is a weed in many places and, as a result, it has been rather well examined from the points of view of biological and agricultural research. It has an exceptional range of adaptation, and therefore deserves additional study and economic development.

Representative Collections

Afghanistan: Griffith, Herbarium of, Herbarium of the late East India Company no. 1485 (distributed 1862–1863) (P). **Algeria:** Maire, R., Dirza supra Aumale, in patis humidis montes, solo calcareo, alt. 1200 m, 31-V-1936, in flower & fruit (P). Maire, R., in incultis, 30-V-1939, in flower & fruit (P). (Type! of *M. hispida* Gaertn. var. *lappacea* (Desr.) Burnat f. *maculifolia* Maire (viewable online at http://www.aluka.org; current June 2010; some evidence of streaking and spotting in leaflets; note that several annuals have been found to exhibit leaf markings under certain cultivated conditions.) Maire, R., in pinetis montis Merdjadjo supra Oran, solo calcareo, alt. 300 m, 31-III-1936, in flower & fruit (P). (Isotype! (at P) of *M. denticulata* Willd. forma *maculata* Maire (viewed online at http://www.aluka.org; current June 2010); holotype (two herbarium sheets so labelled) at MP, viewed online at http://www.aluka.org; current June 2010)) **Argentina:** Pedersen, T.M. 3988, Corrientes, waste ground, 16-IX-1956, in flower & early fruit (P). **Australia:** Goode, R.H. 72, New South Wales, Tamworth, by Peel River between Piallamore and Dungowan, 6-XI-1954, in flower & fruit (BM). Tilden, J.E. 690, Point Lonsdale, Victoria, X-XI, 1912, in flower & early fruit (G). **Belgium:** Vanderborn, environs de St. Trond (Limbg.) bord des chemins, 20-X-1865, in flower & early fruit (K). **Bermuda Islands:** Collins, F.S. 206, roadside, 3-VIII-1913, in fruit (P). **Bolivia:** Mandon, G. 691, Plantae Andium Bolivlensium, vicinis Fornata undique in pratis dumasis, in cultis, alt. 2660–3110 m, 1818, in fruit (P). **Britain:** Townsend, C.C., Arable, Mavlden, wool alien, 23-IX-1959, in fruit (K). **Bulgaria:** Kojoucharov, S. & Kouzmanov, B., les prairies près du village Ljubimetz, la valée de la rivière Maritza, 28-V-1959 in flower & fruit (K). **Canary Islands:** J. Bornmüller, La Palma, in montibus supra Santa Cruz, B-del Rio, alt. 400 m, 30-IV-1901, in flower & early fruit, J. Bormüller Plantae Exsicc. Canariensis 1901 No. 2234 (rather atypical) (B). **Costa Rica:** Khan, Tebbs & Vickery 560, prov. Cartago, Carretera east of San Rafael, Perque Pascador El Cristo Ray, alt. 1600 m, dry roadsides 27-I-1984, in fruit (BM). **Crete:** Townsend, C.C. 69/39, Knossos, local in grassy places amid the ruins of Minos, 5-IV-1969, in flower & fruit (K). **Czechoslovakia:** Kühn, F., Moravia Centr. – Merid., distr. Brno, in ruderatis suburbii cerná Pole dicti in urbe Brno, alt. 260 m planta cum lana adventiva, 10-X-1966, in fruit, Flora Exsiccata Reipublicae Socialisticae Cechoslovacae No. 1532 (K, P). **Egypt:** Keller, A. 44, Cairo, IV-1903, in flower & fruit (P). Simpson, N.D. 934, near Mene village, 6-VIII-1922, dead plant with fruits (HUJ). **Ethiopia:** Jones, S. 28, Eritrea, Sahil Awraja, 20 miles southwest of Naefa, alt. 2400–2650 m, VII-VIII-1987, in flower & early fruit (K). **France:** Foucaud, J. & Gautier, G., route de Villeneuve au col d'Extrem (Pyrenées-Orientales), VI-1894, in fruit, Société Rochelaise 1894 No. 3627 (P). Reau, N., Bord des chemins à Saint-Georges-d'Oleron (Charente-Inférieure), VI-VII-1889, Société Rochelaise 1889, No. 2611 (P). Timbal-Lagrave & Gautier, G., entre Villeneuve et le col d'Extrem, dans les Corbières (Aude), bord des champs, 12-VI-1879, in fruit, Société dauphinoise 1880, No. 2425 (P). **Greece:** Reverchon, E. 29, Crete, Kissamos, rare, 8-IV-1884, in fruit (P). Reverchon, E., Corse, Porto, moissons, 17-V-1885 (E). Reverchon Plantes de Corse 1885 No. 452 (P). **Hawaii:** Faurie 786, insula Hawaii, Hilo, in arenosis littoralis, V-1909, in fruit (BM). Heller, A. 1982, Hawaiian Islands, island of Oahu, Honolulu, 22-III-1895, in fruit (P). **India:** Drummond, J.R. 21499, Punjab, Karnal, 1886, in fruit (P). Thomson, J., Punjab, alt. 1000 ped (sic), Herb. Or. Hook. fil J. Thomson (P). **Iran:** Bunge, M. 1868-13, inter Astrabad & Schchrud, in jugo Elbousensi, V-1858, in flower & fruit (P). Kotschy, Th., pr. Dalechi, in rupestribus collium, mart. m-1842, in flower & early fruit, Th. Kotschy Pl. Pers. Austr. Ed. R.F. Hockenacker 1845 No. 186 (P). Lindsay, N. 820, Lahijan, Guilan, in hedges and on banks, rare, 5-VI-1936, in fruit (BM). **Iraq:** Eig, A. & Zohary, M., plain between Chankula (between Baghdad and Basra), depression, 6-IV-1933, in fruit (HUJ). Meude, E. 199, Rowanduz, alt. "2500", 5-V-1958, in fruit (BM). **Israel:** Feinbrun, N. & Jacobovitch, E., Jerusalem, Mt. Scopus, white calcareous soil, 14-IV-1938, in flower & early fruit, universitatis Hebraicae Hierosolymitanae Flora Palaestina Exsiccata No. 336 (BM, P). Zohary, M., Judean Mtns., University campus, weed in grass, 10-II-1972, in fruit (HUJ). **Italy:** Todaro 850, Palermo a. Mondello, in herbosis arenosis maritimis V (received by P in 1882), Todaro Flora Sicula Exsiccata (P). **Japan:** Faurie 2336, Tokyo, 4-VI-1888, in fruit (P). Faurie 490, Tokyo, 20-VI-1887, in fruit (P). Maximowicz, Yokohoma, 1862, in flower & fruit (P). **Jordan:** Al-Eisawi, D. 1123, Zaiarea, 15 km west Salt, 31-III-1973, in flower & fruit (BM). Al-Eisawi, D. 1399, 10 km east Ash'Shuna, along the road to Irbid, 16-IV-1973, in fruit (BM). **Kenya:** Stewart, R.B. C31, Jats, Jimna, a weed in legume nursery, well-drained, red clay 3-XII-1955, in flower & fruit (BM). **Libya:** Maire, R. & Weiller, M. No. 442, Cyrenaica, in lapidosis calcareis, circa Cyrenem, alt. 600 m, 22-IV-1938, in fruit (type of *M. cyrenaea* Maire & Weiller; isotype P! holotype MP; types online at http://www.aluka.org; current June 2010.) **Madeira:** Press, J.R. & Short, M.J. 526, Quebradas to Funchal, along Levada dos Piorhais, alt. ca. 150 m, 1-IV-1984, in fruit (BM). **Malta:** Bricknell, E.A. 79, central Malta, alt. 275 feet 3-IV-1964, in flower & fruit (K). **Mauritius:** Drennen, D.S.H. 10, Réduit, University of Mauritius Farm, VII-1981, in flower & fruit (K). **Mexico:** Basurto, P. & Durán, G. 200, Puebla, 7 km al southwest de Xicotepek de Juárez 14-III-1979, in fruit (BM). **Morocco:** Maire, R., prope Kasba Tadla, alt. 600 m, 15-IV-1934, in fruit (P). Reading Univ./BM. Exped. 27, 8 km south of Arbaova, west of road to Kenitra (34°50'N, 5°45'W), derelict field, near *Eucalyptus* wood, sandy loam, 31-V-1974, in fruit (BM). **New Zealand:** Bangerten, E.B. 5144, North Island, Bournemouth Terrace (36°44'S, 174°45'E), on roadside verge with viatical weeds, occasional to locally frequent, 22-I-1974, in fruit (K). **Portugal:** Ferreira, M., Coimbra, Lordemao, VI-1910, in fruit Flora Lusitanica Exsiccata Herb. Hort. Bot. Conimbricensis No. 1879 (P). Moller, A., Coimbra: Balea, VI-1889, in fruit, Flora Lusitanica Exsiccata Herb. Hort. Bot. Conimbricensis No. 754 (P). **Rhodesia:** Whellan, J.A. 110, Salisbury, outside Agriculture Experiment Station, weed, 2-VI-1956, in flower & fruit (K). **Saudi Arabia:** Mandaville, J.M. Jr. 7409, Eastern Province, 2 km northwest of Dhahran, at Ramada Hotel, irrigated lawn, 12-IV-1979 in flower & fruit (BM). **Socotra:** Gwynne, M.P. 154, Wadi Klohhen, alt. 3000 feet, grazed land under orange tree, 20-VIII-1956, in flower & early fruit (BM). **South Africa:** Chaberlain, Y.M. 61, Cape Province, Arniston, sand dunes, 25-IX-1956, in flower & fruit (BM). Mogg, A.O.D. 16855, Halfords Union Farm, Pyramids, 12 miles north of Pretoria, common in cultivated ground, 13-VIII-1940, in flower & very early fruit (BM). **Spain:** Gardner, M.F. & S.G. 1631, prov. Almeira, Mojacar, Sierra de Alhamilla, 8 km north-east of Nijar (36°59'N, 1°11'W), alt. 350 m, cultivated terraces, 13-IV-1982, in flower & fruit (BM). **Sweden:** Blom, C., prov. Skane, paroecia Lackalänga ad pagum Lackalänga, adventicia, 20-VIII-1927, in flower & fruit, G. Samuelsson Plantae Suecicae Exsiccatae No. 1052 (K). Fries, H., Göteberg, Kviberg, 16-IX-1925, in flower & fruit (HUJ). **Syria:** Eig, A. & Zohary, M., Soukhne, fields, 1-V-1933, in fruit (HUJ). Zohary, M., Syria bor., Telejin to Abudur, 5-V-1931, in fruit (HUJ). **Tanganyika:** Greenway, P.J. 7938, West Usambaras, Schume, alt. 6500 feet, grassland on moutain slopes, very local, 8-III-1947, in flower & fruit (K). **Tanzania:** Iverron, S.T. & Pocs, T. 85665, Tanga, Lushoto District, west Usambara Mts., alt. 920 m, Shume Forest guesthouse, in the garden, 12-II-1985, in flower & fruit (K). **Tonga:** Soaki, E. 323, Nuku'alok Tonaatapu, alt. 15 feet, grassland near the wireless post, 27-I-1959, in flower & fruit (K). **Tristan de Cunha:** Stableford, H.C., grown from manure from imported fodder, 1953–1954, in flower & fruit (K). **Tunisia:** Cosson, M.M.E., Kelibia, 16-V-1883, in fruit (P). **Turkey:** Balls, E.K. B709, Tarsus, alt. 50 feet, cultivated and waste ground, 10-IV-1934, in flower & fruit (BM). Davis & Hedge D2737, prov. Hatay, Antakya, near St. Peter's Church, alt. 150–300 m, waste places, 27-IV-1957, in fruit (BM). **USA:** Moldenke, H.N. 1111, Florida, Leon Co., Tallahassee, dry sandy soil, 2-V-1930, in fruit (P). Reed, F. & H., Californie du Sud, Riverside, région de San Fernandino, 1933–1934, in flower & fruit (B, P). Vasey, R., San Bernardino, California, V-1880, in flower & fruit (P). **USSR:** Gonczarov, N. & Zaprjagajeva, V., Tadzhikistania, in viciniis vrb. Stalinabad, in gramineto ephemero varie herboso ad pedem montium, 1934, in fruit (BM). Gonczarov, N. & Zaprjagajeva V., Tadzhikistania, in viciniis urb. Stalinabad, in gramineto ephemero varie herboso ad pedem montium, 1934, in fruit (K). **Venezuela:** Alston, A.H.G. 7016, State of Merida, Bailadores, alt. 2000 m, 13-II-1939, in fruit (BM). **Yemen:** Wood, J.R.I. 7/75/246, Jebel Taara, Jibla, bare ground in the village below east summit, 30-V-1975, in flower & fruit (BM). **Yugoslavia:** Jackson, A.K. & Turrill, W.B. 155, Solin, grassy places among Roman ruins, 21-IV-1935, in flower & fruit (K).

Notes

Del Pozo et al. (2002) studied the ecology of *M. polymorpha* in Chile and concluded that, in arid to semi-arid environments where rainfall is erratic and unpredictable, ecotypes are naturally selected that are early flowering and have high growth capacity during the winter, the period of the year when rainfall is most likely to occur in Mediterranean-climate regions. Conversely, in humid areas with higher rainfall predictability and frequency of killing frost, ecotypes are selected with longer periods of growth potential, but lower winter growth

Loi et al. (1993) studied the growth response at different temperatures of eleven collections of *M. polymorpha* from different altitudes in Sardinia. They found that genotypes originating from regions of relatively low temperatures ("Low-temperature Genotypes") grew slowly under cooler experimental conditions but rapidly under warmer experimental conditions. By comparison, genotypes originating from regions of relatively higher temperatures ("High-temperature Genotypes") grew more quickly than the Low-temperature Genotypes under the cooler experimental conditions, but less quickly under the warmer experimental conditions. This observation was interpreted as evidence that the Low-temperature Genotypes are adapted to avoiding frost damage in the winter by having slow growth under cool conditions, but are able to take greater advantage of warm periods by growing quite rapidly.

70. *Medicago popovii* (Korovin) Širj.

M. popovii (Korovin) Širj., Bull. Misc. Information Kew 271 (1928), in obs.

The epithet *popovii* commemorates Mikhail Grigoríevič Popov (1893–1955), a Russian botanist and explorer of the flora of the Asiatic part of the Soviet Union who collected the type specimen of the species.

Subordinate Taxa

None recognized here.

Type Specimen

In describing his new species *Trigonella popovii*, Korovin (1924, the reference given below) provided the following collection information: "Hab. Buchara. In montibus Hissar, ad partem superiorem flpr. Kara-tag, fl. 17, VI. [1]915, M. Popov. Ibidem prope Chakimi fl. 11, 13, VI. [1]915. M. Popov (plantae humillioros depaperatae). Reg. Syrdarja: In valle fl. Angen infra Arasan ad colles argilosos copiosa; fr. VIII (E. Korowin)."

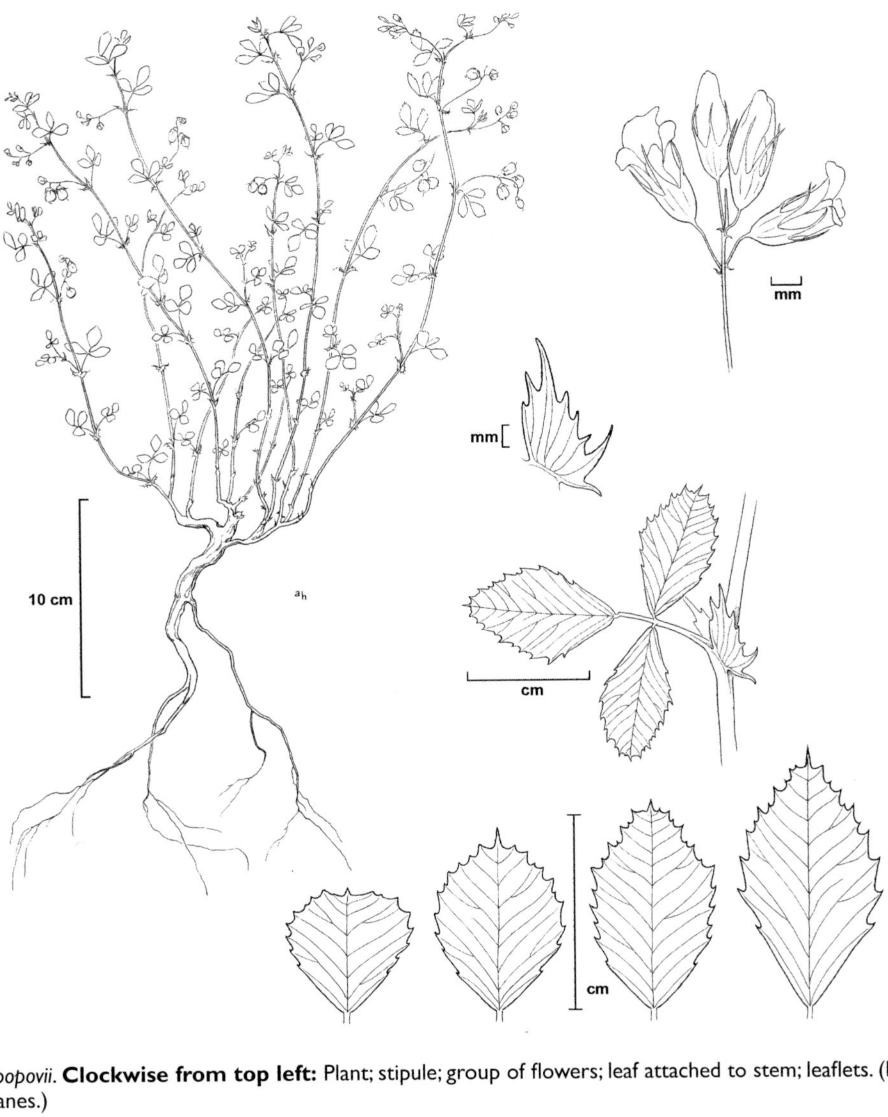

Fig. 70-1. *Medicago popovii.* **Clockwise from top left:** Plant; stipule; group of flowers; leaf attached to stem; leaflets. (Based on collection 727/79, K; artist: A. Hanes.)

Apparent type (information supplied by LE; translated from Russian): M.G. Popov 1129, (Soil Science Expedition of OZU to the basin of the Amu-Daria River, Bukhara Estates, Turkestan), the Gissarskii Ridge, in the mountains at Kosh-Assan(g), in the upper course of the Karataga (Karatag) River, 17-VI-1915. Photo: DAO. (Amudar'ya River is 44°02′N, 59°39′E; this site is in western Uzbekistan. This is consistent with the type information by Vassilczenko (1953), who indicated the type is at LE (although Grossheim (1945) stated "Type in Tashkent").

Synonymy

Homotypic

Trigonella popovii Korovin in Not. Syst. ex Herb. Hort. Petrop. (= Not. Syst. Herb. Hort. Bot. Reipub. Ross. = Botanicheskie Materialy Gerbarija Glavnogo Botaniceskogo Sada R.S.F.S.R. = Botanical Materials of the Herbarium of the Main Botanical Gardens) 5(11-12): 177 (1924), (basionym).

Melissitus popovii (Korovin) Golosk., Fl. Kazakhst. 5: 28 (1961).

Melilotoides popovii (Korovin) Soják, Sborn. Nár. Muz. Praze, 1982(1-2): 104 (1982).

Heterotypic

No names found.

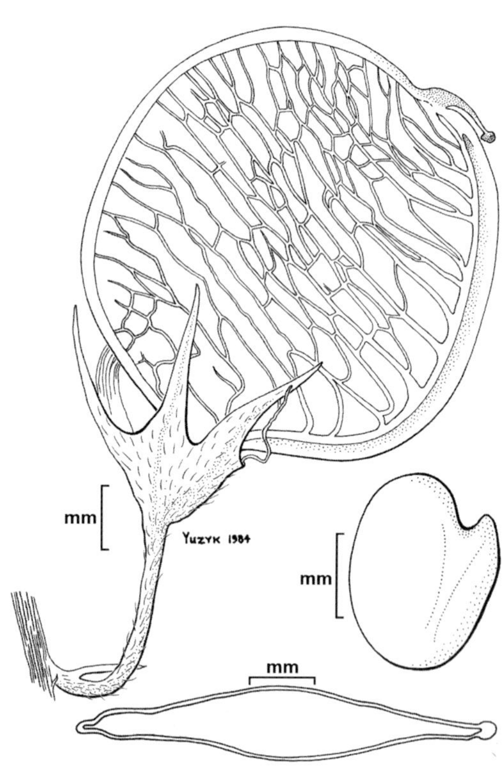

Fig. 70-2. *Medicago popovii.* **Clockwise from top:** Fruit; seed; outline of cross section of fruit. (Based on Popov & Wedensky 253, W; artist: L. Yuzyk.)

Vernacular Names

ENGLISH: No names found.

FRENCH: No names found.

Geographical Distribution

(N = native, I = introduced, U = uncertain status)

ASIA: Kazakhstan (southeastern; N), Kyrgyzstan (N), Tajikistan (N), Turkmenistan (eastern; N), Uzbekistan (N).

This species is endemic to Central Asia, including the middle mountains of western Tian-Shan and western Pamiro-Alai.

Description

Perennial herb with numerous branching stems 10–30(60) cm long arising from a woody base, stems ascending to upright, slightly pubescent with appressed hairs. Leaflets (4)5–15(20) mm long, (4)7–8(12) mm wide, obovate, oval, or sometimes round or cordate, upper side glabrous, lower side sparsely covered with appressed hairs or almost glabrous, up to two-thirds of distal margins serrate. Stipules of lower leaves semi-sagittate, ovate-lanceolate, toothed at base or incised, stipules of upper leaves linear-lanceolate, margin not toothed. Inflorescence of 4–6(10) flowers in a short, almost capitate raceme, the peduncles 10–20 mm long, usually equal to or longer than the subtending petiole, or occasionally shorter. Corolla 6–8 mm long, yellow, becoming reddish with aging and drying to a lilac color. Calyx campanulate, 3–4 mm long, sparsely pubescent with appressed hairs, the teeth lanceolate-subulate, equal to tube or slightly longer. Pod shortly and broadly oval, (5)6–8(10) mm long, 4–6 mm wide, flat, weakly thickened on sutures, glabrous, with thin, oblique to transverse veins, the apex with a curved persistent beak. Seeds (1)2–3 per pod.

Identification Hints

Medicago popovii is distinguishable from the very similar *M. platycarpos* by its smaller fruits (typically 5–10 mm long in *M. popovii* and 15–25 mm long in *M. platycarpos*). Indeed, *M. popovii* in many respects appears to be like *M. platycarpos* but with smaller dimensions. *Medicago popovii* is distinguishable from *M. ruthenica* by its smooth seeds. It differs from the other two species by the notable convexity of both of its sutures.

Habitat and Environmental Adaptation

Habitat

Medicago popovii occurs on slopes of the upper tree–shrub zone, and in meadows in the subalpine zone, extending upwards to 3500 m on southern slopes.

Phenology

Flowering commences in June and continues through July in its native area.

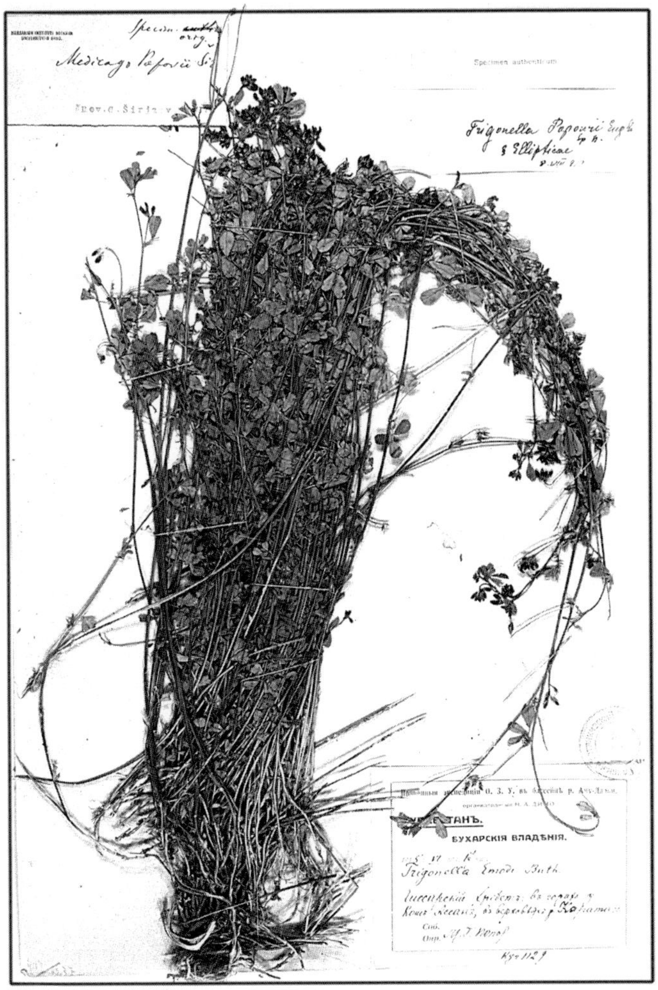

Fig. 70-3. Apparent type of *Trigonella popovii* Korovin, at LE.

Soil

The species has been collected from clayey and stony substrates.

Temperature

Medicago popovii is native to subalpine conditions, so it clearly is exceptionally adapted to cold temperatures.

Water

The species occurs in well-drained substrates and may be adapted to dry conditions.

Biotic Relationships

Pollination

The flowers are of moderate size by the standards of the genus, suggestive of at least some outcrossing.

Dispersal

There are no obvious adaptations for seed dispersal. Given the mountainous habitat frequently occupied, gravity probably plays a significant role in seed distribution.

Herbivory

No information located.

Nodulation and mycorrhizae

USDA (2009) reported positive root nodulation for *M. popovii*.

Disease resistance

Kulik and Dery (1995) examined 27 annual and four perennial species of *Medicago* for resistance to damping-off induced by isolates of *Rhizoctonia solani*, known to be pathogenic to *M. sativa*. While none of the species exhibited resistance to highly pathogenic strains, only *M. popovii* exhibited significantly higher resistance to the moderately pathogenic strains used.

Economic History, Cultivars, and Uses

In its indigenous area the species appears to provide wild forage for livestock. However, *M. popovii* has not yet attracted agricultural development.

Cultivars

No cultivars have been bred.

Forage

According to Vassilczenko (1953), *M. popovii* is "one of the most valuable forage plants, it acquires particular importance in high mountain regions (in the subalpine zone)." However, the species is not cultivated deliberately for forage.

Fodder

No information located.

Green manure, rotation crop, ground cover

No information located.

Other uses (salad sprouts, ornamental, plant products, etc.)

No information located.

Chromosome Number

No information located.

Genetic Relationships and Gene Transferability

Intraspecific

Artificial hybridization of different populations has not been attempted, so it is not known whether there are barriers to hybridization within the species.

Interspecific

No information located.

Exchange by advanced biotechnology

No information located.

Conservation Status

Medicago popovii is endemic to regions of the world where uncontrolled grazing threatens to result in the extinction of plants. Moreover, seed collections in germplasm institutions are almost nonexistent. Accordingly, a conservation assessment and germplasm preservation are urgently required.

Germplasm Sources

Three accessions (perhaps duplicates) have been deposited in major *Medicago* germplasm collections (see Chapter 3).

Research Problems and Priorities

Medicago popovii is very poorly understood at this time. Given that it is a perennial adapted to very cold subalpine conditions, it almost certainly has valuable genes useful for alfalfa improvement, and the species itself has promise as a new crop. The highest priority should be given to acquisition of germplasm of this species.

Representative Collections

USSR: Popov, M. & Wedensky, montes meridionales, Tian-Schan occidentalis, ad declivia argillosa montis Tschimgan Minoris, in fruticetis, 4-VII-1926, in flower & early fruit, Herbarium Florae Asiae Mediae Univ. Asiae Mediae editum Fasc. XI No. 253 (B, G, K, S). (Note that numerous additional collections are cited in Vassilczenko 1953).

Note

I.T. Vassilczenko, an inveterate taxonomic splitter, recorded the following (Vassilczenko 1953, considering the species as *Trigonella popovii*; translated from Russian): "This species requires further study on more extensive material. Its high-mountain forms are lower-growing and possess smaller and less dentate leaflets. One should note the finding in Eastern Tadzhikistan... of plants which belong to *T. popovii* but which, in contrast to the typical forms, have lax elongated (oval) racemes 15–20 mm in length and calyx teeth that are 1.5–2.5 times longer than the tube, the pods are also larger (6)10–15 mm in length, with a pedicel that is equal in length to the calyx tube. It is possible that in this case we have a new species (*T. darvasica* Vass. nom.), though the lack of material impels me, for the time being, to refrain from a description of the latter... Messagetov and Massal'skii on 7 July 1932 in Uzbekistan... collected a plant which was identified by them as "*Medicago popovii* E. Kor." This plant is indeed very similar to *T. popovii*. E. Kor., though it differs from the latter by the strong pubescence of all of its parts... Unfortunately, there are neither flowers nor fruits on the above-mentioned specimen. Provisionally I am assigning it to ...(*T. massagetovii* Vass. nom.)."

71. *Medicago praecox* DC.

M. praecox DC., Adnot. Cat. Pl. Horti Monsp. 123 (1813), (basionym).

The epithet *praecox* is Latin for precocious, a reference to (ostensibly) its early maturation by comparison with other species of the genus.

Subordinate Taxa

None recognized here.

Type Specimen

Heyn and Davis (1970) stated with regard to the type (P!): "(S. France) In locis sterilibus circa Forum-Julian, Balbis (G!)". As noted in Heyn (1963 p. 54), De Candolle wrote with regard to *M. praecox*: "Hab. in locis sterilibus circa Forum-Julium ubi post cl. amic. Balbis reperti." There is a single sheet in Herb. DC. At P, with 3 collections, in 1809, 1810, and 1823, the first two being Balbis collections. The 1809 collection has the information "locis sterilibus..."

Synonymy

Homotypic

Medica praecox (DC.) Bubani, Fl. Pyren. 2: 477 (1900).

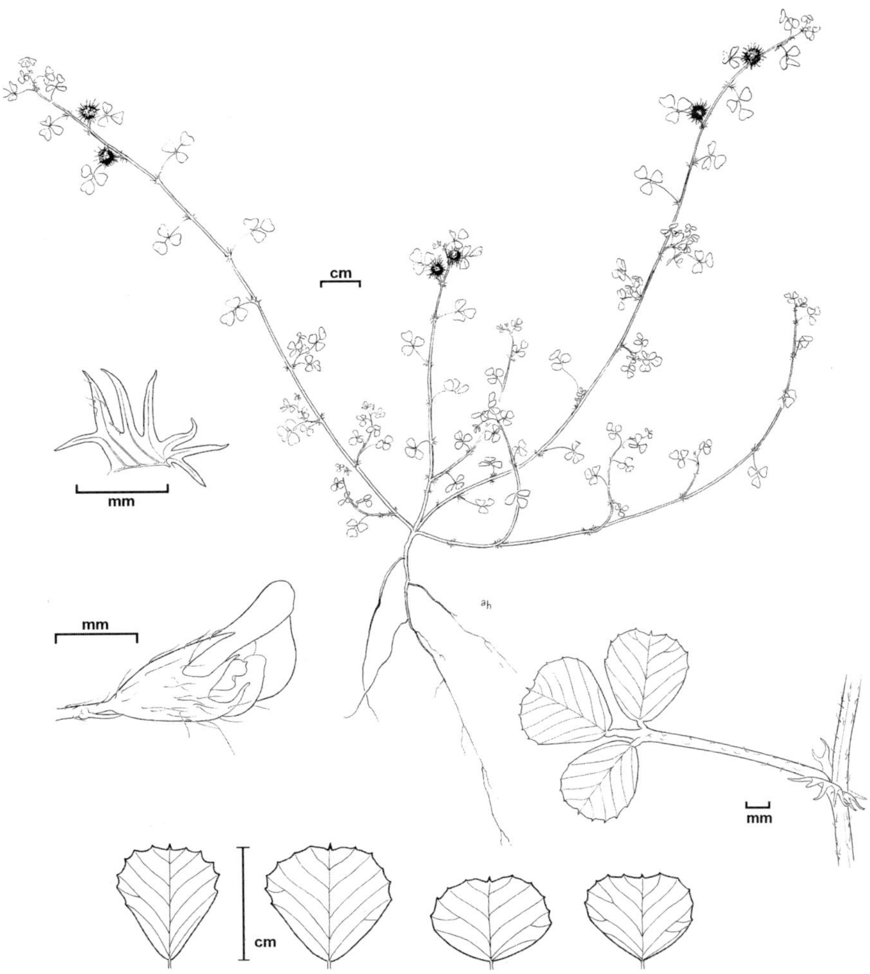

Fig. 71-1. *Medicago praecox.* **Clockwise from top:** Plant; leaf attached to stem; leaflets; side view of flower; stipule. (The plant is based on A. Fiori & A. Béguinot 1310, GH; the remainder is based on E. Small M337, DAO; artist: A. Hanes.)

Heterotypic

M. pontificalis Gennari, Nuovo Giorn. Bot. Ital. 2: 132 (1870).

M. praecox DC. f. *coerulea-venosa* Urb., Verh. Bot. Ver. Brandenb. 15: 75 (1873).

M. ×*casellasii* P. Monts., Collect. Bot., Barcinone 5: 61 (1956) is an alleged hybrid of *M. littoralis* and *M. praecox*.

Vernacular Names

ENGLISH: Early medic, plain trefoil, small leaf bur, small-leaved bur medic (bur is also spelled burr, medic is also spelled medick), early alfalfa (Bolton 1962). The names "Mediterranean medic" and "Mediterranean bur clover" have been encountered, but are too vague to be useful.

FRENCH: La luzerne précoce.

Geographical Distribution

(N = native, I = introduced, U = uncertain status)

ASIA: Cyprus (N), Lebanon (N), Syria (N), Turkey (N).

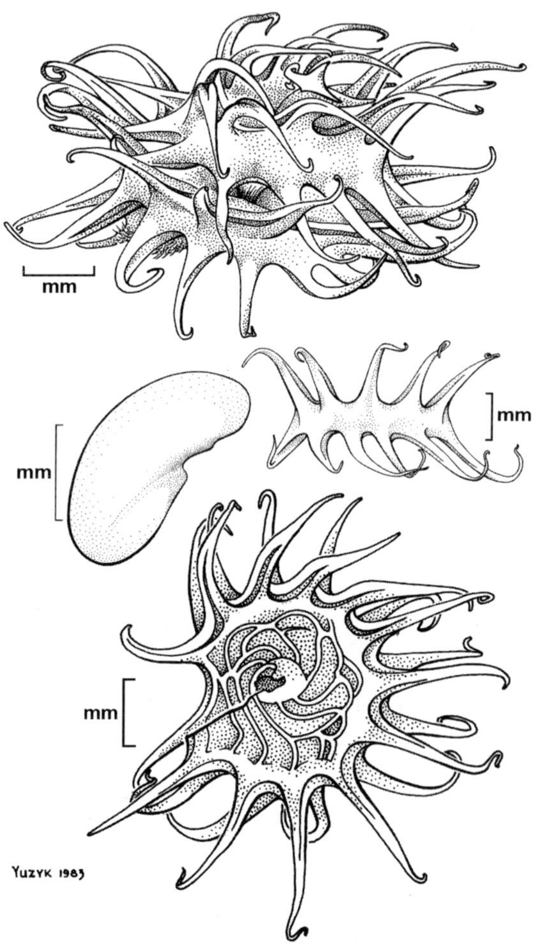

Fig. 71-2. *Medicago praecox.* **Clockwise from top:** Fruit; edge of coil; fruit coil face showing venation; seed. (Based on A. Fiori & A. Béguinot 1310, GH; artist: L. Yuzyk.)

AUSTRALASIA: Australia (I).

EUROPE: Belgium (I), Bulgaria (N), England (I), France (including Corsica; N), Greece (including Crete, Karpathos, and East Aegean Islands; N), Italy (including Sardinia; N), Spain (including Balearic Islands; N), Sweden (I), Ukraine (N), Yugoslavia (N).

NORTHERN AMERICA: USA (California, Massachusetts, Oregon; I).

Medicago praecox is indigenous to the northern and eastern Mediterranean area, and is adventitious elsewhere, primarily in Europe.

Description

Annual herb, branching from the base, branches usually procumbent, sometimes ascending, (5)10–20(30) cm long, sparsely pubescent with simple hairs. Leaflets 2–7(12) mm long, 2–5(10) mm wide, obovate to obcordate, truncate, distal third with serrate margin, glabrous on upper surface, sparsely hairy on lower surface. Stipules ovate-lanceolate to lanceolate, their margins dentate, incised or lacerate. Inflorescence of 1–2 flowers, peduncle shorter than corresponding petiole, usually only one ripe pod remaining on peduncle. Flowers 2–4 mm long, calyx pubescent with simple hairs, teeth unequal, mostly as long as tube, corolla yellow, slightly longer than calyx, 2–3(4) mm long. Young pod protrudes sideways from calyx. Mature pod short-cylindrical, 2–4(5) mm long, pubescent with simple hairs or rarely glabrescent, with 2.5–4(5) loosely appressed coils, 2–4(4.5) mm wide. Grooved spines that are usually hooked present on dorsal suture. Surface of coil with 8–12 very strongly curving radial veins that branch slightly and enter a broad lateral vein near the dorsal suture. Pod with a deep furrow between the lateral vein and dorsal suture (clearly visible only when viewing the pod in cross section, i.e., looking at the coil surface; not clearly visible when viewing the coil edges directly). Seeds 1–2 per coil, 1.7–2.4 mm long, 0.9–1.3 mm wide, reniform, smooth, yellow or brownish-yellow, separated by thin, transverse membranes, radicle usually slightly shorter than half seed length.

Identification Hints

Medicago praecox is similar to *M. coronata* in having small, loosely coiled fruits with a notably wide coil edge, but *M. praecox* normally has less than three flowers per raceme whereas *M. coronata* usually has 6–12.

Habitat and Environmental Adaptation

Habitat

Medicago praecox is an uncommon species found on dry rocky and gravelly slopes, often among shrubs adapted to arid regions of the Mediterranean. Prosperi et al. (1989) found that in Corsica the species occurred in dry and grazed locations, in rangelands and scrublands. With respect to Greece, Thanopoulos (2007) noted its presence in degraded, strongly overgrazed garrique in which weeds occur, and in a variety of weed communities and ruderal places, along roadsides, and beside streams, at altitudes up to 1300 m.

Phenology

The species flowers in early summer, often in April and May.

Soil

Medicago praecox often occupies rocky, sterile soils. Piano and Francis (1992) recorded that in its native area it is found on sandy-loam soils ranging in pH from slightly acidic to slightly basic.

Temperature

Like most annual *Medicago* species, *M. praecox* is a winter annual, completing its life cycle in the middle of the summer when the climate becomes very hot and dry.

Water

Medicago praecox is often found in dry soils. Piano and Francis (1992) recorded that in its native area it requires a minimum annual rainfall of 250 mm.

Biotic Relationships

Pollination

Small (1988) noted that the very small flowers and pollen to ovule ratio (163) are consistent with *M. praecox* being largely self-pollinated. This species is basically an inbreeder.

Dispersal

The indehiscent spiny fruits are adapted to dispersal in animal fur.

Herbivory

No information located.

Nodulation and mycorrhizae

USDA (2009) reported positive root nodulation for *M. praecox*.

Disease resistance

O'Neill et al. (2003) examined the comparative resistance of 36 annual species of *Medicago* to *Phoma medicaginis* (the cause of spring black stem and leaf spot, a major disease of alfalfa) and concluded that *M. praecox* had high resistance to it. In a similar comparative test of 36 annual species of *Medicago*, O'Neill and Bauchan (2000) found that two accessions of *M. praecox* were completely susceptible to anthracnose caused by *Colletotrichum trifolii*. Yaege and Stuteville (2002) tested the resistance of 33 annual species of *Medicago* to powdery mildew (*Erysiphe pisi*) and found that the two collections of *M. praecox* grown were susceptible to the disease (it was observed that often only one of a number of populations of other species tested was resistant).

Economic History, Cultivars, and Uses

Medicago praecox has not attracted agricultural interest to date.

Cultivars

No cultivars have been bred.

Forage

St. John-Sweeting et al. (1996) noted that one accession studied in South Australia had poor dry matter production and so was not of potential commercial interest. Grossheim (1945) described *M. praecox* as "of no interest for forage use", which is probably an accurate description given the small size of the plant. Nevertheless it is probably grazed in the very dry regions where it occurs because there is little else of interest to animals.

Fodder

The species is not harvested for animal feed.

Green manure, rotation crop, ground cover

No information located.

Other uses (salad sprouts, ornamental, plant products, etc.)

No information located.

Chromosome Number

$2n = 14$: Lesins 1952; Lesins and Lesins 1962; Simon and Simon 1965; Gillies 1971; Mariani et al. 1996.

$2n = 16$: Agarwal and Gupta 1983.

Genetic Relationships and Gene Transferability

Intraspecific

Artificial hybridization of different populations has not been attempted, so it is not known whether there are barriers to hybridization within the species.

Interspecific

Lesins (1952) was unable to hybridize *M. praecox* and *M. sativa*, both at the diploid ($2n = 16$) level. Simon and Millington (1967) were unable to produce hybrid seed by crossing *M. praecox* with *M. polymorpha* or *M. rigidula* (*M. rigiduloides*?).

As noted above, a putative hybrid between *M. littoralis* and *M. praecox* was named *M.* ×*casellasii*.

Exchange by advanced biotechnology

No information located.

Conservation Status

Medicago praecox is an uncommon species, but has a fairly large distribution range so it does not seem in danger of genetic erosion.

Germplasm Sources

Approximately 200 accessions of *M. praecox* have been deposited in the principal germplasm banks with large *Medicago* collections (see Chapter 3).

Research Problems and Priorities

Medicago praecox is a small plant of limited vigor, which does not seem to be of much direct use in agriculture, except on very dry, marginal land. Given this adaptation, it clearly has stress-resistant genes that may be useful for improvement of cultivars of *Medicago*, and this possibility requires exploration.

Representative Collections

Belgium: Lousley, J.E., Vesdre, IX-1957, in fruit (K). **Britain:** Lousley, J.E., Bedfordshire, Old Warden, wool alien in onions, 26-IX-1954, in fruit (K). **Bulgaria:** Kuzmanov, B. 80652, Struma valley, grassy places on the hill Malkija Kozuh, 9-VI-1980, in fruit (B). Velcev, V. & Bondev, I., Bulgaria australis, in prope Struma, in herbidis siccis et fruticetis, 24-VI-1953, in fruit, Inst. Bot. Acad. Scientiarium Bulgariae Plantae Bulgaricae Exsiccatae Cent VII No. 670a (B, BM). **Corsica:** Debeaux, O., la Valée du Fanyo, près de Bastia, rochers mica schisteux, 20-IV-1867, in fruit, F. Schultz Herbarium Normale Cent. II No. 1041 (K). **Cyprus:** E. Chapman 296, Malounda, Nicosia district, alt. 1000 feet, roadsides, common, 10-IV-1950, in flower & fruit (K). **France:** Bourgeau, E. 114, env. de Toulon, dans les rochers maritimes aux tablettes, 16-V-1848 (P). Cartier, Le Luc (Var), 5-V-1879, in fruit (BM). Debeaux, O., la vallée de Fango, près de Bastia (Corse), rochers micaschisteux, 20-IV-1867, in fruit, F. Schultz Herbarium Normale Cent. II. No. 104 (P). Foucaud, J., Colliovre (Pyrenées-Orientales), V-1891, in fruit, Société Rochelaise 1891, No. 3052 (P). **Greece:** Rechinger, K.H. 4253, Insula Ikaria (Nikaria), pagum Hag. Kyrikos, 18-24-IV-1934, in fruit (BM, K). Rechinger, K.K. & F. 4253, Insula Ikarai Nikaria, in ditione pagi Hagios Kirykos, ad viam litoream, 18-24-IV-1934, in fruit (K). Tuntas, B., Flora Cycladum, in insula Cythno, IV-1900, in fruit, Dr. Heldreich Plantae Exsicc. Flora Hellenicae (P). **Italy:** Fiori A., Etruria, prov. di Firenze, Lastra a Signa, in adsencu loci dicti Poggio S Romolo, in herbosis silvaticis (ericetis) copiosa, alt. ca. 170 m, solo siliceo-arenoso, 29-V-1910, in fruit, Flora Italica Exsiccata Series II Fiori & Béquinot No. 1310 (BM, K). Joad, G.C., Florence Mt. Rinaldi, 5-V-1872 (K). **Lebanon:** Paul Mouterde, S.J. 9846, Senaiya (entre Gezzine et la mer S. de Saïda, alt. 1000 m, 27-V-1950, in fruit (G). Peyron, E. 1855, Linano, Ain Sofar, 24-V-1894, in fruit (G). **Spain:** Humbert, H., Prov. de Gerone, le long de la playa de las Escudellas près de Culera, 7-VI-1910, in fruit (P). **Sweden:** Blom, C., Gôteborg, Rimgön in ruderatis, 28-X-1958, in fruit (BM). **Turkey:** Davis & Polunin 25850, prov. Antalya, Alanya Promentory, alt. 100 m, rocky limestone slopes, 11-IV-1956, in fruit (BM). Raus, T. 17777, Region C1 (Davis 1970), Prov. Muğla, Osteite des Bafa Gölü bei Kapikiri/Herakleia, alt. 20 m, "5.4.1992", in fruit (B).

72. *Medicago prostrata* Jacq.

M. prostrata Jacq., Hort. Bot. Vindob. 1: 39, tab 89 (1770–1771), (basionym).

The epithet *prostrata* is Latin for low, and is applied botanically to plants of limited stature, often because the stems are decumbent. This species, however, frequently grows erect.

Key to Subordinate Taxa

Pods covered with gland-tipped hairs or rarely glabrous or glabrescent .... *M. prostrata* subsp. *prostrata*

Pods densely white due to a dense covering of simple hairs .... *M. prostrata* subsp. *pseudorupestris*

In reducing the new species that he had recognized in 1917 as *M. pseudorupestris* Hayek to a variety in 1921, *M. prostrata* var. *pseudorupestris* (Hayek) Hayek, Hayek commented that (translated from German): "The specimens with ripe fruits which have now been collected by I. Dörfler on the Pašrik [peak on Yugoslav–Albanian frontier] show that the differences that I assumed to exist in the fruit structure compared with *M. prostrata* are not present; instead, quite apart from the fact that in *M. prostrata* they are glabrous or glandular and dispersed while in *M. pseudorupestris* they bear hairs which are appressed against the surface, the fruits are in all other respects completely identical so that the latter type can only be a form of the other type."

Fig. 72-1. *Medicago prostrata* subsp. *prostrata*. From Hallier (1885, vol. 23, plate 2336).

Fig. 72-2. *Medicago prostrata* subsp. *prostrata*. From (the original species description): Jacquin, N.J. 1770. Hortus botanicus Vindobensis I, Tabula 89. Vienna.

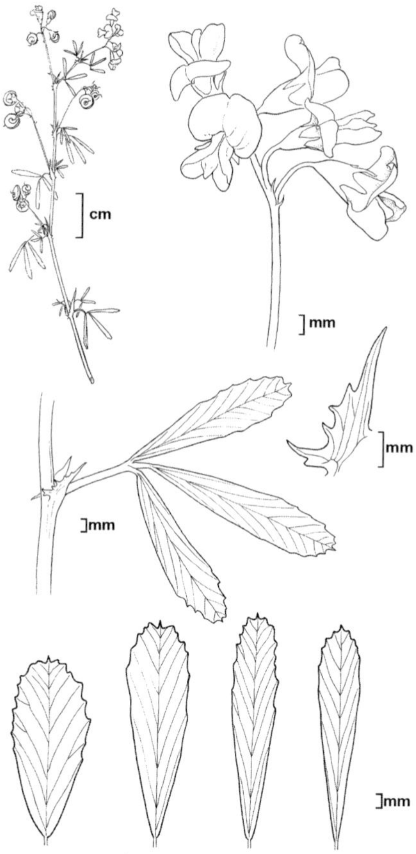

Fig. 72-3. *Medicago prostrata* subsp. *prostrata*. **Clockwise from center left:** Leaf attached to stem; branch with flowers and fruits; group of flowers; stipule; leaflets. (Based on E. Small M279, DAO; artist: A. Hanes.)

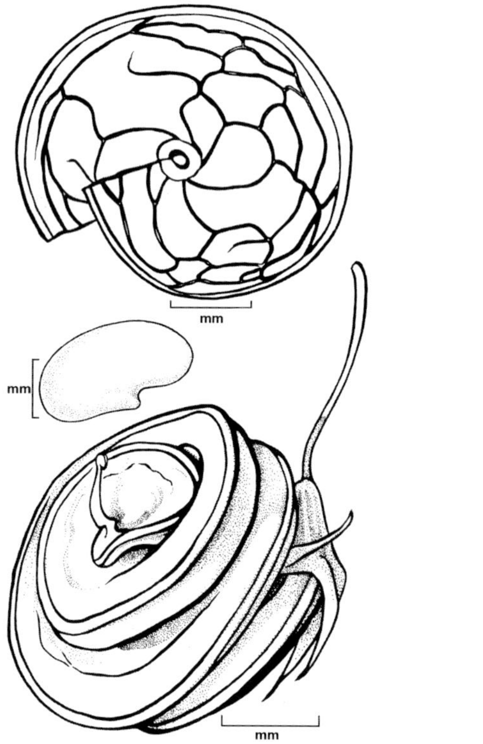

Fig. 72-4. *Medicago prostrata* subsp. *prostrata*. **Top:** Fruit coil face showing venation. **Center:** Seed. **Bottom:** Fruit. (Based on E. Small M279, DAO; artist: A. Hanes.)

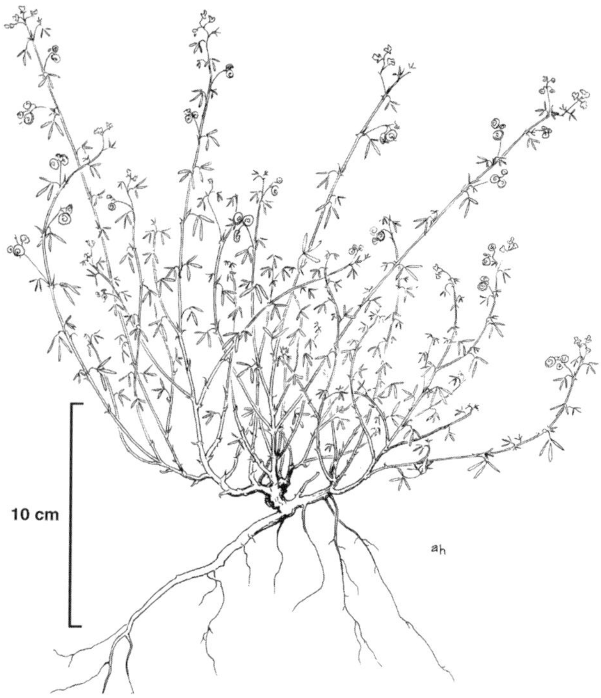

Fig. 72-5. *Medicago prostrata* subsp. *pseudorupestris*. (Plant based on Dorfler 851, A; artist: A. Hanes.)

Bornmüller (1925, pp. 469–470) contended that *M. prostrata* var. *pseudorupestris* is distinguished by having larger flowers (7–8 mm, rarely more than 5 mm in var. *prostrata*), a more abundantly flowered inflorescence (2–4(8) flowers), and pods with fewer coils (1.5–2, versus 2–4 in var. *prostrata*). However, these observations are based on very few plants.

Described as native to Albania, Macedonia, and the Balkans, *M. prostrata* var. *pseudorupestris* was delimited simply by having densely white-pubescent pods. As noted in Chapter 7, there is good reason to believe that dramatic fruit pubescence in part of a species is an adaptive response to local insects.

Type Specimen

The material on which Jacquin based *M. prostrata* was grown in the Vienna Botanical Garden (and based on seed from Italy according to Neibreich (1859: 933).

Synonymy

Homotypic

No names found.

Heterotypic

Note: The extensive synonymy presented in Micevski (1973; often rendered "Mitsevski") should be noted, although the classification system seems very excessive.

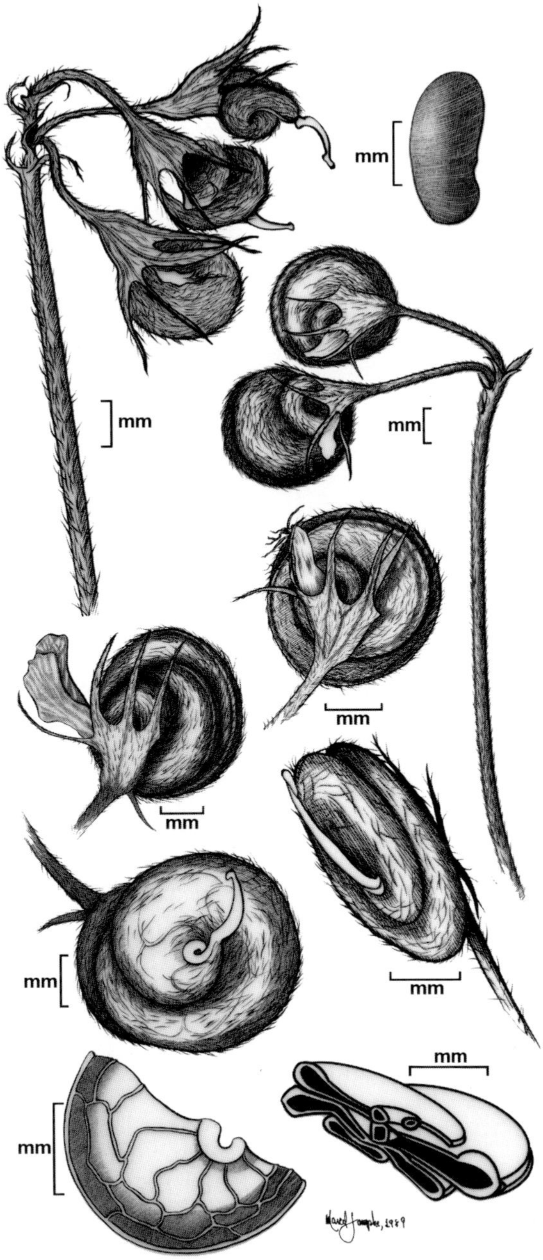

Fig. 72-6. *Medicago prostrata* subsp. *pseudorupestris.* **Clockwise from top left:** Fruiting branch; seed; fruiting branch; fruit viewed from base showing calyx and remnants of staminal column; fruit attached to peduncle; diagram of long-sectioned coiled fruit; part of coil face showing venation; stigma end of fruit; fruit viewed from base showing calyx and remnants of corolla. (Artist: M. Jomphe.)

M. declinata Kit., Linnaea 32: 613 (1863).

M. prostrata Jacq. f. *declinata* (Kit.) Rohlena, Conspectus fl. Montenegr. 178 (1842).

M. prostrata Jacq. f. *declinata* (Kit.) Urb., Verh. Bot. Ver. Prov. Brandenburg 15: 55 (1873), nom. superfl. (Described by Urban (translated from German): "Whole plant short compressed hairy. Pods often interspersed with glandular hairs.")

M. erecta Winterl, Index horti bot. Pest. tab. 4, c. (1788), non. *M. erecta* Kotov (1940).

M. prostrata Jacq. var. *erecta* (Winterl) Borb., Balat. Flor. 422 (1900).

M. prostrata Jacq. f. *glabra* (Kit.) Urb., Verh. Bot. Ver. Prov. Brandenburg 15: 55 (1873). (Described by Urban (translated from German): "Whole plant glabrous or somewhat hairy only on the youngest parts.")

M. prostrata Jacq. f. *glandulifera* Urb., Verh. Bot. Ver. Prov. Brandenburg 15: 56 (1873). (Described by Urban (translated from German): "Whole plant covered with spreading glandular hairs.")

M. prostrata Jacq. f. *glandulifera* (Urb.) Hay., Prodromus Fl. Penins. Balc. 1: 836 (1924).

M. prostrata Jacq. β pubescens Guss., Pl. rar. 313 (1826).

M. teniflora Schur, Enum. Pl. Transsilv. 151 (1866).

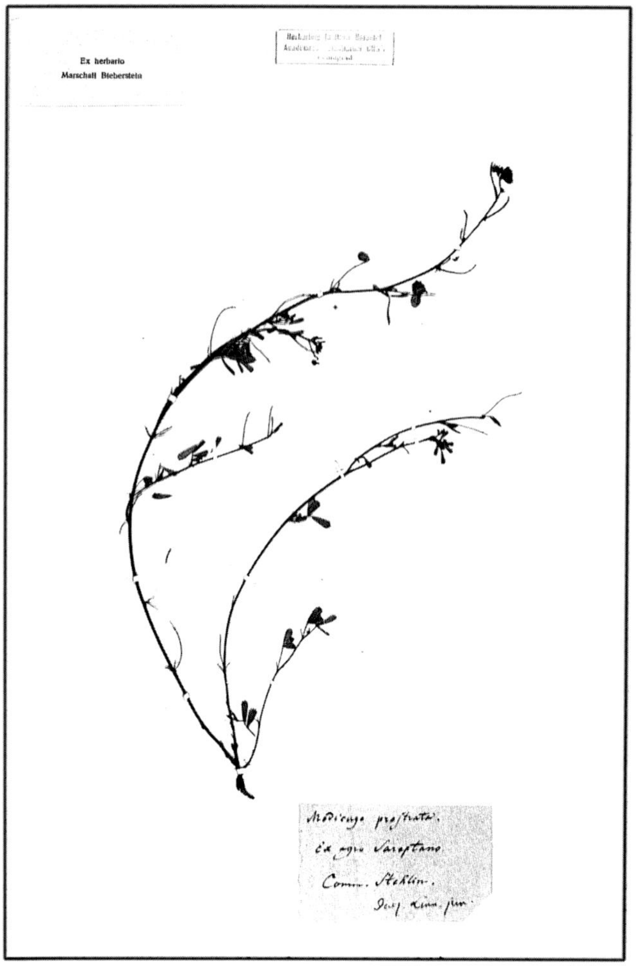

Fig. 72-7. Type specimen of *M. prostrata*. Photograph courtesy of LE. (The label appears to read "Ex herbario Marschall Bieberstein; *Medicago prostrata*, ex agro Sareptans, Comm. Stechlin, letz. Linn. Jun.")

M. tenuifolia C. Presl., Fl. Sic. 1: 20 (1826).

M. prostrata Jacq. var. α *typica* Pospichal, Fl. Oesterr. Küstenl. 1: 356 (1898), nom. inadmiss.

M. prostrata Jacq. var. α *typica* Fiori, Nuova Fl. Anal. d'Italia, 1: 830 (1923/29), nom. inadmiss.

Medicago prostrata subsp. *pseudorupestris* (Hayek) Micevski

M. prostrata Jacq. subsp. *pseudorupestris* (Hayek) Micevski, God. Zborn. Prir.-Mat. Fak. Univ. Kiril & Metodij Skopje, Biol. 25: 140 (1973).

Homotypic

M. pseudorupestris Hayek, Kaiserl. Akad. Wiss. Wien, Math.-Naturwiss. Kl., Denkschr. 94: 172 (1918), (basionym).

M. prostrata var. *pseudorupestris* (Hayek) Hayek, Österr. Botan. Zeitschrift 70(1-2): 12 (1921).

Heterotypic

M. prostrata Jacq. subsp. *pseudorupestris* (Hayek) Hayek f. *glandulosa* Micevski, God. Zborn. Prir.-Mat. Fak. Univ. Kiril & Metodij Skopje, Biol. 25: 141 (1973). (Holotype specified, at SKO.)

M. prostrata Jacq. subsp. *vukovicii* Micevski, God. Zborn. Prir.-Mat. Fak. Univ. Kiril & Metodij Skopje, Biol. 25: 142 (1973). (Holotype specified, at SKO.)

M. prostrata Jacq. subsp. *vukovicii* Micevski var. *vukovicii* f. *pilosa* Micevski, God. Zborn. Prir.-Mat. Fak. Univ. Kiril & Metodij Skopje, Biol. 25: 143 (1973). (Holotype specified, at SKO.)

M. prostrata Jacq. subsp. *vukovicii* Micevski var. *matkae* Micevski, God. Zborn. Prir.-Mat. Fak. Univ. Kiril & Metodij Skopje, Biol. 25: 144 (1973). (Holotype specified, at SKO.)

M. prostrata Jacq. subsp. *vukovicii* Micevski var. *opolkae* Micevski, God. Zborn. Prir.-Mat. Fak. Univ. Kiril & Metodij Skopje, Biol. 25: 144 (1973). (Holotype specified, at SKO.)

Type Specimen

Type: not determined. In his original (1917) description of *M. pseudorupestris*, Hayek cited only one collection: "Nordost-Albanien: Felsige Abhänge in der Gipfelregion der Pašrik, zirka 1800 m (Exs. Nr. 587)." The exsiccata bears the information: "Dörfler 587, Reise im Albanisch Montenegrischen Grenzgebiete 1-VII-1914." The type is thus from northeastern Albania from the region of the Bastrik mountain, 1800 m above sea level. *Taxonomic Literature*, edition II states that the location of the types of Ignaz Dörfler (1866–1950) is unknown. A suitable choice for the type would be Dörfler 587, Reise im Albanisch Montenegrischen Grenzgebiete 1-VII-1914, Nordostalbanien. Felsige Abhänge in der Gipfelregion der Pašrik (cited in Kozuharov 1965). However, Dörfler, I. 851 (see "Representative Collections" below) is cited as a putative type at Zurich (herbarium: Z; http://www.zuerich-herbarien.unizh.ch/datenbanken/details.php?l=&spTypFlg=&spBarCod=Z-000023535; current July 2010).

Vernacular Names

English: No names found.

French: Luzerne couchée.

Geographical Distribution

(N = native, I = introduced, U = uncertain status)

Africa: Mauritius (I).

Europe: Albania (N), Austria (eastern; N), Czechoslovakia (N), Greece (N), Hungary (N), Italy (N), Romania (N), Yugoslavia (N).

Medicago prostrata is native to middle and southeastern Europe.

Description

Perennial herb, branches 10–40(60) cm long, slender, procumbent to ascending, glabrous, glabrescent, or pubescent with either or both of simple and glandular hairs. Leaflets oblong-cuneate or linear-cuneate, distal margin serrate, (3)4–10(15) mm long, (1)2.5–4(5) mm wide. Stipules ovate-lanceolate, the lower ones with dentate margins at base. Inflorescence of 1–8(20) flowers, the peduncle longer than the subtending petiole. Corolla (3)5–7(8) mm long, yellow. Calyx about half length of flower or less, the tube as long as or shorter than the linear teeth, glabrous or pubescent. Pedicels thin, longer than the flowers (unlike most other species of *Medicago*) and recurved. Young pod turns sideways through the calyx teeth. Mature pod (2.5)3–4(5) mm wide, loosely coiled with 2–3(4.5) spirals, lacking dorsal suture spines, glabrous or with simple and/or gland-tipped hairs, the coil face with 5–8 delicate, more or less curved veins that anastomose in the outer half of the face, the coil axis lacking or almost lacking an opening. Seeds 1–2 per coil, yellow or brownish, 2–2.3 mm long, 1–1.3 mm wide, not separated or separated by thin, transverse partitions, radicle about equal to half of seed length.

Identification Hints

Medicago prostrata var. *prostrata* can be confused with other yellow-flowered, montane perennials with coiled, nonspiny fruits that have gland-tipped trichomes. However, *M. sativa* subsp. *glomerata* does not have the long reflexed pedicels of *M. prostrata*; *M. suffruticosa* has much wider leaflets; *M. rupestris* and (the nonmontane) *M. cancellata* are well to the east of *M. prostrata* and the venation of their fruits is much more prominent; and *M. papillosa* is also well to the east and has distinctive interseed fruit partitions.

Habitat and Environmental Adaptation

Habitat

Medicago prostrata occurs on dry rocky hillsides, on grassy sites, in open shrublands, and at the edges of fields. Thanopoulos (2007) noted that in Greece the species has been collected at altitudes from 300 to 2040 m.

Phenology

Flowering occurs from May to August in its indigenous area.

Soil

The species has been collected on rocky limestone slopes and in sandy soils.

"The only known serpentine local endemic in Macedonia is *M. prostrata* subsp. *pseudorupestris*" (http://www.macedoniagiftshop.com/natural_monuments_in_macedonia_1.htm; current July 2007). The Messiano Nero Valley of Greece, where *M. prostrata* subsp. *pseudorupestris* occurs, has half serpentine soil and half calcareous soil (http://eunis.eea.europa.eu/sites-factsheet.jsp?tab=1&idsite=A00040022; current July 2007). Bornmüller (1925) also noted that *M. prostrata* subsp. *pseudorupestris* occurred on serpentine soil in Macedonia. Ritter-Studnička (1970) recorded the presence of *M. prostrata* on serpentine soil in Bosnia. Thanopoulos (2007) described the soils occupied in Greece as "serpentine and calcareous substrates".

Temperature

Medicago prostrata is native to an area where cool, low mountain conditions prevail. However, Nicoloff (1961) stated that Czechoslovakian populations occur in "very dry places with a high temperature."

Water

The species has been recorded as growing in dry soils and is considered to be drought resistant. It appears to grow on well-drained soils. Suksayretrup (1986) studied the xerophytic characters in *M. prostrata*, in comparison with *M. sativa*. Xerophytic characters identified in *M. prostrata* included "thick cuticles and cell walls; small and thick leaves with surface wax, the secretion of which increased under moisture stress; glandular hairs distributed generally but particularly densely on the abaxial side of leaves; and a high cortex:radius ratio in the stems, which accumulated starch granules in the perimedullary zone when the plants were stressed." Additionally, "growth chamber experiments showed that *M. prostrata* had higher stomatal conductance than *M. sativa* under high stress treatment, and a higher shoot water potential under low stress conditions. Plant height of *M. sativa* was reduced more under moisture stress than was that of *M. prostrata*."

Biotic Relationships

Pollination

Small (1988) recorded a pollen to ovule ratio of 221 for *M. prostrata*, which is in the range of that for inbreeding (annual, self-pollinating) species of *Medicago*. Also, the flowers are of relatively small size by comparison with other perennial *Medicago*, suggestive of the same thing. However, no perennial species of the genus is known to be exclusively self-pollinating, and it is likely that *M. prostrata* outbreeds to at least some extent.

Dispersal

The indehiscent fruits have no obvious adaptations to dispersal, but given the typical occupation of slopes, they probably roll downhill fairly well.

Herbivory

Shade and Kitch (1983) noted that the glandular hairs of *M. prostrata* resulted in resistance to the pea aphid (*Acyrthosiphon pisum* (Harris)). Brewer et al. (1986*a*) found that *M. prostrata* was extremely resistant to feeding and oviposition by the adult potato leafhopper *Empoasca fabae*

(Harris). Danielson et al. (1987*a*) found that glandular-haired *M. prostrata* is resistant to oviposition of the alfalfa weevil *Hypera postica* (Gyllenhal); Danielson et al. (1987*b*) found resistance to adult feeding of the weevil; and Danielson et al. (1986) found that the larvae of the weevil did not develop as well as on alfalfa. However, Danielson et al. (1990), in field tests of larval presence and damage from the alfalfa weevil, did not find significant differences between *M. prostrata* and alfalfa cultivars. Lenssen et al. (1988*a, b*) reported that forage quality parameters were not negatively affected by the presence of the glands (both in populations of *M. prostrata* and hybrids with *M. sativa*), and Lenssen et al. (1989) found that sheep foraging was not negatively affected by the presence of glandular hairs on the species (although the animals preferred tetraploid alfalfa).

Nodulation and mycorrhizae

No information located.

Disease resistance

No information located.

Economic History, Cultivars, and Uses

Medicago prostrata has not been employed directly as a crop plant but, as noted below, it has contributed germplasm to experimental cultivars of alfalfa. Vassilczenko (1949) commented that it "may be of value as a pasture-land plant."

Cultivars

No cultivars of *M. prostrata* have been bred (although the species has been used as a parent for cultivars: see the section on Pubescence in Chapter 7).

Forage

Medicago prostrata is not sown for forage.

Fodder

The species is not employed as livestock feed.

Green manure, rotation crop, ground cover

No information located.

Other uses (salad sprouts, ornamental, plant products, etc.)

No information located.

Chromosome Number

$2n = 16$: Nilsson and Lassen 1971; Baltisberger and Baltisberger 1995.

$2n = 16, 32$: Lesins and Lesins 1960; Lesins 1968.

$2n = 32$: Ahuja 1955; Bakšay 1958; Clement 1962; Majovsky et al. 1970*a*.

"*Medicago prostrata* Jacq. has two localities of habitation, a northern in Czechoslovakia–Moravia and in Transylvania, and a southern in Italy–Sicily. At that, the Italian proveniences (sic) are diploids ($2n = 16$), while the Czechoslovak are tetraploids ($2n = 32$)" (Nicoloff 1961).

Gene Transferability and Genetic Relationships

Intraspecific

Artificial hybridization of different populations has not been attempted, so it is not known whether there are barriers to hybridization within the species.

Interspecific

Lesins (1962) hybridized *M. prostrata* with *M. sativa* at both the $2n = 16$ and $2n = 32$ levels, and observed an interspecific barrier when *M. prostrata* was the female parent. Lesins (1968) hybridized $2n = 16$ *M. sativa* (as "*M. glomerata*") with $2n = 16$ *M. prostrata* and the hybrids were quite fertile. Sorensen et al. (1981) crossed diploid *M. prostrata* with tetraploid *M. sativa*, and reported that the hybrid was tetraploid and fertile. A synthetic, glandular-haired alfalfa that included germplasm of *M. prostrata*, with multiple pest resistance, was reported by Sorensen et al. (1994; also see discussion of Pubescence in Chapter 7).

The following taxa are putative hybrids with *M. sativa*:

M. ×*mixta* Sennh. (= "*M. falcata*" × *M. prostrata*), Verh. Zool.-Bot. Ges. Wien 38: 32 (1888).

M. ×*recta* Kit. (= "*M. falcata*" × *M. prostrata*), Linnaea 32: 613 (1863). (Illegitimate homonym.)

Exchange by advanced biotechnology

No information located.

Conservation Status

Given the rarity of the species, there is a need for a conservation assessment of *M. prostrata*.

Germplasm Sources

Twenty accessions are listed among the major germplasm institutions with large *Medicago* collections (see Chapter 3).

Research Problems and Priorities

Medicago prostrata is a rare species that has not received much study to date, except for examination and utilization of its anti-insect glands, which have been transferred to some experimental lines of alfalfa. The species interbreeds with alfalfa and as such is an invaluable resource for transferring its stress-resistant genes. Very little germplasm has been collected to date and increasing germ bank holdings of this species is the most pressing need.

Representative Collections

Albania: Baldacci, A. 164, Alveo fl. Jala ad pontem sub Abate, distr. Teutari, 22-24-VII-1897, in flower & early fruit (BM). Baldacci, A. 416, Cáfa Evozdeca infra Orahovo et Fujepsi, 2-VII-1898, in flower & fruit (BM). **Austria:** Spreitzenhofer, Austria, inferior, in graminosis apricis ad

Felixdorf pr. West-Neustadt, in flower & early fruit (received by P in 1885), Flora Exsiccata Austro-Hungarica No. 25 (BM, K, P). **Czechoslovakia:** Pödpera, J., Moravia Mer., prope urbem Mikulov (Nikolsburg), alt. ca. 380 m, in declivibus stepposo-saxosis collis "Kotelná", 27-VI-1926, in flower & fruit, Flora Exsiccata Reipublicae Bohemicae Slovenicae No. 227 (K, BM). **Hungary:** Bohatsch, F., Ofcn, Adlersberg alt. 200 m, 1-VII-1879 (K). Bohitsch, Budae in saxosis dolomis, 1-VII-1879, in flower & fruit (K). Javorkei, J., Budapest, in saxosis muntium, 3-VI-1922, in flower & very early fruit (K). Sag, G., Hüvosvölgy, near Budapest, 8-VII-1976, in flower & early fruit (BM). **Italy:** Fiori, A., Etruria, Prov. di Grosseto, Montepescali, alt. ca. 320 m, locis saxisis collium, oppidi meridiem versus sitis, Astragalo Muelleri socia, solo argillo-calcareo, 28-V-1911, in flower & early fruit (BM). Groves, H., Montes Sircute, alt. "3000", in saxosis, VII-1876, in fruit (BM). Groves, H., southern Italy, Abruzzi, flanks of Mte. Sircuto, stony ground, VII-1877, fruits apparently shed (BM). Huet du Pavillon, E. & A., ad rupes prope Campo di-Giove, Aprutii 28-VII-1856, in flower, Plantae Neapolitanae No. 299 (K, P). Marchesetti, C., Ager Tergestinus, prope Tergeste (Trieste), in rupestribus apricis montis Spaccato, alt. 400 m, solo calcareo nummulitico, V-1912, in flower & early fruit, Flora Italica Exsiccata Series II Fiori & Béguinot No. 1861 (BM, K). **Romania:** Gergely, I., Turcu, L. & Lörinczi, F., Crisana, distr. Bihor, in rupibus calcareis Dl. Magurii angustiarum Defileul Crisului Repede (Vadul Crisului), alt. 520–560 m, 8-VI-1965, in flower & fruit, Flora Romaniae Exsiccata Herb. Univ. Napucensis No. 3065 (K). **Yugoslavia:** Bierbach, Montenigrina Lovcen, Tvenora, alt. 1350 m, VI-1908, in flower & early fruit (BM). de Tommasini, M., Monte Spaccato, pres de Trieste (Illyrie), lieux herbeux, secs et rocailleux de la montagne calcaire, flowers 19-V, fruits 19-VI-1852, in fruit, F. Schultz, Herbarium Normale Cent. II No. 1038 (BM, K, P, S). Duyfjes, B.E.E., Hengeveld, R., & Van der Voet, V.C.W. 611, Bosna-Hercegovina, Mostar, along Naretva river, 12-VI-1962, in flower (BM). Gilliat-Smith, B. 3461, Dalmatia, Zlakarska, abandoned vineyards, V-1935, in flower K). Justin, Carniolia, in graminosis lapidosis et ad margines viarum prope pagum Vreme in Valle Reka, solo calcareo, in flower & very early fruit, alt. 360 m, VI, Flora Exsiccata Carniolica No. 485 (BM). N.D. Simpson 35.182, Dalmatia, solin near Split, waste ground, 21-IV-1935, in flower (BM). Pichler, T., in monte karst prope Triest, V-1880, in flower & fruit (K). Raap, H., Hostar, an grnsigen alt. 200 m, 20-VI-1895, in flower & fruit, Plantae Hercegovinae exsiccatae a. 1895 a. H. Raap lectae No. 47 (K, P). Stainton, J.D.A. 7844, Matka, west of Skopje, Macedonia, alt. 1000 feet, limestone rocks, 3-VII-1959, in flower & fruit (K). Turrill, W.B. 1145, Flora Karst, near St. Carzian, 20-VIII-1925, in flower & fruit (K). Baenitz, C., Hercegovinae, Velez-Gebirge, Gendarmerie posten Rujiste, auf Wiesen bei Zimlje gornje, alt. 600 m, 6-VI-1898, in flower (BM).

M. prostrata subsp. *pseudorupestris*

Albania: Dörfler, I. 851, Nord-Albanien, Distrikt Hasi Pastrik An Felsen der Sudhange, alt. 1800 m, 24-VII-1918, in flower & fruit (A, BM, M, S; locus classicus of subspecies *pseudorupestris*). Callier, A., Tauria, In rupibus montis cretacei Akkaja prope Karasubar locus classicus, VI-1900, in flower & early fruit, Herbarium Normale editum I Dörfler No. 4227 (BM).

73. *Medicago radiata* L.

M. radiata L., Sp. Pl. 778 (1753), (basionym).

The epithet *radiata* is based on the Latin *radiatus* (rayed, as in a wheel), an allusion to the fruit that is sometimes reminiscent of a wheel with spokes arrayed around an axle.

Subordinate Taxa

None recognized here.

Type Specimen

Lectotypified by Heyn, (1959: 161) (Hb. Cliff., Hort. Sicc. Cliff. 377-8, BM!). Jarvis (2007) noted that Linnaeus referred to "habitat in Italia" and that the lectotype bears the code BM-000646776.

Synonymy

Homotypic

Medica radiata Miller, Gard. Dict. (ed 8) no 3 (1768).

Pocockia radiata (L.) Trautv., Bull. Acad. Petersb. 8: 271 (1841).

Nephromedia radiata (L.) Kostel., Ind. Sem. Hort. Prag. 1844: 92 (1844).

Trigonella radiata (L.) Boiss., Fl. Or. 2: 90 (1872). (The name *Trigonella radiata* Da Costa, coined in 1778, is an old

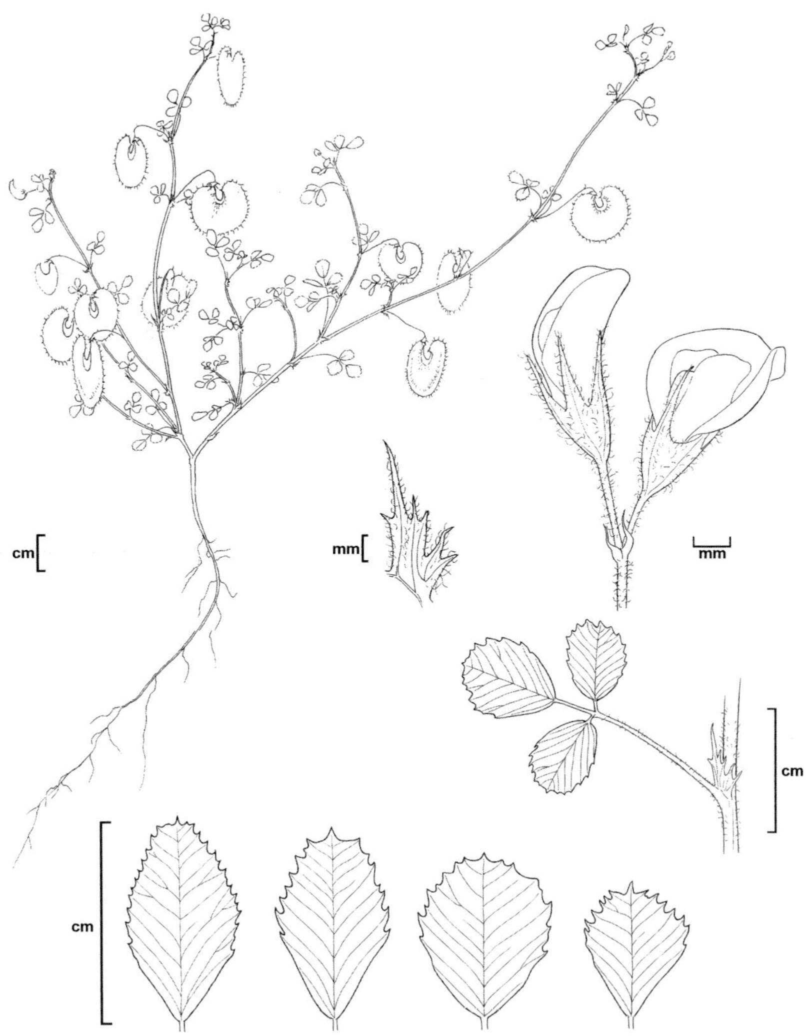

Fig. 73-1. *Medicago radiata.* **Clockwise from top:** Plant; group of flowers; leaf attached to stem; leaflets. **Center:** Stipule. (Based on accession 2531992, US; artist: A. Hanes.)

synonym for a marine mollusc with the current name *Mactra corallina* (L.).)

Melissitus radiatus (L.) Latsch., Trudy Tiflissk. Bot. Inst. 21: 16 (1970).

Pseudomelissitus radiatus (L.) P. N. Ovczinnikov, M.R. Rassulova & G.K. Kinzikaëva in P.N. Ovczinnikov, Fl. Tadzhiksk. SSR 5: 328, 629 (1978).

Heterotypic

M. radiata L. var. *dasycarpa* Ser. in DC., Prod. 2: 172 (1825). (Vassilczenko (1953) stated that the type is at P.)

M. radiata L. f. *dasycarpa* (Ser. in DC.) Urb., Verh. Bot. Ver. Prov. Brandenb. 15: 51 (1873). (Urban (1873) characterized the taxon as "younger parts of the plant somewhat white-tomentose, pods hairy.")

Trigonella dasycarpa (Ser.) Vassilcz., Acta Inst. Bot. Acad. Sci. URSS, Ser. 1, Fasc. 10, 180 (1953).

Melissitus dasycarpus (Ser.) Latsch., Trudy Tiflissk. Bot. Inst. 19: 7 (1958), in obs., sine relat. loc. (listed on exsiccata and in Czerepanov (1995) as "Latsch. ex Golosk.")

Pseudomelissitus dasycarpus (Ser.) Ovcz., M.R. Rassulova & G.K. Kinzikaëva, in P. N. Ovczinnikov, Fl. Tadzhiksk. 5: 329 (1978).

Nephromedia dasycarpa (Ser.) Czerep., Sosud. Rast. SSSR: 238 (1981).

Radiata dasycarpa (Ser.) Ovez., M.R. Rassulova & G.K. Kinzikaëva (op. cit., above).

Radiata dasycarpa (Ser.) Soják in Sborn. Nár. Muz. Praze, 1982(1-2): 103 (1982), comb. superfl.

Radiata glabra Ovcz., M.R. Rasulova & G.K. Kinzikaëva, Fl. Tadzhik. SSR, ed. 2, 6: 688 (1981). (Czerepanov (1995) assigned this to *M. radiata*.)

Radiata leiocarpa Soják, Sborn. Nár. Muz. Praze, 1982(1-2): 103 (1982). (Czerepanov (1995) assigned this to *M. radiata*.)

M. lunata Rchb., Fl. Germ. Excurs. 501 (1832).

Vernacular Names

English: Ray-podded medic (medic is also spelled medick).

French: La luzerne rayonnante, luzerne rayonée.

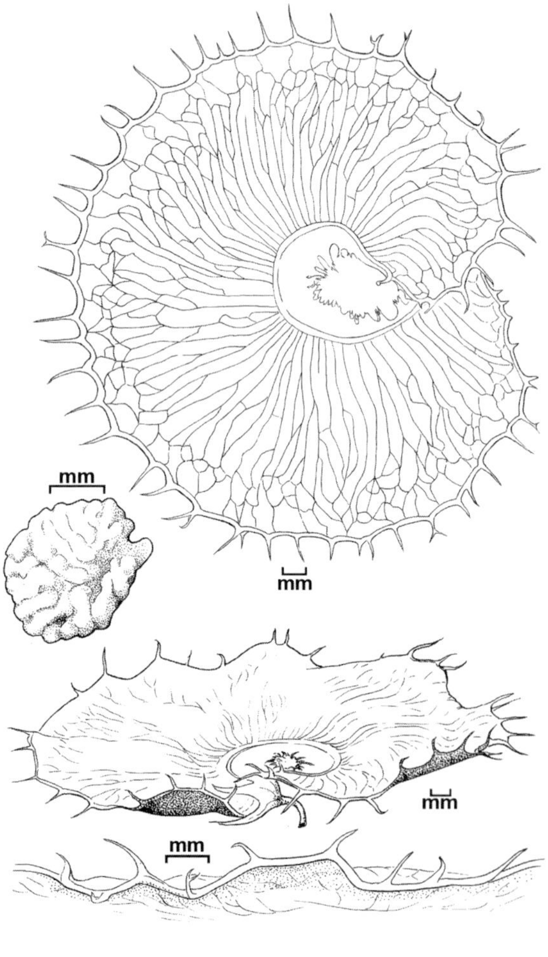

Fig. 73-2. *Medicago radiata.* **Top to bottom:** Fruit; seed; oblique view of fruit; edge of fruit coil. (Based on E. Small M225, DAO; artist: A. Hanes.)

Fig. 73-3. *Medicago radiata.* From Hallier (1885, vol. 23, plate 2338).

Geographical Distribution

(N = native, I = introduced, U = uncertain status)

Asia: Afghanistan (N), Armenia (N), Azerbaijan (N), Iran (N), Iraq (N), Israel (N), Jordan (N), Lebanon (N), Syria (N), Turkey (N), Turkmenistan (U), Uzbekistan (U).

Europe: France (I), Italy (I), Spain (?; not considered present in Greuter et al. 1989).

Medicago radiata is native from the eastern Mediterranean to Central Asia.

Description

Annual herb, branched from base, branches prostrate, procumbent, or ascending, (5)10–25(30) cm long, more or less pilose with simple hairs. Leaflets (4)5–15(20) mm long, (3)4–8(11) mm wide, obovate to oblong or oval or cuneate-obovate, distal margin serrate for about half to two-thirds length of leaflet, margin sometimes incised ("f. *incisa*"), somewhat pubescent (especially on lower surface). Stipules lanceolate-subulate, incised-dentate or with a few teeth at base, pilose. Inflorescence of 1–2(4) flowers, peduncle usually longer than subtending petiole (elongating in fruit). Flowers 4–6 mm long, calyx campanulate, pilose, teeth lanceolate-subulate, about as long as tube, corolla yellow, less than twice length of calyx. Fruit flat, often crescent-shaped, with 0.5–1.5 coils, membranous, pubescent or glabrous, (12)15–20(28) mm at widest point, usually pectinate (spiny-ciliate, some forked) on the convex (dorsal) suture, shortly fimbriate-winged on the concave (ventral) suture, fruit face with a dense net of veins near marginal vein, the fruit wall protruding at the locations of the seeds. Seeds 4–7(8) per pod, (2)2.3–3(3.2) mm long, 2–2.7 mm wide, ovoid-oblong but compressed from sides, yellowish-brown, transversely rugose, radicle about as long as seed length, seeds separated by transverse spongy fruit partitions.

Identification Hints

This very distinctive species is unique in having small spines or fimbriations on both fruit sutures.

Habitat and Environmental Adaptation

Habitat

Medicago radiata occurs on dry, rocky hillsides and in steppes, fields, grasslands, pastures, and deserts.

Phenology

The species flowers from March to June in its Mediterranean native range, the time depending on the local climate.

Soil

Piano and Francis (1992) recorded the species' occurrence in soils with a range of textures that are slightly to moderately basic. Similarly with respect to soil texture, Townsend (1974) reported that in Iraq the species occurs in gravel, sand, silt, and clay.

Temperature

Medicago radiata is found in deserts, indicating heat tolerance, although it is a winter annual and so avoids the hottest period of the year by early completion of its life cycle. However, compared with most other medics, it grows in relatively cool areas (Piano and Francis 1992).

Water

Medicago radiata often occupies dry, well-drained soils in its native area, and indeed seems to be restricted to areas of very low annual rainfall (Gintzburger and Blesing 1979). Piano and Francis noted that in its indigenous regions it requires a minimum annual rainfall of 150 mm.

Biotic Relationships

Pollination

Small (1988) noted that the small flowers and pollen to ovule ratio (123) are consistent with *M. radiata* being largely self-pollinated. This species is basically an inbreeder.

Dispersal

The flat, indehiscent pods of this species may be an adaptation to wind dispersal.

Herbivory

No information located.

Nodulation and mycorrhizae

USDA (2009) reported positive root nodulation for *M. radiata*.

Disease resistance

Yaege and Stuteville (2002) tested the resistance of 33 annual species of *Medicago* to powdery mildew (*Erysiphe pisi*) and found that one collection of *M. radiata* grown was susceptible to the disease whereas three other collections showed slight resistance to it (it was observed that often only one of a number of populations of other species tested was resistant). In a comparative test of 36 annual species of *Medicago*, O'Neill and Bauchan (2000) found that four accessions of *M. radiata* exhibited a range of susceptibility to anthracnose caused by *Colletotrichum trifolii* (one accession was quite susceptible, two were moderately resistant).

Economic History, Cultivars, and Uses

Medicago radiata has attracted little agricultural interest to date.

Cultivars

Azizi et al. (2004) referred to "*Medicago radiata* cv. Radiata". An inquiry regarding this was not answered and there does not seem to be any cultivar of *M. radiata*.

Forage

Medicago radiata is not sown for forage. Townsend (1974) reported that in Iraq "the plant is undoubtedly of some forage value in our northern and western pastures."

Fodder

Medicago radiata is not harvested as animal feed.

Green manure, rotation crop, ground cover

No information located.

Other uses (salad sprouts, ornamental, plant products, etc.)

No information located.

Chromosome Number

2*n* = 16: Tschechow 1932 (as *Trigonella radiata*); Heyn 1956; Clement 1962; Bolkhovskikh et al. 1969, p. 309 and p. 321 (as *T. radiata*); Singh and Singh 1976; Kliphuis and Barkoudah 1977; Lesins and Lesins 1979.

2*n* = 30: Singh and Singh 1976 (as *T. radiata*).

Genetic Relationships and Gene Transferability

Intraspecific

Artificial hybridization of different populations does not seem to have been attempted, so it is not known whether there are barriers to hybridization within the species.

Interspecific

Lesins and Lesins (1979) were unable to hybridize *M. radiata* with *M. heyniana*.

Exchange by advanced biotechnology

No information located.

Conservation Status

Medicago radiata is a fairly widespread species that does not seem in need of conservation attention.

Germplasm Sources

More than 300 accessions are listed among the major germplasm institutions with large collections of *Medicago* seeds (see Chapter 3).

Research Problems and Priorities

Medicago radiata has received limited study to date. This species has drought tolerance and reasonable biomass production for an annual *Medicago*, and so it has some agricultural promise as a crop for arid situations. It likely also harbors genes of potential interest for improvement of the species of *Medicago* currently used in agriculture, but as the species seems to be distant from all other species of the genus (see Chapter 10), this will likely require the use of advanced gene transfer technology.

Representative Collections

France: Herb. L. Rotereau, Corse, Bonifacis, V-1911, in fruit (P). **Iran:** Davis & Bokhari D55704, DS (Times Atlas), ca. 10 km from Bebehan to Omidiye, alt. 350 m, gravelly calcareous hills (steppic), 21-III-1974, in flower & fruit (K). Davis & Bokhari D56316, E6 (Times Atlas), 12 km from Jahrom to Qir, alt. 1200 m, stony calcareous hills, in run-off gullies, 2-IV-1974, in flower & fruit (K). Zohary, M. & Orsshan, G. J002/7, North Iran, Shimran (env. of Teheran), *Artemisia* steppe, on gravelly slope of the Elburs foothills, alt. 1390 m, 5-V-1961, in flower & fruit (HUJ). **Iraq:** Barkley, F.A., & Brahim, J. 2258, 20 km west of Rutba, Western Desert, 6-V-1962, in mature fruit (K). Eig, A. & Zohary, M., 12 km northwest of Kirkuk, alt. 350 m, gypsaceous hills, 16-IV-1933, in late flower & fruit (HUJ). Eig, A., & Zohary, M., env. of Dohuk, slopes of Rubat (the valley of), grazing place, 21-IV-1933, in flower & very early fruit (HUJ). Eig, A. & Zohary, M., 12 km north of Kisick, Keupri, alt 370 m, fallow field, 22-IV-1933, in flower & early fruit (HUJ). Eig, A. & Zohary, M., env. of Tel Afar, hills of Jabal Sasan, stony place, 26-IV-1933, in fruit (HUJ). Polunin, O. 5134, Sharifa near Amadia, alt. 4000 feet, stony ground, 22-IV-1958, in flower & fruit (K). **Israel:** Baum, B., C. Negev, hills around Sde-Boker, 11-IV-1965, in late flower & fruit (HUJ). Eig, A., Zohary, M., & Grizi, A., Judaean Desert, about 3/4 km from Jerusalem, Jericho road (turning to Ein Fuar), Maestrich, gravelly soft steppy soil, northern exposure, 27-III-1935, in fruit (HUJ). Liston, A. 7-84-185/1, Judean Mountains, Jerusalem Forest, dirt path-side in afforested area, population of ca. 30 plants scattered over 100 m length of path, 31-III-1984, in flower & fruit (HUJ). Schmida, A., Luria, M., & Gilboa, M., Har HaNegev, Nahal Elot, "Elot Cliff", northeast exposure, rocky surface, 27-V-1972, in fruit (HUJ). Zohary, M., Judaean Mtns., between Deir-Yassin and Beit-Hakerem, Senon soil, rocks and Batha, eastern exposure, 26-III-1941, in flower & fruit (HUJ). Zohary, M. & Orshan, G., Judaean Desert, km 30 on Jerusalem-Seden road, rocky calcareous hill, 24-III-1961, in late flower & fruit (HUJ). **Israel(–Syria):** Schmida, A., S. Hermon, 4 km NNNW of Majdal Shams, hard Jurassic limestone, alt. 1650 m, 10-VI-1976, in flower & fruit (HUJ). **Jordan:** Boulos, L. & Al-Eisawi, D. 6405, 10–15 km, northwest Zarga, along the road to Alouk, 29-III-1974, in flower & fruit (BM). Gillett, J.B. 15914, Shaubak Experimental Farm (30°30'N, 35°32'E), alt. 1420–1450 m, mean annual rainfall ca. 230 mm, gentle limestone slopes with remnants of red forest soil, grazing animals excluded by fence, locally frequent, 28-IV-1963, in flower & fruit (HUJ). Townsend, C.C. 65/342, between Azraq Druz & Ain el Beida, sandy wadi, 1-V-1965, in fruit (K). **Lebanon:** Meinertzhagen, R., Anti Lebanon, Messalon, alt. 4800 feet, rocky slope, 4-V-1933, in flower & fruit (BM). **Syria:** Bertron, A. 259, vigne sur la crète aride entre Rachauga et la cuvette de Kefer Kouk (1300 m, northeast of Rachaya), alt. "1270", 6-IX-1926, in fruit (P). Eig, A. & Zohary, M., top of Jebel Abd el Aziz, alt 620 m, Cenomanian stones, gray steppic soil, 29-IV-1933, in fruit (HUJ). Eig, A. & Zohary, M., 17 km southwest of Deir ez Zor, gray soil covered by thin gravel, alt. 300 m, 30-IV-1933, in fruit (HUJ). Eig, A. & Zohary, M., 107 km south of Homs, alt. 1430 m, wadi, 2-V-1933, in fruit (HUJ). Samuelsson, G. 3529, Desertum palmyrense, ad radices austr. montis Djebel el Abiad, in valliculo lapidoso ("ouadi"), ca. 15 km ad bor.-occid. a fonte ain el Beida, alt. ca. 500 m, 18-IV-1933, in flower & fruit (K). **Turkey:** Balansa, B. 910, entre le Bosante-Sou et le Kamechly-Tchaï (Cappodoce), collines, 13-VI-1856, in fruit (P). Balls, E.K. B2213, Gaziantep, Kara Tash, alt. 3000 feet, waste land, 3-V-1935, in fruit (BM). Bourgeau, E. 86, Elmalu, in collibus, 13-V & 3-VI-1960, in flower & fruit (G, P). Davis & Hedge D27425, Prov. Maras, Göksun, near Yemis dag, alt. 600 m, eroded shaley slopes in *Pinus brutia* forest, 3-V-1957, in flower & fruit (BM). Davis & Hedge D28094, Prov. Urfa, Akçakale, waste ground at edge of field in plain, alt. 400 m, 17-V-1957, in fruit (BM). Davis & Hedge D29115, Prov. Tunceli, Pertek-Tunceli, 26 miles from Elazig, alt. 1400 m, crystalline limestone ravine, 6-VI-1957, in flower & fruit (BM). Sintensis, P., Armenia turcica, kharput, in saxosis prope Melbé, 28-V-1889, Iter Orientale No. 570 (P). **USSR:** Goloskokov, V., Kazachstania, montes Karatau Syrdariensis, planities in viciniis pag. Aral-Tjube, in declivibus australibus lapidosis, 26-V-1963, in fruit (BM, K, E). Grossheim, A., Transcaucasia prope Erivan, in lapidosis, 15-V-1922, in fruit, Grosseheim & Schischkin Plantae Orientales Exsiccatae No. 88 (BM, K). Vvedensky, Montes meridionales, Tian-Schan occidentalis, ad declivia saxosa-argillosa secus canalem Zach prope pagum Niazbek, 4-V-1920, in flower & fruit, Herbarium Florae Asiae Mediae ab Universitate Asiae Mediae Editum Fasc. XI, Jan. a. 1927, No. 254 (K, P, E).

Note

The taxonomic relationships of *M. radiata* have been controversial. Ingham (1979*a*) found that *M. radiata* produces phytoalexins characteristic of numerous other *Medicago* species. On the basis of flavonoid chemistry, Saleh et al. (1982) concluded that the species was closer to *Medicago* than to *Trigonella*. As reviewed in Chapter 10, phylogenetic studies have shown that the species is distant from other *Medicago* species, and its relationships to them has not been adequately determined.

74. *Medicago retrorsa* (Boiss.) E. Small

M. retrorsa (Boiss.) E. Small in Small, Lassen, and Brookes, Willdenowia 16(2): 434 (1987).

The epithet *retrorsa* is based on the Latin *retrorsus*, turned or bent backward, a reference to the horizontally declinate orientation of the fruits.

Subordinate Taxa

None recognized here.

Type Specimen

Griffith, exsiccata 1069, hab. in Affghania ad Kukou. (Herb. Boiss. G! — merely a single fruit with three immature seeds. Širjaev (1928–1934: 1931, p. 33) presented a photo of a fruit, calling this the type.)

Synonymy

Homotypic

Trigonella retrorsa Boiss., Fl. Orient. 2: 78 (1872), (basionym).

Heterotypic

No names found.

Vernacular Names

English: No names found.

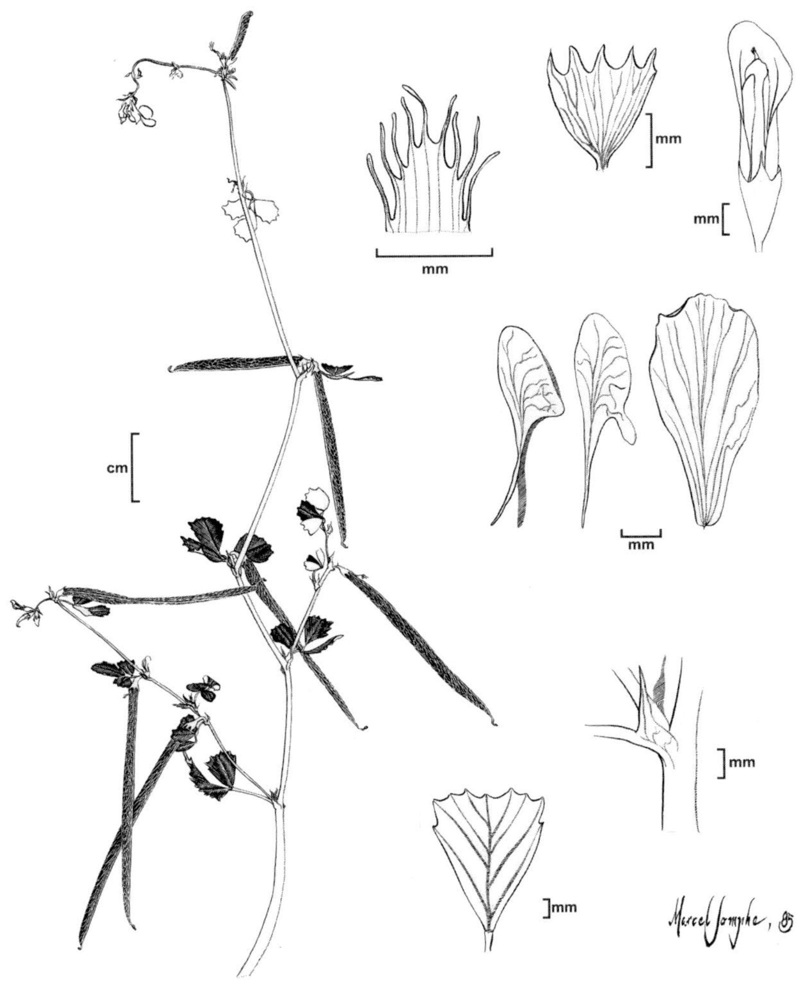

Fig. 74-1. *Medicago retrorsa.* **Left:** Branch with fruits and flowers. **Top right:** Floral parts (clockwise from top left: flattened androecial column (free stamen not shown); calyx; side view of flower; standard petal; wing petal; keel). **Lower right:** Stem node showing stipules; leaflet. (Based on Lemann 1140, K; artist: M. Jomphe.)

FRENCH: No names found.

Geographical Distribution

(N = native, I = introduced, U = uncertain status)

ASIA: Afghanistan (N).

Medicago retrorsa has been collected only from Afghanistan. The single area indicated by Širjaev (1928–1934) is located approximately at 31°24′N, 69°E. Vassilczenko (1953) intimated that the species may also be found in Pakistan.

Description

Annual herb, stems 3–35 cm long, diffusely branched, prostrate, densely pilose. Leaflets obovate or obtriangular, margin of distal half serrate, 6–7 mm long, 4–6 mm wide, glabrous above, sparsely pilose below. Stipules semi-sagittate, incised in lower part. Inflorescence of 1–2 flowers, the peduncle very short (3–5 mm). Flowers ca. 7 mm long, the corolla yellow. Calyx glabrescent, ca. 3 mm long, the teeth lanceolate-triangular and shorter than the tube. Pods horizontally deflexed, glabrescent, straight or slightly curved, linear, 30–50 mm long, 1.75–2 mm wide, the netted venation delimiting elongated areas (alveoli). Seeds smooth(?), radicle about 80% of length of seed.

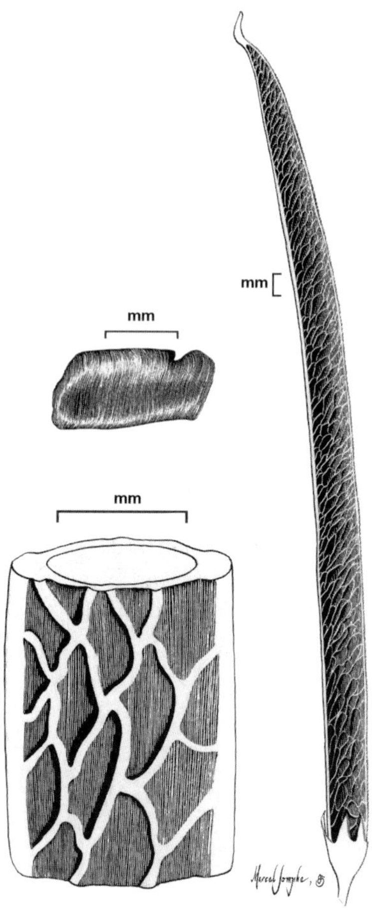

Fig. 74-2. *Medicago retrorsa.* **Clockwise from top left:** Seed; fruit; portion of fruit showing venation. (Based on Lemann 1140, K; artist: M. Jomphe.)

Identification Hints

The deflexed fruits and origin from Afghanistan should serve to identify this rare species.

Habitat and Environmental Adaptation

Habitat

Habitat conditions have not been described for *M. retrorsa*.

Phenology

No information located.

Soil

No information located.

Temperature

No information located.

Water

No information located.

Biotic Relationships

Pollination

The small flowers suggest that *M. retrorsa* is largely an inbreeder.

Dispersal

No obvious adaptations for seed distribution are evident.

Herbivory

No information located.

Nodulation and mycorrhizae

No information located.

Disease resistance

No information located.

Economic History, Cultivars, and Uses

This species has not been employed in agriculture.

Cultivars

No cultivars have been bred.

Forage

Medicago retrorsa is not sown for forage.

Fodder

The species has not been used as a source of dried livestock feed.

Green manure, rotation crop, ground cover

No information located.

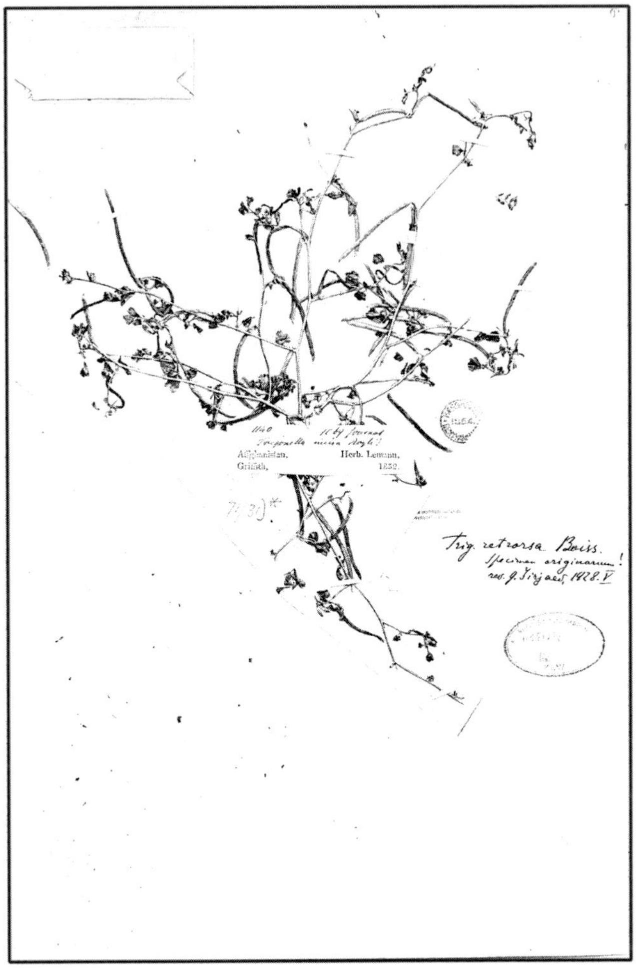

Fig. 74-3. Isotype of *Trigonella retrorsa* at K.

Other uses (salad sprouts, ornamental, plant products, etc.)

No information located.

Chromosome Number

No information located.

Genetic Relationships and Gene Transferability

Intraspecific

Artificial hybridization of different populations has not been attempted, so it is not known whether there are barriers to hybridization within the species.

Interspecific

No information located.

Exchange by advanced biotechnology

No information located.

Conservation Status

Given the rarity of this species, it is in extreme need of a conservation assessment.

Germplasm Sources

No accessions were located.

Research Problems and Priorities

Medicago retrorsa is a distinctive species that has been rarely and inadequately collected. There is almost no information available about its nature and potential. The principal priority is to obtain seeds for germplasm preservation and as a basis for study. Unfortunately, the home of this species, Afghanistan, has been at war for decades, making collection hazardous.

Representative Collections

AFGHANISTAN: Lemann 1140 (K).

75. *Medicago rhodopea* Velen.

M. rhodopea Velen., Sitzungber. Königl. Böhm. Ges. Wiss. Prag. Math.-Naturwiss. Cl. 37: 21 (1893).

The epithet *rhodopea* (often incorrectly rendered "*rhodopaea*") refers to the Rhodope Mountains in the Balkans of southern Bulgaria (where the species has been collected) and northeastern Greece.

Subordinate Taxa

None recognized here.

Type Specimen

See Fig. 75-3.

Synonymy

Homotypic

M. rupestris M.B. var. *rhodopea* (Velen) Stojan. & Stef., Fl. na Bolgaria 12: 631 (1924).

Heterotypic

No names found.

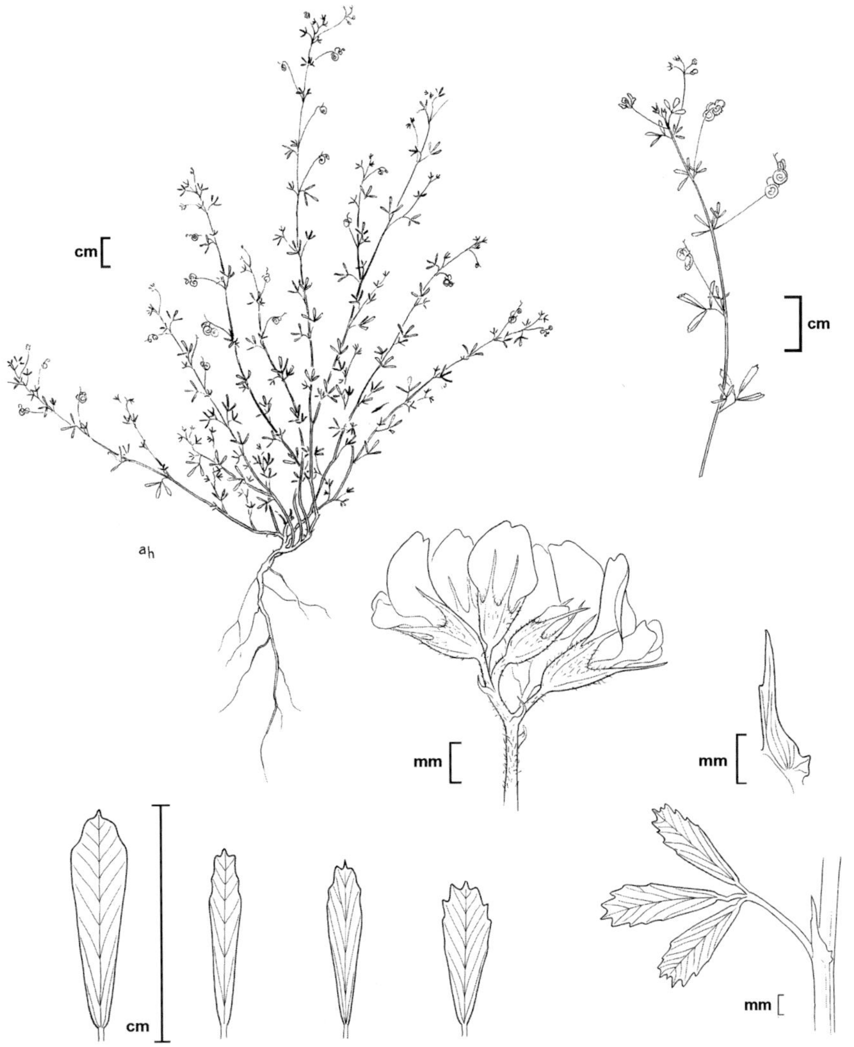

Fig. 75-1. *Medicago rhodopea.* **Clockwise from top left:** Plant; branch with flowers and fruits; stipule; leaf attached to stem; leaflets. **Center:** Group of flowers. (Based on collection 13/2202, US; artist: A. Hanes.)

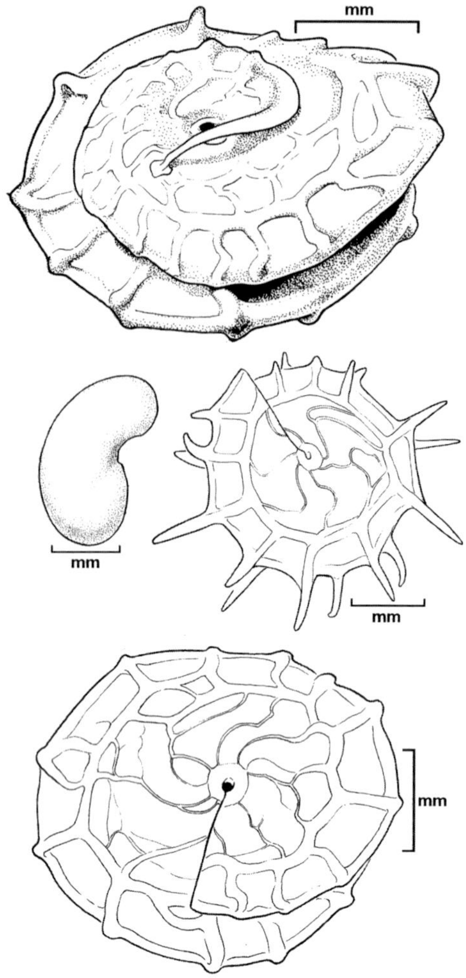

Fig. 75-2. *Medicago rhodopea.* **Clockwise from top:** Tuberculate fruit; spiny fruit coil face showing venation; tuberculate fruit coil face showing venation; seed. (Spiny fruit based on V. Stribrny on 5-VI-1898; artist: A. Hanes.)

Not

"*Medicago saxatilis*", misapplied in the sense of some authors, as in Tutin et al. (1968, p. 155).

Vernacular Names

ENGLISH: Rhodope alfalfa (Nicoloff 1961).

FRENCH: No names found.

Geographical Distribution

(N = native, I = introduced, U = uncertain status)

EUROPE: Bulgaria (N).

Medicago rhodopea is endemic to southeastern Europe, specifically southern Bulgaria. "It is found in the central and eastern Rhodopes, the Thracian plain, the Tundzha valley, and the eastern Stara Planina Mountains" (Nikolov et al. Report of the Green Balkans Movement on the Bulgarian National Biological Diversity Conservation Strategy. http://www.worldwildlife.org/bsp/publications/europe/bulgaria/bulgaria31.html; current July 2010).

Description

Perennial herb, stems arising from crown (woody at base), procumbent to ascending, 5–15(30) cm long, more or less pubescent with appressed simple hairs. Leaflets (3)5–8(10) mm long, 1–3(5) mm wide, linear-cuneate, apex somewhat serrate, appressed-pilose (especially on lower sides). Stipules subulate, usually entire. Inflorescence more or less capitate, of 4–10 flowers, the peduncle longer than the subtending petiole. Flowers 3–5(7) mm long, corolla yellow. Calyx pilose, shorter than half length of flower, the teeth about equal to length of tube. Young pod turns sideways through calyx. Mature pod pubescent, short-cylindrical or lenticular, the dorsal suture with tubercles or spines, 1–2.5 coils, 2.5–4 mm in diameter, the coil face with veins running from the axis to a prominent lateral vein from which elevated ridges extend to the dorsal suture and spines or tubercles, a wide groove between the lateral vein and the dorsal suture. Seeds 1–2 per coil, separated by spongy fruit partitions, ca. 2.3 mm long, 1–1.3 mm wide, reniform, brownish, radicle about half length of seed.

Identification Hints

Medicago rhodopea is distinguished from its close allies *M. prostrata* and *M. rupestris* by having thick pod veins projecting as short spines or tubercles. (Rare spineless forms of the latter two could make identification problematical, but *M. prostrata* and *M. rupestris* occur well to the west of *M. rhodopea*).

Habitat and Environmental Adaptation

Habitat

Medicago rhodopea occurs in rocky and grassy sites in lower mountain regions. Peev et al. commented with respect to *M. rhodopea* in Bulgaria that "Relatively few endemic species and subspecies are found in the lowlands and the low mountain foothills. In the lowlands they are preserved mainly in the lower hills with shrub and herbaceous vegetation, where they were undoubtedly widespread in the past" (Biodiversity of Vascular Plants in Bulgaria; http://www.worldwildlife.org/bsp/publications/europe/bulgaria/bulgaria3.html; current July 2010). Stanev (1977) provides a detailed analysis of a plant community in Bulgaria in which *M. rhodopea* was a constituent.

Phenology

Medicago rhodopea flowers from May to August in its native area.

Soil

This species typically occupies calcareous substrates. McCoy (1987) found that *M. rhodopea* has limited salt tolerance.

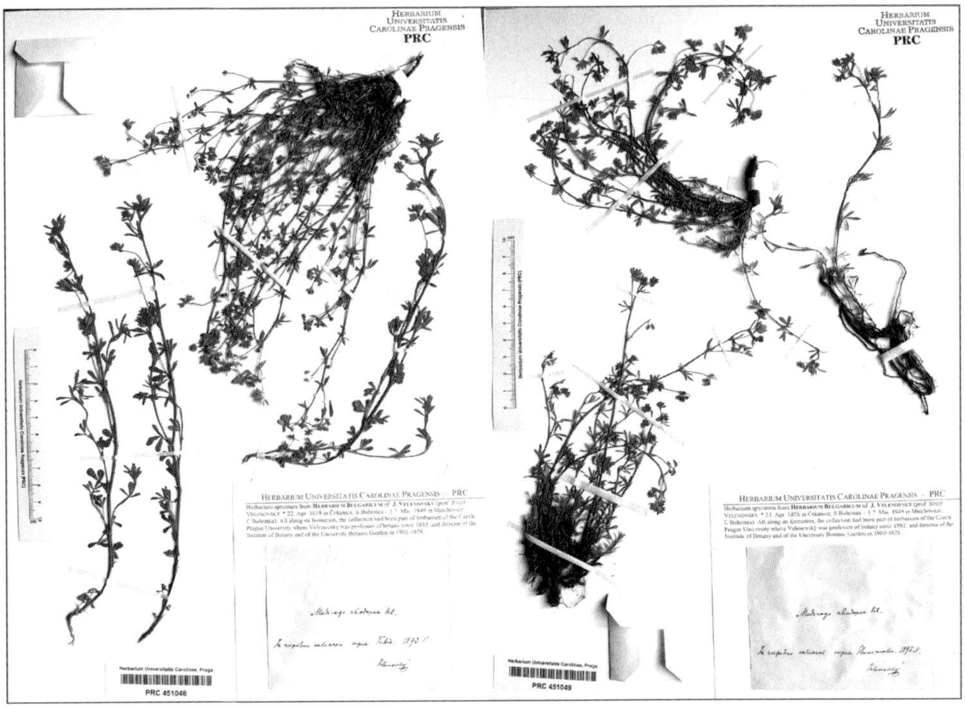

Fig. 75-3. Syntypes of *M. rhodopea* Velen. at PRC. **Left:** "In rupibus calcareis supra Tekir. 1893.V.Velenovsky." This collection is here designated the lectotype. **Right:** "In rupibus calcareis supra Stanimaka. 1893.V. Velenovsky." Photographs courtesy of J. Stepánek, Head Curator, Herbarium, Charles University in Prague, Czech Republic.

Temperature

Most plants grown outdoors in Brandon, Manitoba were winter killed (Lesins and Lesins 1979).

Water

The species typically grows in dry places.

Biotic Relationships

Pollination

The flowers of this species are relatively small for a perennial *Medicago*, suggesting that *M. rhodopea* may inbreed to a considerable extent.

Dispersal

The indehiscent fruits often have spines, but these are not hooked and so would have limited capacity to adhere to fur or feathers. The spines would appear to have value in stopping the fruits from rolling down steep hills or mountain sides.

Herbivory

No information located.

Nodulation and mycorrhizae

No information located.

Disease resistance

No information located.

Economic History, Cultivars, and Uses

Medicago rhodopea has not attracted much agricultural interest as a cultivated crop.

Cultivars

No cultivars have been bred.

Forage

Medicago rhodopea is not sown for forage.

Fodder

This species is not harvested as animal feed.

Green manure, rotation crop, ground cover

No information located.

Other uses (salad sprouts, ornamental, plant products, etc.)

No information located.

Chromosome Number

$2n = 16$: Nicoloff 1961; Gillies 1972*a*; Terzyisku and Poddubnaya-Arnoldi 1973; Lesins and Lesins 1979.

Genetic Relationships and Gene Transferability

Intraspecific

Artificial hybridization of different populations has not been attempted, so it is not known whether there are barriers to hybridization within the species.

Interspecific

Lesins and Gillies (1972*b*) stated that unpublished work by Lesins found that *M. rhodopea* could be hybridized with *M. sativa*, but not at the diploid level. Gillies (1972*a*) studied a triploid hybrid between *M. rhodopea* and *M. sativa* (reported in Lesins 1972). Lesins (1972) had produced the triploid by crossing diploid alfalfa with an artificially doubled tetraploid *M. rhodopea* (he also produced an aneuploid hybrid with $2n$ = 31). Gillies (1972*a*) studied a hybrid between *M. rhodopea* and *M. rupestris* (made by K. Lesins). Lesins (1970) reported a tri-species hybrid of *M. rhodopea*, *M. saxatilis*, and *M. sativa*.

Exchange by advanced biotechnology

Using embryo rescue techniques, McCoy and Smith (1986) obtained hybrids between *M. rhodopea* and (at the same chromosome number) diploid *M. sativa* ($2n$ = 16); such hybrids were obtained relatively frequently. Denton and McCoy (1987) found that the hybrids showed good bivalent pairing at meiosis (indicating considerable chromosomal homology), but were generally infertile.

Conservation Status

The 1997 International Union for Conservation of Nature Red List of Threatened Plants included *M. rhodopea* (http://www.ceu.hu/envsci/soe/ref/nature/threpl11.htm; current July 2007).

Germplasm Sources

Four accessions of *M. rhodopea* are recorded among the major *Medicago* germplasm collections (see Chapter 3). Pederson et al. (1999) made one collection in Bulgaria.

Research Problems and Priorities

Medicago rhodopea has low productivity and does not seem to have much direct potential as a forage crop. However, it has been shown to harbor genes valuable for the improvement of alfalfa and deserves additional development in this respect.

Representative Collections

Bulgaria: Kojoucharov, S. & Kouzmanov, B., sur les collines calcaires près de la ville Asenovgrad, 20-VI-1959, in flower & early fruit (K). Stribrny, V., in collibus, 21-VII-1895, in flower & fruit (BM).

76. *Medicago rhytidiocarpa* (Boiss & Balansa) E. Small

M. rhytidiocarpa (Boiss. & Balansa) E. Small in Small, Lassen & Brookes, Willdenowia 16(2): 434 (1987).

The epithet *rhytidiocarpa* is based on the Greek *rhytido* (wrinkled), plus *karpos* (fruit), an allusion to the moniliform appearance of the fruits.

Subordinate Taxa

None recognized here.

Type Specimen

Balansa, B. 114, Turkey, petits coteaux pierreux bordant la rive droite du Kamechly-Tchaï près de Béréketly

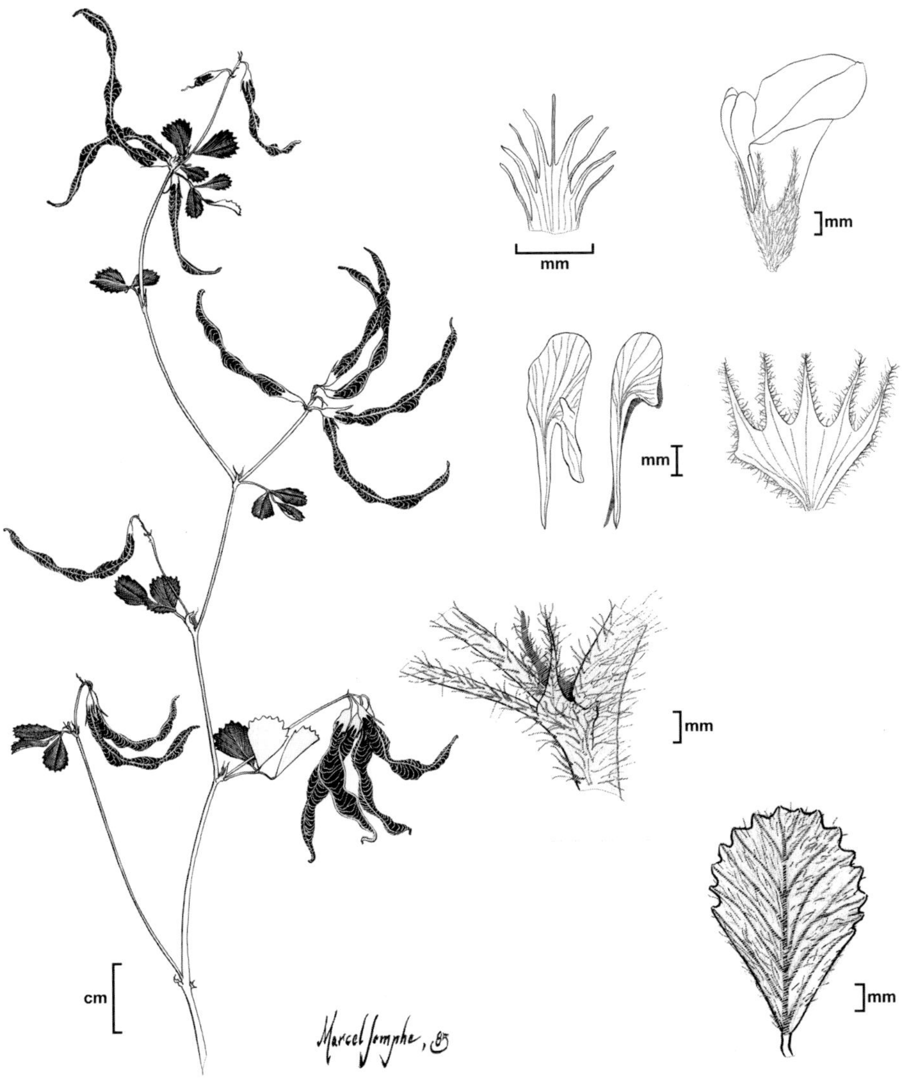

Fig. 76-1. *Medicago rhytidiocarpa.* **Left:** Branch with fruits and flowers. **Top right:** Floral parts (clockwise from top left: flattened androecial column (free stamen not shown); side view of flower; calyx; keel; wing petal). **Lower right:** Leaflet. **Center:** Stem node showing stipules. (Based on holotype, B. Balansa 114, G; artist: M. Jomphe.)

(Cappadoce), 1300 m d'alt., 17-VI-1856 (holotype Herb. Boiss. G! (Fig. 3); isotypes BM!, US!, G!).

Synonymy

Homotypic

Trigonella rhytidiocarpa Boiss. & Balansa in Boiss., Diagn. Pl. Orient., ser. 2, 3 (6): 44 (1859), (basionym).

Heterotypic

T. rhytidiocarpa Boiss. & Balansa f. *typica* Širj., Generis *Trigonella* revisio critica, Spisy Pfir. Fac. Masarykovy Univ. 4: 33 (1931), nom. inadmiss.

T. rhytidiocarpa Boiss. & Balansa f. *glabra* Širj., Generis *Trigonella* revisio critica, Spisy Pfir. Fac. Masarykovy Univ. 4: 33 (1931).

T. rytidocarpa (sic) Boiss., Diagn. Pl. Orient. Ser. 1, 2(9): 12 (1849), in syn.

Vernacular Names

ENGLISH: No names found.

FRENCH: No names found.

Geographical Distribution

(N = native, I = introduced, U = uncertain status)

ASIA: Turkey (N).

Medicago rhytidiocarpa is an endemic of southern Anatolia.

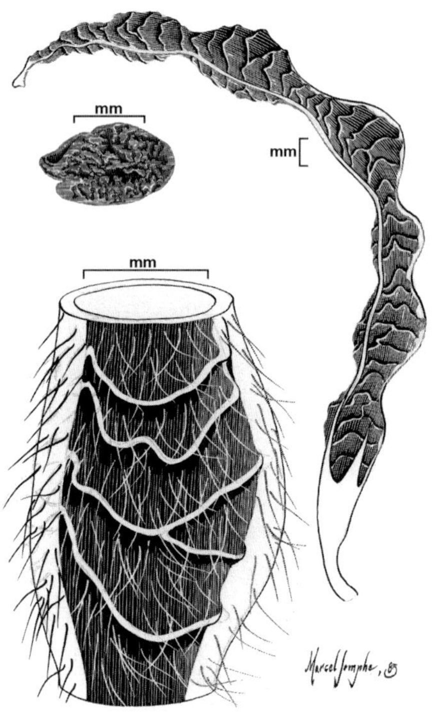

Fig. 76-2. *Medicago rhytidiocarpa.* **Clockwise from top right:** Fruit; portion of fruit showing venation; seed. (Based on isotype, B. Balansa 114, US; artist: M. Jomphe.)

Description

Annual herb, stems procumbent, ascending, or straight, 10–30 cm long, pubescent with appressed hairs or villous. Leaflets obovate or obovate-cuneate, distal half of margin serrate, pubescent on both sides, (4)5–10(12) mm long, (3)4–8(10) mm wide. Stipules semi-sagittate, lanceolate-subulate or lanceolate-acuminate, dentate at base. Inflorescence subumbellate, of (3)6–7(8) flowers, the peduncle 1–3.5 cm long, longer than the subtending petiole. Flowers 6–9 mm long, pedicels at first straight and then reflexed, calyx campanulate, 5–6 mm long, pilose, the teeth lanceolate-subulate and longer than the tube, the corolla yellow. Fruit 2.5–3 cm long, 1–1.5 mm wide, deflexed, linear, falcate to semi-circular, weakly compressed, more or less moniliform (strangulate between the seeds), with a curved short beak at the apex, the fruit surface rugose from the raised, transverse, oblique, nonanastomosing veins, pilose or glabrous. Seeds 4–8 per pod, oblong, 2–2.5 mm long, 1–1.5 mm wide, rugose, yellowish, radicle about 80% of length of seed.

Identification Hints

Medicago isthmocarpa is the only other *Medicago* with fruits so constricted between the seeds that the pods are more or less moniliform. For distinction of these two species, see Chapter 43 on *M. isthmocarpa*. Some species of *Trigonella* have moniliform pods (for generic distinction, see Chapter 9).

Habitat and Environmental Adaptation

Habitat

Medicago rhytidiocarpa occurs on rocky slopes at altitudes of 1300–1450 m.

Phenology

In its native area, *M. rhytidiocarpa* flowers in June and July.

Soil

No information located.

Temperature

No information located.

Water

No information located.

Biotic Relationships

Pollination

Small (1988) recorded a relatively low pollen to ovule ratio of 191, consistent with *M. rhytidiocarpa* being largely self-pollinated and substantially an inbreeder, like essentially all other annual species of *Medicago*. However, the flowers are somewhat larger than most other annual species of the genus, suggesting that some outcrossing may occur.

Dispersal

No obvious adaptations are present for seed distribution.

Herbivory

No information located.

Nodulation and mycorrhizae

No information located.

Disease resistance

No information located.

Economic History, Cultivars, and Uses

Medicago rhytidiocarpa has not been employed in agriculture.

Cultivars

No cultivars have been bred.

Forage

Medicago rhytidiocarpa is not sown for forage.

Fodder

The species is not harvested and dried for hay.

Green manure, rotation crop, ground cover

No information located.

Other uses (salad sprouts, ornamental, plant products, etc.)

No information located.

Chromosome Number

The chromosome number has not been determined.

Genetic Relationships and Gene Transferability

Intraspecific

Artificial hybridization of different populations has not been attempted, and it is not known whether there are barriers to hybridization within the species.

Interspecific

No information located.

Exchange by advanced biotechnology

No information located.

Conservation Status

This species is very rare and grows in an area where the vegetation is subject to disturbance, so it needs the highest category of protection.

Germplasm Sources

No accessions were located.

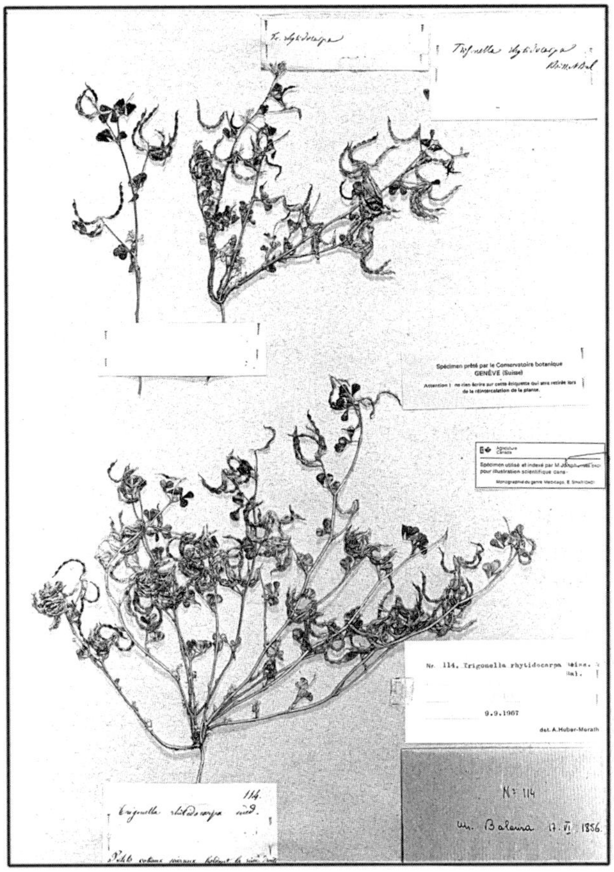

Fig. 76-3. Photograph of holotype of *Trigonella rhytidiocarpa* Boiss. & Balansa, at G.

Research Problems and Priorities

Medicago rhytidiocarpa is a very poorly known species. The greatest need is for collection of seeds for deposition in germplasm banks for basic botanical studies and agricultural assessment.

Representative Collections

TURKEY: Balansa, B. 907, Collines bordant la rive droite du Kamechly-Tchaï près de Béréketly (Cappadoce), 1300 m d'alt., 15-VI-1856 (K, G, E, P). Coode, M.J.E. & Jones, B.M.G. 1248, Niĝde Pozanti to Niĝde, 8–10 m north of ES Junction, alt. 1300 m, fallow fields and roadside, 20-V-1965, in flower (E) Huber-Morath, A. 12045, Niĝde (Cappadocien), Niĝde – Ulukişla, Passhöhe, 3 km nördlich der Abzweigung nach Adana, Eruptivgest, alt. 1450 m, 10-VI-1953, in flower (G). Khan, H. 7G, Karapinas (Konya), on sandy soil, 6-VI-1974, in flower & fruit (G). Reese, H., prov. Niĝde (region C5), district Ulukişla, East von Ulukişla, 10-VI-1937, in flower & fruit (G). Reese, H., prov. Niĝde (region C5), district Ulukişla, 5 km East Ulukişla, 17-VI-1950, in flower & very early fruit (G).

77. *Medicago rigida* (Boiss. & Balansa) E. Small

M. rigida (Boiss. & Balansa) E. Small in Small, Lassen & Brookes, Willdenowia 16(2): 434 (1987).

The epithet *rigida* is based on the Latin *rigidus*, stiff, an allusion to the short, stiff fruits.

Subordinate Taxa

None recognized here.

Type Specimen

Balansa, B. 906, Turkey, à l'ouest du Bosante-Tschaï entre Tarsous et Césarée, collines calcaires, 12-VI-1856, in flower & fruit. (Holotype G!, syntype G!, isotypes P, GH; Huber-Morath (1970) gave "ca. Pozanti" as the type locality.)

Synonymy

Homotypic

Trigonella rigida Boiss. & Balansa in Boiss., Diagn. Pl. Orient. 2(5): 79 (1856), (basionym).

Heterotypic

No names found.

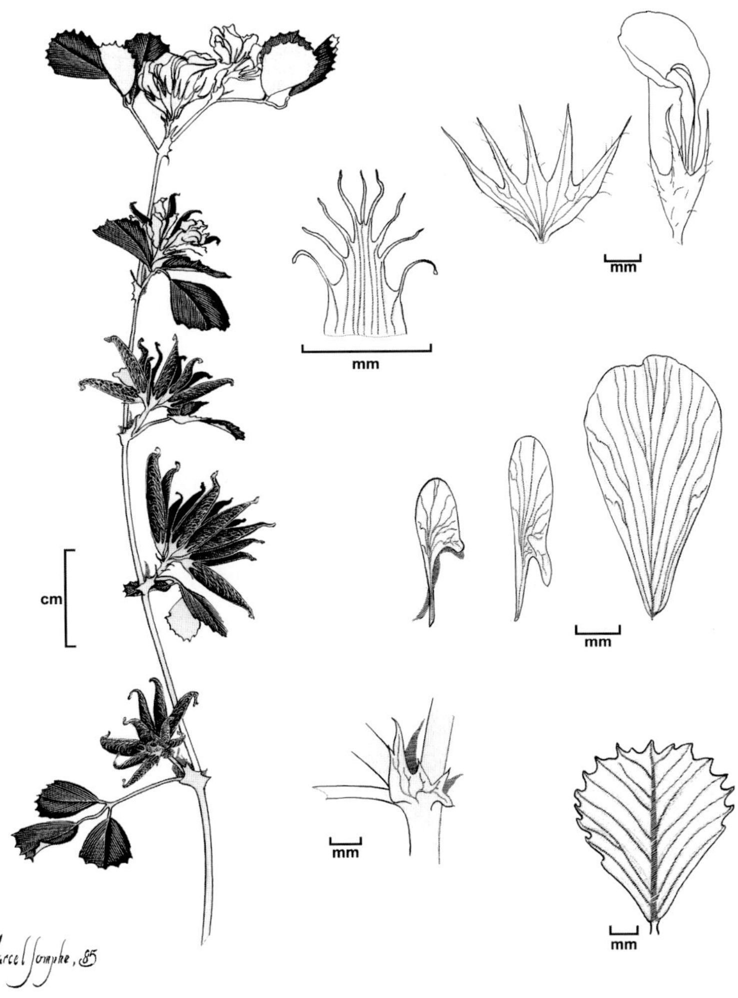

Fig. 77-1. *Medicago rigida.* **Left:** Branch with fruits and flowers. **Top right:** Floral parts (clockwise from left: flattened androecial column (free stamen not shown); calyx; side view of flower; standard petal; wing petal; keel). **Lower right:** Leaflet. **Lower center:** Stem node showing stipules. (Based on specimen annotated "T. Kotschy, Suppl. 61, April 29, 1859", G; artist: M. Jomphe.)

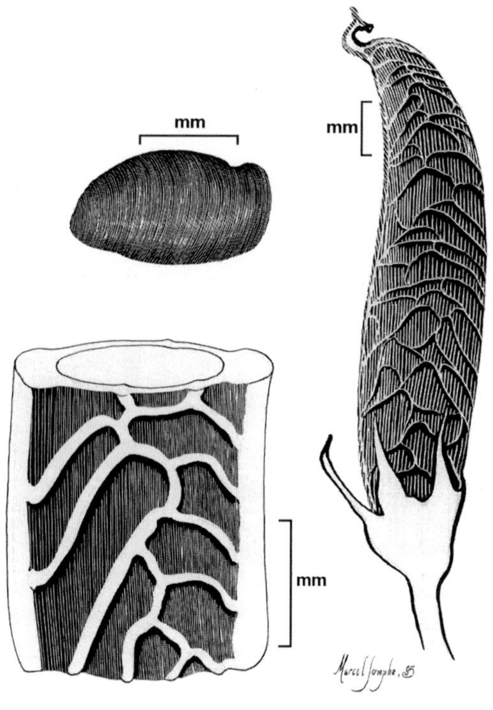

Fig. 77-2. *Medicago rigida.* **Clockwise from top left:** Seed; fruit; portion of fruit showing venation. (Based on specimen annotated "T. Kotschy, Suppl. 61, April 29, 1859", G, detail of fruit venation based on Boissier (?) 248, G; artist: M. Jomphe.)

Not

"*Medicago rigida* Lam." This name has no status. It is a wrong citation (misprint) by Boissier (1872) for "*M. rigidula* Desr. in Lam." (correctly *M. rigidula* (L.) All.). Therefore, the combination *M. rigida* (Boiss. & Balansa) E. Small is not prevented.

Vernacular Names

English: No names found.

French: No names found.

Geographical Distribution

(N = native, I = introduced, U = uncertain status)

Asia: Turkey (N).

Medicago rigida is a local endemic, known from southern Anatolia.

Description

Annual herb, branched from base, stems 5–20(30) cm long, prostrate or ascending, appressed pubescent. Leaflets obovate to obcordate, distal margin serrate, (3)5–8 mm long, (2.5)4–7 mm wide. Stipules lanceolate-acuminate, dentate to incised. Inflorescence of 7–12 flowers in a dense, spherical head, peduncle 2–4 mm long. Flowers 5–6 mm long, corolla yellow, calyx campanulate, 3–4 mm long, teeth linear-subulate, longer than tube. Fruits broadly linear, compressed, erect, slightly curved and hooked towards tip, 6–10 mm long, 2–3 mm wide, reticulate-veined, closely clustered in the infructescence. Seeds ovate to oblong, 2–2.5 mm long, 1–1.75 mm wide, smooth, radicle about 80% of length of seed.

Identification Hints

Medicago rigida is most similar to *M. crassipes*, which has larger fruits (1.2–3.5 cm long versus 0.6–1.0 cm in *M. rigida*).

Habitat and Environmental Adaptation

Habitat

Medicago rigida occurs on dry limestone slopes and sandy river beds at altitudes of 100–900 m.

Phenology

Flowering occurs from April to June in the native area.

Soil

The species occupies calcareous, rocky, gravelly, or sandy substrates.

Temperature

Like most annual species of *Medicago*, *M. rigida* evades an increasingly hot summer by completing its life cycle in early summer.

Water

Medicago rigida occupies dry or at least seasonally dry substrates.

Biotic Relationships

Pollination

Small (1988) noted that the flowers are small and the pollen to ovule ratio is low (327), observations that are consistent with *M. rigida* being largely self-pollinated and substantially an inbreeder.

Dispersal

No obvious morphological adaptations are present that facilitate seed distribution.

Herbivory

No information located.

Nodulation and mycorrhizae

No information located.

Disease resistance

No information located.

Economic History, Cultivars, and Uses

This species has not been developed or used as a crop.

Cultivars

No cultivars have been bred.

Forage

Medicago rigida is not sown for forage.

Fodder

This species is not harvested as dry hay.

Green manure, rotation crop, ground cover

No information located.

Other uses (salad sprouts, ornamental, plant products, etc.)

No information located.

Chromosome Number

$2n = 16$: Singh and Singh 1976.

Genetic Relationships and Gene Transferability

Intraspecific

Artificial hybridization of different populations has not been attempted, so it is not known whether there are barriers to hybridization within the species.

Interspecific

No information located.

Exchange by advanced biotechnology

No information located.

Conservation Status

Medicago rigida is a rare endemic in a region subject to biodiversity loss. Accordingly, a conservation assessment is called for, as well as collection of seeds for deposition in germplasm banks.

Germplasm Sources

No accessions have been located.

Research Problems and Priorities

Medicago rigida is a very poorly documented species. Collection of its seeds and basic botanical and agricultural assessments are required.

Representative Collections

TURKEY: Balansa, B. 116, collines argilo-calcaires situées entre le Bosante-Tchaï et le Kamechly-Tchaï (Cappadoce), 14-VI-1856, in fruit (G). Balansa, B., bords de la route de Tarsous à Césarée à l'ouest du Bosante-Tchaï, terraines calcaires, 9-VIII-1855, in fruit, plant dead when collected and seeds shed (G, P). Balls, E.K. & Gourlay, W.B. B742, Chona near Csmaniye, alt. 400 feet, growing among stones and in sand of stream-bed, 12-IV-1934, in flower & early fruit (K, E). Balls, E.K. 2108, Kurt Tepe, Adana alt. 500 feet, open banks and very stony ground, 11-IV-1935, in flower (BM, E). Davis & Hedge, D.26621, Prov. Adana, Kozan (Anti-taurus), alt. 400 m, igneous slopes, 12-IV-1957, in flower (BM, E). Davis & Hedge, D.27505, Prov. Maras, Maras, Göksun, near Yemis dag. alt. 600 m, shaley slopes, 3-V-1957, in flower & early fruit (BM, E). Davis & Hedge D26759, Prov. Adana dist. Bahce (N. Ananus), Haruniye, alt. 500 m, edge of water course, 17-IV-1957, in flower & early fruit (BM, E). Huber-Morath, A. 12046, Nigde (Pontus), Çiftehan – Pozanti, Urgestein, Geröll 2 km unterhalb Çiftehan, alt. 910 m, 10-VI-1953, in flower (G). Reese, H., prov. Içel (region C5), Gülek – Pozanti, 16-VI-1950, in flower & fruit (G).

78. *Medicago rigidula* (L.) All.

M. rigidula (L.) All., Fl. Pedem. 1: 316 (1785).

The epithet *rigidula* is based on the Latin *rigidus* (stiff, hard, inflexible) plus *ule* (diminutive suffix), a reference to the fairly stiff fruit spines of the species. (The French name *luzerne à tige rigide* ("rigid-stemmed alfalfa") suggests that the stem is stiff; however, this name was coined without insight into just what part of the plant is stiff.)

Warning: As detailed below, *M. rigidula* and *M. rigiduloides* are two very closely related species for which the geographical distribution and morphological separation are not clear. *Medicago rigidula* was split into *M. rigidula* and *M. rigiduloides* in 1990, and so only recent literature has recognized the distinction. Accordingly, it is often difficult to ascertain which species was the subject of much (indeed most) of the literature. The history of the separation of *M. rigidula* and *M. rigiduloides* is briefly given in the following.

Lesins and Lesins (1963*b*) discovered that collections of "*M. rigidula*" from each of the islands of Capri and Corsica had three-pored pollen with a cylindrical shape, whereas collections from Iraq and Turkey had four-pored pollen with a pyramidal-triangular shape. Moreover, in crossing the two kinds, they found that the hybrids had much reduced pollen viability (15% to 25%) compared with the parents (85% to 99%).

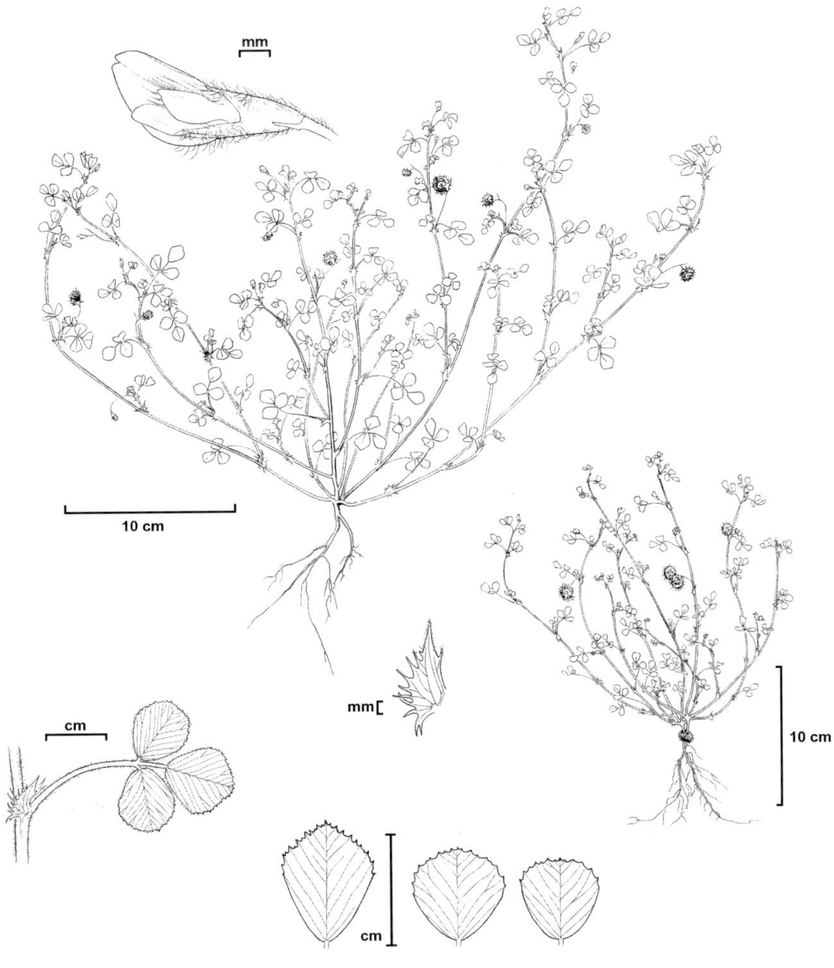

Fig. 78-1. *Medicago rigidula.* **Clockwise from lower left:** Leaf attached to stem; plant; flower; plant emerging from maternal fruit; leaflets. **Center:** Stipule. (Flower and leaf based on E. Small M950, DAO; large plant based on Stevenson 3345, DAO; plant emerging from maternal fruit based on Morariu on 26-V-1963, DAO; artist: A. Hanes.)

Building on the above work, Small et al. (1990*b*) analyzed 30 morphological characteristics and found they could be separated into two groups, one basically of 25 European accessions and the other of 39 Asian accessions. Moreover, the European group proved to have mostly three-pored pollen, whereas the Asian group proved to have mostly four-pored pollen. A small number of African accessions appeared to be similar to the European group. Small (1990*b*) recognized the European and African groups collectively as *M. rigidula sensu stricto* and the Asian group as a new species, *M. rigiduloides* E. Small.

Heft and Groose (reported in a 1996 abstract) examined available accessions previously classified as "*M. rigidula*" in the United States National Plant Germplasm System (NPGS), with the aim of separating them into *M. rigidula* and *M. rigiduloides* as defined by Small (1990*b*) and Small et al. (1990*b*), based on greenhouse-grown plants and examination of the pollen and pods. In total, 307 of the 332 accessions of "*M. rigidula*" then in the NPGS were examined. The vast majority of these produced either predominately three-pored, cylindrical pollen (*M. rigidula*, 103 lines total) or predominately four-pored, pyramidal-triangular pollen (*M. rigiduloides*, 192 lines total). Seven "intermediate" accessions produced pollen typical of both species as well as some intergrade samples of pollen grains, and five accessions proved to be mixed collections (with different species). Four Asian accessions included plants both of *M. rigidula* and *M. rigiduloides*. The authors reported that Asian and European accessions were not as distinct as reported by Small et al. (1990*b*). In the NPGS, 9% of European accessions proved to be *M. rigiduloides* and 15% of Asian accessions were *M. rigidula*. The authors also stated that they found that the pod morphology of European *M. rigiduloides* accessions was essentially identical to Asian accessions of that species and, conversely, Asian *M. rigidula* accessions had pods that were nearly identical to *M. rigiduloides* pods; they concluded that the key to pod morphology presented by Small et al. (1990*b*), and reproduced here, would not be useful for differentiating these species in Asia.

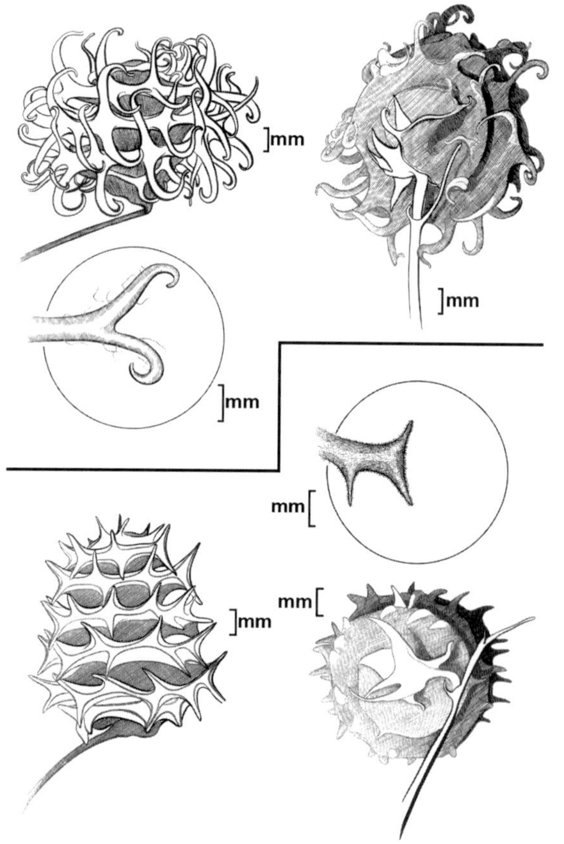

Fig. 78-2. A comparison of fruits of *M. rigidula* (top three figures based on E. Small M1288, DAO) and *M. rigiduloides* (bottom three figures based on E. Small M1320, DAO). Note that the fruits of *M. rigiduloides* tend to be more ovoid in shape, have more coils, and have spines that are shorter and less curved in comparison with the same characters of *M. rigidula*. (Artist: M. Jomphe.)

However, Heft and Groose have assumed that the two species can simply be defined by the one character — whether the pollen grains are predominantly three-pored with a cylindrical shape or predominantly four-pored with a pyramidal-triangular shape. As is well known, one-character classifications are rarely sufficient for defining species (identification is another matter). Moreover, the Heft and Groose data did not appear in a reviewed publication, and the accuracy of the identifications of the accessions has not been verified. Nevertheless, the analysis of Heft and Groose points out a need for a more extensive analysis of character distribution on a geographical basis of the two species (such a study by Small and colleagues is in progress).

In addition, as noted below, Heft and Groose reported that there is a partial sterility barrier between the primarily European plants with three-pored pollen and the primarily Asian plants with four-pored pollen, confirming the earlier work of Lesins and Lesins in this regard.

Small et al. (1990*b*) and Heft and Groose (1996) are in agreement that approximately 90% of European plus African plants comprise one distinctive morphological group and belong to a separate species (*M. rigidula* sens. strict.), whereas approximately 90% of Asian plants comprise a second distinctive morphological group and belong to a separate species. What remains unclear is the extent to which the primarily Asian species occurs in the distribution range of the primarily European plus African species, and vice-versa.

Mehregan et al. (2002), based on material collected in Iran, treated *M. sinskiae*, *M. constricta*, *M. rigidula*, and *M. rigiduloides* as one species — *M. rigidula*. They recognized the first two as subspecies of *M. rigidula* and placed both *M. rigidula* and *M. rigiduloides* in the sense of Small into subsp. *rigidula*. They stated "The study of Small et al. (Small et al. 1990*b*) represents a clinal variation in *M. rigidula* from Europe toward Asia....Regarding this, distinction of Asian populations of *M. rigidula* and placing them in *M. rigiduloides*, does not seem to be logical, the latter species is not accepted in this study."

Subordinate Taxa

None recognized here.

Type Specimen

Heyn (1959: 168) lectotypified the basionym for *M. rigidula* (*M. polymorpha* L. var. κ *rigidula* L.) by the illustration in vol. 2, p. 386 accompanying the phrase "Medica hirsuta echinis rigidioribus" in: Bauhin, J., Cherler, J.H., et al. 1650–1651. Historia Plantarum Universalis, vol. 2. Typographia Caldoriana (printer), Yverdon, Switzerland.

This is accepted in The Linnaean Plant Name Typification Project (http://www.nhm.ac.uk/research-curation/projects/linnaean-typification/; current June 2010; cf. Jarvis 2007), which states the illustration is on p. 385.

For nomenclatural purposes, it is critical to note that the lectotypification is on the basis of an illustration of presumably European material, since the species here interpreted as bearing the name *M. rigidula* is predominantly European (and hence so probably is the lectotype), whereas the very closely related species *M. rigiduloides* is predominantly Asian. (A reviewer of this monograph noted that if Heyn's lectotype cannot be reliably identified as *M. rigidula* or *M. rigiduloides*, an epitype that can be accurately identified could be designated.)

Synonymy

Homotypic

M. polymorpha L. var. κ *rigidula* L. Sp. Pl. 780 (1753), (basionym).

M. rigidula (L.) Desr. in Lam., Encycl. Méthod. 3(2): 634 (1792).

Heterotypic

M. agrestis Ten., Fl. Napol. 1: lxxi (1811–1815). (The publication date was discussed in Sabato (1990). Difficulties with determining the type were discussed by Heyn (1963, p. 107). The label of a type at FI bore the following information (photo provided by FI): 549 (579?), *Medicago agrestis* Ten., Da Tenore in Marzo 1844, Inter segetes Apuliae.).

M. rigidula (L.) All. var. *agrestis* (Ten.) Burnat, Fl. Alpes Marit. 2: 102 (1896).

M. gerardi Waldst. & Kit. ex Willd. subsp. *agrestis* (Ten.) Bonnier & Layens, Fl. Fr. 75 (1894).

M. rigidula (L.) All. subsp. *agrestis* (Ten.) Rouy in Rouy & Fouc., Fl. Fr. 5: 27 (1899).

M. rigidula (L.) All. subsp. *agrestis* (Ten.) Ponert, Feddes Repert. 83: 640 (1973), syn. superfl.

M. bondevii Kozukharov, Mitt. Bot. Inst. Sofia 15: 157 (1965). (Type indicated in publication: Ivan Bondev, 24-IV-1962, SOM).

M. bonofcensis Kit. in Kanitz, Linnaea 32: 615 (1863).

M. rigidula (L.) All. var. *brevispina* Rouy in Rouy & Fouc., Fl. Fr. 5: 26 (1899).

M. cinerascens Jordan in F.W. Schultz, Arch. Fl. France Allemagne 316 (1854). (Type indicated by Heyn (1963): exs. 1161, of Fl. Gall. et Germ. Jordan (BM).)

M. rigidula (L.) All. var. *cinerascens* (Jordan) Rouy, Fl. Fr. 5: 26 (1899).

M. rigidula (L.) All. subsp. *cinerascens* (Jordan) Ponert, Feddes Repert. 83 (9-10): 639 ("1972"; 1973).

M. rigidula (L.) All. f. *cinerascens* (Jordan) Kozuharov, Bulgarian Acad. Sci. 15: 156 (1965).

M. rigidula (L.) All. var. *compacta* Kozuharov, Bulgarian Acad. Sci. 15: 155 (1965) (type indicated in publication).

M. depressa Jordan, Catal. Graines Jard. Dijon (Adnot. Cat. Dijon) 1848: 28 (1848).

M. rigidula (L.) All. f. *discoidaea* Kozuharov, Bulgarian Acad. Sci. 15: 156 (1965) (type indicated in publication).

M. rigidula (L.) All. f. *elegans* Kozuharov, Bulgarian Acad. Sci. 15: 155 (1965) (type indicated in publication).

M. rigidula (L.) All. var. *eriocarpa* Rouy in Rouy & Fouc., Fl. Fr. 5: 26 (1899) (represented as "*M. eriocarpa* Rouy" in various compendia of scientific names).

M. gaditana Pérez Lara ex Willk., Öesterr. Bot. Z. 41(3): 85 (1891).

M. gerardi Waldst. & Kit. ex Willd., Sp. Pl., 3(2): 1415 (1802).

M. rigidula (L.) All. var. *gerardii* (Waldst. & Kit. ex Willd.) Burnat, Fl. Alpes Marit. 2: 101 (1896).

M. germana Jordan in F.W. Schultz, Arch. Fl. France Allemagne 316 (1854).

M. rigidula (L.) All. var. *germana* (Jordan) Rouy in Rouy & Fouc., Fl. Fr. 5: 25 (1899).

Medica incana Moench, Methodus 117 (1794), nom. illeg.

M. rigidula (L.) All. var. *lamottei* Rouy, in Rouy & Fouc., Fl. Fr. 5: 26 (1899), nom. illeg.

M. cinerascens Jordan var. *longiseta* Lamotte, Prodr. Fl. Fr. Plateau Central 192 (1877).

M. gerardi Waldst. et Kit ex Willd. var. ß *minor* Ser. in DC. Prodr. 2: 179 (1825). (A single sheet (which may contain more than one collection) in Herb. DC. has fruits with about 5 coils and extremely curved spines, consistent with *M. rigidula*, not *M. rigiduloides*.)

M. rigidula (L.) All. var. *minor* Ser. in DC., Prodr. 2: 179 (1825).

M. rigidula (L.) All. var. *minor* (Ser.) Morariu, Fl. Rom. 5: 133 (1957), comb. superf.

M. mitis Del. ex Godr. in Gren. & Godr., Fl. France 1: 394 (1848), nom. inval. (name merely cited in synonymy).

M. mitis Del. ex Nyman, Consp. Fl. Europ. 168 (1878).

M. mitis Willd. ex Urb., Ind. Sem. Hort. Berol. 4 (1872).

M. morisiana Jordan, Mém. Acad. Roy. Sci. Lyon, Sect. Sci., Ser 2, 1: 264 (1851).

M. rigidula (L.) All. var. *morisiana* (Jordan) Rouy in Rouy & Fouc., Fl. Fr. 5: 25 (1899).

M. polymorpha L. var *muricata* L., Sp. Pl. 781 (1753). The Linnaean Plant Name Typification Project (http://www.nhm.ac.uk/research-curation/projects/linnaean-typification/; current June 2010; cf. Jarvis 2007) provides the following information regarding the type: "Lectotype = "*Medica cochleata dicarpos capsula spinosa rotunda minore, rarius spinis crassioribus & brevioribus armata*" in Morison, Pl. Hist. Univ., 2: 153, s. 2, t. 15, f. 18, 1680. Designated by: Heyn in Bull. Res. Council Israel, Sect. D, Bot. 7: 168 (1959)."

M. muricata (L.) All., Fl. Pedem. 1: 316 (1785). (Heyn (1963) recognized *M. muricata* (L.) Willd. as *M. italica* and *M. muricata* Benth. as *M. doliata*.)

M. obscura Retz var. *muricata* (L.) Rouy, Fl. Fr. 5: 24 (1899).

M. rigidula (L.) All. f. *pastoralis* Kozuharov, Bulgarian Acad. Sci. 15: 155 (1965) (type indicated in publication).

M. pseudogerardii Schur, Verh. Naturf. Ver. Brünn 15(2): 174 (1877).

M. rigidula Bourg. ex Nyman, Consp. Fl. Eur. 167 (1878)

M. rigidula (L.) All. var. *submitis* Boiss., Fl. Or. 2: 101 (1872). (Type indicated by Heyn (1963): Szovits in prov. Karabagh, Iran (G).)

M. rigidula (L.) All. f. *submitis* (Boiss.) Kozuharov, Bulgarian Acad. Sci. 15: 156 (1965).

M. rigidula (L.) All. f. *subnudalosa* Kozuharov, Bulgarian Acad. Sci. 15: 155 (1965) (type indicated in publication).

M. rigidula (L.) All. var. *suffrutescens* Willk. ex Pérez Lara, Anales Soc. Esp. Hist. Nat. 21: 246 (1892).

M. timeroyi Jordan, Catal. Graines Jard. Dijon (Adnot. Cat. Jord. Dijon), 1848: 29 (1848).

M. rigidula (L.) All. var. *timeroyi* (Jordan) Boreau, Fl. Centr. Fr. d. 3, 2: 149 (1857).

M. tomentosa Chevallier, Fl. Gén. Env. Paris 2: 782 (1827), nom. illeg.

M. rigidula (L.) All. var. *turbiniformis* Kozuharov, Bulgarian Acad. Sci. 15: 156 (1965) (type indicated in publication).

M. villosa DC., Fl. Franç. (ed. 3) iv: 545 (1805), not *M. villosa* Brot., Fl. Lusit. ii: 116 (1804).

M. rigidula (L.) All. var. *compacta* Kozuharov f. *viscosa* Kozuharov, Bulgarian Acad. Sci. 15: 155 (1965) (type indicated in publication).

M. rigidula (L.) All. subsp. *zoharyi* Ponert, Feddes Repert. 83 (9-10): 640 (1973).

(Khassanov (1972, p. 91) cited the following: var. *submitis* Boiss. f. *eusumitis* Abdull. and f. *tuberculata* Abdull.; var. *cinerascens* (Jordan) Rouy f. *indefinata* (no authority cited, perhaps Abdullaeva was intended); and var. *rigidula* f. *vulgaris* (unclear whether Abdullaeva was the intended authority).)

Vernacular Names

English: Tifton burclover, Tifton bur medic, Tifton medic, rigid medic (a name also applied to *M. rigiduloides*), prickly lucerne (bur is also spelled burr, medic is also spelled medick). Bolton (1962) cited the common names "coriaceous alfalfa" and "field alfalfa", but these are very ambiguous.

French: Lampourdette, luzerne raide, luzerne rigide, luzerne à tige rigide, la luzerne de Gérard.

Geographical Distribution

(N = native, I = introduced, U = uncertain status)

Africa: Algeria (N), Egypt (N), Libya (N), Morocco (N), Tunisia (N).

Europe: Albania (N), Balearic Islands (Spain; N), Bulgaria (N), Corsica (France; N), Crete (Greece; N), Czech Republic (I), France (N), Greece (N), Hungary (I), Italy (N), Malta (N), Moldova (N), Portugal (N), Romania (N), Russia (N), Sardinia (N), Sicily (Italy; N), Spain (N), Turkey (N), Ukraine (N), Yugoslavia (N).

Northern America: USA (California, see Riefner and Small 2009; Massachusetts; both of these could be *M. rigiduloides*; I).

As interpreted here, *M. rigidula* is widespread in Europe and Africa. The extent to which it overlaps the Asian range of *M. rigiduloides* is problematical.

Description

Annual herb, more or less pubescent with simple hairs, sometimes also with glandular hairs. Stems procumbent to ascending, typically branched from base, (6)10–40(50) cm long. Mature leaflet blades (6)8–12 mm long, cuneate-obovate to obovate, tips of mature leaflets mostly emarginate (also retuse and obtuse), with terminal tooth, margins serrate from tip to about halfway to base. Stipules dentate to laciniate, often dissected one-third to one-half the width, with 4–8 teeth including a long terminal tooth. Flowers 1–3(6) per peduncle, yellow, (3)4–6(8) mm long, calyx pubescent with simple and/or glandular hairs, length 60% to 80% of corolla length, calyx teeth 40% to 75% of calyx length, peduncle 150% to 300% of length of subtending petiole, pedicel ca. 1 mm long. Pollen grains 3–6-pored, mostly 3-pored. Young pod contracted and concealed within calyx. Mature pod usually ovoid (sometimes cylindrical, discoid, or spherical), usually pubescent with simple and gland-tipped hairs (rarely glabrous), 5–10(12) mm long, 5–10(15) mm wide, with (4)5–6.5(7.5) coils, hardened at maturity, some gaps often present between mature coils (coils usually not tightly adpressed at maturity), facing surfaces of coils with 7–17 strongly curved radial veins that anastomose somewhat towards coil periphery, veins obscure at maturity due to development of proliferating, spongy tissue, edge of coil spiny or spines reduced to tubercles, or spineless, spines when present 0.5–2(2.5) mm long, 10–20(25) per coil, approximately oblique to pod axis, usually straight (sometimes somewhat hooked toward the apex). Seeds (2.5)3–4(4.5) mm long, 1.5–2 mm wide, 1–2 per coil, separated by spongy fruit partitions, smooth, yellow to yellow-brown, radicle about half (or slightly less than half) as long as seed.

Identification Hints

For distinction of *M. rigidula* and *M. rigiduloides*, see the above information as well as Chapter 79 on *M. rigiduloides*. Forms with fruits lacking spines, or with tubercles instead of spines, occur more frequently in *M. rigiduloides* than in *M. rigidula*. Both *M. rigidula* and *M. rigiduloides* can be confused with *M. doliata*, which also often has fruits covered with gland-tipped trichomes. The venation of the coil face of *M. doliata* is less anastomosed peripherally than is the case in the other two species. The species can also be confused with *M. truncatula*, although *M. rigidula* and *M. rigiduloides* rarely have their spines intermeshing between adjacent coils as dramatically as is often the case in *M. truncatula*. In addition, *M. truncatula* rarely has strongly arching radial veins on the pod coil faces or glandular hairs on the pods, both of which are often found in *M. rigidula* and *M. rigiduloides*.

Habitat and Environmental Adaptation

Habitat

Medicago rigidula has been recorded in widely different habitats, including steppes, shrub associations, and open woodlands, and it also occurs as a roadside weed and in fallow ground. Illustrative of the range of habitats occupied, Thanopoulos (2007) recorded that in Greece the species occurs in riparian environments, garrique with rocks and cliffs, as well as in open shrublands, meadows, macchie, degraded forests, meadows, grasslands, as a weed in various crops, on the edges of cereal and fallow fields, and in waste places from sea level to 1800 m.

Phenology

Flowering in much of the Mediterranean area is in March to May, extending to July at higher altitudes and in more northerly locations in parts of Europe.

Soil

The species has been collected on heavy (fine-textured) soils, and also on rocky soils including dry rubble and stony slopes. In Algeria, Abdelguerfi et al. (1988*a*) observed that *M. rigidula* occurs in medium- to fine-textured soils. Piano and Francis (1992) recorded that in its native area it occurs on loam–clay soils ranging in pH from slightly to moderately basic. In Sicily, Piano et al. (1991) observed the species growing in a pH range of 6.5–9.0.

Temperature

Ballard and Charman (2000) noted that *M. rigidula* has "reputed tolerance to cold (below freezing) conditions". Piano and Francis (1992) observed that the species is often one of the dominant medics in the cold environments of West Asia and North Africa. Cocks (1992) compared 12 annual *Medicago* species and observed that *M. rigidula* had high frost tolerance. Azizi et al. (2004) recommended *M. rigidula* for cultivar development for cool regions of Asia and North Africa. Krall et al. (1996*a*) found that 14 of 19 accessions of *M. rigidula* examined for their ability to self-regenerate in Wyoming survived the winter at one or more of the three locations tested. They also noted the initiation of a breeding program based on *M. rigidula* and *M. rigiduloides* to create cultivars suitable for the cold conditions of the United States Western High Plains. Caddel (2007) found that both *M. rigidula* and *M. rigiduloides* exhibited good frost resistance in the Southern Plains of the United States.

Water

Medicago rigidula has been collected on dry hillsides and appears capable of growing in arid areas, but not as arid as several of the more drought-tolerant annual species of *Medicago*. Piano and Francis (1992) stated that it requires a minimum annual rainfall of 250 mm in its native area. (Compare this to the information presented for *M. rigiduloides*.)

Biotic Relationships

Pollination

Small (1988) noted that the small flowers and pollen to ovule ratio (225) are consistent with *M. rigidula* being largely self-pollinated. This species is basically an inbreeder.

Dispersal

The spiny fruits are adapted to dispersal in animal fur, accounting in part for the widespread occurrence of this species.

Herbivory

No information located.

Nodulation and mycorrhizae

USDA (2009) reported positive root nodulation for *M. rigidula*. Groose et al. (1996) examined the comparative *Rhizobium* requirements of five lines each of *M. rigidula* and *M. rigiduloides*. The species differed markedly in *Rhizobium* specificity, and distinct within-species differences in *Rhizobium* specificity among the *Medicago* lines were also evident. All five *M. rigiduloides* (a species that is largely Asian) lines produced more shoot biomass with the two Asian *Rhizobium* strains used than with the four *Rhizobium* strains isolated from Europe and Australia. None of the latter rhizobial strains performed well with any *M. rigiduloides* line. Conversely, the European/Australian *Rhizobium* strains generally performed well with the European *M. rigidula*. Also see the information presented for *M. rigiduloides*. Literature purporting to examine the rhizobial relationships of "*M. rigidula*" (e.g., Brockwell et al. 1988) is difficult to interpret with respect to which species have been investigated, although in cases where accessions are still available it is possible to identify the species. Vâzquez et al. (2001) examined the performance of "*M. rigidula*" inoculated both with *Sinorhizobium meliloti* and arbuscular mycorrhizal fungi of the genus *Glomus* and found that occasional enhancement resulted from the combined use of both micro-organisms, depending on the conditions of growth.

Disease resistance

Yaege and Stuteville (2002) tested the resistance of 33 annual species of *Medicago* to powdery mildew (*Erysiphe pisi*) and found that the six collections of "*M. rigidula*" (the possibility exists that some of these were *M. rigiduloides*) grown were susceptible to the disease (it was observed that often only one of a number of populations of other species tested was resistant). In a comparative test of 36 annual species of *Medicago*, O'Neill and Bauchan (2000) found that an accession of "*M. rigidula*" from Spain (and therefore probably correctly identified as *M. rigidula*) exhibited moderate susceptibility to anthracnose caused by *Colletotrichum trifolii*.

Economic History, Cultivars, and Uses

Medicago rigidula has attracted limited interest to date as an agricultural crop, although it is the subject of experimentation. 'Laramie' is a relatively winter hardy selection intended as a self-regenerating pasture species for the Central High Plains of the United States (Krall et al. 2007). See below for information on the French cultivars 'Ampus' and 'Kalambaka', which are recommended for use in rangelands for sheep rearing. Vitkus (1977) assessed *M. rigidula* as suitable for cultivar selection for Lithuania.

Cultivars

The American cultivar 'Laramie', cited above, was selected by breeders who understood the distinction between *M. rigidula* and *M. rigiduloides*, and it is likely *M. rigidula*. The French cultivars 'Ampus' and 'Kalambaka' were identified by their breeders as *M. rigidula* subsp. *cinerascens* and are probably also *M. rigidula*, not *M. rigiduloides*.

Forage

Medicago rigidula has been sown for forage only to a limited extent. In an experimental test of 17 annual *Medicago* species in southeastern Wyoming, Walsh et al. (2001) found that the accession of *M. rigidula* tested was more productive than any of the other species.

Fodder

This species is not used as a significant source of harvested herbage for animal feed.

Green manure, rotation crop, ground cover

Evans et al. (2004) examined the potential of numerous fodder species to survive winter in North Korea and produce nitrogen, organic matter, and soil stabilization in maize and rice fields. *Medicago rigidula* was judged to have good winter survival and good green biomass production, and to produce fair ground cover. Ghalavand et al. (2003) found that using *M. rigidula* as a cover and/or fallow crop in Iran reduced surface run-off and erosion, and increased water conservation and soil water-holding capacity.

Nair et al. (2006*b*) noted that the French cultivars 'Ampus' and 'Kalambaka' were bred in part "to maintain soils around infrastructures: roadsides, railway lines embankments and disturbed areas; to avoid soil erosion from rainfall and winds, reduce the use of herbicides, limit infiltrations of pesticides in soils and enhance underground life in sloping vineyards."

Other uses (salad sprouts, ornamental, plant products, etc.)

No information located.

Chromosome Number

Before the proposal to split *M. rigidula* into two species, *M. rigidula* and *M. rigiduloides*, the following counts were reported for "*M. rigidula*":

$2n = 14$: Fryer 1930; Clement 1962 (as var. *rigidula* and var. *morisiana*); Simon and Simon 1965; Lesins and Lesins 1962; Gillies 1971*a*; Kliphuis and Barkoudah 1977; Fernandes et al. 1977; Fernandes and Queirós 1978; Abdelguerfi et al. 1988*b*; Mariani et al. 1996; Ibragimova 1999.

$2n = 14, 16$: Heyn 1963; Bolkhovskikh et al. 1969, p. 809; Loon 1980.

$2n = 16$: Ghimpu 1928, 1929*a*, 1929*b* (as *M. gerardi*), 1930; Ferakova and Murin 1978; Abdelguerfi and Guittoneau 1979; Astanova 1981; Agarwal and Gupta 1983*b*; Magulaev 1992.

Small et al. (1990*b*) reported $2n = 14$ for 27 collections of *M. rigidula* s. str. and $2n = 14$ for 28 collections of *M. rigiduloides*, making it clear that this is the predominant count for both species and raising the possibility that the $2n = 16$ counts are erroneous.

Genetic Relationships and Gene Transferability

Intraspecific

See comments for Interspecific, below.

Interspecific

There is a partial barrier to interspecific fertility between *M. rigidula* and *M. rigiduloides* (see information of Lesins, above). Groose et al. (1996) conducted hybridization experiments involving a set of 20 lines of *M. rigidula* and *M. rigiduloides* and found support for the taxonomic segregation of these species. F_1 plants of both intraspecific and interspecific crosses had reduced fertility compared with mid-parent means (all significantly different at P=0.05), but in all cases interspecific F_1 plants exhibited a greater depression of fertility than did intraspecific crosses.

Exchange by advanced biotechnology

No information located.

Conservation Status

This species is fairly widespread, and at this time there is no indication that there are particular forms requiring conservation measures. *Medicago rigidula* has been listed as a protected species in Hungary (http://econets.iucn-ce.org/files/1/060125204354_Species%20Hungary; current July 2010).

Germplasm Sources

There are approximately 3000 accessions in the major germplasm institutions with large *Medicago* collections that are (mostly) identified as *M. rigidula*, although many (perhaps the majority since the majority are from Asia) are *M. rigiduloides*. For research purposes, it is necessary to ascertain the identity of these accessions before employing them.

Research Problems and Priorities

Medicago rigidula has been found to have greater cold tolerance than most other annual species of *Medicago*, and seems to have superior potential for producing useful cultivars for the Mediterranean region and other regions of the world. This potential needs to be more fully examined.

Representative Collections (listed here on the basis that they are from Europe and Africa; these collections were examined before I made the distinction between *M. rigidula* and *M. rigiduloides* and they have not been re-examined on the basis of morphological distinction)

Albania: Rechinger fil. & Scheffer 75, Bertiscus, Alpes boreales albanicae, in collibus circa oppidum Pec (Ipek), in graminosis siccis, alt. 500–700 m, 2-VII-1933, in fruit (US). **Algeria:** Alston, A.H.G. & Simpson, N.D. 37671, 3 miles south of Michelet, Djurdjura Mts, alt. ca. 1170 m, 29-IV-1937, in flower & fruit (BM). Chevallier, L., Birmandreis carrière ecrite de la Colonne Voiral, 27-III-1893, in fruit (P). Durieu, M., Constantine, 26-IV-1860, in fruit (P). Fournier, E., Plain à Batisa, I-1885, in fruit (P). Reboud, V., Constantine, V-1875, in fruit, Société Dauphinoise No. 74 bis (P). **Austria:** Charpin, A. & Greuter, W. 8362, prov. "Basses-Alpes", prope sepulcretum oppidi "Manosque", in rarioribus herbidis dumetorum Quercus cocciferae abundans, alt. 440 m, in flower & fruit, 16-V-1969 (NY). **Bulgaria:** Kojoucharov, S. & Kouzmanov, B., près de village Mazek, la vallée de la rivière Maritza, patûrages secs, 1-V-1959, in fruit (K). Vyhodcevski, N., Bulgaria occidentalis, M. Ljulin, in arenosis saxosis ad cca 650 m s.m., 18-VI-1955, in fruit, Plantae Bulgaricae Exsiccatae Cent. VII No. 672 (UC). **France:** Albert, A., Var, La Farlède, champs arides 1-V-1887, in fruit, Flor Selecta Exsiccata publié par Ch. Magnier, No. 1646 (P). Bouby, H. 9110, Col de Cabrenas, au dessus de Sre. Enimie (Logêre), 18-VII-1969, in fruit (P). Guillet, J. 8838, Pyrenées orientales, Collioure glacis du fort – Miradoux, 27-V-1956, in fruit (P). Jordan, A., Villeurbanne près de Lyon, champs cultivés, vieilles luzernières des terrains tertiaires, fleurs le 14-V, fruits en VI-1850, Flora Galliae et Germaniae, Exsiccata de C.B. No. 535 (P). Jordan, Al, Saint-Alban près de Lyon (Rhone), pâturages secs, lisière de bois, fleurs le 6 mai, fruits le 20 juin 1853, Flora Galliae et Germania Exsiccata de C. Billot No. 1161 (P). Kralik, L., près de Paris, pelouse sablonneuse du bois de Boulogne, 28-V-1843, in flower & fruit, Flora Galliae et Germaniae Exsiccata de C. Billot (P). Mabille, P., Saint-Florent, Corsica, 10-VI-1868, P. Mabille – Flora Corsicum, No. 364 (P). Rotereau, L., Var, Hoyères près l'Ermitage, sur le mica schiste, 4-V-1874 (P). Timbal-Lagrave, Haute-Garonne, Toulouse, canal de Brienne, berge de la rive droite, à moitié, VII-1881, in fruit, Flora selecta Exsiccata, publié par Ch. Magnier, No. 42 (P). **Greece:** De Heldreich, Attica, in vinetis prope Hagin Glykeria, ad radices collis Turcovuni, 30-IV-1884, in fruit; De Heldreich Herbarium Graecum Normale No. 837 (GH, NY, P). Patten, J.B. K59, Crete, Isthmus of Hierapetra, region of Kavoúsi, Tomb Mt., rubble, Phrygana formation, 20-V-1900, in fruit (GH). Tedd, H.G. 1682, Thrace, near Xanthi, alt. 400 feet, rocky hillslopes, 8-V-1936, in flower & fruit (K). **Italy:** Ball, J., south of Tuscany, Chiusi, 5-V-1872, in flower & early fruit (US). Groves, H., prope Florentiam, ad vias et in herboris aridis, V-1874, in flower & fruit (US). Levier, E., circa Florentiam, in herbosis incultis, V-1886, in flower & fruit, F. Schultz Herbarium Normale Nov. Ser. Cent. 22, No. 2156 (GH, P). Schultz, F., circa Florentiam, in herbosis incultis, V-1886, in fruit, F. Schultz Herbarium Normale Nov. Ser. Cent. 22 No. 2156 (P). **Morocco:** de Wilde, J.J.F.E. & P.A.W.J. & Dorgelo, J. 2536, forest between Al-Hajeb & Ifrane, 18 km northwest Ifrane, arable field in forest, 25-V-1961, in fruit (BM). Reading Univ./BM. Exped. 26, 8 km south of Arbaoua, west of road to Kenitra, (30°50′N, 5°45′W), alt. 250 m, derelict field, near Eucalyptus wood, sandy loam, 31-V-1974, in fruit (BM). **Portugal:** Daveau, J. 1061, Provincia da Estremadura, Serra de Monsanto, IV-1884, in flower & fruit (P). Daveau, J., Almada, rampes du Tage, environs de Lisbonne, IV-V-1877, in fruit (P). **Romania:** Borza, Al. & Nyárády, E.I., Dobrogea, distr. Caliacra, ad littorem ponti Euxini (Marea neagra) prope "Farul dela Sabla", alt. 3–5 m, 14-VI-1925, in fruit, Museum Botanicum Universitatis, Cluj Flora Romaniae Exsiccata No. 1267 (P). Cirtu, D. & M., Oltenia, distr. Dolj., in pratis Festucetum valesiacae – Poaetum bulbosae, inter pagum Negoesti et silva Vogna, alt. ca. 190 m, 29-V-1972, in flower & fruit, Flora Olteniae Exsiccata 972 (NY). **Spain:** Ball, J. 331, in Hispania Baetica, juxta Ronda, ex rupibus calcareis montis El Hacho, 20-VI-1851, in flower &

fruit, (GH). Ball, J., ex Hispania Baetica, in collibus juxta Algeciras, 15-IV-1871, in flower & fruit (GH). Jerónimo, Hno, Madrid, Granja de Castilla, lieux incultes, 2-V-1918, in fruit, Plantes d'Espagne – F. Sennen (P). Sennen, F. 968, Catalogne, Barcelone, pentes du Tibidabo, V-1910, in flower & fruit (US). **Turkey:** Fleischer, in collibus Smyrnae, IV-1827, in flower & fruit (MO) (this European taxon is rare in Asia; this location, essentially Izmir on the Adriatic coast, borders Europe). **United Kingdom:** Townsend, C.C. 530625/14, Aronmouth Docks, Glas., 25-VI-1953, in fruit (K). **USSR:** Alexeenko 735, Prov. Dagestan, Distr. Temir-Chan Sclura, A1 urb, l'elrowsk, southwest versus, in campsis arenosis, alt. 100 feet 12-V-1901 (LE). Alexeenko 7982, Prov. Dagestan, Distr. Kurinskij, Pr. St. Diviezi, in decliv. merid, siccis, 6-IV-1902 (LE). Davis 33147, Crimea, Distr. Yalta, Nikita, alt. 80 m, herby dry southwest-facing slopes, 30-V-1959, in flower & fruit (K). Davis 33321,. Crimea, Distr. Yalta, above Nikita, alt. 300 m. sunny bank, 21-VI-1959, in flower & fruit (K). Sintensis, P. 1681, Iter transcaspico-persicum 1900–1901, Regio transcaspica, Kisil Arwat, in collibus, 5-V-1901 (LE). Sintensis, P. 246b, Iter transcaspico-persicum 1900–1901, Regio transcaspica; Aschabad, in montibus supra Nephton, 4-V-1900 (LE). **USA:** Bicknell, E.P., Yonkers Wool Mill, Yonkers, New York, 1-IX-1894, in fruit (NY; collection mixed with *M. minima*). Buchanan, W.F. Jr., Georgia, Clarke Co., Athens, weed in waste places, 21-IV-1937, in fruit & flower (US) (identification confirmed by 3-pored pollen). **USSR:** (confirmed) Tselev, N.N. et al. 363, Crimea, mountain ridge between the town of Aiya and the village of Rezervnoe, 18–20 km southeast of Sevastopol, 15-V-1984, in flower & fruit (A, translated from Russian). Grossheim, A., Gurvitsh, C. & Rzaev, R., Transcaucasia, Azerbajdzhan, distr. Zuvant, prope p. Lerik, in fauce fl. Orant-tihaj, in rupestribus, 15-V-1935, in flower & fruit (A). (rare in Asia, identification confirmed by pollen) Meyer, F.N. 502, prov. Daghestan, between Petrovsk and Temir-Chan-Schura, 15-V-1910, in fruit (GH, NY, UC; note that Meyer 507, collected near here, is identified as *M. rigiduloides*). Schiraevsky, I. 3341, Tauria, In distr. Eupatoria, prope pag. Saki, inter frutices, 3-VII-1910, in fruit (BM, GH, LE, MO, NY, UC, US). **Yugoslavia:** Gilliat-Smith, B. 3243, roadside just outside Sarajevo, along the Miljacka, 4-VII-1933, in flower & fruit (K). Jackson, A.K. & Turrill, W.B. 157, Dalmatia, Solin, grassy places among Roman ruins, 21-IV-1935, in flower & fruit (K). Muller, Fiume, V-1826, in flower & fruit (MO). Pichler, T., Dalmatia, prope Torretto ad Spaleto, in agris, VI-1880, in flower & fruit (K).

79. *Medicago rigiduloides* E. Small

M. rigiduloides E. Small, Can. J. Bot. 68(12): 2614 (1990), (basionym).

The epithet *rigiduloides* is based on *rigidula* (a reference to the species *M. rigidula*) plus *oides* (suffix of Greek origin, indicating similarity of form), an allusion to the very similar appearance of this species to *M. rigidula*.

Note: See Chapter 78 on *M. rigidula* for information regarding difficulties of distinguishing the two species and interpreting the extent to which past literature on "*M. rigidula*" actually refers to *M. rigiduloides*.

Subordinate Taxa

None recognized here.

Type Specimen

E. Small & B. Brookes M1282*c* (holotype, DAO), grown in cultivation in Ottawa, Canada, 21 May to 14 September 1986, using seed obtained from the Parafield Plant Introduction Centre, South Australian Department of Agriculture, Adelaide (accession 15773), collected originally from E. Arbil, Khanzad Range, Iraq, altitude 600 m, in loam soil.

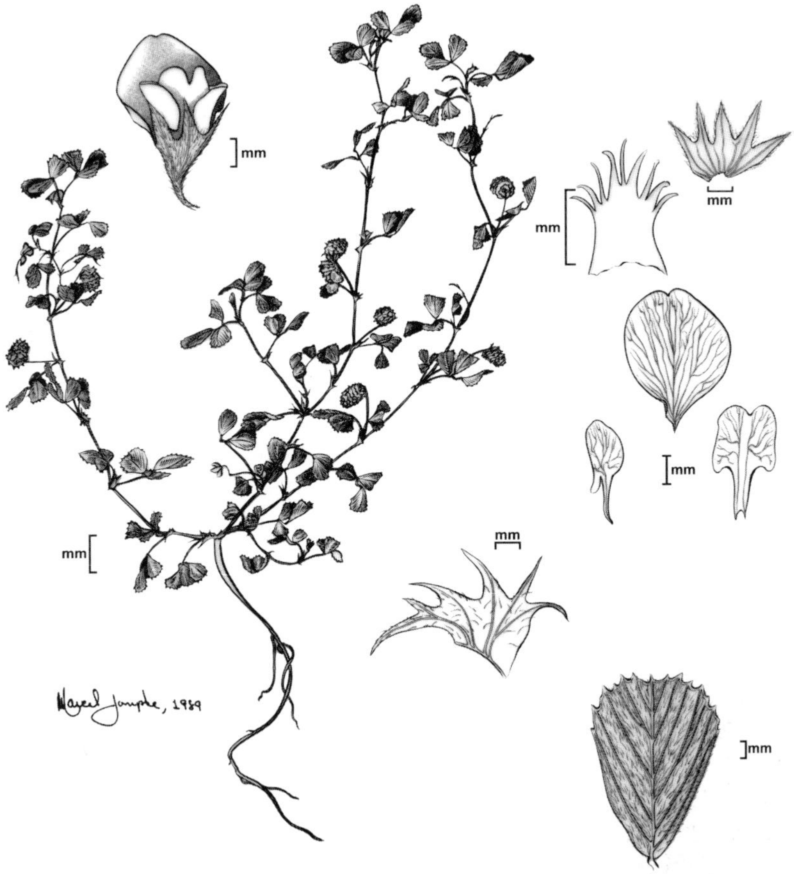

Fig. 79-1. *Medicago rigiduloides*. **Left:** Plant and flower. **Upper right:** Floral parts (clockwise from upper left: flattened androecial column (anthers and free stamen not shown); flattened calyx; standard petal; keel; wing petal. **Lower right:** Leaflet. **Center:** Stipule. (Flower and floral parts based on E. Small M1338, DAO, remainder based on K.H. & F. Rechinger 3763, US; artist: M. Jomphe.)

Synonymy

Homotypic

No names found.

Heterotypic

M. rigidula (L.) All. var. *submitis* Boiss., Fl. Or. 2: 101 (1972) (lectotypifed by Heyn (1963, p. 106): "Lectotype – the specimen collected by Szovits in Iran (Prov. Karabagh) deposited at G".) The specimen is: Szovits 153.)

Vernacular Names

ENGLISH: No names found. *Medicago rigiduloides* and *M. rigidula* collectively have been referred to as "rigid medics" (Caddel 2007).

FRENCH: No names found.

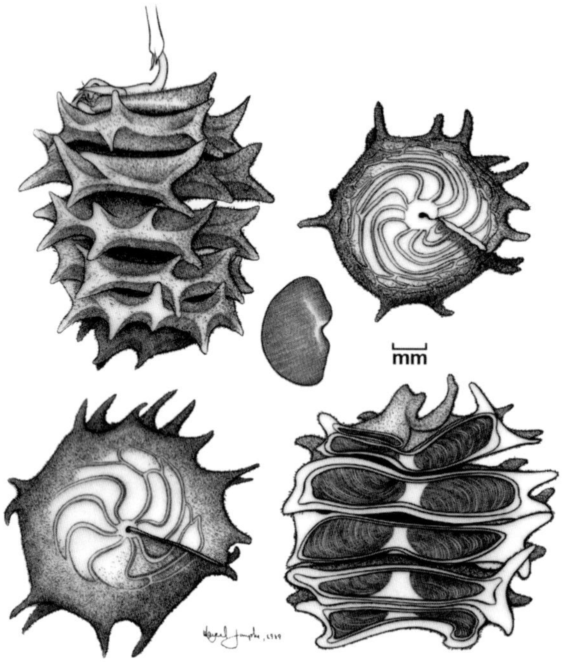

Fig. 79-2. *Medicago rigiduloides*. Fruits (pedicel at top). **Upper left:** Mature fruit with about six coils (based on E. Small M1301, DAO). **Upper right:** Face of a coil from a young fruit (based on E. Small M1338, DAO; note the strongly curving radial veins that extend to periphery). **Lower right:** Longitudinal section through fruit, showing seed cavities (based on E. Small M1275, DAO). **Lower left:** Face of a coil from a mature fruit (based on E. Small M1338, DAO; note appearance of peripheral veinless area because of obscuring of peripheral venation by alveolar (spongy) tissue). **Center:** Seed (based on E. Small M1275, DAO). (Artist: M. Jomphe.)

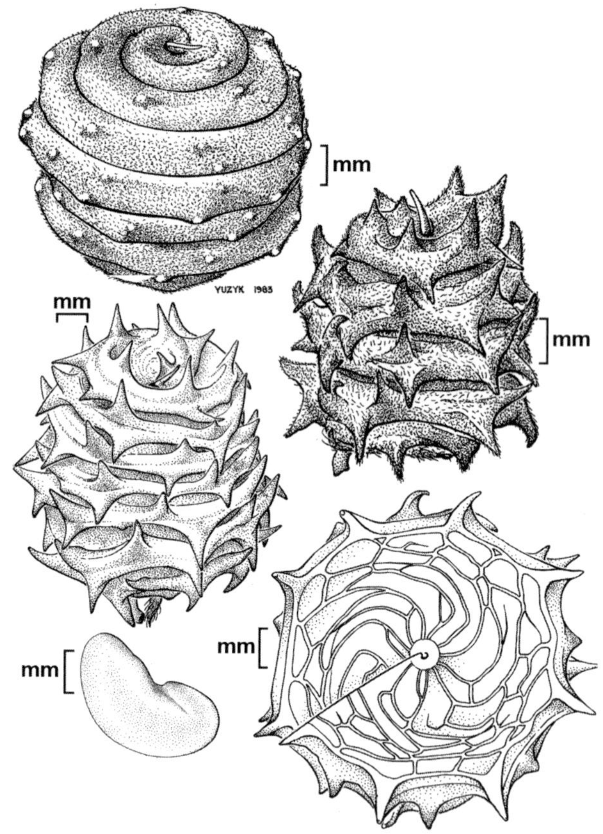

Fig. 79-3. Variation of fruits of *Medicago rigiduloides*. **Clockwise from top left:** Tuberculate (very short-spined) fruit (based on Eig, Zohary & Feinbrun on 6-V-1927, HUJ); spiny fruit (based on Ser Hayek, det. Recl. 47, HUJ); coil face showing venation; seed; spiny fruit (preceding three based on Eig & Zohary on 14-IV-1929, HUJ). (Artist: L. Yuzyk.)

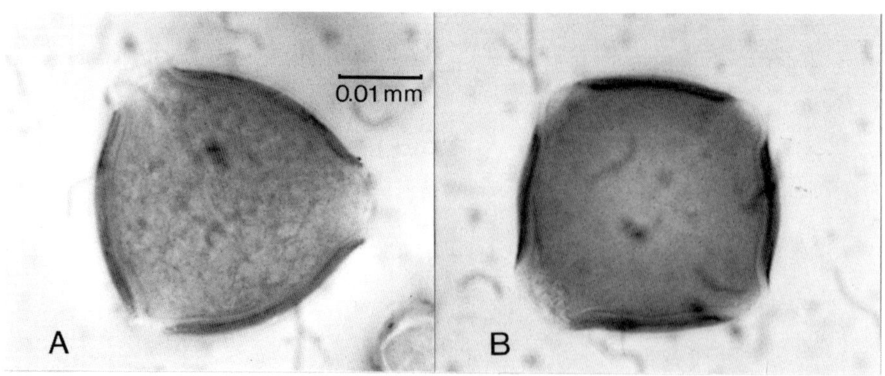

Fig. 79-4. A comparison of pollen grains (in glycerin jelly, polar view) of (A) typical three-pored pollen of the European *Medicago rigidula* (based on E. Small M1269, DAO), and (B) the typical four-pored pollen of Asian *M. rigiduloides* (based on E. Small M1320, DAO).

Geographical Distribution

(N = native, I = introduced, U = uncertain status)

As interpreted here, *M. rigiduloides* is widespread in Asia. The extent to which it overlaps the European and African ranges of *M. rigidula* is uncertain.

Africa: Francis (1981) reported that "*M. rigidula*" (probably *M. rigiduloides*) occurred as small, isolated populations in Tripolitania (northwestern Libya).

Asia: Afghanistan (N), Armenia (N), Azerbaijan (N), Cyprus (N), Dagestan (N), Gruziya (N), Iran (N), Iraq (N), Israel (N), Jordan (N), Kazakhstan (N), Kyrgyzstan (N), Lebanon (N), Syria (N), Russia (N), Tadzhikistan (N), Turkey (N), Turkmenistan (N), Uzbekistan (N).

Description

Annual herb, more or less pubescent with simple hairs, sometimes also with glandular hairs. Stems procumbent to ascending, typically branched from base, 10–40 cm long. Mature leaflet blades (6)8–12 mm long, cuneate to obovate, tips of mature leaflets mostly emarginate (also retuse and obtuse), with terminal tooth, margins serrate from tip halfway to base. Stipules dentate to laciniate, often dissected one-third to one-half of width, with 4–8 teeth including a long terminal tooth. Flowers 1–3(6) per peduncle, yellow, (3)4–6(8) mm long, calyx pubescent with simple and/or glandular hairs, length 60% to 80% of corolla length, calyx teeth 40% to 75% of calyx length, peduncle 150% to 300% of length of subtending petiole, pedicel ca. 1 mm long. Pollen grains 3–6-pored, mostly 4-pored. Young pod contracted and concealed within calyx. Mature pod usually ovoid (sometimes cylindrical, discoid, or spherical), usually pubescent with both simple and gland-tipped hairs (rarely glabrous), 5–10(12) mm long, 5–10(15) mm wide, with (4)5–6.5(7.5) coils, hardened at maturity, some gaps often present between mature coils (coils usually not tightly adpressed at maturity), facing surfaces of coils with 7–17 strongly curved radial veins that anastomose somewhat towards coil periphery, veins obscure at maturity due to developing spongy tissue, edge of coil spiny or spines reduced to tubercles, or spineless, spines when present 0.5–2(2.5) mm long, 10–20(25) per coil, perimeter approximately oblique to pod axis, usually straight (sometimes somewhat hooked toward the apex). Seeds (2.5)3–4(4.5) mm long, 1–2 per coil, separated by spongy fruit partitions, smooth, yellow to yellow-brown, radicle about half (or slightly less) as long as length of seed.

Identification Hints

Medicago rigiduloides is very similar to *M. rigidula*. The pollen of *M. rigiduloides* is almost always four-pored, whereas the pollen of *M. rigidula* is almost always three-pored. See the following key and Chapter 78 on *M. rigidula* for easier characters to employ for distinguishing the two species.

After *M. rigidula*, the species next most likely to be confused with *M. rigiduloides* is *M. doliata*, which also often has fruits covered with gland-tipped trichomes. However, the venation of the coil face of *M. rigiduloides* is less anastomosed peripherally than is the case in *M. doliata*. Still another species with which it might be confused is *M. constricta*, which has much less evident, if any, gaps between the fruit coils at maturity.

Habitat and Environmental Adaptation

Habitat

Medicago rigiduloides has been recorded in a variety of habitats, especially rocky soils.

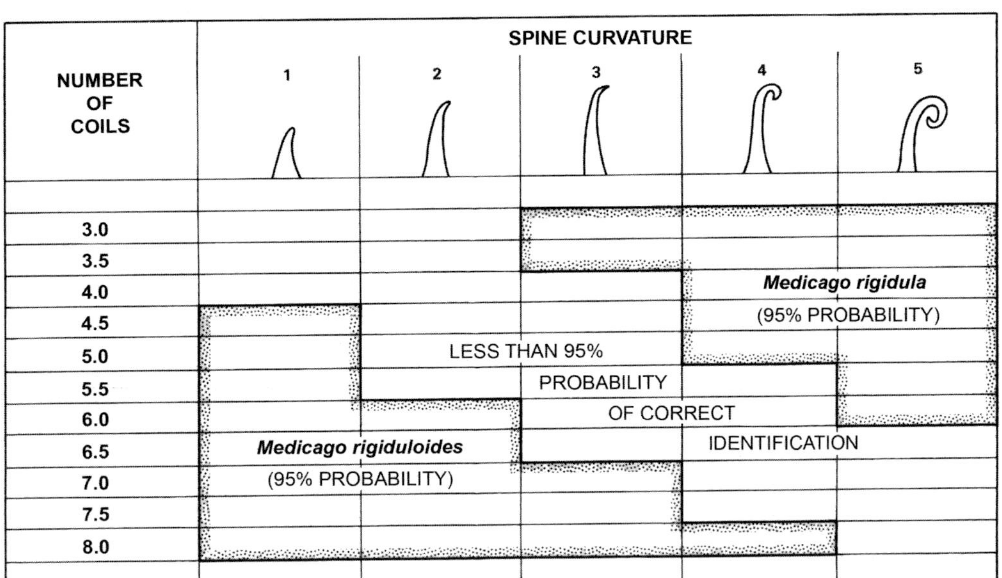

Fig. 79-5. Illustrated key for discriminating *Medicago rigiduloides* and *M. rigidula* (based on Small et al. 1990*b*).

Phenology

Flowering in much of the Mediterranean area occurs from March to May.

Soil

The soils naturally occupied are likely much like those described for *M. rigidula* (see Chapter 78).

Temperature

Ballard and Charman (2000) noted that a South Australian Annual Pasture Legume Collection of *M. rigiduloides* (SA3601) has "reputed tolerance to cold (below freezing) conditions." Cocks and Ehrman (1987) found that, of 22 species of *Medicago* examined in Syria, "*M. rigidula*" (presumably *M. rigiduloides*) was among the most frost tolerant. Mehregan et al. (2003) recorded that "*M. rigidula*" (presumably *M. rigiduloides*) "has good resistance to the hard and cold conditions of the high mountains in Iran", occurring at altitudes up to 3200 m. Krall et al. (1996*a*) found that five of nine accessions of *M. rigiduloides* examined for their ability to self-regenerate in Wyoming survived the winter at one or more of the three locations tested. Krall et al. (1996*a*) pointed out a breeding program based on *M. rigidula* and *M. rigiduloides* to create cultivars suitable for the cold conditions of the United States Western High Plains. Most cultivars of annual *Medicago* have been bred in Australia and lack tolerance to frost (Cocks and Ehrman 1987). In this respect, *M. rigiduloides* (identified as *M. rigidula*) is considered to be a better possibility for West Asia than the available Australian cultivars (Abd El Moneim and Cocks 1986). Caddel (2007) found that *M. rigiduloides* exhibited good frost resistance in the United States Southern Plains.

Water

Abdelguerfi et al. (1988*a*) observed that in Algeria "*M. rigidula*" (probably *M. rigiduloides*) occurs in regions receiving between 450 and 800 mm of precipitation annually, but not in areas receiving less than 450 mm or more than 800 mm.

Biotic Relationships

Pollination

The small flowers indicate that *M. rigiduloides* is largely an inbreeder.

Dispersal

The spiny fruits are adapted to dispersal in animal fur, accounting in part for the widespread occurrence of this species.

Herbivory

No information located.

Nodulation and mycorrhizae

USDA (2009) reported positive root nodulation for *M. rigiduloides*. Very curiously, Ballard and Charman (2000) noted that, in experiments testing the ability of 11 annual *Medicago* species to nodulate and fix nitrogen with naturalized rhizobia from 28 Australian soils, *M. rigiduloides* was prolifically nodulated but failed to fix nitrogen with any of the sources of soil rhizobia, whereas the closely related *M. rigidula* fixed nitrogen with all the rhizobia tested. Lorenzetti et al. (1987) tested the symbiotic response of *Rhizobium meliloti* strains on *M. rigidula* from Syria (presumably *M. rigiduloides*) and observed a high level of specificity, some strains greatly increasing productivity, others having little effect. Also see information presented for *M. rigidula*.

Disease resistance

In a comparative test of 36 annual species of *Medicago*, O'Neill and Bauchan (2000) found that four accessions of "*M. rigidula*" from Asia (and therefore probably *M. rigiduloides*) exhibited strong to moderate susceptibility to anthracnose caused by *Colletotrichum trifolii*.

Economic History, Cultivars, and Uses

Medicago rigiduloides has attracted limited interest to date as an agricultural crop, although it is the subject of experimentation.

Cultivars

No cultivars have been bred.

Forage

Medicago rigiduloides has been sown for forage only to a limited extent.

Fodder

This species is not used as significant source of harvested herbage for animal feed.

Green manure, rotation crop, ground cover

See information for *M. rigidula*. The two species almost certainly have very similar properties and potential.

Other uses (salad sprouts, ornamental, plant products, etc.)

No information located.

Chromosome Number

$2n = 14$ (see information in Chapter 78 on *M. rigidula*).

Genetic Relationships and Gene Transferability

Intraspecific

See information for *M. rigidula*.

Interspecific

See information for *M. rigidula*.

Exchange by advanced biotechnology

No information located.

Conservation Status

This species is fairly widespread and at this time there is no indication that there are particular forms requiring conservation measures.

Germplasm Sources

See information for *M. rigidula*.

Research Problems and Priorities

Medicago rigiduloides, like its very close relative *M. rigidula*, has greater cold tolerance than most other annual species of *Medicago*, and has potential for producing useful cultivars both for the Mediterranean region and other regions of the world. This potential needs to be more fully examined.

Representative Collections (listed here on the basis that they are from Asia; these collections were examined before I made the distinction between *M. rigidula* and *M. rigiduloides*, and have not been re-examined to confirm their identity by using morphology)

Afghanistan: Aitchison, J.E.T., 522, 20-V-1885, in fruit (BM, GH, K). **Azerbaijan:** Moussain, M., Zargami & Tehrani 2-AR, Rezaieh, Ghasemlov, alt. 1350–1650 m, 16-VI-1977, in fruit (HUJ). Moussain, M. & Zargami 1-AF, Rezaieh, Kaloudan Island, alt. 1320–1580 m, 18-VI-1977, in fruit (HUJ). Moussain, M. 4-AK, Kalihar, between Cardane-ye Malauk and Ouch-Ghardash, alt. 1200–1850 m, 29-VI-1977, in fruit (HUJ). **Iran:** Bunge, M. 1868-13, pr. Teheran, V-1859, in fruit (P). Koelz, W. 14435, Takht-i-Jamshed, Fars, on marble mountain, 12-III-1940, in flower & fruit (US). Koelz, W. 15489, Luristan, Durud, alt. 5000 feet, on open slopes, in flower & fruit, 20-V-1940 (US). Lindsay, N. 1081, Tspili, Yehlah, dry places, fairly common, VI-1936 (BM). Lindsay, N. 269, Sultanabad, alt. 5000 feet, dry banks, V-1935, in flower & fruit (BM). Rechinger, K.H. & al. 3763, prov. Kerman, montes Djamal Bariz inter Bam et Djiroft, in delivibus australibus, alt. 2000 m, in flower & fruit, 8-10-V-1948 (US). **Iraq:** Eig, A. & Zohary, M., northeast Iraq, 71 km east of Sulaimani, alt. 720 m, banks of stream, 14-IV-1933, in flower (HUJ). Gillett & Rawi 7217, Jebel Makhul near Ain Diba, steppe, alt. 200 m, 2-IV-1947, in fruit (US). Gillett & Rawi 7521 & 7522, Qaranjir, between Kirkuk & Chem Chemal, alt. 700 m, grassy slopes, 14-IV-1947, in fruit (K). **Israel:** Eig, A., Zohary, M. & Grizi, A., Judaean Desert, about 3/4 km on Jerusalem-Jericho road, soft gravelly soil, northern exposure, 27-III-1935, in fruit (HUJ). Zohary, D., Judaean Mts., Beit Hakerem to Deir-Yassin, Senon soil, rocks, and batha, east slope, 25-III-1941, in fruit (HUJ). Zohary, M., Judean Mts., Jerusalem, Giv'at Ram, roadsides, 20-III-1972, in flower (HUJ). **Israel (–Syria):** Schmida, A., Hermon, Namneman ridge, alt. 1600 m, 27-V-1984, in fruit (HUJ). **Jordan:** Eig, A., Zohary, M. & Feinbrun, N., Ammon, Amman, 7-V-1927, in fruit (HUJ). Eig, A., & Zohary, M., Ammon, Ein Suella, Nubian sandstone, 14-V-1929, in fruit (HUJ). Naftolsky, N., Ammon, Es-Salt, mountains, 20-IV-1929, in fruit (HUJ). **Lebanon:** Bot. Dep., descent to Wadi Joze (east of Wadi Fissen), alt. 1520 m, alluvial gravelly soil, 17-VII-1934, fruits shed (HUJ). **Syria:** Samuelsson, G. 1870, Coelesyria, Baalbek, inter segetes, alt. ca. 1150 m, 29-V-1932, in fruit (NY). Shepard, Dr. & Mrs., Aintab, northern Syria, V-1886, in late flower & fruit (GH). Zohary, M., Syria bor., about Aleppo, 3-V-1931, in flower & fruit (UC). **Turkey:** Coode, M.J.E. & Jones, B.M.G. 720, Yumurtalik (southeast of Adana), near sea level, sandstone cliff overlooking sea, 8-V-1965, in fruit (A). Davis & Hedge D26762, prov. Adana, dist. Bahçe (N. Amanus), Haruniya, alt. 500 m, edge of water course, 17-IV-1957, in flower & fruit (BM). Davis & Hedge, D27008, prov. Hatay, distr. Belen (Amanus), Belen Soguk Oluk, alt. 400 m, roadsides, 23-IV-1957, in flower & fruit (BM). Davis & Hedge D28544, prov. Mardin, Savur, alt. 900 m, vineyards, 24-V-1957, in fruit (BM). Ledingham, G.F. 4295, western edge of Ankara, dry overgrazed areas, alt. ca. 850 m, 10-VII-1965, in fruit (NY). Sintensis, P. 375, Mesopotamia, Biredjik, Djebel Taken, 16-IV-1888, in flower & fruit (P). Sintensis, P. 634, Mesopotamia, Biredjik, Leidun baghlsche, 5-V-1888, in fruit (P). Zohary, M., northeast Turkey, between Mus and Erzurum, 26 km north of Mus, overgrazed steppe with forest remnants, calcareous soil, eastern exposure, alt. 1300–1500 m, V-VII-1963, in flower & fruit (HUJ). **USSR** (confirmed): Fedtschenko, V.A. 1278, Iter Samarcandense Bucharicum, 4-V-1913, in fruit (GH). Frantskeviya, N.A., Kopetdag, 9-VI-1972, in fruit (MO). Karjagin, I., Transcaucasia, Armenia, distr. Migri, contra pag. Legvas, in decliv. austro-occidental. rup. lapidosis sicis cum Paliuro, alt. ca. 900 m, 13-IV-1934, in flower (UC). Goloskokov, V.P.. Kirgizskaya, southwestern tip of the Kirghizean Range, Alman-sai Canyon, southern rocky slopes, 2-VI-1969, in fruit (NY, translated from Russian). Grossheim, A. & Gurvitsh, C., Transcaucasia, respublica Nachitschevan, distr. Shach-buz, inter Kerababa et Aryndzhe, inter segetes et in vineis, 4-VI-1934, in flower & fruit (NY). Meyer, F.N. 507, prov. Daghestan, near Petrovsk, on dry steppe, 15-V-1910, in flower & fruit (MO; note that Meyer 502, not far from this site, is identified as *M. rigidula*). Nepli, G.N. 269, Tadzhikistan, hills in the vicinity of the city of Dushanbe, west slope, herbage-grass association, 26-V-1984, in flower & fruit (NY, translated from Russian).

80. *Medicago rostrata* (Boiss. & Balansa) E. Small

M. rostrata (Boiss. & Balansa) E. Small in Small, Lassen & Brookes, Willdenowia 16(2): 434 (1987).

The epithet *rostrata* is Latin meaning beaked, a reference to the somewhat prominent rostrum or beak (the remnants of the stigma) at the tip of the fruits of this species.

Subordinate Taxa

None recognized here.

Type Specimen

B. Balansa 111, petits coteaux pierreux bordant la rive droite du Kamechly-Tchaï près de Béréketly (Cappodoce), 1300 m d'alt., 17-VI-1856, in flower & fruit (holotype & isotype Herb. Boiss. G!; isotypes K, W).

Synonymy

Homotypic

Pocockia rostrata Boiss. & Balansa in Boiss., Diagn. Pl. Orient. ser. 2, 3(6): 43 (1859), (basionym).

Trigonella rostrata (Boiss. & Balansa) Boiss., Fl. Or. 2: 89 (1872).

Telis rostrata O. Kuntze, Rev., 1: 209 (1891).

Melissitus rostratus (Boiss. & Balansa) Latsch., Trudy Tiflissk. Bot. Inst. 19: 7 (1970).

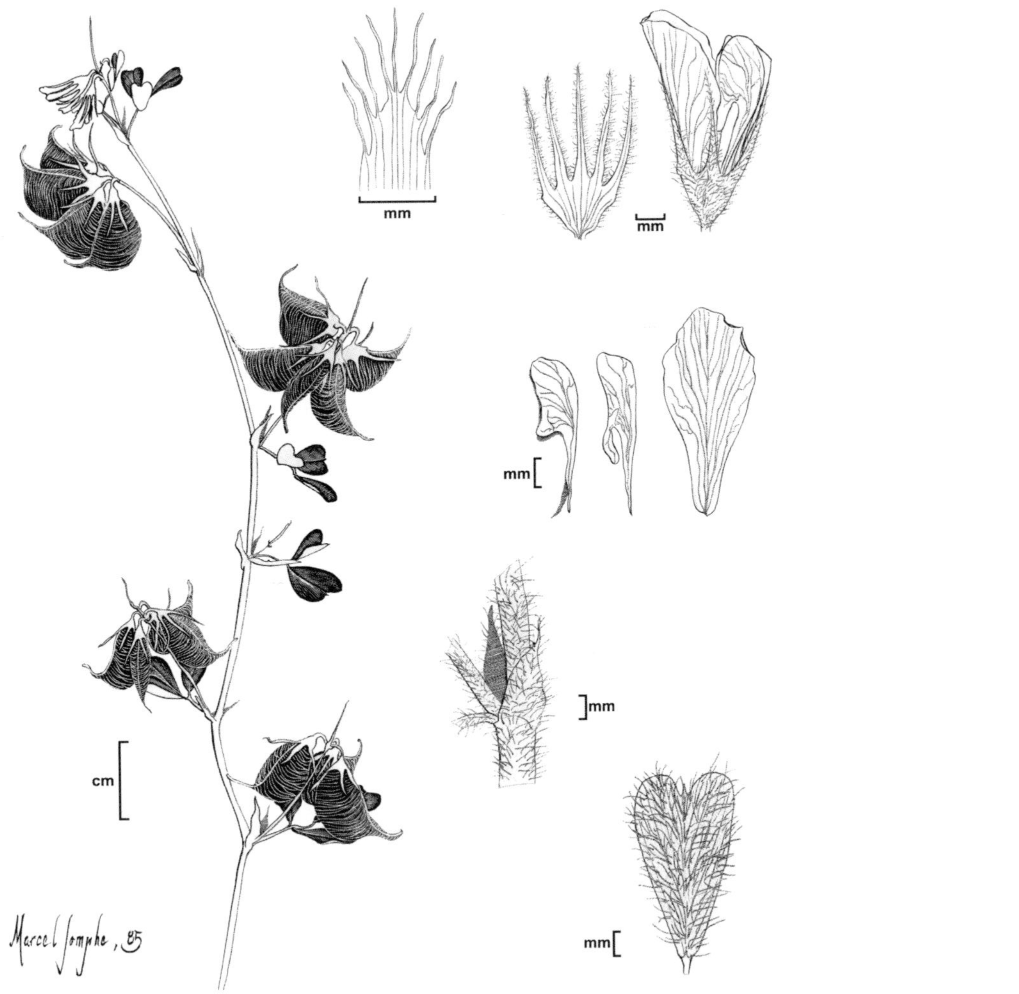

Fig. 80-1. *Medicago rostrata*. **Left:** Fruiting branch. **Upper right:** Floral parts (clockwise from upper left: flattened androecial column (free stamen not shown); calyx; side view of flower; standard petal; wing petal; keel). **Bottom right:** Leaflet; stem node. (Based on holotype, B. Balansa 111, G; artist: M. Jomphe.)

Heterotypic

No names found.

Vernacular Names

English: No names found.

French: No names found.

Geographical Distribution

(N = native, I = introduced, U = uncertain status)

Asia: Turkey (N).

Medicago rostrata is native to central and southern Anatolia.

Description

Annual herb, stems ascending to erect, (3.5)5–20(25) cm long, often branched from base, very pubescent with both simple and gland-tipped hairs. Leaflets obovate or obovate-cuneate, (4)5–8(10) mm long, (2.5)3–(6) mm wide, apically notched but the margin mostly entire, pubescent on both surfaces. Stipules ovate or lanceolate-acuminate, margin entire. Inflorescence umbellate or subumbellate, of 2–6 flowers, the peduncle about as long as the subtending petiole. Flowers about 7 mm long, the pedicels becoming reflexed with development. Corolla yellow. Calyx campanulate, (4)5–6(7) mm long, with simple and gland-tipped hairs, the teeth linear-subulate, much longer than tube. Fruit semi-lunate (dorsal suture straight, ventral suture concave), flat, reflexed, 7–9.5 mm long, 4–5 mm wide, transversely veined, villous with simple and gland-tipped hairs, apex rostrate (the beak bent to the side and about 3 mm long), with 3–6 seeds. Seeds ca. 2.5 mm long, 1.5 mm wide, ovate, the surface rugose, radicle ca. 80% of length of seed.

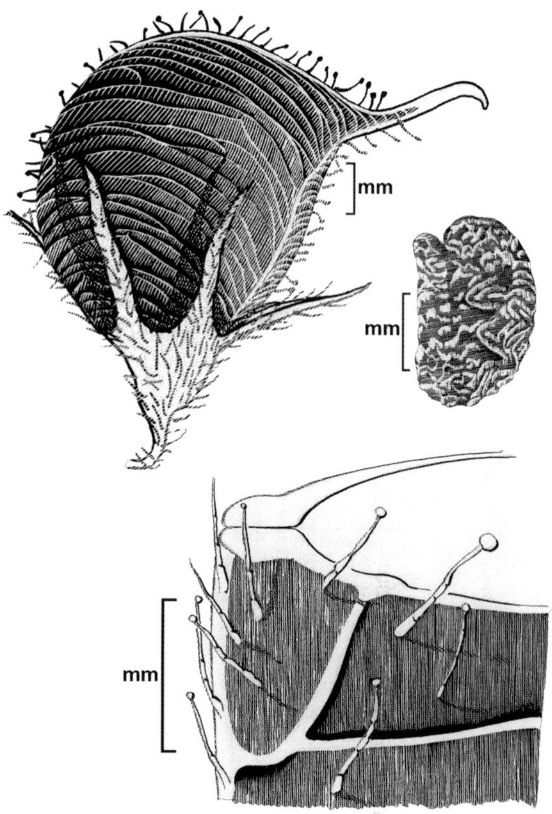

Fig. 80-2. *Medicago rostrata.* **Clockwise from upper left:** Fruit, seed; section of fruit. (Based on holotype, B. Balansa 111, G; artist: M. Jomphe.)

Identification Hints

The only other species of *Medicago* with lunate pods that have gland-tipped hairs is the closely related *M. biflora*. *Medicago biflora* has leaflets with some serrations and fruits that are (10)15–20 mm long with a concave ventral suture. By contrast, *M. rostrata* has entire leaflets (only the tips tend to be notched) and fruits that are 7–9.5 mm long with a straight ventral suture.

Habitat and Environmental Adaptation

Habitat

Medicago rostrata has been collected from stony slopes, steppes, and pine forests at altitudes of 850–1600 m.

Phenology

In its native area *M. rostrata* flowers in May and June.

Soil

The species has often been collected from gravelly and rocky soils.

Temperature

Medicago rostrata occurs in the Mediterranean climate of southern Turkey.

Water

The species often occupies well-drained soils.

Biotic Relationships

Pollination

Small (1988) recorded a pollen to ovule ratio of 588, indicative that the species is an outcrosser. However, the flowers are relatively small, suggesting that *M. rostrata* is largely self-pollinated and substantially an inbreeder, like essentially all other annual species of *Medicago*.

Dispersal

No obvious adaptations of the fruit facilitating seed dispersal are present.

Herbivory

No information located.

Nodulation and mycorrhizae

No information located.

Disease resistance

No information located.

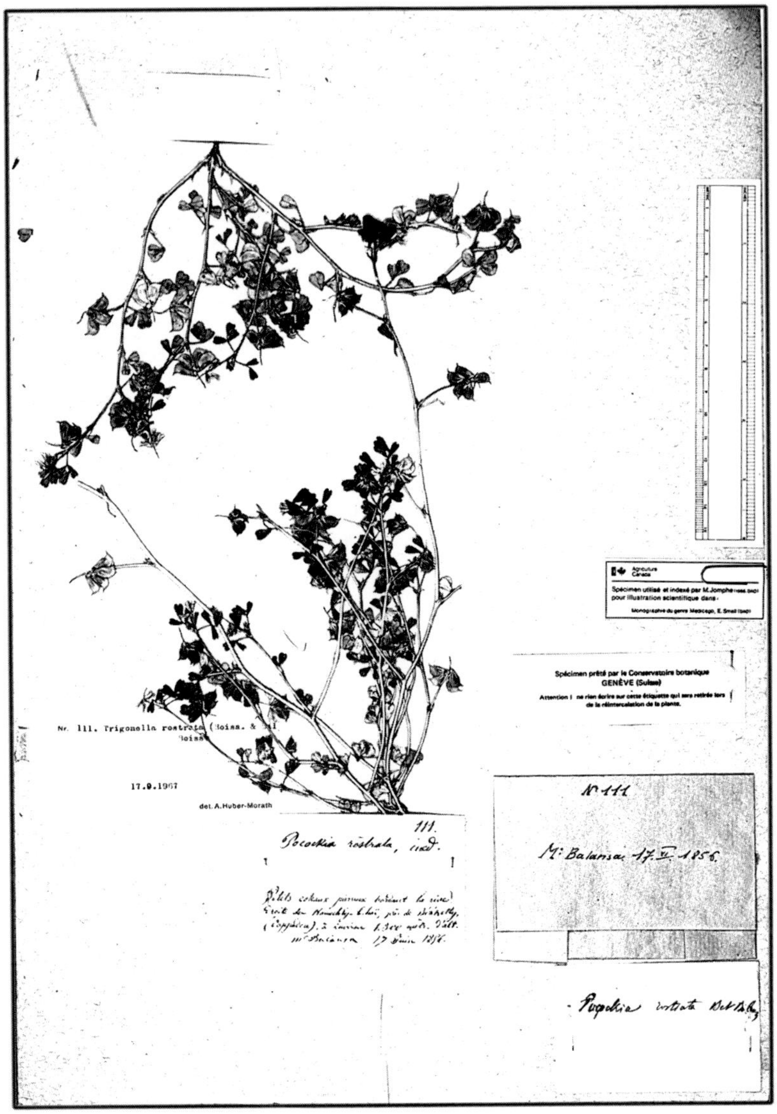

Fig. 80-3. Holotype of *Pocockia rostrata* Boiss. & Balansa, at G.

Economic History, Cultivars, and Uses

Medicago rostrata has not attracted agricultural development to date.

Cultivars

No cultivars have been bred.

Forage

Medicago rostrata is not sown for forage.

Fodder

This species is not harvested for dried animal feed.

Green manure, rotation crop, ground cover

No information located.

Other uses (salad sprouts, ornamental, plant products, etc.)

No information located.

Chromosome Number

No information located.

Genetic Relationships and Gene Transferability

Intraspecific

Artificial hybridization of different populations has not been attempted, so it is not known whether there are barriers to hybridization within the species.

Interspecific

No information located.

Exchange by advanced biotechnology

No information located.

Conservation Status

This endemic species is native to an area of the world

that is quite subject to overgrazing and habitat destruction. A conservation assessment as well as collection and preservation of germplasm need to be carried out.

Germplasm Sources

No accessions have been located.

Research Problems and Priorities

Medicago rostrata has been the subject of very little research. Collection of germplasm for study is the first priority, followed by basic botanical studies and agricultural assessment.

Representative Collections

TURKEY: Akman, Y., (region B4), Ankara – Beypazari – Aladag-radisi, alt. ca. 700 m, 25-V-1971, in flower & fruit (G). Alinoglu, N. OR166, Prov. Ankara, Ankara, Atatürk, Orman Çiftligi, near Dolapdere, 7-VI-1963, in flower & fruit (E). Balansa, B. 276, Phrygie, Kaiagueul-Dérè, à 2 lieus au SSO d'Ouchak, collines incultes, 27-V-(1856?), in flower & fruit (G). Balansa, B. 908, collines bordant la rive droite du Kamechly-Tchaï près de Béréketly (Cappodoce), 1300 m d'alt., 15-VI-1856, in flower & fruit (GH, P). Balansa, B. 1215, Phrygie, Kaia-queul-Dérè, au SO d'Ouchak, vers 850 m d'alt. 14-VI-1857, in flower & fruit (BM, G, P). Bourgeau, E., Lycia, Elmalu, in collibus, 9-VI-1860, in flower & fruit (G, P). Huber-Morath, A. 8185, Kleinasien: Isaurien (region C4), Karaman – Ermenek, Pinus nigra Wald 73 km nach K., alt. 1500 m, 12-VI-1948, in flower & fruit (G). Huber-Morath, A. 8595, Kleinasien: Lycaonien (region C4), Vilajet Konya, Tal hinter dem Elektrizitätswerk 13 km westlich Konya, alt. 1170 m, 6-VI-1948, in flower & early fruit (G). Tobey 2651, Çerum, after Gumushacikëy, alt. 800 m, rocky, scrub, 31-IV-1969, in flower & fruit (E).

81. *Medicago rotata* Boiss.

M. rotata Boiss., Diagn. Pl. Orient. Nov. Ser. 1, 2: 24 (1843), (basionym).

The epithet *rotata* is Latin for wheel-shaped, a reference to the appearance of the pods, which are somewhat reminiscent of miniature wheels.

Key to Subordinate Taxa

Leaflets not laciniate; or if laciniate, most inflorescences with no more than two flowers .... var. *rotata*

At least some leaflets laciniate <u>and</u> inflorescences often with more than two flowers .... var. *eliezeri*

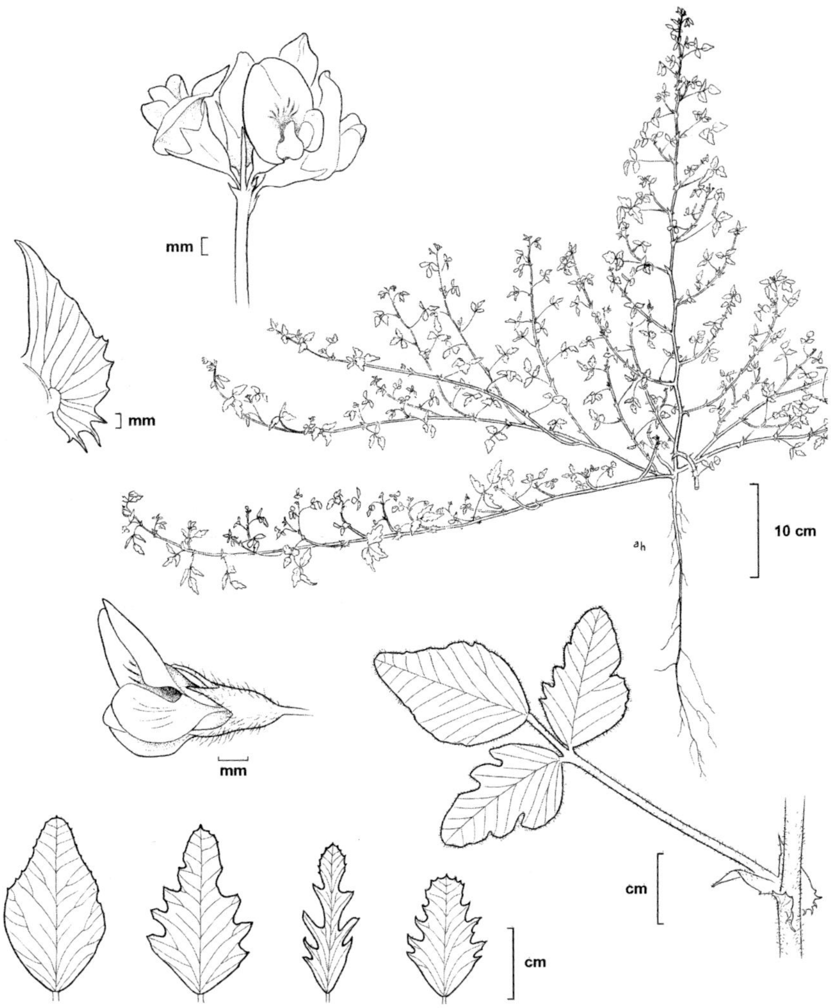

Fig. 81-1. *Medicago rotata* var. *eliezeri*. **Clockwise from upper right:** Plant; leaf attached to stem; leaflets; flower; stipule; group of flowers. (Based on <u>E. Small M335</u>, DAO, except two bottom right leaflets (extreme right based on <u>Dafni</u>, loan 99/238, HUJ; leaflet to the left based on <u>Naftolsky</u>, loan 99/233, HUJ; artist: A. Hanes.)

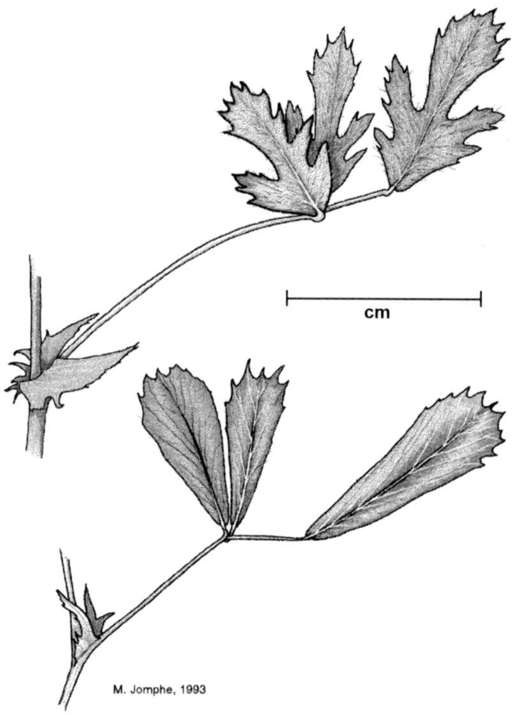

Fig. 81-2. Leaflets of the varieties of *Medicago rotata*. **Top:** Laciniate leaf of *M. rotata* var. *eliezeri* (based on M. Zohary & T. Segal, collected 09 April 1951, DAO). **Bottom:** Regular trifoliolate leaf of *M. rotata* var. *rotata* (based on E. Small M958, DAO). (Artist: M. Jomphe.)

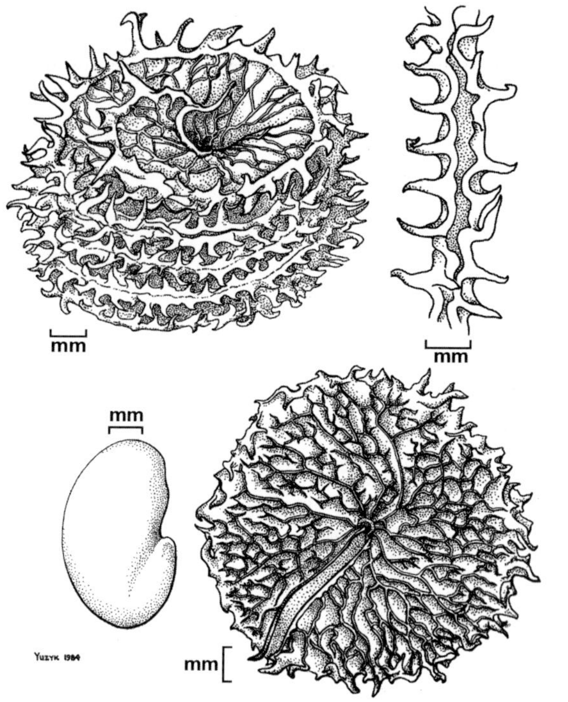

Fig. 81-3. *Medicago rotata* var. *rotata*. **Clockwise from upper left:** Fruit; coil edge; fruit coil face showing venation; seed. (Based on E. Small H2988, DAO; artist: L. Yuzyk.)

Type Specimen

Medicago rotata Boiss. was lectotypified by Heyn (1963: 48): Aucher 937 from Syria (wrongly recorded from Palestine on Aucher & Eloy's sheet, as noted by Heyn and Davis (1970, p. 496)), Herb. Boiss. at G!).

Synonymy (*M. rotata* var. *rotata*)

Homotypic

No names found.

Heterotypic

M. rotata subvar. *laciniata* Nègre, Travaux Inst. Sci. Chérif. Maroc., Sér. Bot. 5: 37 (1956) (holotype at HUJ cited by Heyn (1963); defined by laciniate stipules, not by laciniate leaflets)

M. rotata Boiss. var. *longispina* Evenari in Oppenh. & Evenari, Fl. Cisjord. 283 (1940).

Vernacular Names

English: Wheel medic (medic is also spelled medick).

French: No names found.

Geographical Distribution

(N = native, I = introduced, U = uncertain status)

Africa: Morocco (I).

Asia: Cyprus (N), Iraq (N), Israel (N), Jordan (N), Lebanon (N), Syria (N), Turkey (N).

This species occurs in the Eastern Mediterranean area.

Description

Annual herb, branched from near base, stems (12)15–30(35) cm long, usually ascending, twigs and leaves pubescent or glabrescent with simple hairs. Leaflets 6–15 mm long, 4–9 mm wide, generally about twice as long as broad, usually obovate to elliptical (occasionally ovate or oblanceolate), the tip truncate or apiculate, with a small terminal tooth, distal third of margin serrate. Stipules dentate, with 4–7 teeth, uppermost tooth longer. Inflorescence of 1–5(–6) flowers, peduncle as long to twice as long as subtending petiole. Flowers 6–8 mm, yellow. Calyx 2–4 mm, with simple hairs, teeth as long as tube. Corolla up to twice length of calyx. Young fruit contracted and concealed within calyx. Fruit glabrous (very rarely pubescent), discoid to shortly cylindrical, reminiscent of a miniature tire (all but first and last coils approximately equal in diameter), with 3–5(–7) loosely adpressed coils that are hardened at maturity, 4–10 mm long, 4–10(–12) mm in diameter and essentially lacking a central opening, the coil faces with a dense net of veins, some veins entering the dorsal suture where they join with a prominent vein. Dorsal suture spines inserted at 90°, short (0.5–3 mm), often irregular, some forked. Seeds 3–12 per pod, 3–5 mm long, 2–2.5 mm wide, seedcoat smooth, greenish-brown, yellowish-brown, yellow, or light brown, reniform, occasionally separated by poorly developed transverse, spongy fruit septa, radicle less than half seed length.

Identification Hints

Medicago rotata is easily identified by the short, irregular spines at right angles to the coil face.

Habitat and Environmental Adaptation

Habitat

Medicago rotata is most often observed as a weed in fields, both cultivated and fallow, and occasionally in rocky areas. In presettlement times when human-created fields were not present, the species was presumably a generalist pioneer of open, disturbed habitats.

Phenology

The species flowers from March to May in its native area.

Soil

Piano and Francis (1992) observed that *M. rotata* occurs in basaltic and loam soils that are slightly to moderately basic.

Temperature

Piano and Francis (1992) noted that in its native area *M. rotata* is more or less cold-tolerant compared with other medics. Most cultivars of annual *Medicago* have been bred in Australia and lack tolerance to frost (Cocks and Ehrman 1987). In this respect, *M. rotata* is considered to be a better possibility for west Asia than the available Australian cultivars (Abd El Moneim and Cocks 1986). However, Radwan et al. (1977) found that winter temperature was a critical factor limiting the distribution of *M. rotata* in northern Iraq.

Water

Piano and Francis (1992) concluded that in its native area the species requires a minimum annual rainfall of 350 mm.

Biotic Relationships

Pollination

Small (1988) noted that the small flowers and pollen to ovule ratio (191) are consistent with *M. rotata* being largely self-pollinated. This species is basically an inbreeder.

Dispersal

No information located.

Herbivory

No information located.

Nodulation and mycorrhizae

USDA (2009) reported positive root nodulation for *M. rotata*. Vázquez et al. (2001) examined the performance of *M. rotata* inoculated with *Sinorhizobium meliloti* and arbuscular mycorrhizal fungi of the genus *Glomus* and found occasional enhancement resulted from the combined use of both micro-organisms, depending on conditions of growth.

Disease resistance

O'Neill et al. (2003) examined the comparative resistance of 36 annual species of *Medicago* to *Phoma medicaginis* (the cause of spring black stem and leaf spot, a major disease of alfalfa) and concluded that *M. rotata* is highly susceptible to it. In a similar comparative test of 36 annual species of *Medicago*, O'Neill and Bauchan (2000) found that six accessions of *M. rotata* were very susceptible to anthracnose caused by *Colletotrichum trifolii*. Yaege and Stuteville (2002) tested the resistance of 33 annual species of *Medicago* to powdery mildew (*Erysiphe pisi*) and found that the six collections of *M. rotata* grown were susceptible to the disease (it was observed that often only one of a number of populations of other species tested was resistant).

Economic History, Cultivars, and Uses

This species has not attracted agricultural interest to date.

Cultivars

There are no commercial cultivars. 'Highlander' is an experimental frost-tolerant cultivar from SeedCo Ltd., mentioned in Walsh et al. (2001).

Forage

Medicago rotata is not sown for forage. St. John-Sweeting et al. (1996) noted that one accession studied in South Australia had poor dry matter production and so was not of potential commercial interest. Mohammad and Qamar (1988) grew 14 annual species of *Medicago* in small test plots in Islamabad, Pakistan located in a subtropical, subhumid area that is notably hotter and wetter than the normal habitat of these species. The dry matter yield of *M. rotata* was 5.2 t/ha. Noy-Meir et al. (1989) studied the effects of protecting natural Mediterranean grazing lands by fencing and observed no consistent effects on *M. rotata* populations. Tadmor et al. (1971) conducted studies of ecotypes selected for their ability to survive and persist in the Negev Desert of Israel in an area where rainfall averaged only 90 mm annually, but received an additional 263 mm of water annually in the form of run-off from erratic winter storms. The dry matter production of an ecotype of *M. rotata* was 5.2 t/ha (averaged over 6 years), which is considered rather good given the dry environment.

Fodder

The species is not employed for fodder.

Green manure, rotation crop, ground cover

No information located.

Other uses (salad sprouts, ornamental, plant products, etc.)

No information located.

Chromosome Number

$2n = 16$: Fryer 1930; Clement 1962; Simon and Simon 1965; Kliphuis and Barkoudah 1977; Lesins and Lesins 1979; Agarwal and Gupta 1983*b*; Díaz Lifante et al. 1992; Mariani et al. 1996.

$2n = 16 + 1B$: Heyn 1963.

$2n = 18$: Heyn 1956, 1963.

Genetic Relationships and Gene Transferability

Intraspecific

No information located.

Interspecific

As noted above, *M. rotata* crosses extensively in nature with *M. bonarotiana*. Lesins et al. (1976) found that crossing the two species artificially was easily accomplished. However, Simon and Millington (1967) were unable to produce hybrid seed by crossing "*M. blancheana*" with *M. rotata*, although the two should be easily crossable.

Exchange by advanced biotechnology

No information located.

Conservation Status

This species is fairly widespread and not in need of conservation protection.

Germplasm Sources

Almost 500 accessions are listed among the major germplasm institutions with large holdings of *Medicago* (see Chapter 3).

Research Problems and Priorities

Medicago rotata is one of the many winter-annual species of *Medicago* that has not yet been adequately examined for its agricultural potential. Piano and Francis (1992) described it as having "regional potential".

Note

See Note for *M. bonarotiana* indicating the close morphological relationships of the two species.

MEDICAGO ROTATA BOISS. VAR. *ELIEZERI* EIG

M. rotata Boiss. var. *eliezeri* Eig, Contr. Fl. Pal. (Bull. Inst. Agr. Nat. Hist., Tel-Aviv) 6: 20 (1927), (basionym).

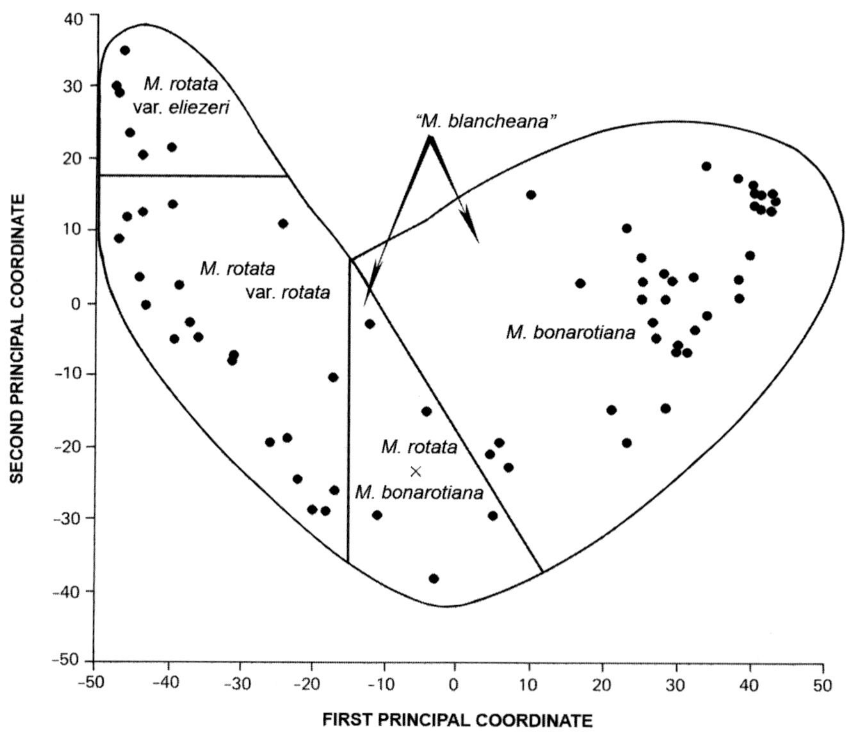

Fig. 81-4. Pattern of variation in the *Medicago blancheana-bonarotiana-rotata* complex (after Small 1984*a*).

The epithet *eliezeri* commemorates Eliezer Factorovsky (1897–1926, a collaborator of Alexander Eig in exploring the flora of Palestine and founding the herbarium of the Hebrew University of Jerusalem (HUJ) (information supplied by A. Danin); Eig also dedicated his genus *Factorovskya* to him, now reduced to *M. hypogaea*.)

Type Specimen

The type is Eig M/405, Esdraelon Plain, env. of Balfouria, Palestine, 7–IV-1924 (HUJ!). A syntype (!) is also at HUJ.

Synonymy

Homotypic

M. rotata subsp. *eliezeri* (Eig) Ponert, Feddes Repert. 83(9/10): 639 (1973).

Heterotypic

No names found.

Vernacular Names

English: No names found.

French: No names found.

Geographical Distribution

(N = native, I = introduced, U = uncertain status)

Middle East: Jordan (N), Israel (N), Syria (N), Turkey (N).

Whether *M. rotata* var. *elizeri* is sporadically distributed throughout the range of *M. rotata* var. *rotata* or occurs in particular areas is unclear.

Description

In recognizing his new variety, Eig (1927) wrote: "Leaflets deeply incised-dentate or lobed; flowerets 2–6, rarely 1–6". Heyn (1963) presented the following key (given here as it presents more distinguishing characters than I have been able to verify):

Peduncles with 1–3(4) flowers. Pods with 4–6 coils, 8–12 mm in diameter; last coil flat. Leaflets serrate, very rarely laciniate ....var. *rotata*
Peduncles with 4–6 flowers. Pods with 3–4(5) coils, 5–7(8) mm in diameter; last coil concave. Leaflets laciniate ....var. *eliezeri*

Although I believe Eig's taxon deserves recognition, many collections of *M. rotata* are difficult to identify to the two varieties recognized. Heyn (1963) noted that intermediate forms between var. *rotata* and var. *eliezeri* sometimes occur with the typical forms, suggesting interbreeding.

Identification Hints

The most prominent characteristic of *M. rotata var. eliezeri* is the laciniation of the leaflets. Some specimens with leaves eaten by insects give a laciniate appearance and have been mislabelled as var. *eliezeri* in collections. Insect damage on the leaf margin shows discoloration, whereas laciniate edges are green. Based on years of examining herbarium specimens of *Medicago*, it is my impression that the foliage of *M. rotata* is exceptionally attacked by insects that feed on the margins of the leaflets. Perhaps the development of leaf laciniation in this species represents leaf damage mimicry to deter herbivores, in accord with the "grazing dispersal hypothesis" (see for example Edwards et al. 1992). The appearance of cut-up leaves may suggest that a herbivore has already attacked the plant, dissuading further attack.

Germplasm Sources

This variety is rarely identified as such in germplasm accessions.

Research Problems and Priorities

The leaf laceration of *M. rotata* var. *eliezeri* is reminiscent of the same phenomenon in some other *Medicago* species (such as *M. laciniata* and *M. monantha*). As suggested above, such leaf dissection may be adaptive, deterring herbivores. Accordingly, it is of interest to attempt to identify specific geographical areas where the leaf character is especially evident, as well as herbivore species that may be responsible for leaf laciniation.

Representative Collections (not distinguished to var. *rotata* or var. *eliezeri*)

Israel: Andursky, J., Judean Mountains, Kiryath-Huarvrim, 2-V-1954, in fruit (G). Eig, A. 803, Esdraelon Plain, env. of Balfourya, basalt soil, fallow, 7-IV-1924, in flower (HUJ) (syntype!). Eig, A., Upper Jordan Valley, Bitaniya, citrus grove, irrigated 22-II-1924, in flower (HUJ). Eig, A., Zohary, M., & Feinbrun, N., Philistean Plain, Qu'beiba to Ramla, 1-IV-1928, in flower (HUJ). Grizi, A., Esdraelon Plain, Megiddo, foot of the hill, 26-III-1956, in flower & early fruit (HUJ). Naftolsky, N., Philistean Plain, Ben-Shemen, wadi, 10-III-1923, in flower & fruit (HUJ). Eig, A., Esdraelon Plain, env. of Balfourya, fallow, 12-IV-1924, in flower & early fruit (HUJ). Naftolsky, N., Upper Jordan Valley, Deganiya B., 30-III-1924, in flower (HUJ). Naftolsky, N., Esdraelon Plain, Tel-Yosef, irrigated, 1925, in flower & early fruit (HUJ). Naftolsky, N., Philistaean Plain, Ben-Shemen, vineyard, 22-IV-1925, in flower & early fruit (HUJ). Segal, T., Shefela, Yaune, wheat field, 5-IV-1950, in flower & early fruit (HUJ). Segal, T., Shefela, Na'an, fields, 10-IV-1950, in flower & early fruit (HUJ). Segal, T., Esdraelon Plain, Sarid, wheat field, heavy soil, 26-IV-1950, in flower & fruit (HUJ). Smoly, E., Shefela,

Gezer, 7-IV-1927, in flower & early fruit (HUJ). Zohary, M., Philistean Plain, Miqve-Yisrael, 15-III-1949, in flower & early fruit (HUJ). Zohary, M., Shefela, Ramla to Gezer km 42, fallow, heavy soil, 9-IV-1951, in flower & fruit (HUJ). Zohary, M., Shefela, about 10 km east of Ramla, fields, heavy and moist soil, 1-V-1952, in flower (HUJ). **Jordan:** Kasapligil, B. 1914, Jordan Valley, Deir Alla Agric. Sta., alt. 200 m, 9-II-1955, in flower (HUJ). **Palestine**: Eig, A., Philistean Plain, south of Gaza, fallow, 4-V-1924, in flower & fruit (HUJ). **Syria:** Dafni, A., Golan, Ein Dib, 31-III-1969, in flower (HUJ). Eig, A. & Zohary, M., Hauran, about 1.5 km west of Souweida, basalt fields, 19-VI-1932, in fruit (HUJ).

82. *Medicago rugosa* Desr.

M. rugosa Desr. in Lam., Encycl. Méthod., Bot. 3: 632 (1792), (basionym).

The epithet *rugosa* is Latin for somewhat wrinkled, descriptive of the edges of the coils of the fruit.

Subordinate Taxa

None recognized here.

Type Specimen

Heyn (1963) commented "Type: *M. rugosa* is described by Desrousseaux from a living plant in the Jardin du Roi. Therefore, it seems as if Morison's fig. 4 in tab 15 of the Plantarum Historiae Oxoniensis 2 (1680), cited by Desrousseaux as a synonym should be accepted as a lectotype. (The possibility cannot be altogether excluded that the living plant serving as the type of Desrousseaux's description was later deposited at the Muséum National d'Histoire Naturelle in Paris)." Jafri (1980) cited as the type "On le cultive au Jardin du Roi" (P)", presumably citing a specimen in Paris.

Synonymy

Homotypic

No names found.

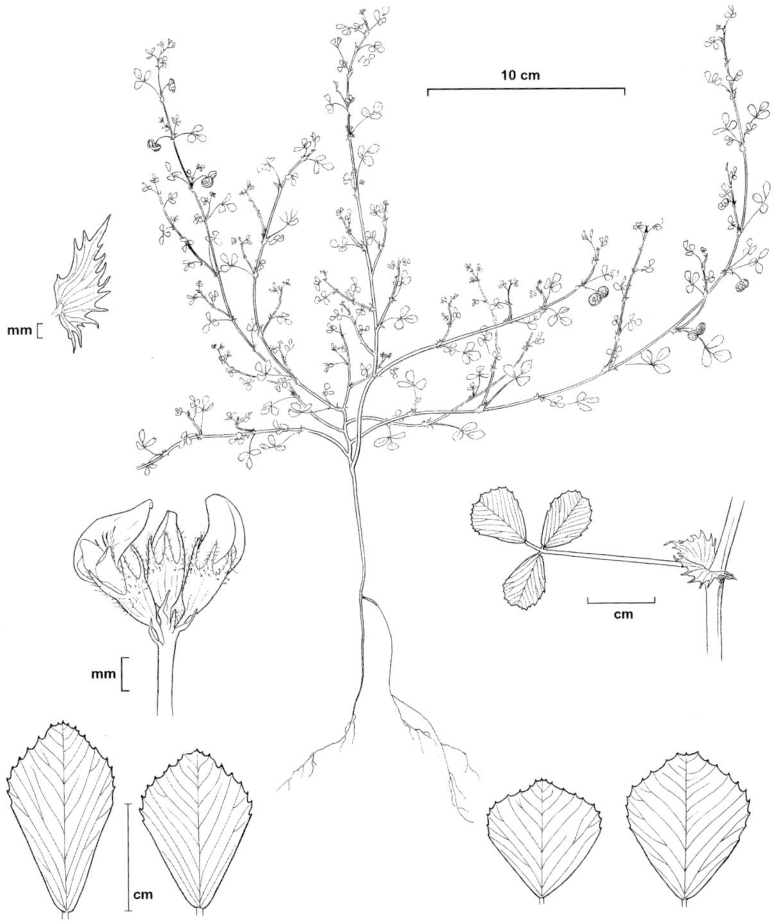

Fig. 82-1. *Medicago rugosa.* **Clockwise from upper left:** Stipule; plant; leaf attached to stem; leaflets; group of flowers. (Based on Fiori 2093, GH, except the leaf and stipule, which are based on E. Small M1018, DAO; artist: A. Hanes.)

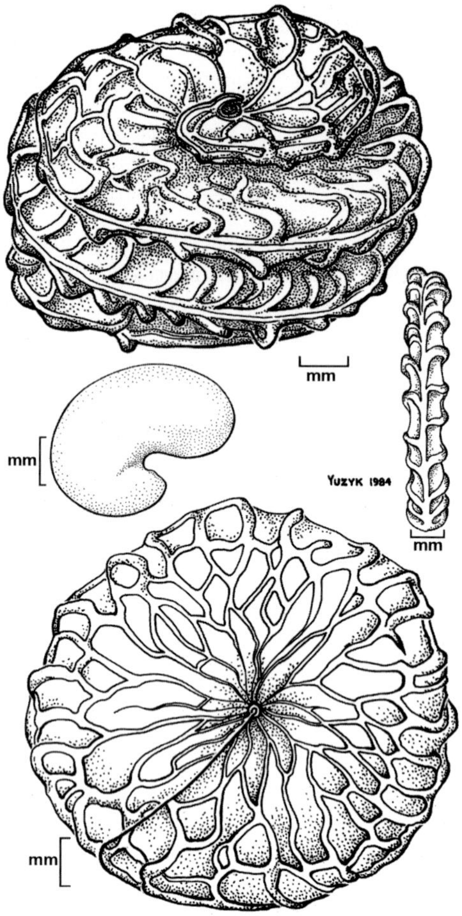

Fig. 82-2. *Medicago rugosa.* **Clockwise from top:** Fruit; fruit coil edge; fruit coil face showing venation; seed. (Based on E. Small M736, DAO; artist: L. Yuzyk.)

Heterotypic

M. elegans Jacq. ex Willd., Sp. Pl. 3(2): 1408 (1802), nom. illeg. (Several authors cite the name as *M. elegans* Pourr. ex Willd.)

M. elegans Pourr. ex Willk. & Lange, Prodr. Fl. Hispan. 3: 384 (1877). (Presumed synonymy.)

M. rugosa Desr. var. *graeca* Heldr. ex Nym., Consp. 167 (1878).

M. elegans Jacq. ex Willd. var. *incisa* Moris, Fl. Sard. 1: 437-438 (1837). (Rarely, laciniate specimens occur, which have inspired this name.)

M. rugosa Desr. f. *incisa* (Moris) Urb., Verh. Bot. Ver. Brandenb. 15: 64 (1873).

M. orbiculariodes P. Candargy, Bull. Soc. Bot. France 44: 161 (1897). (Several authors have indicated that this name may be problematical.)

Vernacular Names

English: Gama medic, rugose medic, wrinkled medic (medic is also spelled medick).

French: La luzerne élégante, luzerne rugueuse, luzerne plisée.

Geographical Distribution

(N = native, I = introduced, U = uncertain status)

Africa: Algeria (N), Libya (N), Tunisia (N).

Asia: Cyprus (N), East Aegean Islands (Greece; N), India (Ghosh and Dutta 1976; I), Iran (Mozzafarian 1988; Mehregan et al. 2002; U), Israel (N), Jordan (N), Lebanon (N), Syria (N), Turkey (N).

Australasia: Australia (I).

Europe: Balearic Islands (Spain; Carreras et al. 2002; U), Corsica (France; N), France (U), Greece (including Crete; N), Ireland (I), Italy (including Sardinia and Sicily; N), Malta (N), Portugal (U; claimed to be naturalized by some authors), Spain (U), United Kingdom (I).

Northern America: USA (I, according to Castroviejo et al. 2001).

Pacific Ocean: Hawaii (I).

Medicago rugosa is omni-Mediterranean.

Description

Annual herb, stems prostrate, ascending or procumbent, 15–30(50) cm long, branched from base, branches densely pubescent with simple and gland-tipped hairs. Leaflets obovate, obovate-cuneate, or oblong to oblanceolate, (7)10–15 mm long, 7–10 mm wide, distal third serrate (rarely laciniate), glabrous above, pubescent below with simple and gland-tipped hairs. Stipules lanceolate to broadly ovate, unequally dentate, with simple and gland-tipped hairs on abaxial surface. Inflorescence of (1)2–3(7) flowers, the peduncle shorter to as long as the corresponding petiole. Flowers (2)2.5–4(5) mm long. Corolla yellow, less than twice as long as calyx. Calyx with simple and gland-tipped hairs, teeth short-triangular, shorter than the tube. Young pod protrudes sideways from the calyx. Mature pod discoid to short-cylindric, 3–4(5) mm long, dorsal suture spineless, with gland-tipped and simple hairs, (2.5)3.5–5 loosely adpressed coils, (4)5–10(12) mm wide. Coil face with (12)18–25(30) slightly curved radial veins that anastomose considerably towards dorsal suture and are thickened at the dorsal suture. Seeds 1 per coil (or absent in some coils), not separated by partitions, curved, (2.5)3–4(4.5) mm long, ca. 1.3 mm wide, smooth, yellow, yellowish, or yellow-brown, radicle less than half length of seed.

Identification Hints

The coil margin of the fruit of this species is distinctive in having strong thickenings. *Medicago noeana* has a similar feature but, unlike *M. rugosa*, its leaves do not have gland-tipped trichomes.

Habitat and Environmental Adaptation

Habitat

Medicago rugosa is often found in fallow fields and along roadsides in its native geographical area. In Greece, Thanopoulos (2007) recorded it in openings of garrique, in other woodlands, fallow fields, and as a weed of cultivated fields, olive groves, and ruderal places, at altitudes up to 1000 m. The species is usually not frequent (Heyn 1963).

Phenology

This species flowers from March to June in its indigenous geographical range, the time depending on its location. In South Africa, Wassermann (1974) found that *M. rugosa* planted in early December showed no sign of flowering up to 20 May 1973, indicating a marked cold requirement. Clarkson and Russell (1975) found that flowering was accelerated by vernalization, long photoperiods (18 and 24 hours), and high temperature.

Soil

In Algeria, Abdelguerfi et al. (1988*a*) noted that *M. rugosa* tends to be in medium- to fine-textured soils, and indeed the species has been observed most often on heavy clay soils (Heyn 1963). Piano and Francis (1992) stated that in its native area the species grows on clay and loam soils that are slightly to moderately basic. In Sicily, Piano et al. (1991) observed the species growing in a pH range of 6.5–9.0. Dunbabin and Slavich (1990) found there was low salt tolerance in *M. rugosa*. The first *M. rugosa* cultivar 'Paragosa' is well adapted to heavy alkaline soils (Anon. 1972) and clay-loam soil (Crawford 1983*b*). Cultivar 'Sapo' has been described as suited to a wide range of soil types from sandy loams to clay loams, with a preference for neutral to alkaline soils ranging in pH from 5.7–8.5, and has a considerable sensitivity to boron.

Temperature

This species is found particularly in Mediterranean-climate areas, growing as a winter annual, tolerating temperatures that often reach 35 °C in the summer. However, winter temperatures need to be relatively warm (Piano and Francis 1992). Cocks and Ehrman (1987) noted that *M. rugosa* in Syria was almost totally intolerant of frost. Christiansen and Svejcar (1989) sowed two accessions of *M. rugosa* in Oklahoma in March 1985 and observed that they were killed by frost and cold in the following winter period. Derkaoui et al. (1991) also found that fall-planted *M. rugosa* in Oklahoma failed to survive over the winter, indicating that the species does not have sufficient cold tolerance for successful self-replication in the southern Great Plains of the United States.

Water

Abdelguerfi et al. (1988*a*) found that in Algeria *M. rugosa* tends to be in the wetter areas of the country (with an annual rainfall of 450–800 mm). Piano and Francis (1992) noted that in its native area the species requires a minimum annual rainfall of 350 mm.

Biotic Relationships

Pollination

Small (1988) noted that the small flowers and pollen to ovule ratio (243) are consistent with *M. rugosa* being largely self-pollinated. This species is basically an inbreeder.

Dispersal

The indehiscent fruits do not appear to have obvious morphological adaptations for seed dispersal. Because the pods lack spines, they would not easily become tangled in animal fur or feathers.

Herbivory

Medicago rugosa has been reported to be resistant to aphids and the alfalfa weevil (*Hypera postica* (Gyllenhal) (Johnson et al. 1980*a*, 1980*b*; Kleitner and Sorensen 1983; Doss et al. 1989). Johnson et al. (1981) found that *M. rugosa* was especially resistant to adult alfalfa weevils compared with other glandular-haired *Medicago* species. Shade and Kitch (1983) noted that the glandular hairs of *M. rugosa* resulted in resistance to the pea aphid (*Acyrthosiphon pisum* (Harris)), and Ridland and Berg (1981) found that seedlings of cultivar 'Paragosa' had a high level of resistance to the pea aphid. MacLean and Byers (1983) observed that the glandular hairs of *M. rugosa* tended to deter the alfalfa blotch leafminer (*Agromyza frontella* (Rondani)) from ovipositing. Shade et al. (1979) and De Haan et al. (2002) found that *M. rugosa* was very resistant to the potato leafhopper (*Empoasca fabae* Harris). Ferguson et al. (1982) found that *M. rugosa* was quite resistant to the spotted alfalfa aphid (*Therioaphis maculata* (Buckton)).

Nodulation and mycorrhizae

USDA (2009) reported positive root nodulation for *M. rugosa*. The species is fairly specific in its rhizobial requirements (Anon. 1972). Brockwell (1971) selected strains of *Rhizobium meliloti* suitable for *M. rugosa*. In Central and Western Spain, it has been shown that, when infected by an effective strain of *Rhizobium*, *M. rugosa* resists freezing temperatures and long periods of winter drought, whereas noninfected plants die (Feuntes 1981).

Disease resistance

O'Neill et al. (2003) examined the comparative resistance of 36 annual species of *Medicago* to *Phoma medicaginis* (the cause of spring black stem and leaf spot, a major disease of alfalfa) and concluded that *M. rugosa* is highly susceptible to it. In a similar comparative test of 36 annual species of *Medicago*, O'Neill and Bauchan (2000) found that nine accessions were very susceptible to anthracnose caused by *Colletotrichum trifolii*, and two accessions showed some resistance to the disease. Yaege and Stuteville (2002) tested the resistance of 33 annual species of *Medicago* to powdery mildew (*Erysiphe pisi*) and found that the 11 collections of *M. rugosa* grown were susceptible to it (it was observed that often only one of a number of populations of other species tested was resistant). Barbetti (1989) found that three cultivars of *M. rugosa* were very highly resistant to Phoma black stem disease (*Phoma medicaginis* Malb. & Roum.).

Economic History, Cultivars, and Uses

Medicago rugosa is one of the seven annual species of *Medicago* for which commercial cultivars have been bred and are often grown. 'Paragosa', the first cultivar, was derived in Australia from an introduction received in 1939 from the Institute of Agronomy in Lisbon, Portugal (Anon. 1972).

Cultivars

For information on the Australian cultivars 'Paragosa', 'Paraponto', and 'Sapo', see the Australian Herbage Plant Cultivars website (http://www.pi.csiro.au/ahpc/legumes/legumes.htm; current June 2010). 'Paragosa' was described in Anon. (1972) and released in Australia in 1966, 'Paraponto' was registered in 1978, and 'Sapo' in 1980. For additional information on cultivar development of *M. rugosa*, see Clarkson (1977) and Crawford (1983*b*).

Forage

Medicago rugosa cultivars are sown for forage in Australia and elsewhere. Mohammad and Qamar (1988) grew 14 annual species of *Medicago* in small test plots in Islamabad, Pakistan located in a subtropical, subhumid area that is notably hotter and wetter than the normal habitat of these species. The dry matter yield of *M. rugosa* was 0.1 t/ha, the poorest of all the species examined. Zhu and Sheaffer (1997) recorded aboveground dry matter yields (in kg/ha) of cultivar 'Sapo' at two sites in Minnesota (compared with the alfalfa cultivar 'Nitro', yields in parentheses): 7106 (3815) and 5724 (4596).

Fodder

Like most annual species of *Medicago*, *M. rugosa* is used primarily as forage (live plants in pastures or rangelands consumed by grazing livestock), and is not harvested for preparation of dried feedstocks.

Green manure, rotation crop, ground cover

No information located.

Other uses (salad sprouts, ornamental, plant products, etc.)

No information located.

Chromosome Number

$2n = 30$: Bauchan and Elgin 1984 (37 accessions evaluated); Piccirilli and Arcioni 1992*a*; Mariani and Falistocco 1991; Mariani et al. 1996.

$2n = 32$: Fryer 1930; Heyn 1956; Clement 1962; Simon and Simon 1965; Fernandes and Queirós 1978; Lesins and Lesins 1979; Abdelguerfi and Guittoneau 1979; Colombo et al. 1980; Aparicio 1986; Abdelguerfi et al. 1988*b*; Slavik et al. 1993.

The $2n = 32$ counts may be inaccurate. Bauchan and Elgin (1984) pointed out that the large chromosomal satellites that are present in this species may have led to "the erroneous chromosome counts of $2n = 32$."

Genetic Relationships and Gene Transferability

Intraspecific

Artificial hybridization of different populations does not seem to have been attempted, so it is not known whether there are barriers to hybridization within the species. *Medicago rugosa* appears not to exhibit the wide variation of many annual species of *Medicago* (Heyn 1963), suggesting that internal barriers to hybridization are unlikely.

Interspecific

Lesins and Lesins (1979) doubled the chromosome number of *M. noeana* to $2n = 32$ and crossed this polyploid with *M. rugosa* (which they considered to also have $2n = 32$), but were unable to obtain hybrids.

Exchange by advanced biotechnology

Piccirilli and Arcioni (1992*a*, *b*) crossed *M. rugosa* ($2n = 30$) with *M. sativa* ($2n = 32$) and obtained 56 viable embryos from which they were able to produce 16 hybrid plants using embryo rescue techniques. Mizukami et al. (2006) produced somatic hybrids of *M. rugosa* and *M. sativa*, of odd chromosomal constitution, and noted that this technology had the potential of transferring insect resistance to alfalfa. See *M. truncatula* for information on somatic hybrids of this species and *M. rugosa*.

Conservation Status

Medicago rugosa is a fairly widespread, if uncommon species, suggesting that it is not in particular need of conservation measures. The species is protected in the French regions of Corsica (http://www.mnhn.fr/mnhn/cbnbp/flore/textes/detail/corse.htm; current July 2010) and Bourgogne (http://perso.orange.fr/bblb/botanica/ListesPlProt/bg.html; current July 2007).

Germplasm Sources

More than 400 accessions of *M. rugosa* are present among the major *Medicago* germplasm collections (see Chapter 3).

Research Problems and Priorities

Medicago rugosa has been developed as a significant crop. It remains a basic resource for future development of improved cultivars. Piano and Francis (1992) described it as having "regional potential".

Representative Collections

Algeria: Durando, Alger, champs près de la Colonne Voirol, 8-V-1865, in fruit, Fragmenta Florae Algeriensis Excissata No. 514 (P). Reboud, V., Constantine, pentes arides du Djebel-Mécid, VI-1878, in fruit, Société Dauphinoise 1879 No. 2018 (K, P). **Cyprus:** Casey, E.C. 1335, Kornos Vovno (Kyrenia Range), alt. 1600 feet, 17-IV-1954, in fruit (K). Casey, M.M.C. 688, near Ay. Hilarion Castle, alt. ca. 1800 feet, 11-V-1949, in fruit (K). **Greece:** De Heldreich, Attica, in arvis prope Athenas, omnibus partibus robustion et praesertim legumine major; quam in typo Florae Mediterraneae,15-IV-1893, in fruit, De Heldreich Herbarium, Graecum Normale No. 1225 (K, P). Gandoger, M. 10943, Crete, Prov. Candia, Gorgolaino, 14-V-1915, in fruit (P). Lewis, G.P. 577, Sardonic Islands, Spetsai, alt. 600–700 feet pathside, meadowland on limestone, 6-V-1989, in fruit (K). Rechinger, K.H. 12844, Crete, distr. Sitia, inter segetes inter Chandras et Maronia, alt. ca 400 m, 11-V-1942, in fruit (G, K). Robson, N.K.B. 2098, Rhodes, Philerimos, 5-IV-1973, in fruit (BM). Townsend, C.C. 69/18, Peloponnese, Mistra, in long grass by the main path, below the Pantenassa, 4-IV-1969, in flower & fruit (K). Townsend, C.C. 70/42, Ins. Patmos immediately behind the port of Scala, in dry grassland grazed by sheep, 4-IV-1970, in fruit (K). Townsend, C.C. 82/314, Attiki, scattered in long grass near the museum, the Acropolis, Athens, 30-IV-1982, in fruit (K). **Israel:** Kasapligil 2357, Bethléem, 3-IV-1955, in flower (old fruit at base of plant) (G). **Italy:** Davis, D. & Sutton, S. 63553, Sicily, Trápany ca. 25 km northeast of Trápani Monte Cofano, alt. 400–600 m, near summit on rocky limestone slopes, 22-V-1979, in fruit (BM). Fiori, A., Sardinia, Prov. di Sassari, Terranova Pausania, in herbosis glareosis secundum viam ferream prope portum, solo granitico, 4-IV-1912, in fruit, Flora Italica Exsiccata Series III Fiori & Béguinot No. 2093 (BM, K). Huet de Pavillon, E. & A., Sicily, prope Panormum, ad radices montis Peregrino, in herbosis, 23-III-1855, in flower & fruit (K). Huter, Porta & Rigo, Calabria I. occid., loc. lapidos. ad camporum margines in mte. Catania supr. Pelluro, sol. argil-granit., alt. 600–700 m, 28-IV-1877, in fruit (K). Stace, C.A. & Cotton, R. 323, Sicily, lower slopes of Madonie Mts. above Isnello south of Cephalu, alt. ca. 700–800 m, 26-V-1972, in fruit (BM). **Lebanon:** Meinertzhagen, R., Beyrouth, among crops, 31-III-1933, in flower & fruit (BM). Pabot, H., Rar Bayada (Liban Sud), 8-IV-1957, in fruit (G). **Malta:** Duthie, J.F., Madonna dela Kala, in campis maritimis, 11-IV-1874, in fruit (K). **Portugal:** Scarlett, Mr. & Mrs. P.W. 13, vicinity of Lisbon, alt. under 200 feet, waste and cultivated places, roadsides, IV-1935, in fruit (K). **Syria:** Peyron, E. 2460, in collibus, 20-IV-1883, in fruit (P). **Tunisia:** Barratte, G., Utique, 4-V-1888, in fruit (P). Townsend, C.C. 83/117, ruins of ancient Dugga, near the temple, on dry stony ground, 8-V-1983, in fruit (K).

83. *Medicago rupestris* M. Bieb.

M. rupestris M. Bieb. Fl. Taur.-Caucas. 2: 225 (1808), (basionym).

The epithet *rupestris* is Latin for "of rocks", descriptive of the often rocky habitat of this species.

Subordinate Taxa

None recognized here.

Type Specimen

Lectotype at LE "from collections of F.A. Marschall von Bieberstein from Crimea" viewable at http://www.mobot.org/MOBOT/Research/LEguide/species/13.html (current July 2010). The presumptive lectotype bears the inscription "ex saxosis Tauriae". Vassilczenko (1949) noted that the original material was described from Crimea (collected by Steven from the city of Ak-Koya near Karasu-Bazar).

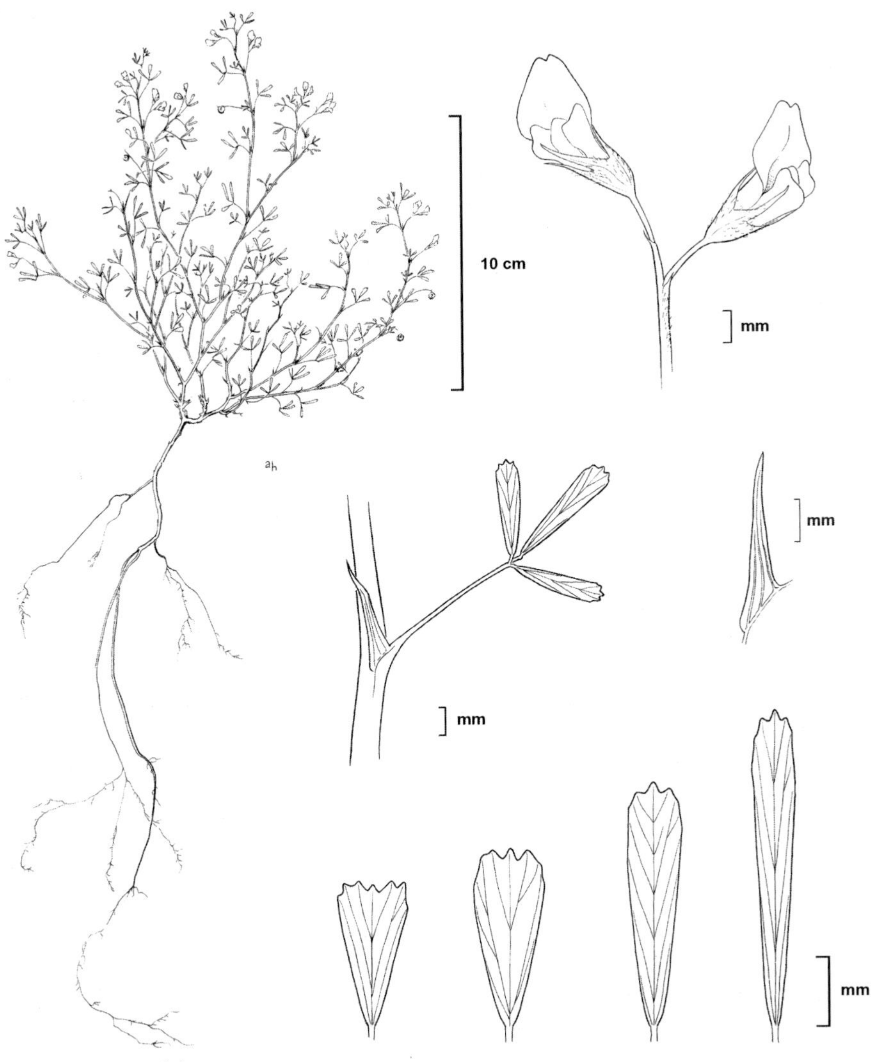

Fig. 83-1. *Medicago rupestris*. **Clockwise from upper left:** Plant; group of flowers; stipule; leaflets. **Center:** Leaf attached to stem. (Based on Stevenson 4115, DAO; artist: A. Hanes.)

Synonymy

Homotypic

No names found.

Heterotypic

No names found.

Not

M. rupestris Pirona (= *M. pironae* Vis.).
M. rupestris Eichv. (= *M. sativa* L.).

Vernacular Names

ENGLISH: Rocky alfalfa (Nicoloff 1961).

FRENCH: No names found.

Geographical Distribution

(N = native, I = introduced, U = uncertain status)

ASIA: Russia (N).

EUROPE: Ukraine (N).

Medicago rupestris is a local endemic of the Lower Crimean Mountains. Vassilczenko (1949) gave the range as the mountainous Crimea, the Taman Peninsula, and the North Black Sea coast as far as Gelendzhik. Didukh (2007) provided the following information (translated from Ukrainian): "Foothills of the Crimean Mountains (between the cities of Simferopol and Belogorsk). The species is also widespread in the northern Caucasus (Taman Peninsula) and the city of Gelendzhik (Krasnodar Province, Russia)."

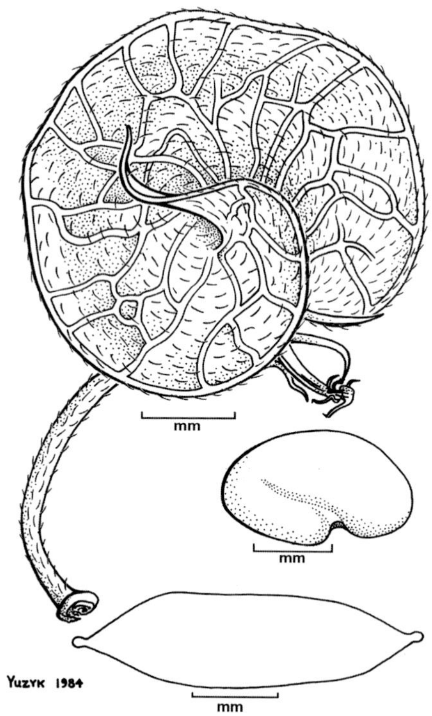

Fig. 83-2. *Medicago rupestris.* **Top:** Fruit. **Center:** Seed. **Bottom:** Outline of cross section of fruit. (Based on Lesins 1847, ALTA; artist: L. Yuzyk.)

Description

Perennial herb, the numerous stems arising from a woody base, stems (5)10–25 cm long, pubescent. Leaflets obovate to narrowly linear-cuneate, the margins entire except for a few teeth at tip, (3)5–7(10) mm long, (1)1.5–2(3) mm wide, the edges often revolute, pubescent especially on lower surface. Stipules linear-subulate, rarely toothed at base, pubescent. Corolla (4)5–6(7) mm long, yellow or orange-yellow. Calyx broadly turbinate-campanulate, pubescent, about half the length of the corolla, teeth about as long as the tube, unequal, subulate. Inflorescence of 2–4(8) flowers, peduncle longer than or equal to subtending petiole. Young pod turns sideways through calyx. Mature pod appressed-pubescent, 3–4(5) mm in diameter, reniform, with 1–1.5 coils, the dorsal margin slightly thickened and lacking spines, prominent thickened veins spreading from ventral suture and anastomosing somewhat, central (axial) hole absent or obscure. Pedicels recurved in fruit. Seeds 1–2(3) per pod, 1.75–2 mm long, 1–1.5 mm wide, ovoid, brown or brownish, not separated by fruit partitions, radicle equal to or slightly longer than half length of seed.

Identification Hints

This species can be confused with the more western *M. prostrata* and the more eastern *M. cancellata*, as well as with *M. rhodopea*. *Medicago rupestris* has fruits with fewer coils than *M. cancellata* and *M. prostrata* (1–1.5 coils in *M. rupestris*, 1.5–3 coils in *M. cancellata*, 2–4.5 coils in *M. prostrata*). *Medicago rhodopea*, by contrast with *M. rupestris*, usually has short spines or tubercles on the dorsal suture.

Habitat and Environmental Adaptation

Habitat

Medicago rupestris occurs in rocky sites, bluffs, and fissures among stones in the lower Crimean Mountains. The plant does not form large populations and is often found among shrub associations, sometimes as a dominant (Didukh 2007).

Phenology

Flowering in its native area is from May to July, with fruit maturing in July and August.

Soil

In its native habitat the species typically occupies calcareous, rocky areas. McCoy (1987) found that *M. rupestris* has limited salt tolerance.

Temperature

Medicago rupestris occurs in the often-cool conditions of mountains and survives over cold winters.

Water

Didukh (2007) termed *M. rupestris* a xerophyte, a description consistent with the species' occupation of well-drained substrates and possession of a deeply penetrating root system.

Biotic Relationships

Pollination

Most perennial species of *Medicago* appear to be facultative outcrossers. The flowers of *M. rupestris* are small, suggestive that at least some selfing occurs.

Dispersal

No obvious morphological adaptations are evident for seed dispersal. The pods lack spines and so would not easily become tangled in fur.

Herbivory

No information located.

Nodulation and mycorrhizae

No information located.

Disease resistance

No information located.

Economic History, Cultivars, and Uses

Medicago rupestris has not attracted agricultural interest, although recently it has been considered as a possible source of genes for alfalfa improvement.

Cultivars

No cultivars have been bred.

Forage

Medicago rupestris is not sown for forage. Grossheim (1945) commented that "*M. rupestris* cannot be recommended as forage." His Russian compatriot Vassilczenko (1949) stated, "this plant may be of some importance as a pasture plant."

Fodder

This species is not harvested for dried livestock feed.

Green manure, rotation crop, ground cover

No information located.

Other uses (salad sprouts, ornamental, plant products, etc.)

The species is used to a small extent as a rock garden ornamental.

Chromosome Number

$2n = 16$: Lesins and Lesins 1966; Gillies 1972*a*.

Genetic Relationships and Gene Transferability

Intraspecific

Artificial hybridization of different populations has not been attempted, so it is not known whether there are barriers to hybridization within the species.

Interspecific

Using embryo rescue techniques, McCoy and Smith (1986) obtained hybrids between *M. rupestris* and (at the same chromosome number) diploid *M. sativa* ($2n = 16$); such hybrids were obtained relatively frequently. Gillies (1972*a*) studied a hybrid between *M. rhodopea* and *M. rupestris* (made by K. Lesins) and observed that chromosome pairing in this hybrid was excellent.

Exchange by advanced biotechnology

No information located.

Conservation Status

Medicago rupestris is a rare endemic. Didukh (2007) provided a brief (Red Book) conservation assessment and noted (translated from Ukrainian): "Protected within the territory of the Ak-Kay National Nature Monument (Belogorsk rayon, Crimea). The status of populations must be monitored". Given the rarity of the plant and its location in an area of the world subject to biodiversity loss, every effort should be made to protect the plant's habitat and seeds should be collected for deposition in germplasm banks.

Germplasm Sources

No accessions are recorded in the major genebanks with large holdings of *M. rupestris* (see Chapter 3). K.A. Lesins' germplasm collection was deposited in the USDA collection, so presumably there are at least some seeds there, although perhaps not yet accessible to the public.

Research Problems and Priorities

Medicago rupestris is sufficiently close to alfalfa (*M. sativa*) that it can be considered to be a continuing source of genes for breeding improved cultivars. Increased experimentation in this regard is very desirable.

Representative Collections

USSR: Callier, A. 62, Felsen in Karakuschbei Karasubazar, 15-VI-1896, in flower (K, P, S). Callier, A. 843, Karasubazar, in rupibus montis cretacei Akkaja, 7-VI-1900, in flower & early fruit (K). Callier, A., in rupibus montis cretacei "Akkaja" prope "Karasubasar" (locus classicus), vi-1900, in flower, Herbarium Normale editium ab I. Dörfler No. 4227 (LE, P). Junge, A. Tauria, in createaceis pr. urb. Karasribarar, 16-VI-1906, in fruit (S). Wankow, J., Flora Taurica, near the village Ak-Kaja, 1-VII-18-VI-1906, in late fruit (K).

84. *Medicago ruthenica* (L.) Trautv.

M. ruthenica (L.) Trautv., Bull. Sci. Acad. Imp. Sci. St. Petersb. 8: 271 (1841).

(Often inappropriately cited as *M. ruthenica* (L.) C.F. Ledebour, Fl. Ross. 1: 523 (1841 and 1842 depending on imprint; actually published in 1843).)

The epithet *ruthenica* is based on *Ruthenica*, *Ruthen* plus the English suffix *ic*, which is derived from the Greek *ikos* (belonging to), denoting the adjectival form of Ruthenia. Ruthenia has had several geographical meanings historically. At the time *M. ruthenica* was named, it referred generally to the Ukraine (the westernmost location).

Subordinate Taxa

None recognized here. Vassilczenko (1953) split *M. ruthenica* into four different species, a treatment that does not seem justified.

Type Specimen

The Linnaean Plant Name Typification Project (http://www.nhm.ac.uk/research-curation/projects/linnaean-typification/; current June 2010) provided the following information: "Linnaean Name: *Trigonella ruthenica* Linnaeus. Reference: *Species Plantarum* 2: 776

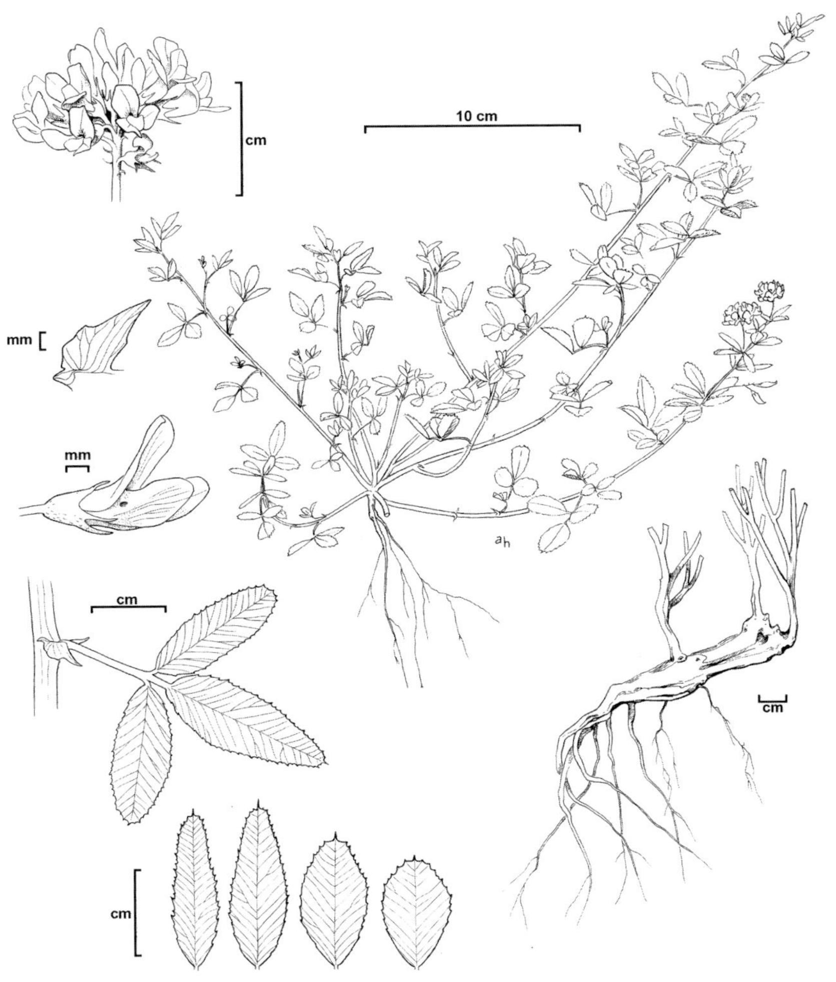

Fig. 84-1. *Medicago ruthenica.* **Clockwise from upper left**: Group of flowers; plant; root; leaflets; leaf attached to stem; side view of flower; stipule. (Based on E. Small M1025, DAO, except root, which is based on USDA Asia 22, GH; artist: A. Hanes.)

(1753). Provenance: 'Habitat in Sibiria.' Type: Lectotype: Herb. Linn. No. 932.1 (LINN). Designated by: Lassen in Turland & Jarvis (ed.), Taxon 46: 483 (1997). Current Name: *Medicago ruthenica* (L.) Trautv."

Synonymy

Homotypic

Pocockia ruthenica (L.) Boiss., Diagn. Pl. Orient., ser. 1, 2(9): 12 (1849).

Trigonella ruthenica L., Sp. Pl. 776 (1753), (basionym), not *T. ruthenica* Patrin in Ledeb. Fl. Alt. 3: 253 (1831), non L. (= *M. platycarpos*, according to Grossheim 1945).

Melissitus ruthenicus (L.) Latsch., Zametski Sist. Geogr. Rast. (Tbilisi) 32: 21 (1976).

Melissitus ruthenicus (L.) C.W. Chang, Fl. Tsinlingensis 1(3): 23 (1981), nom. superfl.

Melilotoides ruthenica (L.) Soják, Sborn. Nár. Muz. Praze, 1982(1-2): 104 (1982).

Turukhania ruthenica (L.) N.S. Pavlova, in Charkevicz, Pl. Vasc. Orient. Extremi Sovietici 4: 321 (1989).

Note: Vassilczenko (1953) referred to "*Medicago ruthenica* Kom. & Alis., Opred. Rast. Dal'nevost. Kr. 2: 662 (1932), non L.", which he equated with *T. korshinskyi* Grossh., assigned below to synonymy of *M. ruthenica* L. Accordingly, the name *M. ruthenica* Kom. & Alis. has no nomenclatural significance.

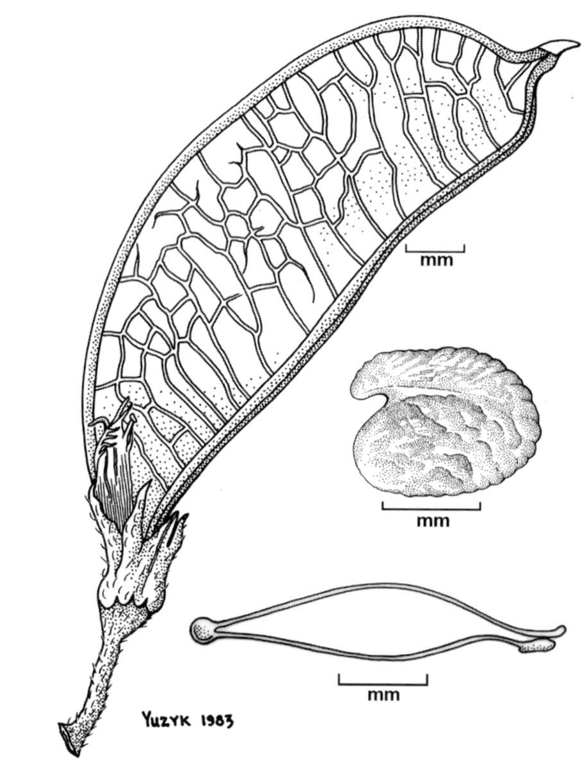

Fig. 84-2. *Medicago ruthenica.* **Upper left:** Fruit. **Center right:** Seed. **Bottom right:** Outline of cross section of fruit. (Based on E. Small M1025, DAO; artist: A. Hanes.)

Heterotypic

M. ruthenica var. *alpina* Usin (status undetermined; at Kew, I noted a type folder for this taxon based on G.N. Potanin's 188 collection from China borealis prov. Kansu) (Vassilczenko (1953: 173) assigned *M. ruthenica* var. *alpina* Maxim. in herb. to *T. archiducis-nicolai* (Širj.) Vassilc.).

M. ruthenica (L.) Trautv. var. *chinensis* Širj., Generis *Trigonella* L. Revisio Critica, Spisy Pfir. Fac. Masarykovy Univ. 1: 17 (1928). (Vassilczenko (1953: 173) suggested that this is based on immature Himalayan (not Chinese) material.)

T. ruthenica L. ß *cuneiloba* Karo ex Fr. (Based on F. Karo, Pl. Dahuricae 249b, plants collected near Perchinsk on the steppes, 1889, LE!)

M. ruthenica (L.) Trautv. var. *genuinam* Širj., Generis *Trigonella* L. revisio critica. Spisy Pfir. Fac. Masarykovy Univ. 1: 17 (1928), nom. inadmiss.

Pocockia ruthenica (L.) Boiss. var. *inschanica* H.C. Fu & Y.C. Tsiang, Intramongolica 3: 287 ("1977", publ. 1978).

Melissitus ruthenicus (L.) Latsch. var. *inschanicus* (H.C. Fu & Y.C. Tsiang) H.C. Fu & Y.C. Tsiang, Acta Sci. Nat. Univ. Intramongolicae 18(2): 391 (1987).

Melilotoides ruthenica (L.) Soják var. *inschanica* (H.C. Fu. & Y.Q. Jiang) H.C. Fu & Y.Q. Jiang, Fl. Intramongolica, ser. 2, 3: 196 (1989).

T. koraiensis Vassilc., Bot. Mat. Gerb. (Leningrad), 14: 234 (1951). Type: V.L. Komarov, 28-VIII-1897, grassy slopes along the valley of the Chashin-Chana River, a tributary of the Yalu (LE! photo DAO; holotype at LE viewable at http://www.mobot.org/MOBOT/Research/LEguide/species/13.html (current July 2010). (This taxon, based on a single Korean collection, seems conspecific with *M. ruthenica*.)

T. korshinskyi Grossh., Fl. USSR 11: 127 (1945). (Type: *S. Korzhinskii*, 19-VII-1891, village of Bakharevoy on the Bureya River, southern stony slope (LE! photo DAO).)

M. korshinskyi (Grossh.) R. Kamelin & Gubanov, Byulleten' Moskovskoe Obshchestvo Ispytatelei Prirody Otdel Biologicheskii 97(5): 63 (1992).

Melissitus korshinskyi (Grossh.) Latsch., Zam. Sist. Geogr. Rast. (Tbilisi) 32: 20 (1976).

T. ruthenica L. var. *latifolia* Karo ex Fr. in sched. (Based on F. Karo, Pl. Dahuricae 249a, plants collected near Perchinsk on the steppes, 1889, flowering, beginning fruiting, LE!)

Pocockia liaosiensis P.Y. Fu & Y.A. Chen, Fl. Plant. Herb. Chinae Bor.-Orient. 5: 176 (1976).

Pocockia ruthenica (L.) Boiss. var. *liaosiensis* (P.Y. Fu & Y.A. Chen) H.C. Fu & Y.C. Tsiang, Fl. Intramongolica 3: 146 (1977 publ. 1978), without basionym ref.

Melilotoides ruthenica (L.) Soják var. *liaosiensis* (P.Y. Fu & Y.A. Chen) H.C. Fu & Y.Q. Jiang, Fl. Intramongolica ser. 2, 3: 195 (1989).

Melissitus ruthenicus (L.) Latsch. var. *liaosiensis* (P.Y. Fu & Y.A. Chen) X.P. Jiang, Fl. Beijing 1: 429 (1984), basionym not cited (so the following combination is correct).

Melissitus ruthenicus (L.) Latsch. var. *liaosiensis* (P.Y. Fu & Y.A. Chen) H.C. Fu & Y.C. Tsiang, Acta. Sci. Nat. Univ. Intramongolicae 18(2): 391 (1987).

Medicago liaosiensis (P.Y. Fu & Y.A. Chen) X.Y. Zhu & Y.F. Du, Legumes of China: a checklist, 603 (2007).

T. ruthenica L. τ *oblongifolia* Karo ex Fr. (based on F. Karo, Pl. Dahuricae 249e, collected 1892, LE!)

Melilotoides ruthenica (L.) Soják var. *oblongifolia* (Fr.) H.C. Fu & Y.Q. Jiang, Fl. Intramongolica ser. 2, 3: 195 (1989), basionym not cited, assumed here.

Melissitus ruthenicus (L.) Latsch. var. *oblongifolius* (Fr.) H.C. Fu & Y.C. Tsiang, Acta. Sci. Nat. Univ. Intramongolicae 18(2): 391 (1987).

T. ruthenica L. var. *petraea* Vassilc., Tr. Bot. Inst. Akad. Nauk SSSR, ser. 1, Flora Sist. Vyss. Rast. Fasc. 14: 170 (1953), nom. illeg.

T. ruthenica L. var. *robusta* Vassilc., Tr. Bot. Inst. Akad. Nauk SSSR, ser. 1, Flora Sist. Vyss. Rast. Fasc. 14: 170 (1953), nom. illeg.

T. schischkinii Vassilc., Bot. Mat. Gerb. (Leningrad), 14: 233 (1951). (Type: No. 23, 31-VIII-1892, southern Primorye, Shkotovo region, Peyshulla River, Zmeinaya, limstone rock, on the peak, LE).

Melissitus schischkinii (Vassilc.) Latsch., Zametski Sist. Geogr. Rast. (Tbilisi) 32: 21 (1976).

Turukhania schischkinii (Vassilc.) N.S. Pavlova in Charkevicz, Plantae Vasculares Orientis Extremi Sovietici: 319 (1989).

M. vassilczenkoi Worosch., Fl. Sov. Dalmn. Vost.: 271 (1966).

Vernacular Names

English: No English common name is available, although the name Russian fenugreek is occasionally found (a name that considers the species to belong to the genus *Trigonella*). The Chinese name is bianxudou (Hua-Zhu and Quing-Chuan 1990). Balabaev (1934) notes that the species has been referred to as a "medic", although this term is normally applied to annual species of *Medicago*.

French: No name found.

Geographical Distribution

(N = native, I = introduced, U = uncertain status)

Europe: Russia (N), Ukraine (N).

Asia: China (N; Manchuria), Mongolia (N), North Korea (N), Russia (N; Siberia).

Medicago ruthenica has been collected mostly from the Transbaikal region to the Pacific Ocean, and south to about 34° N latitude. Hua-Zhu and Quing-Chuan (1990) described the distribution in China as including 10 provinces of northern China, especially in Heilongjiang grasslands, Inner Mongolia dry grasslands, Xinjiang hill grasslands, and the Qinghai-Tibet Plateau grasslands.

Description

Perennial herb, stems 20–40(100) cm long, erect to ascending, few to numerous stems arising from crown atop stout rootstock, stems branched, particularly in upper part, twigs and leaves more or less pubescent with appressed simple hairs. Leaflets (8)12–25(30) mm long, (2)3–4(5) mm wide, elliptical, oblong, obovate, oblong-obovate, oblong-lanceolate, oblong-oblanceloate or linear-oblong, generally 2–6 times as long as broad, cuneate at base, distal four-fifths to half of margin serrate with sharp mostly obtuse teeth, upper surface glabrous, lower surface pubescent to glabrescent, with simple appressed trichomes. Stipules 3–7 mm, lanceolate to lanceolate-subulate, dentate to incised. Inflorescence a short raceme of (3)4–10(15) flowers, peduncles 2–3(5) cm, puberulent, longer than subtending petiole. Flowers 5–6(7) mm, corolla yellow with violet veins, yellow and violet, bluish, or more rarely orange-reddish, brownish-yellow, or a mixture of colors, pedicels 2–4 mm, bracteoles small, subulate. Calyx 2.0–3.0 mm, with simple appressed hairs, teeth shorter than or more or less equal to tube. Corolla usually wider in distal portion than in proximal, more than twice length of calyx, wings a little shorter than standard, keel notably shorter than wings, keel and wings recurving. Fruit glabrous, oblong-oval or oblong-elliptical, flat, straight, or slightly curved, pendulous, often becoming black at maturity, 8–12(20) mm long, 3.5–5.0(7.0) mm wide, dehiscent, ventral suture more or less straight, dorsal suture convex, veins arise from ventral suture, branch, and anastomose on the outer two-thirds to one-half of the coil face, then reach dorsal suture, the beak small and curved. Seeds (1)2–5(6) per pod, 1.5–2.5 mm long, 1.2–1.8 mm wide, rugose, brownish, ovoid, radicle slightly less than seed length.

Identification Hints

This species is closest to *M. platycarpos*. For distinction from the latter, see identification hints for *M. platycarpos* (Chapter 66).

Habitat and Environmental Adaptation

Habitat

Medicago ruthenica occurs on sunny hillsides, dry stony slopes, embankments, and in mixed grass steppes and moist meadows, at times in large numbers. Hua-Zhu and Quing-Chuan (1990) recorded its growth in China on plain and hill areas, in woods, and on roadsides. Campbell et al. (1997) provided ecological notes on 50 collections made in Inner Mongolia; habitats included grasslands, roadsides, near streams, drainage ditches and marshes, and dune areas. Campbell et al. (1999) made collections in Mongolia at altitudes ranging from 175 to 1493 m. The species is only occasionally found as a weed in fields.

Phenology

In much of its native range, the species is in flower June to August (to September), and in fruit from August onward.

Soil

This species most frequently occurs on sandy, gravelly, and stony soils, chiefly chernozem and chernozem-like steppe soils. Some of the collections recognized as *Trigonella schischkinii* come from limestone cliffs. Campbell et al. (1997) observed that the pH of soils where *M. ruthenica* was found ranged from 6.4 to 7.3 in 15 sites in Inner Mongolia. Wang et al. (1989) described *M. ruthenica* as salt–alkali resistant. Lieth and Mochtchenko (2003) listed the species in their halophyte database. Dodd and Orr (1995), in experimental pot culture, found that *M. ruthenica* did not grow well in an acidic soil.

Temperature

Medicago ruthenica is adapted to cool temperate and alpine climates. In the Transbaikal region and northern Manchuria, it withstands extreme continental climates, in some cases with low snowfall and extremely cold winters. In China, Hua-Zhu and Quing-Chuan (1990) noted that *M. ruthenica* was exceptionally cold resistant, growing well in frigid dry areas above 3000 m. Palenski-Brown and Chu (1989) observed that at 3000 m in altitude in Gansu province, China, *M. ruthenica* is exposed to temperatures ranging from –20 °C in winter to 30 °C in summer. Wu et al. (1996) examined the eco-geographical distribution of *M. ruthenica* in Inner Mongolia. The species was distributed mainly in the Central–Eastern region of Inner Mongolia with accumulated temperatures (>10 °C) of 1400 to 3000 hours. In an unpublished (2000) United States Department of Agriculture (USDA) "Notice of Release" concerning *M. ruthenica* germplasm, it was recorded that "approximately 73% of the stand was lost during the winter of 1995–1996 due to poor adaptation to humid winter conditions." Campbell et al. (1997) noted that of the large number of accessions cultivated at Beltsville, Maryland and maintained during the severe winter of 1993 to 1994, more than 50% of the plants died, "suggesting that the species may be less resistant to cold stress under humid conditions than of that encountered in its native habitat."

Water

Medicago ruthenica seems adapted to moderately dry circumstances, although occasionally it also tolerates damp meadows. It withstands dry stony soils well. Wang et al. (1989) and Hua-Zhu and Quing-Chuan (1990) described the species as drought resistant. Campbell et al. (1997) reported that many *M. ruthenica* accessions that were collected in Mongolia may be very drought tolerant. Palenski-Brown and Chu (1989) noted that in Gansu province, China, *M. ruthenica* receives from 200 to 500 mm precipitation annually and exhibits extreme drought resistance during the summer. Wu et al. (1996) examined the eco-geographical distribution of *M. ruthenica* in Inner Mongolia. The species was distributed mainly in the Central–Eastern region of Inner Mongolia with an annual precipitation of 250–450 mm. Gu et al. (1994) found that *M. ruthenica* could be established well in the arid and semi-arid grassland of Inner Mongolia. Dodd and Orr (1995), in experimental pot culture, found that *M. ruthenica* was resistant to drought.

Biotic Relationships

Pollination

Campbell and Bauchan (2002) referred to *M. ruthenica* as "allogamous". Palenski-Brown and Chu (1989) noted that "manual tripping and hand-pollination were necessary for seed set in the glasshouse suggesting insect pollination in the natural habitat." The floral guides (brown streaks) on the petals are an indication of adaptation to insect pollinators. Small (1988) recorded a pollen to ovule ratio of 430, which is consistent with *M. ruthenica* being at least partly cross-pollinated, like other perennial species of *Medicago*. However, the relatively small flowers are suggestive of at least some independence from pollinators. Material for dissemination by the USDA has been replicated using leafcutter bees (*Megachile rotundata*) for pollination.

Dispersal

There are no obvious morphological adaptations of the pod for seed dispersal. Given the mountainous habitat often occupied, gravity probably contributes to dispersal.

Herbivory

Cambell et al. (1999) observed that *M. ruthenica* cultivated in Beltsville, Maryland was significantly more tolerant to potato leafhopper (*Empoasca fabae* Harris) feeding than alfalfa. However, in an unpublished (2000) USDA "Notice of Release" concerning *M. ruthenica* germplasm, it was recorded that "significant variation in ... potato leafhopper (*Empoasca fabae* Harris) damage was noted" (among the 40 accessions grown).

Nodulation and mycorrhizae

USDA (2009) reported positive root nodulation for *M. ruthenica*. Van Berkum et al. (1998) studied rhizobial associates of *M. ruthenica* and concluded that at least three genomic species of *Rhizobium*, including *R. tropici* and *R. mongolense*, formed nitrogen-fixing symbioses with *M. ruthenica*.

Disease resistance

In an unpublished (2000) USDA "Notice of Release" concerning *M. ruthenica* germplasm, it was noted that seed replication had taken place at Prosser, Washington, and "no disease symptoms were noted, however, a detailed pathological examination was not conducted". Campbell and Xia (2002) did not observe a notable occurrence of foliar disease on plants experimentally cultivated at Beltsville, Maryland.

Economic History, Cultivars, and Uses

Medicago ruthenica seems not to have been cultivated as a crop.

Cultivars

No cultivars have been bred.

Forage

Vassilczenko (1953) suggested that this species could be a very valuable forage plant in the USSR. Li (1997) found a protein content of 17.5 % in *M. ruthenica*, which was one of very few wild species of the Tianzhou region, Gansu, China that was a useful fodder plant in the area, and he recommended its domestication. In experimental plantings in China, Hua-Zhu and Quing-Chuan (1990) recorded fresh forage yields of up to 18 000 kg/ha per year and noted that the plants produced supple, tender stems and leaves with 10% to 15% crude protein. They described *M. ruthenica* as "a superior herbage" and noted that two cuttings per year were possible. Zho et al. (2005) studied the phenology and development of *M. ruthenica* on sandy land in China. The crop produced an aboveground, wet-weight biomass of 14.89 t/ha in the second year, the hay yield was 6.10 t/ha and the seed yield was 516 kg/ha. The stem to leaf ratio was 1:0.79. The root system could reach 80 cm in depth in the sowing year and 120 cm in depth in the second year. Li and Wang (1998) recorded that in grasslands of Mongolia, *M. ruthenica* was a preferred species of grazing sheep, and Campbell et al. (1997) also reported that farmers in Inner Mongolia indicated that *M. ruthenica* was preferred by their animals over other indigenous forage. In an unpublished (2000) USDA "Notice of Release" concerning *M. ruthenica* germplasm, it was suggested that "*M. ruthenica* appears to be particularly suitable for use in dry, low-input systems such as grazing land in the western U.S. The new *M. ruthenica* germplasms should be excellent, broad-based sources for selecting vigorous upright or procumbent parents in the development of improved *M. ruthenica* cultivars, or for crossing with alfalfa (*M. sativa*) for improving stress tolerance." Palenski-Brown and Chu (1989) observed a considerable level of hardseedness in *M. ruthenica*, and cultivars selected for more rapid germination may be desirable. Ma et al. (1984) reported that after a natural fire in a grassland in Inner Mongolia, the density and frequency of *M. ruthenica* increased by 200%, and suggested burning the natural grassland might be a method of increasing its forage value.

Fodder

Balabaev (1934) recorded that *M. ruthenica* becomes part of the hay harvested from the steppes and dry valleys of the Transbaikal and Dauriya regions.

Green manure, rotation crop, ground cover

Shao et al. (2006) recommended *M. ruthenica* for controlling erosion in the middle Gansu to Eastern Qinghai-Tibet Plateau loess region of China.

Other uses (salad sprouts, ornamental, plant products, etc.)

Medicago ruthenica has been alleged to have healing properties, and has been listed as a medicinal plant used for heart diseases in the Transbaikal region (Balabaev 1934).

Chromosome Number

$2n$ = 14, 16: Rudyka 1986.

$2n$ = 16: Fryer 1930; Cooper 1935; Lesins 1952; Clement 1962; Měsiček and Soják 1969; Lesins and Lesins 1979; Yan et al. 1989 (as *Melissitus ruthenicus* var. *inschanicus*).

Rudyka (1986) reported counts for "a collection like *Melissitus schischkinii*", which is presumably *M. ruthenica*. The count of $2n$ = 14 is new for flat-podded species of *Medicago*, and whether this number is characteristic of *M. ruthenica* requires exploration.

Genetic Relationships and Gene Transferability

Intraspecific

Studies of natural or artificial hybridization among populations of *M. ruthenica* have not been reported.

Interspecific

See *M. platycarpos* (Chapter 66) regarding the claim that it has been crossed with *M. ruthenica*. Attempts to cross *M. ruthenica* with another of the flat-podded species, *M. popovii*, were unsuccessful (Lesins and Lesins 1979). Oldenmeyer (1956*b*) failed in attempts to cross *M. ruthenica* with *M. sativa*, *M. lupulina*, *M. orbicularis*, or *M. rigidula*. However, Wang et al. (1989) and Li et al. (1990) reported that they had succeeded in crossing *M. ruthenica* with *M. sativa*, although very few pollinations resulted in seed, and meiosis in the F_1 plants was extremely aberrant. By contrast, Lesins (1952) was unable to hybridize *M. ruthenica* and *M. sativa*, both at the diploid ($2n$ = 16) level. Campbell (2003) also cited personal unpublished information that neither classical hybridization techniques nor protoplast fusion techniques had yielded viable hybrid embryos between *M. ruthenica* and *M. sativa*.

Exchange by advanced biotechnology

Campbell (2000) examined variation and relationship among clones of alfalfa (*M. sativa*) and *M. ruthenica* towards the possibility of gene transfer. Campbell (2003) studied genes in *M. ruthenica* that he considered potentially transferable to *M. sativa* for improvement of the latter.

Conservation Status

Medicago ruthenica occurs in remote areas of Asia that are subject to genetic erosion, and so conservation evaluation is desirable. The very recent collections acquired by the USDA will ensure that much of the genetic variation will be available in the future.

Germplasm Sources

As noted in Chapter 3, the following genebanks with large collections of *Medicago* have accessions of *M. ruthenica*: the USDA National Plant Germplasm System/Genetic Resources Information Network has 128 (largely the result of a germplasm collecting expedition, see publications by T.A. Campbell); and the Australian Plant Genetic Resources Information System has 10. Johnson et al. (2006) noted nine USDA collections of *M. ruthenica* from Mongolia acquired during the expedition cited above.

Research Problems and Priorities

Medicago ruthenica appears to offer considerable promise for domestication to cool temperate conditions, although the fact that *M. sativa* presently fills this requirement discourages interest in domestication. Balabaev (1934), however, held that *M. ruthenica* has even greater frost-resistance and hardiness than *M. sativa*. It should also be remembered that *M. ruthenica* seems far more capable of existing on poor stony soils under conditions of the Continental North than *M. sativa*, and so has potential application in low-input agricultural systems. The variation of *M. ruthenica* is poorly understood and basic systematic analyses of the different variants that have been recognized need to be carried out. Although there has been recent acquisition of germplasm (see above), there is still a need for additional germplasm collection of the apparently wide range of ecotypes of this species and evaluation of their agronomic potential. There also needs to be an evaluation of the possibility of gene transfer to alfalfa, particularly those characteristics favoring drought and cold resistance.

Representative Collections

China: Chanet, L. 1358, frontière du Chansi, Kche-Ly, montagnes, VII-1917, in flower & early fruit (K). Chaney, R.W., Inner Mongolia, grasslands, 1925, in flower & fruit (K). Cowdry, N.H. 1512, Kalgan-Chihli, roadsides very common, "Herb. Kew – 18 May 1923", in flower & fruit (K). Cowdry, N.H., Pietaiho, in wooded and damp sand, VII-1919, in flower (K) ("flower red brown on outside, inside yellow"). David, A., Mongolie orientale, Géhol, terrains sabl. en plaine, VI-1864, in flower (P), in flower & fruit (K). Dorsett, P.H. & Morse, W.J. 7079, Chili Province, Ching Lung Chiao, near Great Wall, VIII-IX-1930, in flower & fruit (BM). Kuah & Wang 782, Szechuan, Kanting, 19-VII-1963, in flower & fruit (K). Kuan & Wang 284, Szechuan, Kangting, 23-VII-1963, in flower (K). Licent, E. 207, northern China, Chansi N, montagnes, 18-VII-1914, in flower & early fruit (K). Licent, E. 4112, northern China, vers le Sinlong Chanetle Maho Chan, Kanson, occident, 9-VII-1918, in flower (K). Licent, R.P. 5732, Chine Septentrionale, Plaine du Toumet vers Ou tang tchao, 2-VIII-1919, in flower & early fruit (P). Meyer, F.N. 1099, Prov. Chili between Hwai lai and Tau hua, 31-VII-4-VIII-1913, in flower & early fruit (K, P). Rock, J.F. 12924, southwest Kansu, T'ao River basin, near Choni, banks of river, VII-1925, in flower (K, S). Smith, H. 654, Prov. Chili, Hsiao-wu-tai-shan, inter Sin-pao-an et Hun-ho, in arenosis, alt. ca. 600 m, 14-VIII-1921, in flower (S). Smith, H. 7726, Prov. Shansi, Chieh-hsiuh, Distr. Ch'o mei-shan, in prato herboso, alt. ca. 800 m, 27-IX-1924, in flower (S). Smith, H. 7802, Prov. Shansi: Chieh-hsiuh, Distr. Sung-lin-miao, in prato humido, alt. ca. 800 m, 28-IX-1924, in flower (S). Smith, H. 8222, Prov. Shansi, Lu Yak Shan, 21-VIII-1924, in flower (S). Wang, T.P. 3378, Shansi, Siao-y-hsien, Shangtingshan, 27-VIII-1935, in flower & fruit (K). Wang, T.P. W2210, Kalgan, on sandy river bed, 3-VIII-1934, in flower & fruit (K). **USSR:** Karo, F., Nertschinsk, in steppen, in flower, 1889, F. Karo Plantae Dahuricae No. 249c (K, P). (identified as γ *oblongiloba* Freyn). Karo, F., Nertschinsk, in steppen, in flower, 1889, F. Karo Plantae Dahuricae, No. 249b (K, P) (identified as β *cuyneiloba* Freyn). Karo, F., Nertschinsk, in steppen, in flower, 1889, F. Karo Plantae Dahuricae, No. 249 (K, P) (identified as α *latiloba* Freyn). Komarov, V., Manshuria Rossica, Flum Bina major, 25-VIII-1895, in flower, Komarov Flora Manshuriae No. 934 (K, P).

85. *Medicago sauvagei* Nègre

M. sauvagei Nègre, Compt.-Rend. Séances Soc. Mens. Soc. Sci. Nat. Maroc. 7: 175 (1954).

The epithet *sauvagei*, meaning "of Sauvage", commemorates Charles Philippe Felix Sauvage (1909–1980), a French botanist and apparently the first person to collect this species (in 1948).

Subordinate Taxa

None recognized here.

Type Specimen

Sauvage 6110, Rehamna, Boulaouane, Morocco, 1948 (RAB).

Synonymy

Homotypic

No names found.

Heterotypic

M. sauvagei Nègre var. *recta* Nègre, Luzern. Maroc 180 (1956). (As explained by Heyn (1963), Nègre divided *M. sauvagei* into two varieties, one from each of two localities, and since "var. *recta*" came from the population that included the type of the species, it should have been called var. *sauvagei*. A "syntype" is shown at http://www.aluka.org/action/showMetadata?doi=10.5555/AL.AP.SPECIMEN.MPU009775&pgs=&cookieSet=1; current July 2010.)

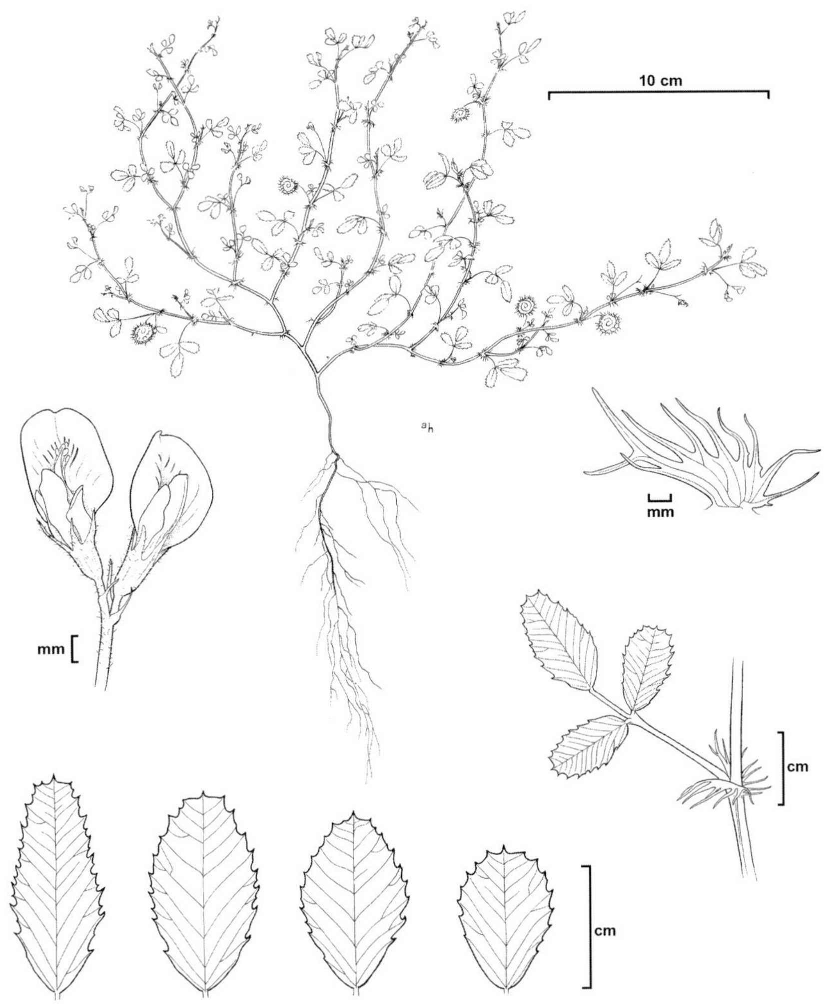

Fig. 85-1. *Medicago sauvagei.* **Clockwise from top:** Plant; stipule; leaf attached to stem; leaflets; group of flowers. (Based on E. Small H2989, DAO; artist: A. Hanes.)

M. sauvagei Nègre var. *uncinata* Nègre, Luzern. Maroc 180 (1956), (type specified in publication).

Vernacular Names

ENGLISH: No names found.

FRENCH: No names found.

Geographical Distribution

(N = native, I = introduced, U = uncertain status)

AFRICA: Morocco (N).

EUROPE: Netherlands (I; specimen recorded by Oostroom and Reichtgelt 1958).

Medicago sauvagei is a rarely collected endemic from Morocco. Bounejmate et al. (1992*b*) did not find it in their extensive collecting of *Medicago* in Morocco.

Description

Annual herb, stems procumbent, decumbent, or ascending, 5–40 cm long, glabrous or glabrescent, branched from base. Leaflets 8–12 mm long, 4–6 mm wide, cuneate to oblanceolate, distal one-third to two-thirds of margin serrate. Stipules laciniate. Inflorescence of 1–3 flowers, the peduncle more or less the length of the subtending petiole. Flowers 2–4(7) mm long. Corolla yellow, about twice as long as calyx. Calyx glabrous or glabrescent with simple hairs, the teeth shorter than the tube. Young pod protrudes sideways through calyx. Mature pod discoid, 5–6 mm long, with 4–6 coils, widest in the middle coils (5)7–8(10) mm in diameter, coil face with ca. 10–15 S-shaped radial veins, some forked, which enter a lateral vein at a distance of 1–2 mm from the dorsal suture. Dorsal sutures with spines that are hooked and grooved, apical coil or part of it spineless. Seeds 1–2(3) per coil, 2.5–4.25 mm long, 1.5–2.5 mm wide, ovoid, radicle slightly more than half (up to two-thirds) as long as the seed, seed smooth, dark yellow to brown-yellow, separated or scarcely separated by fruit partitions.

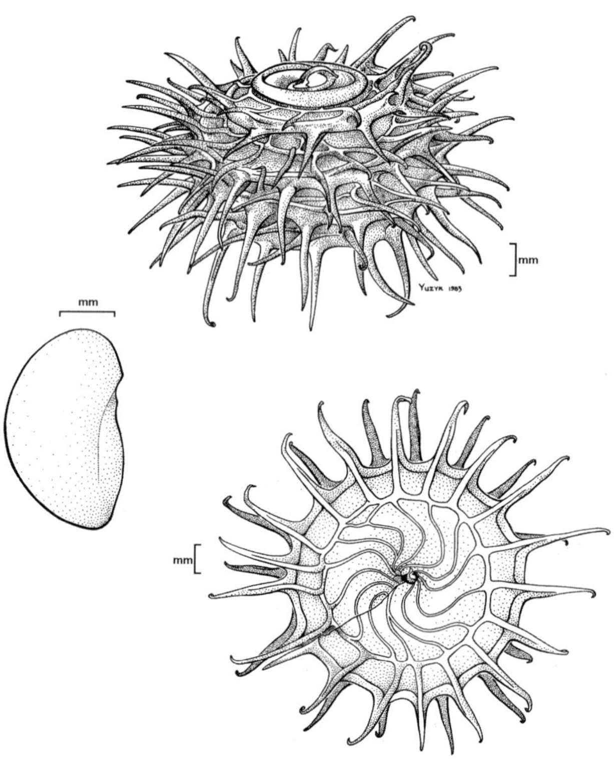

Fig. 85-2. *Medicago sauvagei.* **Top:** Fruit. **Center left:** Seed. **Bottom:** Fruit coil face showing venation. (Based on E. Small H2989, DAO; artist: L. Yuzyk.)

Identification Hints

Medicago sauvagei is similar to *M. minima*, which has much hairier foliage. It is also similar to *M. laciniata* in that the coil face has a submarginal vein that is rather remote from the coil edge. *Medicago sauvagei* has fruits that are usually wider than 5 mm at the widest point, whereas those of *M. laciniata* are rarely this wide; also, the latter species often has laciniate leaves. In addition, the submarginal vein is about one-fifth of the radius of the coil away from the dorsal suture in *M. laciniata*, but one-third to one-quarter of the coil radius away from the dorsal suture in *M. sauvagei*. The extremely rare *M. sauvagei* is most likely to be confused with the ubiquitous *M. polymorpha*, the coil face of which has much more reticulated venation and a submarginal vein much closer to the coil edge.

Habitat and Environmental Adaptation

Habitat

Medicago sauvagei has been collected from two sites in Morocco, both more than 400 m in altitude (Nègre 1956). It has been observed on roadsides (Lesins and Lesins 1979).

Phenology

The species flowers in the spring in its native area.

Soil

Piano and Francis (1992) recorded that *M. sauvagei* occurs on loam-clay soils ranging in pH from slightly to moderately basic. The species has been collected from red and brown clay (Nègre 1956; Lesins and Lesins 1979).

Temperature

The species appears to resemble most annual *Medicago* in being a winter annual in the Mediterranean climate, and therefore adapted to mild winters while setting seed before the hottest time of the summer.

Water

No information located.

Biotic Relationships

Pollination

Small (1988) noted that the small flowers and pollen to ovule ratio (350) are consistent with *M. sauvagei* being largely self-pollinated. This species is basically an inbreeder.

Dispersal

The indehiscent spiny fruits are well adapted to dispersal in animal fur.

Herbivory

No information located.

Nodulation and mycorrhizae

Unlike many of the more "promiscuous" species of *Medicago*, *M. sauvagei* forms symbiotic nodule associations with only a few rhizobia (Rome et al. 1996).

Disease resistance

O'Neill et al. (2003) examined the comparative resistance of 36 annual species of *Medicago* to *Phoma medicaginis* (the cause of spring black stem and leaf spot, a major disease of alfalfa) and concluded that *M. sauvagei* is highly susceptible to it. In a similar comparative test of 36 annual species of *Medicago*, O'Neill and Bauchan (2000) found that two accessions of *M. sauvagei* were completely susceptible to anthracnose caused by *Colletotrichum trifolii*. Yaege and Stuteville (2002) tested the resistance of 33 annual species of *Medicago* to powdery mildew (*Erysiphe pisi*) and found that the two collections of *M. sauvagei* grown were susceptible to the disease (it was observed that often only one of a number of populations of other species tested was resistant).

Economic History, Cultivars, and Uses

Medicago sauvagei is a rare species that has not been employed in agriculture.

Cultivars

No cultivars have been bred.

Forage

Medicago sauvagei is not sown for forage.

Fodder

This species is not employed as dried livestock feed.

Green manure, rotation crop, ground cover

No information located.

Other uses (salad sprouts, ornamental, plant products, etc.)

No information located.

Chromosome Number

$2n = 16$: Lesins and Lesins 1961; Simon and Simon 1965.

Genetic Relationships and Gene Transferability

Intraspecific

Artificial hybridization of different populations has not been attempted, so it is not known whether there are barriers to hybridization within the species.

Interspecific

Singh and Lesins (1972) crossed *M. sauvagei* with *M. laciniata* and observed that the hybrids were fairly sterile and the subsequent progeny suffered from inviability. Lesins and Lesins (1979) reported that they were unable to hybridize *M. sauvagei* with *M. disciformis*.

Exchange by advanced biotechnology

No information located.

Conservation Status

Given the rarity of this plant and its occurrence in an area of the world that is subject to biodiversity loss, this species should be considered to be in great need of conservation measures, especially collecting seeds for deposition in germplasm institutes.

Germplasm Sources

Eighteen accessions of *M. sauvagei* are listed among the holdings of the germplasm institutions with large *Medicago* collections (see Chapter 3).

Research Problems and Priorities

Medicago sauvagei is a rare, hardly examined species that requires germplasm collection and basic botanical and agronomic examination.

Representative Collections

Morocco: See collections cited by Nègre (1954), Heyn (1963), and Lesins and Lesins (1979).

86. *Medicago saxatilis* M. Bieb.

M. saxatilis M. Bieb., Fl. Taur.-Caucas. 2: 225 (1808).

The epithet *saxatilis* is Latin for dwelling or growing among rocks, descriptive of the often rocky habitat of this species.

Subordinate Taxa

None recognized here.

Type Specimen

Lectotype at LE "from collections of F.A. Marschall von Bieberstein from Crimea" viewable at http://www.mobot.org/MOBOT/Research/LEguide/species (current July 2010). Vassilczenko (1949) noted that the type collected by Steven bears the information "In Tauriae saxosis". Vassilczenko (2002*c*) recorded the following regarding the type: "Crimea

Fig. 86-1. *Medicago saxatilis*. **Clockwise from upper left:** Plant; root; stipule; leaf attached to stem; leaflets; side view of flower; group of flowers. **Center:** Flowering branch. (Based on K. Lesins 586, DAO; artist: A. Hanes.)

('in rupestribus calcareis Tauriae, circa Karassubasar et Sympheropolin')."

Synonymy

Homotypic

No names found.

Heterotypic

M. cuspidata Tausch, Flora (Linnaea) 14: 211 (1831).

M. saxatilis M. Bieb. var. *cuspidata* (Tausch) Urb., Verh. Bot. Ver. Brandenb. 15: 59 (1873).

Vernacular Names

English: Rock-inhabiting alfalfa (a name apparently contrived by Bolton 1962).

French: No names found.

Geographical Distribution

(N = native, I = introduced, U = uncertain status)

Europe: Ukraine (N).

Medicago saxatilis is a local endemic of the Crimean Mountains. *Flora Europaea* (Tutin et al. 1968) mistakenly reported it in Bulgaria, the result of confusion with *M. rhodopea*.

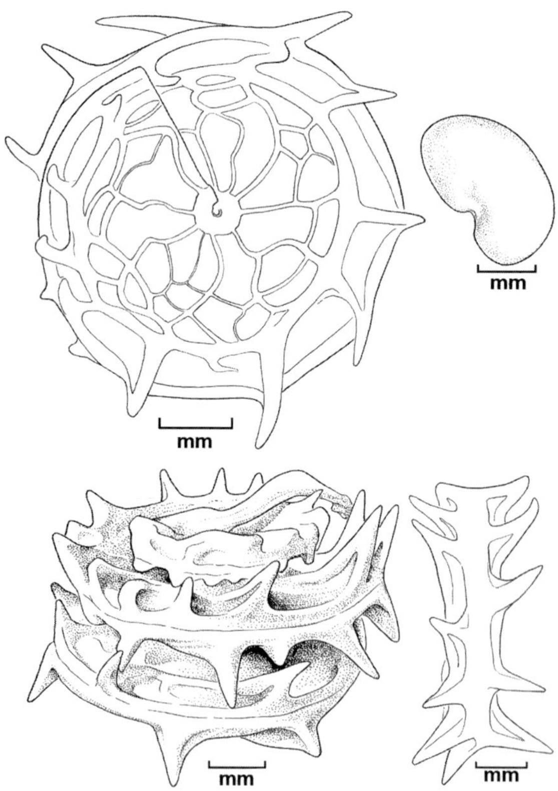

Fig. 86-2. *Medicago saxatilis.* **Clockwise from lower left:** Fruit; fruit coil face showing venation; seed; coil edge. (Based on E. Small M995, DAO; artist: A. Hanes.)

Description

Perennial herb, stems numerous, arising from crown, stems prostrate, decumbent, or ascending, 15–25 cm long, somewhat lignified at base, usually considerably branched, sparingly appressed-pubescent. Leaflets sublinear, linear-cuneate, oblong-cuneate, obovoid, or narrowly obcuneate, 5–10(12) mm long, (2)3–4 mm wide, distal third of margin slightly serrate or dentate, sparsely pubescent beneath and sometimes also above. Stipules narrowly triangular-lanceolate or linear-lanceolate, margin entire or with 1 or 2 teeth at base, sparingly appressed-hairy. Corolla (4)5–9(12) mm long, yellow or orange-yellow. Calyx campanulate, weakly to densely pubescent with white appressed hairs, teeth lanceolate-subulate or subulate-acuminate from a triangular base, equal to about twice as long as tube. Inflorescence capitate, with 7–12 flowers, the peduncle longer than the subtending petiole. Young pod exiting calyx before turning sideways. Mature pod with 2–4(5) coils, 5–6(8) mm in diameter, lenticular to cylindrical, sparsely appressed-hairy or glabrous, the dorsal suture with short, thick, conical, straight spines (rarely tubercles), the coil face with prominent, reticulately anastomosing veins that enter a well-developed lateral vein. Seeds 1–2 per coil, brownish, 3–3.7 mm long, ca. 2 mm wide, separated by spongy partitions, radicle slightly longer than half seed length.

Identification Hints

Medicago saxatilis resembles *M. cancellata*, a species to the northeast, but the latter lacks spines on its pods. *Medicago saxatilis* is also similar to another spiny-fruited species, *M. rhodopea*, which is well to the west of it and has much smaller fruits (the fruit diameter is generally 2–4 mm in *M. rhodopea*, 6–8 mm in *M. saxatilis*). The other closely related spiny-fruited perennial species are also distinctive: *M. daghestanica* usually has violet flowers and *M. pironae* has fruit covered with glandular hairs.

Habitat and Environmental Adaptation

Habitat

Medicago saxatilis occurs on dry, rocky, calcareous mid-mountain, subalpine slopes.

Phenology

This species flowers from May to July in its indigenous area.

Soil

The species typically occupies rocky limestone soils.

Temperature

Medicago saxatilis is native to cool, subalpine conditions. Lesins and Lesins (1979) noted that in Brandon, Manitoba, some plants survived the winter, although regrowth was poor.

Water

The species occurs in dry, well-drained soils.

Biotic Relationships

Pollination

Small (1988) noted that the relatively large flowers and pollen to ovule ratio (508) are consistent with *M. saxatilis* being at least partly cross-pollinated, like other perennial species of *Medicago*.

Dispersal

The indehiscent spiny fruits appear to be adapted to dispersal in animal fur.

Herbivory

No information located.

Nodulation and mycorrhizae

No information located.

Disease resistance

No information located.

Economic History, Cultivars, and Uses

Medicago saxatilis has not been employed as a crop.

Cultivars

No cultivars have been bred.

Forage

Medicago saxatilis is not sown for forage. Vassilczenko (1949) wrote (translated from Russian) that "this species is less xerophilous than *M. rupestris*. The plant tillers abundantly and develops numerous stems. It is of interest for utilization as a pasture-land plant (perhaps also for experiments on cultivation)." Lesins and Lesins (1979) commented that "the species by itself seems to be of little agricultural value. However, due to its crossability with *M. sativa*, it may be of value as a germplasm donor to the cultivated alfalfa, provided some desirable characters are found in *M. saxatilis*."

Fodder

This species is not noted as a source of hay.

Green manure, rotation crop, ground cover

No information located.

Other uses (salad sprouts, ornamental, plant products, etc.)

Medicago saxatilis is grown to a small extent as a rock garden ornamental.

Chromosome Number

$2n = 48$: Lesins and Lesins 1963*c*; Lesins 1970; Lesins and Lesins 1979; Yen and Murphy 1979.

Genetic Relationships and Gene Transferability

Intraspecific

Artificial hybridization of different populations has not been attempted, so it is not known whether there are barriers to hybridization within the species.

Interspecific

Lesins (1970) hybridized *M. saxatilis* (with $2n = 48$ chromosomes) with tetraploid ($2n = 32$) and hexaploid ($2n = 48$) *M. sativa* and reported considerable chromosomal irregularities in the hybrids. Yen and Murphy (1979) crossed *M. saxatilis* with two hexaploid amphiploids ($2n = 6x = 48$) of *M. sativa* and found that while hybrids (both F_1 and F_2) could be established, there was very extensive sterility. Lesins (1970) found that *M. saxatilis* could be easily crossed with *M. cancellata* (also with $2n = 48$) and that the F_1 plants were quite fertile. Lesins (1970) also reported on tri-species hybrids involving *M. rhodopea*, *M. sativa*, and *M. saxatilis*.

Exchange by advanced biotechnology

Using embryo rescue techniques, McCoy and Smith (1986) obtained tri-species hybrids between *M. saxatilis* ($2n = 6x = 48$) and *M. sativa* × *M. papillosa* F_1 hybrids ($2n = 3x = 24$).

Conservation Status

Given the fact that *M. saxatilis* is a quite localized endemic in a region subject to biodiversity loss, its habitat should be afforded protective measures and seeds should be collected for deposition in germplasm institutes.

The "Red Data Book of Ukraine" provides the following exemplary report, translated here from the original Russian (http://www.redbook.iatp.org.ua/1996/k1170.htm; current July 2007):

Scientific value: Endemic species. Promising for selection as a drought-resistant fodder plant.

Status: I Class.

Area: Crimean Mountains foothills (Bahchisaray region).

Habitat: Exposed limestone.

Quantity: Populations are numerous, relatively stable, occur sporadically, occupying areas of insignificant size.

Cause of quantity change: Limestone extraction, excessive cattle grazing.

General parameters: Perennial herbaceous plant 20–30 cm tall. Stalks thin, abundantly ramose. Leaves narrow sphenoid, slightly dentate on edges. Flowers up to 10 mm long. Inflorescence racemose, capitate. Pod diameter 5–6 mm, spirally twisted, almost bare, with short thick straight thornlets and bumps on edge. Blossoming in May to June. Fructification in July. Reproduction by seeds. Petrophyte. Hemixerophyte. Obligate calciphile.

Protection measures: Protected within areas of natural monuments of national importance: Belbek Canyon and Mangup-Kale. The condition of populations ought to be controlled; places of largest localization should be transformed into a sanctuary; possibilities of cultivating should be studied.

Germplasm Sources

Only two accessions are available from the major *Medicago* germplasm collections of the world (see Chapter 3).

Research Problems and Priorities

Germplasm of *Medicago saxatilis* has been in very short supply, limiting examination of its basic botany and agricultural potential. Accordingly, germplasm acquisition is the most pressing need.

Representative Collections

Ukraine: Lesins, K. 586, DAO; Small, E. M995, DAO (cited above).

87. *Medicago scutellata* (L.) Miller

M. scutellata (L.) Miller, Gard. Dict., ed. 8: no. 2 (1768).

The epithet *scutellata* is Latin for "covering of plates or dishes", a reference to the appearance of the fruit coils, which resemble a stack of dishes.

Subordinate Taxa

None recognized here.

Bauchan and Elgin (1984) first observed certain different kinds of *M. scutellata*: (1) nonvigorous plants that flowered about 30 days after planting and had one flower per inflorescence; (2) vigorous plants that flowered about 60 days after planting and had three flowers per inflorescence; and (3) intermediate plants. Bauchan et al. (1986) noted that samples of the early- and late-flowering forms possessed different seed proteins, and that hybrids of these forms have proteins from both parents. Small et al. (1993) examined 233 germplasm accessions of *M. scutellata* grown under greenhouse conditions and confirmed that the early- and late-flowering forms are substantially different morphologically. The early-flowering form is smaller and less vigorous, has fewer flowers in the inflorescence, and has fewer serrations on the leaf blades and

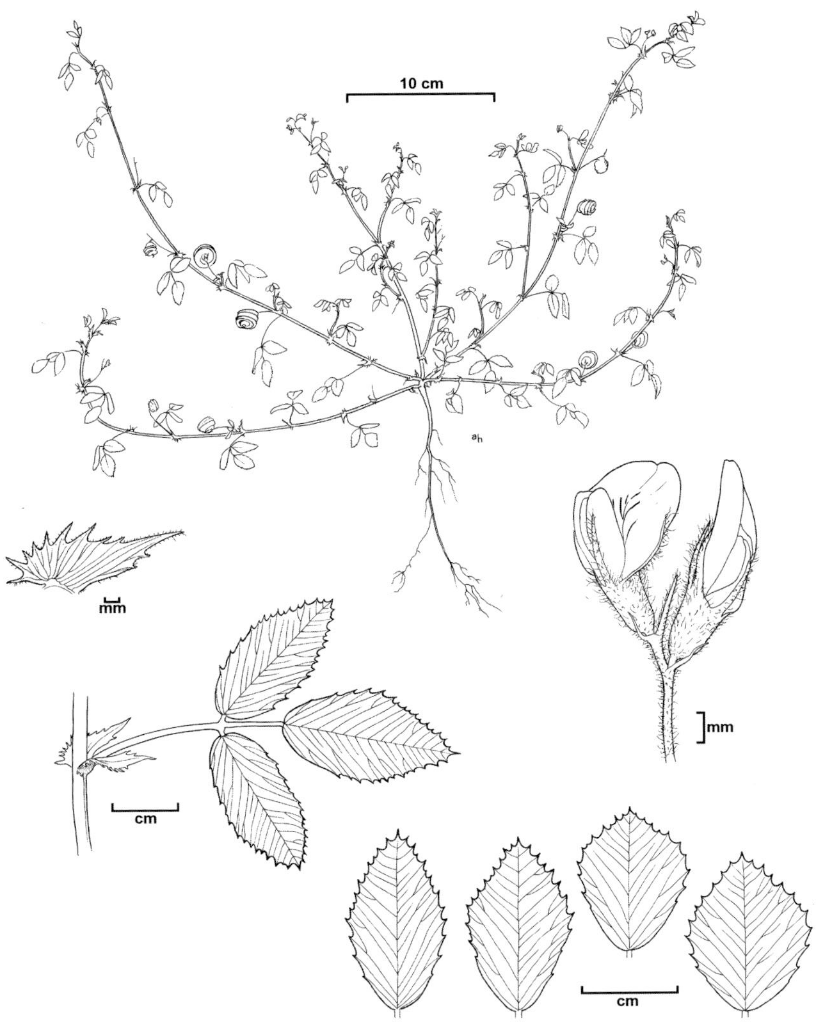

Fig. 87-1. *Medicago scutellata.* **Clockwise from top:** Plant; group of flowers; leaflets; leaf attached to stem; stipule. (Based on E. Small M1020, DAO; artist: A. Hanes.)

stipules than the late-flowering form. By using polyacrylamide gel electrophoresis, Small et al. (1993) identified 11 seed proteins that can be used to distinguish the two forms and putative hybrids. Although both forms appear widespread in the circum-Mediterranean indigenous distribution area, only about a third of the accessions represented the early-flowering form, which seemed to predominate only on isolated Mediterranean islands.

The two basic forms may merit infraspecific taxonomic recognition. However, I examined the large collection of *M. scutellata* at HUJ, and the morphological characters that served to discriminate the two groups grown in the greenhouse study by Small et al. (1993) were generally not adequate for discriminating most plants collected in nature, which were more variable than the greenhouse plants. Since the two kinds of plant co-occur in many countries and can be hybridized experimentally, and since plants of some accessions seem to combine the characteristics of both kinds, they clearly do not merit separate recognition as species. However, it is not yet clear that they merit taxonomic recognition at any rank, given the absence of a reliable means of visual identification. It is desirable to clarify the extent of hybridization between the two forms in nature. Pending a practical way of identifying the two classes of *M. scutellata*, they are not recognized formally here. For choice of future nomenclature, it may be noted that the lectotype seems to be the late-flowering form.

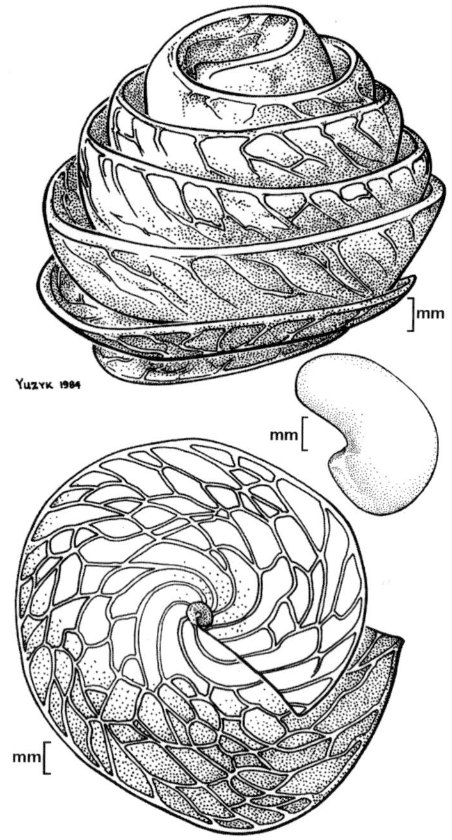

Fig. 87-2. *Medicago scutellata*. **Top:** Fruit. **Center right:** Seed. **Bottom:** Fruit coil face showing venation. (Based on E. Small M284, DAO; artist: L. Yuzyk.)

Type Specimen

Heyn (1959: 164) lectotypified the basionym for *M. scutellata* (L.) Miller (*M. polymorpha* L. var. β *scutellata* L.) by an illustration, tab. 15, f. 3, in: Morison, R. 1680. Plantarum historia universalis Oxoniensis, vol 2: 152, s. 2. Oxford, UK. Accompanying text: "Medica cochleata major dicarpos fructus capsula rotunda globosa scutellata." This lectotypification was disputed by Burtt (1964), but was accepted in The Linnaean Plant Name Typification Project (http://www.nhm.ac.uk/research-curation/projects/linnaean-typification/; current June 2010; cf. Jarvis 2007).

Synonymy

Homotypic

Medicago scutellaris All. in Reichenb. Icon. Fl. Germ. Helv. 22: t. 63 (1867–1886). The epithet "*scutellaris*" is a misprint, as explained in Heyn (1963).

M. polymorpha L. var. β *scutellata* L., Sp. Pl. 779-780 (1753) (basionym).

M. scutellata (L.) All., Fl. Pedem. 1: 315 (1785), nom. superfl.

Medica scutellata (L.) Moench, Meth. 116 (1794).

Medica scutellata Bubani, Fl. Pyren. 2: 574 (1900), nom. superf.

Heterotypic

M. inermis (Lam.) Lam. ex Steud., Nomencl. Bot. 513 (1821), pro syn.

Not

M. scutellata Pall. (Ind. Taur.) ex Bieb. Fl. Taur.-Cauc. 2: 226 (1808). (Listed as a synonym of *M. orbicularis* by Heyn (1963).)

Vernacular Names

English: Shield medic, snail alfalfa, snail clover, snail medic (medic is also spelled medick).

French: La luzerne attrape, la luzerne à écusson(s), luzerne en écusson.

Geographical Distribution

(N = native, I = introduced, U = uncertain status)

Africa: Algeria (N), Libya (N; Francis (1981) reported that *M. scutellata* occurred as small, isolated populations in Tripolitania (northwestern Libya)), Morocco (N), South Africa (I), Tunisia (N).

Asia: Cyprus (N), Israel (N), Japan (I, according to "Alien species recognized to be established in Japan or found in the Japanese wild as of Oct. 27, 2004" http://www.env.go.jp/en/nature/as/041110.pdf; current June 2010), Jordan (N), Lebanon (N), Syria (N), Turkey (N).

Australasia: Australia (I), New Zealand (I).

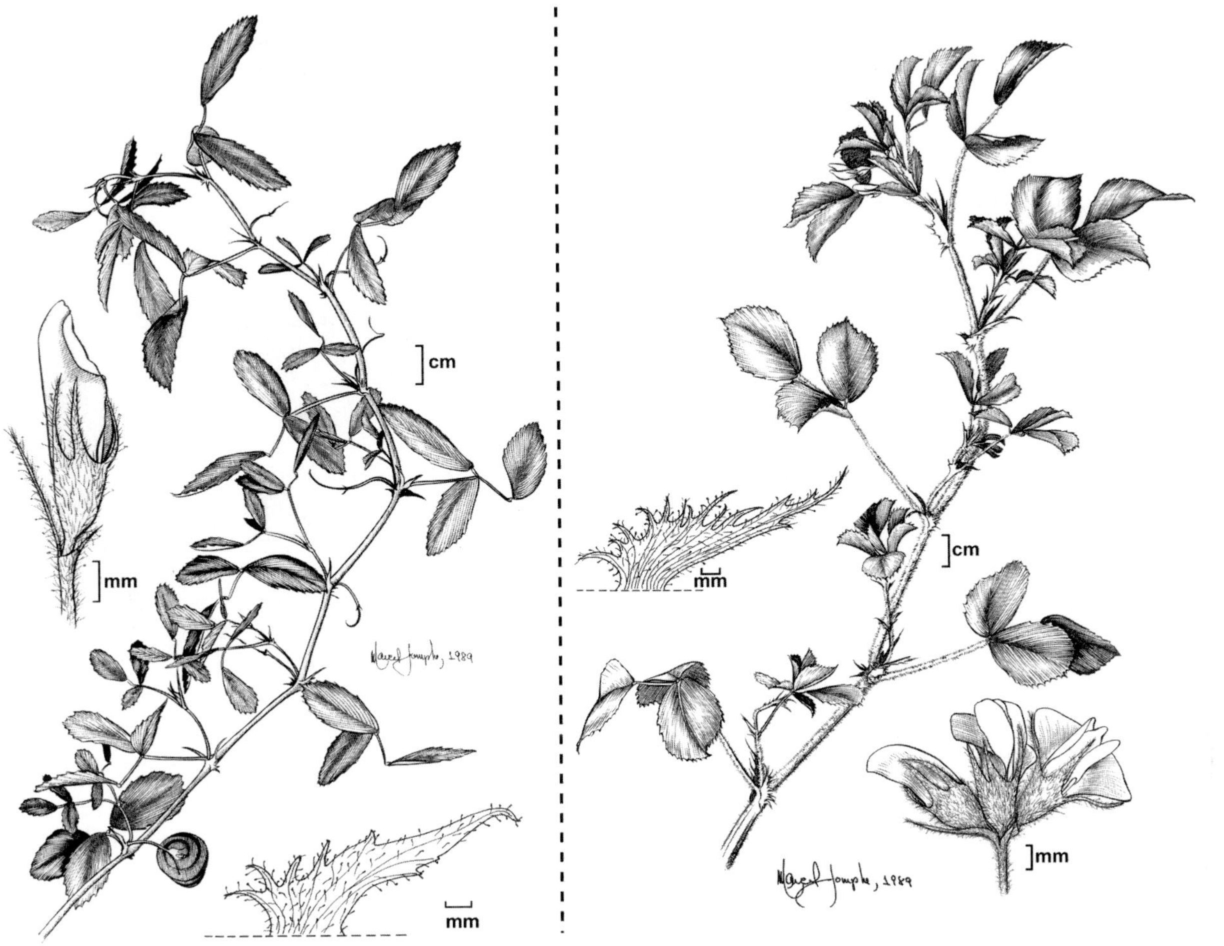

Fig. 87-3. Left: Fruiting branch of the early-flowering form of *Medicago scutellata*. **Center left:** One-flowered inflorescence. **Bottom right:** Stipule (relatively undissected). (Based on a greenhouse-grown plant, E. Small M1728, DAO; artist: M. Jomphe.) **Right:** Fruiting branch of the late-flowering form of *Medicago scutellata*. **Center left:** Stipule (relatively dissected). **Bottom right:** Several-flowered inflorescence. (Based on a greenhouse-grown plant, E. Small M1647, DAO; artist: M. Jomphe.)

Europe: Austria (I), Belgium (I), Czechoslovakia (N), France (including Corsica; N), Germany (I), Greece (including Crete; N), Hungary (N), Karpathos (Greece: East Aegean Islands; N), Italy (including Sicily; N), Malta (N), Netherlands (I), Portugal (U; cf. Malato-Beliz and Guerra 1977), Spain (including Balearic Islands; N), Switzerland (N), Turkey (N), Ukraine (N), Yugoslavia (N).

Northern America: USA (Delaware; I).

Medicago scutellata is omni-Mediterranean, and native also to the Black Sea region and the Crimea. It is widely introduced elsewhere.

Description

Annual herb, usually ascending (sometimes prostrate or decumbent), branched from base, stems (20)25–50(60) cm long, usually densely pubescent with glandular hairs. Leaflets obovate, ovate, ovate-lanceolate, or oblanceolate, the distal two-thirds or less of the margin serrate, glabrous above, very pubescent with glandular hairs and/or simple hairs below, 10–25 mm long, 5–12 mm wide. Stipules ovate-lanceolate to lanceolate, margin dentate, incised or laciniate, abaxial surface densely covered with glandular hairs. Inflorescence of 1–3(4) flowers, the peduncle shorter to as long as the subtending petiole. Flowers 6–9 mm long. Corolla yellow to orange-yellow, less than twice length of calyx. Calyx with glandular hairs, teeth lanceolate, as long as or longer than tube. Young pod protruding sideways from calyx. Mature pod 7–15(19) mm in diameter, the dorsal suture spineless and with glandular hairs, the (4)5–7(8) coils concave (bowl-like) and imbricated (stacked within each other, the convex surface facing the pod base), the coil face with a dense net of veins. Seeds 4–7 mm long, 2.5–4 mm wide, reniform, 1–2 per coil (upper coils usually seedless), separated by thin fruit partitions, the seed surface smooth, light yellow to brown, radicle less than half length of seed.

Identification Hints

Medicago scutellata is unique in having imbricate fruit coils, giving the appearance of a stack of bowls.

Habitat and Environmental Adaptation

Habitat

Medicago scutellata occurs in meadows, cultivated and fallow fields, as a weed of various cultivated crops, and on the margins of woods. De Haan et al. (2002) found that cultivar 'Kelson' was tolerant of shade.

Phenology

In the Mediterranean area, flowering occurs from March to May, and later in more northerly locations. In an experimental trial, Ivory (1976) studied time from germination to first flower in 19 accessions of *M. scutellata* and observed a large range: from 49 to 111 days. Clarkson and Russell (1975) found that flowering was not accelerated by vernalization or long photoperiods (18 and 24 hours), unlike several other species of *Medicago* tested at the same time, but observed high temperature acceleration of flowering.

As observed in Small et al. (1993), cultivar 'Robinson' is a member of the early-flowering group. Indeed, Crawford (1985) recorded its exceptional earliness. It is interesting to note that this early-flowering cultivar is relatively less vigorous compared with the late-flowering group under greenhouse conditions. Apparently under the conditions 'Robinson' is typically grown in Australia its short life cycle is important, and its performance is not inferior. Indeed, "'Robinson' has produced the highest seed yields of all annual medic cultivars tested in experiments in South Australia between 1966 and 1969" (Mackay 1981).

Soil

The species often seems to prefer heavy soil. In Algeria, Abdelguerfi et al. (1988*a*) observed that *M. scutellata* occurs in medium- to very fine-textured soils. Quinlivan et al. (1974) wrote that "in South Australia, the species is suitable for some of the sticky alkaline red clays and clay loams." Piano and Francis (1992) recorded that in its native area it occupies clay and loam soils that are slightly to moderately basic. In Greece, Thanopoulos (2007) recorded that the species occurs near the seashore in sandy places. In Sicily, Piano et al. (1991) observed the species growing in a pH range of 6.5–9.0.

Temperature

The species is a typical Mediterranean winter annual, germinating in the cool (not freezing) temperatures of late winter and setting seed in the middle of the summer to avoid the hottest, driest part of the year. Compared with most other medics, it requires a relatively warm winter climate. Based on growth chamber experiments, Brandsaeter et al. (2002) found that *M. scutellata* 'Kelson' exhibited poor freezing resistance. Krall et al. (1996) found that three accessions of *M. scutellata* examined for their ability to self-regenerate in Wyoming failed to survive the winter at the three locations tested. Christiansen and Svejcar (1989) sowed two accessions of *M. scutellata* in Oklahoma in March 1985 and observed that all the plants were killed by frost and cold in the following winter period. Derkaoui et al. (1991) found that fall-planted *M. scutellata* in Oklahoma failed to survive the winter, indicating that the species does not have sufficient cold tolerance for successful self-replication in the southern Great Plains of the United States.

Water

As with most Mediterranean annual species, *M. scutellata* is native to well-drained soils. Piano and Francis (1992) observed that in its native area the species requires a minimum

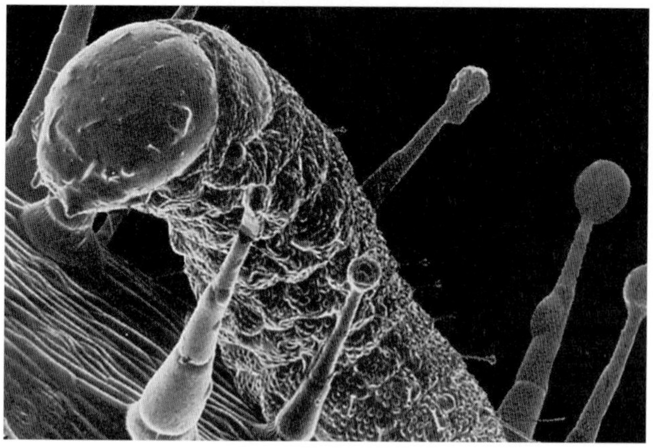

Fig. 87-4. Scanning electron micrograph of a larva of an alfalfa weevil (*Hypera postica*) trapped amidst the glandular hairs on a leaf of *Medicago scutellata*. The larva is about 1.3 mm long. This photo was taken by E. Sorensen and his colleagues at Kansas State University and published in Pierce (1983). It is reproduced here by courtesy of G.R. Bauchan and the USDA Agricultural Research Service.

Fig. 87-5. *Medicago scutellata*. From Hallier (1885, vol. 23, plate 2341).

of 350 mm annual precipitation. In Algeria, it occupies areas with a wide range of annual precipitation, between 350 and 800 mm (Abdelgeurfi et al. 1988*a*). Francis and Poole (1973) observed that it is very susceptible to waterlogging. Conway et al. (2001) studied the performance of *M. scutellata* in Central Queensland, an area that has summer-dominant rainfall and clay soils rather than the rain-fed winter Mediterranean climate and well-drained soils to which the species is adapted (winter rainfall in Central Queensland is typically 80–100 mm); they concluded that *M. scutellata* did not perform well. Cultivar 'Sava' is described as requiring 375–550 mm rainfall in Australia (although claims have been made for a much smaller amount). 'Kelson' is said to require at least 400 mm.

Biotic Relationships

Pollination

Medicago scutellata has relatively large flowers for an annual species of *Medicago*, but nevertheless appears to be primarily an inbreeder.

Dispersal

The adaptive significance of the peculiar fruit shape is unclear, including the question of whether this affects seed distribution. The roundish fruits would likely roll well on the ground, perhaps propelled by wind, and this could be a mechanism for seed distribution. The number of seeds per pod and the size of the fruits and seeds was studied by Abdelguerfi (1989) in Algeria, in relation to rainfall and altitude. He observed that (in a common garden) populations from areas receiving less than 350 mm of rainfall annually produced more seeds per pod than populations receiving 650 mm or more of precipitation, and that populations from higher altitudes produced fewer and smaller seeds than lower-altitude populations. The size of fruits and seeds likely affects dispersal, but whether this is adaptive to particular ecological circumstances is unclear.

Herbivory

The main insect pests of *M. scutellata* include the redlegged earth mite (*Halotydeus destructor* (Tucker)), blue oat mite (*Penthaleus* species), the lucerne flea (*Sminthurus viridis* (L.)), the spotted alfalfa aphid (*Therioaphis maculata* (Buckton)), the bluegreen aphid (*Acyrthosiphon kondoi* Shinji), and the pea aphid (*Acyrthosiphon pisum* (Harris)) (Information obtained in 2007 from the New South Wales Department of Primary Industries, Australia website, no longer current). *Medicago scutellata* has been reported to be resistant to aphids and the alfalfa weevil, *Hypera postica* (Gylh.) (Kleitner and Sorensen 1983; Doss et al. 1989). Ferguson et al. (1982) found that *M. scutellata* was quite resistant to the spotted alfalfa aphid. Shade et al. (1975) examined the resistance of 21 species of *Medicago* to attack by larvae of the alfalfa weevil, and found that *M. scutellata* was 100% resistant. Johnson et al. (1980*a*, 1980*b*) also demonstrated resistance to the alfalfa weevil. Shade and Kitch (1983) noted that the glandular hairs of *M. scutellata* resulted in resistance to the pea aphid. MacLean and Byers (1983) observed that the glandular hairs of *M. scutellata* tended to deter the alfalfa blotch leafminer (*Agromyza frontella* (Rondani)) from ovipositing. Shade et al. (1979) found that *M. scutellata* was very resistant to the potato leafhopper (*Empoasca fabae* Harris).

Nodulation and mycorrhizae

USDA (2009) reported positive root nodulation for *M. scutellata*. In Central and Western Spain, it has been shown that, when infected by an effective strain of *Rhizobium*, *M. scutellata* resists freezing temperatures and long periods of winter drought, whereas noninfected plants die (Feuntes 1981).

Disease resistance

The main diseases of *M. scutellata* include root rot, alfalfa mosaic virus, and phoma (Information obtained in 2007 from the New South Wales Department of Primary Industries, Australia website, no longer current). O'Neill et al. (2003) examined the comparative resistance of 36 annual species of *Medicago* to *Phoma medicaginis* (the cause of spring black stem and leaf spot, a major disease of alfalfa) and concluded that *M. scutellata* is highly susceptible to it. In a similar comparative test of 36 annual species of *Medicago*, O'Neill and Bauchan (2000) found that 17 accessions of *M. scutellata* were extremely susceptible to anthracnose caused by *Colletotrichum trifolii*. Yaege and Stuteville (2002) tested the resistance of 33 annual species of *Medicago* to powdery mildew (*Erysiphe pisi*) and found that the 18 collections of *M. scutellata* grown were susceptible to the disease (it was observed that often only one of a number of populations of other species tested was resistant).

Economic History, Cultivars, and Uses

Medicago scutellata was apparently introduced to Australia by Baron von Mueller in the late 19th century (Baron von Mueller 1895). It is one of the seven principal annual species of *Medicago* from which forage cultivars have been developed, mainly in Australia during the latter half of the 20th century.

Cultivars

Cultivars 'Kelson' (late-flowering), 'Essex' (mid-flowering), and 'Silver' (early-flowering) of *M. scutellata* were released primarily for use in leys in southern Queensland, Australia (Conway et al. 2001). 'Sava' (early-flowering) is another cultivar in use in Australia. 'Robinson', a principal early-flowering cultivar of *M. scutellata*, originated from a ruderal South Australian population (Mackay 1981). It has been noted that mixtures of cultivars with different maturity dates can be sown together as a strategy to address variable climatic conditions.

Forage

Sanadgol et al. (2006) found the unirrigated yield of *M. scutellata* over a 3-year period in the Gorgan plain of Iran

Table 87-1. Forage yields of *M. scutellata* in northern Iraq (kg/ha), after Kernick (1987).

Fertilized with 60 kg/ha P_2O_5		Unfertilized	
Green matter	Dry matter	Green matter	Dry matter
74000	3000	5000	2200

was 2083 kg/ha (dry weight). Ibragimova and Smolenskaya (2001) examined 24 annual species of *Medicago* for their agronomic performance in the Novosibirsk region of Russia, and concluded that only *M. scutellata* and *M. orbicularis* showed promise for the area. Vitkus (1977) assessed *M. scutellata* as suitable for cultivar selection for Lithuania. Belov (1971) examined 37 annual species of *Medicago* for suitability as new crops in the Tashkent region of Uzbekistan and *M. scutellata* was one of the three species recommended for development. Shrestha et al. (1998) concluded from experimental growth of *M. scutellata* in Michigan that its productivity for the year of planting was comparable with that of alfalfa. Zhu and Sheaffer (1997) recorded aboveground dry matter yields (in kg/ha) of cultivar 'Sava' at two sites in Minnesota (compared with the alfalfa cultivar 'Nitro', yields in parentheses): 6888 (3815) and 5706 (4596).

Kernick (1978) provides extensive information on the forage yield of *M. scutellata* in the Middle East. In the littoral zone of Tunisia, green forage yields in the order of 20 000 kg/ha in dryland cultivation have been reported (Kernick 1978). In the steppic zone (220 m elevation) of northern Iraq, during two seasons with rainfall varying from 330 to 475 mm (November to May), fallow fields on light calcareous soil seeded with *M. scutellata* and fertilized with 60 kg/ha P_2O_5 , gave yields varying from 5700–6000 kg/ha green forage (Kernick 1978). In a nursery plot on alluvial soil in the steppic zone (220 m elevation) of northern Iraq, during a season with rainfall of 475 mm (November to May), forage yields were recorded as follows:

Medicago scutellata herbage yield (dry matter) under irrigated conditions in South Africa was 2.43 t/ha (Wasserman 1974). Mohammad and Qamar (1988) grew 14 annual species of *Medicago* in small test plots in Islamabad, Pakistan located in a subtropical, subhumid area that is notably hotter and wetter than the normal habitat of these species. The dry matter yield of *M. scutellata* was 4.6 t/ha. Also in Islamabad, Shaheen and Mohammad (1988) obtained a dry matter yield of 8 t/ha without fertilization, and 14 t/ha with fertilization.

In South East Queensland, Australia, Jones and McLeod (1971) reported that *M. scutellata* yields during peak spring growth exceeded that of alfalfa. They found that digestibility ranged from approximately 80% at early flowering to 50% at the dead stem stage, and stated that the nitrogen content would be adequate for milk production.

Tadmor et al. (1971) conducted studies of ecotypes selected for their ability to survive and persist in the Negev Desert of Israel in an area where rainfall averaged only 90 mm annually, but received an additional 263 mm of water annually in the form of run-off from erratic winter storms. Dry matter production of an ecotype of *M. scutellata* was 5.5 t/ha (averaged over 6 years), which is considered rather good given the dry environment.

Livestock disorders that have been associated with grazing on *M. scutellata* include photosensitization in horses, occasionally red gut in sheep, and frequently bloat in cattle. (Information obtained in 2007 from the New South Wales Department of Primary Industries, Australia website, no longer current.)

Fodder

Like the other annual species of *Medicago* that are used commercially, there may be occasional harvest of herbage for use as dried livestock feed, although the plants are employed principally as live forage.

Green manure, rotation crop, ground cover

This species is recommended as a cover crop for South African vineyards. Stands of *M. scutellata* have been established in Montana to protect soil from wind erosion during the autumn windy season (Sims 1998). De Haan et al. (1997) examined the use of *M. scutellata* cultivars 'Sava' and 'Kelson' as a smother crop used against weeds for an ensuing corn crop, but found that subsequently *M. scutellata* competed against the corn, significantly reducing the crop.

Other uses (salad sprouts, ornamental, plant products, etc.)

In Japan, young seed pods, which taste like young sweet pea pods, are employed as a garnish in salads (G.R. Bauchan, personal communication). *Medicago scutellata* is occasionally used as an ornamental cover for dry, sunny banks. In Victorian times, the plant was popular in gardens under the name "snails" (for its snail-like fruits). Ceclinai et al. (1997) investigated the use of trypsin inhibitors from the seeds as an inhibitor of pest insects. An anticarcinogenic "Bowman-Birk" protease inhibitor has been obtained from the seeds (Catalano et al. 2003).

Chromosome Number

$2n = 14$: Eraç and Tokluoglu 1983. This count was proven erroneous (Small et al. 1993).

$2n = 16$: Ghimpu 1928; Fernandes et al. 1977; Baltisberger et al. 2006.

$2n = 28$: Abdelguerfi and Guittoneau 1979; Abdelguerfi et al. 1988*b*.

$2n = 28, 32$: Colombo et al. 1980.

$2n = 30$: Bauchan and Elgin 1984 (72 accessions evaluated); Mariani and Falistocco 1991; Mariani et al. 1996; Karadağ and Gülcan 1997.

$2n = 16, 32$: Fernandes and Queirós 1978.

$2n = 32$: Ghimpu 1928, 1929*a*, 1929*b*, 1930; Fryer 1930; Senn 1938; Heyn 1956; Clement 1962; Simon and Simon 1965; Bolkhovskikh et al. 1969, p. 309 (2 reports); Dahlgren et al. 1971; Nilsson and Lassen 1971; Lesins and Lesins 1979; Agarwal and Gupta 1983*b*; Zhang et al. 1993; Oberprieler and Vogt 1996*b*.

The $2n = 32$ counts may be inaccurate; Bauchan and Elgin (1984) pointed out that the large chromosomal satellites that are present in this species may have led to "the erroneous chromosome counts of $2n = 32$".

Genetic Relationships and Gene Transferability

Intraspecific

Hybridization between early- and late-flowering types of *M. scutellata* is possible. Hybrids are intermediate in plant form for total plant height, number of flowers per inflorescence, and number of serrations per leaflet, and the F_2 generation was observed to segregate 1:2:1, suggesting a single gene trait with incomplete dominance (Bauchan et al. 1986).

Interspecific

Despite using embryo rescue techniques, Elgin et al. (1977) and Bauchan (1987) were unable to hybridize *M. scutellata* with *M. sativa*. Oldenmeyer (1956*c*) was able to generate pods when the two species were crossed, but these aborted. Sangduen et al. (1982) succeeded in generating a hybrid between *M. sativa* ($2n = 32$) and *M. scutellata* ($2n = 30$). The hybrid was a mixoploid (see Fig. 6-1), with shoots differing in chromosome number, and had very low pollen stainability and would not set seed. Piccirilli and Arcioni (1992*a*, *b*) also attempted to produce hybrids between *M. scutellata* ($2n = 30$) and *M. sativa* ($2n = 32$); they obtained viable embryos, but were unable to mature these into viable plants using embryo rescue techniques.

Exchange by advanced biotechnology

Thomas et al. (1990) fused protoplasts of *M. scutellata* and *M. sativa*, and generated plantlets from resulting calli that proved to be nuclear hybrids. Mizukami et al. (2006) also produced somatic hybrids of *M. scutellata* and *M. sativa*. See Chapter 97 for information on somatic hybrids between *M. truncatula* and *M. scutellata*.

Conservation Status

This species is widespread and does not seem to have developed the considerable variability observed in most of the other common annual *Medicago* (Heyn 1963). It is also well represented in germplasm banks. Accordingly, it does not seem in need of conservation measures.

Germplasm Sources

Almost 700 accessions of *M. scutellata* are listed in the germplasm institutions with large *Medicago* collections (see Chapter 3).

Research Problems and Priorities

Medicago scutellata is an established crop with proven value as forage. The commercial market remains interested in the species for development in the future, and there are no obvious needs that need to be addressed.

Representative Collections

Algeria: Meyer, A., environs d'Alger, cultures du Plateau du Hamma, 21-V-1884, in flower & fruit, Société Dauphinoise 1885 No. 4498 (BM, P). **Balearic Islands:** Bowden, L.F. & Sims, P.A. 1982, Ibiza, Santa Eulalia del Rio, south of town, in long grass, occasional, 12-V-1968, in fruit (BM). Ferguson, L.F. & I.K. 21131, Mallorca, Puerto Pollensa, north of Trl road from Puerta Pollensa to El Villa, arable land, 2-VII-1968, in fruit (BM). **Cyprus:** Cassey, E.O., Kyrenia, alt. 100 feet, on cultivated olive grove, 25-III-1911, in flower & fruit (K). Hansen, A. 503, fields along the Nicosia-Famagusta road, south of Kythrea, 25-III-1970, in flower & fruit (K). **France:** Huet, A., Prudet près Toulon, champs, sur le grès rouge, 13-VI-1871, in fruit, F. Schultz Herbarium Normale Nov. Ser. Cent. 1 No. 39 (P). Réchin, Hérault, Lespignan, champs humides, 21-V-1892, Société Rochelaise 1897 No. 4055 (P). **Greece:** Atchley, S.C. 1129, Crete, Phaestos, alt. 600 (sic), dry hillsides, waste ground, V-1932, in fruit (K). Guiol, F., Argolis, pr. Argos, IV-1931, in flower & fruit (BM). Rechinger, K.H. 16807, insula Euboea septentrionalis, Hagia Anna, ad Versuras, 1-VI-1955, in fruit (K). Reverchon, E., Crete, La Cannée, moissons, 5-V & 2-VII-1883, in fruit, E. Reverchon – Plantes de Crete 1883 No. 3 (P). **Israel:** Davis, P.H. 4223, near Waldheim (d. Haifa), dry bushy places in *Styrax* association, 28-III-1942, in flower & fruit (K). Davis, P.H. 4338, Bingamina Kabara, dry slopes of Carmel facing the sea, 12-IV-1942, in fruit (K). **Italy:** Bicknell, C. & Pollini, L., Liguria occid., S. Remo, frequens in arenosis aridis, olivetorum prope urbem, alt. 150 m, VI-1908, in flower & fruit, Flora Italica Exsiccata Series II Fiori & Béguinot No. 1069 (BM, K). Davis, D. & Suton, S. 63265, Sicily, Caltanisetta, ca. 20 km south of Piazza Armerina, alt. 100 m, banks over limestone rocks, 17-V-1979, in fruit (BM). Davis, D. & Suton, S. 64187, Sicily, Messina, ca. 3 km southeast of Cesarò, alt. ca.1000 m, eroded bank at edge of arable field, 31-V-1979, in fruit (BM). **Lebanon:** Bornmüller, J. & F., ad Berytum (Beirût), ad fluvium Nahr-Beirût, 5-V-1910, in fruit, Bornmüller Iter Syriacum II (1910) No. 11631 (BM) (identified as *M. rugosa*). **Portugal:** Matos, Y. 3696, Lisboa, in agris pr. Stadium, 6-V-1951, in fruit (BM). Scarlett 235, Lisbon area, sea level and upwards, cultivated places, 1935, in fruit (K). **Spain:** Bourgeau, E. 696, Pyrenées Espagnoles, Barcelone, à St. Pédro, 11-V-1847, in fruit (K). Hellman, E. & Hubbard, C. 78, 1 mile northwest of Cordova, alt. 400 feet, in sandy fields, 4-V-1924, in fruit (K). Sennen, F. 1641, Catalogne, Massif du Tibidabo vers Espluga, 24-IV-1913, in flower & fruit (BM). **Syria:** Bertschinger, A., Harasta bei Damaskus, Getreidefeld, alt. 1000 m, 16-V-1937, in fruit (G). Peyron, E. 815, Gadin, segetal, 26-IV-1884, in fruit (P). **Tunisia:** Barratte, G., entre le Belvédère et le Barde, champs, 17-IV-1888, in fruit (P). Cosson, M.M.E. et al., Telourba près Tunis, 6-V-1883, in fruit (P). **Turkey:** Guiol, F., Asia Minor, Ins. Chios, IV-1930, in fruit (BM). **Yugoslavia:** Hauffner, Croatia, Toplitz, 1842, in fruit (K).

88. *Medicago secundiflora* Durieu

M. secundiflora Durieu in Duchartre, Rev. de Bot., Recueil Mens., 1: 365 (1845–1846), (basionym).

The epithet *secundiflora*, based on the Latin *secundus* (one meaning of which is one-sided), plus *flora* (flowers), refers to the secund arrangement of flowers, i.e., on one side of the inflorescence.

Subordinate Taxa

None recognized here.

Type Specimen

Durieu, M. (Durieu de Maisonneuve, M.C.), Vallée du Rummel supérieur à Constantine, Herbier de la Commission scientifique de l'Algérie, 4-V-1840, in fruit (P!; also viewed online at http://www.aluka.org; current June 2010).

Synonymy

Homotypic

M. secundiramea Durieu ex Willk. in Willk & Lange, Prod. Fl. Hispan. 3: 417 (1877), nom. nud.

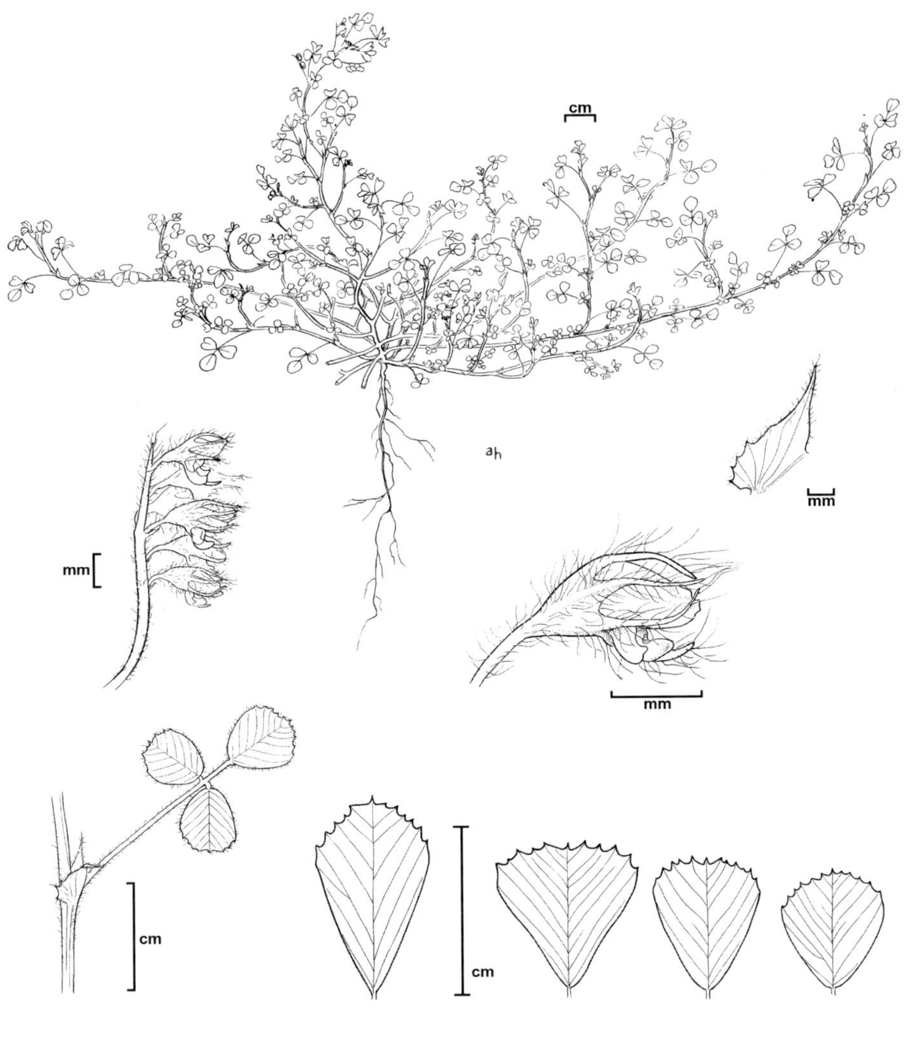

Fig. 88-1. *Medicago secundiflora.* **Clockwise from top:** Plant; stipule; side view of flower; leaflets; leaf attached to stem; flowers on peduncle. (Based on E. Small M339, DAO; artist: A. Hanes.)

M. lupulina L. subsp. *secundiflora* (Durieu) Bonnier & Layens, Fl. Fr. 74 (1894).

Medica secundiflora (Durieu) Bubani, Fl. Pyren. 2: 473 (1900).

M. lupulina L. var. *secundiflora* (Durieu) Firoi, Nuova Fl. Analit. Ital. 1: 828 (1925).

Medicula secundiflora (Durieu) Holub, Folia Geobot. Phytotax. 18(2): 204 (1983).

Heterotypic

No names found.

Vernacular Names

English: Island medic (medic is also spelled medick).

French: Luzerne à fleurs unilatérales.

Geographical Distribution

(N = native, I = introduced, U = uncertain status)

Africa: Algeria (N), Morocco (N), Tunisia (N).

(For Libya, according to Jafri (1980), "reported by Pampanini (Pl. Trip. 146. 1914) from Masellata, Garian etc. but needs confirmation from our area".)

Australasia: Australia (I, according to Castroviejo et al. 2001)

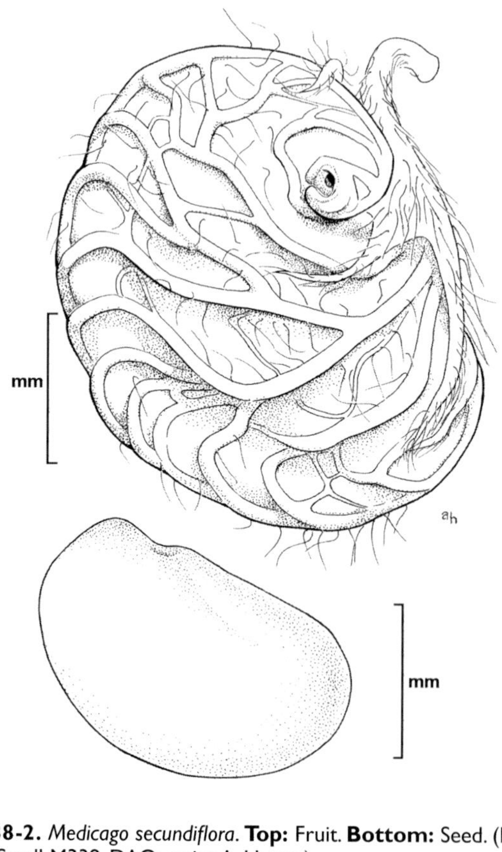

Fig. 88-2. *Medicago secundiflora.* **Top:** Fruit. **Bottom:** Seed. (Based on E. Small M339, DAO; artist: A. Hanes.)

Europe: France (N), Italy (including Sicily; U), Spain (including Balearic Islands; N).

Northern America: USA (I, according to Castroviejo et al. (2001), but not established).

This species is native to North Africa, and is both native and adventive in southern Europe.

Description

Annual herb, branches (2)10–25 cm long, prostrate, procumbent, ascending, or upright, branches arising from base in larger plants, pubescent with simple trichomes producing a grayish or whitish appearance. Leaflets 5–15 mm long, 4–10 mm wide, obovate, oboval, or oblong, pubescent on both surfaces, apical quarter of margin serrate. Stipules narrowly triangular or lanceolate-subulate, entire or toothed at base. Inflorescence of 3–12 flowers in a loose, secund raceme, the peduncle longer than the subtending petiole. Flowers 1.5–2.5 mm long. Calyx campanulate, 1.8–2 mm long, the subulate teeth longer than the tube. Corolla yellow, slightly longer than calyx. Young pod densely covered with long white hairs, the tip protruding through calyx during development and becoming a small coil. Mature pod 2.5–4.0 mm long, somewhat pubescent with simple hairs, grayish with dark anastomosing veins running obliquely from the ventral suture towards the dorsal suture where they curve in a reverse direction. The single seed brownish, reniform (convex on cotyledon side, almost straight on radicle side), 1.5–2.5 mm long, ca. 1.5 mm wide, brown yellow, smooth, radicle somewhat more than half seed length.

Identification Hints

The only species similar to *M. secundiflora* is the much more common and widespread *M. lupulina*, which is easily distinguished by the lack of a secund inflorescence or infructescence (flowers or fruits on one side of the subtending axis).

Habitat and Environmental Adaptation

Habitat

Medicago secundiflora occurs on calcareous slopes, rocky areas, hills, and in dry pastures.

Phenology

The species flowers in late spring and early summer.

Soil

In Algeria, Abdelguerfi et al. (1988*a*) observed that *M. secundiflora* occurs in medium- to fine-textured soils. Nègre (1959) recorded the species in North Africa as growing on calcareous slopes. The plant has often been collected on Mediterranean beach sand.

Temperature

Medicago secundiflora in North Africa withstands cool temperatures in early spring during germination and the hot temperatures of a Mediterranean climate during seed maturation in early summer.

Water

In Algeria, Abdelguerfi et al. (1988*a*) observed that *M. secundiflora* does not show any evident association with annual rainfall. In North Africa, the species has been collected from dry sites, or at least locations that are drying rapidly during seed maturation.

Biotic Relationships

Pollination

Small (1988) noted that the small flowers and pollen to ovule ratio (149) are consistent with *M. secundiflora* being largely self-pollinated. This species is basically an inbreeder.

Dispersal

There are no obvious morphological adaptations of the fruits for seed dispersal. Since the pods are one-seeded achenes, they have become the essential propagules. The species does not have a large distribution area, but this is unlikely to be due to limitations of adaptations for seed dispersal, as evidenced by the worldwide distribution of *M. lupulina*, which also produces achenes.

Herbivory

No information located.

Nodulation and mycorrhizae

No information located.

Disease resistance

No information located.

Economic History, Cultivars, and Uses

Medicago secundiflora has not attracted economic development.

Cultivars

No cultivars have been bred.

Forage

Medicago secundiflora is not sown for forage.

Fodder

This species is not harvested as dried livestock feed. Given its limited frequency, it would not contribute much to grazing on public lands.

Green manure, rotation crop, ground cover

No information located.

Other uses (salad sprouts, ornamental, plant products, etc.)

No information located.

Chromosome Number

2*n* = 16: Lesins and Lesins 1965; Abdelguerfi and Guittoneau 1979; Abdelguerfi et al. 1988*b*; Schlarbaum et al. 1989.

Genetic Relationships and Gene Transferability

Intraspecific

Artificial hybridization of different populations has not been reported, so it is not known whether there are barriers to hybridization within the species.

Interspecific

Lesins and Lesins (1979) were unable to cross *M. lupulina* with its apparent closest relative, *M. secundiflora* (but as noted in Chapter 10, molecular studies indicate that the species are not closely related).

Exchange by advanced biotechnology

No information located.

Conservation Status

This species is relatively rare and, given its occurrence in North Africa where biodiversity and habitat loss are a concern, a conservation assessment is desirable. *Medicago secundiflora* has protected status in regions of France: Languedoc-Roussillon —

http://www.gievpe.org/reglementationfrancaise_loiregionale_languedoc_roussillon.htm; current July 2007) and in National Bassin Parisien —http://cbnbp.mnhn.fr/cbnbp/especeAction.do?action=maille5&cdNom=107713 (current July 2010).

Germplasm Sources

Only four accessions of this species have been listed in the major *Medicago* germplasm collections (see Chapter 3).

Research Problems and Priorities

The small size of *M. secundiflora* suggests that it is unlikely to offer much forage potential, although it may well harbor genes of value for improvement of other species and it deserves to be examined from this perspective.

Representative Collections

Algeria: Balansa, B. 374, Saïda, sur la pente des ravines, 24-V-1852, in flower & fruit (K, P). Balansa, B., Sèrkra, gravier de l'oved-Sèrkra, 21-IV-1853, in flower & fruit (P). Boissier & Reuter, Tlemsen, 1849, in flower & fruit (K). Cosson, E., Constantine, 16-V-1853, in flower & fruit (P). Cosson, E., Lingad cercle de Batna, prov. de Constantine, 14-VI-1853, in fruit (P). D'Alleizette, Ch., Gued Zèrsati, sommet à la pépinière, 16-V-1910, in fruit (P). Davis 52556, halfway between Setif and Kherrata, alt. 1000 m, shaley hillsides (herbaceous flora), 2-V-1971, in flower & early fruit (BM). Letourneux, A., Seybouse près de Bône, sables, 11-IV-1857, in fruit, Fragmenta Florae Algeriensis Exsiccata No. 126 (BM, K, P). Maire, R., Monts de Djelfa, rocailles calcaires dans les forêts de *Pinus halepensis* près Aïn-Bahrara, alt. 1200–1300 m, 16-V-1921, Société Française 1921 Exsicc. Ch. Duffour No. 3734 (P). Reuter, prov. Oran, in montibus supra Tlemsen, IV-1849, in flower & fruit (K, P). **France:** Gautier, C., Ile de la Sidrière de Fitou, dans l'étang de Leucate, près de Narbonne (Aude), nouvelle pour la flore d'Europe, 15-V-1882, in fruit (P). Gautier, G. 2019, Ile de la Sidrière de Leucate, dans l'étang de Leucate, près Narbonne (Aude), sur les sables maritimes, 7-V-1878, in flower & fruit (K, P). Raynal, J. 4194, Juvignac (Hérault), 600 m. au S du Mas de Neussargues, pelouse à *Brachypodium ramosum, Thym. vulg.* et *Asphodel. cerasites* sur calcaires dolomitiques jurassiques, locus novus (intr. par moutons d'Afr. du N probable), 25-IV-1959, in flower & fruit (P). **Libya:** Maire, R. & Weiller, M. 427, Tripolitania, prope Garian in rupestribus calcareis humosis, alt. 600 m. 1-V-1938, in fruit (P). **Spain (Balearic islands):** Ferrer, P. (Cabrera) 122, Baleares, ferra de la Pleta d'es Castell, sobre las peñas meridionales, 10-V-1948, in flower & fruit (G). **Tunisia:** Cuenod, A., Di Rocar (?), 3-IV-1910, in late flower & fruit (G). Pitard, C.J. 97, Djebel-Djeloud, in collibus aridis, V-1907, in flower (G).

89. *Medicago shepardii* Post ex Boiss.

M. shepardii Post ex Boiss., R. Buser, Fl. Orient. Suppl. 163 (Oct. 1888); (also) Post, J. Linn. Soc., Bot. 24: 425 (Dec. 1888).

According to Post (1888), "the name was given in honour of Mrs. Shepard, M.D., an enthusiastic and careful collector." Index Herbariorium, Part II(6), Collectors (1986) lists an F. Shepard with collections in Turkey from 1882–1884. The epithet is often misspelled "*shepardi*".

Subordinate Taxa

None recognized here.

Type Specimen

G. Post, Turkey, Gaziantep, Aintab (Gaziantep), Turkey, VI-1884 (Herb. Boiss. G!, isotype! in general collection at G).

Synonymy

Homotypic

M. tornata (L.) Mill. var. *shepardi* (sic) (Post ex Boiss.) Heyn, Scripta Hierosolymitana 12: 90 (1963). (Post described the taxon in 1896.)

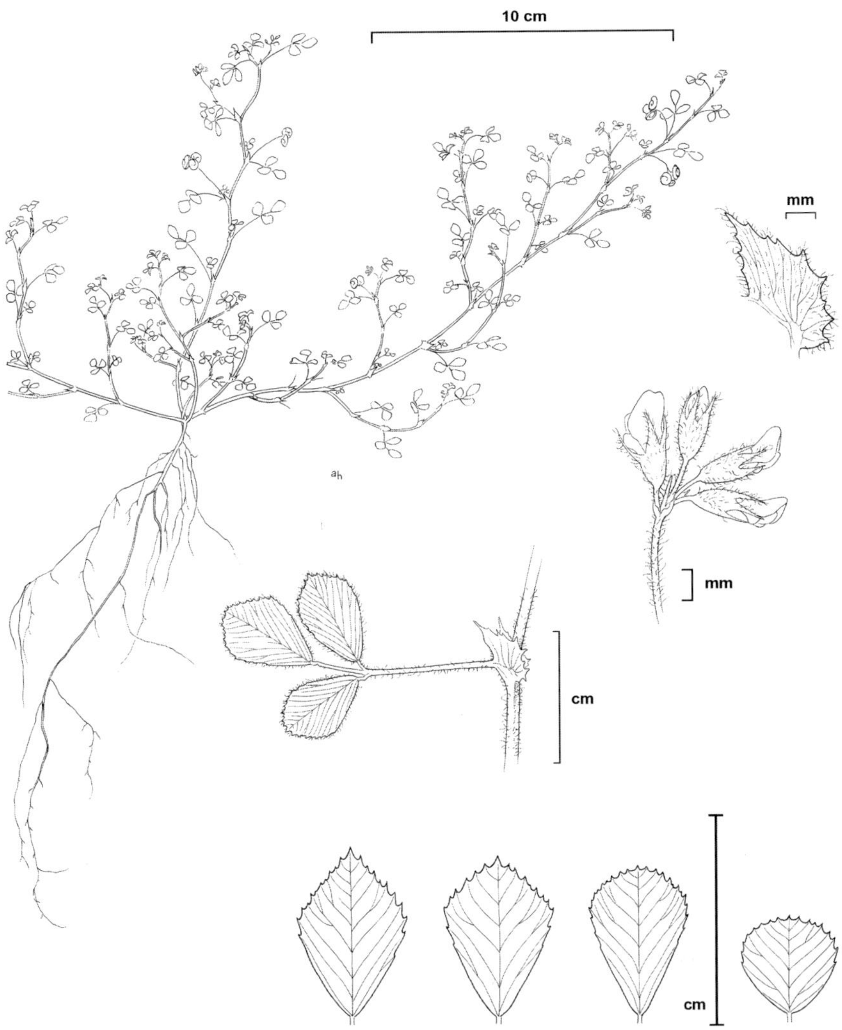

Fig. 89-1. *Medicago shepardii.* **Clockwise from upper left**: Plant; stipule; group of flowers on a peduncle; leaflets; leaf attached to stem. (Based on E. Small M340, DAO; artist A. Hancs.)

Heterotypic

No names found.

Vernacular Names

ENGLISH: No names found.

FRENCH: No names found.

Geographical Distribution

(N = native, I = introduced, U = uncertain status)

ASIA: Turkey (N).

This species has been collected around Gaziantep (formerly Antep or Aintab) and near Ankara, both in south-central Turkey.

Description

Annual herb, stems decumbent to ascending, 15–30 cm long, pubescent with simple hairs. Leaflets 5–10 mm long, 2–6 mm wide, obovate-cuneate, apical third of margin serrate. Stipules lanceolate, nearly entire or with a few small teeth at base. Inflorescence of 2–3(6) flowers, the peduncle longer than the subtending petiole. Flowers ca. 2.5–4 mm long. Corolla yellow, less than twice length of calyx. Calyx pubescent with simple hairs, teeth shorter than or as long as tube. Young fruit contracted within calyx, then turning sideways through it. Mature fruit discoid, spineless, pubescent to glabrescent, with ca. 1.5 coils, 3–5 mm in diameter, the coil face with numerous prominent radial veins that branch and anastomose slightly and enter a submarginal vein that is distant from the dorsal suture, the dorsal suture bearing very short, elevated, wing-like protrusions. Seeds yellow to brown, ca. 3 mm long, 2 mm wide, 1–2(3) per pod, separated by fruit partitions.

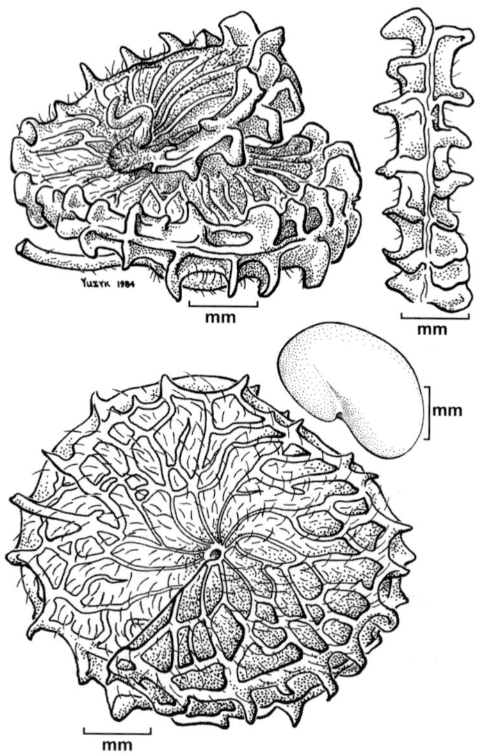

Fig. 89-2. *Medicago shepardii.* **Clockwise from upper left:** Fruit; coil edge; seed; fruit coil face showing venation. (Based on E. Small M340, DAO; artist: L. Yuzyk.)

Identification Hints

The fruits of *M. shepardii* have wing-like protrusions on the coil margins, reminiscent of *M. rugosa*, but are hairy, and the leaves lack the gland-tipped trichomes of the latter species.

Habitat and Environmental Adaptation

Habitat

Medicago shepardii has been collected from roadsides. Presumably in presettlement times it was generally found in open, disturbed habitats.

Phenology

This species flowers in its native area from May to July.

Soil

The species has been collected in rocky soils of volcanic origin.

Temperature

Akan et al. (2005) noted that the annual mean temperature of the native area of the species is 18.7 °C, the maximum mean temperature in July is 46.8 °C, and the minimum mean temperature in February is –6.8 °C. *Medicago shepardii* completes its life cycle between the extreme temperature months.

Water

Akan et al. (2005) observed that the area occupied by *M. shepardii* has a semi-arid Mediterranean climate with an annual rainfall of about 458 mm.

Biotic Relationships

Pollination

Small (1988) noted that the small flowers and pollen to ovule ratio (270) are consistent with *M. shepardii* being largely self-pollinated. This species is basically an inbreeder.

Dispersal

There are no obvious morphological adaptations for seed dispersal in this species. The pod lacks spines serving to attach it to fur or feathers, a dispersal feature of many other annual species of *Medicago*.

Herbivory

No information located.

Nodulation and mycorrhizae

No information located.

Disease resistance

No information located.

Economic History, Cultivars, and Uses

This species is very rare and has not attracted agricultural development.

Cultivars

No cultivars have been bred.

Forage

Medicago shepardii is not sown for forage.

Fodder

This species is not employed for dried livestock feed.

Green manure, rotation crop, ground cover

No information located.

Other uses (salad sprouts, ornamental, plant products, etc.)

No information located.

Chromosome Number

$2n = 16$: Lesins and Lesins 1963*a*.

Genetic Relationships and Gene Transferability

Intraspecific

Artificial hybridization of different populations has not been attempted, so it is not known whether there are barriers to hybridization within the species.

Interspecific

Lesins and Singh (1973) found that *M. shepardii* could not be crossed with *M. italica*.

Exchange by advanced biotechnology

No information located.

Conservation Status

Akan et al. (2005) commented about the area where *M. shepardii* occurs that "The vegetation of the area has been extensively destroyed by anthropogenic effects... the populations of the rare plants should be examined scientifically, and new reports should be prepared for the conservation of plant species." These authors noted that according to Red Data Book categories of the 2000–2001 period, *M. shepardii* had a rating of VU ("vulnerable").

Germplasm Sources

Twelve accessions are listed among the germplasm institutions with large *Medicago* collections (see Chapter 3).

Research Problems and Priorities

Medicago shepardii is a very rare species that has received very little study to date. The first priority is to collect seeds for deposition in germplasm institutes, and for both basic and agricultural studies of the species.

Representative Collections

Turkey: Post, Spring-1891, with the annotation "pour Prof. Urban, Berlin, avec les compliments de George E. Post" (general collection, in type folder, G).

The following collections were cited at a Turkish website (SÜRDÜRÜLEBİILİIR KIRSAL VE KENTSEL KALKINMA DERNEĞİ, KARACADAĞ BİITKİ ÇEŞ İTLİLİĞİ, Doç. Dr. Selçuk ERTEKİIN, Şubat 2002 – Diyarbakır; http://www.surkal.org.tr/tr/karaca.htm; current July 2007): **C7** Diyarbakır: Ergani, Diyarbakır-Siverek, 40 km, 1050 m, taşlık step, 10.5.2001. Diyarbakır-Siverek, 43 km, 1000 m, step, 23.4.2001. Diyarbakır-Siverek, 48 km, 1090 m, taşlık step, 10.5.2001. Küçük Besrek tepesi, 1050 m, kayalık yamaçlar, 23.4.2001. Elvanlı köyü çevresi, 900 m, mera, 15.5.1998. Şanlıurfa: Siverek, Karabahçe-Diyarbakır, 2 km, 1200 m, mezarlık, 31.5.2001. Diyarbakır-Siverek, 78 km, 900 m, taşlık step, 23.4.2001. Siverek-Diyarbakır, 10 km, 850 m, çayır, 23.5.1998. **C8** Diyarbakır: Çınar, Ovabağ yol ayrımı-Ovabağ, 10–12 km, 750 m, taşlık step, 29.4.2001. (**C7** and **C8** are geographical subdivisions of Turkey recognized in the multivolume *Flora of Turkey and the East Aegean Islands*, Edited by P.H. Davis, Edinburgh University Press, UK)

Akan et al. (2005) cited the following collections: Above Aş**l**kköy, 700–800 m, 23.05.2002, steppe, Akan 2017 et al. Above Uyuzpinar K., 600–800 m, 31.05.2002, steppe, Akan 2054 & 2083 et al.

90. *Medicago sinskiae* Uljanova

(a species of uncertain status)

M. sinskiae Uljanova, Novosti Sist. Vyssh. Rast. (= Novit. Syst. Pl. Vasc. Acad. Sci. URSS= "Plant Taxonomy News") 1: 175 (1964), (basionym).

The epithet *sinskiae* commemorates Evgenija Nikolaevna Sinskaya (1889–1965), a Russian botanist and agronomist, noted in part for her work on *Medicago*.

Note

Small and Brookes (1991) concluded that *M. sinskiae* deserves recognition as a separate species. However, this was based only on the type gathering and material subsequently studied from plants grown from it. From this material, I provided samples for the DNA-based phylogenetic study of *Medicago* species by Downie et al. (1998; see Chapter 10). Note the extracted portion of the dendrogram from this study (Fig. 90-3). This shows that at least the type gathering of *M. sinskiae* is distant from *M. rigidula* and *M. rigiduloides*, the two closest species (indeed, many other quite distinctive species of *Medicago* proved to be closer to each other in the Downie et al. study). As indicated below, the type location is in an area of extremely high endemism, a consideration supportive of the concept that "*M. sinskiae*" might be a very localized

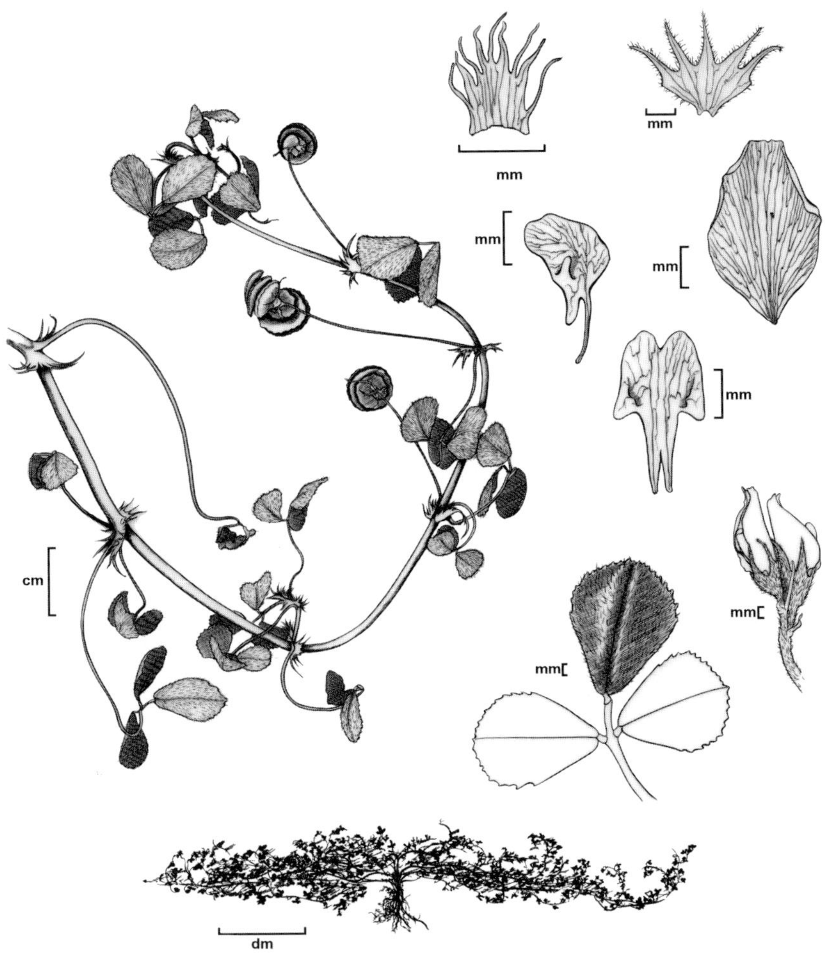

Fig. 90-1. *Medicago sinskiae.* **Clockwise from left:** Fruiting branch; floral parts (clockwise from top left: flattened androecial column (split beside free stamen, which is not shown, anthers also not shown); calyx (split and flattened); standard petal; keel (the two petals separated adaxially and flattened); wing petal); inflorescence with peduncle; leaf; habit drawing of plant. (Based on a greenhouse-grown plant (voucher, E. Small M1578, DAO), grown from seeds from the holotype obtained from the *Medicago* curator, LE; artist: M. Jomphe.)

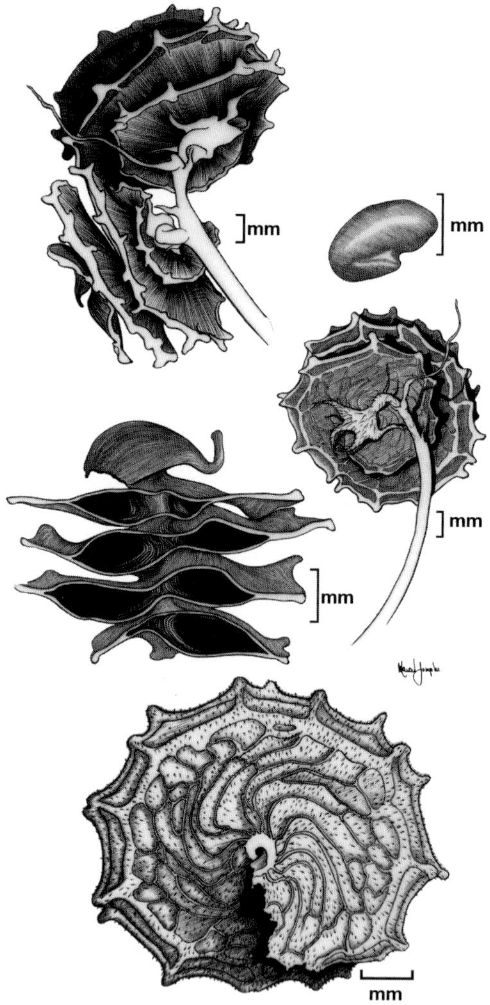

Fig. 90-2. *Medicago sinskiae.* **Clockwise from upper left:** Infructescence; seed; pedicel view of fruit; fruit coil face showing venation; sagittal section of fruit showing seed cavities. (Based on a greenhouse-grown plant (voucher, E. Small M1578, DAO), grown from seeds from the holotype obtained from the *Medicago* curator, LE; artist: M. Jomphe.)

endemic species. Nevertheless, I have not had the opportunity to study populations that others (cited below) have contended are referable to this taxon, and therefore I am now uncertain of what status it deserves. Below, differences are indicated (based on Small and Brookes 1991) between *M. sinskiae* and other species of section *Spirocarpos* subsection *Pachyspirae*. This information on how to discriminate *M. sinskiae* from other species pertains only to the type gathering, and is inappropriate for distinguishing collections referred to this taxon by others (cited below) from *M. rigidula* and *M. rigiduloides*. Because of this uncertainty, *M. sinskiae* is not included in the key to *Medicago* (Chapter 11), where it would key out to *M. rigidula* or *M. rigiduloides*.

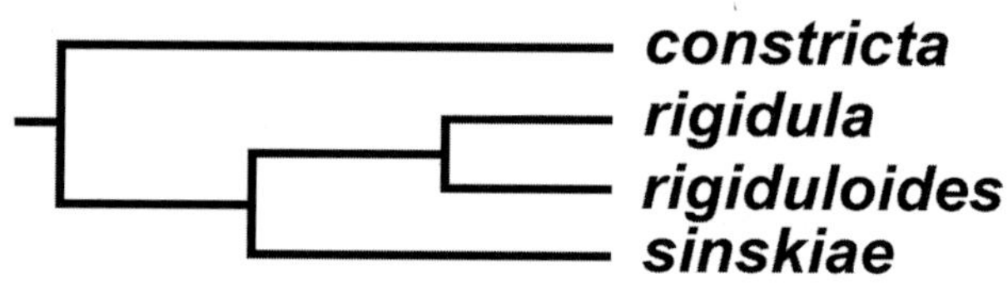

Fig. 90-3. Portion of dendrogram from Downie et al. (1998) showing relationships of *M. sinskiae*.

Subordinate Taxa

None recognized here. Mehregan et al. (2002), based on material collected in Iran, treated *M. sinskiae*, *M. constricta*, *M. rigidula*, and *M. rigiduloides* as one species — *M. rigidula*, recognizing the first two as subspecies of *M. rigidula*. They reported that the material they studied from Iran (most of which I have not had the opportunity to study) included forms with spines up to 3.5 mm long that were usually hooked at the apex, as well as forms with spineless pods (that of the type specimen from Turkmenistan has only tubercles on the pods). Mehregan deposited one of his collections in DAO (Mehregan, I., 445 m (alt.), 70 km Andimeshk to Khorram Abad, Luristan, southwestern Iran, 28-3-2002); I have been unable to distinguish this spiny-fruited collection from *M. rigiduloides* and therefore remain uncertain whether any of the material Mehregan analyzed is conspecific with *M. sinskiae*. As noted below, Khassanov (1972) recognized (but not by valid publication) a spiny-fruited variety.

Type Specimen

Uljanova in 1961 (LE!; see Fig. 90-4).

Synonymy

Homotypic

M. rigidula (L.) All. subsp. *sinskiae* (Uljanova) I. Mehregan & M.R. Rahiminejad, Iran J. Bot. 9: 214 (2002).

Heterotypic

(Reference material has not been examined, and the following nonstatus names of Khassanov are listed provisionally as it is unclear whether the material he had in mind is conspecific with *M. sinskiae*.)

M. sinskiae var. *longispina* Khassan. & M.N. Abdull. in Khassanov, O. Kh., Dikorastushchie lyutserny Srednei Asii. Tashkent, p. 94 (1972). (nom. nud., Latin description not provided. Characterized as having thicker coils and longer spines.)

M. sinskiae var. *vassilczenkoi* Khassan. & M.N. Abdull. in Khassanov, O. Kh., Dikorastushchie lyutserny Srednei Asii. Tashkent, p. 94 (1972). (nom. nud., Latin description not provided. Characterized as having tubercles instead of spines.)

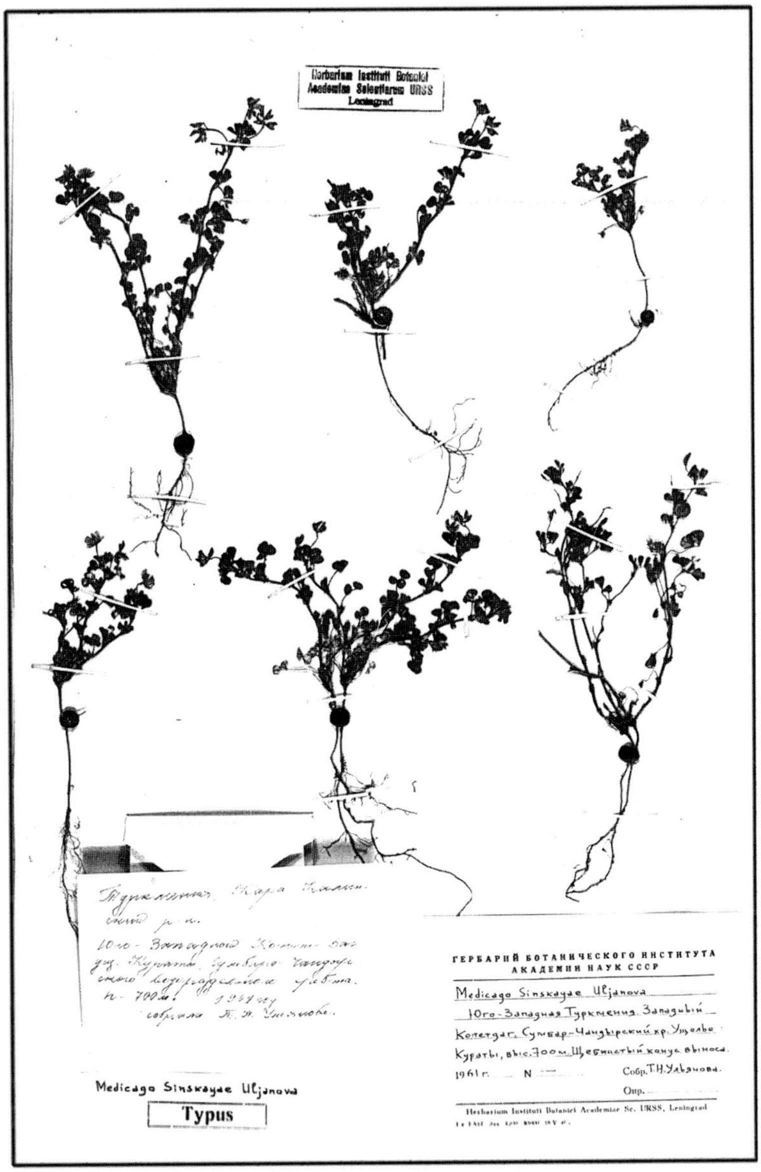

Fig. 90-4. Holotype collection of *Medicago sinskiae* Uljanova at LE. Translation of the label information from the Russian: Southwestern Turkmenia, Karal-Kalin, 10th western Kopet-Dag, Kuraty Canyon of the Sumbar-Chandyrskii watershed ridge, river slope, debris cone, altitude 700 m, collected 1961 by T.N. Uljanova.

M. sinskiae var. *sinskiae* Khassan. & M.N. Abdull. in Khassanov, O. Kh., Dikorastushchie lyutserny Srednei Asii. Tashkent, p. 94 (1972). (comb. inadmiss (this is the autonym). Characterized by lack of spines and presence only of barely visible tubercles.)

Vernacular Names

English: No names found.

French: No names found.

Geographical Distribution

(N = native, I = introduced, U = uncertain status)

Asia: Iran (N), Turkmenistan (N).

This narrow endemic was collected from the type locality in western Kopet-Dag (Turkmenistan) by Uljanova (1964). Mehregan et al. (2003) claimed to have collected it in Iran (Zagrosian region, approximate latitude 34.5°, longitude 47°), as discussed above. Khassanov (1972) stated (translated from Russian) that the "Presence in neighbouring areas of the Iranian Plateau may be presumed."

Description

Annual herb, more or less pubescent with simple hairs. Stems prostrate to ascending, branched from base, (10)15–25(40) cm long. Mature leaflet blades (3)5–8(10) mm long, cuneate to obovate, margins serrate from tip to as much as halfway to base. Stipules(2)3–4(5) mm long, dentate to laciniate, often dissected one-third to one-half width, with 4–8 teeth including a long terminal tooth. Flowers 1–3(5) per peduncle, 3–6 mm long, calyx pubescent with simple hairs, longer than half corolla length. Corolla yellow or orange-yellow. Calyx turbinate to campanulate, teeth longer than tube, peduncle longer than subtending petiole. Young pod concealed within calyx, covered with whitish hairs. Mature pod ovoid, cylindrical, or discoid, pubescent with both simple and gland-tipped hairs, 4–8 mm long, (4)5–6 mm wide, with 2.5–5 coils, hardened at maturity, some gaps often present between mature coils (coils usually not tightly appressed at maturity), facing surfaces of coils with strongly curved radial veins that anastomose somewhat towards coil periphery, veins obscure at maturity because of developing spongy tissue, edge of coil spiny or spines reduced to tubercles. Seeds 1.75–2.5 mm long, 1–2 per coil, separated by spongy fruit partitions, smooth, yellow to yellow-brown, radicle about half as long as length of seed.

Identification Hints

The following table (from Small and Brookes 1991) distinguishes *M. sinskiae* from each of the other species currently placed in *Medicago* sect. *Spirocarpos* subsect. *Pachyspirae*.

Characters best distinguishing *Medicago sinskiae* from other species of *Medicago* sect. *Spirocarpos* subsect. *Pachyspirae*.

Species and its characters	Contrasting characters in *M. sinskiae*
M. constricta Fruit spherical to ovoid, veins on coil face anastomose weakly.	Fruit short-cylindrical, veins on coil face anastomose strongly.
M. doliata Fruit spherical to ovoid.	Fruit short-cylindrical.
M. italica Fruit rarely with glandular hairs, veins on fruit coil face anastomose weakly. Shoot glabrescent.	Fruit with glandular hairs, veins on fruit coil face anastomose strongly. Shoot pubescent.
M. murex Fruit spherical to ovoid, with a peripheral veinless area on coil face, coil face with strong vein branching, glandular hairs absent. Shoot glabrous. Stipules strongly dissected.	Fruit short-cylindrical, lacking peripheral veinless area on coil face, coil face with weak vein branching, glandular hairs present. Shoot pubescent. Stipules weakly dissected.
M. littoralis Fruit lacking glandular hairs, veins on coil face weakly branched.	Fruit with glandular hairs,veins on coil face strongly branched.
M. sphaerocarpos Distinguished like *M. murex*, above.	
M. rigidula* and *M. rigiduloides Fruit discoid, barrel-like, ovoid or spherical; radial veins on coil face strongly curved, moderately branched;fruits mostly spiny (occasionally tubercular).	Fruit short-cylindrical; radial veins moderately curved, strongly branched; fruits tubercular or spiny.
M. soleirolii Fruit lacking tubercles, lacking glandular hairs. Leaflets with gland-tipped hairs.	Fruit with tubercles, with glandular hairs. Leaflets lacking gland-tipped hairs.
M. syriaca Fruit barrel-like, lacking glandular hairs, veins on coil face weakly branching. Shoot glabrescent. Stipules strongly dissected.	Fruit short-cylindrical, with glandular hairs, veins on coil face strongly branching. Shoot pubescent. Stipules weakly dissected.
M. truncatula Fruit barrel-like, veins on coil face weakly branched, glandular hairs absent.	Fruit short-cylindrical, veins on coil face strongly branched, glandular hairs present.
M. turbinata Fruit coil face with peripheral veinless area and weak vein branching, glandular hairs absent.	Fruit coil face lacking peripheral veinless area, with strong vein branching, glandular hairs present.

Habitat and Environmental Adaptation

Habitat

The type collection of *M. sinskiae* is from a river canyon; the species has apparently also been collected from the moist, cold Zagrosian region of Iran.

Phenology

Khassanov (1972) noted that flowering starts in early April and fruits form from late April to early May, maturing in May and June.

Soil

Khassanov (1972; cf. Khassanov and Abdullazhanov 1973) stated (translated from Russian) that *M. sinskiae* was "encountered on stony and crushed rock, and on silty dry mountain slopes of the Western Kopet-Dag."

Temperature

The species occurs in a continental climate.

Water

The native area occupied is generally dry, especially in late summer.

Biotic Relationships

Pollination

This species is largely an inbreeder (Novoselova 2003).

Dispersal

Like many annual Mediterranean species of *Medicago*, it is likely that there are kinds with and without spines, the latter obviously adapted for wider seed dispersion.

Herbivory

No information located.

Nodulation and mycorrhizae

No information located.

Disease resistance

No information located.

Economic History, Cultivars, and Uses

This species has not attracted agricultural attention to date.

Cultivars

No cultivars have been bred.

Forage

Medicago sinskiae is not sown for forage. However, Khassanov (1972) noted that (at least in some forms) the general absence of spines on the fruits makes the species more attractive for forage than spiny-fruited medics, and his studies showed that *M. sinskiae* is an excellent plant for feeding livestock.

Fodder

This species has not been employed as dried livestock feed.

Green manure, rotation crop, ground cover

No information located.

Other uses (salad sprouts, ornamental, plant products, etc.)

No information located.

Chromosome Number

$2n = 14$: Ibragimova 1999.

$2n = 16$: Small and Brookes 1991 (mistakenly stated as $n = 7$ in one place in the paper).

Genetic Relationships and Gene Transferability

Intraspecific

Artificial hybridization of different populations has not been attempted, so it is not known whether there are barriers to hybridization within the species.

Interspecific

No information located.

Exchange by advanced biotechnology

No information located.

Conservation Status

This is a rare, little-studied endemic, the status of which requires clarification. Given its rarity and occurrence in an area of the world subject to biodiversity loss, it should be assumed to be in need of conservation protection.

The Kopet-Dag region, where the type of *M. sinskiae* was collected, is characterized by foothills, dry and sandy slopes, mountain plateaus, and steep ravines (the Kopet-Dag mountain range is on the frontier between Turkmenistan and Iran, extending about 650 km along the border east of the Caspian Sea). Local vegetation includes riparian forests in the river valleys, woodlands, and grasslands. Very high endemism is exhibited among many groups of organisms, both plants and animals. There are 332 endemic plant species, including many species of Fabaceae. This endemism has been attributed to the area being a site of constant dispersal and probably of local isolation and speciation during the Pleistocene Age (less than one million years ago), as well as during the recent glaciation (16 000 to 10 000 years ago). Although the areas of woodland habitat are under protection, heavy logging, collecting of firewood, and overgrazing are occurring and this has extinguished more than a dozen of the rare endemic plant species (http://www.worldwildlife.org/wildworld/profiles/terrestrial/pa/pa1008_full.html; current July 2010). Thus, there is an obvious need to preserve the habitat of the rare endemic *M. sinskiae*.

Germplasm Sources

Not available at present.

Research Problems and Priorities

Medicago sinskiae is a poorly understood species that has not been available for botanical and agricultural studies. Its relationship to *M. rigidula* and *M. rigiduloides* needs to be clarified. The urgent priorities are to document its population biology and distribution and acquire seeds for deposition in germplasm banks.

Representative Collections

Iran: Mehregan et al. (2002) cite the following: W. Reihani 25165 (TARI, spineless form), Mehregan 2.10 (TARI, spiny-fruited form); the identification of these needs to be confirmed. **Turkmenistan:** Type, cited above.

91. *Medicago soleirolii* Duby

M. soleirolii Duby, Bot. Gall. 1: 124 (1828), (basionym).

The epithet *soleirolii* commemorates Joseph François Soleirol (1781–1863), a French engineering officer and botanist, noted for his collections of Corsican plants.

Subordinate Taxa

None recognized here.

Type Specimen

Calvi, Corsica, *Soleirol* misit, III-1827, ex Herb. Gay 1241 (holotype K, isotype BM).

Synonymy

Homotypic

No names found.

Heterotypic

M. plagiospira Durieu in Duchartre, Rev. de Bot., Recueil Mens., 1: 366 (1845–1846) (type indicated by Heyn 1963, p. 38).

Vernacular Names

English: No names found.

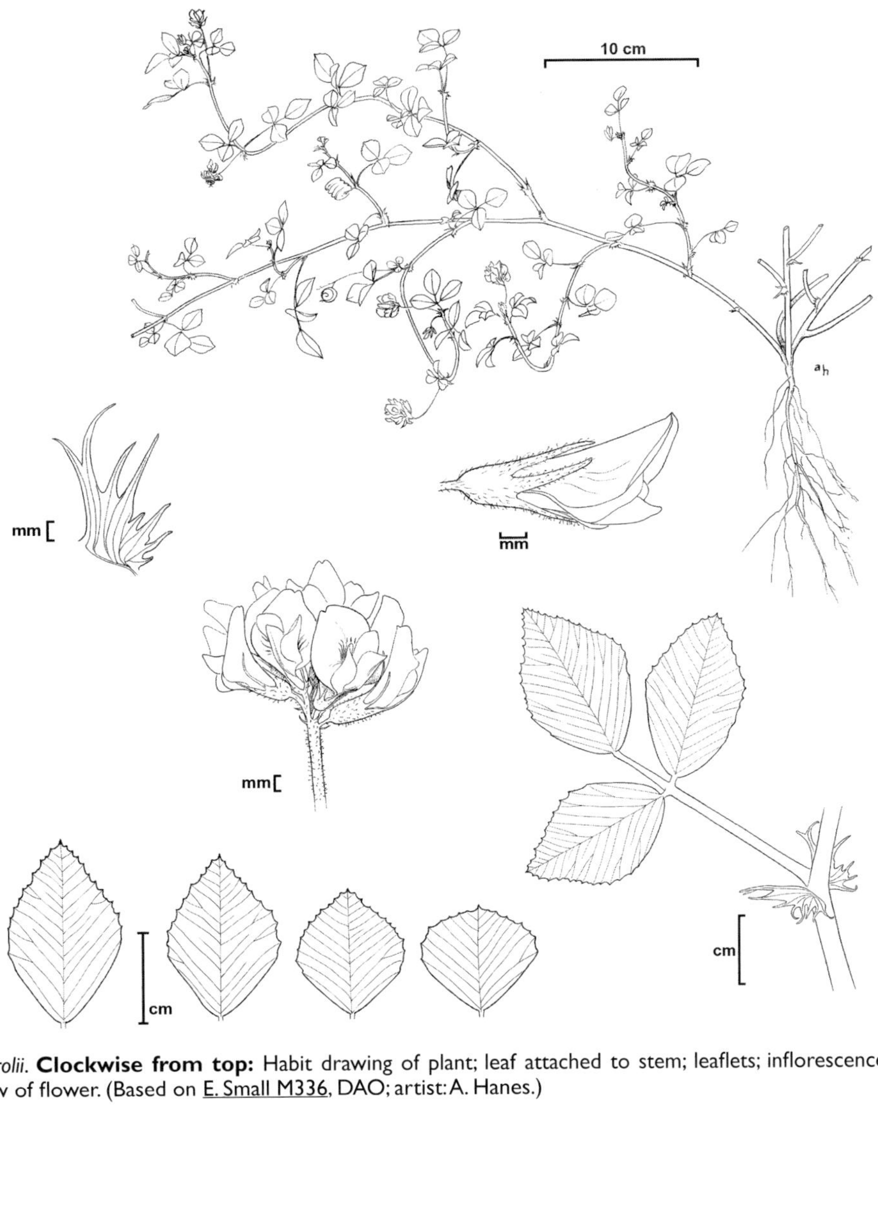

Fig. 91-1. *Medicago soleirolii.* **Clockwise from top:** Habit drawing of plant; leaf attached to stem; leaflets; inflorescence with peduncle; stipule. **Center:** Side view of flower. (Based on E. Small M336, DAO; artist: A. Hanes.)

FRENCH: La luzerne du Soleirol, luzerne de Soleirol.

Geographical Distribution

(N = native, I = introduced, U = uncertain status)

AFRICA: Algeria (N), Tunisia (N).

EUROPE: Balearic Islands (Spain; U), France (including Corsica; U; Jeanmomod and Burdet (1999) commented (translated from French) that "The status of *Medicago soleirolii* in Corsica — whether indigenous or naturalized — remains uncertain but with each new discovery its indigenous nature becomes more likely."), Italy (U; "probably introduced in Italy" according to Castroviejo et al. 2001). (Mistakenly reported as introduced in Crimea.)

Medicago soleirolii is native to the western Mediterranean, occurring mainly in the maritime region of Algeria and Tunisia, but also in France and Italy.

Description

Annual herb, stems ascending or procumbent (rarely erect), 20–50 cm long, usually branched from base. Vegetative aboveground organs covered with simple and glandular hairs, upper surface of leaf usually glabrescent. Leaflets 12–18 mm long, 8–15 mm wide, obovate to ovate (occasionally subrhombic), distal half of margin serrate. Stipules ovate-lanceolate, with margin deeply dentate to laciniate. Inflorescence with (2)3–7 flowers, the peduncle longer than the subtending petiole. Flowers 7–10 mm long. Corolla yellow (orange-yellow also recorded), less than twice length of calyx. Calyx with glandular hairs, teeth slender, subulate-linear, longer than tube. Young pod contracted and concealed within calyx. Mature pod discoid, lenticular, or cylindrical, 2–3 mm long, spineless, glabrous, with (2)3–5(8) coils, middle coils 5–8 mm in diameter, becoming hard at maturity. Coil surface with radial veins anastomosing before entering dorsal suture. Seeds 1–2 per coil, curved and subreniform, 3–5 mm long, smooth, yellow to red-brown, separated by thin fruit membranes.

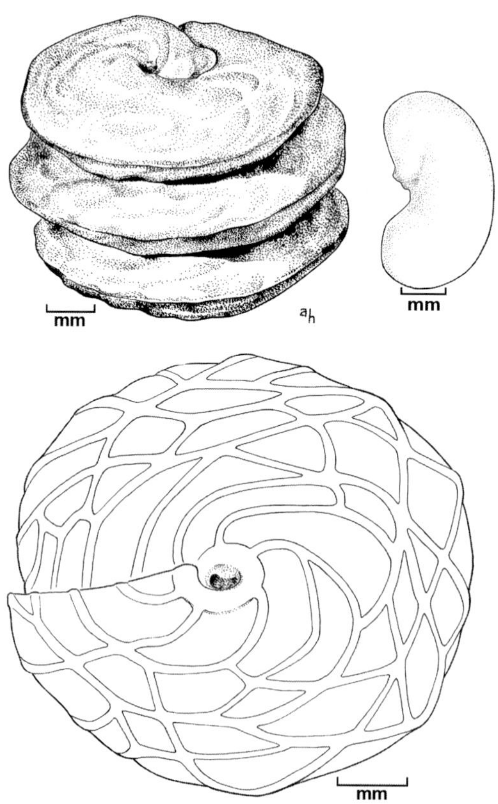

Fig. 91-2. *Medicago soleirolii.* **Clockwise from top left:** Fruit; seed; fruit coil face showing venation. (Based on E. Small M336, DAO; artist: A. Hanes.)

Identification Hints

Heyn (1963) noted that numerous collections of *M. soleirolii* she received were misidentified, and cautioned against uncritically accepting data in the literature concerning this species. *Medicago soleirolii* is easily confused with *M. italica*, but the former is distinguishable by the presence of some gland-tipped hairs on the leaves.

Habitat and Environmental Adaptation

Habitat

Medicago soleirolii has been collected from grassy slopes, prairies, seasides, and ruderal habits.

Phenology

The species flowers from March to May in North Africa, and from April to June in Europe.

Soil

Medicago soleirolii apparently occupies a variety of soils. In Algeria, Abdelguerfi et al. (1988*a*) observed that it occurs in medium- and coarse-textured soils, and Piano and Francis (1992) described it as occupying sands and loams. Lesins and Lesins (1979) collected *M. soleirolii* on sandy soils and fertile hill slopes in Algeria. Species of *Medicago* are generally intolerant of acidic soils, but compared with most other species of the genus, *M. soleirolii* is relatively acid-tolerant (Piano and Francis 1992), although it primarily occupies soils that range from slightly acidic to moderately basic.

Temperature

This species is a typical West Mediterranean winter annual, germinating in later winter and completing its life cycle before the very hot, dry summer period.

Water

Piano and Francis (1992) noted that in its native area the species requires a minimum annual rainfall of 400 mm. In Algeria, *M. soleirolii* occurs in the northeast, the wettest part of the country (Nègre 1959; Abdelguerfi et al. 1988*a*).

Biotic Relationships

Pollination

Small (1988) noted that the low pollen to ovule ratio (154) is consistent with *M. soleirolii* being largely self-pollinated. The species is presumably largely an inbreeder, like essentially all other annual species of *Medicago*, although the exceptionally large flowers suggest that some outcrossing occurs.

Dispersal

No obvious morphological adaptations for seed distribution are present. In particular, the fruits lack spines that would attach the fruits to fur or feathers.

Herbivory

Shade et al. (1979) found that *M. soleirolii* was not resistant to the potato leafhopper (*Empoasca fabae* Harris).

Nodulation and mycorrhizae

USDA (2009) reported positive root nodulation for *M. soleirolii*. Howieson and Ewing (1989) found that *M. soleirolii* was outstanding in its ability to achieve nodulation in acidic soils.

Disease resistance

O'Neill et al. (2003) examined the comparative resistance of 36 annual species of *Medicago* to *Phoma medicaginis* (the cause of spring black stem and leaf spot, a major disease of alfalfa) and concluded that *M. soleirolii* had high resistance to it. Barbetti (2007) identified an accession of *M. soleirolii* with a very high level of resistance to *P. medicaginis*. In a similar comparative test of 36 annual species of *Medicago*, O'Neill and Bauchan (2000) found that one accession of *M. soleirolii* was fairly susceptible to anthracnose caused by *Colletotrichum trifolii,* but a second accession was very resistant. Yaege and Stuteville (2002) tested the resistance of 33 annual species of *Medicago* to powdery mildew (*Erysiphe pisi*) and found that the two collections of *M. soleirolii* grown were susceptible to the disease (it was observed that often only one of a number of populations of other species tested was resistant).

Economic History, Cultivars, and Uses

This species has not attracted agricultural development to date.

Cultivars

No cultivars have been bred.

Forage

Medicago soleirolii is not sown for forage.

Fodder

This species is not harvested as dried feed for livestock.

Green manure, rotation crop, ground cover

No information located.

Other uses (salad sprouts, ornamental, plant products, etc.)

No information located.

Chromosome Number

2*n* = 16: Fryer 1930; Simon and Simon 1965; Lesins and Lesins 1979; Abdelguerfi and Guittoneau 1979; Abdelguerfi et al. 1988*b*; Baltisberger and Widmer 2006.

Genetic Relationships and Gene Transferability

Intraspecific

Artificial hybridization of different populations has not been attempted, so it is not known whether there are barriers to hybridization within the species.

Interspecific

Lesins and Lesins (1979) reported that they crossed *M. soleirolii* with *M. italica* and generated several F_1 plants that were chlorophyll deficient and died within 2 months.

Exchange by advanced biotechnology

No information located.

Conservation Status

This species is uncommon, both in terms of area occupied and frequency of occurrence.

Medicago solerolii is protected in the French regions of Corsica (http://cbnbp.mnhn.fr/cbnbp/protectionAction_IMP.do?action=cdArrete&cdArrete=V94; current July 2010) and Bourgogne (http://perso.orange.fr/bblb/botanica/ListesPlProt/bg.html; current July 2007). The species probably should be given a conservation assessment in other locations.

Germplasm Sources

Forty-one accessions have been listed among the germplasm institutions with large *Medicago* collections (see Chapter 3).

Research Problems and Priorities

Medicago soleirolii is not well documented, and it is in need of basic botanical studies and agricultural assessment for its potential as a source of genes for improvement of the species of *Medicago* currently used as crops. Piano and Francis (1992) described it as having "regional potential".

Representative Collections

Algeria: Durieu, M., La Calle, prairie du Lac Houbera, 31-V-1841, in late fruit (P) (second specimen with "très abondant dans les pâturages autour du lac Houbera près La Calle"). Reboud, V., Le Djebel Ouach près Constantine, prairies, 12-V-1881, in fruit, Société Dauphinoise No. 3591 (P). Redoub, V., Le Djebal Ouach, près Constantine, prairies, 12-V-1881, in fruit, Société dauphinoise, 1882, No. 3591 (P). Trabut & Battandier, Bône, V-1889, in flower & fruit (P). **France:** Delacour, Th. & Gaudefroy, E., semé par les fourrages de l'armée régulière pendant le second siège de Paris, de mars à mai 1871, 23-VI-1872, in flower & fruit (P). Franchet, A., (leg A. Franchet), Orléans, ile Arrault (Loiret), adventice, 6-VII-1871, in fruit (P). Franchet, M., Cheverny, 1872, in flower & fruit, "Herbier de plantes adventices" (P). Franchet, M., Orléans, île Arrault, 6-1871, in flower & fruit, "Herbier de plantes adventices" (P). Franchet, M., Vendôme, la Papeterie, 1873, in late flower & fruit, "Herbier de plantes adventices" (P). Mogguidge, I.T., Alpes Maritimes, near Hôtel Bellevue, Cannes, 10-V-1872 (K). **Italy:** Baglietto, F., in Villa Patrone, infra Vottri et Crevari, in locis graminosis seccus viam provincialem, V-1892, in fruit, Congresso Botanico Internazionale 1892, O. Penzig Selectae Stirpes Liguriae (P). Bicknell, Cl. & Pollini, L., Liguria, in castanetis supra "Sestri Ponente". alt. 100 m. V-(in flower)-VI-(in fruit)-1901, Herbarium Normale Editum ab I Dörfler No. 4228 (BM, P). Fiori, A., Etruria, prov. di Livorno, Ilva (Elba) Insula, in arvis et locis herbosis prope ferrariam officinam (Alti Forni), alt. 10 m, solo calcareo, 23-V-1907, in fruit, Flora Italica Exsiccata Fiori, Béguinot & Pampanini No. 870 (BM, K). Joad, G.C., Sestii (?) di Levante, E. Ligurio, abundant on grassy slopes, east side of bay, 27-IV-1871, in flower & fruit (K). **Tunisia:** Cosson, M.M.E. et al., Fedj El-Saha ou Camp de la Santé, nord de Fernana, Kroumirie, 30-VI-1883, in fruit (P). Cosson, M.M.E. et al., Oued Bou Mouhbal (?), 21-V-1883, in fruit (P).

92. *Medicago sphaerocarpos* Bertol.

M. sphaerocarpos Bertol., Rar. Lig. Pl. (Rariorum Liguriae (Italiae) Plantarum, decas tertia) 3: 60 (1810), (basionym).

The epithet *sphaerocarpos* is based on the Greek for "spherical fruit"; this is an appositive noun, not an adjective, and therefore not Latinized as *sphaerocarpa* to agree with the feminine gender of *Medicago* (nevertheless, one sometimes encounters "*Medicago sphaerocarpa*"). Heyn's (1963) discussion of fruit shape in what was then known as "*M. murex*" (i.e., *M. sphaerocarpos* and *M. murex* collectively) is instructive: "*M. murex* is often known in botanical literature as *M. sphaerocarpos* or *M. sphaerocarpa*. Willdenow (1802) described a form having ovoid pods, Bertolini (1810) another, having spherical pods. Later these two forms were described as two varieties of *M. murex* or *M. sphaerocarpos*. In the present study no varietal rank for these forms was retained, though the shape of the pod is to some extent inherited. The reason for this treatment of the species is that the shape was found to vary from spherical to ovoid even on one plant. This is often, but not always due to the age of the pods. On ripening, the coils gradually thicken, thus gradually changing the shape of the pod from spherical to ovoid... However, it seems that on

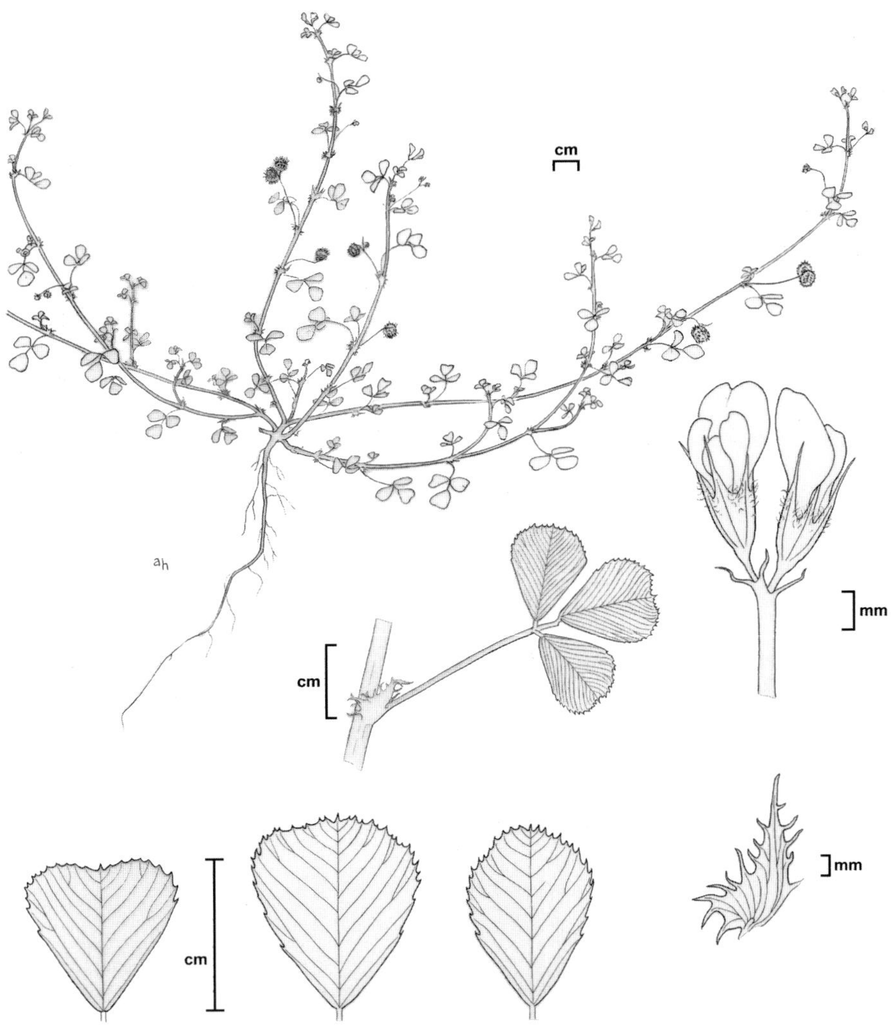

Fig.92-1. *Medicago sphaerocarpos*, wild-collected plant. **Clockwise from top left:** Plant; group of flowers; stipule; leaflets; leaf attached to stem. (Based on F. Raine 30-4-1905, GH; artist: A. Hanes.)

the whole the pods of *M. murex* in the West-Mediterranean are more nearly spherical than those in the East-Mediterranean."

Subordinate Taxa

None recognized here.

Type Specimen

The holotype was apparently lost from BOLO during the Second World War (personal communication, G. Cristofolini.). Future designation of a lectotype needs to carefully take account of the need to ensure that a specimen of this species and not the very closely related *M. murex* is chosen. Bertolini's (1810) description of *M. sphaerocarpos* includes "floribus plerumque 1-2, rario quatuor", i.e., "flowers mostly 1–2, rarely 4", which is diagnostic of this species in contrast to the closely related *M. murex*. Historical collections at FI from the *locus classicus* (at Sarzana, Italy) are this species, not *M. murex* (personal communication, G. Cristofolini). Less significant to the interpretation of Bertolini's species concept is his description of *M. sphaerocarpos* in his *Flora Italica* (1850), in which he noted "leguminibus... margine externo lato, depresso, *bi-trinervi*..." (italicization added), a description apt only for this species, not *M. murex*.

Synonymy

Homotypic

M. murex Willd. var. *aculeata* Urb. f. *sphaerocarpa* (Bertol.) Urb., Verh. Bot. Ver. Brandenb. 15: 72 (1873). (On page

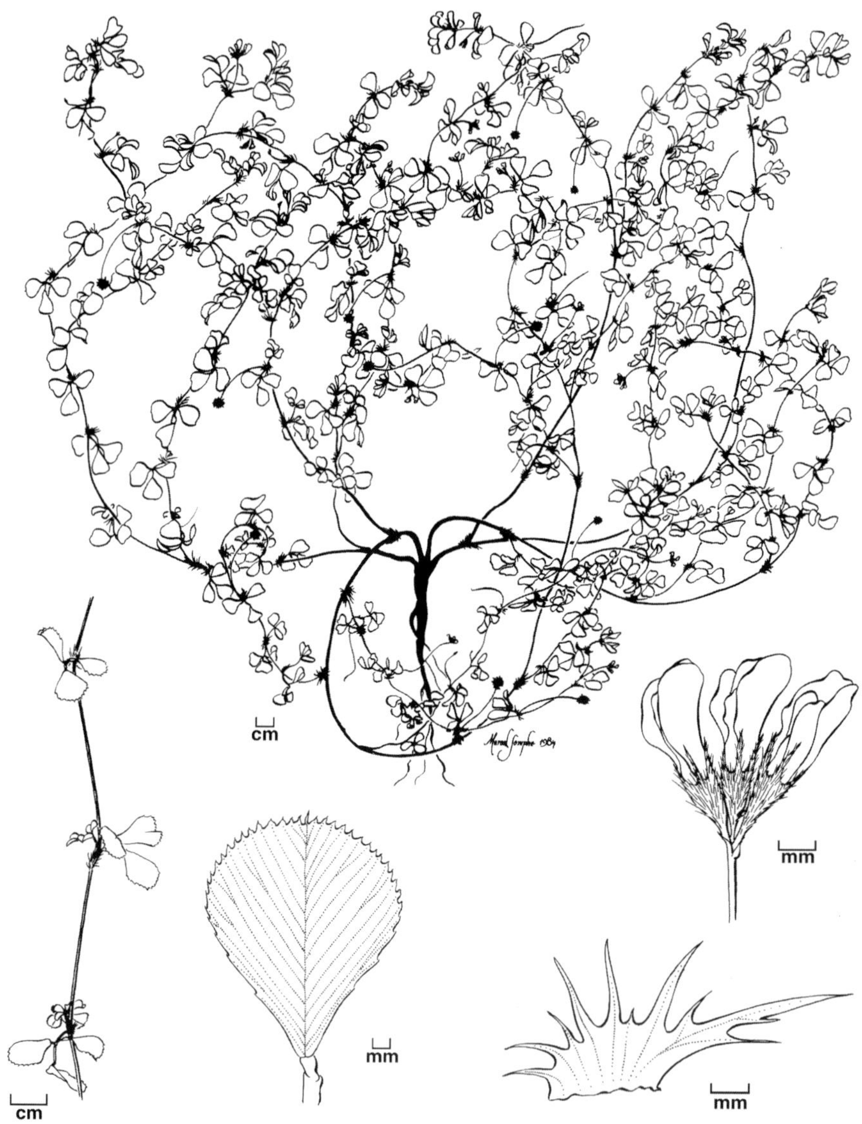

Fig. 92-2. *Medicago sphaerocarpos*, greenhouse-grown plant. **Clockwise from top:** Plant; group of flowers; stipule; leaflet; branch. (Based on E. Small M1147, DAO; artist: M. Jomphe.)

79 of his publication, Urban contrarily appears to have called this a variety. However, rank in Urban's publication is debatable.)

M. murex Willd. var. *sphaerocarpos* (Bertol.) Willk., Willk. & Lange, Prodr. Fl. Hispan. 3: 385 (1877).

Medica sphaerocarpos (Bertol.) Bubani, Fl. Pyren. 2: 480 (1900).

M. murex Willd. subsp. *sphaerocarpos* (Bertol.) K.A. Lesins & I. Lesins, Gen. *Medicago* 175 (1979).

Heterotypic

M. globulifera Desf. ex Poir. in Lam. Encycl. Méthod. Suppl. 3: 526 (1813). (Type!: Herb Poiret (P), collected at "La Garde, 24 Mai." The collection is a mixed gathering with *M. ciliaris*. In Small and Brookes (1985), the left-hand specimen was designated as the type.)

Names undetermined whether synonymous with *M. sphaerocarpos* or *M. murex*

M. sphaerocarpos Bertol. var. *macrocarpa* Moris, Fl. Sardoa 1: 446 (1837).

M. murex Willd. var. *aculeata* Urb. f. *macrocarpa* (Moris) Urb., Verh. Bot. Ver. Brandenb. 15: 73 (1873).

M. murex var. *macrocarpa* (Moris) Willk. in Willk. & Lange, Prodr. Fl. Hispan. 3: 385 (1877).

M. ovata Carmign. in Giorn. Pis. 5, 12: 136 (1810).

M. murex Willd. var. *aculeata* Urb. f. α *ovata* Carmign. in Giorn. Acad. Ital. 1810 ex Urb., Verh. Bot. Ver. Brandenb. 15: 73 (1873).

M. murex Willd. var. *ovata* (Carmign.) Arcangeli, Comp. Fl. Ital. 163 (1882).

M. murex Willd. proles *ovata* (Carmign.) Rouy, Fl. Fr. 5: 33 (1899).

M. ovata hort. ex Steud., Nomencl. Bot. (Steudel) 514.

M. murex var. *rodriguezii* Font Quer ex Casellas, Collect. Bot. (Barcelona) 6: 244 (1962).

M. murex subvar. *rodriguezii* (Font Quer ex Casellas) O. Bolòs & Vigo, Butll. Inst. Catalana Hist. Nat. 38: 69 (1974).

M. sicula Tod., Ind. Sem. Hort. Panorm. 27 (1868. (Type: Todaro Fl. Sic. Exs. 962 (BM, U) (a collection with spineless fruits).) (Abdelguerfi and Guittoneau (1989, p. 426) gave type details as follows: Todaro Fl. Sic. Exs. 962 (BM, U); a collection with spineless fruits.)

M. murex Willd. var. *inermis* (Guss.) Urb. f. *sicula* (Todaro) Urb., Verh. Bot. Ver. Prov. Brandenb. 15: 73 (1873).

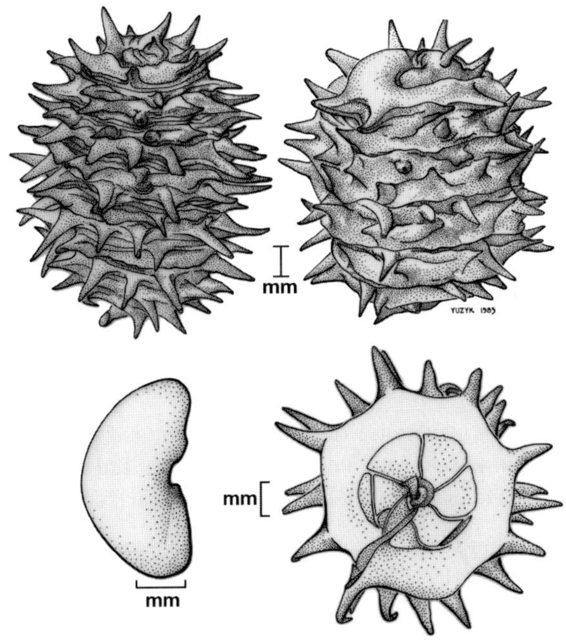

Fig. 92-3. *Medicago sphaerocarpos* and *M. murex*. **Top left:** Fruit of *M. sphaerocarpos* (based on E. Small M1126, DAO). **Top right:** Fruit of *M. murex* (based on E. Small M1102, DAO). Notice extra ridges on coil edges of *M. sphaerocarpos*. **Bottom:** *M. sphaerocarpos* seed and fruit coil face showing venation. (Artist: L. Yuzyk.)

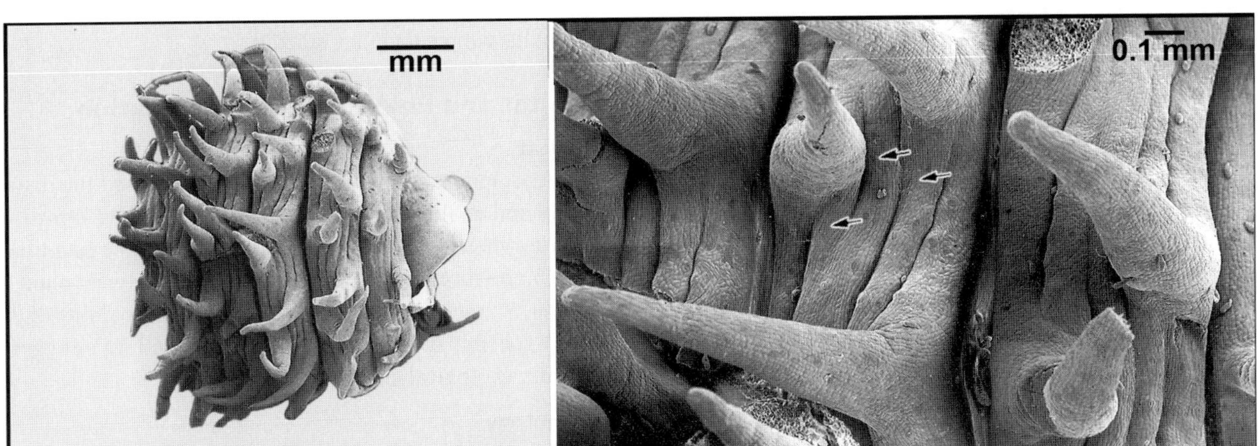

Fig. 92-4. *Medicago sphaerocarpos*. Scanning electron photomicrographs (prepared by B. Brookes). **Left:** Fruit. **Right:** Close-up showing three ridges (arrows) on coil edge. (Based on E. Small M1146, DAO).

(Nègre (1956) referred to *M. murex* var. *ovata* subvar. *genuina* Rouy & subvar. *brevisteta* Rouy; also *M. murex* var. *lonchocarpa* Sennen & Mauricio (Catalogue du Rif, p. 32).)

Not

M. sphaerocarpa Costa, Introduccion a la flora de Cataluña 60 (1864) (cited in The International Plant Names Index (IPNI; http://www.ipni.org/index.html; current July 2010) as a synonym of *M. turbinata*).

Vernacular Names

English: Sphere medic (medic is also spelled medick). (Note: because *M. murex* and *M. sphaerocarpos* have only recently been separated, it is often not possible to distinguish which common name applies to which species.)

French: Luzerne à fruits arrondis (a translation based on the Latin synonym *M. sphaerocarpos*), luzerne à fruits sphériques.

Geographical Distribution

(N = native, I = introduced, U = uncertain status)

Africa: Algeria (N), Morocco (N), Tunisia (N).

Asia: Israel (N), Turkey (N).

Europe: France (presumed present and N), Germany (presumed present and U), Greece (N), Italy (including mainland, Sardinia, and Sicily, N), Portugal (N), Spain (N).

This species is omni-Mediterranean. Gillespie and McComb (1991) noted that while *M. sphaerocarpos* often grows together with the very closely related and similar *M. murex*, in mainland Italy and Sicily it seems to be comparatively more rare than the latter.

Description

See description for *M. murex* (Chapter 55).

Identification Hints

The hard-fruited species of annual *Medicago* (currently assigned to sect. *Spirocarpos* subsect. *Pachyspirae*) are fairly easily distinguished from other *Medicago*, but are often difficult to distinguish from each other. Of this group, *M. sphaerocarpos*, *M. murex*, and *M. turbinata* are the only ones in which the peripheral part of the pod faces of the more or less adpressed coils conspicuously lack venation (*M. syriaca* shows the character, but less conspicuously). However, in all these species a spongy tissue tends to develop on the pods at maturity, obscuring the veins. In many species it appears that this obscuration is more pronounced in the peripheral part of the coil faces than in the proximal parts, leading to the appearance of an absence of peripheral veins in some species that actually possess them. This cause of misidentification can be avoided by examining young pods. *Medicago turbinata* (generally known as *M. tuberculata* Willd.) is quite similar to *M. sphaerocarpos* and *M. murex*, but is conspicuously more pubescent.

Medicago sphaerocarpos and *M. murex* are very similar. The fruits of *M. murex* lack noticeable ridges on the coil edge beside the dorsal suture (ridges are almost always evident in *M. sphaerocarpos*; see Figs. 92-3 and 92-4). (This character should be examined in mature fruits; immature fruits of *M. murex* sometimes seem to have extra coil edge ridges.) Racemes of *M. murex* typically have only one or two flowers, whereas many racemes of *M. sphaerocarpos* have three flowers.

The leaflets of living plants of *M. murex* have white flecks (rarely detectable in dried, herbarium specimens), a feature that is absent in *M. sphaerocarpos*. Abdelguerfi and Guittoneau (1989) were unable to discern a cellular difference in the epidermis between white-flecked areas and normal green areas. This character is temperature sensitive. Gillespie and McComb (1991) found that in midwinter in Perth, Australia (with temperatures averaging 17 °C during the day), the white flecks were always apparent but, at constant temperatures above 20 °C in a glasshouse, the flecks were faint and sometimes difficult to detect. Also, Gillespie and McComb (1991) noted that the leaflets of *M. sphaerocarpos* sometimes had purple (anthocyanin) leaf flecks that were more obvious at lower temperatures, and that sometimes basal, purple shield-like leaf marks were present (characters never observed in *M. murex*).

Small and Brookes (1985) reported that the fully developed lower leaves of *M. murex* tend to be emarginate and those of *M. sphaerocarpos* tend to be rhomboid. However, Gillespie and McComb (1991), on the basis of examining a much larger sample of the two species, concluded that this character was of limited use for discrimination.

Abdelguerfi and Guittoneau (1989) observed a depression in the radicle of the seeds of *M. murex*, but not in *M. sphaerocarpos*. Gillespie and McComb (1991) were unable to confirm this in the sample that they studied.

Habitat and Environmental Adaptation

Habitat

Abdelguerfi and Guittoneau (1989) studied the distribution and ecology of *M. sphaerocarpos* and *M. murex* in Algeria, where *M. sphaerocarpos* was found to be four times as common as *M. murex*. Wherever *M. murex* was found in Algeria, *M. sphaerocarpos* also occurred, but the ability of the latter to occupy drier habitats and more varied soils suggests a greater ecological amplitude.

Phenology

In Australia, cultivar 'Orion' is advertised as a 99-day (mid-flowering) variety with the capacity to stay green for up to 2 weeks longer than many other medic cultivars.

Soil

This species often occurs on light soils. Species of *Medicago* are generally intolerant of acidic soils, but compared with most other species of the genus, *M. sphaerocarpos* is relatively acid-tolerant. In Algeria, where *M. sphaerocarpos* is native, Abdelguerfi and Guittonneau (1989) collected it from 15 sites, with the pH ranging from 5.8 to 8.5. The 'Orion' cultivar is almost as tolerant of soil acidity as is the quite tolerant *M. murex* (Gillespie 1994), growing in a recommended soil pH of 4.8–8.0. 'Orion' is recommended for sandy loams, loamy sands, and clay loams, but has been grown on heavy soils. In Australia it is not recommended for deep sands, where waterlogging is common, or in the infertile red soils of western New South Wales.

Temperature

No information located.

Water

Medicago sphaerocarpos is adapted to low-rainfall, Mediterranean areas that receive most of their precipitation in the winter growing season. In Algeria, where *M. sphaerocarpos* is native, Abdelguerfi and Guittonneau (1989) collected it from 15 sites, the annual rainfall ranging from 750 to 1750 mm. In Australia, 'Orion' is recommended for areas with 275–400 mm (winter-dominant) rainfall. As with other annual species of *Medicago*, waterlogged soils are not tolerated.

Biotic Relationships

Pollination

The small flowers of this species indicate it is a strong inbreeder. The pronounced isozyme uniformity of populations and rarity of heterozygotes (Small et al. 1992) also show that outbreeding is very rare.

Dispersal

Indehiscent spiny fruits such as those that occur in *M. sphaerocarpos* are well adapted to dispersal in animal fur. Gillespie and McComb (1991) reported that only 3 of 135 accessions of *M. sphaerocarpos* they studied had pods that were completely lacking in spines and tubercles, although the presence of accessions with short spines or tubercles was common.

Herbivory

Cultivar 'Orion' is advertised as tolerant of the redlegged earth mite (*Halotydeus destructor* (Tucker)), but not especially tolerant of aphids (it is susceptible to the bluegreen aphid, *Acyrthosiphon kondoi* Shinji).

Nodulation and mycorrhizae

The USDA (2009) reported positive root nodulation for *M. sphaerocarpos*.

Disease resistance

O'Neill et al. (2003) examined the comparative resistance of 36 annual species of *Medicago* to *Phoma medicaginis* (the cause of spring black stem and leaf spot, a major disease of alfalfa) and concluded that *M. sphaerocarpos* (identified as *M. murex*) had high resistance to it. Barbetti (2007) identified two accessions of *M. sphaerocarpos* with very high levels of resistance to *P. medicaginis*. Barbetti (1995) identified several cultivars of *M. sphaerocarpos* that were resistant to pepper spot leaf disease (caused by *Leptosphaerulina trifolii*). In a similar comparative test of 36 annual species of *Medicago*, O'Neill and Bauchan (2000) found that three accessions of *M. sphaerocarpos* (identified as *M. murex*) were quite susceptible to anthracnose caused by *Colletotrichum trifolii*, but one accession was completely resistant. Cultivar 'Orion' has been advertised as resistant to black stem and Phytophthora root rot. Yaege and Stuteville (2002) tested the resistance of 33 annual species of *Medicago* to powdery mildew (*Erysiphe pisi*) and found that the six collections of *M. sphaerocarpos* (identified as *M. murex*) grown were susceptible to the disease (it was observed that often only one of a number of populations of other species tested was resistant).

Economic History, Cultivars, and Uses

As noted below, the species has been used as a forage plant and one cultivar has been selected.

Cultivars

'Orion', the first registered cultivar of *M. sphaerocarpos*, is described in Gillespie (1994), where detailed directions are given for planting and maintenance. This Australian cultivar originated from a collection in Sicily in 1986, and is well adapted to Western Australia and some other regions of the continent.

Forage

Medicago sphaerocarpos is used as a legume component of sowing mixtures for rotational pasture systems in Australia. The hard seed of cultivar 'Orion' results in irregular germination, and accordingly it is not recommended for sowing as a monoculture. Cultivar 'Orion' of *M. sphaerocarpos*, like other species of *Medicago*, is capable of causing digestive problems in livestock, such as bloat in cattle, when consumed in large amounts.

Fodder

In regions of Australia, cultivar 'Orion' has been found to be a good component of pasture mixtures or rotational series that are harvested for fodder.

Green manure, rotation crop, ground cover

'Orion' is employed in Australia as a rotational crop.

Other uses (salad sprouts, ornamental, plant products, etc.)

No information located.

Chromosome Number

$2n = 14$: Lesins et al. 1970; Fernandes and Queirós 1978; Lesins and Lesins 1979; Abdelguerfi and Guittoneau

1979; Agarwal and Gupta 1983*b*; Small and Brookes 1985; Small et al. 1992; Abdelguerfi et al. 1988*b*; Mariani and Falistocco 1990; Gillespie and McComb 1991; Mariani et al. 1996. All the early counts were reported as "*M. murex*".

Genetic Relationships and Gene Transferability

Intraspecific

Artificial hybridization of different populations has not been attempted, so it is not known whether there are barriers to hybridization within the species.

Interspecific

Attempts to experimentally hybridize *M. sphaerocarpos* with *M. murex* have failed (Lesins et al. 1970; Mariani and Falistocco 1990; Gillespie and McComb 1991). Lesins et al. (1970) were unable to hybridize *M. sphaerocarpos* with *M. constricta*, another annual medic with the same chromosome number ($2n = 14$).

Exchange by advanced biotechnology

No information located.

Conservation Status

Medicago sphaerocarpos appears to be widespread, and therefore not in need of special conservation measures.

Germplasm Sources

The South Australia Research and Development Institute *Medicago* seed bank has about 600 collections identified as *M. sphaerocarpos* (see Chapter 3), a result of research having been carried out there on the collection. In other germplasm institutions, it is generally not possible to know from accession data whether accessions are *M. sphaerocarpos* or *M. murex*.

Research Problems and Priorities

The separateness of *Medicago sphaerocarpos* and *M. murex* has only recently come to light, and the comparative geography and ecology of the two species requires more study.

Representative Collections

Algeria: Durando, environs d'Alger, lieux herbeux, V-1866, in fruit, Fragmenta Flora Algeriensis Exsiccata No. 568 (P). Durieu, M., Alger, IV-1840, in early fruit (P). **Corsica:** Mabille, P. 365, Bastia, 20-29-V-1868, in flower & fruit (K). **Greece:** Zohary, M. & Orshan, G. 25404/13, South Crete, 28 km northwest of Tymbakion, the Bwala ridge, a *Poterium Phlomis* "batha" on a brownish-gray soil derived from metamorphic rock, alt. 560 m, 25-IV-1964, in flower & fruit (HUJ). **Israel:** Eig, A., Philistean Plain, Rishon-le-Zion to Nahalet-Yehuda, sandy loam soil, 25-III-1925, in fruit (HUJ). Eig, A., Zohary, M., & Feinbrun, N., Sharon Plain, Wadi Mansur, marsh, 30-V-1929, in fruit (HUJ). Naftolsky, N., Sharon, Petakh-Tikva, vineyard, 24-IV-1927, in fruit (HUJ). Zohary, D., Sharon Plain, east of Ma'abaroth, dried marshes, 10-V-1942, in fruit (HUJ). **Italy:** Bicknell, C., Liguria occid. Bordighera, in olivetis, alt. 2–10 m, solo argilloso, 18-V-1911, in fruit, Flora Italica Exsiccata Series III-Fiori & Béguinot No. 2096. Huter, Porto & Rigo 74, Sicily, Sicilia orient. ad agrorum margines collium arenos supra Messina, sol. granit., alt. 2–300 m, 18-IV-1877, in flower & fruit (K). **Sardinia:** Reverchon, E. 281, Tempio, terrains incultes et granitiques, 10-V-1882, in flower & fruit (K).

Notes

Lesins et al. (1970) and Lesins and Lesins (1979), on the basis of chromosome configurations, suggested that *M. murex* is the direct ancestor of *M. sphaerocarpos*. Mariani and Falistocco (1990), also on the basis of chromosome analysis, suggested that both species could have arisen from a common ancestor. Small et al. (1992) suggested the additional hypothesis that *M. sphaerocarpos* may have given rise to *M. murex*. Falistocco and Moretti (2000) and Falistocco et al. (2002) hybridized the DNA of the two species and noted that their genomes appeared to be very similar.

Brummer et al. (1995) conducted a DNA analysis of six annual species of *Medicago* and noted that "Four accessions belonging to [*M. sphaerocarpos*] were small pubescent plants much different from the large glabrous plants of the other five accessions. Chromosome counts indicated that these plants had $2n = 14$, indicative of [*M. sphaerocarpos*]. These accessions do not appear to be [*Medicago murex*], which has $2n = 16$ (Small and Brookes 1985). However, the PAUP output supports the classification of these four as a different group from the other [*M. sphaerocarpos*] accessions. Further investigation on the true taxonomic placement for these accessions is warranted." I inquired about these and E.C. Brummer replied (on 07 November 2006): "I have not followed up on the [*M. sphaerocarpos*] accessions from my 1995 Genome paper. ... A few years back, I had some discussion with Walter Graves about this. He suggested that [*M. sphaerocarpos*] 1 and 4 in my paper were actually *M. doliata* and that [*M. sphaerocarpos*] 2 and 3 were *M. constricta*. Walt's more knowledgeable than I am (and certainly than I was at the time I did this paper) regarding the annual medics, so I'm guessing he's probably right. However, I've not done anything to confirm that." [In this paragraph the correct names have been substituted and shown in square brackets; as noted in Chapter 55, the name *M. murex* was used until recently for what is now recognized as *M. sphaerocarpos*.]

93. *Medicago strasseri* Greuter, Matthäs & Risse

M. strasseri Greuter, Matthäs & Risse, Willdenowia 12: 201–202 (1982), (basionym).

The epithet *strasseri* honors contemporary Swiss botanist Walter Strasser of Steffisburg, the author of several botanical works, who first collected and recognized the uniqueness of this species.

Subordinate Taxa

None recognized here.

Type Specimen

Greuter, W. & J.J., Matthäs, U., & Risse, H. 19362, Petres Gorge, Crete (35°20′30″N, 24°31′30″E) (Holotype!, B; fuller citation below).

Synonymy

Homotypic

M. arborea L. subsp. *strasseri* (Greuter, Matthäs & Risse) Sobrino et Ceresuela, Portugaliae Acta Biologica 19: 225-237 (2000).

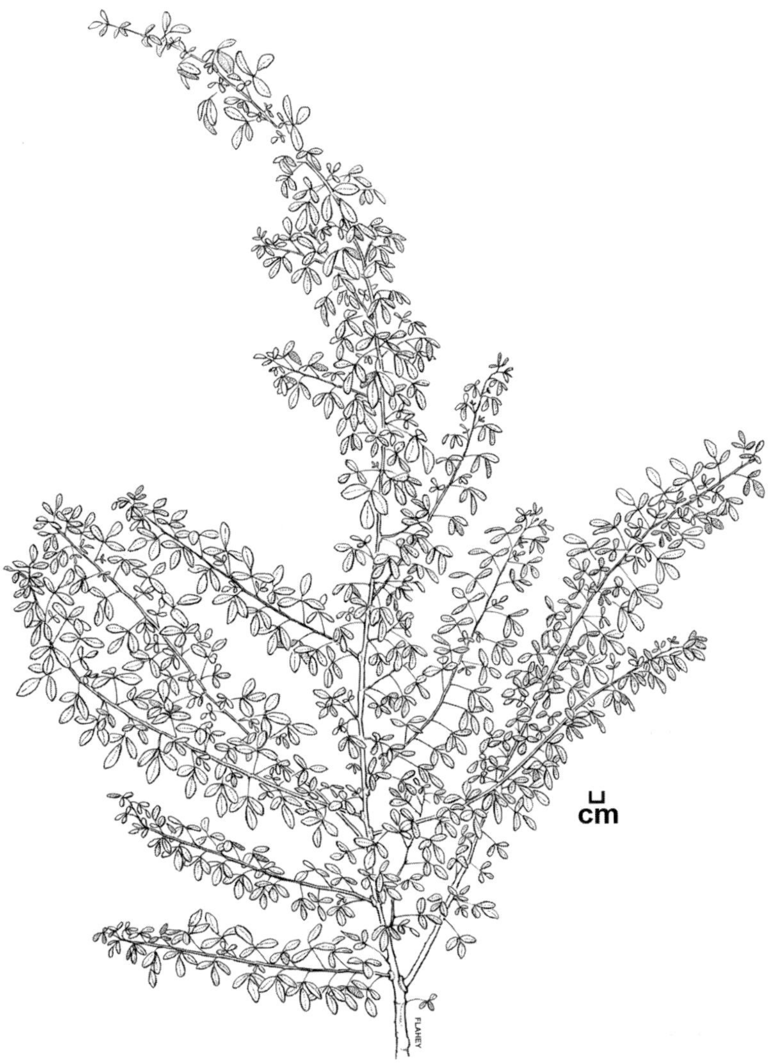

Fig. 93-1. *Medicago strasseri.* Greenhouse-grown, vegetative plant. (Voucher: E. Small M2069, DAO. Grown from seed supplied by R. Vogt, Botanical Garden Berlin-Dahlem; original collection: Griechenland: Kreta, Ep. Pedhias, Felswände am Fuß des Berges Pirjia S von Piskopiano (35°18′N, 25°23′E), Kalkfelsklippen, ca. 350 m, 12.VI.1990, leg W. Greuter; artist: B. Flahey.)

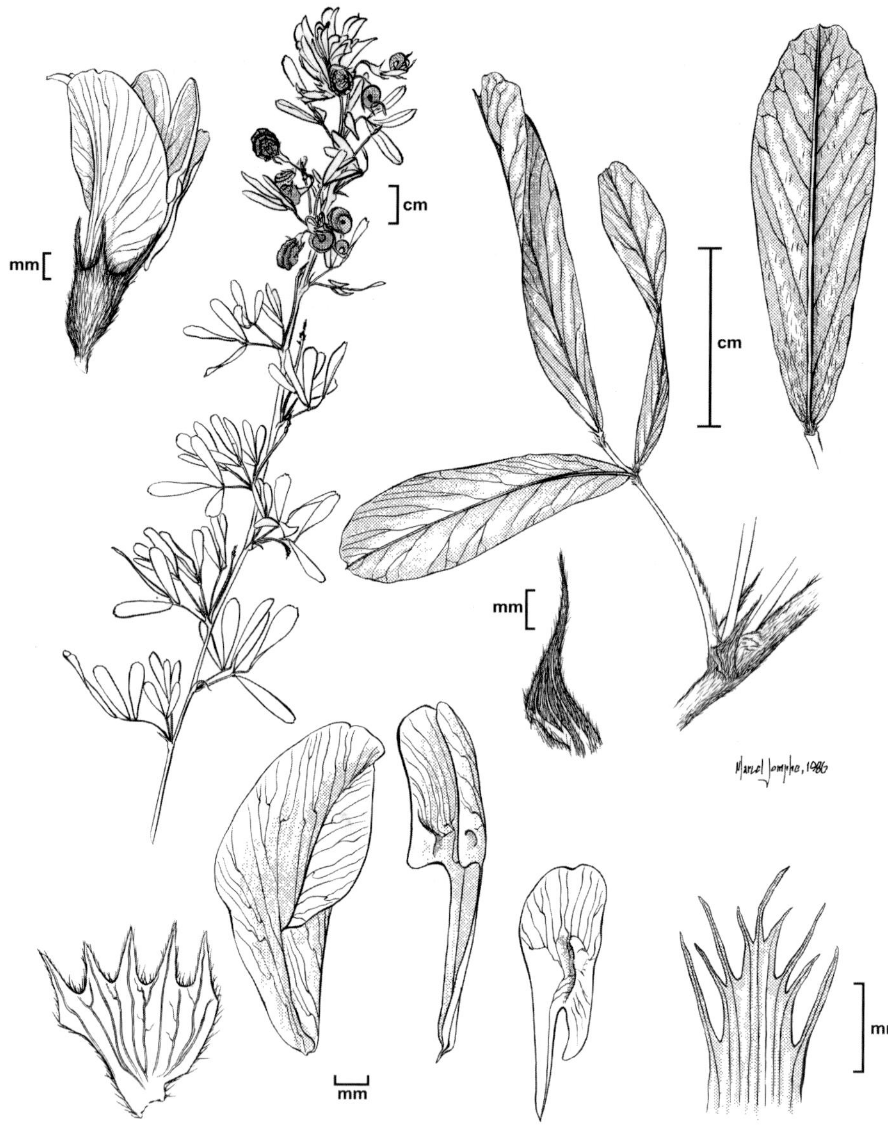

Fig. 93-2. *Medicago strasseri.* **Clockwise from upper left:** Side view of flower; branch with flowers and fruits; leaf at axil of stem; leaflet; stipule; flattened androecial column (split beside free stamen, which is not shown, anthers also not shown); wing petal; keel; standard petal; calyx (split and flattened). (Based on holotype; artist: M. Jomphe.)

Heterotypic

No names found.

Vernacular Names

English: No names found.

French: No names found.

Geographical Distribution

(N = native, I = introduced, U = uncertain status)

Europe: Greece (N).

This species is a narrowly distributed endemic of Crete, where it is known from seven localities (Thanopoulos 2007).

Description

Woody perennial shrub 0.5–1.5 m tall, young twigs and leaves finely and densely pubescent with whitish simple hairs, aging twig surfaces cracking longitudinally to reveal developing bark. Leaflets 10–18 mm long, obovate to obovate-spatulate, generally about three times as long as broad, cuneate at base, more or less entire or slightly denticulate near and at apex. Stipules narrowly triangular-subulate, entire. Inflorescence a very short capitate raceme of 6–12 crowded flowers, peduncles 2–5 cm long. Flowers 9–13 mm long, orange-yellow. Calyx 4–5 mm long, teeth shorter than tube, calyx about as long as fruiting pedicel. Corolla more than twice length of calyx. Fruit with 1.5–2.5(3.0) loosely adpressed coils, disciform or shortly cylindriform, 5–7 mm in diameter with a central opening

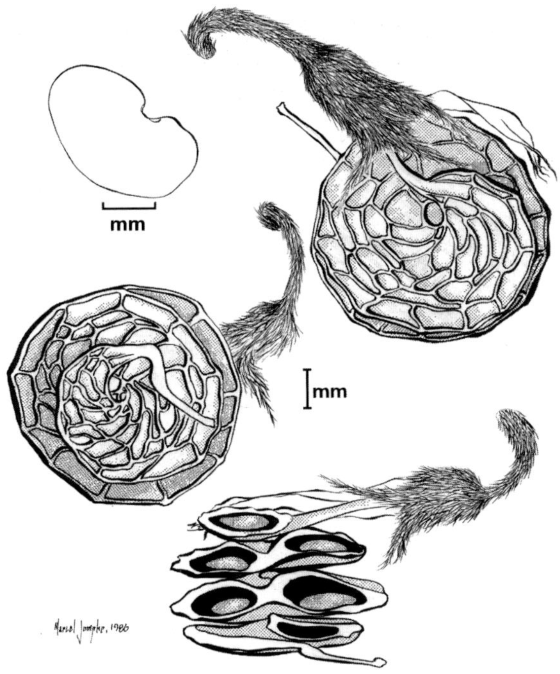

Fig. 93-3. *Medicago strasseri.* **Clockwise from upper left:** Seed; fruit, perspective from style end; sagittally sectioned fruit showing seed cavities; fruit, perspective from basal end. (Based on holotype; artist: M. Jomphe.)

0–0.5 mm wide, without dorsal suture spines (rarely with short, irregular spines), with short, appressed hairs, coil face with curving radial veins that anastomose on the outer two-thirds, then reach dorsal suture. Seeds 3–8 per pod, 2.0–2.5 mm long, yellow or yellow-brown, reniform, separated by a transverse, spongy fruit septum, radicle about equal to half seed length.

Greuter et al. (1982) pointed out that rare spines that develop on the dorsal fruit sutures of *M. strasseri* are fundamentally similar to those of most spiny-fruited annual species of *Medicago* (i.e., with two "roots" from veins from either side of the coil face vasculature).

Identification Hints

Medicago strasseri is easily distinguished from the only similar species, *M. arborea* and *M. citrina*, by having a considerably smaller pod diameter (often 5–7 mm, rarely as wide as 10 mm versus 10–15 mm in the other species) and smaller seed length (2.0–2.5 versus 3.5–4.7 mm in the other species), more numerous coils (1.5–2.5(3) versus 0.5–1.5 in the other species), and a smaller central (axial) opening of the spiral (0–0.5 versus at least 1 mm in diameter but usually larger in the other species). *Medicago strasseri* can be distinguished from *M. arborea* by having a larger fruiting pedicel to calyx length ratio (>1.0 in the former versus ca. 1.0 in the latter), but as *M. citrina* has a ratio of >1.0, this character is not useful for distinguishing the latter from *M. strasseri*.

Additional differences that have been pointed out (Greuter et al. 1982; Strasser 1988) are in stature (up to 4 m in *M. arborea*, up to 1.5 m recorded to date for *M. strasseri*), in flower size (12–15 mm in *M. arborea*, 9–13 in *M. strasseri*), in fruit face venation (less curving radial veins that anastomose more freely in *M. arborea* and present a pattern of relatively small areoli between the dorsal and ventral sutures compared with the areoli near the sutures; cf. Figs. 14-2 and 93-3), in thickening of the dorsal suture (more so in *M. arborea* than in *M. strasseri*), and in leaflet shape (obovate to obcordate, and mostly not more than twice as long as broad in *M. arborea*, obovate-spatulate and about three times as long as broad in *M. strasseri*; cf. Figs. 14-1 and 93-2).

Chebbi et al. (1995) compared a small number of populations of *M. arborea*, *M. citrina*, and *M. strasseri* grown under similar conditions and did not find differences in height (cf. above paragraph), but did observe significant differences in other details, such as more flowers per peduncle, averaging 10.4 in *M. strasseri*, 9.4 in *M. arborea*, and 7.4 in *M. citrina*. The terminal leaflet of *M. strasseri* tended to have a larger length to width ratio (averaging 1.85) compared with about 1.48 for the two varieties of *M. arborea*. However, these observations are based on relatively few plants. Chebbi et al. (1995) did not find differences in flower size (cf. above paragraph) between *M. arborea* and *M. strasseri*.

Habitat and Environmental Adaptation

Habitat

Medicago strasseri colonizes crevices of steep limestone cliffs in gorges of Crete, at altitudes of 30–400 m.

Phenology

The species flowers from February to May in its native sites (Chilton and Turland 1997). The growth rate of *M. strasseri* was found to be significantly higher than that of its close relative *M. arborea* (Alegre et al. 1993*b*).

Soil

This species occurs in areas of schistous, nonkarstic substrates (Greuter et al. 1982).

Temperature

Medicago strasseri was found to be more tolerant of temperatures below 0 °C than its close relative *M. arborea* (Alegre et al. 1993*b*).

Water

"Although by no means moist, the environment is characterized by a fairly high relative humidity of the air" (Greuter et al. 1982). Given the arid climate of Crete, the recorded sites are in areas of relatively high mean annual rainfall and high humidity, where sclerophyllous species such as *Quercus ilex* and *Myrtus communis* are restricted on the island. In experiments comparing the productivity of *M. strasseri* with the two varieties of *M. arborea*, Lefi et al. (2004*b*) found that shoot to root production was highest in *M. strasseri* both under drought and irrigation.

Fig. 93-4. Holotype of *Medicago strasseri* Greuter, Matthäs & Risse.

Biotic Relationships

Pollination

No information available.

Dispersal

Sobrino et al. (2000) noted that two morphotypes could be distinguished, one with spiny pods and one without, and that the spiny feature is associated with explosive dehiscence.

Herbivory

"From the occurrence of *Medicago strasseri* high up on vertical cliff-faces and in a few other places plainly inaccessible to the most intrepid of goats, it must be concluded that the species is absolutely intolerant of grazing" (Greuter et al. 1982).

Nodulation and mycorrhizae

No information available.

Disease resistance

No information available.

Economic History, Cultivars, and Uses

This rare endemic was discovered fairly recently and has not yet been employed economically. Its significance lies in extending the range of genetic variability of the genus *Medicago*, particularly related to the germplasm of *M. arborea*, and potential exploitation is probably best judged in the same perspective as the latter species, which has been employed both as an ornamental, grazing shrub, and soil stabilizer.

Cultivars

No cultivars have been bred.

Forage

This very narrowly distributed species has not been employed as forage, although Bennett et al. (2006) speculated that it has potential as a forage crop. González-Andrés and Ceresuela (1998) found that it has similar crude protein and fiber contents to those of its close relative *M. arborea*, and they suggested that the species would be a good winter forage in winter-cold, semi-arid environments.

Fodder

Medicago strasseri has not been employed as a fodder crop.

Green manure, rotation crop, ground cover

No information reported. The adaptation to vertical cliff faces suggests that, like its close relative *M. arborea*, the species has excellent potential for stabilizing slopes.

Other uses (salad sprouts, ornamental, plant products, etc.)

Bennett et al. (2006) speculated that the species has potential as an ornamental shrub.

Chromosome Number

González-Andrés et al. (1999) reported $2n = 32$ based on unpublished information from the taxonomists who recognized *M. strasseri* (i.e., Greuter, Matthäs, and Risse).

Gene Transferability and Genetic Relationships

Intraspecific

The range of variants included in this species requires clarification before the question of hybridization among intraspecific variants can be addressed.

Interspecific

No information located.

Exchange by advanced biotechnology

No information located.

Conservation Status

This very rare and unique species is deserving of the highest conservation protection. Bennett et al. (2006) stated that the species should be considered as threatened from a conservation viewpoint. *Medicago strasseri* is listed in the Red Data Book of Rare and Threatened Plants of Greece (Phitos et al. 2009).

Germplasm Sources

The authors of the taxon and the South Australia Research and Development Institute (see Chapter 3) appear to be the only potential sources of seeds at present.

Research Problems and Priorities

A related population at Karpathos (Greece) in the South Aegean may deserve taxonomic separation (Greuter et al. 1982, 1984) and requires investigation. The principal immediate need is for the collection and preservation of germplasm of *M. strasseri*.

Representative Collections

Greece: Boratinska, Boratinski, Browicz & Dolatowski 216, Pedhias, mountain above Piskopiano near Hersonisos (35°18′N, 25°23′E), vertical cliffs, 28-V-1983 (herb. Greuter, KOR). Greuter, W. & J.J., Matthäs, U., & Risse, H. 19296, provincia et eparchia Rethimni, in parte meridonali faucium Prasiano in valle fluvii Sfakoriako, alt. 90–100 m, (35°19′N, 24°33′E), 24-V-1982 (B; also at B from Prasiano Gorge: Risse, H. 62, 14-VI-1982). Strasser, W., Creta, provincia et eparchia Rethimni, alt. 30 m, in faucibus Petres a pago Karoti septemtriones versus (35°20′30″N, 24°31′30″E), 22-IV-1982 (B; also at B from Petres Gorge: Risse, H. 98, 26-V-1982; Risse, 575, 12-VI-1982; Greuter, W. & J.J., Matthäs, U., & Risse, H. 19362, 26-V-1982 (holotype); Strasser 22-IV-1982).

94. *Medicago suffruticosa* Ramond ex DC.

M. suffruticosa Ramond ex DC. in Lam. & DC., Fl. Franç., ed. 3, 4(2): 541 (1805), (basionym).

The epithet *suffruticosa* is Latin for subshrubby or somewhat shrubby, from the Latin *sub-* (somewhat less than), plus *fruticosus* (bushy, shrubby), descriptive of the appearance of this low-growing plant. Vassilczenko (1949) made the following comment regarding the appropriateness of the term (translated from Russian): "The name *M. suffruticosa* Ram. is here used incorrectly: among perennial alfalfas there are no real semi-shrubs with buds overwintering at a certain distance from the soil surface. As for the lignification of the bases of stems, it occurs in all the prostrate-ramose types of alfalfa, not only in this species."

Key to Subordinate Taxa

Fruit with simple or gland-tipped trichomes; venation on coil face, if visible, somewhat net-like .... subsp. *suffruticosa*
Fruit glabrous; venation on coil face prominently concentric .... subsp. *leiocarpa*

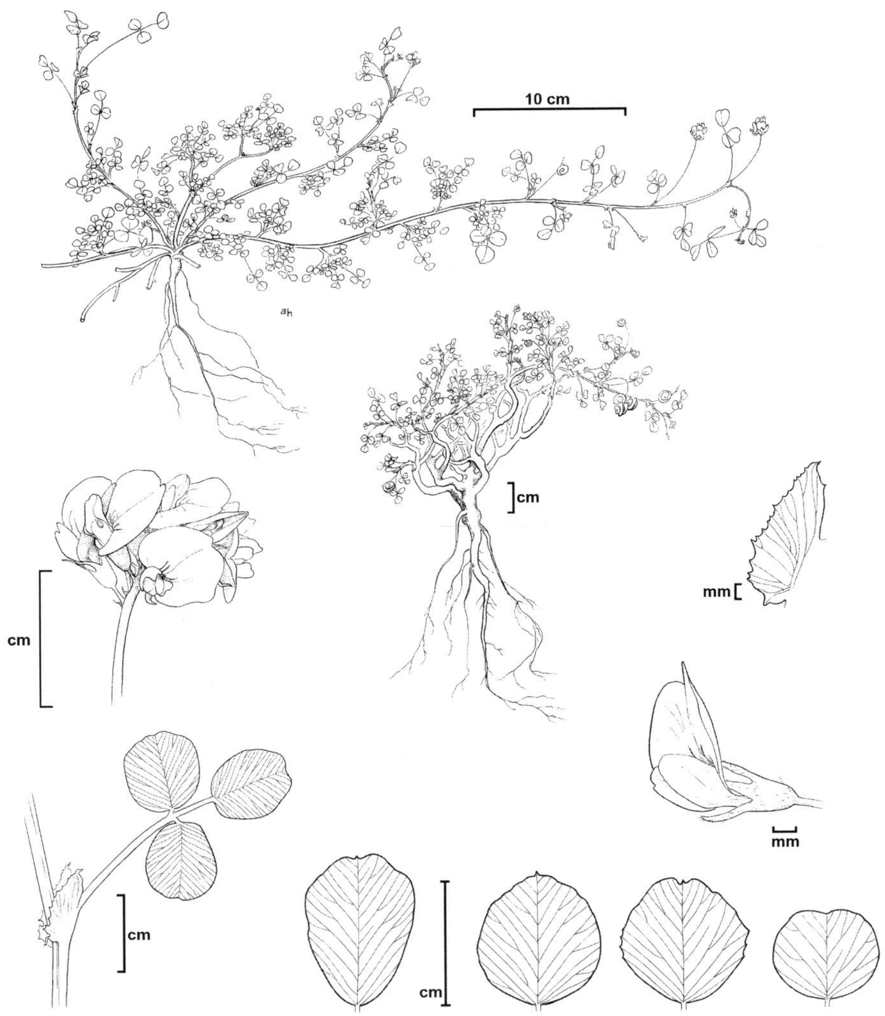

Fig. 94-1. *Medicago suffruticosa* subsp. *leiocarpa*. **Clockwise from top:** Habit drawing of a greenhouse-grown plant; habit drawing of plant collected from the wild; stipule; side view of flower; leaflets; leaf attached to stem; inflorescence with peduncle. (All based on a greenhouse-grown plant, E. Small M199, DAO, except the plant from the wild, which is based on #1156404, US; artist: A. Hanes.)

Medicago suffruticosa subsp. *suffruticosa* is separated from *M. suffruticosa* subsp. *leiocarpa* primarily or exclusively on the basis of fruit pubescence (the former usually has articulated trichomes, or a mixture of these and simple hairs, sometimes simple hairs only). Aside from this very widely accepted taxonomic distinction based on fruit pubescence, various authors have alleged (the exact characters depending on the author) that *M. suffruticosa* subsp. *suffruticosa* has acuminate stipules (versus not acuminate in subsp. *leiocarpa*); larger flowers; fewer fruit coils than *M. suffruticosa* subsp. *leiocarpa* (1.5–3.5 versus 2–4); relatively loose coiling (tighter in subsp. *leiocarpa*); more reticulated coil face venation; a more pronounced fruit coil axis opening; a narrower fruit dorsal suture; and seeds that are irregularly cordate and larger (3.2–3.5 mm × 1.5–1.8 mm in subsp. *leiocarpa* versus 2.5 mm × 1.5 mm in subsp. *suffruticosa* according to Lesins and Lesins 1979), with 3–5 seeds per coil and a radicle almost as long as the cotyledons (versus seeds ovate, 2–3 coil, and a radicle slightly longer than half the length of the seed in subsp. *leiocarpa*). Also, *M. suffruticosa* subsp. *suffruticosa* grows at higher altitudes and in moister locations. However, the nature of the geographical, ecological, morphological, and genetic separation of the two subspecies requires clarification.

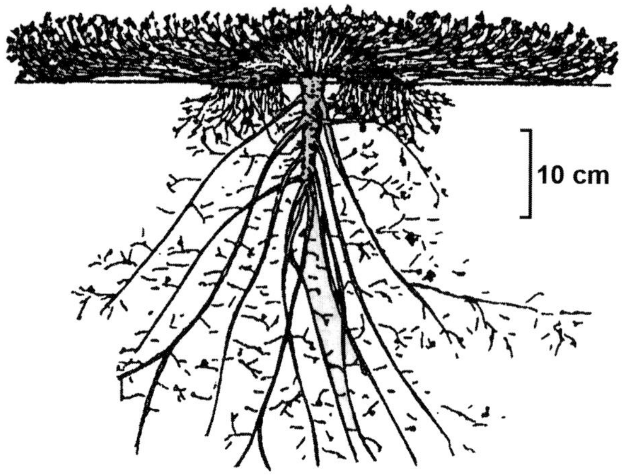

Fig. 94-2. Root system of *M. suffruticosa* subsp. *leiocarpa*, published in Rios et al. (1993). Reproduced with the permission of the authors (drawn by S. Rios).

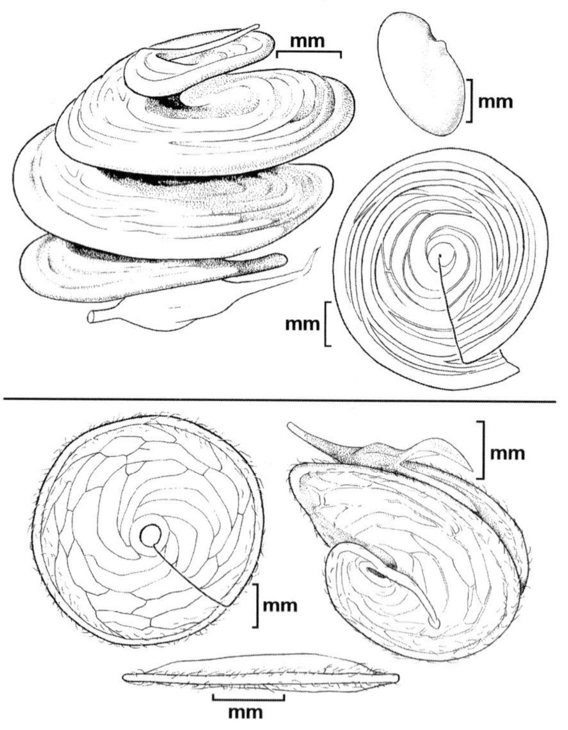

Fig. 94-3. *Medicago suffruticosa*. **Above line:** *M. suffruticosa* subsp. *leiocarpa*. **Clockwise from left:** Fruit; seed; fruit coil face showing venation. (Based on E. Small M199, DAO.) **Below line:** *Medicago suffruticosa* subsp. *suffruticosa*. **Clockwise from right:** Fruit; coil edge; fruit coil face showing venation. (Based on J. Ball in 13–15 May 1871; artist: A. Hanes.)

Nègre (1956) stated (translated from French) that "These characters vary strongly from one individual to another, the only valuable character being the total absence of hairs on the fruits." Such polymorphism in *Medicago* species, with one form having very pubescent (often glandular-haired) fruit and the contrasting form having glabrous fruit, is common, and seems to be related to the presence of herbivorous insects (see Chapters 4 and 7). If indeed glabrous fruit is more common at lower altitudes, one could predict that some insect species attacks the plants at higher altitudes, and this has stimulated the evolution of protective hairs. The pubescence of fruits is the kind of feature that seems to compel taxonomists of *Medicago* to split species into different taxa, but it is possible that future research on *M. suffruticosa* will establish that the character has such a simple genetic base that it does not justify such treatment.

Type Specimen

In his 1805 description of the new species *M. suffruticosa*, De Candolle stated that "Cette plante a été découverte par M. Ramon dans les Pyrenées, aux environs de Barrèges." On 26 July 2007, F. Jacquemoud, Keeper of the Herbaria at G, informed me "that a relevant type specimen is kept in the G-DC herbarium (Prodr. 2: 173, n°11)," and kindly requested that J. Lazare (CECRV, Bayonne, France) check for the possibility that there is another original specimen in the herbarium of Ramond. Geneva provided the illustration of the type, below.

Synonymy

Homotypic

M. suffruticosa* Ramond ex DC. subsp. *suffruticosa

Medica suffruticosa (DC.) Bubani, Fl. Pyren. 2: 472 (1900).

M. suffruticosa* Ramond ex DC. subsp. *leiocarpa (Benth.) Urb.

M. leiocarpa Benth., Cat. Pl. Pyrenées Bas-Languedoc 100 (1826), (basionym). (Vassilczenko (1949) stated the type was described from Narbonne, southern France.)

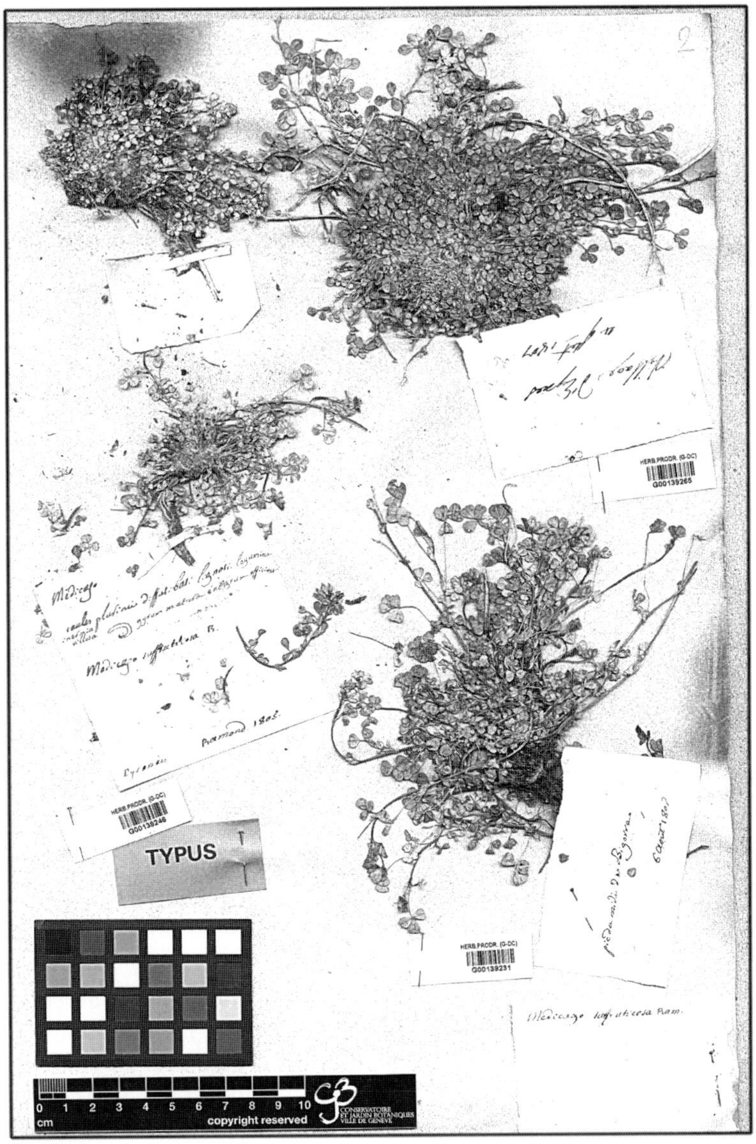

Fig. 94-4. Type of *Medicago suffruticosa* at G-DC. Photograph by P. Riedy, reproduced with the permission of the Conservatoire et Jardin botaniques de la Ville de Genève, Suisse.

M. suffruticosa Ramond ex DC. subsp. *leiocarpa* (Bentham) Urb., Verh. Bot. Vereins Prov. Brandenb. 15: 58 (1873).

M. suffruticosa Ramond ex DC. subsp. *leiocarpa* (Benth.) Rouy in Rouy & Foucaud, Fl. France 5: 16 (1899), comb. superfl.

Heterotypic

M. suffruticosa* subsp. *suffruticosa

M. suffruticosa Ramond ex DC. var. *ballii* Pau, Butl. Inst. Catalana Hist. Nat. 25: 2 (1925).

M. suffruticosa Ramond ex DC. var. *benthamii* Duby, Bot. Gall. 1: 123 (1828).

M. suffruticosa Ramond ex DC. a. *genuina* Willk. & Lange, Prodr. Fl. Hisp. 3: 378 (1878), nom. inadmiss.

M. paillecu Philippe, Fl. Pyren. 1: 211 (1859). (Assignment provisional.)

M. pailleri Philippe ex Nyman, Consp. Fl. Euro. 166 (1878). Merely listed as a synonym.

M. tomentosa Dulac, Fl. Hautes-Pyr. 273 (1867), nom. illeg., non Chevallier (1827).

M. suffruticosa Ramond ex DC. var. *villosa* Benth., Cat. Pl. Pyrenées 100 (1826).

M. suffruticosa Ramond ex DC. f. *villosa* Maire (Alleged type: Faurel, L., 11-VI-1938, pentes rocheuses au nord du Tizi d'Ighid vers 2700 m, Dj Masker, Grand atlas oriental, Algeria (MP), viewed online at http://www.aluka.org; current June 2010)

M. suffruticosa* subsp. *leiocarpa

M. suffruticosa Ramond ex DC. var. *glabra* Costa, Anales Soc. Esp. Hist. Nat. 3: 210 (1874).

M. suffruticosa Ramond ex DC. var. *glabrescens* Nègre, Luzernes du Maroc. 24 (1956).

M. lenticulata Pourret ex Benth., Cat. Pl. Pyrenées 100 (1826), merely listed as a synonym.

M. suffruticosa Ramond ex DC. var. *maroccana* Batt., Bull. Soc. Bot. France 54: 546 (1907). (Type: Ibrahim, 6-VI-1875, Djebel Afougheur, montagne au sud-ouest de la ville de Maroc, Morocco (MP), viewed online at http://www.aluka.org; current June 2010.) (Type inconsistent with the following.) (Battandier distinguished his new variety on the basis that it had long, thin petioles on the lower leaves, obcordate leaflets, and peduncles scarcely with more than two flowers.)

M. suffruticosa Ramond ex DC. subsp. *maroccana* (Batt.) Maire in Jahandiez & Maire, Catal. Pl. Maroc 2: 381 (1932). (Alleged type: Doutté, Tizi-n-Miri (Goundafa), Grand Atlas, Morocco (MP), viewed online at http://www.aluka.org; current June 2010.)

M. leiocarpa var. *segobricensis* Pau, Not. Bot. Fl. Españ. 4: 27 (1891), nom. nud.

Not

M. suffruticosa Grossh., Obz. Krym.-Kavk. *Medicago* 21 (1919), = *M. sativa*.

M. leiocarpa Guss., Syn. Fl. Sic. 2: 362 (1844), = *M. lupulina*.

Vernacular Names

English: Sprawling medic (medic is also spelled medick). Rios et al. (1993) translated the Spanish name *alfalfilla de monte* as "montain little alfalfa" (better, "little montane (or mountain) alfalfa").

French: La luzerne sous-ligneuse, luzerne souffrée. (In his 1805 description of his new species *M. suffruticosa*, De Candolle called it "luserne à souche".)

Geographical Distribution

(N = native, I = introduced, U = uncertain status)

Africa: Morocco (N).

Europe: France (N), Spain (N).

Medicago suffruticosa occurs in southern France (including the Eastern Pyrenées, Corbières, Languedoc), northeastern Spain (from the eastern Pyrenées to Valencia), and Morocco (Atlas Mountains).

With reference to *M. suffruticosa* subsp. *leiocarpa*, Rios et al. (1993) noted that "The taxon is ... distributed along the mountain arch formed by the eastern ranges of the Iberian Peninsula, and its prolongations in Morocco (Rif and Atlas) and France (Rosellon). In Spain, it is present in the mountains of Catalonia (Coastal range from low Llobregat to the Beseit Pass), Valencia (from Morella Pass to Dianique Mts.), Iberian System (ranges of Cuenca, Gúdar, and Javalambre), Andalusian Betic ranges (Alcaraz, Segura, Cazorla, and María) and Murcia (Espuña and Moratalla)."

Description

Perennial herb, stems prostrate to decumbent, woody at base, glabrous, glabrescent, or pubescent, 5–35(50) cm long, branching limited or abundant. A thick taproot develops below, and above a dense crown (numerous crowded stems arising from the top of the taproot). Leaflets obovate or obcordate, length less than twice width, (4)5–10 mm long, 3–5 mm wide, margin serrate (often three-quarters or more), upper surface glabrous, lower surface somewhat pubescent. Stipules ovate-lanceolate or ovate, or slightly dentate (usually at base, sometimes entire margin). Inflorescence with (2)3–9(10) flowers, the peduncle usually longer than the subtending petiole. Corolla 3–7(9) mm long, violet-blue in bud, yellow in open flower. Calyx 3–3.5 mm long, pubescent, the lobes equal to or longer than the tube. Young pod coils within calyx, then turns sideways through it. Mature pod 4–6(7.5) mm wide, with (1)2–4(5) coils, disk-shaped to short-cylindrical, dorsal suture lacking spines, coil face with radial veins that become almost concentric and produce an anastomosing network that enter the dorsal suture, dorsal suture convex, with articulated hairs (in subsp. *suffruticosa*), simple hairs, or glabrous (in subsp. *leiocarpa*). Seeds 2–5 per coil, oval or irregularly cordate, not as reniform as most other *Medicago*, greenish to brownish yellow, smooth, 2–3.25 mm long, 1.5–1.8 mm wide, radicle about half seed length to almost as long as seed, seeds not separated by fruit partitions.

Identification Hints

Medicago suffruticosa subsp. *suffruticosa* most probably would be confused with the possibly sympatric *M. sativa* subsp. *glomerata*. The leaflets of *M. suffruticosa* (both subspecies) are usually broader (length to width ratio usually greater than 3:2) than those of *M. sativa* (length to width ratio usually less than 3:2), and indeed most other perennial species of *Medicago*. *Medicago suffruticosa* also usually has fewer flowers (3–10 as opposed to 4–18), and somewhat longer radicles in the seed (usually about two-thirds of the seed length as opposed to about half). The following other related, somewhat similar perennials that sometimes have fruit covered with gland-tipped trichomes are found well to the east and are in any event distinct: *M. prostrata*, *M. rupestris*, and *M. cancellata* have much narrower leaflets; *M. papillosa* seeds are separated by spongy pads.

Medicago suffrutricosa subsp. *leiocarpa* is easily recognizable by the prominent, concentric venation of the pods. This taxon is unlikely to be confused with other species.

Habitat and Environmental Adaptation

Habitat

Medicago suffruticosa occurs in rocky places at altitudes of 800–3000 m (up to 2000 m in Europe, 3000 m in Morocco). It has been collected from forests, damp pastures, ravines, vineyards, and uncultivated fields. Quarro and Montard (1992) noted that the presence of the species under trees indicates tolerance to shade. Rios et al. (1993) recorded that in southeastern Spain the species is very scarce and usually found in protected sites such as rocky shelves or between thorny shrubs. With reference to Spain and subsp. *leiocarpa*, Rios et al. (1993) observed that it is associated with different plant communities, but frequently occurs in open pine woods with sparse ground cover, and in discontinuous matorrals (xeric shrub associations). It is also present in high mountains forming part of low grasslands and thorny matorrals.

Phenology

Flowering in its native habitat extends from March to August, depending on location (later at higher altitudes). With reference to *M. suffruticosa* subsp. *leiocarpa* in southern Spain, Rios et al. (1993) observed that "Plants flower early in the spring, when stem runners elongate, lasting until the beginning of summer if moisture conditions are good; at high mountain altitudes, flowering starts in May. Fruits ripen during summer, the first fruits maturing in July; later on, with the summer drought, most flowering stems dry off, remaining green only a few short stems around the crown, from which the plant recovers with the first autumn rains."

Soil

The species typically occupies calcareous soils. Rios et al. (1993) with reference to Spain recorded that the species "usually grows on foothills and shallow soils developed from calcareous rocks, but containing some organic matter; exceptionally, it can be found growing on bed-rocks of mountain gullies." McCoy (1987) found that *M. suffruticosa* has limited salt tolerance.

Temperature

The species occupies cool mountains, but does not seem adapted to very cold temperatures. Lesins and Lesins (1979) found that *M. suffruticosa* could not survive the very cold winters in Brandon, Manitoba. However, Rios et al. (1993) noted that the species does tolerate winter frosts in its natural habitat, and in southern Spain where the mean value of the coldest month was 4 °C, the plants did not become dormant during the winter.

Water

Medicago suffruticosa has been collected both from moist and dry substrates and, as noted above, subsp. *suffruticosa* is alleged to occupy moister areas at higher altitudes than subsp. *leiocarpa*. With reference to Spain and subsp. *leiocarpa*, Rios et al. (1993) noted that the taxon "grows at altitudes of 600–1600 m (Mesomediterranean and Supramediterranean bioclimatic belts) in zones where average annual rainfall is 350–400 mm or more (dry and subhumid ombroclimates)." Rios et al. also noted that when moisture conditions are good, the stems root at their nodes.

Biotic Relationships

Pollination

Small (1988) observed that the relatively large flowers and pollen to ovule ratio (346) are consistent with *M. suffruticosa* being at least partly cross-pollinated, like other perennial species of *Medicago*. Olivieri et al. (1991) characterized *M. suffruticosa* as selfing (based on observations of Lesins and Lesins 1979), while characterizing most of the other perennial species of the genus as outcrossing.

Dispersal

Since the pods lack spines that can adhere to fur, there are no obvious adaptations for seed dispersal in *M. suffruticosa*. Given the mountainous sites occupied, gravity doubtless plays a major role, and the lack of spines on the pods may facilitate their rolling downhill. Lesins and Lesins (1979) noted that although the plants did not overwinter in Brandon, Manitoba, they persisted by reseeding, showing a slight tendency to weediness.

Herbivory

No information located.

Nodulation and mycorrhizae

No information located.

Disease resistance

Steuckardt et al. (1972) found that *M. suffruticosa* is susceptible to *Verticillium albo-atrum* R., and Elgin and Ostazeski (1992) observed that *M. suffruticosa* was susceptible to Race 1 of anthracnose caused by *Colletotrichum trifolii* Bain, but had high resistance to Race 2.

Economic History, Cultivars, and Uses

This species has not been developed as an agricultural crop.

Cultivars

No cultivars have been bred.

Forage

Medicago suffruticosa is not currently sown for forage. Lesins and Lesins (1979) observed that in Brandon, Manitoba, spring growth and regrowth, as well as forage yields, were very light. However, they suggested that the species may be of value for establishing pastures in rocky, steeply sloped land at higher altitudes. Rios et al. (1993) described *M. suffruticosa* as a suitable new perennial pasture legume for Mediterranean dry alkaline areas. They observed that it is well grazed by wild fauna in Spain, and that preliminary cultivation studies indicated that it has good agricultural value. Rios et al. also reported a crude protein content of 18.6% and suggested (from growth of a small number of cultivated plants) that dry matter yield under cultivation would be in the order of 5.6 t/ha (about nine times as large as that produced by wild plants in nature). These authors also observed that the plants seem to tolerate grazing very well.

Fodder

The species is not employed as a source of dried animal feed.

Green manure, rotation crop, ground cover

No information located.

Other uses (salad sprouts, ornamental, plant products, etc.)

Medicago suffruticosa is used to a small extent as a rock garden ornamental, with the recommendation that it be grown in rock crevices, with well-drained soil, full sunlight, and protection against winter-wet conditions.

Chromosome Number

$2n = 16$: Clement 1962; Gillies 1972*b*; Lesins 1969; Lesins and Lesins 1979; Galland 1988 (as *M. suffruticosa* subsp. *maroccana*).

Lesins and Lesins (1979) stated that both subspecies have $2n = 16$.

Genetic Relationships and Gene Transferability

Intraspecific

Lesins (1969) crossed the two subspecies of *M. suffruticosa* with each other, noting that good seed set resulted (the pollen fertility of the hybrid was not reported).

Interspecific

Lesins (1969) crossed both subspecies of *M. suffruticosa* with *M. hybrida*. He observed that the F_1 plants had considerable pollen viability and seed set, indicating that the two species are closely related. However, Lesins (1969) was unable to hybridize *M. suffruticosa* with *M. sativa*.

Exchange by advanced biotechnology

No information located.

Conservation Status

Medicago suffruticosa is an uncommon species largely occupying remote mountainous regions. However, sheep grazing can affect many of the populations. There has been limited research conducted on the variation and ecology of the species, and it is poorly represented in germplasm institutions. Under the circumstances it is advisable that conservation assessments be prepared in the various countries in which the species occurs.

Germplasm Sources

Thirty-four accessions are identified as *M. suffruticosa* among the germplasm institutions with large *Medicago* collections, and an additional four are identified as subsp. *leiocarpa* (see Chapter 3).

Research Problems and Priorities

Medicago suffruticosa requires taxonomic, ecological, and agricultural clarification. Given the scarcity of seeds in germplasm collections, expeditions to acquire seeds, as well as ecological data, are recommended.

Representative Collections

Medicago suffruticosa subsp. *suffruticosa*

France: Billiet, P., Couméli-de-Héas (Hautes-Pyrenées), patûrages, alt. 1700 m, terr. granitique, flowers VII, fruits 10-IX-1886, Société dauphinoise, 1887 No. 5214 (P). Bordère, Couméli-de-Héas (Hautes-Pyrenées), alt. 1700 m, terr. granitique, pâturages, VII (fleurs)-10-IX-(fruits)-1886 (one branch in flower, another in fruit), Société dauphinoise 1887 No. 5214 (BM). Bordère, Couméli-de-Héas (Hautes Pyrenées), terrain granitique, alt. 1700 m, Société Rochelaise 1887, No. 2126, VII-IX-1886, flowering & fruiting branches present (BM). Bordère, Gèdre (Hautes Pyrenées), Gavarnie, alt. 1300 m., VII-1878, in flower & early fruit, Dr. C. Baenitz Herbarium Europaeum (P). Bordère, près de Gêdre (Hautes-Pyrenées), alterissements de Sauqué dans les montagnes granitiques, flowers 15-VI, fruits 20-VII-1868, F. Schultz Herbarium Normale, Cent. 11, No. 1039 (P). Bordère, Saugué (haute Pyrenées), VII-1865, in flower & fruit (P). Bouby, H. 3349, route du Tourmalet-Barèges (Pyr.), rocailles, alt. 1300 m, 21-VIII-1955, in flower & fruit. Bouby, H. 5681, rochers au-dessus de Fetges (Pyrenées orientales), alt. 1500 m, 17-VII-1962 (P). Bouby, H. 5715, Col. de la Lloze près de la Llagone (Pyrenées-orientales), alt. 1860 m, 15-VII-1962, in flower & early fruit (P). de Franqueville, A., gazons du pic du Midi (Hautes-Pyrenées), alt. 2000 m., 7-VIII-1854, in flower & fruit, Flora Galliae et Garmaniae Exsiccata de C. Billot No. 1852 (P). Déséglise, A., Pyrenées orientales, Monthouis, 6-VII-1872, in flower & profuse fruit (BM). Ellman, E. & Sandwith, N.Y. 44, Pyrenées Orientales, Vernet-les-bains, alt. 2000–3000 feet common, 4-7-VI-1925, in flower & fruit (K). Mailho, J.B., terrains arides à la Clape (Aude), 10-V-1892, in flower & fruit, Société Rochelaise 1892, No. 3252 (P). Raynal, J. 3121, Col. de Puymorens (Pyrenées orientales), pelouse alpine, alt. vers 2300 m, 11-VII-1958, in fruit (P). **Spain:** Hutchinson, J., Matthews, J.R. and Riley, L.A.M. 169, Flora Pyrenaica, Port de Spain, alt. 7000 feet, 24-VII-1922, in flower & fruit (K). Meinertzhagen, R., Pont de Gavarnie, Hautes Pyrenées, alt. 7000 feet, alpine pasture, 29-VIII-1932, in flower & early fruit (BM). Sandwith, N.Y. 4593, Prepireneos Arayonenses, between Fiscal and Sabiñanigo, Basarán, alt. ca. 1250 m. on bare ground by paths, 3-VII-1956, in flower & fruit (K). Sennen, F. 4378, Cerdagne, Odeillo, talus, alt. 1700 m, 18-VII-1922, in flower & fruit, (K, P). Sennen, F. 5697, Cerdagne, Targasonne, alt. 1600 m, talus, 15-VI-1926, in flower & early fruit (BM). Sennen, F. S698, Cerdagne, Sarèje, lieux piétinés, alt. vers 1300 m, 28-VI-1926, in fruit (BM).

Medicago suffruticosa subsp. *leiocarpa*

France: Biche, B., Hérault, garriques de Castelnau-de-Guers, 3-VI-1886, in flower, Flora Selecta Exsiccata Publié par Ch. Magnier No. 1409 (P). Drake, E., Herbarium of, Aude, Garrigues de la Clappe près Narbonne, 7-VI-1867, in flower & fruit (P). Raynal, J. 4370, Gruissau (Aude), 16-V-1959, in flower & fruit (P). Sennen 5672, Aude – La Nouvelle, garriques, V-1900, in flower & fruit (P). **Morocco:** Chaworth-Musters, J. 176, High Atlas, Djebel Bou Ourloul above Tizi N'Tichka, alt. 2500–3000 m, 19-IV-1937, in fruit (BM).

Davis 55528, GA, J. Toubkal, between Meltner hut & Sidi Chamarouch, alt. 2600–2700 m, rocky igneous slopes, forming dense hard perennial mats, 24-VII-1973, in fruit (BM). Gattefossé, J., Grand-Atlas, berges de l'Oued Reraïa, près Arround, alt. 2200 m, 20-VII-1936, in fruit (K). Hooker, south Morocco, Greater Atlas, Reraia, 7000–8000 feet, V-1871, in flower & early fruit (P). Jahandiez, E. 413, Moyen Atlas, Aïn-Leuh, bois des montagnes calcaires, alt. 1550 m, 23-V-1924, in flower & fruit (P). Jury, S.L., Rejdali, M. & Watson, M.F. 8966, 72 km south from Marrakech, Oukaïmeden, near base of ski lift to mountain summit (31°12′N, 7°51′W), alt. 2650 m, very closely appressed to rocks, 5-VII-1987, in flower & fruit (BM). Lindberg, H. 4291, Atlas medium, Ras-el Ma prope pag. Azrou, alt. ca. 1600 m, 24-VI-1926, in flower & fruit (K). Maire, R. HG1921, Moyen-Atlas, Tiwbadit, rocailles basaltiques au bord du Guigau, alt. 1800–1900 m, 26-VII-1921, in flower & early fruit (P). Reading Univ./B.M. Exped. 744, GA, High Atlas, Oukaïmeden, Jebel Angour (30°12′N, 7°54′W), alt. 2650 m, 15-VI-1974, in flower & early fruit (BM). Balls, E.K. B2820, Djebel Erdous, alt. 11 000 feet, 13-6-1936, in flower (seed in fruit in packet collected Tashdirt to Arround, alt. 7000 feet, 19-VII-1936), growth woody and closely appressed to rocks or ground, no more than 1 foot tall, making mats to 3 feet across, nonlime rocks or open ground, often in partial shade and somewhat moist places (BM). Cosson, E., Herbarium of ("par les soins de M. Beaumier"), Djebel Afougueur, montagnes à l'ouest de la ville de Maroc, Ibrahim, 1-VII-1876 (P). **Spain:** Cannon, P.F. et al., Albacete, Sierra de Segura, 15 km west of Yeste, 1.5 km east of Prados (38°22′N, 2°25′W), alt. 1000 m, open maquis near track, in fruit (BM). Lassen, P. 12-1, prov. Alicante, just north of Bolulla, alt. 200–250 m, moist ground, 11-IV-1979, in flower & fruit (BM). Porta & Rigo 438, Albacete, in glareosis montis mugron, sol. calcareo, alt. 1000–1500 m, 1890, in fruit (K, P). Reverchon, E., Province de Teruel; Balaclocha, dans les vignes incultes, sur le calcaire, alt. 1000 m, VI-1892, in flower & fruit, E. Reverchon Plantes d'Espagne 1892 No. 687 (K, P). Sandwith, N.Y. 4991, prov., Termel, Sierra de Gúdar, lower slopes, above Alcalá de la Silva, alt. ca. 1500 m, 6-VII-1957, in flower & fruit (K). Sennen, F. & Jerónimo 6427, Murcie, Sierra de Espuña, flowers collected 17-V, fruit collected 2-VII-1927 (BM). Sennen, F. 7235, Barcelone, Vallirana, au Lladoner, pâturages calcaires, alt. 4500 m, 20-VI-1929, in flower & fruit (BM). Willkomm, Province de Valance, Sierra de Espadan, lieux herbeux, sur le calcaire triasique, alt. 800 m, VI-1891, in flower & early fruit, E. Reverchon Plantes d'Espagne 1891 No. 687 (K, P). Willmott, A.J., prov. Granada, between Granada and Jaen, rocky roadside near Inquisition, 10-VI-1927, in fruit (BM).

95. *Medicago syriaca* E. Small

M. syriaca E. Small, Canad. J. Bot. 68: 1476 (1990), (basionym).

The epithet *syriaca* denotes the Syrian distribution of this species, first known only from Syria. Subsequently, Mehregan et al. (2002) discovered the species in Iran.

Notes

This species was described in 1990 based on an Australian germplasm accession derived from a west-central Syrian collection, and forwarded for identification. Subsequently, collections were made in southwestern Iran. Because data are very limited, this species requires much study to clarify its characteristics, geography, and ecology.

To check the possibility that the population in question might have arisen recently as an interspecific hybrid, pollen fertility was evaluated by examining stainability with cotton blue (Small 1990*a*). Species currently assigned to *Medicago* section *Spirocarpos* subsection *Pachyspirae* are highly intersterile, and hybrids, when such can be generated, are highly infertile (Lesins and Lesins 1979). The high pollen fertility observed (about 100%) makes recent hybrid origin very unlikely.

Subordinate Taxa

None recognized here.

Type Specimen

E. Small & B. Brookes M1348 (plant 2), grown in an Ottawa (Ontario, Canada) greenhouse, 26 July to 11 November 1988 (holotype: sheet 1 of 3 sheets, all from "plant 2," DAO!).

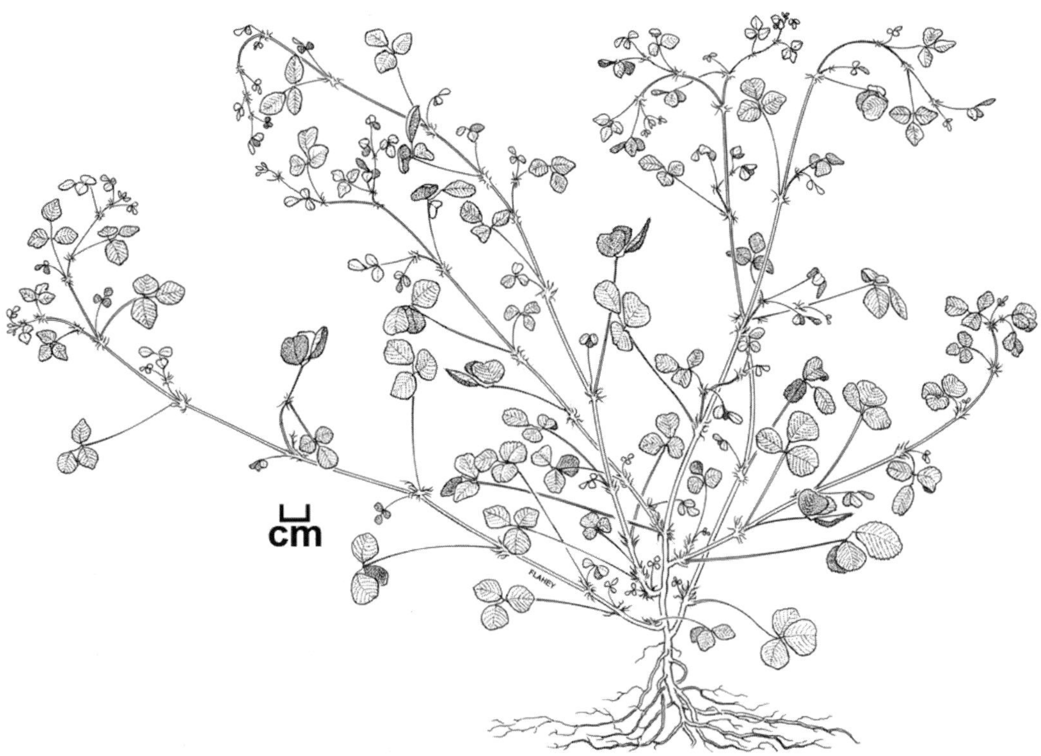

Fig. 95-1. *Medicago syriaca.* Vegetative plant. (Based on E. Small M1348, DAO; artist: B. Flahey.)

Synonymy

Homotypic

No names found.

Heterotypic

No names found.

Vernacular Names

ENGLISH: No names found. Syrian medic is an obvious name (medic is also spelled medick).

FRENCH: No names found.

Geographical Distribution

(N = native, I = introduced, U = uncertain status)

ASIA: Iran (N; Mehregan et al. 2002), Syria (N).

The population from which *M. syriaca* was described was originally collected from Haylin, 37 km northwest of Hama, Syria. Mehregan et al. (2002) stated that the species is "well distributed in SW Iran" and occurs from 20–800 m. A map in Mehregan et al. (2003) shows a collection at approximately latitude 33°, longitude 47°.

Description

Annual herb, more or less glabrous to slightly pubescent. Stems procumbent to ascending, branched near base. Leaflets ovate to obovate, margins serrate 50% to 80% length of leaflet from tip towards base, leaf tip obtuse and with terminal tooth. Stipules strongly laciniate. Flowers 1–4 per peduncle, yellow, 2.5–4 mm long, calyx about two-thirds length of corolla, calyx

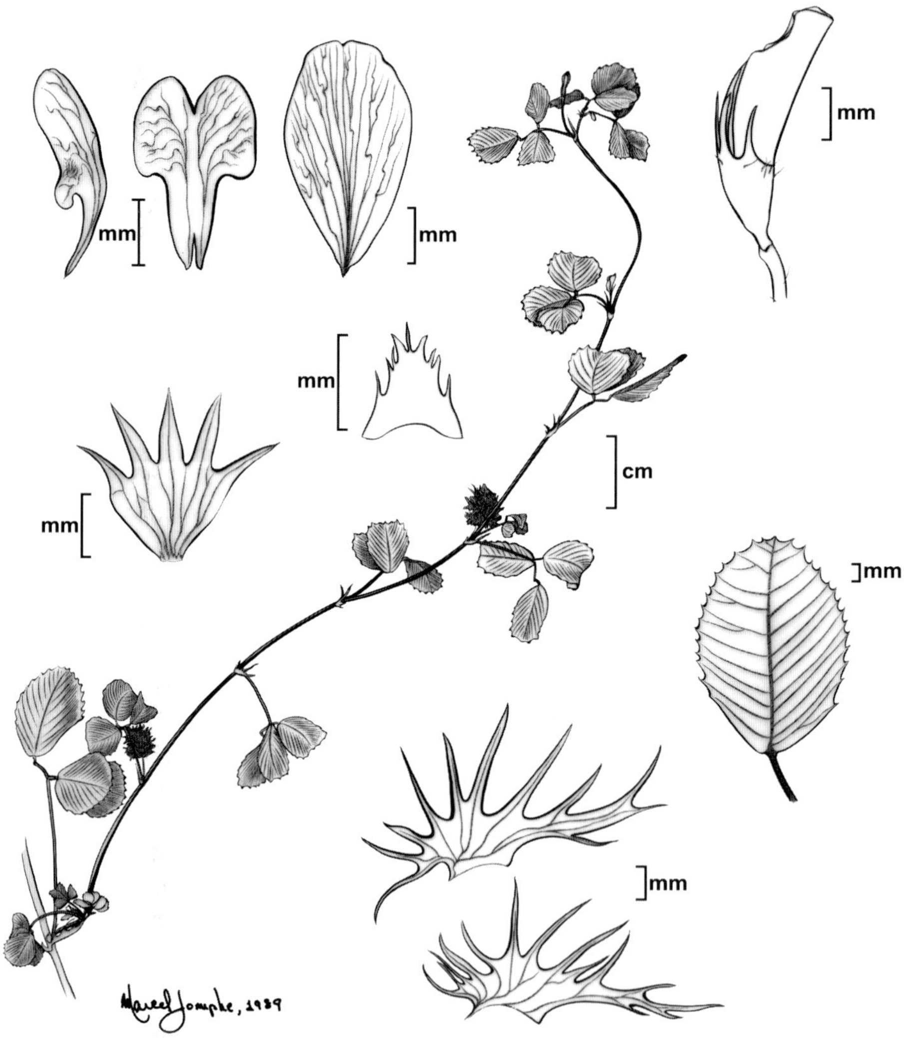

Fig. 95-2. *Medicago syriaca.* **Upper left:** Floral parts (clockwise from upper left: wing petal; keel (the two petals separated adaxially and flattened); standard petal; flattened androecial column (split beside free stamen, which is not shown, anthers also not shown); calyx (split and flattened). **Remainder, clockwise**: Branch; side view of flower; leaflet; stipules. (Based on E. Small M1348, DAO; artist: M. Jomphe.)

lobes about half calyx length. Young fruit emerges sideways through the calyx. Mature pod cylindrical (barrel-like), glabrous, yellow-brown to dark brown at maturity, 3–7 mm long, with 3–6 coils, coils not tightly appressed (gaps usually present between some coils at maturity), hardened at maturity. Facing surfaces of coils show 8–12 very strongly curved veins (almost concentric towards periphery) with few branches and anastomoses, venation obscured at maturity by spongy tissue, peripheral 10% of coil face often more or less veinless, coil edge blunt (lateral veins bordering coil edges and dorsal suture at about same level), coil edge with spines or tubercles (populations with fruits lacking spines or tubercles to be expected), spines when present about perpendicular to fruit axis. Seeds 2–3.5 mm long, 1–2 per coil, smooth, yellow to brown yellow.

This description is based mostly on plants grown under greenhouse conditions in Ottawa, Ontario. Such cultivation invariably has produced plants with more luxuriant vegetative growth than those collected in nature in the mostly arid habitats occupied by annual species of *Medicago*. Accordingly, the size of the shoots and leaves has been omitted. Greenhouse plants had shoots up to 40 cm long, but in nature one could expect plants with shoots 10–30 cm in length (the size range of the Mehregan specimens cited below).

The material grown in the greenhouse produced seeds with a convex (dorsal) surface that was more undulate (see Fig. 95-3) than previously observed in the genus. However, as this variable character may have been influenced by the greenhouse conditions, it is not included as part of the species description.

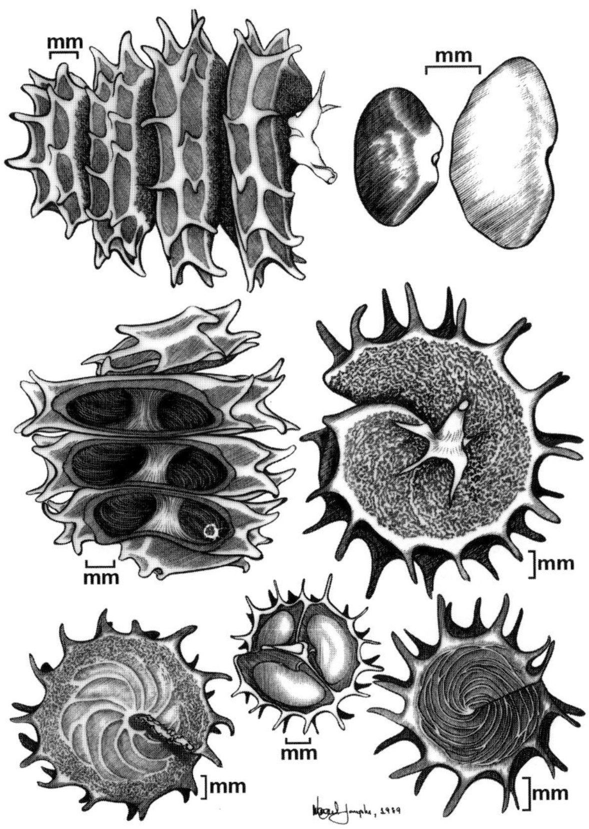

Fig. 95-3. *Medicago syriaca.* **Clockwise from upper left:** Fruit with four spirals, pedicel at right; seeds; end view of mature fruit (pedicel end; note how spongy tissue obscures venation); face view of coil of young fruit (note highly curved radial veins becoming almost concentric near periphery, where a few anastomoses occur; also note irregular and limited veinless area towards periphery); transversely sectioned fruit showing cavities with seeds; end view of young fruit (pedicel end; note peripheral veinless area); sagittally sectioned fruit showing cavities where seeds are located. (Based on E. Small M1348, DAO; artist: M. Jomphe.)

Identification Hints

Medicago syriaca can be distinguished from its most similar relatives by its glabrous, cylindrical fruits with very strongly curving, little-anastomosed radial veins, accompanied by loosely appressed coils with a tendency to show in immature fruits a limited peripheral veinless area on the coil faces.

A useful but variable character for confirming the identification of this species is the proportion of the leaflet margin that is serrated. Most species of subsection *Pachyspirae* have leaflets in which less than three-quarters of the margin is serrate, but leaflets with three-quarters of the margin serrate were common in the observed plants of *M. syriaca* (see Fig. 95-2). The following notes contrast *M. syriaca* with the species most likely to be confused with it.

The closely related pair *M. murex* and *M. sphaerocarpos* are reminiscent of *M. syriaca* in having very limited pubescence on the shoot, fairly dissected stipules, and fruits with coil faces with a peripheral veinless areas. Also, *M. syriaca* and *M. murex* both have a chromosome number of $2n = 16$. However, the veinless zone on the periphery of the fruit is not as well developed in *M. syriaca* as in the other species, the venation on the fruit coil face of *M. syriaca* is much more strongly curved than in the other species, and the fruit is usually barrel-like as opposed to the more frequent spherical or ellipsoid shape in the other species. *Medicago sphaerocarpos* has distinctive grooves on the coil edge by which it can be distinguished with certainty from other species currently placed in section *Spirocarpos* subsection *Pachyspirae*. Interestingly, the DNA phylogenetic analysis of Bena (2001) indicated that the nearest relations of *M. syriaca* are *M. murex*, *M. sphaerocarpos*, and *M. polymorpha*.

Medicago turbinata is the only other species of subsection *Pachyspirae* that develops a peripheral veinless zone on the fruit coil face. It also shares a chromosome number of $2n = 16$ with *M. syriaca*. However, *M. turbinata* is easily distinguished by the greater pubescence of its shoot, and fruits with a wide peripheral veinless zone (up to a third of the coil diameter) with no gaps between the coil faces at maturity, and venation with rather limited curving.

The closely related pair *M. rigidula* and *M. rigiduloides* are quite similar to *M. syriaca*. However, the majority of plants of *M. rigidula* and *M. rigiduloides* have glandular-pubescent fruit, whereas those of *M. syriaca* are glabrous or glabrescent (similarly the shoots of *M. rigidula* and *M. rigiduloides* are generally quite pubescent, whereas those of *M. syriaca* have very limited pubescence). The usual chromosome numbers are also different: $2n = 14$ in *M. rigidula* and *M. rigiduloides* and $2n = 16$ in *M. syriaca*.

Medicago constricta shares with *M. syriaca* the feature of having pods with coil faces with strongly curving, almost concentric venation. However, *M. constricta* shoots are generally much more pubescent than those of *M. syriaca*, the stipules are less dissected, and the pods at maturity show no traces of gaps between the coils, whereas gaps are usually present in the fruits of *M. syriaca*. The chromosome number of *M. constricta* ($2n = 14$) is also different from that of *M. syriaca* ($2n = 16$).

Medicago doliata has the same chromosome number as *M. syriaca*, but is more pubescent, and has fruits distinguishable from those of *M. syriaca* in their ovoid shape, lack of gaps between the coils at maturity, and strong anastomosing of veins toward the periphery of the coil face.

Habitat and Environmental Adaptation

Habitat

No information located.

Phenology

The species flowers from February to March in southwestern Iran.

Soil

No information located.

Temperature

The species appears to germinate in the cool, late-winter period and complete its life cycle in early summer as conditions become hot and dry.

Water

The Hama area of Syria, the type locality of *M. syriaca*, is subject to considerable drought, but is known to have been irrigated locally. Much of southwestern Iran is similarly arid, albeit with local irrigation. It is likely that *M. syriaca* is typical of annual *Medicago* in being a drought-evading, winter annual.

Biotic Relationships

Pollination

The small flowers indicate that this species is largely an inbreeder.

Dispersal

The indehiscent fruits have short spines without pronounced hooks, so they are not ideal for attaching to animal fur, but this may be the most important mechanism for dispersal. However, given that very few collections are known, dispersal seems to be very limited.

Herbivory

No information located.

Nodulation and mycorrhizae

No information located.

Disease resistance

No information located.

Economic History, Cultivars, and Uses

This species has not been the subject of agricultural interest to date.

Cultivars

No cultivars have been bred.

Forage

Medicago syriaca has not been sown for forage.

Fodder

This species has not been harvested for dried animal feed.

Green manure, rotation crop, ground cover

No information located.

Other uses (salad sprouts, ornamental, plant products, etc.)

No information located.

Chromosome Number

$2n = 16$: Small 1990*a*.

Genetic Relationships and Gene Transferability

Intraspecific

Artificial hybridization of different populations has not been attempted to test for the presence of breeding barriers within the species, and indeed few populations of the species are known to date.

Interspecific

No information located.

Exchange by advanced biotechnology

No information located.

Conservation Status

Given the rarity of this species, it needs a conservation assessment as well as measures to protect it.

Germplasm Sources

The germplasm collection from which *M. syriaca* was recognized is SA 14197, collected by the Northfield Research Laboratories and Research Centre, Department of Agriculture, South Australia, which ceased operations in 1993. The accession is housed in the annual *Medicago* genebank of the South Australia Research and Development Institute (SARDI) in Adelaide.

Research Problems and Priorities

Medicago syriaca is a recently discovered species about which very little information is available. Priority should be given to investigating its botanical and agricultural characteristics, studying its ecology in the field, and collecting its germplasm.

Representative Collections

Iran: Mehregan et al. (2002) cite Mehregan 2.15, Mehregan 2.17 and Telvari 372, all from southwestern Iran, and preserved at TARI. Mehregan deposited the following collections at DAO: Mehregan, I., 750 m (alt.), 48 km Gatchsaran to Shiraz, Boustan, southwestern Iran, 1-IV-2002; Mehregan, I. 445 m (alt.), Luristan: 70 km Andimeshk to Khorram Abad, southwestern Iran, 28-III-2002. **Syria:** E. Small & B. Brookes M1348, several collections from greenhouse-grown plants, including the type specimen and two isotype sheets from the same plant, referred to above, and an additional isotype (grown at the same time: three sheets of "plant 3"), and the chromosome voucher (two sheets labelled M1348c, both from the same plant) (DAO).

96. *Medicago tenoreana* Ser.

M. tenoreana Ser. in DC., Prodr. 2: 180 (1825), (basionym).

The epithet *tenoreana*, based on Tenore plus *-ana* (from the Latin *-anus*, "belonging to"), commemorates Michele Tenore (1780–1861), an Italian botanist who authored numerous botanical publications.

Subordinate Taxa

None recognized here.

Type Specimen

Not determined. In her 1963 monograph "The Annual Species of *Medicago*", Heyn noted that "Tenore described (1811) and depicted (1830) a species from Naples under the name of *M. cancellata*. As the epithet "*cancellata*" has earlier been used by Marschall von Bieberstein (1808) for another species of *Medicago*, *M. cancellata* Tenore is a homonym. Seringe in DC. (1825) again described the same species as *M. tenoreana*, this time from a specimen of Balbis from

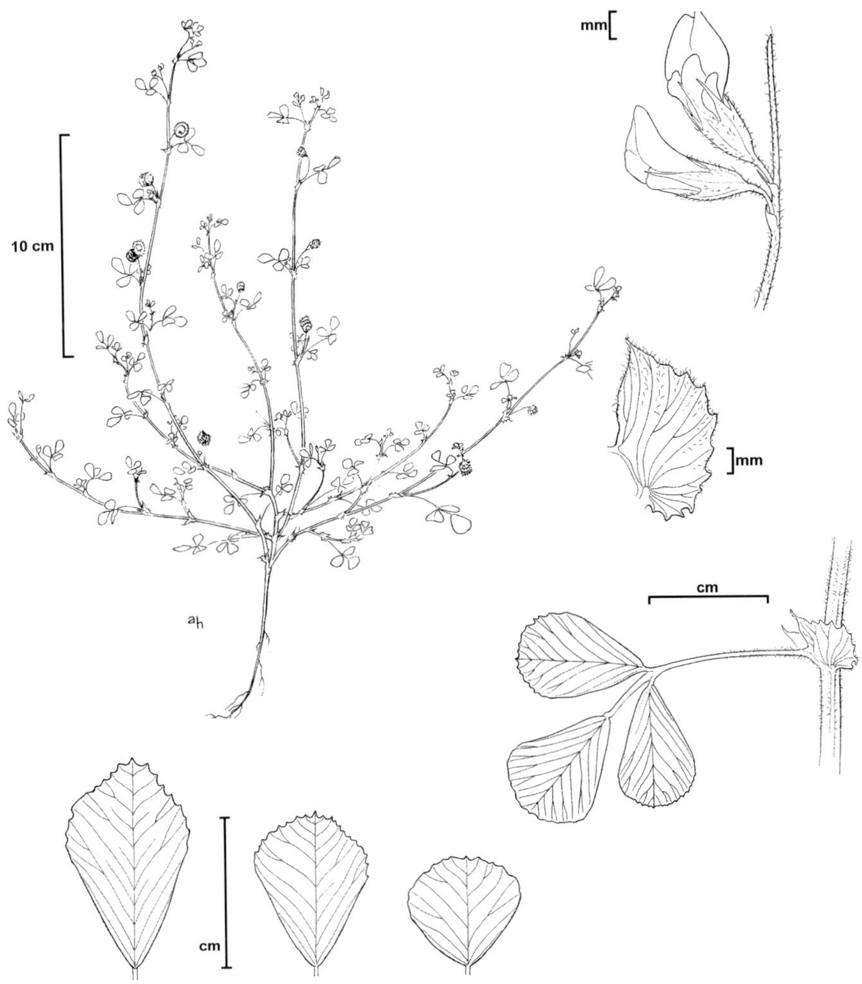

Fig. 96-1. *Medicago tenoreana.* **Clockwise from upper left:** Habit drawing of plant; inflorescence with peduncle (note extended axis of peduncle); stipule; leaf attached to stem; leaflets. (Based on E. Small M 341, DAO, except the habit drawing, which is based on Stevenson 3597, DAO; artist: A. Hanes.)

Naples collected prior to 1811. If such a specimen exists — it should be deposited at the Herbarium of the Botanical Garden of Naples, but unfortunately the author has not been able to obtain authentic material from this herbarium." Seringe (1825) noted "in agris circa Neapolim. *M. cancellata* Ten! cat. 1819. p. 58. non Bieb. (v. s. in herb. Balbis et Reynier nunc Dunant.)" An inquiry to Naples regarding a type did not receive a reply.

Synonymy

Homotypic

No names found.

Heterotypic

M. cancellata Ten., Fl. Nap. XLIV (1811), non M. Bieb. Fl. Taur.-Cauc. 2: 226 (1808).

M. fenstralis Gerh., in herb.

Vernacular Names

ENGLISH: Tenore's medic is an obvious name for this species, based on a literal translation of the Latin name. The name "Italian medic" has been encountered, but is more appropriate for *M. italica* (medic is also spelled medick).

FRENCH: La luzerne de Tenore [Ténore].

Geographical Distribution

(N = native, I = introduced, U = uncertain status)

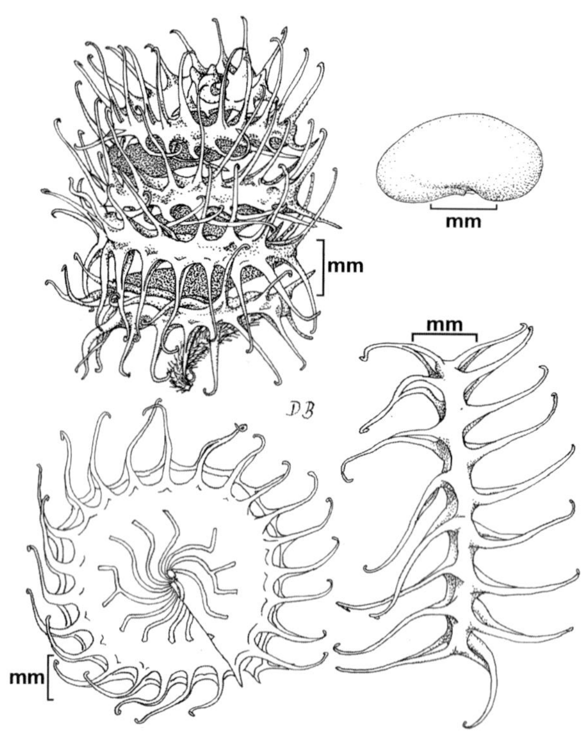

Fig. 96-2. *Medicago tenoreana*. **Clockwise from upper left:** Fruit; seed; edge of coil; fruit coil showing venation. (Based on E. Small M341, DAO; artist: D. Beddoes.)

EUROPE: France (N), Malta (N), Monaco (N), Spain (several authorities question the status), Italy (including Sardinia and Sicily; N), Yugoslavia (N).

Medicago tenoreana is native to the northwestern Mediterranean area.

Description

Annual herb, usually procumbent, branching from base, stems (10)20–35(50) cm long. Vegetative parts pubescent with simple hairs, sometimes also with glandular hairs. Leaflets 6–10 mm long, 3–5 mm wide, obovate to obcordate, distal third (approximately) of margin serrate, pubescent with simple hairs especially on the lower surface, glandular hairs when present are only on the lower surface. Stipules acuminate, margins nearly entire to serrate. Inflorescence with 1–2(4) flowers, the peduncle shorter than the subtending petiole. Flowers 3–5(7) mm long. Corolla yellow, less than twice length of calyx. Calyx pubescent, teeth slender and longer than tube. Young pod with simple and sometimes also glandular hairs, the pod protruding sideways through calyx. Mature pod cylindrical, glabrous or glabrescent, 5–7 mm long, 5–7 mm wide, the 4–6 coils pergamentaceous, loosely appressed (laxly spiralled), the coil face with radial veins that enter a veinless margin at the outer third of the coil diameter. All coils, but sometimes not the first and last with spines, spines ca. 2 mm long, grooved, sometimes hooked, oriented more or less perpendicular to coil face (on each side of the coil) and often slanted somewhat in opposite direction to pod coiling, the spine bases inserted remotely from the dorsal suture. Seeds 1–2 per coil, 2.5–4 mm long, 1–1.3 mm wide, not separated by fruit partitions, radicle somewhat longer than half to two-thirds as long as the seed, seedcoat smooth, yellow to light brown.

Identification Hints

Medicago tenoreana and *M. disciformis* possess some similarities, especially the presence of a peripheral veinless area on the fruit coil face. *Medicago disciformis* has spines slanted towards the fruit base, whereas in *M. tenoreana* the spines point both towards the base and the tip. *Medicago tenoreana* also has thicker coil edges.

Habitat and Environmental Adaptation

Habitat

Medicago tenoreana occurs on dry rocky slopes of hills and has been collected from the edges of fields.

Phenology

Flowering in its indigenous area occurs from April to June.

Soil

Like most other annual species of *Medicago*, *M. tenoreana* appears to prefer calcareous, well-drained soils. Piano and Francis (1992) recorded that in its native area, the species occurs in loam-clay soils that are slightly to moderately basic.

Temperature

No information found. The locations occupied suggest that *M. tenoreana* is a typical winter annual, germinating in the cool conditions of late winter and setting seed by early summer when temperatures are increasing substantially.

Water

This species typically occurs in moderately dry substrates. Piano and Francis (1992) recorded that in its native area the species requires a minimum annual rainfall of 300 mm.

Biotic Relationships

Pollination

The small flowers indicate that *M. tenoreana* is an inbreeder.

Dispersal

The fruits have spines that are sometimes hooked, but the spines appear to be inefficient for attaching the fruits to fur.

Herbivory

No information located.

Nodulation and mycorrhizae

No information located.

Disease resistance

O'Neill et al. (2003) examined the comparative resistance of 36 annual species of *Medicago* to *Phoma medicaginis* (the cause of spring black stem and leaf spot, a major disease of alfalfa) and concluded that *M. tenoreana* had high resistance to it. Elgin and Ostazeski (1992) found that *M. tenoreana* was more than 90% resistant to Races 1 and 2 of anthracnose caused by *Colletotrichum trifolii* Bain. However, in a comparative test of 36 annual species of *Medicago*, O'Neill and Bauchan (2000) found that one accession of *M. tenoreana* was completely susceptible to anthracnose caused by *Colletotrichum trifolii*. Yaege and Stuteville (2002) tested the resistance of 33 annual species of *Medicago* to powdery mildew (*Erysiphe pisi*) and found that the single collection of *M. tenoreana* grown was susceptible to the disease (it was observed that often only one of a number of populations of other species tested was resistant).

Economic History, Cultivars, and Uses

This species has not attracted agricultural interest.

Cultivars

No cultivars have been bred.

Forage

Medicago tenoreana is not sown for forage.

Fodder

This species is not employed as a source of dried feed for livestock.

Green manure, rotation crop, ground cover

No information located.

Other uses (salad sprouts, ornamental, plant products, etc.)

No information found.

Chromosome Number

$2n = 16$: Ghimpu 1928, 1929*a*, 1929*b*, 1930; Simon and Simon 1965; Lesins and Lesins 1979.

Genetic Relationships and Gene Transferability

Intraspecific

Artificial hybridization of different populations has not been attempted, so it is not known whether there are barriers to hybridization within the species.

Interspecific

Lesins and Lesins (1979) were unable to hybridize *M. tenoreana* with *M. disciformis*.

Exchange by advanced biotechnology

No information located.

Conservation Status

Medicago tenoreana is an uncommon species, but is not rare, and does not seem in need of conservation measures.

Germplasm Sources

Sixteen accessions are listed among the germplasm institutions with the largest *Medicago* collections (see Chapter 3).

Research Problems and Priorities

Medicago tenoreana is a rather typical annual species of *Medicago* that does not seem to offer any particular advantages as a special crop. However, it may well have genes of value for improvement of other species, and in this regard deserves additional basic botanical and agricultural study.

Representative Collections

France: Albert, A., Revest, champs cultivés, 10-V-1886 (P). Albert, A., Var, Le Prevest, champs, 8-V-1886, in fruit, Flora Selecta Exsiccata, Publié par Ch. Magnier No. 144 (P). Chambeiron, près de Toulon (Var.), au pied du mont Faron, bord des champs calcaires incultes, 20-VI-1860, in fruit, Flora Galliae et Germaniae Exsiccata de C. Billot No. 3036 (K, P). Reynier, A., Toulon (Var.), pentes herbeuses du Feron au Fort-Rouge, 7-V-1916, in fruit (P). **Italy:** Fiori, A., Apulia, Mons Garganus, in herbosis prope vico Garganico, loco i Cappuccini dicto, alt. 400 m, solo calcareo, 25-V-1913, in flower & fruit, Flora Italica Exsiccata Series III Fiori & Béguinot No. 2098 (BM, K). **Malta:** Borg, J., an alien naturalized in the Botanic Gardens, Malta, IV-1931, in flower & fruit (K). **Yugoslavia:** Pichler, Dalmatia, in agris cefsantil (?), insulae Lesinae, V-1870, in fruit (K).

97. *Medicago truncatula* Gaertn.

M. truncatula Gaertn., De Fruct. Sem. Pl. 2: 350 (1791), (basionym).

The epithet *truncatula* is based on the Latin *truncatus* (truncate), plus *-ulus* (diminutive suffix), a reference to the somewhat truncate ends of the barrel-like fruits (*quasi truncatum* was the Latin phrase used by Gaertner in his original description).

Key to Subordinate Taxa

Pods coiled fairly tightly .... *M. truncatula* var. *truncatula*
Pods coiled very laxly .... *M. truncatula* var. *laxicycla*

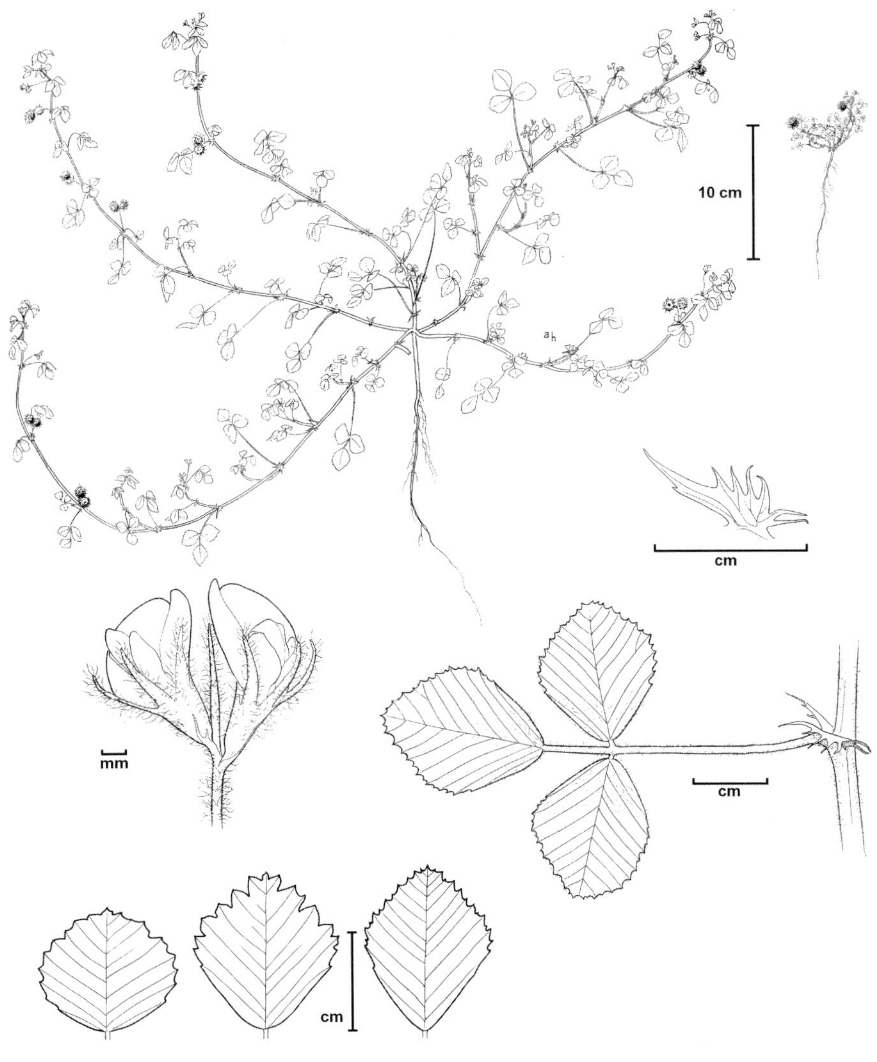

Fig. 97-1. *Medicago truncatula.* **Clockwise from top left:** Habit drawing of a fruiting plant; habit drawing of a diminutive fruiting plant (illustrating ability of the species to reproduce under inhospitable conditions); stipule; leaf attached to stem; leaflets; inflorescence with peduncle. (Based on E. Small M943, DAO, except the diminutive plant, which is based on Eig et al., loan number 99/319, HUJ; artist: A. Hanes.)

Medicago truncatula var. *laxicycla*, easily distinguished from the remainder of the species (*M. truncatula* var. *truncatula*) by its laxly coiled pods, is discussed in detail below.

Ellwood et al. (2006*a*) used simple sequence repeat (SSR) loci to examine representatives of the large *M. truncatula* collection of the South Australian Research and Development Institute (SARDI), and concluded that subspecies *tricycla* (based on 15 collections so labelled) is genetically distinct from subspecies *truncatula*. I was suspicious that because of widespread hybridization and introgression of *M. littoralis* with *M. truncatula* (Small and Brookes 1990*a*), the accessions labelled as subspecies *tricycla* were in fact introgressants, and this explained their genetic distinction from the other samples of *M. truncatula* examined. At my request, S. Ellwood kindly supplied fruits of 11 of the 15 collections identified as subspecies *tricycla*. None of the 11 samples was "pure" *M. truncatula* and all showed background of *M. littoralis*. Five of the samples (10964, 11753, 17498, 18395, 18935; SARDI accession codes) indeed appeared to be *M. littoralis* rather than *M. truncatula*. Two of the samples (2729 and 10962) appeared to be closer to *M. littoralis* than to *M. truncatula*. Sample 2715 seemed to be more or less intermediate between *M. truncatula* and *M. littoralis*. The remaining three samples possessed small pods with small spines: 12564 and 22922 appeared to be *M. littoralis* introgressed with *M. italica* (i.e., they were mostly *M. littoralis*), and 28645 was an odd form that may have been intermediate between *M. littoralis* and *M. truncatula*. Accordingly, it is clear that the group of accessions collectively labeled subspecies *tricycla* appeared to be distinctive because of genetic representation from *M. littoralis*.

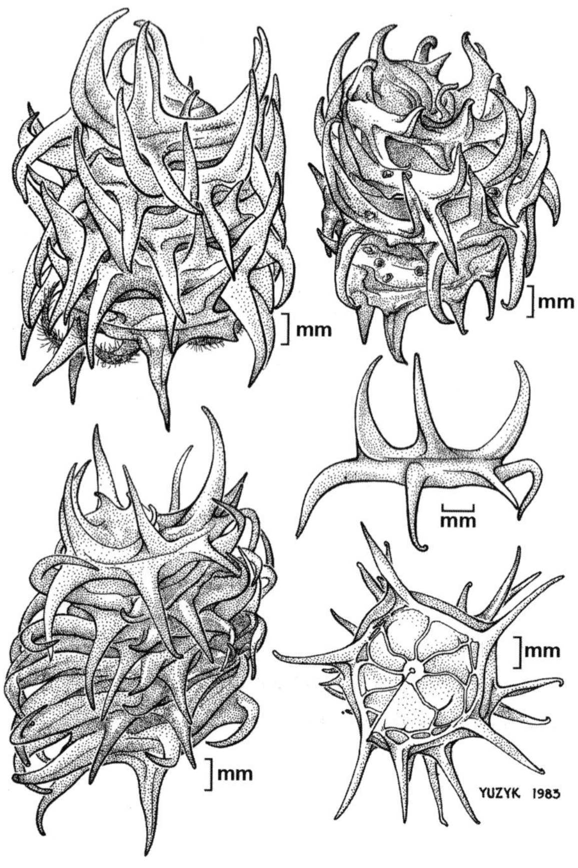

Fig. 97-2. *Medicago truncatula*. Variation in fruit morphology. The fruits clockwise from bottom left were originally identified respectively as *M. truncatula* vars. *longeaculeata, truncatula*, and *tricycla*. **Center right:** Coil edge. **Bottom right:** Fruit coil face showing venation. (Top left based on Eig, Zohary & Feinbrun, 9-5-1927, HUJ; top right based on Eig, Zohary & Feinbrun, 4-4-1927, HUJ; remainder based on Leinkram 24-3-1955, HUJ; artist: L. Yuzyk.)

Additional evidence that subspecies *tricycla* is a doubtful taxon was presented by Quiros and Ostafichuk (1983), who were unable to find any enzyme distinction of an accession identified as subsp. *tricycla* and seven other populations of *M. truncatula* otherwise identified. Given currently available evidence, the concept of subspecies *tricycla* is quite unsupported.

Type Specimen

Lectotypified by Heyn (1963) using an illustration (tab. 155, Gaertn., Fruct. Sem. Pl. 2: 350 (1791)) — a small drawing of a single fruit. (Sales and Hedge (1994) commented that in the absence of a suitable specimen this drawing was a suitable choice as the lectotype.)

Synonymy

Homotypic

M. tribuloides Desr. in Lam. var. *truncatula* (Gaertn.) Koch, Syn. Fl. Germ. 162 (1837).

M. tribuloides Desr. in Lam. subsp. *truncatula* (Gaertn.) Nyman, Consp. Fl. Eur. 167 (1878).

Heterotypic

M. tribuloides Desr. in Lam. var. *breviaculeata* Moris, Fl. Sard. 1: 441 (1837).

M. truncatula Gaertn. var. *breviaculeata* (Moris) Urb., Verh. bot. Ver. Brandenb. 15: 67 (1873).

M. truncatula Gaertn. var. *cosyrensis* (Somm.) Fiori, Fl. Ital. 1: 834 (1925).

M. crassispina Vis., Flora 12(1) Erg. 20 (1829).

M. truncatula Gaertn. var. *breviaculeata* (Moris) Urb. f. *dextrorsa* Urb., Verh. bot. Ver. Brandenb. 15: 67 (1873).

M. truncatula Gaertn. var. *longeaculeata* Urb. f. *dextrorsa* Urb., Verh. Bot. Ver. Brandenb. 15: 68 (1873).

M. tribuloides Desr. in Lam. var. *genuina* Koch Syn. 1: 20 (1837).

M. glabella Presl, Delic. Prag. 44 (1822).

M. hornemaniana Ser. in DC., Prodr. 2: 177 (1825). (Commonly transcribed as "*M. hornemanniana*.")

(*M. truncatula* var. *incinata* was mentioned in Khassanov (1972: 38), but is probably a misprint for *uncinata*.)

M. truncatula Gaertn. var. *longeaculeata* Urb., Verh. Bot. Ver. Brandenb. 15: 67 (1873).

M. truncatula Gaertn. subsp. *longeaculeata* (Urb.) Ponert, Feddes Repert., 83 (9-10): 639 (1973).

M. truncatula Gaertn. subvar. *microcarpa* Maire (Alleged type: Wall, E., 5-V-1934, Cap Ghir, sables, Morocco (MP); information (not verified) viewed online at http://www.aluka.org; current June 2010.)

M. tribuloides Desr. in Lam. var. *narbonensis* Ser. in DC., Prodr 2: 178 (1825). (At Herb. DC. (P) I observed a folder of this taxon with three sheets, two collected on 10 June 1807 and one on 06 June 1807. The last (the most adequate collection) is here designated the lectotype.) ("ssp. *narbonensis*" is mentioned in Neves et al. (2001); the status of this is unclear.)

M. pubescens Hornem. Hort. Reg. Hafn. 2: 726 (1815).

M. truncatula Gaertn. var. *rectiuscula* Rouy in Rouy & Fouc., Fl. Fr. 5: 23 (1899).

M. tribuloides Desr. in Lam. var. *rectiuscula* Rouy in Rouy & Fouc., Fl. France 5: 23 (1899).

M. rigidula Willd. Sp. Pl. 3(2): 1417 (1802), non All.

M. truncatula Gaertn. var. *rugosissima* Maire & Weiller. (Type: Maire, R. 438, Cyrenaica: in rupestribus calcareis humosis ad antrum Lethes, 19-IV-1938 (MP); information (not verified) viewed online at http://www.aluka.org; current June 2010.)

M. truncatula Gaertn. var. *breviaculeata* (Moris) Urb. f. *sinistrorsa* Urb., Verh. bot. Ver. Brandenb. 15: 67 (1873).

M. truncatula Gaertn. var. *longeaculeata* Urb. f. *sinistrorsa* Urb., Verh. Bot. Ver. Brandenb. 15: 68 (1873).

M. tentaculata Willd., Sp. Pl. 3(2): 1413 (1802), nom. illeg. (Heyn (1963) stated that this name is based on a misprint of *M. truncatula*.)

M. truncatula Gaertn. var. *tentaculata* (Willd.) Burnat, Fl. Alp. Marit. 2: 100 (1896).

M. tribuloides Desr. in Lam. proles *tentaculata* (Willd.) Rouy, Fl. Fr. 5: 23 (1899).

M. tribuloides Desr. in Lam. var. *tentaculata* (Willd.) P. Fourn., Quatre Fl. Fr. 547 (1936).

M. tribuloides Desr. in Lam., Encycl. Méth., Bot. 3(2): 635 (1792).

M. truncatula Gaertn. var. *tribuloides* (Desr.) Burnat, Fl. Alp. Marit. 2: 100 (1896).

Medica tribuloides (Desr.) Bubani, Fl. Pyren. 2: 477 (1900).

M. truncatula Gaertn. var. *tricycla* (Nègre) Heyn, Annual Species *Medicago* 99 (1963).

M. truncatula Gaertn. var. *genuina* (Koch) Briq. subvar. *vulgaris* Briq. f. *tricycla* Nègre, C.R. Soc. Sci. Nat. Maroc. 17: 90 (1951).

M. uncinata Willd., Sp. Pl. 3(2): 1417 (1802). (At Herb. Willd. (B), I observed a folder of *M. uncinata* that had 3 specimens. Specimen 1 (in the upper right corner with the label Photo-Nr-893 & IDC Cat. Nr. 14335) is *M. truncatula*, and is here interpreted as the type. The remaining specimens may not be *M. truncatula*.)

M. tribuloides Desr. in Lam. var. *uncinata* (Willd.) Rouy in Rouy & Fouc., Fl. France 5: 22 (1899).

M. tribuloides Desr. in Lam. var. *vulgaris* Rouy in Rouy & Fouc., Fl. France 5: 22 (1899).

Note: The name *M. truncatula* Godr. & Gren., Fl. France Prosp. 1: 395 (recognized in Grenier and Godron 1846) is listed in Tropicos® (Missouri Botanical Garden: http://www.tropicos.org/; current July 2010). Grenier and Godron (1848: 595) recognized the priority of the name *M. truncatula* Gaertn. and its application to the species discussed in this chapter.

Vernacular Names

ENGLISH: Aleppo clover, barrel clover, barrel medic, strong-spined medic (medic is also spelled medick).

FRENCH: La luzerne faux-tribule, luzerne tronquée.

Geographical Distribution

(N = native, I = introduced, U = uncertain status)

AFRICA: Algeria (N), Canary Islands (Spain; N), Egypt (N), Libya (N), Madeira (Portugal; N), Morocco (N), Tunisia (N).

ASIA: Armenia (N), Azerbaijan (N), Cyprus (N), Gruziya (N), Iran (U), Iraq (U), Israel (N), Jordan (N), Lebanon (N), Syria (N), Turkey (N), Turkmenistan (N).

AUSTRALASIA: Australia (I).

EUROPE: Austria (I), France (including Corsica; N), Gibraltar (N), Greece (including Crete, East Aegean islands; N), Italy (including Sardinia and Sicily; N), Malta (N), Portugal (N), Spain (including Balearic Islands; N), Turkey (N), Ukraine (N), Yugoslavia (N).

NORTHERN AMERICA: USA (California; Riefner and Small 2009).

Medicago truncatula is omni-Mediterranean, and widely introduced and adventitious elsewhere.

Description

Annual herb, procumbent to ascending, branched from near base, branches (6)15–50 cm long, vegetative parts pubescent with simple hairs. Leaflets 8–15 cm long, 7–12 mm wide, cuneate-obovate to obovate, distal half with serrate margins, rarely incised or laciniate, stipules deeply dentate to laciniate. Inflorescence of 1–3(5) flowers, peduncle about same length as subtending petiole. Flowers 6–8 mm long, calyx longer than half length of flowers, with simple hairs and subulate teeth longer than the tube, corolla yellow. Young pod contracted inside calyx. Mature pod cylindrical, spiny, the spines divergent and typically intermeshing with those of adjoining coils, pods often pubescent (sometimes glabrescent), 6–12 mm long, with 2.5–8 more or less strongly adpressed coils (5)7–12 mm wide. Coils hard at maturity, the surface with ca. 10 radial veins slightly branching before entering lateral vein (near dorsal suture). Seeds 1–2 per coil, separated by a transverse spongy fruit septum, oblong-ovoid, 2.5–4.5 mm long, 2–2.5 mm wide, radicle less than half length of seed, seedcoat smooth, yellow or brownish-yellow.

Identification Hints

Medicago truncatula is most likely to be confused with the similar but maritime *M. littoralis*, with which it hybridizes and intergrades. See *M. littoralis* (Chapter 48) for information on how the two species differ.

Habitat and Environmental Adaptation

Habitat

Medicago truncatula occurs in inland Mediterranean grasslands and shrublands. The species often is encountered as a weed, and indeed is one of the most common species of *Medicago* found in ruderal situations. Illustrative of the range of possible habitats, in Greece Thanopoulos (2007) noted the species occurs in sandy fields, wet grasslands, wet meadows, strongly overgrazed and degraded garrique, coniferous forests, grasslands, fallow fields, olive groves, and as a weed in cereal crops and waste places.

Phenology

Like all of the Mediterranean-region annual *Medicago* species, *M. truncatula* germinates and develops after the advent of the wet season, and produces seeds before summer drought becomes severe. Because there are more cultivars available of this species than of any other annual *Medicago*, there is a range of maturities, with early maturing cultivars most suited to the driest conditions and shortest growing seasons. Clarkson and Russell (1975) found that flowering was accelerated by vernalization (but not long vernalization, which delayed flowering), long photoperiods (18 and 24 hours), and high temperature. Hochman (1987) also studied the effect of temperature in advancing the flowering time of *M. truncatula* and Van Heerden (1984) also investigated the influence of temperature and day length on phenological development. Bucciarelli et al. (2006), Hanan et al. (2006), Moreau (2006), and Moreau et al. (2006) prepared developmental/ phenotypic frameworks for the species.

On 24 July 2007 I planted seeds of 200 accessions representing 62 species of *Medicago* (every taxon of the genus for which seeds were available) in a greenhouse in Ottawa, Ontario; in early September the first plants to develop flowers were the precocious-flowering *M. hypogaea* (which pushes its closed flowers underground) and the five accessions of *M. truncatula* var. *laxicycla*. At that time the plants of *M. truncatula* var. *laxicycla* were about three-fourths the size of the several kinds of *M. sativa* grown and larger than the majority of the other accessions, indicating that *M. truncatula* var. *laxicycla* is not only extraordinarily early maturing but also moderately rapid in growth rate.

Soil

Medicago truncatula occurs on a variety of soil textures, ranging from sand to clay. Piano and Francis (1992) stated that it generally occurs in loam soils. Zoghlami et al. (1996) observed that in Tunisia *M. truncatula* did not show any apparent preference for soil texture and Abdelguerfi et al. (1988*a*) observed the same for Algeria. Similarly, Bounejmate and Beale (1992) found that in Morocco *M. truncatula* was able to tolerate a wide range of edapho-climatic conditions. In Israel the species typically occupies heavy soils, including terra rossa, basaltic, and alluvial soils (Heyn 1963). *Medicago truncatula* does best on well drained neutral to alkaline soils, and is intolerant of low soil pH (Evans et al. 1990), notably when soluble aluminum and manganese are present (Little et al. 1992). However, populations have been found on acidic soils in Greece (Ewing and Howieson 1987) and ecotypes differ in tolerance to acidic soils (Bounejmate and Robson 1992). Moreover, Sledge et al. (2005) surveyed resistance to aluminum in more than 300 accessions from the United States Department of Agriculture and observed appreciable variation in aluminum tolerance. In Sicily, Piano et al. (1991) observed the species growing in a pH range of 6.0–9.0. In Australia, the pH tolerance range recommended for planting *M. truncatula* cultivars is 5.8–8.0 (Gillespie 1994). Also in Australia, good phosphorus and sulphur levels have proven to be critical to good performance (Clarkson et al. 1989). Veatch et al. (2004) were unable to find salt-tolerant genotypes in *M. truncatula*. Abdelly et al. (1996) observed that although *M. truncatula* was regularly found on the edges of sebkhas (closed depressions with hydromorphic, saline soil, periodically submersed) in Tunisia, the species had limited tolerance of soil salinity. However, *M. truncatula* performed very well in an experimental study of cover crops suitable for saline soils of the Central Valley of California (Mitchell 1996).

Temperature

Medicago truncatula, consistent with its geographical area of origin, is suited to the warm temperate conditions of a Mediterranean climate with dry hot summers and mild moist winters. Annual species of *Medicago* mostly start growing in the Mediterranean winter and so are subject to frost damage. *Medicago truncatula* appears to be one of the more vulnerable species to frost (Bounejmate et al. 1994; Brandsaeter et al. 2002). Based on growth chamber experiments, Brandsaeter et al. (2002) found that *M. truncatula* 'Parabinga' exhibited poor freezing resistance. Krall et al. (1996) found that eight accessions of *M. truncatula* examined for their ability to self-regenerate in Wyoming failed to survive the winter at the three locations tested. Christiansen and Svejcar (1989) sowed six accessions of *M. truncatula* in

Oklahoma in March 1985 and observed that all were killed by frost and cold in the following winter period. Derkaoui et al. (1991) found that fall-planted *M. truncatula* in Oklahoma failed to survive the winter, indicating that the species does not have sufficient cold tolerance for successful self-replication in the southern Great Plains of the United States. Subfreezing temperatures in December 1998 in West Central Texas caused considerable seedling mortality of cultivar 'Jemalong'. Despite the lack of tolerance to cold temperatures usually exhibited by the species, cold-tolerant ecotypes have been collected in Morocco, Algeria, and Sicily (Piano and Francis 1992).

Water

Medicago truncatula is adapted to 250–600 mm of annual rainfall, received mostly during a mild winter and spring. It occurs frequently in dry environments. Piano and Francis (1992) stated that it requires a minimum annual precipitation of 225 mm. Carter (1975) recorded (for Algeria) that *M. truncatula* occurred in regions receiving between 150 and 450 mm of rain annually. In Tunisia, it occupies a semi-arid zone with 250–450 mm of precipitation annually (Zoghlami and Hassen 2004), generally in sites receiving more than 350 mm (Zoghlami et al. 1996). The species is intolerant of soils with a low water table since the plant is not deep rooted. Abdelguerfi and Chebouti (2000) studied the effects of drought on seed production of *M. truncatula* and reported that drought during the flowering period had a more detrimental effect than drought during the vegetative period. Conway et al. (2001) studied the performance of *M. truncatula* in Central Queensland, an area that has summer-dominant rainfall rather than the rain-fed winter Mediterranean climate to which *M. truncatula* is adapted (winter rainfall in Central Queensland is typically 80–100 mm); they concluded that despite relatively unimpressive results there was some potential for persistence of some of the available cultivars. Brahim and Smith (1993) observed that *M. truncatula* in a field plot in Arizona produced seeds after only 125 mm of winter precipitation. Hassen et al. (1998) discuss variants of *M. truncatula* recommended for arid and semi-arid areas.

Senhadri Maamri et al. (2000) studied the ecological distinction of populations of *M. truncatula* in Algeria and observed that, based on growth in a common garden, plants native to regions of high altitude and/or with limited rainfall were not as productive of fruits and seeds as plants native to regions of low altitude and/or high rainfall.

Biotic Relationships

Pollination

Medicago truncatula is largely an inbreeder (Bataillon and Ronfort 2006). Bonnin et al. (1996), on the basis of random amplified polymorphic DNA analysis, and Bonnin et al. (2001) on the basis of DNA (microsatellite loci) variation in a population of *M. truncatula* in France, found that, apparently because of occasional crossing, considerable polymorphism was present and subpopulations only 50 m apart were often distinct.

Dispersal

The indehiscent spiny fruits are adapted to dispersal in animal fur, accounting for the widespread occurrence of this species. However, the characteristic right-angle bending and interlocking of the spines tends to lessen adherence to fur, and an adaptive reason for this spine feature is not clear (the spines thus cover the area between adjacent coils, perhaps making it more difficult for insects to gain access). Kotzé et al. (1995) showed that some intact seeds will pass through the guts of sheep and remain viable.

Herbivory

All the Australian cultivars of *M. truncatula* are susceptible to aphids, particularly the blue-green aphid (*Acyrthosiphon kondoi*) and the spotted alfalfa aphid (*Therioaphis trifolii* f. *maculata*), and to a lesser extent the pea aphid (*A. pisum*) and the cowpea aphid (*A. craccivora*). Klingler et al. (2007) examined genes responsible for resistance to the spotted alfalfa aphid and the blue-green aphid. Other pests include the red-legged earth mite (*Halotydeus destructor*), the lucerne flea (*Sminthurus viridis*) and sitona weevils (*Sitona discoideus*, *S. humeralis*). Cultivars 'Mogul' and 'Sephi' are resistant to blue-green and spotted alfalfa aphids. All the current barrel medic cultivars are susceptible to the lucerne flea and the red-legged earth mite. In Australia, the root lesion nematode (*Pratylenchus neglectus*) can reduce *M. truncatula* productivity, as it does for other pasture medics. In tests of the alfalfa weevil (*Hypera postica*) on 13 annual species of *Medicago*, Pandey et al. (1984) found that *M. truncatula* showed resistance. Resistance of *M. truncatula* to herbivores, based on the release of volatile compounds, was studied by Leitner et al. (2005). Resistance of *M. truncatula* to species of the parasitic plant *Orobanche* was studied by Rodríguez-Conde et al. (2004), and the molecular basis of resistance was reviewed by Pérez-de-Luque et al. (2006). Korth et al. (2006) demonstrated that calcium oxalate crystals in *M. truncatula* are a defence against chewing insects.

Nodulation and mycorrhizae

Medicago truncatula forms symbioses with nitrogen-fixing rhizobia (*Sinorhizobium meliloti*). With respect to *M. truncatula* being a molecular model system, much of the scientific effort has focused on the biology of root–rhizobial interactions and the molecular regulation of nodule development (e.g., Zhu et al. 2002; Tesfaye et al. 2003). Chebouti and Abdelguerfi (2003) compared the root weight, number of nodules, and nodule weight responses of *M. truncatula* when it was grown under either drought or irrigation and found that the three variables decreased more or less proportionately to each other under drought. Zribi et al. (2005) examined the abilities of several populations of *M. truncatula* to trap natural populations of *S. meliloti* and *S. medicae* in Tunisian soils and found no consistent tendencies of given *M. truncatula* populations to interact efficiently with groupings of the 155 *Sinorhizobium* isolates examined. Denton et al. (2007) studied the ecology of nodulation of *M. truncatula* in relation both to rhizobial strain and presence of naturalized rhizobia in the soil, and found that there was an interaction such that some strains nodulated well in *M. truncatula* when there were other rhizobial strains present. In Central and Western Spain it has been shown that, when infected by an effective strain of *Rhizobium*, *M. truncatula* resists freezing temperatures and long periods of winter drought, whereas noninfected plants die (Feuntes 1981).

Medicago truncatula also forms symbioses with arbuscular mycorrhizal fungi, a relationship that (as in other vascular plants) particularly supplies phosphorus to the plant. Gianinazzi-Pearson et al. (2006) presented a detailed analysis of the relationships between arbuscular mycorrhizal fungi and *M. truncatula*. Chen et al. (2005) examined the effects of the mycorrhizal fungus *Glomus intraradices* on uranium uptake by *M. truncatula* from uranium-contaminated soil.

Disease resistance

Seed-borne black stem (also known as leaf spot, *Phoma medicaginis*) is damaging (O'Neill et al. 2003), especially with above average spring rainfall (Crawford et al. 1989). Ellwood et al. (2006*b*) identified genetic resistance to black stem in one germplasm collection. Barbetti (2007) identified an accession of *M. truncatula* with a very high level of resistance to *P. medicaginis*. Leaf rusts (*Uromyces anthyllidis* and *U. striatus*) can reduce hay and seed yields of *M. truncatula* (Crawford et al. 1989). Diseases are most prevalent in dense stands. Alfalfa mosaic virus can significantly reduce growth and productivity (Dall et al. 1989). Several other viruses are also able to infect the plants.

Yaege and Stuteville (2002) tested the resistance of 33 annual species of *Medicago* to powdery mildew (*Erysiphe pisi*) and found a range of resistance from 0% to 100% among the nine collections of *M. truncatula* grown. Tivoli et al. (2006) reviewed "necrotrophic pathogens" affecting *Medicago* species, with special reference to *M. truncatula* as a model plant for molecular exploration of the nature of disease resistance, and Ellwood et al. (2006*c*) did the same for "pathogenic fungi". Raynal (1977) examined the susceptibility of nine species of *Medicago* to *Colletotrichum trifolii* Bain, a principal cause of anthracnose of alfalfa, and found only *M. truncatula* was fully resistant. Elgin and Ostazeski (1992) found that *M. truncatula* was more than 90% resistant to Races 1 and 2 of anthracnose caused by *Colletotrichum trifolii*. In a comparative test of 36 annual species of *Medicago*, O'Neill and Bauchan (2000) found that four accessions of *M. truncatula* were very resistant to anthracnose caused by *Colletotrichum trifolii*, whereas five other accessions exhibited a range of susceptibility to the disease. Torregrosa et al. (2004) examined the molecular (DNA) resistance aspects of this fungal pathogen.

Economic History, Cultivars, and Uses

Medicago truncatula is the most important annual species of *Medicago*, next in agricultural importance only to the perennial *M. sativa* (i.e., alfalfa and its close allies). There are more cultivars of this species than of any other *Medicago* species except *M. sativa*. See below for remarks on the development of *M. truncatula* as a model legume for genome analysis and exploitation. The first cultivar, 'Jemalong', originated from wild plants that were naturalized in southern Australia before 1920 (Clarkson 1977).

Cultivars

Registered cultivars in Australia (with release or registration year in parentheses) include 'Akbar', 'Ascot', 'Cyfield' (1969), 'Borung' (1970), 'Caliph' (1993), 'Cyprus' (1959), 'Ghor' (1967), 'Hannaford' (1939), 'Jemalong' (1955), 'Jester' (2001), 'Mogul' (1992), 'Parabinga' (1986), 'Paraggio' (1982), and 'Sephi' (1984). Detailed information on most of the Australian cultivars is available at the Register of Australian Herbage Plant Cultivars (http://www.pi.csiro.au/ahpc/legumes/legumes.htm; current June 2010). Additional information is available in Clarkson (1977), at the Australian Government (IP Australia) Plant Breeder's Rights (PBR) database of PBR varieties (http://www.ipaustralia.gov.au/pbr/index.shtml; current July 2010), at the New South Wales (Australia) Department of Primary Industries website (http://www.dpi.nsw.gov.au/; current July 2010), and in the online article by Nair et al. (2006*b*). Carneiro et al. (2004) described 'Revilheira', a Portuguese-selected cultivar of *M. truncatula*, as having good adaptation to grazing because of its prostrate to semi-prostrate growth habit, and at least as good levels of dry matter production as available Australian annual *Medicago* cultivars, for purposes of stocking long-duration, rain-fed pastures under Mediterranean conditions. See below for information on the French cultivar 'Salernes', which is recommended for use in rangelands for sheep rearing.

Forage

Medicago truncatula is the most important annual species of *Medicago* sown for forage in the ley farming system of Australia, estimated to be grown there on more than 4.5 million hectares (Pearson et al. 1997). According to St. John-Sweeting et al. (1996), "The predominance of *M. truncatula* with numerous commercialized cultivars is due to its outstanding adaptation to low rainfall environments with soils high in available lime and to long established understanding by farmers of how to manage the species for best results in terms of production and persistence." *Medicago truncatula* has performed well in combination with grasses (*Phalaris aquatica*, *Lolium rigidum*), but does best in pure culture. In Western Australia, dry matter yields of 4.05 to 4.96 t/ha have been reported (*Register of Australian Plant Cultivars*). Among irrigated mixtures with oats (*Avena strigosa*) or perennial ryegrass (*Lolium perenne*), dry matter yields of *M. truncatula* were 4.0 and 4.4 t/ha, respectively

(Lowe and Bowdler 1988). Sanadgol et al. (2006) found the unirrigated yield of *M. truncatula* over a 3-year period in the Gorgan plain of Iran was 1257 kg/ha. *Medicago truncatula* was one of the highest yielding of several annual *Medicago* species evaluated for forage potential in the North Central United States (Zhu et al. 1996). Zhu and Sheaffer (1997) recorded above-ground dry matter yields (in kg/ha) of cultivar 'Sephi' at two sites in Minnesota (compared with the alfalfa cultivar Nitro, yields in parentheses): 6854 (3815) and 5594 (4596). In Alaska, dry matter yields of 3.39 to 3.84 t/ha were obtained for *M. truncatula* (Panciera and Sparrow 1995). Ocumpaugh et al. (2007), in reference to United States productivity, indicated that annual yields were about 4.5 t/ha. In Argentina, Gargano et al. (1990) found yields of annual dry matter in pure stands of 4.08 t/ ha for cultivar 'Jemalong' and 3.77 t/ha for cultivar 'Paraggio'. *Medicago truncatula* has shown potential as an emergency forage in the northern United States when sown in the spring and harvested after 2 months, although regrowth for a second harvest was shown to be poor (Shrestha et al. 1998).

Kernick (1978) provided extensive information on the productivity of *M. truncatula* in the Middle East. He noted that "In the littoral zone of Algeria in a season with 770 mm rainfall (September–May) a fallow pasture comprising seeded *M. truncatula* ('Jemalong') and volunteer annual legumes and grasses yielded on 21 May a green forage yield of 24 400 kg/ha and 70 000 kg/ha DM; of this total *M. truncatula* yielded 7000 kg/ha green forage and 2000 kg/ha DM." In the littoral zone of Tunisia, green forage yields on the order of 20 000 kg/ha in dryland cultivation have been reported for *M. truncatula* (Kernick 1978). In the steppic zone (220 m elevation) of northern Iraq, during two seasons with rainfall varying from 330 to 475 mm (November to May), fallow fields on light calcareous soil seeded with *M. truncatula* cultivar 'Jemalong' and fertilized with 60 kg/ha P_2O_5 gave yields varying from 2000 kg/ha to 3300 kg/ha of green material (Kernick 1978). Tadmor et al. (1971) conducted studies of ecotypes selected for their ability to survive and persist in the Negev Desert of Israel in an area where rainfall averaged only 90 mm annually, but received an additional 263 mm of water annually in the form of run-off from erratic winter storms. Dry matter production of six ecotypes of *M. truncatula* varied from 2.7 to 5.6 t/ha (averaged over 6 years).

In nursery plots on alluvial soil in the steppic zone (220 m elevation) of northern Iraq, during a season with rainfall of 475 mm (November to May), forage yields were recorded as follows:

Table 97-1. Yield of *M. truncatula* cultivars in northern Iraq (kg/ha), after Kernick (1978).

	Fertilized with 60 kg/ha P_2O_5		Unfertilized	
Cultivar	Green matter	Dry matter	Green matter	Dry matter
'Jemalong'	9200	4200	7600	2700
'Borung'	8000	3200	5800	2000
'Cyprus'	4000	1600	3000	1200

However, Kernick (1978) noted that much lower yields — 300 kg/ha of green forage for cultivar 'Jemalong' — were recorded farther south in the steppic zone of Iraq on sandy soils in a season with 340 mm rainfall; the plants were fertilized with 30 kg/ha P_2O_5 and were poorly nodulated because of lack of inoculation. Mohammad and Qamar (1988) grew 14 annual species of *Medicago* in small test plots in Islamabad, Pakistan located in a subtropical, subhumid area that is notably hotter and wetter than the normal habitat of these species. The dry matter yield of *M. truncatula* was 2.20 t/ha.

Seed yields of up to 1330 kg/ha for cultivar 'Jemalong' have been realized in the littoral zone of Algeria (Kernick 1978).

Denney et al. (1979) concluded that *M. truncatula* hay harvested at the flowering stage was similar to a good quality alfalfa hay with 65% digestible dry matter. Denney et al. (1983) and Brand et al. (1991*a*, 1991*b*) examined the nutritional characteristics of *M. truncatula* forage for feeding sheep. Kotzé et al. (1995) examined the nutritional value of *M. truncatula* seeds for feeding sheep. Gallardo et al. (2006) provide detailed analysis of the nutritional chemistry of the seeds. Antiquality factors have been identified in *M. truncatula* and there is risk of bloat in cattle. Photosensitization in horses and occasional red gut in sheep have occasionally been recorded. Some older cultivars have phytoestrogenic potency, mainly from coumestans that inhibit ewe fertility. The coumestans are increased when the plants are damaged by fungal diseases or pest attack. Eyal et al. (1975) found that a combination of seeded *M. truncatula* and *M. polymorpha* provided considerable palatable pasture for sheep in a semi-arid area of Israel, but a fertility disorder apparently due to estrogenic activity was noted. In general, *M. truncatula* provides moderately nutritional forage, but for best production it should not be a major source of feed.

Fodder

In Australia, *M. truncatula* is used to some extent for hay and silage. The more erect-growing cultivars are more easily harvested than the relatively decumbent types.

Green manure, rotation crop, ground cover

This species is recommended as a cover crop for South African vineyards. As noted by Nair et al. (2006*b*), the French cultivar 'Salernes' was bred in part "to maintain soils around infrastructures: roadsides, railway lines embankments and disturbed areas; to avoid soil erosion from rainfall and winds, reduce the use of herbicides, limit infiltrations of pesticides in soils and enhance underground life in sloping vineyards." Ghalavand et al. (2003) found that using *M. truncatula* as a cover and/or fallow crop in Iran reduced surface run-off and erosion, and increased water conservation and soil water-holding capacity.

Other uses (salad sprouts, ornamental, plant products, etc.)

As noted below, *M. truncatula* is considered to have high potential as a molecular factory (i.e., genetically modified to produce desired chemicals, especially by using recombinant proteins).

Chromosome Number

$2n = 14$: Humphries et al. 1978; Vogt and Aparicio 1999.

$2n = 16$: Ghimpu 1929*a*, 1929*b*, Ghimpu 1930 (as both *M. tribuloides* and *M. truncatula*); Clement 1962; Simon and Simon 1965; Dahlgren et al. 1971; Fernandes et al. 1977; Fernandes and Queirós 1978; Loon and Jong 1978; Abdelguerfi and Guittoneau 1979; Lesins et al. 1980; Agarwal and Gupta 1983*b*; Abdelguerfi et al. 1988*b*.

Genetic Relationships and Gene Transferability

Intraspecific

Simon and Millington (1967) observed that intraspecific crosses of *M. truncatula* usually produced fully fertile hybrids (one population distinguished by a chromosomal translocation produced somewhat infertile hybrids). Lilienfeld (1962) crossed plants of different sources in Israel and the fertility of the hybrids was notably reduced.

Interspecific

Simon and Millington (1967) were unable to produce hybrid seed by crossing *M. truncatula* with *M. italica*, *M. rigidula* (*M. rigiduloides*?), *M. murex* (*M. lesinsii*?), *M. constricta*, *M. turbinata*, *M. doliata*, or *M. minima*. Despite interspecific incompatibility, Lake (1989) demonstrated that, for breeding improved cultivars, gene exchange is possible among *M. truncatula*, *M. littoralis*, and *M. italica* (see Chapter 48 on *M. littoralis* and Chapter 44 on *M. italica* for additional information). Jenczewski et al. (1997) crossed *M. truncatula* with *M. italica* and noted very odd segregation in the hybrids. Lesins et al. (1980) hybridized *M. truncatula* with *M. turbinata* — the hybrids were completely sterile, although doubling the chromosome number resulted in some fertility.

Exchange by advanced biotechnology

Tian and Rose (1999) produced asymmetric somatic hybrids (i.e., combining the complete genome of the recipient species with only one or a few transferred chromosomes of the donor species) between *M. truncatula* and *M. scutellata* — a substantial achievement given the sexual incompatibility between these (and most other annual) species, and offering a promising example of how genes can be transferred among all species of *Medicago*. This was the first successful asymmetric somatic hybridization between sexually incompatible annual *Medicago* species. Tian et al. (2002) produced an asymmetric somatic hybrid between *M. truncatula* and *M. rugosa*. In these crosses, the genome of *M. truncatula* predominated.

Conservation Status

Medicago truncatula is a very common species. Moreover, because of the interest in it as cultivars and as a model legume, large germplasm collections are available.

Germplasm Sources

Nair et al. (2006*a*) analyzed the principal gene banks with holdings of *M. truncatula*. (cf. Chapter 3).

At the time the information was collected (see Chapter 3), the Australian *Medicago* Genetic Resource Centre (AMGRC) had 5509 accessions from 40 countries, including 4517 wild accessions, a core collection of 231 accessions, 788 breeding lines, 18 cultivars, 5 mutants, 2 landraces, and 175 recombinant inbred lines. Of these, 4423 lines have up to 64 descriptors recorded, with all lines having at least 30 detailed descriptors, including agronomic traits such as seedling vigor, winter herbage production, flowering time, seed production, pod spininess, changes in seedcoat permeability, herbicide tolerance, and resistance to insects and disease.

The USDA National Plant Germplasm System (NPGS) germplasm collection had 324 accessions originating from 22 countries, especially from Algeria, Morocco, and Tunisia. The entire collection has been evaluated for morphological characters and much of the collection has also received physiological and/or resistance evaluation.

The French National Institute for Agricultural Research (INRA) had 16 cultivated accessions, 638 natural populations, 720 inbred lines, and 1750 mutant lines.

Research Problems and Priorities

As the most important annual species of *Medicago* in agriculture today, there is considerable activity in breeding improved cultivars of *M. truncatula*. Objectives include improved seedling vigor, greater forage yield and seed production, greater resistance to pests and diseases, and reduced estrogenic potency.

Medicago truncatula is being developed as a model legume plant in both classical and molecular genetic studies to elucidate the functions of its genes (e.g., for bacterial and fungal symbiosis, stress resistance, and plant architecture) and exploit its genome (e.g., for improved seed quality and production of specific secondary metabolites). Advantageous in this regard are the plant's small genome (approximately 454 to 526 Mbp, half that of alfalfa), diploid genetics, self-fertility, rapid generation time (the life cycle is short — about 3 months), prolific seed production and good biomass yield, rich natural diversity, amenability to genetic transformation, and synteny with several major legume crops of interest (Huguet and Prosperi 1996; Journet et al. 2001; Colebatch et al. 2002; Young et al. 2003; Choi et al. 2004; Gutierrez et al. 2005; Zhu et al. 2005; Cannon et al. 2006; Mun et al. 2006; Town 2006). Abranches et al. (2005) pointed out that seeds are promising organs for pharmaceutical production because they permit storage of the recombinant product in the seed endosperm for several years without loss of activity; and because *M. truncatula* is autogamous, it offers better containment than the outcrossing *M. sativa* (i.e., it is much easier to prevent gene escape to wild or other cultivated plants). Large-scale, international genomic projects have been initiated for *M. truncatula* (Frugoli and Harris 2001; VandenBosch and Stacey 2003), along with the development of bioinformatics resources (Dénairié 2002). Recent reviews include Thoquet et al. (2002) and Watson et al. (2003). Hundreds of research papers have been generated and there is great prospect of using the information now being produced to utilize *M. truncatula* not just for agricultural purposes, but also for a wide range of biotechnological applications (Abranches et al. 2005). Genetic markers found in *M. truncatula* are useful for determining relationships with other forage species (Eujayl et al. 2004). The entire genome of *M. truncatula* has been sequenced (see http://www.medicago.org; current May 2010) by a consortium of researchers led by the Samuel Noble Foundation, Ardmore, Oklahoma and the University of Oklahoma, with funding by the United States National Science Foundation and the European Union's Sixth Framework Programme.

Representative Collections

Algeria: Alston, A.H.G. & Simpson, N.D. 409, between Tarue & Bourbaki, alt. 960 m, rocky slope, 3-V-1937, in fruit (BM). Bové, Alger, champs, V-1839, in fruit (P). Davis 52436, near Barage, northeast of Biskra, alt. 150 m, near irrigated field, 17-V-1971, in fruit (BM). Meyer, A., environs d'Alger, chemins autour de la ville, IV-V-1884, in fruit, Société Dauphinoise 1885 No. 350 bis (P). Reboud, V., Constantine, 1874, in fruit (collection mixed with *M. rigidula*), Société Dauphinoise No. 350 (P). **Australia:** Aston, H.I. 2364, New South Wales, South Western Plains, 13 km east of Urana, along the Urana to Lockhart Road, (35°19′S, 146°25′E), frequent on hard dry, clay-loam in roadside depression previously inundated, 2-XI-1982, in fruit (K). Keigherty, G.J., & Alford, J.J. 831, Western Australia, 49.5 km northwest of Eucla off track to Reid/Forrest, (31°21′S, 128°36′E), common, on road verge of sandy clay loam, 10-X-1986, in fruit (BRI). **Austria:** Untchj, K., Istria, in litore maris ad Veruda prope Pola, V-1898, in flower & fruit, Herbarium Normale editum ab I. Dörfler No. 4229 (P). **Balearic Islands (Spain):** Bowden, L.F. & Sim, P.A. 561, Mallorca, beside road about 3 km south of Ciudad de Los Lagos towards Arta, abundant, alt. ca. 10 m, 9-VI-1966, in flower & fruit (BM). **Canary Islands (Spain):** Pitard, C.J. 1417, Santa Cruz, ad mare, III-1905, in fruit (P). **Egypt:** Ehranbenz, G., Alexandria (on or before 1871), in fruit (K). **England:** Lousley, J.E., southwest York, Kirkbeaton, 28-IX-1958, in fruit, Wool Adventive Flora of Britain – J.E. Lousley W587 (K). **France:** Fouchard, J., Chatelaillon (Ch.-Inf.), lieux secs, 7-VI-1885, in flower & early fruit, Société Rochelaise 1885 (P). Septimin-Donat 982, Aude, Sallileo Le Gaillaurdi, 1902, in fruit (BM). **Greece:** De Heldreich, Attica: Hagia Glykeria prope Athenas in locis herbosis, ad vias, 30-IV-1884, in fruit, De Heldreich Herbarium Graecum Normale No. 883 (P). Guiol, F., Creta, V-1937, in flower & fruit (BM). Rechinger fil, K.H. 12116, Crete, Distr. Kissamos, Ins. Grabusa Agria, in saxosis calc., 20-IV-1942, in fruit (K). Tuntas, B., Flora Cycladum, in insula Cythno, IV-1900, in fruit, De Heldreich Plantae Exsicc. Flora Hellenicae (P). **Iran:** Bowles Scholarship Bot. Exp. 1356A, western shore L. Urmia, alt. 4700 feet, gravelly foreshore, 7-V-1963, in flower & fruit (K). **Iraq:** Gillett & Rawi 7022, 5 or 6 km above Rawa, alt. 150 m, irrigated cereal fields near river, 27-III-1947, in flower & fruit (K). **Israel:** Naftolsky, N., Shomron, at foot of Mt. Gilbo'a, 9-IV-1927, in fruit (K). **Italy:** Fiori, A., Apulia, Gargano, prope Rodi, loco Coppe di Ceccasole dicto, in herbosis, alt. 280 m. solo calcareo, 24-V-1913, Flora Italia Exsiccata Series III Fiori & Béguinot No. 2095 (BM). Rigo, G., Sicilia, Messina, in herbidis, 3-V-1898, in fruit, G. Rigo, Iter Italicum Quartum Anni 1898 (P). **Jordan:** Boulos, L. & Al-Eisawi, D. 6439, 10–15 km northwest Zarga, along the road to Alouk, 29-III-1974, in fruit (BM). Kasapligil, B., 2151, Tafile, Reshadiye, Jebel El-Gayir, west on the road to Dahl, alt. ca 800 m, 11-III-1955, in fruit (UC). **Lebanon:** Polunin, O. 5281, Harissa, alt. 2300 feet, cultivation terraces, 16-IV-1959, in fruit (K). **Libya:** ben Saud, S. 273, Cyrenaica, el Marj., alt. 450 m, 18-IV-1959, in flower & fruit (K). Sandwith, N.Y. 2264, Cyrenaica El Abiar, alt. ca. 300 m, among rocks in altiplano, 2-IV-1939, in flower & fruit (K). **Madeira:** Bornmüller, J., Funchal, in campis maritimis prope rupem Gurgulho, 23-III-1900, in fruit, J. Bornmüller Plantae Exsiccatae Maderenses No. 476 (P). Bornmüller, J., Funchal, in campis maritimis prope rupem Gurgulho, 23-III-1900, in fruit, J. Bornmüller Plantae Exsiccatae Maderenses No. 476 (P). **Morocco:** Maire, R., prope Skhirat, in arvis arenoso argillaceis, 23-IV-1935, in fruit (P). Sennen, F. 8391, Ulad Settut, Muluya, alluvions, 1-VII-1933, in fruit (BM). **Portugal:** Daveau, J, Serra de Cintra, VI-1882, in fruit, J. Daveau Herbarium Lusitanicum No. 33 (P). Daveau, J., Jachêres près de Serpa, V-1882, in flower & fruit, J. Daveau Herbarium Lusitanicum No. 61 (P).

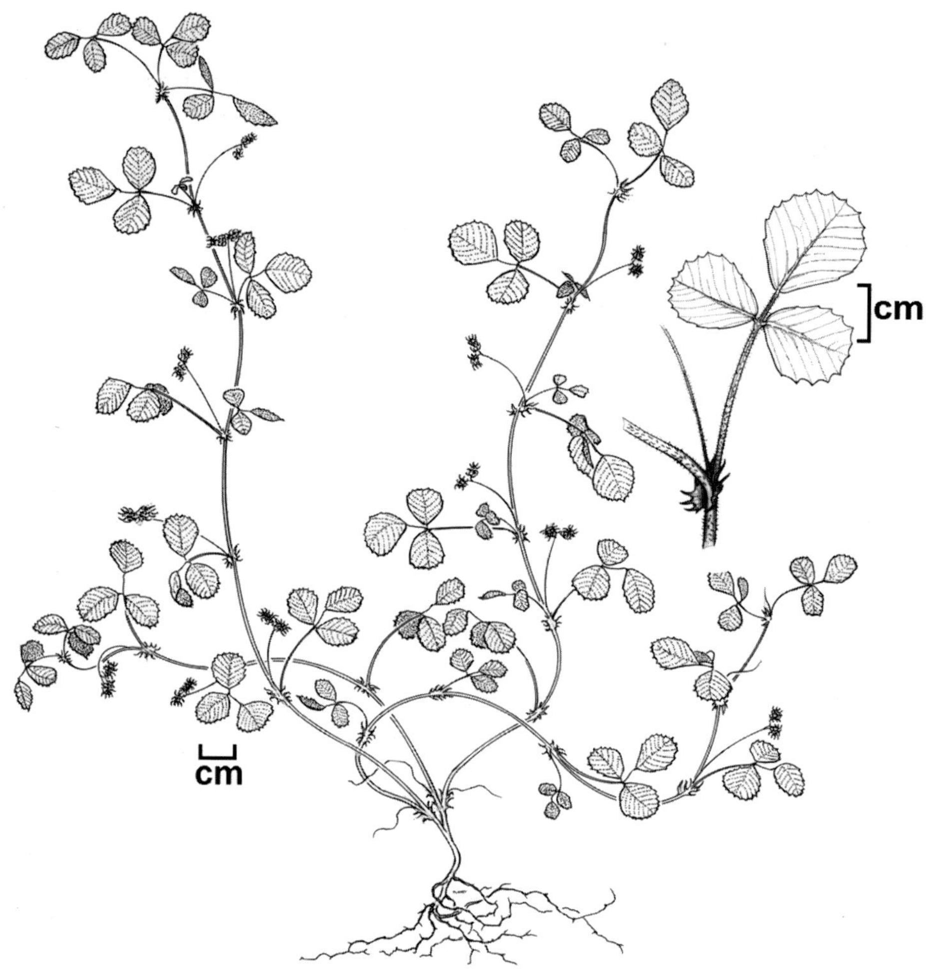

Fig. 97-3. *Medicago truncatula* var. *laxicycla*. Vegetative, greenhouse-grown plant. (Based on E. Small M1762, DAO; artist: B. Flahey.)

Spain: Sennen, F. 540, Catalogne, Barcelone, talus, 20-V-1908, in fruit (BM). Sennen, F., Barcelone, champs sablonneux, 22-V-1921, in fruit (BM). Wolley-Dod, A.H. 1808, Flora Calpensis Governor's cottage, 21-IV-1913, in fruit (K). **Tunisia:** Pitard, C.J., Gabês, in aridis deserti, III-1909, in fruit (BM). Young, R. 122, area around Ae Nouha oasis (ca. 20 miles northwest of Gabes), edge of Oued Crercrara northeast of Bou Hedma in dry gravelly soil, IV-1968, in flower & fruit (BM). **Turkey:** Bourgeau, E., Adalia, sur les pelouses, 2-V-1860, in fruit (P). Brummitt, R.K. 6487, C1 Izmir, near village of Givzel Camli on North coast of Samsun Dagi Peninsula, 12 miles south of Kusadasi, 5-V-1969, in fruit (K). **Yugoslavia:** Jackson, A.K. & Turrill, W.B. 134, Lapad Peninsula, damp place at base of wall, 19-IV-1935, in fruit (K). Jackson, A.K. & Turrill, W.B. 65, above Gravosa (Sergi), waste ground by path, 17-IV-1935, in fruit (K). Petter, Fr., Littoral-Pflanze, Auf Ackerrainen um Torrette bei Spalato IV, (received by P in 1889), Flora Dalmatica Exsiccata No. 245 (P).

Note

There is evidence of geographically based differentiation in *M. truncatula*, although not sufficiently evident to warrant taxonomic recognition. Ronfort et al. (2006) examined DNA (microsatellite) variation in 346 accessions, found evidence of distinction of the Iberian and Moroccan collections, and suggested that this may be because of historical biogeography. Ellwood et al. (2006*a*) examined DNA (SSR loci) variation in 196 accessions, also found evidence of distinctiveness of the Iberian and Moroccan collections, and noted subclusters of accessions from North Africa.

Crawford (1970) studied 206 collections of *M. truncatula* and noted considerable variability in morphology (including foliar markings), vigor, flowering time, weight of seeds, and weight of pods produced. Latta and Quigley (1993) examined the persistence of viable seeds in Australian pastures. Wang and Grusak (2005) examined the structure and development of the fruit wall and seed coat. Benlloch et al. (2003) examined floral development in the species. McComb (1974) studied the inheritance of leaf markings. Neves et al. (2001) described a method for micropropagation.

MEDICAGO TRUNCATULA VAR. *LAXICYCLA* (E. SMALL) E. SMALL

M. truncatula Gaertn. f. *laxicycla* E. Small in Small et al., Pl. Syst. Evol. 174: 38 (1991), basionym.

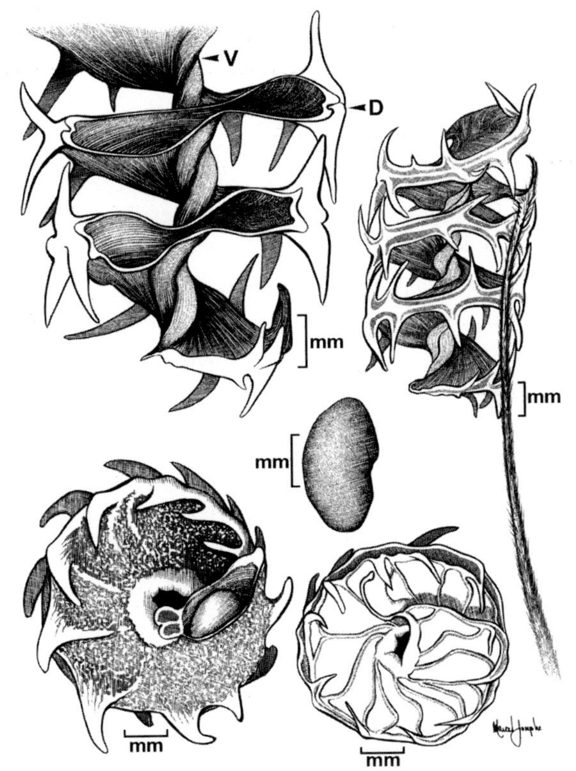

Fig. 97-4. *Medicago truncatula* var. *laxicycla*. **Clockwise from top left:** Sagittal section of portion of fruit (arrows point to gaping ventral suture (V) and dorsal suture (D)); peduncle with fruit; view of stylar end of fruit (venation evident in young fruit); cross section of mature fruit showing a portion of spiral that has been cut to reveal a seed (venation obscured in mature fruit). **Center:** Seed. (Based on E. Small M1353, DAO; artist: M. Jomphe.)

The epithet *laxicycla*, based on the Latin *laxus*, meaning loose, plus *cyclus*, meaning whorled, denotes the loosely coiled fruits that define this taxon.

The elevation of forma *laxicycla* to var. *laxicycla* is proposed here. It was my original intent to publish this as a variety in the 1991 publication cited above, but the editor of the journal at the time, exercising arbitrary taxonomic concepts, would only accept the paper if the new taxon was given the rank of forma.

Type Specimen

E. Small M1353D (DAO). Grown in an Ottawa greenhouse in the winter of 1988 from seeds from the Australian *Medicago* Genetic Resource Centre, South Australia accession 17218. Seeds originally from eastern Libya as L. Blessing C16.

Identification Hints

The coils of the mature fruits of *M. truncatula* var. *truncatula* are adpressed, in obvious contrast to the extremely loosely coiled fruits of *M. truncatula* var. *laxicycla*, so that the latter is easily distinguished. The only other species of *Medicago* with lax fruit coiling is *M. laxispira* (see Chapter 47). However, the coiling in *M. laxispira* is not as loosely spiralled as in *M. truncatula* var. *laxicycla*. Both species are extremely rare and will very seldom be encountered.

Habitat and Environmental Adaptation

Small et al. (1991) conducted germination tests on a number of species of *Medicago* and found that this unusual variant of *M. truncatula* from eastern Libya was the only one in which rapid and simultaneous germination of most of the seeds occurred while still within a fruit. They concluded that the loose coiling of the fruit of the Libyan variety appears to facilitate water entry and seedling escape. Its rapid germination might adapt it to respond quickly to intermittently available moisture, but at the cost of abandoning the strategy of intermittent germination adopted by other populations of annual *Medicago* species. The climate of eastern Libya where *M. truncatula* var. *laxicycla* has been collected is certainly drier than most areas occupied by *M. truncatula*, which makes it plausible that loose fruit coiling evolved as an adaptive response to climate. (For additional discussion, see Chapter 7.)

The fruits of *M. truncatula* var. *laxicycla* are so loosely coiled that they make access to the seeds much easier than in any other known annual population of *Medicago*. This must be disadvantageous in allowing insects to reach the seeds. Presumably this is compensated by the ecological advantage of facilitating access to moisture by the seeds.

Small et al. (1991) reported that mature seeds of *M. truncatula* var. *laxicycla* weighed less than half that of the seeds from other populations of *M. truncatula*. Also (on the basis of cultivated plants) var. *laxicycla* appeared to produce plants that are about 25% smaller than those from other *M. truncatula* populations. The smaller dimensions suggest that *M. truncatula* var. *laxicycla* has evolved the capacity to reproduce with a smaller supply of resources, which is presumably an evolutionary response to its more stressful environment.

Chromosome Number

$2n = 16$, based on population SA17218 (Small et al. (1991); voucher at DAO).

Representative Collections and Germplasm Sources

The seven germplasm collections upon which var. *laxicycla* was based were obtained in 1979 by L. Blessing, an employee of the South Australian Department of Agriculture, during a trip in eastern Libya. From seeds provided from the Australian collections, plants were grown in a greenhouse in Ottawa in 1988. The following presents L. Blessing's collection numbers (prefixed by C), the collection numbers of the Australian *Medicago* Genetic Resource Centre, based in the South Australian Department of Agriculture (prefixed by

SA), and the E. Small voucher collection codes in the DAO herbarium (prefixed by M): C8, SA17217, M1758; C16, SA17218, M1353; C17, SA17219, M1759; C35, SA17222, M1760; C41, SA17224, M1761; C54, SA17225, M1762; and C875, SA17227, M1763.

Notes

Medicago truncatula var. *laxicycla* should not be confused with *M. truncatula* var. *tricycla* (Nègre) Heyn, (Heyn 1963), a dubious taxon, as noted above, that has been more or less defined by having fruits with about three coils, which is similar to *M. truncatula* var. *laxicycla*. Plants assigned to *M. truncatula* var. *tricycla* have been suggested to be typical of relatively arid portions of the distribution range of *M. truncatula*. Plants identified as *M. truncatula* var. *tricycla* that I examined at HUJ and W have fruits that are far more compactly coiled than the variant described in this paper. The same is true for the illustration of *M. truncatula* var. *tricycla* in Nègre (1956, p. 48), referred to by Heyn (1963) in recognizing her new combination.

Ellwood et al. (2006*a*) conducted a DNA-based relationship analysis of 196 accessions from the SARDI, from which the seven accessions here recognized as var. *laxicycla* were obtained. Unfortunately, none of these seven accessions was used in their study.

Small et al. (1991) noted that the seeds from one of the collections of *M. truncatula* var. *laxicycla*, accession SA17217, did not produce plants with fruits of normal appearance. The fruits showed an abnormality of development that has occasionally (but rarely) been observed in the *M. truncatula* – *M. littoralis* complex, in which after each coil is produced there is a (clockwise-anticlockwise) reversal of spiralling (cf. Lesins and Lesins 1979, p. 19, which describes a related but somewhat different phenomenon; another example is E. Small-Spain-265, DAO).

98. *Medicago turbinata* (L.) All.

M. turbinata (L.) All., Fl. Pedem. 1: 315 (1785). (Greuter et al. (1989) considered this to be "nom. confus."; indeed, the name has often been misapplied to *M. doliata* (frequently simply because of misidentification); nevertheless it is adopted here on the basis that *M. turbinata* has become the established name and should be conserved.)

The epithet *turbinata* is Latin for top-shaped, a reference to the fruit. When spines are present, these are sometimes inclined opposite to the direction of fruit coiling, giving the fruit the appearance of a rapidly spinning top.

Subordinate Taxa

None recognized here. Previous taxonomists have split the species into a large number of infraspecific taxa, largely on the basis of the nature or absence of spines or tubercles, despite the fact that these different forms for the most part grow together or in proximity.

Type Specimen

Heyn (1959: 166) lectotypified the basionym for *M. turbinata* (*M. polymorpha* L. var. δ *turbinata* L.) by the illustration

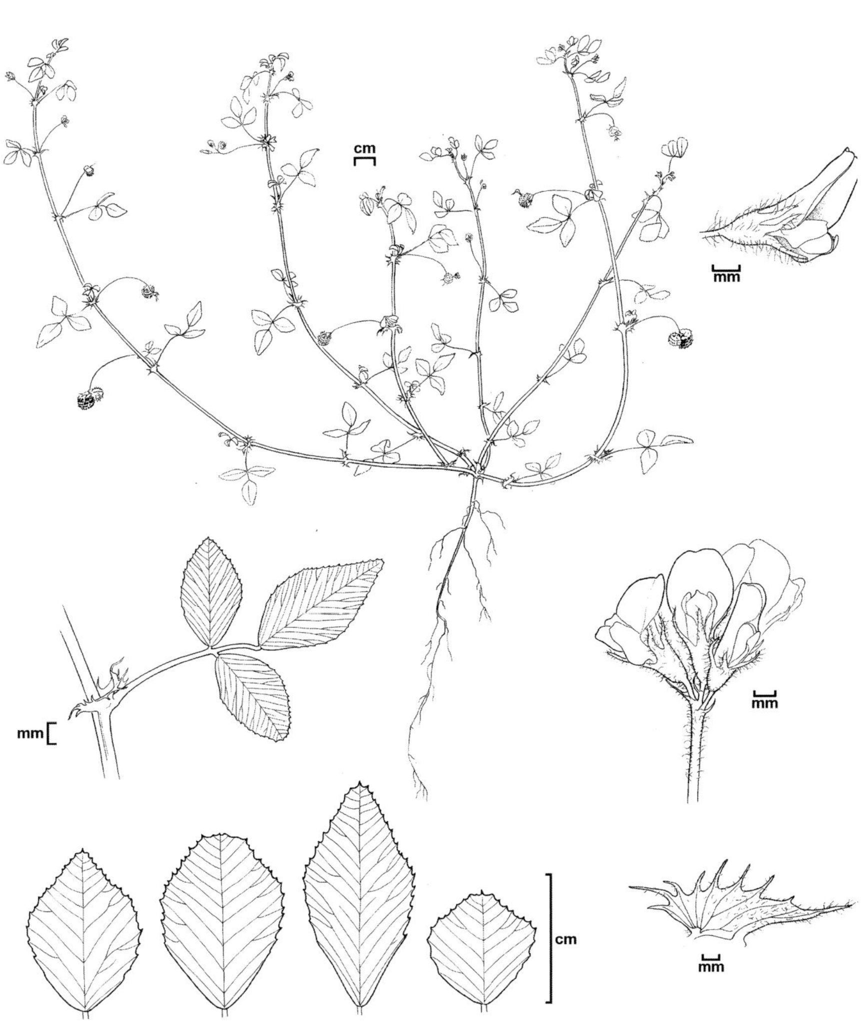

Fig. 98-1. *Medicago turbinata.* **Clockwise from top left:** Habit drawing of plant; side view of flower; inflorescence with peduncle; stipule; leaflets; leaf attached to stem. (Based on E. Small M946, DAO, except the habit drawing, which is based on Andursky, loan number 99/293, HUJ; artist: A. Hanes.)

(accompanied by the text "Medica magna turbinata") in: Bauhin, J., Cherler, J.H., et al. 1650–1651. Historia Plantarum Universalis, vol. 2: 385. Typographia Caldoriana (printer), Yverdon, Switzerland. This lectotypification was accepted in The Linnaean Plant Name Typification Project (http://www.nhm.ac.uk/research-curation/projects/linnaean-typification/; current June 2010; cf. Jarvis 2007).

Synonymy

Homotypic

M. polymorpha L. var. δ *turbinata* L., Sp. Pl. 780 (1753), (basionym).

M. turbinata (L.) Willd., Sp. Pl. 3: 1409 (1802). (As Heyn (1963) pointed out, this is according to synonymy, not according to type.)

Medica tubinata (sic) Medik., Vorles. Churpfälz. Phys.-Öcon. Ges. 2: 386 (1787).

Heterotypic

M. tuberculata (Retz.) Willd. var. *aculeata* Moris in Moris & De Notaris, Fl. Caprar. 36 (1839).

M. turbinata (L.) All. var. *aculeata* (Moris) Urb., Verh. Bot. Ver. Brandenb. 15: 72 (1873).

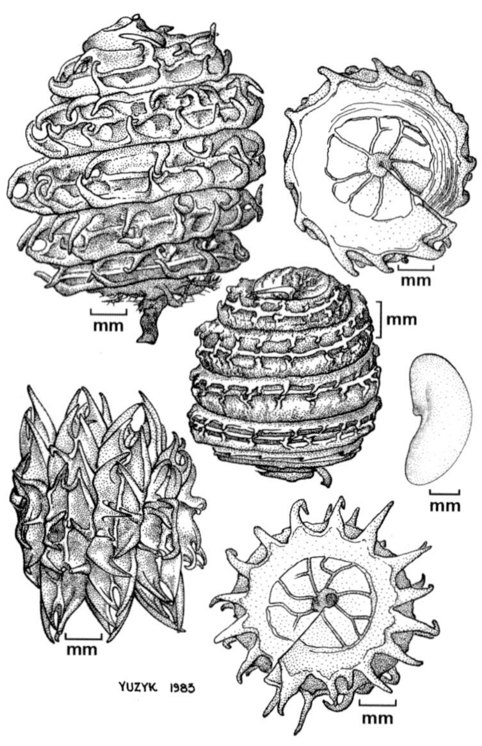

Fig. 98-2. *Medicago turbinata*, showing variation in fruits. **Top:** Fruit (left) and fruit coil face showing venation (right) (based on Zohary 16-5-1927, HUJ; identified as var. *spinulosa*). **Center:** Fruit (based on Zohary 15-4-1924, HUJ; identified as var. *turbinata*). **Center right:** Seed. **Bottom (left to right):** Fruit; fruit coil face showing venation (the preceding three based on Naftolsky 14-5-1927, HUJ; identified as var. *aculeata*). (Artist: L. Yuzyk.)

M. turbinata (L.) All. var. *aculeata* (Moris) Heyn, Scripta Hierosolymitana 12: 119 (1963), nom. superfl.

M. apiculata Bast., Essai Fl. Maine-et-Loire 280 (1809). nom. superf.

M. tuberculata (Retz.) Willd. var. *apiculata* (Bast.) Urb., Verh. Bot. Ver. Brandenb. 15: 72 (1873). (Based on Bast., exsicc. p. 280.)

M. tuberculata (Retz.) Willd. var. *brevispina* Rouy in Rouy & Fouc., Fl. Fr. 5: 32 (1899).

M. catalonica Schrank, Pl. Rar. Hort. Monac. tab. 28 (1819). (Possible type indicated in Heyn (1963 p. 119).)

M. tuberculata (Retz.) Willd. var. *chiotica* Urb., Verh. Bot. Ver. Brandenb. 15: 72 (1873). (Type indicated in Heyn (1963 p. 119).)

M. ciliaris Balb. ex DC., Fl. Franç. (ed. 3, Suppl.) 548 (1815). (Presumed synonymy.)

M. tuberculata (Retz.) Willd. var. *vulgaris* Moris f. *dextrorsa* Urb., Verh. Bot. Ver. Brandenb. 15: 72 (1873).

M. tuberculata (Retz.) Willd. var. *glandulosa* Podp., Verh. Bot.-Zool. Ges. 52: 645 (1902).

M. pubescens DC., Cat. Pl. Horti Monsp. 124 (1813).

M. pubescens DC. & Lam., Fl. Franç. ed. 3, 5: 569 (1815). (In 1993, I observed four specimens without mature fruits in Herb. DC. in P; the one bearing the information "dans un champ...le mai 1807" is particularly suitable as type material and is here designated the lectotype.)

Fig. 98-3. *Medicago turbinata*. From Hallier (1885, vol. 23, plate 244).

M. tuberculata (Retz.) Willd. var. *vulgaris* Moris f. *pubescens* (DC.) Urb., Verh. Bot. Ver. Brandenb. 15: 72 (1873).

M. tuberculata (Retz.) Willd. var. *vulgaris* Moris f. *sinistorsa* Urb., Verh. Bot. Ver. Brandenb. 15: 72 (1873).

M. spinulosa DC., Adnot. Cat. Hort. Monsp. 124 (1813), non Hornem. (*M. spinulosa* DC. in Lam. & DC., Fl. Franç. ed. 3, 5: 569 (1815).)

M. tuberculata (Retz.) Willd. var. *spinulosa* Rouy in Rouy & Fouc., Fl. Fr. 5: 32 (1899).

M. turbinata (L.) All. subsp. *spinulosa* (DC.) Ponert, Feddes Repert. 83 (9-10): 640 (1973), comb. invalid.

M. tuberculata (Retz.) Willd. var. *submutica* Boiss. ap. Evenari in Oppenh. & Evenari, Fl. Cisj. 284 (1940). (Type indicated in Heyn (1963, p. 118).)

M. tuberculata (Retz.) Willd. var. *syriaca* Tod. ex Urb.,Verh. Bot. Ver. Brandenb. 15: 72 (1873). (Based on Todaro, Fl. Sic. Exc. Nr. 1141.)

M. polymorpha L. var. *tuberculata* Retz,, Observ. Bot. 2: 23 (1781). (Specimen with label by P. Lassen in 1977 imprinted lectotype! at LD: Herb. Retzii, Lund, 84/76.5112; photo DAO!)

M. tuberculata (Retz.) Willd., Sp. Pl. 3(2): 1410 (1802).

M. tuberculata (Retz.) Willd. var. *typica* Hal., Consp. Fl. Graec. 1: 364 (1901), nom. inadmiss.

M. tuberculata (Retz.) Willd. var. *vulgaris* Moris in Moris & De Not., Fl. Caprar. 36 (1839), nom. inadmiss.

M. tuberculata (Retz.) Willd. var. *warburgina* Evenari in Oppenh. & Evenari, Fl. Cisj. 284 (1940).

According to the International Plant Names Index (http://www.ipni.org/index.html; current June 2010), the name *M. rugosa* d'Urv., Mém. Soc. Linn. Paris 1: 354 (1822) corresponds to *M. tuberculata* (*M. turbinata*).

Not

M. turbinata sensu Willd. (Sp. Pl. 3(2): 1409 (1802)) = *M. doliata* Carmign. (See *M. doliata* for several infraspecific synonyms of *M. turbinata* that have been adopted in this sense.)

Vernacular Names

English: Tubercled medic. The Latin name of the species translates as "top-shaped medic", a name that could be used. The name could also be "top medic", although this could be taken to mean "the best medic". The name "Southern medic" has been encountered (e.g., United States Department of Agriculture, Natural Resources Conservation Service, Introduced Plants of the United States website at http://plants.usda.gov/java/introduced?startChar; current July 2010), but is too vague to be useful. Kernick (1978) cites the name "cogwheel medick" (Townsend (1974) indicates this is an American name) and Rumbaugh and Johnson (1986) cite the name "cog-wheel burr clover". (Burr is also spelled bur; medic is also spelled medick.)

French: La luzerne en toupie (luzerne entoupie), luzerne à fruits tuberculeux, luzerne turbinée, luzerne à tubercules.

Geographical Distribution

(N = native, I = introduced, U = uncertain status)

Africa: Algeria (N), Egypt (rare; N), Libya (N), Morocco (N), Tunisia (N).

Asia: Cyprus (N), Iraq (N), Israel (N), Jordan (N), Lebanon (N), Syria (N), Turkey (N).

Europe: Bulgaria (U), France (including Corsica; N), Greece (including Crete, Karpathos and East Aegean Islands; N), Italy (including Sardinia and Sicily; N), Malta (N), Portugal (N), Spain (including Balearic Islands; N), Turkey (N), Yugoslavia (N).

Northern America: USA (New Jersey, Oregon; I).

Medicago turbinata is native to Mediterranean Circle countries and is much more common in the northern Mediterranean area. It occurs rarely as an adventive in Europe and elsewhere.

Description

Annual herb, stems 15–40 cm long (up to 150 cm according to Nègre 1956), procumbent or ascending, branched from base. Vegetative parts densely pubescent with simple hairs. Leaflets obovate, ovate, or oblanceolate, 12–16 mm long, 6–8 mm wide, up to three-fourths of distal margin, serrate or doubly serrate. Stipules ovate to lanceolate, margin entire or dentate. Inflorescence of (1)3–8(10) flowers, the peduncle longer than the subtending petiole. Flowers (3)5–8 mm long. Corolla yellow or orange-yellow, less than twice length of calyx. Calyx campanulate, pubescent with simple hairs (also sometimes glandular hairs) or glabrescent, the subulate teeth about equal to or somewhat longer than length of tube. Young pod contracted within calyx. Mature pod spherical to ovoid or short-cylindrical, the apex truncate, pods 6–12(15) mm long, glabrous, with (4)5–7(9) coils, the coil edges with spines, tubercles, or spineless. Coils strongly thickened, lignified, usually adpressed, diameter 5–8 mm, the coil face with radial veins entering a veinless margin occupying the distal quarter or third of the coil. Dorsal suture of coils in young pods protrudes (producing apparent constrictions between coils), this feature often disappearing with pod maturation. Spines obsolete or short (up to ca. 5 mm long); when spines are present, their bases are usually buried in alveolar tissue. Seeds 4–5 mm long, 2–3 mm wide, reniform, 1–2 per coil, separated by transverse fruit partitions, seedcoat smooth, yellow to light brown, radicle less than half length of seed.

Identification Hints

Medicago turbinata can be very difficult to distinguish from other hard-fruited species (currently placed in sect. *Spirocarpos* subsect. *Pachyspirae*), but is most likely to be confused with *M. doliata*. The outer coil face of *M. turbinata* has a veinless border, and this is darker in color than the inner part of the coil where evident veins occur. This veinless border is absent in *M. doliata*, but if only mature fruits are available on which alveolar tissue obscures venation (as is usually the case), one might mistakenly conclude that a veinless zone on the pod face is present. However, when the veinless border appears to be present in *M. doliata*, it is of the same color as the inner part of the coil with evident veins.

Habitat and Environmental Adaptation

Habitat

Medicago turbinata occurs in cultivated fields and waste ground, along roadsides, and on the edges of woods and in open woods.

Phenology

Flowering occurs in the spring (February to June depending on location in Europe; April and May in Turkey).

Soil

Piano and Francis (1992) observed that in its native area the species occurs on clay and loam soils, ranging in pH from slightly acidic to moderately basic. It has also been collected on sandy soils along seashores and in stony places (Thanopoulos 2007). In Israel, the species grows on terra rossa and on alluvial soils (Heyn 1963).

Temperature

The species is a typical Mediterranean winter annual, germinating in late winter and early spring under cool conditions, and completing its life cycle by the increasingly hot days of midsummer. Kurt et al. (1990) found that *M. turbinata* seeds overwintered in Central Anatolia.

Water

Piano and Francis (1992) observed that in its native area the species requires a minimum of 350 mm annual precipitation. The species often occupies dry soils in its native area, although Nègre (1956) commented that in Morocco it prefers soils that remain damp for a long period before drying.

Biotic Relationships

Pollination

Small (1988) observed that the small flowers and pollen to ovule ratio (308) are consistent with *M. turbinata* being largely self-pollinated. This species is basically an inbreeder.

Dispersal

The spiny fruits present in many forms of the species adhere to fur and serve to distribute the seeds.

Herbivory

In preference tests of the alfalfa weevil (*Hypera postica*) for 13 annual species of *Medicago*, Pandey et al. (1984) found that the weevil preferred *M. turbinata* least.

Nodulation and mycorrhizae

USDA (2009) reported positive root nodulation for *M. turbinata*. In Central and Western Spain, it has been shown that, when infected by an effective strain of *Rhizobium*, *M. turbinata* resists freezing temperatures and long periods of winter drought, whereas noninfected plants die (Feuntes 1981).

Disease resistance

Yaege and Stuteville (2002) tested the resistance of 33 annual species of *Medicago* to powdery mildew (*Erysiphe pisi*) and found that the four collections of *M. turbinata* grown were susceptible to the disease (it was observed that often only one of a number of populations of other species tested was resistant). In a comparative test of 36 annual species of *Medicago*, O'Neill and Bauchan (2000) found a range of susceptibility to anthracnose caused by *Colletotrichum trifolii* (from completely susceptible to almost completely resistant) among five accessions of *M. turbinata*.

Economic History, Cultivars, and Uses

This species has attracted little agricultural interest to date.

Cultivars

No cultivars have been bred.

Forage

Medicago turbinata is rarely sown for forage. Henson and Hollowell (1960) recommended *M. turbinata* (as "*M. tuberculata* Willd.") for pasture usage in East Central Texas. Vitkus (1977) assessed *M. turbinata* as suitable for cultivar selection for Lithuania. Belov (1971) examined 37 annual species of *Medicago* for suitability as new crops in the Tashkent region of Uzbekistan, and *M. turbinata* was one of the three species recommended for development. St. John-Sweeting et al. (1996) compared several species in South Australia, and suggested that further study of developing *M. turbinata* commercially was warranted in view of its ability to compete effectively with weeds. Tadmor et al. (1971) conducted studies of ecotypes selected for their ability to survive and persist in the Negev Desert of Israel in an area where rainfall averaged only 90 mm annually, but received an additional 263 mm of water annually in the form of run-off from erratic winter storms. The dry matter production of an ecotype of *M. turbinata* was 6.2 t/ha (averaged over 6 years), which is considered rather good given the dry environment. Mohammad and Qamar (1988) grew 14 annual species of *Medicago* in small test plots in Islamabad, Pakistan located in a subtropical, subhumid area that is notably hotter and wetter than the normal habitat of these species. The dry matter yield of *M. turbinata* was 2 t/ha.

Fodder

Medicago turbinata has not been employed to any significant extent as a source of dried animal feed.

Green manure, rotation crop, ground cover

No information located.

Other uses (salad sprouts, ornamental, plant products, etc.)

Lectins are proteins that selectively bind carbohydrates and carry out several important metabolic functions in cells, and so these have potential medicinal applications. Bird and Wingham (1983) studied the presence of certain lectins in a dozen species of *Medicago* and observed that these were only detectable in *M. disciformis* and *M. turbinata*.

Chromosome Number

$2n$ = 14: Agarwal and Gupta 1983*b* (as "*M. tuberculata* Willd.").

$2n$ = 14, 16: Mariani 1963.

$2n$ = 16: Fryer 1930 (as *M. tuberculata*); Ghimpu 1929*b*, 1930 (as both *M. tuberculata* and *M. turbinata*); Clement 1962 (as *M. tuberculata* and *M. turbinata*); Tschechow 1932 (as *M. tuberculata* and *M. turbinata*); Senn 1938 (as *M. tuberculata*); Heyn 1956; Simon and Simon 1965; Bolkhovskikh et al. 1969, p. 309 (as *M. tuberculata*); Dahlgren et al. 1971; Lesins and Lesins 1979; Agarwal and Gupta 1983*b*.

Genetic Relationships and Gene Transferability

Intraspecific

Artificial hybridization of different populations has not been attempted, so it is not known whether there are barriers to hybridization within the species.

Interspecific

Simon and Millington (1967) were unable to produce hybrid seed by crossing *M. turbinata* with *M. truncatula*, *M. littoralis*, or *M. italica*. Lesins et al. (1980) hybridized *M. truncatula* with *M. turbinata*; the hybrids were completely sterile, although doubling the chromosome number resulted in some fertility. Lesins et al. (1980) and also Quiros and Ostafichuk (1983) observed chlorophyll deficiency in some hybrids between *M. truncatula* and *M. turbinata*.

Exchange by advanced biotechnology

No information located.

Conservation Status

Medicago turbinata is a fairly widespread species and therefore not in apparent need of conservation measures.

Germplasm Sources

More than 500 accessions are available from germplasm institutions with large *Medicago* collections (see Chapter 3).

Research Problems and Priorities

Medicago turbinata is a typical annual *Medicago* and may have some use similar to the seven other annual species that have been the source of pasture cultivars. It deserves to be examined for its agricultural potential.

Representative Collections

Algeria: Azeudt 114, Herbarium of, Alger (BM). **Cyprus:** Chapman, E. 239, Platanisso Karpas, alt. 150 feet, open ground near village, locally common, 7-IV-1950, in flower & fruit (K). Chapman, E. 425, mile 3, Larnaca-F'gusta Road, alt. 10 feet, roadside on edge of marsh, 13-IV-1950, in flower & fruit (K). **England:** Brown, G.G. 748, Hythe Quay, Colchester, on cottonseed refuse, 22-VI-1914, in flower & fruit (K). **Greece:** Reverchon, E., Crete, La Canée, moissons, in flower 5-VI & in fruit 2-VIII-1883, E. Reverchon Plantes de Crête 1883 No. 32 (K, P). Sledge, W.W., Rhodes, Mt. Profitis Elias, bushy places near M. Fountoukli, 20-IV-1981, in flower & early fruit (K) ("var. *chiotica* Urb."). Townsend, C.C. 67/75, Peloponnese, Olympia, barley field at the foot of the Hill of Kronos, 30-IV-1967, in flower & early fruit (K). **Israel:** Davis, P.H. 4243, Tiberias-Kligdal, calcareous slopes, among annuals, 3-IV-1942, in fruit (K). Hepper, F.N. 3462, Upper Galilee, Mt. Meron Nature Reserve, open places on steep hill with some red soil, collected in limestone pockets, 18-IV-1969, in flower & early fruit. Zohary, M. & Amdursky, I., western Upper Galilee, mouth of Wadi Qarn, 24-III-1955, in flower & fruit, Univ. Hebraicae Hierosolymitanae Flora Israëlis Exsiccata No. 650 (BM, K). **Italy:** Fiori, A., Apulia, Hydruntum (Otranto), in arvis circa lacum Alimini piccolo, alt. ca. 10 m solo calcareo, 2-VI-1913, Flora Italica Exsiccata series III – Fiori & Béguinot No. 2094 (K). Huter, Porta & Rigo 74, Sicilia orient. ad agrorum margines collium arenos supra Messina, sol. granit., alt. 2–300 m, 18-IV-1877, in flower & fruit (K) (mixed collection with *M. doliata*). Porta & Rigo 359, Italia austral., Gapygia inter segetes pr. Otranto, sol. calcareo, 14-V-1875, in flower & fruit (K). Rigo, G. 69, prov. Siracusa, Avola, in herbosis collium, solo tuffacea, 21-IV-1898, in flower & fruit (K, P). **Jordan:** Maitland, T.D. 21, Shiekh Jarrah, on edge of wheat field, IV-1963, in flower & fruit (K). **Lebanon:** Meinertzagen, R., Beyrouth, Syria, in crops, 31-III-1933, in flower & fruit (BM). **Portugal:** Scarlett 59, Lisbon district, 1935, in flower & fruit (K). **Spain:** Sennen, Fre., Catalogne, tortiá. talus, 1922 (P). ("*Medicago catalonica* Schrank") **Syria:** Blanche 588, environs de Saïda, IV-1853, in flower & fruit (P). Samuelsson, G. 3289, Bordj prope Djibail, in agro lapidoso, 14-IV-1933, in flower & early fruit (K). **Turkey:** Balansa, B., Village de Tchasusli près de Mersine (Cilicie), 11-VI-1855, in flower & fruit (P). Bourgeau, E., Lycica, Odalia, in graminosis, 21-IV-1860, in fruit (P). Davis 41190, C2 Muğla, d. Marmaris, Sögüt to Bozburan, sea level – 50 m, rocky limestone slopes, 15-IV-1965, in flower & fruit (K). **Yugoslavia:** Huter, Istria, Pola, in agris, 10-V-1867, in flower & fruit (K). Pichler, T., Dalmatia, ad viam ferream pr. Salona, V-1880, in flower & fruit (K).

Literature Cited

Abd El Moneim, A.M., and Cocks, P.S. 1986. Adaptation of *Medicago rigidula* to a cereal-pasture rotation in northwest Syria. J. Agric. Sci. 107: 179–186.

Abdelguerfi, A. 1989. Contribution à l'étude des espèces spontanées du genre *Medicago* en Algérie. Variabilité au niveau des graines et des gousses chez sept populations de *M. scutellata*: relation avec les condition du milieu d'origine. XVI Congrès International des Herbages, Nice, France, 1989. Association Française pour la Production Fourragère, Versailles, France. pp. 219–220.

Abdelguerfi, A., and Guittoneau, G.G. 1979. I.O.P.B. chromosome number reports. Fabaceae. Taxon, 28: 392–393.

Abdelguerfi, A., and Guittoneau, G.G. 1989. Morphologie, écologie et répartition du *Medicago heterocarpa* Spach (= *M. lesinsii* Small) en Algérie, espèce méditerranéenne du complexe *M. murex* Willd. Candollea, 44: 423–433.

Abdelguerfi, A., and Laouar, M. 2000. Effet des conditions bioclimatiques d'origine sur le comportement et la floraison des populations de *Medicago ciliaris* (L.) Krocker. *In* Legumes for Mediterranean forage crops, pastures and alternative uses. *Edited by* L. Sulas. CIHEAM-IAMZ, Zaragoza, Italy. pp. 241–244.

Abdelguerfi, A., Chapot, J.Y., and Conesa, A.P. 1988*a*. Contribution a l'étude de la répartition des luzernes annuelles spontanées en Algérie selon certain facteurs du milieu. Fourrages (France), 113: 89–106.

Abdelguerfi, A., Abdelguerfi-Berrekia, R., and Guittoneau, G.G. 1988*b*. Contribution to the study of wild species of the genus *Medicago* L. in Algeria. III. Chromosome counts in 35 populations of 17 species. Annales Institut National Agronomique El Harrach 12(1, T1): 342–354. (In French.)

Abdelguerfi, A., Laouar, M., Tazi, M., Bounejmate, M., and Gaddes, N.E. 2000. Present et futur des pâturages et des légumineuses fourragères en région méditerranéenne: cas du nord de l'Afrique et de l'ouest de l'Asia. Cahiers Options Méditerranéennes, 45: 461–468.

Abdelguerfi-Laouar, M., Abdelguerfi, A., Bouznad, Z., and Guittoneau, G.G. 2003. Autoecology and distribution of the species complex *Medicago ciliaris* – *Medicago intertexta* in Algeria. Acta Bot. Gallica, 150: 253–265. (In French.)

Abdelkefi, A., Boussaid, M., and Marrakchi, M. 2001. Species complex and reproduction system in the *Intertextae* section: *Medicago ciliaris*, *Medicago intertexta*. Options Méditérraneennes Serie A, Seminaires Méditérraneens, 45: 107–110. (In French.)

Abdelly, C., Zid, E., Hajji, M., and Grignon, C. 1996. Biomass production and nutrition of *Medicago* species associated with halophytes on the edge of a sebkha in Tunisia. *In* Halophytes and biosaline agriculture. *Edited by* R. Choukr-Allah, C.V. Malcolm, and A. Hamdy. Marcel Dekker Inc., New York, NY. pp. 313–324.

Abranches, R., Marcel, S., Arcalis, E., Altmann, F., Fevereiro, P., and Stoger, E. 2005. Plants as bioreactors: a comparative study suggests that *Medicago truncatula* is a promising production system. J. Biotech. 120: 121–134.

Adams, N.R. 1989. Phytoestrogens. *In* Toxicants of plant origin, Vol. 4, Phenolics. *Edited by* P.R. Cheeke. CRC Press Inc., Boca Raton, FL. pp. 23–51.

Agarwal, K., and Gupta, P.K. 1983*a*. Cytological studies in the genus *Trigonella* Linn. Cytologia, 48: 771–780.

Agarwal, K., and Gupta, P.K. 1983*b*. Cytological studies in the genus *Medicago* Linn. Cytologia, 48: 781–793.

Ahuja, M.R. 1955. Chromosome numbers of some plants. Indian J. Genet. Plant Breed. 115: 142–143.

Akan, H., Kaya, Ö.F., Eker, I., and Cevheri, C. 2005. The flora of Kaşmer Daği (Şanliurfa, Turkey). Turk. J. Bot. 29: 291–310.

Al-Bermani, A.-K.K.A., Al-Shammary, K.I.A., Gornall, R.J., and Bailey, J.P. 1993. Contribution to a cytological catalogue of the British and Irish flora, 3. Watsonia, 19: 169–171.

Aldrich, D.T.A. 1984. Lucerne, red clover and sainfoin: herbage production. *In* Forage legumes. *Edited by* D.J. Thomson. Occasional Symposium No. 16, British Grassland Society, Hurley, UK. pp. 126–131.

Alefeld, M. 1859. Zu *Medicago*. Bot. Zeitung (Berlin), 17: 278–279.

Alegre, J., Sancha, J.L., Guía, E., and Agudo, M.A. 1993*a*. Caracterización nutriva de arbustos forrajeros: 1. Composicíon química de leguminosas arbustivas y su evolución estacional. XVIII Jornadas Cientificas de la Sociedad Española para el Estudio de los Pastos. pp. 58–67.

Alegre, J., Navarrete, L., Ceresuela, J.L., and Hornero, J. 1993*b*. La alfalfa leñosa de Creta (*Medicago strasseri* Matthäs, Greuter & Risse): Primeros datos acerca de su potencial interés forrajero. XXXI Reunión Cientifica de la Sociedad Española para el Estudio de los Pastos. pp. 76–80.

Ali, S.I. 1968. Typification of the genus *Medicago* and some of its Linnaean species. Taxon, 17: 540–541.

Ali, S.I. 1972. Typification of the genus *Medicago* and some of its Linnaean species – a reply. Taxon, 21: 665–668.

Ali, S.I. 1977. Flora of West Pakistan. No. 100: Papilionaceae. Department of Botany, University of Karachi, Pakistan (Printed at Ferozsons, Karachi). 390 pp.

Allen, V.G. (Chair) and 32 others (The Forage and Grazing Terminology Committee). 1992. Terminology for grazing lands and grazing animals. J. Prod. Agric. 5: 191–201.

Allison, M.J., Hammond, A.C., and Jones, R.J. 1990. Detection of ruminal bacteria that degrade toxic dihydroxpyridine compounds produced from mimosine. Appl. Environ. Microbiol. 56: 590–594.

Al-Mayah, A.-R.A., and Al-Shehbaz, I.A. 1977. Chromosome numbers for some Leguminosae from Iraq. Bot. Notiser, 130: 437–440.

Al-Yahya, M.A., Mossa, J.S., Al-Badr, A.A., Tariq, M., and Al-Meshai, I.A. 1987. Phytochemical and biological studies on Saudi medicinal plants: part 12. A study on Saudi plants of family "Leguminosae". Int. J. Crude Drug. Res. 25: 65–71.

Amato, G., Stringi, L., and Giambalvo, D. 2004. Productivity and canopy modification of *Medicago arborea* as affected by defoliation management and genotype in a Mediterranean environment. Grass Forage Sci. 59: 20–28.

American Phytopathological Society. 2001. Diseases of legumes. American Phytopathological Society, St. Paul, MN. (Computer optical disc.)

Amich [Amich Garcia], F., and Sánchez [Sánchez Sánchez], J. 1983. Acerca de las subespecies de *Trigonella polyceratia* L. Studia Botanica, 2: 129–132.

Amschler, W. 1935. The oldest pedigree chart. J. Heredity, 26: 233–238.

Anderson, C.R. 1944. Contact dermatitis from alfalfa and bur clover. Archs. Derm. Syph. 5: 201.

Andreu, V., Rubio, J.L., and Cerni, R. 1994. Use of a shrub (*Medicago arborea*) to control water erosion on steep slopes. Soil Use Manage. 10: 95–99.

Andreu, V., Rubio, J.L., Gimeno-Garcia, E., and Llinares, J.V. 1998. Testing three Mediterranean shrub species in runoff reduction and sediment transport. Soil Tillage Res. 45: 441–454.

Andrew, W.D. 1964. The effect of nitrogen stress during flowering on seed production in *Medicago laciniata*. Austral. J. Exp. Agric. Animal Husb. 4: 222–224.

Andrew, W.D., and Hely, F.W. 1960. Frequency of annual species of *Medicago* on the major soil groups of the Macquarie region of New South Wales. Austral. J. Agrc. Res. 2: 705–714.

Anonymous. 1972. *Medicago rugosa* Desr. (gama medic). Register of Australian Herbage Plant Cultivars, 2nd edition. 2 pp.

Anthony, D.W. 1986. The "Kurgan Culture", Indo-European origins, and the domestication of the horse: a reconsideration. Current Anthropology, 27: 291–313.

Anthony, D.W. 1991. The domestication of the horse. *In* Equids in the ancient world, II. *Edited by* R.H. Meadow and H.-P. Uerpmann. Ludwig Reichert Verlag, Wiesbaden, Germany. pp. 250–277.

Anthony, D.W., and Brown, D.R. 1991. The origins of horseback riding. Antiquity (Cambridge, UK.), 65: 22–38.

Aparicio, A. 1986. Números cromosomáticos de plantas occidentales, 422-426. Anales del Jardín Botánico de Madrid, 43: 427–430.

Applebaum, S.W., and Birk, Y. 1979. Saponins. *In* Herbivores, their interaction with secondary plant metabolites. *Edited by* G.A. Rosenthal and D.H. Janzen. Academic Press, New York, NY. pp. 539–566.

Archer, M. 1973. The species preferences of grazing horses. J. Brit. Grassland Soc. 28: 123–128.

Armour, D.J., Mackie, J.M., Musial, J.M., and Irwin, J.A.G. 2008. Transfer of anthracnose resistance and pod coiling traits from *Medicago arborea* to *M. sativa* by sexual reproduction. Theor. Appl. Genet. 117: 149–156.

Armstrong, J.M., and Gibson, D.R. 1941. Inheritance of certain characters in the hybrid of *Medicago media* and *M. glutinosa*. Sci. Agr. 22: 1–10.

Astanova, S.B. 1981. Chromosome numbers of Leguminosae of flora in Tajikistan. Dokl. Akad. Nauk Tadziksk. SSR. 24: 61–63. (In Russian.)

Athar, M., and Mahmood, A. 1978. A qualitative study of the nodulating ability of legumes of Pakistan: list 1. Pakistan J. Bot. 10: 95–100.

Atwood, S.S., and Grunn, P. 1952. Cytogenetics of alfalfa. Bibliogr. Genet. 14: 133–188.

Australian Herbage Plant Cultivars – Legumes. http://www.pi.csiro.au/ahpc/legumes/legumes.htm (current June 2010).

Axelrod, D.I. 1973. History of the Mediterranean ecosystem in California. *In* Mediterranean-type ecosystems. Origin and structure. *Edited by* F. di Castri and H.A. Mooney. Springer-Verlag, Berlin, Germany. pp. 225–277.

Azizi, K., Dehaghi, M.A., and Abbasipour, H. 2004. Effect of different air and root-zone temperatures on nitrogen fixation and nodulation of annual medics. J. Agron. 3: 131–136.

Baburina, O.K., and Shevyakova, N.I. 1987. Salt resistance and the level of quaternary ammonium compounds in wild alfalfa species. Fiziologiya Rastenii (Moscow), 34: 887–893. (In Russian.)

Baburina, O.K., and Shevyakova, N.I. 1995. Betaines as methyl group donors in alfalfa seedlings exposed to salt. Fiziologiya Rastenii (Moscow), 42: 862–870. (In Russian.)

Bailly, X., Olivieri, I., De Mita, S., Cleyet-Marel, J.-C., and Béna, G. 2006. Recombination and selection shape the molecular diversity pattern of nitrogen-fixing *Sinorhizobium* sp. associated to *Medicago*. Molec. Ecol. 15: 2719–2734.

Bailly, X., Olivieri, I., Brunel, B., Cleyet-Marel, J.-C., and Béna, G. 2007. Horizontal gene transfer and homologous recombination drive the evolution of the nitrogen-fixing symbionts of *Medicago* species. J. Bacteriol. 189(14): 5223–5236.

Baker-Brosh, K.F., and Peet, R.K. 1997. The ecological significance of lobed and toothed leaves in temperate forest trees. Ecology, 78: 1250–1255.

Bakšay, L. 1958. The chromosome numbers of Ponto-Mediterranean plant species. Ann. Hist.-Nat. Mus. Nat. Hungarici, 50(Ser. nova, 9): 121–125.

Balabaev, G.A. 1934. Yellow lucernes of Siberia, *Medicago ruthenica* (L.). Ledb. and *M. platycarpos* (L.) Ledb. Bull. App. Bot. Genet. Pl. Breed., Ser. 7(1): 113–123. (In Russian.)

Balbis, G.B. 1801. Elenco delle piante. Torino, Italy. 102 pp.

Ballard, R.A., and Charman, N. 2000. Nodulation and growth of pasture legumes with naturalised soil rhizobia: 1. Annual *Medicago* spp. Austral. J. Exper. Agric. 40: 939–948.

Balodi, B., and Rao, R.R. 1991. The genus *Trigonella* L. Fabaceae in northwest Himalaya. J. Econ. Taxon. Bot. 15: 185–194.

Baltisberger, M. 1987. Chromosomenzahlen einiger Pflanzen aus Albanien Berichte des Geobotanishen Institutes der Eidgenössische Technishe Hochschule Stiftung Rübel, 53: 47–63.

Baltisberger, M. 1990. Numeri cromosomici per la flora Italiana: 1208-1230. Informatore Botanico Italiano, 22: 216–226.

Baltisberger, M. 1991*a*. IOPB chromosome data 3. International Organization of Plant Biosystematists Newsl. 17: 5– 7.

Baltisberger, M. 1991*b*. Cytological investigations of some Greek plants. Flora Mediterranea, 1: 157–173.

Baltisberger, M., and Baltisberger, E. 1995. Cytological data of Albanian plants. Candollea, 50: 457–493.

Baltisberger, M., and Charpin, A. 1989. Chromosomenzählungen von Gilbert Bocquet Berichte des Geobotanishen Institutes der Eidgenössische Technishe Hochschule Stiftung Rübel, 55: 246–251.

Baltisberger, M., and Voelger, M. 2006. IAPT/IOPB chromosome data: Fabaceae. Taxon, 55: 444.

Baltisberger, M., and Widmer A. 2006. Chromosome numbers of plant species from the Canary Islands. Bot. Helv. 116: 9–30.

Baltisberger, M., Mullaj, A., and Tartari, V. 1993. Mediterranean chromosome number reports 3 (185–203). Flora Mediterranea, 3: 348–353.

Baltisberger, M., Aeschimann, D., Charpin, A., and Jeanmonod, D. 2006. Chromosome numbers counted by the late Gilbert Bocquet. Candollea, 61: 61–70.

Barbetti, M.J. 1989. Strategies for control of Phoma black stem in annual *Medicago* species. Austral. J. Exp. Agric. 29: 635–640.

Barbetti, M.J. 1995. Resistance in annual *Medicago* species to *Phoma medicaginis* and *Leptosphaerulina trifolii* under field conditions. Austral. J. Exp. Agric. 35: 209–214.

Barbetti, M.J. 2007. Resistance in annual *Medicago* spp. to *Phoma medicaginis* and *Leptosphaerulina trifolii* and its relationship to induced production of a phytoestrogen. Plant Dis. 91: 239–244.

Barnes, D.K. 1972. A system for visually classifying alfalfa flower color. USDA-ARS Agric. Handbook. No. 424. U.S. Gov. Print. Office, Washington, DC. 12 pp + plates.

Barnes, D.K., and Sheaffer, C.C. 1995. Alfalfa. *In* Forages, 5th edition, Vol. 1, An introduction to grassland agriculture. *Edited by* R.F. Barnes, D.A. Miller, and C.J. Nelson. Iowa State University Press, Ames, IA. pp. 205–216.

Barnes, D.K., Bingham, E.T., Murphy, R.P., Hunt, O.J., Beard, D.F., Skrdla, W.H., and Teuber, L.R. 1977. Alfalfa germplasm in the United States: genetic vulnerability, use, improvement, and maintenance. Agricultural Research Service Tech. Bull. No. 1571. U.S. Dept. Agriculture, Hyattsville, MD. 21 pp.

Barnes, D.K., Goplen, B.P., and Baylor, J.E. 1988. Highlights in the USA and Canada. *In* Alfalfa and alfalfa improvement. *Edited by* A.A. Hanson, D.K. Barnes, and R.R. Hill, Jr. American Society of Agronomy, Crop Science Society of America, Soil Science Society of America, Madison, WI. pp. 1–24.

Barnes, R.F, Miller, D. A., and Nelson, C.J. (Editors). 1995. Forages. Volume 1: An introduction to grassland agriculture. Iowa State Univ. Press, Ames, IA.

Barney, P.A. 1987. The use of *Trifolium repens*, *Trifolium subterraneum* and *Medicago lupulina* as overwintering leguminous green manures. Biol. Agric. Hortic. 4: 225–234.

Baron von Mueller, F. 1895. Select extra-tropical plants. Government Printer, Melbourne, Australia. p. 302.

Basigalup, D.H., Barnes, D.K., and Strucker, R.E. 1995. Development of a core collection for perennial *Medicago* plant introductions. Crop Sci. 35: 1163–1168.

Bass, L.N., Gunn, C.R., Hesterman, O.B., and Roos, E.E. 1988. Seed physiology, seedling performance, and seed sprouting. *In* Alfalfa and alfalfa improvement.

Edited by A.A. Hanson, D.K. Barnes, and R.R. Hill, Jr. American Society of Agronomy, Crop Science Society of America, Soil Science Society of America, Madison, WI. pp. 961–983.

Bataillon, T., and Ronfort, J. 2006. Evolutionary and ecological genetics of *Medicago truncatula*. *In* The *Medicago truncatula* handbook (version 1, 2006). 13 pp. http://www.noble.org/MedicagoHandbook/ (current July 2010).

Bauchan, G.R. 1984. Colchicine doubled *Medicago disciformis* ($2n = 4x = 32$). Agron. Abstracts, p. 58.

Bauchan, G.R. 1987. Embryo culture of *Medicago scutellata* and *M. sativa*. Plant Cell Tiss. Organ Cult. 10: 21–29.

Bauchan, G.R. 2009. Alfalfa (*Medicago sativa* ssp. *sativa* (L.) L. & L.). *In:* Genetic resources, chromosome engineering, and crop improvement. Vol. 5, forage crops. *Edited by* R.L. Singh. CRC Press. Boca Raton, FL. pp. 11–39.

Bauchan, G.R., and Campbell, T.A. 1994. Use of an image analysis system to karyotype diploid alfalfa (*Medicago sativa* L). J. Heredity, 85: 18–22.

Bauchan, G.R., and Elgin, J.H., Jr. 1984. A new chromosome number for the genus *Medicago*. Crop Sci. 24: 193–195.

Bauchan, G., and Greene, S. 2000. National Plant Germplasm System, report on the status of *Medicago* germplasm in the United States, October 2000: http://www.ars-grin.gov/npgs/cgc_reports/alfalfa/alfalfacgc2000.htm (current July 2010).

Bauchan, G., and Greene, S. 2002. Status of the *Medicago* germplasm collection in the United States. Plant Genetic Resources News. 129: 1–8. http://bldg6.arsusda.gov/3.pdf (current July 2010).

Bauchan, G.R., and Hossain, M.A. 1997*a*. Identification of alfalfa chromosomes using Giemsa banding and image anlysis technique. http://www.internationalgrasslands.org/publications/pdfs/1997/1_04_061.PDF (current January 2010).

Bauchan, G.R., and Hossain, M.A. 1997*b*. Karyotypic analysis of C-banded chromosomes of diploid alfalfa: *Medicago sativa* ssp. *coerulea* and ssp. *falcata* and their hybrid. J. Heredity, 89: 533–537.

Bauchan, G.R., and Hossain, M.A. 1998*a*. Karyotypic analysis of N-banded chromosomes of diploid alfalfa: *Medicago sativa* ssp. *caerulea* and ssp. *falcata* and their hybrid. J. Heredity, 89: 191–193.

Bauchan, G.R., and Hossain, M.A. 1998*b*. Cytogenetic studies of the nine germplasm sources of alfalfa. Proc. North Amer. Alfalfa Improvement Conf., Bozeman, MT, Aug. 26, 1998, 36: 21.

Bauchan, G.R., and Hossain, M.A. 2001*a*. A computerized image analysis system to characterize small plant chromosomes. Microscopy Analysis, 48: 9–11.

Bauchan, G.R., and Hossain, M.A. 2001*b*. Distribution and characterization of heterochromatic DNA in the tetraploid African population alfalfa genome. Crop Sci. 41: 1921–1926.

Bauchan, G.R., Salter, R., and Elgin, J.H., Jr. 1986. Intraspecific hybridization among three morphotypes of *Medicago scutellata*. *In* Report of the thirtieth North American alfalfa improvement conference. *Edited by* J.B. Moutray and J.H. Elgin, Jr. North American Alfalfa Improvement Conference, Beltsville, Maryland. p. 48.

Bauchan, G.R., Campbell, T.A., and Hossain, M.A. 2002. Chromosomal polymorphism as detected by C-banding patterns in Chilean alfalfa germplasm. Crop Sci. 42: 1291–1297.

Bauchan, G.R., Campbell, T.A., and Hossain, M.A. 2003. Comparative chromosome banding of nondormant alfalfa germplasm. Crop Sci. 43: 2037–2042.

Baum, B.R. 1968. A clarification of the generic limits of *Trigonella* and *Medicago*. Can. J. Bot. 46: 741–749.

Baum, B.R. 1970. The type of the genus *Medicago* (Leguminosae). Taxon, 19: 80–84.

Belaeva, V.A., and Siplivinsky, V. 1981. I.O.P.B. chromosome number reports LXXIII. Taxon, 30: 857–860.

Belov, A. 1971. Some observations on collections of wild, annual lucerne. Nauchnye Trudy, Tashkentskii Sel'skokhozyaistvennyi Institut, 27: 26-34. (In Russian.)

Bena G. 2001. Molecular phylogeny supports the morphologically based taxonomic transfer of the "medicagoid" *Trigonella* species to the genus *Medicago* L. Pl. Syst. Evol. 229: 217–236. [G. Bena's name is occasionally rendered "Béna" in the literature; for consistency, "Bena" is presented.]

Bena G., Lejeune, B., Prosperi, J.M., and Olivieri, I., 1998*a*. Molecular phylogenetic approach for studying life-history evolution: the ambiguous example of the genus *Medicago* L. Proc. R. Soc., B. Biol. Sci. 265(1401): 1141–1151.

Bena G., Prosperi J.M., Lejeune, B., and Olivieri, I. 1998*b*. Evolution of annual species of the genus *Medicago*: a molecular phylogenetic approach. Molec. Phylogenetics Evol. 9: 552–559.

Bena, G., Lyet, A., Hugeut, T., and Olivieri, I. 2005. *Medicago – Sinorhizobium* symbiotic specificity evolution and the geographic expansion of *Medicago*. J. Evol. Biol. 18: 1547–1558.

Benlloch, R., Navarro, C., Beltran, J.P., and Cañas, L.A. 2003. Floral development of the model legume *Medicago truncatula*: ontogeny studies as a tool to better characterize homeotic mutations. Sexual Plant Reprod. 15: 231–241.

Bennett, S.J. 1999. Using collections to describe ecological relationships. *In* Genetic resources of Mediterranean pasture and forage legumes. *Edited by* S.J. Bennett and P.S. Cocks. Kluwer Academic Publishers, Boston, MA. pp. 41–52.

Bennett, S.J., Broughton, D.A., and Maxted, N. 2006. Ecogeographical analysis of the perennial *Medicago*. CRC Salinity Bulletin (Australia). 62 leaves. (Available online and deposited in the National Library of Australia.)

Bentham, G. 1826. Catalogue des plantes indigènes des Pyrénées et du Bas-Languedoc. Madame Huzard, Paris, France. 128 pp.

Bentham, G. 1865. Leguminosae. *In* G. Bentham and J.D. Hooker. Genera plantarum 1(2). Lovell Reeve, London, UK. pp. 434–600.

Berdahl, J.D., Wilton, A.C., Lorenze, R.J., and Frank, A.B. 1986. Alfalfa survival and vigour in rangeland grazed by sheep. J. Range Manage. 39: 59–62.

Beresnev, E.N., Chekanov, V.I., and Rudetskikh P.A. 1980. The consumption of the grass stand of wormwood ephemeral pastures in the desert zone of the southeastern part of the Kazakh SSR USSR. Izvestiya Akademii Nauk Kazakhskoi SSR Seriya Biologicheskaya, 1: 17–22.

Berggren, H. 1992. Project agro-fibre industrial use for pulp and paper and related products. *In* The future of lucerne–biotechnology, breeding and variety constitution. Proceedings of the X international conference of EUCARPIA, *Medicago* spp. group. *Edited by* P. Rotili and L. Zannone. Istituto Sperimentale per le Colture Foraggere, Lodi, Italy. pp. 55–66.

Bergman, D.K., Dillwith, J.W., Zarrabi, A.A., and Berberet, R.C. 1991. Epicuticular lipids of alfalfa leaves relative to position on the stem and their correlation with aphid (Homoptera: Aphididae) distributions. Environ. Entomol. 20: 470–476.

Bernays, E.A. 1991. Evolution of insect morphology in relation to plants. Phi. Trans. R. Soc. Lond. B 333: 257–264.

Bhaumik, G.H. 1976. Cytological investigations on some members of the tribe Trifolieae (family Papilionaceae). Sci. Cult. 42: 322–324.

Bialy, Z., Jurzysta, M., Mella, M., and Tava, A. 2004. Triterpene saponins from aerial parts of *Medicago arabica* L. J. Agric. Food Chem. 52: 1095–1099.

Bijok, K., Adamkiewicz, E., and Grygorczyk, I. 1971. Studia nad kariologia i rozwojem pylku kilku odmian *Medicago sativa* L. Gz. I. Acta Agrobot. 24: 319–341.

Bilis, J. 2003. Polyploidization and tetraploid black medic selection according to cytomorphological characters. Zemdirbyste, Mokslo Darbai, 84: 141–149. (In Lithuanian.)

Bingham, E.T. 1970. Registration of hexaploid alfalfa germplasm. Crop Sci. 10: 211.

Bingham, E.T., and McCoy, T.J. 1979. Cultivated alfalfa at the diploid level: origin, reproductive stability, and yield of seed and forage. Crop Sci. 19: 97–100.

Bingham, E.T., and Saunders, J.W. 1974. Chromosome manipulations in alfalfa: scaling the cutivated tetraploid to seven ploidy levels. Crop Sci. 14: 474–477.

Bir, S.S., and Kumari, S. 1979. Cytological evolution of the leguminous flora of the Punjab plain. Recent Res. Pl. Sci. (New Delhi), 7: 252–260.

Bir, S.S., and Sidhu, S. 1967. Cytological observations on the North Indian members of family Leguminosae. Nucleus, 10: 47–63.

Bir, S.S., and Sidhu, M. 1980. Cyto-palynological studies on weed flora of cultivable lands of Patiala district (Punjab). J. Palynol. 16: 85–105.

Bird, G.W.G., and Wingham, J. 1983. "New" lectins for the identification of erythrocyte cryptantigens and the classification of erythrocyte polyagglutinability: *Medicago disciformis* and *Medicago turbinata*. J. Clin. Pathol. 36: 196–196.

Birk, Y. 1969. Saponins. *In* Toxic constituents of plant foodstuffs. *Edited by* J.E. Liener. Academic Press, New York, NY. pp. 169–210.

Bjorkqvist, I., Bothmer, R. von, Nillson, O., and Nordenstam, B. 1969. Chromosome numbers in Iberian angiosperms. Bot. Notiser, 122: 271–283.

Blaser, R.E., and Stokes, W.E. 1946. Ecological and morphological characteristics of black medic strains. J. Amer. Soc. Agron. 38: 325–331.

Blumenthal, M.J., and Ison, R.L. 1993*a*. Water use and productivity in sub. clover and murex medic swards, 1. Dry matter production. Austral. J. Agric. Res. 44: 89–107.

Blumenthal, M.J., and Ison, R.L. 1993*b*. Water use and productivity in sub. clover and murex medic swards, 2. Seed production. Austral. J. Agric. Res. 44: 109–119.

Boe, A., Bortnem, R., Higgins, K.F., Kruse, A.D., Kephart, K.D., and Selman, S. 1998. Breeding yellow-flowered alfalfa for combined wildlife habitat and forage purposes. Agricultural Experiment Station, South Dakota State University, Cooperative Fish & Wildlife Research Unit, Brookings, SD. 12 pp.

Bohart, G.E. 1958. Alfalfa pollinators with special reference to species other than honeybees. 10th International Cong. Ent. Proc. Montreal, 4: 929–937.

Bohart, G.E. 1972. Management of wild bees for the pollination of crops. Ann. Rev. Entomol. 17: 287–312.

Boissier, E. 1849. Diagnoses plantarum orientalium novarum. Ser. I. No. 9. Paris, France.

Boissier, E. 1856. Diagnoses plantarum orientalium novarum. Ser. 2, No. 5. Paris, France.

Boissier, E. 1872. Flora orientalis sive enumeratio plantarum in Oriente a Graecia et Aegypto ad India fines hucusque observatarum. H. George, Basel, Switzerland. Vol. 2.

Bokonyi, S. 1978. The earliest waves of domestic horses in east Europe. J. Indo-European Studies, 6: 17–76.

Bokonyi, S. 1984. Horse. *In* Evolution of domesticated animals. *Edited by* I.L. Mason. Longman, London, UK. pp. 162–173.

Bolkhovskikh, Z., Grif, V., Matvejeva, T., and Zakharyeva, O. 1969. Chromosome numbers of flowering plants. Academy of Sciences of the USSR, V.L. Komarov Botanical Institute, 926 pp.

Bolland, M.D.A. 1986. Performance of *M. polymorpha* cv. Serena in comparison with other annual medic species on alkaline soils near Esperance, Western Australia. J. Austral. Inst. Agric. Sci. 52: 163–167.

Bolòs, O., and Vigo, J. 1974. Notes sobre taxonomia i nomenclatura de plants. 1. Bull. Inst. Catalana Hist. Nat. 38 (Sec. Bot., 1): 61–89.

Bolton, J.L. 1962. Alfalfa–botany, cultivation and utilization. Interscience, New York, NY. 474 pp.

Bolton, J.L., and Greenshields, J.E.R. 1950. A diploid form of *Medicago sativa*. Science, 112(2906): 1275–277.

Bolton, J.L., Goplen, B.P., and Baenziger, H. 1972. World distribution and historical developments. *In* Alfalfa science and technology. *Edited by* C.H. Hanson. American Society of Agronomy, Madison, WI. pp. 1–34.

Bondi, A., Birk, Y., and Gestetner, B. 1973. Forage saponins. *In* Chemistry and biochemistry of herbage. *Edited by* G.W. Butler and R.W. Bailey. Academic Press, London, UK. pp. 511–528.

Bonnin, I., Huguet, T., Gherardi, M., Prosperi, J.M., and Olivieri, I. 1996. High level of polymorphism and spatial structure in a selfing plant species, *Medicago truncatula* (Leguminosae), shown using RAPD markers. Am. J. Bot. 83: 843–855.

Bonnin, I., Ronfort, J., Wozniak, F., and Olivieri, I. 2001. Spatial effects and rare outcrossing events in *Medicago truncatula* (Fabaceae). Molec. Ecol. 10: 1371–1383.

Borges, O.L., Stanford, E.H., and Webster, R.K. 1976. Sources and inheritance of resistance to *Stemphylium botryosum* leafspot of alfalfa. Crop Sci.16: 458–461.

Bornmüller, J. 1925. Beiträge zur Flora Mazedoniens. Sammlungen in den Kriegsjahren 1916–1918. Bot. Jahrb. 59: 294–504.

Bósca, I., Sarosi, J., Feher, F., Majko, Z., and Vetesi, M. 1986. Results of breeding a low-saponin lucerne variety as a replacement in the protein requirement of monogastric animals. Novénytermelés, 35: 297–291. (In Hungarian.)

Boscaiu, M., Riera, J., Estrelles, E., and Güemes, J. 1997. Números cromosomáticos de plantas occidentales, 742–745. Anales Jardín Botánico de Madrid, 55: 429–431.

Boulos, L., and Al-Eisawi, D. 1977. Studies on the flora of Jordan, part 8, new and noteworthy plants. Candollea, 32: 269–276.

Bounejmate, M., and Robson, A.D. 1992. Differential tolerance of genotypes of *Medicago truncatula* to low pH. Austral. J. Agric. Res. 43: 731–737.

Bounejmate, M., Robson, A.D., and Beale, P.E. 1992*a*. Annual *Medicago* species in Morocco. II. Distribution in relation to soil and climate. Austral. J. Agric. Res. 43: 751–763.

Bounejmate, M., Beale, P.E., and Robson, A.D. 1992*b*. Annual *Medicago* species in Morocco. I. Species and their abundance. Austral. J. Agric. Res. 43: 739–749.

Bounejmate, M., Loss, S.P., and Robson, A.D. 1994. Effects of temperature and frost on genotypes of *Medicago truncatula* L. and *Medicago aculeata* L. from contrasting climatic origins. J. Agron. Crop Sci. (Germany), 172: 227–236.

Bounejmate, M., Gintzburger, G., Christiansen, S., Robertson, L.D., Abd El Moneim, A., and Konopka, J. 1999. Forage and pasture legume genetic resources at ICARDA. *In* Genetic resources of Mediterranean pasture and forage legumes. *Edited by* S.J. Bennett and P.S. Cocks. Kluwer Academic Publishers, Boston, MA. pp.163–169.

Bousquet-Mélou, A., Louis, S., Robles, C., Greff, S., Dupouyet, S., and Fernandez, C. 2005. Allelopathic potential of *Medicago arborea*, a Mediterranean invasive shrub. Chemoecology, 15: 193–198.

Boutin, J.M., Bretagnolle, V., and Moreau, C. 1994. Use of lucerne by wildlife with special reference to birds. FAO/REUR Technical Series, 36: 47–49.

Bouton, J.H., Smith, S.R., Wood, D., Hoveland, C.S., and Brummer, E.C. 1991. Registration of Alfagraze. Crop. Sci. 31: 479.

Bouzid, S.M., and Papanastasis, V.P. 1996. Effects of seeding rate and fertilizer on establishment and growth of *Atriplex halimus* and *Medicago arborea*. J. Arid Environ. 33: 109–115.

Brahim, K., and Smith, S.E. 1993. Annual medic establishment and the potential for stand persistence in southern Arizona. J. Range Manage. 46: 21–25.

Brand, T.S., Cloete, S.W.P., Villiers, T.T. de, Franck, F., and Coetzee, J. 1991*a*. Nutritive evaluation of *Medicago truncatula* (cv. Jemalong) pasture for sheep. 1. Seasonal influences on the chemical composition and digestibility of pasture. South African J. Animal Sci. 21: 88–94.

Brand, T.S., Cloete, S.W.P., Villiers, T.T. de, and Franck, F. 1991*b*. Nutritive value of *Medicago truncatula* (cv. Jemalong) as pasture for sheep. 2. Voluntary intake and animal production. South African J. Animal Sci. 21: 137–143.

Brandsaeter, L.O., Olsmo, A., Tronsmo, A.M., and Fykse, H. 2002. Freezing resistance of winter annual and biennial legumes at different developmental stages. Crop Sci. 42: 437–443.

Brenac, P., and Sauvaire, Y. 1996. Chemotaxonomic value of sterols and steroidal sapogenins in the genus *Trigonella*. Biochem. Syst. Ecol. 24: 157–164.

Brewer, G.J., Horber, E., and Sorensen, E.L. 1986*a*. Potato leafhopper *Empoasca fabae* (Homoptera Cicadellidae) antixenosis and antibiosis in *Medicago* species. J. Econ. Entomol. 79: 421–425.

Brewer, G.J., Sorensen, E.L., Horber, E., and Kreitner, G.L. 1986*b*. Alfalfa stem anatomy and potato leafhopper (Homoptera: Cicadellidae) resistance. J. Econ. Entomol. 79: 1249–1253.

Brickell, C.D., Alexander, C., David, J.C., Hetterscheid, W.L.A., Leslie, A.C., Malecot, V., Jin, X., Cubey, J.J., and members of the Editorial Committee. 2009. The international code of nomenclature for cultivated plants. 8th edition. International Society for Horticultural Science, Leuven, Belgium. 204 pp.

Britton, N.L., and Brown, A. 1913. An illustrated flora of the northern United States, Canada and the British Possessions, 2nd. ed., vol. 2. Charles Scribner's Sons, New York, NY. p. 350.

Brockwell, J. 1971. Selection of *Rhizobium meliloti* strains for inoculation of *Medicago rugosa* Desr. cv. Paragosa. 1976. Australia CSIRO Division Plant Industry Field Station Record 10(2): 51–58.

Brockwell, J., and Botomley, P.J. 1995. Recent advances in inoculant technology and prospects for the future. Soil Boil. Biochem. 27: 683–697.

Brockwell, J., Gault, R.R., Haile, A., and Rogers, V.E. 1976. Australia CSIRO Division Plant Industry Field Station Record, 15(2): 51–61.

Brockwell, J., Holliday, R.A., Daoud, D.M., and Materon, L.A. 1988. Symbiotic characteristics of a *Rhizobium*-specific annual medic, *Medicago rigidula* (L.) All. Soil Biol. Biochem. 20: 593–600.

Broderick, G.A., and Buxton, D.R. 1991. Genetic varation in alfalfa for ruminal protein degradability. Can. J. Plant Sci. 71: 755–760.

Brookes, B. 1992. Pollination studies in alfalfa, *Medicago sativa* L. (Leguminosae), with special reference to the alfalfa leafcutting bee, *Megachile rotundata* (F.) (Apoidea). Master's thesis, Carleton University Library, Ottawa, ON, Canada. 161 pp.

Brookes, B., and Small, E. 1988. Enhanced floral analysis by low temperature scanning electron microscopy. Scanning Microsc. 2: 247–256.

Brookes, B., Small, E., Damman, H., Fairey, D.T., and Lefkovitch, L.P. 1994. Attractiveness of alfalfa (*Medicago sativa* L.) to wild pollinators in relation to wildflowers. Can. J. Pl. Sci. 74: 779–783.

Brouwer, D.J., and Osborn, T.C. 1997. Identification of RFLP markers linked to the unifoliate leaf, cauliflower head mutation of alfalfa. J. Heredity, 88: 150–152.

Brown, A.H.D. 1989*a*. The case for core collections. *In* The use of plant genetic resources. *Edited by* A.H.D. Brown, O.H. Frankel, D.R. Marshall, and J.T. Williams. Cambridge University Press, Cambridge, UK. pp. 136–156.

Brown, A.H.D. 1989*b*. Core collection: a practical approach to genetic resources management. Genome, 31: 818–824.

Brown, V.K., and Lawton, J.H. 1991. Herbivory and the evolution of leaf size and shape. Phil. Trans. R. Soc. Lond. Ser. B, 333: 265–272.

Brownlee, H. 1985. History of medics in central western New South Wales. *In* The ecology and agronomy of annual medics. Proceedings of a workshop at the Agricultural Research and Advisory Station, Condobolin, 1981. *Edited by* Z. Hochman. Dept. Agriculture New South Wales, Technical Bull. 32. pp. 1–3.

Brummer, E.C., and Bouton, J.H. 1991. Plant traits associated with grazing-tolerant alfalfa. Agron. J. 83: 996–1000.

Brummer, E.C., and Bouton, J.H. 1992. Physiological traits associated with grazing-tolerant alfalfa. Agron. J. 84: 138–143.

Brummer, E.C., Bouton, J.H., and Kochert, G. 1995. Analysis of annual *Medicago* species using RAPD markers. Genome, 38: 362–367.

Brummitt, R.K. 2001. World geographic scheme for recording plant distributions (2nd edition). Pittsburgh, International Working Group on Taxonomic Databases For Plant Sciences (TDWG), Hunt Institute for Botanical Documentation, Carnegie Mellon University, Pittsburgh, PA. 137 pp.

Brummitt, R.K., and Powell, C.E. 1992. Authors of plant names: a list of authors of scientific names of plants, with recommended standard forms of their names, including abbreviations. Royal Botanic Gardens, Kew, UK. 732 pp.

Bucciarelli, B., Hanan, J.M., Palmquist, D., and Vance, C.P. 2006. A standardized method for analysis of *Medicago truncatula* phenotypic development. Plant Physiol. 142: 207–219.

Bullen, K.S. 2002. The lucerne management handbook. 4th ed. Dept. of Primary Industries, Brisbane, Queensland, Australia. 74 pp.

Burton, R.M. 1999. Yet more new plants from the East Aegean island of Samos. Annales Musei Goulandris, 10: 45–51.

Burtt, B.L. 1964. Review: The annual species of *Medicago* by C.C. Heyn. New Phytol. 63: 263–264.

Busso, C.A., Fernandez, O.A., and Fresnillo Fedorenko, D.E. 1998. Dry weight production and partitioning in *Medicago minima* and *Erodium cicutarium* under water stress. Annals Bot. 82: 217–227.

Buttler, K.P. 1989. Chromosomenzahlen von Gefässpflanzen aus Hessen. 4. Folge. Hess. Florist. Briefe, 38: 11–14.

Caddel, J. 2007. Potential for rigid medics in the Southern Plains. *Medicago rigidula* (L.) All. and *Medicago rigiduloides* E. Small. http://forage.okstate.edu/rigidmedics/rigid_medics.htm (current July 2010).

Cain, S.A. 1944. Foundations of plant geography. Hafner, New York, NY. 556 pp.

Çakan, H., Yilmaz, K.T., and Düzenli, A. 2005. First comprehensive assessment of the conservation status of the flora of the Çukurova Deltas, southern Turkey. Oryx, 39: 17–21.

Camp, W.H. 1947. Distribution patterns in modern plants and the problems of ancient dispersal. Ecol. Mongr. 17: 159–183.

Campbell, T.A. 2000. Molecular analysis of genetic variation among alfalfa (*Medicago sativa* L.) and *Medicago ruthenica* clones. Can. J. Plant Sci. 80: 773–779.

Campbell, T.A. 2003. Investigation of variations in NBS motifs in alfalfa (*Medicago sativa*), *M. edgeworthii*, and *M. ruthenica*. Can. J. Plant Sci. 83: 371–376.

Campbell, T.A., and Bauchan, G.R. 2002. Organelle based molecular analyses of the genetic relatedness of cultivated alfalfa (*Medicago sativa* L) to *Medicago edgeworthii* Sirjaev, and *Medicago ruthenica* (L.) Ledebour. Euphytica, 125: 51–58.

Campbell, T.A ., and Xia, Z.L. 2002. Agronomic evaluations of grass and legume forage species collected in the Xinjiang Autonomous Region of the People's Republic of China and Mongolia. Genet. Resourc. Crop Evol. 49: 295–303.

Campbell, T.A., Bao, G., and Xia, Z.L. 1997. Agronomic evaluation of *Medicago ruthenica* collected in Inner Mongolia. Crop Sci. 37: 599–604.

Campbell, T.A., Bao, G., and Xia, Z.L. 1999. Completion of the agronomic evaluations of *Medicago ruthenica* [(L.) Ledebour] germplasm collected in Inner Mongolia. J. Genetic Resources Crop Evolution, 46(5): 477–484.

Candolle, A.L.P.P. de. 1885. Origin of cultivated plants. D. Appleton and Company, New York, NY. 468 pp.

Cannon, S.B., Sterck, L., Rombauts, S., and 24 others. 2006. Legume genome evolution viewed through the *Medicago truncatula* and *Lotus japonicus* genomes. PNAS, 103(40): 14959–14964.

Carneiro, J.P., and Serrao, M.G. 1996. Programs for improvement of *Medicago* sp. at ENMP-Portugal: I. Distribution of annual medics as related to soil properties and climatic conditions. Cahiers Options Mediterraneennes, 18: 23–29.

Carneiro, J.P., de Varennes, A, and Amante , H. 2001. Manganese toxicity in three species of annual medics. J. Plant Nutr. 24: 1957–1964.

Carneiro, J.P., Barradas, A., and Amante, H. 2004. Straight-spined (*Medicago doliata*), burclover (*Medicago polymorpha*) and barrel medick – three new cultivars of annual lucernes (*Medicago* spp.) with a good adaptation to the Mediterranean climate. Melhoramento (Portugal), 39: 28–36. (In Portuguese.)

Carreras, D., Fraga, P., Pons, M., García, Ó., Truyol, M., and Mascaró, C. 2002. Notes and contributions to the knowledge of the flora of Minorca (III). Bolletí de la Societat d'Història Natural de les Balears, 45: 69–80. (In Spanish.)

Carretero, J.L. 1990. *Mollugo cerviana* (L.) Ser. y *Trigonella ovalis* Boiss., novedades para la flora Valenciana. Collecteana Botanica (Barcelona), 18: 156. (In Spanish.)

Carter, E.D. 1975. The potential for increasing cereals and livestock in Algeria. Centro International e mojaramento de Maiz y Trigo. 1–33.

Carter, E.D. 1980. The survival of medic seeds following ingestion of intact pods by sheep. Proceedings Austral. Agron. Conference, Lawes, Queensland. p. 178.

Casellas, J. 1962. El género *Medicago* L. en España. Collectanea Botanica, VI(I-II): 183–291.

Castri, F. di. 1991. An ecological overview of the five regions of the world with a Mediterranean climate. *In* Biogeography of Mediterranean invasions. *Edited by* R.H. Groves and F. di Castri. Cambridge University Press, Cambridge, UK. pp. 3–16.

Castroviejo, S. (Coordinator & Editor) et al. (Editors). 2001. Claves de Flora Iberica. Plantas vasculares de la Península Ibérica e Islas Baleares. Real Jardin Botánico, Madrid, Spain. Vol. 1.

Catalano, M., Ragona, L., Molinari, H., Tava, A., and Zetta, L. 2003. Anticarcinogenic Bowman Birk inhibitor isolated from snail medic seeds (*Medicago scutellata*): Solution structure and analysis of self-association behavior. Biochemistry, 42: 2836–2846.

Cecliani, F., Tava, A., Iiori, R., Mortarino, J., Odoardo, M., and Ronchi, S. 1997. A trypsin inhibitor from snail medic seeds active against pest proteases. Phytochemistry, 44: 393–398.

Cereti, C.F., and Rossini, F. 1994. Sowing and planting of *Medicago arborea* in Mediterranean maquis. Preliminary observations. *In* Fodder trees and shrubs. *Edited by* V. Papanastasis and L. Stringi. CIHEAM-IAMZ (Cahiers Options méditerranéennes No. 4), Zaragoza, Spain. pp. 71–76.

Charman, N., and Ballard, R.A. 2004. Burr medic (*Medicago polymorpha* L.) selections for improved N_2-fixation with naturalised soil rhizobia. Soil Biol. Biochem. 36: 1331–1337.

Chatterton, B., and Chatterton, L. 1989. Fodder for the Near East: annual medic pastures. FAO Plant Production and Protection Paper 97/2. Food and Agriculture Organization of the United Nations, Rome, Italy. 122 pp.

Chaudhri, M.N., Vegter, I.H., and Wal, C.M. de. 1972. Index herbariorum. Part II(3). Collectors. A. Oosthoek's Uitgeversmaatschappij N. V., Uterecht, Netherlands.

Chebbi, H., Pascual-Villalobos, M.J., Cenis, J.L., and Correal, E. 1995. Morphological and molecular characters for the differentiation of woody species within the genus *Medicago*. Fourrages (France), 142: 191–206. (In French.)

Chebouti, A., and Abdelguerfi, A. 2000. Effet du stress hydrique sur la production de gousses et de graines chez quelques populations de *Medicago truncatula* (L.) Gaertn. *In* Legumes for Mediterranean forage crops, pastures and alternative uses. *Edited by* L. Sulas. CIHEAM-IAMZ, Zaragoza, Italy. pp. 237–240.

Chebouti, A., and Abdelguerfi, A. 2003. Effet du stress hydrique sur la nodulation et le système racinaire chez quelques populations de *Medicago aculeata*, *M. orbicularis* et *M. truncatula*. *In* Fixation symbiotique de l'azote et développement durable dans le Bassin méditerranéen. *Edited by* INRA. INRA, Paris, France. pp. 257–262.

Cheeke, P.R. 1971. Nutritional and physiological implications of saponins: a review. Can. J. Anim. Sci. 51: 621–632.

Cheeke, P.R. 1973. Alfalfa: a natural hypocholesteremic agent. Am. J. Clin. Nutr. 26: 133.

Cheeke, P.R., and Shull, L.R. 1985. Natural toxicants in feeds and poisonous plants. Avi Publishing Company, Westport, CT. 492 pp.

Chen, B., Iver, J., Roos, P., and Zhu, Y.-G. 2005. Effects of the mycorrhizal fungus *Glomus intraradices* on uranium uptake and accumulation by *Medicago truncatula* L. from uranium-contaminated soil. Plant Soil, 275: 349–359.

Cheng, Y., Watkin, E.L.J., Ohara, G.W., and Howieson, J.G. 2002. *Medicago sativa* and *Medicago murex* differ in the nodulation response to soil acidity. Plant Soil, 238: 31–39.

Cheng, Y., Howieson, J.G., O'Hara, G.W., Watkin, E.L.J., Souche, G., Jaillard, B., and Hinsinger, P. 2004. Proton release by roots of *Medicago murex* and *Medicago sativa* growing in acidic conditions, and implications for rhizosphere pH changes and nodulation at low pH. Soil Biol. Biochem. 36: 1357–1365.

Chepinoga, V.V., Gnutikov, A.A., Enushchenko, I.V., and Rosbakh, S.A. 2009. IAP T/IOPB chromosome date 8: Fabaceae. Taxon, 58: 1281, E2.

Chesneaux, M.T. 1974. La multiplication végétative chez la luzerne. [Vegetative reproduction in alfalfa.] Ann. Amélior. Plantes, 24(1): 83–91.

Chilton, L., and Turland, N.J. 1997. Flora of Crete. A supplement. Marengo Publications, Retford, Nottinghamshire, UK. 125 pp.

Choi, H.K., Kim, D., Uhm, T., Limpens, E., Lim, H., Mun, J.H., Kalo, P., Penmetsa, R.V., Seres, A., and Kulikova, O. 2004. A sequence-based genetic map of *Medicago truncatula* and comparison of marker colinearity with *M. sativa*. Genetics, 166: 1463–1502.

Chon, S.U., Choi, S.K., Jung, S., Jang, H.G., Pyo, B.S., and Kim, S.M. 2002. Effect of alfalfa leaf extracts and phenolic allelochemicals on early seedling growth and root morphology of alfalfa and barnyard grass. Crop Protect. 21: 1077–1082.

Chow Ben-Shun. 1989. The domestic horse of the pre-Ch'in period in China. *In* The walking larder. *Edited by* J. Clutton-Brock. Unwin Hyman, London, UK. pp. 105–107.

Christiansen, S., and Svejcar, T. 1989. Evaluation of annual medics and subclovers in Oklahoma. Proceedings of the XVI International Grassland Congress, 4–11 October 1989, Nice, France. pp. 239–240.

Christiansen, S., Abd El Moneim, A.M., Cocks, P.S., and Singh, M. 1996. Seed yield and hardseededness of two amphicarpic pasture legumes (*Vicia sativa* ssp. *amphicarpa* and *Lathyrus ciliolatus*) and two annual medics (*Medicago rigidula* and *M. noeana*). J. Agric. Sci. 126: 421–427.

Clarke, W.E., and Burke, H.R. 1977. The curculionid genus *Tychius* Germar: natural history and coevolution with leguminous host plants. Southwest. Entomol. 2: 106–126.

Clarkson, N.M. 1977. Annual medics in Queensland. Queensland Agric. J. 103: 39–60.

Clarkson, N.M., and Russell, J.S. 1975. Flowering responses to vernalization and photoperiod in annual medics (*Medicago* spp.). Austral. J. Agric. Res. 26: 831–838.

Clarkson, N.M., Swann, I.F., and Chaplain, N.P. 1989. Sulfur and phosphorus fertilizers increase the yield of barrel medic (*Medicago truncatula*) five-fold in native pasture on a traprock soil. J. Exp. Agric. 29: 527–532.

Classen, D., Nozzolillo, C., and Small, E. 1982. A phenolic-taxometric study of *Medicago* (Leguminosae). Can. J. Bot. 60: 2477–2495.

Clement, W.M., Jr. 1962. Chromosome numbers and taxonomic relationships in *Medicago*. Crop Sci. 2: 25–28.

Clement, W.M., Jr. 1963. Chromosome relationships in a diploid hybrid between *Medicago sativa* L. and *M. dzhawakhetica* Bordz. Can. J. Genet. Cytol. 5: 427–432.

Clement, E.J., and Smythies, B.E. 1976. On *Trigonella ovalis* Boiss., a forgotten species of the Spanish flora. Lagascalia, 6(2): 243–246.

Clutton-Brock, J. 1992. Horse power. A history of the horse and the donkey in human societies. Harvard University Press, Cambridge, MA. 192 pp.

Cockerell, T.D.A. 1935. Recollections of a naturalist. Bios, 6: 372–385.

Cocks, P.S. 1988. Seed production and seed survival under grazing of annual medics (*Medicago* spp.) in North Syria. J. Agric. Sci. 110: 455–463.

Cocks, P.S. 1990. Dynamics of flower and pod production in annual medics (*Medicago* spp.). I. In spaced plants. Austral. J. Agric. Res. 41: 911–921.

Cocks, P.S. 1992. Plant attributes leading to persistence in grazed annual medics, *Medicago* spp., growing in rotation with wheat. Austral. J. Agric. Res. 43: 1559–1570.

Cocks, P.S. 1993. Seed and seedling dynamics over four consecutive years from a single seed set of six annual medics (*Medicago* spp.) in North Syria. Exp. Agric. 29: 461–472.

Cocks, P.S. 2001. Ecology of herbaceous perennial legumes: a review of characteristics that may provide management options for the control of salinity and waterlogging in dryland cropping systems. Austral. J. Agric. Res. 52: 137–151.

Cocks, P.S., and Bennett, S.J. 1999. Introduction: role of pasture and forage legumes in Mediterranean farming systems. *In* Genetic resources of Mediterranean pasture and forage legumes. *Edited by* S.J. Bennett and P.S. Cocks. Kluwer Academic Publishers, Boston, MA. pp. 9–19.

Cocks, P.S., and Ehrman, T.A.M. 1987. The geographic origin of frost tolerance in Syrian pasture legumes. J. Appl. Ecol. 24: 673–683.

Colebatch, G., Trevaskis, B., and Udvardi, M. 2002. Symbiotic nitrogen fixation research in the postgenomics era. New Phytol. 153: 37–42.

Coley, P.D., Bryant, J.P., and Chapin, F.S. III. 1985. Resource availability and plant antiherbivore defense. Science, 230: 895–899.

Colla, A. 1834. Herbarium Pedemontanum. Vol. 2. Augustae Taurinorum, Italy. (In Latin.)

Colombo, P., and Trapani, S. 1990. Números cromosomáticos de plantas occidentales, 556–567. Anales del Jardín Botánico de Madrid, 47: 179–183.

Colombo, P., Colombo, R., Marceno, C., and Pavone, P. 1978. Numeri cromosomici per la Flora Italiana. 51–7525. Informatore Botanico Italiano, 10: 406–413.

Colombo, P., Marceno, C., and Princiotta, R. 1980. Numeri cromosomici per la Flora Italiana: 794–805. Informatore Botanico Italiano, 12: 333–340.

Colombo, P., Marcenó, C., and Mazza, S. 1987. Números cromosomáticos de plantas occidentales, 436–444. Anales Jardín Botánico de Madrid, 44: 140–142.

Conlan, D.J., Dear, B.S., and Coombes, N.E. 1994. Effect of grazing intensity and number of grazings on herbage production and seed yields of *Trifolium subterraneum*, *Medicago murex*, and *Ornithopus compressus*. Austral. J. Exper. Agric. 34: 181–188.

Constance, L. 1964. Systematic botany—an unending synthesis. Taxon, 13: 257–273.

Conway, M.J., Brandon, N.J., Clem, R.L., Jones, R.M., Robertson, B.A., and Willcocks, J.R. 2001. Growth and persistence of 17 annual medic (*Medicago* spp.) accessions on clay soils in Central Queensland. Trop. Grasslands, 35: 226–234.

Cook, D.R. 1999. *Medicago truncatula*—a model in the making! Curr. Opin. Plant. Biol. 2: 301–304.

Cooper, D.C. 1935. Macrosporogenesis and embryology of *Medicago*. J. Agric. Res. 51: 471–477.

Cooper, D.C. 1936. Chromosome numbers in the Leguminosae. Am. J. Bot. 23: 231–233.

Corleto, A., Cazzato, E., and Laudadio, V. 1993. The influence of cutting frequencies and irrigation on D.M.Y. and forage quality of *Medicago arborea* L. *In* Agriculture: Agrimed research programme: Fodder trees and shrubs in the Mediterranean production systems: objectives and expected results of the EC research contract. *Edited by* V. Papanastasis. Commission of the European Communities, Luxembourg. pp. 125–130.

Correal, E. 1997. Cold stress in high altitude semi-arid Mediterranean areas: cold tolerance improvement potential of *Medicago citrina* and *Bituminaria bituminosa*, woody forage legumes with good drought tolerance. *In* Proceedings of a workshop on improving forage crops for sem-arid areas, 2–4 Oct., Mallorca, Spain. *Edited by* D.G.A. Fernandez, D.R. Martorell, and D.P. Munoz. pp. 242–247.

Coutinho, L. de A., and Santos, A. 1943. Novas contribuições para a cariologia do género *Trigonella* L. Agron. Lusitana, 5: 349–361.

Cramer, H.H. 1967. Plant protection and world crop production. Pflanzenschutznachrischten, 10: 1–524.

Crawford, E.J. 1970. Variability in a large Mediterranean collection of introduced lines of *Medicago truncatula* Gaertn. Proc. 11th Internat. Grassland Congr. 188–192.

Crawford, E.J. 1983*a*. Selecting cultivars from naturally occurring genotypes: evaluating annual *Medicago* species. *In* Genetic resources of forage plants. *Edited by* J.G. McIvor and R.A. Gray. CSIRO, Melbourne, Australia. pp. 135–148.

Crawford, E.J. 1983*b*. Development of gama medic (*Medicago rugosa* Desr.) as an annual leguminous species for dryland farming systems in southern Australia. *In* Proceedings of the XIV International Grassland Congress held at Lexington Kentucky, USA, June 15–24, 1981. *Edited by* J.A. Smith and V.W. Hays. Westview Press, Boulder, Colorado, USA. pp. 224–226.

Crawford, E.J. 1985. Flowering response and centres of origin of annual *Medicago* species. *In* The ecology and agronomy of annual medics. *Edited by* Z. Hochman. Tech. Bull. 32, Department of Agriculture, New South Wales, Australia. pp. 7–11.

Crawford, E.J., Lake, A.W.H., and Boyce, K.G. 1989. Breeding annual *Medicago* species for semiarid conditions in northern Australia. Advances Agron. 42: 399–437.

Cristurean, I. 1972. Species of the genus *Trigonella* L. in the flora of India. Rev. Roum. Biol. – Botanique, 17: 69–77.

Croizat, L., Nelson, G., and Rosen, D.E. 1974. Centers of origin and related concepts. Syst. Zool. 23: 265–287.

Custalow, B. 1991. Protein requirements during exercise in the horse. J. Equine Vet. Sci. 11: 82, 85.

Cymbaluk, N.F. 1990. Comparison of forage digestion by cattle and horses. Can. J. Anim. Sci. 70: 601–610.

Czerepanov, S.K. 1995. Vascular plants of Russia and adjacent states (the former USSR). Cambridge University Press, Cambridge, UK. 516 pp.

Daday, H. 1962. Breeding for creeping root in lucerne (*Medicago sativa* L.). I. The initial response to selection. Austral. J. Agr. Res. 13: 813–820 + 3 plates.

Dahlgren, R., Karlsson, T., and Lassen, P. 1971. Studies on the flora of the Balearic Islands. I. chromosome numbers in Balearic angiosperms. Bot. Notiser, 124: 249–269.

Dakskobler, I. 2005. Floristic novelities from the Soca Valley and adjoining areas in western and northwestern Slovenia. IV. Hacquetia, 4: 173–200. (In Slovenian.)

Dall, D.J., Randles, J.W., and Francki, R.I.B. 1989. The effect of alfalfa mosaic virus on productivity of annual barrel medic (*Medicago truncula*ta). Austral. J. Agric. Res. 40: 807–816.

Dal Vesco, G., Forneris, G., and Pistarino, A. 1987. Loci classici and type in the works and herbaria of Allioni and Balbis. Allionia (Turin), 28: 5–20. (In Italian.)

Damiani, F., and Arcioni, S. 1991. Transformation of *Medicago arborea* with an *Agrobacterium* rhizogenes binary vector carrying the hygromicin resistance gene. Plant Cell Reports, 10: 300–303.

Danielson, S.D., Manglitz, G.R., and Sorensen, E.L. 1986. Development of alfalfa weevil *Hypera postica* (Coleoptera, Curculionidae) larvae when reared on perennial glandular-haired *Medicago* species in the greenhouse. Environmental Entomol. 15: 396–398.

Danielson, S.D., Manglitz, G.R., and Sorensen, E.L. 1987*a*. Resistance of perennial glandular-haired *Medicago* species to oviposition by alfalfa weevils (Coleoptera: Curculionidae). Environmental Entomol. 16: 195–197.

Danielson, S.D., Manglitz, G.R., and Sorensen, E.L. 1987*b*. Resistance in perennial glandular-haired *Medicago* species to feeding by adult alfalfa weevils (Coleoptera: Curculionidae). Environmental Entomol. 16: 708–711.

Danielson, S.D., Mumm, R.F., Manglitz, G.R., and Sorensen, E.L. 1989. Glandular hair densities in three perennial *Medicago* species. Euphytica, 43: 15–19.

Danielson, S.D., Manglitz, G.R., and Sorensen, E.L. 1990. Field resistance of perennial glandular-haired *Medicago* strains and alfalfa cultivars to the alfalfa weevil. Euphytica, 49: 177–182.

Danin, A., and Small, E. 1989. Contributions to the flora of Israel and Sinae. V. *Trigonella sibthorpii* Boiss., a new record from Israel. Israel J. Bot. 38: 121–124.

D'Aoust, M.A., Lerouge, P., Busse, U., Bilodeau, P., Trepanier, S., Gomord, V., Faye, L., and Vezina, L.P. 2004. Efficient and reliable production of pharmaceuticals in alfalfa. *In* Molecular Farming, *Edited by* R. Fischer and S. Schillberg. pp. 1–12.

D'Arcy, W.G. 1970. Jacquin names, some notes on their typification. Taxon, 19: 554–560.

Darwin, C. 1868. The variation of animals and plants under domestication. Orange Judd & Company, New York, NY. 2 vols.

Davis, J.H., Gangstad, E.O., and Hackerott, H.L. 1957. Button clover: a reseeding winter annual legume for north central Texas. Hoblitzelle Agr. Lab., Texas Res. Found. (Renner Texas), Bull. 6. 11 pp.

Davis, P. 2008. Flora of Turkey. Edinburgh University Press, Edinburgh, UK. 11 vols.

Dawson, J.H., Lee, W.O., and Timmon, F.L. 1971. Controlling dodder in alfalfa. Farmers' Bulletin No. 2211. U.S. Dept. Agric., Washington, D.C. 16 pp.

Dear, B.S., and Jenkins, L.J. 1992. Persistence, productivity, and seed yield of *Medicago murex*, *M. truncatula*, *M. aculeata*, and *Trifolium subterranean* on an acid red earth soil in the wheat belt of eastern Australia. Austral. J. Exper. Agric. 32: 319–329.

Dear, B.S., Sandral, G.A., Peoples, M.B., Wilson, B.C.D., Taylor, J.N., and Rodham, C.A. 2003. Growth, seed set and nitrogen fixation of 28 annual legume species on 3 Vertosol soils in southern New South Wales. Austral. J. Exper. Agric. 43: 1101–1115.

De Haan, R.L., and Barnes, D.K. 1998. Inheritance of pod type, stem color, and dwarf growth habit in *Medicago polymorpha*. Crop Sci. 38: 1558–1561.

De Haan, R.L., Sheaffer, C.C., and Barnes, D.K. 1997. Effect of annual medic smother plants on weed control and yield in corn. Agron. J. 89: 813–821.

De Haan, R.L., Sheaffer, C.C., Samac, D.A., Moynihan, M., and Barnes, D.K. 2002. Evaluation of annual *Medicago* for Upper Midwest agroecosystems. J. Agron. Crop Sci. 188: 417–425.

De Koning, C.T., Hughes, S., McLachlan, D., and Duncan, A.J. 2000*a*. Medicago *arborea*: a leguminous fodder shrub for low rainfall farming systems. *In* Proceedings of the 10th Meeting of the Mediterranean Sub-Network of the FAO-CIHEAM Inter-Regional Cooperative Research and Development Network on Pastures and Fodder Crops, Sassari (Italy), 4–9 April 2000. *Edited by* L. Sulas. Centre International de Hautes Études Agronomiques Méditerranéenes, Bari (Italy); Inst. Agronomique Méditerranéen FAO, Rome (Italy); Plant Production and Protection Div., FAO, Rome (Italy); Legumes for Mediterranean forage crops, pastures and alternative uses, Zaragoza (Spain). 4 pp. (pagination irregular).

De Koning, C.T., Hughes, S., McLachlan, D., and Duncan, A.J. 2000*b*. *Medicago arborea* – a leguminous fodder shrub for low rainfall farming systems. Cahiers Options Méditerr. 45: 435–438.

Delay, J. 1969. Halophytes III. Inf. Ann. Caryosty. et Cytogenet. 3: 17–20.

Del Pozo, A., Ovalle, C., Aronson, J., and Avendaño, J. 2000. Developmental responses to temperature and photoperiod in ecotypes of *Medicago polymorpha* L. collected along an environmental gradient in central Chile. Annals Bot. 85: 809–814.

Del Pozo, A., Porqueddu, C., and Ovalle, C. 2001. Ecotypic differentiation of *Medicago polymorpha* in Sardinia and Chile. *In* Quality in lucerne and medics for animal production. Options Méditerranéenes Series A: Séminaires

Mediterranéens no. 45. *Edited by* I. Delgado and J. Lloveras. CIHEAM-IAMN, Zaragoza, Spain. pp. 99–102.

Del Pozo, A., Ovalle, C., Aronson, J., and Avendano, J. 2002. Ecotypic differentiation in *Medicago polymorpha* L. along an environmental gradient in central Chile. II. Winter growth as related to phenology and temperature regime. Plant Ecology, 160: 53–59.

Del Pozo, L.A. Ovalle-M.C., Avendano R.J., Aravena, A.T., and Diaz, M.E. 2001. Combarbala - INIA, an early flowering cultivar of burr medic (*Medicago polymorpha* L.) for Mediterranean dryland areas. Agricultura Tecnica (Chile), 61: 93–96. (In Spanish.)

Dempsey, R. E., Gornall, R.J., and Bailey, J.P. 1994. Contributions to a cytological catalogue of the British and Irish flora, 4. Watsonia, 20: 63–66.

Dénairié, J. 2002. Genomics and genetics of *Medicago truncatula*: resources for legume crop improvement. Grain Legumes 38: 20–21.

Denney, G.D. Hogan, J.P., and Lindsay J.R. 1979. Digestion of barrel medic (*Medicago truncatula*) hay and seed pods by sheep. Austral. J. Agric. Res. 30: 1177–1184.

Denney, G.D., Hogan, J.P., Lindsay, J.R., and Davis, P. 1983. Digestion of *Medicago truncatula* pasture by sheep. South African J. Animal Sci. 13: 11–12.

Denton, M.D., Hilla, C.R., Bellottia, W.D., and Coventrya, D.R. 2007. Nodulation of *Medicago truncatula* and *Medicago polymorpha* in two pastures of contrasting soil pH and rhizobial populations. Appl. Soil Ecol. 35: 441–448.

Denton, C.R., and McCoy, T.J. 1987. Cytogenetic analysis of interspecific hybrids between alfalfa (*Medicago sativa*) L. and *M. rhodopea* Velen. Genome, 29: 853–858.

Derkaoui, M., Caddel, J.L., and Rommann, L.L. 1991. Seed production and reseeding of Mediterranean annual *Medicago* spp. in the southern Great Plains of the USA. Agr. Med. 121: 219–223.

Derkaoui, M., Caddel, J.L., and Rommann, L.L. 1993. Forage quality in annual *Medicago* spp. Agricoltura Mediterranea, 123: 86– 91.

Desborough, P.J. 1976. Evaluation of cool-season, annual legume species on alkaline, cracking-clay soils at Narrabri, New South Wales. 1. Initial Survey. Australian Plant Intro. Rev. 11(3-4): 2–16.

Dhar, A.K., and Dhar, R.S. 2000. Culinary and pot herbs of Jammu and Kashmir. J. Herbs Spices Med. Plants, 7: 7–18.

Díaz Lifante, Z., Luque, T., and Santa Bárbara, C. 1992. Chromosome numbers of plants collected during Iter Mediterraneum II in Israel. Bocconea, Monographiae Herbarii Mediterranei Panormitani, 3: 229–250.

Díaz, S., Noy-Meir, I., and Cabido, M. 2001. Can grazing response of herbaceous plants be predicted from simple vegetative traits? J. Appl. Ecol. 38: 497–508.

Didukh, Ya. P. 2007. *Medicago rupestris*. http://mail.menr.gov.ua/publ/redbook/main.php?num=1169&action=print_ver (current July 2010). (In Ukrainian.)

Diosdado, J.C., Santa-Bárbara, Vioque, C.J., Juan, R., and Pastor, J. 1993. Números cromosomáticos para la flora Española. 691-719. Lagascalia, 17: 173–184.

Diwan, N., Bauchan, G.R., and McIntosh, M.S. 1994. A core germplasm collection for the United States annual *Medicago* germplasm collection. Crop Sci. 34: 279–285.

Diwan, N., McIntosh, M.S., and Bauchan, G.R. 1995. Methods of developing a core collection of annual *Medicago* species. Theor. Appl. Genet. 90: 755–761.

Dixit, V.P., and Jain, P. 1990. The role of castration in atherosclerosis, and prevention with *Medicago sativa* seed extract. Fitoterapia, 61: 37–40.

Dobeš, C., and Hahn, B.1997. Report, IOPB chromosome data 11, International Organization of Plant Biosystematists Newsl. 26/27: 15–18.

Dodd, M.B., and Orr, S.J. 1995. Seasonal growth, phosphate response, and drought tolerance of 11 perennial legume species grown in a hill-country soil. N.Z. J. Agric. Res. 38: 7–20.

Dodd, S. 1916. Trefoil dermatitis. J. Compar. Pathol. Therap. 29: 47–62.

Donald, C.M. 1981. Innovation in Australian agriculture. *In* Agriculture in the Australian economy. *Edited by* D.B. Williams. Sydney University Press, Sydney, Australia. pp. 57–86.

Dorogostaiskaya, E.V. 1978. Anthropophilic flora of Chukotka. Botanicheskii Zhurnal, 63: 246–254. (In Russian.)

Doss, R.P., Gould, S.J., Johnson, K.J.R., Flath, R.A., and Kohner, R.L. 1989. (Z)-Oxacyclotridec-10-en-2-one an alfalfa weevil feeding deterrent from *Medicago rugosa*. Phytochemistry, 28: 3311–3315.

Downie, S.R., Katz-Downie, D.S., Rogers, E.J., Zujewski, H.L., and Small, E. 1998. Multiple independent losses of the plastid *rpo*C1 intron in *Medicago* (Fabaceae) as inferred from phylogenetic analyses of nuclear ribosomal DNA internal transcribed spacer sequences. Can. J. Bot. 76: 791–803.

Droukales, I.L. 1976. Tree lucerne (*Medicago arborea*). Dasika Hronika, 18: 315–321. (In Greek.)

Drower, M.S. 1969. The domestication of the horse. *In* The domestication and exploitation of plants and animals. *Edited by* P.J. Ucko and G.W. Dimbleby. Aldine Publishing Company, Chicago, IL. pp 471–484.

Drozdov, I.P. 1974. Wild-growing legume species in Eastern Siberia, with some importance for cultivation. Rastitel'-nye-Rsursy, 10: 204–208. (In Russian.)

Du, G.Z., and Wang, G, 1990. The classification and ordination of communities in old fields. Acta Prataculturae Sinica, 1(1): 108–116. (In Chinese.)

Duke, J.A. 1981. Handbook of legumes of world economic importance. Plenum Press, New York, NY. 345 pp.

Dunbabin, J.S., and Slavich, P.G. 1990. Relative salt tolerance of annual legumes. *In* Management of Soil Salinity in South East Australia, Proceedings of a symposium held at Albury, New South Wales, Australia, September 18–20, 1989. *Edited by* E. Humphreys, W.A. Muirhead, and A. Van der Lelij. Australian Society of Soil Science Inc., Canberra, Australia. pp. 171–179.

Dunbier, M.W. 1972. Genetic variability in *Medicago lupulina* L. across a valley in the Weka Pass, New Zealand. New Zealand J. Bot. 10: 48–58.

Dunbier, M.W., Eskew, D.L., Bingham, E.T., and Schrader, L.E. 1975. Performance of genetically comparable diploid and tetraploid alfalfa: agronomic and physiological parameters. Crop Sci. 15: 211–214.

Dvorak, F., and Dadakova, B. 1984. Chromosome counts and chromosome morphology of some selected species. Folia Geobot. Phytotax. 19: 41–70.

Dyba, S., Broda, Z., Uczkiewicz, T., and Nawracaa, J. 2003. Saponin content in selected species of *Medicago*. Prace z Zakresu Nauk Rolniczych, 95: 11–14. (In Polish.)

Edwards, P.J., Wratten, S.D., and Parker, E.A. 1992. The ecological significance of rapid wound-induced changes in plants: insect grazing and plant competition. Oecologia, 91: 266–272.

Ehrlich, P.R., and Raven, P.H. 1964. Butterflies and plants: a study in coevolution. Evolution, 18: 586–608.

Ehrman, T.A.M., and Cocks, P.S. 1988. Co-adaptation and geographical distribution of reproductive attributes in Syrian pastures. *In* Pasture, forage and livestock program, annual report 1988. ICARDA, Aleppo, Syria. pp. 66–74.

Ehrman, T., and Cocks, P.S. 1990. Ecogeography of annual legumes in Syria: distribution patterns. J. Appl. Ecol. 27: 578–591.

Ehrman, T., and Cocks, P.S. 1996. Reproductive patterns in annual legumes species on an aridity gradient. Vegetatio, 122: 47–59.

Eig, A. 1926. A contribution to the knowledge of the flora of Palestine. Institute of Agriculture and natural History, Agricultural Experiment Station, Tel Aviv, Israel, Bull. 4. (Leguminosae: pp. 17–20.)

Eig, A. 1927. A second contribution to the knowledge of the flora of Palestine. Institute of Agriculture and natural History, Agricultural Experiment Station, Tel Aviv, Israel, Bull. 6. (Leguminosae: pp. 8-82 + plates.)

El Amry, M.I. 1995. Plant diversity, dispersion, community similarity, and vegetation description in United Arab Emirates. II. Eastern coast. Qatar Univ. Sci. J. 15: 259–266.

Elci, S. 1983. The utilization of genetic resources in fodder crop breeding in Turkey. *In* Conference Proceedings, The utilization of genetic resources in fodder crop breeding, 3– 16 Sept. 1982, Aberystwyth, UK. pp. 322–334.

Elgin, J.H., Jr., and Ostazeski, S.A. 1992. Evaluation of selected alfalfa cultivars and related *Medicago* species for resistance to Race 1 and Race 2 anthracnose. Crop Sci. 22: 39–42.

Elgin, J.H., McMurtry, J.E., and Schaeffer, G.W. 1977. Attempted interspecific hybridization of *Medicago scutellata* and *M. sativa*. Agron. Abst. p. 54.

Ellwood, S.R., D'Souza, N.K., Kamphuis, L.G., Burgess, T.I., Nair, R.M., and Oliver, R.P. 2006*a*. SSR analysis of the *Medicago truncatula* SARDI core collection reveals substantial diversity and unusual genotype dispersal throughout the Mediterranean basin. Theor. Appl. Genet. 112: 977–983.

Ellwood, S.R., Kamphuis, L.G., and Oliver, R.P. 2006*b*. Identification of sources of resistance to *Phoma medicaginis* isolates in *Medicago truncatula* SARDI core collection accessions, and multigene differentiation of isolates. Phytopathology, 96: 1330–1336.

Ellwood, S.R., Lichtenzveig, J., Pfaff, T., Kamphuis, L., and Oliver, R. 2006*c*. Fungi. *In* The *Medicago truncatula* handbook (version 1, 2006). 6 pp. http://www.noble.org/MedicagoHandbook/ (current July 2010).

Epstein, H., and Mason, I.L. 1984. Cattle. *In* Evolution of domesticated animals. *Edited by* I.L. Mason. Longman, New York, NY. pp. 6–27.

Eraç, A., and Tokluoglu, M. 1983. Diploid snail medic (*Medicago scutellata* (L.) Mill.). Z. Pflanzenzucht. 91: 82–85.

Esau, K. 1977. Anatomy of seed plants. 2nd ed. John Wiley & Sons, New York, NY. 550 pp.

Eujayl, I., Sledge, M.K., Wang, L., May, G.D., Chekhovskiy, K., Zwonitzer, J.C., and Mian, M.A.R. 2004. *Medicago truncatula* EST-SSRs reveal cross-species genetic markers for *Medicago* spp. Theor. Appl. Genet. 108: 414–422.

Evans, J., Dear, B., and O'Connor, G.E. 1990. Influence of an acid soil on the herbage yield and nodulation of five annual pasture legumes. Austral. J. Exp. Agric. 30: 55–60.

Evans, J., Wolfe, E., Yun-Guk, M., Yong-Jo, R., and Tong-Ryul, K. 2004. Legumes for cropping system in the Democratic Peoples Republic of Korea. *In* New directions for a diverse planet: proceedings of the 4th international crop science congress, Brisbane, Australia, 26 Sept.–1 Oct. 2004, *Edited by* T. Fischer, N. Turner, J. Angus, L. McIntyre, M. Robertson, A. Borrell, and D. Lloyd. Regional Institute Ltd., Gosford, NSW, Australia. http://www.cropscience.org.au/icsc2004/poster/2/1/2/520_evansj.htm (current July 2010).

Ewing, J.A. 1956. Buttonlcover (*Medicago orbicularis*). Circular No. 104 (revised). University of Tennessee Agricultural Experiment Station, Knoxville, TN. 4 pp.

Ewing, M.A. 1986. Serena and Circle Valley medic establishment. J. Agric., West. Australia, 27: 107–112.

Ewing, M.A., and Howieson, J.G. 1987. A report on the collection of annual species of *Medicago* and associated *Rhizobium meliloti* from acidic regions of mainland Greece and neighbouring islands of the Aegean Sea. Western Australian Dept. Agriculture, Perth. 18 pp.

Ewing, M.A., and Howieson, J.G. 1989. The development of *Medicago polymorpha* L. as an important pasture species for southern Australia. XVI Internat. Grassland Congress, Nice, France, 1989. pp. 197–198.

Eyal, E., Benjamin, R.W., and Tadmor, N.H. 1975. Sheep production on seeded legumes, planted shrubs, and dryland grain in a semiarid region of Israel. J. Range Manage. 28: 100–107.

Fahn, A. 1974. Plant anatomy. 2nd ed. Pergamon Press, Oxford, UK. 611 pp.

Falistocco, E. 1987. Cytogenetic investigations and karyological relationships of two *Medicago*: *M. sativa* L. (alfalfa) and *M. arborea* L. Caryologia, 40: 339–346.

Falistocco, E., and Falcinelli, M. 1991. Cytological and morphological studies in *Medicago hispida* Gaertner (= *M. polymorpha* L.). Annali di Botanica, 49: 13–25.

Falistocco, E., and Falcinelli, M. 1993. Karyotype and C-banding in *Medicago noeana* Boiss., Leguminosae. Cytologia, 58: 151–154.

Falistocco, E., and Moretti, C. 2000. Homologous relationship between *Medicago murex* Willd. and *Medicago lesinsii* E. Small investigated by genomic Southern and in situ hybridization. *In* Proceedings of the XIII Eucarpia *Medicago* spp. group meeting, Perugia, Italy, 13–16 September 1999. pp. 380–384.

Falistocco, E., Torricelli, R., and Falcinelli, M. 2002. Genomic relationships between *Medicago murex* Willd. and *Medicago lesinsii* E. Small investigated by in situ hybridization. Theor. Appl. Genet. 105: 829–833.

Farr, E.R., Leussink, J.A., and Stafleu, F.A. (Editors). 1979. Index nominum genericorum (plantarum), 3 vols. Bohn, Scheltema & Holkeema, Utrecht; Dr. W. Junk, b.v., The Hague, The Netherlands.

Feng, R. 2000. Genetic diversity of rhizobia of *Medicago edgeworthii* by AFLP and RFLP analysis. Acta Microbiologica Sinica, 40(4): 339–345. (In Chinese.)

Fenster, C.B., Armbruster, W.S., Wilson, P., Dudash, M.R., and Thomson, J.D. 2004. Pollination syndromes and floral specialization. Ann. Rev. Ecol. Evol. Syst. 35: 375–403.

Ferakova, V., and Murin, A. 1978. Report in IOPB chromosome number reports LX. Taxon, 27: 223–231.

Ferguson, S., Sorensen, E.L., and Horber, E.K. 1982. Resistance to the spotted alfalfa aphid *Therioaphis maculata* (Homoptera, Aphididae) in glandular haired *Medicago* species. Environ. Entomol. 11: 1229–1232.

Fernandes, A., and Queirós, M. 1978. Contribution à la connaissance cytotaxonomique des Spermatophyta du Portugal. IV. Leguminosae (Suppl. 3). Bol. Soc. Brot. 52: 79–164.

Fernandes, A., and Santos, M.F. 1971. Contribution à la connaissance cytotaxinomique des Spermatophyta du Portugal. IV. Leguminosae. Bol. Soc. Brot., Coimbra, ser. 2A, 45: 177–226.

Fernandes, A., Santos, M.F., and Queirós, M. 1977. Contribution à la connaissance cytotaxonomique des Spermatophyta de Portugal. Leguminosae. Bol. Soc. Brot. 51: 137–186.

Feuntes, S.C. 1981. Grain legumes in unirrigated conditions. Agricultura (Spain), 50(591): 780–782. (In Spanish.)

Figuier, L. 1867. The vegetable world: being a history of plants, with their botanical descriptions and peculiar properties. Chapman, London, UK. 576 pp.

Fiori, B.J., and Dolan, D.D. 1981. Field tests for *Medicago* resistance against the potato leafhopper *Empoasca fabae* Homoptera Cicadellidae. Can. Entomol. 113: 1049–1054.

Fiorini, G., and Raffaelli, M. 1996. Mediterranean chromosome number reports 6 (705-715). Flora Mediterranea, 6: 278–288.

Flannery, K.V. 1969. Origins and ecological effect of early domestication in Iran and the Near East. *In* The domestication and exploitation of plants and animals. *Edited by* P.J. Ucko and G.W. Dimbleby. Aldine Publishing Company, Chicago, IL. pp. 73–100.

Fois, N., Ligios, S., and Sitzia, M. 2000. Stubble management of *Medicago polymorpha* L. and pod consumption by grazing ewes during summer. *In* Legumes for Mediterranean forage crops, pastures and alternative uses. *Edited by* L. Sulas. CIHEAM-IAMZ, Zaragoza, Italy. pp. 365–368.

Fonnesbeck, P.V., and Symons, L.D. 1969. Effect of diet on concentration of protein, urea nitrogen, sugar and cholesterol of blood plasma of horses. J. Animal Sci. 28: 216–219.

Font Quer, P. 1924. New forms of plants. Mem. Mus. Ci. Nat. Barcelona Sér. Bot. 1(2): 7–14. (In Spanish.)

Foster, S., and Duke, J.A. 1990. A field guide to medicinal plants. Eastern and Central North America. Houghton Mifflin Co., Boston, MA. 366 pp.

Fox, F.W., and Wilson, C. 1937. Lucerne as a food for human consumption (laboratory report no. 3 of an inquiry into native diets). South Africa Inst. Medical Res., Johannesburg, South Africa. 28 pp.

Frame, J., and Harkess, R.D. 1987. The productivity of four forage legumes sown alone and with each of five companion grasses. Grass Forage Sci. 42: 213–223.

Frame J., Charlton J.F.L., and Laidlaw A.S. 1998. Temperate forage legumes. CAB International, Wallingford, UK. 327 pp.

Francis, C.M. 1981. The distribution and ecology of annual *Medicago* species in north west Libya. A report based on a plant collection tour. Austral. Plant Intro. Rev. 13: 3–14.

Francis, C.M., and Gillespie, D.J. 1981. Ecology and distribution of subterranean clovers and *Medicago* species in Sardinia. Austral. Plant. Intro. Rev. 13: 15–25.

Francis, C.M., and Katznelson, J.S. 1977. Observations on the distribution and ecology of subterranean clovers, some other clovers, and medics in Greece and Crete. Austral. Plant Intro. Rev. 12: 17–25.

Francis, C.M., and Poole, M.L. 1973. Effect of waterlogging on the growth of annual *Medicago* species. Austral. J. Exp. Agric. Anim. Husband. 13: 711–713.

Frape, D.L., and Boxall, R.C. 1974. Some nutritional problems of the horse and their possible relationship to those of other herbivores. Equine Vet. J. 6: 59–67.

Fresnillo Fedorenko, D.E. 2001. Ten years of ecological research on *Medicago minima* (L.) Bart. *In* Quality in lucerne and medics for animal production. Options Méditerranéenes Series A: Séminaires Mediterranéens no. 45. *Edited by* I. Delgado and J. Lloveras. CIHEAM-IAMN, Zaragoza, Spain. pp. 911–114.

Fresnillo Fedorenko, D.E., Fernandez, O.A., Busso, C.A. 1995. The effect of water stress on top and root growth in *Medicago minima*. J. Arid Environ. 29: 47–54.

Friedman, J., and Elberse, W.T. 1976. Competition between two desert varieties of *Medicago laciniata* (L.) Mill. under controlled conditions. Oecologia, 22: 321–339.

Friedman, J., and Orshan, G. 1974. Allopatric distribution of two varieties of *Medicago laciniata* (L.) Mill. in the Negev Desert. J. Ecol. 62: 107–114.

Frugoli, J., and Harris, J. 2001. *Medicago truncatula* on the move! Plant Cell. 13: 458–463.

Fryer, J.R. 1930. Cytological studies in *Medicago*, *Melilotus* and *Trigonella*. Can. J. Res. 3: 3–50.

Futuyma, D.J., and Slatkin, M. (Editors). 1983. Coevolution. Sinauer Associates Inc., Sunderland, MA. 566 pp.

Fyfe, J.L. 1957. Relational incompatibility in diploid and tetraploid lucerne. Heredity, 11: 286.

Gachet, J.P., and Elmir, A. 1972. Études monographique des *Medicago* annuelles. Annales de l'institute national de la recherches agronomique de Tunisie, 45(1): 3–45.

Gadella, T.W.H., and Kliphuis, E. 1966. Chromosome numbers of flowering plants in the Netherlands. II. Proc. Roy. Netherlands Acad. Sci., Ser C. 69: 541–556.

Gadella, T.W.H., and Kliphuis, E. 1968. Chromosome numbers of flowering plants in the Netherlands. IV. Proc. Roy. Netherlands Acad. Sci., Ser C. 71: 168–183.

Gadella, T.W.H., and Kliphuis, E. 1970*a*. Cytotaxonomic investigations in some angiosperms collected in the Valley of Aosta and in the National Park "Gran Pariso". Carylogia, 23: 363–379.

Gadella, T.W.H., and Kliphuis, E. 1970*b*. Chromosome studies in some flowering plants collected in the French Alps (Hante Savoie). Rev. Gen. Bot. 77: 487–497.

Gadnidze, R.I., Gviniashvili, T.N., Danelia, I.M., and Churadze, M.V. 1998. Chromosome numbers of the species of the Georgian flora. Bot. Zhurn. 83(10): 143–147. (In Russian.)

Galland, N. 1988. Recherche sur l'origine de la flore orophile du Maroc étude caryologique et cytogéographique Travaux de l'Institut Scientifique, Université Mohammed V. Série Botanique, 35: 1–168.

Gallardo, K., Lesignor, C., Darmency, M., Burstin, J., Thompson, R., Rochat, C., Boutin, J.-P., Kuester, H., Buitink, J., Leprince, O., Limami, A., and Grusak, M.A. 2006. Seed biology of *Medicago truncatula*. *In* The *Medicago truncatula* handbook (version 1, 2006). 23 pp. http://www.noble.org/MedicagoHandbook/ (current July 2010).

Galloni, M., Podda, L., Vivarelli, D., and Cristofolini, G. 2007. Pollen presentation, pollen-ovule ratios, and other reproductive traits in Mediterranean legumes (fam. Fabaceae – subfam. Faboideae). Pl. Syst. Evol. 266: 147–164.

Gan G., Li, Z., and Yang, Y. 2007. Studies on the characteristics related to nitrogen fixation in the nodules of *Trigonella arcuata* in Xinjiang arid zone. Arid Zone Research (China), 5(4): 1–6. (In Chinese.)

Garau, G., Reeve, W.G., Brau, L., Deiana, P., Yates, R.J., James, D., Tiwari, R., O'Hara, G.W., and Howieson, J.G. 2005. The symbiotic requirements of different *Medicago* spp. suggest the evolution of *Sinorhizobium meliloti* and *S. medicae* with hosts differentially adapted to soil pH. Plant Soil, 276: 263–277.

Garcia, J., Barker, D.G., and Journet, E.-P. 2006. Seed storage and germination. *In* The *Medicago truncatula* handbook (version 1, 2006). 9 pp. http://www.noble.org/MedicagoHandbook/ (current July 2010).

Gardé, A. 1948. Estudo citológico de algunas espécies do género *Trigonella*. Genetica Iberica, 1: 69–83.

Gardé, A., and Malherios-Gardé, N. 1953. Contribuição para o estudo cariológico de algumas especies de angiospermicas. I. Genetica Iberica, 5: 115–124.

Gardner, C.A., and Bennetts, H.W. 1956. The toxic plants of Western Australia. West Australian Newspapers, Perth, Australia. 253 pp.

Gargano, A.O., Aduriz, M.A., and Saldungaray, M.C. 1990. Forage and seed yields of *Medicago truncatula* cv. Jemalong and cv. Paraggio, and of *Vicia sativa*. Revista Argentina de Produccion Animal, 10: 19–26. (In Spanish.)

Gatehouse, J.A. 2002. Plant resistance towards insect herbivores: a dynamic interaction. New Phytol. 156: 145–169.

Génier, G., Guy, P., and Prosperi, J.M. 1992. Les luzernes. *In* Mieux comprendre: amélioration des espèces végétales cultivées: objectifs et critères de sélection. *Edited by* A. Gallais and H. Bannerot. Institut National de la Recherche Agronomique, Paris, France. pp. 323–338.

Ghalavand, A., Sharifabad, H.H., Azizi, K., and Sanavy, S.A.M. 2003. The effects of cropping systems (annual medics (*Medicago* spp.), fallow and chickpea) on run-off, erosion and soil moisture. Iranian J. Range Desert Res. 10: 429–444, 493.

Ghassali, F., Osman, A.E., and Cocks, P.S. 1998. Rehabilitation of degraded grasslands in north Syria: the use of Awassi sheep to disperse the seeds of annual pasture legumes. Exp. Agric. 34: 391–405.

Ghimpu, M.V. [Often simply V. Ghimpu]. 1928. Contribution à l'étude caryologique du genre *Medicago*. C.R. Acad. Sci. Paris, 187: 245–247.

Ghimpu, M.V. 1929*a*. Sur la caryologie du genre *Medicago*. Bull. Agric. Roumania (Bucarest), 5-6: 175–178.

Ghimpu, M.V. 1929*b*. Recherches chromosomiques sur les luzernes, vignes, chênes et orges. XIV Congr. Internatl. d'Agric. Bucarest: 1–7.

Ghimpu, M.V. 1930. Recherches cytologiques sur les genres: *Hordeum*, *Acacia*, *Medicago*, *Vitis* et *Quercus*. Arch. d'Anat. Micros. 26: 135–250, plates 1-6.

Ghorbani Javid, M., Akbari, G.A., Moradi, F., and Allahdadi, I. 2005. Effects of drought stress on morphological characteristics of two different genotypes of cutleaf medics (*Medicago laciniata* (L.) Mill). The 2nd international conference on integrated approaches to sustain and improve plant production under drought stress. 2005. University of Rome "La Sapienza" Rome, Italy. Abstract P5.28 (p. 216) http://www.plantstress.com/id2/Abstract%20book%20updated.pdf (current July 2010).

Ghorbani Javid, M., Akbari, G.A., Moradi, F, and Allahdadi, I. 2006. The effects of drought stress on proline amino acid and soluble sugars concentration in two cutleaf medic (*Medicago laciniata* (L.) Mill) ecotypes. The first international conference on the theory and practices in biological water saving, 21–25 May 2006, Beijing, China. pp. 60–61. (http://www.conference.ac.cn/ictpb/html/ICTPB5.doc; current March 2007).

Ghosh, R.B., and Dutta, S.C. 1976. Angiosperms not included in the Flora of British India, List V. Bull, Bot. Soc. Bengal, 30: 129–133.

Gianinazzi-Pearson, V., Maldonado-Mendoza, I., Lopez-Meyer, M., Weidmann, S., and Harrison, M.J. 2006. Arbuscular mycorrhiza. *In* The *Medicago truncatula* handbook (version 1, 2006). 19 pp. http://www.noble.org/MedicagoHandbook/ (current July 2010).

Gillespie, D.J. 1987. Murex, a new medic for acid soils. *In* Temperate pastures, their production, use and management. *Edited by* J.L. Wheeler, C.J. Pearson, and G.E. Robards. Australian Wool Corporation/CSIRO, Australia. pp. 172–174.

Gillespie, D.J. 1988. Murex medic – a new pasture species. J. Agric. West Australia, 29: 132–138.

Gillespie, D.J. 1989. *Medicago murex* Willd. (murex medic) cv. Zodiac. Austral. J. Exper. Agric. (Register of Australian Herbage Plant Cultivars), 29: 300–301.

Gillespie, D. 1994. Orion—the first variety of sphere medic. Department of Agriculture and Food, Western Austrlia, Farmnote 49/1994 http://www.agric.wa.gov.au/servlet/page?_pageid=449&_dad=portal30&_schema=PORTAL30&p_start_url=/pls/portal30/docs/FOLDER/IKMP/PAST/PL/MED/FN1994_049.HTM (current October 2006).

Gillespie, D.J., and McComb, J.A. 1991. Morphology and distribution of species in the *Medicago murex* complex. Can. J. Bot. 69: 2655–2662.

Gillies, C.B. 1968. The pachytene chromosomes of a diploid *Medicago sativa*. Can. J. Gent. Cytol. 10: 788–793.

Gillies, C.B. 1970*a*. Alfalfa chromosomes. I. Pachytene karyotype of a diploid *Medicago falcata* L. and its relationship to *M. sativa* L. Crop. Sci. 10:169–171.

Gillies, C.B. 1970*b*. Alfalfa chromosomes. II. Pachytene karyotype of a tetraploid *Medicago sativa* L. Crop. Sci. 10:169–171.

Gillies, C.B. 1971*a*. Pachytene studies in $2n = 14$ species of *Medicago*. Genetica, 42: 278–298.

Gillies, C.B. 1971*b*. Alfalfa chromosomes. III. *Medicago glomerata* Balb. pachytene karyotype. Crop Sci. 11: 463–464.

Gillies, C.B. 1972*a*. Pachytene chromosomes of perennial *Medicago* species. II. Distantly related species whose karyotypes resemble *Medicago sativa*. Hereditas, 72: 289–302.

Gillies, C.B. 1972*b*. Pachytene chromosomes of perennial *Medicago* species. III. Unique karyotypes of *Medicago hybrida* Trautv. and *Medicago suffruticosa* Ramond. Hereditas, 72: 303–310.

Gillies, C.B. 1977. Identification of trisomics in diploid lucerne (*Medicago sativa*). Austral. J. Agric. Res. 28: 309–317.

Gintzburger, G., and Blesing, L. 1979. Genetic conservation in Libya: indigenous forage legumes collection in northern Libya. Distribution and ecology of *Medicago* spp. Part III. FAO_ARC TFUB. FAO, Rome, Italy.

Givnish, T.J. 1979. On the adaptive significance of leaf form. *In* Topics in plant population biology. *Edited by* O.T. Solbrig, S. Jain, G. B. Johnson, and P. H. Raven. Columbia University Press, New York, NY. pp. 375–407.

Givnish, T.J. 1987. Comparative studies of leaf form: assessing the relative roles of selective pressures and phylogenetic constraints. New Phytol. 106 (Suppl): 131–160.

Gledhill, D. 1989. The names of plants. 2nd. ed. Cambridge University Press, Cambridge, UK. 202 pp.

Gmelin, G. 1769. Flora sibirica sive historia plantarum Sibiriae. Tip. Academiae Scientiarum, St. Petersburg, Russia. Vol. 4.

Goertzen, L.R., and Small, E. 1993. The defensive role of trichomes in black medick (*Medicago lupulina* L.). Pl. Syst. Evol. 184: 101–111.

Goldblatt, P. 1981. Chromosome numbers in legumes II. Ann. Missouri Bot. Gard. 68: 551–557.

González-Andrés, F., and Ceresuela, J.L. 1998. Chemical composition of some Iberian Mediterranean leguminous shrubs potentially useful for forage in seasonally dry areas. New Zealand J. Agric. Res. 41: 139–147.

González-Andrés, F., Chávez, J., Montáñez, G., and Ceresuela, J.-L. 1999. Characterisation of woody *Medicago* (sect. *Dendrotelis*) species, on the basis of seed and seedling morphometry. Genet. Resourc. Crop Evol. 46: 505–519.

Good, R. 1974. The geography of the flowering plants. 4th ed. Longman, London, UK. 557 pp.

Goodall, D. 1965. Horses of the world. Country Life Ltd., London, UK. 272 pp.

Grant, W.F., and Small, E. 1996. The origin of the *Lotus corniculatus* (Fabaceae) complex: a synthesis of diverse evidence. Can. J. Bot. 74: 975–989.

Grauman, H.O. 1955. What do we mean by creeping alfalfas? Soil Conservat. 21: 103–104.

Greene, S.L., and Bauchan, G.R. 2000. Status of the U.S. *Medicago* germplasm collection. http://www.naaic.org/Meetings/National/2000meeting/abstracts/Greene.htm (current July 2010).

Greene, S.L., Kisha, T.J., and Dzyubenko, N.I. 2008. Conserving alfalfa wild relatives: is past introgression with Russian varieties evident today? Crop Sci. 48: 1853–1864.

Gregory, M. 1983. A bibliography on the anatomy of organs connected with nyctinastic and seismonastic movements in higher plants. Notes Jodrell Lab., Kew 10: 22–30.

Grenier, J.C.M., and Godron, D.A. 1846. Flore de France. Prospectus. Besançon (De Sainte-Agathe Ainé), France. 8 parts.

Grenier, J.C.M., and Godron, D.A. 1848. Flore de France. J.B. Baillière, Paris, France. Vol. 1(2).

Greuter, W. 1979. The flora and phytogeography of Kastellorizo (Dhodhekanisos, Greece). Willdenowia, 8: 531–611.

Greuter, W., and Raus, T. 1986. Med-Checklist Notulae, 13. Willdenowia, 16: 103–116.

Greuter, W., Matthäs, U., and Risse, H. 1982. Notes on Cardaegean plant. 3. *Medicago strasseri*, a new leguminous shrub from Kriti. Willdenowia, 12: 201–206.

Greuter, W., Matthäs, U., and Risse, H. 1984. Additions to the flora of Crete, 1973–1983 (1984) – II. Willdenowia, 14: 269–297.

Greuter, W., Burdet, H.M., and Long, G. 1989. Med-Checklist: A critical inventory of vascular plants of the circum-Mediterranean countries, Vol. 4: Dicotyledones (Lauraceae–Rhamnaceae). Conservatoire et Jardin botaniques de la Ville de Genève, Switzerland.

Grisebach, A. 1843. Spicilegium florae rumelicae et bithynicae exhibens synopsin plantarum quas in aest. Friedrich Vieweg und Sohn, Braunschweig, Germany. Vol. 1, part 1.

Groose, R.W., Kojis, W.P., and Bingham, E.T. 1988. Combining ability differences between isogenic diploid and tetraploid alfalfa. Crop Sci. 28: 7–10.

Groose, R.W., Ballard, R.A., Charman, N., and Lake, A.W.H. 1996. Cold-tolerant annual medics: *Medicago rigidula* and *M. rigiduloides*. Proc. 35th North American Alfalfa Improvement Conference, Oklahoma, USA. (Abstract) http://www.naaic.org/Publications/1996Proc/alfalfa_breeding.htm #Groose (current July 2010).

Grossheim [= Grossgeim], A.A. 1919. Review of the representatives of the genus *Medicago* in the Crimea and in the Caucasus. Sci. Pap. Appl. Sect. Tiflis Bot. Gard. 1: 1–56. (In Russian, English summary.)

Grossheim [= Grossgeim], A.A. 1945. *Trigonella*, *Medicago*. *In* Komarov, V.L. (ed.). Flora of the U.S.S.R. 11: 99–176; Akademiya Nauk SSSR, Moscow and Leningrad, USSR. (Translated from Russian, 1971, by Israel Program Scientific Translations, Jerusalem.)

Groves, C.P. 1986. The taxonomy, distribution, and adaptation of recent equids. *In* Equids in the ancient world. *Edited by* R.H. Meadow and H.-P. Uerpmann. Ludwig Reichert Verlag, Weisbaden, Germany. pp 11–65.

Gu, A.L., Holzworth, L., and Yun, J.F. 1994. Establishment of legumes in arid and semi-arid grasslands of Inner Mongolia. Grassland of China, 6: 12–15. (In Chinese.)

Gualtieri, G., and Bisseling, T. 2000. The evolution of nodulation. Plant Molec. Biol. 42: 181–194.

Gunn, C.R. 1981. Seed topography in the Fabaceae. Seed Sci. Technol. 9: 737–757.

Gunn, C.R., Skrdla, W.H., and Spencer, H.C. 1978. Classification of *Medicago sativa* L. using legume characters and flower colors. U.S. Dept. Agric., Agric. Res. Serv. Tech. Bull. No. 1574. Agric. Res. Serv., Washington, D.C. 84 pp.

Gupta, M. 1991. Seed coat structure in some species of *Trigonella*. Scanning Microscopy, 5: 787–796.

Gutierrez, M.V., Vaz Patto, M.C., Huguet, T., Cubero, J.I., Moreno, M.T., and Torres, A.M. 2005. Cross-species amplification of *Medicago truncatula* microsatellites across three major pulse crops. Theor. Appl. Genet. 110: 1210–1217.

Guy, P. 1993. Lucerne in Europe: statistical elements. *In* The future of lucerne. Biotechnology breeding and variety constitution. Proceedings of the X international conference of EUCARPIA, *Medicago* spp. Group. *Edited by* P. Rotili and L. Zannone. Istituto Sperimentale per le Colture Fouraggere, Lodi, Italy. pp. 13–17.

Haeckell, E. 1866. Generelle morphologie der organismen. Verlag von Georg Reimer, Berlin, Germany. 462 pp. + plates.

Hall, J.W., and Majak, W. 1989. Plant and animal factors in legume bloat. *In* Toxicants of plant origin, Vol. 3, Proteins and amino acids. *Edited by* P.R. Cheeke. CRC Press Inc., Boca Raton, FL. pp. 93–106.

Hall, M.H., Sheaffer, C.C., and Heichel, G.H. 1988. Partitioning and mobilization of photoassimilate by alfalfa subjected to water deficits. Crop Sci. 28: 964–969.

Hallier, E.H. 1885. Flora von Deutschland, edition 5 of D.F.L. von Schlechtendal et al. F.E. Köhler, Gera-Untermhaus, Germany. Vol. 23.

Halloran, G.M., and Pennell, A.L. 1981. Distribution and variability in development of the genus *Trigonella* in Asia Minor, and its possible use in Australian environments. Austral. J. Agric. Res. 32: 793–800.

Hamly, D.H. 1932. Softening of the seeds of *Melilotus alba*. Bot. Gaz. 93: 345–374.

Han, Y., Zhang, Y., Jiang, G., Sledge, M., Greene, S., Coyne, C., and Monteros, M. 2007. Evaluation of genetic diversity, population structure and identification of a *Medicago truncatula* core collection using SSR markers. ASA-CSSA-SSSA international annual meetings (November 4–8, 2007). http://a-c-s.confex.com/crops/2007am/techprogram/P33905.HTM (current July 2010).

Hanan, J., Bucciarelli, B., and Vance, C.P. 2006. Sharing phenotypic data: A coding system and a developmental model. *In* The *Medicago truncatula* handbook (version 1, 2006). 21 pp. http://www.noble.org/MedicagoHandbook/ (current July 2010).

Hannachi, A.S., Mohamed, B., and Mohamed, M. 1998. Genetic variability organisation and gene flow in natural populations of *Medicago polymorpha* L. prospected in Tunisia. Genetics, Selection, Evolution, 30(Suppl.): S121.

Hansen, N.E. 1909. The wild alfalfas and clovers of Siberia, with a perspective view of the alfalfas of the world. Bull. U.S. Bur. Pl. Ind. 150. Washington, D.C. 31 pp.

Hansen, A. 1977. Some supplementary remarks on a forgotten species of the Spanish flora: *Trigonella ovalis* Boiss. Lagascalia, 7(1): 47–50.

Hanson, C.H. (Editor). 1972. Alfalfa science and technology. American Society of Agronomy, Madison, WI. 812 pp.

Hanson, A.A., Barnes, D.K., and Hill, R.R., Jr. (Editors). 1988. Alfalfa and alfalfa improvement. American Society of Agronomy, Crop Science Society of America, and Soil Science Society of America, Madison, WI. 1084 pp.

Hanson, C.H., Garrison, C.S., and Graumann, H.O. 1960. Alfalfa varieties in the United States. Agriculture Handbook No. 177. U.S. Dept. Agriculture, Washington, D.C. 30 pp.

Hardin, G. 1968. The tragedy of the commons. Science, 162: 1243–1248.

Harper, A.M., Schaber, B.D., Story, T.P., and Entz, T. 1990. Effect of swathing and clear-cutting alfalfa on insect populations in southern Alberta. J. Econ. Entomol. 83: 2050–2057.

Harper, F. 1945. Extinct and vanishing mammals of the Old World. Special Public. 12. American Committee for International Wild Life Protection, New York, NY. 850 pp.

Haslam, S.M., Sell, P.D., and Wolseley, P.A. 1977. A flora of the Maltese Islands. Malta University Press, Msida, Malta. 560 pp.

Hassen, H., Zoghlami, A., Sayari, R., and Kalbousi, R. 1998. Evaluation agronomique d'écotypes spontanés de *Medicago* et de *Scorpiurus* pour la sélection de cultivars adaptés au régions arides et semi-arides. Annales de l'INRAT, 72: 211–235.

Hayek, A. 1917. Beitrag zur Kenntnis der Flora des albanisch-montenegrinischen Grenzgebietes. Denkschriften der math, naturwiss. Klasse der Akademie der Wissenschafften in Wien, Bd. 94, S: 127–210.

Hayek, H. 1921. Diagnosen neuer von J. Dörfler und H. Zerny in den Jahren 1916 und 1918 in Albanien gesammelten. Pflanzenformen. Österr. Bot. Zeitschr. 70(1-2): 12–22.

Hedges, I.C., and Sales, F. 2001. *Trigonella*. Flora Iberica. *Edited by* Consejo Superior de Investigaciones Científicas, Madrid. Real Jardín Botánico de Madrid, Spain, 7(2): 731–741.

Heft, D.C., and Groose, R.W. 1996. Assignment of *Medicago rigidula* accessions in the NPGS into the two species *M. rigidula* and *M. rigiduloides*. Proc. 35th North American Alfalfa Improvement Conference, Oklahoma, USA. [Poster abstract.] http://www.naaic.org/Publications/1996Proc/poster_session.htm (current July 2010.)

Heinrichs, D.H. 1963. Creeping alfalfas. Advances Agron. 15: 317–337.

Heiser, C.B., and Whitaker, T.W. 1948. Chromosome number, polyploidy and growth habit in California weeds. Am. J. Bot. 35: 179–186.

Hekneby, M., Antolin, M.C., and Sanchez-Diaz, M. 2001. Cold response of annual Mediterranean pasture legumes. Options Mediterraneennes Serie A, Seminaires Mediterraneens. 45: 157–161.

Hendry, G.W. 1923. Alfalfa in history. J. Amer. Soc. Agron. 15: 171–176.

Hensen, I. 2002. Seed predation by ants in south-eastern Spain (Desierto de Tabernas, Almería). Anales de Biología, 24: 89–96.

Henson, P.R. and Hollowell, E.A. 1960. Winter annual legumes for the south. USDA Farmer's Bull. 2146. U.S. Dept. Agric., Washington, D.C. 24 pp.

Herms, D.A., and Mattson, W.J. 1992. The dilemma of plants: to grow or defend. Quart. Rev. Biol. 67: 283–335.

Hernandez, A.J., Prieto, N., and Pastor, J. 2001. Management of an olive crop in a semiarid environment using sown or resident leguminous covers. *In* Conservation agriculture, a worldwide challenge; First World Congress on conservation agriculture, Madrid, Spain, 1–5 October, 2001. Volume-2: offered contributions. *Edited by* L Garcia-Torres, J. Benites, and A. Martinez-Vilela. XUL, Cordoba, Spain. pp. 419–423.

Hesterman, O.O., Teuber, L.R., and Livingston, A.L. 1981. Effect of environment and genotype on alfalfa sprout production. Crop Sci. 21: 720–726.

Heyn, C. 1956. Some chromosome counts in the genus *Medicago*. Caryologia, 9: 160–165.

Heyn, C.C. 1959. The genus *Medicago* L. in Linnaeus's Species Plantarum. Bull. Res. Counc. Israel, 7D: 157–174.

Heyn, C. C. 1963. The annual species of *Medicago*. Magnes Press, The Hebrew University, Jerusalem. 154 pp.

Heyn, C.C. 1968. An evolutionary study of fruit morphology in the tribe Trigonelleae (Leguminosae). Phytomorphology, 18: 54–59.

Heyn, C.C. 1969. The typification of *Medicago* (Leguminosae). Taxon, 18: 667–669.

Heyn, C.C. 1981. Trifolieae. *In* Advances in legume systematics, part 1. *Edited by* R.M. Polhill and P.H. Raven. Royal Botanic Gardens, Kew, UK. pp. 383–385.

Heyn, C. C. 1984. *Medicago*. *In* Flora Iranica, Papilionaceae, Vol. 2; Papilionaceae 2 (Tabula), Tab. 156–170. *Edited by* K.H. Rechinger. Akademische Druck-u. Verlagsanstalt, Graz, Austria. pp. 253–271.

Heyn, C.C., and Davis P.H. 1970. *Medicago*. *In* Flora of Turkey and the East Aegean Islands, Vol. 3. *Edited by* P.H. Davis. Edinburgh University Press, UK. pp. 483–511.

Higgs, E., and Jarman, M. 1969. The origins of agriculture: a reconsideration. Antiquity, 43: 31–41.

Hintz, H.F. 1983. Horse nutrition, a practical guide. Arco Publishing Inc., New York, NY. 228 pp.

Hirsch, A.M., Lum, M.R., and Downie, J.A. 2001. What makes the rhizobia-legume symbiosis so special? Plant Physiol. 127: 1484–1492.

Hochman, Z. 1987. Quantifying vernalization and temperature promotion effects on time of flowering of three cultivars of *Medicago truncatula* Gaertn. Austral. J. Agric. Res. 38: 279–286.

Hodgkinson, K.C., Smith, N.G., and Miles, G.E. 1972. The photosynthetic capacity of stubble leaves and their contribution to growth of the lucerne plant after high-level cutting. Austral. J. Agric. Res. 23: 225–238.

Hodgson, J. 1979. Nomenclature and definitions in grazing studies. Grass and Forage Sci. 34: 11–18.

Hojland, J.G., and Poulsen, G.S. 1994. *Medicago sativa* L. ssp. *sativa* (lucerne/alfalfa). *In* Five cultivated plant species: *Brassica napus* L. ssp. *napus* (rape), *Medicago sativa* L. ssp. *sativa* (lucerne/alfalfa), *Pisum sativum* L. ssp. *sativum* (pea), *Populus* spp. (poplars). Dispersal, establishment and interactions with the environment. The National Forest and Nature Agency, Denmark. pp. 29–45.

Hole, F. 1989. A two-part stage model of domestication. *In* The walking larder. *Edited by* J. Clutton-Brock. Unwin Hyman, London, UK. pp. 97–104.

Holland, P.G. 1978. An evolutionary biogeography of the genus *Aloe*. J. Biogeogr. 5: 213–226.

Hollingsworth, P.M., Gornall, R.J., and Bailey, J.P. 1992. Contribution to a cytological catalogue of the British and Irish flora, 2. Watsonia, 19: 134–137.

Holmgren, P.K., and Holmgren, N.H. 1998 [continuously updated]. Index herbariorum: a global directory of public herbaria and associated staff. New York Botanical Garden's Virtual Herbarium. http://sweetgum.nybg.org/ih/ (current July 2010).

Holub, J., and Pouzar, Z. 1967. A nomenclatural analysis of the generic names of phanerogams proposed by F.M. Opiz in his Śeznam Rostlin Květeny České. Folia Geobot. Phytotax. 2: 397–428.

Horjales, M. 1976. Contribución al estudio citotaxonómico de la flora espanola. Trab. Dep. Bot. y F. Veg. (Madrid), 9: 13–18.

Horovitz, A., Meiri, L., and Beiles, A. 1976. Effects of ovule positions in Fabaceous flowers on seed set and outcrossing rates. Bot. Gaz. 137: 250–254.

Hossain, M.A., and Bauchan, G.R. 1999. Identification of B chromosomes using Giemsa banding in *Medicago*. J. Heredity, 90: 428–429.

Hougen-Eitzman, D., and Rausher, M.D. 1994. Interactions between herbivorous insects and plant-insect coevolution. Amer. Nat. 143: 677–697.

Howarth, R.E. 1988. Antiquality factors and nonnutritive chemical components. *In* Alfalfa and alfalfa improvement. *Edited by* A.A. Hanson, D.K. Barnes, and R.R. Hill, Jr. American Society of Agronomy, Crop Science Society of America, and Soil Science Society of America, Madison, WI. pp. 493–514.

Howieson, J.G. 1999. The host-rhizobia relationship. *In* Genetic resources of Mediterranean pasture and forage legumes. *Edited by* S.J. Bennett and P.S. Cocks. Kluwer Academic Publishers, Boston, MA. pp. 96–106.

Howieson, J.G., and Ewing, M.A. 1986. Acid tolerance in the *Rhizobium meliloti* and *Medicago* complex. Austral. J. Agric. Res. 37: 55–64.

Howieson, J.G., and Ewing, M.A. 1989. Annual species of *Medicago* differ greatly in their ability to nodulate on acid soils. Austral. J. Agric. Res. 40: 843–850.

Howieson, J.G., Ewing, M.A., and Dantuono, M.F. 1988. Selection for acid tolerance in *Rhizobium meliloti*. Plant & Soil, 105: 179–188.

Howieson, J.G., Robson, A.D., and Abbott, L.K. 1992. Calcium modifies pH effects on the growth of acid-tolerant and acid-sensitive *Rhizobium meliloti*. Austral. J. Agric. Res. 43: 765–772.

Hrušovská-Osuská, L. 1988. Karyological study of some taxa of the flora of the northern part of Povazský Inovec. Part I. Acta Facultatis Rerum Naturalium Universitatis Comenianae, Botanica, 35: 69–79.

Hua-Zhu, G., and Quing-Chuan, Y. 1990. Studies on eleven wild bianxudou (*Medicago ruthenica*) in China. Report of the 32nd North American alfalfa improvement conference. p. 74.

Huber-Morath, A. 1965. Novitiae florae Anatolicae VII. Bauhinia, 2(3): 295–330.

Huber-Morath, A. 1970. *Trigonella*. *In* Flora of Turkey and the East Aegean Islands, Vol. 3. *Edited by* P.H. Davis. Edinburgh University Press, UK. pp. 452–482.

Huber-Morath, A., and Zohary, M. 1969. Materials for a flora of Turkey. 20 Leguminosae: Trifolieae. Notes Roy. Bot. Gard. Edinb. 29: 321–323.

Huguet, T., and Prosperi, J.M. 1996. *Medicago truncatula*: a legume model-plant. Cahiers Options Méditerranéennes, 18: 171–175.

Humphries, C.J. 1992. Shamrock botany and history of an Irish myth, by E.C. Nelson [book review]. The Linnean, 8: 62–63.

Humphries, C.J., Murray, B.G., Bocquet, G., and Vasudevan, K.N. 1978. Chromosome numbers of phanerogams from Morocco and Algeria. Bot. Not. 131: 391–404.

Hurst, E. 1942. The poison plants of New South Wales. NSW Poison Plants Committee, Sydney, Australia. 498 pp.

Huyghe, C., Bertin, E., and Landry, N. 2007. Medicinal and nutraceutical uses of alfalfa (*Medicago sativa* L.). *In* Advances in Medicinal Plant Research. *Edited by* S.N. Acharya and J.E. Thomas. Research Signpost, Transworld Research Network, Trivandrum, Kerala, India. pp. 147–172.

Ibragimova, S.S. 1999. Chromosome numbers of annual species of the genus *Medicago* L. Tsitologiya, 41(12): 1064. (In Russian.)

Ibragimova, S.S., and Smolenskaya, S.E. 2001. Prospects of introducing annual medics at Novosibirsk region. Doklady Rossiiskoi Akademii Sel'skokhozyaistvennykh Nauk, 1: 9–11. (In Russian.)

Ibrar, M., and Hussein, F. 2003. The effect of salinity on the growth of *Medicago polymorpha* Linn. J. Sci. Technol. (Pakistan), 27: 35–38.

Ignasiak, T., and Lesins, K. 1973. Carotenoids in petals of some perennial *Medicago* species. Biochem. Syst. Ecol. 1: 97–100.

Ingham, J.L. 1979*a*. Isoflavonoid phytoalexins of the genus *Medicago*. Biochem. Syst. Ecol. 7: 29–34.

Ingham, J.L. 1979*b*. Isoflavonoid phytoalexins of *Parochetus communis* and *Factorovskya aschersoniana*. Z. Natursforsch. 31: 290–292.

Ingham, J.L. 1981. Phytoalexin induction and its chemosystematic significance in the genus *Trigonella*. Biochem. Syst. Ecol. 9: 275–281.

Ivanov, A.I. 1977. History, origin, and evolution of the genus *Medicago*, subgenus *Falcago*. Bull. Appl. Bot. Genet. Plant Breed. 59(1): 3–40. (In Russian.)

Ivanov, A.I. 1980. Lucerne. Kolos, Moscow, USSR. (In Russian. Translated by A.K. Dhote, 1988, A.A. Balkema, Rotterdam, The Netherlands. 318 pp.)

Ivory, D.A. 1976. Agronomic variation in *Medicago scutellata* and *M. orbicularis* in south-eastern Queensland. Tropical Grasslands, 10: 165–172.

Jafri, S.M.H. 1980. Flora of Libya. 86. Fabaceae. Al Faateh University, Tripoli, Libya. 304 pp.

Jahan, B., Vahidy, A.A., and Ali, S.I. 1994. Chromosome numbers in some taxa of Fabaceae mostly native to Pakistan. Ann. Missouri Bot. Gard. 81: 792–799.

Jain, P.S. 2003. Development and structure of nodules in leguminous weeds. *In* Advances in microbiology. *Edited by* P.C. Trevedi. Scientific Publishers, Jodhpur, India. pp. 215–225.

Janis, C. 1976. The evolutionary strategy of the Equidae and the origins of rumen and cecal digestion. Evolution, 30: 757–774.

Janzen, D.H. 1980. When is it coevolution? Evolution, 34: 611–612.

Jarvis, C. 2007. Order out of chaos. Linnaean plant names and their types. Linnean Society of London and Natural History Museum, London, UK. 1016 pp.

Jeanmonod, D., and Burdet, H.M. 1999. Notes et contributions à la flore de Corse, XV. Candollea, 54: 385–416.

Jeanrenaud, E. 1974. Water relations in some legume populations on the Black Sea coast. Oecologia, 25: 125–143.

Jeanrenaud, E. 1975. The sublethal water deficit and the degree of drought resistance of some plants from various eco-physiologic groups. Analele Stiintifice ale Universitatii' Al I Cuza' din Iasi, IIa 21: 20–22.

Jelenevskii ["Elenevskii"], A.G., and Kupatadze, G.A. 1968. Systematics of annual *Trigonella* of section *Bucerates*. Biologicheskii Zhurnal Armenii, 21(6): 66–70. (In Armenian.)

Jenczewski, E., Gherardi, M., Bonnin, I., Prosperi, J.M., Olivieri, I., and Huguet, T. 1997. Insight on segregation distortions in two intraspecific crosses between annual species of *Medicago* (Leguminosae). Theor. Appl. Genet. 94: 682–691.

Jenczewski, E., Angevain, M., Charrier, A., Genier, G., Ronfort, J., and Prosperi, J.M. 1998. Contrasting patterns of genetic diversity in neutral markers and agromorphological traits in wild and cultivated populations of *Medicago sativa* L. from Spain. Genetics Selection Evolution (Paris), 30(Suppl. 1): S103–S119.

Jenczewski, E., Prosperi, J.M., and Ronfort, J. 1999*a*. Differentiation between natural and cultivated populations of *Medicago sativa* (Leguminosae) from Spain: analysis with random amplified polymorphic DNA (RAPD) markers and comparison to allozymes. Mol. Ecol. 8: 1317–1330.

Jenczewski, E., Prosperi, J.M., and Ronfort, J. 1999*b*. Evidence for gene flow between wild and cultivated *Medicago sativa* (Leguminosae) based on allozyme markers and quantitative traits. Am. J. Bot. 86: 677–687.

Jewell, P.A. 1969. Wild mammals and their potential for new domestication. *In* The domestication and exploitation of plants and animals. *Edited by* P.J. Ucko and G.W. Dimbleby. Aldine Publishing Company, Chicago, IL. pp. 101–109.

Jha, S.S., and Pal, A. 1992. Seed structure in *Medicago* L. Flora, 187: 209–218.

Johns, T. 1990. With bitter herbs they shall eat it: chemical ecology and the origins of human diet and medicine. Univ. Arizona Press, Tucson, AZ. 356 pp.

Johnson, D.A., Jigjidsuren, S., Sheehy, D.P., Majerus, D.P., Winslow, S.R., and Holzworth, L.K. 2006. Collection and evaluation of forage germplasm indigenous to Mongolia. USDA Forest Service Proceedings RMRS-P-39: 50– 61. (http://www.fs.fed.us/rm/pubs/rmrs_p039/rmrs_p039_050_061.pdf; current July 2010).

Johnson, K.J.R., Sorensen, E.L., and Horber, E.K. 1980*a*. Resistance in glandular-haired annual *Medicago* species to feeding by adult alfalfa weevils (*Hypera postica*). Environ. Entomol. 9: 133–136.

Johnson, K.J.R., Sorensen, E.L., and Horber, E.K. 1980*b*. Effect of temperature and glandular-haired *Medicago* species on development of alfalfa weevil larvae. Crop Sci. 20: 631–633.

Johnson, K.J. R., Sorensen, E.L., and Horber, E. K. 1981. Behavior of adult alfalfa weevils (*Hypera postica*) on resistant and susceptible *Medicago* spp. in free choice preference test. Environ. Entomol.10: 580–585.

Jones, M., Kobayashi, H., Davies, B.W., Taga, M.E., and Walker, G.C. 2007. How rhizobial symbionts invade plants: the *Sinorhizobium–Medicago* model. Nature Reviews/Microbiology, 5: 619–633.

Jones, R.A.C., and Nicholas, D.A. 1992. Studies on alfalfa mosaic virus infection of burr medic (*Medicago polymorpha*) swards: seed-borne infection, persistence, spread and effects on productivity. Austral. J. Agric. Res. 43: 697–715.

Jones, R.J., and Megarrity, R.G. 1986. Successful transfer of DHP-degrading bacteria from Hawaiian goats to Australian ruminants to overcome the toxicity of *Leucaena*. Austral. Vet. J. 63: 259–262.

Jones, R.M., and McLeod, M.N. 1971. Changes in nutritive value throughout the growth cycle of snail medic (*Medicago scutellata*). J. Austral. Inst. Agric. Sci. 37: 63–64.

Jones, W.E., and Bogart, R.B. 1973. Genetics of the horse. 2nd. ed. Caballus Publishers, Fort Collins, CO. 356 pp.

Jordan, D.C. 1981. Reduction of the nodulation barrier in *Medicago laciniata* by alteration of the root temperature. Plant Soil (Netherlands), 61: 93–111.

Journet, E.P., Carreau, V., Gouzy, J., Thoquet, P., Rosenberg, C., Barker, D., Huguet, T., Denarie, J., and Gamas, P. 2001. The model legume *Medicago truncatula*: recent advances and perspectives in genomics. Oléagineux Corps gras Lipides (France), 8: 478–484.

Juan, A., and Crespo, M.B. 1999. Phytosociological behaviour of *Medicago citrina* (Font Quer) (Leguminosae), an Iberolevantine-Mediterranean endemic. Acta Bot. Malacitana, 24: 221-229. (In Spanish.)

Juan, A., Crespo, M.B., Rios, S. 1999. *Medicago citrina* (Font Quer) Greuter (Leguminosae): variability, ecology and habitat conservations. *In* Proceedings of the 39th Scientific Meeting of Spanish Society for Pastures Study. Almeria, Spain. pp. 87–91. (In Spanish.)

Juan, A., Crespo, M.B., Cowan, R.S., Lexer, C., and Fay, M.F. 2004. Patterns of variability and gene flow in *Medicago citrina*, an endangered endemic of islands in the western Mediterranean, as revealed by amplified fragment length polymorphism (AFLP). Molec. Ecol. 13: 2679–2690.

Juan, N.A., Sheaffer, C.C., Barnes, D.K., Swanson, D.R., and Halgerson, J.H. 1993. Leaf and stem traits and herbage quality of multifoliolate alfalfa. Agron. J. 85: 1121–1127.

Julén, G. 1944. Investigations on diploid, triploid and tetraploid lucerne. Hereditas, 30: 567–582.

Jung, G.A., and Larson, K.L. 1972. Cold, drought and heat tolerance. *In* Alfalfa Science and Technology. *Edited by* C.H. Hanson. American Society of Agronomy, Madison, WI. pp. 185–209.

Jurzysta, M., Small, E., and Nozzolillo, C. 1988. Hemolysis, a synapomorphic discriminator of an expanded genus *Medicago* (Leguminosae). Taxon, 37: 354–363.

Jurzysta, M., Burda, S., Oleszek, W., Ploszynski, M., Small, E., and Nozzolillo, C. 1992. Chemical composition of seed saponins as a guide to the classification of *Medicago* species. Can. J. Bot. 70: 1384–1387.

Kamal, R., and Sharma, G.L. 1981. Steroidal analysis of *Trigonella polycerata* L. seeds. Pharmazie, 36(H8): 578.

Kamal, R., Yadav, R., and Mehra, P. 1997. Rotenoid biosynthesis potential of *Trigonella* species in vivo and in vitro. J. Med. Aromat. Plant Sci. 19: 983–993.

Kamari, G., and Matthäs, U. 1986. Cytotaxonomical contributions on the flora of Crete. III. Wildenowia, 15: 515–520.

Kapoor, P., and Ramakrishnan, P.S. 1975. Studies on crop legume behaviour in pure and mixed stands. Agro-Ecosystems, 2: 61–74.

Karadağ, Y., and Gülcan, H. 1997. Research on some cytological characters of some medic species (*Medicago scutellata*, *M. orbicularis*, *M. polymorpha*) occurring in native vegetation of the Çukurova region. Turk. J. Agric. Forest. 21: 121–127. (In Turkish.)

Karshibaev, H.K. 1992. Chromosome numbers of some Fabaceae in the Uzbekistan. Tezisy III Soveshchanie po Kariologii Rastenii. pp. 26–27.

Kaufman, W.H. 1954. Alfalfa seed dermatitis. J. Am. Med. Ass. 155: 1058.

Kawakami, I. 1930. Chromosome numbers in Leguminosae. Bot. Mag. (Tokyo), 44: 319–328.

Kernick, M.D. 1978. Technical data sheet 22. Papilionaceae (= Fabaceae). *Medicago* spp. *In* Ecological management of arid and semi-arid rangelands in Africa and the Near and Middle East (EMASAR) Phase II, Volume 4, Indigenous arid and semi arid forage plants of North Africa, the Near and Middle East. FAO, Rome, Italy. pp. 469–517.

Kerrigan, R.H. 1986. Practical horse nutrition. Griffin Press Limited, Adelaide, Australia. 204 pp.

Keys, R.N., Smith, S.E., Mogensen, H.L., and Small, E. 1995. Microgametophytic plastid nucleoid content and reproductive and life history traits of tribe Trifolieae (Fabaceae). Pl. Syst. Evol. 196: 89–98.

Khan, M.A., and Ahmad, Z. 1993. Ethnobotanical studies of certain rare medicinal plants of Bhopal. Fitoterapia, 64: 516–517.

Khassanov, O. Kh. 1972. Dikorastuschnie lyutserny Srednei Assii [Wild alfalfas of central Asia]. Tashkent, Uzbekistan. 172 pp. (In Russian.)

Khassanov, O. Kh., and Abdullazhanov, E. 1973. Ecology and geography of some wild species of medic in central Asia. Uzbekskii Biologicheskii Zhurnal, 17: 34–36. (In Russian.)

Khatoon, S., and Ali, S.I. 1991. Chromosome numbers in subfamily Papilionoideae (Leguminosae) from Pakistan. Willdenowia, 20: 159–165.

Khatoon, S., and Ali, S.I. 1993. Chromosome atlas of the angiosperms of Pakistan. Dept. Botany, BCC & T Press, Univ. Karachi, Pakistan. 232 pp.

Khatoon, S., and Ali, S.I. 2006. Chromosome numbers and polyploidy in the legumes of Pakistan. Pak. J. Bot. 38: 935–945.

Khoudi, H., Laberge, S., Ferullo, J.M., Bazin, R., Darveau, A., Castonguay, Y., Allard, G., Lemieux, R., and Vezina, L.P. 1999. Production of a diagnostic monoclonal antibody in perennial alfalfa plants. Biotechnol. Bioeng. 64: 135–143.

Kidwell, K.K., Austin, D.F., and Osborn, T.C. 1994. RFLP evaluation of nine *Medicago* accessions representing the original germplasm sources for North American alfalfa cultivars. Crop Sci. 34: 230–236.

Kirschner, J. and Štepánek, J. 1992 *In* Měsíček, J. and Jarolímová, V. List of Chromosome Numbers of the Czech Vascular Plants. Academia, Praha.

Kleitner, G.L., and Sorensen E.L. 1983. Erect glandular trichomes of *Medicago scutellata* (L.) Mill.: gland development and early secretion. Bot. Gaz. 144: 165–174.

Klingler, J.P., Edwards, O.R., and Singh, K.B. 2007. Independent action and contrasting phenotypes of resistance genes against spotted alfalfa aphid and bluegreen aphid in *Medicago truncatula*. New Phytol. 173: 630–640.

Kliphuis, E. 1977. Report in IOPB chromosome number reports LVI. Taxon, 26: 257–274.

Kliphuis, E., and Barkoudah, Y.I. 1977. Chromosome numbers in some Syrian angiosperms. Acta Bot. Neerl. 26: 239–249.

Kliphuis, E., and Wieffering, J.H. 1979. Report in IOPB chromosome number reports LXIV. Taxon, 28: 399.

Knuth, P. 1908. Handbook of flower pollination (J.R. Ainsworth Davis' translation of II. Band, I. Teil of German edition). Clarendon Press, Oxford. 705 pp.

Koç, A., and Tan, M. 1997. Tüylü Yonca (*Medicago papillosa* Boiss.) 'nın Bazı Tarımsal Özellikleri Üzerinde Bir Araştırma. [An investigation on some agronomic characters of "Tüylü Yonca" (*Medicago papillosa* Boiss.)]. Tarla Bitk. Merk. Araş. Enst. Derg. [J. Central Field Crop Res. Inst.], 6(1): 43–48.

Koch, W.D.J. 1837. Synopsis florae Germanicae et Helveticae. F. Wilmans, Frankfurt, Germany. 844 pp.

Kodama, A. 1989. Karyotype analyses of chromosomes in eighteen species belonging to nine tribes in Leguminosae. Bul. Hiroshima Agric. Coll. 8: 691–706.

Korth, K.L., Doege, S.J., Park, S.H., Goggin, F.L., Wang, Q., Gomez, S.K., Liu, G., Jia, L., and Nakata, P.A. 2006. *Medicago truncatula* mutants demonstrate the role of plant calcium oxalate crystals as an effective defense against chewing insects. Plant Physiol. 141: 188–195.

Kotzé, T.N., Brand, T.S., and Agenbag, G.A. 1995. A comparison of pod and seed characteristics, nutritive value, digestibilty and seed survival in cultivars from different *Medicago* species fed to sheep. Afr. J. Range For. Sci. 12: 11–15.

Kovalenko, V.I., Ibragimova, S.S., Shumnyi, V.K., Lapteva, L.S., and Laptev, A.V. 1987. Tripping and evolution of the propagation system of species of the *Medicago* L. genus. Sel'skokhozyaistvennaya Biologiya, 8: 35–40 [Translated from Russian into English by Allerton Press, Inc. as: Soviet Agricultural Biology: Part 1: Plant Biology, 4: 65–70.]

Kozuharov, S. 1965. The species of genus *Medicago* L. (lucerne) in Bulgaria. Bulgarian Academy of Science, Reports (Izvestiya) of the Botanical Institution, 15: 119–188. (In Bulgarian.)

Kozuharov, S., Petrova, A., and Markova, T. 1975. Cytotaxonomic study of Bulgarian Leguminosae species. 1. Bulg. Acad. Sci. [in honor of acad. D. Jordanov], 1975: 31–47.

Kozuharov, S.I., and Kuzmanov, B.A. 1965. Chromosome numbers of flowering plants in Bulgaria. Mitt. Bot. Inst. Bulg. Akad. Wiss., Sofia, 15: 255–258.

Kozuharov, S.I., Kuzmanov, B.A.,and Markova, T. 1972. Report, in IOPB chromosome number reports XXXVI. Taxon, 21: 333–346.

Kozuharov, S.I., Petrova, A.V. and Markova, T. 1973. Report, in IOPB chromosome number reports XL. Taxon, 22: 285–291.

Krall, J., Groose, R.W., and Sobels, J. 1996*a*. Winter survival of Austrian winter pea and annual medic on the western high plains. *In* Progress in new crops. *Edited by* J. Janick. ASHS Press, Alexandria, VA. pp. 237–240.

Krall, J.M., Delaney, R.H., Claypool, D.A., and Groose, R.W. 1996*b*. Evaluation of cold tolerance in annual medics with potential for use in rotation with wheat on the U.S. High Plains. Proc. 35th North American Alfalfa Improvement Conference, Oklahoma, USA. [Poster abstract.] http://www.naaic.org/Publications/1996Proc/poster_session.htm (current July 2010).

Krall, J.M., Groose, R.W., Walsh, M.J., Nayighugu, V., Cecil, J., and Hess, B. 2007. Registration of 'Laramie' annual medic. J. Plant Registr. 1: 32–33.

Kreitner, G.L., and Sorensen, E.L. 1979*a*. Glandular trichomes on *Medicago* species. Crop Sci. 19: 380–384.

Kreitner, G.L., and Sorensen, E.L. 1979*b*. Glandular secretory system of alfalfa species. Crop Sci. 19: 499–502.

Kreitner, G.L., and Sorensen, E.L. 1983. Erect glandular trichomes of *Medicago scutellata* (L.) Mill.: gland development and early secretion. Bot. Gaz. 144: 165–174.

Kreitner, G.L., and Sorensen, E.L. 1985. Structure of the keel-locking mechanism in insect-pollinated and self-pollinated alfalfa species. Crop Sci. 25: 631–634.

Kucherov, E.V., and Mikhailova, T.P. 1978. Feed value of *Trigonella platycarpos* in the Bashkir ASSR USSR. Rastitel'nye Resursy, 14(2): 255–258. (In Russian.)

Kulik, M.M., and Dery, P.D. 1995. Frequency of plants in populations of annual and perennial medics (*Medicago* spp.) that were resistant to damping-off incited by AG 4 isolates of *Rhizoctonia solani*. Zeitschrift Pflanzenkrankheiten Pflanzenschutz, 102: 392–398. (In German.)

Kumari, S., and Bir, S.S. 1990. Karyomorphological evolution in Papilionaceae. J. Cytol. Genetics, 25: 173–219.

Kurt, O., Ozmen, A.T., and Alnoglu, N. 1990. Investigation of annual medic species that can be used on fallow areas and for improvement of rangeland in Turkey. Doga, Turk Tarm ve Ormanclk Dergisi, 14: 306–313. (In Turkish.)

Labadie, J.P. 1976. I.O.P.B. chromosome number report LIV. Taxon, 25: 638.

Labadie, J.P. 1979. Report in IOPB chromosome number reports LV. Taxon, 28: 628–629.

Laberge, S., Castonguay, Y., and Vezina, L.P. 1993. New cold- and drought-regulated gene from *Medicago sativa*. Plant Physiology (USA), 101: 1411–1412.

Lake, W.H. 1989. Forage plant breeding. Transfer of genes for aphid resistance from *Medicago truncatula* (barrel medic) via *Medicago littoralis* (strand medic) to *Medicago tornata* (disc medic). International grassland congress, 16, Nice, France, 4–11 Oct. 1989. pp. 255–256.

Lambert, M.G., Jung, G.A., and Costall, D.A. 1989*a*. Forage shrubs in North Island hill country. 1. Forage production. N. Z. J. Agric. Res. 32: 477–483.

Lambert, M.G., Jung, G.A., Fletcher, R.H., Budding, P.J., and Costall, D.A. 1989*b*. Forage shrubs in North Island hill country. 2. Sheep and goat preferences. N. Z. J. Agric. Res. 32: 485–490.

Lambert, M.G., Jung, G.A., Harpster, H.W., Budding, P.J., and Wewala, G.S. 1989*c*. Forage shrubs in North Island hill country. 3. Forage digestibility. N. Z. J. Agric. Res. 32: 491–497.

Lambert, M.G., Jung, G.A., Harpster, H.W., and Lee, J. 1989*d*. Forage shrubs in North Island New Zealand hill country. 4. Chemical composition and conclusions. N. Z. J. Agric. Res. 32: 499–506.

Lammerink, J. 1968. Genetic variability in commencement of flowering in *Medicago lupulina* L. in the South Island of New Zealand. N. Z. J. Bot. 6: 33–42.

Lamson-Scribner, F., and Kennedy, P.B. 1900. Turkestan alfalfa (*Medicago sativa* var. *Turkestanica*). U.S. Dept. Agric., Division of Agrostology, Circular 25. 20 pp.

Laouar, M, and Abdelguerfi, A. 2000*a*. Ecology and distribution of the *Medicago ciliaris–M. intertexta* complex in Algeria. *In* Lucerne and medics for the XXI century. Proceedings XIII EUCARPIA *Medicago* spp. Group meeting, Perugia, Italy, 13–16 Sept. 1999. *Edited by* F. Veronesi and D. Rosellini. University of Perugia, Perugia, Italy. pp. 162–169. (In French.)

Laouar M., and Abdelguerfi, A. 2000*b*. Étude du complexe d'espèces *Medicago ciliaris–M. intertexta*: caractérisation des différents types d'infrutescences. *In* Legumes for Mediterranean forage crops, pastures and alternative uses. *Edited by* L. Sulas. CIHEAM-IAMZ, Zaragoza, Italy. pp. 39–42.

Laouar, M, and Abdelguerfi, A. 2003. Variabilité morphologique et phénologique chez deux taxa très proches: *Medicago ciliaris* et *M. intertexta*. *In* Fixation symbiotique de l'azote et développement durable dans le Bassin méditerranéen. *Edited by* INRA. INRA, Paris, France. pp. 263–272.

Larkcom, J. 2007. Oriental vegetables: the complete guide for the gardening cook. Revised edition. Kodansha, New York, NY. 232 pp.

Larkin, R.A., and Grauamann, H.O. 1954. Anatomical structure of the alfalfa flower and explanation of the tripping mechanism. Bot. Gaz. 116: 40–52.

Larsen, K. 1955. Cytotaxonomical studies on the Mediterranean flora. Bot. Notiser, 108: 263–275.

Larsen, K. 1956. Chromosome studies in some Mediterranean and South European flowering plants. Bot. Notiser, 109: 293–307.

Lassen, P. 1983. *Medicago heyniana* Greuter. *In* Med-Checklist notulae, 7. *Edited by* W. Greuter and Th. Raus. Willdenowia, 13: 98.

Latta, R.A., and Quigley, P.E. 1993. A comparison of the persistence of *Medicago truncatula* cv. Paraggio with other annual medics in the Victorian Mallee. Austral. J. Exper. Agric. 33: 443–449.

Lattimore, M.-A. 2008. Producing quality lucerne hay: a guide for lucerne hay producers across Australia. Rural Industries Research and Development Corporation (Australia), Barton, Australia. 126 pp.

Laumont, P., and L'Hermite, M. 1953. Note sur la Luzerne arborescente (*Medicago arborea* L.). Inst. Agric. Algérie, 7(7).

Lavrent'-ev, G.F., and Chernenko, E.G. 1975. The results of a study of wild species of *Medicago*. Selekt. Semen. 6: 30. (In Russian.)

Ledingham, G.F. 1940. Cytological and developmental studies of hybrids between *Medicago sativa* and a diploid form of *M. falcata*. Genetics, 24: 1–15.

Lefi, E., Gulias, J., Cifre, J., Ben Younes, M., and Medrano, H. 2004*a*. Drought effects on the dynamics of leaf production and senescence in field-grown *Medicago arborea* and *Medicago citrina*. Ann. Appl. Biol. 144: 169–176.

Lefi, E., Gulias, J., Ribas-Carbo, M., and Medrano, H. 2004*b*. Soil water deficit effects on photosynthesis, water use efficiency and growth of three Mediterranean shrubs: *Medicago arborea*, *Medicago citrina* and *Medicago strasseri*. *In* Rangeland and pasture rehabilitation in Mediterranean areas. *Edited by* Institut des Régions Arides, Medenine (Tunisia), Centre International de Hautes Études Agronomiques Méditerranéennes, Paris (France); FAO, Roma (Italy); Réhabilitation des pâturages et des parcours en milieux méditerranéens, Zaragoza (Spain); and CIHEAM-Institut Agronomique Méditerranéen. pp. 65–68.

Lefi, E., Ben Younes, M., and Medrano, H. 2004*c*. Seasonal variation of leaf production for *Medicago arborea* and *Medicago citrina* under irrigation and drought conditions. *In* Rangeland and pasture rehabilitation in Mediterranean areas. *Edited by* Institut des Régions Arides, Medenine (Tunisia), Centre International de Hautes Études Agronomiques Méditerranéennes, Paris (France); FAO, Roma (Italy); Réhabilitation des pâturages et des parcours en milieux méditerranéens, Zaragoza (Spain); and CIHEAM-Institut Agronomique Méditerranéen. pp. 327–330. http://ressources.ciheam.org/om/pdf/c62/04600181.pdf (current July 2010).

Lefi, E., Medrano, H., and Cifre, J. 2004*d*. Water uptake dynamics, photosynthesis and water use efficiency in field-grown *Medicago arborea* and *Medicago citrina* under prolonged Mediterranean drought conditions. Ann. Appl. Biol. 144: 299–307.

Lehman, W.F., and Marble, V.L. 1986. Nondormant alfalfa in the United States and the world. Report of the Thirtieth North American Alfalfa Improvement Conference. p. 60.

Leitner, M., Boland, W., and Mithöfer, A. 2005. Direct and indirect defences induced by piercing-sucking and chewing herbivores in *Medicago truncatula*. New Phytol. 167: 597–606.

Lenssen, A.W., Sorensen, E.L., Posler, G.L., and Harbers, L.H. 1988*a*. Forage quality of perennial glandular-haired and eglandular *Medicago* populations. Crop Sci. 28: 168–171.

Lenssen, A.W., Sorensen, E.L., Posler, G.L., and Harbers, L.H. 1988*b*. Total cell wall and fiber concentrations of perennial glandular-haired and eglandular *Medicago* populations. Can. J. Plant Sci. 68: 439–448.

Lenssen, A.W., Sorensen, E.L., Posler, G.L., and Harbers, L.H. 1989. Sheep preference for perennial glandular-haired and eglandular *Medicago* populations. Crop Sci. 29: 65–68.

Lenssen, A.W., Banfield, J.D., and Cash, S.D. 2001. The influence of trichome density on the drying rate of alfalfa forage. Grass Forage Sci. 56: 1–9.

Lesins, K. [= Lesins, K.A.] 1952. Some data on the cytogenetics of alfalfa. J. Heredity, 43: 287–291.

Lesins, K. 1956. Interspecific hybrids between alfalfa, *Medicago sativa* L. and *M. dzhawakhetica*. Agron. J. 48: 583.

Lesins, K. 1958. Progenies from a cross *Medicago dzhawakhetica* × *M. sativa*. Proc. X Internatl. Congr. Genetics, 2: 165.

Lesins, K. 1959. Note on a hexaploid *Medicago*: *M. cancellata* M. B. Can. J. Genet. Cytol. 1: 133–134.

Lesins, K. 1961*a*. Interspecific crosses involving alfalfa. I. *Medicago dzhawakhetica* (Bordz) Vass. × *M. sativa* L. and its peculiarities. Can. J. Genet. Cytol. 3: 135–152.

Lesins, K. 1961*b*. Interspecific crosses involving alfalfa. II. *Medicago cancellata* M. B × *M. sativa* L. Can. J. Genet. Cytol. 3: 316–324.

Lesins, K. 1962. Interspecific crosses involving alfalfa. III. *Medicago sativa* L. × *M. prostrata* Jacq. Can. J. Genet. Cytol 4: 14–23.

Lesins, K. 1968. Interspecific crosses involving alfalfa. IV. *Medicago glomerata* × *M. sativa* with reference to *M. prostrata*. Can. J. Genet. Cytol. 10: 536–544.

Lesins, K.A. 1969. Relationship of taxa in the genus *Medicago* as revealed by hybridization. IV. *M. hybrida* × *M. suffruticosa*. Can. J. Genet. Cytol. 11: 340–345.

Lesins, K. 1970. Interspecific crosses involving alfalfa. V. *Medicago saxatilis* × *M. sativa* with reference to *M. cancellata* and *M. rhodopaea*. Can. J. Genet. Cytol. 12: 80–86.

Lesins, K. 1971. Interspecific hybrids involving alfalfa. Part 6. Ineffectiveness of alloploidy in induction fertility in *Medicago pironae* × *Medicago daghestanica* hybrids. Can. J. Genet. Cytol. 13: 437–442.

Lesins, K. 1972. Interspecific crosses involving alfalfa. VII. *Medicago sativa* × *M. rhodopea*. Can. J. Genet. Cytol. 14: 221–226.

Lesins, K. 1976. Alfalfa, lucerne. *In* Evolution of crop plants. *Edited by* N.W. Simmonds. Longman, London, UK. pp. 165–168.

Lesins, K. and Erac, A. 1968*a*. Relationship of taxa in the genus *Medicago* as revealed by hybridization. I. *M. striata* × *M. littoralis*. Can. J. Genet. Cytol. 10: 263–275.

Lesins, K. and Erac, A. 1968*b*. Relationship of taxa in the genus *Medicago* as revealed by hybridization. III. *M. aschersoniana* × *M. laciniata*. Can. J. Genet. Cytol. 10: 777–781.

Lesins, K., and Gillies, C.B. 1968. Relationship of taxa in genus *Medicago* as revealed by hybridization. II. *Medicago pironae* × *Medicago daghestanica* with reference to *Medicago sativa*. Can. J. Genet. Cytol. 10: 454–459.

Lesins, K., and Gillies, C.B. 1972. Taxonomy and cytogenetics of *Medicago*. *In* Alfalfa science and technology. *Edited* by C.H. Hanson. American society of Agronomy, Madison, WI. pp. 53–86.

Lesins, K.A., and Lesins, I. 1960. Sibling species in *Medicago prostrata*. Jacq. Can. J. Genet. Cytol. 2: 416–417.

Lesins, K. A., and Lesins, I. 1961. Some little-known *Medicago* species and their chromosome complements. Can. J. Genet. Cytol. 3: 7–9.

Lesins, K.A., and Lesins, I. 1962. Trueness-to-species in seed samples of *Medicago*, with a note on $2n$=14 species. Can. J. Genet. Cytol. 4: 337–339.

Lesins, K. A., and Lesins, I. 1963*a*. Some little-known Medicagos and their chromosome complements. II. Species from Turkey. Can. J. Genet. Cytol. 5: 133–137.

Lesins, K.A., and Lesins, I. 1963*b*. Pollen morphology and species relationship in *Medicago* L. Can. J. Genet. Cytol. 5: 270–280.

Lesins, K.A., and Lesins, I. 1963*c*. *Medicago saxatilis* M.B., a second high-ploidy species in the genus. Can. J. Genet. Cytol. 5: 348–350.

Lesins, K.A. and Lesins, I. 1964. Diploid *Medicago falcata* L. Can. J. Genet. Cytol. 6: 152–163.

Lesins, K. A., and Lesins, I. 1965. Some little-known Medicagos and their chromosome complements. III. Some Mediterranean species. Can. J. Genet. Cytol. 7: 97–102.

Lesins, K. A., and Lesins, I. 1966. Little-known Medicagos and their chromosome complements. IV. Some mountain species. Can. J. Genet. Cytol. 8: 8–13.

Lesins, K. A., and Lesins, I. 1979. Genus *Medicago* (Leguminosae). A taxogenetic study. Dr. W. Junk bv, The Hague, The Netherlands. 228 pp.

Lesins, K., and Singh, S.M. 1973. Relationships of taxa in the genus *Medicago* as revealed by hybridization. VII. *M. tornata* complex. Can. J. Genet. Cytol. 15: 321–325.

Lesins, K., Lesins, I., and Gillies, C.B. 1970. *Medicago murex* with $2n$=16 and $2n$=14 chromosome complements. Chromosoma, 30: 109–122.

Lesins, K., Singh, S.M., and Erac, A. 1971. Relationship of taxa in genus *Medicago* as revealed by hybridization. V. Section *Intertextae*. Can. J. Genet. Cytol. 13: 335–346.

Lesins, K., Sadasivaiah, R.S., and Singh, S.M. 1976. Relationship of taxa in genus *Medicago* as revealed by hybridization. VIII. Section *Rotatae*. Can. J. Genet. Cytol. 18: 345–355.

Lesins, K., Dickson, J., and Ostafichuk, L.1980. Relationship of taxa in *Medicago* as revealed by hybridization. IX. *Medicago turbinata* × *Medicago truncatula*. Can. J. Genet. Cytol. 22: 137–142.

Levin, D.A. 1973. The role of trichomes in plant defense. Quart. Rev. Biol. 48: 3–15.

Lev-Yadun, S. 2000. Why are underground flowering and fruiting more common in Israel than anywhere else in the world? Current Sci. 79: 289.

Lewis, H. 1966. Speciation in flowering plants. Science, 152: 167–172.

Li, C.T. 1997. Wild legumes resources and their utilization in alpine meadow of Tianzhou area, Gansu. Grassland of China 1: 23–24, 28. (In Chinese.)

Li, H., Luo, X.Y., and Wang, D.K. 1990. Study on cross-breeding of *Pocockia ruthenica* and *Medicago sativa*. II. Observations on the meiosis of F_1 pollen mother cells. Grassland of China, 1990(6): 20–23.

Li, Y.H., and Wang, S.P. 1998. Grazing intake of Inner Mongolian fine-wool sheep and their preference for grassland forage species. Acta Prataculturae Sinica, 7: 50–53.

Lieth, H., and Mochtchenko, M. 2003. Cash crop halophytes: recent studies. 10 years after the Al Ain meeting. Kluwer Academic Publishers, Dordrech, The Netherlands. 250 pp. + database in attached CD.

Lila, M., Genier, G., and Pietraszek, W. 1989. Study of an anti-nutritional factor in lucerne (*Medicago sativa*): the saponins. *In* Proc. 16 Int. Grassland Congress, 1989, Nice, Versailles, France. Assoc. Française pour la Production Fourragère. pp. 779–780.

Lilienfeld, F.A. 1962. Plastid behavior in reciprocally different crosses between two races of *Medicago truncatula* Gaertn. Seiken Zihô, 13: 3–38.

Linnaeus, C. 1753. Species plantarum. Facsimile edition (1957) by Ray Society, London, UK. 2 vols.

Little, I.P., Chartres, C.J. and Young, R.R. 1992. The relationship of soil properties to the growth of barrel medic at Condobolin, New South Wales. Austral. J. Soil Res. 30: 371–382.

Livingstone, A.L., Knuckles, B.E., Teuber, L.R., Hesterman, O.B., and Tsai, L.S. 1985. Minimising the saponin content of alfalfa sprouts and leaf protein concentrates. *In* Nutritional and toxicological aspects of food safety. *Edited by* M. Friedman. Plenum Press, New York, NY. pp. 253–268.

Lloyd, D.L., Taylor, G.B., Johnson, B., and Teasdale, K.C. 1997. Patterns of seed softening and seedling emergence of nineteen annual medics during three years after a single seed crop in southern Queensland. Austral. J. Exper. Agric. 37: 767–778.

Loader, L.R. 1979. Effect of sulphur on the productivity of a native pasture containing *Medicago minima*, on the eastern Darling Downs, Queensland. Queensland J. Agric. Anim. Sci. 36: 25–29.

Loi, A., Howieson, J.G., Cocks, P.S., and Caredda, S. 1993. The adaptation of *Medicago polymorpha* to a range of edaphic and environmental conditions: effect of temperature on growth, and acidity stress on nodulation and nod gene induction. Austral. J. Exper. Agric. 33: 25–30.

Loi, A., Porqueddu, C., Veronesi, F., and Cocks, P.S. 1995. Distribution, diversity and potential agronomic value of *Medicago polymorpha* in Sardinia. J. Agric. Sci. 124: 419–426.

Lolicato, S., and Mary-Anne Lattimore, M.-A. 1998. Irrigated lucerne: a guide to profitable irrigated lucerne hay production in northern Victoria and southern NSW. 2nd ed. New South Wales Agriculture,Australia. 46 pp.

Loon, J.C. Van. 1980. *In* Chromosome number reports LXIX. Taxon, 29: 718–720.

Loon, J.C. Van, and Jong, H. de. 1978. Report in IOPB chromosome number reports LIX. Taxon, 27: 53–61.

Loon, J.C. Van, and Kieft, B. 1980. Report in IOPB chromosome number reports LXVIII. Taxon, 29: 538–542.

Loon, J.C., and Setten, A.K. van. 1982. I.O.P.B. chromosome number reports LXXVI. Taxon, 31: 589–592

Lorenzetti, S., Falcinelli, M.,and Rosini, G. 1987. Use of selected strains of *Rhizobium meliloti* in the cultivation of *Medicago rigidula* L. and *Medicago noeana* Boiss. Annali della Facolta di Agraria Universita degli Studi di Perugia, 41: 173–180. (In Italian.)

Löve, A., and Kjellqvist, E. 1974. Cytotaxonomy of Spanish plants. IV. Dicotyledons: Caesalpiniaceae – Asteraceae. Lagascalia, 4: 153–211.

Löve, A., and Löve, D. 1956. Cytotaxonomical conspectus of the Icelandic flora. Acta Horti Gothoburgensis, 20: 65–291.

Löve, A., and Löve, D. 1982. I.O.P.B. chromosome number reports LXXVI. Taxon, 31: 574–598.

Lovka, M., Susnik, F., Löve, A., and Löve, D. 1972. Report, in IOPB chromosome number reports XXXVI. Taxon, 21: 333–346.

Lövkvist, B., and Hultgård, U.-M. 1999. Chromosome numbers in south Swedish vascular plants. Opera Botanica, 137: 1–42.

Lowe, C.C., Marble, V.L., and Rumbaugh, M.D. 1972. Adaptation, varieties, and usage. *In* Alfalfa science and Technology. *Edited by* C.H. Hanson. American Society of Agronomy, Inc., Madison, WI. pp. 391–413.

Lowe, K.F., and Bowdler, T.M. 1988. Effects of height and frequency of defoliation on the productivity of irrigated oats (*Avena strigosa*) cultivar Saia and perennial ryegrass (*Lolium perenne*) cultivar Kangaroo Valley grown alone or with barrel medic (*Medicago truncatula*) cultivar Jemalong. Austral. J. Exp. Agric. 28: 57–68.

Lubenets, P.A. 1953. Species composition and breeding assessment of cultivated and wild alfalfas. Bull. Appl. Bot. Genet. Sel. Forage Crops, 30(2): 3–155. (In Russian.)

Lubenets, P.A. 1972. Alfalfa–*Medicago* L. (A brief survey of the genus and the classification of the subgenus *Falcago* (Rchb.) Grossh.) Bull. Appl. Bot. Genet. Sel. Forage Crops, 47(3): 1–68. (In Russian.)

Lubenets, P.A., and Vitkus, A.A. 1978. Growth and developmental features of perennial alfalfa species in the Lithuanian SSR USSR and their protein amino-acid ascorbic-acid carotene thiamine and riboflavine content, part 1, growth and development. Lietuvos TSR Mokslu Akademijos Darbai Serija C Biologijos Mokslai, 3: 21–30. (In Russian.)

Luque, T. and Díaz Lifante, Z. 1991. Chromosome numbers of plants collected during Iter Mediterraneum I in the SE of Spain. Bocconea, Monographiae Herbarii Mediterranei Panormitani, 1: 303–364.

Luque, T., Mejías, J.A., and Díaz Lifante, Z. 1988. Números cromosómicos para la flora Española, 544–550. Lagascalia, 15: 130–133.

Ma, S.J., Shi, Y.H., and Xia, M. 1984. Effect of burn-cultivation in spring on the plant community structure and aboveground biomass in *Stipa grandis* grassland. Grassland of China, 21: 47–51. (In Chinese.)

Mackay, J.H.E. 1981. Register of Australian Herbage Plant Cultivars (suppl. to 1972 ed.). C.S.I.R.O., Melbourne, Australia. 122 pp.

MacLean, P.S., and Byers, R.A. 1983. Ovipositional preferences of the alfalfa blotch leafminer (Diptera:

Agromyzidae) among some simple and glandular-haired *Medicago* species. Environ. Entomol. 12: 1083–1086.

Magulaev, A.Y. 1980. Chromosome numbers of some Fabaceae in the North Caucasus. Bot. Zh. 6: 836–843.

Magulaev, A.Y. 1992. About cytogeography of *Medicago* L. species in the northern Caucasus flora. Tezisy III Soveshchanie po Kariologii Rastenii, pp. 33–34.

Magulaev, A.Y. 2001. On the karyology of *Trigonella* of Caucasus. 3d. Int. conf. biodiversity of Caucasus. Nalchik. p. 32. (In Russian.)

Majak, W., Howarth, R.E., Fesser, A.C., Goplen, B.P., and Pedersen, M.W. 1980. Relationships between ruminant bloat and the composition of alfalfa herbage. II. Saponins. Can. J. Animal Sci. 60: 699–708.

Major, D.J., Hanna, M.R., and Beasley, B.W. 1991. Photoperiod response characteristics of alfalfa (*Medicago sativa* L.) cultivars. Can. J. Plant Sci. 71: 87–93.

Majovsky, J. et al. 1974. Index of chromosome numbers of Slovakian flora. Part 4. Acta Fac. Rerum Nat. Univ. Comenianae Bot. 22: 1–20.

Majovsky, J. et al. 1978. Index of chromosome numbers of Slovakian flora. Part 6. Acta Fac. Rerum Nat. Univ. Comenianae Bot. 26: 1–42.

Malallah, G.A., and Brown, G. 1999. Determination of chromosome number of Kuwaiti flora I. Cytologia, 64: 181–196.

Malato-Beliz, J., and Guerra, J.A. 1977. Notas de floristica [Floristic Notes] – 10. Lagascalia, 7(1): 55–75. (In Portuguese.)

Manglitz, G.R., and Sorensen, E.L. 1998. Alfalfa germplasm resources and insect resistance. *In* Global plant genetic resources for insect-resistant crops. *Edited by* S.L. Clement and S.S. Quisenberry. CRC Press, Boca Raton, FL. pp. 149–168.

Marble, V.L. 1989. Fodders for the Near East: alfalfa. FAO Plant Production and Protection Paper 97/1. Food and Agriculture Organization of the United Nations, Rome, Italy. 207 pp.

Maremmani, A., Bedini, S., Matoševic, I., Tomei, P.E., and Giovannetti, M. 2003. Type of mycorrhizal associations in two coastal nature reserves of the Mediterranean basin. Mycorrhiza, 13: 1432–1890.

Mariani, A. 1963. Determinazione del numero cromosomico di alcune specie di *Medicago* con particolare ruguardo alla *Medicago tuberculata* Willd. Caryologia, 16: 139–142.

Mariani, A. 1975. Cytogenetic research on hexaploid alfalfa, *Medicago sativa* L. Caryologia, 28: 359–373.

Mariani, A., and Falistocco, E. 1990. Chromosome studies in $2n = 14$ and $2n = 16$ types of *Medicago murex*. Genome, 33: 159–163.

Mariani, A., and Falistocco, E. 1991. Cytogenetic analysis of *Medicago rugosa* and *Medicago scutellata*. J. Genet. Breed. 45: 111–116.

Mariani A., Pupilli, F. and Calderini, O. 1996. Cytological and molecular analysis of annual species of the genus *Medicago*. Can. J. Bot. 74: 299–307.

Martyn, R.S., and Oram, R.N. 1990. *Medicago tornata* (L.) Mill. var. *spinulosa* (disc medic) cv. Rivoli. Register of Australian Herbage Plant Cultivars, Austral. J. Exp. Agric. 30: 442.

Mason, I.L. 1984. Goat. *In* Evolution of domesticated animals. *Edited by* I.L. Mason. Longman, London, UK. pp. 85–99.

Masson, P., and Gintzburger, G. 2000. Les légumineuses fourragères dan les systèmes de production méditerranéens: utilisations alternatives. Cahiers Options Méditerranéennes, 45: 395–405.

Mastrorilli, M., and Colucci, R. 1997. Role of *Medicago arborea* (L.) in pastures in southern Italy. Murgia. Rivista di Agronomia, 31(1, Suppl.): 250–253. (In Italian.)

Mastrorilli, M., Colucci, R., and Marrone, G. 1999. *Medicago arborea* in a forage system of southern Italy: fodder resource and soil erosion protection. Rivista di Agronomia, 33: 240–245. (In Italian.)

Materon, L.A., and Zibilske, L. 2001. Delayed inoculation and competition of nitrogen-fixing strains in *Medicago noeana* (Boiss.) and *Medicago polymorpha* (L.). Agriculture Ecosystems Environment, 17: 175–181.

Mathews, F.P. 1937. Photosensitization and the photodynamic diseases of man and the lower animals. Arch. Pathol. 23: 399.

Maureira-Butler, I.J., Pfeil, B.E., Muangprom, A., Osborn, T.C., and Doyle, J.F. 2008. The reticulate history of *Medicago* (Fabaceae). Syst. Biol. 57: 466–482.

Maxted, N., Bennett, S.J., and Cocks, P.S. 1998. Ecogeography and genetic conservation. *In* Genetic resources of Mediterranean pasture and forage legumes. Proceedings of an international workshop, York, Western Australia, Australia, 14–18 Oct. 1996. pp. 53–66.

Mazin, V.V., Ostretsova, I.N., Agafodorova, M.N., and Pogorilaya, E.V. 1994. Cytokaryological characterization of the lucerne species, *Medicago lupulina* L., *Medicago borealis* Grossch. and their somatic hybrid, as an evaluation of source material for breeding for fodder value. Sel'skokhozyajstvennaya Biologiya, No. 4: 101–106. (In Russian, English abstract.)

McComb, J.A. 1974. Leaf marks in *Medicago*, with special reference to their inheritance in *Medicago truncatula*. Austral. J. Bot. 22: 67–80.

McCoy, T.J. 1985. Interspecific hybridization of *Medicago sativa* L. and *M. rupestris* M.B. using ovule–embryo culture. Can. J. Genet. Cytol. 27: 238–245.

McCoy, T.J. 1987. Tissue culture evaluation of sodium chloride tolerance in *Medicago* sp. cellular versus whole plant response. Plant Cell Reports, 6(1): 31–34.

McCoy, T.J., and Bingham, E.T. 1988. Cytology and cytogenetics of alfalfa. *In* Alfalfa and Alfalfa Improvement. *Edited by* A.A. Hanson, D.K. Barnes, and R.R. Hill, Jr. American Society of Agronomy, Crop Science Society of America, and Soil Science Society of America, Madison, WI. pp. 737–776.

McCoy, T.J., and Bingham, E.T. 1991. Alfalfa cytogenetics. *In* Chromosome engineering in plants: genetics, breeding, evolution. Part B. *Edited by* T. Tsuchiya and P.K. Gupta. Elsevier, Amsterdam, The Netherlands. pp. 399–418.

McCoy, T.J., and Echt, C.S. 1993. Potential of trispecies bridge crosses and random amplified polymorphic DNA markers for introgression of *Medicago daghestanica* and *M. pironae* germplasm into alfalfa (*M. sativa*). Genome, 36: 594–601.

McCoy, T.J., and Quarisa, G.L. 1989. Allotetraploid behavior of hybrids of *Medicago sativa* L. and *M. papillosa* Boiss. Genome, 32: 6–11.

McCoy, T.J., and Smith, L.Y. 1983. Genetics, cytology, and crossing behavior of an alfalfa (*Medicago sativa*) mutant resulting in failure of the post-meiotic cytokinesis. Can. J. Genet. Cytol. 25: 390–397.

McCoy, T.J., and Smith, L.Y. 1984. Uneven ploidy levels and a reproductive mutant required for interspecific hybridization of *Medicago sativa* L. and *M. dzhawakhetica* Bordz. Can. J. Genet. Cytol. 26: 511–518.

McCoy, T.J., and Smith, L.Y. 1986. Interspecific hybridization of perennial *Medicago* species using ovule–embryo culture. Theor. Appl. Genet. 71: 772–783.

McCoy, T.J., Echt, C.S., and Mancino, L.C. 1991. Segregation of molecular markers supports an allotetraploid structure for *Medicago sativa* × *Medicago papillosa* interspecific hybrid. Genome, 34: 574–578.

McDonald, W.J., Goodyer, G.J., and Nikandrow, A. 1995. Lucerne for pasture and fodder. 2nd ed. New South Wales Agriculture, Orange, N.S.W., Australia. 44 pp.

McGregor, S.E. 1976. Insect pollination of cultivated crop plants. Agric. Res. Service, U.S. Dept. Agric. Handbook No. 496. Washington, D.C. 411 pp.

McKee, R. 1916. Button clover. Farmers' Bull. No. 730. U.S. Dept. Agriculture, Washington, D.C. 11 pp.

McKenzie, J.S., Paquin, R., and Duke, S.H. 1988. Cold and heat tolerance. *In* Alfalfa and Alfalfa Improvement. *Edited by* A.A. Hanson, D.K. Barnes, and R.R. Hill, Jr. American Society of Agronomy, Crop Science Society of America, and Soil Science Society of America, Madison, WI. pp. 259–302.

McNeill, J., Barrie, F.R., Burdet, H.M., Demoulin, V., Hawksworth, D.L., Marhold, K., Nicolson, D.H., Prado, J., Silva, P.C., Skog, J.E., Wiersema, J., and Turland, N.J. 2006. International code of botanical nomenclature (Vienna code). A.R.G. Gantner Verlag, Ruggell, Liechenstein. 568 pp. Available online at http://ibot.sav.sk/icbn/main.htm. (current July 2010).

Medicus, F.K. 1789. Philosophische Botanik, mit kritischen Bemerkungen, Heft 1. Mannheim, p. 209.

Meglič, V., Pelikán, J., and Babinec, J. 2003. Current state of the alfalfa genetic resources in Europe. 25th Eucarpia Meeting, Brno 2003 (Section 1). p. 17.

Mehregan, I., Moussavi, M., and Nasrabadi, N. 2003. The genus *Medicago* in Iran: biodiversity and variation centers. Rostaniha, 4: 5–18.

Mehregan, I., Rahiminejad, M.R., and Azizian, D. 2002. A taxonomic revision of the genus *Medicago* L. (Fabaceae) in Iran. Iran. J. Bot. 9: 207–221.

Merabet, C., Bekki, A., Benrabah, N., Bey, M. Baba-Hamed, Bouchentouf, L., Ameziane, H., Rezki, M.A., Domergue, O., Cleyet-Marel, J.-C., Avarre, J.-C., Béna, G., Bailly, X., and de Lajudie, P. 2006. Distribution of *Medicago* species and their microsymbionts in a saline region of Algeria. Arid Land Res. Manage. 20: 219–231.

Měsiček, J. 1992. *In* J. Měsíček and V.J. Jarolímová. List of Chromosome Numbers of the Czech Vascular Plants. Academia, Praha.

Měsiček, J., and Soják, J. 1969. Chromosome counts of some Mongolian plants. Folia Geobot. Phytotax., Praha, 4: 55–86.

Měsiček, J., and Soják, J. 1992. Chromosome numbers of Mongolia angiosperms. I. Preslia, 64: 193–206.

Metawi H.R., Kandil, H.M., and El-Shaer H.M. 1999. Sheep productivity under extensive and semi-intensive production systems in Egypt. *In* Systems of sheep and goat production: organization of husbandry and role of extension services. *Edited by* R. Rubino and P. Mohrand-Fehr. CIHEAM-IAMZ, Zaragoza, Italy. pp. 239–243.

Micevski, K. 1973. Contribution to the taxonomy and chorology [regional geography] of the *Medicago prostrata* Jacq. agg. in Macedonia. God. Zborn. Prir.-Mat. Fak. Univ. Kiril & Metodij Skopje, Biol. 25: 137–147 + 4 plates. (In Macedonian, German summary.)

Michaud, R., W.F. Lehman, and Rumbaugh, M.D. 1988. World distribution and historical development. *In* Alfalfa and alfalfa improvement. *Edited by* A.A. Hanson, D.K. Barnes, and R.R. Hill, Jr. American Society of Agronomy, Crop Science Society of America, and Soil Science Society of America, Madison, WI. pp. 25–91.

Michener, C.D. 1979. Biogeography of the bees. Ann. Missouri Bot. Gard. 66: 277–347.

Miersch, J., Schlee, D., and Rosche, I. 1994. Ecological significance of allelochemicals: occurrence of canavanine in *Medicago sativa* L. *In* Allelopathy in agriculture and forestry. *Edited by* S.S. Narwal and P. Tauro. Scientific Publishers, Jodhpur, India. pp. 153–166.

Mikhailova, T.P. 1979. Structural and morphological peculiarities of *Trigonella platycarpos* under cultured conditions. Poleznye Rastenii Dikorastushchoi Flory Bashkirii I Puti Ikh Ratsional' Noe Ispol' Zovanie, 1979: 122–137. (In Russian.)

Mikhailova, T.P. 1984. Morphogenesis of *Trigonella platycarpos*. Biulleten' Glavnogo Botanicheskogo Sada. 133: 57–62. (In Russian.)

Miller, D.R. 2005. Past and future milestones in the genetic improvement of alfalfa, "the queen of forages". J. Idaho Acad. Sci. 41(1): 10–11.

Mitchell, J. 1996. Cover crops for saline soils. (http://www.sarep.ucdavis.edu/ccrop/ccres/11.HTM; current July 2008).

Mizianty, M., Frey, L., and Mirek, Z. 1981. Contribution to the knowledge of the chromosome numbers of Polish vascular plants. Fragmenta Floristica Geobotanica, 27: 19–30.

Mizukami, Y., Kato, M., Takamizo, T., Kanbe, M., Inami, S., and Hattori, K. 2006. Interspecific hybrids between *Medicago sativa* L. and annual *Medicago* containing alfalfa weevil resistance. Plant Cell Tissue Organ Culture, 84: 80–89.

Mohamed, M.K. 1997. Chromosome counts in some flowering plants from Egypt, Egyptian J. Bot. 37: 129–156.

Mohammad, N., and Qamar, I.A. 1988. Dry matter yield, root nodulation and nitrogen fixation in 37 varieties of Australian annual medics. Pakistan J. Agric. Res. 9: 390–395.

Mohr, E. 1971. The Asiatic wild horse, *Equus przevalskii* Poliakoff, 1881. J.A. Allen & Co. Ltd., London, UK. Translation by D.M. Goodall. 124 pp.

Mole, S. 1994. Trade-offs and constraints in plant-herbivore defense theory: a life-history perspective. Oikos, 71: 3–12.

Monteiro, M. 2000. Plant regeneration from protoplasts of alfalfa (*Medicago sativa* L.) and optimization of the chemical of fusion technique between *Medicago sativa* and *Medicago arborea* L. Piracicaba, SP (Brazil). 83 pp. (In Portuguese.)

Montford, S., and Small, E. 1999*a*. A comparison of the biodiversity friendliness of crops with special reference to hemp (*Cannabis sativa*). J. Internat. Hemp Assoc. 6: 53–63.

Montford, S., and Small, E. 1999*b*. Measuring harm and benefit: the biodiversity friendliness of *Cannabis sativa*. Global Biodiversity, 8(4): 2–13. [Also published in French as: Montford, S. et E. Small. 1999. Comment mesurer les dommages et les avantages des cultures: le *Cannabis sativa* et le respect de la biodiversité. La biodiversité mondiale, 8(4): 2–15.]

Montmollin, B. de. 1986. Étude cytotaxonomique de la flore de la Crète. III. Nombres chromosomiques. Candollea, 41: 431–439.

Moore, P.D. 1995. How plants deal with what's eating them. Science Watch, 6(4): 7–8.

Moradi, F., M. Ghorbani Javid, M., Akbari, G.A., Allahdadi, I., and Khoshkholgh Sima, N.A. 2005. Drought tolerance in cutleaf medic (*Medicago laciniata* L. Mill): a morphological and physiological perspective. The 2nd international conference on integrated approaches to sustain and improve plant production under drought stress. 2005. University of Rome "La Sapienza" Rome, Italy. Abstract P5.59 (p. 233). http://www.plantstress.com/id2/Abstract%20book%20updated.pdf (current July 2010).

Moreau, D. 2006. Morphology, development and plant architecture of *M. truncatula*. *In* The *Medicago truncatula* handbook (version 1, 2006). 6 pp. http://www.noble.org/MedicagoHandbook/ (current July 2010).

Moreau, D., Salon, C., and Munier-Jolain, N. 2006. Using a standard framework for the phenotypic analysis of *Medicago truncatula*: an effective method for characterizing the plant material used for functional genomics approaches. Plant Cell Environ. 29: 1087–1098.

Mozaffarian, V. 1988. New species and new plant records from Iran. Iran. J. Bot. 4: 61–70. (In Persian.)

Muir, J.P., Ocumpaugh, W.R., and Butler, T.J. 2005. Trade-offs in forage and seed parameters of annual *Medicago* and *Trifolium* species in north-central Texas as affected by harvest intensity. Agron. J. 97: 118–124.

Muller, M.H., Balsera, C., Genier, G., Prosperi, J.M., Roussel, S., Santoni, S., Soudiere, O., Tauzin, D., Vabre, M., and Ronfort, J. 2001*a*. Mitochondrial DNA diversity and phenotypic variation in wild and cultivated populations of *Medicago sativa*: insights into the dynamics of contact zones between the two related forms. Genetics Selection Evolution (Paris), 33(Suppl. 1): S381–S398.

Muller, M.H., Prosperi, J.M., Santoni, S., and Ronfort, J. 2001*b*. How mitochondrial DNA diversity can help to understand the dynamics of wild-cultivated complexes. The case of *Medicago sativa* in Spain. Molec. Ecol. 10: 2753–2763.

Muller, M.H., Prosperi, M., Santoni, S., and Ronfort, J. 2003. Inferences from mitochondrial DNA patterns on the domestication history of alfalfa (*Medicago sativa*). Molec. Ecol. 12: 2187–2199.

Muller, M.H., Poncet, C., Prosperi, J.M., Santoni, S., and Ronfort, J. 2006. Domestication history in the *Medicago sativa* species complex: inferences from nuclear sequence polymorphism. Molec. Ecol. 15: 1589–1602.

Mulligan, G.A. 1957. Chromosome numbers in Canadian weeds. I. Can. J. Bot. 35: 779–789.

Mulligan, G.A. 1984. Chromosome numbers of some plants native and naturalized in Canada. Naturaliste Can. 111: 447–449.

Mun, J.-H., Kim, D.-J., Choi, H.-K., and 11 others. 2006. Distribution of microsatellites in the genome of *Medicago truncatula*: a resource of genetic markers that integrate genetic and physical maps. Genetics, 172: 2541–2555.

Murin, A. 1978. Report, in Index of chromosome numbers of Slovakian flora. Part 6. Acta Fac. Rerum Nat. Univ. Comenianae, Bot. 26: 1–42.

Murin, A., and Ferakova, V. 1981. Acta Fac. Rerum Nat. Univ. Comenianae, Bot. 28: 59–62.

Murin, A., and Sheikh, M.Y. 1971. Report, in IOPB chromosome reports XXXII. Taxon, 20: 349–356.

Murr, J. 1986. Two rare forms from Upper Austria. Deutsche botanische Monatsschrift, 15: 199–200. (In German.)

Musselman, L.J., and Saoud, N.S. 2004. The type specimens of George Edward Post in Beirut and Geneva. Turk. J. Bot. 28: 155–160.

Nair, R.M., Howie, J., Ballard, R., Hutton, R., Charman, N., and Preston, C. 2004. Genetic improvement of strand medic (*Medicago littoralis* Rohde ex Lois.) for Australian farming systems. *In* New directions for a diverse planet: proceedings of the 4th International Crop Science Congress, Brisbane, Australia, 26 Sept.–1 Oct. 2004. *Edited by* T. Fischer, N. Turner, J. Angus, L. McIntyre, M. Robertson, A. Borrell, and D. Lloyd. The Regional Institute Ltd., Gosford, NSW, Australia. http://www.cropscience.org.au/icsc2004/symposia/6/3/716_nairrm.htm (current July 2010).

Nair, R.M., Hughes, S.J., Ellwood, S., Oliver, R., Green, S.K., Delalande, M., Wen, J., and Oldroyd, G. 2006*a*. *M. truncatula* stock centres. *In* The *Medicago truncatula* handbook (version 1, 2006). 18 pp. http://www.noble.org/MedicagoHandbook/ (current July 2010).

Nair, R., Howie, J., and Delalande, M. 2006*b*. *Medicago truncatula* cultivars. *In* The *Medicago truncatula* handbook (version 1, 2006). 14 pp. http://www.noble.org/MedicagoHandbook/ (current July 2010).

Nasirzadeh, A.R., Rabbani, M., and Joukar, L. 2006. Evaluation of yield quality in five annual *Medicago* species under irrigated and dry conditions. Iran. J. Rangelands Forests Plant Breed. Genet. Res. 13: 401–416. (In Persian.)

Natarajan, G. 1978. Report in IOPB chromosome number reports LXII. Taxon, 27: 519–535.

Naveh, Z., and Vernet, J.L. 1991. The palaeohistory of the Mediterranean biota. *In* Biogeography of Mediterranean invasions. *Edited by* R.H. Groves and F. di Castri. Cambridge University Press, Cambridge, UK. pp. 19–32.

Nègre, R. 1956. Les luzernes du Maroc. Travaux Inst. Sci. Chérif. Maroc., Sér. Bot. 5. 119 pp. + map.

Nègre, R. 1959. Révision des *Medicago* d'Afrique du Nord. Bull. Soc. Hist. Nat. Afr. Nord, 5: 267–314.

Neilreicht, A. 1859. Flora von Nieder-Oesterreich. Vol. 2. Vienna, Austria.

Nenz, E., Pupilli, F., Damiani, F., and Arcioni, S. 1996. Somatic hybrid plants between the forage legumes *Medicago sativa* L. and *Medicago arborea* L. Theor. Appl. Genet. 93: 183–189.

Neves, L.O., Tomaz, L., and Fevereiro, M.P.S. 2001. Micropropagation of *Medicago truncatula* Gaertn. cv. Jemalong and *Medicago truncatula* ssp. *Narbonensis*. Plant Cell Tissue Organ Cult. 67: 81–84.

Newcomer, A.D. 1978. Milk, lactase, and gene distribution. Digestive Diseases, 23: 961–962.

Nichols, P.G.H., Loi, A., Nutt, B.J., and 23 others. 2007. New annual and short-lived perennial pasture legumes for Australian agriculture—15 years of revolution. Field Crops Res. 104: 10–23.

Nicoloff, H. 1961. Cytological study of *Medicago rhodopaea* Velen. Comptes Rendus de l'Acad. Bulg. Sci. 14: 847–850. (In English and Bulgarian.)

Nieddu, G., Graviano, O., Lostia, M., and Porqueddu C. 2000. Effects of *Medicago polymorpha* L. cover cropping in Sardinia vineyards. *In* Legumes for Mediterranean forage crops, pastures and alternative uses. *Edited by* L. Sulas. CIHEAM-IAMZ, Zaragoza, Italy pp. 449–452.

Nilsson, Q., and Lassen, P. 1971. Chromosome numbers of vascular plants from Austria, Mallorca and Yugoslavia. Bot. Notiser, 124: 270–276.

Novoselova, L.V. 2003. Bud cleistogamy at annual species of genus *Medicago*. Czech. J. Genet. Plant Breed. 39(Special Issue): 285–287.

Novosyelova, [sic; cf. above] L.V. 1998. The bud cleistogamy of some annual species of the genus *Medicago* (Fabaceae). Botanicheskii Zhurnal, 83(11): 82–87. (In Russian.)

Noy-Meir, I.,Gutman, M., and Kaplan, Y. 1989. Responses of Mediterranean grassland plants to grazing and protection. J. Ecol. 77: 290–310.

Oakenfull, D. 1981. Saponins in food—a review. Food Chem. 6: 19–40.

Oakenfull, D., and Sidhu, G.S. 1989. Saponins. *In* Toxicants of plant origin, Vol. 2, glycosides. *Edited by* P.R. Cheeke. CRC Press Inc., Boca Raton, FL. pp. 97–141.

Oakley, R.A., and Garver, S. 1917. *Medicago falcata*, a yellow-flowered alfalfa. U.S. Dept. Agric. Bull. 428. Washington, D.C. 70 pp.

Oberprieler, C. and Vogt, R. 1996*a*. Chromosome number of North African phanerogams. VI. Some counts in Leguminosae. Willdenowia, 25: 669–680.

Oberprieler, C. and Vogt, R. 1996*b*. Mediterranean chromosome number reports 6 (692-699). Flora Mediterranea, 6: 262–266.

O'Connor, P.J., Smith, S.E., and Smith, E.A. 2002. Arbuscular mycorrhizas influence plant diversity and community structure in a semiarid herbland. New Phytol. 154: 209–218.

Ocumpaugh, W.R. 1999. Armadillo bur medic: a new legume variety for central and south Texas. Texas Forage and Grassland Council Advances (Texas Forage and Grassland Council, Georgetown, Texas), 12(1): 1–3.

Ocumpaugh, W.R., Hussey, M.A., Grichar, W.J. Jr., Read, J.C., Bade, D.H., Pinchak, W.E., Smith, G.R., Lane, R.A., Muir, J.P., and Coleman, S.W. 2004. Registration of 'Armadillo' burr medic. Crop Sci. 44: 1023–1024.

Ocumpaugh, W.R., Ueckert, D.N., Muir, J.P., Butler, T.J., and Reed, R.L. 2007. Registration of 'Devine' little burr medic. J. Plant Registr. 1: 1–2.

Oeder, G.C., Müller, O.F., and Vahl, M. (*Editors*). 1761–1883. Flora Danica, Copenhagen, Denmark. 17 vols.

Okumura, K., Takai, T., Kanbe, M., and Gau, M. 1995. Variations in pollen production and seed set of male sterile alfalfa somaclones. Breeding Science, 45: 25–29.

Oldach K.H., Peck, D.M., Cheong, J., Williams, K.J., and Nair, R.M. 2008. Identification of a chemically induced point mutation mediating herbicide tolerance in annual medics (*Medicago* spp.). Ann. Bot. 101: 997–1005.

Oldenmeyer, R.K. 1956*a*. Interspecific hybrids between alfalfa, *Medicago sativa* L., and *M. dzhawakhetica* Bordz. Agron. J. 48: 583.

Oldenmeyer, R.K. 1956*b*. Distant relatives of cultivated alfalfa, *Medicago ruthenica* and *M. platycarpa*. Agron. J. 48: 583–584.

Oldenmeyer, R.K. 1956*c*. Interspecific hybridization in *Medicago*. Agron. J. 48: 584–585.

Oleszek, W., Jurzysta, M., and Górski, P.M. 1992. Alfalfa saponins—the allelopathic agents. *In* Allelopathy: basic and applied aspects. *Edited by* S.J.H. Rizvi and V. Rizvi. Chapman & Hall, London, UK. pp. 151–167.

Oleszek, W.A., Hoagland, R.E., and Zablotowicz, R.M. 1999. Ecological significance of plant saponins. *In* Principles and practices in plant ecology. Allelochemical interactions. *Edited by* K.M.M. Inderjit Dakshini and C.L. Foy. CRC Press, Boca Raton, FL. pp. 451–465.

Oliva, R., Fernandez, C.J., Pajaron, S., Rodriguez P.M.L., Pueche, J., Ortiz, A., Ruiz, R.M., Oliver, J.J.L., Posse, H.M.F., Fernandez, P.J., Ramos, R., and Valdes, B. 1978. Chromosome numbers of the Spanish flora. 45-83. Lagascalia, 8: 105–125. (In Spanish.)

Olivieri, I., Gouyon, P.H., and Prosperi, J.M. 1991. Life cycles of some Mediterranean invasive plants. *In* Biogeography of Mediterranean invasions. *Edited by* R.H. Groves and F. di Castri. Cambridge University Press, Cambridge, UK. pp. 145–157.

Olvera-Novoa, M.A., Campos G., S., Sabido, M.G., and Martínez Palacios, C.A. 1990. The use of alfalfa leaf protein concentrates as a protein source in diets for tilapia (*Oreochromis mossambicus*). Aquaculture, 90: 291–302.

O'Neill, N.R., and Bauchan, G.R. 2000. Sources of resistance to anthracnose in the annual *Medicago* core collection. Plant Disease, 84: 261–267.

O'Neill, N.R., Bauchan, G.R., and Samac, D.A. 2003. Reactions in the annual *Medicago* spp. core germ plasm collection to *Phoma medicaginis*. Plant Disease, 87: 557–562.

Oostroom, S.J. van, and Reichgelt, T.J. 1958. Het geslacht *Medicago* in Nederland en België. Acta Bot. Neerl. 7: 90–123.

Orloff, S.B., Carlson, H.L., and Teuber, L.R. (Editors.) 1997. Intermountain alfalfa management. University of California Division of Agriculture and Natural Resources, Oakland, CA. 102 pp. (online at http://alfalfa.ucdavis.edu/; current July 2010).

Orpin, C.G. 1981. Isolation of cellulolytic phycomycete fungi from the caecum of the horse. J. Gen. Microbiol. 123: 287–296.

Orpin, C.G. 1988. Nutrition and biochemistry of anaerobic Chytridiomycetes. BioSystems, 21: 365–370.

Ørskov, E.R., and Kay, R.N.B. 1987. Non-microbial digestion of forages by herbivores. *In* The nutrition of herbivores. *Edited by* J.B. Hacker and J.H. Ternouth. Academic Press, Toronto, Canada. pp. 267–280.

Othman, R.B., Sorensen, E.L., Liang, G.H., and Horber, E.K. 1981. Density and distribution of erect glandular hairs on annual *Medicago* species. Bot. Gaz. 142: 237–241.

Ovalle, M.C., Pozo, L.A., del Avendano, R.J., Aravena, A.T., and Diaz, M.E. 2001. Cauquenes – INIA, a new Chilean cultivar of burr medic (*Medicago polymorpha*) for Mediterranean dryland areas. Agricultura Tecnica (Chile), 61: 89–92. (In Spanish.)

Packer, J.S. 1970. The flight and foraging behavior of the alkali bee *Nomia melanderi* (Ckll.) and the alfalfa leafcutter bee *Megachile rotundata* (F.). Ph.D. thesis, Utah State University, Logan, UT. 119 pp.

Palenski-Brown, F.J., and Chu, A.C.P. 1989. Effects of presowing treatments on germination of *Medicago ruthenica* L. (Trautv.) Lebed. synonym *Trigonella ruthenica* L. New Zealand J. Agric. Res. 32: 441–446.

Panciera, M.T., and Sparrow, S.D. 1995. Effects of nitrogen fertilizer on dry matter and nitrogen yields of herbaceous legumes in interior Alaska. Can. J. Plant Sci. 75: 129–134.

Pandey, K.C., Faruqui, S.A., Amar, S., and Patil, B.D. 1984. Screening of *Medicago* species for resistance to alfalfa weevil. Indian J. Agric. Sci. 54: 196–199.

Panitsa, M., and Tzanoudakis, D. 2000. Flora, vegetation and important habitat types of Tilos island group (East Aegean): a management approach. *In* Proc. of the 8th Scientific Congress of Greek Botanical Society, Patras. *Edited by* G. Kamari, G. Psaras, and Th. Costantinidis. pp. 342–351. (In Greek.)

Paquin, R. 1986. Effect of soil moisture on the free proline content and percentage of total sugars in cold and drought-hardened alfalfa (*Medicago media* cultivar Rambler). Can. J. Plant Sci. 66: 95–102. (In French.)

Parish, R., and Turkington, R. 1990. The influence of dung pats and molehills on pasture composition. Can. J. Bot. 68: 1698–1705.

Parkhurst, D.F., and Loucks, O.L. 1972. Optimal leaf size in relation to environment. J. Ecology, 60: 505–537.

Parra, R., Valdés, B., Fernández, I., Holguín, F.A., Rubio, L.M., and Ruiz, O. 1999. Números cromosómicos de plantas de Marruecos, III. Lagascalia, 21: 149–154.

Patanè, C., and Gresta, F. 2006. Germination of *Astragalus hamosus* and *Medicago orbicularis* as affected by seed-coat dormancy breaking techniques. J. Arid Environ. 67: 165–173.

Pathipanawat, W., Jones, R.A.C., and Sivasithamparam, K. 1996. Resistance to alfalfa mosaic virus in button medic (*Medicago orbicularis*). Austral. J. Agric. Res. 47: 1157–1167.

Pavlova, D. 1996. Mediterranean chromosome number reports 6 (762–766). Flora Mediterranea, 6: 323–328.

Pavone, L.V., and Reader, R.J. 1982. The dynamics of seed bank size and seed state of *Medicago lupulina*. J. Ecol. 70: 537–547.

Pearson, C.J., Brown, R., Collins, W.J., Archer, K.A., Wood, M.S., Petersen, C., and Bootle, B. 1997. An Australian temperate pastures database. Austral. J. Agric. Res. 48: 453–466.

Pederson, G.A., Quesenberry, K.H., Smith, G.R., and Guteva, Y.K. 1999. Collection of *Trifolium* sp. and other forage legumes in Bulgaria. Genetic Resourc. Crop Evol. 46: 325–330.

Pederson, M.W. 1978. Low-saponin alfalfa for non-ruminants. *In* Proc. Second Internat. Green Crop Drying Congress, Saskatoon. pp. 148–155.

Pederson, M.W., Berrong, B., Wall, M.E., and Davis, K.H., Jr. 1973. Modification of saponin characteristics of alfalfa by selection. Crop Sci. 13: 731–735.

Peel, M.D., Asay, K.H., Waldron, B.L., Jensen, K.B., Robins, J.G., and Mott, I.W. 2009. 'Don', a diploid falcata alfalfa for Western U.S. Rangelands. J. Plant Registrations, 3: 115–118.

Pérez-Bañon, C., Juan, A., Petanidou, T., Marcos-García, M., and Crespo, M.B. 2003. The reproductive ecology of *Medicago citrina* (Font Quer) Greuter (Leguminosae): a bee-pollinated plant in Mediterranean islands where bees are absent. Plant Syst. Evol. 241: 29–46.

Pérez-de-Luque, A., Lozano, M.D., Maldonado, A.M., Jorrín, J.V., Dita, M.A., Die, J., Román, B., and Rubiales, D. 2006. *Medicago truncatula* as a model for studying interactions between root parasitic plants and legumes. *In* The *Medicago truncatula* handbook (version 1, 2006). 31 pp. http://www.noble.org/MedicagoHandbook/ (current July 2010).

Perez-Garcia, F., Ceresuela, J.L., Gonzalez, A.E., and Aguinagalde, I. 1992. Flavonoids in seed coats of *Medicago arborea* and *Medicago strasseri* (Leguminosae): Ecophysiological aspects. J. Basic Microbiol. 32: 241–248.

Pérez-Mellado, V., Ortega, F., Martín-García, S., Perera, A., and Cortázar, G. 2000. Pollen load and transport by the insular lizard, *Podarcis lilfordi* (Squamata, Lacertidae) in coastal islets of Menorca (Balearic Islands, Spain). Israel J. Zool. 46: 193–2000.

Peterson, T.A., and Russelle, M.P. 1991. Alfalfa and the nitrogen cycle in the Corn Belt. J. Soil Water Conserv. 46: 229–235.

Phillips, D.A., and DeJong, T.M. 1984. Dinitrogen fixation in leguminous crop plants. *In* Nitrogen in crop production. *Edited by* R.D. Hauck. American Society of Agronomy, Madison, WI. pp. 121–132.

Phillips, E.P. 1917. A contribution to the flora of the Leribe Plateau and environs: with a discussion of the relationships of the floras of Basutoland, the Kalahari, and the south-eastern regions. Ann. S. Afr. Mus. 16(1): 1–377.

Phitos, D., Constandinidis, T., and Kamari, G. (Editors). 2009. The red data book of rare and threatened plants of Greece. Vol. 2, E–Z. Hellenic Botanical Society, Patras, Greece. (In Greek. The treatment of *Medicago* is on pages 168 to 183.)

Piano, E. 1987. Perennial and annual herbaceous legumes in the fodder and pastoral Mediterranean systems. FAO, Eur. Coop. Network on Past. and Fodder Crop Prod. Bull. No. 5 (Additional). pp. 5–34.

Piano, E., and Francis, C.M. 1992. The annual species of *Medicago* in the Mediterranean region. Ecogeography and related aspects of plant introduction and breeding. *In* Proceedings of the 10th International Conference of Eucarpia *Medicago* spp. Group. Istotuto Sperimentale per la Colture Foraggere, Lodi, Italy. pp. 373–385.

Piano, E., Sardara, M., and Pusceddu, S. 1982. Observations on the distribution and ecology of subterranean clover and other annual legumes in Sardinia. Rivista di Agonomia, 18: 273–283.

Piano, E., Pusceddu, S., and Spanu, F. 1991. Observations on the distribution and ecology of annual medics and subterranean clovers in Sicily (Italy). I[e] Congrès International des Terres de Parcours, Montpellier, France, 1991. pp. 405–309.

Pías, B., and Guitián, P. 2001. Flowering phenology and pollen-to-ovule ratio in coastal dune communities near Eurosiberian-Mediterranean border in the NW Iberian Peninsula. Flora (Jena), 196: 475–482.

Piccirilli, M., and Arcioni, S. 1992*a*. Production of interspecific hybrid plants in the genus *Medicago* through embryo rescue. *In* Ploidy and chromosome manipulation in forage breeding. Proceedings of the 17th meeting of the Fodder Crops Section of EUCARPIA, Alghero, Italy, 14–18 October 1991. *Edited by* F. Veronesi, S. Bullitta, and S. Caredda. pp. 174–176.

Piccirilli, M., and Arcioni, S. 1992*b*. New interspecific hybrids in the genus *Medicago* through in vitro culture of fertilized ovules. *In* Angiosperm pollen and ovules. *Edited by* E. Ottaviano, D.L. Mulcahy, M. Sari-Gorla, and G.B. Mulcahy. Springer-Verlag, New York, NY. pp. 325–330.

Pierce, R. 1983. Hairy, sticky alfalfa fights off insects. Agricultural Research (USDA, Washington, D.C.), 32(2): 6–7.

Pieters, A.J. 1927. Green manuring principles and practices. John Wiley & Sons, New York, NY. 356 pp.

Pinkerton, A., and Randall, P.J. 1993. A comparison of the potassium requirements during early growth of *Lotus pedunculatus*, *Medicago murex*, *M. polymorpha*, *M. truncatula*, *Ornithopus compressus*, *Trifolium balansae*, *T. resupinatum*, *Pennisetum clandestinum*, and *Phalaris aquatica*. Austral. J. Exper. Agric. 33: 31–39.

Pistoia, A., Balestri, G., Ferruzzi, G., and Secchiari, P. 2003. Productive trend and nutritional characteristics in *Medicago arborea* L. J. Agric. Environ. Internat. Develop. 97: 177–184.

Plowden, C.C. 1972. A manual of plant names. George Allen & Unwin Ltd., London, UK. 260 pp.

Podlech, D., and Dieterle, A. 1969. Chromosomenstudien an afghanischen Pflanzen. Candollea, 24: 185–243.

Podol'naia, L.P., and Efimova, E.V. 1986. Development of *Trigonella cretacea* (Bieb.) Taliev (family Fabaceae Lindl.) fruit. Vestn. Leningr. Univ. (Leningrad), 1986(1): 28–35. (In Russian, English summary.)

Pogan, E., Czapik, R., Jankun, A., and Kuta, E. 1982. Further studies in chromosome numbers of Polish angiosperms. Part XV. Acta Biologica Cracoviensia, Series Botanica, 24: 113–126.

Pogan, E., Czapik, R., and Jankun, A. 1985. Further studies in chromosome numbers of Polish angiosperms. Part XVIII. Acta Biologica Cracoviensia, Series Botanica, 27: 57–74.

Pólya, L. 1948. Chromosome numbers of certain alkali plants. Arch. Biol. Hungarica, 18: 145–146.

Ponert, J. 1973. Combinationes novae, stati novi et taxa nova non tantum specierum turcicarum. Feddes Repert. Z. Bot. Taxon. Geobot. 83: 617–644.

Poole, M.L. 1970. An agronomic evaluation of *Medicago tornata* cv. Tornafield, a new medic cultivar. Austral. J. Exp. Agric. Animal Husb. 10: 405–409.

Popenko, A.K., and Pershina, E.P. 1979. Use of various forms of lactic-acid bacteria in ensilage of plants from the semi arid zone of the Kazakh-SSR USSR. Izvestiya Akademii Nauk Kazakhskoi SSR Seriya Biologicheskaya, 6: 61–65. (In Russian.)

Porqueddu, C. 2001. Screening germplasm and varieties for forage quality: constraints and potentials in annual medics. Options Méditerranéens Serie A, Seminaires Méditerranéens, 45: 89–98.

Porqueddu, C., Loi, A., and Fara, G. 1998. Anglona: the first Italian variety of annual medic. *In* IV Congresso Nazional Bidiversità: Germoplasma locale e sua valorizzazione. Alghero, Italy, 8–11 Sept. 1998. pp. 891–894. (In Italian.)

Porqueddu, C., Nieddu, S., and Fiori, P.P. 2000. Use of subterranean clover and burr medic as cover crops in vineyards. *In* Legumes for Mediterranean forage crops, pastures and alternative uses. *Edited by* L. Sulas. CIHEAM-IAMZ, Zaragoza, Italy. pp. 365–368.

Post, G.E. 1888. Diagnoses plantarum novarum orientalium. J. Linn. Soc., Bot. 24: 419–441.

Price, K.R., Johnson, I.T., and Fenwick, G.R. 1987. The chemistry and biological significance of saponins in foods and feedingstuffs. CRC Crit. Rev. Food Sci. Nutr. 26: 27–135.

Prosperi, J., Ronfort, J., and Genier, G 1999. Constraints to the introduction of medics in French Mediterranean farming systems. *In* Proceedings of the 13th Eucarpia *Medicago* spp. Group Meeting, Perugia, Italy, 13–16 September 1999. pp. 154–161.

Prosperi, J.M., Gensollen, V., Olivieri, I., and Mansat, P. 1989. Observations sur la répartition et l'écologie de luzernes annuelles et de trèfle souterrain en Corse. *In* Proceedings of the XVI International Grassland Congress, 4–11 October 1989, Nice, France. Association Française pour la Production Fourragere, Centre National de Recherche Agronomique, Versailles, France. pp. 295–296.

Prosperi, J.M., Angevain, M., Bonnin, I., Chaulet, E., Génier, G., Jenczewski, E., Olivieri, I., and Ronfort, J. 1996. Genetic diversity, preservation and use of genetic resources of Mediterranean legumes: alfalfa and medics [part of workshop: The genus *Medicago* in the Mediterranean region: current situation and prospects in research; 1995 October, Hammamet, Tunisie]. Options Méditerranéennes, 18: 71–89.

Prosperi, J.-M., Auricht, G.C., Genier, G., and Johnson, R.C. 2001. Medics (*Medicago* L.). *In* Plant Genetic Resources of Legumes in the Mediterranean. *Edited by* N. Maxted and S. Benett. Kluwer, Dordrecht, the Netherlands. pp. 99–114.

Puckridge, D.W., and French, R.J. 1983. The annual legume pasture in cereal ley farming systems of southern Australia: a review. Agric. Ecosyst. Environ. 9: 229–267.

Pupilli, F., Arcioni, S., and Damiani, F. 1991. Protoplast fusion in the genus *Medicago* and isoenzyme analysis of parental and somatic hybrid cell lines. Plant Breeding, 106: 122–131.

Puri, H.S. 1979. Evolutionary tendencies in the fruit and seed morphology of *Trigonella* spp. *In* Recent Researches in Plant Sciences. *Edited by* S.S. Bir. Kalyani Publishers, New Delhi, India. pp. 154–158.

Putnam, D. 1998. Contribution of alfalfa to wildlife and the environment. 28th National Alfalfa Symposium, Bowling, Green, KY, Feb. 26–27. http://www.calhay.org/paper.html. (current September 2008).

Putnam, D., Russelle, M., Orloff, S., Kuhn, J., Fitzhugh, L., Godrey, L., Kiess, A., and Long, R. undated. Alfalfa, wildlife and the environment. The importance and benefits of alfalfa in the 21st century. The California Alfalfa and Forage Association, Novato, CA. 24 pp. http://www.calhay.org/pdf/BrochureFINAL.pdf. (current September 2008).

Qi, R.-H. 1997. A comparative study on three species of *Trigonella* L. Grassland of China [Zhongguo caoyuan], 4: 52–54. (In Chinese.)

Quarro, M., and Montard, F.X. de. 1992. Étude de la productivité des parcours de la zone d'Ain-Leuh (Moyen Atlas, plateau Central). II. Modélisation de la production d'herbe: interactions climats × sol × recouvrement arboré. Agronomie, 12: 489–501.

Queirós, M., and Ormonde, J. 1984. Contribução para o conhecimento citotaxonómico da florados Açores – I. Boletim da Sociedade Broteriana, 57: 77–85.

Quinlivan, B.J., McComb, J.A., and Devitt, A.C. 1974. Annual medics in Western Australia. West. Austral. Dept. Agric. Bull. 3874. 20 pp.

Quiros, C.F., and Bauchan, G.R. 1988. The genus *Medicago* and the origin of the *Medicago sativa* complex. *In* Alfalfa and alfalfa improvement. *Edited by* A.A. Hanson, D.K. Barnes and R.R. Hill, Jr. American Society of Agronomy, Crop Science Society of America, Soil Science Society of America, Madison, WI. pp. 93–124.

Quiros, C.F., and Ostafichuk, L. 1983. Allozymes and genetic variability in *Medicago turbinata*, *Medicago truncatula*, and their hybrids. Can. J. Genet. Cytol. 25: 286–291.

Rabeler, R.A. 1984. An unnecessary infraspecific combination in *Medicago* (Fabaceae). Taxon, 33: 321–322.

Radcliffe, J.E. 1983. Fodder trees: an option for dry hill country. Special Publication, Centre of Resources Management, Lincoln College, Canterbury, New Zealand, No. 26. pp. 49–57.

Radwan, M.S., Al Hasan, A.M., and Al Fakhry, A.M. 1977. Annual species of *Medicago* in northern Iraq. Egypt. J. Bot. 20: 141–144.

Radwan, M.S., Al Fakhry, A.K., and Al Hasan, A.M. 1978. Some observations on the performance of annual medics in northern Iraq. Mesopotamia J. Agric. 13: 55–67. (In Arabic.)

Radwan, M.S., Al Fakhry, A.K., and Al Hasan, A.M. 1980. Variation between and within species of annual *Medicago* from northern Iraq. Egypt. J. Agron. 5: 153–159.

Ramert, B. 1993. Mulching with grass and bark and intercropping with *Medicago littoralis* against carrot fly (*Psila rosae* (F.)). Biol. Agric. Hortic. 9: 125–135.

Ramos, M.E., Robles, A.B., Ruiz-Mirazo, J., Cardoso, J.A., González-Rebollar, J.L. 2006. Effect of gut passage on viability and seed germination of legumes adapted to semiarid environments. *In* Sustainable grassland productivity: proceedings of the 21st general meeting of the European Grassland Federation [= Grassland science in Europe, vol. 11], Badajoz, Spain, 3–6 April, 2006. *Edited by* J. Lloveras, A. Gonzalez-Rodriguez, O. Vazquez-Yanez, J. Pineiro, O. Santamaria, L. Olea, and M.J. Poblaciones. Sociedad Espanola para el Estudio de los Pastos, Madrid, Spain. pp. 315–317.

Ranger, C.M., and Hower, A.A. 2001*a*. Glandular morphology from a perennial alfalfa clone resistant to the potato leafhopper. Crop Sci. 41: 1427–1434.

Ranger, C., and Hower, A.A. 2001*b*. Role of the glandular trichomes in resistance of perennial alfalfa to the potato leafhopper (Homoptera: Cicadellidae). J. Econ. Entomol. 94: 950–957.

Ransom, K. 2006. Making lucerne pay: integrating crops and lucerne on mixed farms. Dept. of Primary Industries, Bendigo, Vic., Australia. 94 pp.

Raynal, G. 1977. Comparison in artificial contaminations of pathogenicities of *Colletotrichum* spp. isolated in France on lucerne. Annales Phytopathol. 9: 193–204. (In French.)

Reasat, M., Karapetyan, J., and Nasirzadeh, A. 2003. Karyotypic analysis of *Trigonella* genus of Fars Province. Iran. J. Rangelands Forests Plant Breed. Genet. Res. 11: 127–145, 162. (In Persian.)

Rechinger, K.H. 1984. *Trigonella*. *In* Flora Iranica, Papilionaceae Vol. 2. *Edited by* K.H. Rechinger. Akademische Druck-u. Verlagsanstalt, Graz, Austria. pp. 207–253.

Reed, C.A. 1959. Animal domestication in the prehistoric Near East. Science, 130: 1629–1639.

Reed, C.A. 1969. The pattern of animal domestication in the prehistoric Near East. *In* The domestication and exploitation of plants and animals. *Edited by* P.J. Ucko and G.W. Dimbleby. Aldine Publishing Company, Chicago, IL. pp. 361–380.

Reese, G. 1957. Über die Polyploidiespektren in der nordsaharischen Wüstenpflanzen. Flora, 144: 598–634.

Reeve, W.G., Bräu, L., Castelli, J., Garau, G., Sohlenkamp, C., Geiger, O., Dilworth, M.J., Glenn, A.R., Howieson, J.G., and Tiwari, R.P. 2006. The *Sinorhizobium medicae* WSM419 lpiA gene is transcriptionally activated by FsrR and required to enhance survival in lethal acid conditions. Microbiology, 152: 3049–3059.

Reeves, R.G. 1930. Nuclear and cytoplasmic division in the microsporogenesis of alfalfa. Am. J. Bot. 17: 29–41.

Reeves, T.G., and Ewing, M.A. 1993. Is ley farming in Mediterranean zones just a passing phase? *In* Grasslands for our world. *Edited by* M.J. Baker. SIR Publishing, Wellington, New Zealand. pp. 810–818.

Riasat, M. [cf. "Reasat," above], Nasirzadeh, A., and Heidari, M. 2005. Determination of the best methods of seed germination and growth index in some species of *Trigonella* in Fars province. Iran. J. Rangelands Forests Plant Breed. Genet. Res. 13(3): Pe247–Pe256. (In Persian.)

Ricciardelli-D'Albore, G.C. 1984. Observations on the pollinating insects of some Leguminosae (*Onobrychis vicifolia* Scop., *Lotus corniculatus* L., *Medicago arborea* L., *Medicago sativa* L.) in a specialized area. Redia (Italy), 67: 145–155. (In Italian.)

Richards, K.W. 1984. Alfalfa leafcutter bee management in Western Canada. Publication 1495/E. Agriculture Canada, Ottawa, Canada. 53 pp.

Riday H., and Brummer, E.C. 2004. Morphological variation of *Medicago sativa* subsp. *falcata* genotypes and their hybrid progeny. Euphytica, 138: 1–12.

Riday H., Brummer, E.C., Campbell, T.A., Luth, D., and Cazcarro P.M., 2003. Comparisons of genetic and morphological distance with heterosis between *Medicago sativa* subsp. *sativa* and subsp. *falcata*. Euphytica, 131: 37–45.

Ridland, P.M., and Berg, G.N. 1981. Seedling resistance to pea aphid (*Acyrthosiphon pisum*) of lucerne, annual medic and clover species in Victoria Australia. Austral. J. Exp. Agric. 21: 506–511.

Riefner, R.E., Jr., and Small, E. 2009 *Medicago rigidula* and *M. truncatula* (Fabaceae): new to the California flora. J. Bot. Res. Inst. Texas, 3: 325–331.

Rios, S., Robledo, A., and Correal, E. 1993. *Medicago suffruticosa*, a new perennial pasture legume for Mediterranean dry alkaline areas. *In* Management of Mediterranean shrublands and related forage resources (7th meeting FAO-European Sub-Network on Mediterranean Pastures and Fodder Crops). REUR Technical Series (FAO) no. 28. Food and Agricultural Institute, Rome, Italy. pp. 12–15.

Ritter-Studnička, H. 1970. Die vegetation der serpentinvorkommen in Bosnien. Vegetatio, 21: 75–156.

Rizk, T.Y., Al-Hasan, A.M., El-Tekriti, R.A., and Alawi, B.J. 1978. Effect of salinity on germination and seedling vigor of some annual medics, *Medicago* spp. Mesopotamia J. Agric. 13: 105–122.

Robertson, M. 2006. Lucerne prospects: drivers for widespread adoption of lucerne for profit and salinity management. CRC for Plant-Based Management of Dryland Salinity, Perth, W.A., Australia. 64 pp.

Robertson, S.M., and Smith, J.G. 2006. Do some naturalised legumes have the potential to improve pasture production in the Victorian Mallee? Austral. J. Exper. Agric. 46: 615–625.

Robledo, A., Rios, S., and Correal, E. 1991. Weeds in cereal fallow systems of Northwestern Murcia: their importance as a feed resource for sheep, 1: botanical composition. Proceedings of 1991 meeting of Weed Control Spanish Society, Cordoba, December 11–12, 1991. Sociedad Espanola de Malherbologia, Madrid, Spain. pp. 70–73. (In Spanish.)

Robledo, A., Rios, S., Correal, E. 1993. The *Medicago arborea* group in the Mediterranean Basin, 1: Origin, distribution and morphology. Revista-Pastos, 23(2): 55–67. (In Spanish.)

Rodríguez-Conde, M.F., Moreno, M.T., Cubero, J.I., and Rubiales, D. 2004. Characterization of the *Orobanche-Medicago truncatula* association for studying early stages of the parasite-host interaction. Weed Res. 44: 218–223.

Rogers, M.E., Craig, A.D., Munns, R.E., and 14 others. 2005. The potential for developing fodder plants for the salt-affected areas of southern and eastern Australia: an overview. Austral. J. Exp. Agric. 45: 301–329.

Roggero, P.P., and Porqueddu, C. 1999. The agronomic purpose of evaluation: relating nurseries to field situations. *In* Genetic resources of Mediterranean pasture and forage legumes. *Edited by* S.J. Bennett and P.S. Cocks. Kluwer Academic Publishers, Boston, MA. pp. 107–119.

Rome, S., Fernandez, M.P., Brunel, B., Normand, P., and Cleyet-Marel, J.-C. 1996. *Sinorhizobium medicae* sp. nov., isolated from annual *Medicago* spp. Internat. J. Syst. Bacteriol. 46: 972–980.

Romic, S. 1978. The dietary and protective value of alfalfa in feeding horses. Poljoprivredna Znanstvena Smotra, 45(55): 5–16. (In Russian.)

Ronfort, J., Bataillon, T., Santoni, S., Delalande, M., David, J.L., and Prosperi, J.M. 2006. Microsatellite diversity and broad scale geographic structure in a model legume: building a set of nested core collection for studying naturally occurring variation in *Medicago truncatula*. BMC Plant Biology 6(28). http://www.biomedcentral.com/content/pdf/1471-2229-6-28.pdf. (Web accessible only; current July 2010).

Rose, R.J., Schlarbaum, S.E., Small, E., and Johnson, L.B. 1988. Chloroplast genomic variation and phylogeny in *Medicago* section *Intertextae*. Can. J. Bot. 66: 1352–1358.

Rosenthal, J.P., and Kotanen, P.M. 1994. Terrestrial plant tolerance to herbivory. TREE, 9: 145–148.

Rossiter, R.C. 1966. Ecology of the Mediterranean annual-type pasture. Advances Agronomy, 18: 1–56.

Rostovtseva, T.S. 1977. Chromosome numbers of some plant species from the south of Siberia. II. Bot. Zurn. SSSR, 62: 1034–1042. (In Russian.)

Royle, J.F. 1839. Illustrations of the botany and other branches of the natural history of the Himalayan Mountains, Vol. II, Plates. W.H. Allen and Co., London, UK.

Rudyka, E.G. 1986. Chromosome numbers in some representatives of the Alliaceae, Fabaceae, Malvaceae, Poaceae families. Bot. Zurn. SSSR, 1426–1427. (In Russian.)

Ruíz de Clavijo, E. 1993. Números cromosomáticos para la flora Española 664-690. Lagascalia, 17: 161–172.

Ruiz, T., and Devesa, J.A. 1998. Seed germination in wild medics (*Medicago*, Leguminosae) from southwestern Europe (Spain). Acta Botanica Gallica, 145: 81–91.

Rumbaugh, M.D., and Johnson, D.A. 1986. Annual medics and related species as reseeding legumes for northern Utah pastures. J. Range Manage. 39: 52–58.

Rumbaugh, M.D., and Pendery, B.M. 1990. Germination salt resistance of alfalfa (*Medicago sativa* L.) germplasm in relation to subspecies and centers of diversity. Plant and Soil, 124: 47–51.

Russelle, M.P. 2001. Alfalfa. Amer. Sci. 89: 252–261.

Russi, L., Cocks, P.S., and Roberts, E.H. 1992. Coat thickness and hardseededness in some *Medicago* and *Trifolium* species. Seed Sci. Res. 2: 243–249.

Ryakhova, D.K., and Kucherov, E.V. 1975. Wild-growing species of grasses and legumes promising for introduction into cultivation. Povyshenie Produktivnosti Senokosov I Pastbishch. 1975: 35–40. (In Russian.)

Ryder, O.A. 1988. Przewalski's horse—putting the wild horse back in the wild. Oryx, 22: 154–157.

Saadani, Y., Kayouli, C., and Narjisse, H. 1989. Valeur nutritive d'un parcours mixte à *Acacia cyanophylla*, *Atriplex nummularia*, et *Medicago arborea*. Proc. 16 Internat. Grassland Congress, Nice, France, Asssoc. Française pour la Production Fourragère. pp. 943–944.

Sabato, S. 1990. Remarks on the publication dates of Tenore's Flora Napolitana. Taxon, 39: 409–416.

Sagatelyan, A.A. 1980. New and rare species for the flora of the Megri region of the Armenian SSR USSR. Bot. Zhurn. 65: 1230–1234. (In Russian.)

Şakiroğlu, M., and Brummer, E.C. 2007. Little heterosis between alfalfa populations derived from the Midwestern and Southwestern United States. Crop Sci. 47: 2364–2371.

Saleh, N.A.M., Boulos, L., El-Negoumy, S.I., and Abdalla, M.F. 1982. A comparative study of the flavonoids of *Medicago radiata* with other *Medicago* and related *Trigonella* species. Biochem. Syst. Ecol. 10: 33–36.

Salensky, W. 1907. Prjevalsky's horse. (Translation by M.H. Hayes and O.C. Bradley). Hurst and Blackett Limited, London, UK. 67 pp.

Sales, F., and Hedge, I.C. 1994. *Medicago* L. (Leguminosae): typification and nomenclature. Anal. Jard. Bot. Madrid, 51: 321–323.

Samac, D., Jung, H.-J., and Lamb, J. 2006. Development of alfalfa (*Medicago sativa* L.) as a feedstock for production of ethanol and other bioproducts. *In* Alcoholic fuels. *Edited by* S. Minteer. CRC Press, Boca Raton, Fl. pp. 79–98.

Samuel Roberts Noble Foundation. 1955. Button clover. Agricultural Division, Samuel Roberts Noble Foundation, Ardmore, Oklahoma. 2 pp. (Typescript.)

Samuel Roberts Noble Foundation. 1959. Button clover. Agricultural Division, Samuel Roberts Noble Foundation, Ardmore, Oklahoma. 9 pp.

Sanadgol, A.A., Chaichi, M., and Kalagari, A.B. 2006. Comparison of forage yields of 5 annual medics in Gorgan region. Iran. J. Range Desert Res. 13: 62–68. (In Persian.)

Sangduen, N., Sorensen, E.L., and Liang, G.H.O. 1982. A perennial × annual *Medicago* cross *Medicago sativa*, *Medicago scutellata*. Can. J. Genet. Cytol. 24: 361–365.

Saniewska, A., Jarecka, A., Bialy, Z., and Jurzysta. 2005. Antifungal activity of saponins from *Medicago arabica* L. shoots against some pathogens. Allelopathy, J. 16: 105–111.

Sareen, T.S., and Singh, P.D. 1976. Cytological studies in some north Indian Leguminosae. Proc. Indian Sci. Congr. Assoc. 63: 122–123.

Sareen, T.S., and Trehan, R. 1976. Report in IOPB chromosome number reports LIV. Taxon, 25: 631–649.

Sareen, T.S., and Trehan, R. 1977. Cytological studies in some Papilionaceae. Proc. Indian Sci. Congr. Assoc. 64: 156.

Sattaur, O. 1991. Rare horses ready for return to Mongolian home. New Scientist, 129(1751): 26.

Saunders, R.J., Howie, J.H., and Auricht, G.C. 1993. *Medicago orbicularis* (L.) Bart – a potential new medic for the cereal belt of southern Australia. Proc. 6th Australian Society of Agronomy conference, Armidale 1992. p. 614.

Savova, D., Al-Mazyad, P.R., and Felber, F. 1996. Cytogeography of *Medicago falcata* L. and *M. sativa* L. in Switzerland. Botanica Helvetica, 106: 197–207.

Sawada, H., Kuykendall, D., and Young, J.M. 2003. Changing concepts in the systematics of bacterial nitrogen-fixing legume symbionts. J. Gen. Appl. Microbiol. 49: 155–179.

Schlarbaum, S.E. 1988. Characterization of somatic chromosome morphology in alfalfa, *Medicago sativa* L.: comparison of donor plant with regenerated protoclone. Cytologia, 53: 499–507.

Schlarbaum, S.E., Small. E., and Johnson, L.B. 1984. Karyotypic evolution, morphological variability and phylogeny in *Medicago* sect. *Intertextae*. Pl. Syst. Evol. 145: 203–222.

Schlarbaum, S.E., Rose, R.J., Small, E., and Johnson, L.B. 1989. Karyotypic and chloroplast genomic diversity in *Medicago* section *Lupularia* (Leguminosae). Pl. Syst. Evol. 166: 69–78.

Schmiedeknecht, M., and Lesins, K. 1968. Weitere Untersuchungen zur Resistenz von *Medicago*-Arten gegen *Pseudopeziza medicaginis* (Lib.) Sacc. Theoret. Appl. Genet. 38: 188–194.

Schröck, O. 1943. Beobachtungen an einem Bastard zwischen Luzerne und Gelbklee. Züchter, 15: 4–10.

Schulz, O.E. 1901. Monographie der gattung *Melilotus*. Bot. Jahrb. Syst. Pflanzgesch. Pflanzengeog. 29: 660–735.

Schwarz, O. von, and Klinkowsi, M. 1933. *Medicago falcata* L. ssp. *Urumovii* Degen em. Schw. et Klink., a subspecies of the sickle alfalfa which has remained unnoticed in Germany. Verh. Bot. Ver. Prov. Brandenb. 74(2): 180–184.

Scofield, C.S. 1908. The botanical history and classification of alfalfa. U.S. Bureau Plant Industr. Bull. No. 131(2): 11–19.

Scrugli, A., and Bocchieri, E. 1977. Numeri cromosomici per la flora Italiana: 348–357. Inform. Bot. Ital. 9: 127–133.

Semerenko, L.V. 1989. Chromosome numbers in some members of the families Asteraceae, Fabaceae, Orchidaceae and Poaceae from the Berezinsky Biosphere Reservation (Byelorussia) flora. Bot. Žurn. 745: 1671–1673.

Semerenko, L.V., and Shvets, I.V. 1989. Karyologicheskoe isuchenie bobovykh (Fabaceae) Belorussii. *In* Tesizy II Symp. Plant Karyology. Novosibirsk. pp. 66–68.

Senhadri Maamri, F., Rahal Bouziane, H., Rahmani, D.H., and Abdelguerfi, A. 2000. Étude biométrique des gousses et des graines de populations de deux générations de *Medicago truncatula* Gaert. *In* Legumes for Mediterranean forage crops, pastures and alternative uses. *Edited by* L. Sulas. CIHEAM-IAMZ, Zaragoza, Italy. pp. 249–253.

Senn, H.A. 1938. Chromosome number relationships in the Leguminosae. Bibliogr. Genetica, 12: 175–345.

Seringe, N.C. 1825. *Medicago*, *Trigonella*, *Pocockia*. *In* Prodromus systematis naturalis regni vegetabilis. Vol. 2. *Edited by* A.P. de Candolle. Treuttel and Würtz, Paris, France. pp. 171–185.

Serra Laliga, L., Perez Botella, J., and Izquierdo, J. 2001. *Medicago citrina* (Font Quer) Greuter (Leguminosae), on the Iberian Peninsula. Anal. Jard. Bot. Madrid, 59(1): 158–159. (In Spanish.)

Sessitsch, A., Howieson, J.G., Perret, X., Antoun, H., and Martínez-Romero, E. 2002. Advances in *Rhizobium* research. Crit. Rev. Plant Sci. 21: 323–378.

Shade, R.E., and Kitch, L.W. 1983. Pea aphid (Homoptera Aphididae) biology on glandular-haired *Medicago* species. Environ. Entomol. 12: 237–240.

Shade, R.E., Thompson, T.E., and Campbell, W.R. 1975. An alfalfa weevil larval resistance mechanism detected in *Medicago*. J. Econ. Entomol. 68: 399–404.

Shade, R.E., Doskocil, M.J., and Maxon, N.P. 1979. Potato leafhopper resistance in glandular-haired alfalfa species. Crop Sci. 19: 287–289.

Shah, G.L., and Kothari, M.J. 1975. Observations on stomata and hairs on vegetative and floral organs in the tribe Trifolieae (family Papilionaceae). Austral. J. Bot. 23: 111–122.

Shaheen, A, and Mohammad, N. 1988. Effect of fertilization and inoculation on nitrogen fixation and fodder yield of snail medic (*Medicago scutellata*). Biologia (Pakistan), 34: 133–140.

Shao, X.W.K., Dong, S.-K., Huang, X.-X., and Kang, M.-Y. 2006. Regionalisation of suitable herbages for grassland reconstruction in agro-pastoral transition zone of northern China. New Zealand J. Agric. Res. 49: 73–84.

Sheaffer, C.C., Tanner, C.B., and Kirkham, M.B. 1988. Alfalfa water relations and irrigation. *In* Alfalfa and alfalfa Improvement. *Edited by* A.A. Hanson, D.K. Barnes, and R.R. Hill, Jr. Agronomy Monograph No. 29, ASA/CSSA/SSSA, Madison, WI. pp. 373–409.

Sheaffer, C.C., Gunsolus, J.L., Jewett, J.G,, and Lee, S.H. 1991. Annual *Medicago* as a smother crop in soybean. J. Agron. Crop Sci. 188: 408–416.

Sherratt, A. 1983. The secondary exploitation of animals in the Old World. World Archaeology, 15: 90–104.

Shi, Z.Y., Zhang, L.Y., Li, X.L., Feng, G., Tian, C.Y., and Christie, P. 2007. Diversity of arbuscular mycorrhizal fungi associated with desert ephemerals in plant communities of Junggar Basin, northwest China. Applied Soil Ecol. 35: 10–20.

Shinners, R.H. 1956. Authorship and nomenclature of bur clovers (*Medicago*) found wild in the United States. Rhodora, 58: 1–13.

Short, R.V., Chandley, A.C., Jones, R.C., and Allen, W.R. 1974. Meiosis in interspecific equine hybrids. II. The Przewalski's horse/domestic horse hybrid. Cytogenet. Cell Genet. 13: 465–478.

Shrestha, A., Hesterman, O.B., Squire, J.M., Fisk, J.W. and Sheaffer, C.G. 1998. Annual medics and berseem clover as emergency forages. Agron. J. 90: 197–201.

Sibole, J., Montero, E., Bennàssar, A., Cabot, C., Barceló, J., and Poschenrieder, C. 1994. Bases fisiológicas, bioquímicas y moleculares de la nutrición mineral de las plantas. V Simposium Nacional y I Ibérico de Nutrición Mineral de las Plantas. pp. 275–283.

Sidhu, M.K. 1979. Distributional and cytological studies of the weed flora of cultivable fields of Patiala district (Panjab). Ph.D. Thesis, Patiala, India. 230 pp.

Sidhu, M., and Bir, S.S. 1983. Karyological studies on weeds on cultivable lands in Punjab, India. Trop. Plant Sci. Res. 1: 1–13.

Sidhu, S.S., and Cavers, P.B. 1977. Maturity dormancy relationships in attached and detached seeds of *Medicago lupulina* L. (black medick). Bot Gaz. 138: 174–182.

Silici, S., and Gökceoglu, M. 2007. Pollen analysis of honeys from Mediterranean region of Anatolia. Grana, 46: 57–64.

Simon, J.P. 1965. Relationship in annual species of *Medicago*. II. Interspecific crosses between *M. tornata* (L.) Mill. and *M. littoralis* Rhode. Austral. J. Agr. Res. 16: 51–60.

Simon, J.P. 1979. Serological studies in *Medicago, Melilotus*, and *Trigonella* (Leguminosae). II. Relationships of critical taxa of sections *Medicago* and *Lupularia* of the genus *Medicago*. Bot. Gaz. 140: 452–460.

Simon, J.P., and Millington, A.J. 1967. Relationship in annual species of *Medicago*. III. The complex *M. littoralis* Rhode – *M. truncatula* Gaertn. Austral. J. Bot. 15: 35–73.

Simon, J.P., and Simon, A. 1965. Relationship in annual species of *Medicago*. I. Number and morphology of chromosomes. Austral. J. Agric. Res. 16: 37–50.

Simpson, G.G. 1936. Horses and history. Nat. Hist. 38: 277–288.

Simpson, G.G. 1951. Horses. Oxford University Press, New York, NY. 247 pp.

Sims, J.R. 1998. Snail medic for green manure, saline seep and wind erosion and grazing black medic with cattle in a medic pasture small grain rotation. *In* Proceedings 36th North American Alfalfa Improvement Conference, Bozeman, Montana 2–6 August 1998. p. 17.

Sims, J.R., Koala, S., Ditterline, R.L., and Wiesner, L.E. 1985. Registration of 'George' black medic. Crop Sci. 25: 709–710.

Sims, J.R., Solum, D.J., Westcott, M.P., Jackson, G.D., Kushnak, G.D., Wichman, D.M., et al. 1991. Yield and bloat hazard of berseem clover and other forage legumes in Montana. Mont. AgRes. 8: 4–10.

Sinclair, T.R. 1998. Historical changes in harvest index and crop nitrogen accumulation. Crop Sci. 38: 638–643.

Singh, A., and Roy, R.P. 1970. Karyological studies in *Trigonella*, *Indigofera* and *Phaseolus*. Nucleus (Calcutta), 13: 41–54.

Singh, A., and Singh, D. 1976. Karyotype studies in *Trigonella*. Nucleus (Calcutta), 19: 13–16.

Singh, B.P. 1989. Invasion of plants in Ganganagar district of Rajasthan India. J. Econ. Taxon. Bot. 13: 281–283.

Singh, S.M., and Lesins, K. 1972. Relationship of taxa in the genus *Medicago* as revealed by hybridization. VI. *M. laciniata* × *M. sauvagei*. Can. J. Genet. Cytol. 14: 823–828.

Sinskaya, E.N. 1948. Wild alfalfas of the Caucasus and the prospects of utilizing them in selection and seed production. Bull. Appl. Bot. Plant Breed. 28((1): 11–47. (In Russian.)

Sinskaya, E.N. 1950. Medic–*Medicago* L. em. Brief review of the genus. *In* Flora of cultivated plants of the U.S.S.R. Vol. 13(1). *Edited by* E.N. Sinskaya. pp. 7–195. (Translated from Russian in 1961 by the Israel Program for Scientific Translations, Jerusalem.)

Širjaev [= Sirjaev], G. 1928. A new *Medicago* from China. Kew Bull. Misc. Inform. 7: 270.

Širjaev, G. 1928–1934. Generis *Trigonella* L. revisio critica. Spisy Přir. Fac. Masarykovy Univ. 102, 110, 128, 136, 148, 170, 192.

Sitzia, M., and Fois, N. 1999. Seeds and pods of *Medicago polymorpha* L. as summer feed resource for sheep in Mediterranean conditions. Rivista di Agronomia (Italy), 33: 185–188. (In Italian.)

Sitzia, M., Ligios, S., and Fois, N. 2000. *Medicago polymorpha* L. forage production and its quality when grazed by ewes. *In* Legumes for Mediterranean forage crops, pastures and alternative uses. *Edited by* L. Sulas. CIHEAM-IAMZ, Zaragoza, Italy. pp. 191–194.

Skalińska, M., Pogan, E., et al. 1966. Further studies in chromosome numbers of Polish angiosperms. Acta. Biol. Cracov. Ser. Bot. 9: 31–58.

Skinner, D.Z. 1992. A PCR-based method of identifying species-specific repeated DNAs. Biotechniques, 13: 210–214.

Skinner, D.Z., Bauchan, G.R., Auricht, G., and Hughes, S. 1999. Developing a core collection from a large annual *Medicago* germplasm collection. *In* Core collections for today and tomorrow. *Edited by* R.C. Johnson and T. Hodgkin. International Plant Genetic Resources Italy, Rome, Italy. pp. 61–67. http://www2.bioversityinternational.org/publications/Web_version/43/ (current Sept. 2010.)

Skovsted, A. 1939. Cytological studies in twin plants. Compt. Rend. Trav. Lab. Carlsberg, Sér. Physiol. 22: 427–446.

Slavik, B., Jarolimova, V., and Chrtek, J. 1993. Chromosome counts of some plants from Cyprus. Candollea, 48: 221–230.

Sledge, M.K., Pechter, P., and Payton, M.E. 2005. Aluminum tolerance in *Medicago truncatula* germplasm. Crop Sci. 45: 2001–2004.

Small, E. 1972. Adaptation in *Clarkia*, section *Myxocarpa*. Ecology, 53: 808–818.

Small, E. 1973. Xeromorphy in plants as a possible basis for migration between arid and nutritionally-deficient environments. Botaniska Notiser, 126: 534–539.

Small, E. 1981*a*. A numerical analysis of major groupings in *Medicago* employing traditionally used characters. Can. J. Bot. 59: 1553–1557.

Small, E. 1981*b*. An alfalfa germ plasm expedition in Turkey. Forage Notes Res. Branch Agric. Canada, 25(2): 56–66.

Small, E. 1982. *Medicago* collecting in Turkey. Pl. Genet. Resources Newslett. 49(March): 11–12.

Small, E. 1983. Pollen ploidy-prediction in the *Medicago sativa* complex. Pollen & Spores, 25: 305–320.

Small, E. 1984*a*. A clarification of the *Medicago blancheana-bonarotiana-rotata* complex. Can. J. Bot. 62: 1693–1696.

Small, E. 1984*b*. Hybridization in the domesticated-weed-wild complex. *In* Plant biosystematics. *Edited by* W.F. Grant. Academic Press, Toronto, Canada. pp. 195–210.

Small, E. 1985*a*. Morphological differentiation in *Medicago sativa* s.l. in relation to ploidy. Can. J. Bot. 63: 1747–1752.

Small, E. 1985*b*. Insect pests and the evolution of defensive glandular trichomes in alfalfa. Can. J. Pl. Sci. 65: 589–596.

Small, E. 1985*c*. Taxonomic discoveries of interest to the alfalfa agronomist. Forage Notes Res. Branch Dep. Agic. Canada, 28(for 1984): 24–29.

Small, E. 1986*a*. Taxonomy of glandular wild alfalfa (*Medicago sativa* L.). Can. J. Bot. 64: 2125–2129.

Small, E. 1986*b*. A clarification of the *Medicago papillosa* complex. Can. J. Bot. 64: 2800–2806.

Small, E. 1987*a*. A taxonomic study of the "medicagoid" *Trigonella*. Can. J. Bot. 65: 1199–1211.

Small, E. 1987*b*. Reduction of *Ursia* to *Trifolium*. Taxon, 36: 578–583.

Small, E. 1987*c*. Generic changes in Trifolieae subtribe Trigonellinae. *In* Advances in legume systematics, part 3. *Edited by* C.H. Stirton. Royal Botanic Gardens, Kew, UK. pp. 159–181.

Small, E. 1988. Pollen-ovule patterns in tribe Trifolieae (Leguminosae). Pl. Syst. Evol. 160: 195–205.

Small, E. 1989*a*. Polythetic generic separation in tribe Trifolieae subtribe Trigonellinae (Leguminosae). Can. J. Bot. 67: 1480–1492.

Small, E. 1989*b*. The evolution of genera in the Leguminosae. *In* Advances in legume biology. *Edited by* C.H. Stirton and J.L. Zarucchi. Monogr. Syst. Bot. Missouri Bot. Gard. 29: 467–486.

Small, E. 1990*a*. *Medicago syriaca*, a new species. Can. J. Bot. 68: 1473–1478.

Small, E. 1990*b*. *Medicago rigiduloides*, a new species segregated from *M. rigidula*. Can. J. Bot. 68: 2614–2617.

Small, E. 1992. Anik alfalfa and the horse. *In* Alfalfa seed production in the Peace River region, update 1992. *Edited by* D. Fairey. Joint Publ. 92-03 Alberta Alfalfa Seed Producers' Association Peace Branch and Fairview College, Alberta. pp. 1–3.

Small, E. 1994. Additional clarification of the *Medicago blancheana-bonarotiana-rotata* complex. Can. J. Bot. 72: 829–831.

Small, E. 1996. 1995 George Lawson Medal Review: Adaptations to herbivory in alfalfa (*Medicago sativa*). Can. J. Bot. 74: 807–822.

Small, E. 1997. *Medicago* (alfalfa). *In* Munro, D.B., and Small, E. 1997. Vegetables of Canada. National Research Council Press, Ottawa, Canada. pp. 270–278.

Small, E. 2003. Distribution of perennial *Medicago* with particular reference to agronomic potential for the semiarid Mediterranean climate. *In* New perennial legumes for sustainable agriculture. *Edited by* S.J. Bennett. University of Western Australia Press, Crawley, Australia. pp. 57–80.

Small, E., and Bauchan, G.R. 1984. Chromosome numbers of the *Medicago sativa* complex in Turkey. Can. J. Bot. 62: 749–752.

Small, E., and Brookes, B.S. 1982. Coiling of alfalfa pods in relation to resistance against seed chalcids. Can. J. Pl. Sci. 62: 131–135.

Small, E., and Brookes, B. 1983. The systematic value of stigma morphology in the legume tribe Trifolieae with particular reference to *Medicago*. Can. J. Bot. 61: 2388–2404.

Small, E., and Brookes, B. 1984*a*. Coiling of alfalfa pods in relation to resistance against seed chalcids: additional observations. Can. J. Pl. Sci. 64: 659–665.

Small, E., and Brookes, B. 1984*b*. Taxonomic circumscription and identification in the *Medicago sativa-falcata* (alfalfa) continuum. Econ. Bot. 38: 83–96.

Small, E., and Brookes, B. 1984*c*. Reduction of the geocarpic *Factorovskya* to *Medicago*. Taxon, 33: 622–635.

Small, E., and Brookes, B. 1985. *Medicago lesinsii*, a new Mediterranean species. Can. J. Bot. 63: 728–734.

Small, E., and Brookes, B. 1986. Glandular trichomes on cotyledonary petioles of Leguminosae tribe Trifolieae. Can. J. Pl. Sci. 66: 1019–1023.

Small, E., and Brookes, B. 1990*a*. A numerical taxonomic analysis of the *Medicago littoralis – M. truncatula* complex. Can. J. Bot. 68: 1667–1674.

Small, E., and Brookes, B. 1990*b*. A taxonomic simplification of *Medicago italica*. Can. J. Bot. 68: 2103–2111.

Small, E., and Brookes, B. 1991. A clarification of *Medicago sinskiae* Uljan. Can. J. Bot. 69: 100–106.

Small, E., and Catling, P.M. 2007. Blossoming treasures of biodiversity: 24. Alfalfa—making agriculture friendly. Biodiversity, 8(2): 15–24.

Small, E., and Fawzy, M. 1991. A clarification of the *Medicago polyceratia – Medicago orthoceras* complex. Can. J. Bot. 69: 1907–1912.

Small, E., and Fawzy, M. 1992. Morphogeographic variation in the *Medicago monantha* complex. Can. J. Bot. 69: 1907–1912.

Small, E., and Jomphe, M. 1989*a*. A note on *Trigonella (Melilotus) bicolor*. Can. J. Bot. 67: 1604–1606.

Small, E., and Jomphe, M. 1989*b*. A synopsis of the genus *Medicago* (Leguminosae). Can. J. Bot. 67: 3260–3294.

Small, E., and Lefkovitch, L.P. 1982. Agrochemotaxometry of alfalfa. Can. J. Pl. Sci. 62: 919–928.

Small, E., and Lefkovitch, L.P. 1986. Relationships among morphology, geography, and interfertility in *Medicago*. Can. J. Bot. 64: 45–52.

Small, E., Bassett, I.J., and Crompton, C.W. 1981*a*. Pollen variation in tribe Trigonelleae (Leguminosae) with special reference to *Medicago*. Pollen & Spores, 23: 295–320.

Small, E., Crompton, C.W., and Brookes, B. 1981*b*. The taxonomic value of floral characters in tribe Trigonelleae (Leguminosae), with special reference to *Medicago*. Can. J. Bot. 59: 1578–1598.

Small, E., Lefkovitch, L.P., and Brookes, B. 1981*c*. Remarkable asymmetries in trifoliolate leaves with particular reference to *Medicago*. Can. J. Bot. 59: 662–671.

Small, E., Lefkovitch, L.P., and Classen, D. 1982. Character set incongruence in *Medicago*. Can. J. Bot. 60: 2505–2510.

Small, E., Grant, W.F., and Crompton, C.W. 1984. A taxonomic study of the *Lotus corniculatus* complex in Turkey. Can. J. Bot. 62: 1044–1053.

Small, E., Lassen, P., and Brookes, B. 1987. An expanded circumscription of *Medicago* (Leguminosae, Trifolieae) based on explosive flower tripping. Willdenowia, 16: 415–437.

Small, E., Jomphe, M., and Heyn, C.C. 1988. Reduction of *Medicago makranika* (Leguminosae tribe Trifolieae) to *Indigofera nephrocarpa* (tribe Indigoferae). Taxon, 37: 964–966.

Small, E., Lefkovitch, L.P., and Brookes, B. 1989. Preferential infestation of fruits within the *Medicago truncatula* – *M. littoralis* complex by the bean weevil *Bruchidius bimaculatus*. Pl. Syst. Evol. 163: 35–41.

Small, E., Brookes, B., and Lassen, P. 1990*a*. Circumscription of the genus *Medicago* (Leguminosae) by seed characters. Can. J. Bot. 68: 613–629.

Small, E., Brookes, B., and Crawford, E.J. 1990*b*. Intercontinental differentiation in *Medicago rigidula*. Can. J. Bot. 68: 2607–2613.

Small, E., Jurzysta, M., and Nozzolillo, C. 1990*c*. The evolution of hemolytic saponin content in wild and cultivated alfalfa. Econ. Bot. 44: 226–235.

Small, E., Jomphe, M., and Brookes, B. 1991. *Medicago truncatula* forma *laxicycla*, a new taxon with loose fruit coiling promoting rapid germination of the fruit-retained seeds. Pl. Syst. Evol. 174: 37–46.

Small, E., Warwick, S.I., and Brookes, B. 1992. Isozyme variation and alleged progenitor-derivative relationships in the *Medicago murex* complex. Pl. Syst. Evol. 181: 33–43.

Small, E., Bauchan, G.R., Salter, R., Brookes, B., and Auricht, G.C. 1993. A systematic comparison of morphology and seed proteins of early- and late-flowering forms of *Medicago scutellata*. Can. J. Bot. 71: 183–192.

Small, E., Brookes, B., and Jomphe, M. 1994. A bee that loves herbs. Herbarist, 61: 57–62.

Small, E., Brookes, B., Lefkovitch, L.P., and Fairey, D.T. 1997. A preliminary analysis of the floral preferences of the alfalfa leafcutting bee (*Megachile rotundata*). Can. Field-Nat. 111: 445–453.

Small, E., Warwick, S.I., and Brookes, B. 1999. Allozyme variation in relation to morphology and taxonomy in *Medicago* section *Spirocarpos* subsection *Intertextae* (Fabaceae). Pl. Syst. Evol. 214: 29–47.

Small, E., Marcus, D. McElroy, A. and Butler, G. 2007. Apparent increase in biomass and seed productivity in hemp (*Cannabis sativa*) resulting from branch proliferation caused by the European corn borer (*Ostrinia nubilalis*). J. Industrial Hemp, 12: 15–26.

Smith, D., and Nelson, C.J. 1967. Growth of birdsfoot trefoil and alfalfa. 1. Responses to height and frequency of cutting. Crop Sci. 7: 130–133.

Smith, M.A., and Baltensperger, A.A. 1984. Field performance of three annual medics. New Mexico Agric. Exp. Stat. Res. Report 525: 1004.

Smith, S.R., Bouton, G.H., and Hoveland, C.S. 1992. Persistence of alfalfa under continuous grazing in pure stands and in mixtures with tall fescue. Crop Sci. 32: 1259–1264.

Sobrino, E., Hervella, A., Ceresuela, J.L., Barbado, A., Viviani, A., De Andrés, F., and Tenorio, J.L. 2000. Morfología y taxonomía de la sección *Dendrotrelis* del género *Medicago* (Fabaceae). Portulagaliae Acta Biologica, 19: 225–237.

Soják, J., and Pouzar, Z. 1984. Proposal to conserve *Medicago* L. with *M. sativa* L. as its type (Fabaceae). Taxon, 33: 519–521.

Sokolovskaya, A.P., Probatova, N.S., and Rudyka, E.G. 1989. Chromosome numbers in some species of the flora of the Soviet far east from the families Actinidiaceae, Aristolochiaceae, Fabaceae, Ranunculaceae, Saxifragaceae. Bot. Žurn. 74: 268–271.

Song, J.S., Sorensen, E.L., and Liang, G.H. 1988. A new method to prepare root tip chromosomes in alfalfa. Cytologia, 53: 641–645.

Šopova, M., and Sekovski, Ž. 1989. Chromosome atlas of some Macedonian angiosperms. V. God. Zborn. Biol. Prir.-Mat. Fak. Univ. Kiril & Metodij, 39–40: 353–365.

Šopova, M., Sekovski, Ž., and Jovanovska, M. 1983. Chromosome atlas of some Macedonian angiosperms. IV. Ann. Fac. Biol. Univ. Skopje, 36: 73–86.

Sorensen, E.L., Sangduen, N., and Liang, G.H. 1981. Transfer of glandular hairs from diploid *Medicago prostrata* to tetraploid *M. sativa*. *In* Report of the twenty-seventh alfalfa improvement conference. July 8–10, 1980, University of Wisconsin, Madison, WI. Page 53.

Sorensen, E.L., Horber, E.K., and Stuteville, D.L. 1985. Registration of KS-108GH-5 glandular-haired alfalfa germplasm with multiple pest resistance. Crop Sci. 25: 1132.

Sorensen, E.L., Byers, R.A., and Horber, E.K. 1988. Breeding for insect resistance. *In* Alfalfa and alfalfa improvement. *Edited by* A.A. Hanson, D.K. Barnes, and R.R. Hill. American Society of Agronomy, Inc., Crop Science Society of America, Inc., Soil Science Society of America, Inc., Madison, WI. pp. 859–902.

Sorensen, E.L., Stuteville, D.L., Horber, E.K., and Skinner, D. 1994. Registration of KS224 glandular-haired alfalfa germplasm with multiple pest resistance. Crop Sci. 34: 544.

Southworth, W. 1914. Alfalfa hybridization. J. Hered. 5: 448–457.

Southworth, W. 1928. Influences which tend to affect seed production in alfalfa and an attempt to raise a high seed producing strain by hybridization. Scient. Agric. 9: 1–28.

Spedding, C.R.W., and Diekmahns, E.C. 1972. Grasses and legumes in British agriculture. Bull. 49, Commonwealth Bureau of Pastures and Field Crops. Commonwealth Agriculture Bureaux, Farnham Royal, Bucks, England. 511 pp.

Spira, T.P., and Wagner, L.K. 1983. Viability of seeds up to 211 years old extracted from adobe brick buildings of California and northern Mexico. Am. J. Bot. 70: 303–307.

Sprent, J.I. 2007. Evolving ideas of legume evolution and diversity: a taxonomic pespective on the occurrence of nodulation. New Phytol. 174: 11–25.

Sprent, J.I., and James, E.K. 2007. Legume evolution: where do nodules and mycorrhizas fit in? Plant Physiology, 144: 575–581.

St. John-Sweeting, R.S., Auricht, G.C., Shaw, C.L., and Mohamed, A.A. 1996. Comparisons of new *Medicago* spp. in South Australia. Proceeding 8th Australian Agronomy conference Toowoomba 1996. pp. 514–517.

Stafleu, F.A., and Cowan, R.S. 1976. Taxonomic literature, ed. 2, 1. Regnum Veg. Vol. 94.

Stanev, S. 1977. Geobotanical characteristics of some phytocenoses specific for the Besaparski Ridges Bulgaria. Fitologija, 7: 25–50. (In Bulgarian.)

Stanford, E.H., Clement, W.M., Jr., and Bingham, E.T. 1972. Cytology and evolution of the *Medicago sativa–falcata* complex. *In* Alfalfa science and technology. *Edited* by C.H. Hanson. American Society of Agronomy, Madison, WI. pp. 87–101.

Staszewski, Z., and Jakubowska, B. 1992. The possibility of hybridization [between] *Medicago varia* Mart. and annual species of lucerne. *In* Ploidy and chromosome manipulation in forage breeding, proceedings of the 17th meeting of the Fodder Crops Section of EUCARPIA, Alghero, Italy, 14–18 October 1991. *Edited by* F. Veronesi, S. Bullitta, and S. Caredda. Consiglio Nazionale delle Ricerche, Rome, Italy. pp. 171–173.

Staszewski, Z., Jakubowska, B., Veronesi, F., Bullitta, S., and Caredda, S. 1992. Ploidy and chromosome manipulation in forage breeding. Proceedings of the 17th meeting of the Fodder Crops Section of EUCARPIA, Alghero, Italy, 14–18 October 1991. *Edited by* F. Veronesi and S. Bullitta. Consiglio Nazionale delle Ricerche, Rome, Italy. pp. 171–173.

Stearn, W.T. 1960. Notes on Linnaeus's "Genera Plantarum". Introduction to 1960 facsimile of Carl Linnaeus's Genera Plantarum, 5th ed. 1754. Hafner, New York, NY. pp. v–xxiv.

Stebbins, G.L. 1947. Types of polyploids. Their classification and significance. Advances Genet. 1: 14–18.

Stebbins, G.L. 1981. Coevolution of grasses and herbivores. Ann. Missouri Bot. Gard. 68: 75–86.

Stebbins, G.L. 1982. Darwin to DNA, molecules to humanity. W.H. Freeman and Company, San Francisco, CA. 491 pp.

Stebbins, G.L. 1986. The origin and success of polyploids in the boreal circumpolar flora: a new analysis. Trans. Bot. Soc. Edinb. 150th Anniversary Suppl.: 17–31.

Steele, K.P., and Wojciechowski, M.F. 2003. Phylogenetic analyses of tribes Trifolieae and Vicieae, based on sequences of the plastid gene, *mat*K (Papilionoideae: Leguminosae). *In* Advances in legume systematics, part 10, higher level systematics. *Edited by* B.B. Klitgaard and A. Bruneau. Royal Botanic Gardens, Kew, UK. pp. 355–370.

Steele, K.P., Ickert-Bond, S.M., Zarre, S., and Wojciechowski, M.F. 2010. Phylogeny and character evolution in *Medicago* (Leguminosae): evidence from analyses of plastid *trn*K/*mat*K and nuclear *GA3ox1* sequences. Am. J. Bot. 97: 1142 –1155.

Stepanov, N.V. 1994. Chromosome numbers of some higher plant taxa of the flora of the Krasnoyarsk region. Botanicheskii Zhurnal, 79(2): 135–139. (In Russian.)

Stepanova, G.V., Atlakova, E.N., and Zatchina, G.P. 1996. Breeding *Medicago lupulina* for pasture and green manure. Proc. 35th North American Alfalfa Improvement Conference, Oklahoma, USA. [Poster abstract.] http://www.naaic.org/Publications/1996Proc/poster_session.htm (current July 2010).

Steuckardt, R., Hempel, K., and Kuhfuss, K.H. 1972. Studies of resistance to *Verticillium albo-atrum* R. et B. in lucerne collections from the Hungarian People's Republic, the USSR and the German Democratic Republic. Agrobotanika, 13: 31–44. (In German.)

Stevens, M. 2005. The role of eyespots as anti-predator mechanisms, principally demonstrated in the Lepidoptera. Biol. Rev. 80: 573–588.

Steyn D.G. 1934. The toxicology of plants in South Africa together with a consideration of poisonous foodstuffs and fungi. Central News Agency, Johannesburg, South Africa. 631 pp.

Stirton, C.H. 1982. The genus *Medicago* (Leguminosae) in Southern Africa. Bothalia, 14: 27–35.

Stock, C., Howard H., and J. Everest. 1944. The ascent of *Equus*. Los Angeles County Museum Sciences Series No. 8, Paleontology Publication No. 5. 38 pp.

Stoyanova, S.D. 2001. *Ex situ* conservation in the Bulgarian genebank.1. Effect of storage. PGR Newsl. 128: 68–76.

Strasser, W. 1988. *Medicago strasseri* and *Draba strasseri*, two new species from Greece. Mitt. Naturforsch. Ges. Bern. 45: 168–172.

Strid, A. 1971. Chromosome numbers in some Albanian angiosperms. Bot. Notiser, 124: 490–496.

Strid, A. 2000. New taxa described in Grisebach's "Spicilegium Florae Rumelicae et Bithynicae" (1843-46). Preslia, 72: 241–321.

Strid, A., and Franzen, R. 1981. I.O.P.B. chromosome number reports LXXIII. Taxon, 30: 829–842.

Stringi, L., Gristina, L., Giambalvo, D., and Accardo, A. 1993. Productivity of *Medicago arborea* subjected to a simulated sheep grazing system in a semi-arid environment. *In* Agriculture: Agrimed research programme: Fodder trees and shrubs in the Mediterranean production systems: objectives and expected results of the EC research contract. *Edited by* V. Papanastasis. Commission of the European Communities, Luxembourg. pp. 131–137.

Stringi, L., Amato, G., Giambalvo, D., and Accardo, A. 1994*a*. Behaviour and phenotypic variability of some *Medicago arborea* populations in Sicily. *In* Fodder trees and shrubs. *Edited by* V. Papanastasis and L. Stringi. CIHEAM-IAMZ (Cahiers Options méditerranéennes No. 4), Zaragoza, Spain. pp. 51–60.

Stringi, L. Giambalvo, D., Accardo, A., and Amato, G. 1994*b*. Breeding system of *Medicago arborea* in open pollinated and in caged conditions. *In* Fodder trees and shrubs. *Edited by* V. Papanastasis and L. Stringi. CIHEAM-IAMZ (Cahiers Options méditerranéennes No. 4), Zaragoza, Spain. pp. 103–109.

Stuteville, D.L., and Erwin, D.C. 1990. Compendium of alfalfa diseases, 2nd edition. APS Press, St. Paul, MI. 84 pp.

Suksayretrup, K. 1986. Xerophytism in *Medicago*. Dissertation Abstracts International, B Sciences and Engineering, 47(5): 1802B. 98 pp. (Thesis.)

Sulas, L. 2005. The future role of forage legumes in Mediterranean-climate areas. *In* Grasslands: developments, opportunities, perspectives. *Edited by* S.G. Reynolds and J. Frame. Food and Agriculture Organization of the United Nations, Rome, Italy/Science Publishers, Inc., Enfield, NH. pp. 29–54.

Sulc, R.M., Johnson, K.D., Sheaffer, C.C., Undersander, D.J., and van Santen, E. 2004. Forage quality of potato leafhopper resistant and susceptible alfalfa cultivars. Agron. J. 96: 337–343.

Sulc, R.M., van Santen, E., Johnson, K.D., Sheaffer, C.C., Undersander, D.J., Bledsoe, L.W., Hogg, D.B., and Willson, H.R. 2001. Glandular-haired cultivars reduce potato leafhopper damage in alfalfa. Agron. J. 93: 1287–1296.

Summers, C., and Putnam, D. (Editors). 2008. Irrigated alfalfa management for Mediterranean and desert zones. University of California, Division of Agriculture and Natural Resources, Publication 8305. http://alfalfa.ucdavis.edu/IrrigatedAlfalfa/ (24 chapters, individually paginated. Current July 2010).

Sumnevich, G.P. 1932. Onekotorykh aziatshikh vidakh roda *Medicago* L. [Some Asiatic species of the genus *Medicago* L.] Sistematicheskie Zametki Gerbariya Tomskogo Un-ta, Nos. 1-2: 1–14 + plate.

Tadmor, N.H., Shanan, L., and Evenari, M. 1971. "Runoff farming" in the desert. V. Persistence and yields of annual range species. Agron. J. 63: 91–95.

Taylor, G.B., and Ewing, M.A. 1988. Effect of depth of burial on the longevity of hard seeds of subterranean clover and annual medics. Austral. J. Exper. Agric. 28: 77–81.

Tazima, Y. 1984. Silkworm moths. *In* Evolution of domesticated animals. *Edited by* I.L. Mason. Longman, London, UK. pp. 416–424.

Terzyisku, D.P., and Poddubnaya-Arnoldi, V.A. 1973. Cytoembriological experiments on Rhodopian lucerne (*Medicago rhodopaea* Ull.) Bull. Glav. Bot. Sad. 89: 20–29. (In Russian.)

Tesfaye, M., Samac, D.A., and Vance, C.P. 2006. Insights into symbiotic nitrogen fixation in *Medicago truncatula*. Molec. Plant Microbe Interact. 19: 330–341.

Thanopoulos, R. 2007. The genus *Medicago* in Greece. 1. A review of species diversity, geographical distribution and ecological adaptation. Fl. Medit. 17: 217–276.

Thies, J.A., Barnes, D.K., and Smith, D.M. 1985. Associations between geographic origin and resistance to bacterial wilt and fusarium wilt among alfalfa plant introductions. Proc. 19th Central Alfalfa Improvement Conference. p. 30.

Thomas, H. 1980. Terminology and definitions in studies of grassland plants. Grass Forage Sci. 35: 13–23.

Thomas, M.R., Johnson, L.B., and White, F.F. 1990. Selection of interspecific somatic hybrids of *Medicago* by using *Agrobacterium*-transformed tissues. Plant Sci. (Ireland), 69: 189–198.

Thomas, P.K. 1989. Utilization of domestic animals in pre- and protohistoric India. *In* The walking larder. *Edited by* J. Clutton-Brock. Unwin Hyman, London, UK. pp. 108–112.

Thompson, J.L. 1892. The Clydesdale horse. Agricultural Gazette of New South Wales (Australia), 3: 608–616.

Thompson, J.L. 1895. Rotation of crops. Agricultural Gazette of New South Wales (Australia), 6: 479–486.

Thompson, J.N. 1982. Interaction and coevolution. John Wiley & Sons, New York, NY. 179 pp.

Thompson, J.N. 1989. Concepts of coevolution. TREE, 4: 179–183.

Thomson, E.F., Rihawi, S., Cocks, P.S., Osman, A.E., and Russi, L. 1990. Recovery and germination rates of seeds of Mediterranean medics and clovers offered to sheep at a single meal or continuously. J. Agric. Sci. 114: 295–299.

Thoquet, P., Ghérardi, M., Journet, E.-P., Kereszt, A., Ané, J.-M., Prosperi, J.-M., and Huguet, T. 2002. The molecular genetic linkage map of the model legume *Medicago truncatula*: an essential tool for comparative legume genomics and the isolation of agronomically important genes. BMC Plant Biol. 2(1). http://www.biomedcentral.com/1471-2229/2/1 (current July 2010).

Tian, D., and Rose, R.J. 1999. Asymmetric somatic hybridisation between the annual legumes *Medicago truncatula* and *Medicago scutellata*. Plant Cell Reports, 18: 989–996.

Tian, D., Niu, C., and Rose, R.J. 2002. DNA transfer by highly asymmetric somatic hybridisation in *Medicago truncatula* (+) *Medicago rugosa* and *Medicago truncatula* (+) *Medicago scutellata*. Theor. Appl. Genet. 104: 9–16.

Tischler, G. 1934. Die Bedeutungen der Polyploidie für die Verbreitung der Angiospermen, erläutert an den Arten Schleswig-Hosteins, mit Ausblicken auf andere Florengebiete. Bot. Jahrb. 67: 1–36.

Tivoli, B., Baranger, A., Sivasithamparam, K., and Barbetti, M.J. 2006. Annual *Medicago*: from a model crop challenged by a spectrum of necrotrophic pathogens to a model plant to explore the nature of disease resistance. Annals Bot. 98: 1117–1128.

Tomé, G.A. 1947. Das casos de poliploidia en alfalfa (*Medicago sativa* L.). Rev. Fac. Agron. Vet. Univ. Buenos Aires, 11: 299–307.

Torregrosa, C., Cluzet, S., Fournier, J., Huguet, T., Gamas, P., Prosperi, J.M., Esquerre-Tugaye, M.T., Dumas, B., and Jacquet, C. 2004. Cytological, genetic, and molecular analysis to characterize compatible and incompatible interactions between *Medicago truncatula* and *Colletotrichum trifolii*. Molecular Plant-Microbe Interactions MPMI. 17(8): 909–920.

Town, C.D. 2006. Annotating the genome of *Medicago truncatula*. Curr. Opin. Plant. Biol. 9: 122–127.

Townsend, C.C. 1968. Contribution to the flora of Iraq. V. Notes on the Leguminales. Kew Bull. 21: 433–438.

Townsend, C.C. 1974. *Trigonella*. *In* Flora of Iraq, vol. 3, Leguminales. *Edited by* C.C. Townsend and E. Guest. Ministry of Agriculture, Baghdad, Iraq. pp. 87–111.

Townsend, M.S., Henning, J.A., and Currier, C.G. 1994. The alfalfa catalog software package. Agron. J. 86: 337–339.

Trautvetter, E.R. 1841. Über die mit *Trifolium* verwandten Pflanzengattungen. Bull. Sci. Acad. Imp. Sci. St.-Petersbourg, 8: 267–272.

Travis, R.L., and Reed, R. 1990. Phototropic leaflet reorientation in alfalfa germplasm selected for leaf color and size. J. Agron. Crop Sci. 165: 281–286.

Trinajstic, I., and Pandza, M. 2002. Moon trefoil – *Medicago arborea* L. in the flora of Croatia. Agronomski Glasnik, 64: 137–143. (In Croatian.)

Troitzky, N. 1923. *Medicago dzhawakhetica* var. *timofeewii*. Rep. (Vestn.) Bot. Gard. Tiflis, New Ser. 1: 90–91. (In Russian.)

Trumble, J.T., Kolodny-Hirsch, D.M., and Ting, I.P. 1993. Plant compensation for arthropod herbivory. Annu. Rev. Entomol. 38: 93–119.

Tschechow, W. 1932. Karyo-systematic analysis of the tribe Trifolieae D.C. [(Family Leguminosae Juss.)]. Bull. Appl. Bot. Pl. Breed. Ser. 2, 1: 119–143. (In Russian.)

Turkington, R., and Cavers, P.B. 1979. The biology of Canadian weeds. 33. *Medicago lupulina* L. Can. J. Plant Sci. 59: 99–110.

Turland N.J. 1995. *Medicago heyniana* Greuter. *In* The red data book of rare and threatened plants of Greece. *Edited by* D. Phitos, A. Strid, A.S. Snogerup, and W. Greuter. World Wildlife Fund, Athens, Greece. pp. 360–361.

Turland, N.J., and Jarvis, C.E. (Editors). 1997. Typification of Linnaean specific and varietal names in the Leguminosae (Fabaceae). Taxon, 46: 457–485.

Turland, N.J., Chilton, L., and Press, J.R. 1993. Flora of the Cretan area. Annotated checklist & atlas. The Natural History Museum, London, UK. 439 pp.

Tutin, T.G. 1968. *Medicago*. *In* Flora Europaea. Vol. 2. *Edited by* T.G. Tutin, V.H. Heywood, N.A. Burges, D.M. Moore, D.H. Valentine, S.M. Walters, and D.A. Webb. Cambridge University Press, Cambridge, UK. pp. 153–157.

Twamley, B.E. 1974. Recurrent selection for seedling vigor in birdsfoot trefoil. Crop Sci. 14: 87–90.

Tyler, V.E. 1993. The honest herbal. 3d ed. Pharmaceutical Products Press (Haworth Press), Binghamton, NY. 375 pp.

Ueckert, D.N., Petersen, J.L., and Ocumpaugh, W.R. 2003. Production and natural regeneration of annual medics in west-central Texas. Texas J. Agric. Nat. Res. 16: 86–92.

Uerpmann, H.-P. 1987. The ancient distribution of ungulate mammals in the Middle East. Beihefte zum Tübinger Atlas des Vorderen Orients, Reihe A, Nr. 27, Dr. Ludgwig Reichert, Wiesbaden, Germany. 173 pp.

Uhrikova, A. 1978. Report in Index of chromosome numbers of Slovakian flora. Part 6. Acta Fac. Rerum Nat. Univ. Comenianae, Bot. 26: 1–42.

Uljanova, T. 1964. A new species of medic from Turkmenia. Novit. Syst. Pl. Vasc. Acad. Sci. URSS, 1964: 175–177. (In Russian.)

Undersander, D.J. 2004. Alfalfa management guide. American Soc. Agronomy, Madison, WI. 58 pp.

Urban, I. 1873. Prodromus einer Monographie der gattung *Medicago* L. Verh. Bot. Vereins Prov. Brandenburg, 15: 1–85.

Urban, I. 1882 ("1881"). Herr I. Urban sprach über einige für die Flora Aegyptens neue Arten der Gattung *Trigonella* L. Sitz. Verh. Bot. Vereins Prov. Brandenberg, 23: 66–72.

USDA. 2009. Nodulation Plants in GRIN Taxonomy. http://www.ars-grin.gov/~sbmljw/cgi-bin/taxnodul.pl?language=en (National Genetic Resource Program, Germplasm Resources Information Network, Online Database. Current July 2010).

Vacek, V. 1997. Grigorij Ivanovič Širjajev. Folia Historica (Markyk Univ., Brno) 55 (Jan.). 3 pp.

Vaitsis, T.A. 1993. Breeding *Medicago arborea* for Mediterranean conditions. *In* Agriculture: Agrimed research programme: Fodder trees and shrubs in the Mediterranean production systems: objectives and expected results of the EC research contract. *Edited by* V. Papanastasis. Commission of the European Communities, Luxembourg. pp. 171–175.

Vaitsis, T., and Konstantakis, A. 1994. Collecting and breeding *Medicago arborea* indigenous germplasm in Greece. *In* Fodder trees and shrubs (Cahiers Options méditerranéennes No. 4). *Edited by* V. Papanastasis and L. Stringi. CIHEAM-IAMZ, Zaragoza, Spain. pp. 61–70.

Valdenegro, M., Barea, J.M., and Azcon, R. 2001. Influence of arbuscular-mycorrhizal fungi, *Rhizobium meliloti* strains and PGPR inoculation on the growth of *Medicago arborea* used as model legume for re-vegetation and biological reactivation in a semi-arid Mediterranean area. Plant Growth Regul. 34: 233–240.

Van Berkum, P., Beyene, D., Bao, G., Campbell, T.A., and Eardly, B.D. 1998. *Rhizobium mongolense* sp. nov., is one of three rhizobial genotypes identified which nodulate and form nitrogen-fixing symbioses with *Medicago ruthenica* [(L.) Ledebour]. Internat. J. Syst. Bacteriol. 48: 13–22.

Van Deynze, A.E., Fitzpatrick, S., Hammon, B., McCaslin, M.H., Putnam, D.H., Teuber, L.R., and Undersander, D.J. 2008. Gene flow in alfalfa: biology, mitigation, and potential impact on production. Council for Agricultural Science and Technology, Ames, IA. 30 pp. http://voyager.noble.org/books/CAST_Alfalfa_Gene_Flow157.pdf (current July 2010).

Van Heerden, J.M. 1984. Influence of temperature and daylength on the phenological development of annual *Medicago* species with particular reference to *Medicago truncatula* cultivar Jemalong. South African J. Plant Soil, 1: 73–78.

Van Loon, J. 1974. A cytological investigation of flowering plants from the Canary Islands. Acta Bot. Neerl. 23: 113–124.

Vance, C.P., Heichel, G.H., and Phillips, D.A. 1988. Nodulation and symbiotic dinitrogen fixation. *In* Alfalfa and alfalfa improvement. *Edited by* A.A. Hanson, D.K. Barnes, and R.R. Hill, Jr. American Society of Agronomy, Crop Science Society of America, Soil Science Society of America, Madison, WI. pp. 229–257.

VandenBosch, K.A., and Stacey, G. 2003. Summaries of legume genomics projects from around the globe. Community resources for crops and models. Plant Physiol. 131: 840–865.

Vassilczenko, I.T. 1949. Alfalfa—the best forage plant. Trudy botanicheskogo instituta imenu V.L. Komarova [Transactions of the V.L. Komarov Botanical Institute] Ser. 1 (Fasc. 8): 1–240. [USSR Academy of Sciences Publishing House, Moscow–Leningrad, USSR; cited in text as: Acta Inst. Bot. Acad. Sci. URSS, Ser. 1, Fasc. 8 (1949).]

Vassilczenko, I.T. 1951. New species and sections of *Trigonella*. Bot. Mat. (Leningrad), 14: 229–235. (In Latin and Russian.)

Vassilczenko, I.T. 1953. A review of the species of the genus *Trigonella* L. Tr. Bot. Inst. Akad. Nauk SSSR, Ser. 1, Flora Sist. Vyss. Rast. Fasc. 14. pp. 124–269. (In Russian.)

Vassilczenko, I.T. 1972. Genus novum *Rhodusia* Vass. e. familia Papilionacearum. Nov. Sist. Vyss. Rast. 9: 203–206. (Translated from Russian, 1978, Amerind Publishing Co., New Delhi.)

Vassilczenko, I.T. 1979. On the systematic position of the genera *Medicago* L. and *Trigonella* L. Nov. Syst. Pl. Vasc., Acad. Sci. URSS, Inst. Komarovii 16: 130–138. (In Russian.)

Vassilczenko, I.T. 1984. *Medicago sativa* L. s.l. *In* Flora Iranica, Papilionaceae Vol. 2. *Edited by* K.H. Rechinger. Akademische Druck-u. Verlagsanstalt, Graz, Austria. pp. 271–274.

Vassilczenko. I.T. 2002*a*. *Trigonella*. *In* Flora of Russia: the European part and surrounding regions. [Translation of Fedorov, A.A. (Editor). 1987. Flora Evropeisko Chasti SSRS, Vol. 6, Nauka, Leningrad, USSR.] *Edited by* A.A. Fedorov. Vol. 6. CRC Press, Boca Raton, FL. pp. 270–278.

Vassilczenko. I.T. 2002*b*. *Turukhania*. *In* Flora of Russia: the European part and surrounding regions. [Translation of Fedorov, A.A. (Editor). 1987. Flora Evropeisko Chasti SSRS, Vol. 6, Nauka, Leningrad, USSR.] *Edited by* A.A. Fedorov. Vol. 6. CRC Press, Boca Raton, FL. p. 279.

Vassilczenko. I.T. 2002*c*. *Medicago*. *In* Flora of Russia: the European part and surrounding regions. [Translation of Fedorov, A.A. (Editor). 1987. Flora Evropeisko Chasti SSRS, Vol. 6, Nauka, Leningrad, USSR.] *Edited by* A.A. Fedorov. Vol. 6. CRC Press, Boca Raton, FL. pp. 279–291.

Vavilov, N.I. 1951. The origin, variation, immunity, and breeding of cultivated plants. Transl. from Russian by K.S. Chester. Chronica Botanica Vr. 1/6. 364 pp.

Vázquez, M.M., Azcón, R., and Barea, J.M. 2001. Compatibility of a wild type and its genetically modified *Sinorhizobium* strain with two mycorrhizal fungi on *Medicago* species as affected by drought stress. Plant Sci. 161: 347–358.

Veatch, M.E., Smith, S.E., and Vandemark, G. 2004. Shoot biomass production among accessions of *Medicago truncatula* exposed to NaCl. Crop Sci. 44: 1008–1013.

Vereshchagin, N.K. 1959. The mammals of the Caucasus. Izdatel'stvo Akademii Nauk SSSR, Moscow-Leningrad. Translated from Russian by A. Lerman and B. Rabinovich, Israel Program for Scientific Translations, Jerusalem, 1967. 816 pp.

Verkaar, H.J. 1986. When does grazing benefit plants? TREE, 1: 168–169.

Verloove, F., and Lambinon, J. 2006. The non-native vascular flora of Belgium: a new nothospecies and three new combinations. Syst. Geog. Plants, 76(2): 217–220.

Veronesi, F., and Rosellini, D. 2000. Screening germplasm of *Medicago polymorpha* suitable as a cover crop in Mediterranean orchards. *In* Lucerne and medics for the XXI Century, Proceedings XIII Eucarpia *Medicago* spp. Group Meeting, Perugia, Italy, 13–16 September 1999. *Edited by* L. Russi and S. Lorenzetti. Universita di Perugia, Perugia, Italy. pp. 170–175.

Vickery, R. 1995. A dictionary of plant-lore. Oxford University Press, Oxford, UK. 437 pp.

Villegas, M. Del C., Rome, S., Maure, L., Domergue, O., Gardan, L., Bailly, X., Cleyet-Marel, J. C., and Brunel, B. 2006. Nitrogen-fixing sinorhizobia with *Medicago laciniata* constitute a novel biovar (bv. *medicaginis*) of *S. meliloti*. Syst. Appl. Microbiol. 29: 526–538.

Vitale, M., Pupilli, F., Labombarda, P., and Arcioni, S. 1998. RAPD analysis reveals a low rate of outcrossing in burr medic (*Medicago polymorpha* L.). Genetic Res. Crop Evol. 45: 337–342.

Vitek, E., Kiehn, M., Pascher, K., Starlinger, F., Greimler, J., Stocker, U., Lehner, S., Beinhofer, P., and Blaha, A. 1992. Beiträge zur Flora von Österreich-weitere Chromosomenzählungen Verhandlungen der Zoologisch-botanischen Gesellschaft in Wien, 129: 215–226.

Vitkus, A.A. 1977. Growth development and productivity characteristics of 19 1-year old alfalfa species in the Lithuanian-SSR, USSR. Lietuvos TSR Mokslu Akademijos Darbai Serija C Biologijos Mokslai, 3: 27–32. (In Russian.)

Vitkus, A.A. 1987. Biological features and chemical composition of a local population of *Medicago lupulina* in the southeastern Lithuanian SSR USSR 6. Growth and development of annual plants as a function of the time of sowing. Lietuvos TSR Mokslu Akademijos Darbai-Serija C Biologijos Mokslai, 1: 14–20. (In Russian.)

Vitkus, A.A., Lapinskene, N.A., and Chiapaite, A.B.A. 1977. Peculiarities of growth and development of *Medicago carstiensis* Wulf. [and] lucerne species in the Lithuanian SSR in 1973–1975. Lietuvos TSR Mokslu Akademijos Darbai Serija C Biologijos Mokslai, 4: 13–20. (In Russian.)

Vivant, J. 2000. *Medicago orbicularis* subsp. *castellana* (J. Casellas) P. Monts. observé dans les Hautes-Pyrénées en vallée d'Aure. Monde Pl. 95(470): 8.

Vogt, R., and Aparicio, A. 1999. Chromosome numbers of plants collected during Iter Mediterraneum IV in Cyprus Bocconea. Monographiae Herbarii Mediterranei Panormitani, 11: 117–169.

Volenec, J.J., and Cherney, J.H. 1990. Yield components, morphology, and forage quality of multifoliolate alfalfa phenotypes. Crop Sci. 30: 1234–1238.

Wagner, L.K., and Spira, T.P. 1994. Germination, recruitment and survival in the weedy annual *Medicago polymorpha* in successive wet and dry years. Amer. Midl. Nat. 131: 98–108.

Walsh, M.J., Delaney, R.H., Groose, R.W., and Krall, J.M. 2001. Performance of annual medic species (*Medicago* spp.) in southeastern Wyoming. Agron. J. 93: 1249–1256.

Walters, C., Wheeler, L.M., and Grotenhuis, J.M. 2005. Longevity of seeds stored in a genebank: species characteristics. Seed Sci. Res. 15: 1–20.

Wang, K., Li, H., and Luo, X. 1989. Research on distant hybridization between *Melissutus* [sic] *ruthenicus* C.W. Chang and *Medicago sativa* [cv.] Zhaodong. *In* XVI international Grassland Congress, Nice, France, 4–11 Oct., 1989. *Chaired by* R. Jarrige. Association Française pour la Production Fourragère, France. pp. 333–334.

Wang, H., and Li, D.-Z. 2005. Pollination biology of four *Pedicularis* species (Scrophulariaceae) in Northwestern Yunnan, China. Ann. Missouri Bot. Gard. 92: 127–138.

Wang, H.L., and Grusak, M.A. 2005. Structure and development of *Medicago truncatula* pod wall and seed coat. Annals Bot. 95: 737–747.

Ward, D.E. 1983. Chromosome counts from New Mexico and southern Colorado. Phytologia, 54: 302–309.

Wassermann, V.D. 1974. The possibilities of annual medics in summer rainfall situations. Proc. Grassland Soc. Southern Africa, 9: 179–184. (In Afrikaans.)

Watson, B.S., Asirvatham, V.S., Wang, L., and Sumner, L.W. 2003. Mapping the proteome of barrel medic (*Medicago truncatula*). Plant Physiol. 131: 1104–1123.

Watson, D.P. 1948. Structure of the testa and its relation to germination in the Papilionacaeae tribes Trifolieae and Loteae. Ann. Bot. (London), 12: 385–409.

Watt, J.M., and Breyer-Brandwijk, M.G. 1962 The medicinal and poisonous plants of Southern and Eastern Africa. Being an account of their medicinal and other uses, chemical composition, pharmacological effects and toxicology in man and animal. 2nd edition. E. & S. Livingstone, Edinburgh, UK. 1457 pp.

Weaver, J.E. 1926. Root development of field crops. McGraw-Hill, New York, NY. 291 pp.

Weir, B.S. 2008. The current taxonomy of rhizobia. http://www.rhizobia.co.nz/taxonomy/rhizobia.html (current July 2009).

Wentworth, J.E., Bailey, J.P., and Gornall, J. 1991. Contributions to a cytological catalogue of the British and Irish flora, 1. Watsonia, 18: 415–17.

Wheeler, W.A. 1950. Forage and pasture crops. D. Van Nostrand Co., New York, NY. 752 pp.

White, J.G.H., Meijer, G., and Douglas, J.A. 1975. College Glutinosa Lucerne. Proceedings New Zealand Grassland Association Conference, 36(2): 172–179.

White, M.J.D. 1978. Modes of speciation. W.H. Freeman and Co., San Francisco, CA. 455 pp.

White, R. 1989. Husbandry and herd control in the Upper Paleolithic. Current Anthropology, 30: 609–632.

Wiley, E.O. 1981. Phylogenetics. John Wiley & Sons, New York, NY. 439 pp.

Wilkinson, H.P. 1983. The anatomy of the pulvinus in various Mimosoideae in relation to function. Notes Jodrell Lab., Kew, 10: 1–11.

Willoughby, D.P. 1974. The empire of *Equus*. Thomas Yoseloff Ltd., London, UK. 475 pp.

Winter, K., Troughton, J.H., Evenari, M., Lauchli, A., and Luttge, U. 1976. Mineral ion composition and occurrence of CAM-like diurnal malate fluctuations in plants of coastal and desert habitats of Israel and the Sinai. Oecologia, 25: 125–143.

Wodehouse, R.P. 1971. Hayfever Plants. 2nd edition. Hafner, New York, NY. 280 pp.

Wojciechowski, M.F. 2003. Reconstructing the phylogeny of legumes (Leguminosae): an early 21st century perspective. *In* Advances in legume systematics, part 10. *Edited by* B.B. Klitgaard and A. Bruneau. Royal Botanic Gardens, Kew, UK. pp. 5–35.

Wojciechowski, M.F., Sanderson, M.J., Steele, K.P., and Liston, A. 2000. Molecular phylogeny of the "temperate herbaceous tribes" of papilionoid legumes: a supertree approach. *In* Advances in legume systematics, part 9. *Edited by* P.S. Herendeen and A. Bruneau. Royal Botanic Gardens, Kew, UK. pp. 277–298.

Woodfield, D.R., and Bingham, E.T. 1995. Improvement in two-allele autotetraploid populations of alfalfa explained by accumulation of favourable alleles. Crop Sci. 35: 988–994.

Wright, N.C. 1954. The ecology of domesticated animals. *In* Progress in the physiology of farm animals, Vol. 1. *Edited by* J. Hammond. Butterworths Scientific Publications, London, UK. pp. 191–251.

Wu, R., Bao, G.P., and Yong, S.P. 1996. Eco-geographical distribution of *Melilotoides* plants in Inner Mongolia. Grassland of China [Zhongguo caoyuan], 2: 5–6. (In Chinese.)

Wu, Y.-F., Ge, R.-D., and Zhang, S.-Z. 1986. Analysis concerning the types of *Trigonella* chromosome groups. Grassland of China [Zhongguo caoyuan], 3: 48–50. (In Chinese.)

Xu, B.-S., Weng, R.-F., and Zhang, M.-Z. 1992. Chromosome numbers of Shanghai plants I. Investigatio et Studium Naturae, 12: 48–65.

Yaege, J.R., and Stuteville, D.L. 2002. Reactions of accessions in the annual *Medicago* core germ plasm collection to *Erysiphe pisi*. Plant Disease, 86: 312–315.

Yahiaoui S., and Abdelguerfi, A. 1999. Growth and phenology of some *Medicago ciliaris* populations: interaction with the site of origin. *In* Dynamics and sustainability of Mediterranean pastoral systems, *Edited by* M. Etienne. (= Cahiers Options Méditerranéennes 39; Meeting of the Mediterranean Sub-Network of the FAO-CIHEAM Inter-Regional Cooperative Research and Development Network on Pastures and Fodder Crops, 1997/11/26-29, Badajoz, Spain.). pp. 107–110. (In French.)

Yakovlev, G.P. 1988. Plants of central Asia. From the material of the V.L. Komarov Botanical Institute, Academy of Sciences of the USSR. Vol. 8a, Leguminosae. Nauka, Leningrad, USSR. 125 pp. (In Russian. Translated as: Plants of central Asia. Plant collections from China and Mongolia. Science Publishers, Enfield, NH. 170 pp.)

Yakovlev, G.P., Sytin, A.K., and Roskov, Yu.R. 1996. Legumes of northern Eurasia. A checklist. Royal Botanic Gardens, Kew, UK. 724 pp.

Yan, G.-S., Zhang, S.-Z., Xue, F.-H., Wang, L.-Y., Yun, J.-F., and Fu, X.-Q. 2000. The chromosome numbers and natural distribution of 38 forage plants in north China. Grassland of China [Zhongguo caoyuan], 5: 1–5. (In Chinese.)

Yan, G.-X., Zhang, S.-Z., Yan, J.-F., Fu, X.-Q., and Wang, L.-Y. 1989. Chromosome numbers and geographical distribution of 68 species of forage plants. Grassland of China [Zhongguo caoyuan], 4: 53–60. (In Chinese.)

Yang, G.-F., and Zhao, C.-X. 1986. Karyotype analysis of *Medicago sativa* and *Vicia villosa* chromosomes. Grassland of China [Zhongguo caoyuan], 2: 58–60. (In Chinese.)

Yao, H., and Tan, D.H. 2005. Size-dependent reproductive output and life-history strategies in four ephemeral species of *Trigonella*. Zhiwu Shengtai Xuebao, 29: 954–960. (In Chinese.)

Yen, S.-T., and Murphy, R.P. 1979. Cytology and breeding of hexaploid alfalfa. I. Stability of chromosome number. Crop Sci. 19: 389–393.

Young, N., Mudge, J., and Ellis, T. 2003. Legume genomics: more than peas in a pod. Curr. Opin. Plant. Biol. 6: 199–204.

Young, R.R., and Lagudah, E.S. 1994. Ribosomal DNA, peroxidase and extensin variation in genomic DNA of *Medicago laciniata* (L.) Miller from Australia and Africa. Austral. J. Agric. Res. 45: 1013–1023.

Young, R.R., Croft, P.H., and Sandral, G.A. 1992. Variation in flowering times and agronomic characteristics of *Medicago laciniata* (L.) Miller collected from diverse locations in New South Wales. Austral. J. Exper. Agric. 32: 59–63.

Young, R.R., Lloyd, D.L., O'Neill, M., Teasdale, K., and Johnson, B. 1996. Variation in naturalised *Medicago polymorpha* collected from inland eastern Australia: 11 – Queensland. Proc. 8th Aust. Agronomy Conf., Toowoomba. http://www.regional.org.au/au/asa/1996/poster/680lloyd.htm (current July 2010).

Yu, Y., Fujimoto, F., and Oba, S. 1999. Variation of local populations of *Medicago arabica* (L.) Huds. and strains of *Rhizobium meliloti* and their symbiotic compatibility in nitrogen fixation capacity. J. Jpn. Soc. Grassland Sci. 44: 320–327.

Zarrabi, A.A., Caddel, J.L., and Berberet, R.C. 1985. Blue alfalfa aphid resistance evaluation as affected by seedling age. Proc. Central Alf. Imp. Conf. 19: 26–27.

Zary, A.M.S., and Hickenbick, M.C.M. 1986. Estudos citogenéticos em "alfafa crioula" (*Medicago sativa* L.) – Leguminosae. Ciencia e Cultura (Sao Paulo), 38: 940–941.

Zeng, L.W., Cocks, P.S., Kailis, S.G., and Kuo, J. 2005. The role of fractures and lipids in the seed coat in the loss of hardseededness of six Mediterranean legume species. J. Agric. Sci. 143: 43–55.

Zeuner, F.E. 1963. A history of domesticated animals. Hutchinson, London, UK. 560 pp.

Zeven, A.C. 1998. Landraces: A review of definitions and classifications. Euphytica, 104: 127–139.

Zhang, Z.-P., Wu, L.-H., and Kang, Y.-F. 1993. Analyses of karyotypes in four forage species of *Medicago*. Botanical Research, 1: 55–59.

Zho, S.F., Sun, Q.Z., Han, J.G., and Wang, Y.Q. 2005. Studies on biomass of *Melilotoides ruthenica* on Kerqin sandy land. Grassland of China [Zhongguo caoyuan], 27: 26–30. (In Chinese.)

Zhu, B., Li, K., and Li, Y. 1985. Karyotype analysis of two species of *Medicago*. Chinese Bull. Bot. 3(5): 47–49. (In Chinese.)

Zhu, H., Cannon, S.B., Young, N.D., and Cook, D.R. 2002. Phylogeny and genomic organization of the TIR and non-TIR NBS-LRR resistance gene family in *Medicago truncatula*. Molec. Plant Microbe Interact. 15: 529–539.

Zhu, H., Choi, H.K., Cook, D.R., and Shoemaker, R.C. 2005. Bridging model and crop legumes through comparative genomics. Plant Physiol. 137: 1189–1196.

Zhu, Y., and Sheaffer, C.C. 1997. Growth analysis of spring and summer seeded annual *Medicago* spp. Crop Sci. 37: 1514–1519.

Zhu, Y., Sheaffer, C.C., and Barnes, D.K. 1996. Forage yield and quality of six annual *Medicago* species in the north-central USA. Agron. J. 88: 955–960.

Zoghlami A., and Hassen H. 1999. Observation on the distribution and ecology of annual *Medicago* species in northern Tunisia. *In* Genetic resources of Mediterranean pasture and forage legumes. *Edited by* S.J. Bennett and P.S. Cocks. Kluwer Academic Publishers, Boston, MA. pp. 235–238.

Zoghlami A., and Hassen H. 2004. Ressources génétiques des espèces spontanées de légumineuses fourragères et pastorales en Tunisie. *In* A. Ferchichi (compiler). Réhabilitation des pâturages et des parcours en milieux méditerranéens. CIHEAM-IAMZ (Zaragoza, Spain). pp. 375–377.

Zoghlami, A., Hassen, H., Seklani, H., Robertson, L., and Salkini, A.K. 1996. Distribution des luzernes annuelles en Tunisie centrale en fonction des facteurs édaphiques et climatiques. Fourrages (France), 145: 5–16.

Zohary, M. 1972. Flora Palaestina. Part 2. Platanaceae to Umbelliferae. Israel Academy of Sciences and Humanities, Jerusalem, Israel. Text & plates in separate volumes.

Zohary, M., and Heller, D. 1984. The genus *Trifolium*. The Israel Academy of Sciences and Humanities, Jerusalem, Israel. 606 pp.

Zribi, K., Mhamdi, R., Hugeut, T., and Aouani, M.E. 2005. Diversity of *Sinorhizobium meliloti* and *S. medicae* nodulating *Medicago truncatula* according to host and soil origins. World J. Microbiol. Biotech. 21: 1009–1015.

Glossary

Abaxial – the surface or part of a lateral organ turned or facing away from the axis and toward the plant's base; thus, the underside of a leaf or petal. cf. Adaxial.

Achene – a small, dry, indehiscent, one-seeded fruit with a tight, thin pericarp (fruit wall developed from the ovary wall).

Acuminate – (as applied to leaves, sepals, etc.) with the tip (sometimes applied to the base) tapering gradually to a point, usually with somewhat concave sides.

Acute – (generally applied to tips, sometimes also to bases, of leaves and petals) with two almost straight or slightly convex sides converging to terminate in a sharp point (the point shorter and usually broader than in an acuminate tip).

Adaxial – turned or facing toward the axis and apex; thus the upper, if not always the uppermost, surface of an organ (sometimes used as equivalent to ventral). cf. abaxial.

Adpressed – see appressed.

Adventitious – adventive (which see); (also, a reference to an organ or part of a plant or body developing in an unusual place, as adventitious roots that arise elsewhere than from the primary root).

Adventive – locally or temporarily naturalized.

Ala (plural alae) – see Wing (petal).

Allopolyploidy – possession of more than two sets of chromosomes derived by hybridization from different parental species (or in some contexts, simply from parents with notably different genomes). cf. autopolyploidy.

Ambiguous name – a name that has been applied to different taxa by different authors and has become difficult to interpret.

Anastomosing – netted, interveined.

Aneuploidy – Presence or lack of one or a few chromosomes in an individual. Sometimes species have been produced by aneuploidy.

Annual – of one season's duration, from seed to maturity and death.

Anthesis – the period during which flowers of a plant have matured to the stage that they are shedding and/or acquiring pollen for sexual fertilization.

Apomorphy – a derived characteristic that is shared by a group of related organisms and inherited from their most recent common ancestor.

Appressed – lying close to or flat against.

Arcuate – curved or bowed.

Areole – the open space formed and bounded by anastomosing veins. (In Cactaceae, the point at which a cluster of spines arises; a node.)

Ascending – rising up; produced somewhat obliquely or indirectly upward.

Autonym – when an infraspecific taxon is proposed and named for the first time, or if only one infraspecific taxon is separately recognized from the remaining plants of a species at a given rank, the remaining plants are automatically placed in a parallel taxon with an epithet that repeats the species epithet, but the taxon so named does not have an authority. For example, in this monograph, in the species *M. papillosa* Boiss. a subspecies is recognized with the name *M. papillosa* subsp. *macrocarpa* (Boiss.) Urb. Automatically, a parallel subspecies is recognized with the name *M. papillosa* subsp. *papillosa*, inclusive of those plants not in the other subspecies. Where one or more infraspecific taxa are accepted at a given rank, one must be named by the autonym.

Autopolyploidy – possession of more than two sets of chromosomes derived from within the same species; having a chromosome complement with more than two complete copies of the genome of a single, ancestral species. (Definitions differ somewhat depending on whether a genetic, cytogenetic, or taxonomic perspective is being taken. cf. Allopolyploidy. Autopolypoidy and allopolyploidy are ends of a continuum between duplicate sets of genetically homologous and nonhomologous chromosomes. The multiple sets of genomes are at least all notably similar in autopolyploids, notably different in allopolyploids.)

Awn – a bristle-like part or appendage.

Baleage – see Haylage.

Banner – the uppermost (generally the largest) petal of a papilionaceous corolla; = standard, vexillum.

Base number – the lowest common denominator of haploid chromosome numbers among the species of a genus (or taxa of some group). This is represented by x (haploid or gametic number) or $2x$ (diploid or somatic number). The base number is believed to be characteristic of the ancestral state, with divergence to other numbers by aneuploidy (which see) and polyploidy (which see). The base chromosome number in *Medicago* is $x = 8$.

Basionym – the first name published for a taxon. If the group labeled by that name is transferred to another taxon (e.g., a species moved to a different genus), the epithet remains the same, unless there is a conflict. (e.g., when *Trigonella carica* was transferred to *Medicago*, the correct name became *Medicago carica*; however, when *Trigonella aurantiaca* was transferred to *Medicago*, it became *Medicago phrygia* because there was a conflict with a preexisting name, *Medicago aurantiaca*.)

Beak – a long, prominent, and substantial tapered or pointed ending; applied particularly to prolongations of fruits and pistils.

Biennial – growing for two seasons from seed to maturity and death.

Bivoltine – having two broods in a season (referring to insect reproduction).

Blade – the thin, expanded part of a leaf or petal (also known as the lamina); the leaf or similar organ excluding the petiole (stalk), stipe, or claw.

Bract – a modified, small, protective leaf, typically associated with the inflorescence or buds.

ca. – from the Latin circa, about or approximately.

Calcareous – in reference to soil: chalky or alkaline with considerable calcium carbonate.

Campanulate – bell shaped.

Capitate – descriptive of an inflorescence with the flowers in a head, or aggregated into a very dense, compact cluster.

Carina – keel, which see.

cf. – from the Latin confer, compare with.

Chernozem – a very black topsoil, rich in humus, typical of cool to temperate, semi-arid regions such as the grasslands of European Russia.

Circumscription – the characters and other information used to describe and define a given taxon that separate that taxon from all other taxa.

Clade – a group of organisms or taxa that includes the most recent common ancestor of all of its members and all of the descendants of the most recent common ancestor.

Claw – the long, narrow (or narrowing), petiole-like base of petals or sepals in some flowers. cf. limb.

comb. – from the Latin combinatio, meaning combination.

comb. illeg. – an illegitimate combination.

comb invalid. – an invalid or not valid combination.

comb. nov. – from the Latin combination nova, new combination. When a species is transferred from one genus to another and the same specific epithet is used, the new name is a new combination (and similarly for transfers at other ranks).

cons. – from the Latin, conservandus, to be conserved, to be kept.

Coppice – a thicket of small trees or shrubs (especially one cut back regularly).

Cordate – heart-shaped; often descriptive of leaves with a sinus and two rounded lobes at the base.

Correct name – the name that must be adopted for a taxon with a particular rank, position, and circumscription, under the rules of botanical nomenclature (McNeill et al. 2006).

Cotyledon – seed leaf; the primary leaf or one of them (if several are present) of the embryo.

Crown – A prominent part of the central axis in some plants, located where the stem and roots merge, near ground level. In *Medicago* species with crowns, branches are produced from the top and sides of the crown, roots below. In many perennial *Medicago*, the crown functions as dormant tissue (in addition to the roots) to survive winter and store reserves, and also functions as the center of regrowth following severe herbivory. In alfalfa, form, size, and depth with respect to the soil surface are subject to genetic and environmental influences. As noted in Chapter 12, there is considerable variation in *Medicago sativa* with respect to crown development, some kinds lacking crowns, others producing large crowns. In most alfalfa cultivars, crown characteristics are critical to survival and productivity. A separate botanical meaning of crown, not relevant to *Medicago*, is the upper part of a tree where considerable branching occurs.

Cultigen – from the Latin cultus, cultivated, and gens, kind, refers to a group of plants that have been domesticated (genetically altered for agricultural or horticultural purposes by humans). The term has sometimes been applied to entire plant species that are known only in cultivation (sometimes also as escapes), but also sometimes to the domesticated phase of a species which exists both in the wild and in cultivation. Also, different kinds of domesticates may be known as cultigens (e.g., oilseed, fiber, and narcotic forms of *Cannabis sativa* L. have been referred to as different cultigens). The term cultigen almost always designates a much broader assemblage of plants that the term cultivar (which see), and often cultigens include several cultivars.

cultivar – (also cv., from the Latin cultivarietas, meaning cultivar) a genetically distinctive, named (with a nonscientific name) assemblage of plants, usually selected by plant breeders or horticulturists, and maintained in cultivation for its usefulness to people. Naming of cultivars is governed by the *International Code of Nomenclature for Cultivated Plants* (Brickell et al. 2009).

Cuneate (cuneiform) – wedge-shaped; triangular with the narrow end at point of attachment; often descriptive of leaves and petals.

Cuspidate – with an apex somewhat abruptly and sharply concavely constricted in an elongated, sharp-pointed tip.

Declinate – bent downward or forward, the tip(s) often recurved.

Decumbent – reclining or lying on the ground, but with the tip ascending.

Deflexed – reflexed (which see).

Dehiscence – the method or process of opening of a seed pod (also of an anther, although not used in this context in this book). In the Fabaceae, dehiscence is generally by opening of one or both sutures to release the seeds.

Dentate – with sharp, spreading, rather coarse indentations or teeth that are perpendicular to the margin.

Denticulate – minutely or finely dentate.

Depressed – more or less pressed down (in reference to a shoot or branch).

det. – for the Latin determinavit, meaning he, she, or it identified, and indicating that someone (usually with a measure of expertise) has provided an identification.

Diadelphous – pertaining to stamens, borne in two distinct bundles, or with several stamens united and an additional solitary stamen apart from them, as in many Fabaceae.

Diagnosis – a brief description of essential characteristics of a taxon that in the opinion of the author of the taxon distinguishes it from other taxa.

Diploid – pertaining to the number of chromosomes in the nonreproductive (somatic) cells of an organism or species, and designated by $2n$. cf. Haploid. Also, where there are related populations or species differing in numbers of chromosome complements, those with the lower numbers may be referred to as diploids in contrast with those with higher (polyploid) numbers of chromosome sets.

Dissected – divided into many slender segments.

Doubly serrate – with coarse serrations bearing minute teeth on their margins (i.e., the coarse teeth have fine teeth).

ead. comb. – from the Latin eadem combinatio, same combination (as previously cited).

Emarginate – with a shallow notch at the apex.

emend. – from the Latin emendavit, referring to a modification of the circumscription of a taxon. "Smith emend. Jones" means that Jones revised the inclusiveness of the taxon previously recognized by Smith.

Endemic – a species (or infraspecific taxon) with a native distribution naturally restricted geographically to a very limited area (although perhaps of much wider distribution in the distant past). The word is employed both as a noun and an adjective.

Entire – with a margin that is continuous, lacking indentations or teeth.

Ephemeral – a plant that completes its life cycle very rapidly.

Epithet – the last term of the binomial Latin name of a species, or of the trinomial name of a subspecies or variety (etc.).

Escape – a plant that has escaped from cultivation and is found growing in natural vegetation.

et al. – from the Latin et aliores, meaning "and others", the standard way of citing the authors of given publications when there are more authors than one wishes to cite for purposes of brevity.

ex – from the Latin ex, after, and sometimes used in citing the author of a Latin name when that author has taken up the same name of another author who coined the name but did not validly publish the name according to the rules of nomenclature. When so used, e.g., for the combination *M. suffrutricosa* Ramond ex DC., the second author (DC.) cited, validly published the name (or the epithet) indicated or suggested by the first (Ramond).

excl. – from the Latin exclusus, excluded.

Exsiccata (plural: exsiccatae) – from the Latin exsiccatus, dried. Usually one of multiple copies of a herbarium specimen, generally with an imprinted label giving considerable information, distributed to various herbaria and (once) considered to be a form of publication (for nomenclatural purposes, after 1953 the publication of a new botanical name in exsiccata is valid only if a separate printed publication is distributed independently of the exsiccatum itself.).

f. – forma, from the Latin forma, one of the lowest taxonomic ranks (below a variety); or, when following an author, filius, Latin for son (e.g., L. f.: son of Linnaeus).

Fabaceae – alternatively known as the Leguminosae, the third largest family of flowering plants (after Orchidaceae, orchids; and Asteraceae, sunflowers), with about 20 000 species. The family is second only to Poaceae (grasses) in economic importance. For a general account, see Wojciechowsky, M.F., Mahn, J., and Jones, B. Fabaceae. http://tolweb.org/Fabaceae/21093/2006.06.14 (current July 2010).

Falcate – strongly curved sideways, sickle-shaped.

Fodder – in this monograph: harvested plant material fed (usually in a dried or processed form) to animals (usually livestock). As discussed in the text (Chapter 2: definitions of forage, fodder, pastures, and related terms), a term also commonly used in other senses.

Forage – in this monograph: living plants (normally near ground level) consumed by free-ranging animals (usually livestock in pastures or rangelands). As discussed in the text (Chapter 2: definitions of forage, fodder, pastures, and related terms), a term also commonly used in other senses.

Forb – (variously and usually imprecisely defined) a nonwoody or soft-stemmed, upright, vascular plant (excluding the graminoid, i.e., grass or grass-like families, i.e. grasses, sedges, and rushes, sometimes excluding all monocots) of "limited" stature. A forb is most widely defined as a herb (which see) excluding the graminoids (and sometimes also excluding vascular plant groups that lack flowers), but some definitions also exclude annuals, defining a forb as a perennial that dies back to ground level at the end of the season.

Funiculus – a structure attaching an ovule to the placental wall of the ovary (equivalent to the umbilical cord of an animal). A subsequent scar (funicular scar = hilum) frequently remains on the seed (reminiscent of a belly button).

gr. – grex (or the Latin grupo): a group ("series") of related species within a genus. The word has been applied particularly to orchid hybrids, but is occasionally applied to other plants.

Garrique – low, soft-leaved scrubland found on limestone soils around the Mediterranean Basin, generally near the seacoast.

Glabrate – nearly glabrous, or becoming glabrous with maturity or age.

Glabrescent – nearly glabrous, or becoming glabrous with age; minutely and not visibly (to the naked eye) pubescent.

Glabrous – not hairy.

Halophyte – a plant particularly tolerant of various mineral salts in the soil solution, usually of sodium chloride. Halophytes habitually grow in saline or semi-saline substrates.

Haploid – pertaining to the number of chromosomes in the gamete of an organism or species, and designated by *n*. Normally this is half the chromosome number in most (somatic) cells of the organism. cf. Diploid.

Hard seed – seed in which a proportion germinates irregularly and late (typically after several months or years), usually because the seed coat is impermeable and will not take up water until the coat is broken down. Hardseedness or seed dormancy is undesirable in crops that need to be started all at once, but is desirable in self-regenerating pastures since planting is unnecessary for several years.

Haylage – a preserved form of forage, intermediate between hay and silage. Of these three forms of feed, hay has the lowest moisture content — no more than 20% when harvested (i.e., after drying; freshly cut forage typically contains 75% to 85% water) and ca. 12% when stored. Haylage is usually prepared by baling semi-wilted herbage at a higher moisture content than hay (baled haylage is sometimes called "baleage") and storing it in a sealed plastic wrap that produces an oxygen-free environment as the harvested plant material respires. Because of the higher moisture level and airtight environment, the forage ferments and is preserved by lactic acid production during fermentation, resulting in a slight acidic product (about pH 5.0). Compared with hay, the forms of ensilage have several advantages, including greater preservation of nutrients, more consistent forage quality when properly stored, better adaptation to mechanized harvesting and feeding, and less rain-free weather required for harvesting. Haylage has a lower water content (typically 35% to 55%) than silage (60% to 70%), so the fermentation process is weaker and slower compared with wet silage preparation, which results in a pH of 3.5–4.0. Unless properly prepared, haylage can develop moulds and bacteria that can cause botulism. Silage is fed to ruminant animals such as cattle and sheep; haylage is also widely used to feed cattle, and also horses to some extent.

Head – pertaining to an inflorescence, a short, crowded cluster of flowers.

Herb – a soft-stemmed plant (or at least one that lacks definitely woody stems). If perennial, herbs naturally die back to ground level at the end of the growing season. Most herbs have limited stature (although banana plants are often stated to be herbs). cf. Forb.

Heterotypic synonym (taxonomic synonym) – a synonym (which see) based on a type (which see) that is different from that of the accepted scientific name. cf. Homotypic synonym.

Hilum – in the seed, a scar or mark indicating where the ovule was attached.

Holotype – the one permanently preserved specimen or illustration used by the author or designated by the author as the nomenclatural type, i.e., the reference specimen which permanently is attached to the name and fixes its application. cf. Lectotype.

Homotypic synonym (nomenclatural synonym) – a synonym (which see) based on a type (which see) that is based on the same type as that of another name (especially the accepted scientific name). cf. Heterotypic synonym.

hort. (without an initial capital H, although often this rule is not followed) – from the Latin hortulanorum, of gardens, and indicating that the name has been employed in horticulture, but does not have standing in Latin botanical nomenclature (hort ex following by an author means that the author based his or her new name on an epithet used previously in horticulture). ("Hort" with a capital H and not followed by a period indicating abbreviation refers to the name of an author, Fenton John Anthony Hort, 1828–1892.)

Imbricate – overlapping like shingles.

in adnot. – from the Latin in adnotatione, meaning in a note or annotation.

in – citation of a name in the article or book authored by another author or other authors (e.g. "*M. coronata* (L.) Desr. in Lam." indicates that Desrousseaux published the combination in a publication authored by Lamarck). Sometimes this information is desirable to avoid confusion regarding exactly who is the author or authors of a name (i.e., the author of the name or the author of the publication). At one time it was required to include this in full citation of such names, but now it is optional.

Incised – cut, slashed irregularly, more or less deeply and sharply.

Indehiscent – pertaining to a fruit that does not regularly open to release the seeds. The opposite of dehiscent. (While fruits are generally one or the other, some fruits are "tardily dehiscent.")

Indigenous – native (which see), not exotic or introduced.

ined. – from the Latin ineditus, meaning unpublished.

Infraspecific – referring to any unit of classification below the species level (e.g., subspecies, variety). Sometimes (undesirably) used as a synonym of intraspecific, which see.

in herb – from the Latin in herbarium, for a name recorded on a herbarium collection and without status (unless subsequently validly published).

in obs. – from the Latin in observatio, recorded simply as an observation.

in sched. (plural: in schedae) – from the Latin in scheda, indicating that a name was observed on a herbarium specimen or label. Such names have often not been validly published, but the information that an investigator had at least contemplated recognizing it may be of interest. However, valid publication could be effected when such names were distributed on sets of exsiccatae specimens (100 or more specimens distributed to as many herbaria) accompanied by printed diagnoses (a situation often indicated in print by "cum notula").

in syn. – from the Latin in synonymia, in synonymy; such names have been listed as synonyms by authors who merely recognized the names as equivalents without validly publishing them.

incl. – from the Latin inclusus, including.

ined. – from the Latin ineditus, unpublished.

Inflorescence – the flowering part of a plant, or more specifically a cluster of flowers on a single stem.

Intraspecific – arising or occurring within a species; contrasts with interspecific, referring to the same phenomena between or among species.

Introduction – a plant introduced to a locality (where it is not native), almost always as a result of human influence (it is arguable whether long-distance dispersal by birds or by other natural means should be included).

Isotype – a herbarium collection (specimen) that is a duplicate of the holotype (a collection of the same taxon made by the same collector(s) from the same site at the same time).

Karyotype – in reference to the chromosome complement, the number and shape (morphology) of the chromosomes.

Keel – the two front, united petals of a papilionaceous flower (= carina).

Key – a written system of character choices, used for the identification of organisms. Dichotomous keys present a series of couplets of mutually exclusive statements, each of which leads to a further pair of choices or (ultimately in all cases) to an identification.

Lacerate – irregularly and more or less broadly and shallowly cut, as if torn.

Laciniate – irregularly slashed into narrow, pointed lobes.

Lanceolate – lance-shaped, narrowly ovate, but 3–6 times as long as broad and with the broad point below the middle, tapering to a spear-like apex.

Landrace – Variously defined (Zeven 1998). Crop landraces are plant populations (representing a part of the overall variation pattern of a cultivated species) that were domesticated by people before the age of modern plant breeding. They are typically adapted to the local environment (including climate, soil, herbivores, and disease-causing organisms) in which they were selected and are more variable than modern cultivars. Landraces are invaluable for breeding cultivars. For an example of a landrace in *Medicago*, see the section in Chapter 12 titled "Evolution of Purple-Flowered Alfalfa in Historical Times" (landrace Kayseri).

Lectotype – When a holotype (which see) was not designated at the time of publication of a scientific name, or is missing or destroyed, or is found to belong to more than one taxon, an appropriate lectotype is desirably designated as a substitute. This is a specimen or illustration designated from the original material used by the author of a name at the time he or she described the new taxon.

leg. – from the Latin legit, he gathered, and referring to the collector of a specimen.

Limb – the flat, expanded portion of a segment, such as a petal. cf. Claw.

Marcescent – withering but the remains persisting.

Macchie (= maquis) – based on the French *maquis* or Italian *macchia*, a Mediterranean-region shrubland typically consisting of densely growing evergreen shrubs such as sage, juniper, or myrtle. A similar type in North America is chaparral. The drought-resistant plants found in macchie tend to grow in arid, rocky areas.

Micropyle – a minute opening in the ovule of a seed plant through which the pollen tube usually enters. A pore may remain when the seed develops and may serve as a point of water entry to enable germination.

Misspelling – in nomenclature, a name that has been incorrectly spelled.

Moniliform – constricted at regular intervals so as to resemble a string of beads.

Montane – pertaining to lower mountain areas, typically with moist, cool slopes and often dominated by evergreens.

Morphology – the form and structure of an organism.

Mycorrhiza (plural mycorrhizae, mycorrhizas) – a symbiotic (generally mutualistic, occasionally weakly pathogenic) association between a fungus and the roots of a plant, either intracellularly (as in arbuscular mycorrhizal fungi) or extracellularly as in ectomycorrhizal fungi.

Mycorrhizal – a term describing roots that associate with a fungal mycelium, the plant deriving a benefit through symbiosis with the fungus.

Native – an organism that occurs naturally in a country or region. Native species are indigenous to the country or region.

Naturalized – a species that has either been introduced (see Introduction) or has escaped (see Escape, cf. Adventive) and reproduces in its new location.

Neotype – a specimen or illustration selected as the type when a holotype, lectotype, or syntype are unavailable (neotypes are designated normally when no candidates as type are available, and there seems no reasonable prospect that such will be discovered).

Nodule (root nodule) – a knoblike outgrowth, especially on the roots of species of the legume family (Fabaceae), containing bacteria involved in nitrogen fixation.

nom. (nomen) – from the Latin nomen, name.

nom. ambig. – from the Latin nomen ambiguum, ambiguous name, i.e., a name used in different senses that has become a long-persistent source of error.

nom. confus. – from the Latin nomen confusum, i.e., a name causing confusion that is not used because it cannot be assigned unambiguously to a particular taxon.

nom. cons. – from the Latin nomen conservandum, conserved name, i.e., a name conserved according to the *International Code of Botanical Nomenclature* (McNeill et al. 2006).

nom. illeg. (nom. illegit.) – from the Latin nomen illegitimum, an illegitimate name, i.e., a name that is illegitimate according to the *International Code of Botanical Nomenclature* (McNeill et al. 2006). ("Illegitimacy" results from several possible actions, for example, publishing a name for a group when a legitimate name, which should have been used, already exists for it.)

nom. inadmiss. – a name of a kind that cannot be accepted under the rules of nomenclature.

nom. inval. – from the Latin nomen invalidum, i.e., a name that is invalid or not valid according to the *International Code of Botanical Nomenclature* (McNeill et al. 2006). ("Invalidity" results from several possible actions, for example failing to provide a description in Latin.)

nom. nov. (n.n.; n. nov) – from the Latin nomen novum, meaning new name.

nom. nud. – (sometimes n.n.) from the Latin nomen nudum (and equivalent to "nomen"), literally a naked name, one published without description or reference to a previously published description, and accordingly not afforded any taxonomic status.

nom. superfl. – from the Latin nomen superfluum, meaning superfluous name, i.e., one that should not have been published because a suitable name already existed.

Nomenclature – the branch of biology concerned with the naming of organisms.

Nomenclatural synonym – a name based on the same type specimen.

non – the Latin non, not.

Obcordate – deeply lobed at the apex; the opposite of cordate.

Oblanceolate – the reverse of lanceolate; with respect to a leaf, broader at the distal third than at the middle and tapering toward the base.

Obligate – essential, unable to exist without (contrasts with facultative, meaning optional or able to exist with or without).

Oblong – longer than broad, with the sides nearly or quite parallel most of their length.

Obovate – the reverse of ovate, the terminal half broader than the basal half.

Obtuse – blunt, rounded; generally applied to an apex or base terminating gradually in a blunt or rounded end.

op. cit. – from the Latin opera citato: in the work cited.

Orbiculate – circular or disc-shaped.

orth. var. – orthographic variant, an incorrect or alternative spelling of the same name.

Ovate – egg-like in outline, with the broader end below the middle.

Palearctic – designating or of the biogeographical realm that includes Europe, parts of North Africa, and most of Asia.

Papillate (papillose) – bearing minute, pimple-like protuberances.

Papilionaceous – pertaining to a corolla characteristic of the papilionoid legumes; "shaped like a butterfly", having a large standard petal, two lateral wings (wing petals), and two lower petals united along the lower margin to form a carinate structure (keel) enclosing the stamens and pistil.

Pasture – an area with vegetation used by grazing animals, especially livestock. See Chapter 2 (definitions of forage, fodder, and pastures) for additional information.

Pedicel – the stalk of a single flower or fruit.

Peduncle – the stalk of an inflorescence.

Perennial – living three or more seasons.

Pergamentaceous – parchmentlike.

pers. comm. – an oral or written personal communication.

Petiolule – stalk of a leaflet.

Podzolic– pertaining to a leached soil formed mainly in cool, humid climates.

Polyploidy – presence of extra sets of chromosomes compared with ancestral groups. cf. Autopolyploidy, Allopolyploidy, Haploid, Diploid.

Polytomy (polychotomy) – a section of a phylogeny in which the evolutionary relationships can not be fully resolved to dichotomies. Soft polytomies result from data that are insufficient to fully resolve a phylogeny, hard polytomies when several lineages split essentially simultaneously.

Pontic – referring to the region of Pontus (an ancient country of northeastern Asia Minor).

p.p. – from the Latin pro parte, which see.

Procumbent – trailing or lying flat.

pro parte (p.p.) – Latin, in part, partly. Indicates that a taxon in the sense of one author includes a part only of the taxon with the same name as circumscribed by an earlier author.

Prostrate – lying flat on the ground.

pro syn. – from the Latin pro synonymo, as a synonym, i.e., denoting a nomenclatural change in the synonymy published by an author or authors. Reflecting the fact that the author(s), by relegating the name to synonymy while publishing it did not accept it as a good name; such names are not afforded taxonomic recognition, and are recorded here simply to avoid confusion should they occasionally be encountered.

proles – an intercalated rank between subspecies and variety.

Provisional name – a name given in anticipation of a valid description (see nom. provis.).

Puberulent – minutely pubescent, the hairs scarcely visible to the unaided eye.

Pubescent – hairy.

Pulvinus – a minute gland or swelling at the petiole base producing leaf movement in response to environmental conditions such as light, vibrations, and heat.

quoad – Latin, as to, as regards, with respect to; used in taxonomic citations to indicate which part of a taxon as circumscribed by an earlier author is being referred to by the present author.

quoad syn. – from the Latin quoad synonymum, according to the synonymy presented.

Raceme – a simple, elongated, indeterminate inflorescence with pedicellate or stalked flowers.

Racemose – having flowers in a raceme-like inflorescence.

Radicle – the embryonic root of a seed.

Recurved – bent or curved downward or backward.

Reflexed – abruptly recurved or bent downward or backward; abruptly deflexed at more than a 90° angle.

Reniform – kidney-shaped.

Rhizome – underground, horizontal stem (generally spreading for a distance).

Rootstalk – slow-growing, woody, upright, underground base of a perennial herbaceous plant that annually produces new stems and/or leaves.

Rostrate – having a beak or beaklike projection.

Ruderal – pertaining to a plant: one that is usually weedy and introduced to a region (i.e., of foreign origin), associated by virtue of adaptation to disturbed sites with human dwellings, construction, agriculture, roadways, etc. Ruderal plants often manifest rapidly growing seedlings, early flowering, self-pollination, and rapid maturation and release of numerous seeds (associated with a short life), with seeds ranging widely in dormancy; pertaining to a habitat: one with disturbed (and often fertile) soil, often with weeds that have replaced native plants, and frequently open (and sunny).

Rugose – wrinkled.

Satellite – a portion of a chromosome, generally at its end and separated by a constriction. Counting the number of chromosomes in a cell can be confounded when the observer mistakenly evaluates satellites as separate chromosomes.

Secund – one-sided; said of inflorescences when the flowers appear as if borne from only one side.

sens. – from the Latin sensu, in the sense of; this may (when followed by an author) refer to the manner in which that author interpreted or used a name.

sensu lato (sens. lat., s.l.) – from the Latin sensu lato, a taxon interpreted in the enlarged, broad, or comprehensive sense (including groups that are sometimes considered to belong to other taxa).

sensu stricto (s. str., or s.s.; sometimes italicized) – Latin for "in the strict sense", referring to the use of a name in a narrow sense, i.e., including only a restricted range of the variation pattern denoted by others who have used the term in a broad sense. cf. sensu lato.

Serpentine – (as applied to soils, rocks, and outcrops) dominated by certain (serpentine) minerals. Serpentine soils are toxic to most plant species because of the high concentrations of nickel, chromium and cobalt; and growth is further inhibited by low concentrations of potassium and phosphorus and a low ratio of calcium to magnesium. The flora growing on serpentine outcrops is usually distinctive, with slow-growing, specialized plants. Serpentine soils are widely distributed on Earth and are often occupied by rare or endangered endemic plants.

Serrate – a description of the margin of a leaf with (often apically directed) teeth resembling those of a saw.

Setiform – bristle-shaped.

sic – Latin, used after a word that looks wrong or absurd to show that it has been quoted correctly.

Sessile – without a stalk or essentially so.

sine auct. – from the Latin sine auctore, without authority given for the name in question.

Sinuate – wavy-margined.

Sister group – the closest relative to the group mentioned, sharing a common ancestor with it.

ssp. – subspecies; equivalent to subsp.

sp. n; sp. nov. – from the Latin species nova, meaning new species.

Standard (petal) – see Banner (also known as vexillum).

Steppe – a vast, semi-arid, grass-covered plain, as found in south-eastern Europe, Siberia, and central North America.

Strangulate – contracted or constricted at various intervals (cf. Moniliform, which is an extreme form of strangulation).

subsp. – subspecies; equivalent to ssp.

Sulcate – with a long, narrow groove, channel, or furrow (applied in the text to spines with a groove along the length).

Supernumerary – also known as B chromosomes, supernumerary chromosomes are extra chromosomes found in some (but not all) individuals of a species, not considered essential for life, characteristically heterochromatic (i.e., with certain biological stains of different appearance from the main or A chromosomes, and typically deficient in genes), and controversially of benefit or harm to the individual or population depending on circumstances.

Synonym – a name that was applied to a taxon in the past, but which should not be used because it is not the one correct name according to the *International Code of Botanical Nomenclature* (McNeill et al. 2006). cf. Heterotypic, Homotypic.

Syntype – (1) Any specimen cited in the protologue (published information accompanying the original description of a new taxon) when there is no holotype. (2) Any of two or more specimens simultaneously designated as types.

t; tab – from the Latin tabula, meaning plate.

t; teste – from the Latin teste, witness, meaning on the evidence of, according to (usually a person).

Taxa – plural of taxon.

Taxon – a named unit (group) of classification, e.g., genus, species, subspecies.

Taxonomy – the science of classification, identification, and nomenclature of organisms.

Terete – cylindrical but usually tapered somewhat towards one or both ends; round in cross section.

Terra rossa – a reddish-brown soil overlapping limestone bedrock.

Testa (plural testae) – a hard protective outer layer of the seeds of flowering plants; seedcoat.

Teste – see t; teste.

Topotype – a specimen collected from the type locality. This does not normally bear on the nomenclature of the taxon if it was not cited at the time of taxon description by the taxonomist who named it.

Tortuous – bent or twisted in various directions.

Translocations – a chromosomal abnormality in which a part of a chromosome has broken off one chromosome and attached to another. Such transfer of segments is normally harmful, but has often led to the evolution of new species of plants.

Trifoliolate – descriptive of a compound leaf with three leaflets. (This term is often confused with "trifoliate", which means having three leaves.)

typ. cons. – from the Latin typus conservandus, meaning conserved type, i.e., by special provision the type has been conserved (whereas normally it would have been rejected).

Type (nomenclatural type) – the element to which the name of a taxon is permanently attached. This is usually a collection on a herbarium sheet that is permanently preserved in a herbarium.

Umbel – an indeterminate, often flat-topped inflorescence with the pedicels and peduncles arising from a common point, resembling the stays of an umbrella.

Uncinate – hooked obtusely at the tip.

Undulate – wavy (up and down, not in and out).

var. – variety (varietas, i.e., a formal category of classification).

Vernalization – exposure of seeds, seedlings, or plants to low temperature (sometimes also moisture) to initiate or accelerate germination, development, or flowering.

Verrucose – having a wartlike or nodular surface.

Vexillum – see Banner.

Villous – having long, soft, not matted hairs.

Wing – pertaining to a papilionaceous flower, one of the two lateral petals (ala, plural alae).

× – the (mathematical) symbol used to indicate a cross between two species (e.g., *Medicago bonarotiana* × *M. rotata*), a binomial (scientific name) for a hybrid species (e.g., *Medicago* ×*varia*), or an intergeneric hybrid (e.g., ×*Triticosecale*, a hybrid between *Triticum* and *Secale*)

Xerophyte – a plant of a dry, arid habitat.

Guide to Scientific Synonyms of Species

- Thousands of scientific names have been proposed for *Medicago*. The following list of species synonyms, in conjunction with the lists of infraspecific synonyms presented in the text of this monograph for every species recognized, is the first comprehensive guide to the identity of the thousands of names that have been recorded for the genus. As noted below, several internet-based databases present extensive lists of names, mostly junior (i.e., incorrect) species names, without an indication of the correct name (because expert evaluation of this was not verified or even known). Most of the names have been available in specialist publications, but have not been previously collected and evaluated, so that their applicability and merit has been unclear.
- This list includes all species names that are given as synonyms of the 87 species of *Medicago* recognized in this monograph, as well as species belonging to other genera that were once recognized as species of *Medicago*. Names are included here regardless of nomenclatural status (i.e., whether or not the names were validly published or are legitimate).
- Names in bold are the accepted names for the 87 species recognized in this monograph. The numbers in parentheses after the names are the chapters for the 87 species in which the names that are not bolded also appear.
- The symbol ≡ means that the names of the two species are based on (or presumed to be based on) the same type specimen (and will therefore be listed in the synonymy section of the corresponding chapter under *Homotypic* names).
- The symbol = means that the names of the two species are based on (or presumed to be based on) different type specimens (and will therefore be listed in the synonymy section of the corresponding chapter under *Heterotypic* names).
- The symbol * means that the name is present in widely used plant nomenclatural lists (e.g., International Plant Names Index: http://www.ipni.org/ (the successor of Index Kewensis, referred to below); Tropicos® (Missouri Botanical Garden): http://www.tropicos.org/; International Legume Database & Information Service: http://www.ildis.org/), but synonymy of the name has not been determined (these names are old and obscure and will almost never be encountered). In these cases, publication details are provided. Several of the names are by Steudel (Steud.), who merely listed names, which are accordingly often problematical.

Note: With relatively few exceptions, the thousands of names of infraspecific taxa are found in the synonymy of the species of *Medicago* indicated below (e.g., the name *Medicago intertexta* (L.) Mill. subsp. *aculeata* Urb. is found in Chapter 42 on *Medicago intertexta*). In exceptional cases, an infraspecific name will be found only in the synonymy of a species that is not listed below (because the taxonomist who coined that name did not correctly identify the species in which the new name should be placed). For example, the name *M. intertexta* (L.) Mill. var. *decandollei* (Tin.) Urb. is found in the chapter on *M. ciliaris* since it is a synonym of *M. ciliaris*, not of *M. intertexta*. An index to these exceptional names is not provided here because they are simply taxonomic errors, of academic interest only.

Buceras elliptica Moench = *Medicago monspeliaca* (54)
Buceras monspeliacum (L.) All. ≡ *Medicago monspeliaca* (54)
Buceras mutica Moench = *Medicago polyceratia* (68)
Buceras polyceration All. = *Medicago polyceratia* (68)
Buceras torosa Moench = *Medicago medicaginoides* (51)
Buceras tortuosa Steud. = *Medicago medicaginoides* (51)

Crimaea cretacea (M. Bieb.) Vassilcz. ≡ *Medicago cretacea* (29)

Diploprion medicaginoides Viv. = *Medicago coronata* (27)

Factorovskya aschersoniana (Urb.) Eig = *Medicago hypogaea* (41)

Falcatula monspeliaca (L.) Fourr. ≡ *Medicago mospeliaca* (54)

Kamiella archiducis-nicolai (Širj.) Vassilcz. ≡ *Medicago archiducis-nicolai* (15)

Lotus medicaginoides Retz. ≡ *Medicago medicaginoides* (51)

Lupularia parviflora Opiz ≡ *Medicago lupulina* (49)

Lupulina aurata Noulet ≡ *Medicago lupulina* (49)

Medica apiculata (Willd.) E.L. Greene = *Medicago polymorpha* (69)
Medica arabica (L.) Medik. ≡ *Medicago arabica* (13)
Medica arborea (L.) Mill. ≡ *Medicago arborea* (14)

Medica bimuricata Moench = *Medicago polymorpha* (69)

Medica ciliaris Medik. ≡ *Medicago ciliaris* (24)

Medica coronata (L.) Bubani ≡ *Medicago coronata* (27)

Medica coronata (L.) Medik. ≡ *Medicago coronata* (27)

Medica denticulata (Willd.) Greene = *Medicago polymorpha* (69)

Medica disciformis (DC.) Bubani = *Medicago disciformis* (31)

Medica echinata Lam. = *Medicago arabica* (13)

Medica falcata (L.) Miller ≡ *Medicago sativa* subsp. *falcata* (12)

Medica hirsuta (L.) Bubani = *Medicago minima* (52)

Medica incana Moench = *Medicago rigidula* (78)

Medica inermis Lam. ≡ *Medicago orbicularis* (58)

Medica intertexta (L.) Medik. ≡ *Medicago intertexta* (42)

Medica intertexta (L.) Moench ≡ *Medicago intertexta* (42)

Medica italica Mill. ≡ *Medicago italica* (44)

Medica legitima Clus ex E.L. Greene = *Medicago sativa* subsp. *sativa* (12)

Medica littoralis (Loisel.) Bubani ≡ *Medicago littoralis* (48)

Medica lupulina (L.) Scop. ≡ *Medicago lupulina* (49)

Medica maculata Bubani = *Medicago arabica* (13)

Medica marina (L.) Lam. ≡ *Medicago marina* (50)

Medica maritima Bubani = *Medicago marina* (50)

Medica media Fourr. = *Medicago sativa* subsp. ×*varia* (12)

Medica muricata (Willd.) Bubani = *Medicago polymorpha* (69)

Medica orbicularis (L.) Medikus ≡ *Medicago orbicularis* (58)

Medica orbiculata Bubani ≡ *Medicago orbicularis* (58)

Medica pentacycla (DC.) Bubani = *Medicago polymorpha* (69)

Medica polymorpha (L.) Scop. ≡ *Medicago polymorpha* (69)

Medica pourretii (Noulet) Bubani = *Medicago hybrida* (40)

Medica praecox (DC.) Bubani ≡ *Medicago praecox* (71)

Medica radiata Mill. ≡ *Medicago radiata* (73)

Medica reticulata (Benth.) Bubani = *Medicago polymorpha* (69)

Medica sativa (L.) Lam. ≡ *Medicago sativa* (12)

Medica sativa (L.) Mill. ≡ *Medicago sativa* (12)

Medica scutellata Bubani ≡ *Medicago scutellata* (87)

Medica scutellata (L.) Moench ≡ *Medicago scutellata* (87)

Medica secundiflora (Durieu) Bubani ≡ *Medicago secundiflora* (88)

Medica sphaerocarpos (Bertol.) Bubani ≡ *Medicago sphaerocarpos* (92)

Medica suffruticosa (DC.) Bubani ≡ *Medicago suffruticosa* (94)

Medica sylvestris Bubani = *Medicago sativa* subsp. *falcata* var. *falcata* (12)

Medica tomentosa Moench = *Medicago marina* (50)

Medica tornata (L.) Medik. = *Medicago italica* (44)

Medica tribuloides (Desr.) Bubani = *Medicago truncatula* (97)

Medica tubinata (sic, often transcribed as *turbinata*) Medik ≡ *Medicago turbinata* (98)

Medica villosa Bubani, Fl. Pyren. 2: 478 (1900) =? *Medicago rigidula* (78)

Medicago aculeata Gaertn. = nomen confusum (see 32)

Medicago aculeata Rchb., Fl. Germ. Excurs. 503 (1832), description vague, applicable to species of section *Spirocarpos* subsect. *Pachyspirae* (type, if extant, not examined)

Medicago aculeata Willd. = *Medicago doliata* (32)

Medicago aegagropila Desv. = *Medicago granadensis* (35)

Medicago afghanica (Bordzil.) Vassilcz. = *Medicago sativa* subsp. *sativa* (12)

Medicago agrestis Ten. = *Medicago rigidula* (78)

Medicago agropyretorum Vassilcz. = *Medicago sativa* subsp. *caerulea* (12)

Medicago alaschanica Vassilcz. = *Medicago sativa* subsp. ×varia (12)

Medicago alatavica Vassilcz. = *Medicago sativa* subsp. ×varia (12)

Medicago alba (Medik.) E.H.L. Krause, Fl. Deutschland (Sturm – ed. 2) 9: 128 (1901) ≡ *Melilotus alba* Medik.

Medicago altissima (Thuill.) E.H.L. Krause, Fl. Deutschland (Sturm – ed. 2) 9: 128 (1901) ≡ *Melilotus altissima* Thuill.

Medicago ambigua Jordan ex Boreau = *Medicago orbicularis* (58)

Medicago americana Larrañaga, Escritos de Don Dámaso Antonio Larrañaga, 2: 233 (1923). The description is applicable to several spiny-fruited, annual *Medicago* species, but given the location of Uruguay and the fact that *M. polymorpha* is mentioned separately, this is probably *M. minima* (52).

**Medicago amethystina* C.L. Gmelin ex Steud., Nomencl. Bot. 513 (1821)

Medicago anguina (Delile) C. Romero Zarco ≡ *Trigonella anguina* Delile

Medicago angulata Láng (Adolf Franz Láng, 1795–1863), Enum. Pl. Hung. 2 (1822?). An early version of Index Kewensis identified this as *Medicago sativa* (12)

Medicago angustifolia Chevallier = *Medicago sativa* subsp. *falcata* var. *falcata* (12)

Medicago annularis Besser = *Medicago sativa* subsp. ×varia (12)

Medicago apennina J. Woods = *Medicago lupulina* (49)

Medicago apiculata Bast. = *Medicago turbinata* (98)

Medicago apiculata Willd. = *Medicago polymorpha* (69)

Medicago applanata Willd. ex Hornem. = *Medicago orbicularis* (58)

Medicago applanata Willd. ex Schlecht. = *Medicago orbicularis* (58)

Medicago arabica (L.) All. ≡ *Medicago arabica* (13)

***Medicago arabica* (L.) Huds. (13)**

Medicago arborea Aubl. ≡ *Medicago arborea* L. (14)

***Medicago arborea* L. (14)**

Medicago arborescens C. Presl = *Medicago arborea* (14)

***Medicago archiducis-nicolai* Širj. (15)**

Medicago arcuata (C.A. Mey.) Trautv. = *Medicago medicaginoides* (51)

Medicago arenaria Ten. = *Medicago littoralis* (48)

***Medicago arenicola* (Hub.-Mor.) E. Small (16)**

Medicago aschersoniana Urb. = *Medicago laciniata* (45)

Medicago asiatica Sinskaya = *Medicago sativa* subsp. *sativa* (12)

Medicago astroites Bertol. = *Medicago italica* (44)

***Medicago astroites* (Fisch. & C.A. Mey.) Trautv. (17)**

Medicago aurantiaca Godr. = *Medicago sativa* subsp. *falcata* var. *falcata* (12)

**Medicago banofoensis* Kit., Linnaea 32: 615 (1863). Description quite inadequate for identification.

Medicago beipinensis Vassilcz. = *Medicago sativa* subsp. *caerulea* (12)

Medicago berteroana Moris = *Medicago polymorpha* (69)

Medicago berteroana Nyman, Consp. Fl. Eur. 169 (1878), listed in various nomenclatural compendia, but merely incidentally mentioned as a synonym of "*M. apicularis* W." (*Medicago polymorpha*, 69)

Medicago besseriana (Ser.) E.H.L. Krause ≡ *Trigonella besseriana* Ser.

Medicago biancae Todaro ex Urban, Urban (1873), listed as a species in various nomenclatural compendia, but Urban recognized only a form of *Medicago orbicularis* (58)

Medicago biancae (Urb.) Pinto da Silva = *Medicago orbicularis* (58)

***Medicago biflora* (Griseb.) E. Small (18)**

Medicago bimuricata Retz. ex Steud., Nomencl. Bot. 513 (1821) =? *Medicago polymorpha* (69)

Medicago blancheana Boiss. = *Medicago bonarotiana* × *M. rotata* (19)

***Medicago bonarotiana* Arcang. (19)**

Medicago bondevii Kozukharov = *Medicago rigidula* (78)

Medicago bonofcensis Kit. = *Medicago rigidula* (78)

Medicago borealis Grossh. = *Medicago sativa* subsp. *falcata* var. *falcata* (12)

Medicago brachyacantha A. Kerner = *Medicago minima* (52)

***Medicago brachycarpa* Fisch. ex M. Bieb. (20)**

"*Medicago brachycarpa* Fisch." ≡ *Medicago brachycarpa* (20)

Medicago brachyloba Fisch. ex Moris = *Medicago brachycarpa* (20)

Medicago braunii Godron in Gren. & Godron = *Medicago littoralis* (48)

Medicago breviflora Gilib. = *Medicago lupulina* (49)

Medicago caerulea Less. ex Ledeb. = *Medicago sativa* subsp. *caerulea* (12)

Medicago calcar Lowe = *Medicago italica* × *Medicago littoralis* (44)

Medicago canariensis Benth. ex Webb = *Medicago italica* × *Medicago littoralis* (44)

Medicago cancellata (Desf. ex Pers.) Kryl. = *Medicago medicaginoides* (51)

Medicago cancellata Kryl. = *Medicago medicaginoides* (51)

***Medicago cancellata* M. Bieb. (21)**

Medicago cancellata Ten. = *Medicago tenoreana* (96)

Medicago canescens Menyharth = *Medicago lupulina* (49)

Medicago canescens R. Graham, Numer. List n. 5946 (1831); mentioned as a synonym of *M. denticulata* in an early edition of Index Kewensis, so possibly = *Medicago polymorpha* (69)

***Medicago carica* (Hub.-Mor.) E. Small (22)**

***Medicago carstiensis* Wulfen (23)**

Medicago ×*casellasii* Montserrat, an alleged hybrid of *Medicago littoralis* (48) and *M. praecox* (71)

Medicago caspia Grunes ex Boiss. = *Medicago polymorpha* (69)

Medicago caspica Jacq. ex Spreng. = *Medicago polymorpha* (69)

Medicago catalonica hort ex Urb. = *Medicago polymorpha* (69)

Medicago catalonica Schrank = *Medicago turbinata* (98)

Medicago caucasica Vassilcz. = *Medicago sativa* subsp. ×*varia* (12)

Medicago cibinensis Lerchenfeld ex Schur = *Medicago sativa* subsp. ×*varia* (12)

Medicago. *ciliaris* Balb. ex DC. = *Medicago turbinata* (98)

***Medicago ciliaris* (L.) All. (24)**

Medicago ciliaris (L.) Krock. ≡ *Medicago ciliaris* (24)

Medicago ciliaris Luce = *Medicago lupulina* (49)

Medicago ciliaris Savi = *Medicago polymorpha* (69)

Medicago cinerascens Jord. = *Medicago rigidula* (78)

Medicago circinnata L. ≡ *Hymenocaros circinatus* (L.) Savi var. *nummularius* (DC.) Bornm.

Medicago cirtensis Rouy = *Medicago sativa* subsp. *glomerata* (12)

Medicago ciscaucasica B. Fedtsch. = *Medicago cancellata* (21)

***Medicago citrina* (Font Quer) (25)**

Medicago clypeata Lindl. = *Medicago orbicularis* (58)

**Medicago cochleata* Pall., Reise Suedl. Statthaltersch. Russ. Reich. 2: 336 (1801)

Medicago coelesyriaca E.H.L. Krause, Fl. Deutschland (Sturm – ed.2) 9: 121 (190) ≡ *Trigonella caelesyriaca* Boiss.

Medicago coerulea Less. ex Nyman = *Medicago sativa* subsp. *caerulea* (12)

Medicago commutata Tod. = *Medicago italica* × *Medicago littoralis* (44)

**Medicago compressa* Biroli ex Colla, Herb. Pedem. 2: 100 (1834). Colla (1834) suggested the affinities of this are with *Medicago tuberculata*, i.e., *Medicago turbinata* (98)

Medicago confinis W.D.J. Koch = *Medicago polymorpha* (69)

Medicago confinis (W.D.J. Koch) Black = *Medicago polymorpha* (69)

Medicago conica Schkuhr = *Medicago polymorpha* (69)

Medicago connivens Trautv. = *Medicago medicaginoides* (51)

***Medicago constricta* Durieu (26)**

**Medicago contorta* Gilib., Fl. Lit. Inch. 2: 94 ("1781" = 1782). An early version of Index Kewensis identified this as *Medicago sativa* (12)

Medicago cordata Desr. in Lam. = *Medicago arabica* (13)

Medicago corniculata (L.) Trautv. ≡ *Trigonella corniculata* L.

***Medicago coronata* (L.) Bartal. (27)**

Medicago coronata (L.) Desr. in Lam. ≡ *Medicago coronata* (27)

Medicago coronata (L.) Gaertn. = *Medicago polymorpha* (69)?; = *Medicago turbinata* (98)?

Medicago coronata Pall. = *Medicago arabica* (13)

Medicago corrugata Durieu in Duchartre = *Medicago italica* (44)

Medicago corymbifera W.L.E. Schmidt ex Schltdl. = *Medicago lupulina* (49)

**Medicago crassa* Porta, Nuovo Giorn. Bot. Ital. 11: 249 (1879)

***Medicago crassipes* (Boiss.) E. Small (28)**

Medicago crassispina Vis. = *Medicago truncatula* (97)

***Medicago cretacea* M. Bieb. (29)**

**Medicago cretica* Steud., Nomencl. Bot. 512 (1821)

Medicago crinita C. Presl = *Medicago intertexta* (42)

Medicago cuneata Woods = *Medicago orbicularis* (58)

Medicago cupaniana Guss. = *Medicago lupulina* (49)

Medicago cuspidata Tausch = *Medicago saxatilis* (86)

Medicago cyclocarpa Hy = *Medicago sativa* subsp. ×*varia* (12)

Medicago cylindracea DC. = *Medicago littoralis* (48)

Medicago cyrenaica Maire & Weiller = *Medicago polymorpha* (69)

***Medicago daghestanica* Rupr. ex Boiss. (30)**

**Medicago decaisneana* Tod., Ind. Sem. Hort. Panorm. 35 (1866)

Medicago decandollei Tineo ex Guss. = *Medicago muricoleptis* (56)

Medicago declinata Kit. = *Medicago prostrata* (72)

Medicago dentata E.H.L. Krause, Fl. Deutschland (Sturm – ed. 2), 9: 121 (1901) ≡ *Melilotus dentatus* Pers.

Medicago denticulata Willd. = *Medicago polymorpha* (69)

Medicago denticulata (Willd.) Boiss. = *Medicago polymorpha* (69)

Medicago depressa Jordan = *Medicago rigidula* (78)

Medicago dicarpa Trautv. = *Medicago monantha* (53)

Medicago difalcata Sinskaya = *Medicago sativa* subsp. *falcata* var. *falcata* (12)

Medicago diffusa Poiret = *Medicago laciniata* (45)

Medicago diffusa Steud., Nomenc. Bot. 513 (1821). An early version of Index Kewensis identified this as *Medicago prostrata* (72)

***Medicago disciformis* DC. (31)**

Medicago distans Poir. = *Medicago polymorpha* (69)

***Medicago doliata* Carmign. (32)**

**Medicago dubia* E.H.L. Krause, Fl. Deutschland (Sturm – ed. 2) 9: 142 (1901)

Medicago dzhawakhetica Bordz. = *Medicago papillosa* (62)

Medicago echinata Banks & Sol., Nat. Hist. Aleppo (ed. 2) 2: 261 (1794) = *Onobrychis echinata* (Banks & Sol.) Eig.

Medicago echinata Boucher , Extr. Fl. Abbeville 56 (1803). An early version of Index Kewensis identified this as *Medicago hispida*, i.e. *Medicago polymorpha* (12)

Medicago echinata St. Lag. = *Medicago intertexta* (42)

Medicago echinata Steud. = *Medicago orbicularis* (58)

Medicago echinus DC. = *Medicago intertexta* (42)

***Medicago edgeworthii* Širj. (33)**

Medicago elegans Jacq. ex Willd. = *Medicago rugosa* (82)

Medicago elegans Pourr. ex Willd. = *Medicago rugosa* (82)

Medicago elegans Pourr. ex Willk. & Lange = *Medicago rugosa* (82)

Medicago elongata (Roch.) Simonk. ex A. Kerner = *Medicago minima* (52)

Medicago erecta Kotov = *Medicago sativa* subsp. *falcata* var. *falcata* (12)

Medicago erecta Winterl = *Medicago prostrata* (72)

"*Medicago eriocarpa* Rouy" = *Medicago rigidula* (78)

Medicago esculenta Rottler ex Willd. = *Trigonella corniculata* L.

Medicago eusativa Sinsk. = *Medicago sativa* subsp. *sativa* (12)

Medicago falcata L. = *Medicago sativa* subsp. *falcata* (12)

Medicago falcata Lam. = *Medicago sativa* subsp. *falcata* (12)

Medicago falcata-sativa Rchb. = *Medicago sativa* subsp. ×*varia* (12)

Medicago fenstralis Gerh. = *Medicago tenoreana* (96)

***Medicago fischeriana* (Ser.) Trautv. (34)**

Medicago fissa Trautv. = *Medicago polyceratia* (68)

Medicago flexuosa Ten. (= ? See *Medicago polymorpha* (69))

Medicago foenum-graecum (L.) E.H.L. Krause ≡ *Trigonella foenum-graecum* L.

Medicago fructicans Hornem., Cat. Hort. Hafin. 24 (1807); an early version of Index Kewensis identified this as *Medicago cretacea* (29)

Medicago gaditana Pérez Lara ex Willk. = *Medicago rigidula* (78)

Medicago galilaea Boiss. = *Medicago granadensis* (33)

Medicago gerardi Waldst. & Kit. ex Willd. = *Medicago rigidula* (78)

Medicago germana Jordan = *Medicago rigidula* (78)

Medicago glabella Presl = *Medicago truncatula* (97)

Medicago glabra Aubl., Hist. Pl. Guiane 778-779 (1775). Not identified, not a *Medicago* species.

Medicago gladiata (Steven ex M. Bieb.) E.H.L. Krause ≡ *Trigonella gladiata* Steven ex M. Bieb.

Medicago glandulosa Davidov = *Medicago sativa* subsp. *falcata* var. *viscosa* (12)

Medicago glandulosa (Mert. & W.D.J. Koch) Davidov = *Medicago sativa* subsp. *falcata* var. *viscosa* (12)

Medicago globosa Presl = *Medicago constricta* (26)

Medicago globulifera Desf. = *Medicago sphaerocarpos* (92)

Medicago globulosa Desv. = *Medicago doliata* (32)

Medicago glomerata Balb. = *Medicago sativa* subsp. *glomerata* (12)

Medicago glomerata Fisch. ex Moris = *Medicago brachycarpa* (20)

Medicago glutinosa M. Bieb. = *Medicago sativa* subsp. *glomerata* (12)

Medicago gordejevi Komarov, Bot. Mater. Gerb. Glavn. Bot. Sada R.S.F.S.R. 2: 133 (1921) ≡ *Trifolium gordejevi* (Kom.) Z. Wei.

Medicago gordejevi Komarov, Kom. 11: 129 (1941) i Alis. Opred. Rast. Dal'nevost. Kraya, (1932) ≡ *Trifolium gordejevi* (Kom.) Wei Zhi.

Medicago gracilis Biv. = *Medicago littoralis* (48)

Medicago gracilis E.H.L. Krause, Fl. Deutschland (Sturm – ed. 2) 9: 129 (1901) = *Melilotus gracilis* DC.

Medicago gracillima Tineo ex. Urb. = *Medicago polymorpha* (69)

Medicago graeca Hornem. = *Medicago minima* (52)

***Medicago granadensis* Willd. (35)**

Medicago granatensis Willd. ex. Ser. in DC. = *Medicago granadensis* (35)

Medicago grandiflora (Grossh.) Vassilcz. = *Medicago sativa* subsp. ×*varia* (12)

Medicago grossheimii Vassilcz. = *Medicago sativa* subsp. *sativa* × subsp. *glomerata* (12)

Medicago gunibica Vassilcz. = *Medicago sativa* subsp. *glomerata* (12)

***Medicago halophila* (Boiss.) E. Small (36)**

Medicago hamata Presl = *Medicago polymorpha* (69)

Medicago hamosa (L.) Trautv. ≡ *Trigonella hamosa* L.

***Medicago heldreichii* E. Small (37)**

Medicago helix Willd. = *Medicago italica* (44)

Medicago hemicoerulea Sinskaya = *Medicago sativa* subsp. *caerulea* (12)

Medicago hemicycla Grossh. = *Medicago sativa* subsp. ×*varia* (12)

Medicago heterocarpa Duerrnb. ex Murr, Deutsch. Bot. Monatsschr. 15: 199 (1897). Allegedly (and extremely unlikely to be) a hybrid of *Medicago minima* and *Medicago sativa* subsp. *falcata*.

Medicago heterocarpa hort ex Urb. = *Medicago polymorpha* (69)

Medicago heterocarpa Spach ex Cosson & Durieu = *Medicago murex* (55)

***Medicago heyniana* Greuter (38)**

Medicago hirsuta (L.) All., Fl. Pedem. = *Medicago minima* (52)

Medicago hirsuta (L.) Bartal. = *Medicago minima* (52)

Medicago hirsuta Schkuhr, Botanisches Handb. 2: 420 (1796). An early version of Index Kewensis identified this as *Medicago denticulata*, i.e., *Medicago polymorpha* (69)

Medicago hirsuta Thuill., Fl. Env. Paris, 390 (1799). An early version of Index Kewensis identified this as *Medicago rigidula* (78), but the short description is consistent with this and several other species.

Medicago hirsuta Vill., Hist. Pl. dauph. 3: 403 (1789). Villars introduced this as "*Medicago hirsuta* Linn.", a name listed above. Nevertheless, the name has been listed in Index Kewensis (an early version identified it as *Medicago hispida*, i.e., *Medicago polymorpha* (69)). The description is quite imprecise, even if interpreted as a new taxon of Villars.

Medicago hirsutissima Cyrillo ex Steud. = *Medicago minima* (52)

Medicago hispanica Steud., Nomencl. Bot. (ed. 2) 2(9): 109 (1841) = *Hymenocaros circinatus* (L.) Savi var. *nummularius* (DC.) Bornm.

Medicago hispida Gaertn. = *Medicago polymorpha* (69)

Medicago hornemaniana ("*M. hornemanniana*") Ser. in DC. = *Medicago truncatula* (97)

***Medicago huberi* E. Small (39)**

Medicago hybrida Boiss., Diagn. Pl. Orient., ser 1, 2(9): 12 (1849). Merely an incidental mention without information as to status, a name without status that has nevertheless been listed in nomenclatural databanks.

**Medicago hybrida* E.H.L. Krause, Fl. Deutschland (Sturm – ed. 2) 9: 141 (1901)

***Medicago hybrida* (Pourr.) Trautv. (40)**

***Medicago hypogaea* E. Small (41)**

Medicago hystrix Ten. = *Medicago polymorpha* (69)

Medicago ignatzii Lesins and Lesins = *Medicago papillosa* (62)

Medicago immaculata Schur = *Medicago arabica* (13)

Medicago inconspicua Nevski = *Medicago minima* (52)

Medicago inermis (Lam.) Lam. ex Steud., Nomencl. Bot. 513 (1821) = *Medicago scutellata* (87)). (An early version of Index Kewensis equated this to *M. orbicularis*, *M. scutellata*, *M. tornata*, and *M. turbinata*).

Medicago inermis Urb., Ind. Sem. Hort. Berol. (Berlin) 4 (1872) = *Medicago polymorpha* (69). (One of several of Urban's names in this publication interpreted in nomenclatural lists as a "species", although Urban simply presented an infraspecific classification of "*M. hispida*" (*M. polymorpha*).)

Medicago intermedia DC. ex Ser. in DC. = *Medicago italica* (44)

Medicago intermedia DC. ex G. Don = *Medicago polymorpha* (69)

Medicago intermedia Schultes = *Medicago sativa* subsp. *falcata* var. *falcata* (12)

Medicago intertexta (L.) All. ≡ *Medicago intertexta* (42)

Medicago intertexta (L.) Gaertn. ≡ *Medicago intertexta* (42)

***Medicago intertexta* (L.) Mill. (42)**

Medicago intertexta Willd. = *Medicago ciliaris* (24)

***Medicago isthmocarpa* (Boiss. & Balansa) E. Small (43)**

Medicago italica (L.) E.H.L. Krause ≡ *Melilotus italicus* (*italica*) (L.) Lam.

***Medicago italica* (Mill.) Grande (44)**

Medicago italica (Mill.) Steud. ≡ *Medicago italica* (44)

Medicago italica (Mill.) Steud. ex Fiori ≡ *Medicago italica* (44)

Medicago jemenensis Sinsk. = *Medicago sativa* subsp. *sativa* (12)

Medicago juliae Sennen, Bull. Soc. Bot. France 78: 186 (1931). (Name without status, merely published with incidental mention, but often included in nomenclatural lists.)

**Medicago justischmidtii* E.H.L. Krause, Fl. Deutschland (Sturm – ed. 2) 9: 121 (1901)

Medicago karatschaica Latsch. = *Medicago sativa* subsp. *sativa* (12)

Medicago komarovii Vassilcz. = *Medicago sativa* subsp. ×*varia* (12)

Medicago kopetdaghi Vassilcz. = *Medicago sativa* subsp. *sativa* (12)

Medicago korshinskyi (Grossh.) R. Kamelin & Gubanov = *Medicago ruthenica* (84)

Medicago kotovii O.D. Wissjul. = *Medicago sativa* subsp. *falcata* var. *falcata* (12)

Medicago kultiassovii Vassilc. = *Medicago sativa* subsp. ×*varia* (12)

***Medicago laciniata* (L.) Mill. (45)**

Medicago ladak Vassilcz. = *Medicago sativa* subsp. ×*varia* (12)

Medicago laevis Desf. = *Medicago italica* (44)

**Medicago langeana* Tod., Ind. Sem. Hort. Panorm. 23, 31 (1867)

***Medicago lanigera* E. Winkl. & B. Fedtsch. (46)**

Medicago lappacea Desr. = *Medicago polymorpha* (69)

Medicago lavrenkoi Vassilcz. = *Medicago sativa* subsp. ×*varia* (12)

Medicago laxiflora Schur = *Medicago sativa* subsp. *falcata* var. *falcata* (12)

***Medicago laxispira* Heyn (47)**

Medicago leiocarpa Benth. = *Medicago sufruticosa* (94)

Medicago leiocarpa Guss. = *Medicago lupulina* (49)

Medicago lenticularis Desr. in Lam. = *Medicago italica* (44)

Medicago lenticulata Pourret ex Benth. = *Medicago suffruticosa* (94)

Medicago leonis Sennen & Mauricio =? *Medicago laciniata* (45)

Medicago lesinsii E. Small = *Medicago murex* (55)

Medicago lessingii Fisch. & Mey. ex Kar. = *Medicago sativa* subsp. *caerulea* (12)

Medicago liaosiensis (P.Y. Fu & Y.A. Chen) X.Y. Zhu & Y.F. Du = *Medicago ruthenica* (84)

Medicago libyca Spreng. = *Medicago coronata* (27)

Medicago lilacea Hy = *Medicago sativa* subsp. ×*varia* (12)

Medicago lineata Tausch = *Medicago minima* (52)

***Medicago littoralis* Rohde ex Loisel. (48)**

Medicago loreti Albert = *Medicago polymorpha* (69)

Medicago lunata (Boiss.) Boiss. = *Medicago biflora* (18)

Medicago lunata Rchb. = *Medicago radiata* (73)

***Medicago lupulina* L. (49)**

Medicago macrocarpa Moris = *Medicago sativa* subsp. *falcata* var. *falcata* (12)

Medicago macrocarpa (Urb.) Goldschmidt = *Medicago sativa* subsp. *sativa* (12)

Medicago maculata Hook. & Arn., Bot. Beechey Voy. 330 (1794). Mistakenly listed as a new species name in nomenclatural lists; Hooker and Arnott merely called this "*Medicago maculata* Sibth.", noted next.

Medicago maculata Sibth. = *Medicago arabica* (13)

Medicago maculata Willd. = *Medicago arabica* (13)

Medicago marginata Willd. = *Medicago orbicularis* (58)

***Medicago marina* L. (50)**

Medicago marina Pall., Reise Russ. Reich. 3: 590 (1776). Listed as a new species name in nomenclatural databanks, but merely mentioned by P.S. Pallas as a species observed during an expedition.

Medicago maritima Pall. ex Georgi = *Medicago marina* (50)

Medicago media Pers. = *Medicago sativa* subsp. ×*varia* (12)

***Medicago medicaginoides* (Retz.) E. Small (51)**

Medicago mesopotamica Vassilcz. = *Medicago sativa* subsp. *sativa* (12)

**Medicago mexicana* Sweet, Hort. Brit. (ed. 2) 133 (1830)

Medicago meyeri Gruner = *Medicago minima* (52)

Medicago microcarpa Balbis in Bernh. = *Trifolium filiforme* L.

Medicago microdon Ehrenb., = *Medicago polymorpha* (69)

***Medicago minima* (L.) Bartal. (52)**

Medicago minima (L.) Grufb. ≡ *Medicago minima* (52)

Medicago minima (L.) L. ≡ *Medicago minima* (52)

Medicago minima (L.) L. ex Bartal. ≡ *Medicago minima* (52)

Medicago mirandana Sennen & Elias = *Medicago sativa* subsp. ×*varia* (12)

Medicago mitis Del. ex Godr. = *Medicago rigidula* (78)

Medicago mitis Del. ex Nyman = *Medicago rigidula* (78)

Medicago mitis Willd. ex Urb. = *Medicago rigidula* (78)

Medicago mitra Willd. ex Urb. = *Medicago polymorpha* (69)

Medicago ×*mixta* Sennh. = *Medicago falcata* × *Medicago prostrata* (see 72)

Medicago mniocarpa Wallr. ex Ser. in DC. = *Medicago lupulina* (49)

Medicago mollissima Roth = *Medicago minima* (52)

Medicago monantha Blatt. = *Medicago laciniata* (45)

***Medicago monantha* (C.A. Mey.) Trautv. (53)**

***Medicago monspeliaca* (L.) Trautv. (54)**

Medicago morisiana Jord. = *Medicago rigidula* (78)

Medicago munnularia Steven = *Medicago cretacea* (29)

***Medicago murex* Willd. (55)**

**Medicago muricata* Desr. In Lam., Encycl. 3(2) 634 (1792) =? *Medicago rigidula* (78)

Medicago muricata Griseb., Spic. Fl. Rumel. 1(1): 17 (1843). The description suggests this is *Medicago littoralis* × *Medicago truncatula* (see 48)

Medicago muricata (L.) All. = *Medicago rigidula* (78)

Medicago muricata (L.) Willd. = *Medicago italica* × *Medicago littoralis* (44)

Medicago muricata Salzm. ex Ball, J. Linn. Soc. Bot. 16: 416 (1878). A name listed in nomenclatural databanks, but this was merely mentioned in synonymy of *Medicago intertexta* (42)

Medicago muricoleptis DC. = *Medicago muricoleptis* (56)

Medicago muricoleptis hort ex Urb., Ind. Sem. Hort. Berol. 4 (1872). Urban stated this was not Tineo's *M. muricoleptis* (below), but the name is merely listed as a variant of *M. hispida*, i.e., *Medicago polymorpha* (69)

***Medicago muricoleptis* Tineo (56)**

Medicago muricula St. Lag. = *Medicago murex* (55)

Medicago neglecta Guss. = *Medicago doliata* (32)

Medicago neurocarpos Kunze ex. Urb. = *Medicago polymorpha* (69)

Medicago nigra (L.) Krock. = *Medicago polymorpha* (69)

**Medicago noccae* Balb., Giorn. Fis. 8: 65, pl. 2 (1815). Description too vague to assess identity.

***Medicago noeana* Boiss. (57)**

Medicago nummularia DC. ≡ *Hymenocaros circinatus* (L.) Savi var. *nummularius* (DC.) Bornm.

Medicago nummularia Stev. = *Medicago cretacea* (12)

Medicago obscura Retz. = *Medicago italica* (44)

Medicago obscura Vis. = *Medicago polymorpha* (49)

Medicago ochroleuca M. Kult. = *Medicago sativa* subsp. ×*varia* (12)

Medicago officinalis (L.) E.H.L. Krause ≡ *Melilotus officinalis* (L.) Lam.

Medicago oligocarpa Corbiere = *Medicago polymorpha* (69)

Medicago oliviformis ("*olivaeformis*") Guss. = *Medicago doliata* (32)

Medicago ononidea De Coincy ex Rouy & Fouc. = *Medicago minima* (52)

Medicago ononidea (Rouy) A.W. Hill = *Medicago minima* (52)

Medicago orbiculariodes P. Candargy = *Medicago rugosa* (82)

Medicago orbicularis (L.) All. ≡ *Medicago orbicularis* (58)

***Medicago orbicularis* (L.) Bartal. (58)**

Medicago orientalis Vassilcz. = *Medicago sativa* subsp. *sativa* (12)

Medicago ornithopodioides (L.) Trautv. ≡ *Trifolium ornithopodioides* (L.) Sm.

***Medicago orthoceras* (Kar. & Kir.) Trautv. (59)**

***Medicago ovalis* (Boiss.) Urb. (60)**

Medicago ovata Carmign. = *Medicago murex* (55) or *Medicago sphaerocarpos* (92)

Medicago ovata hort. ex Steud., Nomencl. Bot. 514 (1821). cf. *Medicago ovata* Carmign.

Medicago oxalioides Schur = *Medicago arabica* (13)

Medicago paillecu Philippe = *Medicago suffruticosa* (94)

Medicago paillerii Philippe ex Nyman = *Medicago suffruticosa* (94)

***Medicago pamphylica* (Huber-Mor. & Širj.) E. Small (61)**

Medicago panormitana Tineo ex Urb. = *Medicago intertexta* (42)

***Medicago papillosa* Boiss. (62)**

**Medicago paradoxa* Sprengel ex Steud., Nomencl. Bot. 514 (1821)

Medicago parviflora (Desf.) E.H.L. Krause = *Melilotus indica* (L.) All.

Medicago pauciflora Ledeb. = *Medicago sativa* subsp. ×*varia* (12)

**Medicago pedunculata* Ehrenb. ex Sweet, Hort. Brit. (ed. 2) 134 (1830)

Medicago pekinensis Vassilcz. = *Medicago sativa* subsp. *caerulea* (12)

**Medicago pennata* Schwaegrichen ex Steud., Nomencl. Bot. 514 (1821)

Medicago pentacycla DC. = *Medicago polymorpha* (69)

***Medicago persica* (Boiss.) E. Small (63)**

Medicago persica Stev. ex Fisch. = *Medicago polymorpha* (69)

***Medicago phrygia* (Boiss. & Balansa) E. Small (64)**

Medicago pilosa Fisch. ex Moris = *Medicago brachycarpa* (20)

**Medicago pilosa* Schleicher, Cat. Pl. Helv. (ed. 3) 46 (1815)

**Medicago pinnatifida* Steud., Nomencl. Bot. 514 (1821)

***Medicago pironae* Vis. (65)**

Medicago plagiospira Durieu = *Medicago soleirolii* (91)

Medicago platycarpos (L.) Ledeb. = *Medicago platycarpos* (66)

***Medicago platycarpos* ("*platycarpa*") (L.) Trautv. (66)**

***Medicago plicata* (Boiss. & Balansa) Širj. (67)**

Medicago plumbea Bertol. = *Medicago italica* (44)

Medicago polia (C.J. Brand) Vassilcz. = *Medicago sativa* subsp. *sativa* (12)

**Medicago polonica* Bellardi ex Balbis, Herb. Pedem. 2: 100 (1834). Colla (1834) suggested this is *Medicago minima* (52)

Medicago polyantha Dulac = *Medicago polymorpha* (69)

**Medicago polycarpa* Schur, Verh. Naturf. Vereins Brünn 15(2): 175 (1877). Description somewhat vague, but probably *Medicago polymorpha* (69)

Medicago polycarpa Willd. = *Medicago polymorpha* (69)

Medicago polycarpa Willd. ex Godr. = *Medicago polymorpha* (69)

***Medicago polyceratia* (L.) Trautv. (68)**

Medicago polychroa Grossh. = *Medicago sativa* subsp. *sativa* × subsp. *glomerata* (12)

Medicago polycycla Moris = *Medicago intertexta* (42)

***Medicago polymorpha* L. (69)**

Medicago polymorpha Roxb., Hort. Bengal. 58 (1814). Name listed in nomencatural databanks, but Roxsburgh merely published a list of names, not attributed to himself.

Medicago polyphorma Larrañaga = *Medicago polymorpha* (69)

Medicago pontificalis Gennari = *Medicago praecox* (71)

***Medicago popovii* (Korovin) Širj. (70)**

Medicago pourretii Noulet = *Medicago hybrida* (40)

***Medicago praecox* DC. (71)**

Medicago praesativa Sinskaya = *Medicago sativa* subsp. *sativa* (12)

Medicago procumbens Besser = *Medicago sativa* subsp. *falcata* var. *falcata* (12)

***Medicago prostrata* Jacq. (72)**

Medicago pseudo-gerardi Nyman, Consp. Fl. Eur. 168 (1878). A name included in nomenclatural databanks, but Nyman merely listed it in synonymy of *Medicago rigidula* (78)

Medicago pseudogerardii Schur = *Medicago rigidula* (78)

Medicago pseudogranatensis Blatt. = *Medicago laciniata* (45)

Medicago pseudorupestris Hayek = *Medicago prostrata* (72)

Medicago psilocarpa Trautv. = *Medicago monantha* (53)

Medicago pubescens DC. = *Medicago turbinata* (98)

Medicago pubescens DC. & Lam. = *Medicago turbinata* (98)

Medicago pubescens (Edgew. ex Baker) Širj. ≡ *Medicago edgeworthii* (33)

Medicago pubescens Guss., Pl. Rar. 313 (1826). Mistakenly listed in nomenclatural databanks as a species; Gussone recognized a variety, *pubescens*, of *Medicago prostrata* (72)

Medicago pubescens Hornem. = *Medicago truncatula* (97)

**Medicago pugionata* Trautv., Bull. Sci. Acad. Imp. Sci. St. Petersb. 8: 272 (1841)

Medicago pulchella Lowe = *Medicago minima* (52)

Medicago pulchella Tod. = *Medicago minima* (52)

Medicago pungens Trautv. = *Trigonella maritima* Delile

Medicago purpurea Gueldenstaedt, Reise 1: 195 (1787). Merely a published name without any accompanying information.

Medicago pusilla Viv. = *Medicago littoralis* (48)

Medicago quasifalcata Sinskaya = *Medicago sativa* subsp. *falcata* var. *falcata* (12)

***Medicago radiata* L. (73)**

Medicago recta (Desf.) Willd. = *Medicago minima* (52)

Medicago recta Kit. = *Medicago falcata* × *Medicago prostrata* (see 72)

Medicago recurvata Post (unpublished name on herbarium specimen, noted by Heyn 1963) = *Medicago bonarotiana* (19)

Medicago reniformis Dulac = *Medicago lupulina* (49)

Medicago reticulata Benth. = *Medicago polymorpha* (69)

***Medicago retrorsa* (Boiss.) E. Small (74)**

"*Medicago ×revolii*" Coste & Soulei (Soulié), Le monde des plantes, 22(129): 6 (1921) = *Medicago lupulina* (49)

Medicago reynieri Albert = *Medicago polymorpha* (69)

***Medicago rhodopea* Velen. (75)**

***Medicago rhytidiocarpa* (Boiss. & Balansa) E. Small (76)**

***Medicago rigida* (Boiss. & Balansa) E. Small (77)**

Medicago rigidula Bourg. ex Nyman = *Medicago rigidula* (78)

Medicago rigidula Hoffm. ex Steud. = *Medicago minima* (52)

***Medicago rigidula* (L.) All. (78)**

Medicago rigidula (L.) Desr. ≡ *Medicago rigidula* (78)

Medicago rigidula Willd. = *Medicago truncatula* (97)

***Medicago rigiduloides* E. Small (79)**

Medicago rivularis Vassilcz. = *Medicago sativa* subsp. ×*varia* (12)

Medicago roborovskii Vassilcz. = *Medicago sativa* subsp. ×*varia* (12)

Medicago romanica D. Brândza = *Medicago sativa* subsp. *falcata* var. *falcata* (12)

Medicago romanica Prodán = *Medicago sativa* subsp. *falcata* var. *falcata* (12)

***Medicago rostrata* (Boiss. & Balansa) E. Small (80)**

***Medicago rotata* Boiss. (81)**

Medicago rugosa d'Urv. = *Medicago turbinata* (98)

***Medicago rugosa* Desr. (82)**

Medicago rugulosa Batt. = *Medicago littoralis* (48)

Medicago rupestris Eichv. = *Medicago sativa* subsp. *caerulea* (12)

***Medicago rupestris* M. Bieb. (83)**

Medicago rupestris Pirona = *Medicago pironae* (65)

**Medicago ruthenica* E.H.L. Krause, Fl. Deutschland (Sturm – ed. 2) 9: 129 (1901)

Medicago ruthenica Kom. & Alis. = *Medicago ruthenica* (84)

***Medicago ruthenica* (L.) Trautv. (84)**

Medicago sabolensis H. Lév. = misspelling of *Medicago ×sabulensis* H. Lév.

Medicago × sabulensis H. Lév., alleged hybrid of *Medicago marina* and *Medicago littoralis* (see 50)

Medicago sagarrae Sennen = *Medicago sativa* subsp. *falcata* var. *falcata* (12)

Medicago salsuginosa Dumort = *Medicago arabica* (13)

Medicago sardoa Moris ex G. Don = *Medicago polymorpha* (69)

***Medicago sativa* L. (12)**

Medicago sativa-falcata Ritschl. = *Medicago sativa* subsp. ×*varia* (12)

Medicago sativoides Léveillé = *Medicago sativa* subsp. ×*varia* (12)

***Medicago sauvagei* Nègre (85)**

***Medicago saxatilis* M. Bieb. (86)**

Medicago schimperiana Hochst. ex Boiss. = *Medicago laciniata* (45)

Medicago schirjaewii Hand.-Mazz. ≡ *Medicago archiducis-nicolai* (15)

Medicago schischkinii Sumnev. = *Medicago sativa* subsp. ×*varia* (12)

Medicago scutellaris All. ≡ *Medicago scutellata* (87)

Medicago scutellata (L.) All. ≡ *Medicago scutellata* (87)

***Medicago scutellata* (L.) Mill. (87)**

Medicago scutellata Pall. ex Bieb. = *Medicago orbicularis* (58)

***Medicago secundiflora* Durieu (88)**

Medicago secundiramea Durieu ex Willk. ≡ *Medicago secundiflora* (88)

Medicago sessilis Peyron ex Post = *Medicago minima* (52)

***Medicago shepardii* Post ex Boiss. (89)**

**Medicago sibirica* DC., Prodr. 2: 173 (1825). (De Candolle (1825) equated this with *Melilotus sibirica* Poir.)

Medicago sicula Tod. = *Medicago murex* (55) or *Medicago sphaerocarpos* (92)

Medicago silvestris = orthographic variant of *Medicago sylvestris*.

***Medicago sinskiae* Uljanova (90)**

Medicago sirjeaewii (*schirjaewii, schirjaëwii*) Hand.-Mazz. ≡ *Medicago archiducis-nicolai* (15)

Medicago sogdiana Vassilc. = *Medicago sativa* subsp. *sativa* (12)

***Medicago soleirolii* Duby (91)**

Medicago sorrentinii Tineo = *Medicago murex* (55)

Medicago sphaerocarpa Costa = *Medicago turbinata* (98)

***Medicago sphaerocarpos* ("*sphaerocarpa*") Bertol. (92)**

Medicago spinulosa DC. = *Medicago turbinata* (98)

Medicago spuria Hy = *Medicago sativa* subsp. ×*varia* (12)

**Medicago stellata* Trautv., Bull. Sci. Acad. Imp. Sci. St. Petersb. 8: 272 (1841)

Medicago stipularis Wallr. = *Medicago lupulina* (49)

***Medicago strasseri* Greuter, Matthäs & Risse (93)**

Medicago striata Bast. = *Medicago italica* (44)

Medicago stricta Wallr. ex Urb. = *Medicago polymorpha* (69)

**Medicago strumaria* hort. ex Steud., Nomencl. Bot. 514 (1821)

Medicago subdicycla (Trautv.) Vassilcz. = *Medicago sativa* subsp. ×*varia* (12)

Medicago subfalcata Schur = *Medicago sativa* subsp. ×*varia* (12)

Medicago subfalcata Sinskaya = *Medicago sativa* subsp. *falcata* var. *falcata* (12)

Medicago subinermis Bertol. = *Medicago littoralis* (48)

Medicago suffruticosa Grossh. = *Medicago sativa* subsp. ×*varia* (12)

***Medicago suffruticosa* Ramond ex DC. (94)**

Medicago sulcata (Desf.) E.H.L. Krause ≡ *Melilotus sulcata* Desf.

Medicago sylvestris ("silvestris") Fr. = *Medicago sativa* subsp. *falcata* var. *falcata* (12)

***Medicago syriaca* E. Small (95)**

Medicago syriaco-palestinica (Bordakov) Sinsk. = *Medicago sativa* subsp. *sativa* (12)

Medicago tadzhicorum Vassilcz. = *Medicago sativa* subsp. *sativa* (12)

Medicago talyschenisis Latsch. (Lachashvili) = *Medicago arabica* (13)

Medicago tenderiensis Opperman ex Klokov = *Medicago sativa* subsp. *falcata* var. *falcata* (12)

***Medicago tenoreana* Ser. (96)**

Medicago teniflora Schur = *Medicago prostrata* (72)

Medicago tentaculata Willd. = *Medicago truncatula* (97)

Medicago tenuifolia C. Presl = *Medicago prostrata* (72)

Medicago tenuis (Fisch. ex M. Bieb.) Trautv. = *Medicago medicaginoides* (51)

Medicago terebella St.-Lag. = *Medicago polymorpha* (69)

Medicago terebellum Willd. = *Medicago polymorpha* (69)

Medicago tetracycla Presl = *Medicago littoralis* (48)

Medicago tetrahemicycla Sinskaya = *Medicago sativa* subsp. ×*varia* (12)

Medicago tianschanica Vassilc. = *Medicago sativa* subsp. ×*varia* (12)

Medicago tibetana (Alef.) Vassilcz. = *Medicago sativa* subsp. ×*varia* (12)

Medicago timeroyii Jordan = *Medicago rigidula* (78)

Medicago tinei Loj. & Poj. = *Medicago arabica* (13)

**Medicago todaroana* Lojac., Malpighia 20: 202 (1906)

Medicago tomentosa Chevallier = *Medicago rigidula* (78)

Medicago tomentosa Dulac = *Medicago suffruticosa* (94)

Medicago tornata (L.) Mill. = *Medicago italica* (44)

Medicago tornata Webb ex Ball, J. Linn. Soc., Bot. 16: 412 (1878). An inaccurate citation that has been repeated in some nomenclatural databases; Ball did not recognize a new species.

Medicago tornata Willd. ex Ces., Passer, & Gibelli = *Medicago italica* × *Medicago littoralis* (44)

**Medicago tortuosa* Schwaegrichen ex Steud., Nomencl. Bot. 514 (1821)

Medicago transoxana Vassilcz. = *Medicago sativa* subsp. *sativa* (12)

Medicago trautvetterii Sumnev. = *Medicago sativa* subsp. ×*varia* (12)

Medicago tribuloides Desr. = *Medicago truncatula* (97)

Medicago tricycla DC. = *Medicago italica* (44)

Medicago tricycla Senn. = *Medicago littoralis* (48)

Medicago trigonella E.H.L. Krause = *Trigonella laciniata* L.

Medicago trigyra Senn. = *Medicago littoralis* (48)

Medicago tripolitanica (Bordakov) Sinskaya = *Medicago sativa* subsp. *sativa* (12)

***Medicago truncatula* Gaertn. (97)**

Medicago truncatulata Godr. & Gren.: see *Medicago truncatula* Gaertn.

Medicago tuberculata (Retz.) Willd. = *Medicago turbinata* (98)

Medicago tuberculata Urb. = *Medicago polymorpha* (69)

Medicago tunetana (Murb.) Vassilcz. = *Medicago sativa* subsp. *sativa* × subsp. *glomerata* (12)

Medicago tu(r)binata Carmign., Giorn. Pisano Lett. 12(1): 13 (1810): see "*Medicago turbinata* sensu Willd."

***Medicago turbinata* (L.) All. (98)**

"*Medicago turbinata* sensu Willd." = *Medicago doliata* (32)

Medicago umbellata hort. ex DC. = *Medicago carstiensis* (23)

Medicago uncinata Nym. = *Medicago polymorpha* (69)

Medicago uncinata Urb. = *Medicago polymorpha* (69)

Medicago uncinata Willd. = *Medicago truncatula* (97)

Medicago urbani Heldr. ex Boiss. = *Medicago coronata* (27)

Medicago vardanis Vassilcz. = *Medicago sativa* subsp. *falcata* var. *falcata* (12)

Medicago ×*varia* T. Martyn = *Medicago sativa* subsp. ×*varia* (12)

Medicago vassilczenkoi Vorosch. = *Medicago ruthenica* (84)

Medicago vera Kirschleger = *Medicago sativa* subsp. *sativa* (12)

**Medicago verloti* Songeon & E.P. Perrier, Billotia 1: 74 (1864)

Medicago vermicularis Cesati = *Medicago coronata* (27)

**Medicago villosa* Brot., Fl. Lusit. 2: 116 (1804). The description is most consistent with *M. doliata* (32)

Medicago villosa DC. = *Medicago rigidula* (78)

**Medicago villosula* Baumg., Enum. Stirp. Transsilv. 2: 382 (1816); the description is more or less consistent with *M. rigidula* (78)

Medicago virescens Grossh. = *Medicago sativa* subsp. *glomerata* (12)

Medicago virginica L. ≡ *Lespedeza virginica* (L.) Britton

"*Medicago viscida* Schur", Verh. Naturf. Vereins Brünn 15(2): 173 (1877) (a name incorrectly stated to exist in some nomenclatural databases, although Schur merely ranked this as a variety) = *Medicago minima* (52)

**Medicago viscidula* Bellardi ex Colla, Herb. Pedem. 2: 100 (1834). Colla (1834) suggested the affinities of this are with *Medicago orbicularis* (58)

Medicago vulgaris Alef., a name represented as published as a species in nomenclatural lists, but merely published as a variety, and illegitimate (see 12)

Medicago willdenowii Boenn. = *Medicago lupulina* (49)

Medicago willdenowii Mérat = *Medicago lupulina* (49)

Medicula lupulina (L.) Medik. ≡ *Medicago lupulina* (49)

Medicula secundiflora (Durieu) Holub ≡ *Medicago secundiflora* (88)

Melilotoides archiducis-nicolai (Širj.) Yakovlev ≡ *Medicago archiducis-nicolai* (15)

Melilotoides biflora (Griseb.) Czerep. ≡ *Medicago biflora* (18)

Melilotoides brachycarpa (Fisch. ex M. Bieb.) Soják ≡ *Medicago brachycarpa* (20)

Melilotoides cretacea (M. Bieb.) Soják ≡ *Medicago cretacea* (29)

Melilotoides karkarensis (Semen. ex Vassilcz.) Soják = *Medicago platycarpos* (66)

Melilotoides platycarpos (L.) Soják ≡ *Medicago platycarpos* (66)

Melilotoides popovii (Korovin) Soják ≡ *Medicago popovii* (70)

Melilotoides pubescens (Edgew. ex Baker) Yakovlev ≡ *Medicago edgeworthii* (33)

Melilotoides ruthenica (L.) Soják ≡ *Medicago ruthenica* (84)

Melilotoides tibetica (Vassilcz.) Yakovlev = *Medicago edgeworthii* (33)

Melilotus lupulinus (L.) Trautv. ≡ *Medicago lupulina* (49)

Melilotus polyceratoides= *Medicago polyceratia* (68)

Melissitus archiducis-nicolai (Širj.). ≡ *Medicago archiducis-nicolai* (15)

Melissitus biflorus (Griseb.) Latsch. ≡ *Medicago biflora* (18)

Melissitus brachycarpus (Fisch. ex M. Bieb.) Latsch. ≡ *Medicago brachycarpa* (20)

Melissitus cretaceus (M. Bieb.) Latsch. ≡ *Medicago cretacea* (29)

Melissitus dasycarpus (Ser.) Latsch. = *Medicago radiata* (73)

Melissitus karkarensis (Semen. ex Vassilcz.) Golosk. = *Medicago platycarpos* (66)

Melissitus korshinskyi (Grossh.) Latsch. = *Medicago ruthenica* (84)

Melissitus ovalis (Boiss.) Latsch. = *Medicago ovalis* (60)

Melissitus platycarpus (L.) Golosk. ≡ *Medicago platycarpos* (66)

Melissitus popovii (Korovin) Golosk. ≡ *Medicago popovii* (70)

Melissitus radiatus (L.) Latsch. ≡ *Medicago radiata* (73)

Melissitus rostratus (Boiss. & Balansa) Latsch. ≡ *Medicago rostrata* (80)

Melissitus ruthenicus (L.) C.W. Chang ≡ *Medicago ruthenica* (84)

Melissitus ruthenicus (L.) Latsch. ≡ *Medicago ruthenica* (84)

Melissitus schischkinii (Vassilc.) Latsch. = *Medicago ruthenica* (84)

Nephromedia dasycarpa (Ser.) Czerep. = *Medicago radiata* (73)

Nephromedia radiata (L.) Kostel. ≡ *Medicago radiata* (73)

Pocockia biflora (Griseb.) Walp. ≡ *Medicago biflora* (18)

Pocockia cilicica Boiss. & Balansa = *Medicago brachycarpa* (20)

Pocockia glomerata (Fisch.) Boiss. = *Medicago brachycarpa* (20)

Pocockia liaosiensis P.Y. Fu & Y.A. Chen = *Medicago ruthenica* (84)

Pocockia lunata (Boiss.) Boiss. = *Medicago biflora* (18)

Pocockia ovalis (Boiss.) Boiss. = *Medicago ovalis* (60)

Pocockia platycarpos Boiss. ≡ *Medicago platycarpos* (66)

Pocockia plicata Boiss. & Balansa ≡ *Medicago plicata* (67)

Pocockia radiata (L.) Trautv. ≡ *Medicago radiata* (73)

Pocockia rostrata Boiss. & Balansa ≡ *Medicago rostrata* (80)

Pocockia ruthenica (L.) Boiss. ≡ *Medicago ruthenica* (84)

Pseudomelissitus dasycarpus (Ser.) Ovcz., Rasulova & Kinzikaëva = *Medicago radiata* (73)

Pseudomelissitus radiatus (L.) Ovcz., Rassulova & Kinzikaëva ≡ *Medicago radiata* (73)

Radiata dasycarpa (Ser.) Ovez., Rassulova & Kinzikaëva = *Medicago radiata* (73)

Radiata dasycarpa (Ser.) Soják = *Medicago radiata* (73)

Radiata glabra Ovcz., Rasulova & Kinzikaëva = *Medicago radiata* (73)

Radiata leiocarpa Soják = *Medicago radiata* (73)

Rhodusia arborea (L.) Vassilcz. ≡ *Medicago arborea* L. (14)

Spirocarpus minimus (L.) Opiz ≡ *Medicago minima* (52)

Telis glomerata (Fisch.) O. Kuntze = *Medicago brachycarpa* (20)

Telis incisa Kuntze = *Medicago polyceratia* (68)

Telis lunata (Boiss.) O. Kuntze = *Medicago biflora* (18)

Telis monspeliaca (L.) Kuntze ≡ *Medicago monspeliaca* (54)

Telis rostrata O. Kuntze ≡ *Medicago rostrata* (80)

Trifolium lupulinum Savi ≡ *Medicago lupulina* (49)

Trigonella alindana Huber-Mor. = *Medicago carica* (22)

Trigonella amandiana Samp. = *Medicago polyceratia* (68)

Trigonella ×ambigua Samp. = *Medicago polyceratia* (68)

Trigonella anatolica Boiss. & Balansa = *Medicago orthoceras* (59)
Trigonella arborea (L.) Vassilcz ≡ *Medicago arborea* (14)
Trigonella archiducis-nicolai (Širj.) Vassilcz. ≡ *Medicago archiducis-nicolai* (15)
Trigonella arcuata C.A. Mey. = *Medicago medicaginoides* (51)
Trigonella arenicola Huber-Mor. ≡ *Medicago arenicola* (16)
Trigonella aschersoniana Urb. = *Medicago hypogaea* (41)
Trigonella astroites Fisch. & C.A. Mey. = *Medicago astroites* (17)
Trigonella aucheri Boiss. = *Medicago brachycarpa* (20)
Trigonella aurantiaca Boiss. = *Medicago phrygia* (64)
Trigonella biflora Griseb. ≡ *Medicago biflora* (18)
Trigonella brachycarpa (Fisch. ex M. Bieb.) Moris ≡ *Medicago brachycarpa* (20)
Trigonella brahuica Boiss. = *Medicago monantha* (53)
Trigonella cancellata Desf. ex Pers. = *Medicago medicaginoides* (51)
Trigonella carica Hub.-Mor. ≡ *Medicago carica* (22)
Trigonella crassipes Boiss. ≡ *Medicago crassipes* (28)
Trigonella cretacea (M. Bieb.) Taliev. ≡ *Medicago cretacea* (29)
Trigonella dasycarpa (Ser.) Vassilcz. = *Medicago radiata* (73)
Trigonella divaricata Clairv. = *Medicago monspeliaca* (54)
Trigonella ×*elmaluensis* = *Medicago monspeliaca* × *Medicago crassipes* (54)
Trigonella filiformis Fisch. ex Ledeb. = *Medicago medicaginoides* (51)
Trigonella fischeriana Ser. ≡ *Medicago fischeriana* (34)
Trigonella flexuosa Fisch. ex M. Bieb. = *Medicago fischeriana* (34)
Trigonella geminiflora Bunge = *Medicago monantha* (53)
Trigonella glomerata (Fisch.) Hort. Paris ex Fisch. & Mey. = *Medicago brachcarpa* (20)
Trigonella halophila Boiss. ≡ *Medicago halophila* (36)
Trigonella hybrida Pourr. ≡ *Medicago hybrida* (40)
Trigonella incisa Benth. = *Medicago monantha* (53)
Trigonella isthmocarpa Boiss. & Balansa = *Medicago isthmocarpa* (43)
Trigonella karkarensis Semen. ex Vassilcz. = *Medicago platycarpos* (66)
Trigonella koraiensis Vassilc. = *Medicago ruthenica* (84)
Trigonella korshinskyi Grossh. = *Medicago ruthenica* (84)
Trigonella kotschyi (non Fenzl.) Rawi = *Medicago phrygia* (64)
Trigonella lasia Blatt. = *Medicago edgeworthii* (33)
Trigonella lunata Boiss. = *Medicago biflora* (18)
Trigonella macroglochin Dur. = *Medicago monantha* (53)
Trigonella mareschiana Hand.-Mazz. = *Medicago phrygia* (64)
Trigonella mniocarpa Wallr. ex Ser. in DC. = *Medicago lupulina* (49)
Trigonella monantha C.A. Mey. ≡ *Medicago monantha* (53)
Trigonella monspeliaca L. ≡ *Medicago monspeliaca* (54)
Trigonella noeana Boiss. = *Medicago monantha* (53)
Trigonella orthoceras Kar. & Kir. ≡ *Medicago orthoceras* (59)
Trigonella ovalis Boiss. ≡ *Medicago ovalis* (60)
Trigonella pamphylica Hub.-Mor. & Širj. ≡ *Medicago pamphylica* (61)
Trigonella persica Boiss. = *Medicago persica* (63)
Trigonella phrygia Boiss. & Balansa ≡ *Medicago phrygia* (64)
Trigonella pinnatifida Cav. = *Medicago polyceratia* (68)
Trigonella platycarpos L. ≡ *Medicago platycarpos* (66)
Trigonella plicata (Boiss. & Balansa) Boiss. ≡ *Medicago plicata* (67)
Trigonella polycarpa Boiss. & Heldr. ≡ *Medicago heldreichii* (37)
Trigonella polycerata Bak. = *Medicago monantha* (53)
Trigonella polyceratia Anth. = *Medicago monantha* (53)
Trigonella polyceratia L. ≡ *Medicago polyceratia* (68)
Trigonella polyceratoides Lange = *Medicago polyceratia* (68)
Trigonella popovii Korovin ≡ *Medicago popovii* (70)
Trigonella pubescens Edgew. ex Baker ≡ *Medicago edgeworthii* (33)
Trigonella radiata (L.) Boiss. ≡ *Medicago radiata* (73)
Trigonella retrorsa Boiss. ≡ *Medicago retrorsa* (74)
Trigonella retzii Spreng. = *Medicago medicaginoides* (51)
Trigonella rhytidiocarpa Boiss. & Balansa ≡ *Medicago rhytidiocarpa* (76)
Trigonella rytidocarpa Boiss. = *Medicago rhytidiocarpa* (76)
Trigonella rigida Boiss. & Bal. ≡ *Medicago rigida* (77)
Trigonella rostrata (Boiss. & Balansa) Boiss. ≡ *Medicago rostrata* (80)
Trigonella ruthenica L. ≡ *Medicago ruthenica* (84)
Trigonella ruthenica Patrin ex Ledeb. = *Medicago platycarpos* (66)
Trigonella schischkinii Vassilc. = *Medicago ruthenica* (84)
Trigonella sinuata Boiss. = *Medicago astroites* (17)
Trigonella sinuata Boiss. var. *brachycarpa* Boiss. = *Medicago halophila* (36)

Trigonella sirjaevii Hub.-Mor. = *Medicago huberi* (39)

Trigonella striata L. fil. = *Medicago medicaginoides* (51)

Trigonella tenuis Fisch. ex M. Bieb. = *Medicago medicaginoides* (51)

Trigonella tibetica Vassilcz. = *Medicago edgeworthii* (33)

Trigonella uncinata Eig = *Medicago monantha* (53)

Turukhania ruthenica (L.) N.S. Pavlova ≡ *Medicago ruthenica* (84)

Turukhania platycarpos (L.) Vassilcz. ≡ *Medicago platycarpos* (66)

Turukhania schischkinii (Vassilcz.) N.S. Pavlova = *Medicago ruthenica* (84)

Comprehensive Index of English Common Names

Notes: Medic (the spelling consistently used below) is also spelled medick; bur (the spelling consistently used below) is also spelled burr. The numbers in parentheses after the scientific names are the chapters in which the common names are presented.

Lunate medic – *Medicago biflora* (18)
Luserne – *Medicago sativa* (12)
Luzern – *Medicago sativa* (12)
Marine alfalfa – *Medicago marina* (50)
Mountain little alfalfa – *Medicago suffruticosa* (94)
Murex medic – *Medicago murex* (55)
Moon trefoil – *Medicago arborea* (14)
Nicholas's fenugreek – *Medicago archiducis-nicolai* (15)
Nicholas's medic – *Medicago archiducis-nicolai* (15)
Non(e)such – *Medicago lupulina* (49)
Orbicular alfalfa – *Medicago orbicularis* (58)
Piron's medic – *Medicago pironae* (65)
Plain trefoil – *Medicago praecox* (71)
Prickly lucerne – *Medicago rigidula* (78)
Purple alfalfa – *Medicago sativa* subsp. *sativa* (12)
Purple lucerne – *Medicago sativa* subsp. *sativa* (12)
Purple medic – *Medicago sativa* subsp. *sativa* (12)
Pyrenean medic – *Medicago hybrida* (40)
Ray-podded medic – *Medicago radiata* (73)
Rhodope alfalfa – *Medicago rhodopea* (75)
Rigid medic – *Medicago rigidula* (78)
Rock-inhabiting alfalfa – *Medicago saxatilis* (86)
Rocky alfalfa – *Medicago rupestris* (83)
Romanian alfalfa – *Medicago sativa* subsp. *falcata* var. *falcata* (12)
Round-fruited medic – *Medicago orbicularis* (58)
Rugose medic – *Medicago rugosa* (82)
Salt trefoil – *Medicago halophila* (36)
Sand alfalfa – *Medicago sativa* subsp. ×*varia* (12)
Sand fenugreek – *Medicago hypogaea* (41)
Sea alfalfa – *Medicago marina* (50)
Sea medic – *Medicago marina* (50)
Semi-twisted alfalfa – *Medicago sativa* subsp. ×*varia* (12)
Shield medic – *Medicago scutellata* (87)
Shore medic – *Medicago littoralis* (48)
Short-fruited fenugreek – *Medicago brachycarpa* (20)
Shrub medic – *Medicago arborea* (14)
Shrubby medic – *Medicago arborea* (14)
Siberian alfalfa – *Medicago sativa* subsp. *falcata* var. *falcata* (12)
Sickle medic – *Medicago sativa* subsp. *falcata* var. *falcata* (12)
Sickle-podded alfalfa – *Medicago sativa* subsp. *falcata* var. *falcata* (12)
Sickle-podded medic – *Medicago sativa* subsp. *falcata* var. *falcata* (12)
Širjaev's fenugreek – *Medicago huberi* (39)
Small alfalfa – *Medicago minima* (52)
Small bur clover (small burclover) – *Medicago minima* (52)
Small bur medic – *Medicago minima* (52)
Small leaf bur medic – *Medicago praecox* (71)
Small-leaved bur medic – *Medicago praecox* (71)
Small medic – *Medicago minima* (52)
Small wooly bur medic – *Medicago minima* (52)
Snail alfalfa – *Medicago scutellata* (87)
Snail clover – *Medicago sativa* (12), *Medicago scutellata* (87)
Snail medic – *Medicago scutellata* (87)
Southern bur clover – *Medicago arabica* (13), *Medicago polymorpha* (69)
Sphere medic – *Medicago sphaerocarpos* (92)
Spiny medic – *Medicago murex* (55)
Spotted bur medic – *Medicago arabica* (13)
Spotted burclover (spotted bur-clover, spotted bur clover) – *Medicago arabica* (13)
Spotted medic – *Medicago arabica* (13)
Sprawling medic – *Medicago suffruticosa* (94)
Star-fruited fenugreek – *Medicago monspeliaca* (54)
Steppe alfalfa – *Medicago sativa* subsp. *falcata* var. *falcata* (12)
Stickyweed – *Medicago minima* (52)
Strand medic – *Medicago littoralis* (48)
Strong-spined medic – *Medicago truncatula* (97)
Swedish alfalfa – *Medicago sativa* subsp. *falcata* var. *falcata* (12)
Tattered medic – *Medicago laciniata* (45)
Tifton bur medic – *Medicago rigidula* (78)
Tifton burclover – *Medicago rigidula* (78)
Tifton medic – *Medicago rigidula* (78)
Toothed bur clover – *Medicago polymorpha* (69)
Toothed medic – *Medicago polymorpha* (69)
Tree alfalfa – *Medicago arborea* (14)
Tree lucerne – *Medicago arborea* (14)
Tree medic – *Medicago arborea* (14)
Trefoil – *Medicago lupulina* (49)
Tubercled medic – *Medicago turbinata* (98)
Variegated alfalfa – *Medicago sativa* subsp. ×*varia* (12)
Variegated lucerne – *Medicago sativa* subsp. ×*varia* (12)
Water medic – *Medicago littoralis* (48)
Wheel medic – *Medicago bonarotiana*, *M. rotata* × *M. bonarotiana* (19), *Medicago rotata* (81)
Wooly alfalfa – *Medicago lanigera* (46)
Wool(l)y bur medic – *Medicago minima* (52)
Wrinkled medic – *Medicago rugosa* (82)
Yellow-flowered alfalfa – *Medicago sativa* subsp. *falcata* var. *falcata* (12)
Yellow-flowered medic – *Medicago sativa* subsp. *falcata* var. *falcata* (12)
Yellow medic – *Medicago sativa* subsp. *falcata* var. *falcata* (12)
Yellow trefoil – *Medicago lupulina* (49)

Comprehensive Index of French Common Names

Note: The numbers in parentheses after the scientific names are the chapters in which the French names are presented.

Comprehensive Index of Cultivars Cited in the Monograph[1]

Note: The numbers in parentheses after the scientific names are the chapters in which the cultivar names are presented.

[1] Some of these entries are experimental rather than commercial cultivars, and some would not meet the criteria currently required to satisfy cultivar status according to the *International Code of Nomenclature for Cultivated Plants*. Although Grimm and Kayseri are sometimes considered to be cultivars, they are not enclosed by the conventional single quotes that indicate cultivars because they are landraces, thus considerably more heterogeneous than typical cultivars.